SpringerWienNewYork

Bernardete Ribeiro
Rudolf F. Albrecht
Andrej Dobnikar
David W. Pearson
Nigel C. Steele (eds.)

Adaptive and Natural Computing Algorithms

Proceedings of the International Conference in Coimbra, Portugal, 2005

SpringerWienNewYork

Dr. Bernardete Ribeiro
Departamento de Engenharia Informática
Universidade de Coimbra, Coimbra, Portugal

Dr. Rudolf F. Albrecht
Institut für Informatik
Universität Innsbruck, Innsbruck, Austria

Dr. Andrej Dobnikar
Fakulteta za Računalništvo in Informatiko
Univerza v Ljubljani, Ljubljana, Slovenia

Dr. David W. Pearson
Equipe Universitaire de Recherche en Informatique
de Saint-Etienne (Groupe de Recherche de Roanne)
Institut Universitaire de Technologie de Roanne
Université Jean Monnet, Saint-Etienne, France

Dr. Nigel C. Steele
Division of Mathematics
School of Mathematical and Information Sciences
Coventry University, Coventry, U.K.

Printed in Austria
SpringerWienNewYork is part of Springer Science + Business Media,
springeronline.com

Camera-ready copies provided by authors and editors
Printed by Novographic Druck G.m.b.H., 1230 Wien, Austria
Printed on acid-free and chlorine-free bleached paper
SPIN 11393269

With 433 Figures

ISBN 3-211-24934-6 SpringerWienNewYork

Preface

The ICANNGA series of Conferences has been organised since 1993 and has a long history of promoting the principles and understanding of computational intelligence paradigms within the scientific community and is a reference for established workers in this area. Starting in Innsbruck, in Austria (1993), then to Alès in France (1995), Norwich in England (1997), Portoroz in Slovenia (1999), Prague in the Czech Republic (2001) and finally Roanne, in France (2003), the ICANNGA series has established itself for experienced workers in the field. The series has also been of value to young researchers wishing both to extend their knowledge and experience and also to meet internationally renowned experts.

The 2005 Conference, the seventh in the ICANNGA series, will take place at the University of Coimbra in Portugal, drawing on the experience of previous events, and following the same general model, combining technical sessions, including plenary lectures by renowned scientists, with tutorials.

In a rapidly advancing world, where technology and engineering change dramatically, new challenges in modern informatics compel us to broaden the Conference themes in order to take into account new developments. Nevertheless we have kept the acronym ICANNGA, which now stands for International Conference on Adaptive and Natural ComputiNG Algorithms

It is interesting to note that the themes of this conference have their origins in attempts to obtain understanding of natural processes. Indeed, Nature itself seems capable of providing a source of inspiration both for new research areas and new methodologies. Thus the focus of ICANNGA is on both the theoretical aspects and also on the practical applications of computational paradigms inspired by adaptive and natural processes, with special emphasis on adaptive and machine learning systems, neural networks, evolutionary algorithms, fuzzy systems, support vector machines, biological computation and quantum computing.

Amongst the papers accepted for presentation, there are contributions reporting theoretical developments as well as practical applications both for neural networks, genetic algorithms and evolutionary computation. Unsupervised and clustering techniques as well as self-organising maps are also the subjects of a substantial number of papers. Also, the relevance of computational neuroscience with a strong biological trend is evidenced by the number of papers submitted. It is pleasing to observe the number of contributions reporting successful applications of the technology to the solution of industrial problems. Predominantly, there are applications of neural networks to process engineering and to robotics and control which may reflect the maturity reached by the methodology. Contributions also abound in the field of evolutionary computation particularly in combinatorial and optimisation problems. They range from evolutionary computation theory to real-world applications, from global numerical optimisation to combinatorial optimisation, from evolutionary learning to multi-agent systems. The topics covered by the accepted papers reflect the breadth and depth of these fields. It is worth noting that they are inherently interdisciplinary. They interact closely with computer science, engineering, biology, mathematics, economics, and many papers at this conference illustrate such interdisciplinarity quite clearly.

Contributed papers in the field of natural computing, swarm optimisation and bioinformatics and computational biology are no less compelling. The awareness in the academic communities that hybrid approaches of intelligent systems will be necessary if the real challenges in computational intelligence are to be tackled is a fundamental stimulus to their investigation and has been acknowledged for the many papers with cross fertilising ideas submitted. Many papers are dedicated to machine learning and heuristics and others to soft computing applications. Some papers are devoted to the theory of computation in which traditional information is replaced by its quantum physical counterpart showing that it is an interesting challenge to study quantum computation. In addition, kernel based algorithms, able to solve tasks other than classification, represent a revolution in pattern recognition bridging existing gaps. Some papers either theoretical or application oriented are dedicated to pattern recognition, intelligent signal processing and computer vision.

We would like to express our gratitude to everyone who contributed in any way to the completion of this volume. In particular, we thank the members of the Programme Committee for reviewing the submissions and making the final decisions on the acceptance of papers, and the additional reviewers Mika Hirvensalo, Jorge Tavares, Pino Buzzanca, José Neto, Armando Vieira and Miguel Rocha for their work. We thank also

Jorge Tavares and Pedro Martins (University of Coimbra) for their difficult and time consuming task of checking paper source files for correctness and preparing the LaTeXvolume initial pages, Silvia Schilgerius (Springer-Verlag) and finally to all authors for their valuable research work and for their submissions.

B. Ribeiro	University of Coimbra, Portugal
R. Albrecht	University of Innsbruck, Austria
A. Dobnikar	University of Ljubljana, Slovenia
D. W. Pearson	Jean Monnet University of Saint-Etienne, France
N. C. Steele	Coventry University, England

Contents

Neural Networks in Process Engineering

Neural Networks in Robotics and Control

Clustering and Unsupervised Learning

Self-Organising Maps

Part II – Evolutionary Computation

Evolutionary Computation I

Evolutionary Computation II

Genetic Algorithms Theory

Complex Systems and Genetic Programming

Swarm Optimization

Molecular and Quantum Computing

Part III – Adaptive and Natural Computing

BioInformatics and Computational Biology I

BioInformatics and Computational Biology II

Machine Learning and Heuristics III

Feature Selection and Kernel Methods

Computer Security

Part V – Computer Vision and Pattern Recognition

Computer Vision and Image Processing

Signal Processing and Pattern Recognition

Part VI – Hybrid Methods and Tools

Hybrid Intelligent Systems

High Performance and Parallel Computing Tools

ICANNGA
International Conference on Adaptive and Natural Computing Algorithms
Coimbra, Portugal, March 21-23, 2005

International Advisory Committee

R. Albrecht, University of Innsbruck, Austria
A. Dobnikar, University of Ljubljana, Slovenia
V. Kůrková, Czech Academy of Science, Czech Republic
D. Pearson, University Jean Monnet, Saint-Etienne, France
B. Ribeiro, University of Coimbra, Portugal
N. Steele, Coventry University, UK

Programme Committee

B. Ribeiro (Chair), University of Coimbra, Portugal
R. Albrecht, University of Innsbruck, Austria
G. Andrejkova, P.J. Šafarík University, Košice, Slovakia
B. Beliczynski, Warsaw University of Technology, Poland
H. Bischof, Institute for Computer Graphics and Vision, Graz, Austria
L. Borges de Almeida, Technical University of Lisbon, Portugal
W. Brauer, Technische Universitaet München, Germany
P. Carvalho, University of Coimbra, Portugal
K. Cios, University of Colorado, USA
E. Costa, University of Coimbra, Portugal
A. Dobnikar, University of Ljubljana, Slovenia
M. Dorigo, Université Libre de Bruxelles, Belgium
A. Dourado, University of Coimbra, Portugal
G. Dray, École des Mines d'Alès, France
E. Gaura, Coventry University, UK
C. Giraud-Carrier, ELCA Informatique SA, Lausanne, Switzerland
J. Henriques, University of Coimbra, Portugal
K. Hlavackova-Schindler, Inst. of Computer Aided Automation, Vienna, Austria
O. Hoshino, Oita University, Japan
P. C. Kainen, Georgetown University, USA
H. Karatza, Aristotle University of Thessaloniki, Greece
M. Kárný, Inst. of Information Theory and Automation, Czech Republic
V. Kůrková, Institute of Computer Science, Czech Republic
P. Larranaga, University of the Basque Country, Spain
A. Leonardis, University of Ljubljana, Slovenia
T. Martinez, Brigham Young University, USA
F. Masulli, University of Pisa, Italy
L. Mokhnache, University of Batna, Algeria
R. Neruda, Institute of Computer Science, Czech Republic
N. Oliveira, University of Coimbra, Portugal
N. Pavesic, University of Ljubljana, Slovenia
D. W. Pearson, University Jean Monnet, Saint-Etienne, France
E. Pelikan, Institute of Computer Science, Czech Republic
A. Pinkus, Technion Israel Institute of Technology, Israel
J. Principe, University of Florida, USA

ADFUNN: An Adaptive Function Neural Network

Dominic Palmer-Brown, Miao Kang

Computational Intelligence Research Group, School of Computing, Leeds Metropolitan University, UK
E-mail: {d.palmer-brown, m.kang}@leedsmet.ac.uk

Abstract

An adaptive function neural network (ADFUNN) is introduced. It is based on a linear piecewise artificial neuron activation function that is modified by a novel gradient descent supervised learning algorithm. This Δf process is carried out in parallel with the traditional Δw process. Linearly inseparable problems can be learned with ADFUNN, rapidly and without hidden neurons. The Iris dataset classification problem is learned as an example. An additional benefit of ADFUNN is that the learned functions can support intelligent data analysis.

1 Motivation

True to its name the artificial neuron derives from a joint biological-computational perspective. Taking the sum of weighted inputs is biologically plausible, and adapting a weight is a reasonable model for modifiable synaptic efficacy. But the common assumption that the output activation function is fixed is for computational rather than biological reasons. A fixed analytical function facilitates mathematical analysis to a greater degree than an empirical one. Nonetheless, there are some computational benefits to modifiable activation functions, and they may be biologically plausible as well.

Recent work in computational neuroscience is beginning to suggest that neuromodulators play a role in learning by modifying the neuron's activation function [1,2]. From a computational point of view, it is surprising that real neurons could be essentially fixed entities with no adaptive aspect, except at their synapses, since such a restriction leads to non-linear responses typically requiring many neurons.

MLPs with an appropriate number of nodes are very effective, but if the activation function is not optimal, neither is the number of hiddens which in turn depends on the function. Training, which is typically slow on linearly inseparable data, always requires hidden nodes. In some cases it may help to adapt a slope-related parameter of the activation function, but not if the analytic shape of the function is unsuited to the problem, in which case many hidden nodes may be required. In contrast, with an adaptive function approach it should be possible to learn linear inseparable problems fast, even without hidden nodes. The simplest test case is the two input exclusive-OR. A single classic artificial neuron is incapable of solving it, yet by choosing or learning the appropriate activation function, the solution is readily learnable with one neuron.

2 Related Approaches to Iris Dataset

There are many neural network methods for solving linearly inseparable problems, such as the popular Iris plants dataset. Most of them use a network of multi-player perceptions (MLPs), so they require hidden neurons.

A reinforcement learning method called SANE [3] (Symbiotic, Adaptive Neuro-Evolution) uses an evolutionary algorithm. It searches for effective connections and/or connection weights within a neural network. The dynamic components (i.e. weights and/or architecture) of the neural networks are encoded in structures that form the genetic chromosomes. With the Iris dataset, SANE constructs a network of 5 inputs, 3 hidden and 3 output units, and the transfer function of the network is the sigmoid. The level (averaged over 50 runs) of the learning and generalisation abilities is about 90%. Cantu-Paz [5] also achieved about 90% accuracy with 5 hidden unites, and a spiking neural network with 4 hiddens has achieved more than 96% accuracy [6].

It is possible to solve the problem quite well without a hidden layer, by transforming the input space, but this involves many extra input neurons. For example, Eldracher [4] use several units for each of the four measurements, in a form of coarse coding, and with 90 training and 60 test patterns they achieve an overall misclassification rate of 2.5%.

3 A Single Layer Adaptive Function Network

In this paper, we describe how to solve linearly inseparable problems using a simple adaptive function neural network (ADFUNN) which is based on linear piecewise function neurons, and has no hidden layer.

3.1 XOR Experiment

The XOR is a simple binary example of linear inseparability, and therefore serves as a good basic test to establish that linearly inseparable problems can be solved by ADFUNN. Two weights are needed for the two inputs and there is one output. Weights are initialized randomly between -1 and 1, they are then normalized. F point values are also initialized randomly between -1 and 1. Each F point is simply the value of the activation function for a given input sum. F points are equally spaced, and the function value between points is on the straight line between them. This network is adapted as follows:

A = input node activation, E = output node error.
Step 1: calculate output error
Step 2: adapt weights:
    Δw = learning rate * A * E
    w' = w + Δw
    weights normalization
Step 3: adapt function:
    Δf = learning rate * E
    f' = f + Δf
Step 4: repeat step 1 to step 3 until the output error is close to 0.

The ADFUNN learns the problem very fast with a learning rate of 0.5. Too low a learning rate makes the network learn very slowly, but too high a learning rate makes the weights and objective function diverge.
An example of the weights after learning is: w1 = 0. 60, w2 = 0.80, and therefore, the summed inputs are:

input pattern (0, 0): w1 * 0 + w2 * 0 = 0,
input pattern (0, 1): w1 * 0 + w2 * 1 = 0.80,
input pattern (1, 0): w1 * 1 + w2 * 0 = 0.60,
input pattern (1, 1): w1 * 1 + w2 * 1 = 1.40.

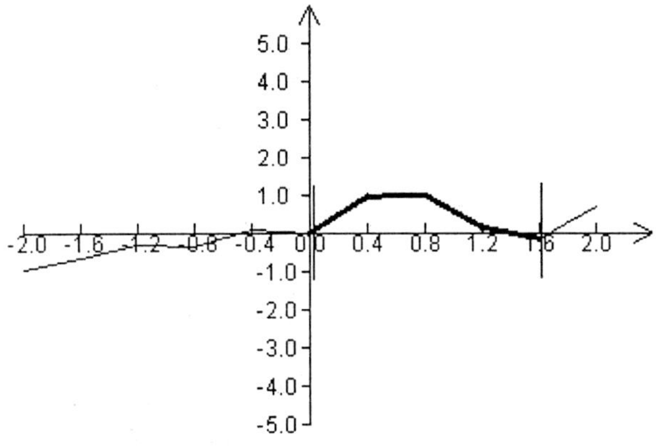

Fig. 1. XOR problem solved using ADFUNN

As can be seen in Fig 1, a characteristic curve of XOR is learned. The line between two borders of the curve marks the learned region, within which adaptation has occurred. The data all projects onto this range, so beyond it none of the points are relevant in the final analysis.

From the above curve, we can see that when input pattern is (0, 1), the corresponding f(x) = 1.0 where x = 0. 80. Similarly, the other three inputs give the expected correct answer. In the region projected onto between (0, 1) and (1, 0), the slope of the activation is nearly 0 and f = 1, and beyond this region, the function slopes down towards 0. Thus, we have a continuous, or fuzzy, XOR.

3.2 Iris Experiments

The iris dataset [7] consists of 150 four dimensional data. Four measurements: sepal length, sepal width, petal length, and petal width, were made by Fisher on 50 different plants from each of three species of Iris (Iris Setosa, Iris Versicolor and Iris Virginica). One class is linearly separable from the other two, but the other two are not linearly separable from each other.

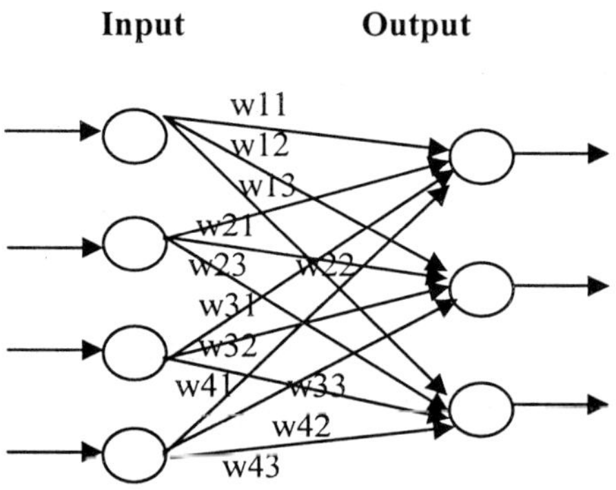

Fig. 2. ADFUNN for the Iris dataset experiment

Because the weights are initialized randomly between [-1, 1], the $\sum aw_j$ has a known range [-11, 11]. The iris dataset has a precision of 0.1, so this range can be coded with a resolution of 0.1 as the F-point interval. Therefore, 221 points ((11+11)/0.1+1=221) are sufficient to encode the data precisely.
Clearly, every dataset will have its own range. To cope with this, we need two learning constants, WL and FL. WL depends on input data range and the F point interval, whereas FL just depends on the required output range. We use WL = 0.01, FL = 0.1 to adapt weights and functions respectively.
Although we did not do so for the simple XOR case, from basic calculus, we should take into account the fact that we have a variable function slope. Hence, Δw = WL*Fslope*A*error, Thus, the algorithm becomes:

Step 1: calculate output error
Step 2: adapt weights for each output neuron:
    Δw = WL*Fslope*A*E
    w' = w + Δw
    weight normalization
Step 3: adapt function for each output neuron:

$\Delta f = FL * E$

$f' = f + \Delta f$

Step 4: randomly select a pattern in Iris-dataset to train

Step 5: repeat step 1 to step 4 until the output error goes to 0 or tends to a steady state.

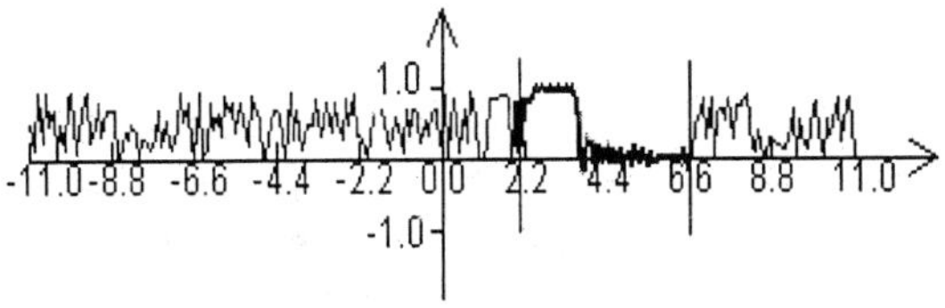

Fig. 3. Iris-Setosa function using ADFUNN

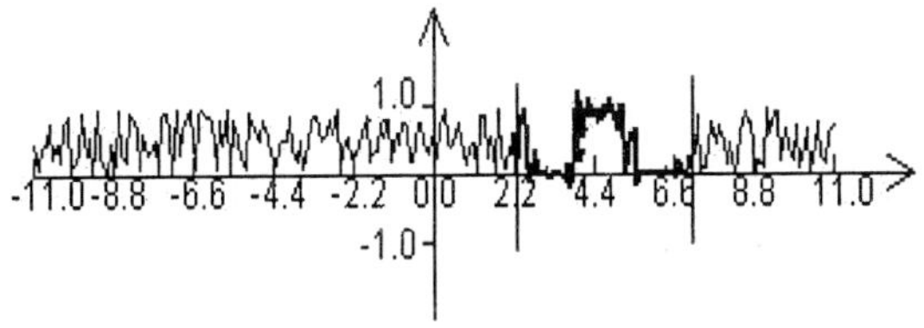

Fig. 4. Iris- Versicolor function using ADFUNN

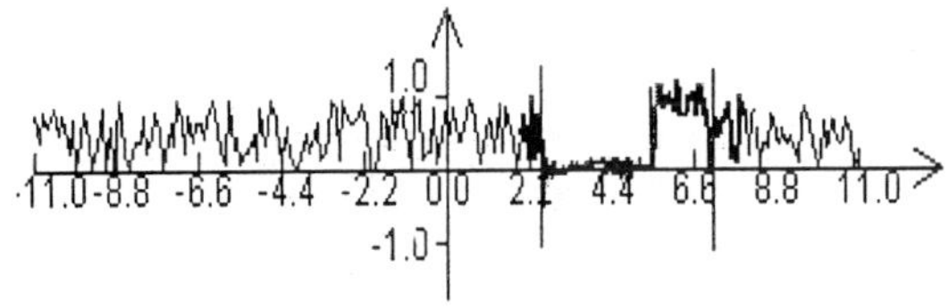

Fig. 5. Iris- Virginica function using ADFUNN

4 Results

In this case, after 25 epochs, the average error for these three classes tends to a steady state, which is always around 0.065. We ran more than 100 simulations (it learns the task in no more than 30 epochs in each run). The average final error is consistently in the range [0.05, 0.15]. In addition, from the weights for each class, which are consistent across the simulations, we can conclude that, for Iris-Setosa, petal length and petal width are the two most important measurements, whilst for Iris-Versicolor, petal length is the most important measurement. The same for Iris-Virginica, petal length is the most important measurement. After many comparisons between the weights for the three classes, we found sepal width is the least important measurement.

The problem is learned with 100% successful classification when all patterns are used in training, and also with only 120, whilst 90 patterns were insufficient to achieve complete generalization. Results are tabulated in figure 9. Comparing figures 3, 4, 5 and 6, 7, 8 we can see that there are only slight differences in the functions, showing that they have solved the problem in approximately the same manner, irrespective of the number of training patterns. The functions are clearly non-linear and non-monotonic, illustrating the value of the adaptive function method which is able to acquire these functions.

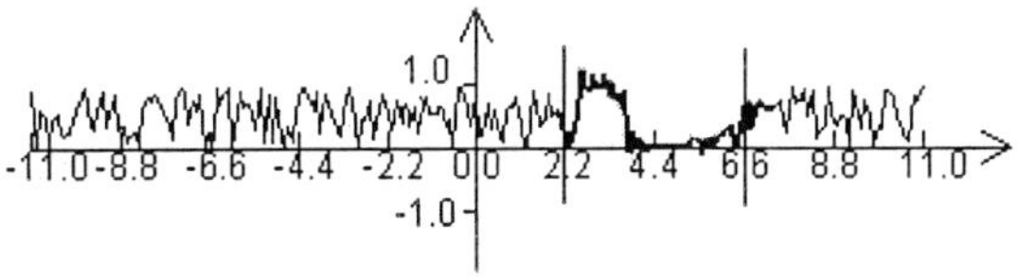

Fig. 6. Iris-Setosa function with 120 training patterns.

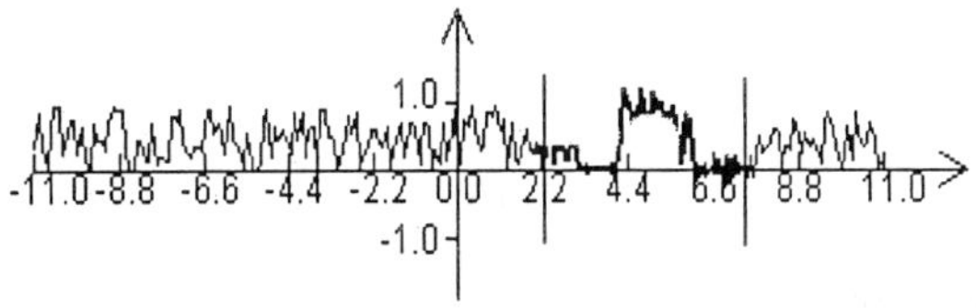

Fig. 7. Iris-Versicolor function with 120 training patterns.

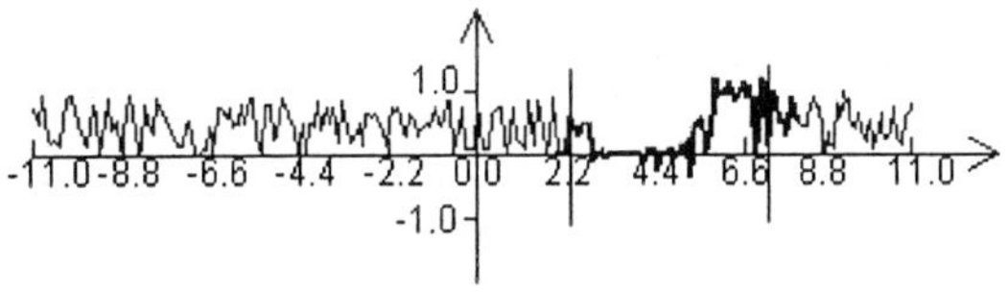

Fig. 8. Iris-Virginica function with 120 training patterns.

Performance	**Best**	**Average**	**Worst**
Average error	0.05	0.10	0.15
Generalisation 90 train 60 test	(55/60) 93.33%	89.17%	(51/60) 85%
Generalisation 120 train 30 test	(30/30) 100%	95%	(27/30) 90%

Fig. 9. Evaluation of ADFUNN for Iris dataset (100 runs)

We can interpret the data by considering weights and functions together. To identify Setosa, both petal width and petal length must be taken into account as they both have equally strong weights. In contrast, Iris-Versicolor and Iris-Virginica are most dependent on petal length, but it is their activation functions which show that Virginica has longer petals.

	Sepal length	Sepal width	Petal length	Petal width
Setosa	0.32	0.19	0.42	0.40
Versicolor	0.31	0.30	0.52	0.39
Virginica	0.43	0.38	0.45	0.31

Fig. 10. Typical weight values for the iris dataset

5 Conclusion and Future Work

One of the main conclusions is that an effective ADFUNN can be realized using the following general learning rules:

$$\Delta w = WL * Fslope * A * E$$
$$w' = w + \Delta w \text{ (normalized)}$$
$$\Delta f = FL * E$$
$$f' = f + \Delta f$$

Thus, ADFUNN performs supervised learning. In related work, we are exploring unsupervised and reinforcement learning regimes that combine complimentary adaptation rules within a single network: namely, the snap-drift algorithm [8]. It is proposed that snap-drift be followed by a supervised phase acting on the activation functions alone, to perform classification.

Related work on connectionist natural language parsers [9, 10] will allow ADFUNN to be compared to MLPs and SRNs on phrase segmentation and recognition tasks, and this may lead to the creation of multilayer ADFUNN.

Intelligent data analysis with neural networks requires analysis of the weights to establish the most important factors and generate simplified equations to explain network decisions [11]. In ADFUNN, the learned functions offer insights into the data, in addition to the weights. We will investigate algorithms for replacing learned functions with matched analytical ones to automatically generate mechanistic models from trained networks.

In complex domains, there will be a need to use hidden nodes in a ML-ADFUNN to represent arbitrary concave decision functions. However, just as the single layer ADFUNN is more powerful than the SLP, so the multilayer ADFUNN will be more powerful in learning than MLPs, and may well require fewer hidden neurons.

References

[1] Scheler, G. (2004) Regulation of Neuromodulator Efficacy - Implications for Whole-Neuron and Synaptic Plasticity. Progress in Neurobiology 72(6)

[2] Scheler, G. (2004) Memorization in a neural network with adjustable transfer function and conditional gating. Quantitative Biology

[3] Moriarty, D.E. (1997) Symbiotic Evolution of Neural Networks in Sequential Decision Tasks. Ph.D. thesis, Department of Computer Science. The University of Texas at Austin

[4] Eldracher, M. (1992) Classification of Non-Linear-Separable Real-World-Problems Using Δ-Rule, Perceptions, and Topologically Distributed Encoding. In Proceedings of the 1992 ACM/SIGAPP Symposium on Applied Computing. Vol. I, ACM Press, pp 1098-1104

[5] Cantu-Paz, E. (2003) Pruning Neural Networks with Distribution Estimation Algorithms. GECCO 2003. USA, pp. 790-800

[6] Pavlidis, N.G., Tasoulis, D.K., Plagianakos, V.P., Nikiforidis, G. and Vrahatis, M.N. (2004) Spiking Neural Network training Using Evolutionary Algorithms. International Joint Conference on Neural Networks (IJCNN 2004). Budapest, Hungary

[7] Fisher, R.A. (1936) The Use of Multiple Measurements in Taxonomic Problems. Annals of Eugenics 7. pp.179-188

[8] Lee. S.W., Palmer-Brown. D. and Roadknight, C. M. (2004). Performance-guided Neural Network for Rapidly Self-Organising Active Network Management. Neurocomputing, 2004 (in press)

[9] Palmer- Brown, D., Tepper, J. and Powell, H. (2002). Connectionist Natural Language Parsing, Trends in Cognitive Sciences, 6(10) pp.437 – 442

[10] Tepper, J., Powell, H. and Palmer-Brown D. (2002). A Corpus-based Connectionist Architecture for Large-scale Natural Language Parsing, Connection Science, 14(2)

[11] Roadknight, C.M., Balls, G., Mills, G. and Palmer-Brown, D. (1997). Modelling Complex Environmental Data. IEEE Transactions on Neural Networks, 8(4): pp.856 – 862

Certain comments on data preparation for neural networks based modelling

Bartlomiej Beliczynski
Warsaw University of Technology
Institute of Control and Industrial Electronics
Koszykowa 75, 00-662 Warsaw
Email: b.beliczynski@isep.pw.edu.pl

Abstract

The process of data preparation for neural networks based modelling is examined. We are discussing sampling, preprocessing and decimation, finally urguing for orthonormal input preprocessing.

1 Introduction

In a very first step of input/output modelling of a physical plant one selects several variables named inputs and others outputs. Those variables are functions of time and they are usually called signals. The process of modelling requires expressing of dependencies between the input and output variables. Usual building blocks for modelling consist of specially selected static or linear dynamical systems and universal function approximators. Contemporary, the role of funtion approximators is usually fulfilled by neural networks. Those mentioned preselected systems may be viewed as preprocessors for neural networks [1].

The neural feedforward architecture is know to be universal function approximator. Even with only one-hidden layer, it can approximate any function (from a class) with any degree of accuracy, provided that sufficiently large number of hidden units is used, [2]. So a natural question arises: why input preprocessing is needed when the feedforward neural architecture is used? There are several answers to it.

Traditionally input preprocessing was used to scale input signals. There were some intuitive and formal arguments for that, but practical results was that with scaled signals, the learning process was significantly faster. This kind of preprocessing is widely used and is implemented via linear transformation.

But really important input preprocessing has to be done if the input and output signals have significantly different spectrum and they cannot be effectively sampled with the same sampling time [3].

The third role of input preprocessing is to transform a set of input signals which cannot span desired output (the output is not described by a function of inputs), into a new set of signals which exactly or approximately can. The input signals are preprocessed thus by a dynamical system. The simplest and the most widely used is a line of delays, others could be considered [1], [4], [5] as well.

In this paper data preparation for neural based modelling is considered. We will be examining sampling, preprocessing and decimation processes. The concept of almost linearly dependent signals will be introduced and used. Finally some arguments for orthonormal input preprocessing will be presented.

This paper is organized as follows. In Section 2 general scheme of data preparation for neural modelling is presented. In Section 3 almost linearly dependent signals concept is formulated and discussed, then in Section 4 sampling process is examined. Dynamic preprocessing is briefly discussed in Section 5. Finally conclusions are drawn in Section 6.

2 Neural network teaching data preparation

Majority of processes modelled by using neural networks are continuous in time. Process variables carrying information about their inputs $u(t)$ and outputs $y(t)$ are continuous-time signals. Throughout this paper we will be considering only one output signal. Continuous-time signals before being used must be sampled with sufficiently high sampling rate. Since that a finite number of samples represent the signals. The process of data preparation for neural networks based input/output modelling is shown in Fig. 1. The $U^{'}$, $U^{''}$, $Y^{'}$, U and Y are matrices of numbers. Along each column of $U^{''}$ and $Y^{'}$ matrices the samples of input and output signals are located. Thus, the preprocessing can be applied to the matrix $U^{''}$ columns resulting with $U^{'}$.

The signal samples contained in the columns of $U^{'}$ and $Y^{'}$ matrices are characterized by the sampling time T. For U and Y the sampling time is n times larger, where n is a decimation constant. Number of columns in U' is usually larger than in $U^{''}$. This is due to pre-

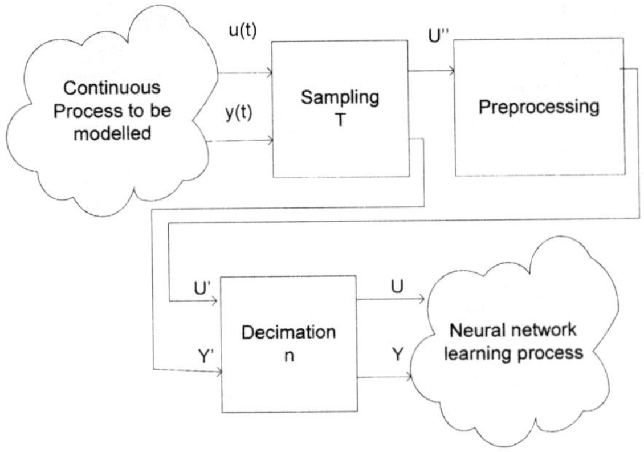

Fig. 1. The process of data preparation for neural modelling.

processing. In order to be able to approximate the relationship between U' and Y' columns by a feedforward neural networks the relationship must be static i.e. be described by a function $U' \rightarrow Y'$. More precisely let $U' \in \Re^{N+1,m}$, $Y' \in \Re^{N+1,1}$ such that

$$U' = \begin{bmatrix} u'(0) \\ u'(1) \\ \cdots \\ u'(N) \end{bmatrix} \text{ and } Y' = \begin{bmatrix} y'(0) \\ y'(1) \\ \cdots \\ y'(N) \end{bmatrix} \quad (1)$$

then the set of input/output pairs $\left\{(u'(i), y'(i))\right\}_{i=0}^{N}$ should be a function. The role of preprocessing block is to ensure that. At least for some type of preprocessing (for instance line of delays), allowed sampling time is bounded from both sides [6], [7]. Too large and too small sampling time should be avoided. So usually after preprocessing, the samples are taken with much lower sampling rate. This is done via so called decimation process. An important property of this process is that, it does not destroy static relationship between the data set as demonstrated in the following Proposition.

Proposition 1 *If the relationship between U' and Y' is static then the relationship between U and Y is also static.*

Proof: The U and Y matrices are created from U' and Y' by deleting certain rows from both or in other words deleting certain pairs from the set $\left\{(u'(i), y'(i))\right\}_{i=0}^{N}$. If relationship between U' and Y' is static i.e. it is a function, it means that for any i and j, $0 \leq i, j \leq N$, there does not exist such situation that $u'(i) \neq u'(j) \rightarrow y'(i) = y'(j)$, so if some pairs from the set $\left\{(u'(i), y'(i))\right\}_{i=0}^{N}$ are deleted, such situation does not exist either. ■

3 Continuous-time signals: almost linearly dependent signals

Let assume that the system to be modelled is excited by the input continuous signals denoted as $u(t)$, $t \in [0, t_{\max}]$ being collection of m measurable signals

$$u(t) = (u_1(t), ..., u_m(t)) \quad (2)$$

$u_i(t) \in \mathcal{L}^2$, $i = 1, ..., m$.Similarly assume that the output signal is $y(t)$, $y(t) \in \mathcal{L}^2$.

Let construct the following matrix

$$\Gamma_c = [< u_i, u_j >]_{i,j=1,...,m} \quad (3)$$

where $< .,. >$ denotes scalar product which for $\mathcal{L}^2$ space is usually defined as

$$< u_i, u_j >= \frac{1}{t_{\max}} \int_0^{t_{\max}} u_i(t)u_j(t)dt, \; i, j = 1, ..., m \quad (4)$$

Matrix Γ_c property is used to define linear independence of signals. Is said that the signals $u_1(t), ..., u_m(t)$ are linearly independent if matrix (3) is nonsingular i.e. $rank(\Gamma_c) = m$.

In practice signals are measured with errors and computer operations are error prone. If two signal samples are located well within measurement error they can be treated as indistinguishable [6] reducing rank of the matrix Γ_c. Here it is a certain generalization of that concept and further development.

Let select only two input signals. Two opposite situations might be distinguished. The first one takes place when two signals are very different in their amplitudes. We may put such case into mathematical framework $||u_2|| << ||u_1||$. In engineering practice u_2 is then often neglected. In the second case the signals are the same magnitude say $||u_2|| = ||u_1|| = 1$ and thy are very closed one to another, say $||u_1 - u_2|| \leq \varepsilon$, where ε is a small number.

If $||u_1 - u_2|| = 0$, then $u_1 = u_2$ and both of signals are carrying the same information and the one is useless. If however distance between two signals is not zero but a very small, in practice the two signals could be treated as indistinguishable from measurement point of view. One may say $\varepsilon-$indistinguishable [8]. For tapped delay neural architecture these two signals fulfill the following condition $u_2(t) = u_1(t+T)$ where T is a sampling time. Thus for small sampling time $u_1(t)$ and $u_1(t+T)$ become indistinguishable.

If one considers general input preprocessing involving various dynamic transformations, also $u_2(t) = -u_1(t)$

should be avoided. In such case of course the two signals are distinguishable, but linearly dependent. Let note that

$$||u_1 - u_2||^2 = \begin{cases} 0 \text{ when } u_1 = u_2, < u_1, u_2 >= 1 \\ 2 \text{ when } < u_1, u_2 >= 0 \\ 4 \text{ when } u_1 = -u_2, < u_1, u_2 >= -1 \end{cases}$$

We need to place our observation in a more formal framework.

Definition 1 *Two normalized signals* u_1, u_2, $||u_1|| = ||u_2|| = 1$ *are almost linearly dependent with accuracy* δ, $0 < \delta < 1$, *iff* $|< u_1, u_2 >| \geq 1 - \delta$.

Note that for normalized signals u_1, u_2, the scalar product $< u_1, u_2 >$ expresses cosines of an angle between u_1 and u_2. So if two signals are approximately linearly dependent with accuracy δ, it means that the angle φ between the two fulfills condition $|\cos\varphi| \geq 1 - \delta$

If one denotes

$$\varphi_0 = \arccos(1 - \delta) \tag{5}$$

then φ_0 could be interpreted as "safety margin". If δ is a small number then $\varphi_0 \approx \delta$. To avoid liner dependence the angle φ, $\varphi \in [0, \pi]$ between two signals should be

$$\varphi_0 \leq \varphi \leq \pi - \varphi_0 \tag{6}$$

Proposition 2 *Two signals* $u_1, u_2 \in \mathcal{L}^2$ *and* $||u_1|| = ||u_2|| = 1$ *are almost linearly dependent with accuracy* δ, $0 < \delta < 1$,*iff*

1. $||u_1 - u_2||^2 \leq 2\delta$ *or* $||u_1 - u_2||^2 \geq 4 - 2\delta$, *or*
2. $cond(\Gamma_c) \geq \frac{2-\delta}{\delta}$

Proof: ad 1) Because $||u_1 - u_2||^2 = ||u_1||^2 - 2 < u_1, u_2 > + ||u_2||^2 = 2 - 2 < u_1, u_2 >$, then $< u_1, u_2 >= 1 - ||u_1 - u_2||^2 /2$ and $|< u_1, u_2 >| \geq 1 - \delta$ iff $||u_1 - u_2||^2 \leq 2\delta$ or $||u_1 - u_2||^2 \geq 4 - 2\delta$.

ad 2) Because Γ_c matrix is symmetric positive semidefinite then its singular values are the same as its eigenvalues. Eigenvalues of Γ_c matrix in this case are the following $\lambda_{1,2} = 1 \pm |< u_1, u_2 >|$ and $\lambda_{\max} = 1 + |< u_1, u_2 >|$ and $\lambda_{\min} = 1 - |< u_1, u_2 >|$. Thus $\lambda_{\max} \geq 2 - \delta$, $\lambda_{\min} \leq \delta$ and $cond(\Gamma_c) \geq \frac{2-\delta}{\delta}$. And vice versa, starting with the last condition we obtain easily definition property. ■

One can acknowledge from Proposition 2 that approximately linearly dependent signals will give large condition number of Γ_c matrix.

An interesting and important case is when two signals are shifted by sampling time i.e. $u_2(t) = u_1(t + T)$. Of course, when T is equal to zero both signals are indistinguishable.

Example 1 *Let* $u_1(t) = \frac{1}{\sqrt{2}} \sin(\frac{2\pi}{T_p} t)$ *and* $u_2(t) = \frac{1}{\sqrt{2}} \sin(\frac{2\pi}{T_p}(t + T))$.*where* T_p *denotes obserwational period and* T *is the sampling time. Let denote also* $f = \frac{1}{T_p}$. *Thus* $||u_1(t)|| = ||u_2(t)|| = 1$, $< u_1(t), u_2(t) >= \cos(2\pi fT)$ *and the two signals will be linearly independent with margin* $\varphi_0 = \arccos(1 - \delta)$, *if*

$$\frac{1}{2\pi} \arccos(1 - \delta) \leq fT \leq \frac{1}{2} - \frac{1}{2\pi} \arccos(1 - \delta) \tag{7}$$

4 Sampled signals

Contemporary measurement and information processing technology requires that the signals are sampled. Correctness and quality of modelling of a physical plant is also influenced by choice of sampling time.

Suppose that within range of time $[0, t_{\max}]$ we collected $u(k)$ and $y(k)$, $k = 0, 1, ...$signals, sampling continuous signals u and y with period T and achieving the following vectors

$$U = [U_1, ..., U_m] = \begin{bmatrix} u(0) \\ u(1) \\ ... \\ u(N) \end{bmatrix} \in \Re^{N+1,m} \tag{8}$$

where

$$U_i = \begin{bmatrix} u_i(0) \\ u_i(1) \\ ... \\ u_i(N) \end{bmatrix} \in \Re^{N+1,1}$$

$$u(k) = [u_1(k), ..., u_m(k)] \in \Re^{1,m}$$

$i = 1, ..., m$ and $k = 0, ..., N$.

Similarly

$$Y = \begin{bmatrix} y(0) \\ y(1) \\ ... \\ y(N) \end{bmatrix} \in \Re^{N+1,1} \tag{9}$$

Let define Γ_d matrix

$$\Gamma_d = [< U_i, U_j >]_{i,j=1,...,m} \tag{10}$$

where $< U_i, U_j >= \frac{1}{N+1} \sum_{k=0}^{N} u_i(k) u_j(k)$.

Note that $\Gamma_d = \frac{1}{N+1} U^T U$ and $rank(\Gamma_d) = rank(U)$.

The following two Propositions demonstrate that while sampling one can loose linear independence of continuous-time variables.

Proposition 3 *[8]* $rank(\Gamma_d) \leq rank(\Gamma_c)$

Proposition 4 $\lim_{T \to 0} \Gamma_d = \Gamma_c$

Proof: Let distinguish $N+1$ points in the range of time $[0, t_{\max}]$ enumerated $0, 1, ..., N$ as in (8). Thus the scalar product (4) can be approximated through sampled signals $< u_i, u_j >= \lim_{N\to\infty} \frac{1}{N+1} \sum_{k=0}^{N} u_i(k\frac{t_{\max}}{N+1}) u_j(k\frac{t_{\max}}{N+1})$ and note that $T \to 0$ as $N \to \infty$. ■

5 Dynamic preprocessing

A feedforwad neural network being the universal approximator may approximate well some complicated nonlinear functions. But a simple, single input/ single output lag to be approximated by such network - it is an impossible task. In such case a dynamic transformation of the input signals has to be used. This is shown in Fig. 2

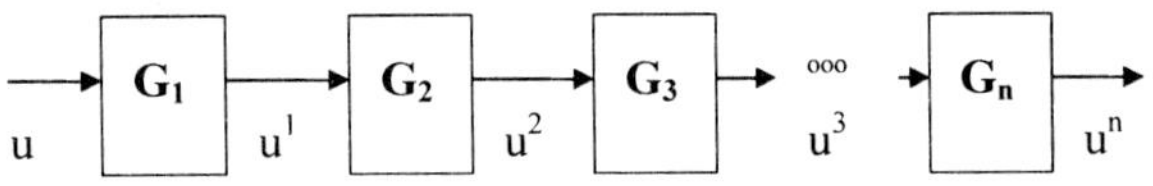

Fig. 2. A line of dynamic transformations of the input signal.

Instead of the input signal u only, the signal $X = (u, u^1, u^2, ..., u^n)$ will be an input to the neural net. We hope that now y could well be described as a function of X. In each case however it must be checked that a static relationship between given X and y exists [9].

As an input preprocessing for dynamic systems a line of delays could be used which together with the neural network form so called tapped delay neural architecture. In this case G_i from Fig. 2 is

$$G_i(j\omega) = e^{-j\omega T}, \quad i = 1, ..., n$$

and T denotes sampling time. Such tapped delay neural architecture has number of performance limitations [6].

A better way of input preprocessing is to introduce allpass transfer functions chain with the following cells

$$G_i(j\omega) = \frac{1 - j\omega T_i}{1 + j\omega T_i} = e^{-j2\arctan(\omega T_i)}$$

(or its discrete time version). Note that the magnitude of $G_i(j\omega)$ is always equal to 1.The time delay of each cell could now be controlled.

There are several reasons why to use input signals orthonormalization, so the signals from the blocks shown in Fig. 2 are mutually orthogonal i.e. $< u^i, u^j >= 0$.But from our considerations, it comes out directly that according to formula (6), for two signals orthonormality ensures the largest angular distance (i.e. $\frac{\pi}{2} - \varphi_0$) between two signals. It gives also the largest distance between the two signals i.e. $||u_1 - u_2|| = \sqrt{2}$.

A good preprocessing will generate a number of orthonormal signals inputting the neural net [5]. One has to noticed that the output signals are orthonormal only for pariculary particular inputs.

6 Conclusions

The process of data preparation for neural based modelling was presented and discussed. We have distinguished and examined some properties of sampling, input preprocessing and decimation, demonstrating particular usefulness of almost linear dependence signals concept. Finally our arguments for orthonormal preprocessing were presented.

References

[1] A H Chung Tsoi and A Back. Static and dynamic preprocessing methods in neural networks. *Engng. Applic. Artif. Intell.*, 8(6):633–642, 1995.

[2] A.Pinkus M.Leshno, V.Lin and S.Schocken. Multilayer feedforward networks with a nonpolynomial activation function can approximate any function. *Neural Networks*, 13:350–373, 1993.

[3] L M Grzesiak, B Beliczynski, and B Ufnalski. Input preprocessing in tapped delay neural architecture for induction motor speed estimation. In *Proc. EPE Conf.* European Power Electronics Conference, Toulouse 2003., EPE, 2003.

[4] P S C Heuberger and P M J Van Den Hof. A genelized orthonormal basis for linear dynamical systems. *IEEE Trans. on Autom. Contr.*, 40:451–464, 1995.

[5] B Nines and F Gustafsson. A unifying construction of orthonormal bases for system identification. *IEEE Trans. Automat. Contr.*, 42:515–521, 1997.

[6] B Beliczynski and L Grzesiak. Induction motor speed estimation: Limitations of certain neural schemes based on time instances of periodic signals. In *Proc. IECON'01*, pages 653–657. IEEE Service Center, 2001.

[7] B Beliczynski and L Grzesiak. Induction motor speed estimation: Neural versus phenomenological model approach. *Neurocomputing*, 43:17–36, 2002.

[8] B Beliczynski. On input discretisation processes for tapped delay neural architecture. In *Proc. ICANNGA'03*, pages 28–32. Springer, 2003.

[9] B Beliczynski and K Kubowicz. Appoximately static relationship between model variables. *Przeglad Elektrotechniczny*, 53(4):385–389, 2004.

A simple method for selection of inputs and structure of feedforward neural networks

H. Saxén and F. Pettersson

Heat Engineering Laboratory, Åbo Akademi University, Finland
E-mail: hsaxen@abo.fi

Abstract

When feedforward neural networks of multi-layer perceptron (MLP) type are used as black-box models of complex processes, a common problem is how to select relevant inputs from a large set of potential variables that affect the outputs to be modeled. If, furthermore, the observations of the input-output tuples are scarce, the degrees of freedom may not allow for the use of a fully connected layer between the inputs and the hidden nodes. This paper presents a systematic method for selection of both input variables and a constrained connectivity of the lower-layer weights in MLPs. The method, which can also be used as a means to provide initial guesses for the weights prior to the final training phase of the MLPs, is illustrated on a class of test problems.

1 Introduction

When neural networks of multi-layer perceptron (MLP) type are used in black box modeling, e.g., for prediction of variables in a complex industrial process, a frequently occurring problem is that there are numerous factors that may potentially influence the variables to be modeled (predicted). Furthermore, it is also common that the number of observations is limited. The second problem has the consequence that the number of inputs and/or the number of hidden nodes has to be restricted to yield a meaningful parameter estimation problem (with, say, the number of network weights not exceeding a tenth of the number of output residuals [1]). This also means that all potential inputs cannot be included in the model. A remedy would be to limit the dimension of the input space by considering only meaningful variables. If knowledge of the system studied exists, it is possible to use such for the elimination of superfluous inputs or for preprocessing of the inputs, thus reducing the input variable dimension. In cases where there is no such knowledge, one has to resort to automatic methods for selection of relevant input variables. In nonlinear modeling, there are no general criteria for making such choices, even though papers have been published on how to tackle the problem [2,3]; an excellent review on pitfalls in measuring the importance of inputs is provided in [4]. A way to tackle the problem is to use constructive methods based on a growing neural network [5,6], but these techniques are often not well suited for problems where the input variables are correlated. Furthermore, their ability to solve problems with a low signal-to-noise ratio is also limited. On the other hand, pruning of large trained networks have also been proposed as a remedy [7,8], and in more recent papers the possibility to use genetic algorithms for a simultaneous optimization of network weights and structure has been explored [9-11]. However, most pruning methods are hampered by a laborious retraining process. Furthermore, the risk of getting stuck in local minima, which is always present in MLP training, may result in "wrong decisions" concerning the usefulness of certain network connectivities.

The present paper proposes a simple pruning approach, where a large single-layer sigmoid network with random initial weights in the lower layer of connections is used as a starting point. The complexity of this network part is gradually decreased by removing, on each iteration, the least significant connection. In comparing the networks with each other, the upper layer weights are determined by linear least squares. This makes the method simple, efficient, rapid and robust. The method is described in the next subsection, followed by an illustration of it in Section 3. The final section presents some concluding remarks.

2 The Method

The method proposed in this paper is based on the following assumptions: We confine our study to feedforward neural networks of MLP type with a single layer of hidden nonlinear (typically sigmoidal) units and a single linear output node. The former limitation is motivated by the fact that such networks have been shown to be able to approximate any continuous differentiable function to arbitrary accuracy, if the number of hidden nodes is large enough, while an extension to networks with multiple outputs is obvious.

The approach is based on the practical observation that for an arbitrary, but known, choice of weights in the lower layer of connections, **W**, (cf. Fig. 1) there are generally a corresponding set of weights, **w**, to the output node that will lead to a relatively good solution of the approximation problem at hand. After propagating the K input vectors through the first layer of connections and through the hidden nodes, the corresponding outputs of each of the n hidden nodes, can be determined. Collecting these together in a matrix, $\mathbf{Z}$, where a first column of ones (for the bias) is included, the upper-layer weights can be easily determined by solving the linear problem

$$\min_{\mathbf{w}}\left\{F=\sqrt{\tfrac{1}{K}\left\|\mathbf{f}-\hat{\mathbf{f}}\right\|^2}\right\} \quad (1)$$

where

$$\hat{\mathbf{f}}=\begin{bmatrix}\hat{f}_1\\ \hat{f}_2\\ \hat{f}_3\\ \vdots\\ \hat{f}_K\end{bmatrix}=\begin{bmatrix}1 & z_{1,1} & z_{1,2} & \cdots & z_{1,n}\\ 1 & z_{2,1} & z_{2,2} & \cdots & z_{2,n}\\ 1 & z_{3,1} & z_{3,2} & \cdots & z_{3,n}\\ \vdots & \vdots & \vdots & & \vdots\\ 1 & z_{K,1} & z_{K,2} & & z_{K,n}\end{bmatrix}\begin{bmatrix}w_0\\ w_1\\ w_2\\ \vdots\\ w_n\end{bmatrix}=\mathbf{Zw} \quad (2)$$

by, e.g., Householder reflections using an orthogonal-triangular factorization [12].

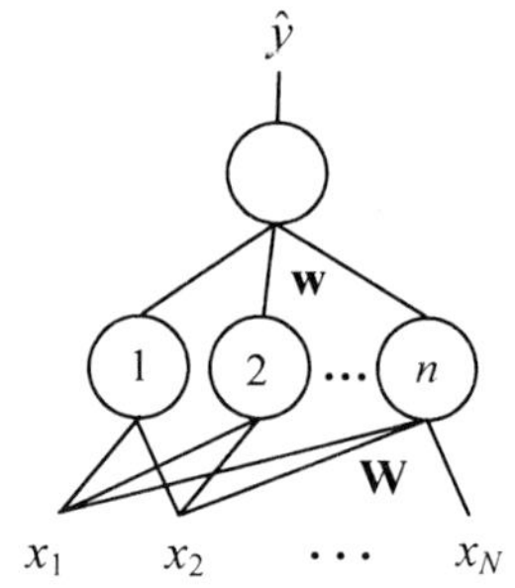

Figure 1. Schematic of the networks used in the study.

The fact that an arbitrary (but well scaled) weight matrix $\mathbf{W}$ is sufficient is also in agreement with the seemingly odd observation, reported by numerous investigators, that a large network trains more rapidly than a small network, despite its larger weight space. The reason is that in a large network the likelihood increases of initially finding hidden nodes that operate in a proper region for solving the problem at hand. Obviously, such a large network is over-sized in a parametric sense, so it should be possible to remove superfluous connections in its lower part without major loss of accuracy of the fit. This brings us into the basic pruning step of the method that in a nutshell can be condensed into the following algorithm:

1. Select a (sufficient) maximum number of hidden nodes, n, and generate a random weight matrix, $\mathbf{W}_0$, for the lower layer of connections. Introduce a set of indices to its weights, I_0 (excluding the biases), and set the iteration index to $i = 1$.
2. Reset, in turn, a weight j in the set I_{i-1}, and determine the corresponding optimal upper-layer weights $\mathbf{w}_j$ by Eqs. (1) and (2) and determine the value of the objective function, $F_{i,j}$.
3. Set $\mathbf{W}_i = \mathbf{W}_{i-1}$ and $\mathbf{I}_i = \mathbf{I}_{i-1}$. Remove the weight, j_i^* that corresponds to the minimum value of the objective function, i.e., $j_i^* = \arg\min_{\mathbf{w}}\{F_{i,j}\}$ from both $\mathbf{W}_i$ and $\mathbf{I}_i$.
4. Set $i = i+1$. If $i < n{\cdot}N$, go to 2. Else, end.

3 Illustration of the Method

In order to illustrate the method outlined above, the following function with $N = 3$ inputs, (x_1 , x_2 and x_3) and one output (y) is used.

$$y = a\left(x_1^2 + 0.5x_1x_2\right) + (1-a)\left(0.5x_2x_3 + x_3^2\right) + b\varepsilon \quad (3)$$

Obviously, the function is designed to yield a varying and nonlinear dependence between inputs and the output for different values of the parameter $a \in (0,1)$: The output, y, is independent of x_3 for $a = 1$, for $a = 0$ it is independent of x_1, while for $0 < a < 1$ it depends on all input variables.

A few runs of the algorithm next illustrate its possible use. In the analysis the inputs as well as the noise term, ε, are taken to be normally distributed random variables with zero mean and unit variance, i.e., ε, x_i = $\mathcal{N}(0,1)$, while the non-negative parameter b is used to control the signal-to-noise ratio. The analysis is started from a network with six hidden nodes, which was considered sufficient for the task at hand.

3.1 Reference case

The model was first run with $K_{tr} = 100$ observations in the training set, using an additional $K_{te} = 100$ observations in a test set for verification of the resulting models, using the parameter values $a = 0.5$ and $b = 0.2$. Figure 2 shows an example of the inputs and the output of a training set.

Figure 3 illustrates a typical evolution of the method on the training (solid lines) and test (dashed lines) sets, where the root-mean square errors F_{i,j^*} (cf. Eq. (1)) have been depicted as a function of the remaining weights (excluding biases) in the lower part of the networks. Note that the "iterations" progress from right to left in the figure. In summary, five to seven non-zero weights (to five active hidden nodes) are sufficient to produce an acceptable fit, and the corresponding networks generalize well on the test set. This is also seen in Fig. 4, which shows the resulting fit on the test set for the network with five remaining lower-layer weights.

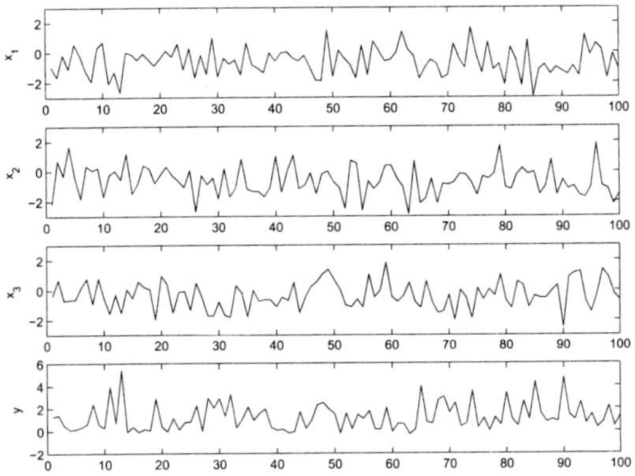

Figure 2. Inputs and output in a training set of the reference case.

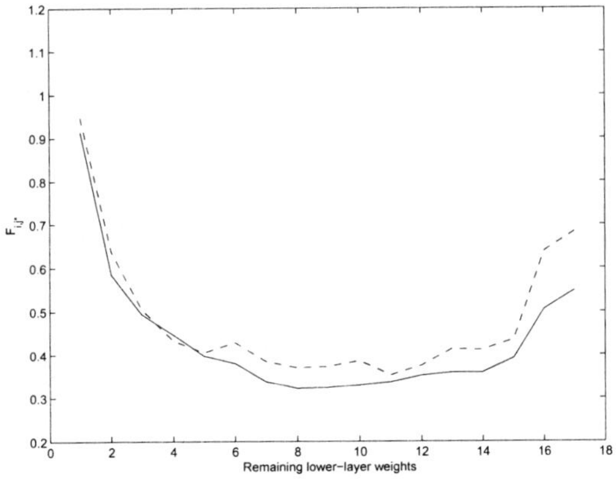

Figure 3. Evolution of the errors with the number of remaining weights (excluding biases) in the lower part of the network (—— training set, - - - test set).

Figure 5 shows the errors on the training set for networks evolved from four random weight matrices. Even though the levels of the errors are seen to differ between the networks, the general features are very similar, especially as far as the minimum connectivity for achieving a good fit is concerned. A noteworthy fact is also that the fits for the largest networks are worse than those of the somewhat pruned networks, even on the training set. Thus, superfluous lower-layer connections can be directly detrimental for the model. This is the result of the fact that the lower-layer weights are not retrained, but it still serves to illustrate the impact of the initial weights on the conditioning of the training problem. Therefore, the proposed technique can also be seen as a method for initializing a (sparse) network.

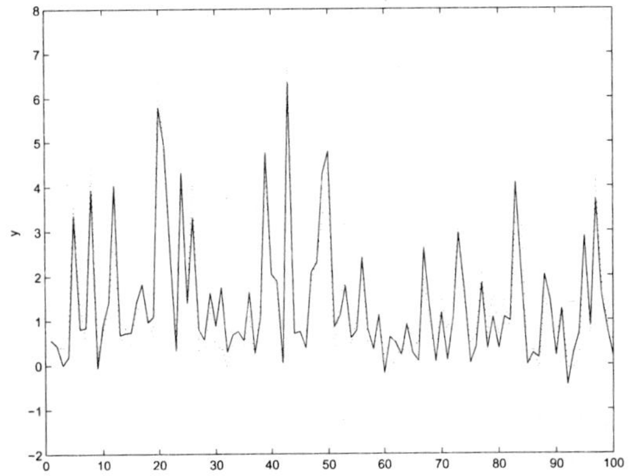

Figure 4 Example on the model fit on the test data using five non-zero weights in the lower part of the network (—— observations, ······ model).

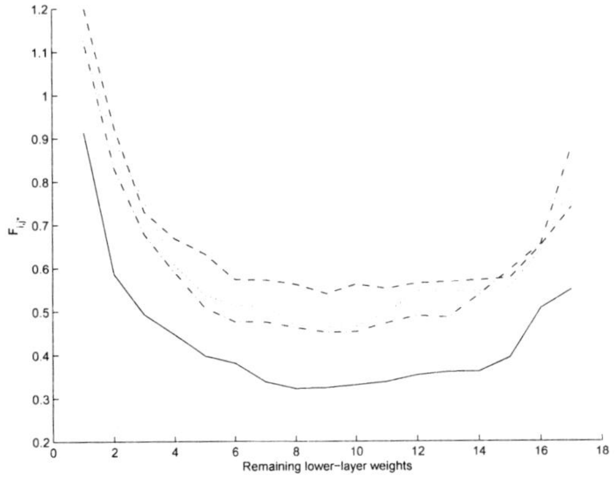

Figure 5. Evolution of the training errors with the number of remaining weights (excluding biases) in the lower part of the networks from four random weight matrices.

3.2 Detection of relevant inputs

The capability of the model to detect and remove irrelevant inputs is next evaluated by creating a data set with identical parameters to those of the previous subsection, except $a = 0.1$. This gives rise to a function y that only slightly depends on x_1. The algorithm was used to detect the one of the inputs that would be first eliminated, i.e., would lose all its connections to the hidden layer.

Starting from 20 random initial weight matrices, $\mathbf{W}_0$, the first input variable to be eliminated was always x_1. Table 1 shows the "frequency" of the number of remaining weight connections at the point where the final connection to x_1 was excluded. Even though there is some scattering, it is interesting to note that the required lower-layer complexity at this point corresponds quite well to the one required for solving the task with $a = 0.5$. Another interesting observation is that the generalization ability of the networks at these very points turns out to be close to optimal for many of the runs. Figure 6 illustrates this behavior for four of the runs, where the test set errors have been depicted. The arrows in the figure indicate the network complexity where the first input variable (x_1) was excluded from the model. Moving from left to right, these are the points where a further added complexity does not improve the function approximation provided by the network.

Table 1. Frequency of remaining lower-layer weights at the point where the first input (x_1) was eliminated by the pruning method for the function (3) with $a = 0.1$.

Remaining weights	Frequency
3	2
4	5
5	4
6	6
7	1
8	1
9	1

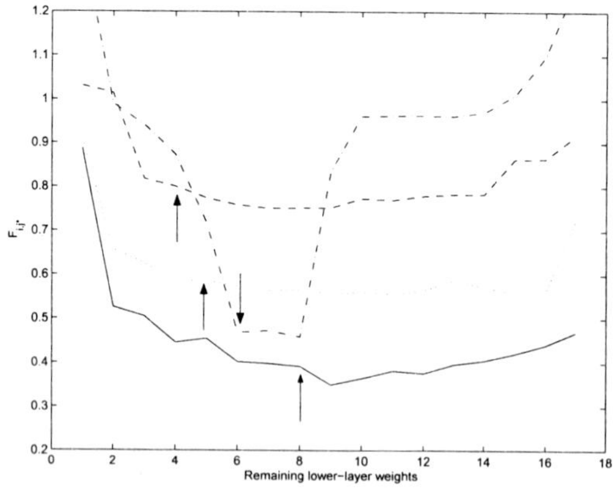

Figure 6. Test set errors of four networks trained on y generated by Eq. (3) with $a = 0.1$. Arrows denote the points where x_1 was excluded from the model.

4 Conclusions

The paper has described a systematic method for selection of both input variables and a constrained connectivity of the lower-layer weights in MLPs. It also provides initial guesses for a gradient based training. The method has been illustrated on a class of test problems, where it has shown promising performance.

REFERENCES

[1] Principe J. C., N. R. Euliano and W. C. Lefebvre, (1999) *Neural and adaptive systems: Fundamentals through simulations*, John Wiley & Sons, New York.

[2] Sridhar, D.V., E. B. Bartlett and R. C. Seagrave, (1998) "Information theoretic subset selection for neural networks", *Comput. Chem. Engng.* **22**, 613-626.

[3] Bogler, Z., (2003) "Selection of quasi-optimal inputs in chemometrics modeling by artificial neural network analysis", *Analytical Chimica Acta* **490**, 31-40.

[4] Sarle, W.S., (2000) "How to measure importance of inputs", ftp://ftp.sas.com/pub/neural/importance.html

[5] Frean, M., (1989) "The Upstart Algorithm. A method for Constructing and Training Feed-forward Neural Networks", Edinburgh Physics Department, Preprint 89/469, Scotland.

[6] Fahlman, S.E. and C. Lebiere, (1990) "The Cascade-Correlation Learning Architecture", in *Advances in Neural Information Processing Systems* **II**, (Ed. D.S. Touretzky), pp. 524-532.

[7] Le Chun, Y., J. S. Denker and S. A. Solla, (1990) "Optimal Brain Damage", in *Advances in Neural Information Processing Systems* **2**, ed. D.S. Touretzky, pp. 598–605, (Morgan

[8] Thimm, G. and E. Fiesler, (1995) "Evaluating pruning methods", Proc. of the *1995 International Symposium on Artificial Neural Networks (ISANN'95)*, Hsinchu, Taiwan, ROC.

[9] Maniezzo, V., (1994) "Genetic Evolution of the Topology and Weight Distribution of Neural Networks", *IEEE Transactions on Neural Networks* **5**, 39–53.

[10] Gao, F., M. Li, F. Wang, B. Wang and P. Yue, (1999) "Genetic Algorithms and Evolutionary Programming Hybrid Strategy for Structure and Weight Learning for Multilayer Feedforward Neural Networks", *Ind. Eng. Chem. Res.* **38**, 4330-4336.

[11] Pettersson, F. and H. Saxén, (2003) "A hybrid algorithm for weight and connectivity optimization in feedforward neural networks", in *Artficial Neural Nets and Genetic Algorithms* (eds. Pearson, D. et al.), pp. 47-52, Springer-Verlag.

[12] Golub, G., (1965) "Numerical methods for solving linear least squares problems", *Numer. Math.* **7**, 206–216.

The Concept and Properties of Sigma-if Neural Network

M. Huk, H. Kwasnicka
Department of Computer Science, Wroclaw University of Technology, Poland
E-mail: {maciej.huk, halina.kwasnicka}@pwr.wroc.pl

Abstract

Our recent works on artificial neural networks point to the possibility of extending the activation function of a standard artificial neuron model using the conditional signal accumulation technique, thus significantly enhancing the capabilities of neural networks. We present a new artificial neuron model, called Sigma-if, with the ability to dynamically tune the size of the decision space under consideration, resulting from a novel activation function. The paper discusses construction of the proposed neuron as well as training Sigma-if feedforward neural networks for well known sample classification problems.

1 Introduction

The basic constituent of a classic artificial neural network (ANN) is the neuron, used to process signals presented as input, through an activation function and a nonlinear output (threshold) function. The former function, hereafter referred to as *A*, determines the activation level of the neuron, while the latter (*F*) bases on the result returned by *A* to construct the neuron's output value. [1,2] The importance of the threshold function as well as of the weights attached to individual interneural connections is well discussed in numerous publications. [1,3,4,5] However almost all of them assume the activation function to be a simple weighted sum of neuron input values and connections weights. In fact, analyzing the behavior of artificial neurons with nonstandard activation functions may lead us to develop new classification structures, with interesting and useful properties. [6-13]

2 Sigma-if Neuron

The authors propose a special type of neuron, whose activation function can be clearly interpreted from a biological perspective. In the case of real neurons, individual dendrites differ in length, allowing a biological neural network to associate incoming signals with particular connections and processing areas. Such a property can be incorporated into classic feedforward neural network models through connections grouping and conditional signals accumulation technique. [11,14]

More specifically, the *M* dendrites of a Sigma-if neuron are divided into *K* distinct groups, by complementing each *i*-th input connection with an additional integer parameter $\theta_i \in \{0,1,..,K\!-\!1\}$, determining membership in one of the groups. This allows us to divide the process of signals accumulation into *K* steps, where *K* is a function of neuron's grouping vector $\theta^T=[\theta_1, \theta_2, ..., \theta_M]$:

$$K(\theta) = \max_{i=1}^{M}(\theta_i). \quad (1)$$

During each step *k* (from 0 to *K*-1) the neuron accumulates data belonging to one selected group, such that

$$\theta_i = k. \quad (2)$$

Within each *k*-th group, partial activation $\Delta g(k)$ is determined as a weighted sum of input signals and the appropriate Kronecker's delta:

$$\Delta g(k, w, x, \theta) = \sum_{i=1}^{M} w_i x_i \delta(k, \theta_i), \quad (3)$$

where w_i and x_i are coefficients of the neuron's weight vector *w* and input vector *x*. This process is repeated until the activation derived from respective groups exceeds a preselected activation threshold net^*. It can be described by the following recursive formula (vectors *w*, *x* and θ are omitted for clearness):

$$net(k)=\begin{cases}\Delta g(k)H(net^*-net(k-1))+net(k-1) & : k\geq 0\\ 0 & : k<0\end{cases} \quad (4)$$

where *H* is Heaviside's function. This sum is then treated as the neuronal activation value. Input from remaining (heretofore unconsidered) groups is neglected. Thus, the proposed form of activation function *A* is:

$$A(w, x, \theta) = net(K, w, x, \theta). \quad (5)$$

For completeness, it is also important to note that in the final stages of determining the output value *Y* of the neuron, function (5) serves as a parameter of the nonlinear threshold function *F*:

$$Y(w, x, \theta) = F(A(w, x, \theta)). \quad (6)$$

Neurons of the presented type can easily be used for building network structures. One can choose architectures similar to classic synchronous feedforward neural networks, but recurrent realizations are also possible. While there are no special restrictions regarding network architecture, Sigma-if neurons can work in fully-connected as well as in sparse structures.

3 Training the Sigma-if Neural Network

The Sigma-if network requires a suitable training procedure. To characterize the influence of neuron modifications on network capabilities, we have separated the problem of grouping vector coefficients selection from that of connections weights calculation.

While in our study we use fully-connected feedforward Sigma-if networks with one hidden layer, the weights vector can be determined through a slightly modified back propagation algorithm. Its modification involves adding an assertion stating that during each weight adjustment cycle, only those weights can be changed, which have recently influenced the output values of the network. The reason for such an approach is as follows: the training algorithm should limit the influence of immediately-recognizable patterns on network structures used in categorizing data which rely on connections with greater θ_i values.

The selection of coefficients of the grouping vector should also be adjusted to the problem considered by the network. Yet, looking for an optimal grouping vector is generally a very difficult task. In practice it involves computationally expensive multidimensional and multimodal optimization. Thus use of reasonable heuristics is hence justified.

Following preliminary experiments we have decided to use the random walk technique. When the random grouping vector selected at the beginning of the learning process doesn't enable the backpropagation algorithm to reduce the network classification error below an assumed target level, we randomly select all theta values again. This process is repeated for every hundred backpropagation cycles until the network is successfully trained. Experience shows that the typical number of grouping vector selection attempts required to properly train the network is on the order of 10. Considering that, and remembering that our main goal was to check the basic properties of the Sigma-if model, the proposed solution seems to be acceptable.

4 Properties of the Sigma-if network

It is worth mentioning that proper selection of connection θ_i parameters between the input and hidden layers is very important. If the selection of grouping vector coefficients is highly disadvantageous, some highly important data attributes served through connections with high θ_i values, may end up not being considered at all. This is a particularly pressing issue when the low-theta connections carry strong noise, exceeding the activation threshold net^* of the neuron. Nevertheless, it is not a hopeless situation, due to another property of the Sigma-if network. It can perform reactivation of inactive and important attributes through minimization of active connection weights. This is achieved through the back propagation algorithm by lowering the weights of dendrites which distort the classification process.

The above property suggests that diversification of theta values may have other positive consequences. It enables neurons to separate disruptive input signals from those that carry useful information, thus increasing the signal-to-noise ratio. This can act as very effective noise filter, but only when noised connections have greater theta values than inputs required for proper classification

The next consequence of neuronal inputs grouping is that simple analysis of the activity of input to hidden layer connections in a properly trained network may yield information about subsets of data attributes important for the classifier. This analysis is reduced to observing which neuronal inputs are considered when establishing the network output as a response to a particular test input patterns.

However the most important feature of the proposed network is the ability to discriminate the input space in an adaptive manner. The conditional signal accumulation technique enables it to partition the data space with hypersurfaces using an increasing number of dimensions, where the number of attempts is determined by the number of distinct theta values assigned to neuronal inputs. This can easily be observed in the case of a single Sigma-if neuron. Despite the use of a sigmoid threshold function in its body, the Sigma-if neuron - unlike the classic neuron - can solve simple linearly inseparable problems, since it is able to use (depending on the circumstances) all or just some of the information present on its inputs.

It should, however, be noted that the presented approach fails for the XOR function. In this case, each straight line perpendicular to one of the dimensions of the data space and passing through a selected point corresponding to a training pattern from class zero contains a point belonging to a class different than zero. It is therefore impossible to separate both points by means of straight lines perpendicular to selected normal vectors of data spaces (hypersurfaces reduced to one dimension). In this case, a different solution may be utilized: rotating the coordinate set by a preselected acute angle. Following such a transformation, a single Sigma-if neuron will be able to properly classify points defining the XOR function.

5 Results of Experiments

The main goal of our experimental study was to find evidence for the basic theoretical expectations concerning the presented Sigma-if neuron model. The second goal was to compare classification capabilities of Sigma-if and classic neural networks, and to show that conditional signal accumulation technique can be useful in data mining applications. At the end we wanted to check if there are any differences between classic and Sigma-if networks in the context of knowledge extraction.

The tested networks were trained using the previously described modified back propagation algorithm, with randomly selected theta values, controlled by a separate algorithm which oversaw the space of convergence of the learning process. Behind tests with single neuron, the number of neurons in the hidden layer was set to the value assuring best training results for the classic network model. All neurons used a bipolar sigmoid threshold function and the activation threshold level was set to a value of net^*=0.4.

5.1 Single Sigma-if Neuron

The basic functionality test of the Sigma-if network involved gauging the properties of a single neuron. According to theoretical analysis, this neuron is not able to properly dissect the data space of the XOR problem (over the real number space). However, even a slight repositioning on one of the points defining the data space enables rapid training of the neuron. Such a modification allows the Sigma-if neuron to dissect the decision space with two different hypersurfaces, one of which must be a straight line perpendicular to one of the dimensions.

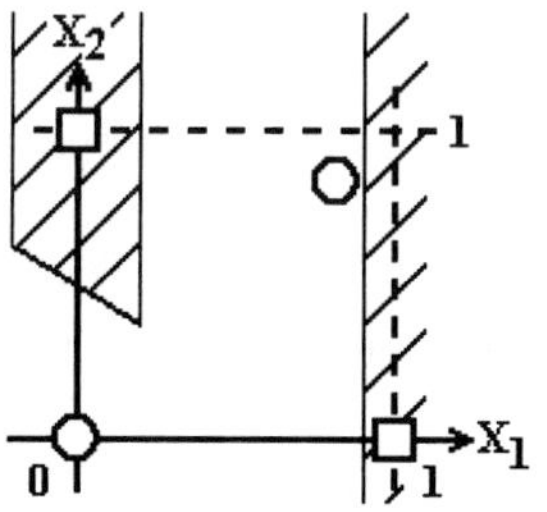

Fig. 1. Sample shape of trained Sigma-if neuron decision borders for the modified XOR function.

Fig. 1 presents a fragment of the decision space of a sample Sigma-if neuron trained to properly classify points belonging to the following function:

$$D(x_1, x_2) = \begin{cases} 0 & : (x_1, x_2) \in \{(0,0), (0.8, 0.8)\} \\ 1 & : (x_1, x_2) \in \{(0,1), (1,0)\} \end{cases} \qquad (7)$$

The parameters of this trained neuron are as follows: weight vector w =[1.3, 2.5] and grouping vector θ=[0, 1]. Thus since θ_1 is less than θ_2, input x_2 will be considered only if the partial neuron activation $\Delta g(0)$ (equal to the product of w_1 and x_1) is less than the activation threshold net^*. It can be seen that when the neuron's output values less or equal to 0.5 and greater than 0.5 are treated as classes 0 and 1 respectively, the Sigma-if neuron with the presented parameters correctly solves the linearly inseparable problem defined by (eq. 7).

This experimental confirmation of the theoretical potential of the Sigma-if neuron is further strengthened by achieving positive results for a training set which includes XOR function points, rotated by 45° around point (1,0). It is , however, important to note that training results depend on the activation threshold level. For the presented problem positive results were obtained only for net^* between 0 and 1. Outside that range, successful training was not possible.

5.2 Sigma-if performance for artificial problems

The promising results described in the previous section have led us to check the performance of the proposed model for more complicated artificial problems. For this purpose we have chosen the well known nested spirals testbed, and the two-class 10x10 checkerboard classification task. [15,16] Both problems have been used to train a fully-connected classic neural network with two inputs, 50 neurons in one hidden layer and two outputs. The obtained results have been compared with the outcome of analogous tests of a Sigma-if neural network with an identical architecture.

Test results have shown that in the case of the nested-spirals testbed, the Sigma-if network acts very similarly to the standard feedforward neural network. Both networks have been able to properly solve the two-spirals problem at similar computational cost. The differences between both types of neural networks only became apparent during the second test. While the standard model was unable to reduce the classification error below 50%, the Sigma-if network reached a stage where almost 70% of patterns were classified correctly.

5.3 Minimization of the number of active attributes

Another experiment involved training the Sigma-if network to minimize the number of attributes required for proper classification of selected data sets from the UCI Machine Learning Repository. This necessitated extending the mechanism which controlled the random selection of grouping vector parameters with facilities for analyzing changes in the number of active input connections. Thus, when the number of active input attributes exceeds 50%, the training algorithm forces random selection of the grouping vector.

Such a mode of operation, for a limited number of neurons in the hidden layer (2 to 4, depending on the training set), enabled us to extract subsets of data attributes which had the greatest impact on the classification process. Table 1 presents the capabilities of the process, assuming at least 80% classification accuracy with the limited attribute set. The best results have been achieved for those training sets which can be expected to contain a lot of redundant data (i.e. Sonar and Breast-Cancer-W). Surprisingly, for the latter set it was possible to reduce the number of active attributes to just two, while retaining a 92% accuracy (the two relevant attributes are Uniformity-of-Cell-Shape and Single-Epithelial-Cell-Size).

Table 1. Trimming data attribute sets with Sigma-if network.

Training set	No. of attributes (total)	No. of attributes (limited set)	Gain [%]
Heart	13	9	30.7
Hypothyroid	29	19	34.5
Iris	4	3	25.0
Breast-Cancer-W	9	2	77.7
Monk1	6	6	0
Monk2	6	6	0
Monk3	6	6	0
Sonar	60	10	83.3
Vote	16	9	43.7
		Mean	32.7

On average, the conducted tests permitted a 30% reduction in the number of attributes used in classification. We can therefore conclude that the proposed solution is well adapted to real-life applications, where data gathering can often be costly.

5.4 Extraction of knowledge from the Sigma-if network

The promising results of the above experiments raise questions about whether such networks process and store knowledge in a way that differs from classic neural networks. To shed some light on that problem, it is necessary to use a knowledge extraction method. The task of extracting knowledge from such an atypical neural network as the Sigma-if network essentially limits the selection of applicable algorithms to black-box-type methods. Such methods are scarce, so we have come to rely on the Trepan algorithm, since the available sources strongly recommend it. [17, 18]

Just like in the previous experiments, the results obtained are very interesting. It appears that, when compared to decision trees extracted from classic neural networks, the Sigma-if trees are much more readable - they are less complex and use a smaller number of decision attributes. They also more accurately represent the functioning of the network from which they have been derived as well as the properties of data being classified. It seems advisable to further study the possibility of applying Sigma-if networks in data mining applications.

6 Summary

The theoretical considerations presented in this paper and confirmed by experimental results clearly point to the fact that conditional signal accumulation is a useful technique in the area of neuronal data processing. In light of its potential benefits, the proposed modifications of classic artificial neural networks require further extensive study. It is, for example, difficult to explain why Monk training sets have proven more difficult to classify, even though they also include redundant data attributes.

It is also worth to underline that full exploitation of the Sigma-if model's potential is possible only when a suitable method of selecting grouping vector coefficients is used. The heuristic approach, mentioned earlier, has yielded some interesting results, but no definitive solution can yet be presented. This problem determines further directions of Sigma-if network research.

While the Sigma-if neuron is only a simple computational model, not designed for modeling biological neurons, it would nevertheless be interesting to research the relations between the properties of Sigma-if networks and those of biological processing systems.

7 Acknowledgements

The above research in the area of intelligent information systems and machine learning is funded by the Polish Ministry of Scientific Research and Information Technology within the 4 T11C 02425 research grant. This support is gratefully acknowledged.

References

[1] Duch, W., Jankowski, N. (1999) Survey of neural transfer functions. Neural Computing Surveys 2: 163-212

[2] Anthony, M., Bartlett, P.L. (1999) Neural networks learning: Theoretical foundations, 1st ed. Cambridge University Press, Cambridge

[3] Hammer, B., Tino, P. (2003) Recurrent neural networks with small weights implement definite memory machines. Neural Computation 15(8): 1897-1929

[4] Murray, A.F., Edwards, P.J. (1992) Synaptic weight noise during MLP learning enhances fault-tolerance,

generalisation and learning trajectory. Advances in Neural Information Processing Systems 5: 491-498

[5] Hornik, K., Stinchcombe, M., White, H. (1989) Multilayer feedforward networks are universal approximators. Neural Networks 2: 359-366

[6] Durbin, R., Rumelhart, D. (1989) Product units: A computationally powerful and biologically plausible extension to backpropagation networks. Neural Computation 1: 133-142

[7] Schmitt, M. (2002) On the complexity of computing and learning with multiplicative neural networks. Neural Computation 14: 241-301

[8] Duch, W., Jankowski, N. (2001) Transfer functions: hidden possibilities for better neural networks. In Verleysen, M. (ed.) 9th European Symposium on Artificial Neural Networks. D-Facto, Brugge, pp. 81-94

[9] Cohen, S., Intrator, N. (2002) A hybrid projection based and radial basis function architecture: initial values and global optimization. Pattern Analysis & Applications 5: 113-120

[10] Schmitt, M. (2002) Neural networks with local receptive fields and superlinear VC dimension. Neural Computation 14(4): 919-956

[11] Huk, M. (2004) The Sigma-if neural network as a method of dynamic selection of decision subspaces for medical reasoning systems. Journal of Medical Informatics & Technologies 7: 65-73

[12] Ridella, S., Rovetta, S., Zunino, R. (1999) Representation and generalization properties of Class-Entropy Networks. IEEE Transactions on Neural Networks 10(1): 31-47

[13] Banarer, V., Perwass, C., Sommer, G. (2003) The hypersphere neuron. In: Verleysen, M. (ed.) 11th European Symposium on Artificial Neural Networks. D-Side, Brugge, pp. 469-474

[14] Huk, M. (2003) Determining the relevancy of attributes in classification tasks through nondestructive elimination of interneuronal connections in a neural network (in Polish). Pozyskiwanie Wiedzy z Baz Danych 975: 138-147

[15] Fahlman, S.E., Lebiere, C. (1990) The Cascade-Correlation learning architecture. Advances in Neural Information Processing Systems 2: 524-532

[16] Kaufman, L., (1999) Solving the quadratic programming problem arising in support vector classification. In: Scholpkopf, B., Burges, C.J.C., Smola, A.J. (eds.) Advances in Kernel Methods - Support Vector Learning. MIT Press, Boston, pp. 146-167

[17] Tickle, A., Andrews, R., Golea, M., Diederich, J. (1998) The truth will come to light: directions and challenges in extracting the knowledge embedded within trained artificial neural networks. IEEE Transactions on Neural Networks 9(6): 1057-1068.

[18] Craven, M.W., Shavlik, J.W. (1996) Extracting tree-structured representations of trained networks. Advances in Neural Information Processing Systems 8: 24-30

Beta wavelet networks for function approximation

Wajdi BELLIL[1], Chokri BEN AMAR[2] and Adel M. ALIMI[2]

REGIM: Research Group on Intelligent Machines

[1] Department of Electrical Engineering, High Institute of Technological Studies of Gafsa, Tunisia

[2] Department of Electrical Engineering, ENIS, University of Sfax, Tunisia

E-mail: {Wajdi.bellil, chokri.benamar, adel.alimi}@ieee.org

Abstract

Wavelet neural networks (WNN) have recently attracted great interest, because of their advantages over radial basis function networks (RBFN) as they are universal approximators. In this paper we present a novel wavelet neural network, based on Beta wavelets, for 1-D and 2-D function approximation. Our purpose is to approximate an unknown function f: $R^n \rightarrow R$ from scattered samples (x_i; $y_i = f(x)$) i=1....n, where:

- ✓ we have little a priori knowledge on the unknown function f which lives in some infinite dimensional smooth function space,
- ✓ the function approximation process is performed iteratively: each new measure on the function (x_i; $f(x_i)$) is used to compute a new estimate $\hat{f}$ as an approximation of the function f.

Simulation results are demonstrated to validate the generalization ability and efficiency of the proposed Beta wavelet network.

1 Introduction

Combining the wavelet transform theory with the basic concept of neural networks [1-3], a new mapping network called wavelet neural network or wavenets (WNN) is proposed as an alternative to feedforward neural networks for approximating arbitrary nonlinear functions. Kreinovich proved in [14] that if we use a special type of neurons (wavelet neurons), then the resulting neural networks are *optimal* approximators in the following sense: as $\varepsilon \rightarrow 0$, the number of bits that is necessary to store the results of a 3-layer wavelet neural network approximation, increases slower than for any other approximation scheme.

Wavelets occur in a family of functions and each is defined by dilation a_i which controls the scaling parameter and translation t_i which controls the position of a single function, named the mother wavelet $\psi(x)$. Mapping functions to a time-frequency phase space, WNN can reflect the time-frequency properties of function more accurately than the RBFNN. Given an *n*-element training set, the overall response of a WNN is:

$$\hat{y}(w) = w_0 + \sum_{i=1}^{N_p} w_i \Psi_i\left(\frac{x - t_i}{a_i}\right) \tag{1}$$

where *Np* is the number of wavelet nodes in the hidden layer and w_i is the synaptic weight of WNN. A WNN can be regarded as a function approximator which estimates an unknown functional mapping:

$$y = f(x) + \varepsilon \tag{2}$$

where *f* is the regression function and the error term ε is a zero-mean random variable of disturbance. There are a number of approaches to WNN construction (a brief survey is provided in [9-12]), we pay special attention on the model proposed by Zhang [1, 6, 8].

2 The Beta wavelet

The Beta function [13] is defined as:

if p>0, q>0, (p, q) ∈ IN

$$\beta(x) = \begin{cases} \left(\frac{x - x_0}{x_c - x_0}\right)^p \left(\frac{x_1 - x}{x_1 - x_c}\right)^q & if\ x \in [x_0, x_1] \\ 0 & \text{else} \end{cases} \tag{3}$$

where, $x_c = \dfrac{px_1 + qx_0}{p + q}$

2.1. The derivatives of Beta function

We proved in [4, 5] that all derivatives of Beta function $\in L^2(IR)$ and are of class C^∞. The general form of the nth derivative of Beta function is:

$$\Psi_n(x) = \frac{d^n \beta(x)}{dx^n}$$

$$= \left[(-1)^n \frac{n!p}{(x - x_0)^{n+1}} + \frac{n!q}{(x_1 - x)^{n+1}}\right]\beta(x) + P_n(x)P_1\beta(x)$$

$$+ \sum C_n^i \left[(-1)^n \frac{(n-i)!\,p}{(x - x_0)^{n+1-i}} + \frac{(n-i)!\,q}{(x_1 - x)^{n+1-i}}\right] P_1(x)\beta(x) \tag{4}$$

where: $P_1(x) = \dfrac{p}{x - x_0} - \dfrac{q}{x_1 - x}$

and $P_n(x) = (-1)^n \dfrac{n!p}{(x - x_0)^{n+1}} - \dfrac{n!q}{(x_1 - x)^{n+1}}$

The first (BW1), second (BW2) and third (BW3) derivatives of Beta wavelet are shown graphically in Figure 1.

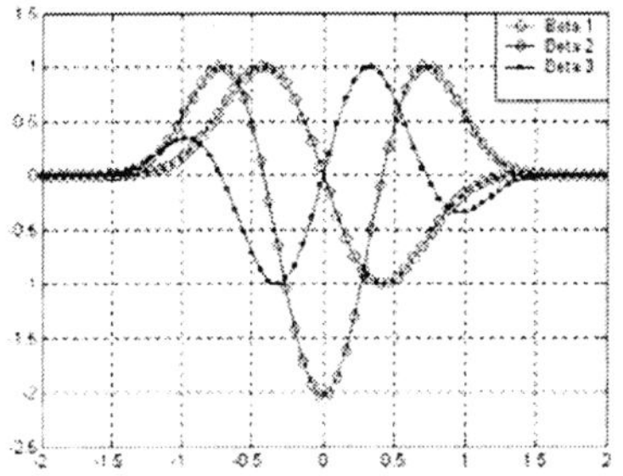

Fig.1. First, second and third derivatives of Beta function.

2.2. Proposition

if p = q , for all n $\in$ IN and $0 < n < p$ the functions

$$\Psi_n(x) = \frac{d^n\beta(x)}{dx^n} \quad \text{are wavelets [4, 5].} \qquad (5)$$

3 Experiments

In this section, we present two experimental results of the proposed Beta Wavelet Neural Networks (BWNN) on approximating two functions using the *Stepwise selection by orthogonalization* training algorithm. First, simulations on the 1-D function approximation $f(x)=0.5x\sin(2x)+\cos^2(2x)$ are conducted to validate and compare the proposed BWNN with some other wavelets. The input x is constructed by the uniform distribution on [-2.5, 2.5], and the corresponding output y is functional of $y = f(x)$ and is artificially contaminated by random errors. The training and test data are composed of 50 points and 500 points, respectively. Beta wavelet is chosen as the mother wavelet for training network. Second, the two-dimension function:

$$f(x_1, x_2) = \frac{e^{-\frac{81}{16}[(x1 - 0.5)^2 + (x2 - 0.5)^2]}}{3} \quad \text{is}$$

approximated to illustrate the robustness of the proposed wavelets family. The training set D contains 11x11 uniform spaced points, and 11x11 stochastic points. The test set V is constructed by evenly spaced 21x21 grid on [-1, 1]x[-1, 1].

3.1 1-D interpolation using the stepwise selection by orthogonalization algorithm

These results are given, using the Stepwise selection by orthogonalization algorithm, on a Neural Wavelet Networks using 9 wavelets, 4 levels decomposition, 500 iterations, 50 points for training and a uniform spaced points. $f(x)=0.5x\sin(2x)+\cos^2(2x)$.

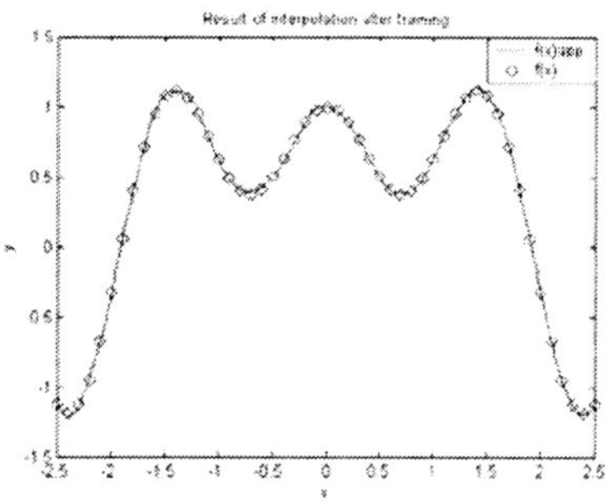

Fig.2. Result of interpolation by Mexican hat.

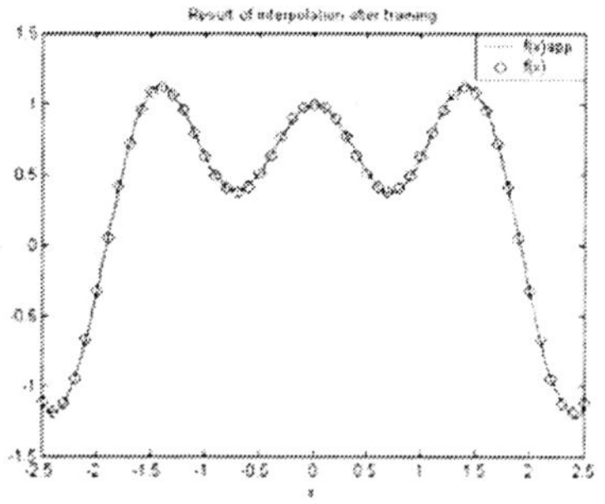

Fig.3. Result of interpolation by BW2.

Approximated functions are displayed in Figures 2 and 3. The Normalized Root Mean Square Error *NRMSE* of the Mexican hat WNN is 0.0138434 compared to 0.009344716 the BW2 WNN achieved. From these simulations we can deduce the efficiency of Beta wavelet in term of function interpolation. The table below gives the normalized square root mean square error and mean square error using traditional wavelets and Beta wavelet:

	Stepwise selection by orthogonalization algorithm	
Wavelets	NSRMSE	MSE (e-005)
Mexican hat	0.0138434	10.7061
Beta 1	0.0295078	48.6430
Beta 2	*0.00934716*	*4.88097*

Table.1 Comparison of NSRMSE and MSE for Beta wavelets and some others in term of 1-D approximation.

3.2 1-D interpolation of noisy data using the stepwise selection by orthogonalization algorithm

These results are given, using on a Neural Wavelet Network using 9 wavelets, 4 levels decomposition, 500 iterations, 50 points for training, uniform spaced points.

$$f(x)=0.5x\sin(2x)+\cos^2(2x)+\varepsilon(x) \qquad (6)$$

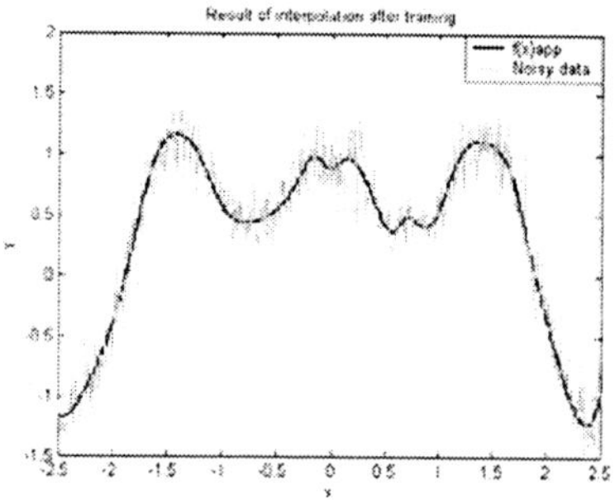

Fig.5. Result of interpolation of noisy data by Mexican hat wavelet.

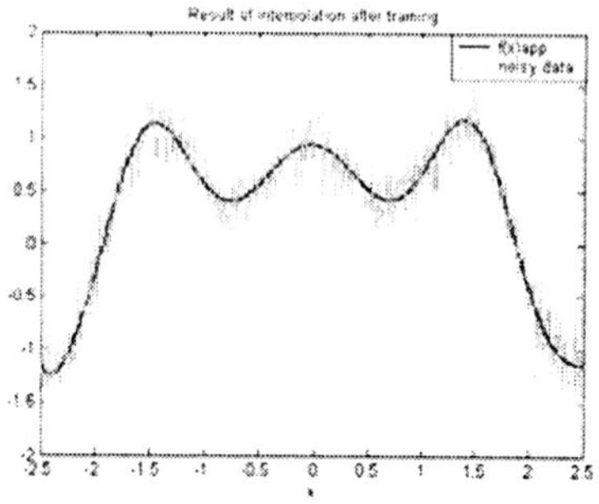

Fig.6. Result of interpolation of noisy data by BW2 wavelet.

We display in figure 5 the result of approximation of a noisy signal using Mexican hat WNN and in figure 6 the Beta WNN one. From these simulations we can scc that Beta 2 WNN is more efficient than the Mexican hat wavelet on noisy data approximation.

3.3 2-D interpolation using the stepwise selection by orthogonalization algorithm

These results are given, using the Stepwise selection by orthogonalization algorithm on a Neural Wavelet Networks using 4 wavelets, 4 levels decomposition, 200 iterations, 11x11 points for training.

$$f(x_1\ x_2) = \frac{e^{-\frac{81}{16}[(x1-0.5)^2 + (x2-0.5)^2]}}{3} \qquad (7)$$

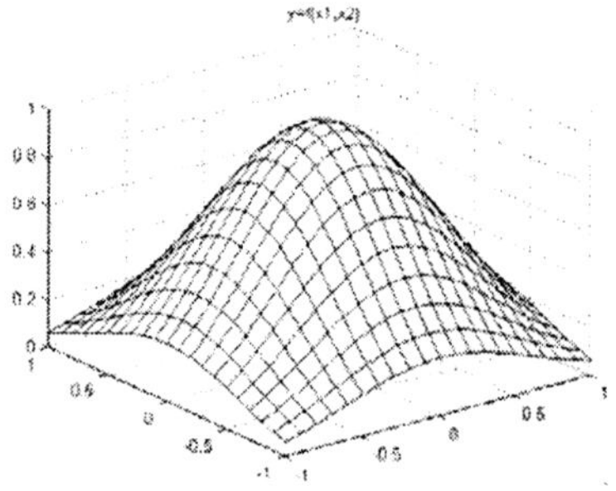

Fig.7. 2-D data to be interpolated.

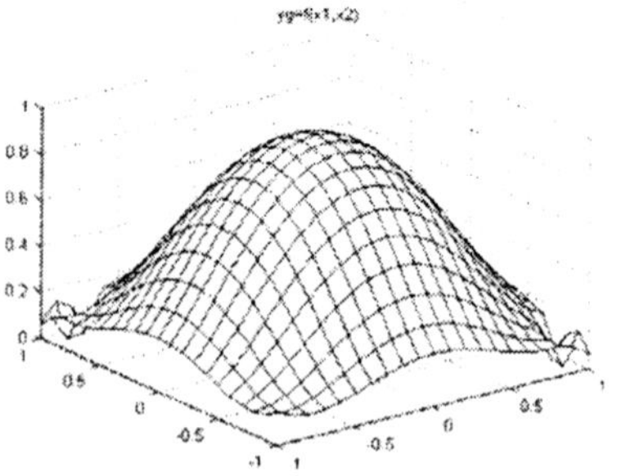

Fig.8. Result of 2-D interpolation by Mexican hat wavelet after training using uniform spaced input patterns.

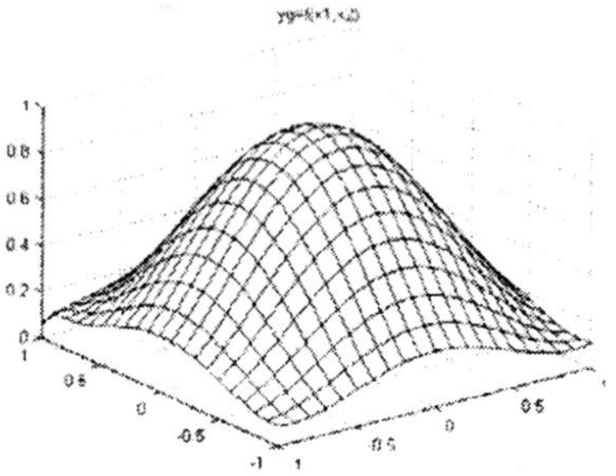

Fig.9. Result of 2-D interpolation by BW2 wavelet after training using uniform spaced input patterns.

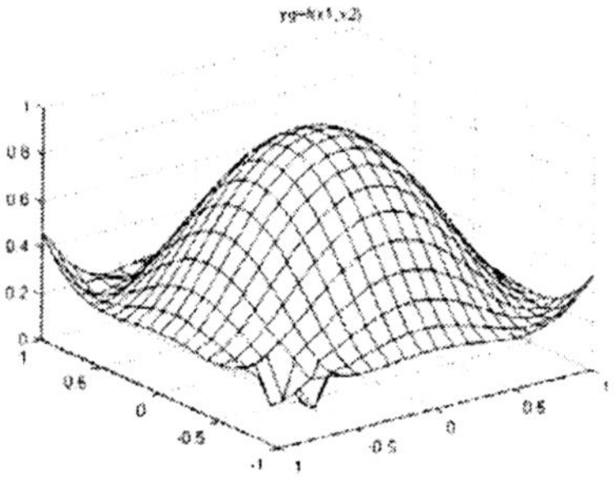

Fig.10. Result of 2-D interpolation by Mexican hat wavelet after training using randomly input patterns.

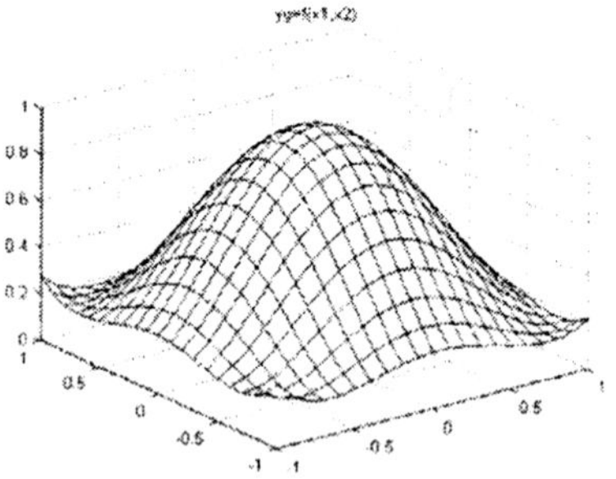

Fig.11. Result of 2-D interpolation by BW2 wavelet after training using randomly input patterns.

We display in Figure 7 the 2-D data we use in our tests, in figure 8 the result of 2-D interpolation using Mexican hat WNN on which we see some distortion in amplitude and at the edges. In figure 9 using Beta 2 WNN we reduce the amplitude distortion. In figure 10 we display the result of interpolation using Mexican hat WNN on which distortion becomes greater than its homolog using Beta 2 WNN displayed in figure 11.

4 Conclusion

We present two experimental results of the proposed Beta Wavelet Neural Networks (BWNN) on approximating two functions using the *Stepwise selection by orthogonalization* training algorithm. First, simulations on the 1-D function approximation on which we prove the superiority of Beta wavelets in term of NSRMSE. Second, the two-dimension function is approximated with the second derivative of Beta wavelet and the Mexican hat wavelet to illustrate the robustness of the proposed wavelets family. The training set D contains 11x11 uniform spaced points, and 11x11 random points, the test set V is constructed by evenly spaced 21x21 grid on [-1, 1]x[-1, 1]. So the new Beta wavelets family has the superiority of approximation in the 1-D and the 2-D case. This new wavelet family can be used to approximate volume using the 2-D 1-D 2-D technique.

Acknowledgment

The authors would like to acknowledge the financial support of this work by grants from the General Direction of Scientific Research and Technological Renovation (DGRSRT), Tunisia, under the ARUB program 01/UR/11/02.

References

[1] Q. Zhang, (1997) Using Wavelet Network in Nonparametric Estimation, IEEE Trans. Neural Network, Vol. 8, pp.227-236.

[2] S.T. Li and E. Leiss, (2001) On Noise-immune RBF Networks, in Radial Basis Function Neural Networks: Recent Developments in Theory and Applications, Editors: Howlett, R. J. and Jain, L.C., Springer Verlag, pp. 95-124.

[3] P. Cristea, R. Tuduce, and A. Cristea, (2000) Time Series Prediction with Wavelet Neural Networks, Proceedings of IEEE Neural Network Applications in Electrical Engineering, pp. 5-10.

[4] W. Bellil, C. Ben Amar et M. A. Alimi, (2003) Beta Wavelet Based Image Compression, International Conference on Signal, System and Design, SSD03, Tunisia, Mars, vol. 1, pp. 77-82.

[5] W. Bellil, C. Ben Amar, M. Zaied and M. A. Alimi, (2004) La fonction Beta et ses derivees: vers une nouvelle famille d'ondelettes, First International Conference on Signal, System and Design, SCS'04, Tunisia, Mars, vol. 1, PP. 201-207.

[6] Q. Zhang and A. Benveniste, (1992) Wavelet Networks, IEEE Trans. on Neural Networks 3 (6),PP 889-898.

[7] J. Zhang, G. G. Walter, Y. Miao and W. N. Wayne Lee, (1995) Wavelet Neural Networks For Function Learning, IEEE Trans. on Signal Processing 43 (6), PP1485-1497.

[8] Q. Zhang (1997) Using Wavelet Network in Nonparametric Estimation. IEEE Trans. on Neural Networks, Vol. 8, No. 2, pp. 227-236.

[9] Y. Oussar, I. Rivals, L. Personnaz (1998) Training Wavelet Networks for Nonlinear Dynamic Input-Output Modeling Neurocomputing.

[10] C.S.Chang, Weihui Fu, Minjun Yi, (1998) Short term load forecasting using wavelet networks Engineering Intelligent Systems for Electrical Engineering and Communications 6, 217-23.

[11] J.Echauz, (1998) Strategies for Fast Training of Wavelet Neural Networks, 2nd International Symposium on Soft Computing for Industry, 3rd World Automation Congress, Anchorage, Alaska, May 10-14, 1-6.

[12] V.Kruger, Happe, A., Sommer, G (1999) Affine real-time face tracking using a wavelet network Proc. Workshop on Recognition, Analysis, and Tracking of Faces and Gestures in Real-Time Systems, IEEE, 141-8.

[13] M. A. Alimi, (2000) The Beta System: Toward a Change in Our Use of Neuro-Fuzzy Systems. International Journal of Management, Invited Paper, pp. 15-19.

[14] V. Kreinovich, O. Sirisaengtaksin. S. Cabrera, (1992) Wavelets compress better than all other methods: a 1-dimensional theorem. University of Texas at El Paso, Computer Science Department, Technical Report.

Speeding up backpropagation with Multiplicative Batch Update Step

Pedro Cruz
Department of Mathematics, University of Aveiro, Portugal
E-mail: jpedro@mat.ua.pt

Abstract

Updating steps in a backpropagation neural network with multiplicative factors $u > 1$ and $d < 1$ has been presented by several authors. The istatistics field of Stochastic Approximation has a close relation with backpropagation algorithms. Recent theoretical results in this field show that for functions of one variable, different values of u and d can produce very different results: fast convergence at the cost of a poor solution, slow convergence with a better solution, or produce a fast move towards a solution but without converging. To speed up backpropagation in a simple manner we propose a batch step adaptation technique for the online backpropagation algorithm based on theoretical results on simple cases.

1 Introduction

Speeding up backpropagation has been a constant challenge and several techniques has been applied like using second order information [4].

Also several works on multiplicative step update where developed [5,6,8] and each of them uses the common update algorithm

$$\gamma_t = \begin{cases} \gamma_{t-1} \times u & \text{if condition C} \\ \gamma_{t-1} \times d & \text{otherwise} \end{cases} \tag{1}$$

where $u > 1$ and $0 < d < 1$ are real parameters. Possibly variable γ_t is guaranteed to be limited $0 < \Gamma_d < \gamma_t < \Gamma_u$.

Condition C could be a gradient condition [5] or a condition on the quadratic error [6,8]. In both cases the quadratic error

$$E_t = E(W_t, x, d) = (1/2) \sum_{i=1}^{n} (\text{NET}_i(W_t, x) - d)^2, \tag{2}$$

is to be minimized, with (x, d) being the pattern presented at the iteration, W_t is the weight vector that describes net state, and NET_i is the ith output of the neural network.

Gradient condition uses change of signs: if

$$\frac{\delta E_t}{\delta w} \times \frac{\delta E_{t-1}}{\delta w} > 0 \tag{3}$$

then step must be incremented otherwise reduced. Different updates are going to occur for each weight w in the net parameters $W = (w_0, w_1, \ldots, w_n)$.

Another type of condition is the error condition which is based on increases or decreases of global error ΔE_t causing the step to change.

We propose an algorithm that uses a gradient condition for step update. Next we present theoretic results that are guides to the proposed algorithm.

2 Stochastic Approximation and New results

Many problems in Stochastic Approximations are described as the following. Consider the problem of searching for the zero point of a function, according to the stochastic approximation procedure

$$x_{t+1} = x_t - \gamma_t y_t, \tag{4}$$

$$y_t = \varphi(x_t) + \xi_t \tag{5}$$

where φ is the function under consideration, x_t means the tth approximation of the zero x^* of φ, y_t is the tth disturbed observation of φ with random disturbance ξ_t.

If step size values γ_t of the procedure (4), (5) are deterministic and satisfy $\sum \gamma_t = \infty$, $\sum \gamma_t^2 < \infty$, the sequence of x_t is proved to converge to x^* [2]. There are various versions of (4) and (5) aiming at accelerating convergence of x_t. We are concentrating here on the methods using step size adaptation in the course of algorithm, decreasing γ_t every time that the two consecutive differences $\Delta x_{t-1} = \gamma_{t-1} y_{t-1}$ and $\Delta x_t = \gamma_t y_t$ have the same sign, and increasing γ_t, otherwise.

Consider the following update rule

$$\gamma_t = \begin{cases} \min\{u\,\gamma_{t-1}, \Gamma\} & \text{if} \quad y_{t-1} y_t > 0, \\ d\,\gamma_{t-1} & \text{if} \quad y_{t-1} y_t \le 0, \end{cases} \tag{6}$$

$t = 2, 3, \ldots$ is used. Here $0 < d < 1 < u$, $0 < \gamma_0 \le \Gamma_u$, $\gamma_1 \in \{\min\{u\gamma_0, \Gamma\}, d\gamma_0\}$, Γ_u is a positive constant. Let us point out the main differences with standard algorithm. Suppose that $\{\xi_t\}$ is a sequence of i.i.d.r.v. with zero mean, besides $\text{P}(\xi_t > 0) = \text{P}(\xi_t < 0)$. Under some additional assumptions on φ, ξ_t, and Γ_u, stated below, the process defined by (4), (5), (6) a.s. diverges if

$ud > 1$, and converges if $ud < 1$. In fact this algorithm can converge to a region near one of the zeros of φ. See [7] for details.

3 Batch Step Update and Backpropagation

Using the above algorithm with feed forward neural networks is not possible since partial derivatives changes sign very frequently indicating a non smooth surface and this cause a fast step decrease and a very poor solution is reached.

However, using constant step, one can observe two facts about a weight w. We use w_t to indicate the sequence of values of some predetermined weight of the vector of parameters W. First fact is that during pattern presentation w_t oscilates frequently, causing the training to freeze. The second fact is that observing w_t values after the full training set has been used one notice that w_t doesn't oscilate, doing some progression on the same direction each batch time. Change of weight direction occurs but not frequently as after each pattern presentation.

Consider the following 'time' definitions: time T indicate the batch number and is incremented after B pattern presentations; time t is incremented after each pattern presentation.

We propose the following batch update rule for step

$$\gamma_T = \begin{cases} \min\{\gamma_{T-1} \times u, \Gamma_u\} & \text{if } \Delta w_T \times \Delta w_{T-1} > 0 \\ \max\{\gamma_{T-1} \times d, \Gamma_d\} & \text{if } \Delta w_T \times \Delta w_{T-1} \leq 0 \end{cases} \tag{7}$$

for $T = 1, 2, \ldots$ with

1. $u > 1$ and $0 < d < 1$;
2. γ_T is bounded by Γ_d and Γ_u constants;
3. γ_0 some constant that could be much less than Γ_u since step can grow.

After the step update rule we define the weight update rule. It is known that online training has been shown to produce better solutions than weight batch update. So, each weight is going to be updated after each pattern presentation. We propose the following update algorithm for a single weight w

$$w_t = w_{t-1} - \gamma_{t/B} \frac{\delta E(W_{t-1}, x_t, d_t)}{\delta w} \tag{8}$$

with $t = 1, \ldots$,and where t/B uses integer division, and $E(W, x, d)$ is defined in (2).

4 A numerical study

We use the MNIST digit database [3] to study the new algorithm performance but using only digits $\{0, 1, 2, 3\}$.

Four identical shape neural networks where used, one for each digit. Each net has the following structure: 28×28 inputs, 10 hidden units and two outputs (first output is 1 when a good pattern is presented, otherwise is valued -1, and bad patterns are classifiyed 1 in the second output, otherwise -1).

Each digit has aproximatly 6000 exemplars (called *good patterns*). The remaning digits are 54000 *bad patterns*). So, the training set has 60000 patterns. The test set has 1000 patterns (approximately 1000 samples of each digit).

Each net was trained using the 6000 'good patterns' against other 6000 randomly choosen 'bad patterns'. So, total training set is 12000. We used these parameters: the number of t iterations was 5×12000, the step was updated every $B = 3000$ (1/4 of training set dimension) and $\Gamma_u = 0.001$ (that is $1/(\text{large fan in})$, and $\Gamma_d = 10^{-7}$. Initial step was $\gamma_0 = 0.0001$. This two values where randomly set.

Figure 1 describes the minimum wrong values on the test set in all the 5×12000 iterations. The step schedules where: **Bar 1** $u = 1.1$ and $d = 0.9/1.1$ ($ud < 1$), **Bar 2** $u = 1.1$ and $d = 1/1.1$ ($ud = 1$), **Bar 3** $u = 1.1$ and $d = 1.08/1.1$ ($ud > 1$) and last **Bar 4** with constant step $\gamma = \gamma_0 = 0.0001$ ($u = d = ud = 1$).

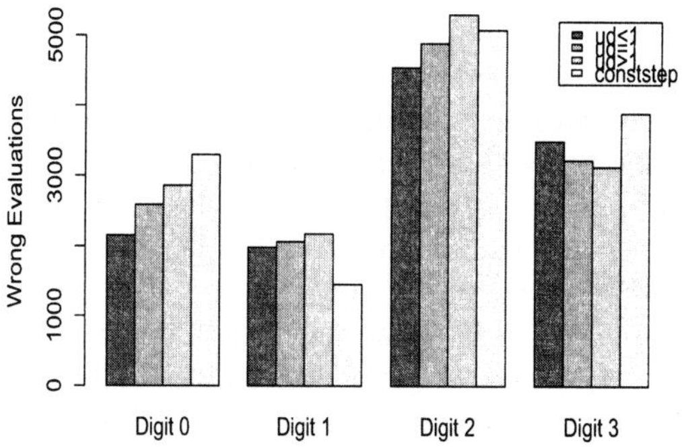

Fig. 1. Number of wrong classifications vs. step schedule in a short period of training steps. The dark bar is $ud < 1$ case, then $ud = 1$, $ud > 1$ and finally constant step.

The main observation is that the algorithm with step update rule $ud < 1$ behaves better in 3 cases: digits $\{0, 2, 3\}$.

The number of wrong classifications are very high for two reasons: small number of iterations and the described method of training makes harder for the net to classify with the simple structure.

As a conclusion of this work we can say that a very

simple familly of algorithms for step update was presented and these simple results are encouraging. However, more work on these methods should be done.

5 Acknowledgements

The author wishes to thank the financial support by the Portuguese program PRODEP 'Medida 5 - Acção 5.3 - Formação Avançada de Docentes do Ensino Superior - Concurso nr. 2/5.3/PRODEP/2001'.

References

[1] Hertz, J., Krogh, A., Palmer, R. G., (1991) Introduction to the theory of neural computation, ISBN 0-201-50395-6, Addison-Wesley Longman Publishing Co., Inc.

[2] Kushner, Harold J., Yin, G.George (1997) Stochastic approximation algorithms and applications, Applications of Mathematics. 35. Berlin: Springer. xxi, 417 p.

[3] LeCun, Y., Bottou, L., Bengio Y., Haffner, P. (1998) Gradient-based learning applied to document recognition. Proc. of the IEEE, vol. 86, nr. 11, 1998.

[4] Spall, J. C., (2000) Adaptive stochastic approximation by the simultaneous perturbation method., IEEE Trans. Autom. Control, vol.45, nr.10: 1839–1853.

[5] Silva, F. M., Almeida, L. B. (1990),Speeding up backpropagation, Advanced Neural Computers, ed. R. Eckmiller, Elsevier Science Publishers, Amsterdam, pp151-158.

[6] Salomon, R., Hemmen, van J. L., (1996), Accelerating Backpropagation through Dynamic Self-Adaptation, Neural Networks, vol. 9, nr. 4: 589–601.

[7] Plakhov, A., Cruz, P. (to be published), A stochastic approximation algorithm with multiplicative step size adaptation.

[8] Battiti, R., (1992) First and second order methods for learning: between steepest descent and Newton's method, Neural Computation, vol. 4, nr. 2: 141–166, The MIT Press, Cambridge–Massachusetts.

Generating Sequential Triangle Strips by Using Hopfield Nets

Jiří Šíma[1]
Institute of Computer Science, Academy of Sciences of the Czech Republic
P. O. Box 5, 18207 Prague 8, Czech Republic, E-mail: sima@cs.cas.cz

Abstract

The important task of generating the minimum number of sequential triangle strips (tristrips) for a given triangulated surface model is motived by applications in computer graphics. This hard combinatorial optimization problem is reduced to the minimum energy problem in Hopfield nets by a linear-size construction. The Hopfield network powered by simulated annealing (i.e. Boltzmann machine) which is implemented in a program HTGEN can be used for computing the semi-optimal stripifications. Practical experiments confirm that one can obtain much better results using HTGEN than by a leading stripification program FTSG although the running time of simulated annealing grows rapidly near the global optimum.

1 Sequential triangular strips

Piecewise-linear surfaces defined by sets of triangles (triangulation) are widely used representations for geometric models. Computing a succinct encoding of a triangulated surface model represents an important problem in graphics and visualization. Current 3D graphics rendering hardware often faces a memory bus bandwidth bottleneck in the processor-to-graphics pipeline. Apart from reducing the number of triangles that must be transmitted it is also important to encode the triangulated surface efficiently. A common encoding scheme is based on sequential triangle strips which avoid repeating the vertex coordinates of shared triangle edges. Triangle strips are supported by several graphics libraries (e.g. IGL, PHIGS, Inventor, OpenGL).

In particular, a *sequential triangle strip* (hereafter briefly *tristrip*) of length $m-2$ is an ordered sequence of $m \geq 3$ vertices $\sigma = (v_1, \ldots, v_m)$ which encodes $m-2$ different triangles $\{v_p, v_{p+1}, v_{p+2}\}$ for $1 \leq p \leq m-2$ so that their shared edges follow alternating left and right turns as indicated in Figure 1 by a dashed line. Thus a triangulation consisting of a single tristrip with n triangles allows transmitting of only $n+2$ (rather than $3n$) vertices. In general, a triangulated surface model T with n triangles that is decomposed into k tristrips $\Sigma = \{\sigma_1, \ldots, \sigma_k\}$ requires only $n + 2k$ vertices to be transmitted. A crucial problem is to decompose a triangulated surface model into the fewest tristrips. This stripification problem has recently been proved to be NP-complete in article [1] which also contains relevant references. In the present paper a new method of generating tristrips Σ for a given triangulated surface model T with n triangles is proposed which is based on a linear-time reduction to the minimum energy problem in Hopfield network $\mathcal{H}_T$ having $O(n)$ units and connections. This approach has been inspired by a more complicated and incomplete reduction (sequential cycles were not excluded) introduced in [2].

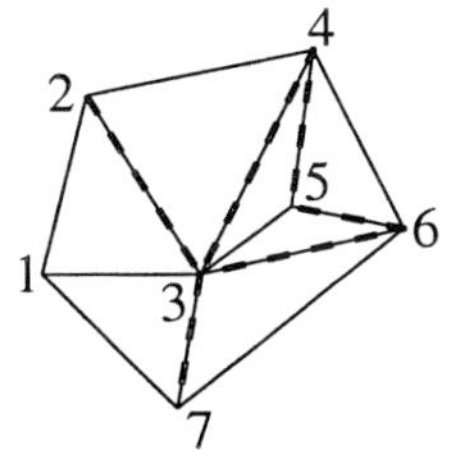

Fig. 1. Tristrip (1,2,3,4,5,6,3,7,1).

The paper is organized as follows. After a brief review of the basic definitions concerning Hopfield nets in Section 2, the main construction of Hopfield network $\mathcal{H}_T$ for a given triangulation T is described in Section 3. The correctness of this reduction has formally been verified [3] by proving a one-to-one correspondence between the optimal stripifications of T and the minimum energy states in $\mathcal{H}_T$. Thus, $\mathcal{H}_T$ combined with simulated annealing (i.e. Boltzmann machine) has been implemented in a program HTGEN which is compared against a leading stripification program FTSG in Section 4. Practical experiments show that HTGEN can compute much better stripifications than FTSG although the running time of HTGEN grows rapidly when the global optimum is being approached.

2 The minimum energy problem

Hopfield networks [4] having well-constrained convergence behavior represent a very influential associative

[1]Research partially supported by projects 1M0021620808, LN00A056 of The Ministry of Education of the Czech Republic.

memory model which is connected to the much-studied Ising spin glass model in statistical physics [5]. Part of the appeal of Hopfield nets also stems from their natural hardware implementations using electrical networks or optical computers, and their application in combinatorial optimization [6].

Formally, a Hopfield network is composed of s computational *units* or *neurons*, indexed as $1, \ldots, s$, that are connected into undirected graph or *architecture*, in which each connection between unit i and j is labeled with an integer *symmetric weight* $w(i,j) = w(i,j)$. The absence of a connection within the architecture indicates a zero weight between the respective neurons, and vice versa. For example, $w(j,j) = 0$ is assumed for $j = 1, \ldots, s$. The *sequential discrete* dynamics of such a network is here considered, in which the evolution of the network *state* $\mathbf{y}^{(t)} = (y_1^{(t)}, \ldots, y_s^{(t)}) \in \{0,1\}^s$ is determined for discrete time instants $t = 0, 1, \ldots$, as follows. The *initial state* $\mathbf{y}^{(0)}$ may be chosen arbitrarily, e.g. $\mathbf{y}^{(0)} = (0, \ldots, 0)$. At discrete time $t \geq 0$, the *excitation* of any neuron j is defined as $\xi_j^{(t)} = \sum_{i=1}^{s} w(i,j) y_i^{(t)} - h(j)$ including an integer *threshold* $h(j)$ local to unit j. At the next instant $t+1$, one (e.g. randomly) selected neuron j computes its new output $y_j^{(t+1)} = H(\xi_j^{(t)})$ by applying the Heaviside activation function H, that is, j is *active* when $H(\xi) = 1$ for $\xi \geq 0$ while j is *passive* when $H(\xi) = 0$ for $\xi < 0$. The remaining units do not change their states, i.e. $y_i^{(t+1)} = y_i^{(t)}$ for $i \neq j$. In this way the new network state $\mathbf{y}^{(t+1)}$ at time $t+1$ is determined.

Also *macroscopic time* $\tau = 0, 1, 2, \ldots$ is introduced during which all the units in the network are updated. A computation of a Hopfield net *converges* or *reaches a stable state* $\mathbf{y}^{(\tau^*)}$ at macroscopic time $\tau^* \geq 0$ if $\mathbf{y}^{(\tau^*)} = \mathbf{y}^{(\tau^*+1)}$. The well-known fundamental property of a symmetric Hopfield net is that its dynamics is constrained by *energy* function $E(\mathbf{y}) = -\frac{1}{2}\sum_{j=1}^{s}\sum_{i=1}^{s} w(i,j) y_i y_j + \sum_{j=1}^{s} h(j) y_j$ which is a bounded function defined on its state space whose value decreases along any nonconstant computation path ($\xi_j^{(t)} \neq 0$ is assumed without loss of generality). It follows from the existence of such a function that starting from any initial state the network converges towards some stable state corresponding to a local minimum of E [4]. Thus the cost function of a hard combinatorial optimization problem can be encoded into the energy function of a Hopfield net which is then minimized in the course of computation. Hence, the *minimum energy problem* of finding a network state with minimum energy is of special interest. Nevertheless, this problem is in general NP-complete [5] (see [7] for related results).

A stochastic variant of Hopfield model called the *Boltzmann machine* [8] is also considered in which randomly selected unit j becomes active at time $t+1$, i.e. $y_j^{(t+1)} = 1$, with probability $P(\xi_j^{(t)})$ which is computed by applying the probabilistic activation function $P : \mathbf{R} \longrightarrow (0,1)$ defined as $P(\xi) = 1/(1 + e^{-2\xi/T^{(\tau)}})$ where $T^{(\tau)} > 0$ is a so-called *temperature* at microscopic time $\tau \geq 0$. This parameter is controlled by *simulated annealing*, e.g. $T^{(\tau)} = T^{(0)}/\log(1+\tau)$ for sufficiently high initial temperature $T^{(0)}$. The simulated annealing is a powerful heuristic method for avoiding the local minima in combinatorial optimization.

3 The reduction

For the purpose of reduction the following definitions are introduced. Let T be a set of n triangles that represents a triangulated surface model homeomorphic to a sphere in which each edge is incident to at most two triangles. An edge is said to be *internal* if it is shared by exactly two triangles; otherwise it is a *boundary* edge. Denote by I the set of internal edges in triangulation T. Furthermore, a *sequential cycle* is a "cycled tristrip", that is, an ordered sequence of vertices $C = (v_1, \ldots, v_m)$ where $m \geq 4$ is even, which encodes $m - 2$ different triangles $\{v_p, v_{p+1}, v_{p+2}\}$ for $1 \leq p \leq m-2$ so that $v_{m-1} = v_1$ and $v_m = v_2$. Also denote by I_C and B_C the sets of internal and boundary edges of sequential cycle C, respectively, that is $I_C = \{\{v_p, v_{p+1}\}\,;\, 1 \leq p \leq m-2\}$ and $B_C = \{\{v_p, v_{p+2}\}\,;\, 1 \leq p \leq m-2\}$. An example of the sequential cycle is depicted in Figure 2 where its internal and boundary edges are indicated by dashed and dotted lines, respectively. In addition, let $\mathcal{C}$ be the set of all sequential cycles in T.

For each sequential cycle $C \in \mathcal{C}$ one unique *representative* internal edge $e_C \in I_C$ can be chosen as follows. Start with any cycle $C \in \mathcal{C}$ and choose any edge from I_C to be its representative edge e_C. Observe that for a fixed orientation of triangulated surface any internal edge follows either left or right turn corresponding to at most two sequential cycles. Thus denote by C' the sequential cycle having no representative edge so far which shares its internal edge $e_C \in I_C \cap I_{C'}$ with C if such C' exists; otherwise let C' be any sequential cycle with no repre-

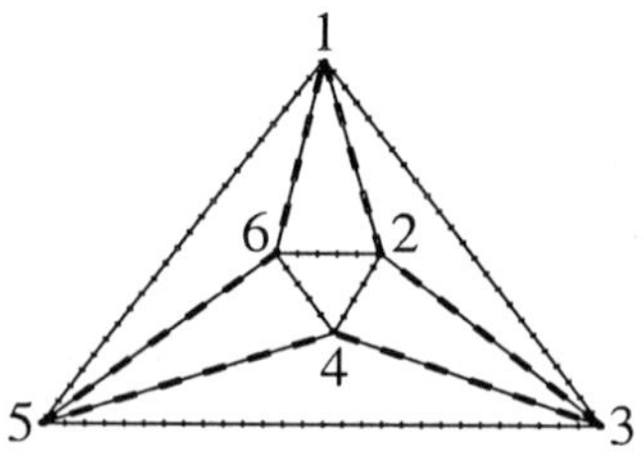

Fig. 2. Sequential cycle (1,2,3,4,5,6,1,2).

sentative internal edge or stop if all the sequential cycles do have their representative edges. Further choose any edge from $I_{C'} \setminus \{e_C\}$ to be the representative edge $e_{C'}$ of C' and repeat the previous step with C replaced by C'. Clearly, each edge represents at most one cycle because set $I_{C'} \setminus \{e_C\} \neq \emptyset$ always contains only edges that do not represent any cycle so far. If it were not the case then another sequential cycle C'' different from C would obtain its representative edge $e_{C''}$ from $I_{C'} \cap I_{C''}$ and hence a representative edge would already be assigned to C' before C is considered.

Hopfield network $\mathcal{H}_T$ corresponding to triangulation T will now be constructed. With each internal edge $e = \{v_1, v_2\} \in I$ two neurons ℓ_e and r_e are associated whose states either $y_{\ell_e} = 1$ or $y_{r_e} = 1$ indicate that e follows the left or right turn, respectively, along a tristrip according to the chosen orientation of triangulated surface. Let $L_e = \{e, e_1, e_2, e_3, e_4\}$ with $e_1 = \{v_1, v_3\}$, $e_2 = \{v_2, v_3\}$, $e_3 = \{v_2, v_4\}$, and $e_4 = \{v_1, v_4\}$ be the set of edges of the two triangles $\{v_1, v_2, v_3\}$, $\{v_1, v_2, v_4\}$ that share edge e. Denote by $J_e = \{\ell_f, r_f \,;\, f \in L_e \cap I\}$ the set of neurons local to e that are associated with the internal edges from L_e. Unit ℓ_e is connected with all neurons from J_e via negative weights except for units r_{e_2} (if $e_2 \in I$), ℓ_e, and r_{e_4} (if $e_4 \in I$) whose states may encode a tristrip that traverses edge e by the left turn. Such a situation (for $L_e \subseteq I$) is depicted in Figure 3 where the edges shared by triangles within the tristrip together with associated active neurons r_{e_2}, ℓ_e, r_{e_4} are marked. Similarly, unit r_e is connected with neurons from J_e except for units ℓ_{e_1} (if $e_1 \in I$), r_e, and ℓ_{e_3} (if $e_3 \in I$) corresponding to the right turn. Thus for each internal edge $e \in I$ define weights $w(i, \ell_e) = -7$ for $i \in J_e \setminus \{r_{e_2}, \ell_e, r_{e_4}\}$ and $w(i, r_e) = -7$ for $i \in J_e \setminus \{\ell_{e_1}, r_e, \ell_{e_3}\}$. Hence, the states of Hopfield network $\mathcal{H}_T$ with the negative symmetric weights which enforce locally the alternation of left and right turns encode tristrips. Furthermore, for each representative edge e_C ($C \in \mathcal{C}$) define $j_C = \ell_{e_C}$ if e_C follows the left turn along sequential cycle C or $j_C = r_{e_C}$ if e_C follows the right turn along C. Let $J = \{j_C \,;\, C \in \mathcal{C}\}$ be the set containing all such neurons whereas $J' = \{\ell_e, r_e \notin J \,;\, e \in I\}$ denotes its complement. The thresholds of neurons associated with internal edges are defined as $h(j) = -5 + 2b_{e(j)}$ for $j \in J'$ and $h(j) = 1 + 2b_{e(j)}$ for $j \in J$ where $e(j) = e$ for $j \in \{\ell_e, r_e\}$ and $b_e = |\{C \in \mathcal{C} \,;\, e \in B'_C\}| \leq 2$ for $B'_C = B_C \setminus L_{e_C}$.

Nevertheless, Hopfield network $\mathcal{H}_T$ must also avoid the states encoding cycled strips of triangles around sequential cycles [1]. Such infeasible states would have less energy E than those encoding the optimal stripifications [3]. For this purpose, two auxiliary neurons d_C, a_C are introduced for each sequential cycle $C \in \mathcal{C}$. Unit d_C computes the disjunction of outputs from all neurons i associated with boundary edges $e(i) \in B'_C$ of C which, being active, enables the activation of unit j_C associated with representative edge e_C. Hence, any tristrip may pass through edge e_C along the direction of C only if a boundary edge of C is is a part of another tristrip crossing the sequential cycle C. This ensures that the states of Hopfield network $\mathcal{H}_T$ do not encode sequential cycles. In addition, unit a_C balances the contribution of d_C to energy E when j_C is passive. As depicted in Figure 4 this is implemented by thresholds $h(d_C) = h(a_C) = 1$ and symmetric weights $w(i, d_C) = w(d_C, i) = 2$ for $e(i) \in B'_C$, $w(d_C, j_C) = w(j_C, d_C) = 7$, $w(d_C, a_C) = w(a_C, d_C) = 2$, and $w(j_C, a_C) = w(a_C, j_C) = -2$, for each sequential cycle $C \in \mathcal{C}$. This completes the construction of Hopfield network $\mathcal{H}_T$.

Moreover, observe that the number of units $s = 2|I| + 2|\mathcal{C}|$ (similarly the number of connections) in $\mathcal{H}_T$ is linear in terms of triangulation size $n = |T|$ because the number of sequential cycles $|\mathcal{C}|$ can be upper bounded by $2|I| = O(n)$ since each internal edge can belong to

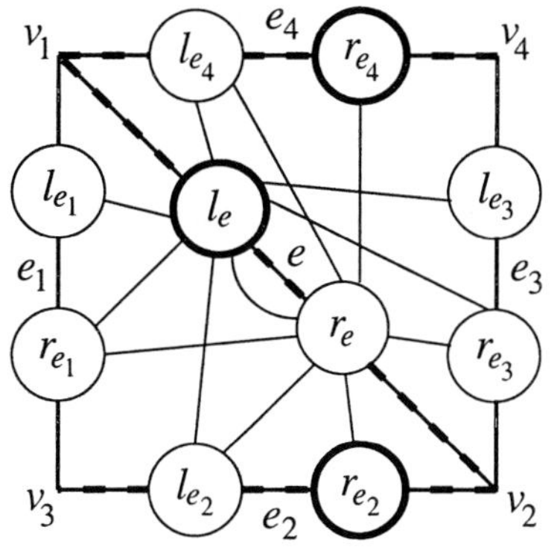

Fig. 3. The construction of $\mathcal{H}_T$ related to $e \in I$.

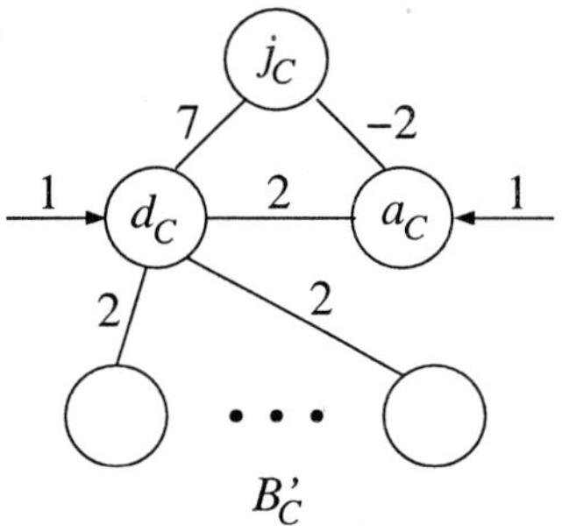

Fig. 4. The construction of $\mathcal{H}_T$ related to $C \in \mathcal{C}$.

Table 1. The average number of tristrips for "grid" models obtained by HTGEN and FTSG

n	HTGEN			FTSG
392	$T^{(0)} = 5, \varepsilon = 0.3$ 88 $\tau^* = 23$ (0.10s)	$T^{(0)} = 10, \varepsilon = 0.05$ 63 $\tau^* = 166$ (0.72s)	$T^{(0)} = 18, \varepsilon = 0.01$ 53 $\tau^* = 1648$ (7.21s)	67
1152	$T^{(0)} = 5, \varepsilon = 0.1$ 243 $\tau^* = 442$ (0.76s)	$T^{(0)} = 10, \varepsilon = 0.05$ 172 $\tau^* = 347$ (6.01s)	$T^{(0)} = 15, \varepsilon = 0.1$ 151 $\tau^* = 1107$ (18.99s)	187
2312	$T^{(0)} = 7, \varepsilon = 0.1$ 404 $\tau^* = 117$ (5.31s)	$T^{(0)} = 10, \varepsilon = 0.05$ 337 $\tau^* = 489$ (21.29s)	$T^{(0)} = 15, \varepsilon = 0.1$ 297 $\tau^* = 1967$ (86.28s)	373

at most two cycles. In addition, it has been proved [3] that the classes of equivalent optimal stripifications of T are mapped one to one to the minimum energy states that are reached during any sequential computation by $\mathcal{H}_T$ starting at the zero initial state (or $\mathcal{H}_T$ can be initialized arbitrarily if one asymmetric weight is introduced).

4 Experiments

A C++ program HTGEN has been created to automate the reduction from Section 3 including the simulation of Hopfield network $\mathcal{H}_T$ using simulated annealing (see Section 2). The input for HTGEN is an object file (in the Wavefront .obj format) describing triangulated surface model T by a list of geometric vertices with their coordinates followed by a list of triangular faces each composed of three vertex reference numbers. The program generates corresponding $\mathcal{H}_T$ which then computes stripification of T. This is extracted from final stable state $\mathbf{y}^{(\tau^*)}$ of $\mathcal{H}_T$ at microscopic time τ^* into an output .obj file containing a list of tristrips together with vertex data. The user may control the Boltzmann machine by specifying the initial temperature $T^{(0)}$ and the stopping criterion ε given as the maximum percentage of unstable units at the end of stochastic computation.

Program HTGEN has been compared against a leading practical system FTSG that computes stripifications [1]. Apart from other data, experiments have been conducted using "grid" models which are generated by randomly triangulating each square in a $b \times b$ regular grid containing of $n = 2(b-1)^2$ triangles. The average number of tristrips obtained by HTGEN and FTSG are summarized in Table 1 where 10 random models were used for each grid size $b = 15, 25, 35$. The results from HTGEN were further averaged for each model over 10 trials of simulated annealing applied for three different initial temperatures $T^{(0)}$ and stopping criteria ε. The corresponding average convergence times τ^* together with the running times in seconds (on common PC) increase as $T^{(0)}$ increases (and ε decreases). Thus $T^{(0)}$ controls the trade-off between the running time and the quality of stripification. One can achieve much better results by HTGEN than by using FTSG with its most successful options (-dfs, -alt) although the running time of HTGEN grows rapidly when the global optimum is being approached. As concerns the time complexity, system HTGEN cannot compete with real-time program FTSG providing the stripifications within a few milliseconds. Nevertheless, HTGEN can be useful if one is interested in the stripification with a small number of tristrips at a preprocessing stage.

References

[1] Estowski, R., Mitchell, J.S.B., Xiang, X. (2002) Optimal decomposition of polygonal models into triangle strips. In: *Proceedings of the 18th Annual Symposium on Computational Geometry*. ACM Press, pp. 254–263.

[2] Pospíšil, D. (2002) Generating triangle strips by Hopfield network. Student's project (in Czech), Faculty of Informatics, Masaryk University, Czech Republic.

[3] Šíma, J. (2004) Tristrips on Hopfield networks. Technical report V-908, Institute of Computer Science, Academy of Sciences of the Czech Republic.

[4] Hopfield, J.J. (1982) Neural networks and physical systems with emergent collective computational abilities. *Proceedings of the National Academy of Sciences USA 79*(8): 2554–2558.

[5] Barahona, F. (1982) On the computational complexity of Ising spin glass models. *Journal of Physics A: Mathematical and General 15*(10): 3241–3253.

[6] Hopfield, J.J., Tank, D.W. (1985) "Neural" computation of decision in optimization problems. *Biological Cybernetics 52*(3): 141–152.

[7] Šíma, J., Orponen, P. (2003) General-purpose computation with neural networks: A survey of complexity theoretic results. *Neural Computation 15*(12): 2727–2778.

[8] Ackley, D.H., Hinton, G.E., Sejnowski, T.J. (1985) A learning algorithm for Boltzmann machines. *Cognitive Science 9*(1): 147–169.

The Linear Approximation Method to the Modified Hopfield Neural Network Parameters Analysis

S. I. Bauk[1], S. M. Perovich[2], A. Lompar

[1]Faculty of Maritime Studies, University of Montenegro
[2]Department of Electrical Engineering, University of Montenegro

Abstract

The dynamic of Hopfield network is usually described by the system of differential equations. Our idea is to modify Hopfield network in aim to allow its behavior description by the system of transcendental exponential equations solvable analytically by the Special Trans Function Theory (STFT). Furthermore, the linear approximation method to the system of transcendental exponential equations describing the modified Hopfield network, based upon the STFT, has been discussed in some details.

1 Introduction

The Hopfield type neural network is an important one due to its applicability in solving associative memory, pattern recognition and optimization problems [1]-[3]. The segment of the nonlinear active electronic circuit in Figure 1 represents classical Hopfield neural network consisting of n neurons. This circuit contains resistors, capacitors, ideal current sources and amplifiers with activation functions, which are often differentiable monotonically increasing ones [4].

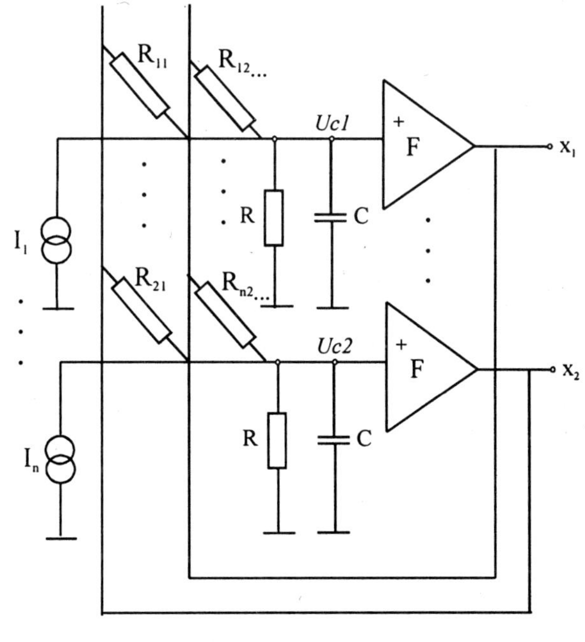

Fig. 1. The classical Hopfield network electronic model consisting of n neurons

In the paper capacitors at the amplifiers inputs have been replaced with the inverse polarized diodes with large capacity of approximately $0.5 \div 0.9\mu F$ (Figure 2). By these replacements, the classical Hopfield model analysis come to be moved from the field of linear differential equations to the field of exponential equations solvable easily by the linear approximation method proposed throughout the next sections.

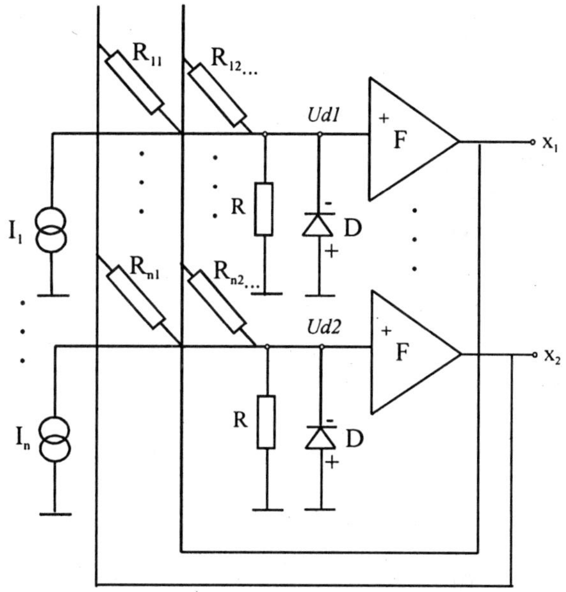

Fig. 2. The modified Hopfield network architecture consisting of n neurons

2 The differential equations transformation into the exponential ones

The electronic circuit segment of Hopfield network with n neurons is given in Figure 1 and its dynamic can be expressed with the linear differential equation of type:

$$C\frac{du_{c_i}}{dt}+\frac{u_{c_i}}{R}=\sum_{j=1}^{n}\frac{1}{R_{ij}}\left(f\left(u_{c_i}\right)-u_{c_i}\right)+I_i\,; \quad i=\overline{1,n} \quad (1)$$

where I_i is an external input signal (or bias) gained from the ideal current source, u_{c_i} is i-th capacitor voltage, $f\left(u_{c_i}\right)$ is a neuron output voltage, R is the same resistance for each amplifier, while R_{ij} are the resistors representing network weights. By replacing capacitors (Figure 1) with the inverse polarized diodes (Figure 2) with large capacity, at the amplifiers inputs, becomes possible to replace the system of the linear differential equations (1) with a system of exponential equations:

$$i_s\left(1-e^{\frac{-u_{d_i}}{V_T}}\right)+\frac{u_{d_i}}{R}=\sum_{j=1}^{n}\frac{f(u_{d_i})-u_{d_i}}{R_{ij}}+I_i;\quad i=\overline{1,n}\cdot \tag{2}$$

where I_i is i-th external current or bias, i_s is diode saturation current, u_{d_i} is i-th neuron input signal, that is diode voltage, V_T is the thermal voltage, and $f(u_{d_i})$ is i-th neuron activation function. The system of exponential equations (2) is to be linear approximated by the Special Trans Function Theory (STFT) approach, in the manner being described in the next section.

3 The linear approximation method based on STFT

Let us suppose that i-th neuron activation function, that is i-th amplifier transfer function, has the simplest form $f(u_{d_i})=C_{I_i}=const.$ Under this assumption equation (2) can be rewritten as follows:

$$i_s\left(1-e^{\frac{-u_{d_i}}{V_T}}\right)+\frac{u_{d_i}}{R}=\sum_{j=1}^{n}\frac{C_{I_i}-u_{d_i}}{R_{ij}}+I_i;\quad i=\overline{1,n}\cdot \tag{3}$$

Now, the term $e^{-\frac{u_{d_i}}{V_T}}$ can be approximated with:

$$e^{\frac{-u_{d_i}}{V_T}}=1-\frac{u_{d_i}}{V_T}+\delta_i;\quad i=\overline{1,n} \tag{4}$$

where δ_i is a linear approximation error. By taking into account this linear approximation, equation (3) takes the form:

$$i_s\left(\frac{-u_{d_i}}{V_T}-\delta_i\right)+\frac{u_{d_i}}{R}=\sum_{j=1}^{n}\frac{C_{I_i}-u_{d_i}}{R_{ij}}+I_i;\quad i=\overline{1,n}\cdot \tag{5}$$

It is to be pointed out that equation (4) can be solved analytically by the Special Trans Function Theory [5]-[7]. Namely, for given δ_i and V_T, we can obtain value of u_{d_i} in analytical closed-form:

$$u_{d_i}=V_T\cdot trans(\delta_i,V_T);\quad i=\overline{1,n} \tag{6}$$

where $trans(\delta_i,V_T)$ is new trans function defined as follows:

$$trans(\delta_i,V_T)=\lim_{x\to\infty}\left[\ln\left(\frac{\sum_{k=0}^{[x+1]}(-1)^k e^{-k\xi}\frac{(x+1-k)^k}{k!}}{\sum_{k=0}^{[x]}(-1)^k e^{-k\xi}\frac{(x-k)^k}{k!}}\right)\right];$$

$$\xi=1+\delta_i;\ i=\overline{1,n} \tag{7}$$

The equation (7) is obtained by classical STFT approach [8]. By using in this manner obtained u_{d_i}, we can now determine the new value of δ_i, that is $^{(1)}\delta_i$, as remain part of MacLaurin series:

$$^{(1)}\delta_i=\frac{(u_{d_i}/V_T)^2}{2!}-\frac{(u_{d_i}/V_T)^3}{3!}+\frac{(u_{d_i}/V_T)^4}{4!}-\frac{(u_{d_i}/V_T)^5}{5!}+\ldots \tag{8}$$

By means of this new value for δ_i, i.e. $^{(1)}\delta_i$, the new one for u_{d_i}, that is for $^{(1)}u_{d_i}$, can be calculated on the base of transformed equation (5), which after some elementary transformations takes the form:

$$^{(1)}u_{d_i}=\frac{\sum_{j=1}^{n}\frac{C_{I_i}}{R_{ij}}+I_i+i_s\,^{(1)}\delta_i}{\frac{i_s}{V_T}+\frac{1}{R}+\sum_{j=1}^{n}\frac{1}{R_{ij}}};\ i=\overline{1,n} \tag{9}$$

The above procedure is to be continued until not the inequalities $\left|^{(1)}\delta_i-\delta_i\right|\le\varepsilon$ and $\left|^{(1)}u_{d_i}-u_{d_i}\right|\le\zeta$, for enough small real ε and ζ, be satisfied.

3.1 The numerical example

Some numerical results related to previously described method to the u_{d_i} estimation, for given values of R, R_I, i_s, V_T, C_I and I, in the case of one neuron, are presented in Table 1.

The same method could be applied to two or more neurons forming the network. The iterative procedures should be performed independently, for each neuron, until all δ_i and u_{d_i} $(i=\overline{1,n})$ be obtained with the appropriate accuracy.

Table 1. The numerical results obtained through 15 iterations, for $R = 0.7M\Omega; R_1 = 0.5M\Omega; i_s = 1\mu A; I = 1nA; C_1 = 0.5V$ and $V_T = 25mV$

It.	δ	${}^{(1)}\delta$	$\left\|{}^{(1)}\delta - \delta\right\|$
1.	0.300000	0.299393	0.606883 E-03
2.	0.299393	0.298790	0.602752 E-03
3.	0.298790	0.298191	0.598671 E-03
4.	0.298191	0.297597	0.594636 E-03
5.	0.297597	0.297006	0.590649 E-03
6.	0.297006	0.296419	0.586707 E-03
7.	0.296419	0.295836	0.582809 E-03
8.	0.295836	0.295257	0.578957 E-03
9.	0.295257	0.294682	0.575148 E-03
10.	0.294682	0.294111	0.571383 E-03
11.	0.294111	0.293543	0.567660 E-03
12.	0.293543	0.292979	0.563978 E-03
13.	0.292979	0.292419	0.560337 E-03
14.	0.292419	0.291862	0.556736 E-03
15.	0.291862	0.291309	0.553175 E-03
It.	u_{d_i}	${}^{(1)}u_{d_i}$	$\left\|{}^{(1)}u_{d_i} - u_{d_i}\right\|$
1.	0.022222	0.019729	2.492305 E-03
2.	0.022196	0.019718	2.478252 E-03
3.	0.022170	0.019706	2.464277 E-03
4.	0.022145	0.019694	2.450378 E-03
5.	0.022120	0.019683	2.436556 E-03
6.	0.022094	0.019672	2.422808 E-03
7.	0.022069	0.019660	2.409135 E-03
8.	0.022044	0.019649	2.395536 E-03
9.	0.022020	0.019638	2.382010 E-03
10.	0.021995	0.019627	2.368556 E-03
11.	0.021971	0.019616	2.355173 E-03
12.	0.021947	0.019605	2.341861 E-03
13.	0.021922	0.019594	2.328619 E-03
14.	0.021898	0.019583	2.315447 E-03
15.	0.021875	0.019572	2.302343 E-03

4 The capacitor and the diode equalization

The basic question related to the proposed Hopfield neural network modification is undoubtedly: under which condition(s) the capacitor can be replaced with an inverse polarized diode? The equalization can be realized under the assumption that the current $i(t)$ (Figure 3) takes the form:

$$i(t) \approx \gamma_i e^{\beta_i t} \tag{10}$$

where

$$\gamma_i = \frac{\alpha_i i_s}{1-\alpha_i e^{-\beta_i t}};\ \alpha_i = \frac{i(0)}{i(0)+i_s} \text{ and } \beta_i = \frac{i_s}{CV_T}.$$

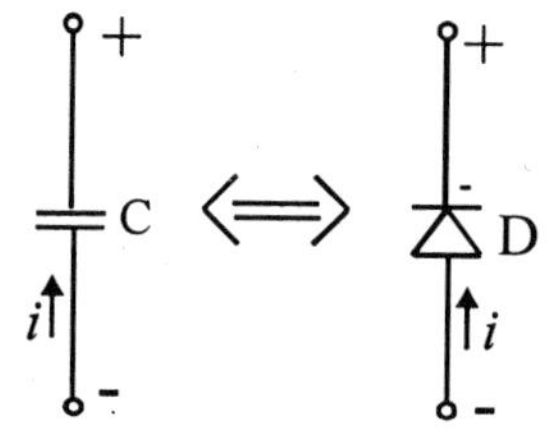

Fig. 3. The capacitor and the inverse polarized diode equalization

The expression (10) for current $i(t)$ is obtained through following steps:

(a) by equalizing capacity (C) voltage u_C with an inverse polarized diode (D) voltage u_d:

$$u_C = \frac{1}{C}\int_0^t i(t)dt \Leftrightarrow u_d = -V_T \ln\left(\frac{i(t)+i_s}{i_s}\right), \tag{11}$$

that is

$$\frac{1}{C}\int_0^t i(t)dt = -V_T \ln\left(\frac{i(t)+i_s}{i_s}\right); \tag{12}$$

(b) by differentiation the above equalization, we get:

$$\frac{i(t)}{C} = -V_T \frac{i'(t)}{i(t)+i_s}; \tag{13}$$

(c) now, we are performing some elementary transformations (formulae (14)-(17)):

$$\frac{1}{CV_T} = -\frac{i'(t)}{i(t)}\cdot\frac{1}{(i(t)+i_s)} = -\frac{i'(t)}{i(t)(i(t)+i_s)}; \tag{14}$$

$$\frac{1}{CV_T} = -\frac{i'(t)}{i_s}\left(\frac{1}{i(t)} - \frac{1}{i(t)+i_s}\right); \tag{15}$$

$$\frac{i_s}{CV_T} = -\frac{i'(t)}{i(t)} + \frac{i'(t)}{i(t)+i_s} \Big/ \int_0^t; \tag{16}$$

$$\frac{i_s}{CV_T}t = -\ln\frac{i(t)}{i(0)} + \ln\left(\frac{i(t)+i_s}{i(0)+i_s}\right) = \ln\frac{(i(t)+i_s)i(0)}{(i(0)+i_s)i(t)} \tag{17}$$

d) by using replacements $\alpha_i = \dfrac{i(0)}{i(0)+i_s}$, as well as $\beta_i = \dfrac{i_s}{CV_T}$, we finally obtain expression for $i(t)$ in the following form:

$$i(t) = \frac{\alpha_i i_s e^{-\beta_i t}}{1-\alpha_i e^{-\beta t}} \approx \gamma_i e^{-\beta_i t} \quad (18)$$

which is identically the same as (10). By supplying the current in the circuit representing Hopfield modified structure, changing in time in accordance with (10), the capacitor and inverse polarized diode equalization can be performed. In other words, the condition of the capacitor and the diode equalization is provided in this manner rather theoretically. This equalization is of up-most importance in stability and network energy function analysis. It is to be noted that the stability and energy function analyses of the modified Hopfield-type neural network are to be the subjects of further more rigorous investigation.

5 Conclusions

The idea of Hopfield neural network electronic model modification by the capacitor replacement with the inverse polarized diode with great PN junction capacity has been proposed in the paper. It has been done in aim to enable Hopfield network architecture analysis in the domain of exponential equations instead of in the domain of differential ones. Besides, the exponential equations, describing modified Hopfield architecture, have been linear approximated and solved analytically by STFT approach. The obtained numerical results confirm in a way the validity and applicability of proposed method.

The equalization between the capacitor and the diode at the neuron input has been examined in the paper, as well. It is shown that this equalization can be established under the condition that the current in the modified Hopfield structure has the appropriate continuous-time form.

The further investigation should be oriented toward modified Hopfield architecture stability and energy function analysis, which are of prior importance in improving system performance and in solving some difficult problems, like some of the associative memory, the pattern recognition and the optimization problems are.

References

[1] Juang J-C. (1999), Stability Analysis of Hopfield-Type Neural Networks, *IEEE Transactions on Neural Networks*, vol. 10, no. 6, pp 1336-1374

[2] Quiao H., et al. (1999), Nonlinear Measures: A New Approach to Exponential Stability Analysis for Hopfield-Type Neural Networks, *IEEE Transactions on Neural Networks*, vol. 12, no. 2, pp 360-369

[3] Kakeya H., Okabe Y. (2000), Fast combinatorial optimization with parallel digital computers, *IEEE Transactions on Neural Networks*, vol. 11, no. 6, pp 1323-1331

[4] Hassoun M. H. (1995), *Fundamentals of Artificial Neural Networks*, The MIT Press, London

[5] Bauk S., Perovich S. M. (2003) STFT iterative procedure for estimating probability of clients arrivals in case of M/M/1 queue, *SYM-OP-IS*, S&M, pp 735-737

[6] Bauk S., Perovich S. M. (2003), Procjena vjerovatnoće nastupanja otkaza primjenom teorije specijalnih trans funkcija, *NSS-OMO*, S&M, pp 60-63

[7] Bauk S., Perovich S. (2004) Considering the special trans function theory approach to the probability estimation (submitted to the *Statistics and Probability Letters*)

[8] Perovich S. (2004) Research monograph – The Special Trans Function Theory, University of Montenegro

The Analytical Analysis of Hopfield Neuron Parameters by the Application of Special Trans Function Theory

S. M. Perovich[1], S. I. Bauk[2], N. Konjevic[2]

[1]Department of Electrical Engineering, University of Montenegro
[2]Faculty of Maritime Studies, University of Montenegro

Abstract

The subject of the theoretical analysis presented in the paper is Hopfield neuron electronic model modification based upon capacitor replacement with an inverse polarized diode. The modified neuron parameters have been analytically analyzed by application of the Special Trans Function Theory (STFT). The obtained results are presented numerically and graphically.

1 Introduction

Well known Hopfield network can be thought of as a single-layer neural network of continuous nonlinear units mutually completely connected with feedback [1]. The nonlinear active electronic circuit shown in Fig. 1 represents this network active unit, that is one neuron. As obvious this circuit consists of resistors, capacitor, an ideal current source and a nonlinear amplifier. Our idea is to replace the capacitor (C) with an inverse polarized diode (D), as shown in Fig. 2. Namely, in this manner, our intention is to shift in a way Hopfield neuron parameters analysis from the domain of linear differential equation to the domain of exponential transcendental equation solvable analytically by the application of Special Trans Function Theory (STFT) [2]-[13].

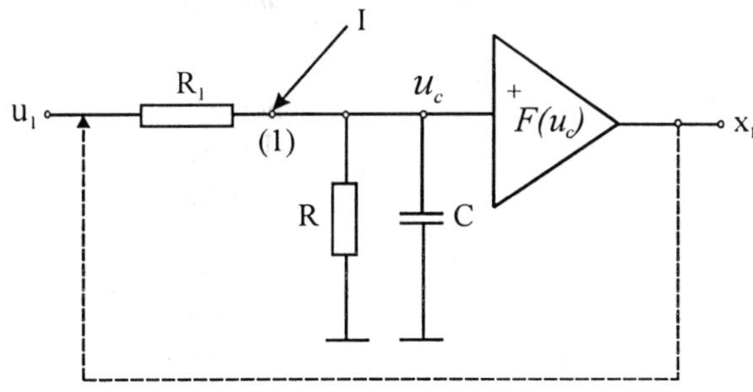

Fig. 1. Hopfield neuron model

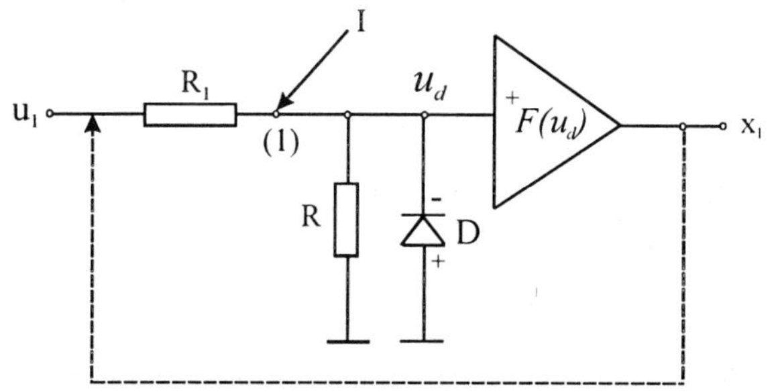

Fig. 2. Modified Hopfield neuron model

2 On Hopfield neuron model transformation

The electronic circuit segment of Hopfield network representing one neuron is given in Fig. 1. The equation describing the evolution of its output state can be obtained by applying Kirchoff's current law to the node (1) in the form:

$$C\frac{du_C}{dt}+\frac{u_C}{R}=\frac{u_1}{R_1}+I \qquad (1)$$

where I is an external input signal (or bias) gained from an ideal current source, u_c is capacitor voltage and $F(u_C)$ is an activation function (amplifier transfer function), which is usually of type: $F(u_C)=\frac{1}{1+e^{-u_C}}$.

By replacing capacitor (Fig. 1) with an inverse polarized diode (Fig. 2) with large capacity of approximately $0.9\mu F$, we are in position to replace differential equation (1) with an exponential equation of type:

$$i_s\left(1-e^{\frac{-u_d}{V_T}}\right)+\frac{u_d}{R}=\frac{u_1}{R_1}+I \qquad (2)$$

where i_s is the saturation current; u_d is the neuron input signal, i.e. diode voltage; V_T is the thermal voltage; I is an external current or bias.

The multiplication of equation (2) by the term $\frac{R}{V_T}$ gives the following equation:

$$\frac{Ri_s}{V_T} - \frac{Ri_s}{V_T} \cdot e^{-\frac{u_d}{V_T}} + \frac{u_d}{V_T} = \frac{R}{V_T} \cdot \frac{u_1}{R_1} + \frac{RI}{V_T} \quad (3)$$

By introducing several transformations of the form:

$$y = \frac{u_d}{V_T}; \alpha = -\frac{Ri_s}{V_T}; \Omega = \frac{R}{V_T} \cdot \frac{u_1}{R_1}; \beta = \frac{RI}{V_T}$$

we get transcendental equation of type:

$$\alpha e^{-y} - \alpha + y = \Omega + \beta . \quad (4)$$

Finally, by the replacement: $\alpha_1 = \alpha + \Omega$, the transcendental equation:

$$\alpha e^{-y} + y = \alpha_1 + \beta \quad (5)$$

appears.

This equation is directly solvable by application of the Special Trans Function Theory (STFT). Thus, by solving (5) in analytical closed form becomes possible to analyze analytically diode voltage u_d with respect to values u_1, R_1, R and (or) I. Besides, by STFT application it becomes possible to obtain analytically *gradients:* $\frac{\Delta u_d}{\Delta u_1}; \frac{\Delta u_d}{\Delta R_1}; \frac{\Delta u_d}{\Delta R}$ and (or) $\frac{\Delta u_d}{\Delta I}$.

3 Genesis of the analytical closed form solution to the equation (5)

According to the Special Trans Function Theory [2]-[13], in aim to solve equation (5) analytically in closed-form, first of all, an appropriate equation for identification (EQID), to (5), is to be found. The theoretical method to the genesis of EQID is the intuitive or analogy. Thus, in this case, EQID is linear differential equation and takes the following form:

$$\varphi'(z) + \alpha\varphi(z-1) - \alpha_2\varphi(z) = 0 \quad (6)$$

where $\alpha_2 = \alpha_1 + \beta$. The asymptotic solution of this equation has the form

$$\varphi_{as} = e^{yz} \quad (7)$$

and it is essential to the EQID genesis. Consequently, by the replacement of this asymptotic solution into the equation (6), the equation (5) is to be obtained. On the other hand, equation (6) is solvable analytically by Laplace transform. Thus, after applying Laplace transform equation (6) takes the form

$$\Phi(s) \cdot \left(s + \alpha e^{-s} - \alpha_2\right) = \varphi(0) \quad (8)$$

where

$$\Phi(s) = L(\varphi(z)) \quad (9)$$

or

$$\Phi(s) = \frac{\varphi(0)}{s - \alpha_2 + \alpha e^{-s}} . \quad (10)$$

Finally,

$$\Phi(s + \alpha_2) = \frac{\varphi(0)}{s + \alpha e^{-\alpha_2} \cdot e^{-s}} = \frac{\varphi(0)}{s\left(1 + \frac{\alpha e^{-\alpha_2}}{s} e^{-s}\right)}$$

or

$$\frac{\varphi(0)}{s} \sum_{n=0}^{\infty} (-1)^n \left(\alpha e^{-\alpha_2}\right)^n \cdot \frac{e^{-ns}}{s^n} \quad (11)$$

under the condition $\frac{\alpha e^{-\alpha_2}}{s} e^{-s} << 1$.

Now, inverting term by term, in original z *domain,* we obtain

$$\varphi(z) e^{-\alpha_2 z} = \varphi(0) \sum_{n=0}^{[z]} (-1)^n \left(\alpha e^{-\alpha_2}\right)^n \cdot \frac{(z-n)^n}{n!} \quad (12)$$

and finally,

$$\varphi(z) = \varphi(0) e^{\alpha_2 z} \sum_{n=0}^{[z]} (-1)^n \left(\alpha e^{-\alpha_2}\right)^n \cdot \frac{(z-n)^n}{n!} . \quad (13)$$

According to the unique solution principle from equations (7) and (13) follows:

$$\lim_{z\to\infty} \left(\frac{\varphi(z)}{\varphi_{as}(z)} \right) = 1 \quad (14)$$

or

$$\lim_{z\to\infty} \frac{\varphi(z+1)}{\varphi(z)} = \frac{\varphi_{0as} e^{y(z+1)}}{\varphi_{0as} e^{yz}} = e^y . \quad (15)$$

From the above equalization we have:

$$y = \lim_{z\to\infty} \left[ln\left(\frac{\varphi(z+1)}{\varphi(z)} \right) \right] = \lim_{z\to\infty} \left[ln\left(\frac{e^{\alpha_2(z+1)}}{e^{\alpha_2 z}} \cdot \frac{\varphi_0(z+1)}{\varphi_0(z)} \right) \right] \quad (16)$$

so, the final expression for y becomes:

$$y = \alpha_2 + \lim_{z\to\infty} \left[ln\left(\frac{\varphi_0(z+1)}{\varphi_0(z)} \right) \right] = \alpha_2 + trans_N(\alpha, \Omega, \beta) \quad (17)$$

where

$$\varphi_0(z) = \sum_{n=0}^{[z]} (-1)^n \left(\alpha e^{-\alpha_2}\right)^n \cdot \frac{(z-n)^n}{n!} \qquad (18)$$

and

$$\varphi_0(z+1) = \sum_{n=0}^{[z+1]} (-1)^n \left(\alpha e^{-\alpha_2}\right)^n \cdot \frac{(z+1-n)^n}{n!}. \qquad (19)$$

The error function is defined as:

$$G = \alpha e^{-y} + y - \alpha_1 - \beta \qquad (20)$$

for y derived by (17), and it must satisfy the inequality $G \le g_0$, where g_0 is enough small real positive number. Therefore, the diode voltage u_d, as neuron input signal, could be obtained by using special trans function, instead of linear differential equation (1) (applicable to the classical Hopfield neuron model) in the following manner:

$$u_d = V_T\left(\alpha_2 + trans_N(\alpha,\alpha_2)\right) \qquad (21)$$

where $trans_N(\alpha,\alpha_2)$ is a new neuron trans function defined as:

$$trans_N(\alpha,\alpha_2) = \lim_{z\to\infty}\left[\ln \left(\frac{\sum_{n=0}^{[z+1]} (-1)^n \left(\alpha e^{-\alpha_2}\right)^n \cdot \frac{(z+1-n)^n}{n!}}{\sum_{n=0}^{[z]} (-1)^n \left(\alpha e^{-\alpha_2}\right)^n \cdot \frac{(z-n)^n}{n!}} \right)\right].$$

4 The neuron input signal derivation function

The partial derivation function of the neuron input signal u_d upon voltage u_1 value, can be defined on the base of the equation (21). Let us note, that α_2 is a function of voltage u_1, so:

$$\frac{\Delta u_d}{\Delta u_1} = V_T\left(\frac{\Delta\alpha_2}{\Delta u_1} + \frac{\Delta trans_N(\alpha,\alpha_2)}{\Delta\alpha_2}\cdot\frac{\Delta\alpha_2}{\Delta u_1}\right) \qquad (22)$$

since, $\frac{\Delta\alpha_2}{\Delta u_1} = \frac{R}{V_T R_1}$ follows that:

$$\frac{\Delta trans_N(\alpha,\alpha_2)}{\Delta\alpha_2}\cdot\frac{\Delta\alpha_2}{\Delta u_1} = \lim_{z\to\infty}\left(\frac{\sum_{n=1}^{[z+1]} (-1)^{n+1}\left(\alpha e^{-\alpha_2}\right)^n \cdot \frac{(z+1-n)^n}{(n-1)!}}{\sum_{n=0}^{[z+1]} (-1)^n\left(\alpha e^{-\alpha_2}\right)^n \cdot \frac{(z+1-n)^n}{n!}} - \frac{\sum_{n=1}^{[z]} (-1)^{n+1}\left(\alpha e^{-\alpha_2}\right)^n \cdot \frac{(z-n)^n}{(n-1)!}}{\sum_{n=0}^{[z]} (-1)^n\left(\alpha e^{-\alpha_2}\right)^n \cdot \frac{(z-n)^n}{n!}} \right)\cdot\frac{R}{R_1} \qquad (23)$$

and finally, it is observed that

$$\frac{\Delta u_d}{\Delta u_1} = V_T \cdot \frac{\Delta\alpha_2}{\Delta u_1}\left(1 + \frac{\Delta trans_N(\alpha,\alpha_2)}{\Delta\alpha_2}\right) = \left[1 + \lim_{z\to\infty}\left(\frac{\sum_{n=1}^{[z+1]} (-1)^{n+1}\left(\alpha e^{-\alpha_2}\right)^n \cdot \frac{(z+1-n)^n}{(n-1)!}}{\sum_{n=0}^{[z+1]} (-1)^n\left(\alpha e^{-\alpha_2}\right)^n \cdot \frac{(z+1-n)^n}{n!}} - \frac{\sum_{n=1}^{[z]} (-1)^{n+1}\left(\alpha e^{-\alpha_2}\right)^n \cdot \frac{(z-n)^n}{(n-1)!}}{\sum_{n=0}^{[z]} (-1)^n\left(\alpha e^{-\alpha_2}\right)^n \cdot \frac{(z-n)^n}{n!}} \right)\right]\cdot\frac{R}{R_1}. \qquad (24)$$

Consequently, it is possible to track changes of u_d analytically with respect to the voltage u_1, changing for small arbitrary chosen positive values. The appropriate numerical results and graphical presentations are given in the next section.

5 The numerical results analysis

In this section some simulation results obtained by previously performed theoretical analysis to the equation (5) are presented. Thus, in the Table 1 are given some numerical results of y, that is, of diode voltage u_d and error $|G|$ for different values of integer $[z]$ controlling number of the solution accurate digits. The other values being involved in proposed Hopfield neuron model analysis are given:

$$R = 0.1M\Omega;\ R_1 = 0.5M\Omega;\ i_s = 1\mu A;$$
$$V_T = 25mV;\ u_1 = 0.5V \text{ and } \ I = 1nA.$$

Table 1. Values for y, u_d and $|G|$ obtained by STFT for various $[z]$

$[z]$	y	$u_d[mV]$
5	1.20502516381988	30.125629095497
10	1.20398526249551	30.099631562388
15	1.20398516473426	30.099629118356

20	1.20398516603802	30.099629150950
25	1.20398516603764	30.099629150941
$[\Xi]$	$\lvert G \rvert$	
5	2.28733096933276 E-03	
10	2.12205883757394 E-07	
15	2.86742806929197 E-09	
20	8.27563712002544 E-13	
25	2.89004931097736 E-15	

Some other results for u_d and error $|G|$, obtained for various values of voltage u_I and resistance R_I, while parameters R, i_s, V_T and *I take the same values as in* the previous example, are shown in Table 2.

The essential advantage of the STFT approach is the possibility of the partial derivation of the diode voltage u_d.

Namely, by using a fixed value of u_I, we are in position to calculate $\Delta u_d / \Delta u_1$ on the basis of equation (24). Some derivation $\Delta u_d / \Delta u_1$ results, for u_I taking discrete values from the interval ($0.1 \div 0.5V$), are presented in Figure 3.

Table 2. The values of u_d and $|G|$ for different u_I and R_I ($[\Xi] = 25$)

$R_I = 0.5[M\Omega]$		
$u_I[V]$	$u_d[mV]$	$\lvert G \rvert$
0.1	4.29953528	2.11292764263141 E-06
0.2	9.22965817	3.86432836806117 E-08
0.3	14.99410535	7.57895527586649 E-10
0.4	21.84172434	5.53645670975378 E-12
0.5	30.09962915	2.89004931097736 E-15
$u_I = 0.5[V]$		
$R_I[M\Omega]$	$u_d[mV]$	$\lvert G \rvert$
0.50	30.0996291	2.89004931097736 E-15
0.55	26.1471403	8.21911982917811 E-15
0.60	23.1101060	2.19401927181728 E-12
0.65	20.7055690	6.46926609504384 E-12
0.70	18.7552700	5.86914995770460 E-11

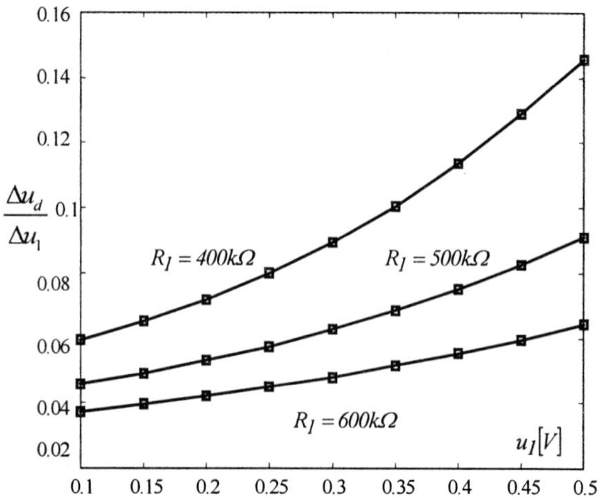

Fig. 3. The partial derivation $|\Delta u_d / \Delta u_1|$ graphical presentation (obtained by equation (24))

6 Conclusions

The paper has proposed an original approach to the modified Hopfield neuron electronic model parameters analytical analysis by the application of recently developed Special Trans Function Theory (STFT). The obtained numerical results confirm applicability and validity of the proposed method. Namely, the architecture of Hopfield electronic neuron model has been changed by replacing capacitor at the amplifier input (representing neuron body) with an inverse polarized diode whose capacity is to be similar to biological neuron membrane capacity (order of μF). This modification comes to remove Hopfield neuron parameters analysis from the field of the linear differential equation to the field of the exponential transcendental equation solvable analytically by STFT approach.

References

[1] Hassoun M. H. (1995), *Fundamentals of Artificial Neural Networks*, The MIT Press, London

[2] Perovich S.M. (2004), *The Special Trans Function Theory*, Research monograph, University of Montenegro

[3] Perovich S. M. (1992), The Transcendental Method in the Theory of Neutron Slowing Down, *Journal of Physics A: Mathematical and General 25*, pp 2969

[4] Perovich S. M. (1995), Reply to the Comment on the Transcendental Method, *Journal of Physics A: Mathematical and General 28*, pp 4191
[5] Perovich S. M. (2001), Concerning Contribution to the Special Trans Function Theory, *ETF Journals of Electrical Engineering, University of Montenegro 5*, pp 27
[6] Perovich S. M. (1996), Transcendental Method in Nonlinear Circuit Theory, *Electronics Letters 16*, pp 1433
[7] Perovich S. M. (1997), Concerning the Analytical Solution of the Disperse Equation in the Linear Transport Theory, *Transport Theory and Statistical Physics 6*, pp 705
[8] Perovich S. M., et al. (1999), Some Comments Concerning the Discrete Eigenvalue, *Nucl. Sci. Engng. 131*, pp 439
[9] Perovich S. M., Bauk S. I., et al. (2002), The Special Trans Function Theory of the AM Detector Transfer Factor Analytical Analysis, *ICECS 9th IEEE Conference Proceedings*, Dubrovnik, Croatia, pp 991-994.
[10] Perovich S. M., Bauk S. I. (2003), The STFT Time Signal Analysis to the Nonlinear Greatz's Circuits, *7th ISSPA/IEEE Conference*, Paris, France
[11] Perovich S. M., Bauk S. I., et al. (2002), Concerning the Improvement of the Error Function of the AM Detector Transfer Factor Numerical Calculations by Special Trans Function Theory, *ISSAT Conference*, Anaheim, California
[12] Perovich S. M., Bauk S.I., et al. (2004), Calculating Conductive Fluid High Level with Special Trans Function Theory, *46th IEEE International Symposium Electronics in Marine-ELMAR*, Zadar, Croatia, pp 519-523

Time-Oriented Hierarchical Method for Computation of Minor Components

M. Jankovic[1], H. Ogawa[2]

[1]Control Department, EE Institute "Nikola Tesla", Serbia and Montenegro
[2]Department of Computer Science, Tokyo Institute of Technology, Japan
E-mail: elmarkoni@ieent.org, ogawa@og.cs.titech.ac.jp

Abstract

This paper proposes a general method that transforms known neural network MSA algorithms, into MCA algorithms. The method uses two distinct time scales. A given MSA algorithm is responsible, on a faster time scale, for the "behavior" of all output neurons. On this scale minor subspace is obtained. On a slower time scale, output neurons compete to fulfill their "own interests". On this scale, basis vectors in the minor subspace are rotated toward the minor eigenvectors. Actually, time-oriented hierarchical method is proposed. Some simplified mathematical analysis, as well as simulation results are presented.

1 Introduction

Neural networks provide a way for parallel on-line computations of the principal/minor component analysis (PCA/MCA) or principal/minor subspace analysis (PSA/MSA). Due to their parallelism and adaptivity to input data, such algorithms and their implementations in neural networks are potentially useful in feature extraction and data compression. Generally speaking, the purpose of PCA is to derive a relatively small number of deccorelated linear combinations (principal components) of a set of random zero-mean variables while retaining as much of the information from the original variables as possible. Among the objectives of PCA are: dimensionality reduction, determination of linear combinations of variables, feature selection, visualization of multidimensional data, identification of underlying variables of identification of groups of objects or outliers [1].

Nevertheless, it has been clearly shown that computing the last principal components of a data sequence, i.e. those principal components endowed with the smallest (non-zero) powers, may be very useful as well, for instance in moving target following [2], frequency estimation [3], adaptive array processing, emitter location and signal parameter estimation [4], biological data analysis and understanding [5], noise reduction problems, function approximation like curve fitting and surface fitting [6] or robust constrained beamforming [7].

A layer of parallel linear artificial neurons shown in Fig. 1.

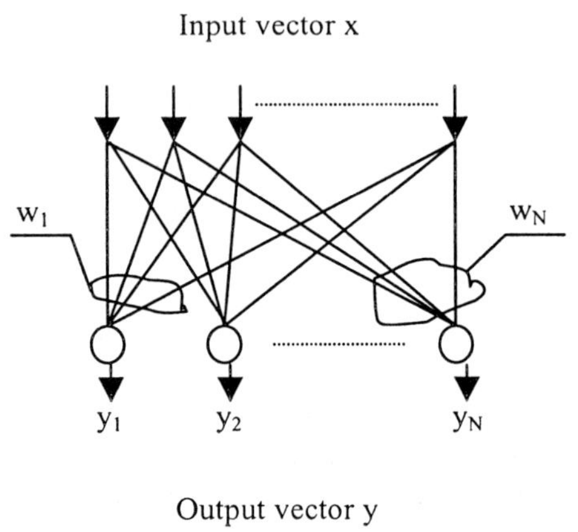

Fig.1. The linear layer of artificial neurons

The output of the n-th unit (n=1, 2, ..., N) is $y_n = w_n^T x$, with x denoting a K-dimensional input vector of the network and w_n denoting a weight vector of the n-th unit. This network, together with appropriate learning rule, could be used as a powerful technique for learning and tracking principal/minor information in time series.

Within last years various MCA and MSA learning algorithms have been proposed and mathematically investigated [1-12]. Most of them are based on local Hebbian learning. Due to their locality it has been argued that these algorithms are biologically plausible. MSA algorithms are useful for the problems in which only a minor subspace identification is of interest and not the decorellation property. In this paper we propose a simple method for converting MSA algorithms to MCA algorithms. It is named Time-Oriented Hierarchical Method (TOHM). Of course, all the time we are talking about parallel algorithms for estimation of MCA and related MSA. Algorithms which discuss sequential extraction of minor components are not considered (for review of those algorithms see e.g. [1]).

2 Time-Oriented Hierarchical Method

A general method for transformation of MSA algorithms to MCA algorithms will be introduced. The main idea is that

Each neuron tries to do what is the best for his family, and then what is the best for himself.

We shall call this idea "the family principle". In other words, the algorithm consists of two parts: the first part is responsible for the family-desirable feature learning and the second part is responsible for the individual-neuron-desirable feature learning. The second part is taken with a weight coefficient which is, by absolute value, smaller than 1. This means that some time-oriented hierarchy in realization of the family and individual parts of the learning rules is made.

In order to realize "the family principle", we propose the following general method, which transforms MSA algorithm, denoted by FP_{MSA} (defines Δw_{PSA}) to a MCA algorithm, denoted by LA_{MCA} (defines Δw_{PCA}):

$$LA_{MCA} = FP_{MSA} + IP\left(\max\left(E\left\{\left((Dy)^T y\right) \middle| \begin{array}{l} w_k^T w_k = 1, \\ k = 1,2\ldots N \end{array}\right\}\right)\right) \quad (1)$$

where D is a diagonal matrix with nonzero elements d_n and such that $|d_n|<1$. IP denotes an individual part of the learning rule (defines Δw_{IP}). This is an algorithm for achieving maximization of $E((Dy)^T y)$ under the constraints $w_k^T w_k=1$ for $k=1,2,\ldots,N$. If all d_n are equal to α, we have the homogenous case. It is not difficult to see that if homogenous MSA algorithm is used and all d_n are equal, then we have fully homogenous MCA algorithm.

3 Neural Learning on Grassman/Stiefel Submanifold

In this paragraph it will be explained how proposed learning method can be related to the neural learning on a Grassman minor submanifold (it will be defined later in the section). Only homogeneous case (all d_n are equal to α) will be analyzed. In the inhomogeneous case, algorithm can be related to a neural learning on a Stiefel minor submanifold. All definitions can be made analogously.

First, we define a Grassman on-submanifold (similar to definition of Grassman manifold):

The space of matrices $W \in O^{KxN} \subset R^{KxN}$ ($N \le K$) such that $W^T W=I$ and a function J: $O^{KxN} \to R$ such that $J(W)=J(WQ)$ for any NxN orthonormal matrix Q is called the Grassman on-submanifold.

Neural network algorithms frequently can be seen as algorithms that maximize/minimize cost function under some constraint, which is usually orthogonality constraint. In other words it can be written as:

Find W_{eks} such that: $J(W_{eks})$ =max/min $J(W)$, $W \in O^{KxN}$.

A standard way to obtain desired solutions is to define the Lagrangian function:

$$J_l(W) = J(W) + l\,\mathrm{tr}\left(W^T W - I\right) \quad (2)$$

where l is so-called Lagrangian multiplier, and to look for free extremes of the function J_l(W), for instance by means of gradient ascent/descent technique, that is

$$\frac{dW}{dt} = \eta \nabla J_l. \quad (3)$$

In order to ensure orthonormality, the iterative orthogonalization of the columns of W could be employed as well, for instance by the Gram-Schmidt orthogonalization of the matrix updated by gradient optimisation of $J(W)$ or projection onto orthogonal group [13]. However, imposing the orthonormality constraint iteratively may be problematic in practice (see e.g. [8,9]). That is the reason why the researchers started to study learning paradigms that keep the weight matrix orthogonal at any time. Such algorithms are known as SOC (Orthonormal Strongly-Constrained) algorithms.

First a SOC MCA algorithm will be introduced. Let's analyze the following system of equations:

$$y^2 = \max, \quad (4)$$

$$y = \frac{w^T}{\sqrt{w^T w}} x, \quad (5)$$

where x represents input vector for the single layer single output (y) neural network and w represents weight vector. Now, we can easily see that equation (5) can be written as

$$y^2 = \frac{w^T x x^T w}{w^T w} = \max. \quad (6)$$

If we directly construct gradient ascent algorithm taking the gradient of y^2 with respect to w, we have the following learning algorithm:

$$w(i+1) = w(i) + \gamma(i)\frac{xy\sqrt{w(i)^T w(i)} - y^2 w(i)}{w(i)^T w(i)}. \quad (7)$$

Time index i, for x and y is omitted in order to shorten equations. It is not difficult to see that if $w(i)^T w(i)=1$ we directly have famous Oja's learning rule. If we extend this to multiple output case we have

$$tr(yy^T) = \max$$

$$y_k = \frac{w_k(i)^T}{\sqrt{w_k(i)^T w_k(i)}} x, \quad (8)$$

where y_k represents the k-th output neuron and w_k represents the k-th column of W. Learning rule for weight vectors is given as

$$w_k(i+1) = w_k(i) + \gamma(i)\frac{xy_k\sqrt{w_k(i)^T w_k(i)} - y_k^2 w_k(i)}{w_k(i)^T w_k(i)}. \quad (9)$$

Now, we define Grassman minor submanifold:
The space of matrices $W\in O^{KxN} \subset R^{KxN}$ $(N\leq K)$ such that $W^TW=I$ and W spans the minor subspace defined by the N minor eigenvectors of matrix C, and a function J: $O^{KxN}\to R$ such that $J(W)=J(WQ)$ for any NxN orthonormal matrix Q will be called the Grassman minor submanifold.
If algorithm (9) is applied under strict constraint for W – W is such that $W^TW=I$ and W spans the minor subspace defined by matrix C, we actually have learning algorithm

$$w_k(i+1) = w_k(i) + \gamma(i)\left(xy_k - y_k^2 w_k(i)\right), \quad (10)$$

and

$$y_k = w_k^T x. \quad (11)$$

Now, our cost function in compact notation is

$$J = E(tr(yy^T)) = E(tr(W^T xx^T W) = tr(W^T CW) = \max. \quad (12)$$

It is not difficult to see that function J satisfies $J(W)=J(WQ)$ for any NxN orthonormal matrix Q. In other words we can say that our algorithm (10) performs neural learning on Grassman minor submanifold. So, we have one SOC algorithm. Using stochastic approximation [14] we can relate (10) to following ordinary differential equation (in compact notation)

$$\frac{dW}{dt} = \left(CW - W\mathrm{diag}\left(W^T CW\right)\right), \quad (13)$$

where diag(W^TCW) is a diagonal matrix which consists of diagonal elements of W^TCW. If we write this equation for each column w_k, we have

$$\frac{dw_k}{dt} = \left(Cw_k - \lambda_k w_k\right), \quad (14)$$

where λ_k is the k-th element of diag(W^TCW). We can easily conclude that the stationary points of these equations are minor eigenvectors of the matrix C. Since we are performing neural learning on minor subspace the resulting w_k will be equal to minor eigenvectors of matrix C. If the $w_k(i)$ of the corresponding discrete algorithm visits infinitely often a compact subset of the domain of attraction of the solution of (14), then the solution of (14) is also a solution for the corresponding discrete algorithm (10). The proof is lengthy and won't be presented here. It can be done by the approach used in [15].
Proposed method is not easy to implement in practice. The reason is that in online methods for MCA, minor Grasmman submanifold is not known. So, we have to apply Lagraingian method to our learning rule which results in the following algorithm:

$$w_k(i+1) = w_k(i) + \gamma(i)\left(xy_k - y_k^2 w_k(i)\right) + \beta\gamma(i)\Delta W_{MSA}, \quad (15)$$

where ΔW_{MSA} represents part of the learning rule that is contributed by the adopted MSA learning rule which makes "weak" constraint. Equation (15) can be written as

$$w_k(i+1) = w_k(i) + \alpha\gamma(i)\left(xy_k - y_k^2 w_k(i)\right) + \gamma(i)\Delta W_{MSA}. \quad (16)$$

If α is small enough we can consider that part multiplied by α does not affect the MSA learning part and appoximately we have a learning on Grassman minor submanifold. Now, we can see that equation (16) actually represents a method proposed in this paper, in the homogenouus case. If α is small enough, we can assume that analysis performed for SOC algorithm (10) can be valid for (16). It must be said that it is necessary to perform stability analysis for any particular selection of MSA. Only after such analysis it is possible to select proper α.

4 Simulation results

In order to illustrate effectiveness of the proposed TOHM we shall consider the small-scale numerical simulations whose results are given in Table 1-4 . The number of inputs was $K = 5$ and the number of output neurons was $N = 3$. Artificial zero-mean vectors with uncorrelated elements were generated by the following equations:

```
x(1,1) = sin(i/2);
x(2,1) = (rem(i,23) - 11)/9)^5;
x(3,1) = (rem(i,27) - 13)/9);
x(4,1) = ((rand(1,1) < .5) * 2 - 1) * log(rand(1,1) + .5);
x(5,1) = -.5 + rand(1,1).
```

In such case, eigenvectors are $c_1 = (01000)^T$, $c_2 = (00100)^T$, $c_3 = (10000)^T$, $c_4 = (00010)^T$ and $c_5 = (00001)^T$ (sorted in such way that c_1 corresponds to the largest eigenvalue and c_5 corresponds to the smallest eigenvalue). Let d be the vector which consists of diagonal elements d_k of matrix D in (1).
Tables 1-4 contain simulation results for MCA derived by implementation of TOHM on some of the stable MSA. MCA algorithm derived from some of the MSAx (x∈{1,2,3}) by the TOHM is denoted by the TOHM MSAx.

Table 1 Weight vectors of the TOHM MLA1 after 66000 iterations; d={0.36, 0.09, 0.018}

W		
-1.0001	-0.0337	-0.0905
0.0191	0.0326	-0.0461
-0.0737	-0.0496	-0.0395
-0.0217	-1.0043	0.2019
-0.0621	0.0644	0.9972

Tables 1, 2 and 3 contain simulation results for inhomogeneous TOHM-MSA1, TOHM-MSA2 and

TOHM-MSA3, respectively. MSA1-3 are defined in refs. 10, 11 and 12, respectively. Table 4 contains results for TOHM MSA2 algorithm in homogeneous case.

Table 2 Weight vectors of the TOHM MLA2 after 5000 iterations; d={0.36, 0.09, 0.018}

W		
0.0177	1.0032	0.0437
0.0103	-0.0232	-0.0227
-0.0199	-0.0188	-0.0068
1.0015	-0.0135	0.0028
0.0059	0.0393	-1.0042

Table 3 Weight vectors of the TOHM MLA3 after 20000 iterations; d={-0.005, -0.0025,- 0.00125}

W		
-0.0067	0.9971	0.0031
0.0957	-0.0728	-0.0507
0.0385	0.0177	-0.0293
-0.0651	0.0057	-0.9966
0.9925	0.0147	-0.0577

Table 4 Weight vectors of the TOHM MLA2 after 4500 iterations; d={0.36, 0.36, 0.36}

W		
0.0040	0.9967	0.0440
-0.0069	0.0778	-0.0179
0.0300	-0.0659	-0.0076
1.0004	0.0035	0.0715
-0.0621	-0.0416	0.9989

The simulation results show that the TOHM is useful.

5 Conclusion

In this paper, a general method (named time-oriented hierarchical method – TOHM) that transforms the MSA learning rules for a single layer linear neural network into MCA learning rules is analyzed. Introduction of the two distinct time bases is the novelty of the proposed algorithm. This indirectly means that possible biological implementation of the network requires two types of the neurotransmitters. On a faster time scale, MSA algorithm is responsible for the "behavior" of the all output neurons (family). On a slower scale, output neurons will compete for "fulfillment of their own interests". On this scale, basis vectors in the minor subspace are rotated toward the minor eigenvectors. Some simplified mathematical analysis, as well as simulation results are presented.

References

[1] A. Chichocki, S.–I. Amari (2003) Adaptive Blind Signal and Image Processing – Learning Algorithms and Applications. John Wiley and Sons, New York

[2] R. Klemm (1987) Adaptive airborne MTI: an auxiliary channel approach. IEE Proceedings 134: 269-276

[3] G. Mathew and V. Reddy (1994) Orthogonal Eigensubspace estimation using neural networks. IEEE Trans. On Signal Processing: 42: 1803-1811

[4] R. Schmidt (1986) Multiple emitter location and signal parameter estimation. IEEE Trans. On Antennas and Propagation 34: 276-280

[5] L. Wiscott (1998) Learning invariance manifolds. International Conference on Artificial Neural Networks: 555-560

[6] L. Xu, E. Oja, C.Y. Suen (1992) Modified Hebbian learning for curve and surface fitting. Neural Networks 5: 441-457

[7] S. Fiori (2003) A Neural Minor Component Analysis Approach to Robust Constrained Beamforming. IEE Proceedings - Vision, Image and Signal Processing 150: 205-218

[8] T.-P. Chen and S. Amari (2001) Unified stabilization approach to principal and minor components. Neural Networks 14: 1377-1387

[9] T.-P. Chen, S. Amari, and Q. Lin (1998) A unified algorithm for principal and minor components extraction. Neural Networks 11: 385-390

[10] S. Fiori (2002) A minor subspace algorithm based on neural Stiefel dynamics. International Journal of Neural Systems 12: 339 – 350

[11] S.C. Douglas, S.Y. Kung and S. Amari (1998) A self-stabilized minor subspace rule. IEEE Signal Processing Letters 5: 328-330

[12] K. Abed-Meraim, S. Attallah, A. Ckheif and Y. Hua (2000) Orthogonal Oja algorithm. IEEE Signal Processing Letters 7: 116-119

[13] S. Fiori (2001) A theory for learning by weight flow on Stiefel-Grassman Manifold. Neural Computation 13: 1625-1647

[14] L. Ljung (1977) Analysis of recursive stochastic algorithms. IEEE Trans. Automat. Contr. 22, 551-575

[15] E. Oja, J. Karhunen (1985) On stochastic approximation of the eigenvectors and eigenvalues of the expectation of a random matrix. J. Math. Anal., Appl. 106: 69-84

Evolution versus Learning in Temporal Neural Networks

Hédi Soula, Guillaume Beslon, Joël Favrel *

*PRISMa Lab., National Institute of Applied Sciences (INSA), Lyon, France, e-mail:hed@prisma.insa-lyon.fr

Abstract

In this paper, we study the difference between two ways of setting synaptic weights in a "temporal" neural network. Used as a controller of a simulated mobile robot, the neural network is alternatively evolved through an evolutionary algorithm or trained via an hebbian reinforcement learning rule. We compare both approaches and argue that in the last instance only the learning paradigm is able to exploit *meaningfully* the temporal features of the neural network.

1 Introduction

When trying to control a robot with a neural network, the evolutionist paradigm is one of the most widespread approaches [1]. It becomes even more important when dealing with temporal neural networks[1] [2, 3, 4]. Indeed, lacking an obvious learning paradigm for such kind of networks, the evolutionary approach is generally considered as the only alternative.

We argue that looking for a learning paradigm for this kind of neurons can be very fruitful. By putting its efficiency aside, we can find it more interesting in its *qualitative* expressions (although not in its quantitative ones).

Indeed, evolutionist robotic is based on the robot's ability to do a task *globally*. It has to learn a global policy losing thereby the local (here temporal) properties of the network. Hence it becomes unclear that a genetic algorithm is able to make an actual exploitation of this temporal feature.

2 Temporal Artificial Neural Networks

In the last decade, many discoveries have changed our views about the neural treatment of information. The "firing rates" models of neural nets are widespread among the neural networks community. However, while what mattered in these models is the firing frequency of the neurons, contemporary neurobiology suggests that the temporal factor (i.e. the exact time of a spike) is used by real neurons in a brain.

Indeed, temporal synchronizations between far neurons have been detected in most of cerebral areas (see [5] for a review). Shortly after, other authors ([6, 7]) have showed that the exact time of firing of a neuron can code precisely the nature of a stimulus. It is then useless to wait for more spikes to transmit the neural information to the remaining of the net (for a review of all these arguments see [8]). All this strongly supports the need to introduce time in the network. In the next section we propose a very simple model that allows the time to appear.

3 The Scheduled Mode

We present here briefly (see [9, 10] for more details) the Scheduled Mode. Relying on a specific asynchronous update rule in a network of Boolean neurons, we define the neural imbalance d_i as the difference between the state of a neuron i and its incoming stimulation.

$$d_i(t) = (1 - 2x_i(t)) \left(\sum_{k=1}^{n} W_{ik} x_k(t) + \sigma(s^2, i, t) \right)$$

where $x_i(t)$ is the state of the neuron i ($i \in [1..n]$) at a given time t ($x_i(t) \in \{0, 1\}$, $1 \leq i \leq n$), W_{ik} is the synaptic weight between neuron k and neuron i and $\sigma(s^2, i, t)$ is a centred Gaussian noise of variance s^2.

The imbalance characterizes the network's influence on the neuron. A high value indicates a high pressure for the neuron to change its state. In order to take this into account, only the most imbalanced neuron is updated at given time step. That is :

$$d_i(t) = \mathrm{MAX}_{j=1}^{n}(d_j(t))$$

The main difference between scheduled mode and classical update modes in neural networks (synchronous/asynchronous modes) is that, in general, the cells state are not all checked. Thus, the neural activity quickly propagates into the network through most imbalanced neurons and an output can be computed without taking into account lower imbalances. Sensorial inputs can thereby change while only some neurons have changed their state.

Then, a neural flow is created whose relative speed depends on the strength of the synaptic connexions. Moreover, the neurons involved in this flow will change their state in order (starting from the most stimulated ones).

[1] By temporal networks we mean networks of neurons that use the time of firing as a mean of coding and not only the firing rates (e.g. spiking neurons)

As such, the Scheduled Mode is one of the simplest possible implementation of a temporal neural network.

4 The Task and Architecture

A "food and nest" task (i.e. moving to catch a moving prey and bring it back to its nest) will serve as test.

The robot has to learn to step over a prey in order to capture it and deliver it to its nest (see figure 1). At each catch a new prey is created but the robot cannot carry more than one. When it steps over the nest, its captured prey is destroyed and the process starts again. The prey is moving (slower than the robot) sideways and bounces off the wall.

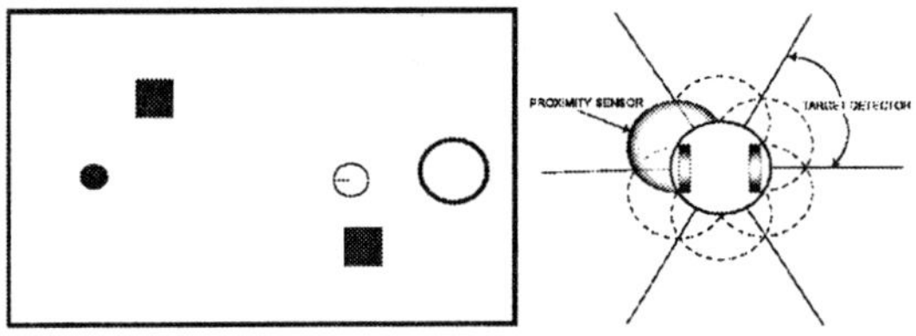

Figure 1. *Left* : The simulated environment. The big circle on the right is the nest. The robot is displayed besides the nest and the small black circle is the prey. The two squares are fixed obstacles. *Right*: Agent's perceptions

The robot is round shaped with six perception sectors. Each of them contains a binary proximity sensor and a target detector that indicates which kind of object faces this sector (see figure 1). It means that it can distinguish the nest and the prey but it can actually "see" them only with a 60° precision. Moreover, the robot is aware of an obstacle at a very short distance. However, neither the prey nor the nest are considered as obstacles. Two binary "proprioceptive" inputs are added and inform the robot of its own state (empty or full). The robot has two propulsion wheels (differential propulsion) whose straight and angular speed are noised.

Lastly, the neural controller is consisting of 20 input neurons (6 neurons are set to code for the proximity sensors value, 6 neurons for the prey's direction, 6 neurons for the nest's direction and remaining 2 neurons for the proprioceptive sensors). This input layer is connected to an all-to-all connected hidden layer made of 40 neurons. The 2 motoneurons are also connected to this hidden layer.

If both motoneurons are in the same state the robot move forward, else it turns left or right depending on which neurons is "on". The motor occurrence is updated at *each time step* whatever the neuron state is (remember that the update rule is asynchronous).

5 Learning algorithms

5.1 Evolutionary synaptic modification

We use a simple genetic algorithm to evolve the neural controller. Each population is composed of 100 genotypes. Each genotype codes for the $(20 \times 40 + 40 \times 40 + 2 \times 40 = 2480)$ synaptic weights of the neural net. The other states variables are constant through the adaptation process. The synaptic weights are chosen between -1.0 and 1.0 in the initial genotype set.

At each generation, we evaluate all the robots during 100000 time steps according to a fitness function which depends on the number of capture and homing and on the average distance made after the last capture (or homing).

The first 20 individuals are selected and reproduced to obtain the next generation. At this stage, all the genotypes are altered through mutation and crossover.

5.2 Hebbian Learning

In order to use the temporal aspect of the Scheduled Mode we used a local learning paradigm with reinforcement. As such, the Hebbian [11] paradigm rests mainly on temporal correlations between pairs of neurons. Thus, by tracing the neurons activities, we are able to measure the influence of a neuron over one another. Let T_{ji} be a trace of the connection between pre-synaptic neuron i and post-synaptic neuron j. In order to record the correlation between i and j, we set :

$$T_{ji}(t) = \gamma T_{ji}(t-1) + (x_i(t) - x_i(t'_i))\beta^{t-t'_i}\Delta x_j$$

Where t'_i was the last time step where the neuron $x_i(t)$ changed its state. The weights are updated via :

$$\Delta W_{ji} = r(t)\alpha\sigma(s^2, j, t)(1 - |W_{ji}|)T_{ji}$$

Where $r(t)$ is the reinforcement received by the controller at the current time t and α a learning factor. Note that when a neuron is in the same state for a long time, its influence in the learning equation vanishes – it has no influence on the variation of the postsynaptic weights anymore. $\sigma(s^2, j, t)$ is the same value that was calculated for imbalance for neuron j at time t (the noise). We set $r = 1$ whenever the robot catches a prey or delivers it in the nest, $r = -1$ when it hits an obstacle and $r = 0$ otherwise. α, s and the values of the reinforcement were constant during the learning trials of 20 random networks – the distribution of the initial weights followed an uniform law between -0.1 and 0.1 for the 2480 weights.

6 Results and comparison

6.1 Quantitative success of evolutionary algorithm

We conducted the evolutionary process over a hundred generations (that is a billion step time in all). Figure 2 shows the fitness results.

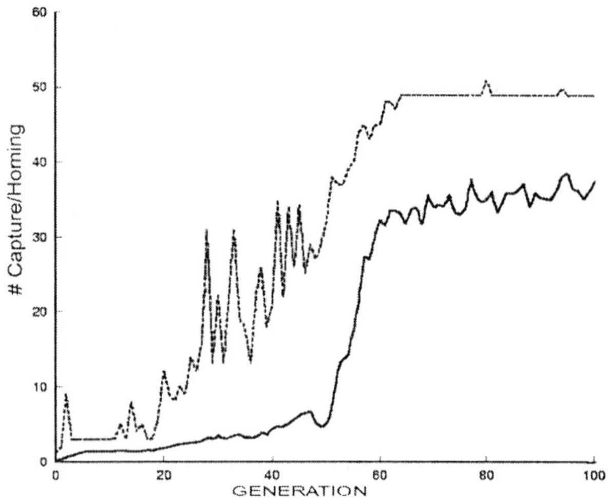

Figure 2. Average (plain line) and maximum (dashed line) fitness.

As usual in this kind of task, the first generations (here 50) are used to learn the basic task (i.e. one capture and one return). A "hunting" behaviour is then consolidated during the second period since the selection pressure is directed toward an optimization of the task (more and more capture/return pairs). The fitness is more or less equal to the sum of capture plus homing. At the end, the best individual is able to make 50 capture and homing within 100 000 time steps. It is very near the optimal. Indeed, the small difference with optimality comes from the fact that all the robots turned in the same direction. The evolution converges to a local maximum. Forced to *optimize* a behaviour, the robots didn't actually manage to turn in both direction making u-turns to reach the preys.

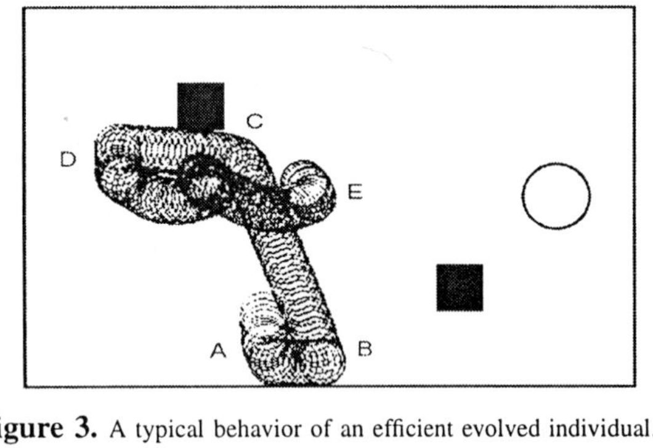

Figure 3. A typical behavior of an efficient evolved individual. It moves from "A" to "E"

Figure 3 displays a typical behavior of an efficient individual at the end of the adaptation process. The prey (not showed) was taken in "D". As it was said before, the robot always turns in the same direction (left in this case). The prey approach is made via straight lines. Since the target sensor's precision is only 60 degrees, this kind of approach is made possible by moving very quickly. Moreover its obstacle avoidance strategy seems stereotypical (90 degrees turn whether in "A", "B" or "C"). It shows the same stereotypical feature during its curious nest approach (the loops from "D" to "E" however far it is from the nest).

This entire means that the evolution algorithm did not enable a temporal learning but a "macro" learning. It was able to extract the invariant dynamics of the *whole* environment.

6.2 Qualitative result of hebbian learning

The resulting behaviours of the learning experiments are strikingly different. For instance, the efficiency is weak compared to the evolution experiment. Figure 4 shows the score of an average individual after a learning session. The session runs over 1 million time steps and the final results are the same (50) as the genetically created best individual (over 100 000 time steps)[2]!

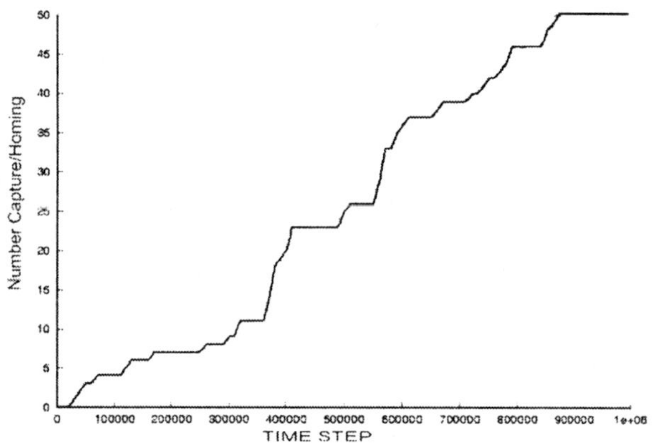

Figure 4. Learning curve of an average robot. The number of capture or homing vs time

At the start of learning process, the robot is to roam randomly inside the environment. So the first 370 000 time steps are devoted to hit obstacles and fall, by chance, on a prey then a nest. After that period it learns and quickly manage to make the task (between 370 000 and 430 000 for example). At this moment, it is able to make an average of 4615 time steps for one capture or one return. That is a bit more than twice the time needed by the best genetically created individual.

Unfortunately, if the learning process goes on, the scores decreased. This is mostly due to weight saturation and a flaw in the hebbian algorithm. This question will be handled in the next section. However it is obvious that the task is learned. Moreover, the robot managed to be quite efficient – during a part of its "life" at least.

[2]Here we don't take into account here that to obtain that best individual we needed 1 billion step time of simulation

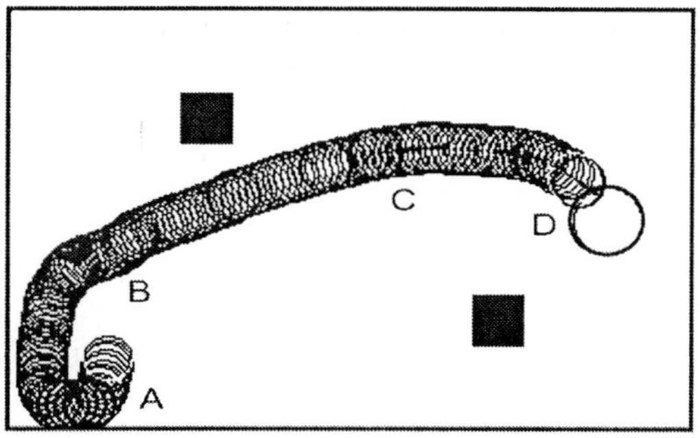

Figure 5. A typical behavior during the learning process. It moves from "A" to "D".

But the most striking difference appears on the behaviours. Indeed, in the case of learning, the robots did not move in straight lines and their behaviours were not stereotypical. They did turn in both directions. Figure 5 shows an example of how they behaved. The robot takes the prey (not showed) in "B", avoids the walls in two different ways (between "A" and "B"). It begins an elliptic nest approach between "B" and "C".

Furthermore, almost all the robots made this kind of strange approach toward the prey and the nest – we named it the "leaf" approach.

In doing this, the robots made micro-movements (small variations around its average "head" direction). These micro-movements were not constant. They appeared at the beginning of an approach and their rates increased at the end. All were to keep the target (prey or nest) on the limit of is "front eye". To evade the combined limitations of the controller (the Scheduled Mode imposes *a time of reaction*) and of the actual precision of the sensors, the robots learned to keep the target around the border of their inputs.

An indirect consequence is that the robots are much slower than with a straight line approach. This impedes significantly the overall performance but leaves more room for an eventual adaptiveness. Since nothing has been done to make such behaviours appear, they *emerged* from this temporal learning.

7 Discussion

As said before, the scheduled mode is not a perfect implementation of temporal neural networks. Its fundamental flaw lies in the fact that it induces just *one* flow of update sequence of neurons. The hebbian learning reinforces indefinitely this sequence then leads to a saturation and a degeneration of the sequence itself. Thus, the robots cannot stabilize what they had already learned.

Nevertheless, the results are still interesting when compared with the evolutionary approach. Micromovements and non-stereotypical behaviours could not be observed in the latter. The fundamental fact lies in the external relation of a robot and its environment. Indeed the fitness driven evolution emphasizes a global and optimized behavior while reinforcement learning emphasizes a local and temporal behaviour. Thus, while the overall performance seems poor, the temporal coherence of this learning enables the robot to extract a temporal relation between itself, its environment and the task. Because learning and optimizing can be two different things, we showed in this paper that a non-optimal but coherent policy could give some adaptation results that may have not arisen if we had just looked at the performance.

References

[1] D. Floreano and F. Mondada. Evolutionary neurocontrollers for autonomous mobile robots. *Neural Networks*, (11):1461–1478, 1998.

[2] D. Floreano and C. Mattiusi. Evolution of spiking neural controllers for autonomous vision-based robots. In T. Gomi, editor, *Evolutionnary Robotics*. Berlin: Springer-Verlag, 2001.

[3] E. Di Paolo. Spike timing dependent plasticity for evolved robots. *Adaptive Behavior*, 10(3):243–263, 2002.

[4] H. Soula, G. Beslon, and J. Favrel. Evolving spiking neurons nets to control an animat. In *Proc. of ICANNGA 20003, Roanne*, pages 193–197. Springer Verlag, 2003.

[5] F. Varela, J-P. Lachaux, E. Rodriguez, and J. Martinerie. The brainweb: Phase synchronization an large-scale integration. *Nature Neuroscience Review*, 2:229–239, 2001.

[6] S. Thorpe, D. Fize, and C. Marlot. Speed of processing in the human visual system. *Nature*, 381:520–522, 1996.

[7] J.J. Hopfield. Pattern recognition computation using action potential timing for stimulus representation. *Nature*, 376(6535):33–36, July 1995.

[8] W. Maas and C. Bishop, editors. *Pulsed Neural Networks*. MIT Press, Cambridge, Massachusetts, 2001.

[9] G. Beslon, H. Soula, and J. Favrel. A neural model for animats brain. In *Proc of ICANNGA 2001, Prague*, pages 352–355, 2001.

[10] H. Soula, G. Beslon, and J. Favrel. Controlling an animat with a self-organized modular neural network. In *Proc. of EWLR'2001, Prague*, pages 39–46, 2001.

[11] D.O. Hebb. *The Organization of Behavior*. J. Wiley and Sons, New York, 1949.

Minimization of empirical error over perceptron networks

Věra Kůrková [1]
Institute of Computer Science, Academy of Sciences of the Czech Republic
E-mail: vera@cs.cas.cz

Abstract

Supervised learning by perceptron networks is investigated a minimization of empirical error functional. Input/output functions minimizing this functional require the same number m of hidden units as the size of the training set. Upper bounds on rates of convergence to zero of infima over networks with n hidden units (where n is smaller than m) are derived in terms of a variational norm. It is shown that fast rates are guaranteed when the sample of data defining the empirical error can be interpolated by a function, which may have a rather large Sobolev-type seminorm. Fast convergence is possible even when the seminorm depends exponentially on the input dimension.

1 Introduction

The goal of supervised learning is to adjust parameters of a neural network so that it approximates with a sufficient accuracy a functional relationship between inputs and outputs known only by a sample of input/output pairs. It is desirable that the system also generalizes well, i.e., it satisfactorily processes new data that were not used for training. Learning from data with generalization capability was studied theoretically in the framework of regularized optimization [4], [14], [10]. Theoretical results describing optimal solutions can be applied to kernel models, a special case of which are radial-basis function networks with constant width. But the most common neural networks built from perceptrons cannot be represented as kernel models.

In this paper, we investigate minimization of empirical error functionals over sets of functions computable by perceptron networks. We estimate rates of convergence of infima over networks with n hidden units to the global infimum achievable by a network with the same number of hidden units as the size of the training set.

2 Approximate minimization of empirical error

Let $\mathcal{R}$ denote the set of real numbers, Ω be a nonempty set and $z = \{(u_i, v_i) \in \Omega \times \mathcal{R}, i = 1, \ldots, m\}$ be a sample of input/output pairs of data. A standard approach to learning from empirical data used, e.g., in back-propagation, is based on minimization of the *empirical error* functional defined as $\mathcal{E}_{z,V}(f) = \frac{1}{m}\sum_{i=1}^{m} V(f(u_i), v_i)$, where $V : \mathcal{R} \times \mathcal{R} \to [0, \infty)$, satisfying for all $y \in \mathcal{R}$, $V(y, y) = 0$, is called a *loss function*. The most common loss function is the *square loss* $V(f(u), v) = (f(u) - v)^2$, we denote by $\mathcal{E}_z$ the empirical error functional with this loss function, i.e., $\mathcal{E}_z(f) = \frac{1}{m}\sum_{i=1}^{m}(f(u_i) - v_i)^2$.

Let M be a subset of a normed linear space $(X, \|.\|)$ and $\Phi : X \to \mathcal{R}$ be a functional. Using standard notation from optimization theory, we denote by (M, Φ) the problem of minimization of Φ over M; M is called the *hypothesis set*. Elements of the set $argmin\,(M, \Phi) = \{g \in M : \Phi(g) = \inf_{g \in M} \Phi(g)\}$ are called *solutions* (or *minimum points*) of the problem (M, Φ). For $\varepsilon > 0$, elements of the set $argmin_{\varepsilon}(M, \Phi) = \{g \in M : \Phi(g) < \inf_{g \in M} \Phi(g) + \varepsilon\}$ are called *ε-near minimum points* of (M, Φ). A sequence $\{g_n\}$ of elements of M is called Φ-minimizing if $\lim_{n\to\infty} \Phi(g_n) = \inf_{g \in M} \Phi(g)$.

Typical hypothesis sets used in neurocomputing are sets of functions computable by *neural networks with n hidden units* of a given type *and one linear output*. Such sets are of the form $span_n\, G = \{\sum_{i=1}^{n} w_i g_i : w_i \in \mathcal{R},\, g_i \in G\}$, where G is the set of functions computable by the computational units.

Standard hidden units are *perceptrons*. For $\Omega \subseteq \mathcal{R}^d$ and $\psi : \mathcal{R} \to \mathcal{R}$ we denote by $P_d(\psi, \Omega) = P_d(\psi) = \{f : \Omega \to \mathcal{R} \mid f(x) = \psi(a_i \cdot x + b_i), a_i \in \mathcal{R}^d, b_i \in \mathcal{R}\}$ the set of functions on Ω computable by perceptrons with the activation function ψ (we write $P_d(\psi)$ when Ω is clear from context). The most common activation functions are *sigmoidals*, which are monotonic increasing functions $\sigma : \mathcal{R} \to \mathcal{R}$ (i.e., for all $t_1, t_2 \in \mathcal{R}, t_1 \leq t_2$ implies $\sigma(t_1) \leq \sigma(t_2)$) satisfying $\lim_{t\to-\infty} \sigma(t) = 0$ and $\lim_{t\to\infty} \sigma(t) = 1$.

An important type of a sigmoidal is the *Heaviside function* $\vartheta(t) = 0$ for $t < 0$ and $\vartheta(t) = 1$ for $t \geq 0$. To shorten notation, we write $H_d(\Omega)$ instead of $P_d(\vartheta, \Omega)$. Note that $H_d(\Omega)$ is the *set of it characteristic functions of closed half-spaces of $\mathcal{R}^d$ intersected with Ω.*

Ito [6, p.73] proved that any function defined on a finite subset of $\mathcal{R}^d$ can be exactly represented as a function computable by a perceptron network with any sigmoidal activation function. The following theorem is a reformu-

[1] This work was partially supported by GA ČR grant 201/05/0557.

lation of Ito's result in terms of optimization theory.

Theorem 2.1 (Ito) *For all positive integers d, m, all samples of data $z = \{(u_i, v_i) \in \mathcal{R}^d \times \mathcal{R}, i = 1, \dots, m\}$ with all u_i distinct and all sigmoidal functions σ, the problem $(span_m \mathcal{P}_d(\sigma), \mathcal{E}_z)$ has a solution and* $\min_{f \in span_m P_d(\sigma)} \mathcal{E}_z(f) = 0$.

A drawback of this result is that the number of hidden units in the network interpolating the sample of data is equal to the size of the sample. For large samples, such networks might not be implementable. Moreover in typical applications of neural networks, a number of hidden units much smaller than the size of the training set is chosen before learning. Using such networks, only suboptimal solutions can be achieved. To compare such suboptimal solutions with the optimal one given by Theorem 2.1 we estimate rates of convergence of $\{\inf_{f \in span_n H_d(\Omega)} \mathcal{E}_z(f)\}$ to zero as n goes to m.

A useful property that allows application of several tools for investigation of approximate minimization of a functional is its continuity. The next proposition shows that continuity of the empirical error defined by a sample z follows from continuity of the evaluation functionals at the input data $u_1, \dots, u_m$. For a normed linear space $(X, \|.\|)$ of functions on some set Ω and $x \in \Omega$, the *evaluation functional at* x, denoted by $\mathcal{F}_x$, is defined for all $f \in X$ as $\mathcal{F}_x(f) = f(x)$.

Proposition 2.2 *Let $(X, \|.\|)$ be a normed linear space of functions on a nonempty set Ω, m a positive integer, $z = \{(u_i, v_i) \in \Omega \times \mathcal{R}, i = 1, \dots, m\}$, $V : \Omega \times \Omega \to \mathcal{R}$ a loss function and $f \in X$. If for all $i = 1, \dots, m$, $\mathcal{F}_{u_i}$ is continuous at f and V is continuous at $(f(u_i), v_i)$, then $\mathcal{E}_{z,V}$ is continuous at f.*

Proof. By continuity of V we get for every $\varepsilon > 0$ some $\eta > 0$ such that $\|f(u_i) - g(u_i)\| \leq \eta$ implies $|V(f(u_i), v_i) - V(g(u_i), v_i)| \leq \varepsilon$. As all $\mathcal{F}_{u_i}$ are continuous at f, there exists $\delta > 0$ such that $\|g - f\| \leq \delta$ implies $\|f(u_i) - g(u_i)| \leq \eta$ and hence $|\mathcal{E}_{z,V}(f) - \mathcal{E}_{z,V}(g)| = \left|\frac{1}{m}\sum_{i=1}^{m}(V(f(u_i), v_i) - V(g(u_i), v_i))\right| \leq \varepsilon$. □

It is easy to show that $\mathcal{E}_z$ is continuous on the space $\mathcal{M}(\Omega)$ of bounded measurable functions on $\Omega \subseteq \mathcal{R}^d$ with the supremum norm $\|.\|_{\sup}$ and that it is convex.

Proposition 2.3 *Let $\Omega \subseteq \mathcal{R}^d$, $z = \{(u_i, v_i) \in \Omega \times \mathcal{R}, i = 1, \dots, m\}$ and $v_{\max} = \max\{|v_1|, \dots, |v_m|\}$. Then $\mathcal{E}_z : (\mathcal{M}(\Omega), \|.\|_{\sup}) \to \mathcal{R}$ is continuous and its modulus of continuity at any $f \in \mathcal{M}(\Omega)$ satisfies $\omega_f(t) \leq t(t + 2\|f\|_{\sup} + 2v_{\max})$.*

In contrast to the supremum norm, in the $\mathcal{L}_2$-norm evaluation functionals need not to be continuous. Lack of continuity can be easily demonstrated for $span_2 H_d([0,1]^d)$: denoting by χ_n the characteristic function of $[0, \frac{1}{n}] \times [0,1]^{d-1}$, we get a sequence $\{n\chi_n\} \subset span_2 H_d([0,1]^d)$ with all elements having $\mathcal{L}_2$-norms equal to 1, on which the evaluation functional at zero is not bounded (and thus it is not continuous). Similarly, one can disprove continuity of evaluation functionals on $span_2 P_d(\sigma, \Omega)$ for any sigmoidal function σ.

Recently, a class of Hilbert spaces of point-wise defined functions, on which all evaluation functionals are continuous, became popular in learning theory. Such spaces are called reproducing kernel Hilbert spaces (RKHS) and they can be represented as completions of linear combinations of "translations" of kernels (symmetric positive semidefinite functions) [1], [4], [14]. Continuity of empirical error functionals on RKHSs allows one to apply theory of generalized inversion and regularization in infinite dimensional Hilbert spaces to describe solutions of the problem of minimization of the empirical error over RKHSs [10]. Moreover, continuity of empirical errors is essential for derivation of estimates of rates of approximate optimization over kernel models of bounded complexity [11], [12].

However, it is not clear, whether there exist kernels, for which sets $P_d(\sigma)$ are contained in the corresponding RKHSs. So in investigation of minimization of empirical errors over perceptron networks, we focus on the space of bounded measurable functions with the supremum norm. To derive rates of approximate optimization over $span_n H_d(\Omega)$ we take advantage of a result from nonlinear approximation theory [2] giving an upper bound on supremum norm rates of approximation by $span_n H_d(\Omega)$ in terms of a norm called *variation with respect to half-spaces*. It is a special case of *G-variation* [9] defined for any bounded nonempty subset G of a normed linear space $(X, \|.\|)$ as the Minkowski functional of the closed convex symmetric hull of G, i.e.,

$$\|f\|_G = \inf\left\{c > 0 : c^{-1}f \in cl\, conv\,(G \cup -G)\right\},$$

where the closure cl is taken with respect to the topology generated by the norm $\|.\|$. Note that G-variation can be infinite (when the set on the right-hand side is empty).

Here we consider H_d-variations with respect to the topology generated by the supremum and $\mathcal{L}_2$-norm (we indicate the norm by the notation $\|.\|_{H_d,\sup}$, $\|.\|_{H_d,\mathcal{L}_2}$, resp. It is easy to check that $\|.\|_{H_d,\mathcal{L}_2} \leq \|.\|_{H_d,\sup}$ as well as $\|.\|_{P_d(\sigma),\mathcal{L}_2} \leq \|.\|_{P_d(\sigma),\sup}$ and that $\|.\|_{H_d,\mathcal{L}_2} = \|.\|_{P_d(\sigma),\mathcal{L}_2}$ for any sigmoidal σ.

Barron [2] estimated rates of approximation by $span_n H_d$ in the supremum norm (see also [3, p. 201] and [5, p. 25]).

Theorem 2.4 (Barron) *For all positive integers d, n, every compact $\Omega \subset \mathcal{R}^d$ and every $f \in \mathcal{M}(\Omega)$,*

$$\|f - span_n H_d(\Omega)\|_{\sup} \leq c\sqrt{\tfrac{d+1}{n}}\|f\|_{H_d(\Omega)\sup},$$
where c is an absolute constant.

The next theorem estimates rates of convergence of infima of a continuous functional over $span_n H_d$ to the global minimum.

Theorem 2.5 *Let* $\Omega \subset \mathcal{R}^d$ *be compact,* $\Phi : (\mathcal{M}(\Omega), \|.\|_{\sup}) \to \mathcal{R}$ *a functional such that there exists a solution* f^o *of the problem* $(\mathcal{M}(\Omega), \Phi)$ *at which* Φ *is continuous with the modulus of continuity* ω, $\{\varepsilon_n\}$ *a sequence of positive real numbers converging to zero, and* $\{f_n\}$ *a sequence of* ε_n*-minimum points of* $(span_n H_d(\Omega), \Phi)$*. Then for every positive integer n:*
(i) $\inf_{f \in span_n H_d(\Omega)} \Phi(f) - \Phi(f^o) \leq \omega\left(c\sqrt{\tfrac{d+1}{n}}\|f^o\|_{H_d(\Omega),\sup}\right)$;
(ii) if $\|f^o\|_{H_d(\Omega),\sup} < \infty$*, then* $\{f_n\}$ *is a* Φ*-minimizing sequence and* $\Phi(f_n) - \Phi(f^o) \leq \omega\left(c\sqrt{\tfrac{d+1}{n}}\|f^o\|_{H_d(\Omega),\sup}\right) + \varepsilon_n$.

Proof. (i) For every $\varepsilon > 0$, let $f_n^\varepsilon \in span_n H_d$ be such that $\|f^o - f_n^\varepsilon\|_{\sup} < \|f^o - span_n H_d\|_{\sup} + \varepsilon$. Then $\inf_{f \in span_n H_d} \Phi(f) - \Phi(f^o) \leq \Phi(f_n^\varepsilon) - \Phi(f^o) \leq \omega(\|f_n^\varepsilon - f^o) \leq \omega(\|f^o - span_n H_d\| + \varepsilon)$. By Theorem 2.4, infimizing over ε we get $\inf_{f \in span_n H_d} \Phi(f) - \Phi(f^o) \leq \omega\left(c\sqrt{\tfrac{d+1}{n}}\|f^o\|_{H_d,\sup}\right)$.

(ii) By the definition of ε_n-minimum point, $\Phi(f_n) - \Phi(f^o) \leq \inf_{f \in span_n H_d} \Phi(f) - \Phi(f^o) + \varepsilon_n$. So by (i), $\Phi(f_n) - \Phi(f^o) \leq \omega\left(c\sqrt{\tfrac{d+1}{n}}\|f^o\|_{H_d,\sup}\right) + \varepsilon_n$. As $\lim_{n\to\infty} \varepsilon_n = 0$ and $\|f^o\|_{H_d,\sup}$ is finite, $\{f_n\}$ is Φ-minimizing. □

Combining Theorems 2.1, 2.5 and Proposition 2.3 we get the following upper bound on rates of approximate minimization of $\mathcal{E}_z$ over $span_n H_d(\Omega)$.

Corollary 2.6 *Let* $\Omega \subset \mathcal{R}^d$ *be compact,* $z = \{(u_i, v_i) \in \Omega \times \mathcal{R}, i = 1, \ldots, m\}$, $f^o \in \mathcal{M}(\Omega)$ *such that* $\mathcal{E}_z(f^o) = 0$, $\{\varepsilon_n\}$ *a sequence of positive reals converging to zero,* $\{f_n\}$ *a sequence of* ε_n*-minimum points of* $(span_n H_d(\Omega), \mathcal{E}_z)$*. Then for every n:*
(i) $\inf_{f \in span_n H_d(\Omega)} \mathcal{E}_z(f) \leq \frac{(d+1)c^2\|f^o\|^2_{H_d,\sup}}{n} + \frac{\sqrt{d+1}(2c\|f^o\|^2_{H_d,\sup} + v_{\max}\|f^o\|_{H_d,\sup})}{\sqrt{n}}$;
(ii) $\{f_n\}$ *is* $\mathcal{E}_z$*-minimizing and* $\mathcal{E}_z(f_n) \leq \frac{(d+1)c^2\|f^o\|^2_{H_d,\sup}}{n} + \frac{\sqrt{d+1}(2c\|f^o\|^2_{H_d,\sup} + v_{\max}\|f^o\|_{H_d,\sup})}{\sqrt{n}} + \varepsilon_n$*, where c is an absolute constant.*

3 Estimates of variation with respect to half-spaces

Corollary 2.6 shows that the speed of convergence of suboptimal solutions of the problem of minimization of $\mathcal{E}_z$ over the set of functions computable by networks with n Heaviside perceptrons depends on the smallest value of H_d-variation on the set of functions interpolating the data z.

To estimate H_d-variations of smooth elements of this set we take an advantage of a result from [8] bounding from above H_d-variation of a smooth function by a product of its certain Sobolev-type seminorm with
$$k_d \sim \left(\frac{e^d}{d2^{d-2}\pi^{d-1}}\right)^{1/2},$$
which as a function of d is *decreasing exponentially fast.*

For a function $f \in \mathcal{C}^d(\mathcal{R}^d)$ define
$$\|f\|_{d,1,\infty} = \max_{|\alpha|=d} \|D^\alpha f\|_{\mathcal{L}_1(\mathcal{R}^d)}.$$
For d odd and f sufficiently rapidly vanishing at infinity, an upper bound
$$\|f\|_{H_d(\mathcal{R}^d),\sup} \leq k_d\|f\|_{d,1,\infty} \qquad (1)$$
was derived in [8].

Thus by Corollary 2.6, for any sample of data z, which can be interpolated by a function f^o satisfying
$$\|f^o\|_{1,d,\infty} \leq \frac{1}{k_d} \sim \left(\frac{d2^{d-2}\pi^{d-1}}{e^d}\right)^{1/2},$$
infima of $\mathcal{E}_z$ over $span_n H_d$ converge to zero with rate
$$c^2\sqrt{\frac{d+1}{n}} + (2c + v_{\max})\frac{d+1}{n}$$
as by (1) f^o has H_d-variation at most 1.

However, there exist samples of data, which cannot be interpolated by functions with small H_d-variations. Such samples $z = \{(u_i, v_i), i = 1, \ldots, m\}$ can be obtained from real-valued Boolean functions $h : \{0,1\}^d \to \mathcal{R}$ by setting $\{0,1\}^d = \{u_1, \ldots, u_{2^d}\}$ and $v_i = h(u_i)$. If $f : \Omega \to \mathcal{R}$ is an extension of h, then $\|f\|_{H_d(\Omega),\sup} \geq \|h\|_{H_d(\{0,1\}^d),\sup}$.

To show that there exist functions on $\{0,1\}^d$ with $H_d(\{0,1\}^d)$-variations depending on d exponentially, we use a geometric characterization of G-variation from [13]
$$\|f\|_G \geq \frac{\|f\|^2}{\sup_{g \in G} |g \cdot f|}. \qquad (2)$$
So functions that have small inner products with all elements of G (are "almost orthogonal" to G) have large G-variations.

For a Hilbert space $(X, \|.\|)$ we define on its unit ball S_1 a pseudometrics $\rho_X(f,g) = \arccos|f \cdot g|$, which measures the distance as the minimum of the two angles between f and g and between f and $-g$ (it is a pseudometrics as the distance of antipodal vectors is zero). For $\alpha > 0$, let $\mathcal{N}_\alpha(S_1)$ denote the α-covering number of S_1 with respect to ρ_X, i.e., the size of the smallest α-net in S_1. The next proposition shows that when for some α close to $\pi/2$, the cardinality of G is smaller than $\mathcal{N}_\alpha(S_1)$, then in S_1 there exists a function with a "large" G-variation. It also guarantees existence a function with G-variation at least $\frac{1}{\varepsilon}$ for any subset G of the unit sphere S^{m-1} in $\mathcal{R}^m$ of smaller cardinality than the ε-*quasiorthogonal dimension* $dim_\varepsilon m$ of $\mathcal{R}^m$. For $\varepsilon > 0$, $dim_\varepsilon m$ was defined in [7] as the maximal number of vectors which are pairwise ε-*quasiorthogonal*, i.e., $|u \cdot v| \leq \varepsilon \|u\| \|v\|$.

Proposition 3.1 *(i) If G is a subset of the unit sphere S_1 in a Hilbert space X and $\alpha \in [0, \pi/2]$ is such that $card\, G < \mathcal{N}_\alpha(S_1)$ with respect to the pseudometrics ρ_X, then there exists $f \in S_1$ with $\|f\|_G \geq 1/\cos\alpha$.*
(ii) If $G \subset S^{m-1} \subset \mathcal{R}^m$ such that $card\, G < dim_\varepsilon m$, then there exists $f \in S^{m-1}$ with $\|f\|_G \geq \frac{1}{\varepsilon}$.

Proof. (i) If $card\, G < \mathcal{N}_\alpha(S_1)$, then there exists $f \in S_1$ such that for all $g \in G$, $\rho_X(f, g) \geq \alpha$ and hence $|f \cdot g| \leq \cos\alpha$. Then by (2) $\|f\|_G \geq \frac{1}{\sup_{g \in G} |f \cdot g|} \geq 1/\cos\alpha$.
(ii) follows from (i) as $dim_\varepsilon m \leq \mathcal{N}_{\arccos(\varepsilon)}(S^{m-1})$. □

Theorem 3.2 *For every positive integer d there exists a sample $z = \{(u_i, v_i) : i = 1, \ldots, 2^d\} \subset \{0,1\}^d \times \mathcal{R}$ such that for every $\Omega \supseteq \{0,1\}^d$ and every $f : \mathcal{R}^d \to \mathcal{R}$ such that $\mathcal{E}_z(f) = 0$, $\|f\|_{H_d(\Omega),\sup} \geq \frac{2^{(d-1)/2}}{d\sqrt{ln 2}}$.*

Proof. It was shown in [7] that $dim_\varepsilon m \geq e^{m\varepsilon^2/2}$. On the other hand, $card\, H_d(\{0,1\}^d) = 2^{d^2 - d\log_2 d + \mathcal{O}(d)}$ [15]. Denoting $H_d^o(\{0,1\}^d)$ the set of normalized elements of $H_d(\{0,1\}^d)$ with respect to l_2-norm on $\mathcal{R}^{2^d}$, we get $\|.\|_{H_d(\{0,1\}^d),\sup} \geq \|.\|_{H_d(\{0,1\}^d),l_2} \geq \|.\|_{H_d^o(\{0,1\}^d),l_2}$. As $card\, H_d^o(\{0,1\}^d) < 2^{d^2}$, for $\varepsilon = \frac{\sqrt{d ln 2}}{2^{(d-1)/2}}$, $card\, H_d^o(\{0,1\}^d) < e^{(2^d\varepsilon^2)/2}$ and hence by Proposition 3.1 (ii) there exists a function $h \in S^{2^d-1}$ with $\|h\|_{H_d(\{0,1\}^d),\sup} \geq \|h\|_{H_d^o(\{0,1\}),l_2} \geq \frac{2^{(d-1)/2}}{d\sqrt{ln 2}}$. Let $(u_1, \ldots, u_{2^d}) = \{0,1\}^d$, $v_i = f(u_i)$ and $z = \{(u_i, v_i) : i = 1, \ldots, m\}$. Then for every $f : \Omega \to \mathcal{R}$, for which $\mathcal{E}_z(f) = 0$, $\|f\|_{H_d(\Omega),\sup} \geq \frac{2^{(d-1)/2}}{d\sqrt{ln 2}}$. □

Note that by (1) for every d odd and $f^o \in \mathcal{C}^d(\mathcal{R}^d)$ sufficiently rapidly vanishing at infinity interpolating the sample described in Theorem 3.2, $\|f^o\|_{1,d,\infty} \geq \left(\frac{2^{2d-3}\pi^{d-1}}{d e^d ln 2}\right)^{1/2}$.

4 Discussion

We have shown that fast convergence of infima of the empirical error functional $\mathcal{E}_z$ over networks with n Heaviside perceptrons to zero can be achieved for samples that can be interpolated by functions f^o with the Sobolev seminorm $\|f^o\|_{d,1,\infty} = \max_{|\alpha|=d} \|D^\alpha f^o\|_{\mathcal{L}_1(\mathcal{R}^d)}$ depending exponentially on the input dimension. Note that the seminorm $\|f^o\|_{1,d,\infty}$ is much smaller than the Sobolev norm $\|f^o\|_{d,1} = \sum_{|\alpha| \leq d} \|D^\alpha f^o\|_{\mathcal{L}_1(\mathcal{R}^d)}$ as instead of summation of iterated partial derivatives of f over all α with $|\alpha| \leq d$ only their maximum over α with $|\alpha| = d$ is taken.

We have also shown that there exist samples of data constructed using special Boolean functions, for which the Sobolev seminorms of interpolating functions are even larger than the exponential size allowed for fast convergence described in Corollary 2.6.

The proof of Proposition 3.1 is existential, but in [13] a lower bound $\mathcal{O}(2^{d/6})$ on $H_d(\{0,1\}^d)$-variation was derived for a concrete function, namely the "inner product modulo 2".

References

[1] Aronszajn, N. (1950). Theory of reproducing kernels. Transactions of AMS, 68: 337-404.

[2] Barron, A. R. (1992). Neural net approximation. In: Proceedings of the 7th Yale Workshop on Adaptive and Learning Systems (pp. 69–72).

[3] Cheang, G. H. L., Barron, A. R. (2000). A better approximation for balls. Journal of Approximation Theory 104: 183–203.

[4] Cucker, F. and Smale, S. (2001). On the mathematical foundations of learning. Bulletin of AMS 39: 1-49.

[5] Donahue, M. J., Gurvits, L., Darken, C., Sontag, E. (1997). Rates of convex approximation in non-Hilbert spaces. Constructive Approximation 13: 187–220.

[6] Ito, Y. (1992). Finite mapping by neural networks and truth functions. Math. Scientist 17: 69–77.

[7] Kainen, P. C., Kůrková, V. (1993). Quasiorthogonal dimension of Euclidean spaces. Applied Math. Letters 6: 7–10, 1993.

[8] Kainen, P. C., Kůrková, V., Vogt, A. (2004). A Sobolev-type upper bound for rates of approximation by linear combinations of plane waves. Research Report ICS-2003-900, Institute of Computer Science, Prague.

[9] Kůrková, V. (2003). High-dimensional approximation and optimization by neural networks. Chapter 4 In: Advances in Learning Theory: Methods, Models and Applications. (Eds. J. Suykens et al.) (pp. 69-88). IOS Press, Amsterdam.

[10] Kůrková, V. (2004). Learning from data as an inverse problem. In: Proceedings of COMPSTAT 2004 (Ed. J. Antoch) (pp. 1377–1384). Physica-Verlag, Heidelberg.

[11] Kůrková, V., Sanguineti, M. (2004). Error estimates for approximate optimization by the extended Ritz method. SIAM Journal on Optimization (to appear).

[12] Kůrková, V., Sanguineti, M. (2004). Learning with generalization capability by kernel methods of bounded complexity. Journal of Complexity (to appear).

[13] Kůrková, V., Savický, P., Hlaváčková, K. (1998). Representations and rates of approximation of real-valued Boolean functions by neural networks. Neural Networks 11: 651–659.

[14] Poggio T., Smale, S. (2003). The mathematics of learning: dealing with data. Notices of AMS 50: 536-544.

[15] Shläfli, L. (1950). Gesamelte mathematische abhandlungen. Band 1. Basel, Verlag Birkhäuzer.

Interval Basis Neural Networks

A. Horzyk[1]

[1]Department of Automatics, University of Science and Technology, Poland
E-mail: horzyk@agh.edu.pl

Abstract

The paper introduces a new type of ontogenic neural networks called Interval Basis Neural Networks (IBNNs). The IBNN configures the whole topology and computes all weights after *a priori* knowledge collected form training data. The training patterns are grouped together producing intervals separately for all input features for each class after statistical analyses of the training data. This IBNNs feature make possible to computed all network parameters without training. Moreover, the IBNN takes into account the distances between patterns of the same classes and builds the well-approximating model especially on the borders between the classes. Furthermore, the IBNNs are insensitive for quantity differences in patterns representation of classes. The IBNNs always correctly classify training data and very good generalize other data.

1 Introduction

Though neural networks are very modern computational tool they are not always easy to use because of many unknown configuration parameters. Many ontogenic neural networks [1,3-8] can help to solve problems of architecture construction but many times their generalization is poor. The generalization is a fundamental problem of ontogenic neural networks use.

This paper introduces a new type of ontogenic neural networks – called Interval Basis Neural Networks (IBNNs). The IBNNs are non-linear three-layers partially connected ontogenic neural networks constructed after the statistical analyses of the training data. The construction of IBNNs is deterministic. Training data are analyzed separately for each input feature searching for intervals characteristic for each class. Such separable intervals are completed with non-linear concave functions with controlled shape of descend that make possible to achieve good approximation especially on the borders between different classes. The shape of these concave functions (fig. 2) are computed after the quantity and distances of other training data of the same class from the hyperregions and sub-hyperregions borders. The hyperregions and sub-hyperregions are produced after the intervals computed for all training patterns of the same class.

The adaptation of RBF networks by Lowe [6] also enables to change the shape of radial basis functions. Lowe describes the importance of the adaptation of RBF function in view of generalization. The shape of slopes can be regulated also for bicentral functions [2].

2 IBNN Construction Process

For any given training data $U=\left\{\left(u^1,C^{m_1}\right),\dots,\left(u^N,C^{m_N}\right)\right\}$ consisting of the pairs: the input vector $u^n=\left[u_1^n,\dots,u_K^n\right]$ ($u_k^n\in\Re$), the adequate class $C^m\in\left\{C^1,\dots,C^M\right\}$ of this input vector $u^n\in C^{m_n}$ and input features *1, ..., K*, the intervals (for each feature) of patterns of the same class can be computed (fig. 1). Different intervals are characteristic for different groups of patterns of each class and can partially identify them. The model of each class can be defined by combining such intervals for all input features into hyperregions and sub-hyperregions in the input space. In order to find out such intervals effectively training data have to be sorted after each input feature.

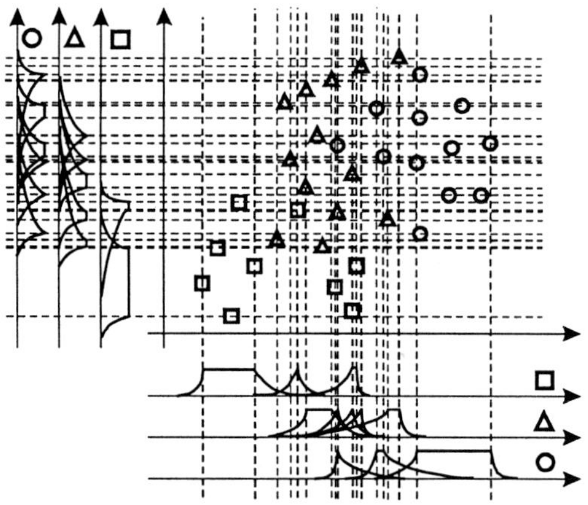

Fig. 1. Extraction of intervals and computation of transfer function for these intervals in view of hyperregions.

In order to generalize correctly especially outside the computed intervals the concave slopes (fig. 1.-2.) are added at the borders of such intervals. These concave slopes are functions that project the quantity and the distances of other input patterns of the same class from the considered intervals. The training patterns are classified always as 100% similar to their classes. Each interval defines a single neuron in the 1-st layer of the constructed neural network (fig. 3.). Such a neuron is connected to the input related with feature of its interval. The transfer function f_T of the 1-st layer neurons is defined as follows:

$$f_T(x) = \begin{cases} 1 - tgh\dfrac{x_k - x_k^R}{1+\alpha_k^R} & if\ x_k > x_k^R \\ 1 & if\ x_k^L \le x_k \le x_k^R \\ 1 - tgh\dfrac{x_k^L - x_k}{1+\alpha_k^L} & if\ x_k < x_k^L \end{cases} \quad (1)$$

where

x_L - is a left limit of the concerned interval,

x_R - is a right limit of the concerned interval,

α_L - is a left slope parameter of the transfer function,

α_R - is a right slope parameter of the transfer function,

The slope parameters of the transfer function (fig. 2) are defined as follows:

$$\alpha_k^L = q^m \sum_{i \in L} \frac{1}{1+\left(x_k^L - u_k^i\right)^2} \quad (2)$$

$$\alpha_k^R = q^m \sum_{i \in R} \frac{1}{1+\left(x_k^R - u_k^i\right)^2} \quad (3)$$

where

$$L = \left\{u^n : u^n \in U \cap C^m \ \&\ u_k^n < x_k^L\right\} \quad (4)$$

$$R = \left\{u^n : u^n \in U \cap C^m \ \&\ u_k^n > x_k^R\right\} \quad (5)$$

$$q^m = \frac{N}{K \cdot N_m} \quad (6)$$

N – is a number of training data,

N_m – is a number of training data that define class C^m,

K – is a number of classes defined in the training data.

The slope parameters (2) and (3) define the slopes of the transfer function (1). They depend on other training patterns of the same class. The more training patterns in the left (L) or in the right (R) direction of the interval are the less sharp the suitable slope is. This feature of the IBNN is very important in view of generalization. The specifically computed slopes of transfer functions contain information about other training patterns that define the same classes. Such intervals of each class define models of classes for each feature separately.

Coefficient q^m makes the computations of slope parameters (fig. 2) insensitive for differences in patterns quantity representating classes, for quantity of training data and for quantity of classes defined in the training data.

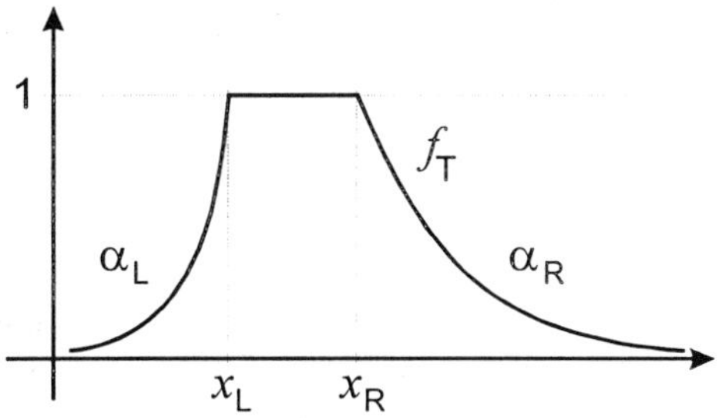

Fig. 2. The transfer function for the 1-th layer neurons.

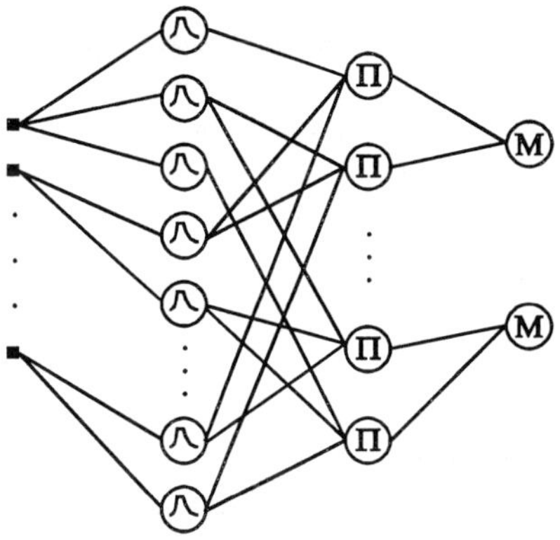

Fig. 3. The 3-layer topology of the IBNN for hyperregions.

Each neuron of 1-st layer is always connected to a single input corresponding with a certain input feature. The weights of 1-st layer neurons are always equal 1. The 2-nd layer contains neurons (fig. 3) that compute (limited – for sub-hyperregions) products of some combinations of output values of the 1-st layer neurons producing hyperregions or sub-hyperregions. Only these combinations given by training data that are related to intervals of different existing input features are transformed to connections. Each 2-nd layer neuron represents a group of training patterns of the same class. Such neurons can correctly approximate the values of classification inside the intervals (hyperregions or sub-hyperregions) and outside them after the specific information came from the computed value of slope parameters. Moreover, the advanced analyses of the sorted input data for all features lead to find out the separate sub-hyperregions for each class. The sub-hyperregions in data subspace can not include patterns of other classes. The sub-hyperregions are easy to find because they are convex. The sub-hyperregions are finally grouped together creating different sub-hyperfigures as shown in the figures 4-5. Theoretically, there could exist many possible sub-hyperfigures in the hyperspace that consist of the different sub-hyperregions covering the training data in different ways as shown in the figure 6. Each training pattern has to be closed in at least one sub-hyperregion. The smallest sub-hyperregions can consist of single point in the input space representing single pattern. In order to avoid the problem of ambiguity of sub-hyperfigures there are used two criteria:

1. Criterion of the density of the patterns of the same class: the longer distances between patterns the lower density in the space and the lower probability to create the sub-hyperregion.
2. Criterion of maximum quantity of training samples closed in sub-hyperregions taking into account the quantity of patterns representing different classes. The sub-hyperregions for different classes are compared regarding the quantity of patterns they contain divided by the quantity of patterns representing appropriate classes they represent in order to make the IBNN insensitive for differences in quantity representation of different classes.

The described criteria are used to define the k-th sub-hyperregion coefficient computed as follows:

$$RC_m^k = \frac{Q^k}{V \cdot N_m} \tag{7}$$

where

Q – quantity of patterns included in the given sub-hyperregion,

V – volume of the sub-hyperregion.

where k-th sub-hyperregion consists of patterns of m-th class.

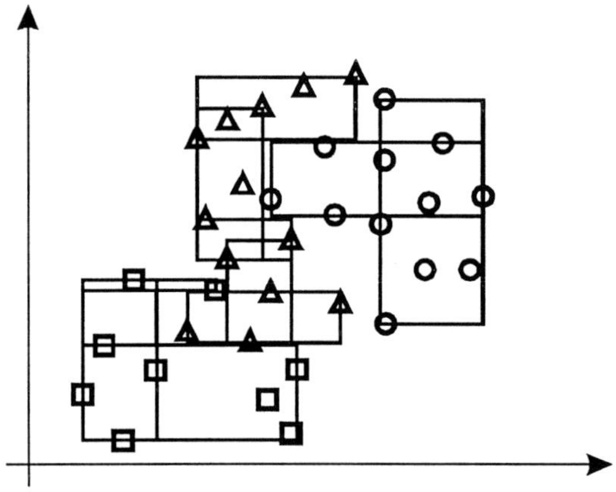

Fig. 4. Producing multiple sub-hyperregions after the intervals.

In order to IBNN classify properly using concave functions at boarders of sub-hyperregions forming sub-hyperfigures representing classes two important conditions have to be true:

1. The sub-hyperregions for the patterns of the same class should overlap if only possible.
2. The sub-hyperregions for the patterns of different classes have to be always separate.

Finally, the overlapped sub-hyperregions of different classes are compared together using coefficients computed for them after the described criteria in order to choose this one which has the maximal value of this coefficient. Other overlapping sub-hyperregions should be omitted or cut off by the winning sub-hyperregion.

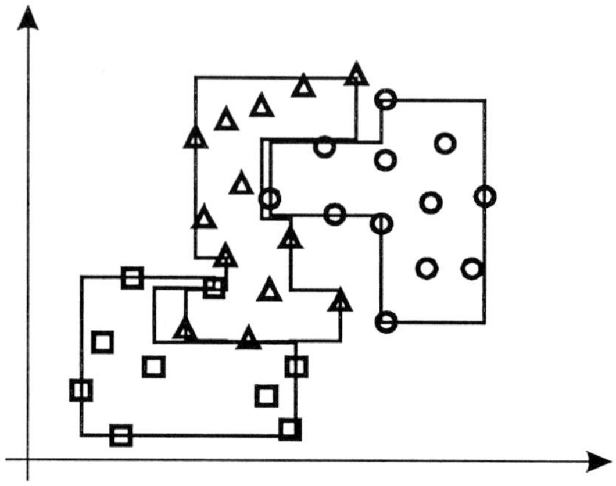

Fig. 5. Connected sub-hyperregions producing sub-hyperfigures for the presented classes.

The transfer function of the 2-nd layer neurons is defined as follows:

$$f_{\Pi}(y) = \prod_{k=1}^{K} y_k \tag{8}$$

The weights of 2-nd layer neurons are always set to 1.

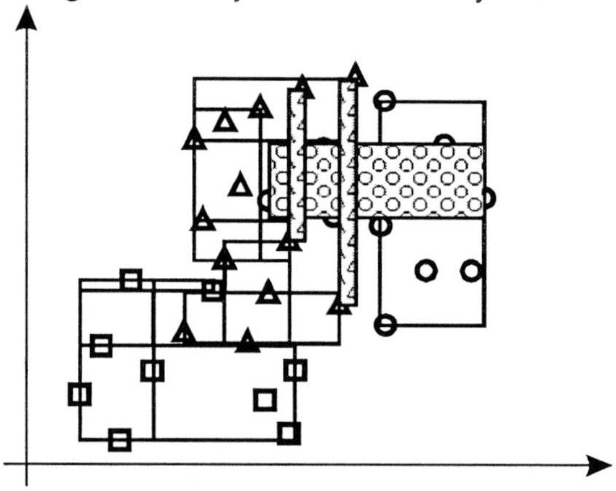

Fig. 6. The use of the criterion (7) prevent the possible collisions of sub-hyperregions.

The 3-rd layer neurons choose the maximum value of the 2-nd layer neurons of the same class. The output values of the 3-rd layer neurons measure the similarities to the classes defined in the training data set. The transfer function of the 3-rd layer neurons is defined as follows:

$$f_M^m(z) = \max_{j \in J}\left\{z_j^i : z_j^i \in C^m\right\} \quad (9)$$

The result classification is defined by the function:

$$f_C(s) = \arg \max_{m=1,\dots,M}\left\{s_m : s_m = f_M^m(z)\right\} \quad (10)$$

This function can be optionally transformed to the neuron of 4-th layer of IBNN. If there is no maximum value for 3-rd layer neurons the output can be defined as zero. The zero output suggest that the network can not univocally qualify the input vector to any of the trained classes.

3 Comparison to other methods

The IBNNs are easy to use ontogenic neural networks. They don't need to be trained because all network parameters are computed in the configuration process automatically. They work similarly to RBF networks (KNN), bicentral based networks and PNN [1, 4, 10, 15] creating a specific hyperregions in the input data space. In comparison to well-known RBF networks which are partially configured and partially trained the IBNNs are only configured after *a priori* knowledge about training data. Only some statistical analyses are needed to find out specific intervals for each class and each feature. For big training data sets computation of IBNN could be sometimes time consuming because the training data have to be sorted for each input feature in order to separate intervals. Intervals as well as RBF neurons group together input data of the same class. In MLP networks the hard-limiter neuron the boundary is a hyperplane, in RBF networks the boundary is the circumference of a hypervolume (hypersphere with Euclidean distance) centered around class samples, while in IBNNs the boundary is hyperregion outspreaded on some subset of class samples. The hyperregions are completed with specific slope functions that are necessary to approximate and generalize outside the IBNN hyperregions and sub-hyperregions. The IBNN is also a good alternative for automatically configured ontogenic SONN [4-5] because SONNs demand binary data for configuration of the networks and computation of weights parameters. IBNN as well as SONN are deterministic and fully automatic. They do not demand any configuration parameters.

4 Conclusions

The described IBNNs can be used as an alternative method to other kernel neural networks (KNN). The deterministic IBNNs configuration process is based on some specific statistical analyses of the training data. The introduced method finds out intervals specific for each training class and constructs specific sub-hyperregions for them. Each hyperregion and sub-hyperregion define the data of one of the trained class and can be used to approximate and generalize data in-between the boundaries of a considered hyperregion. There are defined new specific slope functions that help to generalize outside the hyperregions. The training data are always correctly classified. Moreover, the IBNNs are not sensitive for differences in patterns quantity representation of classes, for quantity of classes. While the IBNNs need to sort training data for each input feature the computations can be time-consuming for big training data sets. On the other hand, big data sets can be better optimized and the IBNN model can be better created.

Acknowledgements: Support from research funds of Polish Committee for Scientific Research is gratefully acknowledged.

References

[1] Duch, W., Korbicz, J., Rutkowski, L., Tadeusiewicz, R. (eds) (2000) Biocybernetics and biomedical engineering, Neural networks, vol. 6, EXIT, Warsaw.

[2] Duch, W., Jankowski N. (1997) New neural transfer functions, Journal of Applied Mathematics and Computer Science, 7(3): 639-658.

[3] Fiesler, E., Beale, R. (eds) (1997) Handbook of neural computation, IOP Publishing Ltd and Oxford University Press, Bristol & New York.

[4] Horzyk, A., Tadeusiewicz, R., (2004) Self-Optimizing Neural Networks, Advances in Neural Networks – ISNN 2004, Proc. of International Symposium on Neural Networks, Dalian, China, Springer-Verlag, Berlin - Heidelberg, pp. 150-155.

[5] Horzyk, A. (2003) Introduction to Self-Optimising Neural Networks for Classification Tasks, Advanced Computer Systems, Sołdek J., Drobiazbiewicz L. (eds), Szczecin, INFORMA, pp.127-135.

[6] Lowe, D. (1989) Adaptive radial basis function nonlinearities and the problem of generalization, 1st IEEE International Conference on Artificial Neural Networks, London, UK, pp. 171-175.

[7] Specht, D. F. (1990) Probabilistic neural networks. Neural Networks, 3: 109-118.

[8] Tadeusiewicz, R., Mikrut, Z., (1998) Neural-Based Object Recognition Support - From Classical Preprocessing to Space-Variant Sensing. Invited paper. In M. Heiss (ed.): Proceedings of the International ICSC/IFAC Symposium on Neural Computation NC '98, Vienna University of Technology, ICSC Academic Press, Canada/Switzerland, pp. 463-468.

Learning from Randomly-Distributed Inaccurate Measurements

John Eidson, Bruce Hamilton, Valery Kanevsky
Agilent Technologies
E-mail: {john_eidson, bruce_hamilton, valery_kanevsky}@agilent.com

Abstract

Traditional measurement systems are designed with tight control over the time and place of measurement of the device or environment under test. This is true whether the measurement system uses a centralized or a distributed architecture. Currently there is considerable interest in using mobile consumer devices as measurement platforms for testing large dispersed systems. There is also growing activity in developing concepts of ubiquitous measurement, such as "smart dust." Under these conditions the times and places of measurement are random, which raises the question of the validity and interpretation of the acquired data. This paper presents a mathematical analysis that shows it is possible under certain conditions to establish dependence between error bounds and confidence probability on models built using data acquired in this manner.

1 Introduction

Traditionally measurements are made on systems to gain insight into their behavior, as inputs to optimize operation, and for fault detection. As systems become physically large and dispersed traditional measurement techniques become more difficult and expensive. Examples of such systems are cellular phone systems, the internet, and epidemic control. For large dispersed systems, traditional measurement techniques, both centralized and distributed, typically spatially undersample due to the cost and difficulty in installing and communicating with a sufficient number of probes.

Currently there is considerable interest in using mobile consumer devices such as cell phones, as measurement platforms for testing and monitoring these large dispersed systems. Such devices are appealing because of their large numbers, physical dispersion, and inherent computing and communication capabilities. Increasingly these devices are also becoming suitable platforms for sensors. For example many of today's cell phones can measure their location, characteristics of the cellular infrastructure, and have the ability to take a picture or to access external devices.

However unlike a traditional measurement system, a measurement system based on mobile consumer devices will at best have limited ability to specify the time and place of measurement. Instead these systems will depend on random spatial and temporal distribution of these devices relative to the phenomenon being measured. As a result we have a large number of randomly distributed inherently inaccurate measurements. This raises interesting questions as to the validity and interpretation of data gathered using these devices.

This paper presents a mathematical analysis that shows it is possible under certain conditions to place error bounds and confidence probability on mathematical models built using data acquired in this manner. The measurement of the spatial variation of the RF fields associated with the cellular phone infrastructure is used as an example. The main body of this paper is focused on the theorem that allows the specification of accuracy and confidence probability for appropriate spatial models constructed from the experimental data, given certain constraints. This theorem will be a *probably approximately correct* (PAC) model of the underlying data. Following this development, the theorem is applied to the RF field example to illustrate its practical use.

2 Mathematical Setting

Consider a space variable $z \in D$, where D is a domain in $\mathbb{R}^k$. Assume there is a family of random variables (measurements) $\xi(z)$, and they are conditionally independent upon any finite subset of the z's where measurements take place. We suppose that D can be represented as a union of open balls of a fixed radius r. Let $r^*>0$ be the exact upper bound of values of r for which the representation is possible. Hence, r^* is a characteristic of "smoothness" of the boundary of D. The space variable z, as viewed in the distributed measurements environment, is also random with a distribution function $F(b)=P(z\in b)$, where b is a measurable subset of D in Borel's sense [1]. In the context of RF field measurement $k = 2$. $\xi(z)$ represents the value of an individual observation, i.e. a measurement result at point z. Let mean value $E\xi(x) = m(x)$ be an unknown function of x, to be estimated, based on spatially distributed observations $\xi(z_i)$ at n points, represented by the set: $A = \{z_1,\ldots, z_n\}$. A is assumed to be a set of independent random points, generated by a certain underlying random process as, e.g. in case of mobile measurement. A framework to our approach to "learn" m(x) from random observations $\xi(z_i)$ is influenced by Statistical Learning Theory (*SLT*) [3], [6]. In *SLT*, however, a *hypothesis space* [6] of candidate functions to approximate the unknown m(x) does not depend on sample $A = \{z_1,\ldots, z_n\}$ as, e.g. in polynomial regression, where a typical *hypothesis space* is the set of polynomials of a fixed degree. Instead, we deal with sets of functions each of which is sample dependent as, e.g. in the case of histograms. Due to sample dependency our approach is essentially non-parametric and the *hypothesis space* is infinite dimensional. This sample dependency leads to first deriving bounds for conditional, upon sample A, probabilities of all events of interest and later averaging them to achieve the final result.

Assumption 1: Measurement Points Distribution Regularity

If d, $d \subset D$ is a ball of a radius r then:

- The probability distribution of measurement locations $F(\bullet)$ is absolutely continuous with respect to a planar Lebesgue measure on $\mathbb{R}^2$. In particular, support of F cannot have $\mathbb{R}^2$ Lebesgue measure zero.
- $F(d) > \gamma > 0$
  where $\gamma = \gamma(r)$ depends only on the radius of a ball but not on the location of its center.

Assumption 2: Smoothness

We assume that unknown underlying, subject to measurement, function m(z) is sufficiently smooth, i.e. we exclude from consideration all excessively volatile functions with, e.g. unbounded change over bounded domain.

$$|m(x) - m(y)| < K\rho^{\alpha}(x,y)$$

for all $x,y \in D$; where K and α are positive constants, $\rho(x,y)$ is the Euclidean distance between x and y.

Assumption 3: Bounded Measurement Range

The random variable $\xi(z)$ that represents a measurement result at point z has a limited range:

$$a \le \xi(z) \le b$$

2.1 Problem Statement

We want to construct such an *estimator* M(x) for m(x), i.e. a function $M(x) = M(x; \xi(z_1),\ldots, \xi(z_n))$ of $x \in D$ and measurements $\xi(z_i)$ such that large deviations of M(x) from m(x) with respect to a given metric are unlikely if n is sufficiently large. In the case of a uniform metric this can be specified as follows: for any $\epsilon>0$, $\delta>0$ there is n such that

$$P\{ \sup_{x\in D} | M(x; \xi(z_1),\ldots, \xi(z_n)) - m(x) | < \epsilon\} > 1 - \delta \quad (1)$$

If (1) is satisfied for an estimator M(x), it enables us to build a *Confidence Blanket(CB)* for m(x). In case of RF strength measurements *CB* at the confidence level $1-\delta$ is defined as a subset in the functional space of functions over D with the uniform metric, i.e.:

$$\{f: M(x; \xi(z_1),\ldots, \xi(z_n)) - \epsilon < f(x) < M(x; \xi(z_1),\ldots, \xi(z_n)) + \epsilon\}$$

and (1) holds. The latter can be interpreted as a claim that probability that *CB contains* an unknown function m(x) is at least $1-\delta$. Alternatively, in the functional space with integral metric the subset is defined as: $\{f: \int_D |M(x) - f(x)| \Phi(dx) < \varepsilon\}$ and $P\{ \int_D |M(x) - m(x)| \Phi(dx) < \varepsilon\} > 1 - \delta$. Here $\Phi(\bullet)$ is a probability measure/weight on a class of Borel's measurable subsets of D i.e. $\int_D \Phi(dx) = 1$. Weight Φ reflects a relative importance of the level of approximation of m(x) by M(x) in different parts of D. In this formulation the choice of M(x) is similar to one in the density estimation problem [2].

3 Confidence Blanket

We will derive a confidence blanket for two different metrics: uniform and integral.

Estimator M(x) for m(x). For every subset s of A of cardinality $|s|$ we define

$$G(\xi, s) = \begin{cases} \frac{1}{|s|} \sum_{w \in s} \xi(w), & \text{if } |s| \neq 0 \\ 0, & \text{if } |s| = 0 \end{cases} \quad (2)$$

Let $d(x) = d(x;r)$ be an open ball of radius r centered at point x and $s = s(x;r)$ be defined as $A \cap d(x;r)$, i.e. a subset of measurement points within d(x;r). Let $S = S(r)$ be a class of all such subsets. One simple candidate for an estimator of m(x) can be obtained as an average value of observations that happened at points inside d(x;r), i.e.

$$M(x)=G(\xi, A\cap d(x;r)) \quad (3)$$

where $k_n=k_n(x;r) = |s|$, i.e. the cardinality of measurement points inside d(x;r). Due to the nature of the set A, $k_n(x,r)$ can be viewed as the number of successes in n Bernoulli's trials ($p=F(d(x))$) for every $x \in D$ and r. Notice that by (3), M(x) is effectively defined as a function on S, i.e. $M(x) = M(s(x;r)) = M(s)$. There is an inconvenience in the definition of M(x) when $k_n = 0$. Let us correct it by breaking the probability of the complement to the event in (1) into summands related to two, properly chosen, mutually exclusive events, i.e. if $\Omega=\{ \sup_{x\in D} | M(x; \xi(z_1),\ldots, \xi(z_n)) - m(x) | \ge \epsilon\}$, then

$$P\{\Omega\}= P\{\Omega; \sup_{x\in D} | k_n(x)/n-F(d(x)) | < \beta\} + P\{\Omega; \sup_{x\in D} | k_n(x)/n-F(d(x)) | \ge \beta\} \quad (4)$$

Here semicolon between events stands for intersection and β is an arbitrary non-negative constant. If β is small enough, e.g. $\beta < \gamma$, event $k_n(x)=0$ may affect only the second summand. Noticing that $E(k_n(x)/n) = F(d(x))$ the second term in (4) is bounded as in the Vapnik-Chervonenkis theorem [3], [4]:

$$P\{\sup_{x\in D} | k_n(x)/n-F(d(x)) | \geq \beta\} \leq 8|S| \exp(-n\beta^2/32) \quad (5)$$

We will see later (Lemma 1) that $|S| = n^2-n+2$.

Remark: Generally speaking, set Ω, as uncountable union of events: $\Omega = \bigcup_{x\in D} \{| M(x; \xi(z_1),\ldots, \xi(z_n)) - m(x) | \geq \epsilon\}$, may not be a measurable set, i.e. not an event, itself. However, function $|M(x; \bullet ,\ldots, \bullet) - m(x)|$, as defined by (3), is a random map from separable subset D^{n+1} of $\mathbb{R}^{2(n+1)}$ (two coordinates for every argument) into compact $[a,b]\subset\mathbb{R}$ and, therefore, has a stochastically equivalent separable modification [7]. The latter will be used instead without a change in the notation. The existence of such modification means, in particular, that Ω can be represented as a union over an everywhere dense countable subset of x's in D and, thus, become an element of the corresponding sigma-algebra, i.e., a measurable subset, i.e. an event [7].

Hereafter, until stated otherwise, all expectations and probabilities are conditional upon $A=\{z_1,\ldots, z_n\}$, i.e. $E(\bullet)$ and $P\{\bullet\}$ will stand for $E(\bullet /A)$ and $P\{\bullet /A\}$ respectively. We will return to unconditional probabilities at the end of the derivation for *CB* bounds based on the *Total Expectation/Probability Formula(TEF)*: $E(\bullet)=E(E(\bullet/A))$ [1], which is transformed into $P\{\bullet\}=E(P\{\bullet/A\})$, when applied to indicators of events. It follows from Assumption 2 that $k_n\to\infty$ along with n, for any x and r, i.e. measurements visit every sub domain of D infinitely often when $n\to\infty$. If $\hat{m}(x)$ is the mean value of M(x) then (3) yields:

$$\hat{m}(x) = E(M(x)) = \begin{cases} \frac{1}{k_n} \sum_{w\in d(x)} m(w), & \text{if } k_n \neq 0 \\ 0, & \text{if } k_n = 0 \end{cases} \quad (6)$$

The natural way to assess $P\{\Omega; \sup_{x\in D} | k_n(x)/n-F(d(x)) | \leq \beta\}$ is to approximate the difference $M(x) - m(x)$ by $M(x) - \hat{m}(x)$ and then deal with the latter which is a centralized sum of independent random variables (see (3), (6)).

Let $B_\beta = \{ \sup_{x\in D} |k_n(x)/n-F(d(x))| \leq \beta\}$. Straightforward manipulations result in:

$$P\{ \sup_{x\in D} |M(x) - m(x)| \geq \epsilon; B_\beta \}$$
$$= P\{ \sup_{x\in D} |M(x)\, \hat{m}(x) + \hat{m}(x) - m(x)| \geq \epsilon; B_\beta \}$$
$$\leq P\{ \sup_{x\in D} |M(x) - \hat{m}(x)| + \sup_{x\in D} | \hat{m}(x) - m(x)| \geq \epsilon; B_\beta \}$$
$$= P\{ \sup_{x\in D} |M(x) - \hat{m}(x)| \geq \epsilon - \sup_{x\in D} | \hat{m}(x) - m(x)|; B_\beta \} \quad (7)$$

It follows from the definition of M(x) and $\hat{m}(x)$, which also can be viewed as functions on S, that

$$\sup_{x\in D} |M(x) - \hat{m}(x)| = \sup_{s\in S} |M(s) - \hat{m}(s)| \quad (8)$$

This way we are able to reduce the original definition of "sup" from over an infinite number of points x in D to over a finite number of different subsets of S which can be obtained by intersecting it with the set $B(r) = \{d(x;r)\}$ of all balls of fixed radius r. The upper bound on the cardinality $|S|$ of S will play a critical role in the accuracy of the estimate of $P\{\Omega\}$.

4 Main Results

We establish confidence bounds for the uniform and integral metrics.

4.1 Uniform metric

We first find the maximum number $|S|$ of different subsets of a set of n points in $\mathbb{R}^2$ that can be cut off by circles of the same radius.

Bounds on $|S|$**.** We can derive an upper bound for $|S|$ from *Sauer's Lemma* [2]. Let V be the *Vapnic-Chervonenkis* (*VC*) [3] dimension of the set of balls of the same radius and U be the *VC* dimension of the set of all balls. Then obviously $V \leq U= k+1=3$, in the case of the Euclidean plane [5]. On the other hand, for example, a set of three points at the vertices of the equilateral triangle can be shattered by a set of balls with a fixed radius r greater than $\sqrt{3}/3$, if the distance between vertices is 1. Therefore $V\geq 3$ and, hence, V=3. *Sauer's Lemma* now implies: $|S| \leq (ne/3)^3 < n^3$. However, we can do better. The maximal achievable value of $|S|$ over all sets of n points is called the *shatter coefficient* [2]. Certainly, this value does not depend on r. Let N(n) stand for this value.

Lemma 1. $N(n) = n^2-n+2$

Proof. We start with a simple construction to prove a lower bound. Let n points be equidistantly placed on the circumference of a ball of radius h. Then, for any integer $1\leq k<n$, all n subsets of k consecutive points can be cut off by balls of a fixed radius r>h. In addition, if we account for an empty set and the entire set of n points, the total number of thus generated subsets is $n(n-1)+2 = n^2-n+2$.

For the idea of the following proof of the upper bound we are indebted to professor Shakhar Smorodinsky [8].

At each point of A place a ball centered at this point with the same radius, say one, as is used to achieve N(n) distinct subsets. This arrangement of balls partitions the plane into a number of components, usually referred to as faces. The claim is that the quantity of components is an upper bound for N(n). Let c be one of the centers of unit balls, which generate S. Generally speaking $c \notin A$. A corresponding subset $s(c) \subset S$ consists of all points whose distance to c is less than 1. By construction those are exactly ones whose corresponding balls bound the component to which c belongs. It means that every component in the arrangement contains no more than one c. Therefore $N(n) \le$ {maximum number of components}$= n^2-n+2$. The latter can be proven, e.g. by induction arguments.

We can now state the bounds under the uniform metric:

Theorem 1. Under assumptions 1–3,

$$P\{\sup_{x\in D} | M(x; \xi(z_1),\ldots, \xi(z_n)) - m(x) | < \epsilon\} \ge 1-2(n^2-n+2)(\exp(-\lambda^2 h^2(n))+4\exp(-n\beta^2/32)) \quad (9)$$

where $\lambda=\sqrt{2}/(b-a)$, $\beta < \gamma$ is an arbitrary positive number and $h(n) = (n(\gamma-\beta))^{1/2}(\epsilon - Kr^\alpha)$.

Proof. Sensible bounds for $P\{\sup_{x\in D} |M(x) - \hat{m}(x)| \ge \epsilon - \sup_{x\in D} |\hat{m}(x) - m(x)|; B_\beta\}$ require $\sup_{x\in D} |\hat{m}(x) - m(x)|$ to be limited by ϵ. Given B_β has occurred,

$$\sup_{x\in D} |\hat{m}(x) - m(x)| = \sup_{x\in D} \left|\frac{1}{k_n} \sum_{w\in s} (m(w) - m(x))\right| \quad (10)$$

can be assessed, due to the definition (6) and Assumption 2 (smoothness). All terms $|m(w)-m(x)|$ in (10) are bounded uniformly, since w and x are inside a certain ball d(x;r) of radius r: $|m(w)-m(x)| \le K\rho^\alpha(x,w) \le Kr^\alpha$. This leads straight to:

$$|\hat{m}(x) - m(x)| = \left|\frac{1}{k_n} \sum_{w\in s} m(w) - m(x)\right|$$

$$\le \frac{1}{k_n} \sum_{w\in s} |m(w)-m(x)| \le Kr^\alpha$$

Therefore, a choice for r such that $Kr^\alpha < \epsilon$ is feasible. Two more observations are needed to bound the probability in (7). First, the reduction in (8) allows for representing probabilities in (7) in terms of a new variable, s. Second, the event $\{\sup_{s\in S} |M(s) - \hat{m}(s)| \ge t; B_\beta\}$ is equivalent to the union $\bigcup_{s\in S} \{|M(s) - \hat{m}(s)| \ge t; B_\beta\}$. These observations yield the chain of inequalities:

$$P\{\sup_{s\in S} |M(s) - \hat{m}(s)| \ge \epsilon - \sup_{z\in D} |\hat{m}(z) - m(z)|; B_\beta\}$$

$$\le P\{\sup_{s\in S} |M(s) - \hat{m}(s)| \ge \epsilon - Kr^\alpha; B_\beta\}$$

$$\le \sum_{s\in S} P\{|M(s) - \hat{m}(s)| \ge \varepsilon - Kr^\alpha; B_\beta\} \quad (11)$$

The number of summands in (11) is not greater than $|S| \le n^2-n+2$, hence Hoeffding's inequality applied to every term in (11) implies:

$$P\{\sup_{s\in S} |M(s) - \hat{m}(s)| > \varepsilon - Kr^\alpha; B_\beta\}$$

$$\le 2(n^2-n+2) \max_{s\in S} \{\exp(-2\frac{k_n}{(b-a)^2}(\epsilon - Kr^\alpha)^2)\}$$

$$= 2(n^2-n+2)\exp(-\lambda^2(\epsilon - Kr^\alpha)^2 k_n) \quad (12)$$

where k_n is such that $B_\beta = \{\sup_{x\in D} | k_n(x)/n-F(d(x)) | \le \beta\}$ holds. Let us return now to unconditional probabilities in (12) via *TEF*, i.e.

$$P\{\sup_{s\in S} |M(s) - \hat{m}(s)| \ge \varepsilon - Kr^\alpha; B_\beta\}$$

$$\le 2(n^2-n+2)E\{\exp(-\lambda^2(\epsilon - Kr^\alpha)^2 \min_{s\in S} k_n); B_\beta\} \quad (13)$$

where $E\{X(\omega); A\}$ stands for $\int_A X(\omega(\omega)P(d, X(\omega)$ is a random variable, A is a measurable subset of a probability space and $P(\bullet)$ is a probability measure on this space. Remember that random variable k_n is a function of x. Since

$$E\{\exp(-\lambda^2(\epsilon - Kr^\alpha)^2 \min_{s\in S} k_n); B_\beta\}$$

$$= E\{\exp(-\lambda^2(\epsilon - Kr^\alpha)^2 \min_{x\in D} k_n); B_\beta\}$$

on B_β the value of $\frac{k_n}{n} \ge \gamma-\beta$ and consequently,

$$E\{\exp(-\lambda^2(\epsilon - Kr^\alpha)^2 \min_{x\in D} k_n); B_\beta\}$$

$$\le P(B_\beta)\exp(-n\lambda^2(\epsilon - Kr^\alpha)^2(\gamma-\beta))$$

$$< \exp(-\lambda^2 h^2(n)).$$

Finally (4), (5) and (11) yield the following bound: $P\{\Omega\} \le 2(n^2-n+2)(\exp(-\lambda^2 h^2(n))+4\exp(-n\beta^2/32))$.

Now the confidence probability for *CB* can be estimated as:

$$P\{\sup_{x\in D} | M(x; \xi(z_1),\ldots, \xi(z_n)) - m(x) | < \epsilon\} \ge 1-2(n^2-n+2)(\exp(-\lambda^2 h^2(n))+4\exp(-n\beta^2/32)) \quad (14)$$

Remark 1. This bound holds for an arbitrary r and β within the interval $0<\beta<\gamma$. Recall that γ is a function of r and therefore there are optimal values for r and β that maximize the bound in (14).

Remark 2. Assumption 3 can be relaxed in favor of any condition on the distribution of $\xi(z)$ that preserves, for example, an exponential decay of the density's tail. The finiteness of $E(\exp(\nu\xi(z)))$ for some $\nu > 0$, uniformly over z, for instance, is sufficient. It permits using Hoeffding's inequality to estimate the probability in (7).

4.2 Integral Metric

We start with some preliminary preparations similar to the previous case. As before, a set $A = \{z_1,\ldots, z_n\}$ of measurement points' locations is treated as fixed and all expectations / probabilities are conditional upon A unless stated otherwise. Let us centralize M(x) with

$$\hat{m}(x): \; P\{\int_D |M(x) - m(x)|\,\Phi(dx) \geq \varepsilon\}$$

$$\leq P\{\int_D |M(x) - \hat{m}(x)|\,\Phi(dx) + \int_D |\hat{m}(x) - m(x)|\,\Phi(dx) \geq \varepsilon\} \quad (15)$$

and estimate $|\hat{m}(x) - m(x)|$ as:

$$|\hat{m}(x) - m(x)| \leq \begin{cases} Kr^{\alpha}, & \text{if } k_n \neq 0 \\ m(x) \leq a, & \text{if } k_n = 0 \end{cases} \quad (16)$$

Since

$\int_D |\hat{m}(x) - m(x)|\,\Phi(dx) = \int_{\{k_n(\bullet) \neq 0\}} |\hat{m}(x) - m(x)|\,\Phi(dx) + \int_{\{k(\bullet)=0\}} |\hat{m}(x) - m(x)|\,\Phi(dx)$, the inequality (16) leads to the bound

$\int_D |\hat{m}(x) - m(x)|\Phi(dx)$
$\leq Kr^{\alpha}\Phi(x \mid k_n(x)\neq 0) + (b-a)\Phi(x \mid k_n(x)=0)$

With the last remark, Markov's inequality $P(Y\geq t)\leq \frac{EY}{t}(t, Y\geq 0)$ [1], applied to the random variable $\int_D |M(x) - \hat{m}(x)|\,\Phi(dx) + a\Phi\Phi(: k_n(x) = 0)$ implies:

$$P\{\int_D |M(x) - \hat{m}(x)|\,\Phi(dx) + u_2 \geq \varepsilon - u_1\}$$

$$\leq \frac{E\int_D |M(x) - \hat{m}(x)|\,\Phi(dx) + u_2}{\varepsilon - u_1} \quad (17)$$

where $u_1 = Kr^{\alpha}\Phi(x \mid k_n(x)\neq 0)$ and $u_2 = (b-a)\Phi(x \mid k_n(x)=0)$. Naturally, r should be such that $Kr^{\alpha}\Phi(x: k_n(x)\neq 0) < \epsilon$.

By Fubini's theorem one can swap $\int$ and E in (17). To treat $E|M(s) - \hat{m}(s)|$ we use the fact $EY = \int_0^{\infty} P\{Y \geq t\}dt$, which holds for any nonnegative random variable Y if EY exists. That way the assessment of $E|M(s) - \hat{m}(s)|$ is reduced to finding a bound for the corresponding probability. Hoeffding's inequality applied to $P\{|M(x) - \hat{m}(x)| \geq t\}$ bounds the numerator of the right hand side in (17) by

$$\int_D \Big(\int_0^{\infty} 2\exp(-2t^2 k_n(x)/(b-a)^2)dt\Big)\,\Phi(dx) + u_2$$

$$= (b-a)\sqrt{\frac{\pi}{2}} \int_{\{k_n(\bullet)\neq 0\}} \frac{\Phi(dx)}{\sqrt{k_n(x)}} + u_2$$

and, consequently inequality (17) now looks as follows:

$$P\{\int_D |M(x) - \hat{m}(x)|\,\Phi(dx) + u_2 \geq \varepsilon - u_1\}$$

$$\leq \frac{1}{\varepsilon - u_1}\Big((b-a)\sqrt{\frac{\pi}{2}} \int_{\{k_n(\bullet)\neq 0\}} \frac{\Phi(dx)}{\sqrt{k_n(x)}} + u_2\Big) \quad (18)$$

Remark: This derivation is true only for those x's where $k_n(x) \neq 0$. Such x's do not contribute to the value of the integral on the right hand side of (17) due to the definition of M(x) and $\hat{m}(x)$. The latter is reflected on of the domain of integration in (17) and (18). In order not to have these reservations in the future and for the convenience in applying Fubini's theorem again, let us define random, with respect to set $A = \{z_1,\ldots, z_n\}$, variable $j_n(x)$ as: $j_n(x) = \begin{cases} 1/\sqrt{k_n(x)}, & \text{if } k_n(x) \neq 0 \\ 0, & \text{if } k_n(x) = 0 \end{cases}$. We can now return to actual probabilities via *TEF* by taking expectations over all possible sets $A = \{z_1,\ldots, z_n\}$ from both sides of the inequality

$$P\{\int_D |M(x) - m(x)|\,\Phi(dx) \geq \varepsilon\}$$

$$\leq \frac{(b-a)\sqrt{\frac{\pi}{2}} \int_D j_n(x)\Phi x)\Phi(+u_2}{\varepsilon - u_1} \quad (19)$$

which follows from (15) and 18). Indeed if Fubini's theorem is applied to the right hand side of (19), we can evaluate $Ej_n(x)$ first and then estimate the integral over D. Following the definition of j_n, the calculation of Ej_n can be represented as :

$Ej_n = E\{j_n; k_n\neq 0\} + E\{j_n; k_n=0\} = E\{j_n; k_n\neq 0\}$
$= E\{1/\sqrt{k_n}; k_n\neq 0\}$

In its own order, set $\{k_n\neq 0\}$can be further split into two subsets

$\{k_n(x)\neq 0\} = \{k_n(x)/n - F(d(x)) > -\beta;\ k_n(x)\neq 0\}$
$\cup\{k_n(x)/n - F(d(x)) \leq -\beta;\ k_n(x)\neq 0\}$

and therefore

$E\{1/\sqrt{k_n};\ k_n(x)\neq 0\}$
$= E\{1/\sqrt{k_n};\ \{k_n(x)/n - F(d(x)) > -\beta\} \cap \{k_n \neq 0\}\}$
$+ E\{1/\sqrt{k_n};\ \{k_n(x)/n - F(d(x)) \leq -\beta\} \cap \{k_n \neq 0\}\}$

Let us work with each of two terms separately. Since $\{k_n(x)/n-F(d(x))>\beta\} \subset \{k_n(x)\neq 0\}$, for the first term, we have

$$E\{1/\sqrt{k_n};\ \{k_n(x)/n-F(d(x))>-\beta\}\cap\{k_n\neq 0\}\}$$
$$= E\{1/\sqrt{k_n};\ k_n(x)/n-F(d(x))>-\beta\}$$
$$< \frac{1}{\sqrt{n(F(d(x))-\beta)}} P(k_n(x)/n - F(d(x)) > -\beta\beta$$
$$< \frac{1}{\sqrt{n(F(d(x))-\beta)}}$$
$$\leq \frac{1}{\sqrt{n(\gamma-\beta)}}$$

and for the second

$$E\{1/\sqrt{k_n};\ \{k_n(x)/n-F(d(x))\leq -\beta\}\cap\{k_n\neq 0\}\}$$
$$\leq 1\cdot P(k_n(x)/n-F(d(x))\leq -\beta)$$
$$\leq \exp(-2n\beta^2)$$

The last is Hoeffding's bound for Bernoulli's random variable $k_n(x)$. Here $\beta<\gamma$ is an otherwise arbitrary constant. As the result of this observation we have:

$$Ej_n < \frac{1}{\sqrt{n(\gamma-\beta)}} + \exp(-2n\beta^2) \tag{20}$$

Notice that this bound depends on r, since $\gamma=\gamma(r)$, but does not depend on x. The next step is to estimate $u_1 = Kr^{\alpha}E\Phi(x;\ k_n(x)\neq 0)$. Since Φ is a probability measure $u_1 \leq Kr^{\alpha}$. In order to estimate Eu_2 we introduce the random indicator function $I_{[k_n(\bullet)=0]}(x)$ and by applying Fubini's theorem we obtain:

$$E\Phi(x;\ k_n(x)=0)$$
$$= E\int_D I_{[k_n(\bullet)=0]}(x)\Phi)\Phi(d$$
$$= \int_D (E(I_{[k_n(\bullet)=0]}(x)))\Phi)))\Phi \tag{21}$$

By the definition of indicator function $E(I_{[k_n(\bullet)=0]}(x)) = P(k_n(x)=0)=(1-F(d(x)))^n$. Finally, $\int_D (1-F(d(x)))^n\,\Phi(dx) \leq (1-\gamma)^n$. The latter can now be combined with (17), (19) and (20) to enable the bound

$$P\{\int_D |M(x)-m(x)|\,\Phi(dx) \geq \varepsilon\}$$
$$\leq \frac{(b-a)\left(\sqrt{\frac{\pi}{2n(\gamma-\beta)}} + \exp(-2n\beta^2) + (1-\gamma)^n\right)}{\varepsilon - Kr^{\alpha}}$$
$$= (\sqrt{\pi}/\lambda)h^{-1}(n) + \frac{(b-a)(\exp(-2n\beta^2)+(1-\gamma)^n)}{\varepsilon - Kr^{\alpha}} \tag{22}$$

We can summarize this derivation as:

Theorem 2. For arbitrary $\epsilon>0$, $r<(\epsilon/K)^{1/\alpha}$ and $\beta<\gamma$, if conditions 1–3 are true the following bound holds:

$$P\{\int_D |M(x)-m(x)|\,\Phi(dx) < \varepsilon\} \geq 1-\left((\sqrt{\pi}/\lambda)h^{-1}(n) + \frac{(b-a)(\exp(-2n\beta^2)+(1-\gamma)^n)}{\varepsilon - Kr^{\alpha}}\right) \tag{23}$$

Remark 1. To bring the formulation of *CB* to the form similar to the uniform case, let us observe the inclusion of events $E_1=\{\Phi\{x:\ |M(x)-m(x)| \geq t\} \geq \epsilon/t\} \subseteq E_2 = \{\int_D |M(x)-m(x)|\,\Phi(dx) \geq \varepsilon\}$ and consequently $P(E_1) \leq P(E_2)$.

In other words, under the conditions of Theorem 2, one is unlikely to observe large ($\geq$ t) absolute deviations of the model M(x) from an unknown phenomenon m(x) over a significant (in terms of Φ) portion of the domain D.

Remark 2. The bounds in both cases of uniform (14) and integral (23) metrics are functions of r, β, ϵ and n. Since the left hand side of these inequalities depends on ϵ as well, a value for ϵ is usually picked beforehand by some practical concerns. It is obvious that only by raising n can bounds be made arbitrarily close to 1. For the two other variables there exist optimal values $r = r(n)$ and $\beta = \beta(n)$ to tighten the bound in (22). On the other hand, that is the "slowest" term $\sqrt{\frac{\pi}{2n(\gamma-\beta)}}$ which sets the rate of convergence in (22) and (23) as a function of n. Therefore it makes sense to set $\beta(n)$ such that $\exp(-2n\beta^2) = O(n^{1/2})$, e.g. $\beta = (\ln(n)/n)^{1/2}$.

4.3 Rate comparison for two metrics

We compare the convergence rates for bounds using the integral metric and uniform metric. To simplify calculations we set $K=\alpha=1$. With constant values for ϵ, r, and β, convergence of the bound is faster for the uniform metric case. However, a scenario when all parameters ϵ, r, and β are functions of n can produce an asymptotic feature with the opposite tendency, i.e. convergence is faster with integral metric. Because the rate of convergence to zero of a sum is not faster then that of a summand, it is sufficient, for instance, to show that with proper choice of $\epsilon(n)$, r(n), and $\beta(n)$ every term in (5) goes to zero faster than, say, the first term, $n^2\exp(-\lambda^2h^2(n))$, up to the $O(\bullet)$ equivalence, in (4). The choice of h(n) should strike a balance: h(n) must grow fast enough to offset a second degree multiplier $n^2 = \exp(2\ln(n))$, but not too fast, so that all terms in (5) converge to zero faster than $\exp(2\ln(n)-\lambda^2h^2(n))$. If, for example, r(n) is chosen such that $\gamma(n) = (2\ln(n)+\ln\ln\ln(n))/n)^{1/3}$, $\epsilon(n)$ is set as $r(n) + \sqrt{2\lambda^{-1}\gamma(n)}$ and $\beta(n) = (1/2)\gamma(n)$ then it can be verified that the bound in (5) is $O(1/\ln(n))^{-1/2}$ which is $o((\ln\ln(n))^{-1})$, whereas the asymptotic behavior of the first term in (4) is like $O((\ln\ln(n))^{-1})$.

5 Numerical Simulation

We performed an analysis using simulated measurements of RF energy at varying distances from a single tower, following the Lee model of RF propagation [9]. Because of space constraints, we present results only for the uniform metric.

5.1 Choice of parameter values

We set the signal strength to range between a minimum and maximum which we have observed experimentally, and set the received signal level to $r^{-3.4}$, a common value for even terrain and moderate clutter. The distance from the tower was varied from 1 mile to 15 miles. Simulated measurements were made at uniformly distributed points. Their values were obtained by perturbing the true function with normally distributed noise having zero mean and σ of 0.15 times the range of actual values. A representative set of measurements is shown in Figure 1.

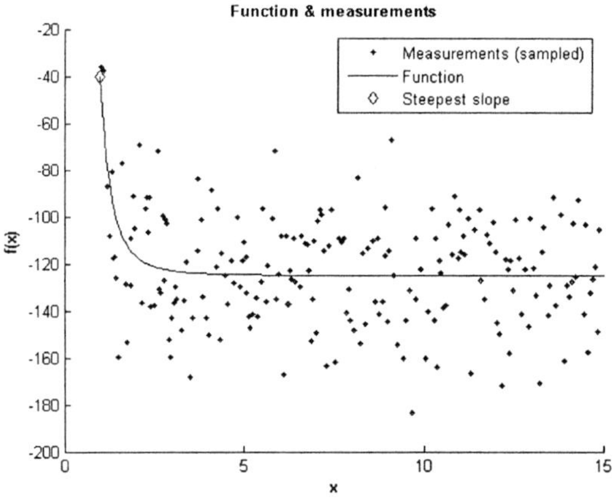

Figure 1: Function and simulated measurements

The effect of dimension k of D appears in the upper bound on |S| (see Lemma 1). Consider an interval I of fixed width r, initially positioned to the left of all n measurement points, and sliding to the right until it is to the right of all the points. Each point will make a transition from the right of I to inside it, and another from inside to the left of it; and it can make each transition only once. There is a different set of measurements only when a point makes a transition. There can be at most 2n transitions, and thus at most 2n+1 states (counting the initial state). Thus |S| is 2n+1 in the 1-dimensional case rather than n^2-n+2. We have made that substitution in what follows.

In evaluating equation (4) we chose ϵ = 8 dB, an error bound commonly accepted for drive tests and model optimizations. Parameters a and b are -40 dBm and -125 dBm, respectively, as described above. We chose the rate of signal decay as 34 dB/decade, consistent with our choice of received signal level = $r^{-3.4}$ above. In choosing a value for γ we make use of the fact that measurement locations are uniformly distributed: γ = (xmax-xmin)/2r. Once γ is known, we choose the value for β which minimizes an upper bound for δ (see (1)), given all the other parameters. We are interested in the behavior of δ for different n, in particular, what value of n is required to give acceptable confidence (say, 95%) for realistic values of the other parameters. We will thus vary n over a range of values. It remains to choose r.

When all other parameters are fixed, there is a value of r which minimizes an upper bound for δ: If r is small, γ and β both approach 0. Thus $\exp(-n\beta^2)$ and $\exp(-n\lambda(\gamma-\beta))$ both approach 1 and (4) is approximately n^2. If r is large, it approaches a limit imposed by the Hoeffding inequality, so $\epsilon \approx K\rho^{\alpha}$, λ approaches 0, $\exp(-n\lambda(\gamma-\beta))$ approaches 1, and again (4) is approximately n^2. For intermediate values the exp(...) terms balance the n^2 terms and (4) can be <1.

We swept a ball of optimum radius over the range of distances from the tower, and observed that the values estimated from (3) are stepwise-constant over varying intervals, as shown in Figure 2.

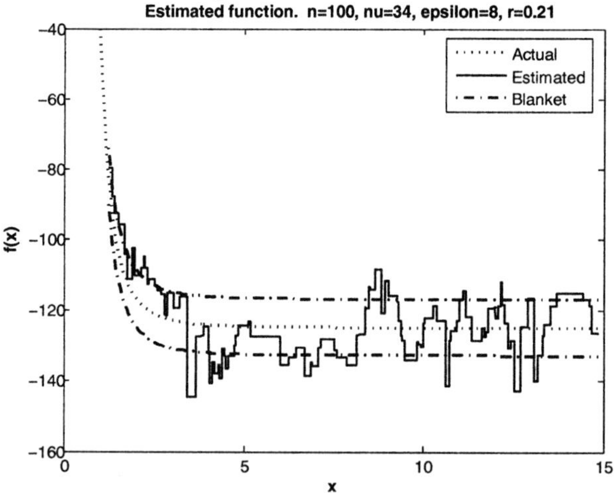

Figure 2: Estimated values are stepwise-constant

To estimate the actual confidence, we ran 100 trials with fixed n, randomly generated measurements, and optimum r. In each trial we compared the modeled signal level with M(x) as given in (3), and observed the fraction of the time that the error exceeded ϵ. We repeated this for a variety of values of n. The results are given in Table 1.

	Uniform metric		Integral metric	
N	δ	Actual	δ	Actual
1×10^2	6.2×10^5	0.7	12.06	0.99
1×10^3	4.6×10^7	0.5	3.94	0
1×10^4	2.4×10^8	0.4	1.37	0
1×10^5	1.6×10^5	0	0.41	0
5×10^5	2.0×10^{-1}	0	0.18	0
1×10^6	5.9×10^{-11}	0	0.12	0
5×10^6	5.0×10^{-91}	0	0.05	0

Table 1: Confidence bound vs. actual confidence

6 Conclusion

The Confidence Blanket appears to be an adequate mathematical concept to make statistical inferences from distributed inaccurate data. Though not tight, explicit lower bounds for confidence probability were established for two different metrics. These bounds were confirmed by numerical simulation. This approach can be extended in a variety of directions, e.g. for metrics in different functional spaces such as Sobolev's, to include infinite dimensional measurements such as images, and to use a different local approximation method.

7 Acknowledgments

We thank Eujine Pechersky, Vladimir Vapnik and Felix Lazebnik for helpful comments.

8 References

[1] A.N. Shiryaev(1996) Probability, Springer-Verlag, New York

[2] L. Devroy and G. Lugosi (2001) Combinatorial Methods in Density Estimation, Springer-Verlag, New York

[3] V. Vapnik(1998) Statistical Learning Theory, John Willey & Sons, New York

[4] L. Devroy, L. Györfi and G. Lugosi (1997) A Probabilistic Theory of Pattern Recognition, Springer-Verlag, New York

[5] R.M. Dudley (1983) Balls in $\mathbb{R}$k do not cut all subsets of k+2 points, Advances in Mathematics, vol. 34, pp. 41-45

[6] F. Cucker and S. Smale, On the Mathematical Foundation of Learning, Bulletin of the American Mathematical Society (2001), vol. 39, Number 1, pp.1-49, 2001, New York

[7] I.I. Gihman and A.V. Skorohod (1983) Theory of Stochastic Processes, Springer-Verlag, New York

[8] Personal communication (2004)

[9] G. Evans, B. Joslin, L. Vinson, B. Foose (1997) The Optimization and Application of the W. C. Y. Lee Propagation Model in the 1900 MHz Frequency Band, IEEE Vehicular Technology Conference, Phoenix, Arizona

Combining Topological and Cardinal Directional Relation Information in Qualitative Spatial Reasoning

Haibin Sun, Wenhui Li

School of Computer Science and Technology
Jilin University, Changchun,China 130012
Offer_sun@hotmail.com, liwh@public.cc.jl.cn

Abstract

Combining different knowledge representation languages is one of the main topics in Qualitative Spatial Reasoning (QSR). In this paper, we combine well known RCC8 calculus (RCC8) and cardinal direction calculus (CDC) based on regions and give the interaction tables for the two calculi. The interaction tables can be used as a tool in solving constraint satisfaction problems (CSP) and consistency checking procedure of QSR for combined spatial knowledge.

1 Introduction

Qualitative spatial reasoning is very useful in improving the reasoning efficiency in answering spatial queries, and can avoid time-consuming quantitative geometry computation. But reasoning with only one aspect of spatial knowledge is not realistic, combining and integrating different kinds of knowledge is an emerging and challenging issue in QSR. [1] has dealt with the combination of topological knowledge and metric size knowledge in QSR, and [2] has combined the cardinal direction knowledge and the relative orientation knowledge.

We first introduce the two formalisms of topological and cardinal directional relations, respectively. The region considered in this paper is non-empty, connected point-set homeomorphic to a unit disk in $\Re^2$.

1.1 RCC8 Calculus (RCC8)

Topology is perhaps the most fundamental aspect of space. Topological relationships are invariant under topological transformations, such as translation, scaling, and rotation. Examples are terms like *neighbor* and *disjoint* [3]. The RCC-8 (Figure 1) is a set of eight jointly exhaustive and pairwise disjoint (JEPD) relations, called base relations, denoted as DC, EC, PO, EQ, TPP, NTPP, TPPi, NTPPi, with the meaning of DisConnected, Extensionally Connected, Partial Overlap, Equal, Tangential Proper Part, Non-Tangential Proper Part, and their converses. Exactly one of these relations holds between any two spatial regions. A reasoning system has been derived according to the property that a situation involving a number of topological relations is possible if and only if the set of model-constraints associated with all of the relations does not entail any of the entailment constraint formulae. A composition table has been obtained by a spatial reasoner using the above technique. For details about the reasoner and composition table, please refer to [4].

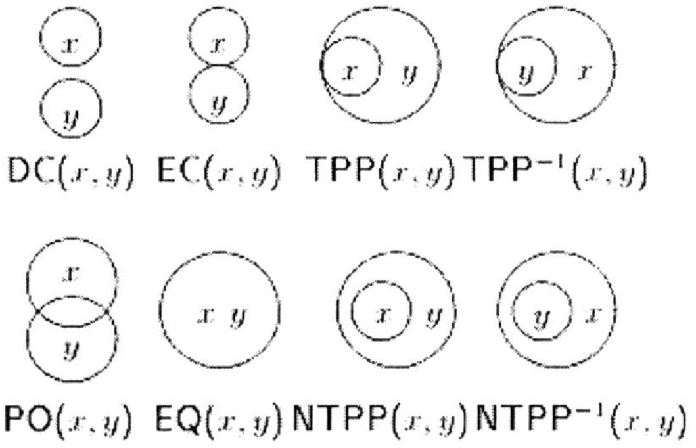

Figure 1. Two-dimensional examples for the eight basic relations of RCC-8

An alternative approach to representing and reasoning about topological relations has been put forward [5], which is based on point-set topology.

1.2 Cardinal Direction Calculus (CDC)

Direction-also called orientation-relationships are important and common-sense linguistic and qualitative properties used in everyday situations and qualitative spatial reasoning.

[6] introduced a direction-relation model for extended spatial objects that considers the influence of the objects' shapes. It uses the projection-based direction partitions and an extrinsic reference system, and considers the exact representation of the target object with respect to the reference frame. The reference frame with a polygon as reference object has nine direction tiles: north (N_A),

northeast (NE_A), east (E_A), southeast (SE_A), south (S_A), southwest (SW_A), west (W_A), northwest (NW_A), and same (O_A). The cardinal direction from the reference object to a target is described by recording those tiles into which at least one part of the target object falls (Figure 2).

$$dir_{RR}(A,B)=\begin{bmatrix} NW_A \cap B & N_A \cap B & NE_A \cap B \\ W_A \cap B & O_A \cap B & E_A \cap B \\ SW_A \cap B & S_A \cap B & SE_A \cap B \end{bmatrix}$$

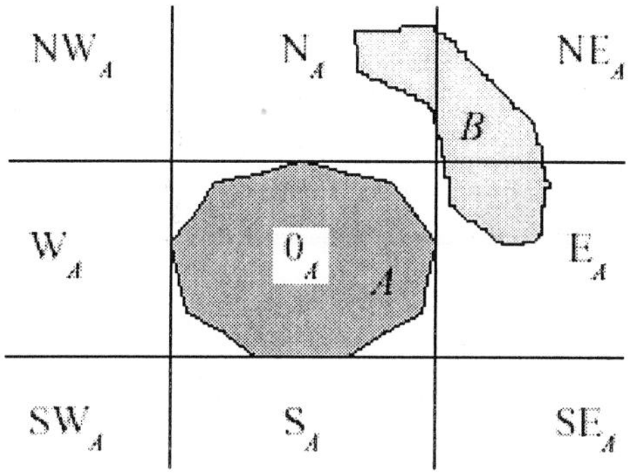

Figure 2. Capturing the cardinal direction relation between two polygons, A and B, through the projection-based partitions around A as the reference object.

For cardinal directions between two polygons, a 3×3 matrix captures the neighborhood of the partition around the reference object and registers the intersections between the target and the tiles around the reference object (Equation 1). The elements in the direction-relation matrix have the same topological organization as the partitions around the reference object. We consider the emptiness and non-emptiness of the nine intersections between the nine tiles formed around the reference object and the exact representation of the target object to describe coarse cardinal directions.

2.Reasoning about combined knowledge of RCC8 and CDC relations

Our main goal is to combine the topological and cardinal direction relation knowledge, which is motivated and inspired by the work of [2]. They try to combine the cardinal direction relations and relative orientation relations based on 2D points. We think our work is more practical, because in the real world we are often faced with regions, not points. And our work is different from theirs in the following aspects:

(1) we investigate two kinds of binary region-based relations; they focus on combining binary and ternary relations, which are all point-based.

(2) We derive additional interaction rules between RCC8 and CDC.

(3) In our procedure, we use two queues instead of one queue used by their work, which can be computed parallel.

Reasoning within the combined knowledge of RCC8 and CDC can be divided into two parts: internal reasoning within the RCC8 and CDC and interaction reasoning between them.

To facilitate the representation of the interaction rules, we denote a basic cardinal direction relation by a set *SB* including at most nine elements, i.e. the nine single-tile [7] cardinal direction relations. For example, *B:S:SE:SN* can be denoted by {*B,S,SE,SN*}. The general cardinal direction relation can be regarded as a set *GB*, whose element is the set *SB*. So we have the relation: $SB \in GB$. The universal relation is the set *BIN*={*B, N, NE, E, SE, S, SW, W, NW*}, and the universe, i.e. the set of all possible cardinal relations, is denoted by *U*.

For two arbitrary regions X and Y, we use *RCC8*(X, Y) to denote the topological relation of X to Y, and we stipulate that *GB*(X, Y) represents the general cardinal direction relation of the primary region X to the reference region Y, and that *SB*(X, Y) represents any basic cardinal relation that belongs to *GB*(X, Y).

2.1 Reasoning within RCC8 and the interaction from RCC8 to CDC (RCC8-To-CDC)

For the internal reasoning within the RCC8, the composition table for pairs of atoms can be found in [8] and [9].

The interaction from the atomic relations of RCC8 to CDC relations has been described in Table 1, where the left column is the eight atomic RCC8 relations (see Fig. 1) and the right column is the cardinal direction relations induced by the left column RCC8 relations. We now give interpretation for this table.

For arbitrary regions X and Y, the table is divided by the value of *RCC8*(X, Y) into eight case as follows:

(1) DC(X, Y): the induced cardinal direction relation *GB*(X, Y) is the universe, i.e. *U*.

(2) EC(X, Y): every basic cardinal direction relation *SB*(X, Y) that belongs to *GB*(X, Y) must includes the relation *B*.

(3) PO(X, Y): every basic cardinal direction relation *SB*(X, Y) that belongs to *GB*(X, Y) must include the relation *B* and an atomic relation that belongs to *BIN*.

(4) TPP(X, Y): the induced cardinal direction relation *GB*(X, Y) includes only one element *B*.

(5) TPPi(X, Y): every basic cardinal direction relation *SB*(X, Y) that belongs to *GB*(X, Y) must include the relation *B* and an atomic relation that belongs to *BIN*.

(6) NTPP: the induced cardinal direction relation *GB*(X, Y) includes only one element *B*.
(7) NTPPi(X, Y): the induced cardinal direction relation *GB*(X, Y) includes only one element *BIN*, i.e. the tiles formed by reference object Y are all occupied by the primary object X.
(8) EQ(X, Y): it is obvious that the induced cardinal direction relation *GB*(X, Y) can only include the relation *B*.
We use *R-to-C* to denote the operation that captures the above interaction between RCC8 knowledge and CDC knowledge, in the direction RCC8-To-CDC, by inferring CDC knowledge from given RCC8 knowledge.
The induced cardinal direction relation by a general RCC8 relation (i.e. the disjunction of atomic RCC8 relations), say Q (X, Y), is the union of cardinal direction relations induced by all atomic RCC8 relations that belong to the general RCC8 relation; namely:

$$GB(X,Y) = \bigcup_{r \in Q} R - to - C(r(X,Y)) .$$

Table 1. The interaction table from the basic relations of CDF to RCC8 relations

Atomic RCC8 relation	Induced cardinal direction relation
DC	U
EC	$\forall SB \in GB : B \in SB$
PO	$\forall SB \in GB$, $\exists R \in BIN : R \in SB \cap B \in SB$
TPP	$\{\{B\}\}$
TPPi	$\forall SB \in GB$, $\exists R \in BIN : R \in SB \cap B \in SB$
NTPP	$\{\{B\}\}$
NTPPi	$\{BIN\}$
EQ	$\{\{B\}\}$

2.2 Reasoning within CDC and the interaction from CDC to RCC8 (CDC-To-RCC8)

The converse operation and the composition operation have been investigated in [10] and [7] respectively.
The RCC8 knowledge induced from the basic cardinal direction relations, denoted by *SB*, is presented in the table 2, which describes the interaction between RCC8 knowledge and CDC knowledge, in the direction CDC - To- RCC8, in four cases:
(1)*SB*={*B*}: the induced RCC8A relation is the disjunction of *DC*, *EC*, *PO*, *TPP*, *NTPP*, *EQ* and *TPPi* .
(2) $\exists R \in BIN : R \in SB \cap B \in SB$: the induced RCC8A relation is the disjunction of *DC*, *EC*, *PO* *TPPi* and *NTPPi*.
(3) Universal relation: the induced RCC8A relation is the disjunction of *DC*, *EC*, *PO*, *TPPi* and *NTPPi*.
$B \notin SB$: the induced RCC8A relation is only *DC*.
We use *C-to-R* to denote the operation that captures the above interaction between RCC8 knowledge and CDC knowledge, in the direction CDC -To- RCC8, by inferring RCC8 knowledge from given CDC knowledge.
The induced RCC8 relation by a general CDC relation (i.e. the disjunction of basic cardinal direction relations), say Q (X, Y), is the union of RCC8 relations induced by all basic cardinal direction relations that belong to the general CDC relation; namely:

$$RCC8(X,Y) = \bigcup_{r \in Q} C - to - R(r(X,Y)) .$$

Table 2. The interaction table from the basic relations of CDF to RCC8 relations

Basic cardinal direction relation (SB)	*RCC8 relation*
{B}	$DC \vee EC \vee PO \vee TPP \vee NTPP \vee EQ \vee TPP^{-1}$
$\exists R \in BIN : R \in SB \cap B \in S$ or *BIN*	$DC \vee EC \vee PO \vee TPP^{-1} \vee NTPP^{-1}$
$B \notin SB$	DC

3 Summary

We have presented the combination of two calculi of spatial relations well-known in Qualitative Spatial Reasoning (QSR): RCC8 calculus and R. Goyal and M. Egenhofer's cardinal direction calculus. In this paper, the interaction between the two kinds of knowledge has been handled, and we have also given two interaction tables. The work in this paper can be applied to the research field of Geographic Information System (GIS), image understanding and computer vision, etc. The fuzzy spatial reasoning with fuzzy knowledge of topological and directional relations can be of very interest in the future.

References

[1] A. Gerevini, J. Renz. (2002) Combining Topological and Size Constraints for Spatial Reasoning, *Artificial Intelligence (AIJ)*, 137(1-2):1-42.

[2] A Isli, V Haarslev and R Möller. (2001) Combining cardinal direction relations and relative orientation relations in Qualitative Spatial Reasoning, Technical report FBI-HH-M-304/01, Fachbereich Informatik, Universität Hamburg.

[3] M. Egenhofer. (1989) A Formal Definition of Binary Topological Relationships, *Third International Conference on Foundations of Data Organization and Algorithms (FODO)*, Paris, France.

[4] B. Bennett. (1994) Spatial Reasoning with Propositional Logics Principles of Knowledge Representation and Reasoning: Proceedings of the 4th International Conference (KR94), edited by Doyle, J and Sandewall, E and Torasso, P, Morgan Kaufmann, San Francisco, CA..

[5] M. Egenhofer and R. Franzosa. (1991) Point-Set Topological Spatial Relations, *International Journal of Geographical Information Systems* 5 (2): 161-174.

[6] R. Goyal and M. Egenhofer. (2000) Cardinal Directions between Extended Spatial Objects, *IEEE Transactions on Knowledge and Data Engineering*, (in press).

[7] S. Skiadopoulos and M. Koubarakis. (2004) *Composing cardinal direction relations*, Artificial Intelligence, 152(2) pp. 143--171.

[8] J.Renz. (2002) Qualitative Spatial Reasoning with Topological Information, LNCS 2293, Springer-Verlag, Berlin.

[9] D. A. Randell, A. G. Cohn and Z. Cui. (1992) Computing Transitivity Tables: A Challenge For Automated Theorem Provers. Proceedings CADE 11, Springer Verlag, Berlin.

[10] Serafino Cicerone, Paolino Di Felice. (2004) Cardinal directions between spatial objects: the pairwise-consistency problem, *Information Sciences* 164:165-188.

An Evidence Theoretic Ensemble Design Technique

H. Altınçay
Department of Computer Engineering, Eastern Mediterranean University
KKTC, Mersin 10, Turkey
E-mail:hakan.altincay@emu.edu.tr

Abstract

Ensemble design techniques based on resampling the training set are successfully used to improve the classification accuracies of the base classifiers. In Boosting technique, each training set is obtained by drawing samples with replacement from the available training set according to a weighted distribution which is iteratively updated for generating new classifiers for the ensemble. The resultant classifiers are accurate in different parts of the input space mainly specified the sample weights. In this study, a dynamic integration of boosting based ensembles is proposed so as to take into account the heterogeneity of the input sets. In this approach, a Dempster-Shafer theory based framework is developed to consider the training sample distribution in the restricted input space of each test sample. The effectiveness of the proposed technique is compared to AdaBoost algorithm using nearest mean type base classifier.

1 Introduction

Boosting is a popular ensemble creation technique which takes into account the classification results of the previous classifiers to construct additional ones. The sequential structure of the algorithm allows to create new classifiers which are more effective on the training samples that the current ensemble has a poor performance. In order to achieve this, weighting is applied on the training samples where a training sample with a high weight has a larger probability of being used in the training set of the next classifier. The weights are updated in an iterative manner so that new classifiers mainly focus on the samples difficult to classify. AdaBoost is the most popular boosting algorithm.

In AdaBoost technique, the reliability of the classifier outputs is dependent on the input due to the weighted resampling [1]. For instance, the decision of a classifier for an input lying in a restricted space that is resampled by a large number of times is expected to be reliable. However, this may not be true for the input spaces which have no representatives in the resampled training set. Hence, the fact that the classifiers developed using AdaBoost may be accurate in some regions of the input space should be considered during the combination operation. The weighted majority voting rule used in the AdaBoost algorithm does not take into account this input dependent information. Dynamic combination schemes that take into account the distribution of the training samples in the restricted space where the given test sample lies may provide better results.

In order to take into account the distribution of the training samples in different parts input space, a Dempster-Shafer theory based (evidential) pattern classification technique is proposed by Denæux [2]. In that approach, each neighbor of a given test sample in the training set is considered as a piece of evidence supporting the class that the training sample belongs. The basic probability assignments from all neighbors are then combined to predict the class of the tested sample. In boosting technique, weights for different samples are naturally available where the ultimate aim is analogous; each training sample has a different influence on the decision depending on the difficulty of its classification since, more replicas of difficult samples are used than easier ones in classifier training. However, the sample weights are explicitly considered only in the training phase. In this study, an evidence-theoretic framework for boosting is proposed so as to take into account the weights and distances of the neighboring training samples in both training and testing boosting based ensembles. In the proposed approach, the weight update mechanism of AdaBoost is preserved where a weight and distance dependent belief structure assignment is developed. The proposed approach is used for boosting nearest mean classifier (NMC) where better accuracies than AdaBoost are obtained.

2 Evidential Pattern Classification

Let Ω denote the set of class labels and $\mathcal{S} = \{(x_n, y_n)\}$, $n = 1, \ldots, N$ be the set of training samples where x_n denotes the nth input sample and $y_n \in \Omega$ is its label. Given a test sample x, each training sample is considered to provide a piece of evidence about the class label of x. In other words, each $x_n \in \mathcal{S}$ induces a belief structure m_n with two focal elements, $\{w_q\}$ and

Ω where $w_q \in \Omega$ is the class that the training sample x_n belongs as,

$$m_n(\{w_q\}) = \alpha\phi_q(||x_n - x||) = \alpha e^{-\gamma_q ||x_n - x||^2}$$
$$m_n(\Omega) = 1 - m_n(\{w_q\}) \tag{1}$$

where $||x_n - x||$ is the Euclidean distance between x_n and x. The class-independent design parameter α and the class-dependent parameter γ_q determine the way basic probability values are assigned to $\{w_q\}$ and Ω. As α increases, the evidence provided is considered to be more certain. The influence of the distance on basic probability assignments is class dependent and it is adjusted by γ_q. As the distance increases, more probability mass is assigned to Ω. These belief structures are combined using Dempster's rule of combination [3]. In order to make a joint decision, pignistic probabilities can be computed as $P_{Bet}(w_i) = \sum_{w_i \subseteq B} \frac{m(B)}{|B|}$, $\forall w_i \in \Omega$, where B denotes the focal elements getting nonzero basic probability value [4]. Then, the class getting the maximal $P_{Bet}(.)$ value can be selected as the joint decision.

In summary, the main idea is to treat each neighboring sample as a piece of evidence for the class label of the tested sample. In this paper, this idea is used to propose an evidence-theoretic framework for boosting where the sample weights are explicitly used to compute the measurement level classifier outputs.

3 Evidence Theoretic Framework for Boosting

Let $W_t(n)$ denote the weight of the nth training sample in $\mathcal{S}$ initialized to $1/N$ and t denotes the iteration count. $\Omega = \{w_a, w_b\}$ since a 2-class problem is considered. Let $\mathcal{S}_t$ denote bootstrap sample set obtained by drawing with replacement N samples from $\mathcal{S}$ using distribution W_t. $d_n(j)$ is defined as the Euclidean distance of x_n to its jth nearest neighbor in $\mathcal{S}$ denoted by $neig_n(j)$. $d_n(1)$ is zero which is the distance of the sample with itself and $d_n(2)$ is the distance from the closest different sample.

The proposed algorithm named as E-Boost is given below. The initial classifier makes use of equal sample weights. In each iteration, the weights of the correctly classified training samples are decreased. Given the current weight vector W_t, the training set is resampled to generate a new ensemble member. Then, each training sample x_n is classified by taking into account its k-nearest neighbors. For this purpose, each neighboring sample having the *same* class label as the classifier decision (totally, $k' \leqslant k$) induces a belief structure m_j with focal elements $\{w_q\}$ being the decision of the trained classifier and Ω. The main idea behind this is to compute the total support on the output of the classifier. It should also be noted that k' is different for each training sample.

for $t = 1, ..., T$
  Build classifier $\mathcal{C}_t$ using sample set $\mathcal{S}_t$ resampled from $\mathcal{S}$ using distribution W_t.
  for n = 1 to N
    Compute the most likely class, $w_q = \mathcal{C}_t(x_n)$
    Compute the k-nearest neighbors of x_n in $\mathcal{S}$ using $d_n(j), j = 1, \ldots, k$.
    for j = 1 to k' // *consider neighbors of the same class as* w_q
      $m_j(\{w_q\}|neig_n(j)) = f(q, W_t(neig_n(j)), d_n(j))$
      $m_j(\Omega|neig_n(j)) = 1 - m_j(\{w_q\}|neig_n(j))$
    end
    $m^n_{comb}(.) = m_1(.|neig_n(1)) \oplus \ldots \oplus m_{k'}(.|neig_n(k'))$
    Compute pignistic probability as,

$$P_{Bet}(w_q) = m^n_{comb}(w_q) + m^n_{comb}(\Omega)/2$$

    The combined decision is computed as,

$$C_t(x_n) = \begin{cases} w_q & if\ |P_{Bet}(w_q) - P_{Bet}(w'_q)| > \tau \\ rand(\{w_a, w_b\}) & otherwise \end{cases} \tag{2}$$

    Calculate the weighted error using $\epsilon_t = \frac{1}{N}\sum_{n=1}^{N} W_t(n)(1 - q_{n,t})$ where $q_{n,t} = 1$ if x_n is correctly classified and zero otherwise.
    Compute $\alpha_t = \frac{1}{2}\ln(\frac{1-\epsilon_t}{\epsilon_t})$, $\epsilon_t \in (0, 0.5)$ and update the weights using
    $W_{t+1}(n) = \frac{W_t(n)}{Z_t} e^{-\alpha_t}$ if $C_t(x_n) = y_n$, where Z_t is a normalization factor so that W_{t+1} is a distribution.
  end
end

The basic probability assignment is defined as,

$$m_j(\{w_q\}|neig_n(j)) = \frac{(W_t(neig_n(j))/W_{t,max})}{1 + e^{\frac{1}{d^q_{avg}}(d_n(j) - d^q_{avg})}} \tag{3}$$

where $W_{t,max}$ is the maximum weight value at the tth iteration. d^q_{avg} used in the denominator is the within-class average of k-nearest neighbor Euclidean distances for the class w_q. As seen in the equation, the basic probability value is proportional to the weight of the neighbor; higher weights correspond to stronger evidence. The weights are normalized by their maximum values so that $m_j(\{w_q\}|neig_n(j)) < 1$. The denominator also depends on the decided class. As the distance increases above the average d^q_{avg}, the evidence provided is considered to decrease whereas a smaller distance corresponds to an increasing evidence.

k' belief structures are then combined using Dempster's rule of combination to compute the combined belief structure, $m^n_{comb}(.)$. The pignistic probabilities obtained from the combined belief structure are used to select the most likely pattern class. If the decision of the classifier is w_a, it is expected that $P_{Bet}(w_a) >> P_{Bet}(w_b)$ where the support on w_b comes from the basic probability assigned to Ω. In the proposed algorithm, the threshold τ is used to make sure that there is enough support on the decision of the classifier. If $P_{Bet}(w_a) \approx P_{Bet}(w_b)$, the classifier is not considered to provide reliable information about the class of the sample under concern. Also, it may be the case that $P_{Bet}(w_a) = P_{Bet}(w_b) = 0$ due to no local support. In both of these cases, the decision is randomly selected.

The decisions generated by the tth classifier after combining the evidence from each k'-nearest neighbors are used to compute the weighted error, ϵ_t as in AdaBoost. If $\epsilon_t \in (0, 0.5)$, the weights of the correctly classified samples are updated. In the proposed approach, the weights of misclassified samples are not increased since, it is observed that this may easily lead to classifiers with higher error rates than 0.5. Such an update was also considered for AdaBoost and named as conservative AdaBoost. When the condition that $\epsilon_t \in (0, 0.5)$ is not satisfied, the algorithm is terminated. The output of the algorithm is a set of T different weight vectors, T classifiers and the α_t values obtained for each weight vector.

During testing an unseen input vector, combined belief structures $m_{comb}(.)$ are computed for each different member classifier. The input vector is firstly tested by each classifier and then each k' nearest neighbors having the same class label as the most likely class generated by the classifier induces a belief structure. k' belief structures are then combined using Dempster's rule of combination to compute the combined belief structure, $m_{comb}(.)$. The pignistic probabilities obtained from the combined belief structures, $\{P_{Bet}^t\}_{t=1}^T$ are aggregated to compute the resultant pignistic probabilities using weighted averaging as,

$$P_{Bet}^{avg}(w_i) = \frac{1}{T}\sum_{t=1}^{T} \alpha_t P_{Bet}^t(w_i) \qquad (4)$$

Then, the class assigned to the tested pattern is the one getting the highest $P_{Bet}^{avg}(.)$ value. Since different belief structures are obtained using resampling from the same training set, Dempster's rule of combination cannot be used. Due to the commutativity property of averaging and the linear relationship between credal level information (using basic probability values) and pignistic probabilities, averaging is considered to be a good candidate for bagging evidential k-NN classifiers [5]. Following this reasoning, the weighted form of averaging as given above is used in this study.

4 Experiments

In order to investigate the benefits of the proposed evidential framework for boosting, experiments are conducted on artificial and real data sets from UCI machine learning repository and ELENA database.

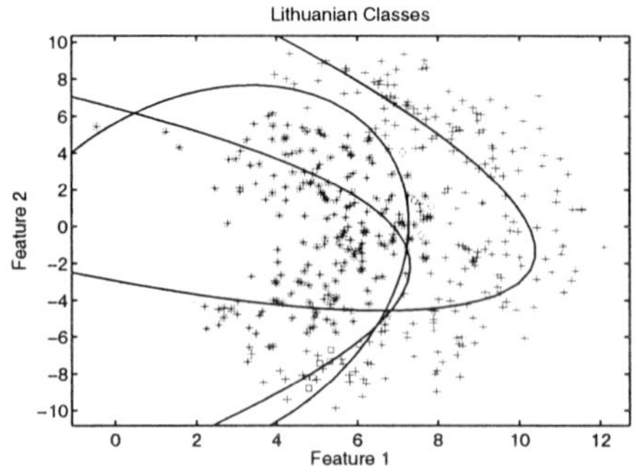

Fig. 1. The scatter plot of training samples for 2-D Lithuanin classes.

4.1 Experiments on Artificial Data Sets

Two different artificial data sets are generated using the PRTOOLS toolbox [6]. In both of these experiments, we set $k = 5$ and $\tau = 0$. For the first 2-D classification problem, Lithuanin classes are generated. For each class, 200 samples are used for training and 200 for testing. AdaBoost is run to generate an ensemble of three normal densities based quadratic classifiers. The developed classifiers are illustrated on the scatter plot of the training data in Figure 1. The first class is represented using +'s and the second class samples are represented using *'s. The test samples that are correctly classified by the proposed algorithm but misclassified by AdaBoost are also marked on the figure. □'s represent such test patterns from the first class and ◇'s represent those belonging to the second class. As seen in the figure, the proposed technique is more effective on the difficult samples lying on the border of the classes. Three of the differently classified test samples belonging to the first class and five belonging to the second class are misclassified by two of the three classifiers. This is the main reason for AdaBoost to be unsuccessful for these samples. However, they are correctly classified by E-boost due to the support from their neighbors. In this experiment the classification accuracy of the base classifier is 91.25%, where 93.50% and 95.75% accuracies are achieved by AdaBoost and proposed algorithm respectively.

For the second 2-D classification problem, 'banana' classes are generated where the same number of training and test samples are considered. Figure 2 illustrates ten NMC type ensemble members obtained using AdaBoost. Since the NMC type classifier is a weak one, it is less sensitive to changes in the training set providing decision boundaries that may be close to some others leading to correlated decisions and only 1% improvement. However, making use of the different weights of the training

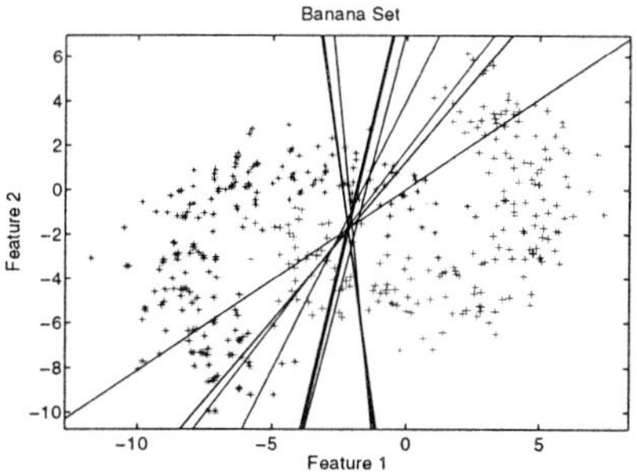

Fig. 2. Ten NMC type ensemble members obtained for 'banana' data set using AdaBoost algorithm.

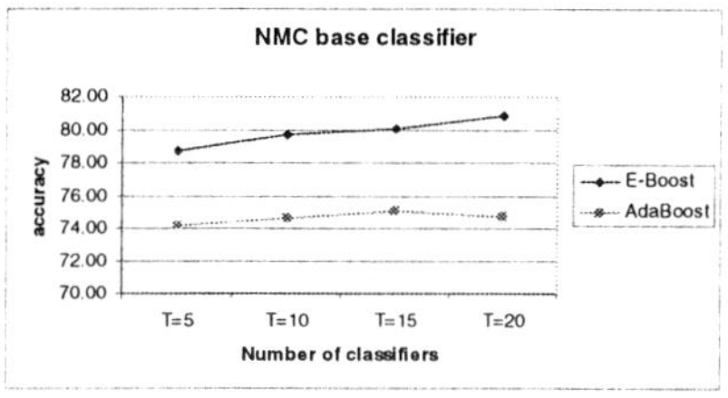

Fig. 3. Average accuracies achieved by E-boost and AdaBoost for $T = 5$, 10, 15 and 20 over eleven data sets.

samples explicitly during testing reduces the correlations providing 14.75% improvement over the base classifier.

4.2 Experiments on Real Data Sets

In order to evaluate the proposed algorithm, experiments are conducted on nine different data sets in the UCI machine learning repository and two in the ELENA data set. The experiments are repeated ten times and the average accuracies are computed. In order to evaluate the proposed framework on real data sets, nearest mean classifier (NMC) is selected as the base classifier. The experimental results for $T = 10$, $k = 5$ and $\tau = 0$ are presented in Table 1. As seen in the table, E-Boost provided much better accuracies compared to AdaBoost. The experiments are repeated also for $T = 10$, 15 and 20, and the average accuracies over eleven data sets are presented in Figure 3. The figure show that better ensembles are obtained as the number of classifiers is increased up to 20 which is not the case in AdaBoost.

5 Conclusions

In this study, an evidence theoretic framework is proposed for boosting. Experimental results have shown that the use of local information is highly useful for

Table 1. Experimental results for $T = 10$ classifiers.

Data set	Base	AdaBoost	E-Boost
monks-1	65.97	65.74	70.83
monks-2	53.47	54.17	64.81
monks-3	80.56	82.64	89.35
wdbc	88.30	88.42	92.28
breast-cancer-w	95.76	95.71	96.54
sonar	65.32	71.77	80.48
ionosphere	74.29	87.05	90.67
liver	55.53	61.75	62.82
heart	61.87	66.37	67.47
phoneme	71.73	71.91	83.73
clouds	75.25	75.27	77.51
Average	*71.64*	*74.62*	*79.68*

boosting based classifier ensembles. The proposed approach should also be evaluated for other types of classifiers such as fisher's linear discriminant. Also, the effect of k and τ should be investigated. Our preliminary experiments have shown that the average accuracy increases to 80.82% for $\tau = 0.15$. Also, the performances achieved are observed to be better for smaller k value in some of the data sets. Analysis of the proposed algorithm in terms of these parameters and estimation of their best fitting values are the main topics our current research.

References

[1] A. Tsymbal and S. Puuronen. Bagging and boosting with dynamic integration of classifiers. *Principles of Data Mining and Knowledge Discovery, Proceedings of PKDD 2000*, pages 116–125, 2000.

[2] T. Denæux. A k-nearest neighbor classification rule based on dempster-shafer theory. *IEEE Transactions on Systems, Man, and Cybernetics*, 25(5):804–813, 1995.

[3] G. Shafer. *A Mathematical Theory of Evidence.* Princeton University Press, 1976.

[4] P. Smets and R. Kennes. The transferrable belief model. *Artificial Intelligence*, 66:191–234, 1994.

[5] J. François, Y. Grandvalet, T. Denæux, and J. M. Roger. Resample and combine: an approach to improving uncertainty representation in evidential pattern classification. *Information Fusion*, 4:75–85, 2003.

[6] R. P. W. Duin. PRTOOLS (version 4). A Matlab toolbox for pattern recognition. *Pattern Recognition Group, Delft University, Netherlands*, 2004.

Cortical Modulation of Synaptic Efficacies through Norepinephrine

O. Hoshino

Department of Human Welfare Engineering, Oita University, Japan
E-mail: hoshino@cc.oita-u.ac.jp

Abstract

I propose a norepinephrine- (NE-) neuromodulatory system, which I call "enhanced-excitatory and enhanced-inhibitory (E-E/E-I) system". The E-E/E-I system enhanced excitatory and inhibitory synaptic connections between cortical cells, modified their ongoing background activity, and influenced subsequent cognitive neuronal processing. When stimulated with sensory features, cognitive performance of neurons, signal-to-noise (S/N) ratio, was greatly enhanced, for which one of the three possible S/N enhancement schemes operated under the E-E/E-I system, namely; i) signal enhancement more than noise increase, ii) signal enhancement and noise reduction, and iii) noise reduction more than signal decrease. When a weaker (or subthreshold) stimulus was presented, the scheme (ii) effectively enhanced S/N ratio, whereas the scheme (iii) was effective for enhancing stronger stimuli. I suggest that a release of NE into cortical areas may modify their background neuronal activity, whereby cortical neurons can effectively respond to a variety of external sensory stimuli.

1 Introduction

It is well known that a release of norepinephrine (NE) into target brain areas through noradrenergic (e.g., locus coeruleus (LC)) pathways facilitates the efficacies of excitatory and inhibitory synaptic transmissions within the targeted neuronal circuits [1]. NE binds to α and β adrenoceptors of neurons, activates second messenger systems, and augments the efficacies of excitatory (e.g., glutamatergic) and inhibitory (e.g., GABAergic) synaptic transmissions [2]. Such NE-induced neuromodulation has been well demonstrated for cortical pyramidal cells [3]. Although many experiments have demonstrated that NE-release in certain cortical areas modifies neuronal excitation and/or inhibition, little is known about how these neuronal modulations affect the cognitive performance of the cortices [2,3].

The purpose of the present study is to propose a neural network model whose dynamic behavior is altered when dosed with NE. By simulating the model, I investigate how NE modulates the dynamic behavior of neurons and what neural mechanisms are essential for NE-mediated cognitive enhancement. We use "signal-to-noise (S/N)" ratio as a cognitive performance measure.

2 Neural Network Model

I construct a neural network model for the cortex (Figure 1a). The model consists of an input (IP) and an output (OP) network. Feature stimuli Fn (n = 1, 2, 3, 4, 5) activate their corresponding groups of IP neurons ("ellipses"), whose action potentials are sent to the OP network via divergent/convergent feedforward projections ("solid lines") and activate corresponding cell assemblies ("circles"). As shown in Figure 1b, the OP network consists of neuron units, each of which contains a pyramidal cell (PYC) ("large triangle"), a small basket cell (SBC) ("small circle") and a large basket cell (LBC) ("large circle"). In each unit, the PYC and the SBC are reciprocally connected via a positive (PYC-to-SBC) and a negative (SBC-to-PYC) synapse. The PYC positively synapses on the LBC. Groups of PYCs form cell assemblies. PYCs within cell assemblies are connected with each other via positive synapses, and there is no

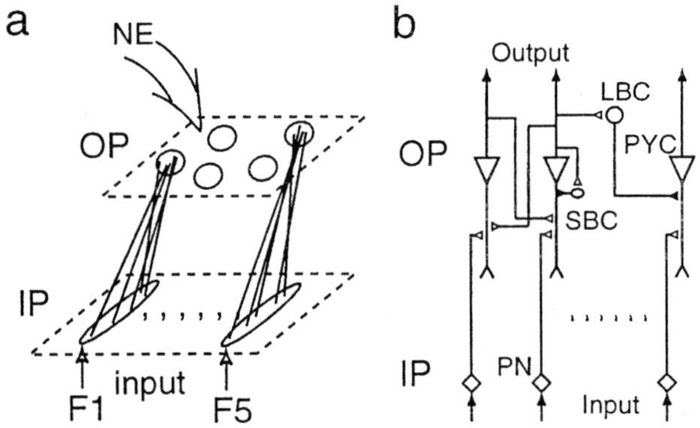

Fig. 1. Structure of the neural network model. (a) Feature stimuli Fn (n = 1, 2, 3, 4, 5) are applied to corresponding groups of IP neurons ("ellipses"), whose action potentials are sent to the OP network via divergent/convergent feedforward projections ("solid lines") and activate corresponding cell assemblies ("circles"). NE (norepinephrine) is dosed into the OP network. (b) PYC, SBC and LBC denote, respectively, pyramidal, small basket and large basket cell. "Open" and "filled" small triangles denote excitatory and inhibitory synapses, respectively.

connection between PYCs across cell assemblies. LBCs negatively synapse on the PYCs of the other cell assemblies through lateral (LBC-to-PYC) connectios.

I assume here a primary cortical area whose neurons have tuning properties to specific sensory features. To make the PYCs feature-selective, I create in the output (OP) network multiple dynamic cell assemblies that are spatially separated from each other (see Figure 1a). Due to such separable property, the dynamics of the OP network allows a given cell assembly to be selectively activated against others when its corresponding feature stimulus is presented to the input (IP) network. For simplicity, the IP network contains only projection neurons (PNs) between which there is no connection. That is, the IP network works exclusively as an input layer. These neurons are an integrate-and-fire type of neurons (for detail, see ref. [4]).

The efficacies of both excitatory and inhibitory synapses are enhanced as a function of a dose level of norepinephrine, or concentration of NE ([NE]). Neuromodulation of excitatory ($w_{ij}^{PY,PY}$: PYC-to-PYC) and inhibitory ($w_{ij}^{PY,LB}$: LBC-to-PYC, $w_{ij}^{PY,SB}$: SBC-to-PYC) synaptic efficacies are described by the following equations.

$$\frac{dw_{ij}^{PY,PY}}{dt} = \alpha_{PY}([NE]_0 - [NE])[NE] - \beta_{PY}(w_{ij}^{PY,PY} - w_0^{PY,PY}) \tag{1}$$

$$\frac{dw_{ij}^{PY,LB}}{dt} = \alpha_{LB}[NE] - \beta_{LB}(w_{ij}^{PY,LB} - w_0^{PY,LB}) \tag{2}$$

$$\frac{dw_{ij}^{PY,SB}}{dt} = \alpha_{SB}[NE] - \beta_{SB}(w_{ij}^{PY,SB} - w_0^{PY,SB}) \tag{3}$$

Equation 1 defines the excitatory synaptic modulation between PYCs, which is based on observed results [5]. Equations 2 and 3 define the inhibitory synaptic modulation from LBC to PYC and from SBC to PYC, respectively, which are simple hypothetical representations based on observed results [2]. In the present study, I focused especially on the postsynaptic (PYC-to-PYC, LBC-to-PYC and SBC-to-PYC) actions of NE on the activities of PYCs that play, in general, major roles in cognitive information processing in the cortex. For simplicity, I did not modulate the other excitatory synapses, PYC-to-LBC and PYC-to-SBC. Unless otherwise stated elsewhere, $w_0^{PY,PY} = 7.0$, $w_0^{PY,LB} = 0.1$, $w_0^{PY,SB} = 30.0$, $\alpha_{PY} = 3.0$, $\alpha_{LB} = 0.8$, $\alpha_{SB} = 60.0$, $\beta_{PY} = \beta_{LB} = \beta_{LB} = 1.0$ and $[NE]_0 = 2.0$.

3 Results

As shown by the raster plots of action potentials in Figure 2a, the PYCs have ongoing (background) activity, when no external stimulus and no dose of NE are applied. The random and brief emergence of the five (F1-5) dynamic cell assemblies, or population activation of PYCs, characterizes the present ongoing neuronal activity. The temporal formation of each dynamic cell assembly arises from mutual excitation between the PYCs within cell assemblies. The brief nature of the dynamic cell assemblies arises largely from the self-inhibition mediated through their accompanying SBCs. Due to such a self-inhibitory mechanism, the more the PYCs emit action potentials, the greater the activities of the PYCs tend to be suppressed.

When the IP network is stimulated with a sensory feature (F2), whose duration is indicated by a "horizontal bar" in Figure 2a, the PYCs of the cell assembly

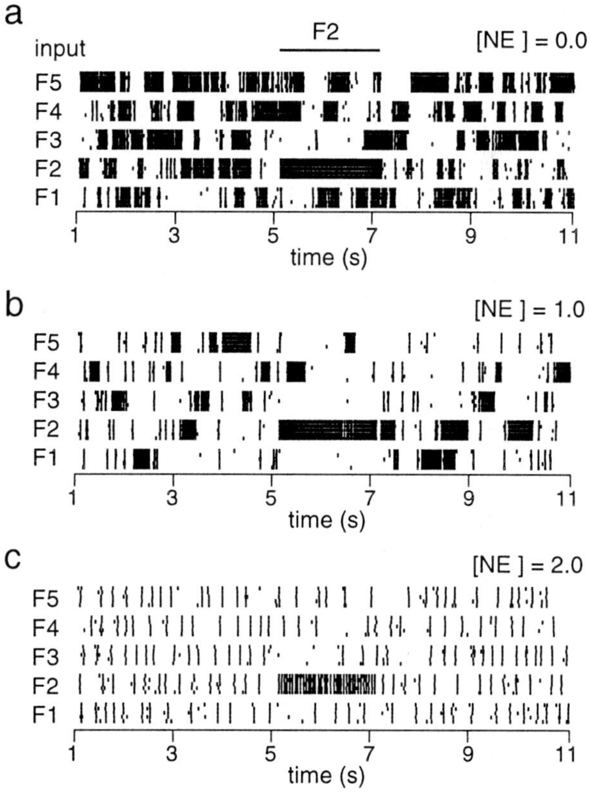

Fig. 2. Dependence of the dynamic behavior of the OP network on dose levels of NE ([NE]). Raster plots of PYC action potentials of cell assemblies that are sensitive to features F1-5 are shown. (a) NE is not dosed, or [NE] = 0.0. A "horizontal bar" indicates a stimulation (F2) presentation period. (b)-(c) NE-induced neuromodulation operated under the E-E/E-I system.

corresponding to the stimulus are activated and emit a long burst of action potentials. After switching off the input, the state of the OP network returns to the ongoing state. Note that the other dynamic cell assemblies (F1, F3, F4 and F5) tend to frequently emerge during the stimulation period. This indicates that the lateral inhibition across dynamic cell assemblies, which is mediated through LBC-to-PYC inhibitory connections, is not so strong under the original condition, or at [NE] = 0.0.

Figure 2b-c shows how the dynamic behavior of the network is modulated by the E-E/E-I system. The period of each brief burst under the ongoing state is deceased as the dose level of NE ([NE]) increases (Figure 2a → 2b → 2c), which is due largely to the enhanced self-inhibition of PYCs through SBC-to-PYC feedback connections. Note that the activation of the dynamic cell assemblies tends to be temporally separated from each other as [NE] increases, that is, they are not likely to overlap in the time course. This is due largely to the enhanced lateral inhibition through LBC-to-PYC connections.

Such temporal segregation of dynamic cell assemblies is essential for processing the applied feature stimulus (F2) in that as "feature-detection neurons" of an early sensory cortex the PYCs must respond selectively to a specific feature stimulus, while the other PYCs are not allowed to respond, or emit fewer action potentials. Note that although the ongoing PYC activity is decreased as [NE] increases, the synchronous PYC activity within cell assemblies is well preserved (e.g., see Figure 2c). The term, "synchronous activity", implies that the PYCs within cell assemblies generate action potentials almost at the same time.

I evaluated the cognitive performance of the network in terms of "evoked-to-background" PYC activity ratio, or [stimulus-induced firing rate of PYCs]/[ongoing firing rate of PYCs]. I applied the same feature (F2) stimulus with various stimulus intensities; $\varepsilon = 0.3$ (strong: Figure 3a), $\varepsilon = 0.05$ (weak: Figure 3b) and $\varepsilon = 0.02$ (too weak: Figure 3c). In Figure 3a-c, the ongoing ("circles") and stimulus-induced ("triangles") PYC activities are shown ("left"). The evoked-to-background activity is shown ("right"), which I call here signal-to-noise (S/N) ratio in a practical sense.

Background neuronal activity itself could contain significant information as internal representations. Hence, we cannot straightforwardly call the background activity as "noise", and therefore should use the term "evoked-to-background" activity ratio rather than "signal-to-noise" ratio (S/N ratio) in a strict sense. The reason why I used the term S/N ratio instead of evoked-to-background ratio is to evaluate the present simulation results in relation to experimental (neurophysiological) observations [1,2,3,5], in which evoked-to-background activity ratio was preferentially called S/N ratio. Nevertheless, such a use of S/N ratio is unusual in such a field of engineering, because noise is not allowed to involve any significant information (or signal). I understand that the term "signal-to-noise" ratio in experimental neuroscience might be used in a more practical sense than in engineering. In the present study, I employed these terms (signal, noise and S/N ratio) based on a neurophysiological use but not on an engineering use.

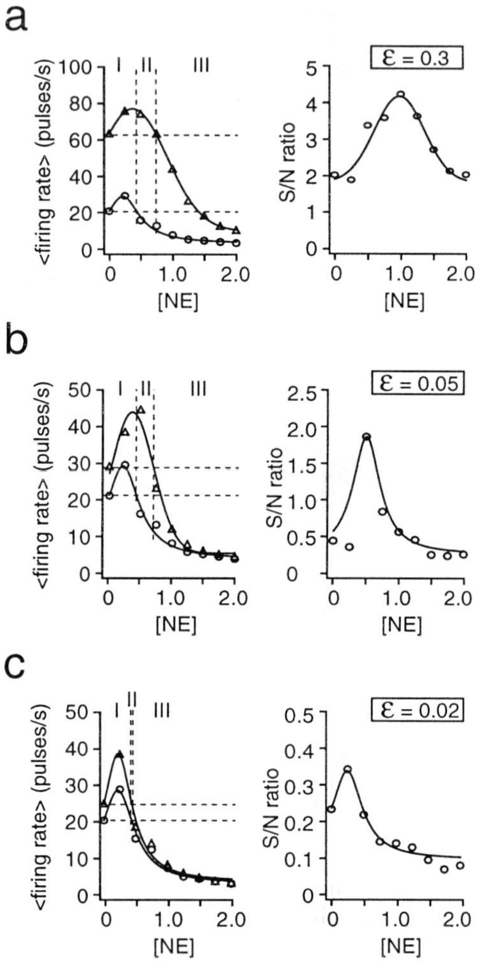

Fig. 3. Neuronal behavior (left) and S/N ratio (right) of PYCs. The model is presented with a feature stimulus with strong (a), weak (b) and too weak (c) intensity. In each figure, the left shows the ongoing firing rate ("circles") and stimulus-induced firing rate ("triangles") of a PYC. Regions marked by "I", "II" and "III" indicate that three distinct types of S/N enhancements take place.

For stronger stimuli (see the right of Figure 3a), S/N ratio is enhanced at an intermediate level of [NE] ([NE] = ~1.0). The stimulus-induced PYC-activity is progressively depressed at [NE] = ~1.0 (see the "triangles" of Figure 3a). This implies that S/N enhancement is possible provided that noise (or background PYC activity) is reduced more than signal (or evoked PYC activity). That is, noise reduction is as fairly effective as signal enhancement for improving S/N ratio. For weaker stimuli (see the right of Figure 3b), S/N ratio is enhanced at lower levels of [NE]. Figure 3c (right) shows fewer S/N enhancements for too weak stimuli.

Fig. 4 shows NE-mediated S/N enhancement for the same stimulus whose intensity is changed between 0.02 and 1.0. There are few S/N enhancements for too weak intensities ($\varepsilon < 0.03$). A greater S/N enhancement occurs for weaker intensities ($0.03 < \varepsilon < 0.1$) at lower levels of NE ("arrow" of Fig. 4). For strong intensities ($\varepsilon = \sim 1.0$), significant S/N enhancement occurs at higher levels of NE, [NE] = ~1.5.

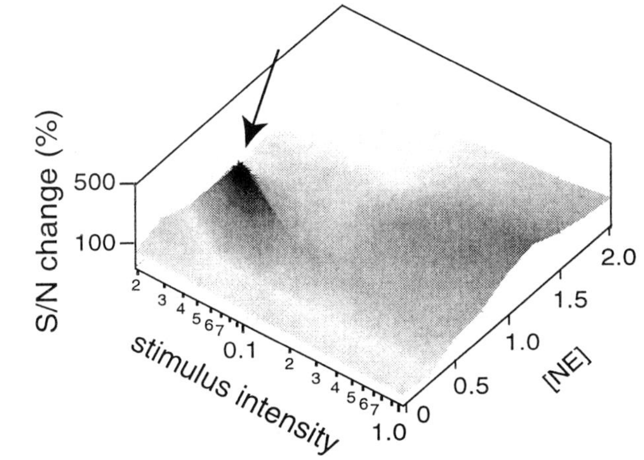

Fig. 4. Dependence of S/N enhancement on stimulus intensity and the dose level of NE, [NE]. An "arrow" indicates a peak in S/N enhancement.

It might be that scheme (ii), or signal enhancement and noise reduction, is quite effective for improving the S/N ratio for the weak stimulus. For the strong stimulus, scheme iii), or noise reduction that surpasses signal decrease, might work. Although the level of S/N enhancement is low for the too weak stimulus (see Fig. 2a), scheme (i) (or signal enhancement that surpasses noise increase) contributes to S/N enhancement.

4 Conclusions

I have proposed here a NE neuromodulatory system, investigated how NE alters ongoing background cortical activity and influences subsequent cognitive performance. The efficacies of the excitatory and inhibitory synaptic connections among pyramidal cells, small basket cells and large basket cells were modulated depending on the concentration of NE. I have found three possible schemes for S/N enhancement, namely; i) signal enhancement that surpasses noise increase, ii) signal enhancement and noise reduction, and iii) noise reduction that surpasses signal decrease. For weaker (or subthreshold) stimuli, signal-enhancement and noise-reduction scheme worked well, where NE application at lower concentration effectively improved the cognitive performance (S/N ratio) of the cortical network.

In these schemes, noise reduction played an essential role for the enhancement of S/N ratio. The reduction of noise could be established through neuromodulation of the inhibitory synaptic (SBC-to-PYC and/or LBC-to-PYC) efficacies. The SBC-to-PYC synapses contribute to suppressing PYC activity through feedback inhibition, and therefore reduce the ongoing PYC activity as [NE] increases. The LSBC-to-PYC synapses contribute to suppressing other PYCs through lateral inhibitory connections. This means that when one PYC assembly is active, the other assemblies tend to be suppressed, whereby the overall firing rate of the PYCs (or the ongoing PYC activity) can be reduced.

References

[1] Woodward, D.J., Moises, H.C., Waterhouse, B.D., Yeh, H.H., Cheun, J.E. (1991) Modulatory actions of norepinephrine on neural circuits. In: Kito, S. et al. (eds.) Neuroreceptor Mechanisms in Brain. Plenum Press, New York, pp. 193-209

[2] Mouradian, R., Sessler, F.M., Waterhouse, B.D. (1991) Noradrenergic potentiation of excitatory transmitter action in cerebrocortical slices: evidence for mediation by an a-1 receptor-linked second messenger pathway. Brain Research 546: 83-95

[3] Waterhouse, B.D., Mouradian, R., Sessler, F.M., Lin, R.C.S. (2000) Differential modulatory effects of norepinephrine on synaptically driven responses on layer V barrel field cortical neurons. Brain Research 868: 39-47

[4] Hoshino, O. (2004) Neuronal bases of perceptual learning revealed by a synaptic balance scheme. Neural Computation 16: 563-594

[5] Devilbiss, D.M., Waterhouse, B.D. (2000) Norepinephrine exhibits two distinct profiles of action on sensory cortical neuron responses to excitatory synaptic stimuli. Synapse 37: 273-282

Associative Memories with Small World Connectivity

Neil Davey, Lee Calcraft, Bruce Christianson, Rod Adams

Department of Computer Science, University of Hertfordshire, UK
E-mail: {N.Davey, L.Calcraft, B.Christianson, R.G.Adams}@herts.ac.uk

Abstract

In this paper we report experiments designed to find the relationship between the different parameters of sparsely connected networks of perceptrons with small world connectivity patterns, acting as associative memories.

1 Introduction

It is possible to build associative memory models from networks of simple perceptrons. These networks perform much better than the canonical Hopfield model, both in terms of capacity and pattern completion. It is also possible to use such networks of perceptrons with sparse or diluted connectivity, and the performance is relatively robust, even at high rates of dilution. Of course real neural networks have sparse connectivity (for example in the cortex of the mouse each neuron is connected to about 0.1% of the other neurons [1]), which motivates the investigation undertaken here. It is also known that in biological systems the networks have a *small world* characteristic [2, 3]. That is they exhibit short path lengths (the minimum number of nodes on a path) between any pair of neurons, as in a random network, but also show a cliquish behaviour, with locally clustered connections. The advantage of such clustering is apparent in the mean connection length (the average wiring length), which is far smaller than in an equivalent random network, see Figure 1. A further benefit is that the wiring complexity is much reduced in networks with predominantly local connections. Recent research has shown that a small world Hopfield network could be a relatively effective associative memory [4]. In our earlier paper [5] we showed how the high capacity model could benefit from small world connectivity. Here we investigate the detailed relationship between the nature of the connection graph, the size of the network and the resulting performance. Sections 2 and 3 describe the background and the computational model. After the performance measure is explained in Section 4 the results are given in Section 5. The paper finishes with a discussion.

2 Background

The simple *small world* model of Watts and Strogatz [2] consists of a regular N-node ring lattice. Each node is connected to $k/2$ neighbours on either side, where k is typically small compared to N. The mean path length between any pair of random points is therefore high. A fraction, p, of these local connections is then *rewired* to randomly selected nodes, see right diagram in Figure 1. They showed that at surprisingly low values of p, the mean path length in the network dropped dramatically, resulting in a small world regime: highly clustered but with low path lengths. Many real networks have been shown to have a small world architecture, including the internet, human acquaintance networks and real networks of neurons [2]. Theoretical work [6] has now shown the detailed relationship between the characteristics of such networks.

Much is known about the effect of sparse connectivity on the standard Hopfield neural network. Some work has also been undertaken on diluted Hopfield networks with modular and small world connectivity [4, 7-9].

For the higher capacity version of the Hopfield network, trained using perceptron learning much less is known about the effect of connectivity patterns. It was shown in [10] that capacity falls linearly with dilution and in [11] that structured local connectivity could help in storing locally correlated data.

The sparse network with only local connectivity can be considered as a simple example of a *Cellular Neural Network* and it has been proposed [12] that such networks can be used as associative memories.

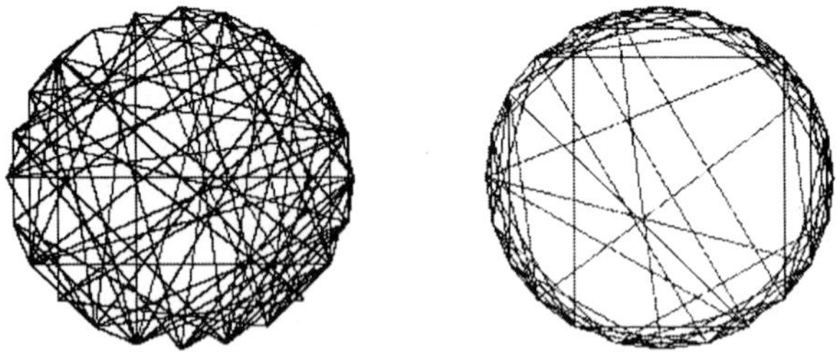

Fig. 1. A ring, with random connectivity on the left and small world connectivity on the right.

3 Network Model

The high capacity models studied here are a modification of the standard Hopfield network. The net input, or *local field*, of a unit, is given by: $h_i = \sum_{j \neq i} w_{ij} S_j$ where S (± 1) is the current state and w_{ij} is the weight on the connection from unit *j* to unit *i*. The dynamics of the network is given by the standard update: $S_i' = \Theta(h_i)$, where Θ is the heaviside function. Unit states may be updated synchronously or asynchronously. Here we use asynchronous, random order updates. If a training pattern ξ^{μ} is a fixed point of the dynamics then it is successfully stored, and is said to be a *fundamental memory*. A network state is stable if, and only if, all the local fields are of the same sign as their corresponding unit, equivalently the *aligned local fields*, $h_i S_i$, should be positive.

We examine sparse networks with small world connectivity. The network topology is similar to the original Watts and Strogatz model. We start with an *N*-ring regular lattice, with each unit connected to its *k* nearest neighbours, and then rewire with probability *p*. However as a network of perceptrons is not necessarily constrained to have symmetric connections we can generalize the network to a weighted directed graph (as is the case for real neural networks). The rewiring does not therefore maintain the symmetry of connectivity of the original regular lattice. In fact it has been shown [5] that for sparse networks of this type, symmetric weights give rise to poor performance.

The networks are trained using the normal perceptron training rule:

```
Begin with zero weights
Repeat until all local fields are correct
 Set state of network to one of the ξ^p
 For each unit, i, in turn:
   Calculate h_i^p ξ_i^p.  If this is less than T
   then change the weights to unit i
   according to:
```

$$\forall j \neq i \quad w'_{ij} = w_{ij} + \frac{\xi_i^p \xi_j^p}{k}$$

Where ξ^p denotes the training patterns, and *T* is the learning threshold which here has the value of 10.

4 Performance Measure

We are interested in how well the small world networks and random networks, trained using the perceptron style learning rule described above, perform as associative memories. The capacity of such networks is determined by the number of incoming connections (*k*) that each perceptron has. For random pattern sets a perceptron can learn up to *2k* patterns [13]. Assuming roughly regular connectivity graphs (as is the case here) the capacity will be determined by the level of dilution and not the specific pattern of connections, and hence is not subject to empirical investigation.

We use *R*, the normalised mean radius of the basins of attraction, as a measure of attractor performance in these networks. It is defined as:

$$R = \left\langle \left\langle \frac{1 - m_0}{1 - m_1} \right\rangle \right\rangle$$

where m_0 is the minimum overlap an initial state must have with a fundamental memory for the network to converge on that fundamental memory, and m_1 is the largest overlap of the initial state with the rest of the fundamental memories. The angled braces denote a double average over sets of training patterns and initial states. Details of the algorithm used can be found in [10]. A value of $R = 1$ implies perfect performance and a value of $R = 0$ implies no pattern correction.

5 Results

In this paper we report experiments designed to find the relationship between the different parameters of the model. We summarise these parameters in Table 1.

Table 1: Parameters of the Model

N	The size of the network
k	The number of connections each unit makes
α	The loading per connection: *size of training set / k*
p	The proportion of rewired connections

5.1 Fixed Size Networks

The first set of experiments fixes the network size, *N*, at 1000 units, arranged in a ring as described earlier. Initially each unit is connected to its $k = 20$, 40 or 60 neighbours. Random training sets of 1000-ary vectors are created. The number of vectors in the training set is determined by the specific values of α and *k*. For example with $\alpha = 0.3$, $k = 20$ implies a training set of 6 vectors, $k = 40$ 12 vectors and $k = 60$ 18 vectors (as used in Figure 3). The attractor performance (*R* value) is then measured as the network is progressively rewired, as described above. All results presented are averages over 10 runs.

Figure 2 gives the results. Considering first the overall pattern, it can be seen that, in all cases, as the amount of rewiring is increased the performance of the network is also improved. In fact all the networks reach a point at which pattern correction behavior is perfect. It is also apparent that higher levels of rewiring are required as the loading, α, of the networks is raised.

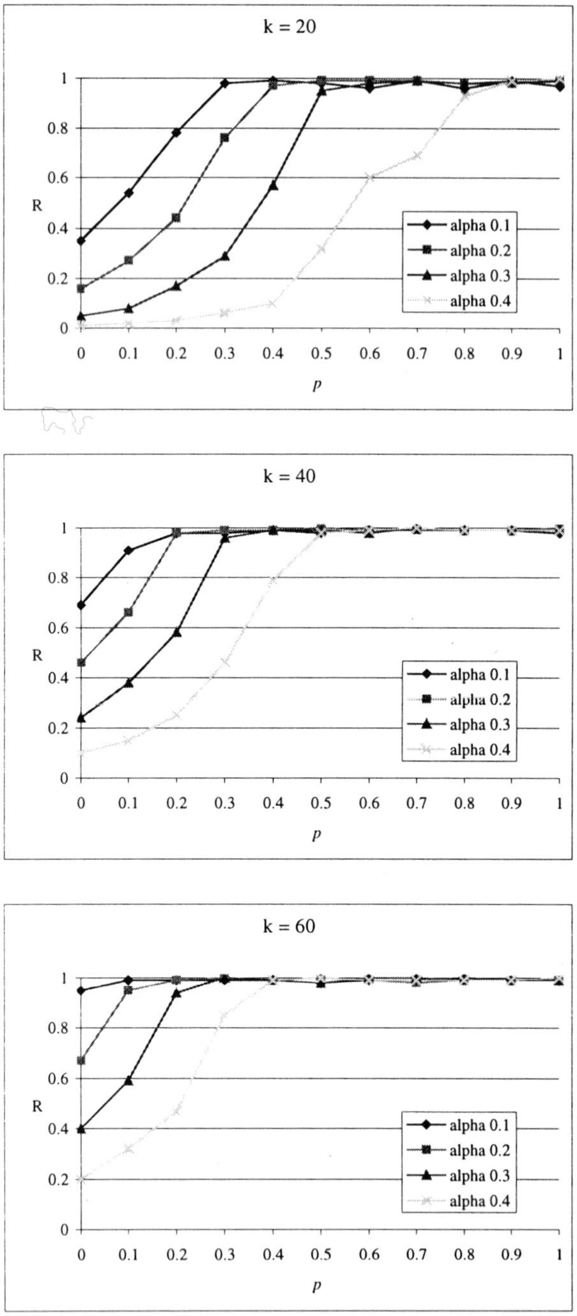

Fig. 2. Networks of size $N = 1000$. The attractor performance R is reported for different values of connectivity k, and at varying loadings, α, as the networks are progressively rewired, by amount p. Results are averages over 10 runs.

Now examining the comparative performance at different levels of connectivity it can be seen that, for a given loading, the larger the value of k the better the network performs. Figure 3 gives a different view of this, fixing the loading at 0.3, and plotting R for different k values on the same graph.

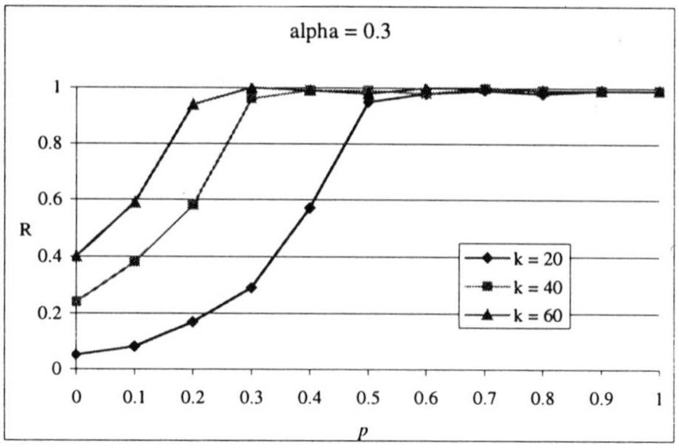

Fig. 3. A different view of the results in Figure 2. α is fixed at 0.3. It can be seen how the higher values of k give better performance.

5.2 Size Effects

The next set of results examines how increasing the size of the network by an order of magnitude affects the performance of the network. Figure 4 summarises the results, here the loading on the networks is fixed at $\alpha = 0.3$. The lower two plots on the graph show the performance of networks of sizes $N = 1000$ and $N = 10{,}000$ with purely local connectivity ($p = 0$). As observed above increasing the level of (local) connectivity improves performance. The larger network has a poorer attractor performance than the $N = 1000$ version, although when k reaches 800 a point of good performance is reached. The explanation for both these observations is probably that if the size of the network is relatively large, when compared to the local neighbourhood of each unit, then it is progressively harder for information to propagate through the network. In other words the more local the connectivity relative to the size of the network the more difficult is it to make a successful global computation such as pattern correction.

The upper pair of lines shows the dramatic effect of just 10% rewiring. Once again the smaller network initially has better performance, but by $k = 180$ the $N = 10{,}000$ network is performing almost perfectly. In this configuration the network is storing 54 10,000-ary vectors, with each unit having only 180, mostly local, connections and is still able to perform almost perfect pattern completion. This architecture is extremely parsimonious. The mean wire length is only about 12% of that of a random network with the same level of connectivity.

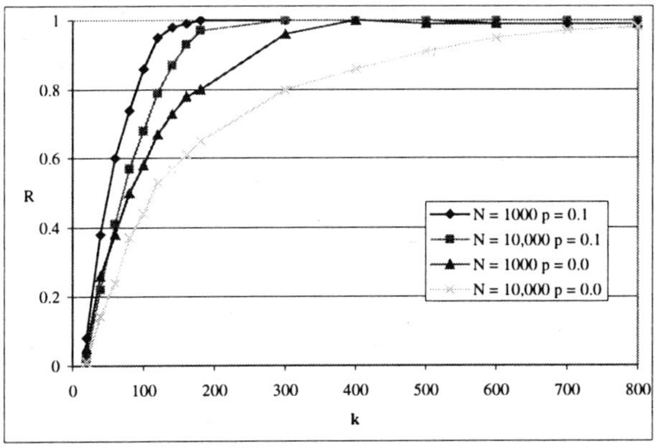

Fig. 4. Networks of size $N = 1000$ and $N = 10,000$, either with local connectivity only $p = 0$ (lower two plots), or with $p = 0.1$, (upper two plots). The attractor performance R is reported for different values of connectivity k. In all cases the loading of the network is $\alpha = 0.3$. Results are averages over 10 runs.

6 Discussion

The fully connected models of associative memory, such as the standard Hopfield network are difficult to realize in VLSI implementations and are not viable as abstract models of real neural networks, which have sparse connectivity. Locally connected associative memories have been proposed as possible solutions to the first of these issues [12]. However as our results show the pattern correction ability of networks with only local connections is compromised. In order to restore the performance of locally connected networks to that of an equivalent random network it is necessary to introduce some rewiring – to change the pattern of connectivity to one of a *small world graph*. The results show that the amount of rewiring required is dependant upon both the relative level of connectivity and the loading on the network. More rewiring is required with either increased loading or decreased relative connectivity. However the very large networks of 10,000 units performed very well with only 10% rewiring at a loading of 0.3. In fact looking at relative connectivity the 10,000 unit network can be thought of as performing better than the 1000 unit network, since it produces perfect pattern completion with a much lower ratio of connections to network size.

Perhaps the most interesting feature of these small world networks is that they suggest how global computation is accomplished in real networks of neurons where distal connectivity comes at considerable cost.

Further work will explore other types of small world connectivity patterns, such as *scale free* networks [8]. We are also interested in finding the pattern of connectivity that maximizes performance whilst minimizing wiring cost.

References

1. Braitenberg, V. and A. Schüz, *Cortex: Statistics and Geometry of Neuronal Connectivity.* 1998, Berlin: Springer-Verlag.
2. Watts, D. and S. Strogatz, *Collective Dynamics of 'small-world' networks.* Nature, 1998. **393**: p. 440-442.
3. Shefi, O., et al., *Morphological characterization of in vitro neuronal networks.* Physical Review E, 2002. **66**(021905).
4. Bohland, J. and A. Minai *Efficient Associative Memory Using Small-World Architecture.* Neurocomputing, 2001. **38-40**: p. 489-496.
5. Davey, N., B. Christianson, and R. Adams. *High Capacity Associative Memories and Small World Networks.* in *IJCNN.* 2004. Budapest.
6. Newman, M.E.J., *Models of the Small World.* Journal of Statistical Physics, 2000. **101**(3/4): p. 819-841.
7. McGraw, P. and M. Menzinger, *Topology and computational performance of attractor neural networks.* Physical Review E, 2003. **68**: p. 047102.
8. Stauffer, D., et al., *Efficient Hopfield pattern recognition on a scale-free neural network.* European Physical Journal B, 2003. **32**(3): p. 395-399.
9. Torres, J.J., et al., *Influence of topology on the performance of a neural network.* Neurocomputing, 2004. **58-60**: p. 229-234.
10. Davey, N. and R. Adams, *High Capacity Associative Memories and Connection Constraints.* Connection Science, 2004. **16**(1): p. 47-66.
11. Turvey, S.P., *PhD Thesis: Analysing and Enhancing the Performance of Associative Memory Architectures*, in *Computer Science.* 2003, University of Hertfordshire: Hatfield.
12. Brucoli, M., L. Carnimeo, and G. Grassi. *Discrete-time cellular neural networks for associative memories: a new design method via iterative learning and forgetting algorithms.* in *38th Midwest Symposium on Circuits and Systems.* 1995.
13. Hertz, J., A. Krogh, and R.G. Palmer, *Introduction to the Theory of Neural Computation.* 1991, Redwood City, CA: Addison-Wesley Publishing Company. 327.

A Memory-Based Reinforcement Learning Model Utilizing Macro-Actions

Makoto Murata and Seiichi Ozawa
Graduate School of Science and Technology, Kobe University, Japan
E-mail: ozawasei@kobe-u.ac.jp

Abstract

One of the difficulties in reinforcement learning (RL) is that an optimal policy is acquired through enormous trials. As a solution to reduce waste explorations in learning, recently the exploitation of macro-actions has been focused. In this paper, we propose a memory-based reinforcement learning model in which macro-actions are generated and exploited effectively. Through the experiments for two standard tasks, we confirmed that our proposed method could decrease waste explorations especially in the early training stage. This property contributes to enhancing training efficiency in RL tasks.

1 Introduction

Recently, in order to avoid waste explorations in reinforcement learning (RL) tasks, several researchers have studied learning algorithms that allow agents to find a series of useful actions called *macro-actions* [1, 2]. For example, when we stand in front of a door we have never seen, first we recognize the knob, reach out our hand for it, and then try to push and pull the door to open. We often attempt to apply some acquired macro-actions to a new task, instead of searching for all combinations of possible actions. Clearly, even when we search for appropriate actions, we exploit some knowledge on actions learned from the past experiences. This fact indicates that we should introduce a mechanism of generating and exploiting macro-actions into reinforcement learning.

On the other hand, we have proposed an RL agent model based on function approximation, in which a value function is approximated by Resource Allocating Network with Long-Term Memory (RAN-LTM) [4]. In RAN-LTM, several representative input-output pairs (called *memory items*) on the approximated value function are stored in long-term memory. Although this learning scheme can be categorized into memory-based learning, the differences from the conventional approaches lie in the number of memory items and their embedded information. In RAN-LTM, memory items are restricted to the minimum number that can suppress the interference caused by incremental learning, and the memory items can store not only the information on an approximated value function but also some other information (e.g., complexity of approximated functions).

In this paper, we propose a new version of RAN-LTM which has a mechanism of generating and exploiting macro-actions to enhance the learning efficiency in RL tasks. To realize this mechanism, new information on the usefulness of state-action pairs as a component of a macro-action is added to the memory items. This information is accumulated in all of the retrieved memory items during an episode. The degree of usefulness is decided depending on how much a series of retrieved memory items contribute to receiving high reward; if this usefulness is high, the corresponding state-action is regarded as a component of a macro-action. Thus, this information is exploited to control the randomness of action selection; that is, when a memory item with high usefulness is retrieved, the probability function of action selection is changed such that a greedy action is taken.

This paper is organized as follows. In Section 2, we present an Actor-Critic model in which function approximators are implemented by RAN-LTM. Section 3 describes how to introduce a mechanism of generating and exploiting macro-actions into RAN-LTM. In Section 4, several experiments are carried out in the two standard tasks. Finally, we state conclusions in Section 5.

2 Actor-Critic Model Using RAN-LTM

Actor-Critic model is one of the most frequently used RL models [3]. In Actor, an action is selected based on a preference value $P_k(\boldsymbol{x}(t), a(t))$, while Critic is used to estimate a state value $V(\boldsymbol{x}(t))$. After selecting an action, Critic evaluates a new state $\boldsymbol{x}(t+1)$ based on TD error $e(t)$ shown below:

$$e(t) \leftarrow r(t+1) + \gamma V(\boldsymbol{x}(t+1)) - V(\boldsymbol{x}(t)) \quad (1)$$

where γ is a discount factor. The preference value for a selected action is updated as follows:

$$P_{k'}(\boldsymbol{x}(t), a(t)) \leftarrow P_{k'}(\boldsymbol{x}(t), a(t)) + \beta e(t) \quad (2)$$

where k' is the index of the action selected in Actor and β is a positive constant.

To approximate a value function $V(\boldsymbol{x}(t))$ and a preference $P_k(\boldsymbol{x}(t), a(t))$, RAN-LTM is adopted as a function

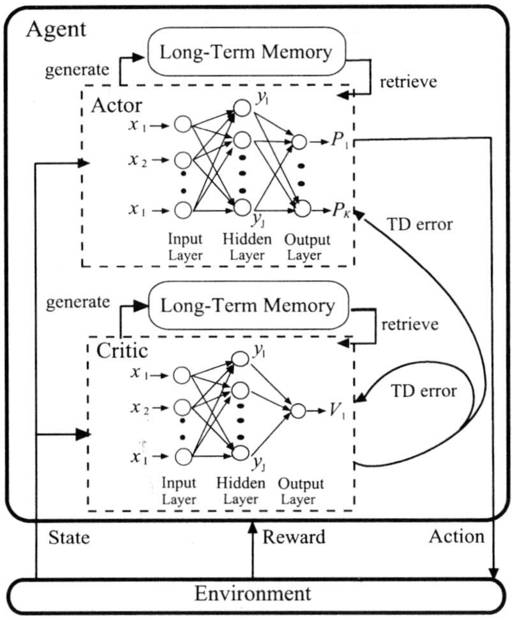

Fig. 1. Structure of Actor-Critic model using RAN-LTM.

approximator in both Actor and Critic. The structure of the Actor-Critic model is shown in Fig. 1. The outputs of the two RAN-LTMs correspond to $V(\boldsymbol{x}(t))$ in Critic and $P_k(\boldsymbol{x}(t), a(t))$ in Actor.

The inputs of RAN-LTM in Actor and Critic at time t is denoted as $\boldsymbol{x}(t) = \{x_1(t), \cdots, x_I(t)\}'$ and they correspond to the agent's states $\boldsymbol{s}(t) = \{s_1(t), \cdots, s_I(t)\}'$. Here I is the number of input units (states). As shown in Eq. (3), the hidden outputs $\hat{\boldsymbol{y}}(t) = \{\hat{y}_1(t), \cdots, \hat{y}_J(t)\}'$ are given by

$$\hat{y}_j(t) = \exp(-\frac{\| \boldsymbol{x}(t) - \boldsymbol{c}_j \|^2}{{\sigma_j}^2}) \quad (j = 1, \cdots, J) \quad (3)$$

where $\boldsymbol{c}_j$ and σ_j^2 are a center vector and a width parameter of the jth radial basis function; J is the number of hidden units. The hidden outputs $\hat{\boldsymbol{y}}(t) = \{\hat{y}_1(t), \cdots, \hat{y}_J(t)\}'$ are normalized by the sum of hidden outputs. Then, the network outputs $\boldsymbol{z}(t)$ are obtained from

$$z_k(t) = \sum_{j=1}^{J} w_{kj} y_j(t) + \gamma_k \quad (k = 1, \cdots, K) \quad (4)$$

where y_j, w_{kj} and γ_k are a normalized hidden output, a connection weight from the jth hidden unit to the kth output unit and a bias of the kth output unit, respectively; K is the number of output units. As stated before, the outputs of Critic correspond to state values $V(\boldsymbol{x}(t))$, while the outputs of Actor correspond to the preference of actions $P_k(\boldsymbol{x}(t), a(t))$. In the softmax strategy, the agent actions are selected based on the following probability:

$$Pr\{a(t) = a\} = \frac{e^{P(\boldsymbol{x}(t),a)/\tau(t)}}{\sum_{b=1}^{K} e^{P(\boldsymbol{x}(t),a_b)/\tau(t)}} \quad (5)$$

where $\tau(t)$ is a temperature to control the randomness of agent actions. Note that K is equal to the number of actions to be selected by agents.

In reinforcement learning, a training sample is given incrementally. In such a situation, neural networks can easily forget the knowledge acquired before. To suppress unlearning, we utilize memory items that are extracted from the approximated function. When learning a new training sample, some memory items are simultaneously trained with the new sample in RAN-LTM to keep input-output relations as much as possible. The readers who want to know the detail training procedures in RAN-LTM can refer to [4].

3 Generating and Exploiting Macro-Actions

In this section, we introduce a mechanism of generating and exploiting macro-actions into Actor-Critic model described in the previous section. As stated in Section 1, we define macro-actions as a deterministic series of useful actions that lead to large rewards with high possibility. To generate such macro-actions, an agent has to retain all the state-action pairs leading to high rewards in some way. However, if the agent's states are continuous, it is obvious that all the possible pairs cannot be held in memory. Considering that memory items in RAN-LTM are distributed over the state space, we easily come upon to add extra information on successful experiences to these memory items.

Let us introduce a new index L_m in the mth memory item in Actor. When an agent receives a reward r_T at time T, L_m of all memory items retrieved during the current episode are updated as follows:

$$L_m^{NEW} \leftarrow \eta L_m^{OLD} + \eta^{T-t_m-1}(r_T - \bar{r}) \quad (6)$$

where $\bar{r}$, t_m and η are the average reward, the time to recall the mth memory item, and a decay constant, respectively. This update is based on the profit sharing method. Note that L_m of the mth memory item becomes large when an action taken in the corresponding state leads to a high reward to the end and an agent frequently encounters the state. This suggests that a state with large L_m should be considered as the component of a macro-action. On the other hand, if an agent encounters a state with large L_m, the action with the highest preference should be taken deterministically.

Based on this idea, we propose the following algorithm of selecting agent's actions which gives a mechanism of generating and exploiting macro-actions for the agent model shown in Fig. 1. Let us denote this agent model RAN-LTM-MA. Note that RAN-LTM-MA does not generate and exploit macro-actions explicitly; macro-actions are generated as a result of controlling the temperature $\tau(t)$ in the action selector that is determined by L_m of the nearest memory item.

[Algorithm]

1. Recall a memory item $\tilde{M}_m : (\tilde{\boldsymbol{x}}_m, L_m, t_m)$ whose $\tilde{\boldsymbol{x}}_m$ has the nearest distance to the state $\boldsymbol{x}_t$.
2. If the distance between $\boldsymbol{x}_t$ and $\tilde{\boldsymbol{x}}_m$ is smaller than a threshold, set $t_m = t$.
3. Update $\tau(t)$ as follows: $\tau(t) \leftarrow \tau(t)/L_m$. Select action a_{t+1} according to the probability in Eq. (5). If the agent receives no reward, go back to Step 1. Otherwise, go to Step 4.
4. Update L_m for all retrieved memory items $\tilde{M}_m$ based on Eq. (6).
5. Reset t, then go back to Step 1.

4 Experiments

To see the usefulness of introducing macro-actions in RAN-LTM, RAN-LTM-MA is applied to the following two standard problems: Mountain-Car Task and Grid-World Task [3].

4.1 Mountain-Car Task

Mountain-Car Task is a task in which a car driver (agent) learns an efficient policy to reach a goal located on a hill. In this task, the goal of a car agent is to successfully drive up a steep slope and to reach the goal as soon as possible. There are three actions to be selected: *full throttle to the goal, zero throttle, full throttle in the opposite direction*. These actions are represented by the following values: $a(t) = \langle +1, 0, -1 \rangle$. A car agent is positioned at random when starting an episode. The position $u(t)$ and velocity $\dot{u}(t)$ are updated based on the following equations:

$$\begin{aligned} u(t+1) &= B[u + \dot{u}(t)] \\ \dot{u}(t+1) &= B[\dot{u}(t) + 0.07a(t) - 0.0025\cos(3u(t))] \end{aligned}$$

where B is a function to restrict the working area, $\{u \mid -1.2 \leq u \leq 0.5\}$, $\{\dot{u} \mid -0.07 \leq \dot{u} \leq 0.07\}$. $u(t)$ and $\dot{u}(t)$ are given as inputs of Actor and Critic, and the value of -1 is given at every step as rewards to all actions taken.

The training is carried out through 200 episodes, which are divided into the 4 learning stages: 1-50, 51-100, 101-150 and 151-200. Since we expect that the introduction of macro-actions contribute to enhancing the training speed, the average steps needed to reach the goal is examined at the above 4 stages. For comparative purposes, the average steps are evaluated for the Actor-Critic models in which Resource Allocating Network (RAN) [5] and RAN-LTM are adopted as function approximators. The results are shown in Table 1.

Table 1. The average steps at the 4 learning stages.

Episodes	RAN	RAN-LTM	RAN-LTM-MA
1-50	980	814	490
51-100	586	451	206
101-150	480	324	210
151-200	327	245	258

As seen from Table 1, at the early stage (1-100 episodes), the average steps in RAN-LTM-MA are greatly reduced as compared with those in the other two models. This result suggests that RAN-LTM-MA can find a proper policy with less experiences by generating and exploiting macro-actions. However, at the last stage (151-200 episodes), the differences in the performances of RAN-LTM and RAN-LTM-MA are not clear. This shows that there are no significant differences in the accuracy as a function approximator; thus one can say that introducing macro-actions contribute to avoiding waste exploration in the early stage of learning.

4.2 Grid-World Task

In this task, the purpose of an agent is to find the shortest path from a start point to a goal in a 10×10 grid-world. Here we adopt 6 grid-world tasks as shown in Fig. 2. In these tasks, an agent can move in 4 directions (east, south, west, north). At the beginning of an episode, an agent is placed at each start point. The reward (+1) is only given to the agent when reaching the corresponding goal.

The training is carried out through 100 episodes, and in each episode the agent's policy is evaluated by the number of steps to the goal. To remove the dependency on random seeds, the training is repeated 30 times and the average steps are evaluated.

Figure 3 shows the time courses of the average steps taken by an agent and the action probability in Task 1. This action probability is obtained by averaging the probability of selecting optimal actions on the three typical shortest paths. If this value is high, one can say that the agent takes optimal actions deterministically. As seen from Fig. 3, the average steps of RAN-LTM-MA rapidly decrease from the 20th episode; at the same time, the action probability rapidly increases. This result suggests that some useful macro-actions are generated on the optimal paths in RAN-LTM-MA and the exploitation of these macro-actions leads the decrease of the average steps. To verify this, we examine the actions selected with 99% and 90% probability. The result is shown in Fig. 4 where black and white arrows correspond to the

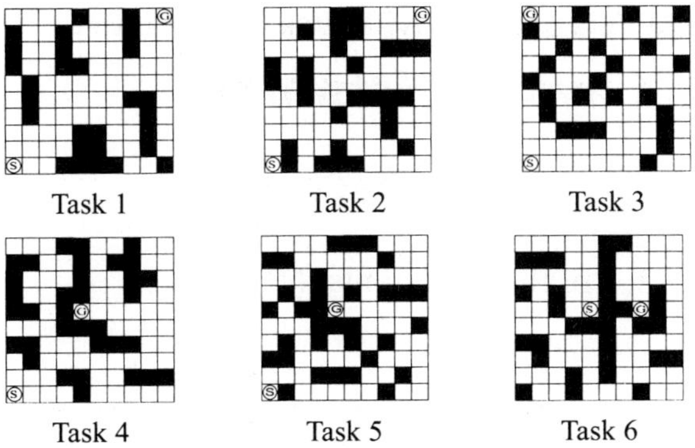

Fig. 2. Evaluated grid-world tasks.

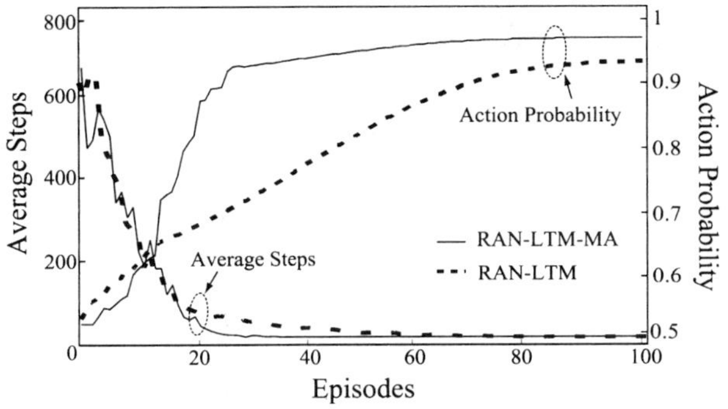

Fig. 3. Time courses of average steps and action probability in Task 1.

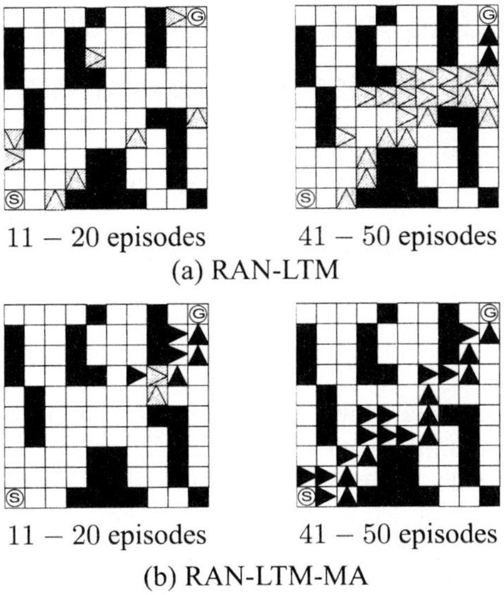

Fig. 4. Selected actions with 99% and 90% in Task 1. Black and white arrows correspond to the directions of selected actions with 99% and 90%, respectively.

directions of selected actions with 99% and 90%, respectively. In Fig. 4, we can see that some series of deterministic actions are generated in RAN-LTM-MA after the 20th episode.

Table 2 shows the number of needed episodes to get a proper policy in the 6 grid-world tasks. Here we define that an agent acquire a proper policy when the steps to a goal is less than the following value: (*steps of the shortest path*) × 1.2. From the results in Table 2, we can conclude that RAN-LTM-MA can get a proper policy promptly in all tasks.

Table 2. The number of needed episodes to get a proper policy in the 6 grid-world tasks.

Task	RAN	RAN-LTM	RAN-LTM-MA
1	64	64	33
2	66	65	33
3	70	70	41
4	59	59	26
5	54	54	27
6	86	86	39

5 Conclusions

In this paper, we proposed a new agent model in which macro-actions were generated and exploited effectively. Through the experiments, we verified that our proposed model could decrease waste explorations especially at the early stage of training. Further experiments are needed to ensure whether such macro-actions really contributes to enhancing the training efficiency in any RL tasks.

Acknowledgment

This research was partially supported by the Ministry of Education, Science, Sports and Culture, Grant-in-Aid for Scientific Research (C) 16500130, and by the Okawa Foundation for Information and Telecommunications.

References

[1] Mcgovern, A., Sutton, R. S., Fagg, A. H. (1997) Roles of macro-actions in accelerating reinforcement learning. Proceedings of the 1997 Grace Celebration of Women in Computing: 13-18

[2] Precup, D., Sutton, R. S. and Singh, S. P. (1997) Planning with closed-loop macro actions. In Working Notes of the 1997 AAAI Fall Symposium on Model-directed Autonomous Systems: 70-76

[3] Sutton, R. S. and Barto, A. G. (1998) Reinforcement learning - An introduction. The MIT Press

[4] Ozawa, S. and Abe, S. (2004) A memory-based reinforcement learning algorithm to prevent unlearning in neural networks. In Neural Information Processing: Research and Development: Rajapakse, J. C. and Wang, L., Eds., Springer-Verlag: 238-255

[5] Platt, J. (1991) A resource allocating network for function interpolation. Neural Computations, 3: 213-225

A Biologically Motivated Classifier that Preserves Implicit Relationship Information in Layered Networks

Charles C. Peck James Kozloski Guillermo A. Cecchi
A. Ravishankar Rao
IBM T.J. Watson Research Center, P.O. Box 218, Yorktown Heights, NY 10598

Abstract

A fundamental problem with layered neural networks is the loss of information about the relationships among features in the input space and relationships inferred by higher order classifiers. Information about these relationships is required to solve problems such as discrimination of simultaneously presented objects and discrimination of feature components. We propose a biologically motivated model for a classifier that preserves this information. When composed into classification networks, we show that the classifier propagates and aggregates information about feature relationships. We discuss how the model should be capable of segregating this information for the purpose of object discrimination and aggregating multiple feature components for the purpose of feature component discrimination.

1 Introduction

Classical Artificial Neural Networks (ANNs) are typically arrayed in layers, with each layer sending its responses to higher layers. This arrangement permits each successive layer to respond to increasingly complex combinations of features or attributes in the input space.

These layered networks, however, have fundamental limitations. Consider an array of ANN's that have been trained with visual inputs to realize an orientation-selective, topographically organized map, as in the primary visual cortex [4]. Each ANN effectively encodes two pieces of information: edge location and edge orientation. These two pieces of information are implicitly related to each other by their shared classifier response. Now consider four additional ANN classifiers that respond to: 1) a vertical edge at any location, 2) horizontal edge at any location, 3) an edge of any orientation in the top half of the image, and 4) an edge of any orientation in the bottom half of the image. Finally, assume that a final ANN classifier exists that responds when a horizontal edge is present in the top half of the image based on the responses of the four classifiers.

If edges can only be horizontal or vertical, this example network will respond properly in each of the four cases where a single edge is present in the image. When two edges are present, however, an erroneous response can occur. In particular, when a vertical edge is present in the top of the image and a horizontal edge is present in the bottom, both attributes will be simultaneously present and the response will be ambiguous. This is Rosenblatt's "superposition catastrophe" [5].

The root of the superposition catastrophe is that the implicit relationship between an edge's orientation and location is lost when these two attributes are independently classified by the set of four classifiers. Since this lost information is not propagated forward through the network, an ambiguity regarding the sources of attributes can exist at the output.

A related problem is the "component discrimination" problem. This problem involves generating an output response specific to an initial classifier when that classifier contributes to a second classifier giving rise to the final output response. For example, let us modify the previous network such that the horizontal and vertical edge classifiers instead recognize squares and triangles in the image. Let us also modify the output objective such that it is to respond only when a particular edge is active and the edge contributed to the recognition of a square. In this case, implicit information is also lost: the relationship between the edge and the other edges giving rise to the square. This information is subsumed into the "square" classifier. To respond properly, the relationship between the output response and the square classifier must be propagated back to each contributing edge and then forward to the output response. In this manner, the relationship between the edge and the square classifier can be tested by the output response.

2 Solution Requirements

Both the superposition catastrophe and the component discrimination problems result from the loss of information about the relationships among features in the input space or the relationships inferred by higher order classifiers. To solve these problems, therefore, this information must be preserved and made available for classification tasks throughout the network.

One approach for preserving this relationship information is to represent it explicitly as a relationship value and couple it with the classifier responses. To solve the superposition catastrophe and component discrimination problems with these values, the following requirements must be met: 1) *Uniqueness:* relationship values must be sufficiently distinct to avoid erroneous relationship interpretations, 2) *Propagation:* relationship values must propagate forward and backward, 3) *Aggregation:* classifiers must take the disparate, but sufficiently similar relationship values of its feedforward and feedback inputs and produce a unified relationship value; and 4) *Selectivity:* relationship values must be used to modulate classifier responses to inputs.

3 A Biologically Motivated Solution

Since the 1989 discovery that neural synchronization correlates with global visual input properties [3], neural oscillation and synchronization have been viewed as a possible means for conveying this implicit relationship information. While both Choe [2] and Seth [6] have modelled neural synchronization, neither effort satisfies the objectives of this paper or operates at the desired level of abstraction. Choe's work uses a spiking model and it operates at a lower level of abstraction. Seth's work does not use self-organizing classifiers and may not provide a sufficient basis for classifier learning.

Here, we propose a classifier model that operates on the neocortical minicolumn level of abstraction and uses minicolumn synchronization as the means to manage relationship information. At this level of abstraction, hundreds of neurons are modelled as a single computational system or unit. Furthermore, communication between minicolumns is modelled as an aggregation of the hundreds or thousands of axons that connect them. Thalamo-cortical inputs are modelled as a firing rate, ρ. Cortico-cortical inputs are modelled as pulses up to 12.5ms wide and are represented as a tuple (γ, T), where γ and T are a pulse's amplitude and reference time, respectively. While not shown here, this model is based on an analysis of the cortical microcircuit. The model explains observations of neural oscillations as the "chopping" of minicolumn inputs. The unique time required for a minicolumn input to propagate through a specific minicolumn, j, and "chop" subsequent inputs determines the natural period, τ_j of that minicolumn.

In this model, the classification response, ψ_j, of minicolumn j is a weighted sum modulated by the coincidence of each input pulse with the pulse cycle of the minicolumn. This coincidence-based modulation addresses the selectivity requirement.

$$\psi_j = \sum_{i \in \mathrm{FF}_j} w_{ij} \gamma_i e^{-\left(\frac{T_i - T_j}{2c_1}\right)^2} + \sum_{k \in \mathrm{TH}_j} w_{kj} \rho_k, \qquad (1)$$

where T_j is the reference time of the current pulse of minicolumn j, FF_j is the set of feedforward cortical inputs to minicolumn j, γ_i and T_i are the parameters of the current pulse for minicolumn i, w_{ij} is the weight from minicolumn i to j, TH_j is the set of thalamic inputs to minicolumn j, ρ_k is the current firing rate of thalamic relay cell k, w_{kj} is the weight from thalamic relay cell k to minicolumn j, and c_1 is the standard deviation of the Gaussian. The final amplitude response of minicolumn j is: $\gamma_j = \sigma\left(\psi_j, \frac{2.2}{\sqrt{N_{\mathrm{FF}_j} + N_{\mathrm{TH}_j}}}, 0\right)$, where $\sigma(x, \alpha, \beta) = \frac{1}{1+e^{-\alpha(x-\beta)}}$, the sigmoid function.

Computationally, the aggregation requirement is addressed with synchronization, and synchronization is achieved by adjusting each minicolumn's reference time, T_j, relative to the reference time's of other minicolumns using a convex mapping (a modulated weighted sum of reference time differences). The next reference time, T_j', for a minicolumn's pulse is computed as follows:

$$\begin{aligned} T_j' &= T_j + \tau_j \\ &+ \frac{1}{c_2} \sum_{i \in \mathrm{FF}_j \cup \mathrm{FB}_j} (T_i - T_j)\, w_{ij} \gamma_i e^{-\left(\frac{T_i - T_j}{2c_1}\right)^2}, \end{aligned} \qquad (2)$$

where FB_j is the set of feedback cortical inputs to minicolumn j and c_2 is a constant corresponding to the maximum contribution from inputs affecting the timing. The use of feedforward and feedback inputs for timing adjustment addresses the propagation requirement. Finally, the uniqueness requirement is addressed by using the natural period, τ_j, as a bias that competes with the pull toward synchronized reference times.

This model provides a basis for learning, but it is independent of the particular method for updating the thalamo-cortical and cortico-cortical weights.

4 Results

Our experimental plan was designed to determine how well the propagation and aggregation requirements are met while simultaneously supporting classification and self-organized learning. We will not explore the uniqueness and selectivity requirements in this paper because those requirements require multiple inputs and an entirely different experimental set up.

The experimental network we used is shown in Figure 1. This network consists of three hypercolumn-like arrays of minicolumns arranged in a hierarchy. At the lowest hierarchical level, two 10x10 networks, denoted

M1L and M1R, each receive feedforward inputs from 2 real values (x, y) in the range [0,1]. At the highest hierarchical level, one 10x10 network, denoted M2, receives inputs from the M1L and M1R networks. Each M2 minicolumn receives feedforward inputs from every M1L and M1R minicolumn and each M1L and M1R minicolumn also receives feedback connections from every M2 minicolumn. This connectivity mimics slowly varying projections between hypercolumns. Within each array, every minicolumn makes a center-excitatory/surround-inhibitory pattern of connections with its neighbors.

Our first experiment was to determine whether we could achieve self-organization, both with and without pulsatile inputs. The M1L and M1R results, shown in Figure 1, demonstrate that we were successful in achieving self-organization without pulsatile inputs. Each minicolumn is shaded to indicate the angular location of inputs to which it responds best. The shading key in shown in the bottom of the figure. Similarly, the M2 results demonstrate self-organization based on high dimensionality pulsatile inputs.

Our second set of experiments explored the propagation and synchronization-based aggregation capabilities of our model. The experiments were executed for 30,288 iterations and each stimulus was presented for 24 iterations. Figures 2A–D illustrates activation and synchronization that occurs when the network is presented with a stimulus after self-organization has begun, but not yet completed (completed self-organization is shown in Figure 1). At this stage, we observed that half of the network became sensitive to half of the input space in all three networks. Categories were further subdivided at later stages. Note that synchronization and self-organization are learned concomitantly. Figure 2A shows the classification response of each minicolumn in the three networks. Minicolumns in the bottom half of M1L and M1R are strongly activated by the input, as they are in the right half of M2. The remaining minicolumns are weakly activated.

Figure 2B shows the inter-pulse period for each minicolumn in the three networks. The strongly activated minicolumns, rendered from light gray to white in Figure 2A, share very similar periods. Even minicolumns receiving large numbers of weak inputs, such as the left half of M2, converge to a shared period. Due to the averaging effects in Equation 2, all shared periods derived from large numbers of inputs have similar values. Weakly activated minicolumns in M1L and M1R have diverse periods derived primarily from their unique natural periods. They do not achieve a shared period due to weak feedback from the left half of M2. This diversity also shows the basis for satisfying the uniqueness requirement in our model.

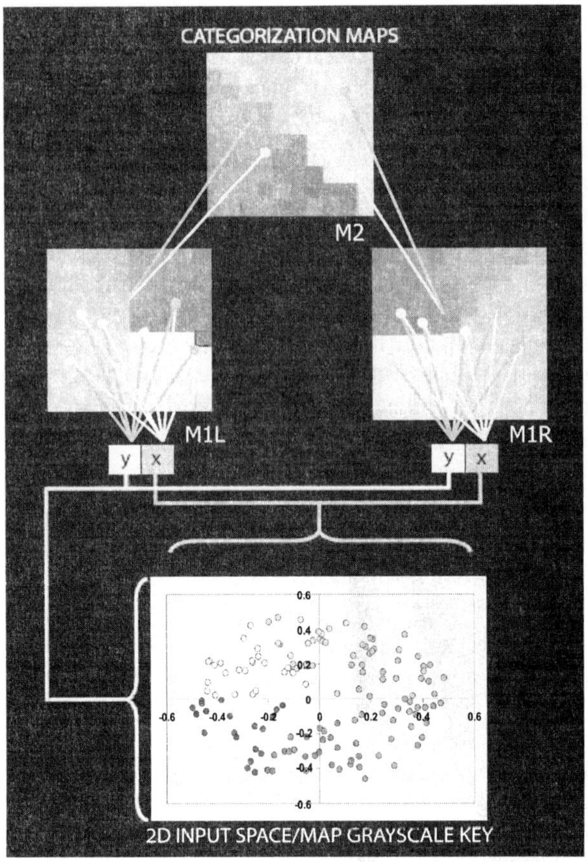

Fig. 1. Experimental Setup. Input space (bottom) and two levels of minicolumn classifiers' (middle, top) preferred-stimulus maps demonstrating self-organization.

The period alone is insufficient to demonstrate synchronization. Figure 2C depicts the pairwise coincidence of pulse reference times. The shades of the lines between two minicolumn conveys the percentage of times their pulse reference times varied by less than 4% during a stimulus. The results demonstrate that minicolumns within strongly activated areas were highly synchronized with each other. In addition, minicolumns in weakly activated areas of M2 were synchronized with each other, but out of phase with the strongly activated M2 minicolumns.

While Figure 2C shows intra-network synchrony, Figure 2D similarly shows inter-network synchrony. In this visualization, the networks are arranged in two planes and all pairs of minicolumns in the network analyzed. The shaded connections show that minicolumns in the strongly activated areas of M1L, M1R, and M2 are synchronized within and *between* their networks. Fi-

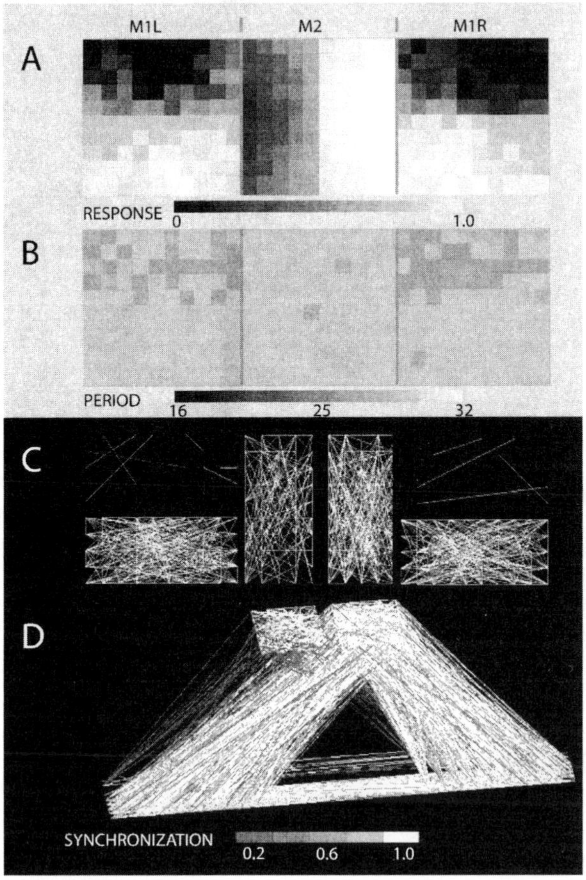

Fig. 2. Results. A. Minicolumn classification response. B. Minicolumn inter-pulse interval in response to a stimulus; C. Minicolumn synchronization within self-organized networks; D. Minicolumn synchronization within and between self-organized networks, dark gray is weak, white is strong.

nally, minicolumns in the weakly activated area of M2 are strongly synchronized with each other and weakly synchronized with some minicolumns in the weakly activated areas of M1L and M1R. This demonstrates that weakly responding minicolumns do not aggregate, which is important for achieving uniqueness.

5 Conclusions

Our results demonstrate that we have solved two key requirements for relationship information preservation: propagation and aggregation of relationship values across a classification network. However, the requirements of selectivity and uniqueness were not fully demonstrated because multiple inputs were not simultaneously presented.

While the forced interdependence between classification and synchronization is not exploited for selectivity and uniqueness in these experiments, the achievement of learning with dynamic inputs is noteworthy and nontrivial. Purely static learning rules are not easily adapted to dynamic inputs [7]. We have solved this problem by learning over those inputs which arrive within a time interval corresponding to a particular phase of an internal oscillation, and by explicitly controlling the relationship between timing and amplitude. We have achieved self-organization, with classification contingent upon synchronization, using both a correlation-based learning approach (Figure 1) and learning rules derived from Kohonen map algorithms [1] (results not shown).

The model also allows for the classification of a feature during a single presentation to become progressively more dependent on only those inputs that are related to the emerging classification. Such a mechanism could allow for more robust classification in the presence of noise, as the mechanism dynamically ignores inputs that do not contribute to the classification. The mechanism should also allow for segregation of nearby features that are derived from different objects, because it allows for classification of a feature to become progressively less a function of those inputs that become more strongly synchronized with other emerging classifications. This selectivity is a requirement for solving Rosenblatt's "superposition catastrophe." By propagating relationship information inferred by classification, this model establishes a basis for grouping features with objects and object discrimination. Experiments using simultaneously presented objects to test the full range of capabilities of our model are ongoing.

References

[1] Bednar, J. A. & Miikkulainen, R. (2003) Neurocomputing, 52-54:473-480.

[2] Choe, Y. & Mikkulainen, R. (2004) Biol Cybern., 90(2):75-88.

[3] Gray C. M. & Singer W. (1989) Proc Natl Acad Sci U S A 86: 1698-1702.

[4] Grinvald,A. & Lieke, E. & Frostig, R. D. & Gilbert, C. D. & Weisel, T. N. (1986) Nature, 324(6095):361-4

[5] Rosenblatt, F. , (1962) Principles of Neurodynamics: Perception and the Theory of Brain Mechanisms. Washington: Spartan Books

[6] Seth, A.K. & McKinstry, J.L. & Edelman, G.M. & Krichmar, J.L. (2004) Cerebral Cortex May 13 Advanced access.

[7] Song S., & Abbott L. F. (2001) Neuron Oct 25;32(2):339-50.

Large Scale Hetero-Associative Networks with Very High Classification Ability and Attractor Discrimination Consisting of Cumulative-Learned 3-Layer Neural Networks

Yohtaro Yatsuzuka[1], Yo Ho[2]

[1]KDDI R&D Laboratories Inc., Japan
[2]July System Co., Japan
E-mail: yohtaro@kddilabs.jp

Abstract

Auto-Associative neural networks have limited memory capacity and no classification capability. We propose a hetero-associative network consisting of a cumulative-learned forward 3-layer neural network and a backward 3-layer neural network, and a hetero-tandem associative network. The hetero-tandem associative network has a spindle type single cyclic-associative network with cumulative learning and is connected in tandem with the subsequent hetero-associative network. These hetero-associative networks with classification ability have high recognition performance as well as rapid attractor absorption.

Consecutive codification of outputs in the forward network was found to produce no spurious attractors, and coarse codification of converged attractors can be easily identified as training or spurious attractors.

Cumulative learning with prototypes and additive training data adjacent to prototypes can also drastically improve associative performance of both the spindle type single cyclic- and hetero-associative networks, allowing them to effectively be connected in tandem.

1 Introduction

Studies have been made of various associative memories. These studies mainly focused on memory capacity for storing associative states and recognition performance in terms of the directly attracted state [1,2]. Conventional associative neural network memories have a limited and relatively small memory capacity [2,3,7]. Moreover, they have no classification capability to allow wide expansion of applications.

This paper proposes neural network architectures for hetero-associative networks with classification and attractor discrimination abilities. This paper further describes approaches for attaining large memory capacity and superior performance through cumulative learning of a 3-layer neural network with a large number of hidden layer units. A cumulative-learned hetero-tandem associative network with spindle type single cyclic- and hetero-associative networks in tandem is also proposed. The spindle type associative network is different from the sand glass type in that the number of hidden layer unit is much larger than that of input layer units.

This paper is organized as follows. In Section 2 we briefly describe the basic configuration of the proposed hetero-associative networks. In Section 3, we discuss hetero-associative network models. In Section 4 simulation results are then discussed for various types of associative networks. In Section 6 we summarize our findings and draw some conclusions.

2 Basic Configuration of Proposed Hetero-Associative Networks

2.1 Hetero-associative networks with classification ability and attractor discrimination

Figure 1 shows the basic configuration of the proposed hetero-associative network. This network has a large memory capacity as well as category classification capability and attractor discrimination. The associative network consists of a forward network for category classification, a backward network to extract an attractor for an associative input X(0) in the associative loop process, and a discriminator for converged attractors. The forward and backward networks are respectively comprised of a 3-layer neural network having one hidden layer. A set of prototypes assigned to categories is initially learned by the forward network. Cumulative learning is then conducted for the prototypes and additive training data applied to improve recognition performance. The backward network can explicitly recall the prototype as a training attractor corresponding to the category in the form of its stable equilibrium state. However, additive training data having a strong correlation to the prototypes cannot be directly recalled due to its implicit embedding in the forward neural network.

A ring memory is therefore prepared to extract these additive training data by using the corresponding training attractor. A cyclic-associative network is used for the ring memory to store a number of sequences.

Each sequence includes the prototype and the additive training data adjacent to the prototype.

The attractor discriminator is also prepared for the forward network with coarse codification of binary outputs such as winner-takes-all codes to reliably perform processes in the subsequent higher hierarchical network by utilizing the category output [4].

A hetero-associative network with classification capability for abstraction or generalization and recall capability of prototypes for instantiation or specialization is an attractive means for allowing complex information to proceed along paths through higher and/or lower hierarchical neural networks having top-down and bottom-up directions.

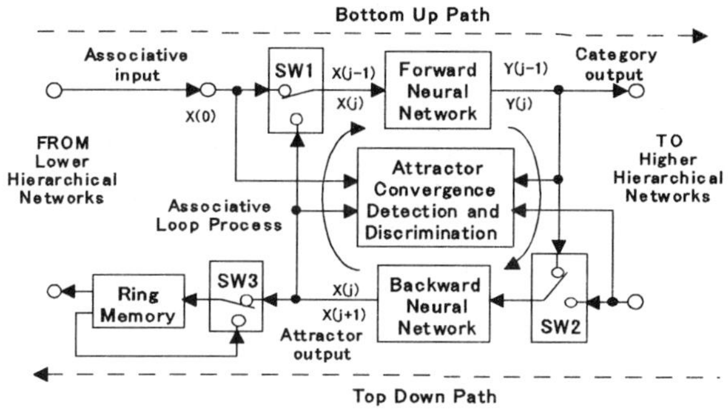

Fig.1. Basic configuration of proposed hetero-associative network.

2.2 Tandem associative network

In the hetero-tandem associative network, a spindle type single cyclic-associative network is connected in tandem with the subsequent hetero-associative network as shown in Fig.2. Both the spindle type associative network and the forward network in the hetero-associative network are cumulative-learned to drastically improve recognition performance. Both the attracted and distracted outputs of the cyclic-associative network through the associative loop process are fed into the forward network. The classified category and attractor outputs are respectively derived from the forward and backward networks.

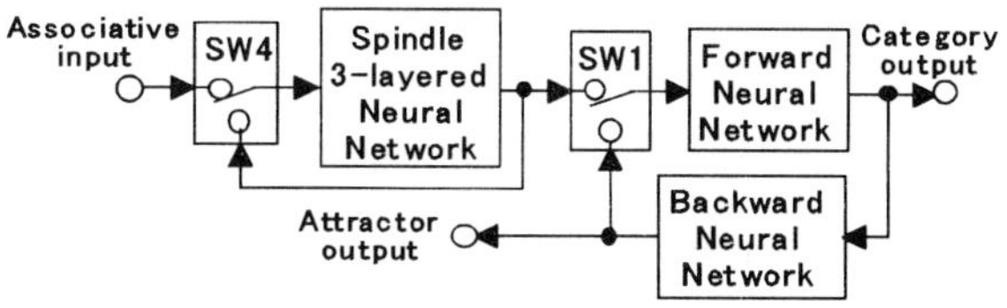

Fig.2. Hetero-tandem associative network with spindle type single cyclic- and hetero-associative networks.

3 Design Aspects of Hetero- and Spindle type Associative Networks

3.1 Learning of Forward and Backward 3-layer Neural Networks

In supervised learning of the forward neural network, binary codes are assigned as targets to categories in classifications. M pairs of a training input vector corresponding to a prototype as a training attractor and a target vector for classification, given by $\{\mathbf{U}(1),\mathbf{T}(1)\},..,\{\mathbf{U}(M),\mathbf{T}(M)$, m=1,2,.. M, are stored in the forward network through initial learning and the cumulative learning. Here, $\mathbf{U}(m)$ and $\mathbf{T}(m)$ are vectors whose elements $U_r(m)$, r=1,..R, and $T_k(m)$, k=1,..K take a binary value {0,1}, where R and K are respectively the number of input and output units. In the backward network, M pairs of training data, $\{\mathbf{T}(1),\mathbf{U}(1)\},..,\{\mathbf{T}(M),\mathbf{U}(M)\}$, m=1,.., M, consisting of components inversely arranged from those for the forward network are also used for initial learning.

3.1.1 Modified error back-propagation algorithm:

Storage of training attractors as prototypes requires complete learning by the forward and backward networks with rapid and stable convergence in binary space. A modified error back-propagation algorithm can meet these conditions.

The modified portions in the error back-propagation algorithm are given in A.

A. Modified error back-propagation algorithm

The modified portions in the error back-propagation algorithm are only given as follows.

The change of $\Delta W^s_{k,n}(m)$ to be made to the weight of $W^s_{k,n}(m)$ between n-th unit of the hidden layer (v) and k-th unit of output layer (s) for the m-th training input data is given by

$$\Delta W^s_{k,n}(m)=\alpha(1-\eta)\,\delta^s_k(m)O^v_n(m)+\eta\,\Delta W^s_{k,n}(m-1), \quad (1)$$

where α is the learning rate, η is the momentum rate, and $O^v_n(m)$ is the unit output of n-th unit of the hidden layer. When the binary unit output of the k-th unit of the output layer is correct, compared with the target, $T_k(m)$ {0,1},

$$\delta^s_k(m) = (T_k(m) - O^s_k(m))(0.25)^{(1-g)}\{O^s_k(m)(1 - O^s_k(m))\}^g, \quad (2)$$

where g is the emphasis rate in the range 0 to 1, $O^s_k(m)$ is the unit output of k-th unit of the output layer for the m-th training input data.

When the binary output of the k-th unit the output layer is false, compared with the target, $T_k(m)$, then

$$\delta^s_k(m) = (T_k(m) - O^s_k(m))\{O^s_k(m)(1 - O^s_k(m))\}, \quad \text{if } O^s_k(m)(1 - O^s_k(m)) \geqq D_{limit}, \quad (3)$$

$$= (T_k(m) - O^s_k(m))\{D_{limit}\}, \quad \text{if } O^s_k(m)(1 - O^s_k(m)) < D_{limit}, \quad (4)$$

where D_{limit} is the threshold value for the limiter in the range 0 to 0.25. The sigmoid function was used as an activation function.

The limited value of $\delta^s_k(m)$ given by Eq.3 and 4 for the false unit output can easily prevent for the network to be trapped into a state having tenacious local minima. The emphasis rate of g given by Eq.2 also accelerates the convergence speed for the correct unit output.

The modified back-propagation algorithm in the 3-layer neural network always achieves further more rapid and ensure stable convergence of very large numbers of prototypes and additive training data than does the error-perturbation algorithm [5,6]. The 3-layer neural network can successfully lean more than 8192 prototypes by using the modified updating algorithm.

3.1.2 Cumulative learning: After the initial learning of prototypes (prototype-learning), adjacent data to the prototypes that provide false outputs due to the generalization property are cumulatively added to the prototypes as additive training data in the repetitive cumulative-learning of the forward network. The cumulative learning is a very attractive method for significantly improving the rate of generalization [5], and results in higher performance in hetero-associative networks. Here, HD1 test data providing false outputs are extracted by utilizing the generalization property and cumulatively added as the additive training data. The cumulative learning is repetitively proceeded until the HD1 test data becomes entirely correct through the 3-layer neural network.

3.2 Detection of Attractor Convergence

Table 1 lists the relations of inputs, X(j) and outputs, Y(j) of the forward network for the associative input X(0) in the j-th associative loop process, and for discrimination of the converged attractors.

Table 1 Relations of inputs and outputs of forward and backward networks for associative input in the associative loop process and for discriminating the attractor.

Number of associative loops j	States of inputs and outputs of forward and backward nets	Index of code Y(j)	Attractor
j = 1	Converged state X(0) =X(1), Y(0) =Y(1), X(1) =X(2)	Allocated	One-Shot Training
		Non-Allocated	One-Shot Spurious
$J \geqq j \geqq 1$ Where J=10	Generalized output of the forward net Just before convergence X(j-1) ≠ X(j), Y(j-1) = Y(j), X(j) =X(j+1)	Allocated	Training
		Non-Allocated	Spurious
	Generalized output of the backward net Just before convergence X(j-1)≠X(j), Y(j-1)≠Y(j), X(j) =X(j+1)	Allocated	Training
		Non-Allocated	Spurious
	Non-converged state at J X(J-1) ≠X(J), Y(J-1)≠Y(J), X(J)≠X(J+1)	Non-Allocated	Diverged

At a stable equilibrium state,

$$X(j-1)=X(j),\ Y(j-1)=Y(j),\ \text{and}\ X(j)=X(j+1) \quad (5)$$

are respectively satisfied, indicating an attracted state. Just before converging to this state, either of the forward or backward network always have the generalized state at (j-1)th, as shown in Table 1.

If the converged Y(j) is a binary code allocated to the target, then the output of the backward network, X(j+1) is discriminated to a one-shot training attractor at j=1 or a training attractor generalized at j>1. If not, the output is discriminated to a one-shot spurious attractor at j=1 or a spurious attractor generalized at j>1.

The associative input X(0) is training input data only when the one-shot training attractor is detected. In the other case, X(0) is always unseen input data.

3.3 Discrimination of Detected Attractor

Spurious attractors generally appear, depending on the coarse codification for targets in the forward network and the single cyclic-associative network. The literal index of binary output of the forward network is therefore utilized to discriminate whether the binary output is allocated to a target, or not by the code index discrimination algorithms given in B.

B. Code Index Discrimination Algorithms

I. Distributed binary codes

The index of a distributed binary code allocated to a target, Cidx, is generally given by

$$\text{Cidx} = \text{Lidx} + \text{CidxMin}, \quad (6)$$

where Lidx is the literal index of a pure-binary code, and CidxMin and CidxMax are the respective minimum and maximum numbers among Cidxes.

For distributed binary codification having the discontinuous index step Q, where Q is a integer larger than 0,

IF $\text{CidxMin} \leqq \text{Oidx}(j) \leqq \text{CidxMax}$, (7)

Then IF $(\text{Oidx}(j) - \text{CidxMin}) \bmod Q = 0$, (8)

Then the output is one of the allocated binary codes,

Else the output is a non-allocated binary code,

Else the output is a non-allocated binary code,

where Oidx(j) is the index derived from the binary output of the forward network in the j-th associative loop process.

II. Winner-takes-All codes (Place codes)

IF Hamming weight = 1, (9)

Then the output is one of allocated the binary codes,

Else the output is a non-allocated binary code.

If the binary output of Y(j) is one of binary codes allocated to the targets, then the converged attractor is a training attractor. If not, then this is a spurious attractor as summarized in Table 1.

Allocation of binary codes having consecutive literal indexes to the targets corresponding to categories, such as pure-binary distributed codes with a compact density, are especially effective in preventing the appearance of spurious attractors, whereas allocatiing of binary codes with discontinuous indexes produces many spurious attractors. A coarse codification called Winner-takes-all or Place Code (One out of M codes) is an attractive means for achieving higher generalization performance, though spurious attractors still appear.

4. Simulations

4.1 Simulation Models

In the forward network having R=75 input units, 140 hidden units and K=7 output units, a pure-binary output representation with the index step Q=1 and CidxMin=0 was prepared for a dense codification. The backward network has K=7 input units, 140 hidden units and R=75 output units.

To make performance comparisons, 8 bit codes (K=8) having Q=1 with CidxMin=10, and Q=2 with CidxMin=0 were also prepared for the equivalently consecutive codification. Furthermore, 9 bit codes (K=9) of Q=3, and 10 bits of O=5 with CidxMin=0 were prepared for the discontinuous codification.

A forward network with Place codes was also prepared as a coarse codification.

Sand glass type and spindle type single cyclic-associative networks that have the output layer directly connected to the input layer in the 3-layer neural network with R=75 input and output units were used. The spindle type network has 200 hidden units, whereas the sand glass type network has 75 hidden units.

4.2 Data sets for training, test and validation

Data sets for training, test and validation were prepared. The training data set had prototypes composed from feature vectors having M=128 input samples. A test data set HD1adjacent to the training data set was composed from the input data elements having a Hamming distance of 1 from the corresponding prototype. Validation data sets of HDp having a Hamming distance p, where p is from 2 to 10, were also prepared to validate performance of the hetero-associative networks. The sizes of these sets were respectively 9,598 in HD1, 355,040 in HD2 and 128,000 in HDp, $3 \leqq p \leqq 10$. Data sets referenced to prototypes of 16 Chinese characters given in [3] were also prepared.

5 Simulation Results and Discussions

5.1 Prototype-learning for associative networks

Retrieval dynamics of hetero-associative networks for the test and validation data are described in Figures

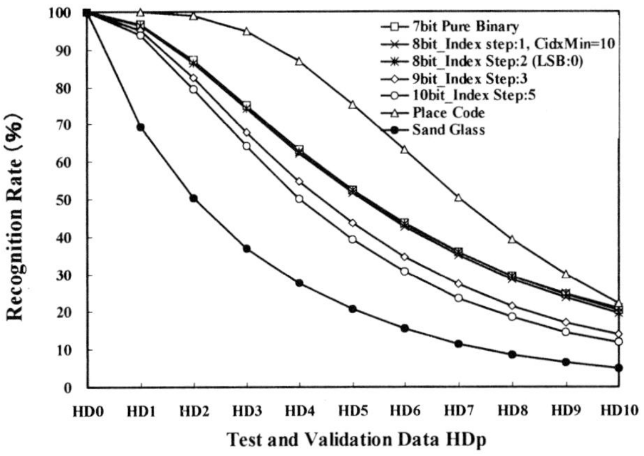

Fig.3. Recognition rate of hetero-associative networks for test and validation data. Here, HD0 indicates the set of prototypes.

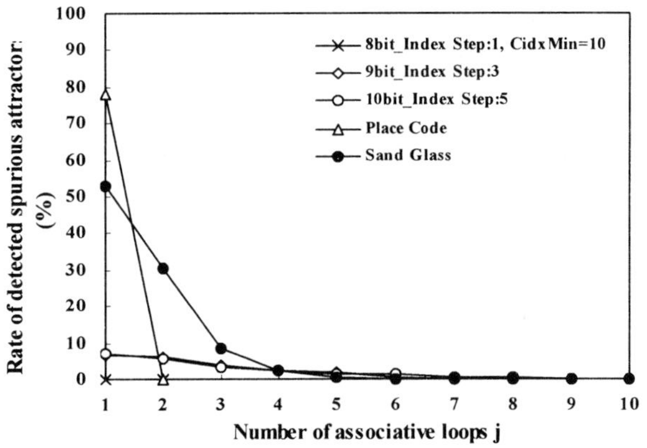

Fig.4. Rate of detected spurious attractors in HD10 validation data versus number of associative loops.

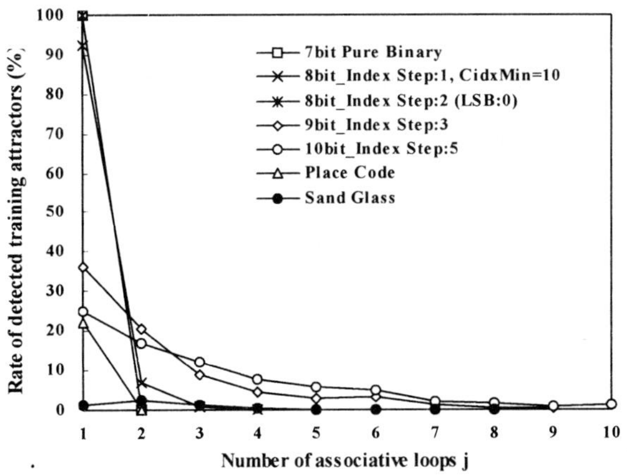

Fig.5. Rate of detected training attractors in HD10 validation data versus number of associative loops.

3 to 5. The recognition rate includes spurious and diverged attractors as the false recognition and therefore corresponds to the generalization rate. Various types of consecutive codification provide almost the same recognition rate as shown in Fig.3.

The coarse codification of the Place Code provided a recognition rate superior to those from consecutive codification due to the robust mapping from the hidden layer output to the coarse binary output of the output layer.

The coarseness however usually produces the spurious attractors, as shown in Fig.4. Discrimination of the spurious attractors providing a false category output can resolve the defect to process them in the higher-hierarchical network.

The consecutive codification provided very rapid attracting ability, and produced only training attractors with no spurious attractors, as shown in Fig.5. The discontinuous codification on the other hand produced a large amount of spurious attractors, resulting in degradation of the recognition rate. Discontinuous codification of Q=5 largely degraded the attractor absorption, resulted in divergence at J=10.

The Place Code always achieved a very high percentage of correctness of one-shot training attractors at j=1. This percentage was as hihg as 99% as a result of drawing almost all ambiguous associative inputs into spurious attractors.

The sand glass type single cyclic-associative network storing 128 training attractors yielded poor recognition rats, even though an optimum number of 75 hidden units was used for the generalization rate, as shown in Fig.3. This is a reason why the number of 128 of the memorized states is larger than the number of 75 of input layer units, providing a storage greater than the random memory capacity for the prototypes [7].

The input feature space was not widely and separately mapped into the space formed by the hidden layer to provide coarse binary outputs through the prototype-learning, even if a larger number of 200 of the hidden units were used in the spindle type network.

5.2 Cumulative learning of associative networks

Table 2 summarizes the simulated results of the cumulative learning. The 200 hidden units in the spindle type cyclic-associative network were required in order for the cumulative learning to achieve relatively rapid convergence and higher generalization.

Figure 6 shows the effect of cumulative learning on the recognition rates. The cumulative learning yielded a drastic improvement in recognition rates of the spindle type single cyclic-associative network.

Both the successfully separable mapping from the input feature space to the huge space of the hidden layer and the robust mapping from the hidden layer space to the coarse output space were performed by cumulative learning of ambiguous data adjacent to prototypes as the additive training data. The huge amount of additive training data adjusted the category boundaries in the hidden layer under the modified error

Table 2 Simulated results of the cumulative learning.

Networks	Total number of epochs	Prototypes M / Additive training data
Forward net: Consecutive codes, Prototype-learning	65	128
		0
Forward net: Consecutive codes, Cumulative learning	108	128
		560
Forward net: Place Codes, Prototype-learning	157	128
		0
Forward net: Place Codes, Cumulative learning	198	128
		12
Sand glass type: Prototype-learning	40	128
		0
Spindle type: Cumulative learning	1066	128
		3653

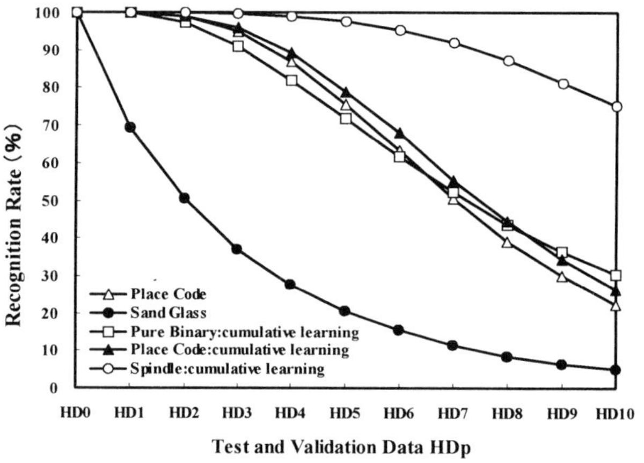

Fig.6. Recognition rate of hetero- and spindle type associative networks having cumulative learning for test and validation data.

back-propagation algorithm with error perturbation [6] to expand the prototype data storage capacity, providing superior associative performance. The improved generalization property also effectively reduces spurious attractors.

In the hetero-tandem associative network, all outputs providing spurious attractors and diverged outputs of the spindle and sand glass type networks were absorbed into one-shot training attractors by the subsequent hetero-associative network with consecutive codification. These results occurred with 68.1% correct category outputs in spurious attractors and diverged outputs of the HD2 (Hamming distance) validation data and 14.1% in them of the HD1 data. Here, the significantly highest recognition rate was achieved, while maintaining almost the same properties as in the spindle type associative network with cumulative learning.

Recognition rates of hetero-associative networks are also described in Fig.7 for a Chinese character set given by Y. Wu [2]. The numbers of prototypes of M and input elements of R are respectively 16 and 63. Sand glass type associative networks by LII methods by Y. Wu have no feedback loop and utilize optimally computed weights to provide close winner-takes-all patterns in the hidden layer (17 units) and eliminate all spurious attractors. An excellent recognition rate was obtained for data in the vicinity of the prototypes whereas proportional rapid degradation was however observed as the Hamming distance from the prototype increased.

The sand glass type single cyclic-associative network with 50 units of the hidden layer achieves a very high recognition rate due to the small number of prototypes in comparison with the number of input units, and so requires only a small memory capacity. It is therefore not necessary to use a spindle type network.

Hetero- and hetero-tandem associative networks with cumulative learning again achieved a large improvement and superior performance over a wide range of Hamming distances from the prototypes. In the hetero-tandem associative network with cumulative learning, only 40 additive training data in the sand glass type network and 91 in the forward network of the subsequent hetero-associative network were used respectively.

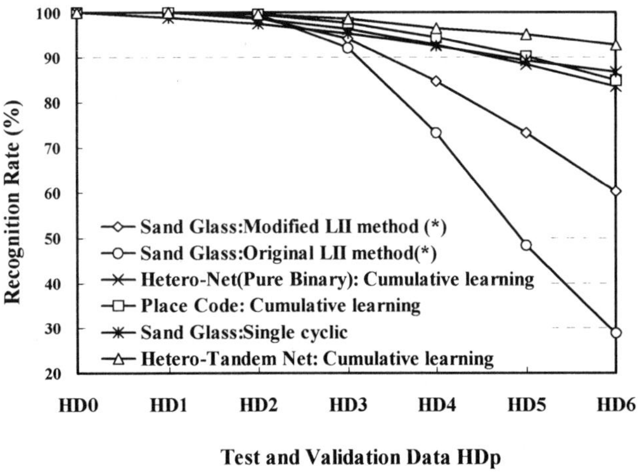

Fig.7. Recognition rate of hetero-associative networks for Chinese characters given in [2], and sand glass type associative networks in [2] are also given by *.

7 Conclusion

We propose hetero-associative networks consisting of 3-layer neural networks with classification, recall of prototypes and attractor discrimination abilities. These networks include a hetero-tandem associative network having a spindle type single cycle-associative network connected in tandem with the subsequent hetero-associative network.

Consecutive codification with a compact code density of outputs of a forward neural network in the hetero-associative network provides very rapid attracting ability and produces only training attractors. Discontinuous codification on the other hand produces a large amount of spurious attractors that lower the recognition rate. These spurious and diverged outputs are successfully discriminated by the literal indices of the distributed binary output codes.

A spindle type single cyclic-associative network and hetero-associative networks having coarseness in binary codification of 3-layer networks with a large amount of hidden layer units offer large capacity memory storage and also achieve very high recognition rate through cumulative learning with a modified error back-propagation algorithm.

Simulations also indicate that the hetero-tandem associative network produces only training attractors with excellent associative properties, along with the cumulative learning and the consecutive codification.

8 References

[1] Hassoun, M. H. (1993) Dynamic Associative Neural Memories. In: Hassoun, M. H.(eds.) Associative Neural Memories, Theory and Implementation 1993. Oxford University Press, pp.3-27.

[2] Yingquan Wu and Stella N. Batalama. (2001) Improved One-Shot Learning for Feedforward Associative Memories with Application to Composite Pattern Association. IEEE Trans. on Systems, Man, and Cybernetics-Part B: Cybernetics, 31, 1:119-125.

[3] Amari, S. Yanai, H. (1993) Statistical Neurodynamics of Various Types of Associative Nets. In: Hassoun, M. H.(eds.) Associative Neural Memories, Theory and Implementation 1993. Oxford University Press, pp.169-183.

[4] deCallatay, A. M. (1986) Natural and Artificial Intelligence Processor Systems Compared to the Human Brain 1986. Elservier Science Publishers B. V.

[5] Yatsuzuka, Y., Enomoto, M. (1996) A Binary Three Layer neural Network with Switched Error Perturbation and Reiterative Learning Utilizing the Generalization Property. ANNIE'96, 6:35-44.

[6] Yatsuzuka, Y., Sugiyama, T. (2002) Parallel Neural Networks by Actively Repetitive Learning with Autonomous Label Assignment and Majority Voting. Proc. of the First International AISO Congress on Autonomous Intelligent Systems, 10028-02-YY-111.

[7] Amari, S. and Maginu, K., (1988) Statistical neurodynamics of associative memory. Neural Networks, 1:63-73.

Crack width prediction of RC structures by Artificial Neural Networks

Carlos AVILA[1], Yukikazu TSUJI[1], Yoichi SHIRAISHI[2]

[1]Department of Civil Engineering, Gunma University, Japan
[2]Department. of Computer Sciences, Gunma University, Japan
{cavila67, tsuji}@civil.ce.gunma-u.ac.jp, siraisi@keim.cs.gunma-u.ac.jp

Abstract

This paper proposes the use of Artificial Neural Networks (ANN) for the prediction of the maximum surface crack width of precast reinforced concrete beams joined by steel coupler connectors and anchor bars (jointed beams). Two different training algorithms are used in this study and their performances are compared. The first approach used Back propagation (BPANN) and the second one includes Genetic Algorithms (GANN) during the training process. Input and output vectors are designed on the basis of empirical equations available in the literature to estimate crack widths in common reinforced concrete (RC) structures and parameters that characterize the mechanical behavior of RC beams with overlapped reinforcement. Two well-defined points of loading are considered in this study to demonstrate the suitability of this approach in both, a linear and a highly nonlinear stage of the mechanical response of this type of structures. Remarkable results were obtained, however, in all cases the combined Genetic Artificial Neural Network approach resulted in improved prediction performance over networks trained by error back propagation.

1 Introduction

Previous studies show applications of artificial neural networks in the civil engineering field However, as far as the authors know, this is the first time that this type of approach is applied to the nonlinear cracking phenomenon of reinforced concrete structures and particularly to those structures joined by mechanical connectors.

This paper briefly describes the experimental program. It then describes the modeling methodology that is used to forecast the maximum crack width of jointed beams.

2 Experimental program

2.1 Materials and specimens description

Twenty four full scale precast RC beams joined by sets of a cast iron coupler and anchor bars which hereinafter is called AC-system were cast "ref.1". The specimens have different geometrical properties, reinforcement and anchor bars configuration according with Fig.1 (a), Fig.1 (b). Mechanical properties of concrete and AC-System are specified in "ref.1".

The specimens were incrementally, vertically and statically loaded to failure. Cycles of loading and unloading were applied to simulate a service load condition in an attempt to study the external cracking response of jointed beams. Maximum surface crack width is evaluated at an anchor bars tensile stress of 200 MPa and 300 MPa as well. Herein after, called 200 MPa cycle and 300 MPa cycle respectively.

A summary of the experimental results is shown in Table 1.

3 Neural network modeling

The first approach makes use of a feed forward artificial neural network with error back propagation as training algorithm, chosen primarily for its simplicity. Genetic Algorithms are used in the second approach to select the parameters that control learning in the Artificial Neural Network models. GA's were implemented to increase the rate of learning "ref.2" taking into account the complexity of the problem.

3.1 Artificial Neural Network's topology

Similar procedures were run in order to establish an appropriate ANN topology in both approaches. This process consists of evaluating the performance of the network when the number of hidden neurons is modified. Fig.2 shows the relation between the number of hidden neurons and BPANN's response evaluated for the testing stage in terms of the correlation coefficient(r). As was expected, the above-mentioned figure demonstrates that not necessarily a larger number of neurons in the hidden layer gives as a result a much more accurate neural network response since over fitting can be caused by a large number of connections. Table 2 shows the topology of the BPANN's and GANN's for 200 MPa and 300 MPa cycles adopted in this study because of the above-mentioned process.

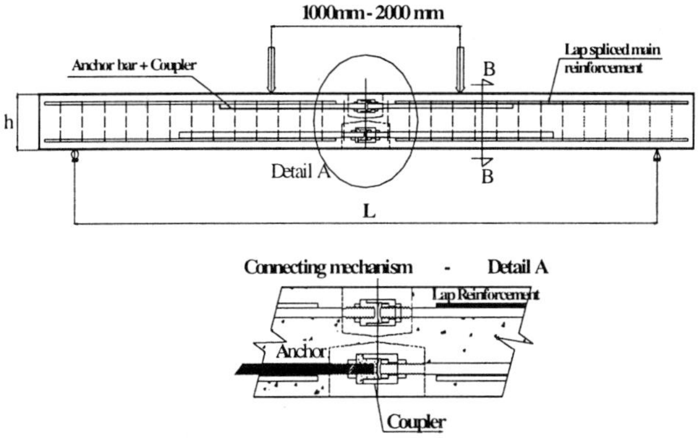

Fig. 1(a) Details of jointed beams

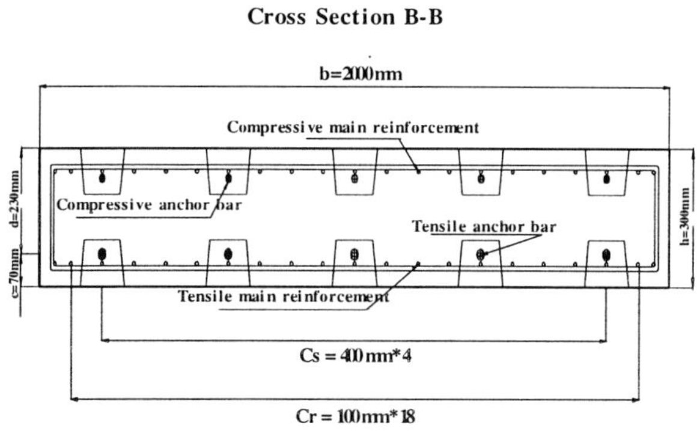

Fig. 1(b) Cross section beam No.01 – S1

Table 1 Data sets summary

Beam	EXPERIMENTAL RESULTS			CRACKS ANN I/O VECTORS					
	Mu [kN.m]	W_{200MPa} [mm]	W_{300MPa} [mm]	200 MPa cycle			300 MPa cycle		
				INPUT		OUTPUT	INPUT		OUTPUT
				Eq. (1)	Eq. (2)	Exp.	Eq. (1)	Eq. (11)	Exp.
				w JSCE	Cs / Cr	w_{200MPa}	w JSCE	Cs / Cr	w_{300MPa} / c
No.01	235	0.151	0.600	0.1068	0.2375	**0.151**	0.4270	0.2375	**0.0104**
No.02	326	0.279	0.717	0.1061	0.1782	**0.279**	0.3902	0.1782	**0.0122**
No.03	215	0.181	0.414	0.1292	0.3166	**0.181**	0.4403	0.3167	**0.0072**
No.04	197	0.236	0.834	0.1180	0.2500	**0.236**	0.9107	0.2500	**0.0145**
No.05	231	0.161	0.546	0.0879	0.1900	**0.161**	0.3535	0.1900	**0.0093**
No.06	128	0.038	0.120	0.0315	0.3350	**0.038**	0.1157	0.3351	**0.0022**
No.07	219	0.354	1.051	0.2140	0.2010	**0.354**	0.7328	0.2011	**0.0178**
No.08	204	0.456	1.024	0.3361	0.2231	**0.456**	0.8748	0.2232	**0.0174**
No.09	230	0.483	1.045	0.2910	0.2510	**0.483**	0.7974	0.251	**0.0177**
No.10	198	0.132	0.608	0.1159	0.4068	**0.132**	0.4661	0.4069	**0.0106**
No.11	212	0.274	0.744	0.1310	0.2380	**0.274**	0.5624	0.2375	**0.0126**
No.12	139	0.031	0.575	0.0548	0.1900	**0.031**	0.5098	0.1900	**0.0097**
No.13	179	0.458	1.273	0.2952	0.2010	**0.458**	0.9172	0.2011	**0.0209**
No.14	177	0.512	1.292	0.3770	0.2375	**0.512**	0.9156	0.2375	**0.0212**
No.15	168	0.421	1.035	0.2096	0.4021	**0.421**	0.6713	0.4021	**0.0180**
No.16	188	0.142	0.574	0.1210	0.4020	**0.142**	0.4507	0.4021	**0.0100**
No.17	129	0.013	0.723	0.0463	0.2010	**0.013**	0.8087	0.2011	**0.0123**
No.18	231	0.227	0.641	0.113	0.258	**0.227**	0.286	0.258	**0.0085**
No.19	239	0.132	0.496	0.110	0.300	**0.132**	0.235	0.300	**0.0066**
No.20	220	0.196	0.533	0.112	0.360	**0.196**	0.297	0.360	**0.0071**
No.21	306	0.315	1.008	0.224	0.300	**0.315**	0.256	0.300	**0.0126**
No.22	320	0.180	0.588	0.110	0.292	**0.180**	0.287	0.292	**0.0078**
No.23	179	0.194	0.709	0.103	0.212	**0.194**	0.267	0.212	**0.0094**
No.24	173	0.249	0.667	0.126	0.243	**0.249**	0.281	0.243	**0.0088**

Notation: **Mu**: Ultimate bending moment; **w**: Maximum surface crack width; **I/O**: Input/Output vectors
w_{200MPa}, w_{300MPa}: Crack width at the 200MPa and 300MPa cycle, **c:** Concrete cover thickness **Cs,Cr** as Fig.1(b)

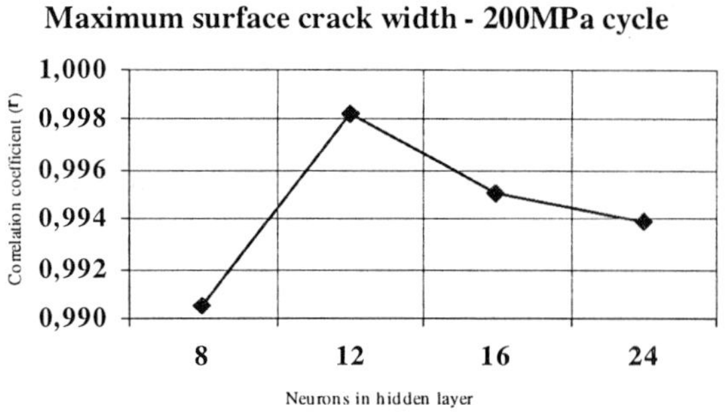

Fig. 2 r vs. neurons in the hidden layer

Table 2 Neural Network Topology

	200 MPa	300 MPa
BPANN	2-12-1	2-24-1
GANN	2-60-1	2-72-1

Here, 2-12-1: Neurons in input layer = 2,
Neurons in hidden layer = 12,
Neurons in output layer = 1.

3.2 Data set division

Table 1 reports the summary of experimental results of ultimate bending moment and maximum surface crack width at service load conditions. This study follows the criteria given by Shahin "ref.3" for data set division. The application of BPANN's and GANN's to study the flexural response of jointed beams uses seventeen beams for training and seven beams for verification. It is about 70% of the available data for training and about 30% of the data set for testing.

3.3 Mapping relationships

Identical mapping relationships are used for both models. Due to the nature of the experimental data, certain differences have been established for the input and output vectors of the 300MPa cycle.

3.3.1 Crack width prediction – 200 MPa cycle

The design of input and output vectors is based on Eq. (1) and Eq. (2). Eq. (1) is recommended by the Japan Society of Civil Engineers (JSCE) "ref.4" to evaluate maximum surface crack width of conventional RC beams undergoing flexural actions. The left term of this equation and the relation between the anchor bars and main reinforcement spacing evaluated by Eq. (2) are used to feed the input layer. Eq. (2) is proposed in this paper to include in the ANN model the effect of reinforcement spacing in noncontact lap splices to the cracking response of RC structures. The experimental value of maximum surface crack width of jointed beams is utilized as output to train the network. The aim is to find a mapping relationship between the input vector and the desired output during training before forecasting the maximum surface crack width for an untrained input instance during verification.

$$\{4.c + 0.7(c_s - \phi)\}.\varepsilon_s \Rightarrow w \quad (1)$$

$$S = \frac{C_r}{C_s} \quad (2)$$

Where, S: Reinforcement spacing relation,
w: maximum surface crack width,
c: concrete cover thickness,
C_s: anchor bars spacing,
C_r: main reinforcement spacing,
ϕ: anchor bars diameter,
ε_s: anchor bars strain at the middle span

Table 3 reports the neural network response for the testing data set of both training methods. 30000 epochs where used as stopping criteria for the standard back propagation algorithm and it took 34 seconds on a 1GHz personal computer. On the other hand, the GA training obtains its better population of connection weights at 29000 epochs. However, it took 41 seconds, to achieve this result. As it could have been expected, longer computer processing time is required. Nevertheless, the GANN shows a significant improvement (Average = 3%) with respect to its counterpart the standard back propagation ANN regarding the high level of nonlinearity present in this phenomenon.

3.3.2 Crack width prediction – 300 MPa cycle

The experimental value of maximum surface crack width for the 300 MPa cycle reported cracks larger than 1.0mm. This fact introduces an additional step into the data set processing stage. As can be observed in Table 3 and Table 5 the output vector was normalized in terms of the cover thickness (**c**). It responds to the fact that the sigmoid function used for each neuron as transference function modulates their outputs into the close interval [0,1] "ref.5". The learning process (Fig. 3) uses the left term of Eq. (1) evaluated at 300MPa and the value given by Eq. (2) as input. On the other hand, the experimental value of maximum surface crack width normalized with respect to the concrete cover thickness (**c**) is used as output vector during training.

In the case of the standard back propagation training, 30000 epochs were used as stopping criteria for the training process, which took approximately 13 seconds. Its counterpart the ANN with genetic training needed 30000 epochs and 30 seconds to give its results. Table 4 shows the predicted values of maximum surface

Table 3 Crack width prediction at 200MPa cycle

	Eq. (1) Left term	Eq. (2) Cr/Cs	W_{200MPa} [mm] Exp	Prediction ANN	GANN
No.03	0.129	0.317	**0.181**	**0.232**	**0.222**
No.04	0.118	0.250	**0.236**	**0.217**	**0.221**
No.05	0.088	0.190	**0.161**	**0.158**	**0.160**
No.13	0.295	0.201	**0.458**	**0.439**	**0.448**
No.17	0.113	0.258	**0.227**	**0.200**	**0.202**
No.18	0.110	0.300	**0.132**	**0.171**	**0.171**
No.24	0.103	0.212	**0.194**	**0.191**	**0.194**

W_{200MPa}: Maximum surface crack width at 200MPa cycle

Table 4 Crack width prediction at 300MPa

	Eq. (1) Left term	Eq. (2) Cr/Cs	W_{300MPa}/C experiment	prediction ANN	GANN
No.04	0.9107	0.250	**0.0145**	**0.0145**	**0.0145**
No.08	0.8784	0.223	**0.0174**	**0.0151**	**0.0187**
No.12	0.5098	0.190	**0.0097**	**0.0126**	**0.0122**
No.16	0.4507	0.402	**0.0100**	**0.0107**	**0.0100**
No.19	0.235	0.300	**0.0066**	**0.0068**	**0.0071**
No.23	0.267	0.212	**0.0094**	**0.0084**	**0.0092**
No.24	0.2810	0.243	**0.0088**	**0.0082**	**0.0088**

W_{300MPa}: Maximum surface crack width at 300MPa cycle;
C: concrete cover thickness

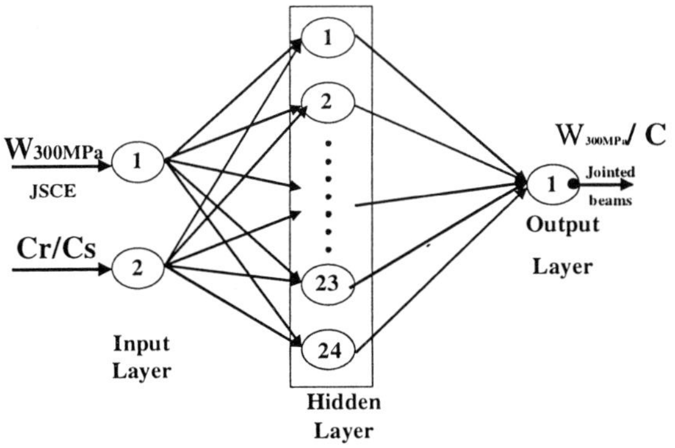

Fig. 3 ANN for forecasting crack width at 300MPa

crack width for the 300 MPa cycle, where an average improvement of 7% achieved with the GANN can be clearly observed.

The 300MPa case represents a stage of loading close to the yielding of reinforcement. It presents a much higher level of nonlinearity than the case of 200MPa, so that both cases have been treated separately and the larger number of neurons reported for the 300MPa cycle may be a consequence of this fact.

4 Conclusions

1) The genetic training resulted in improved prediction performance over the predictions of its counterpart the network trained by standard error backpropagation. An average of 7% for the 300MPa is significant considering the high level of nonlinearity present at this level of stresses.

2) A combination of ANN's and equations proposed for normal beams proved to be suitable to predict a nonlinear phenomenon such as the cracking behaviour of jointed beams in spite of the limitations imposed by those equations.

3) The case of 300MPa requires more neurons than those required for the 200MPa. It is probably due to the higher level of nonlinearity presented at the 300MPa case with respect to the 200MPa case.

Acknowledgement

The research presented in this paper is based on the experimental work conducted in KAIEI-TECHNO Co. Ltd. Japan.

References

[1] Avila, C., Tsuji, Y., Morita, T., and Iizuka, Y. (2003) Flexural behavior of precast RC beams joined by cast iron couplers. Cement Science and Concrete Technology, No.57: 683-690

[2] Montana, D., Davis, L. (1989) Training Feed forward Neural Networks using Genetic Algorithms. Proceedings of the International Joint Conference on Artificial Intelligence: 762-767.

[3] Shahin, M., Maier H. and Jaksa, M. (2000) Evolutionary data division methods for developing artificial neural network models in geotechnical engineering, Department of Civil & Environmental Engineering. The University of Adelaide. Research report No. R.171.

[4] JSCE (2002) Standard Specifications for Concrete Structures - Structural Performance Verification JSCE, Japan (in Japanese).

[5] http://www.compapp.dcu.ie, The backpropagation learning algorithm, School of Computer Applications, Dublin City University

A neural network system for modelling of coagulant dosage used in drinking water treatment

B. Lamrini[1], A. Benhammou[1], A. Karama[1] and M-V. Le Lann[2]

[1] Laboratoire d'Automatique et d'Etude des Procédés, Faculty of Sciences Semlalia, POB:2390, 40000-Marrakech, Morocco
E-mail : {lamrini.b, benhammou, karama}@ucam.ac.ma
[2] LAAS/CNRS, 7, Avenue du Colonel Roche, 31077-Toulouse cedex 4, France
and INSA, DGEI, 135, Avenue de Rangueil, 31077-Toulouse cedex 4, France
E-mail: mvlelann@laas.fr

Abstract

This paper presents the elaboration and validation of "soft sensor" using neural networks for on-line estimation of the coagulation dose from raw water characteristics. The main parameters influencing the coagulant dosage are firstly determined via a PCA. A brief description of the methodology used for the synthesis of neural model is given and experimental results are included. The training of the neural network is performed using the Weight Decay regularization in combination with Levenberg-Marquardt method. The performance of this soft sensor is illustrated with real data.

1 Introduction

Drinking water is the most essential element for mankind. The coagulation process which requires the addition of chemical coagulant is the critical process in the drinking water treatment. The control of a good coagulation is essential for maintenance of satisfactory treated water quality and economic plant operation. Basically, coagulant dosage is chosen empirically by operators based on their past experience, laboratory jar-testing and various information on water quality parameters. The jar-test apparatus simulates mixing, flocculation, setting, and a single test may take about one hour to be performed. Disadvantages associated with jar-testing are that regular samples have to be taken requiring manual intervention and operators can make manually in raw water quality. There is no mechanistic model describing the coagulant dosage related to the different variables affecting the process. Consequently, there is a need for a fast and reliable method for determining the required coagulant rate which can be used instead of the jar-test analysis.

Artificial neural networks are ideal solution for our problem because of their features. They take data and learn underlying relationships, even if the relationships are difficult to find and describe. They do not require a priori precise knowledge on the relationships of the process variables. Various applications of these models have been recently reported in the drinking water treatment industry (as examples: the forecasting of drinking water [1,2], the prediction of the coagulant dosing [3, 4, 5]).

This investigation aims to develop a soft sensor for the on-line estimation of optimal coagulant dosage from raw water characteristics. Previous research show the efficiency of a such approach using neural networks [3, 4].

This paper is organised as follows: First, a brief description of neural network is presented. Then the drinking water treatment process is described. In the subsequent section, the neural software sensor for the on-line estimation of the coagulant dosage is developed. Finally, experimental results are presented.

2 Neural Networks

Neural networks are known to be able to successfully represent complex functions in various fields. There are a wide variety of artificial neural networks existing in the literature, of which the feedforward structure is one of the most commonly used in modelling and control. Feedforward neural networks, such as the multilayer perceptron, consist usually of many simple processing elements arranged in layers as shown in Figure 1. Each element takes its input from the weighted sum of the outputs of the elements of the previous layer. This input is then passed through a nonlinear function, often called the activation function, to form the element's output .

In this study, a three layer feedforward neural network has been adopted : it has been shown that is a network with a single hidden layer (as shown in figure. 1) can simulate any continuous function.

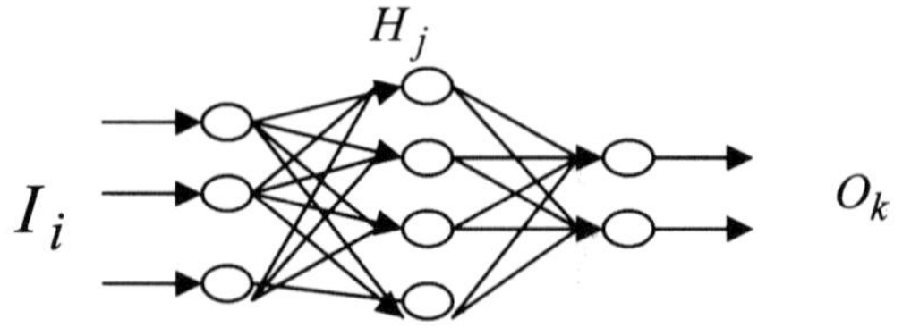

Fig 1. Feedforward neural network

The neural network models consist of the following set of equations:

$$O_k = f_k\left(\sum_j w_{jk} H_j\right) \text{ and } H_j = f_j\left(\sum_i w_{ij} I_i\right) \tag{1}$$

where O_k denotes the network outputs, H_j the hidden neurons outputs, I_i the network inputs, w_{ij} the weights between the input layer and the hidden layer and w_{jk} the weights between the hidden layer and the output layer.

Training of the neural network involves adjusting the weights w_{ij} and w_{jk} by using the backpropagation learning algorithm [6], so that the network emulates the non-linear function underlying the training data set. The network weights are adjusted by minimising the following criteria derived from the difference between real and neural outputs respectively t and O.

$$C = \frac{1}{N}\sum_{i=1}^{N}(t_i - O_i)^2 \tag{2}$$

3 Water treatment process

The plant of drinking water treatment concerned by this study is the drinking water treatment plant Rocade located in Marrakech. It provides water to more than 1,5 millions inhabitants. Raw water is extracted from the channel Rocade. In case of resource failure (raw, pollution...), the treatment plant takes the raw water from a pumping plant Takerkoust. 60% of city needs are assured by the treatment plant, the complement is brought by the underground resources (well, drilling…). It has a nominal capacity to process 1400 l/sec of water. The treated water is stored in two tanks and transported toward the water supply network.

The drinking treatment plant involves physical and chemical processes. The figure 2 presents a schematic overview of the various operations necessary to treat the raw water at the Rocade water treatment plant of Marrakech. The treatment consists in essentially of first disinfection, coagulation-flocculation, settling, filtration and final disinfection.

4 Prediction of Coagulant Dosage

The coagulation process involves many complex physical and chemical phenomena which are difficult to model using mechanistic and chemical phenomena traditional description. The coagulant dose ensuring optimal treatment efficiency has been shown experimentally to be non-linearly correlated to raw water characteristics which are usually available on line.

A neural soft sensor for the prediction of coagulant dosage is developed in too stages. The Factorial Analysis in Principal Component method (APC) is firstly applied to determine the main parameters affecting the prediction of the optimal coagulant dosage. These parameters will be then considered as the input variables of the neural model for which the training algorithm will be performed.

As far as the dimensional analysis is concerned, 9 describers of the raw water quality (temperature, pH, turbidity, total carbonates, total suspend solids (TSS), oxydability, dissolved oxygen, conductivity and the coagulant dosage (passive parameter)) are used. A number of 89 samples have been used like individuals. Every sample underwent different physical and chemical analysis as well as to the jar-testing to determine the coagulant rate. The process of simplification would allowed in the plan to extract the following variables: temperature, pH, TSS, dissolved oxygen and conductivity.

4.1 Neural soft sensor training

The neural network used for modelling of the coagulant dosage is a MLP type. Note that the network inputs are the observed values of the retained raw water quality parameters.

For the determination of the architecture network, the pruning approach "*Weight Decay*" [7, 8] is used, starting from a relatively large network then removing connections in order to arrive at a suitable network architecture [9, 10]. This approach, allowing to eliminate the weak weights, consists in adjusting the weights using the new performance function instead of C defined in (2):

$$C'(w) = C(w) + \alpha\,\Omega(w) \tag{3}$$

Where $$\Omega(w) = \frac{1}{n}\sum_{i=1}^{n} w_i^2 \tag{4}$$

n is the number of weight network, α is a parameter that determines the importance of the two terms in the new performance function $C'(w)$. Using this performance functions will cause the network to have smaller weights and biases, and this will force the network response to be smoother and less likely to overfit. This method presents the advantage to be simple to implement, since the gradient of C' can be very easily calculated from the gradient of C and from network weights. It is sufficient to add the quantity αw to the gradient vector ∇C calculated by the Back-propagation algorithm:

$$\nabla C' = \nabla C + \alpha\, w \tag{5}$$

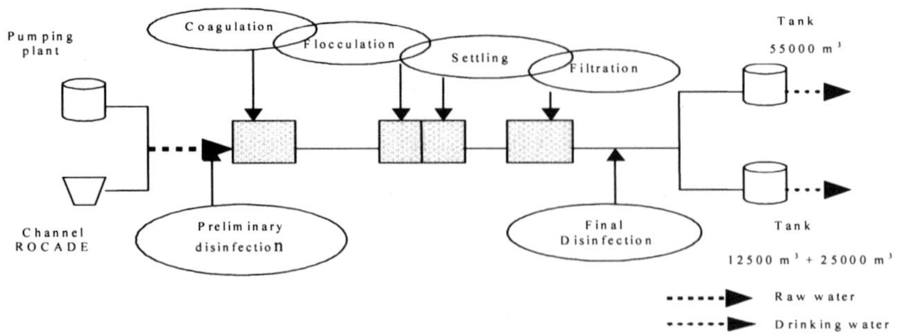

Fig. 2. Simplified synopsis of water treatment plant

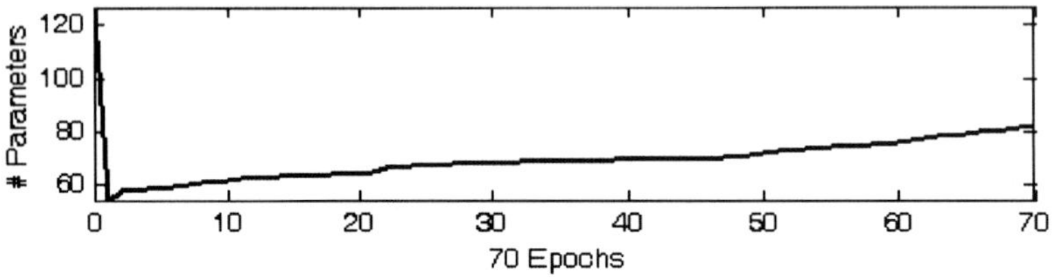

Fig. 3. Effective number of parameters = 81.9283

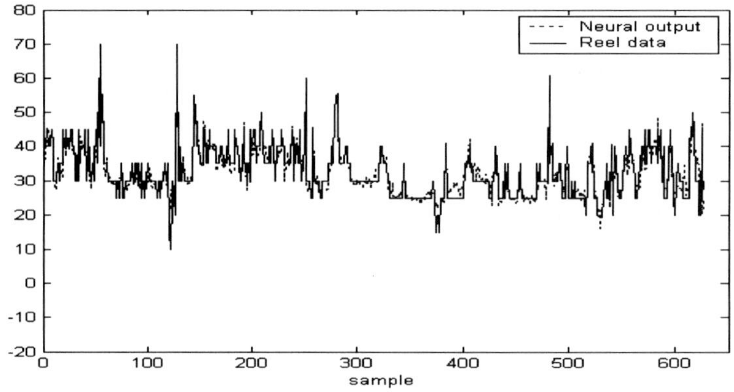

Fig. 4. Neural coagulant dosing rate with respect to the reel data on the validation data.

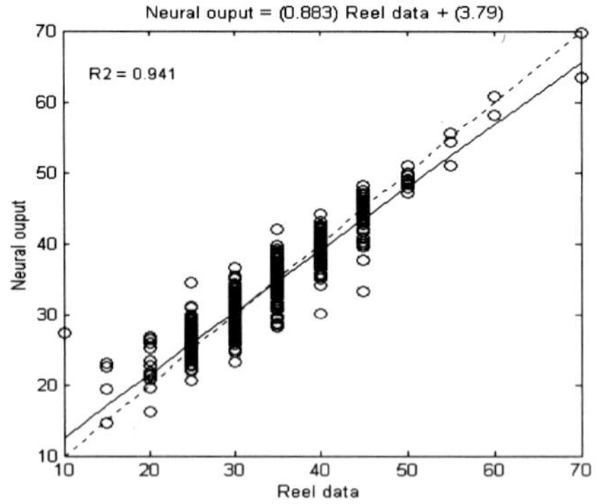

Fig. 5. Correlation between neural coagulant dosing rate and real data.

5 Results and discussion

The experimental data of four years (2511 samples, from January 2000 to July 2003) have been used to establish the neural model as the basis of a coagulant dosing estimation software. For the conductivity and the oxygen dissolved, invalid data have been removed and missing ones have been replaced.

75% of the global data is used for the network training and the remainder (25%) for validating the neural model. The obtained networks consist of five inputs, a single hidden layer with 18 sigmoidal neurones and one linear output. The neural model have been built using the regularisation method in combination with Levenberg-Marquardt training algorithm. It has 127 connections in the beginning of training. The advantage of this algorithm is that it provides a measure of how many weights of the network are efficiently used by the network. In our case, the final qualified network uses approximately 82 weights (fig 3), out of the 127 total weights in the 5-18-1 network.

The figure 4 shows the validation of the neural model. We notice that the coagulant dosage computed with the neural network model is very smoothed to the real data. Consequently, the neural network generalizes well to new data. The obtained sum of the squared error is 0.007. Figure 5 shows the correlation between neural output and real data. The correlation coefficient computed on the validation set is equal to 0.94.

6 Conclusion

A software sensor based on neural network model has been described. We showed in this paper that the coagulant dosing is non-linearly correlated to the raw water characteristics such as TSS, temperature, pH, conductivity and dissolved oxygen. Experimental results using the data raw water plant showed the efficiency and soundness of this approach. The performance of the network depend on the quality and the completeness of data provided for training the system. To take into account of the uncertainty bound, we expect to apply the Bootstrap sampling approach for the evaluation of uncertainty measure related to the neural prediction.

References

[1] Canu, S., Ding, X., Granvalet, Y (1997) Statistiques et Méthodes Neuronales : Une application des réseaux de neurones pour la prévision à un pas de temps. Thiria, S., Lechevallier, Y., Gascul, O., Canu, S. (eds) Chapitre 7, Dunod, Paris, pp. 120-131.

[2] Canu, S., Sobral, R., Lengellé, R (1990) Formel neural network as an adaptative model for water demand. Proceedings of INNC'90. Kluwer Academic Publishers 1: 131-136.

[3] Gagnon, C., Grandjean, B.P.A, Thibault, J (1997) Modelling of coagulant dosage in a water treatment plant. Artificial Intelligence in Engineering 11: 401- 404.

[4] Adgar, A., Cox, C.S., Daniel, P.R., Lowdon, A (1996) Chemical dosing philosophies for a water treatment plant: results of some pilot plant experimentation. In Proceeding of the IEE Conference Publication 427(2): 1052-1057.

[5] Evans, J., Enoch, C., Johnson, M., Williams, P (1998) Intelligent based auto-coagulation control applied to a water treatment works. In Proceedings of International Conference on Control: 141-145.

[6] Rumelhart, D.E., McClelland, J.L (1986) Parallel distribution processing: exploration in the microstructure of cognition. Cambridge. MA: MIT Press 1.

[7] Mackay, D.J.C (1992a) Bayesian interpolation. Neural Computation 4: 415-447.

[8] Mackay, D.J.C (1992b) A pratical Bayesian Framework for backpropagation Networks. Neural Computation 4: 448-472.

[9] Hinton, G.E (1987) Learning Translation Invariant Recognition in Massively Parallel Networks. Proceedings PARLE Conference on Parallel Architectures and Languages Europe. Bakker, J.W., Nijman, A.J., Treleaven, P.C. (Eds.), Springer-Verlag, Berlin, pp.1-13.

[10] Gallinari, P., Cibas, T (1999) Practical complexity control in multilayer perceptrons. Signal Processing 74: 29-46.

ANN modeling applied to NO_x reduction with octane. A^{nn} future in personal vehicles

Mats Rönnholm[1], Kalle Arve[1], Kari Eränen[1], Fredrik Klingstedt[1], Tapio Salmi[1], Henrik Saxén[2]

[1]Laboratory of Industrial Chemistry, Process Chemistry Centre,
[2]Heat Engineering Laboratory
Åbo Akademi University, FIN-20500 Åbo/Turku, Finland
Tel: +358-2-2154431, Fax: +358-2-2154479, e-mail: mats.ronnholm@abo.fi

Abstract

A silver/alumina catalyst was tested for its NO_x reduction activity during oxygen-rich conditions and during variation in the input parameters (nitric oxide, octane and oxygen). A multi-bed approach was tested where the initial bed was divided into four beds acting in different temperature rages. The experimental data were investigated by means of artificial neural networks that were demonstrated to be able to model the process.

1 Introduction

Combustion of fossil fuels in vehicle engines produces environmentally harmful or unwanted exhaust gases, such as nitrogen and carbon oxides. Fuel-efficient engines operate under lean conditions, i.e. with a large excess of oxygen. This results in an enhanced formation of nitrogen oxides (NO_x), mainly nitrogen monoxide (NO) and nitrogen dioxide (NO_2), which should be converted to nitrogen (N_2). We are investigating a potential method of *selective reduction of NO by hydrocarbons* (HC-SCR) in the presence of excess oxygen, which is capable of removing NO_x from different exhausts.

Among all environmental problems, NO_x elimination from the exhaust of diesel and lean-burn otto engines is one of the most urgent. As the HC-SCR-method, in contrast to the alternative, i.e. complicated NO_x-storage technique, can be considered an ultimate solution to the catalytic after-treatment of exhaust gases from these lean-burn engines, this is the subject of our present investigation. We have used artificial neural networks to model the still complex behavior of the HC-SCR, some results of which are shown in Figure 1.

2 Experiments and ANN approach

2.1 Setup of the experiments

Initially, separate activity tests were performed with one catalyst bed and with a multi-bed reactor concept, where one bed was divided into four beds in the temperature range 300 °C - 550 °C with steps of 50 °C. In these experiments data points were obtained by 1) varying either nitrogen monoxide or octane concentration, keeping the concentration of O_2 constant and 2) by changing the C_8H_{18}/NO concentrations pair-wise, keeping the O_2 concentration constant and 3) by keeping both NO and C_8H_{18} constant but varying the incoming O_2 concentration. The following gas concentrations were applied: 375 – 500 – 1000 ppm NO, 134 – 187.5 – 375 – 500 – 750 ppm C_8H_{18} and 1 – 12 vol.% O_2.

2.2 Reaction mechanism

On the basis of earlier studies [1,2], the simplified heterogeneous-homogeneous reaction mechanism of Fig. 3 is proposed for the system at hand. In this mechanism different surface species, such as nitro-compounds, R-NCO and nitriles, are formed through reaction of hydrocarbons with NO and O_2 over the Ag/alumina, and are further transformed into amine and/or ammonia, which desorb into the gas phase. The ammonia and amine species react in the gas phase together with activated NO_x species to form N_2. The activation of NO takes place over the catalyst and is accelerated by suitable reducing agents. Thus, the gas phase reaction is a crucial part of the complete HC-SCR mechanism.

2.3 Neural Network Modeling

The results obtained from the experiments were investigated using a neural network model. Training of the networks were performed by means of the NNDT software [3], using cross validation to verify the models. The experiments with four beds were divided into one group with 83 experiments for training and another group with eight experiments for testing. All networks used had four inputs and four outputs and sigmoidal activation of the nodes in the hidden layer. The inputs were the temperature and the incoming concentrations of NO, octane and O_2, while the outputs to be predicted were the outgoing concentrations of N_2, CO, O_2 and CO_2.

In an earlier study with the single-bed concept [4] neural networks were found to provide a good description of the one-bed reactor concept. The present study focuses on the possibilities to further optimise the reactor by using a more accurate description of the process. By an accurate (neural) model it is possible to monitor the measurements during an ongoing experimental plan, to discover outliers in the data and to improve and revile reaction structures. The last issue is important since the interacting gas-phase and the heterogeneous chemical reaction network can be continuously improved using different concepts of structure and chemical properties of the bed.

3 Results and Discussion

Training was performed on data for the bed temperatures 300, 400, 450, 500 and 550°C, while the test series were carried out at 350°C using the four bed concept. The problem was tackled by an architecture consisting of two parallel networks, one for simple nitrogen conversion estimation and another more complex for all the final products. A typical training result is reported in Table 1, where the weights are given for the simple network model with nitrogen conversion as its single output. The fit is good, as can be seen in Figure 4. Overfitting, which appeared in the model applied for the single bed concept (cf. Fig. 2) has been reduced and the novel model is able to explain also the low- and high-temperature regions well, as is seen in Fig. 5. However, more experiments should be made by varying the distance between the beds of catalyst, before the model is used for a final optimal reactor construction. Since the area between two beds of catalyst is of utmost importance for increasing the low temperature efficiency this must be included in the model. Another new feature to be considered is to vary the type of intermediate beds to get the optimum conditions at each stage.

References

[1] Eränen, K., Klingstedt, F., Arve, K., Murzin, D. Yu., (2004). On the mechanism of the selective catalytic reduction of NO with higher hydrocarbons over a silver/alumina catalyst. Journal of Catalysis 227: 328-343.

[2] Arve, K., Popov, E. A., Rönnholm, M., Klingstedt, F., Eloranta, J., Eränen, K., Murzin D. Yu., (2004). From a fixed bed Ag-alumina catalyst to a modified reactor design: how to enhance the crucial heterogeneous-homogeneous reactions in HC-SCR. Chemical Engineering Science 59: 5277-5282.

[3] Saxén, B. Saxén, H. (1995). NNDT – A Neural Network Development Tool, Proc. Intern. Conf. on Artificial Neural Networks and Genetic Algorithms (ICANNGA'95), Alés, France, 325-328.

[4] Rönnholm, M., Klingstedt, F., Eränen, K., Lindfors, L-E. (2003). Artificial neural network modelling applied on NO_x reduction with octane in excess oxygen over Ag/Al_2O_3. Reaction Kinetics and Catalysis Letters, Vol. 78, No 2, 331-340.

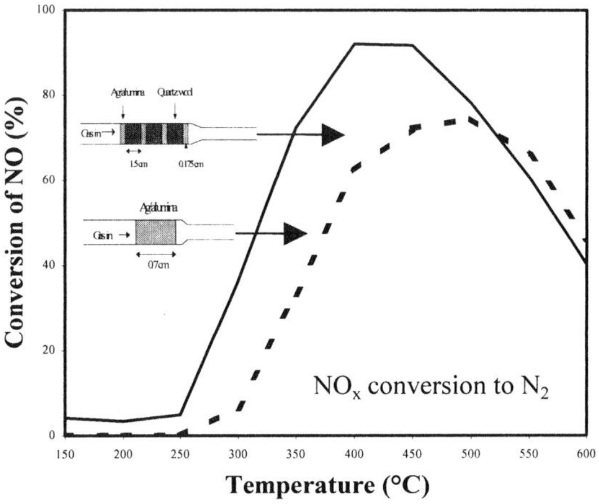

Fig. 1. Four-beds of Ag/alumina vs. single bed. Catalyst mass = 0.4 g, $C_1/NO_x = 6$.

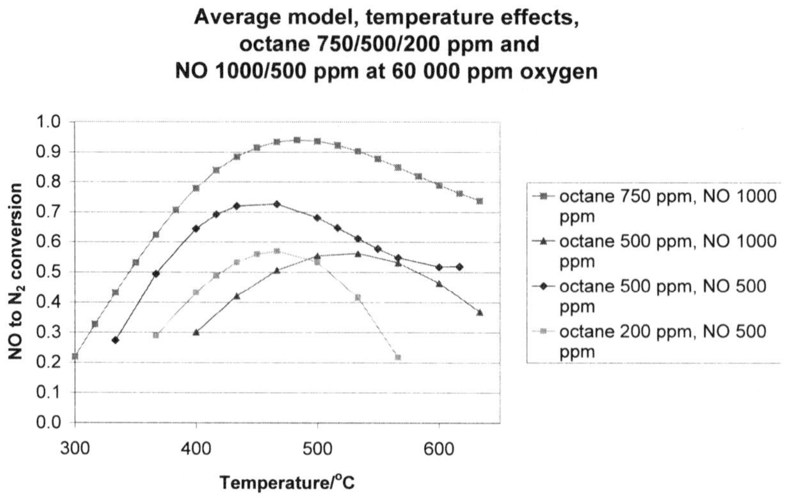

Fig.2. Conversion of NO to N_2, using the average value model for the one-catalyst bed.

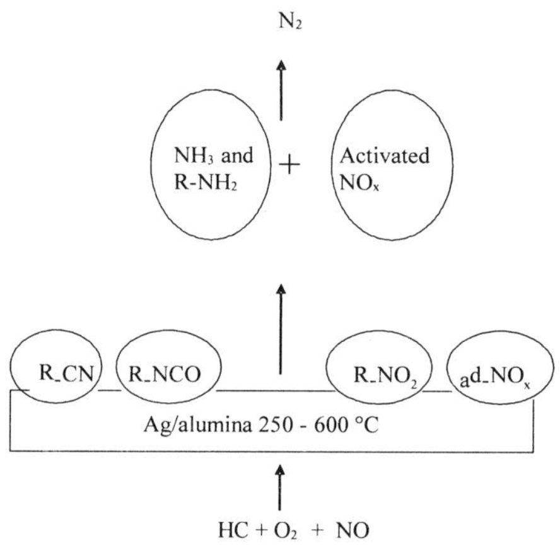

Fig. 3. Simplified reaction scheme for C_8H_{18}-SCR over Ag/alumina catalyst.

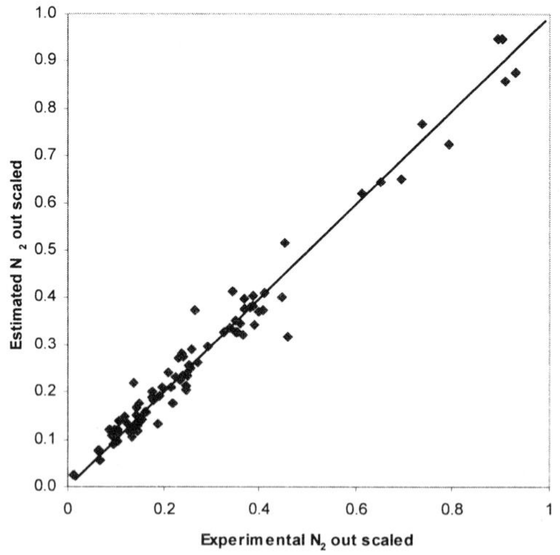

Fig. 4. Comparison between measured and estimated outgoing concentration of N_2 in the four-bed concept.

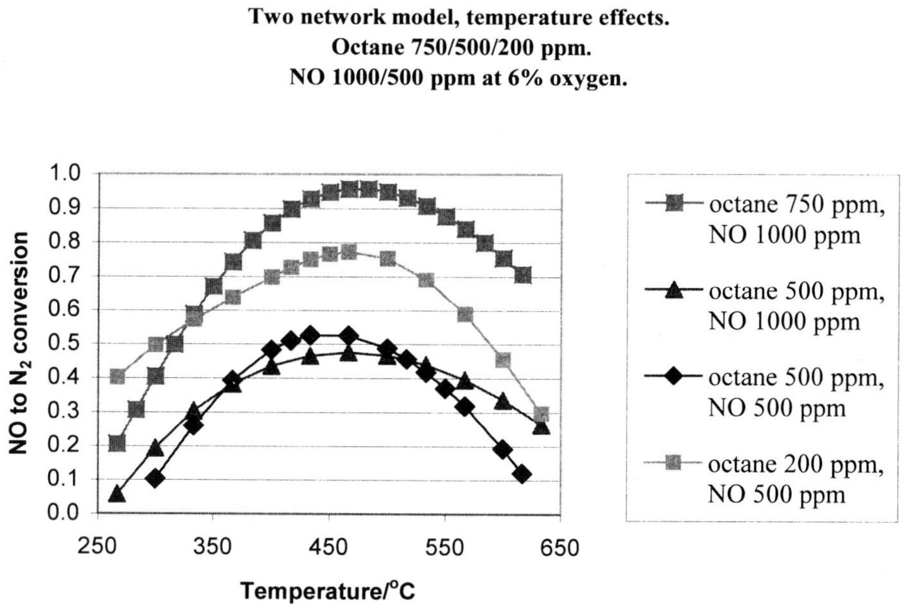

Figure 5. Results of the model of the conversion of nitrogen oxide at different temperatures.

Table 1 Weights of a 4-5-1 network predicting nitrogen release only. The first weigh "matrix" gives the lower layer weights, where each row corresponds to the weights to a hidden node (and the first column to biases), while the last row gives the corresponding upper layer weights to the single output node.

-21.1129	17.93223	2.607068	7.955934	11.82626	
-0.72449	0.189899	1.649559	0.339092	-7.02734	
2.292501	-2.68317	-1.90314	0.758894	6.357047	
4.33179	-4.58947	2.790449	-7.51199	6.831674	
-0.13595	-0.76213	1.68122	0.359621	-7.32542	
-11.1784	0.24837	43.01443	9.796465	1.452536	-33.1646

Acknowledgements

The financial support from the European Union through the Ammonore project (GRD-CT 2001-00595) is gratefully acknowledged. This work is part of the activities at the Åbo Akademi Process Chemistry Centre within the Finnish Centre of Excellence Programme (2000-2005) by the Academy of Finland.

A method for detecting cause-effects in data from complex processes

M. Helle & H. Saxén

Heat Engineering Laboratory, Abo Akademi University, Finland
E-mail: mikko.helle@abo.fi

Abstract

When models are developed to aid the decision making in the operation of industrial processes, lack of understanding of the underlying mechanisms can make a first-principles modeling approach infeasible. An alternative is to develop a black-box model on the basis of historical data, and neural networks can be used for this purpose to cope with nonlinearities. Since numerous factors may influence the variables to be modeled, and all potential inputs cannot be considered, one may instead solely focus on occasions where the (input or output) variables exhibit larger changes. The paper describes a modeling method by which historical data can be interpreted with respect to changes in key variables, yielding a model that is well suited for analysis of how changes in the input variables affect the outputs.

1 Introduction

Industrial processes are generally complex and the number of variables that potentially affect the product quality is often large. Mathematical modeling can be used as a tool for throwing light on the interrelationships between measured variables in order to improve the process performance. However, it is not always possible to develop a first principles model, because the underlying mechanisms might not be known well enough to make this approach feasible. On the other hand, perturbation experiments of the input variables that would shed light on their relations to the outputs are seldom allowed in industrial plants. If historical data from the process is available, an alternative is to make a data-driven approach where a black-box model is developed; today, neural networks have become popular tools for such nonlinear modeling. A drawback is that there are numerous factors that potentially influence the variables to be modeled (e.g., product quality) and the number of available observations ("data points") may, furthermore, be limited. The low degree of freedom can be tackled by constraining the input variable dimension and therefore it is no longer desired to capture all changes in the outputs by the model. An alternative is to focus the data analysis on occasions where the (input or output) variables exhibit larger changes or upsets, and use only such information in the modeling. This approach is also justified by the fact that process variables often drift in time, so their absolute values are not necessary important, but, instead, their relative changes should be considered. The method described in this paper is designed to use historical data from complex processes to interpret changes in key variables and to use the information in a consecutive modeling stage. The resulting model is therefore expected to be well suited for what-if analysis of how changes in the "input" variables affect the outputs.

2 The Problem

2.1 The Blast Furnace Process

Data from an industrial iron making blast furnace is studied in this work [1]. Due to the complexity of the process it is very difficult, if not impossible, to isolate the effect of changes in a single (input) variable on the product quality. In general, many variables change more or less simultaneously so the influence of one variable is easily lost in the noise of the output signals. The process is, therefore, well suited to be tackled using the proposed technique. In this paper, data from a Finnish blast furnace are used in a study where the effect of the properties of the main energy source in the process, metallurgical coke, as well as the injection rate of the auxiliary fuel, oil, on the quality of the product, pig iron, is analyzed. The inputs are variables that characterize the energy input, such as the alkali content, size and strength of coke, while the outputs are the key alloying components in pig iron, i.e., carbon, silicon and sulfur.

2.2 Formulation of the problem

Because of the relatively slow dynamics of the blast furnace process with respect to changes in the charged material, a change in a variable was expressed as the difference between three-week mean values separated by a two-week period. These mean values were considered sufficient to filter out single odd measurements and also to make sure that the values reflected a relatively stable state of the blast furnace. The two-week delay between the periods was introduced to give time for possible changes to influence the process. The above filtering method was also governed by the fact that coke properties were reported weekly. The blast furnace measurement variables, in turn, are daily mean values, which were aggregated to weekly mean values. Denoting the weekly mean value of the jth variable at week i by $\bar{x}_j(i)$, the change is given by

$$\Delta x_j(i) = \left(\sum_{k=i+1}^{i+3} \bar{x}_j(k) - \sum_{k=i-4}^{i-2} \bar{x}_j(k) \right) \Big/ \sum_{k=i-4}^{i-2} \bar{x}_j(k) \quad (1)$$

Figure 1 illustrates the how the three-week mean values (depicted by horizontal solid lines) around the present point, i (=184), are considered in expressing the change in the coke strength after reaction (CSR).

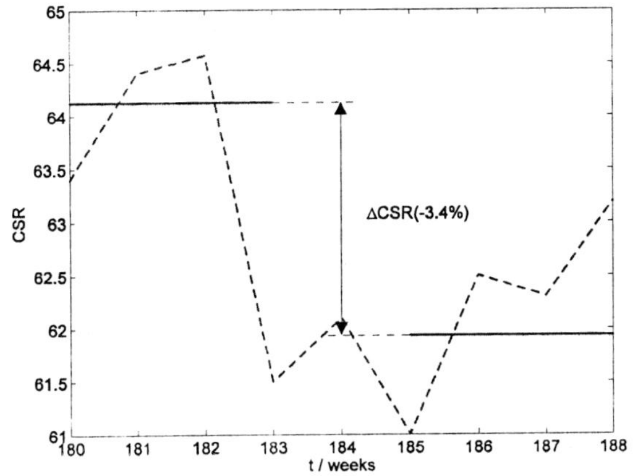

Figure 1. Changes in coke strength after reaction, CSR.

A change is calculated for every week in the data for both the model inputs and outputs. For the available data period spanning roughly seven years of operation of the blast furnace, the total number of weekly observations was about 360. The resulting changes in the values of the coke strength are illustrated in Fig. 2. Clearly, the variable exhibits a number of smaller and larger changes in both directions, characterized by spikes in the figure.

As mentioned earlier, the idea was to concentrate on major changes only. Therefore, every local minimum and maximum of the changes in the input variables is selected for further analysis. At this stage limits can be set for the accepted minima and maxima, e.g., if $\Delta CSR \in (-1\%,+1\%)$ it could be considered as insignificant and rejected from the analysis. Application of this rule (cf. dashed lines in Fig. 2) yields about 70 major changes in the CSR values.

After the extreme values have been selected, the corresponding time instants are noted and combined with the time instants detected by the same procedure applied on the other inputs variables. Finally, a data set of changes of the inputs and outputs for all these time instants can be created. Figure 3 shows an example of the evolution of the changes in the five inputs during a period of 30 weeks, and a time instant (i=196) where the coke strength (CSR) shows a local maximum. The values of the changes in all five inputs variables, as well as the changes in the outputs, correspond to one row in the data series created for the forthcoming analysis.

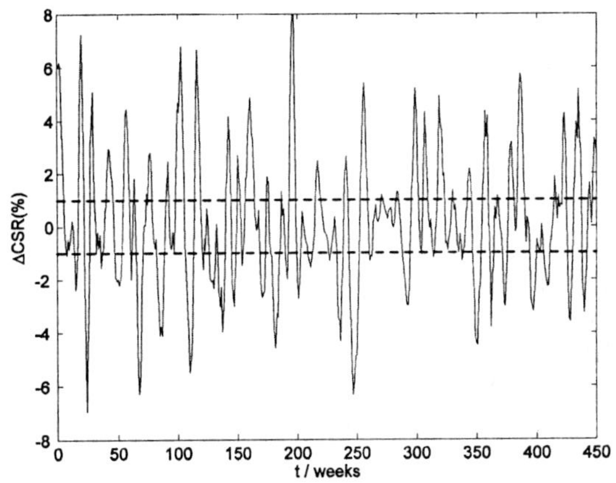

Figure 2. Changes in CSR values during the data period.

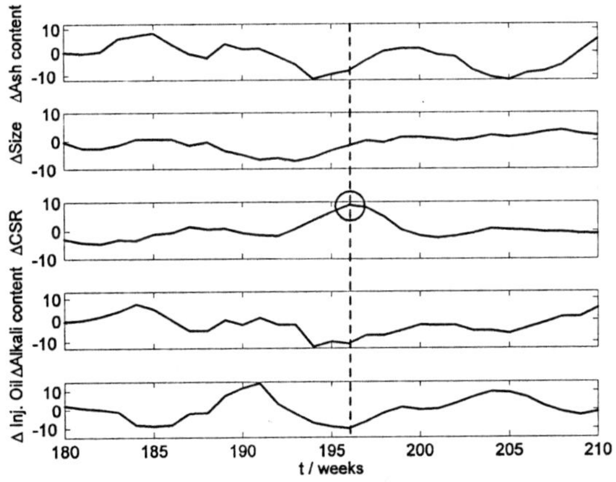

Figure 3. Example on Δ-values for five variables (inputs).

Before the modeling step, each variable (i.e., column of the data file) is normalized by dividing the values by the estimated standard deviation, s, of the variable in question: Thus, the new value of the jth variable at week k is given by

$$\Delta \tilde{x}_j(k) = \Delta x_j(k) / s_{\bar{x}_j} \quad (2)$$

After normalization, any row (i.e., week) containing at least one value $\notin (-3,+3)$ was removed, because such values were considered extreme and not meaningful to include in the model. For the system studied, this implied a rejection of about 7 % of the observations

3 Modeling

The information extracted by the procedure outlined in the previous section was used for developing black-box models of the relations between changes in fuel characteristics and in pig iron composition. The modeling was carried out using feedforward neural networks, with five inputs (cf. Fig. 3) and three outputs (cf. Fig. 4). Because of the restricted number of observations in the training set, the number of hidden neurons was limited to four. The networks entertained were trained with a hybrid algorithm that used a genetic algorithm (GA) to optimize the weights and connections between the input layer and the hidden layer, while linear least squares was applied to optimize the weights to the output nodes [2]. A penalty on the complexity of the network was added to the objective function, so the algorithm prefers simple networks (i.e., with few connections). Therefore, it also takes a stand on whether inputs are relevant for the model or not. Five different data sets of training and testing were created, by excluding a given share of the data from the training set and using it in the test set instead. The points in this test set were chosen by random. A neural model was developed for every such choice of training/test sets, partly to tackle the problem caused by the limited size of the available data set, which makes the performance of the models sensitive to the choice of test data. Another reason for developing multiple models was that the GA training of the lower layer weights was expected to yield quite different models. Figure 4 shows the results of one of the models on the training data for the three (normalized) outputs. Clearly, the model has not even been able to capture the major changes in the outputs, but this is, in fact, a desired feature, since numerous other factors (shutdowns, internal disturbances, etc.) also affect the hot metal chemistry. The fit in the lowest subpanel is poor, for reasons to be discussed in the next section.

4 Model analysis

The trained networks can next be used to analyze how the inputs influence the outputs. The analysis was done by allowing one (normalized) input at a time to change from −3 to 3 and registering the behavior of the outputs, keeping all other inputs at zero.

Since both inputs and outputs express (normalized) changes in the corresponding variables, this analysis will reveal how a change in, e.g., coke particle size is reflected in, e.g., pig iron carbon content.

4.1 General results

Figure 5 shows an example of the outcome of such an analysis, where the five curves in each subpanel correspond to the results of the models developed on the five different data sets; the consensus between the different models is seen to be acceptable. In studying the results one can note that a horizontal line may imply that the current input is of little importance or that it has been excluded from the model by the GA training. Another noteworthy fact is that behavior at the edges (i.e., close to ±3 for the inputs) is rather uncertain, because such values are scarce in the training data.

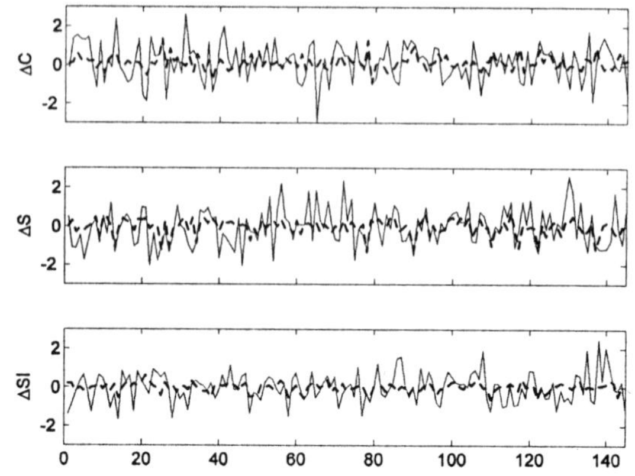

Figure 4. Example on the model's fit on the training data (—— desired output, - - - - model output).

An analysis of the figure also shows that in some cases the resulting change is either negative or positive independent of the value of the input variable. This would imply that whatever happens to the input, there would be a transition in the outputs. This is an undesired feature that arises due to the assumption of independence of the input variables. A simple way to resolve the problem is to shift the relations so that no change in the fuel characteristics (i.e., an input of zero) will yield a constant composition (i.e., a zero change in the output). After this correction, the results of Fig. 6 were obtained, and these will be briefly analyzed in the following.

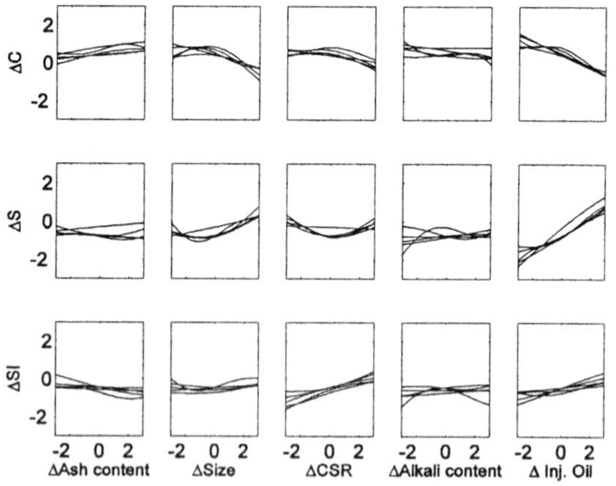

Figure 5. Interpretation of the interrelations between changes in the inputs and changes in the outputs provided by the neural models.

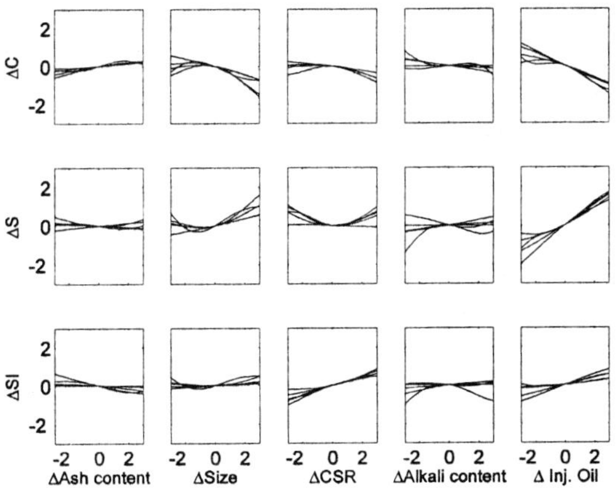

Figure 6. Interpretation of the interrelations between changes in the inputs and outputs with removed bias.

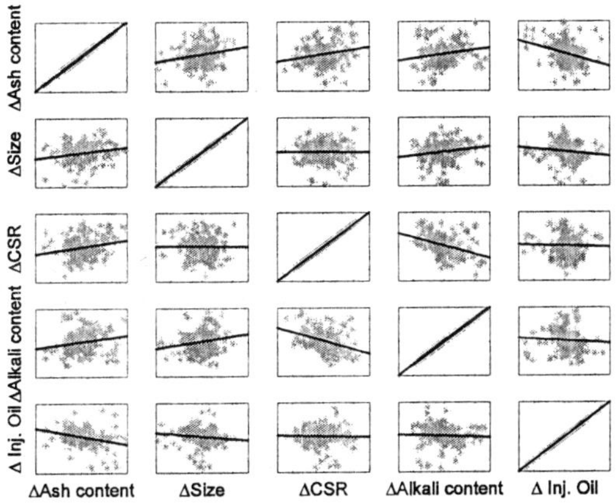

Figure 7. Interrelations between changes in the inputs, with lines representing linear relations.

Since the inputs are not independent linear correlations between them were determined and considered in the analysis. Using the correlations depicted by the lines in Fig. 7, the outcome was, however, almost identical with the one presented in Fig. 6. The obvious reason for this is the relatively weak correlations between the inputs (cf. Fig. 7).

4.2 Specific results

The ash content of the coke is seen to have little influence on the product quality. Since the ash content changes only gradually in the plant, the blast furnace operators have possibilities to consider these changes in time. Another observation is that the changes in the carbon and sulfur contents are strongly negatively correlated. This is in line with practical knowledge of the system at hand [3,4]. It should be pointed out that the carbon and sulfur content of pig iron are strongly dependent on the physical conditions in the lower part of the blast furnace, where coke is the only remaining solid. Yet another feature is that the silicon content is not much affected by the inputs (except CSR), which is in agreement with the fact that the silicon content is mainly determined by the fuel rate and not so much by the coke properties. The clear effect of a change in injected oil on the change in pig iron carbon content can be explained by the fact that when oil injection is increased the amount of coke charged into the furnace decreases, and the contact between liquid iron and coke is therefore also decreased.

5 Conclusions

The paper has described a modeling method by which historical data from complex processes can be interpreted with respect to changes in key variables. Neural networks have been used as modeling tools and the technique has been illustrated on process data from ironmaking industry. The findings of the study are in general agreement with practical experience, even though some of the detected interrelations, e.g. the one between ΔCSR and ΔC, call for a deeper analysis. Forthcoming work will include a more detailed analysis of how coke properties affect other central blast furnace variables.

Acknowledgments

The authors are grateful for financial support provided by the National Technology Agency of Finland (Tekes), Rautaruukki Oyj, the Academy of Finland and the Foundation of Waldemar von Frenckell.

References

[1] Omori, Y. (ed.) (1987) Blast Furnace Phenomena and Modelling, The Iron and Steel Institute of Japan, Elsevier, London

[2] Petterson, F. and Saxén, H. (2003) A hybrid algorithm for weight and connectivity optimization in feedforward neural networks. In: Pearson, D. et al. (eds) Artificial Neural Nets and Genetic Algorithms. Springer, Wien, pp. 47-52

[3] Raipala, K. (2003) On Hearth Phenomena and Hot Metal Carbon Content in Blast Furnace, Doctoral Thesis, Helsinki University of Technology, Helsinki

[4] Husslage, W. (2004) Dynamic distributions – Sulphur transfer and flow in a high temperature packed coke bed, Doctoral Thesis, Delft University of Technology, Febodruk, Enschede.

Predictive data mining on rubber compound database

M. Trebar, U. Lotrič[1]
Faculty of Computer and Information Science, University of Ljubljana, Slovenia
E-mail: {mira.trebar, uros.lotric}@fri.uni-lj.si

Abstract

Neural network based predictive data mining techniques are used to find relationships between rubber compound parameters obtained by rheological and mechanical tests. The preprocessing methods appropriate to the problem are also introduced. Good prediction of different rubber compound parameters evidently indicate that the majority of rubber compounds' mechanical properties can be devised from the rheological measurements of cross-linking process.

1 Introduction

Data mining, also known as knowledge discovery in databases (KDD), is an automated extraction of knowledge implicitly stored in large databases [1]. Very often the goal of data mining is to predict future trends and behaviours, allowing businesses to make knowledge-driven decisions. Predictive data mining tools search databases for hidden parameters, finding predictive information that experts may miss. Lately, neural networks as general nonparametric and robust models, designed to find nonlinear relations between specified input-output pairs [2], are becoming more and more popular in pure data mining applications. Especially when the prediction is the ultimate goal of data mining, neural networks can be seriously considered as an appropriate model [3], mainly due to their good generalization properties.

In rubber industry, the quality of rubber products significantly depends on the quality of rubber compounds. Therefore, instantly after mixing, the rubber compounds are rigorously tested in order to determine whether they are good enough to be released for further production. The most important parameters that determine the quality of rubber compounds are measured by variety of rheological and mechanical tests. While rheological tests are quickly performed on raw compounds, mechanical tests performed on vulcanizates are very time consuming. Thus, the main idea of this work is to quest for strong relationships between rheological and mechanical tests in order to reduce the latter to the minimum or even completely abandon them.

[1] The work is sponsored in part by Slovenian Ministry of Education, Science and Sport by grants V2-0886, L2-6460 and L2-6143.

A brief overview of predictive data mining and neural networks is given in the next section. In the third section the data mining on rubber compound database is outlined. In section four the experimental setup and results in terms of prediction accuracy are given. The main findings are drawn in the last section.

2 Predictive data mining and neural networks

The process of predictive data mining consists of three stages [4]: the initial data exploration, model building with validation/verification, and deployment, outlined in Fig. 1. Exploration, as the first stage of data mining, may

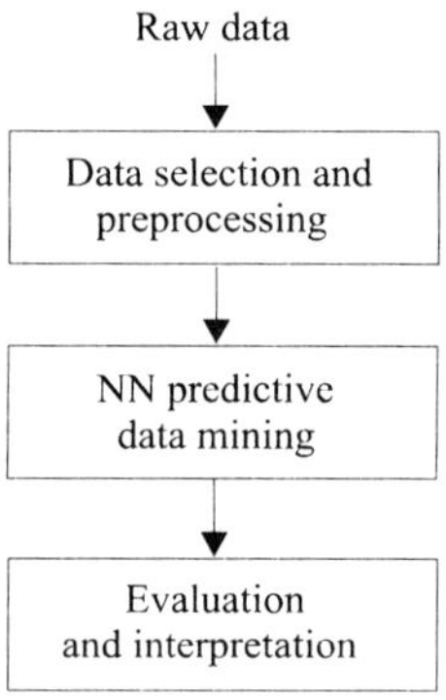

Fig. 1. Data mining process

involve data cleaning to remove noise and correct inconsistencies in the data [1], extraction of the target data set or even more target data subsets and data transformations, such as normalization. The second stage involves a model building and validation. Depending on the nature of the problem, it may include also a choice of appropriate prediction models. Various models are tested in this phase and the best one, based on specified criterion function, is chosen. In the last stage, the obtained model enables prediction and consequent interpretation of results from new input data.

In our prediction modeling, feed-forward neural networks, multilayered perceptrons (MLP), are used. Feedforward neural networks [5] are designed to find nonlinear relations between specified input-output pairs $\{\mathbf{p}(q), \mathbf{d}(q)\}$. Input vectors $\mathbf{p}(q)$ are layer by layer propagated through the neural network in order to obtain corresponding output vectors $\mathbf{y}^L(q)$, where L is the number of layers. The objective of training algorithm is to find such set of weights that minimizes the performance function. The second-order derivative based Levenberg-Marquardt algorithm [6] is used, where the weights are iteratively updated in batch mode, i.e., only after applying the entire set of input-output pairs to the neural network.

3 Rubber compound database

The rheological properties of rubber compounds, affecting its deformation and flowing capabilities, are extremely important for ascertainment of their behavior during the production process [7]. In vulcanization phase of production process, a plastic rubber compound becomes elastic. In order to estimate the quality of final rubber products, the rubber compound is vulcanized in laboratory and some mechanical test are performed on obtained vulcanizates. As the vulcanization phase, lasting about 20 minutes, and the following mechanical tests, performed on hardness tester and tensile testing machine (Fig. 2) by versed operator, taking additional 5 minutes, are time consuming, the rubber producers are trying to reduce this process to a minimum.

In order to meet the foregoing and still obtain enough information about a compound, a rheological instrument called moving dye rheometer (MDR) is used (Fig. 2), performing tests directly on raw rubber compounds in about 2 minutes. It measures a degree of cross-linking, manifested in torque, as a function of time during vulcanization. From the torque curve, presented in Fig 3, some important parameters like ML, MH, t_{10}, t_{50}, t_{90} and t_{s1} are determined.

Fig. 2. Instruments used to measure rubber compound properties: moving dye rheometer (left), hardness tester (top right) and tensile tester (bottom right).

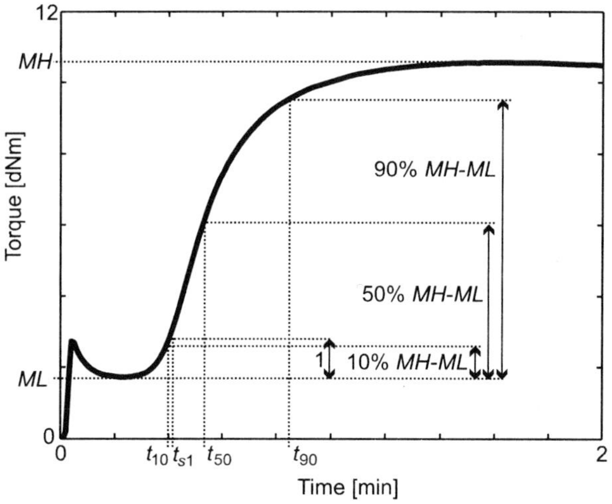

Fig. 3. Torque curve, measured is the effect of cross-linking during vulcanization process, with indicated meaning of parameters ML, MH, t_{10}, t_{50}, t_{90} and t_{s1}.

Although the information about vulcanizates is inherently present in this curve, it is not easy to interpret it in terms of mechanical tests used by rubber engineers. Thus, an attempt was made to relate the rheologycal parameters to the hardness, measured on hardness tester, and parameters measured by tensile tester, like: $M50$, $M100$ and $M200$ - stresses required to elongate a test sample for additional 50%, 100% and 200%, respectively, E - the elongation at break and TS - tensile stress at sample break.

4 Experimental results

Prior to the modeling, the data cleaning was performed to remove inconsistent data. Additionally, rarely mixed rubber compounds were excluded from modeling – only those mixed at least once weekly and having at least 150 measurements were kept. Besides, when the production process of a new rubber compound is restrained in quality, the tests on rheological instruments are shortened. In this case the last maximum on the momentum curve is missing (Fig. 3), thus substantially changing the meaning of other parameters. Therefore, we have used the basic data set (B) and the one from which the tests having longer test times were removed (T). Both types of data sets were further used in their original form with $n = 7$ input parameters (ML, MH, t_{10}, t_{50}, t_{90}, t_{s1} and test time) or preprocessed in a special way, having rubber compound testing database in mind with $n = 2 \times 7 = 14$ input parameters (data sets

BP and TP, respectively). Namely, there are many different rubber compounds in a database and the average values of their parameters differ substantially. Therefore, values of each parameter in a chosen input-output sample were replaced by two values: the average value for a compound it belongs to, and its relative deviation from the average. In this case, only the relative deviation from the average values was predicted.

For each data set the input-output pairs were divided into two sets: the multilayered perceptron was trained with input-output pairs included in the training set (80%) and the performance of the models was tested on the input-output pairs in test set (20%). The last part of training set (20%) was used for early stopping in order to reduce the effect of overfitting. Proportions of input-output pairs belonging to each rubber compound were equal in all three sets.

Multilayered perceptron with one or two hidden layers was used in our experiments. In the case of multilayered perceptron with one hidden layer, the number of neurons was altered from n to $2\,n$ and in the case of two hidden layers, the number of neurons was varied from n to $2\,n$ on first layer and from $n/2$ to n on the second layer. To evaluate the models three criteria were used: (i) the root mean squared error, normalized to the standard deviation NRMSE, (ii) the mean absolute percentage error MAPE and (iii) the percentage of correctly classified samples %OK. Each measurement and/or predicted result can fall into one of three classes: inside, above or below the specified limit range. A given sample is correctly classified when the measurement and the prediction fall in the same class.

In Table 1 the prediction results for hardness are shown. In this case, the total number of samples is 11592 for data sets B and BP and 11016 for data sets T and TP. The criteria used for the selection of MLP topologies was the mean absolute percentage error MAPE obtained on the training set. Considering the NRMSE, the results with preprocessed input data (P) are at least 15% better in both cases. Even though the results on basic data set with preprocessing (BP) are better than in data set B, the effect of variable testing time cannot be completely overcome.

Table 1. Prediction results for hardness

Data set	MLP	NRMSE		MAPE	
		train	test	train	test
B	7-7-1	0.18	0.28	0.03	0.04
BP	14-14-1	0.15	0.23	0.03	0.04
T	7-7-1	0.17	0.20	0.02	0.03
TP	14-21-1	0.13	0.17	0.02	0.02

In prediction of tensile tester parameters the number of data samples in data sets B and BP was 3612 and in the case of data sets T and TP 3531. The best results in prediction of parameters $M50$, $M100$, TS and E are summarized in Tables 2, 3, 4 and 5, respectively.

Table 2. Prediction results for stress ($M50$)

Data set	MLP	NRMSE		MAPE	
		train	test	train	test
B	7-10-1	0.46	0.56	0.11	0.12
BP	14-21-1	0.43	0.42	0.09	0.09
T	7-7-1	0.39	0.31	0.09	0.05
TP	14-14-1	0.39	0.32	0.09	0.06

Table 3. Prediction results for stress ($M100$)

Data set	MLP	NRMSE		MAPE	
		train	test	train	test
B	7-7-3-1	0.42	0.56	0.15	0.16
BP	14-14-1	0.41	0.43	0.12	0.13
T	7-7-3-1	0.31	0.31	0.11	0.09
TP	14-21-1	0.32	0.33	0.10	0.09

Table 4. Prediction results for tensile stress (TS)

Data set	MLP	NRMSE		MAPE	
		train	test	train	test
B	7-12-6-1	0.22	0.27	0.13	0.19
BP	14-21-1	0.21	0.23	0.11	0.12
T	7-7-3-1	0.21	0.25	0.12	0.17
TP	14-21-1	0.21	0.23	0.11	0.11

Table 5. Prediction results for elongation (E)

Data set	MLP	NRMSE		MAPE	
		train	test	train	test
B	7-9-1	0.73	0.41	0.09	0.10
BP	14-16-1	0.73	0.40	0.09	0.10
T	7-8-1	0.72	0.32	0.09	0.09
TP	14-21-1	0.70	0.35	0.09	0.09

Similarly as in the case of hardness, the prediction of all tensile tester parameters is improved by preprocessing of the basic data set B, which is not always the case when preprocessing is applied on data set T. In case of parameters $M50$, $M100$ and TS the effect of preprocessing on basic data set B is far more expressed than in the case of data set T. In case of parameter E there was no significant difference between the results on data sets B and BP in comparison to the data sets T and TP, respectively. The better prediction results on data set T in comparison with data set B show a significant role of input parameter test time.

The third criterion, the percentage of correctly classified samples %OK, was calculated only for compounds having the limit range defined due to the quality issues. In Table 6 the prediction of hardness for some compounds is given. The architecture of multilayered perceptron is the same as in the Table 1. In most cases the percentage of correctly classified samples %OK for the presented compounds is very high. Further analysis of incorrectly classified samples has confirmed our thesis that majority of these samples were just inside or outside boundary lines of specified limit range.

Table 6. Prediction results of hardness for some compounds.

Data set	Compound	MAPE		%OK	
		train	test	train	test
B	20704	0.06	0.08	93	81
	20965	0.02	0.02	98	100
	21367	0.01	0.01	98	100
	21588	0.02	0.03	99	100
	RM1	0.02	0.01	66	98
BP	20704	0.03	0.05	98	97
	20965	0.02	0.02	99	100
	21367	0.01	0.01	98	100
	21588	0.02	0.02	99	100
	RM1	0.02	0.01	67	98
T	20704	0.02	0.02	98	100
	20965	0.03	0.02	97	100
	21367	0.01	0.01	97	100
	21588	0.02	0.03	98	100
	RM1	0.02	0.01	65	97
TP	20704	0.03	0.02	99	100
	20965	0.02	0.02	99	100
	21367	0.01	0.01	98	100
	21588	0.01	0.02	99	100
	RM1	0.02	0.01	67	98

Table 7 shows similar results in prediction of elongation parameter E. As we have already seen in Table 5, the preprocessing in this case does not improve prediction performance. Again, the alteration of %OK criterion can be addressed to the prediction results being very close to the specified limit boundary. The results of parameters $M50$, $M100$ and TS are very similar for all types of data sets.

5 Conclusions

The neural network predictive models yielded very good results in prediction of mechanical parameters of rubber compounds from their rheological properties. Two types of preprocessing, i.e. modification of input parameters and additional cleaning of database, were used to outline the necessity of proper data preparation.

Table 7. Prediction results of elongation (E) for some compounds.

Data set	Compound	MAPE		%OK	
		train	test	train	test
B	21585	0.10	0.09	99	99
	215851	0.08	0.06	100	100
	21588	0.18	0.27	99	100
	21594	0.13	0.09	99	100
	RM1	0.06	0.07	99	100
BP	21585	0.11	0.10	99	99
	215851	0.08	0.06	100	100
	21588	0.13	0.17	99	100
	21594	0.11	0.11	98	100
	RM1	0.05	0.06	99	99
T	21585	0.10	0.08	99	99
	215851	0.08	0.05	100	100
	21588	0.16	0.23	99	100
	21594	0.11	0.08	98	100
	RM1	0.05	0.06	100	99
TP	21585	0.10	0.11	99	99
	215851	0.07	0.06	100	100
	21588	0.14	0.19	99	100
	21594	0.11	0.12	100	100
	RM1	0.08	0.09	99	100

To improve prediction and (above all) the classification of predicted results, some additional parameters should be retrieved from the last part of momentum curves, where the properties of rubber compound are close to the properties of vulcanizates used in mechanical tests.

References

[1] Han, J., Kamber, M., (2001) Data Mining: Concepts and Techniques, Academic Press, London

[2] Mitra, S., (2002) Data Mining in Soft Computing Framework: A survay, IEEE Trans. On Neural Networks, Vol. 13, No. 1: 3-14

[3] Pham, D. T., Xing,Liu (1995) Neural networks for Identification, Prediction and Control, Springer - Verlag, London

[4] Data Mining Techniques, http://www.statsoft.com/textbook/stdatmin.html

[5] Haykin, S. (1999) Neural networks: a comprehensive foundation, 2nd ed., Prentice-Hall, New Jersey

[6] Hagan, M. T., Menhaj, M. B. (1994) Training feedforward networks with the marquardt algorithm, IEEE Trans. Neural Netw 5(6): 989-993

[7] Painter, P. C., Coleman, M. M. (1997) Fundamentals of Polymer Science, Technomic, Lancaster

Applying Neural Network to Inverse Kinematic Problem for 6R Robot Manipulator with Offset Wrist

Z. Bingul, H. M. Ertunc, C. Oysu

Department of Mechatronics Engineering, Kocaeli University, Turkey
E-mail: {zaferb, hmertunc, coysu}@kou.edu.tr

Abstract

An Artificial Neural Network (ANN) using backpropagation algorithm is applied to solve inverse kinematics problems of industrial robot manipulator. 6R robot manipulator with offset wrist was chosen as industrial robot manipulator because geometric feature of this robot does not allow to solve inverse kinematics problems analytically. In other words, there is no closed form solution for this problem. As the number of neurons at hidden layer is varied between 4 and 32, the robot joint angles ($\theta_1\theta_2\ldots\ \theta_6$) were predicted with average errors of 8.9°, 7.8°, 8.3°, 13°, 8.5°, and 10.5° for the 1st, 2nd, 3rd,4th and 6th joint, respectively.

1 Introduction

ANN uses data sets to obtain the models of systems in fields such as robotics, factory automation, and autonomous vehicles. Their ability to learn by example makes artificial neural networks very flexible and powerful. Therefore, neural networks have been intensively used for solving regression and classification problems in many fields. In short, neural networks are nonlinear processes that perform learning and classification. Recently neural networks have been used in many areas that require computational techniques such as pattern recognition, optical character recognition, outcome prediction and problem classification. The current focus in learning research lies on increasingly more sophisticated algorithms for the off-line analysis of finite data sets, without severe constraints on the computational complexity of the algorithms.

In robot inverse kinematics learning, however, special constraints need to be taken into account when approaching a learning task. The complexity in the inverse kinematics problem of industrial robot manipulators arises from their geometry and nonlinear equations (trigonometric equations) occurring between Cartesian space and joint space. Some other difficulties in inverse kinematics problem are : i) kinematic equations are coupled, ii) multiple solutions and singularities may exist. Matematical solutions for inverse kinematics problems may not always correspond to physical solutions and method of its solution depends on the robot configuration. Conventional numerical approaches to the inverse calibration of robots are time-consuming and suffer from numerical problems of ill-conditioning and singularities.

The conversion of the position and orientation of a robot manipulator end-effector from Cartesian space to joint space is called as inverse kinematics problem. This relationship between joint space and Cartesian space is illustrated in Figure 1.

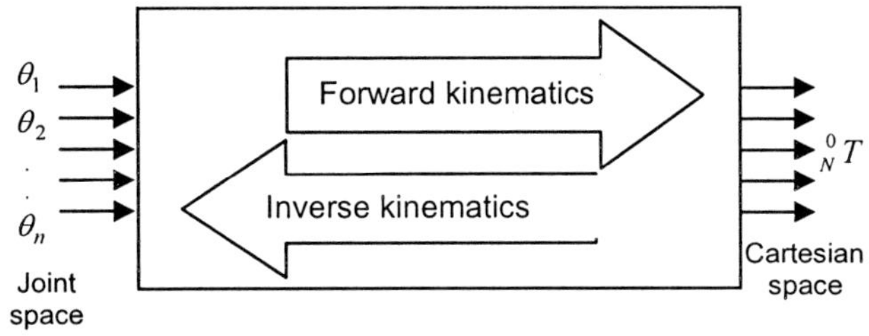

Fig. 1. The schematic representation of forward and inverse kinematics.

There are three types of inverse kinematics solution: complete analytical solution (closed form solution), numerical solutions and semi-analitical solutions. In the first type, all of the joint variables are solved analytically according to given configuration data. Closed form solution is preferable because in many applications where the manipulator supports or is to be supported by a sensory system, the results from kinematic computations need to be supplied rapidly in order to have control actions. In the second type of solution, all of the joint variables are obtained iterative computational procedures. There are four disadvantages in these: a) incorrect initial estimations, b) before executing the inverse kinematics algorithms, convergence to the correct solution can not be guarantied, c) multiple solutions are not known, d) there is no solution, if the Jacobian matrix is singular. In the third type, some of the joint variables are determined analytically in terms of two or three joints variables and these joint variables computed numerically. Disadvantage of numerical approaches to inverse kinematics problems is also heavy computational calculation and big computational time.

When coupling of the position and orientation kinematics occurs, there may be no exist efficient closed form solutions. In this case, it is better to use Neural Networks to solve the inverse kinematics problems since trigonometric equations in inverse kinematics problems can not be solved analytically. In other words, it is not possible to formulate a mathematical model that has a clear mapping between Cartesian space and Joint space for inverse kinematics problem. To overcome this problem, ANN uses the samples to obtain the nonlinear model of such systems. Their ability to learn by example makes artificial neural networks very flexible and powerful when the traditional model-based modeling techniques break down. Many researchers have experimented with this approach by applying it to several robot configurations [1-12]. However, this approach has not applied to robot manipulator with offset wrist whose inverse kinematics solution is no exist in efficient closed form.

In this paper, a neural network using backpropagation algorithm in Matlab Neural Network Toolbox is employed to predict the joint angles. Simulation results show that prediction performance from the approximation accuracy point of view is satisfactory with over 90% correlation.

2 Robot Inverse kinematics

For a six jointed robot manipulator, the position and orientation of the end-effector with respect to the base is given by

$$ {}_{end-effector}^{base}T = \begin{bmatrix} r_{11} & r_{12} & r_{13} & p_x \\ r_{21} & r_{22} & r_{23} & p_y \\ r_{31} & r_{32} & r_{33} & p_z \\ 0 & 0 & 0 & 1 \end{bmatrix} \quad (1)$$

where r_{ij}'s represent the rotational elements of transformation matrix (i and j= 1, 2 and 3) and p_x, p_y and p_z are the elements of position vector. The product of the link transformations yield forward kinematic of the robot manipulator.

$$ {}_6^0T = {}_1^0T(q_1)\,{}_2^1T(q_2)\,{}_3^2T(q_3)\,{}_4^3T(q_4)\,{}_5^4T(q_5)\,{}_6^5T(q_6) \quad (2)$$

where q_i is the joint variable (revolute joint or prismatic joint) for joint i. This forward transformation matrix is equated with end-effector transformation matrix. To find the inverse kinematics solution, it should be solved for q_i as function of the known elements of ${}_{end-effector}^{base}T$.

Denavit-Hartenberg (D-H) kinematic parameters of the robot manipulator used in this study is listed in Table 1 and kinematic structure of this robot is shown in Figure 2.

Table 1. D-H kinematic parameters for robot manipulator.

i	θ_i	α_{i-1}	a_{i-1}	d_i
1	θ_1	0	0	h_1
2	θ_2	90	0	d_2
3	θ_3	-90	0	l_2
4	θ_4	0	l_3	0
5	θ_5	-90	l_4	0
6	θ_6	90	0	d_6

3 The Structure of Backpropagation Neural Network

The neural network built in this study was shown in Figure 3. There is a total of 12 inputs to the network and there are six outputs from the network. The 12 inputs to the network are the nine elements of the rotation matrix (r_{ij} $i,j=1:3$) and the Cartesian position of the arm (P_x, P_y, P_z). The inputs are arranged in vector form from the Rotation matrix. On the other hand, the outputs are three joint angles θ_1, θ_2, θ_3 and three wrist joint angles θ_4, θ_5, and θ_6.

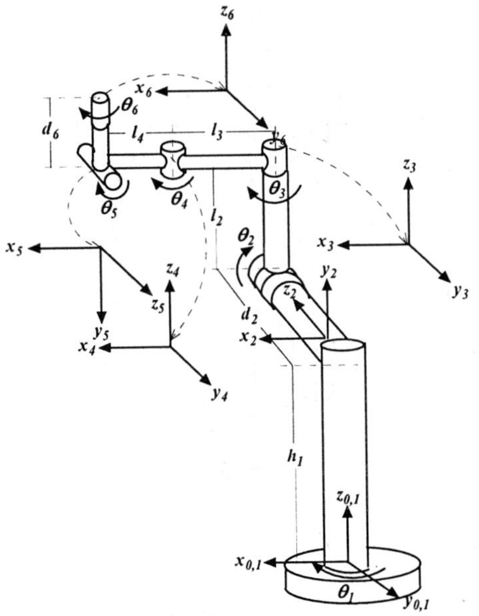

Fig. 2. Coordinate frame attached to the rigid body of the robot manipulator.

A simulation data set for training and testing of neural network was generated by forward kinematic equation 2. This data set includes 10^6 input-output pairs for every 10° joint angle. A working data set 4000 data points that was composed of input-output vector pairs was selected randomly from the data set. Again, the input vector consists of 12 input variables from Rotation matrix and the output vector, or also called as target vector, includes six joint angles. While 60% of the data set was randomly assigned as training set, the remaining 40% was employed for testing. Both the input and output variables were normalized to the [-1,1] range.

As seen from Fig.3, the neural network consists of an input layer with 12 neurons, a hidden layer with 20 neurons and an output layer with 6 neurons. While the neurons at input and hidden layer have sigmoid activation functions, the output layer has linear neurons so that the network could produce values outside the range -1 to +1. Input vectors and the corresponding target vectors from training set are used to train the network until it can approximate a function between the input and output variables. Training procedure adjusted the weighting coefficients using Levenberg-Marquardt algorithm. In this procedure, a termination criterion was set as 10^{-2} and all the weighting coefficients were initially assigned randomly. Then input vectors from the test data set are presented to the trained backpropagation network. The responses of the network, i.e. joint angles, are compared to targets in the test data.

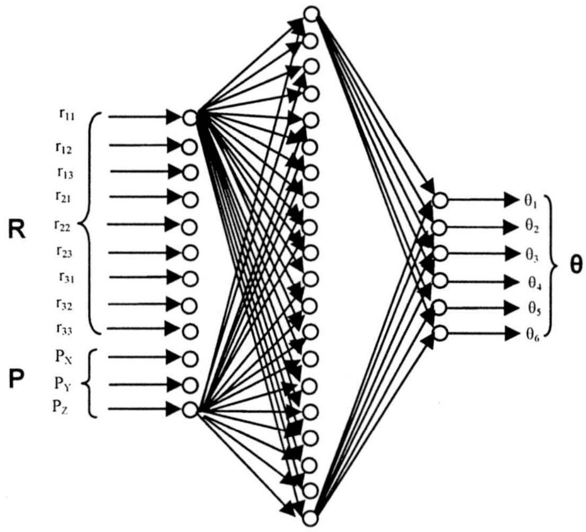

Fig. 3. The neural network structure.

4 The Results and Discussion

The main drawback of using neural networks to approximate the inverse kinematics of robot arms is the high number of training samples required to attain an acceptable precision. After training 60 % of the input-output vector pairs of the data set, the network was tested on the 40 % of the pairs of data set that are not included in the training set. Figure 4 shows 10 points of network output represented by 'x' and corresponding targets represented by 'o' for each angle. As seen from the figure, the neural network predicted the actual joint angle, target, successfully at some angle values. To obtain a better idea for the performance of the neural network prediction, the errors of joint angles for 10 test points was plotted in Fig. 5. Since error values fluctuate around the origin, effective error, root mean square error, is calculated for each angle as follows: 8.9°, 7.8°, 8.3°, 13°, 8.5°, and 10.5° for θ_1, θ_2, θ_3, θ_4, θ_5, and θ_6, respectively. Note that, the biggest effective error is found for θ_4. In fact, when Fig. 4 and Fig. 5 are examined, the worst prediction performance and bigger error amplitudes are observed for fourth joint angle.

In order to examine the effects of the number of neurons at hidden layer on inverse kinematics problem, the number of neurons at hidden layer is varied between 4 and 32. The results of this are illustrated in Figure 6. The more hidden neurons are added to hidden layer, the more powerful networks are obtained. But computation time gets longer with increasing of hidden neurons.

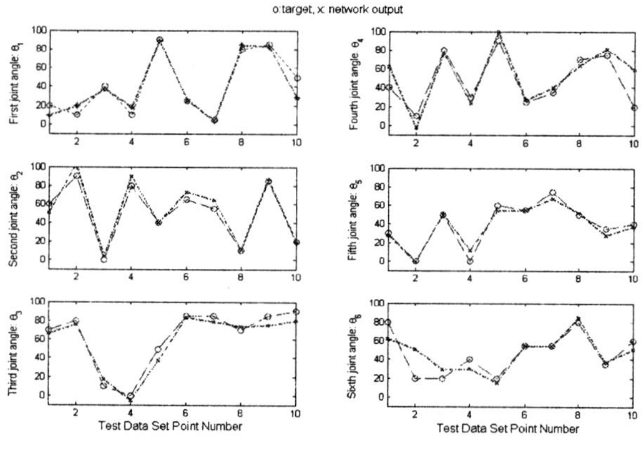

Fig. 4. The output of the neural network and target values.

5 Conclusion

Neural Networks using backpropagation algorithm has been applied to inverse kinematics problem for 6R robot manipulator with an offset wrist which does not have closed form solution. Inverse kinematics computation using an artificial neural network that learns the inverse kinematics of a robot manipulator has been employed by many researchers. The inverse kinematics function of the robot manipulator with an offset wrist, much more difficult robot structure, is a multi-valued and discontinuous function. Therefore, it is difficult for a well-known multi-layer neural network to approximate such a function. A coarse mapping can be obtained easily, but an accurate representation of the true mapping is often not feasible or extremely difficult. In this work, joint angles θ_1, θ_2, θ_3, θ_4, θ_5, and θ_6, were predicted with root mean square errors of 8.9°, 7.8°, 8.3°, 13°, 8.5°, and 10.5°, respectively.

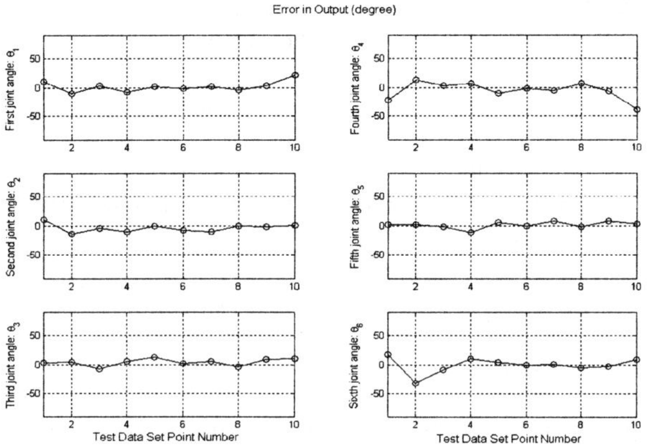

Fig. 5. Errors of joint angles in Test Data set.

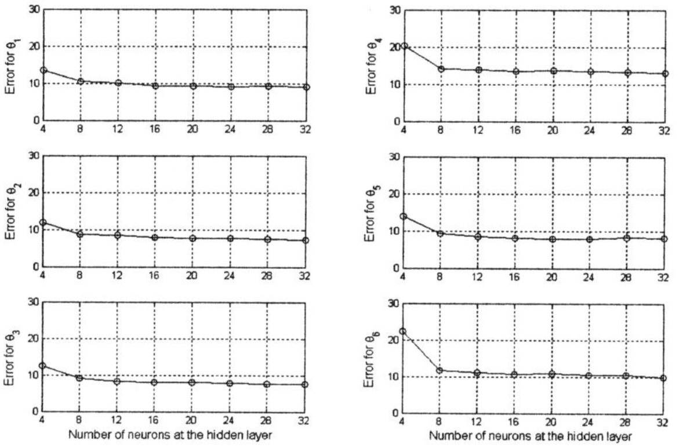

Fig. 6. The effects of the number of neurons at hidden layer on the absolute average errors.

References

[1] Guez, A., Ahmad, Z., (1989) Accelerated Convergence in the Inverse Kinematics via Multilayer Feedforward Networks. IJCNN89. vol. II, 341–344

[2] Guez, A., Ahmad, Z. (1989) A Solution to the Inverse Kinematic in Robotics Using Neural Network Processing. IJCNN89. vol. II, 299–304

[3] Nguyen, L., Patel, RV., Khorasani, K. (1990) Neural Network Architectures for the Forward Kinematics Problem in Robotics. IJCNN90. vol. III, 393–399

[4] Lee, S., Kil, RM. (1990) Robot Kinematic Control based on Bidirectional mapping Neural Network. IJCNN90. vol. III, 327–335

[5] Torras, C. (1993) Symbolic planning versus neural control in robots. From Neural Networks to Artifical Intelligence, Research Notes in Neural Computing 4, Springer-Verlag: Berlin Heidelberg New-York, 509-523.

[6] Krose, BJA. and van der Smagt, PP. (1993) An Introduction to Neural Networks. 5th ed. Chap. 7 Robot Control, University of Amsterdam.

[7] Xiaolin, Z., Lewis, J., N-Nagy, FL. (1996) Inverse Robot Calibration Using Artificial Neural Networks. Engineering Applications of Artificial Intelligence, 9: 83-93.

[8] Lou, YF., Brunn, P. (1999) A Hybrid Artificial Neural Network Inverse Kinematic Solution for Accurate Robot Path Control. Proceedings of the I MECH E Part I Journal of Systems & Control in Engineering, 213: 23-32.

[9] Martn, P., Millan, JDR. (2000) Robot arm reaching through neural inversions and reinforcement learning. Robotics and Autonomous Systems, 31: 227-246.

[10] Oyama, E., et.al, (2001) Inverse Kinematics Learning by Modular Architecture Neural Networks with Performance Prediction Networks. Proc. IEEE Int. Conf. on Robotics and Automation, 1006-1012

[11] Vijayakumar, S., D'souza, A., Shibata, T., Conradt, J., Schaal, S., (2002) Statistical Learning for Humanoid Robots. Autonomous Robots, 12: 55-69

[12] Torras C. (2003) Handbook of Brain Theory and Neural Networks, 2nd ed. MIT Press, Cambridge, Massachusetts, 979-983

Local Cluster Neural Network Chip for Control

Liang Zhang[1], Joaquin Sitte[1], Ulrich Rueckert[2]
[1]School of Software Engineering and Data Communications,
Queensland University of Technology, Australia
[2]University of Paderborn, Germany
E-mail: {lj.zhang, j.sitte}@qut.edu.au, rueckert@hni.upb.de

Abstract

The local cluster neural network (LCNN) is an alternative to RBF networks that performs well in digital simulation. The LCNN is suitable for an analog VLSI implementation that is attractive for a wide range of embedded neural net applications. In this paper, we present the input-output characterisation of LCNN analog chip. The effect of manufacturing variations on the chip's function is investigated and analyzed.

1 Introduction

The most common use of feedforward neural nets is for finding a multivariate function that best interpolates from a given set of sample points. For this purpose the neural net designer has a choice of two main architectures: the multilayer perceptron (MLP) with sigmoidal activation functions, and the localised activation functions, such as radial basis functions (RBF) [1] [2]. In this paper we describe the *Local Cluster neural net* (LCNN) architecture (1) proposed by Geva and Sitte [3] and its analogue VLSI implementation [4] [5]. The LCNN is a special kind of multilayer perceptron where the sigmoidal neurons combine in clusters that have a localised response in input space. LCNN are more general than RBF neural nets and have all the computational advantages of the latter, while retaining the analog VLSI implementation advantages of weighted sums and sigmoids.

2 Local Cluster Neural Network definition

Like a Multilayer Perceptron (MLP) LCNN uses sigmoidal neurones in two hidden layers as Fig. 1. Clusters of sigmoids form functions localised in input space. Each neuron in the second hidden layer outputs such a local reponse function. The LCNN output is a linear combination of localised scalar functions in n-dimensional input space:

$$y(\vec{x}) = \sum_{\mu=1}^{m} v_\mu L_\mu(\mathbf{W}_\mu, \vec{r}_\mu, k, \vec{x}) \quad (1)$$

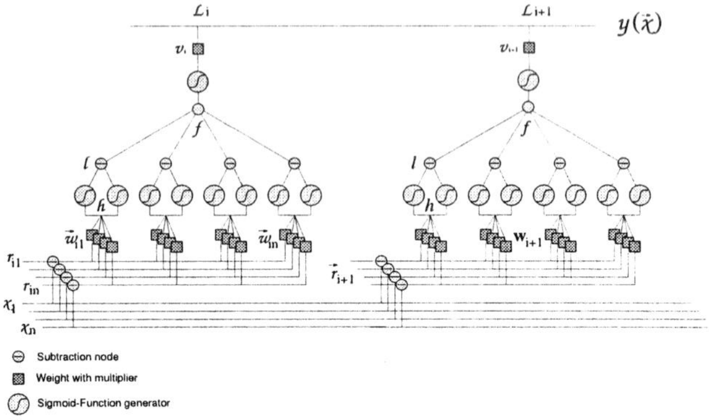

Fig. 1. Network structure of LC net

where v_μ is the output weight. The local function L_μ is the result of applying a sigmoidal *window* to a sum $f(\mathbf{W}, \vec{r}, k, \vec{x})$ of n-dimensional *ridge* functions.

$$L(\mathbf{W}, \vec{r}, k, \vec{x}) = \sigma_0(f(\mathbf{W}, \vec{r}, k, \vec{x}) - b) \quad (2)$$

The constant b allows shifting the function with respect to the window. The ridge functions $l(\vec{w}, \vec{r}, k, \vec{x})$ in

$$f(\mathbf{W}, \vec{r}, k, \vec{x}) = \sum_{i=1}^{n} l(\vec{w_i}, \vec{r_i}, k, \vec{x_i}) \quad (3)$$

are built out of 2 opposing n-dimensional sigmoid functions:

$$l(\vec{w}, \vec{r}, k, \vec{x}) = \sigma(k, h^+) - \sigma(k, h^-) \quad (4)$$

Where

$$\begin{aligned} h^+ &= \vec{w}^T(\vec{x} - \vec{r} + 1) \\ h^- &= \vec{w}^T(\vec{x} - \vec{r} - 1) \end{aligned} \quad (5)$$

For $\sigma(k, h)$ we chose the logistic sigmoid function:

$$\sigma(k, h) = \frac{1}{1 + e^{-kh}} \quad (6)$$

where k is the slope of the sigmoid.

The generation of a local cluster in analogue electronics requires modules for the following operations:

(i) Subtraction of the position vector $\vec{r}$ of the cluster from the input vector subtract.

(ii) Calculation of 2 displaced sigmoid functions (6) for each dot product eq.(5) as shown in Fig. 2(a).

(iii) Subtraction of the two sigmoids to get the ridge function (4) as shown in Fig. 2(b).

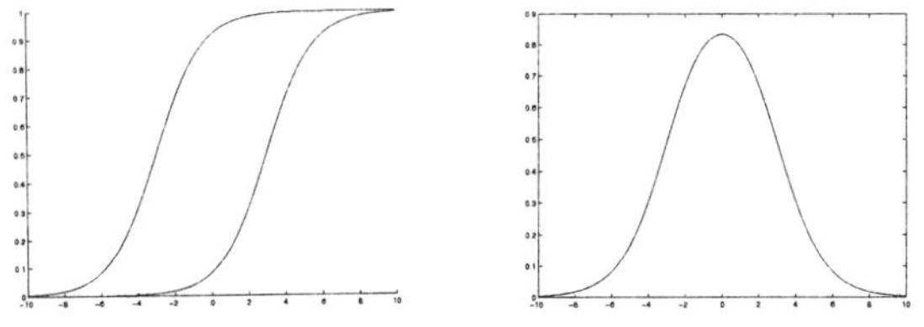

(a) $\sigma(k, h^+), \sigma(k, h^-)$ (b) Difference of sigmoids

Fig. 2. Ridge function $l(\vec{w}, \vec{r}, k, \vec{x})$

(iv) Summation of the ridge functions (3). Fig. 3(a) shows function $f(\mathbf{W}, \vec{r}, k, \vec{x})$ in two dimensions.

(v) Removal of the ridges by application of output sigmoids as per eq.(2). Fig. 3(b) shows a local function $L(\mathbf{W}, \vec{r}, k, \vec{x})$ in two dimensions.

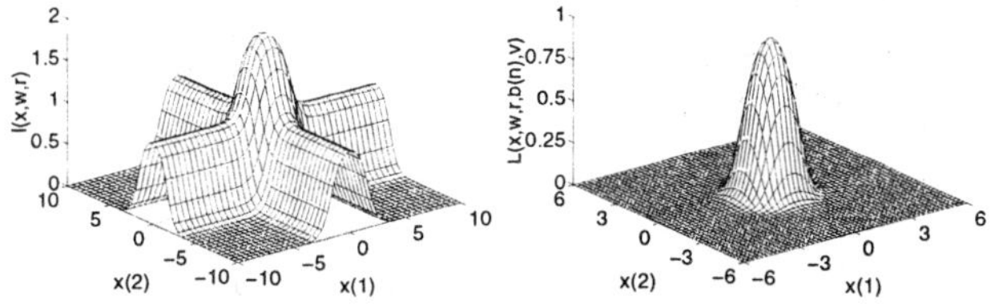

(a) $f(\vec{w}, \vec{r}, k, \vec{x})$ (b) $L(\vec{w}, \vec{r}, k, \vec{x})$

Fig. 3. Two dimensional combination of ridges (a) and output of a local function (b)

3 Hardware implementation of LCNN

The *LCX* is an analogue electronic implementation of the LCNN. All mathematical functions (1 - 6) used in the LCNN are realized in analog current mode CMOS VLSI circuits. *LCX* consists of 8 equal clusters which are sufficient for small applications, 6 inputs, one output and 8-bit digital weight storages. Fig. 4 shows the structure of a cluster. Each cluster has the same functionality. Firstly the position of the local function is determined by $\vec{x} - \vec{r}$. Then dot product $\vec{w}^T \cdot (\vec{x} - \vec{r})$ is computed. The next step is generating ridges and summing them into function $f(\vec{w}, \vec{r}, k, \vec{x})$ and subtracting b. The last step is getting the sigmoid output $L(\vec{w}, \vec{r}, k, \vec{x})$. The chip output is the weighted sum of all cluster outputs.

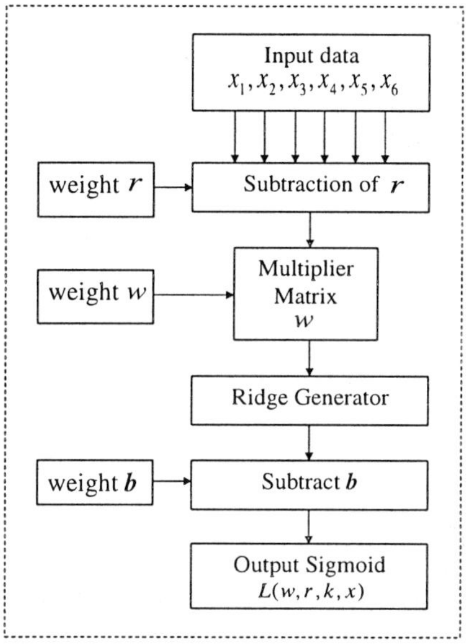

Fig. 4. The structure of LC cluster

The LCNN output is determined by the digital weights, refer to LC function (1), which are downloaded to the 8-bit weight storages. Each cluster has 45 weights, hence the *LCX* chip has 360 weights (8 clusters×45 weights) in total.

4 Test procedures and results for LCNN chip

Analog circuits are much more susceptible to manufacturing inaccuracies and fluctuations than digital circuits. These circuits will not only deviate from their expected design behavior but each one will deviate differently. Offsets, deviations and distortions are the result. Comprehensive test were carried out to determine the absolute deviation from design specification as well as the relative intra-chip and inter-chip fluctuations. The main results of these tests are summarized and discussed in this section.

4.1 Test procedures for LCNN chip

We used a PC to automate the measurements of the test process. The personal computer is connected to a LCNN test board. This board provided DA-converters for the output of analog voltages, AD-converters for measuring input of analogue voltages and digital output channels for loading digital weights onto the chips.

The software managed and loaded inputs and the digital weights onto the chip via the digital output channels. Analog stimuli patterns were generated and the response of the chip was measured. The measured data is saved to a file.

4.2 Effect of weight variations on cluster outputs

To analyze the transfer functions we record the outputs while varying specific inputs across their range for different sets of weights. Fig. 5 shows the output function shape for diagonal weight matrix with value *w* in the interval [64, 127] in steps of 8. Only one input is changed while all other inputs are zero. The width of the ridge decreases as the weight w increases. Fig. 6 shows the outputs with weight *v* variation in [8, 127] in steps of 8. The height of The ridge increases linearly with weight v. Fig. 7 shows the outputs with weight *b* variation in [0, 1920] in steps of 128. The height of ridge is varied with the variation of weight b. Fig. 8 shows the outputs with weight *r* variations in [-127, 127] in steps of 32. The ridge position from input center is varied with the variation of weight r.

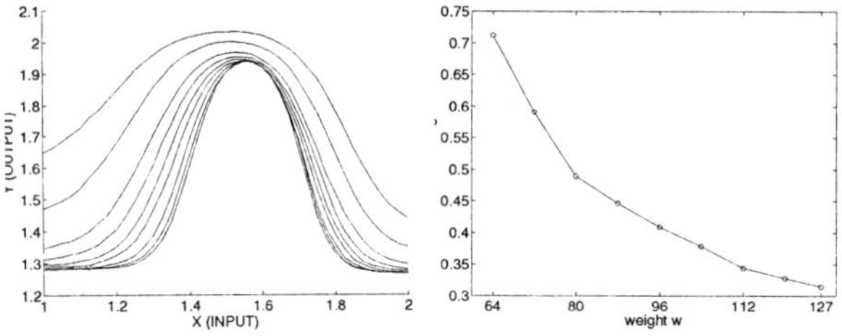

Fig. 5. Cluster 7 output shape variation for different weight w.

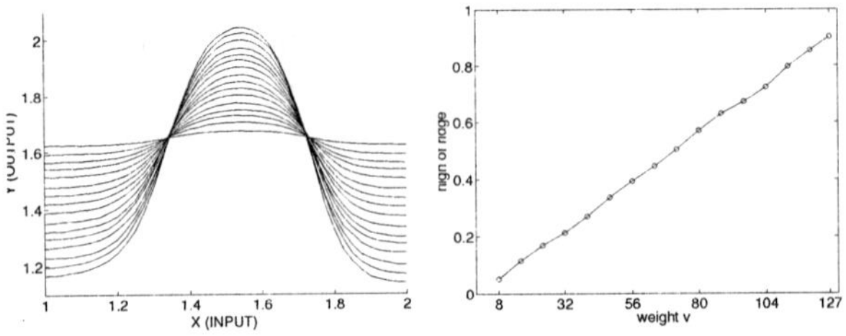

Fig. 6. Cluster 7 output shape variation for different weight v.

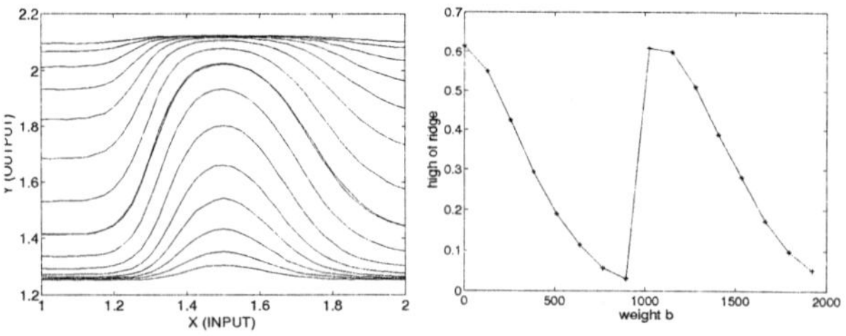

Fig. 7. Cluster 4 output shape variation for different weight b.

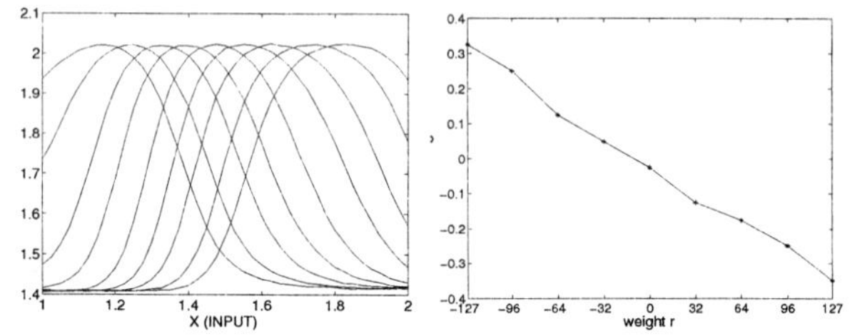

Fig. 8. Cluster 4 output shape variation for different weight r.

4.3 Deviations between clusters

Deviation test is in two aspects: the deviations in LCNN analog output against LCNN theoretical output and the output fluctuations in different clusters and different input channels.

Fig. 9 shows the fluctuation of ridge height in 4 different clusters (cluster 4 to cluster 7) and in different inputs (No. 0 to No. 5) with the same set of weight (w is 96, v is 96 and b is 1664).

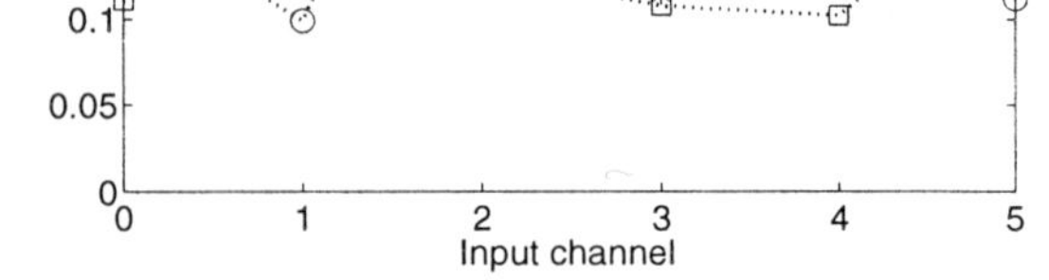

Fig. 9. Output fluctuations from height of ridge in 4 different clusters and 6 different channels.

Table 1 presents the ridge center values. Ridge centers are determined by weight r ∈ [-127, 127]. The LCNN chips are designed to generate center ridge when r is set to 0. The analog circuits produce deviations on ridge center position in different clusters and different inputs.

Table 1. Deviation of ridge centers

	cluster4	cluster5	cluster6	cluster7
channel0	-13	-127	-17	13
channel1	-6	-44	96	-6
channel2	46	-20	95	63
channel3	25	59	63	-47
channel4	88	93	70	27
channel5	42	-60	-70	-45

4.4 Two-dimensional visualisation

The previous section showed the effect of varying one parameter at a time. The output is a function of 6 inputs. Fig 10 shows the output of the LCNN chip when 2 inputs are scanned, with different b values. The measured outputs are in good accordance with the mathematical model.

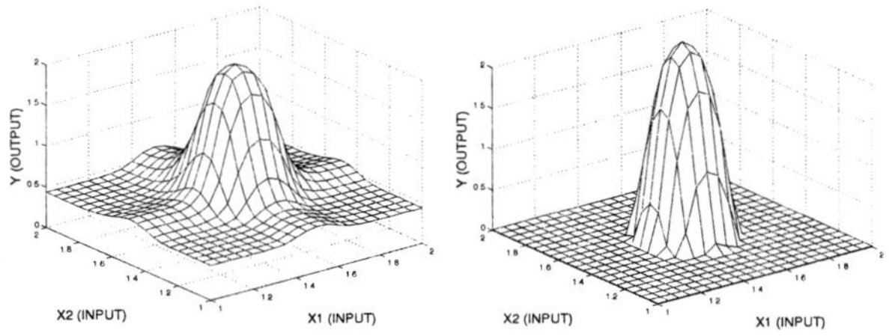

Fig. 10. Two dimensional localised cluster output.

5 Training of Local Cluster Neural Network

LCNN training is the method to seek the minimal error in the output by comparing to a desired response. We use gradient decent algorithm for LCNN training.

LCNN training has been demonstrated its good performance in digital computer simulation [3].

The weights obtained by training the mathematical model (LCNN) on a data set will produce a different function on the chip due to fluctuations. The lack of weight transferability can be overcome by in-loop training as shown in fig. 11. The in-loop training is to connect a computer with LCNN chip and use neural net software to train the chip. The computer writes input and weights into the chip, reads output from the chip, and calculates and updates the weights in the software. The updated weights are downloaded to the chip for the next training iteration. The training is stopped when current error is less than objective error.

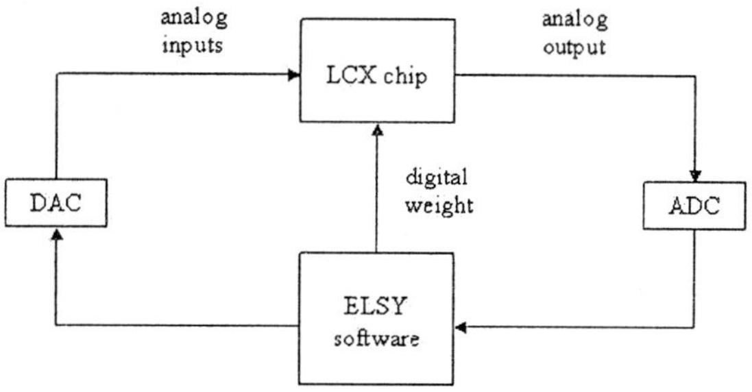

Fig. 11. LCX chip in-loop training

6 Conclusions

The Local Cluster Neural Network (LCNN) has good performance for function approximation in digital computer simulation. The mathematical operations can be implemented in analog neural net hardware as well. However, in contrast to digital implementations analog circuits implementations deviate from the ideal mathematical model. The functionality of LCNN and its accuracy are influenced by the fluctuations as shown in this paper.

In-loop training is the first step to overcome these fluctuations partly. A better solution could be in-circuit learning or on-chip training. Therefore our research at the next stage is the implementation of LCNN on-chip training.

References

[1] Tomaso Poggio and Federico Girosi. Networks for approximation and learning. *IEEE*, 78:1481–1497, 1990.

[2] J. Park and Sandberg I. W. Universal approximation using radial-basis-functions networks. *Neural Computation*, (3):246–257, 1991.

[3] Shlomo Geva, Kurt Malmstrom, and Joaquin Sitte. Local cluster neural net: Architecture, training and applications. *Neurocomputing*, (20):35–56, March 1998.

[4] Tim Körner. *Analog VLSI Implementation of a Local Cluster Neural Net.* PhD thesis, University of Paderborn, 2000.

[5] Joaquin Sitte, Tim Körner, and Ulrich Rückert. Local cluster neural net analog vlsi design. *Neurocomputing*, (19):185–197, August 1997.

A Switching Controller for Nonlinear Systems via Fuzzy Models

M. Boumehraz[1], K. Benmahammed[2]

[1]Department of Automatic, University of Biskra, Algeria
[2]Department of Electronics, University of Setif, Algeria
E-mail: medboumehraz@netcourrier.com, Khierben@ieee.org

Abstract

A Lyapunov based switching control design method for non linear systems using fuzzy models is proposed. The switching controller consists of several linear state feedback controllers; only one of the linear controllers is employed at each moment according to a switching scheme. The gains of the linear state feedback controllers are derived based on Lyapunov stability theory. The fuzzy design model is represented as a set of uncertain linear subsystems and then sufficiency conditions for the system to be globally stabilisable by the switching controller are given. The proposed design method is illustrated trough numerical simulations on the chaotic Lorenz system.

1 Introduction

Fuzzy techniques have been widely adapted to model complex non linear plants. By using a Takagi-Sugeno fuzzy model, a non linear system can be expressed as a weighted sum of simple subsystems. This model gives a fixed structure to some non linear systems and thus facilitates their analysis. There are two ways to obtain the fuzzy model: 1) - by applying identification methods with input-output data from the plant [1] [2], 2)-or directly from the mathematical model of the non linear plant [3].

In this paper, we propose a Lyapunov based design of a switching linear controller for a class of fuzzy models.

The rest of the paper is organized as follows. Section 2 reviews the continuous T-S fuzzy models. Section 3 gives the structure of the switching controller and the controller design method is proposed in section 4. Then we provide an application example; the control problem of the chaotic Lorenz system. Finally, we present our conclusions.

2 Fuzzy model

The continuous-time Takagi-Sugeno fuzzy dynamic model is a piecewise interpolation of several linear models through membership functions. The fuzzy model is described by fuzzy if-then rules. The ith rule of the fuzzy model for the non linear system is of the form:

Plant rule i:

if $z_1(t)$ *is* F_{i1} *and* ... $z_g(t)$ *is* F_{ig} *then*

$$\dot{x}(t) = A_i x(t) + B_i u(t) \quad \text{for } i = 1, 2, \ldots, r \tag{1}$$

where $x(t) = [x_1(t), x_2(t), \ldots, x_n(t)]^T \in R^n$ denotes the state vector, $u(t) = [u_1(t), \ldots, u_m(t)]^T \in R^m$ is the input vector. F_{ij} is the fuzzy set, $A_i \in R^{nxn}$, $B_i \in R^{nxm}$, r is the number of if-then rules, and $z_1(t), z_2(t), \ldots, z_g(t)$ are some measurable system variables, i.e. the premise variables.

The output of the fuzzy model can be expressed as:

$$\dot{x}(t) = \frac{\sum_{i=1}^{r} \omega_i(z(t))[A_i x(t) + B_i u(t)]}{\sum_{i=1}^{r} \omega_i(z(t))}$$

$$= \sum_{i=1}^{r} \alpha_i(z(t))[A_i x(t) + B_i u(t)] \tag{2}$$

where $\omega_i(z(t)) = \prod_{j=1}^{g} F_{ij}(z(t))$

$F_{ij}(z_j(t))$ is the grade of membership of $z_j(t)$ in F_{ij}.

The TS fuzzy model (2) is a general non linear time varying equation and has been used to model the behaviour of complex non linear dynamic systems.

The TS fuzzy model (2) has strong nonlinear interactions among the fuzzy rules, which complicates the analysis and the control of the system. In order to overcome these difficulties, the TS fuzzy model (2) is represented as a set of uncertain linear systems. Each sub-space is defined as:

$$S_l = \left\{ X \middle| \alpha_l(x) \geq \alpha_l^{min} \right\}, \quad l = 1, 2, \ldots, r \tag{3}$$

where $0 \leq \alpha_i^{min} \leq 1$ is a scalar to be determined.

The characteristic function of S_l is defined by:

$$\zeta_i = \begin{cases} 1 & x \in S_l \\ 0 & x \notin S_l \end{cases} \qquad \sum_{i=1}^{r} \zeta_i = 1 \tag{4}$$

In each subspace S_l, the fuzzy model (2) can be represented as:

$$\dot{x}(t) = \left(A_l + \sum_{\substack{i=1 \\ i \neq l}}^{r_l} \alpha_i(t)(A_i - A_l) \right) x(t) + \left(B_l + \sum_{\substack{i=1 \\ i \neq l}}^{r_l} \alpha_i(t)(B_i - B_l) \right) u(t) \tag{5}$$

$$\dot{x}(t) = \left(A_l + (1 - \alpha_l(t)) \sum_{\substack{i=1 \\ i \neq l}}^{r} \alpha_i'(t)(A_i - A_l) \right) x(t) + \left(B_l + (1 - \alpha_l(t)) \sum_{\substack{i=1 \\ i \neq l}}^{r} \alpha_i'(t)(B_i - B_l) \right) u(t)$$

$$\sum_{\substack{i=1 \\ i \neq l}}^{r} \alpha_i'(t) = \sum_{\substack{i=1 \\ i \neq l}}^{r} \frac{\alpha_i(t)}{1 - \alpha_l(t)} = 1$$

$$\dot{x}(t) = (A_l + (1 - \alpha_l(t)) \Delta A_l) x(t) + (B_l + (1 - \alpha_l(t)) \Delta B_l) u(t) \tag{6}$$

where

$$\Delta A_l(\alpha'(t)) = \sum_{\substack{i=1 \\ i \neq l}}^{r} \alpha_i'(t)(A_i - A_l)$$

$$\Delta B_l(\alpha'(t)) = \sum_{\substack{i=1 \\ i \neq l}}^{r} \alpha_i'(t)(B_i - B_l) \tag{7}$$

The fuzzy system in subspace S_l consists of a dominant nominal system with matrix A_l and a set of interacting systems determining the effect of the control law on the other non-dominant systems. The fuzzy model (6) can be viewed as an uncertain linear dynamical system model [4].

We assume that the matrices ΔA_l and ΔB_l can be written as:

$$\Delta A_l(\alpha'(t)) = B_l . D_l(\alpha'(t)) + E_l(\alpha'(t))$$
$$\Delta B_l(\alpha'(t)) = B_l . F_l(\alpha'(t)) \tag{8}$$

The matrices $B_l . D_l(\alpha'(t))$ and $B_l . F_l(\alpha'(t))$ model the matched uncertainties whereas the term $E_l(\alpha'(t))$ represent the mismatched uncertainties. We assume that:

$$F_l(\alpha'(t)) + F_l^T(\alpha'(t)) + \frac{2}{1 - \alpha_l(t)} I_m > 0 \tag{9}$$

We assume also that the matrix functions $D_l(\alpha'(t))$ and $E_l(\alpha'(t))$ are bounded by:

$$\|D_l(\alpha'(t))\| \leq \overline{D_l}, \quad \|E_l(\alpha'(t))\| \leq \overline{E_l}$$

$$\sigma_l = \min_{\alpha(t)} \left[\frac{1}{2} \lambda_{min} \left(F_l(\alpha'(t)) + F_l^T(\alpha'(t)) \right) \right] \tag{10}$$

for all $\alpha'(t) \in [0,1]^{r-1}$.

3 Switching controller approach

A switching controller is employed to control the fuzzy system (2). The switching controller consists of some linear state feedback controllers that will be switched from one to another to control the system. The switching controller is described by:

$$u(t) = -\sum_{l=1}^{r} \zeta_l K_l x(t) \tag{11}$$

with

$$K_l = \frac{\gamma_l}{2} B_l^T P_l \tag{12}$$

and

$$\sum_{l=1}^{r} \zeta_l = 1, \quad \zeta_l \in \{0,1\} \tag{13}$$

K_l is the local state feedback gain in subspace S_l to be designed. The parameter $\gamma_l > 0$ is a scalar and the matrix P_l is the positive definite solution of the following algebraic Ricatti equation:

$$A_l^T P_l + P_l A_l - \eta_l P_l B_l B_l^T P_l = -2 Q_l \tag{14}$$

where $Q_l \in R^{n \times n}$ is a symmetric positive definite matrix and η_l is any given positive constant.

It can be seen that (11) is a linear combination of r linear state-feedback controllers.

4 Controller design

In this section, the switching controller will be designed to guarantee the system stability.

Theorem:

The state feedback controller given by (12) where:

$$\gamma_l = \frac{\eta_l + \delta_l^2 \overline{D_l}^2}{1 + \sigma_l} \tag{15}$$

globally asymptotically stabilise the uncertain sub-system (6) for arbitrary $D_l(\alpha'(t))$, $E_l(\alpha'(t))$ and $F_l(\alpha'(t))$ that satisfy the norm bounds (10) and the conditions (8) and (9) if :

$$\frac{1}{\delta_l^2} < 2 \left(\frac{1}{1 - \alpha_l(t)} \lambda_{min}(Q_l) - \overline{E_l} \lambda_{max}(P_l) \right) \tag{16}$$

and

$$\alpha_l(t) > 1 - \frac{\lambda_{min}(Q_l)}{\overline{E}_l\lambda_{max}(P_l)} \tag{17}$$

Proof: Let define the positive definite function V_l as:

$$V_l(t) = x^T P_l x \tag{18}$$

where $P_l \in R^{nxn}$ is the solution of algebraic Ricatti equation (14).

$$\dot{V}_l = \dot{x}^T P_l x + x^T P_l \dot{x}$$

$$= x^T\left(A_l^T P_l + P_l A_l\right)x + 2(1-\alpha_l)x^T P_l B_l D_l x + 2(1-\alpha_l)x^T P_l E_l x$$

$$-\gamma_l x^T P_l B_l\left[I_m + \frac{1}{2}(1-\alpha_l)\left(F_l^T + F_l\right)\right]B_l^T P_l x$$

$$\leq x^T\left(A_l^T P_l + P_l A_l - \eta_l P_l B_l B_l^T P_l\right)x + 2(1-\alpha_l)x^T P_l B_l D_l x$$

$$+ 2(1-\alpha_l)x^T P_l E_l x - \delta_l^2 \overline{D}_l^2 x^T P_l B_l B_l^T P_l x$$

$$\leq -2\lambda_{min}(Q_l)\|x\|^2 - \delta_l^2\overline{D}_l^2\left\|B_l^T P_l x\right\|^2 +$$

$$2(1-\alpha_l)\left\|B_l^T P_l x\right\|\left\|D_l\right\|\|x\| + 2(1-\alpha_l)\left\|P_l E_l x\right\|\|x\|$$

$$\leq -2\lambda_{min}(Q_l)\|x\|^2 + 2(1-\alpha_l)\left\|P_l E_l x\right\|\|x\| + \frac{1}{\delta_l^2}(1-\alpha_l)\|x\|^2$$

$$\leq \left[-2\lambda_{min}(Q_l) + 2(1-\alpha_l)\overline{E}_l\lambda_{max}(P_l) + \frac{1}{\delta_l^2}(1-\alpha_l)\right]\|x\|^2 \tag{19}$$

$$\dot{V}_l < 0 \Rightarrow -2\lambda_{min}(Q_l) + 2(1-\alpha_l)\overline{E}_l\lambda_{max}(P_l) + \frac{1}{\delta_l^2}(1-\alpha_l) < 0$$

$$\dot{V}_l < 0 \Rightarrow \frac{1}{\delta_l^2} < \frac{2}{1-\alpha_l}\lambda_{min}(Q_l) - 2\overline{E}_l\lambda_{max}(P_l)$$

$$\frac{2}{1-\alpha_l}\lambda_{min}(Q_l) - 2\overline{E}_l\lambda_{max}(P_l) > 0 \Rightarrow$$

$$2\overline{E}_l\lambda_{max}(P_l) < \frac{2}{1-\alpha_l}\lambda_{min}(Q_l)$$

$$\alpha_l > 1 - \frac{\lambda_{min}(Q_l)}{\overline{E}_l\lambda_{max}(P_l)} \tag{20}$$

We define:

$$\alpha_l^{min} = max\left(0, 1 - \frac{\lambda_{min}(Q_l)}{\overline{E}_l\lambda_{max}(P_l)}\right) \tag{21}$$

α_l^{min} is the minimum value of α_l that guarantee the stability of the global system using the local subsystem S_l state-feedback gain K_l.

In each subspace, the command is given by:

$$u(t) = -K_l x(t) \tag{22}$$

The boundary of the sub-region S_l is determined by the minimal value that guarantees its stability α_l^{min}.

Lemma:

The global system is asymptotically stable if there exists, at each moment *t*, at least one value $\alpha_k(t)$ satisfying:

$$\alpha_k(t) \geq \alpha_k^{min},\ k = 1, 2, \ldots, r \tag{23}$$

or $\bigcup_{l=1}^{r} S_l = S$ where S is the global state space.

In overlapping regions many subsystems may satisfy this condition. In this case the control is inferred by selecting the control of the dominant system whose membership function is of maximum distance from the boundary of its stability region determined by α_l^{min}:

$$u(t) = -K_l x(t) \tag{24}$$

$$l = \underset{i=1,r}{arg\ max}\left\{\alpha_i(t) - \alpha_i^{min}, i = 1, 2, \ldots, r\right\} \tag{25}$$

The design procedure of the switching controller is summarized in the following steps:

- *Step 1:* Obtain the fuzzy plant model of the non linear plant by means of the methods in [1],[2], [4], or other suitable ways.
- *Step 2:* Determine the subsystems S_l matrices A_l, B_l, ΔA_l and ΔB_l for $l = 1, \ldots, r$ and check if condition (9) is verified for each subsystem.
- *Step 3:* Design the state-feedback gain K_l for each subsystem S_l according to (14), (15) and (16). And determine the value of α_l^{min} for $l = 1, \ldots, r$.
- *Step 4:* Check if the condition (23) is satisfied, otherwise go to *Step 3* and choose other values for the free design parameters.

5 Simulation example

To show the effectiveness of the proposed method, we simulate the control of the chaotic Lorenz system. The control objective is to drive its chaotic trajectory to the origin. The Lorenz equations are as follows [5]:

$$\frac{d}{dt}\begin{bmatrix} x_1(t) \\ x_2(t) \\ x_3(t) \end{bmatrix} = \begin{bmatrix} -\sigma x_1(t) + \sigma x_2(t) \\ r x_1(t) - x_2(t) - x_1(t)x_3(t) \\ x_1(t)x_2(t) - b x_3(t) \end{bmatrix} \tag{26}$$

The nominal values of (σ, r, b) are $(10, 28, 8/3)$ for chaos to emerge. The system can be described by the following T-S fuzzy model [5]:

Rule 1: If $x_1(t)$ is about M_1 then $\dot{x}(t) = A_1 x(t)$

Rule 2: If $x_1(t)$ is about M_2 then $\dot{x}(t) = A_2 x(t)$

Where

$$A_1 = \begin{bmatrix} -\sigma & \sigma & 0 \\ r & -1 & -M_1 \\ 0 & M_1 & -b \end{bmatrix}, \; A_2 = \begin{bmatrix} -\sigma & \sigma & 0 \\ r & -1 & -M_2 \\ 0 & M_2 & -b \end{bmatrix}$$

$M_1 = -20$ and $M_2 = 30$ (27)

The membership functions are given by:

$$\mu_1(x(t)) = \frac{-x_1(t) + M_2}{M_2 - M_1}, \; \mu_2(x(t)) = \frac{x_1(t) - M_1}{M_2 - M_1} \quad (28)$$

The input matrices B_1 and B_2 are chosen as:

$$B_1 = B_2 = I_3 \quad (29)$$

The fuzzy model can be decomposed into two subsystems:

Subsystem 1: $\dot{x}(t) = (A_1 + (1 - \mu_1)\Delta A_1)x(t) + B_1 u(t)$

Subsystem 2:

$\dot{x}(t) = (A_2 + (1 - \mu_2)\Delta A_2)x(t) + B_2 u(t)$

Where: $\Delta A_1 = A_2 - A_1$, $\Delta A_2 = A_1 - A_2$,

$\Delta B_1 = \Delta B_2 = 0$

For $Q_1 = Q_2 = 50 I_3$ and $\eta_1 = \eta_2 = 0.1$ the resolution of the Ricatti equation (14) gives:

$$P_1 = \begin{bmatrix} 3.1211 & 1.5040 & 0.4201 \\ 1.5040 & 3.1526 & 0.0437 \\ 0.4201 & 0.0437 & 2.9063 \end{bmatrix}, \; P_2 = \begin{bmatrix} 3.0589 & 1.3771 & -0.5838 \\ 1.3771 & 3.1339 & -0.0675 \\ -0.5838 & -0.0675 & 2.9162 \end{bmatrix}$$

and

$\alpha_1^{min} = \max(0, -0.4181) = 0, \alpha_2^{min} = \max(0, -0.4490) = 0$

The state feedback gains:

$$K_1 = \begin{bmatrix} 96.433 & 46.467 & 12.981 \\ 46.467 & 97.404 & 1.350 \\ 12.981 & 1.350 & 89.794 \end{bmatrix} \; K_2 = \begin{bmatrix} 91.425 & 41.160 & -17.447 \\ 41.160 & 93.668 & -2.016 \\ -17.447 & -2.016 & 87.160 \end{bmatrix}$$

The initial values of states are $x(0)^T = [20, 20, 20]$. The simulation time is 40 s. The control input is activated at t=20s. Before the activation of the control the phase trajectory of the Lorenz system was chaotic. However, after the activation of the command the phase trajectory is quickly directed to the origin as shown in figures 1 and 2. In this example the boundary of the two sub-spaces are determined by $\alpha_1^{min} = 0$ and $\alpha_2^{min} = 0$ which means that the two sub-spaces are equal to the global state space and the chaotic system can be controlled using only one controller.

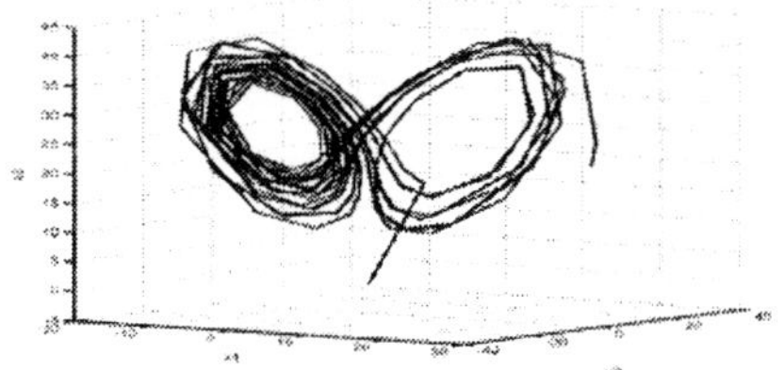

Fig. 1. The phase trajectory of the controlled Lorenz system.

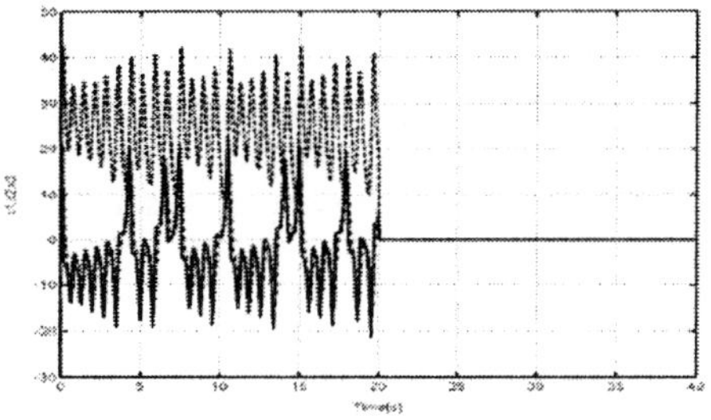

Fig. 2. States of the Lorenz system.

6 Conclusion

In this paper a Lyapunov based method has been proposed to design a fuzzy model based switching controller for non linear systems. Under some conditions this switching controller has the ability to stabilize the non linear system. The control of the chaotic Lorenz system has been used demonstrate the effectiveness of this approach.

References

[1] Takagi, T., Sugeno M., (1985) Fuzzy identification of systems an its application to modelling and control, IEEE Transactions on Systems, men and Cybernitics 15: 116-132.

[2] Sugeno M., Kang, G. T. (1988) Structure identification of fuzzy models, Fuzzy Sets and Systems, 28:15-33.

[3] Kim E., Park M., Ji S., Park M. (1999) A new approach to fuzzy modeling, IEEE Transactions on fuzzy systems, 7: 120-132.

[4]- Zak S. H. (1999) Stabilizing fuzzy system models using linear controllers, IEEE Transactions on Fuzzy Systems, 7: 236-240.

[5] Lee H. J., Park J. B., Chen G. (2001) Robust fuzzy control of nonlinear systems with parametric uncertainties, IEEE Transactions on fuzzy systems, 9: 369-380.

Competitive Decentralized Autonomous Neural Net Controllers

Takehiro Ohba and Masaru Ishida

Chemical Resources Laboratory, Tokyo Institute of Technology
R1-17 4259 Nagatsuta, Midori-ku, Yokohama 226-8503, Japan
E-mail: ohba@res.titech.ac.jp

Abstract

A simple and effective method is proposed for controlling a system consists of small processes. Each process is controlled by a decentralized autonomous neural network controller. These controllers compete with each other in order to increase their performances. As a result of the competition, the performance of whole system is kept at a suboptimal level. A control of example system consist of lots of processes is performed.

1 Introduction

A control of a distributed system is essential for the field of manufacturing, power generation and robotics. A well-designed distributed autonomous control system can attain tolerance for malfunction and quick adaptability for various changes during operation. In previous studies in [1], [2], a large-scale system that consists of a number of subsystems is controlled by decentralized control methods. The decentralized control of each subsystem has been performed by using only its local information of each subsystem. Sometimes it is assumed that only local output feedback is available [3].

Neural networks have been used for decentralized control by papers [4], [5]. In those studies, learning features of the neural network have been molded in the known framework of decentralized control method. The genetic algorithm is also used for decentralized control on [6].

In this paper, a new strategy to use neural network for the decentralized control is proposed. This strategy is based on two simple policies: First, each output of the small process is calculated by using only the output of whole system. By adopting this policy, each controller needs not any common rules or data. Second, the demand for higher performance of each individual process leads the higher performance of the whole system. In this strategy, each controller behaves like stockholder in the stock exchange market and pursues their benefit.

An example system is adopted, and the characteristics of the proposed method are discussed based on the control results.

2 Controlled System

An example controlled system consists of three small processes in which the output value y_i is equal to its input value u as Eq.(1). The output value of whole system Y is the sum of the output values of the processes as Eq.(2). We assumed that the performance index of each process p_i depends on its output value y_i as Eq.(3). In Fig.1, the profiles of the performance index of the three processes are shown. For Process 1 and 2, their performances take the largest value at $y_1 = y_2 = 0.25$, and for Process 3, at $y_3 = 1.0$. The sum of the both performance indexes is the performance of the whole system in Eq.(4).

The objective of the control is to control the output of the system Y at a specified target value R, and to let the performance index of the whole system P as high as possible.

$$y_i = u_i \ , \ u_i = [0,1] \tag{1}$$

$$Y = \sum_i^3 y_i \ , \ Y = [0,3] \tag{2}$$

$$p_i = f_i(y_i) \ , p_i = [0,1] \tag{3}$$

$$P = \sum_i^3 p_i \ , \ P = [0,3] \tag{4}$$

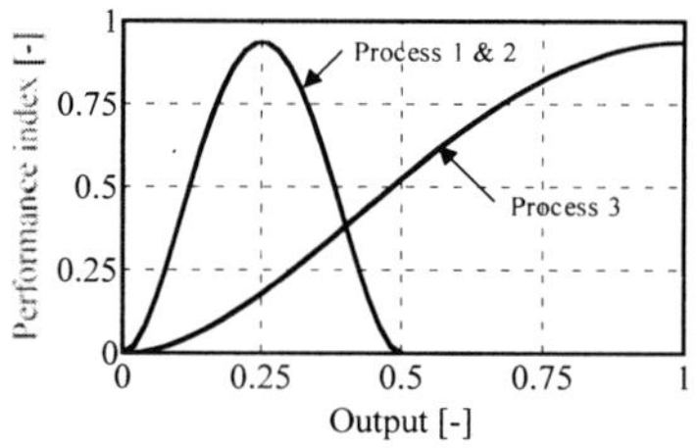

Fig. 1. Performance index profiles of 3 processes

3 Control Strategy

3.1 Control of output of whole system Y

The control system diagram is shown as Fig.2. The control of the output of whole system Y is performed by using the neural net controllers shown in Fig. 3. The network has two inputs ΔY, a_i, and an output Δu_i and two weights of the connections $\omega_{Y,i}$, $\omega_{a,i}$. ΔY is the difference between the target value and the present value, and Δu_i is the change in the ith manipulated variable. Since ΔY is used as input and Δu_i as output, this control system takes velocity form. a_i is a preference factor which is used for the competition with other controllers and mentioned later.

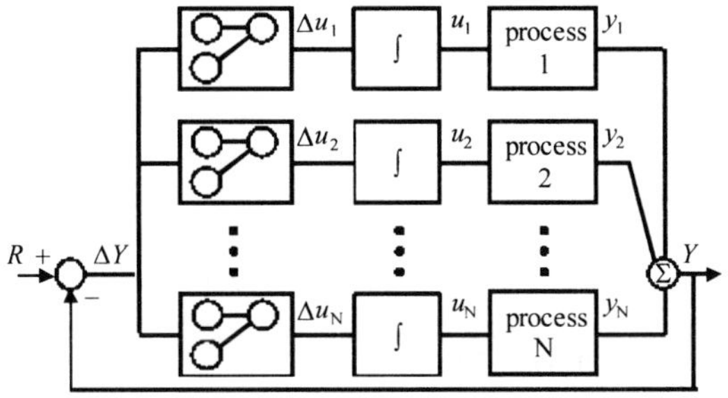

Fig. 2. Control system structure

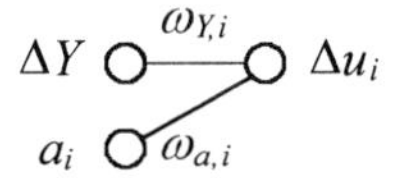

Fig. 3. Structure of neural net controller

For the set-point control, the input variable ΔY_j at sampling step j is the error of the controlled variable Y_j from the target value R.

$$\Delta Y_j = R - Y_j \tag{5}$$

The change in the value $\Delta u_{i,j}$ is calculated by the neural network as following manner.

The product of each input variable and each weight of the connection is inputted into a sigmoidal function f shown by (6) and (7).

$$\Delta u_{i,j} = f\left(\omega_{Y,i,j}\Delta Y_j + \omega_{a,i,j}\Delta a_{i,j}\right) \tag{6}$$

$$f(x) = \frac{1-e^{-x}}{1+e^{-x}} \tag{7}$$

The new value for the input of the process $u_{i,j+1}$ is obtained by adding the value of the output $\Delta u_{i,j}$ to the present value $u_{i,j}$.

$$u_{i,j+1} = u_{i,j} + \Delta u_{i,j} \tag{8}$$

Then the output of each process $y_{i,j+1}$ and the output of the whole system Y_{j+1} are obtained by Eq.(1) and Eq.(2), respectively.

The values of the weights of the connections $\omega_{a,i}$ are fixed as 1.0 and only the weights $\omega_{Y,i}$ are used as modeling parameters. The initial values of the weights of the connections, $\omega_{Y,i}$, are given randomly, and they are updated at every control action by the following direct learning method. We use the values, $\Delta Y_{j-1,\text{learning}}$ and $\Delta u_{i,j-1}$ as the data set for learning of the responses of the process. Since the previous control action $\Delta u_{i,j-1}$ changed the previous output Y_{j-1} to the present value Y_j, the weight of the connection $\omega_{Y,i}$ should be updated to output $\Delta u_{i,j-1}$ when the inputs are $\Delta Y_{j-1,\text{learning}}$ shown by (9) [7].

$$\Delta Y_{j-1,\text{learning}} = Y_j - Y_{j-1} \tag{9}$$

The output values of the neural network $\Delta u_{i,j-1,\text{learning}}$ are calculated by using $\Delta Y_{j-1,\text{learning}}$

$$\Delta u_{i,j-1,learning} = f\left(\omega_{Y,i,j}\Delta Y_{j-1,\text{learning}} + \omega_a a_{i,j-1}\right) \tag{10}$$

Each weight of the connection is updated by the following delta rule. The weight of connection $\omega_{Y,i}$ is slightly changed to the direction that the difference between $\Delta u_{i,j-1}$ and $\Delta u_{i,j-1,\text{learning}}$ decrease. A variable α is the update coefficient and it is specified 0.85.

$$\omega_{Y,i,j+1} = \omega_{Y,i,j} - \alpha\frac{d(\Delta u_{i,j-1} - \Delta u_{i,j-1,learning})}{d\omega_{Y,i,j}} \tag{11}$$

Figure 4 shows the control result when R=2.25 when the initial outputs of processes are zero. The output of whole system Y reached the target value after 6 iterations. Since the initial value of weight of 3rd process $\omega_{Y,3}$ is larger than others, the output y_3 increased faster and finally, y_3 became the maximum value 1.0 and y_1 became 0.9. The control results for whole range of output in Fig. 5 show that the proposed neural network can control the output of whole system.

When the learning is not performed, the outputs of processes oscillated and the output of whole system cannot be controlled

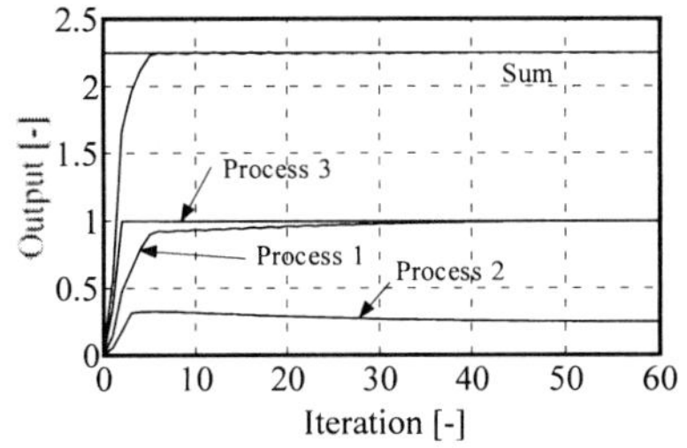

(a) Changes in the output

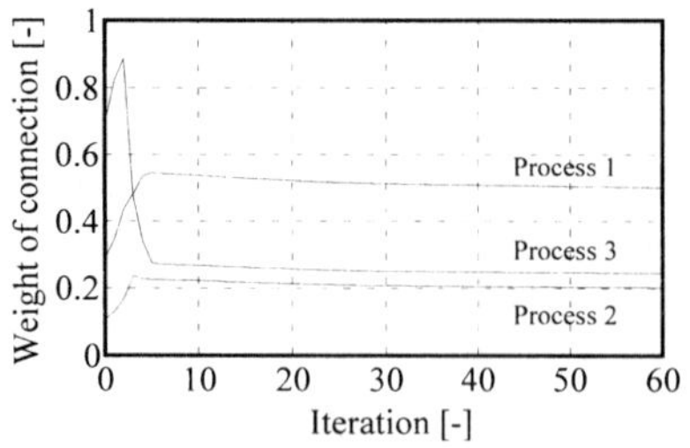

(b) Changes in weights of connections

Fig. 4. Control result for R=2.25

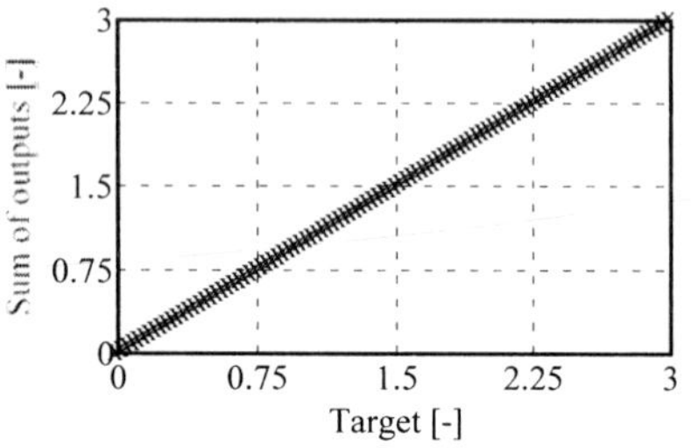

Fig. 5. Control results for 101 targets

3.2 Increase in performance index

The second input for the neural network is introduced to increase the performance index p_i and P.

In Fig.4, since $\omega_{Y,2}$ is smaller than $\omega_{Y,1}$, the rate of change in y_1 is faster than that of y_2. We can use this characteristic feature to control the competition among the neural net controllers. When a_i is positive, the change in the output of the controller Δu_i is increased on similar occasions when the weight $\omega_{Y,i}$ increased, and when a_i is negative, Δu_i is decreased. With this property, we can also change the output u_i to increase the performance of process by setting appropriate value for a_i. The value of $\omega_{a,i}$ is fixed at 1.0. The variable a_i is proposed as Eq.(12).

$$a_{i,j} = f\left(\frac{dp_{i,j}}{dy_{i,j}}\right) p_{i,j} \left|\Delta Y_j\right| + \left| f\left(\frac{dp_{i,j}}{dy_{i,j}}\right) \right| C_a \quad (12)$$

This second input a_i is called the preference factor. The first term is the product of three variables. The slope of the performance index profile against the output of the process is used. This value indicates the direction for increase in the performance of the process. Since the value of the slope often takes huge value, the sigmoidal function in (7) is used for standardization. The value of the performance index p_i is used in order to introduce the competition with other processes. By employing $p_{i,j}$, priority is given to the process that has higher performance. The absolute value of the error of the whole output $|\Delta Y|$ will prevent the process from deviating from the target value. By using this absolute value, the first term is kept at zero when the output of whole system has reached the target value. The second term is a bias factor used for optimization after the target value is achieved. Since C_a has constant small value, the output gradually moves to the optimal point.

3.3 Control results

The control result for R=2.25 is shown in Fig. 6. As mentioned before, the output of the whole system is controlled satisfactorily. After reached the target value at 6th iteration, the performance of whole system increased gradually. Since the process 1 increased and the process 2 decreased in Fig. 4(a), the total performance index increased. This optimization is obtained by the competition between the process 1 and 2.

The 101 sets for the target value between 0 and 3 are examined and each examination was performed five times. The initial values of weights of connections are specified randomly between 0 and 1. The control results are shown in Fig.5. Both cases with and without the preference factor were tested and both gave satisfactory control of the output of the whole system Y.

Figure 7 shows the achieved values of performance index of whole system P. The solid line indicates the theoretical maximum values and the dashed line indicates the theoretical minimum values. It is found that the performance became high when the preference factor is used. When the value of C_a takes 0.0005, the performance index of the whole system became highest.

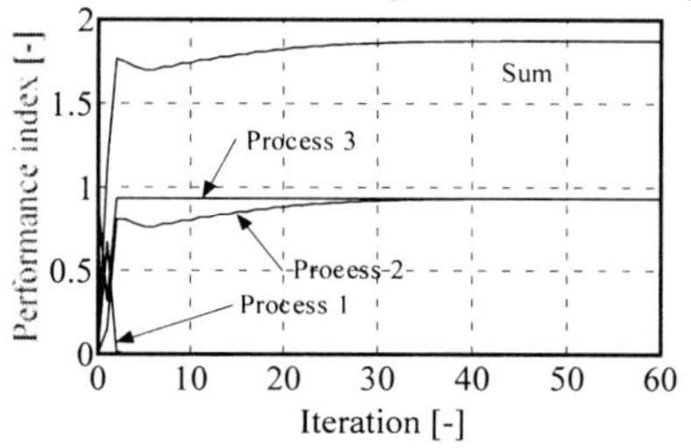

(a) Changes in performance indexes

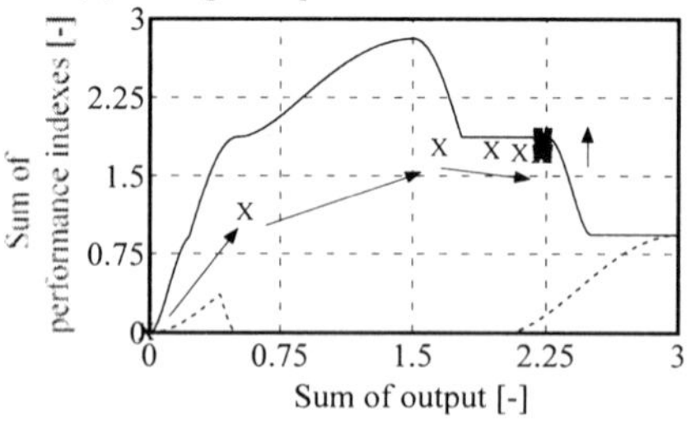

(b) Changes in sum of performance indexes

Fig. 6. Control result for R=2.25

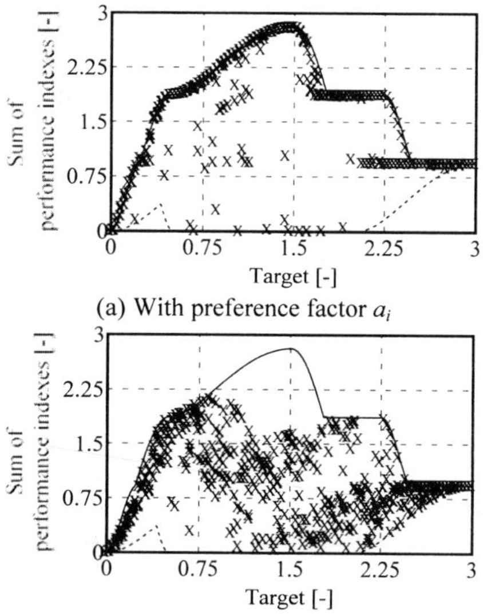

(a) With preference factor a_i

(b) Without preference factor a_i

Fig. 7. Results of performance indexes

4 Application for larger system

The control of the system consists of 50 processes are performed. From the 1st to 49th process has same performance index profile as process 1 and 2 shown in Fig. 1. The 50th process has the profile same as process 3 in Fig. 1. The control of whole output Y is also accomplished and Fig. 8 shows the results of performance indexes achieved.

From these results, the proposed method achieved sufficient result for such a large system.

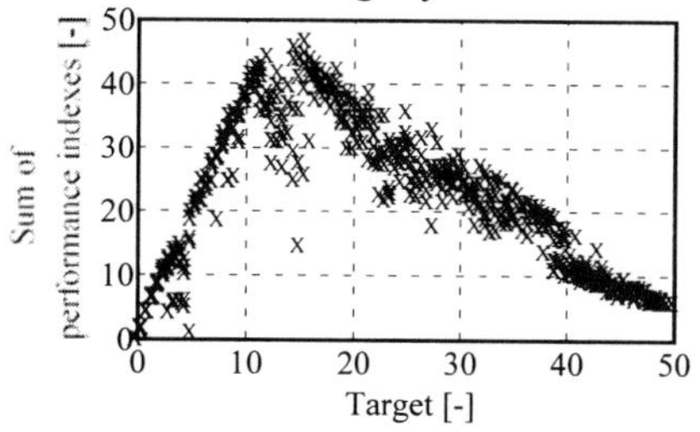

Fig. 8. Control results for 50 processes system

5 Conclusion

Proposed competitive decentralized autonomous neural network controllers are very simple but effective for controlling the system consists of small processes. Each neural net controller uses only the error of the whole system and the preference factor for increasing its performance. The preference factor is the key to achieve efficient and low-cost operation and the bias factor for optimization makes remarkable progress.

The example control results shows that the proposed method is effective. This method can be used in various field of science.

References

[1] Lin Shi and Sunil K. Singh (2003) Decentralized adaptive controller design for large scale systems with higher order interconnections. IEEE Trans. Autom. Contr. 37:1106-1118

[2] Wen Changyun and Y. C. Soh (1999) Decentralized Model Reference Adaptive Control Without Restriction on Subsystem Relative Degrees. IEEE Trans. Autom. Contr. 44: 1464-1469

[3] Jain Sandeep and F. Khorrami (1997) Decentralized adaptive output feedback design for large-scale nonlinear systems. IEEE Trans. Autom. Contr. 42:729-735

[4] Chen, Chyi-Tsong and Yen Jia-Hwang (1998) Multivariable process control using decentralized single neural controllers. J. of Chem. Eng. of Japan 31:14-20

[5] Nardi, F., Hovakimyan, N. and Calise A J. (2001) Decentralized control of large-scale systems using single Hidden layer neural networks. Proceedings of the American Control Conference 4:3122-3127

[6] Lewin D.R. and A. Parag (2003) A constrained genetic algorithm for decentralized control system structure selection and optimization. Automatica 39:1801-1807

[7] Ohba, T. and M. Ishida (1998) Application of a Neural Network to Evaluation of Interactions in a MIMO Process. A.I.Ch.E. Journal 44:2018-2024

Improved hierarchical fuzzy control scheme

Taher M. Jelleli & Adel M. Alimi

Research Group on Intelligent Machines (REGIM)
Department of Electrical Engineering, National Engineering School of Sfax, BP W, Sfax 3038, Tunisia
E-mail: {jellelitaher@uk2.net, adel.alimi@ieee.org}

Abstract

New modifications in mapping hierarchical fuzzy control scheme are proposed, to get effective and optimized control. The scheme was developed so that one can easily understand and modify fuzzy rules in different levels of the hierarchy. The presented approach ensures the universal approximation of functions in a compact domain. To validate this conceptual approach, we consider a textile data base as a non linear, multivariable and dynamic system.

1. Introduction

The design of fuzzy system architecture requires an addition of available information on the system behaviour, certain human knowledge or expertise, to define the controller's structure, the fuzzy rules and theirs necessary parameters. Currently, the problem of conception of fuzzy controller is treated to reach optimised computational time, reduced capacity of storage and number of fuzzy rules.

Introduced in 1974, the fuzzy control like a technology targeting the industrial applications has in fact, added a promising dimension to the classics existing control methods. In his vaguest term, the fuzzy control methodology benefits from some merits face to other approaches. The existing fuzzy control Methods can be classified among those described like intelligent control methods, as soon as the fuzzy control inserts some notions of the human expertise.

The human knowledge used in the conception of controllers is necessary and even advantageous. It is relatively the new axis of fuzzy technology that has the tendency to take place in existing control techniques as neural networks, expert systems...etc. Indeed, the theory of the fuzzy control can be rigorous while processing some interpretable analytic structures. However, the difficulty of this type of control appears in the development of controller structures.

In fact, the need to minimise the complexity of fuzzy rules base in order to reduce the computational time and the number of rules is not new. In this field, Mamdani has been the first to lead for computational time reduction [1]. His algorithm is based on a cylindrical fuzzy relation approximation. Other algorithms, as Sugeno, Takagi-Sugeno, Larsen, Product-Sum-Gravity...etc, have been developed respectively according to the same idea that is based on simple applied operations to the fuzzy sets of every input output parameters. However, these algorithms endure the exponential growth of computational time and the storage space demonstrated by Koczy and Hirota [2] through the expression of the uniform complexity. Indeed, the number of fuzzy rules of a rule base increases exponentially according to the number of antecedent's parameters as fuzzy terms (fuzzy sets) and the number of input-output system variables. It is the reason for that the number of inputs doesn't pass the number of five in general fuzzy applications case. In this context, Moser, Klement and Tikks demonstrate that for the more fuzzy algorithms part, the property of universal approximation of function is not verified in the case where the number of antecedent terms was limited. Therefore, the conception of a rule base has two important objectives, the first is to accomplish a good function approximation, and the second is to reduce the number of fuzzy rules [3]. The major difficulty resides in the contradiction generated by these two constraints. Besides, the hierarchical fuzzy systems have been created in order to remedy this big sudden problem by standard multi-variables fuzzy systems. This problem gives back the implementation of some theoretical approaches impassable [4][5][6][7].

We are going to present our contribution in the following parameters improvement: the computational time of an algorithm based on a TSK description, as well as the reduction of the rule base size of a fuzzy description generated by the choice of a certain configuration.

This paper is organised as follows: In section 1, basic idea for the concept foundation was presented. In section 2, we are giving the eventual problems of this approach, followed in section 3 by a suggested solution. The corresponding algorithm steps are shown in section 4. At least, the conception was tested with a chosen type of application and results are given. At last, we give conclusions in the rest of the paper.

2. Hierarchical fuzzy system formulation

The idea is to conceive a simple system while analysing the problem in levels and meaningful behaviours. Otherwise, these behaviours that are functions of input system variables are characterised by different levels of complexity. Therefore, we bring the fuzzy description from his most complex state towards the least complex even the simplest one. We can define in fact, a complex behaviour by the association of several simple behaviours, from which we use a behavioural decomposition while associating variables linearly by two in order to construct an elementary behaviour. Of this fact, the complexity of behaviour decreases from one level to another. Theoretically, this convergence is desirable in order to satisfy the property of rule number minimisation.

The proposed hierarchy proceeds by use of two successive variables of an input vector $(x_1,x_2,x_3,\ldots x_n)^T$, two by two by alternation [7][8]. For a level Li, every input couple is associated by two intermediate variables that are the Y_{ij} as $i\in[1\ L]$, $j\in[1\ n]$. These intermediate variables are used in rules consequences in order to preserve the rules physical meaning [3]. Of this fact, we analyse the system so that, if n as the number of input variables is pair, we don't risk to have a remaining variable, if no we subdivide variables by pair and the remaining variable will be added to the last level us adopted in [4].

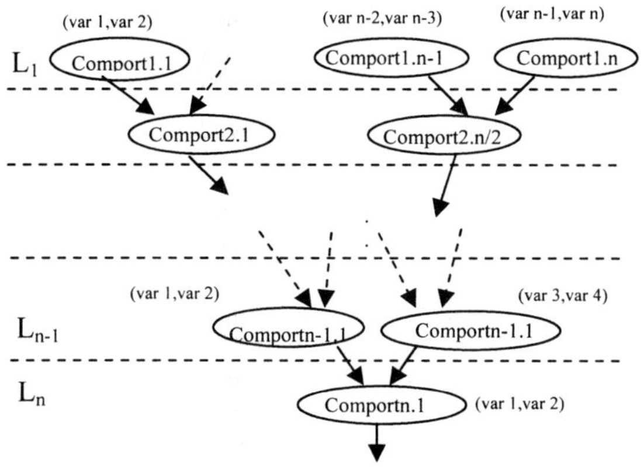

Fig. 1. General behavioral decomposition

The formulation of rules given by the proposed hierarchy for a given Li level is presented as follows:

$$R_{L,nss}: IF\ x_{L,i}\ is\ A^{i'}_{L,i}\ and\ x_{L,i+1}\ is\ A^{j}_{L,i+1}\ then$$
$$Y_{L+1,k} = Q\ (x_{L,i}, x_{L,i+1}, Y_{L,i}, Y_{L,i+1}),\ \forall (i', j) \in [1, m],$$
$$\{l,k\} \in \{n/2^{L-1}-1,\ n/2^{L-1}\}.$$

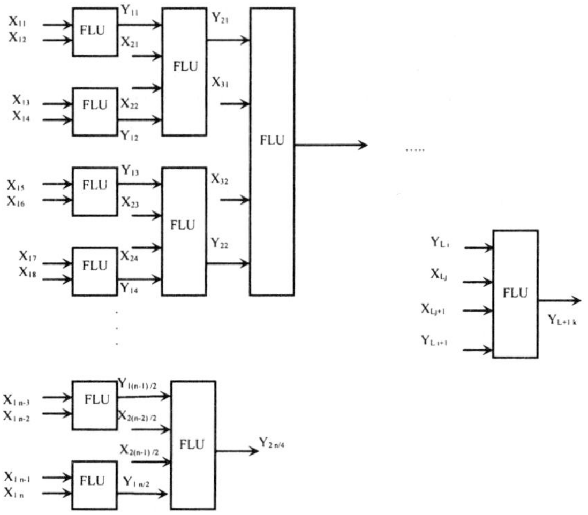

Fig. 2. General proposed conception

3. Existing implementation problems

In [9] we are presenting a reduced representation of fuzzy rule base systems by proposing a theoretical formulation of the rule base. This proposal was encountering problems and presenting disadvantages in some elaborate points. First, the case of odd number of input variables, where the last variable was inserted in the last level as intermediate variable in the consequences parts. Here, the problem of the parameter evaluation appeared when we are modelling with Sugeno fuzzy model. Suggested solution, is to evaluate the rest of parameters with an artificial added variable.

Second, the operation of inserting intermediate variables in the consequences. Does it really preserve the rules physical meaning? Third, if we are modelling with Mamdani fuzzy representation, what are the criteria for adopting the universe of discourse of these parameters, their fuzzy sets and can we justify these choices?

4. Improved hierarchical fuzzy scheme

The scheme so given doesn't apparently present changes, but we are eliminating the inserted intermediate variables from the rules consequences in order to evaluate them separately from the whole rules bases. As a result, all fuzzy logic units conserve their rules physical meaning, with no presence of intermediate or artificial variables. In a first level, all inputs are evaluated in association to extract outputs or behaviours. These output parameters would be evaluated after been associated in a chosen universe of discourse with a given membership functions configuration. In fact, to overcome the problem of physical meaning, of intermediate layers outputs, we propose separating them in rule blocs in order to check them with a multi criteria fuzzy decision maker for deciding which behaviour is dominating. This operation is redone level to other while deciding in last one which is the most dominant behavior. From a level to other, all dominant behaviour benefit from the revaluation of its input variables in their universe of discourse. The following scheme was presenting all behavioural fuzzy units separately in the controller structure.

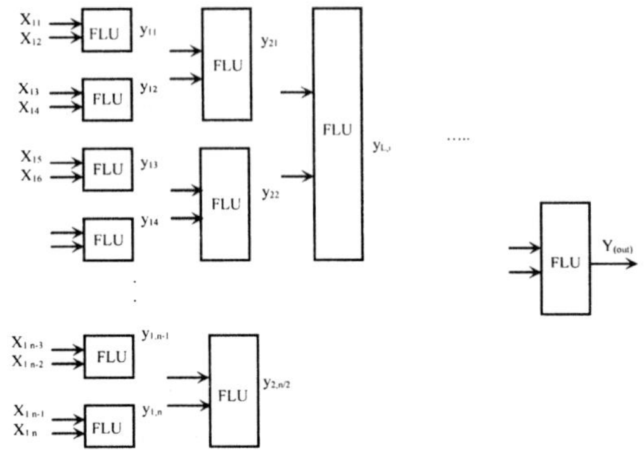

Fig. 3. Multi level fuzzy control scheme

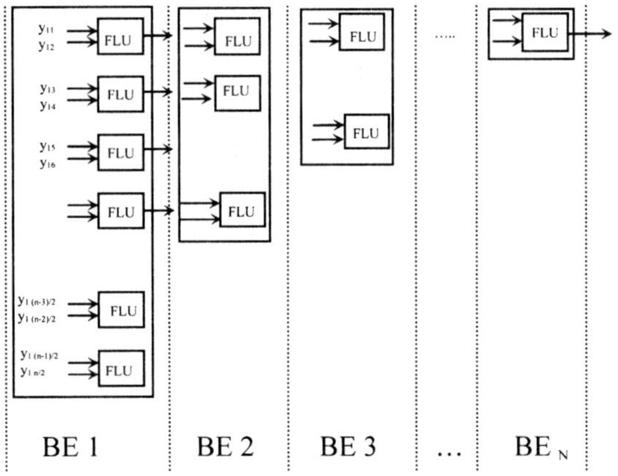

Fig. 4. Behaviors enhancement

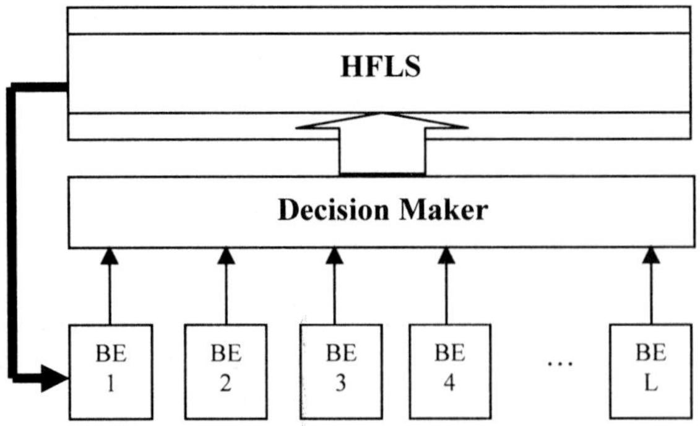

Fig. 5. Supervised fuzzy control scheme

Corresponding Algorithm:

Step 1 : First level behavioural evaluation
Theoretical formulation:

$$\textit{If } x_{Li} \textit{ is } A_i^{\alpha} \textit{ and } x_{Li+1} \textit{ is } A_{i+1}^{\beta}$$
$$\textit{then } Y_{LK} \textit{ is } f(x_{Li}, x_{Li+1})$$

Where, (x_{Li}, x_{Li+1}) are input system associated by two and $(A_i^{\alpha}, A_{i+1}^{\beta})$ are fuzzy input sets. We are adopting *Sugeno* fuzzy model representation for all existing consequences. In this first step, after associating variables, we generate behaviours in order to make dominant one.

Step 2 : First level artificial behavioural enhancement
Theoretical formulation:

$$\textit{If } Y_{Lnss} \textit{ is } B_i^{\alpha} \textit{ and } Y_{Lnss+1} \textit{ is } B_{i+1}^{\beta}.$$
$$\textit{then } D_{nss,nss+1} \textit{ is } g(Y_{Lnss}, Y_{Lnss+1})$$

where, (Y_{Lnss}, Y_{Lnss+1}) are intermediate variables, $(B_i^{\alpha}, B_{i+1}^{\beta})$ are vague input sets, $D_{nss,nss+1}$ represents the decision made by the two parameter evaluation and *nss* is the subsystem number.

Here, we are manipulating intermediate variables corresponding to two successive behaviours outputs to extract the one that is dominating by a choice of certain evaluation criteria without loss of behavioral interactions. This operation is totally separated from physical system behaviours.

Step 3 : Decision Making
Theoretical formulation:

$$\textit{If } D_{nss,nss+1} \textit{ is } \psi_D$$
$$\textit{then } (x_{L+1,i} \textit{ is } A_i^{\alpha} \textit{ and } x_{L+1,i+1} \textit{ is } A_{i+1}^{\beta'})$$

ψ_D ; represent input fuzzy sets for decisions.

This formulation means that, regarding the decision; we reselect variables that formulate the dominant behaviour in order to second re-evaluation. This stage looks like an eliminatory game, so all non-dominant behaviours should skip to let place to those that are winners. These three steps are executed in a looped manner, until sweeping all hierarchical levels.

5. Simulation and Results

The results are given as follows with a unified number of fuzzy sets m=5. The mentioned number of fuzzy rules could be retrieved by a standard fuzzy system representation based on an association of input variables by two. Between those results and those calculated in [4], we mention for the given configuration (data base with 5 inputs, 3 outputs) these points:

Gap of:
Rules number : 125
Membership Functions : 15

Rules Block : 1

For the computational time, if we don't take into account the time for decision or enhancement, the effective period to calculate output behaviour for one HFLU is equal to 2.2min, so we get a 2.2min gap time.

The MSE calculated with the improved structure on the outputs are: 8.84 % for output 1, 0.0544 % for output 2, 2.247 % for output 3. The presented solution ensures efficient optimization of the algorithm parameters as well as rules number, time of computation, algorithm bloc decomposition (decomposed structure). Consequently, the modification of an algorithm part does not affect any other without loss of dependencies.

6. Conclusion

In this paper, we gave improved structure to model a fuzzy system. We denote the conservation of the physical meaning of the system fuzzy rules. The modified algorithm preserves the propriety of function approximation in a compact domain. It ensure the reduction of effective rule base dimension

7. References

[1] Baranyi, P., Yam, Y. (2000) Fuzzy rule base reduction Chapter 7 of Fuzzy IF-THEN Rules in Computational Intelligence: Theory and Applications Eds, D.Ruan and E.E. Kerre, Kluwer, pp.135-160

[2] Kóczy, L.T., Hirota, K. (1997) Size Reduction by Interpolation in Fuzzy Rule Bases, IEEE Trans. S. M. C. 27: 14-25

[3] Taher, M.J., Alimi, M.A. (2003) Les systèmes flous hiérarchiques: états de l'arts, Proc of the fourth Tunisian conference on Automatics STA'03, 77-83

[4] Huwendiek, O., Brockmann W. (1999) Function approximation with decomposed fuzzy systems, Fuzzy Sets and Systems, 101: 273-286

[5] Joo, M.G., Lee, J.S. (2002) Universal approximation by hierarchical fuzzy system with constraints on the fuzzy rule, Fuzzy Sets and Systems 130: 175-188

[6] Raju, G.V.S., Zhou, J., Kisner, R.A. (1991) Hierarchical Fuzzy Control, Int. J. Control, 54-4:1201-1216

[7] Wang, L.X. (1998) Universal approximation by hierarchical fuzzy systems, Fuzzy Set and Systems 93:223-230

[8] Wei, C., Wang, L.X. (2000) A note on universal approximation by hierarchical fuzzy systems, Information Sciences, 123: 241-248

[9] Taher, M.J., Alimi, M.A. (2004) Proved proprieties for hierarchical fuzzy control scheme, the Eighth IASTED International Conference on Artificial Intelligence and Soft Computing, 451-058:120-126.

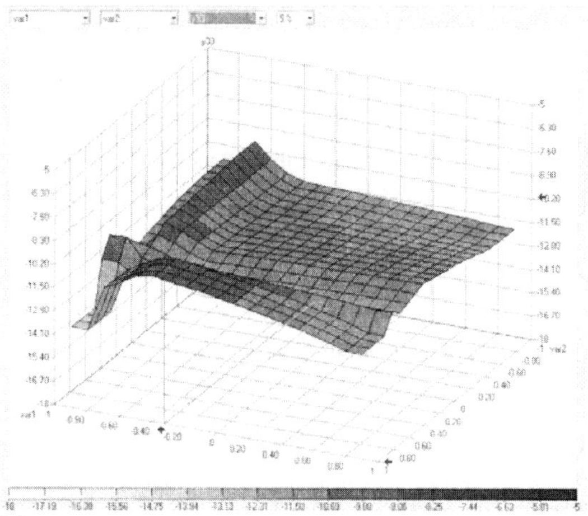

(a)

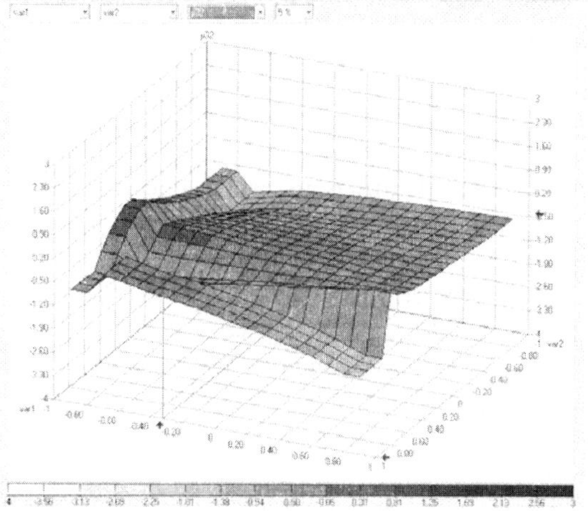

(b)

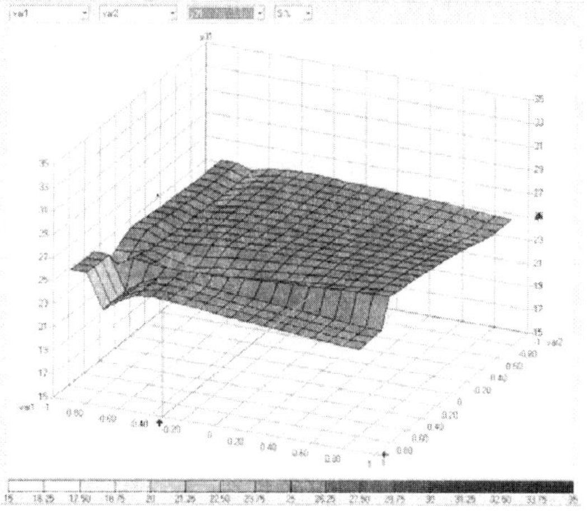

(c)

Fig. 6. (a),(b),(c) are fuzzy outputs system. These outputs present a large constant area, a thing that reflect the possibility of reducing again the rules data bases by reformulating rules in different levels of the hierarchy.

On-line Inference of Finite Automata in Noisy Environments

Ivan Gabrijel [1], Andrej Dobnikar [2]

[1] Private Researcher, Obrtniška ulica 3, SI-8210 Trebnje, Slovenia, E-mail: *Ivan.Gabrijel@guest.arnes.si*
[2] University of Ljubljana, Faculty of Computer and Information Science, Tržaška 25, SI-1001 Ljubljana, Slovenia

Abstract

The most common type of noise in continuous systems of the real world is Gaussian noise, whereas discrete environments are usually subject to noise of a discrete type. The established, original solution for on-line inference of finite automata that is based on generalized recurrent neural networks is evaluated in the presence of noise of both types. It showed quite good performance and robustness.

1 Introduction

An interesting and successful approach to the on-line identification and extraction of finite automata (FAs) is introduced in recent years [1], [2], [3]. It is shown that a FA can be identified in an on-line manner by a sufficiently large generalized recurrent neural network completely reliably. The on-line identification of FAs that is based on neural networks and the corresponding on-line extraction are original according to our knowledge. Real environments are often subject to unexpected situations and to unpredictable phenomena that are usually denoted as noise. The most natural type of noise is normally distributed, white, Gaussian noise with a zero mean and a certain standard deviation. It has a continuous nature and is superimposed on a pure signal. On the other hand, a noise in discrete environments is usually of a discrete nature.

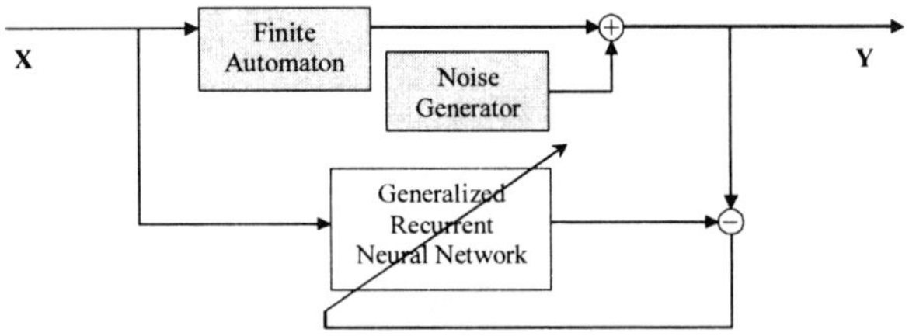

Fig. 1. The process of on-line identification of a FA with a generalized recurrent neural network in noisy environments.

The generalized architecture of recurrent neural networks (GARNN) [1], [2], [3] consists of arbitrary connected neurons that are divided into several groups. The neurons within each group process simultaneously and the groups are processed sequentially at each time moment. A special case of this architecture that consists of two groups of neurons is selected and the recurrent neural networks of this type are successfully applied to the problems of on-line inference of FAs [1], [2], [3] and to the trajectory tracking problems [4].

It is well known that neural networks possess a nice property to generalize. Of interest here is how the proposed solution for the on-line identification and reconstruction of FAs is resistant to the presence of noise in the environment. Fig. 1 illustrates how noise is introduced in the process of on-line identification of FAs. A particular type of noise is superimposed at the output of identified FA. This is the most realistic case. A FA is a discrete system and it is identified on-line by a GARNN, which has a continuous nature. Therefore, the robustness of the conceived solution should be assessed according to both discrete and continuous types of noise.

The paper is organized as follows. Section 2 defines the selected noise types. A revision of the on-line extraction process is given in Section 3. Experiments are described together with results and comments in Section 4. The conclusion summarizes the paper.

2 Definition of selected noise types

This paper deals with binary automata. Therefore, the only possible values at the output of a FA are 0 and 1. Discrete and continuous types of noise were used to represent conditions in a noisy environment. Let us define the applied noise types more specifically.

2.1 Discrete type of noise

A discrete noise event is defined as the realization of value 1 instead of the correct value 0 or vise versa. Additionally, a level of discrete, binary noise is defined as the portion of discrete noise events in the digital data stream. It is obvious that there is no uncertainty connected with noise levels 0.0 or 1.0. The most uncertain binary data contain discrete, binary noise at level 0.5. These noise levels can be described in terms of information theory. The entropy of discrete, binary noise levels 0.0 and 1.0 is equal to 0, which means that corresponding data contain complete information. On the other hand, the entropy of discrete, binary noise level 0.5 is equal to 1, which means that corresponding data contain no information at all.

2.2 Continuous type of noise

The normally distributed white noise with a zero mean is applied as the noise of continuous type. The probability density function of the normal or Gaussian distribution function is described by the well known

Gaussian function as

$$N(x,\mu,\sigma)=\frac{1}{\sqrt{2\pi}\sigma}e^{\frac{-(x-\mu)^2}{2\sigma^2}} \quad , \qquad (1)$$

where symbol μ denotes its mean value and σ is its standard deviation. The zero mean, $\mu = 0$, is selected and the noise level is defined as the standard deviation σ. This type of noise can be explained in terms of information theory. There is no noise at noise level 0 because standard deviation of the distribution equals zero. The entropy of such a data stream is equal to 0. When the noise level is monotonically increased, the entropy of the corresponding data approaches one. It reaches one theoretically at infinite noise level.

3 Extraction process revised

The on-line extraction process was originally proposed for use in certain environments [1], [2], [3]. It consists of on-line clustering, on-line extraction and state minimization stages. It depends on one parameter named the threshold attraction force, $F^{\diamond}$. The refs. 1, 2 and 3 are recommended for the detailed description of the procedure and for a deeper understanding of this revision. The process has shown very good performance in many experiments. However, is it possible for this method to be successfully applied in the environments with the presence of noise as well?

A brief inspection of the algorithm reveals its potential pitfall. Namely, the established on-line extraction method relies on correct desired output values. However, the desired output values are uncertain in the noisy environments. This would lead to the extraction of an incorrect structure of the underlying automaton in such environments. Fortunately, the basic extraction algorithm [1], [2], [3] can be easily generalized to be applicable in noisy environments.

Information on center c_i and mass m_i is maintained for each cluster during the on-line clustering stage of the extraction process. The required information is extended on each cluster i with a vector

$$\mathbf{f}_i^* = \left[f_{i1}^*, f_{i2}^*, \ldots, f_{im}^*\right]^T \quad , \qquad (2)$$

where m is number of desired output symbols. This vector keeps a record of desired output symbol frequencies. Each time when the cluster i is selected for updating and when the desired output symbol is $\mathbf{d}_j$, the j-th component of the vector $\mathbf{f}_i^*$ is updated as follows

$$f_{ij}^* = f_{ij}^* + 1 \quad . \qquad (3)$$

The value of each component actually counts how many times the corresponding desired output symbol was encountered during the process of extraction. When creating a new state of an automaton in the extraction stage of the extraction process by using information on the cluster i, vector $\mathbf{f}_i^*$ is used instead of the current desired value to assign an output symbol to the newly created state. The assigned output symbol is the one with the highest frequency[1] from the vector $\mathbf{f}_i^*$. This ensures correct output function of the extracted automaton. This extension is sufficient for successful extraction in the presence of noise of the discrete type. However, there are infinitely many output symbols possible in the case of continuous noise and this extension alone is not sufficient in such environments.

Further generalization of the extraction algorithm is very natural and it also ensures its applicability in the environments with continuous noise. Another attraction force based clustering process is started and performed in parallel with the clustering of the state space activities. It clusters the desired output patterns and is controlled by its own attraction force threshold parameter. Instead of using the desired output patterns in the extraction stage, indices of the corresponding clusters from the desired output space clustering process are used.

Suitable values for parameters of the extraction process depend on a structure of the trained neural network and on input and output sequences. It would be beneficial to determine these parameters automatically. An extension to the extraction process was devised in order to automate it. Records are kept in the attraction force clustering process on a minimal force between a data pattern and an existing cluster that have led to the adaptation of the cluster and on a maximal attraction force that result in forming a new cluster. Let these values be described by F^+ and F^- respectively. One can start with an arbitrary small (large) value of the attraction force threshold. If the extraction is found to be unsuccessful, the process is restarted now with the selection of parameter $F^{\diamond} > F^+$ ($F^{\diamond} < F^-$). This procedure is then repeated until the underlying automaton is successfully extracted.

These generalizations and extension result in a very sophisticated extraction algorithm that is able to extract the underlying automaton in an on-line manner in noise free and in noisy environments of both discrete and continuous types of noise.

4 Experimental work

4.1 Description of experiments

The robustness of the identification process regarding discrete noise is assessed in the first place. Each successful experiment of each problem of all

[1] The output symbol can be randomly selected in the case of equal highest frequencies.

Table 1. Results of on-line identification and extraction in the presence of discrete and continuous noise. The trained neural networks are tested on data without noise and the numbers of successful identifications and extractions in ten trials at various noise levels are shown. The second number in some cells describes a possible difference in number of successful trials between extraction and identification processes. Note that only results at selected noise levels are represented.

Problem	GARNN size	Success at different noise levels									
		Discrete noise					Continuous noise				
		0.05	0.10	0.15	0.20	0.45	0.1	0.2	0.3	0.4	1.0
Tomita 1'	1–3	10	10	10	10	9	10	10	10	10	10
Tomita 2'	2–4	10	10	10	10	5+2	10	10	10	10	10
Tomita 3'	3–5	10	7+2	10	10	0	10	10	10	10	0
Tomita 4'	1–3	9	9+1	10	8	0	9	10	6+3	6+1	0
Tomita 5	9–11	7	6	2	2	0	3	0	0	0	0
Tomita 6	7–9	10	10	9	5	0	5	0	0	0	0
Tomita 7'	4–6	10	7+1	6	1	0	9+1	9	9	8	0
T-XOR 0	5–7	10	9	10	10	2+2	10	10	7	6	9
T-XOR 1	7–9	10	9	10	10	2	10	10	10	9	0
T-XOR 2	11–13	10	9	9	8	0	10	8	3	0	0
T-XOR 3	16–18	9	7	7	3	0	4	0	1	1	0
T-XOR 4	16–18	9	3	2	2	0	2	0	0	0	0
Rivest & Zuckerman M1-1	1–3	10	9	9	7	0	10	10	10	10	10
Rivest & Zuckerman M2-1	2–4	10	10	10	10	1	10	10	10	10	10
Rivest & Zuckerman M1-2	4–6	10	10	10	9	0	10	10	10	10	4
Rivest & Zuckerman M1-3	4–6	9	1	0	0	0	9	9	9	8+2	0
Rivest & Zuckerman M2-3	5–7	6	3	2	0	0	10	10	9+1	5-1	0
Rivest &Zuckerman M1-4	6–8	5	1	0	0	0	10	9	2	0	0
Rivest & Zuckerman M1-5	9–11	4	0	0	0	0	9	0	0	0	0
Rivest & Zuckerman M2-5	7–9	1	0	0	0	0	9	0	0	0	0
Rivest & Zuckerman M1-6	9–11	0	0	0	0	0	3+1	0	0	0	0

problem domains as described in refs. 2 and 1 is repeated in the presence of discrete noise of different levels. Discrete noise starts at level 0.05, followed by experiments at noise levels 0.10, 0.15, until noise level 0.45, increasing the noise level each time by 0.05.

The successful experiments on all problems in all problem domains are then reconstructed in the uncertain environments that introduce continuous noise into the learning process. The noise levels considered range from 0.1 to 1.0, increasing by steps of 0.1.

The extraction process is again applied at the end of identification in all these noisy environments in all experiments that produce successful trials in the same way as in refs. 2 and 1, where noise was not introduced in the output function of underlying automaton.

The results of all these experiments and comments are given in the following subsections. Average hit ratio is used as the measure of success. The success of trivial identificators is calculated for each problem of each problem domain in order that the results of identification can be interpreted with greater relevance. A trivial identificator always responds with the most probable output pattern or symbol for a particular problem. The overall hit ratios of trivial identificators for each problem domain can be calculated by averaging their hit ratios on corresponding problems, and are: 69.05% for the Tomita problem domain, 50.00% for the temporal XOR problem domain and 82.31% for the Rivest & Zuckerman problem domain.

4.2 Results of experiments

The results obtained in experiments in noisy environments are given in this subsection. The numbers of completely successful trials out of ten performed in each experiment at various noise levels are shown in Table 1. The extraction succeeded in obtaining the correct structures of the underlying automata from all successfully learned GARNNs. Moreover, the extraction manages to extract a correct automaton even in many cases where the neural networks are not completely successful. Differences in number of successful trials between extraction and identification processes are shown in the same table. Average hit ratios of on-line identification in the presence of discrete and continuous noise of all problems for each problem domain are compared with the overall success of trivial identificators and depicted in Fig. 2.

4.3 Comments on results

It can be seen from Table 1 that at least one successful trial exists for each problem of the first two problem domains and for the first three problems of the last problem domain despite the presence of discrete

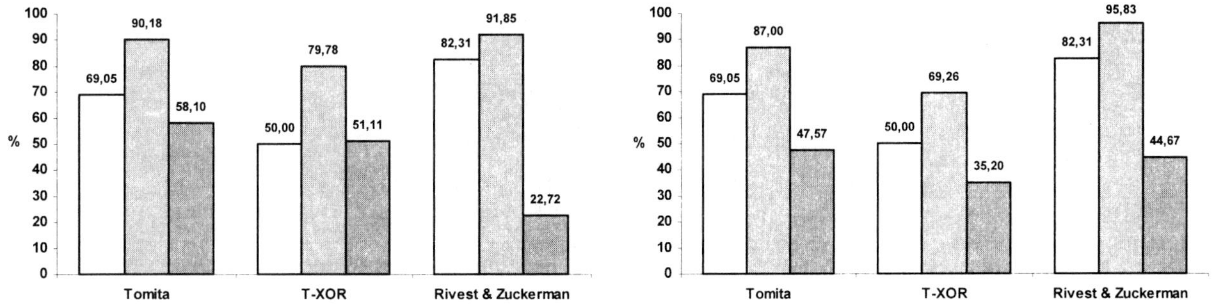

Fig. 2. Success of on-line identification of FAs of each problem domain in environments with the presence of: discrete noise (the left chart) and continuous noise (the right chart). The white columns represent the average success of the trivial identificators, the light gray columns describe the average success of all experiments in a particular problem domain and the dark gray columns describe percentages of completely successful trials.

noise at levels 0.05, 0.10, 0.15 and 0.20. All trials at any noise level are determined to be unsuccessful only in the case of the most complex problem of the third domain. It is obvious from the left chart of Fig. 2 that the most robust problem domain regarding the presence of discrete noise are Tomita automata. It is followed by the T-XOR problem domain, which is slightly less robust. The least robust problem domain is the Rivest & Zuckerman automata [5], [2].

Similarly, it can be seen from Table 1 that at least two successful trials exist for each problem of each problem domain despite the presence of continuous noise at level 0.1. It is obvious from the right chart of Fig. 2 that the first problem domain is also the most robust regarding the presence of continuous noise. The Rivest and Zuckerman automata problem domain is now, by a small amount, less robust. The least robust is the T-XOR problem domain. However, this robustness is far closer to the performances of the other two problem domains than the robustness of the least robust problem domain in the experiments with discrete noise. It looks like continuous noise events can be easier predicted on a statistical basis.

It should be noted that the GARNNs are evaluated that perform successfully in the noiseless environment. Although more complex problems express smaller robustness, it is expected on the basis of Cover's theorem on separability of patterns [1], [2] that larger structures would produce more reliable solutions in noisy environments.

5 Conclusion

Our on-line identification and extraction procedures are evaluated in this paper to assess their robustness regarding the presence of noise. Discrete and continuous types of noise are considered. The results clearly show that on-line identification is quite a robust process and can successfully identify a FA even in environments with significant disturbances of noise of both types. Additionally, it was shown that the generalized on-line extraction reliably interprets successfully trained GARNNs, which identify FAs, in terms of internal structures of the automata.

References

[1] Gabrijel, I., Dobnikar, A. (2003) On-line identification and reconstruction of finite automata with generalized recurrent neural networks. *Neural networks* 16 (1): 101-120.

[2] Gabrijel, I. (2002) *Generalized Architecture of Recurrent Neural Networks and On-Line Identification of Finite Automata – Ph.D. Thesis*. University of Ljubljana, Faculty of Computer and Information Science, Ljubljana, Slovenia.

[3] Gabrijel, I., Dobnikar, A. (2001) On-line identification and rule extraction of finite state automata with recurrent neural networks. In: Kurkova, V., Steele, N. C., Neruda, R., Karny, M. (eds.) *Proceedings of the International Conference on Artificial Neural Nets and Genetic Algorithms*. Springer-Verlag, Vienna, Austria, pp. 78–81.

[4] Gabrijel, I., Dobnikar, A. (2003) Generalized recurrent neural networks and continuous dynamic systems. In: Pearson, D. W., Steele, N. C., Albrecht, R. (eds.) *Proceedings of the International Conference on Artificial Neural Nets and Genetic Algorithms*. Springer-Verlag, Vienna, Austria, pp. 9–12.

[5] Ron, D., Rubinfeld, R. (1997). Exactly learning automata of small cover time. *Machine Learning*, 27 (1), 69–96.

Improved Clustering by Rotation of Cluster Centres

D. W. Pearson[1], M. Batton-Hubert[2]
[1]EURISE, Jean Monnet University of Saint-Etienne, Saint-Etienne, France
E-mail: david.pearson@univ-st-etienne.fr
[2]Centre SITE, Ecole des Mines de Saint-Etienne, Saint-Etienne, France
E-mail: batton@emse.fr

Abstract

In this paper we present a method that leads to the improvement of a subtractive clustering model by modifying the centres. In order to keep within certain bounds, a centre is modified by rotating it.

1 Introduction

One of the most successful methods of subtractive clustering is the one that was proposed by Chiu in 1994 [1]. It is based on fuzzy logic and uses potentials calculated for each data point. A data point is described by a pair of values (x, y) where x is a vector input and y is a vector output in the general case, in our specific case y is a scalar. The point with the highest potential is chosen as a centre, the potentials are modified and the process repeats until the potentials are all lower than a pre-specified threshold. When the algorithm terminates, the set of centres provides the model.

The basic algorithm, with no improvements, attempts to fit the model to a set of training data using only the provided data points. We are interested in the case of supervised learning and so "fitting" the model means that we want the model output to be as close as possible to the actual output for each data point in the training set. The model output is a weighted sum of the outputs associated to the centres. Improvements to the algorithm can be made by calculating new outputs and/or moving centres to provide a better model fit.

Various improvements of the basic algorithm were proposed by Chiu in [1], these are based on linear least squares calculations. We have also looked at how other sorts of improvements can be made [2], also based on linear least squares ideas. In this paper we present another type of improvement based on ideas coming from differential geometry and control theory.

The method that we present in this paper is selective in that it operates on one centre that has to be user specified. This means that the user should visualise the results and choose a centre that has an influence on a badly fitted data point.

2 Subtractive clustering

We have'nt the space here to present the Chiu algorithm in its entirety, the interested reader is invited to consult [1]. Basically the model is the following where $x \in \mathbb{R}^n$ is a vector of inputs, the $c_k \in \mathbb{R}^n$ are the centres for $k = 1, \ldots, p$, the ω^k are weights, α is a parameter determining the range of influence of each cluster centre and $y \in \mathbb{R}$ is the output:

$$y(x) = \phi(x) = \frac{\sum_{k=1}^{p} \omega^k \phi_k(x)}{\sum_{k=1}^{p} \phi_k(x)} \tag{1}$$

where

$$\phi_k(x) = e^{-\alpha \|x - c_k\|^2}$$

In general, the parameter α is chosen by the user, usually by trial and error and then the Chiu algorithm is put to work to find the centres and the weights. In some cases, depending on the data, considerable data reduction can take place when the algorithm is applied i.e. only a small proportion of the data points are needed as cluster centres.

3 Rotation

Using the model (1) we choose a centre, c_k say, and let this centre depend upon a parameter t. Then fix x and let y be a function of $c_k(t)$, we indicate this by the following

$$y_{c_k(t)}(x) = \phi(x) \tag{2}$$

We now think of $c_k(t)$ tracing out some sort of trajectory which causes the output of (2) to vary. This is inspired by the problem of output tracking in control theory [3]. In other words we let $c_k(t)$ satisfy the following differential equation

$$\dot{c}_k(t) = Ac_k(t)$$
$$c_k(0) = c_k \quad (3)$$

where $\dot{c}_k(t) = \frac{dc_k(t)}{dt}$ and the matrix $A : \mathbb{R}^n \to \mathbb{R}^n$ is to be calculated. For notational convenience we let $\phi_\star(x) = \frac{\partial\phi(x)}{\partial c_k}$ then we can differentiate (2) with respect to t and obtain the following expression

$$\dot{y}_{c_k(t)}(x) = \phi_\star(x)\dot{c}_k(t)$$

and upon substituting (3) we obtain

$$\dot{y}_{c_k(t)}(x) = \phi_\star(x)Ac_k(t) \quad (4)$$

Now choose a data point of interest, x_i say, for which the corresponding output, $y(x_i)$, is erroneous. Let y_i be the desired output at x_i then if we can calculate the matrix A in (4) such that

$$\phi_\star(x_i)Ac_k = y_i - y(x_i)$$

and integrate (3) over $[0,1]$ then, theoretically, we will have

$$y_{c_k(0)}(x_i) = y(x_i)$$
$$y_{c_k(1)}(x_i) = y_i$$

We will not obtain this result exactly, of course, because the calculation of A is only local and we have no control over the second and higher order derivatives. The change of centre will also affect other outputs. In order to maximise desired effects and minimise undesirable effects we place a constraint on A. We don't want $c_k(t)$ to change drastically because the elements usually correspond to physical quantities which are themselves naturally constrained. Therefore we require that $\|c_k(t)\| = \|c_k(0)\|$ i.e. that the modification is just a rotation. The solution to (3) is $c_k(t) = e^{tA}c_k(0)$ and so we need $Q(t) = e^{tA}$ to be an orthogonal matrix.

Let $Q^TQ = 1$ where Q is an $n \times n$ orthogonal matrix that satisfies the differential equation $\dot{Q} = AQ$ and 1 is the $n \times n$ identity matrix. Now differentiate the original constraint to get $\dot{Q}^TQ + Q^T\dot{Q} = 0$ and substitute $\dot{Q} = AQ$ to get $Q^T(A^T + A)Q = 0$ which, for nonzero Q is satisfied by $A^T = -A$ i.e. A is antisymmetric. In general, an antisymmetric matrix will generate an orthogonal matrix via the corresponding differential equation.

Thus, we need to calculate A to satisfy the above constraint and so that A is also antisymmetrical. Generically, this can be done as follows. Let $A = U - U^T$ where U is a strictly upper diagonal matrix with zeros on the diagonal, A is easily seen to be antisymmetric. Substitute this into the above constraint and we get

$$\phi_\star(x_i)(U - U^T)c_k = y_i - y(x_i)$$

but $\phi_\star(x_i)U^Tc_k$ is a scalar and so we can transpose it to $c_k^TU\phi_\star(x_i)^T$ and so the above becomes

$$\phi_\star(x_i)Uc_k - c_k^TU\phi_\star(x_i)^T = y_i - y(x_i) \quad (5)$$

Let the vector $u \in \mathbb{R}^{n^2}$ denote the elements of the matrix U listed in lexicographic order (row by row), then (5) can be rewritten, by making use of the Kronecker product $(A \otimes B)$ of two matrices [4], as follows

$$(\phi_\star(x_i) \otimes c_k^T)u - (c_k^T \otimes \phi_\star(x_i))u = y_i - y(x_i) \quad (6)$$

Due to the particular structure of the matrix U it will contain $\frac{n(n+1)}{2}$ zeros. The columns of (6) corresponding to these zeros can be eliminated and then, provided at least one of the remaining columns has a non zero entry, a solution can easily be found for u. Following that, the solution of (3) will provide the new centre.

4 Example

We tried our method on some benchmark data easily accessible via Internet [5]. The data are continuous and concern wine characteristics. There are 13 input variables (all positive) and 1 output variable taking the value of 1, 2 or 3 depending on the class of the particular wine. There are 178 data samples in total, 59 in class 1, 71 in class 2 and 48 in class 3. Our aim is to develop a method of centre modification and so we used all the data when we applied Chiu's method. We normalised all the input data to be inside the interval $[0,5]$. After some experiments we found that $\alpha = 1$ gave good results. After clustering we were left with 26 clusters. The correlation between the desired outputs and the actual outputs was 0.985, the results are presented in figure 1.

Looking at figure 1 we see that one of the biggest errors is produced for data sample 84 where the desired output is 2 but the achieved output is 2.83 and so we decided to modify this value. We chose the cluster centre with the minimum distance from data input number 84 to operate on, the 5th cluster found by Chiu's algorithm. After calculating u

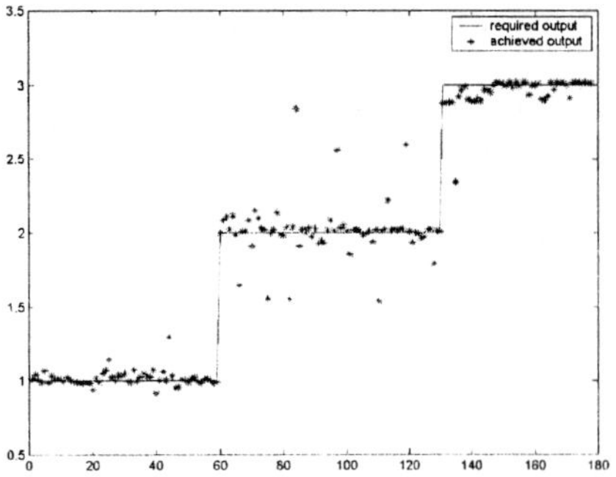

Fig. 1. Results without rotation.

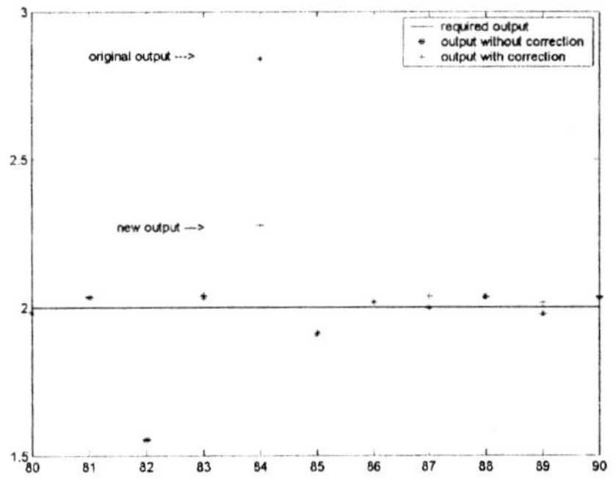

Fig. 2. Results with rotation, close up.

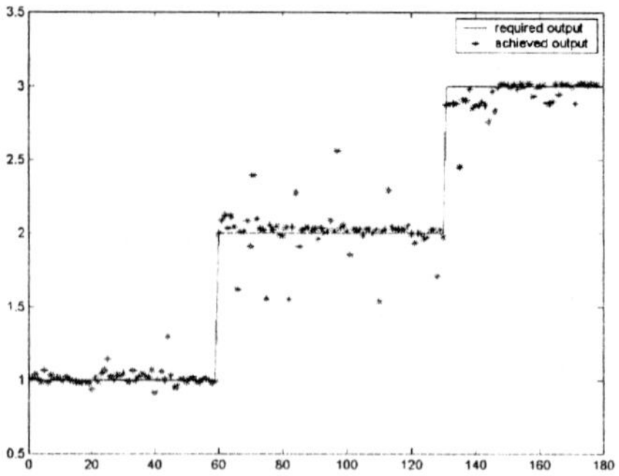

Fig. 3. Results with rotation.

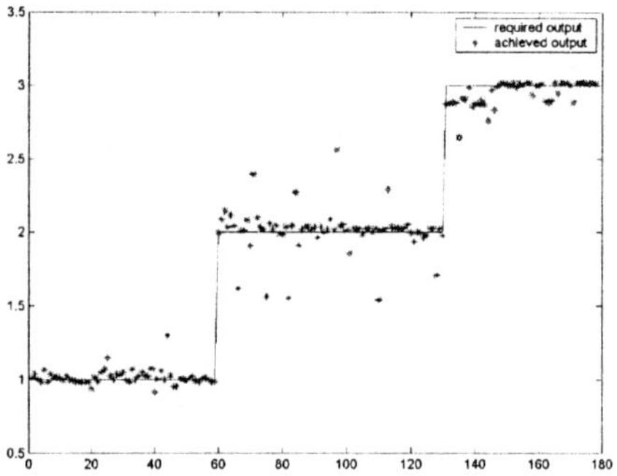

Fig. 4. Results after second rotation.

we integrated (3) over $[0, 1]$ which resulted in the following change of centre

$$\begin{bmatrix} c_5(0) & \rightarrow & c_5(1) \\ 4.0728254 & & 3.7966833 \\ 1.5775862 & & 2.9639282 \\ 3.5913313 & & 3.2256754 \\ 3.0833333 & & 3.2013357 \\ 2.5000000 & & 2.4897982 \\ 2.0618557 & & 1.8922485 \\ 1.4763780 & & 1.3869867 \\ 3.9393939 & & 3.9921567 \\ 2.2905028 & & 2.0351556 \\ 0.92307692 & & 1.2345569 \\ 3.1578947 & & 2.9666360 \\ 2.8375000 & & 2.7482188 \\ 1.4285714 & & 1.5265881 \end{bmatrix}$$

In figure 2 we present a close up of the data between samples 80 and 90 where the effect on number 84 is clearly seen going from a value of 2.83 at the output to a value of 2.28 which is a lot closer to the desired value of 2. As expected, other outputs have also been changed. The correlation between desired and achieved outputs changed from 0.985 to 0.989.

In figure 3 we present all the results after modifying the 5th cluster centre. We can see that one of the largest remaining errors is for the class 3 and corresponds to the point number 119.

We applied the algorithm a second time, this time it was the 14th centre that was closest to the data point. The original and modified centres are presented below.

$$\begin{bmatrix} c_{14}(0) & \rightarrow & c_{14}(1) \\ 4.4976399 & & 4.4849561 \\ 0.81034483 & & 0.72355217 \\ 3.6532508 & & 3.6972663 \\ 2.8333333 & & 2.8146892 \\ 3.3950617 & & 3.4944491 \\ 3.2603093 & & 3.3023906 \\ 1.2795276 & & 1.3911103 \\ 4.1666667 & & 4.1117278 \\ 0.58659218 & & 0.46353234 \\ 1.2192308 & & 1.0425569 \\ 2.9824561 & & 2.9788016 \\ 2.4125000 & & 2.4049777 \\ 2.2321429 & & 2.1895329 \end{bmatrix}$$

With the second modification the correlation between desired and achieved output becomes 0.99 and in figure 4 we present the results. Comparing figures 1 and 4 we can see a slight overall improvement has been obtained with the two centre rotations.

5 Conclusion

Chiu's algorithm, in its original form, has proved to be very reliable and in general gives good results. We have looked at how to make a local improvement of the model by rotating cluster centres. We are now looking at how to improve our method by adding constraints to our calculations.

Our ultimate aim, beyond our reach at the moment, is to tackle the data streaming problem. This is where a global model has been trained on some data coming from a database and the model is then put online for various reasons. For example the model could be for controlling a plant, or for forecasting certain variables. In our case we are particularly interested in forecasting levels of atmospheric pollution. When the model is put online it uses current data in order to carry out its tasks. But these new data are usually stored in a database and could themselves be used for training the model. Just how to make the best use of new data when they become available is the data streaming problem. We believe that our method could be used as a local adjustment of the global model and this is one of the directions we are looking into.

6 References

[1] Chiu, S. (1994) Fuzzy Model Identification Based on Cluster Estimation. Journal of Intelligent and Fuzzy systems, Vol. 2, No. 3.

[2] Dray, G., Peton, N. and Pearson, D.W. (1998) Centre Influence Modification in Subtractive Clustering. Control98, Coimbra, Portugal.

[3] Isidori, A. (1995) Nonlinear Control Systems. Springer-Verlag.

[4] Wonham W.M. (1979) Linear Multivariable Control: a Geometric Approach, Springer-Verlag.

[5] "UCI machine learning repository" http://www.ics.uci.edu/ mlearn/MLRepository.html

Hierarchical Growing Neural Gas

K.A.J. Doherty, R.G. Adams, N. Davey
Department of Computer Science, University of Hertfordshire, United Kingdom
E-mail: {K.A.J.Doherty, R.G.Adams, N.Davey}@herts.ac.uk

Abstract

This paper describes TreeGNG, a top-down unsupervised learning method that produces hierarchical classification schemes. TreeGNG is an extension to the Growing Neural Gas algorithm that maintains a time history of the learned topological mapping. TreeGNG is able to correct poor decisions made during the early phases of the construction of the tree, and provides the novel ability to influence the general shape and form of the learned hierarchy.

1 Introduction

The discovery of hierarchical structure through statistical methods is generally referred to as cluster analysis or numerical taxonomy, and these techniques are well established. The statistical generation of a hierarchical clustering can be achieved by either agglomerative or divisive methods. However, the divisive methods suffer from the inability to recover from a poor decision in the construction of the dendrogram [1].

Unsupervised Competitive Learning is the artificial neural network foil to cluster analysis. A sub-set of the family of Unsupervised Competitive Learning methods are the Growing Self-organising networks. Each node of the network has a position vector in the input space, and nodes are connected by edges to form graphs. The competitive Hebbian rule [2] produces the edges that form a sub-set of the Delaunay triangulation for the nodes [3]. The neighbourhood of a node is defined by the edges incident to the node. The positions of the nodes are altered in response to each input, and the structure of the network is modified by the insertion and deletion of nodes and edges. This dynamic behaviour can result in disjoint graph structures. The procedure used to form the graph and the resultant graph structure are called topology representing networks (TRN) [4].

TRN research has mainly focused on the discovery of concepts without hierarchical structure, but there have been recent attempts to discover and learn the taxonomy of concepts contained in an unlabelled set of data. A family of hierarchical neural clusterers has emerged based on Fritzke's Growing Cell Structures (GCS) [5]. However, the GCS algorithm has some inherent problems, which we will describe later in this document.

In this paper we propose a new unsupervised hierarchical, top-down classifier. Our model uses the Growing Neural Gas (GNG) algorithm [6], removing the reliance on the sometimes less-than-successful partitioning produced by GCS, and provides the ability to alter the general shape and form of the tree structure. The remainder of this paper is organised as follows: In the next two sections, we describe the dynamics and performance of the GCS and GNG networks, and the known hierarchical variants. In section 4, we present our model and the results of our experiments, and in the final section, section 5, we draw our conclusions.

2 Topology Representing Networks

In this section, we very briefly describe the dynamics, and comment on the performance, of the GCS and GNG networks.

2.1 Growing Cell Structures

The GCS algorithm grows a network composed of k-dimensional network construction units, for which k is user-defined (fig. 1). Since k is generally less than that of the arity of the input data, the GCS model performs a dimensionality-reducing mapping from the (possibly high dimensional) input space into a (generally lower) user-defined dimensional output space.

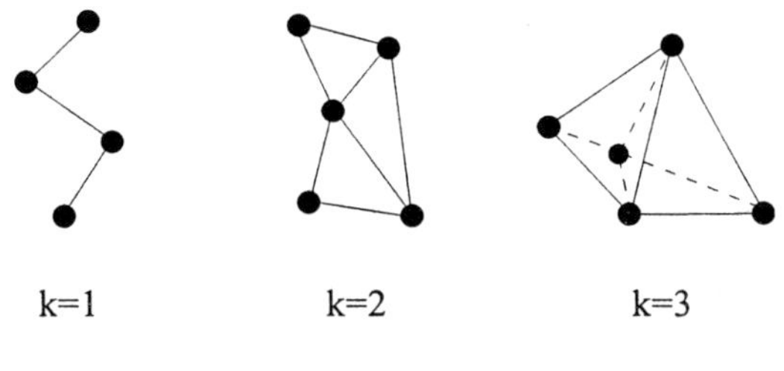

Fig. 1. Examples of GCS network topologies for k-dimensional network construction units [5].

Periodic node deletion occurs based on a measure of node activity and the volume of the input space classified by the node. The calculation of the Voronoi volume required for node deletion is difficult in dimensions greater than 2, and algorithm implementations have resorted to estimates [7, 5].

2.2 Growing Neural Gas

Growing Neural Gas is similar to GCS but does not impose the strict network-topology preservation rule, and edges are deleted based on an age criterion. The network incrementally learns the Delaunay triangulation corresponding to the topological relationships inherent in the data set, and continues until a user defined stopping criterion is satisfied.

2.3 GCS and GNG Comparison

Fritzke claimed that GCS "automatically finds meaningful partitions of the data", and that GCS was able to estimate the probability density of the input "under a wide range of parameter settings" [5]. In later work, Köhle and Merkl used GCS for document classification. Their results suggested that the GCS automatic cluster boundary generation aided the identification of cluster structure. However, they also noted that the algorithm is very sensitive to parameter selection, and for a wide range of parameters, GCS was "unable to produce semantically meaningful classification results" [8]. The strict topological preservation rule of GCS can result in massive purges in the GCS network, causing much of the accumulated learning to be lost [9]. The results of our own investigations also suggest that GCS clustering results are heavily dependent on the network parameter settings.

In a performance comparison of three incremental networks (including GNG and GCS) and the multilayer perceptron, the networks were benchmarked on four datasets, and scored for classification error, convergence rate and parameter sensitivity [10]. The GNG algorithm returned the superior benchmark score. The algorithm converged rapidly and showed little dependence on the network parameter settings. Again, the results of our own experiments agree with these findings.

3 TreeGCS

Various hierarchical variants to GCS and GNG have been proposed [9, 11]. TreeGCS [7] is an interesting variant. TreeGCS is a top-down, incremental learning hierarchical classifier, that maintains a time history tree of the graph connectivity of a standard GCS algorithm. As a part of the normal GCS dynamics, periodic node deletion takes place, occasionally resulting in graphs splitting into two or more disjoint graphs. Every disjoint GCS graph structure is represented by a leaf node in the tree representation. Every epoch, the tree representation is examined, and if a GCS graph structure represented by a leaf node has split, then a new child for every new graph is inserted into the hierarchy beneath the old leaf node. Similarly, if a disjoint sub-graph is deleted from the network, then the leaf node associated with the sub-graph is removed from the tree, and the hierarchy is updated to remove any inconsistent structure e.g. nodes with a single child are removed.

4 TreeGNG

Based on the results of our own experiments and the work of others, we consider that the relatively poor performance of GCS should preclude its use as the underlying algorithm. We believe that an improved basis for this time mapping approach would be the GNG algorithm. We propose the TreeGNG algorithm (fig. 2), which follows the TreeGCS algorithm but uses GNG as the underlying clustering algorithm, has a user-defined graph generation rate and replaces the epoch count stopping criterion with a more flexible user-defined stopping criterion.

```
Until stopping criterion is satisfied
 For each input
  Run GNG, generate graph structures
  If (Tree generation time)
   If (No. of clusters increased)
    Identify the tree node that now
    points to multiple clusters,
    create new children for this node
    and associate the new children
    with the clusters
   Elseif (No. of clusters decreased)
    Remove the associated tree node
    from the hierarchy and ensure the
    tree structure contains no
    singletons
   End if
  End if
 End for
End program
```

Fig. 2. The TreeGNG Algorithm

Within the GCS algorithm, the rate of node deletion is a user-defined iteration count; within the TreeGCS algorithm, the generation of the tree structure is based on an epoch-by-epoch examination of the graph connectivity. The same approach could be followed for TreeGNG. However, we decided that we could make use of the fact that the shape of the final tree will be dependent on the relationship of these two key periods. If *tree generation* occurs very infrequently in relation to the frequency of *node deletion*, then the tree will be very shallow and possibly have a large branching factor. As tree generation becomes more frequent, the tree will tend to get deeper with a smaller branching factor, and taken to the limit, if the tree is generated every time a change occurs in the number of graphs, then the tree will be binary. This aspect was not examined by Hodge and Austin, but we believe it provides a useful tool, as the the general shape and form of the tree can be specified.

Fig. 3 illustrates the TreeGNG graph splitting and the resultant tree growth.

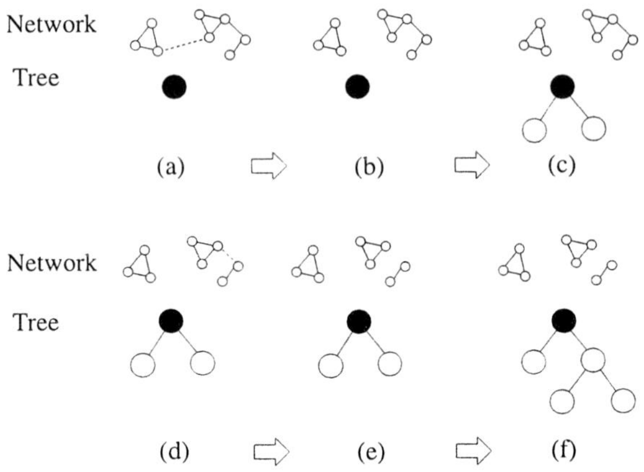

Fig. 3. GNG graph splitting and tree growth. In network (a), following the standard GNG ageing dynamics, the dashed edge is marked for deletion. Edge deletion results in two disjoint networks (b). At tree generation time, the tree is updated with new subordinate nodes (c) to reflect the the splitting of the graph. The edge ageing and deletion process is repeated in graphs (d) and (e) producing further tree growth (f) at the next tree generation time.

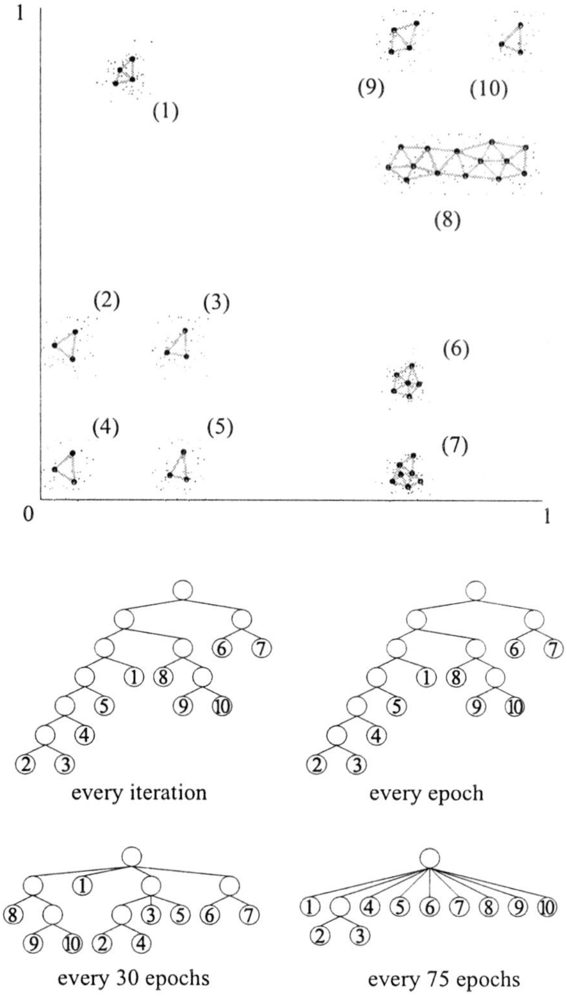

Fig. 4. GNG clustering of the MIX10 data set (upper fig.) with 50 nodes, a GNG edge deletion age of 1 epoch and a node insertion rate of 2 per epoch. The time history trees for a range of tree generation intervals are shown in the lower figures. With the tree generation occurring every 30 epochs, the tree structure was broadly in-line with our expectations of an appropriate hierarchical representation.

4.1 Experimental Results

We tested the utility of our algorithm on a range of data, and report the results for two synthetic data sets. The data sets comprised of 675 and 900 elements in $\mathbb{R}^2$. For both data sets, we ran the GNG algorithm for 5000 epochs with a range of parameters. For all the parameters considered, GNG satisfactorily clustered these data in-line with our expectations, and the induced Delaunay triangulation indicated the appropriate number of discrete clusters.

Figs. 4 and 5 (upper) shows the results of clustering with 50 codebook vectors, for a GNG edge deletion age of 1 epoch and 2 nodes insertions per epoch. It should be noted that the overall clustering time can be reduced by using a smaller edge deletion age, with little (if any) impact on the quality of the final clustering. We recorded the time history of the graph connectivity for tree generation intervals ranging from 1 iteration to 100 epochs, and typical resultant tree structures are shown (figs. 4 and 5 lower). The trees confirm that frequent tree generation results in binary tree representations, whilst an extended period between tree generations results in wider, shallower trees; but of more importance is that between these two extremes, the tree structure is in-line with our expectation of the most appropriate tree structure. In addition, the TreeGNG tree structures do no exhibit the instability noted with TreeGCS where "for many parameter settings" the network repeatedly deletes and reinstates the

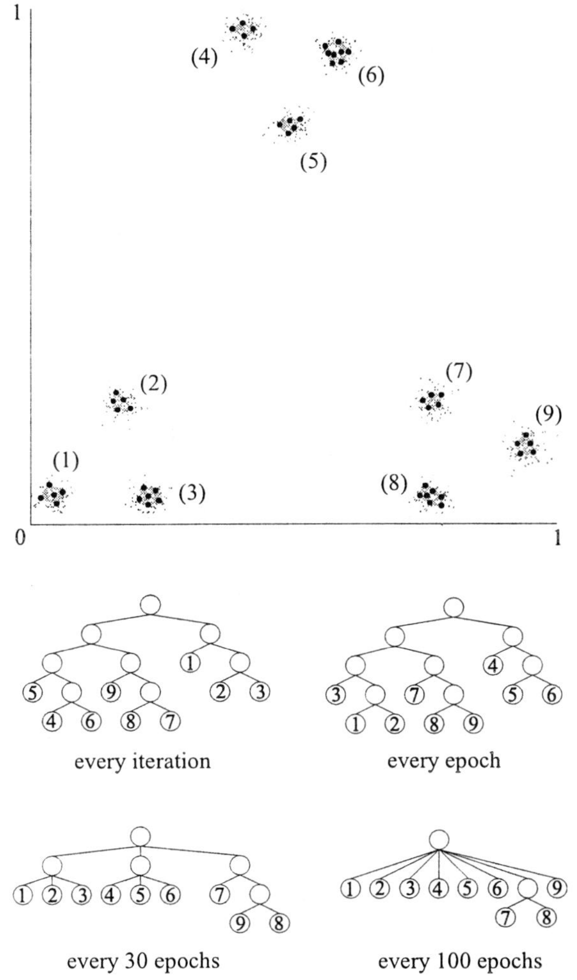

Fig. 5. GNG clustering of the GAUS9 data set (upper fig.) with 50 nodes, a GNG edge deletion age of 1 epoch and a node insertion rate of 2 per epoch. The resultant time history trees for a range of tree generation intervals are shown in the lower figures. With the tree generation occurring every 30 epochs, the tree structure was broadly in-line with our expectations of an appropriate hierarchical representation.

same clusters [7]. The results of tests on other synthetic data sets have indicated that TreeGNG is able to produce trees in-line with our expectations of an approriate structure.

5 Conclusions

We have developed an unsupervised top-down hierarchical classification tool based on Fritzke's GNG algorithm and Hodge and Austin's TreeGCS. Our method dynamically learns and adjusts the tree structure in response to the input data, and produces stable hierarchical representations for a broad range of network parameters. The periodic edge removal and competitive Hebbian learning of the GNG algorithm allows the network to recover from poor decisions in the generation of the hierarchy and thus overcome one of the major problems with TreeGCS and the divisive statistical methods. The algorithm also provides the novel ability to influence the general shape and form of the learned hierarchy.

References

[1] Everitt, B. (1993), Cluster Analysis, Edward Arnold, London, 3rd edition

[2] Hertz, J., Krogh, A., Palmer, R. (1991), Introduction to the Theory of Neural Computation, Addison-Wesley, Redwood City, CA

[3] Fritzke, B. (1996), Growing Self-Organizing Networks-Why?, In: Proc. European Symposium on Artificial Neural Networks, pp. 61–72

[4] Martinetz, T. M., Schulten, K. J. (1994), Topology Representing Networks, Neural Networks, 7(3):507–522

[5] Fritzke, B. (1994), Growing Cell Structures - A Self-Organising Network for Unsupervised and Supervised Learning, Neural Networks, 7(9):1441–1460

[6] Fritzke, B. (1995), A Growing Neural Gas Network Learns Topologies, Advances in Neural Information Processing Systems, pp. 625–632

[7] Hodge, V. J., Austin, J. (2001), Hierarchical Growing Cell Structures: TreeGCS, IEEE Trans. Knowledge and Data Engineering, 13(2):207–218

[8] Köhle, M., Merkl, D. (1996), Visualising Similarities in High Dimensional Input Spaces with a Growing and Splitting Neural Network, Lecture Notes in Computer Science, 1112:581–586

[9] Burzevski, V., Mohan, C. K. (1996), Hierarchical Growing Cell Structures, In: Proc. IEEE Int. Conf. Neural Networks, Washington D.C., volume 3, pp. 1658–1663

[10] Heinke, D., Hamker, F. H. (1998), Comparing Neural Networks: A Benchmark on Growing Neural Gas, Growing Cell Structures, and Fuzzy ARTMAP, IEEE Trans. Neural Networks, 9(6):1279–1291

[11] Cao, X., Suganthan, P. (2002), Hierarchical Overlapped Growing Neural Gas Networks with Applications to Video Shot Detection and Motion Characteristics, In: Proc. Int. Joint Conf. Neural Networks, IEEE, Hawaii, USA, volume 2, pp. 1069–1074

A Clustering Algorithm using Cellular Learning Automata based Evolutionary Algorithm

R. Rastegar, M. Rahmati, M. R. Meybodi
Computer Eng. Department, Amirkabir University, Tehran, Iran
{rrastegar, rahmati, meybodi}@ce.aut.ac.ir

Abstract

In this paper, a new clustering algorithm based on CLA-EC is proposed. The CLA-EC is a model obtained by combining the concepts of cellular learning automata and evolutionary algorithms. The CLA-EC is used to search for cluster centers in such a way that minimizes the squared-error criterion. The simulation results indicate that the proposed algorithm produces clusters with acceptable quality with respect to squared-error criteria and provides a performance that is significantly superior to that of the K-means algorithm.

1 Introduction

Clustering is an important unsupervised classification method used in identifying some inherent structure present in a set of data. The purpose of clustering is to group data into subsets that have useful meaning in the context of a particular problem [1]. Various clustering methods have been developed which may be classified into the following categories: hierarchical clustering, learning network clustering, mixture model clustering, objective-function-based clustering, and partition clustering, etc [5][16]. The clustering problem can be stated as finding the clusters such that the between-group scatter is maximized and within-group scatter is minimized. Many heuristic techniques for clustering exist in the literatures, which address the global minimization of squared-error criterion function, Genetic Algorithms (GA) [1][3][4][7] and Simulated Annealing (SA) [6] are two of these techniques.

The Cellular Learning Automata (CLA), which is introduced for the first time in [11], is a mathematical model for modeling dynamical complex systems that consists of large number of simple components. The simple components of CLA, which have learning capabilities, act together to solve a particular problem. This model has been applied to several problems such as image processing [10], channel assignment in cellular mobile system [2], function optimization [13], modeling of rumor diffusion [8], VLSI Placement [19], and modeling commerce networks [9]. In [13], CLA and evolutionary computing are combined to obtain a new model called CLA-EC for optimization problems. This model is capable of performing search in complex, large and multimodal landscape. In this paper, a new clustering algorithm based on CLA-EC is proposed. The CLA-EC is used to search for cluster centers in such a way that minimizes the squared-error criterion. Due to parallel nature of CLA-EC model, the proposed algorithm is appropriate for clustering large data set. In order to demonstrate the effectiveness of the proposed CLA-EC-clustering, 6 different two-dimensional data sets and IRIS data set are considered. Our experimental results of clustering indicate that the CLA-EC based clustering algorithm provides a performance that is significantly superior to that of the K-means algorithm.

The rest of the paper is organized as follows. Section 0II briefly presents the clustering problem. Section III gives a brief review of learning automata, cellular learning automata and CLA-EC model. The proposed clustering algorithm is described in section IV. Section V presents the simulation results for different data sets and the last section (VI) is the conclusion.

2 Clustering

In a clustering problem, a data set, in N-dimensional Euclidean space, $S=\{x_1,\ldots,x_M\}$ is given, where $x_i \in R^N$ and M is the number of data. Considering K clusters, represented by $C_1,\ldots,C_K$,the clusters should satisfy the following conditions,

$$C_i \neq \phi \ , \ i=1,\ldots,K \ , \ \bigcup_{i=1}^{K} C_i = S \ ,$$

$$C_i \cap C_j = \phi \ , \ i \neq j \ , \ i,j = 1,\ldots,K \ .$$

Among various clustering methods, the K-means method is more attractive than others in practical applications [15]. The K-means clustering algorithm is one of the well-known clustering methods, which is based on an iterative hillclimb-ing algorithm. One of the most important disadvantages of this algorithm is that it is very sensitive to the initial configuration and may be trapped in a local minimum [15]. Therefore, several approximate methods such as Simulated Annealing (SA) [6] and Genetic Algorithm (GA) [1][3][4][7] have been developed to solve the above problem.

3 The CLA-EC model

Learning Automata: Learning Automata are adaptive decision-making devices operating on unknown random environments. The Learning

Automaton has a finite set of actions and each action has a certain probability (unknown for the automaton) of getting rewarded by the environment of the automaton. The aim is to learn choosing the optimal action (i.e. the action with the highest probability of being rewarded) through repeated interaction on the system. If the learning algorithm is chosen properly, then the iterative process of interacting on the environment can be made to result in selection of the optimal action. Figure 1 illustrates how a stochastic automaton works in feedback connection with a random environment. Learning Automata can be classified into two main families: fixed structure learning automata and variable structure learning automata (VSLA). In the following the variable structure learning automata is described.

A VSLA is a quintuple $< \alpha, \beta, p, T(\alpha,\beta,p) >$, where α, β, p are an action set with s actions, an environment response set and the probability set p containing s probabilities, each being the probability of performing every action in the current internal automaton state, respectively. Function T is the reinforcement algorithm, which modifies the action probability vector p with respect to the performed action and the received response. Let a VSLA operate in an environment with $\beta=\{0,1\}$. Let $n\in N$ be the set of nonnegative integers that represent instance of iterations. A general linear schema for updating action probabilities can be represented as follows. Let action i be performed at instance n. If $\beta(n)=0$ *(reward)*,

$$p_i(n+1) = p_i(n) + a[1 - p_i(n)]$$

$$p_j(n+1) = (1-a)p_j(n) \qquad \forall j \quad j \neq i$$

If $\beta(n)=1$ *(penalty)*,

$$p_i(n+1) = (1-b)p_i(n)$$

$$p_j(n+1) = (b/s-1) + (1-b)p_j(n) \quad \forall j \quad j \neq i$$

Where a and b are reward and penalty parameters. When $a=b$, the automaton is called L_{RP}. If $b=0$ the automaton is called L_{RI} and if $0<b<<a<1$, the automaton is called $L_{R\varepsilon P}$. For more Information about learning automata the reader may refer to [12] [17].

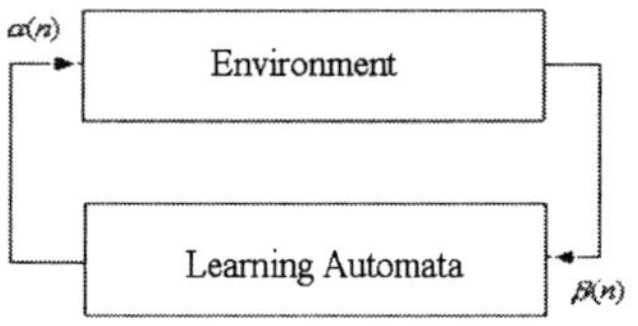

Fig. 1. The interaction between learning automata and environment

Cellular Learning Automata: The Cellular Learning Automata (CLA) [11][20] is a mathematical model for dynamical complex systems that consists of large number of simple components. The simple components, which have learning capabilities, act together to produce complicated behavioral patterns. A CLA is a cellular automata in which learning automaton (or multiple learning automaton) is assigned to its every cell. The learning automaton residing in a particular cell determines its state (action) on the basis of its action probability vector. There is a rule that CLA operate under it. The rule of CLA and the actions selected by the neighboring LAs of any particular LA determines the reinforcement signal to that LA (multiple LA). In CLA, the neighboring LAs of any particular LA constitute its local environment. The local environment is non-stationary because it varies as action probability vector of neighboring LAs vary. The operation of cellular learning automata can be described as follows: At the first step, the internal state of every cell is specified. The internal state of every cell is determined on the basis of action probability vectors of learning automaton (automata) residing in that cell. The initial value may be chosen on the basis of past experience or at random. In the second step, the rule of cellular learning automata determines the reinforcement signal to each learning automaton (automata) residing in that cell. Finally, each learning automaton updates its action probability vector on the basis of the supplied reinforcement signal and the chosen action. This process continues until the desired result is obtained.

The model of CLA-EC (Cellular Learning Automata based Evolutionary Computing) [13]: The CLA-EC model [13] is obtained by combining cellular learning automata and evolutionary computing. This model is capable of performing search in complex, large and multimodal landscape. In CLA-EC, similar to other evolutionary algorithms, the parameters of the search space are encoded in the form of genomes. Each genome has two components, model genome and string genome. Model genome is a set of learning automata. The set of actions selected by this set of learning automata determines the second component of genome (string genome). Based on a local rule, a reinforcement signal vector is generated and given to the set of learning automata. According to the learning algorithm, each learning automaton in the set of learning automata updates its internal state according to a learning algorithm. Then each learning automata in a cell chooses one of its actions using its probability vector. The set of actions chosen by the set of automata residing in a cell determines a candidate string genome that may replace the current string genome. The fitness of this string genome is then compared with the fitness of the string genome residing in that cell. If the fitness of the generated genome is better than the quality of the sting genome of the cell, the generated string genome becomes the string genome of that cell. The process of generating string genome by the cells of the CLA-EC is repeated until a termination condition is satisfied. In order to have an effective algorithm, the designer of the algorithm must be careful about determining a suitable genome representation, fitness function for the problem at hand, the parameters of CLA such as the number of cells (population size), the topology, and the type of the learning automata for each cell. Assume f be a real function that is to be minimized.

$$\min\{f(\xi) \mid \xi \in B^m\}$$

where B^m is $\{0,1\}^m$. In order to use CLA-EC for optimization of function f, first a set of learning automata will be associated to each cell of CLA-EC.

The number of learning automata associated to a cell of CLA-EC is the number of bits in the string genome representing points of the search space of *f*. Each automaton has two actions: 0 and 1. The CLA-EC iterates the following steps until the termination condition is met.

Step1: every automaton in cell *i* chooses one of its actions using its action probability vector.

Step 2: cell *i* generates a new string genome, η^i, by combining the actions chosen by the set of learning automata of cell *i*. The newly generated string genome is obtained by concatenating the actions of the automata (0 or 1) associated to that cell.

Step 3: Every cell *i* computes the fitness value of string genome η^i; if the fitness of this string genome is better than the one in the cell, then the new string genome η^i becomes the string genome of that cell. That is

$$\xi^i_{n+1} = \begin{cases} \xi^i_n & f(\xi^i_n) \le f(\eta^i_{n+1}) \\ \eta^i_{n+1} & f(\xi^i_n) > f(\eta^i_{n+1}) \end{cases}$$

where ξ^i_n and η^i_n present the string genome and the new string genome of cell *i* at instance *n*.

Step 4: Se cells of the neighboring cells of the cell *i* are selected. This selection is based on the fitness value of the neighboring cells according to truncation strategy [18].

Step 5: Based on the selected neighboring cells a reinforcement vector is generated. This vector becomes the input to the set of learning automata associated to the cell. Let *Ne(i)* be set selected neighbors of cell *i*. Define,

$$N_{i,j}(k) = \sum\nolimits_{l \in Ne(i)} \delta(\xi^{l,j}_n = k),$$

where,

$$\delta(\exp) = \begin{cases} 1 & \text{if exp is true} \\ 0 & \textit{otherwise} \end{cases}$$

$\beta^{i,j}$, the reinforcement signal given to learning automaton *j* of cell *i*, is computed as follows,

$$\beta^{i,j}_n = \begin{cases} u(N_{i,j}(1) - N_{i,j}(0)) & \textit{if } \xi^{i,j}_t = 0 \\ u(N_{i,j}(0) - N_{i,j}(1)) & \textit{if } \xi^{i,j}_t = 1 \end{cases}$$

where *u*(.) is a step function. The overall operation of CLA-EC is summarized in the algorithm of figure 2.

```
While not done do
  For each cell i in CLA do in parallel
      Generate a new string genome;
      Evaluate the new string genome;
      If f(new string genome) < f(old string genome) then
          Accept the new string genome
      End if
      Select Se cells from neighbors of cell i ;
      Generate the reinforcement signal vector;
      Update internal state LAs of cell i
  End parallel for
End while
```

Fig. 2. Pseudocode of CLA-EC

4 A CLA-EC based clustering algorithm

We propose to use the CLA-EC model to determine the *K* cluster centers of the data set in R^N; thereby clustering the set of *M* points of $S=\{x_1,\ldots,x_M\}$. The sum of the squared distances of the points from their respective cluster centers is taken as the clustering metric, and denotes it by *f*. The aim is to search the cluster centers in such a way that function *f* be minimized. The proposed algorithm consists of three phases: preprocessing phase, the CLA-EC phase and the clustering phase.

4.1 Preprocessing Phase

The purpose of preprocessing phase is to reduce the size of the search space on which CLA-EC will operate. To reduce the size of the search space, at first the largest and the smallest values of each dimension of data set is found as follows:

$$\min_j = \min_{1 \le i \le M} \{\mathbf{x}_{i,j}\}, \ \max_j = \max_{1 \le i \le M} \{\mathbf{x}_{i,j}\}$$

$$\Delta_j = \max_j - \min_j$$

where $x_{i,j}$ is the *j*th components of x_i. Second, a new search space where we denote it by R' which is $R' = [0,\Delta_1] \times \ldots \times [0,\Delta_N]$ is defined. where $\times$ is Cartesian product sign.

4.2 The CLA-EC Phase

In the CLA-EC phase, clusters are optimized with respect to the squared error criterion. The characteristics of the applied CLA-EC are as follows.

String genome representation: Each string genome is represented by a binary string consisting of $M \times N$ parts where each part is a representation of an encoded real number. Let $\lambda'_{i,j}$ be $(i \times N + j)^{\text{th}}$ part of string genome where *j* is the dimension of the center of cluster *i* in R'. If binary representation of $\lambda'_{i,j}$ has w_{ij} bits then in a *N*-dimensional space with *K* clusters, the length of a string genome is $m=\Sigma\Sigma w_{ij}$.

Fitness function (total sum of distances): To compute the fitness value of ξ, at first, we compute $\lambda'_{i,j}$ by decoding ξ, and set $\lambda_{i,j}$ to be $(\lambda'_{i,j}+min_j)$. The fitness value of genome is computed as follows:

$$f(\xi) = f(\boldsymbol{\lambda}_1,\ldots,\boldsymbol{\lambda}_K) = \sum_{i=1}^{M} (\mathrm{A}^*_i)^2,$$

where,

$$\mathrm{A}_{i,j} = \left\| \mathbf{x}_i - \boldsymbol{\lambda}_j \right\| \text{ And } \mathrm{A}^*_i = \min_{1 \le j \le K} \mathrm{A}_{i,j}$$

Parameters of CLA: A one dimensional CLA with wrap around connection and with the neighborhood shown in figure 3a is used. The neighbors of cell *i* are cell *i-1* and cell *i+1*. The architecture of each cell is shown in figure 3b. Each cell is equipped with *m* learning automata. The string genome determiner compares the new string genome with the string genome residing in the cell. The string with the higher quality replaces the string genome of the cell. Depending on the neighboring string genomes and the string genome of the cell, a reinforcement signal will be generated by the signal generator.

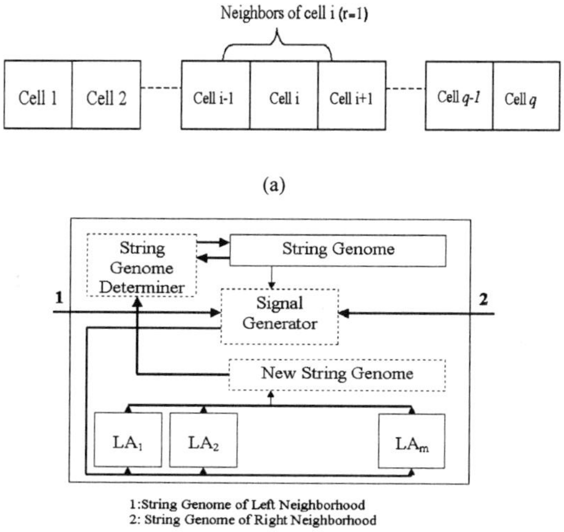

Fig. 3. The topology of the CLA-EC used in this paper.

Termination Criteria: CLA-EC stops after a pre-specified number of iterations. The best string genome found in the last iteration is the solution to the clustering problem. For the experimentations that follow the maximum number of iteration is set to 200.

4.3 The Clustering Phase

In this phase, the clusters are created using their centers, which are encoded in the best string genome reported by the pervious phase. This is done by assigning each point x_i, $i=1...M$, to one of the clusters C_k with center λ_k such that,

$$C_k = \arg\min_{1 \le j \le K} A_{i,j},$$

where

$$A_{i,j} = \left\| \mathbf{x}_i - \boldsymbol{\lambda}_j \right\|$$

All ties are resolved arbitrary.

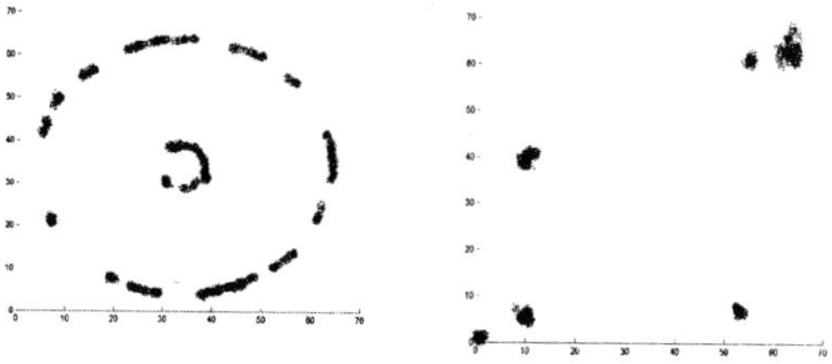

(a) (b)
Fig. 4. (a) Data 1 (b) Data 2.

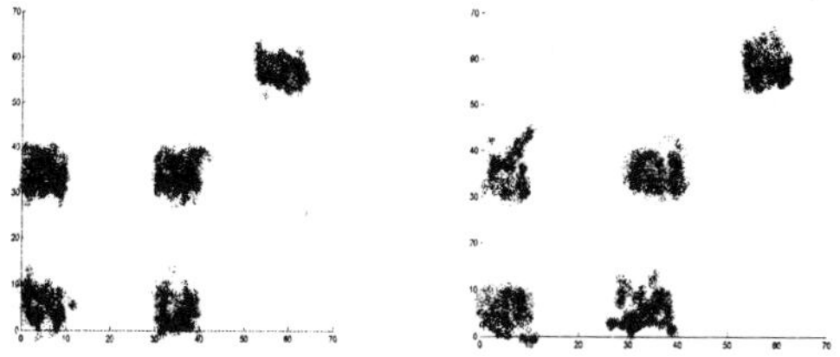

(a) (b)
Fig. 5. (a) Data 3 (b) Data 4.

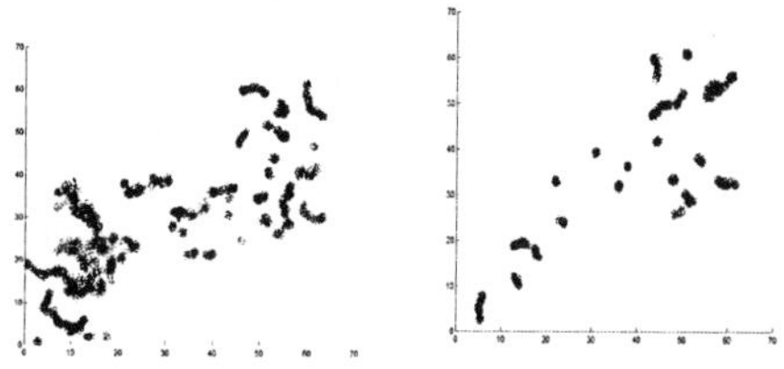

(a) (b)
Fig. 6. (a) Data 5 (b) Data 6.

5 Simulation results

Several simulations are conducted in order to evaluate the effectiveness of the proposed method. The results are then compared with the results obtained for K-means algorithm. Simulations are conducted for seven different data sets, which we call them Data 1, Data 2, Data 3, Data 4, Data 5, Data 6 and IRIS Data set. The characteristics of these data sets are given below.

Data 1: a two-dimensional data set with 4 clusters and 59 points as shown in figure 4a.

Data 2: a two-dimensional data set with 4 clusters and 25 points as shown in figure 4b.

Data 3: a two-dimensional data set with 4 clusters and 170 points as shown in figure 5a.

Data 4: a two-dimensional data set with 5 clusters and 128 points as shown in figure 5b.

Data 5: a two-dimensional data set with 5 clusters and 100 points as shown in figure 6a.

Data 6: a two-dimensional data set with 3 clusters and 35 points as shown in figure 6b.

Iris data: This data set represents different categories of irises having four feature values. The four feature values represent the sepal length, sepal width, petal length and the petal width in centimeter. It has 3 clusters with 150 points.

For the sake of convenience in presentation, we use *CLA-EC(automata(a,b), se ,q)* to refer to the CLA-EC algorithm with *q* cells, the number of selected cell *Se* and when using learning automata *automata* with reward parameter *a* and penalty parameter *b*.

Experiment I: In this experiment we study the effectiveness (quality of clusters found) of the proposed clustering algorithm with respect to CLA-EC parameters such as penalty and reward parameters, the number of cells, and the number of selected cells. Tables 1 of [21] shows the results of the CLA-EC-clustering algorithm for data set Data 1 for different values of the parameters of the CLA-EC such as the type of the learning automata, the penalty and the reward parameters of the learning automata, the number of selected cells and the number of cells. For each simulation maximum number of iterations of CLA-EC is taken to be 200. It is clear that as the mean and the standard deviation decrease the quality of the clustering becomes better. By careful inspection of the results reported in Table 1 [21] it is found that as the number of cells increases, the mean and the standard deviation of the result decreases. Also, it has been found that, better results are obtained when each automaton uses L_{RP} or $L_{R\varepsilon P}$ learning algorithm and when *Se* is set to 1. Figure 7 shows the effect of the number of cells in CLA-EC(L_{RP}(0.01,0.01),1,-) on clustering IRIS data set. It is shown that as the number of cells increases the quality of clustering becomes better. Figure 8 shows the fitness of the best genome (solid line) and the mean of the fitness of genomes (dashed line) for each iteration when using CLA-EC(L_{RP}(0.01,0.01),1,5) for Data 1 and Data 4.

Experiment 2: In this experimentation we compare the results of the proposed algorithm with that of K-means algorithm. For this experimentation CLA-EC has 5 cells, each automaton uses L_{RP} learning algorithm with $a=b=0.1$, *Se* is *1* and the maximum number of iterations is set to be 200. The results of 50 simulations for Data 2 and Data 3 are shown in figure 9. For Data 1 it is found that the CLA-EC-clustering algorithm provides the optimal value of 9502.44 in 28% of the runs whereas K-means algorithm attains this value in 8% of the runs. Both algorithms get trapped at a local minimum for the other runs. For Data 2, CLA-EC-clustering attains the best value of 239.10 in all the runs. K-means, on the other hand, attains this value for 28% of the runs, while in other runs it gets stuck at some of its local minima (such as 3433.77, 3497.16 and 5551.52). For Data 3, Data 4, Data 5, Data 6, and IRIS data set the CLA-EC-clustering attains the best values of 15545.09, 1873.71, 5525.34, 3000.43, and 46.44 in 30%, 100%, 10%, 40%, %30 of the runs, respectively. The best values attained by the K-means algorithm for these data sets are 15545.09, 1873.71, 5515.34, 3000.43, and 46.44 in 20%, 30%, 2%, 30%, and %24 of runs, respectively. Table 2 shows the summary of results of this experiment. By careful inspection of the results it is found that the CLA-EC(L_{RP}(0.1,0.1),1, 5) performs better than the K-means method for Data 1, Data 2, Data 3, Data 4, Data 4, Data5, Data 6, and IRIS data set.

6 Conclusions

In this paper, a new clustering algorithm based on CLA-EC, called CLA-EC-clustering, was developed. The CLA-EC finds the cluster centers, in such a way that the squared-error criterion be minimized. In order to demonstrate the effectiveness of the CLA-EC-clustering, 6 different two-dimensional data sets and IRIS data set were considered. The results of simulations showed that the CLA-EC-clustering algorithm provides a performance that is significantly superior to that of the K-means algorithm. Due to the parallel nature of CLA-EC, the proposed algorithm is very suitable for clustering large data set.

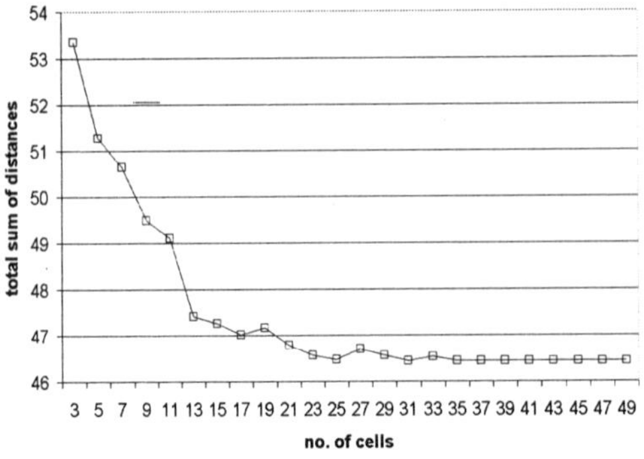

(a)

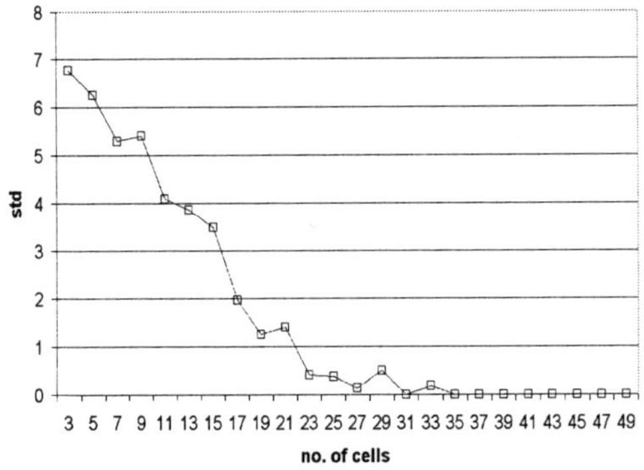

(b)

Fig. 7. Effect of the number of cells on the quality of clustering for IRIS data set when using CLA-EC(L_{RP}(0.01,0.01),1, -) -- (a) shows mean and (b) shows the standard deviation over 50 runs.

7 References

[1] Bandyopadhyay, S., and Maulik, U., "An Evolutionary Technique based on K-means Algorithm for Optimal Clustering in R^N ", Information Sciences, No. 146, PP. 221-237, 2002.

[2] Beigy, H., and Meybodi, M. R., "A Self-Organizing Channel Assignment Algorithm: A Cellular Leaning Automata Approach", Vol. 2690 of Springer-Verlag Lecture Notes in Computer Science, PP. 119-126, Springer-Verlag, 2003.

[3] Bhuyan, J. N., Raghavan, V. V., and Elayavalli, V. K., "Genetic Algorithm for Clustering with an Ordered Representation", International Conference on Genetic Algorithms'91, PP. 408-415, 1991.

[4] Garai, G., and Ghaudhuri, B. B., "A Novel Genetic Algorithm for Automatic Clustering", Pattern Recognition Letters, No. 25, PP. 173-1187, 2004.

[5] Jain, A. K., Duin, R. P. W., and Mao, J., "Statistical Pattern Recognition: A Review", IEEE Transaction on Pattern Analysis and Machine Intelligence, Vol 22, PP. 4-37, 2000.

[6] Klein, R., and Dubes, R., "Experiments in Projection and Clustering by Simulated Annealing", Pattern Recognition, Vol 22, PP. 213-220, 1989.

[7] Maulik, U., and Bandyopadhyay, S., "Genetic Algorithm-Based Clustering Technique", Pattern Recognition, No. 33, PP. 1455-1465, 2000.

[8] Meybodi, M. R., and Taherkhani, M., "Application of Cellular Learning Automata to Modeling of Rumor Diffusion", in Proceedings of 9th Conference on Electrical Engineering, Power and Water institute of Technology, Tehran, Iran, PP. 102-110, May 2001.

[9] Meybodi, M. R., and Khojaste, M. R., "Application of Cellular Learning Automata in Modeling of Commerce Networks", in Proceedings of 6th Annual International Computer Society of Iran Computer Conference CSICC2001, Isfahan, Iran, PP. 284-295, 2001.

[10] Meybodi, M. R., and, Kharazmi, M. R., "Image Restoration Using Cellular Learning Automata", in Proceedings of the Second Iranian Conference on Machine Vision, Image Processing and Applications, Tehran, Iran, PP. 261-270, 2003.

[11] Meybodi, M. R., Beygi, H., and Taherkhani, M., "Cellular Learning Automata", in Proceedings of 6th Annual International Computer Society of Iran Computer Conference CSICC2001, Isfahan, Iran, PP. 153-163, 2001.

[12] Narendra, K. S., and Thathachar, M. A. L., *Learning Automata: An Introduction*, Printice-Hall Inc, 1989.

[13] Rastegar, R., and Meybodi, M. R., "A New Evolutionary Computing Model based on Cellular Learning Automata", Accepted in IEEE CIS 2004, Singapore, 2004.

[14] Saheb Zamani, M., Mehdipour, M., and Meybodi, M. R., "Implementation of Cellular Learning Automata on Reconfigurable Computing Systems", IEEE CCGEI 2003 Conference, Montreal, Canada, May 2003.

[15] Selim, S. Z., and Ismail, M. A., "K-means-type Algorithm: Generalized Convergence Theorem and Characterization of Local Optimality", IEEE Transaction on Pattern Analysis and Machine Intelligence, Vol 6, PP 81-87, 1984.

[16] Yang, M. S., and Wu, K. L., "A Similarity-Based Robust Clustering Method", IEEE Transaction on Pattern Analysis and Machine Intelligence, Vol. 26, No. 4, April 2004.

[17] Thathachar, M. A. L., Sastry, P. S., "Varieties of Learning Automata: An Overview", IEEE Transaction on Systems, Man, and Cybernetics-Part B: Cybernetics, Vol. 32, No. 6, PP. 711-722, 2002.

[18] Goldberg, D. E., *Genetic Algorithms in Search, Optimization and Machine Learning*, Addison-Wesley, New York, 1989.

[19] Meybodi, M. R., and Mehdipour, F., "VLSI Placement Using Cellular Learning Automata", in Proceedings of 8th Annual International Computer Society of Iran Computer Conference CSICC2001, Mashhad, Iran, PP. 195-203, 2003.

[20] Beigy, H. and M. R. Meybodi, "A Mathematical Framework for Cellular Learning Automata", Advances in Complex Systems, to appear.

[21] Rastegar, R., Rahmati, M., and Meybodi, M. R., "A New Clustering Algorithm based On CLA-EC", Technical Report, Computer Eng. Department, Amirkabir University, 2004.

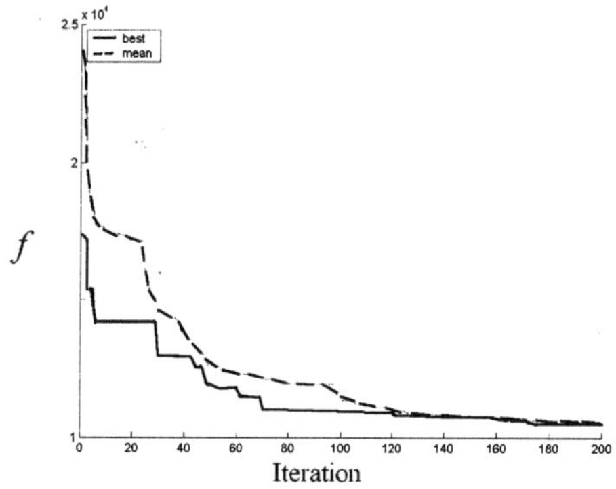

(a)

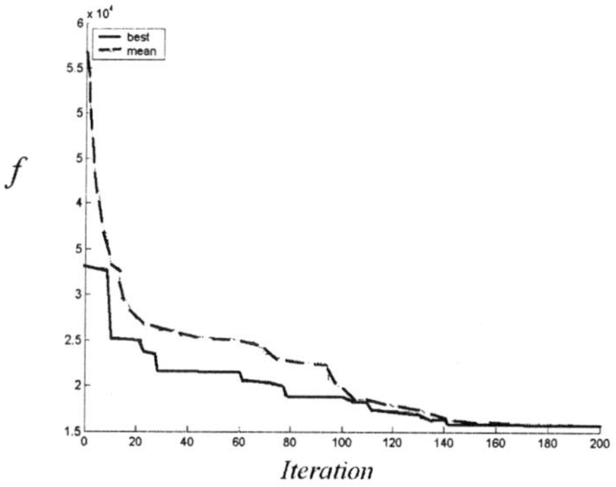

(b)

Fig. 8. The fitness of the best genome (solid line) and the mean of the fitness of genomes (dashed line) for each iteration when using CLA-EC(L_{RP}(0.01,0.01),1,5) for a) Data 1 b) Data 4

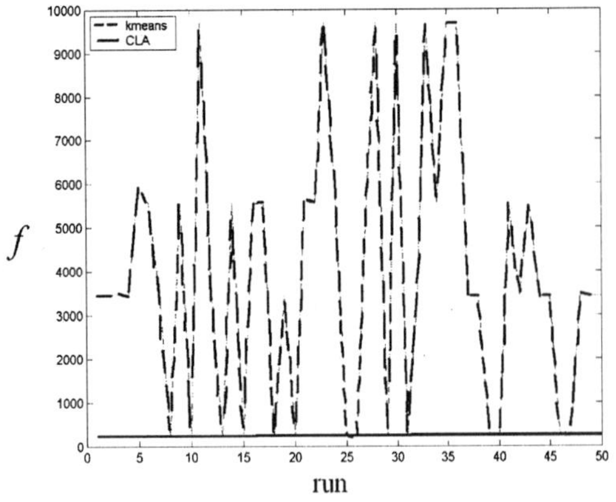

(a)

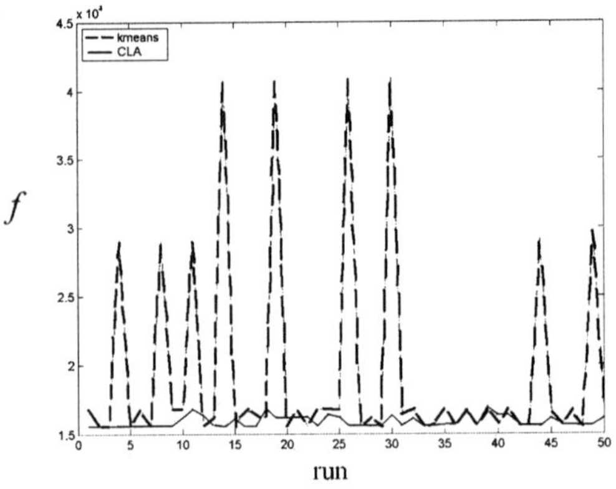

(b)

Fig. 9. Comparison of the K-means and the CLA-EC(L_{RP}(0.1,0.1),1, 5) (a) shows the total sum of distances obtained for K-means and CLA-EC(L_{RP}(0.1,0.1),1, 5) over 50 different runs for Data 2- (b) shows the total sum of distances obtained for K-means and CLA-EC(L_{RP}(0.1,0.1),1, 5) over 50 different runs for Data 3. The solid line is for the CLA-EC(L_{RP}(0.1,0.1),1, 5) and the dashed line is for the K-means algorithm.

Tables 1. The results of the CLA-EC(L_{RP}(0.1,0.1),1, 5) algorithm (*maximum 200 iterations*) and the K-means algorithm for Data 1,2,3,4,5,6, IRIS - Columns 'Mean' and 'Std' show the mean and standard deviation over 50 runs.

Data Set	(CLA-EC-clustering) Mean	(CLA-EC-clustering) Std	(Kmeans) Mean	(Kmeans) Std
1	10067.92	656.10	10373.33	694.78
2	239.10	0	3980.59	3073.99
3	15889.02	413.71	19457.38	7526.98
4	1873.71	0	8473.58	5424.68
5	5683.50	229.75	5920.9	817.05
6	3078.0	118.17	3206.67	469.77
Iris	51.27	6.25	52.96	8.37

Estimating the number of clusters from distributional results of partitioning a given data set

U. Möller[1]

[1] Bioinformatics - Pattern Recognition Group, Hans Knöll Institute for Natural Products Research Jena, Germany
E-mail: Ulrich.Moeller@hki-jena.de

Abstract

When estimating the optimal value of the number of clusters, C, of a given data set, one typically uses, for each candidate value of C, a single (final) result of the clustering algorithm. If distributional data of size T are used, these data come from T data sets obtained, e.g., by a bootstrapping technique. Here a new approach is introduced that utilizes distributional data generated by clustering the original data T times in the framework of cost function optimization and cluster validity indices. Results of this method are reported for model data (100 realizations) and gene expression data. The probability of correctly estimating the number of clusters was often higher compared to recently published results of several classical methods and a new statistical approach (Clest).

1 Introduction

Data clustering is very common. However, rather than 'recognizing' the true structure, a clustering algorithm merely superimposes a structure on a data set, even if this structure does not exist. In order to validate clustering results, different techniques are considered. Whereas relative criteria (in particular, the large family of validity indices) do not involve statistical tests, other techniques are based, e.g., on bootstrapping and hypothesis testing [1]. Biostatisticians currently introduce these techniques into the challenging field of gene expression data analysis in order to allow for statistical inference.

Nevertheless, the singularity of a sample data set may not be the only source of randomness of a clustering result. The perhaps most frequently used and least heuristic clustering algorithms are based on function optimization. For the global optimization, combinatorial, rule-based or analytical solutions are usually infeasible or unknown. To avoid the bias due to heuristic decisions made by the analyst, stochastic methods are often used that contain randomness explicitly, like random initialization [2], and random search [3]. This leads to random output of the clustering algorithm, but as a result of the algorithm. The analysis of this second type of randomness is not only desirable for an adequate data interpretation, but also useful for the previous task of cluster validation.

When clustering data by using an optimization scheme, one typically 'validates' some *final* result, for example, the partition obtained after a number of iterations of an algorithm, such as simulated annealing. If the algorithm only provides a local minimum (such as K-means or fuzzy C-means), it is also common to let the algorithm run a number of times and take the best result with respect to an objective function. If these methods worked well, they would have found the global optimum, i.e., a result that is (usually) unique and not random. If the search effort was specified a priori, it remains unclear how closely the final result matches the optimum. Möller et al. [3, 4] have investigated the uniqueness of obtained objective function values and validity index curves. However, this requires high computational costs, and rules that define when to stop the search.

The validation strategy presented below is based on distributional results of multiple (re-)partitioning rather than a single (final) partition. Furthermore, it relies on distributional data of the algorithmic search effort. One principle of this strategy is comparable to that used in bootstrapping and hypothesis testing. Information about the degree of randomness in a large set of results are utilized for estimating the numbers of clusters. Robust results that are less depending on random settings of the algorithm indicate non-random structure in the data.

It corresponds with experience that distinct clusters are easily identified, even by relatively simple methods. As a conclusion, one can expect the following if the specified and true number of clusters are equal. i) The objective function has clear (global) minimum with a wide and smooth basin of attraction. Hence, the bottom can be reached by descending from many starting points. ii) The distinctness of the true clusters yields quite a low cost function value, i.e. the 'valley' of the function will have a strong gradient. Thus, rather few iterations might be required to reach the minimum.

Conversely, if consistent results of multiple clustering and/or relatively few iterations are observed, this can be interpreted as an indication of existing clusters. A conceptual advantage is that both consistency and effort are general features that can be defined independently of the heuristically chosen objective and validity functions.

2 Methods

Clustering is viewed as an operator that transforms a set of unlabeled items into subsets of equally labeled items. This is only one step in the flow chart of cluster analysis [5]. We consider a clustering algorithm based on function optimization that requires the number of clusters, C, to be specified beforehand. Hence, estimating the number of clusters is another step, and the most important cluster validation problem. Many cluster validity indices have been proposed [1, 5]. The use of both an objective function and a validity index is justified as follows. Common objective functions can "lead to very unrealistic partitions", and "some of the intuitively desirable properties that we want a partition to have may not be captured by a functional that is easily optimized." [5]

2.1 Clustering-and-cluster validation procedure

1. For $C = 2,.., C_{max}$ clusters,
   a) apply the clustering algorithm T times to the data set using $t = 1,.., T$ random initial partitions and/or random sequences of partitions thereby minimizing a cost function, E. Calculate the validity index value for each partition, $V(C,t)$, as well as the minimum values $E_{min}(C) = \min_t E(C,t)$, and $V_{min}(C) = \min_t V(C,t)$.
   b) Re-scale: $E(C,t) = E(C,t) - E_{min}(C)$, $t = 1,.., T$.
2. Re-scale $V(2,.., C_{max})$ globally between 0 and 1.
3. For $C = 2,.., C_{max}$,
   a) calculate the percentage, $P(C)$, of the subset of partitions for which holds: $E(C,t) \leq E_{min}(C)+\varepsilon_E$, and $V(C,t) \leq V_{min}(C)+\varepsilon_V$, where V is a validity index to be minimized. ε_E and ε_V are small positive values.
   b) Calculate the average number of iterations, $I(C)$, performed by the clustering algorithm.
4. Set a threshold value $0 < \pi \leq 100$ (%).
   Calculate four estimates in the number of clusters:
   C_1 is the largest value of C with $P(C) \geq \pi$.
   C_2 is that value of the set $\{C \mid P(C) \geq \pi\}$ with the smallest validity index score $V_{min}(C)$.
   C_3 is the value of C, where the graph connecting the π% percentiles $V_\pi(2),.., V_\pi(C_{max})$ shows its strongest increase, Δ_{max}, from C to $C+1$, and where $P(C) \geq \pi$. If $\Delta_{max} < 0$ set $C_3 = 1$.
   C_4 is found analogously to C_3, but $I(2),.., I(C_{max})$ are used rather than V_π, and $C_4 \in \{2,.., C_{max}\}$

If V is an index which is to be maximized (e.g., Dunn's index [5]), the maximum principle is applied accordingly in steps 1a, 3, and for the calculation of C_2 and C_3.

2.2 Parameters for the analysis procedure

- *Clustering algorithm:* fuzzy C-means (FCM) (cf. [1])
- *Number of clustering trials:* $T = 100$
- *FCM parameters:* squared Euclidean distances, fuzzy exponent: 1.2, maximum number of iterations: 300; 100^*, minimum objective function improvement: 10^{-10}; $10^{-6\,*}$, * typical default values, used here for the microarray data
- *Maximum number of clusters:* $C_{max} = 7$ for the model data and $C_{max} = 9$ for the microarray data.
- *Cluster validity index:* Davis-Bouldin index (cf. ref. 5) *Tolerance values:* $\varepsilon_E = 0$, $\varepsilon_V = 0.01\Delta V$, where ΔV is the range of $V(C,t)$ over all values of C and t.
- *Threshold value:* $\pi = 50$ %.

Step 3a quantifies the robustness of candidate partitions that received the best scores for the quality criteria (E and V). The tolerance values ε_E and ε_V were included to allow negligible (e.g. rounding) differences between almost identical results. Stable estimates, C_i, were obtained for a wide range of parameter π. The rule for setting $C_3 = 1$ was introduced based on results of this procedure for uniformly distributed one-cluster data sets.

3 Data

Model data: 100 realizations (data sets) were generated for a model by Tibshirani et al., extended by Dudoit and Fridlyand [6] when testing analysis tools for microarray data. The model consists of four clusters, 10 variables, and the size of each cluster is random, either 25 or 50. The cluster means of three variables are also randomly chosen, from a $N(\mathbf{O}_3, 25\mathbf{I}_3)$ distribution, where $\mathbf{I}_p$ is the $p \times p$ identity matrix. The cluster members are multivariate normally distributed. The cluster means of seven variables are zero. These variables represent noise. Only those simulations are used, where members of different clusters have a minimum Euclidean distance of 1.
Microarray data: expression time courses of five groups of genes which have been characterized as showing an expression peak in one of five phases of the yeast cell cycle ([7], Table 1). The 214 time courses selected here can be expected to contain five clusters associated with the cell cycle phases, early G_1, late G_1, S, G_2, M.

4 Results

4.1 Model data

Table 1 contains the votes of the estimators $C_1,.., C_4$ for 100 realizations of the model. Three estimators provided the correct number of clusters ($C = 4$) with probabilities of 84 to 87%. C_2 falsely suggested two clusters based on a majority decision (45%). The reason for this is indicated by the example in Table 2: the validity index used rated the coarse structure (two clusters) as being a little clearer than the finer structure (four clusters). This situation may often occur and is not contradictory to the case, that the finer structure is the true structure one is interested in. Hence, it may be more successful to find the largest number of clusters above which the cluster validity, in the literal sense, decreases clearly (i.e. increases if validity is described by small values). This transition is recognized to occur from $C = 4$ to $C = 5$ in Fig. 1, where the results underlying the estimated numbers C_i were plotted for the same data set characterized in Table 2.

Table 1 Probabilities (in %) of correct estimates in the number of clusters, C, for the 100 realizations of the four-cluster model.

$C \rightarrow$	1	2	3	4	5	6	7
C_1	–	0	9	**89**	2	0	0
C_2	–	**45**	33	22	0	0	0
C_3	0	4	11	**84**	1	0	0
C_4	–	3	9	**87**	1	0	0

Table 2 Results of 100 clustering trials for one data set of the four-cluster model. See section 2.1 for the definitions.

$C \rightarrow$	2	3	4	5	6	7
$P(C)$	**54**	**95**	**100**	0	0	0
$I(C)$	25	29	**11**	125	144	155
$V_\pi(C)$	0	0.14	**0.06**	**0.52**	0.78	0.74

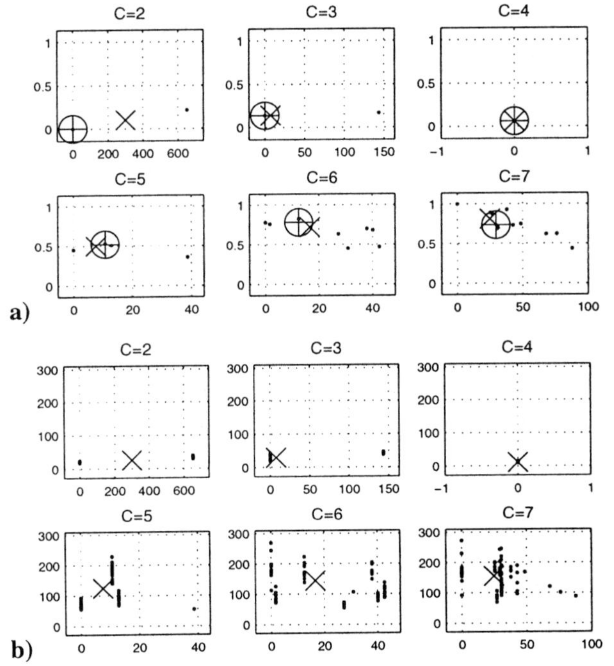

Fig. 1 Results of 100 runs of the FCM clustering algorithm thus partitioning one model data set into C clusters, respectively. Rescaled cost function, $E(C,t)$, on the x-axis a) against the (0,1) normalized Davis-Bouldin cluster validity index, and b) against the number of FCM iterations. ⊕ V_π = median, × mean

4.2 Gene expression data

Based on the obtained estimates, $C_1 = 5$, $C_2 = 4$, $C_3 = 5$, $C_4 = 4$, the number of clusters may be assumed to be four or five. More insight is obtained from the underlying results (Table 3, Fig. 2).

Table 3 Results of 100 clustering trials for the yeast cell cycle gene expression data. See section 2.1 for the definitions.

$C \rightarrow$	2	3	4	5	6	7	8	9
$P(C)$	**100**	0	**100**	**94**	0	0	0	0
$I(C)$	18	39	**39**	**63**	76	63	67	72
$V_\pi(C)$	0.05	0.26	0	**0.32**	**0.65**	0.69	0.73	0.83

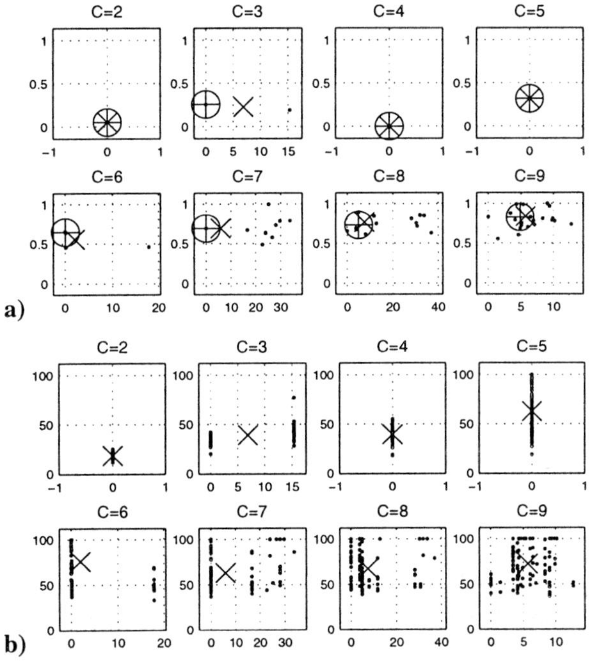

Fig. 2 Results analogous to those in Fig. 1 for the yeast cell cycle gene expression data.

Robust cost function and validity index results, $P(2)$, and small values of both the index, $V_\pi(2)$, and the number of iterations, $I(2)$, indicate a coarse structure of two distinct clusters. The Davis-Bouldin index, used here, rated the four-cluster partition as being the best. This rating is supported by the optimal robustness score, $P(4) = 100\%$. The fact that the four rather than two clusters required more iterations to be found is natural, because the clustering problem is more complex. In this sense, $C \geq 4$ is indicated by the result that, on average, finding the four clusters did not require more iterations than generating three clusters. Moreover, two minima were found equally frequently for $C = 3$ instead of a unique solution (see Fig. 2). Nevertheless, apart from the four-cluster solution, the five-cluster partition also seems to be non-random as suggested by robust cost function and index results, $P(5) = 94\%$. The fact that the index does not exhibit its minimum at $C = 5$ may have the reason that some of the five clusters are closer to each other than those of the four-cluster partition. Indeed one cluster at $C = 4$ was subdivided at $C = 5$. Perhaps this separation required an increased number of iterations from several (unfavorable) starting points of the cost function minimization. As expected, the five-cluster partition most strongly agreed with the five clusters biologically characterized by Cho et al. (ref. 7, table 1).

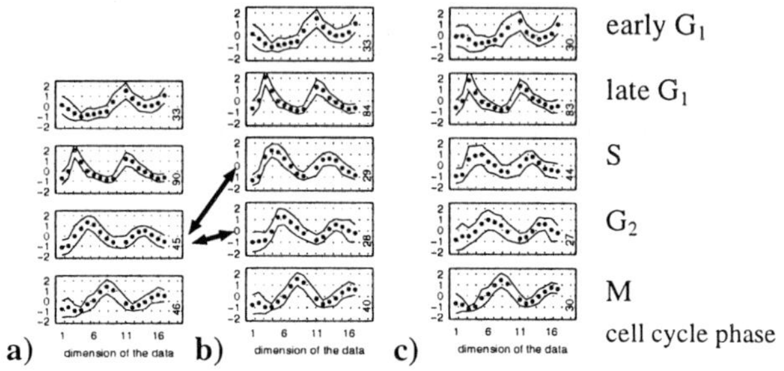

Fig. 3 Means and standard deviations of the gene expression time course clusters. a-b) four and five-cluster partitions obtained by clustering, c) biologically characterized clusters

5 Discussion

It was demonstrated that distributional rather than single clustering results for a particular data set give improved insight into existing data structures at different levels of resolution. In addition to the automated estimation of numbers, such as C_1 to C_4, plots of distributional results may be a valuable tool. Analysts often let the clustering algorithm run a multiple number of times, anyway (and retain the best result). In these cases the evaluation of all results rather than the best only is an advantage that can be obtained with almost no extra costs.

In addition to the model data considered here, the seven other models used in ref. 6 were also analyzed, each based on 100 realizations. Moreover, 39 other cluster validity indices (see ref. 4, 5) were tested replacing the Davis-Bouldin index in the above procedure. Because these results are too comprehensible, they cannot be presented here. The usefulness of the approach introduced in section 1 and demonstrated by an ad hoc implementation, was confirmed also for the other model data and validity indices. However, it seems that the objective function of the clustering algorithm and the validity indices should be adapted accordingly if, for example, elongated clusters are expected.

There is much room for developments. One possibility is the application of the Gustafson-Kessel algorithm that can deal with elongated clusters by computing a set of cluster covariance matrices. Furthermore, different sample data sets might be analyzed in real applications, thereby obtaining probability scores of correct estimation, similar to those provided in Table 1 for the different model realizations. In the absence of sufficient real data, bootstrapping and re-sampling techniques may be helpful. A striking feature of the above model is that few variables represent clusters, and a multiple of variables represent noise. This simulates conditions which are typical when learning from gene expression data in the unsupervised mode. Regarding the identification rates of 84–89% for the model data (Table 1), the above procedure was successful in comparison to seven other methods (Clest, gap statistics with and without principal components, mean silhouette width, and two other indices) which achieved rates of 18 to 74% for the same model [6]. Hence, the presented approach may be promising for microarray data analysis, because it can find cluster characteristics in a relatively large amount of unspecific information. Future work will be directed to the search or identification of estimators that combine advantages of different validity measures and methods.

Acknowledgement: This work has been supported by the German BMBF, grant no. 0312704 D.

References

[1] Theodoridis, S., Koutroumbas, K. (1999) Pattern Recognition. Academic Press, San Diego

[2] Peña, J.M., Lozano, J.A., Larrañaga, P. (1999) An empirical comparison of four initialization methods for the K-Means algorithm. Pattern Recognition Letters 20: 1027–1040

[3] Möller, U., Galicki, M., Barešová, E., Witte, H. (1998) An efficient vector quantizer providing globally optimal solutions. IEEE Trans. Signal Processing 46: 2515–2529

[4] Möller, U., Ligges, M., Georgiewa, P., Grünling, C., Kaiser, W.A., Witte, H., Blanz, B. (2002) How to avoid spurious cluster validation? A methodological investigation on simulated and fMRI data. NeuroImage 17: 431–446

[5] Bezdek, J.C., Pal, N.R. (1998) Some new indexes of cluster validity. IEEE Trans. Syst., Man and Cybern. B28: 301–315

[6] Dudoit, S., Fridlyand, J. (2002) A prediction-based resampling method for estimating the number of clusters in a dataset. Genome Biology 3: 0036.1-0036.21

[7] Cho, R.J., Campbell, M.J., Winzeler, et al. (1998) A genome-wide transcriptional analysis of the mitotic cell cycle. Molecular Cell 2: 65–73

AUDyC Neural Network using a new Gaussian Densities Merge Mechanism

Habiboulaye Amadou Boubacar[1,2], Stéphane Lecoeuche[1,2], Salah Maouche[1]

1 Laboratoire Automatique, & Génie Informatique et Signal,
Université des Sciences et Technologies de Lille, Bâtiment P2, 59655 Villeneuve d'Ascq, France
2 Département Génie Informatique et Productique,
Ecole des Mines de Douai, 941, Rue Charles Bourseul, BP838, 59 508 Douai, France
E-mail: {amadou,lecoeuche}@ensm-douai.fr,salah.maouche@univ-lille.fr

Abstract

In the context of evolutionary data classification, dynamical modeling techniques are useful to continuously learn clusters models. Dedicated to on-line clustering, the AUDyC (Auto-adaptive and Dynamical Clustering) algorithm is an unsupervised neural network with auto-adaptive abilities in non-stationary environment. These particular abilities are based on specific learning rules that are developed into three stages: "Classification", "Evaluation" and "Fusion". In this paper, we propose a new densities merge mechanism to improve the "Fusion" stage in order to avoid some local optima drawbacks of Gaussian fitting. The novelty of our approach is to use an ambiguity rule of fuzzy modelling with new merge acceptance criteria. Our approach can be generalized to any type of fuzzy classification method using Gaussian models. Some experiments are presented to show the efficiency of our approach to circumvent to AUDyC NN local optima problems.

1 Introduction

Numerous classification techniques using leaning processes have been developed for data clustering in pattern recognition. But, most of them are not enough accurate in long term applications, In fact, in many real-life applications, non-stationeries data are common and class models have to evolve in time. For example, in voice and face recognition models undergo variations with ageing. In diagnosis application, the system functioning modes would change in the time due to the wear of its parts and failures. It is thus necessary to refine the class models using adaptive rules to take account of the evolutions. In case of evolutionary data, by using a suitable similarity measure, unsupervised learning processes have to create class models and to modify them in time according to the structure and the evolution of unlabeled data. Within the framework of unsupervised classification, neural network algorithms are widely used and confer to the classifier auto-adaptive learning abilities [1]. For examples, the Fuzzy Min-Max Clustering algorithm [2] uses hyperbox prototypes models to represent classes, the Cluster Detection and Labelling [3] neural network is based on a Euclidian similarity measure. Although their constructive architectures, most of neural network algorithms are not adapted to online applications requiring real dynamical modeling of classes.

To achieve dynamical classification, Lecoeuche and Lurette [4] have proposed a new architecture called AUDyC Auto-Adaptive and Dynamical Clustering. It correctly defines data densities with Gaussian prototypes models by using membership function based on Zadeh theory. AUDyC NN has specific online learning rules that allow the creation of new prototypes and complex classes, adaptation and elimination of existent classes in order to incorporate new information. However, in spite of its performances, the AUDyC fusion procedure is not robust and could induce some local optimum problem.

In probabilistic model construction, the drawbacks of local convergence problems are well-posed for finite mixture models [5]. Some solutions based on split and merge mechanisms are proposed to improve the widely used Expectation-Maximisation (EM) algorithm, [5]. In order to circumvent local maxima, Ueda and Nakano [6] present Split and Merge EM (SMEM) to operate in overpopulate and underpopulate regions. In the same way, Zhang and Yi [7] proposed the Competitive EM algorithm that is the unsupervised version of EM by incorporating efficient split and merge operations. However, the previous techniques used for estimation of mixtures model parameters are based on (prior and posterior) probability conditions and are not useable for online dynamical model in non-stationery environment.

In this paper, we proposed a new Gaussian densities merge mechanism that consists in an ambiguity rule and acceptance criteria. Before presenting it, the next section describes briefly the AUDyC Neural Network.

2 Description of the AUDyC neural network

The AUDyC network [4] consists of a feed-forward architecture with three layers. It is very similar to many networks in Pattern Recognition with constructive structure. Contrarily to its input layer (D neurons), the Hidden and the Output layers (resp. J neurons and I neurons) are totally free of evolution: connections and nodes. The connections matrix W_1 between the input layer and the hidden layer memorizes the Gaussian prototypes centers M_j and the connections matrix W_2 between the Hidden layer and the Output layer defines the membership of prototypes to classes. In the AUDyC NN, a prototype is defined as a single Gaussian density and a class is a complex structure formed by one or more Gaussian prototypes that are enough close to each other. The AUDyC neural network connections weights are defined by:

$$W_1 = [M_j]_{J\times D}, \quad j = 1...J \quad (1)$$

$$W_2 = [\delta_{ji}]_{I\times J}, \quad j = 1...J, i = 1...I \quad (2)$$

$$\text{With} \quad \delta_{ji} = \begin{cases} 1 & P_j \in C_i \\ 0 & Overwise \end{cases} \quad (3)$$

The membership degree μ_k^j of a sample X_k to any Gaussian prototype P_j corresponding to the neuron j is calculated in the following way:

$$\mu_k^j = \exp\left(-\frac{1}{2}(X_k - M_j)'\Sigma_j^{-1}(X_k - M_j)\right) \quad (4)$$

With M_j and Σ_j respectively the center and the covariance matrix of the prototype P_j.

μ_k^j is the decision function established to satisfy fuzzy conditions in Zadeh theory [9] :

$$\begin{aligned} &\forall k = 1,...,N \quad \forall j = 1,...,J \quad 0 \le \mu_k^j \le 1 \\ &\forall k = 1,...,N \quad \forall j = 1,...,J \quad 0 < \sum_{k=1}^{N} \mu_k^j < N \end{aligned} \quad (5)$$

N is the cardinality of the data sample $\chi = \{X_k\}_{k=1,...,N}$.

The AUDyC structure is completely adaptive. The initialization of the network is done at the first data acquisition X_1 with $J = 1$ and $I = 1$. The first Gaussian prototype is characterized by its center $M_1 = X_1$ and a parameter S_{ini} of the initial covariance matrix of the prototype. The construction of the network is done on-line in an iterative way thanks to specific learning rules [4]. Its auto-adaptive properties allow to develop connections and neurons of the Hidden layer and the Output layer and thus to modify the structure of the network. The unsupervised leaning process (shown on Fig.1) consists in three stages:

1st stage: "Classification" makes possible the creation and the adaptation of Gaussian prototypes and classes according to the decision function.

2nd stage: "Evaluation" is used to eliminate aberrant prototypes and classes due to the noise. This phase is based on some rules of distance rejection.

3rd stage: "Fusion" is needed to merge Gaussian prototypes or/and complex classes that are close. It based on some rules of ambiguous rejection [4].

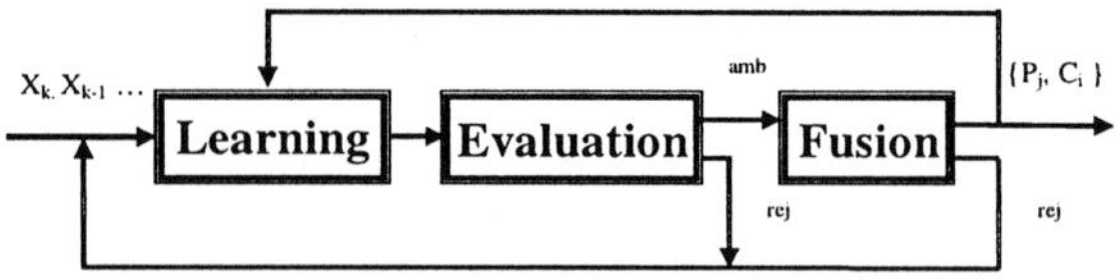

Fig.1: AUDyC NN unsupervised learning process

AUDyC learning strategy self-adapts its architecture in order to incorporate new information and to take into account evolutions of data distributions. Fig.2 shows an illustration on synthetic data.

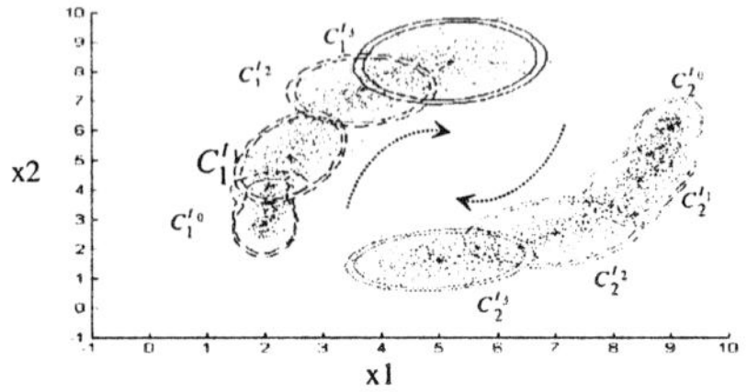

Fig.2: AUDyC dynamical classification on two evolutionary Gaussian distributions (of 2000 dated samples)

More information on the AUDyC NN and its performances for dynamical classification are described in [4][8]. However, in spite of its performances, the AUDyC network has some limits in terms of decrease of its qualities of classification. We notice non-optimal densities modeling in some situations of Gaussian prototype fusion. Indeed, ambiguity rejection rules of this fusion stage lead to some local maxima. The next section presents new density merge mechanism to improve the AUDyC classifier.

3 A new densities merge mechanism

3.1 Ambiguity problematic

In the context of fuzzy classification of incomplete information, ambiguous data shared by clusters decrease the reliability of classifiers. This is due to uncertainties and inaccuracies inherent to data information. To overcome such problems, it is thus useful to define suitable decision rules of ambiguity reject. Many of them lead exclusively to fusion or adaptation of the related prototypes. Note for example ambiguity rules in [9]. The AUDyC NN uses four kinds of ambiguity decision rules. But only one can be selected at the same time and fixed for the whole clustering process. Indeed, when AUDyC classifier encounters ambiguity data shared by two or several Gaussian prototypes, the first and the second decision rule come to adapt only one of them because it has the highest membership degree or because it's the biggest in term of population. The inconvenience of this initiative is that a unique Gaussian density could be represented by overlapped prototypes models (fig.3-a). The third ambiguity rule consists in adapting all prototypes of the ambiguity. This technique induces the same inconvenience as the 2 first rules and in addition assigns a unique data to more prototypes in the same time. Finally, the fourth decision rule comes to merge all prototypes which share an ambiguity data. This solution is widely used to handle with ambiguous information in classification. But, in many situations of Gaussian data classification, it induces errors in modeling as illustrated on Fig.3-b (not aligned densities) or in Fig.3-c (high and weak densities). The succession of such phenomena could more affect the quality of the global clustering.

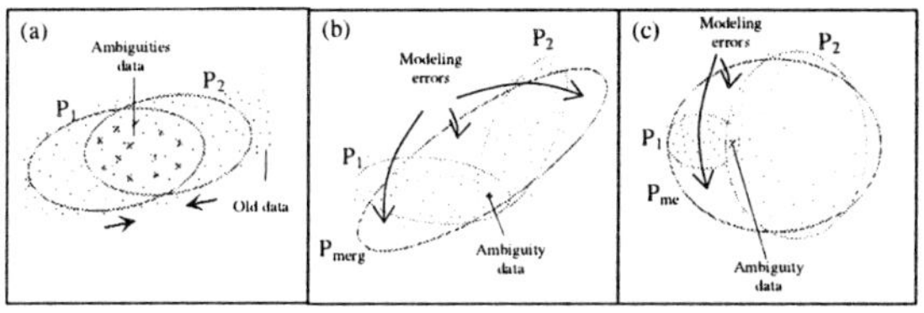

Fig.3: Modeling errors induced by ambiguity rejection criteria

3.2 New criteria for Gaussian densities merge

To solve local optima problems discussed above, we propose a new densities' merge criterion. Our main idea consists at first in detecting Gaussian prototypes candidates by using an ambiguity rule; and secondly, to examine the densities fits according to their relative orientations and their shapes. We deal with this by setting an additive condition based on densities patterns analysis. This will be the merge acceptance criterion. We propose the following densities measure as:

$$A_c(P_j, P_g) = tr\left(\Sigma_j \Sigma_g^{-1} + \Sigma_g \Sigma_j^{-1}\right) - 2.D \tag{6}$$

A_c comes from the well-known Kullback-Leibler Distance [10] for Gaussian densities. By studying the KLD, A_c corresponds to the sensitive term of densities form, orientation discordance. Numerous experiments have shown that A_c is a well-established measure of two local data densities fitting. Consider Gaussian prototypes, the new proposed densities merge mechanism is set as follow:

- First: Detect Gaussian prototypes candidates for fusion thanks to ambiguity criterion with the threshold N_{amb}:
  $Card\{X \in P_g \;\&\; X \in P_j\} \geq N_{amb} \;\Rightarrow\; P_j, P_g$ candidates
- Second : Evaluate the merge acceptance criterion

$$\begin{cases} A_c(P_j,P_g) < th_{ac} \Rightarrow \; merge\_P(P_j,P_g) = P_m, \quad \& \; adapt \; P_m \\ A_c(P_j,P_g) \geq th_{ac} \Rightarrow \; adapt \begin{cases} P_j, & \mu^j(X) > \mu^g(X) \\ P_g, & otherwise \end{cases} \end{cases}$$

- Third : Fusion of the classes

$$P_j \in C_1, P_g \in C_2, \quad C_1 \neq C_2 \quad \Rightarrow \quad merge\_C(C_1, C_2) = C_m$$

The first step consists in analyzing all the X data which belong to two different prototypes. If their cardinality is higher than N_{amb}, the concerned prototypes are selected. In the second step, in case of the acceptance criterion comes to merge the two prototypes, the new merged prototype is allocated to the class of its predecessors and all X data are used to adapt this prototype [4].
The Gaussian prototypes merge operation is done by using the recursive equations [10]. This increases the algorithm speed by avoiding relearning of all data:

$$n_m = n_j + n_g$$

$$M_{P_m} = \frac{1}{n_m}\left(n_j.M_{P_j} + n_g.M_{P_g}\right) \tag{7}$$

$$\Sigma_m = \frac{1}{n_m - 1}\left[\begin{array}{l}(n_j - 1).\Sigma_j + (n_g - 1).\Sigma_g \\ + \frac{n_j.n_g}{n_m}.\left(M_{P_j} - M_{P_g}\right)\left(M_{P_j} - M_{P_g}\right)^T\end{array}\right]$$

In case of the criterion is higher than the threshold, the prototype, which presents the highest membership degree with regard to the ambiguity datum, is adapted with the new information using recursive equations in [4]. After several test, the suitable threshold of acceptance merge is fixed at $th_{ac} = 3{,}2$.

In the third step, if the prototype candidates did not belong to the same class their classes are merged.

4 Experiments

To illustrate how the AUDyC algorithm performs with the new merge mechanism, 3 synthetic datasets have been created in non-stationary environment. The classification quality is appreciated by comparing the results of the modified AUDyC classifier with its previous version. To generate dataset, let consider t a uniform distribution in $[0,1]$ and rnd_1, rnd_2, two iid $N(0,1)$. Two evolutionary Gaussian modes are created with 1000 samples generated by the following equations:

$$[Gauss1] = \begin{bmatrix} f_1(t) \\ g_1(t) \end{bmatrix} + a_1 \begin{bmatrix} rnd_1 \\ rnd_2 \end{bmatrix}$$
$$[Gauss2] = \begin{bmatrix} f_2(t) \\ g_2(t) \end{bmatrix} + a_2 \begin{bmatrix} rnd_1 \\ rnd_2 \end{bmatrix} \tag{8}$$

For all the experiments in this paper the AUDyC NN is initialized with the parameters: $\Sigma_{ini} = 0.4 * I_D$ (initial covariance matrix), $\mu_{\max} = 0.02$ (membership threshold of sample to prototype), $\mu_{\min} = 0.01$ (membership threshold of sample to class), $NP_{\max} = 300$ (maximal cardinality of any Gaussian prototype) and $N_{amb} = 1$.

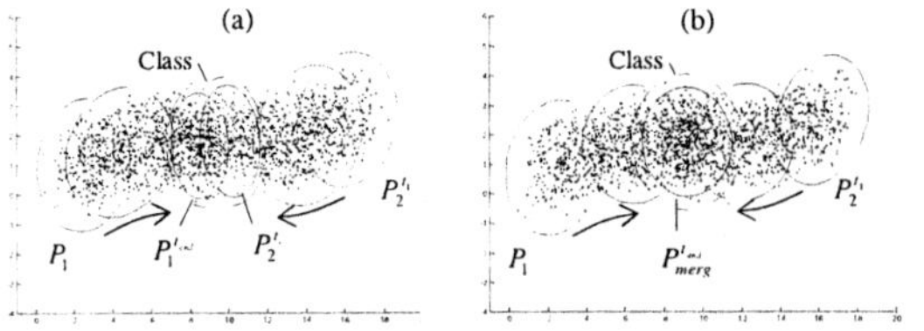

Fig.4: Evolution of two Gaussian modes coming to merge
$f_1(t) = 10t; f_2(t) = 15 - 10t; g_1(t) = g_2(t) = \sin(\pi t) - 1; a_1 = a_2 = 3$

(4-a) Dynamical dataset modeling is done by using the AUDyC classifier with its second ambiguity rule. Consequently, the single natural density resulting of modes evolution is not properly clustered; as it is representing by two Gaussian models. (4-b) The classification of the data evolution is done by using the modified AUDyC algorithm with our proposed merge mechanism. The classifier merges correctly the two evolutionary Gaussian modes in one density mode.

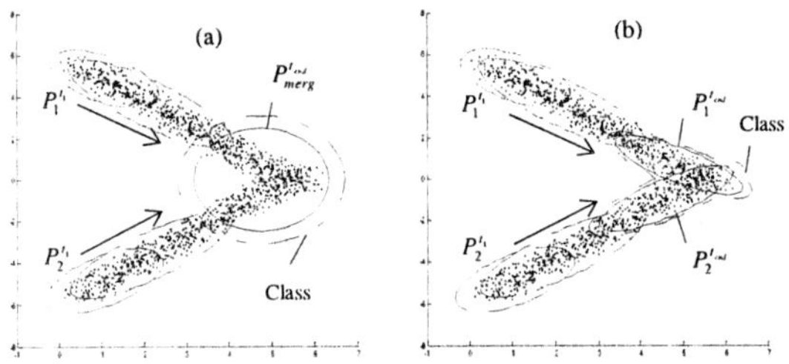

Fig.5: Evolution and crossover of two non-alignment densities.
$f_1(t) = f_2(t) = 9t; g_1(t) = -6 + 10t; g_2(t) = 5 - 10t; a_1 = a_2 = 1$

(5-a) When the AUDyC classifier is used with its fourth ambiguity rule, the two Gaussian models are merged to a single one. This induces modeling errors and local optima. (5-b) The dynamical classification is obtained by using the modified AUDyC. The two densities are not merged but are enough close to belong to the same class. Then we obtain a better classification of data.

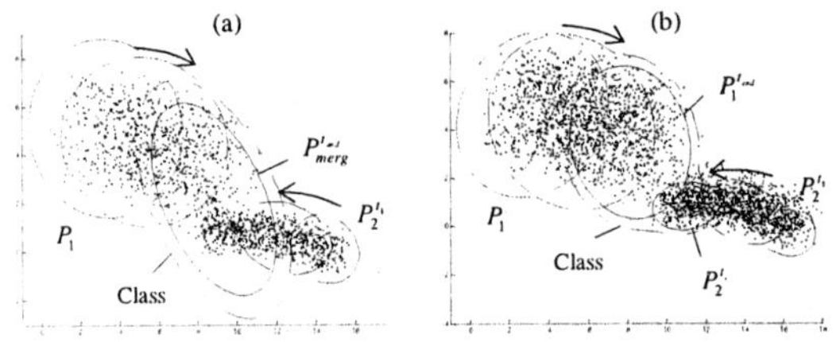

Fig.6: Evolution and crossover of weak and high densities
$f_1(t) = 10t; f_2(t) = 14 - 10t; a_1 = 5;$
$g_1(t) = 1.2 + \cos(\pi t); g_2(t) = \cos(\pi t) - 1.2; a_2 = 1$

(6-a) As in the previous experiment, the AUDyC classifier is used with the fourth ambiguity rule. In this experiment too, we notice modeling error and local optima problem. (6-b) The experiment with the new merge mechanism gives a better modeling by avoiding the prototypes fusion; they are assigned to the same multiprototypes class.

5 Conclusion

Our main contribution consists in: at first establishing a new approach of Gaussian densities criteria for the well-posed ambiguity rejection problem in fuzzy classification; secondly circumventing local optima drawbacks by selecting Gaussian densities fitting to merge. The proposed densities merge mechanism improved the AUDyC NN learning rules for dynamical classification of evolving data. The new mechanism is tested on some examples chosen to illustrate the main difficulties that the merge process has to deal with. By comparison with the previous AUDyC NN that encounters local maxima, the modified version gives a better dynamical classification.
To reach the algorithm global convergence, later works will involve the design of a suitable density split operations for dynamical models in fuzzy classification.

References

[1] Deng, D., Kasabov, N. (2003) On-line pattern analysis by evolving self organizing maps. Neurocomputing 51: 87-103.

[2].Simpson, P.K. (1993) "Fuzzy min-max neural networks - Part2: Classification" IEEE Trans. Fuzzy systems, 11) : 32-45

[3] Eltoft T. (1998) A new Neural Network for Cluster-Detection-and-Labeling. IEEE Trans. NN, 9(5): 1021-1035.

[4] Lecoeuche, S., Lurette C. (2003) Auto-adaptive and Dynamical Clustering Neural Network. ICANN03: 350-358.

[5] McLachlan G. J., Peel D. (2000) Finite Mixture Models. Eds Wiley, ISBN 0-471-00626-2

[6] Ueda, N., Nakano, R.. (2000) SMEM algorithm for mixture models, Neural Computing 12: 2109–2128.

[7] Zhang, B. Xing, Y. (2003) Competitive EM algorithm for finite mixture models, Pattern recognition 37: 131-144

[8] Lecoeuche, S., Lurette C. (2004) New supervision architecture based on on line modelization of non stationary data, Neural Computing and Applications, To be published.

[9] Mouchaweh S. (2003) Conception d'un système de diagnostic adaptatif et prédictif basé sur la méthode Fuzzy Pattern Matching pour la surveillance en ligne des systèmes évolutifs, PhD Univ RCA France

[10] Shaohua, K., Shellappa, R. (2004) Kullback-Leibler Distance between Two Gaussian Densities in Reproducing Kernel Hilbert Space. Chicago DC, *ISIT*

[11] P. M. Kelly (1994). An Algorithm for Merge Hyper-ellipsoidal Clusters. Los Alamos National Laboratory.

The Growing Hierarchical Self-Organizing Feature Maps And Genetic Algorithms for Large Scale Power System Security

M. BOUDOUR[1] , A. HELLAL[2]

[1]Electrical Engineering Department , University of Sciences & Technology, Algiers, Algeria
[2] Electrical Engineering Department, Polytechnic Institute , Algiers, Algeria
E-mail : (mboudour, ahellal)@yahoo.com

Abstract

This paper proposes a new methodology which combines supervised learning, unsupervised learning and genetic algorithm for evaluating power system dynamic security. Based on the concept of stability margin, pre-fault power system conditions are assigned to the output neurons on the two-dimensional grid with the growing hierarchical self-organizing map technique (GHSOM) via supervised ANNs which perform an estimation of post-fault power system state. The technique estimates the dynamic stability index that corresponds to the most critical value of synchronizing and damping torques of multimachine power systems. ANN-based pattern recognition is carried out with the growing hierarchical self-organizing feature mapping in order to provide an adaptive neural net architecture during its unsupervised training process. Numerical tests, carried out on a IEEE 9 bus power system are presented and discussed. The analysis using such method provides accurate results and improves the effectiveness of system security evaluation.

1. Introduction

Power system spontaneous oscillations, due mainly to the development of interconnection of large electric power systems, are due to the lack of damping of the mechanical mode of the interconnected system [1,2]. An important task in power system operation is to decide whether the system is currently operating safely, critically or unsafely. The conventional eigenvalue analysis method requires a lot of computational efforts as the size of power systems become larger. The investigation of the synchronizing and damping torques method [3-6] has shown that it gives efficient means for determining the dynamic stability with a great reduction of the on-line computational requirements. They can also provide a clear picture about the dynamic performance of the individual machines using the security contour visualization.
ANN-based methods have been recently developed for power system security assessment [8-14]. The advantage of these methods is the reduction of the on-line computational requirements. The disadvantages are: (1) elaborate off-line computations are required, and (2) if some components of the pattern vector are strongly correlated, the methods become inaccurate, and a feature extraction must be performed to hopefully yield an uncorrelated set of components. Feature extraction is the process of mapping all available features into a composite feature set of lower dimension. A new feature extraction technique based on genetic algorithm is presented and applied to power system security assessment [15].
One of the most important aspects of achieving good neural network performance has proven to be the proper selection of training features. The curse of dimensionality states, as a rule of thumb, the required cardinality of the training set for accurate training increases exponentially with the input dimension [16]. Feature extraction techniques applied to power system security assessment have previously been presented in [17].
In this paper, pre-fault power system conditions are assigned to the output neurons on the two-dimensional grid with the growing hierarchical self-organizing map technique (GHSOM) via supervised Artificial Neural Networks (ANNs) which perform an estimation of post-fault power system state. The Kohonen SOM memory runs an unsupervised clustering algorithm. It is easily trained and has attractive properties such as topological ordering and good generalization. However, two drawbacks remain untouched: firstly it uses fixed network architecture in terms of number and arrangement of neural processing elements, which has to be defined prior to training. Secondly, hierarchical relations between the input data are not mirrored in a straightforward fashion [18]. To resolve both limitations, the GHSOM composed of SOM-like neural networks with adaptive architecture was used for data mining applications in [19] to improved representation of the hierarchical structure of the analysed data collection.
In the present work, the synchronizing and damping torques analysis which utilizes a linearized small perturbation model [4,5,20] is considered and its application generalized to multimachine power systems for the dynamic security assessment where a new criterion is defined. A GHSOM-based method is presented to calculate the estimate security index so that an output neuron calls the index corresponding to an input pattern. The effect of the control parameter for the (depth / shallowness) of the GHSOM is examined and the effectiveness of the proposed technique demonstrated in a IEEE 9 bus power system.

2 Dynamic Security Evaluation

2.1 Synchronizing and Damping Torques Analysis

The concept of damping and synchronizing torques is extended to calculate the corresponding torque components in a multimachine power system: obtained by breaking the torque of the machine with respect to its own rotor angle and speed responses according to the equation:

$$\Delta T_e = T_s \Delta\delta + T_d \Delta\omega \qquad (1)$$

The time responses of ΔT_e,, $\Delta\delta$ and $\Delta\omega$ include the effect of the various modes and so are the computed damping (T_d) and synchronizing (T_s) torques. The linearized model of the subsystem in a n-machine power system is depicted in Fig.1.

The self damping and synchronizing torques of machine i are:

$$T_{sii} = T_{seii} + T_{smii}$$
$$T_{dii} = T_{deii} + T_{dmii} \qquad (2)$$

where "e" stands for electrical and "m" for mechanical.

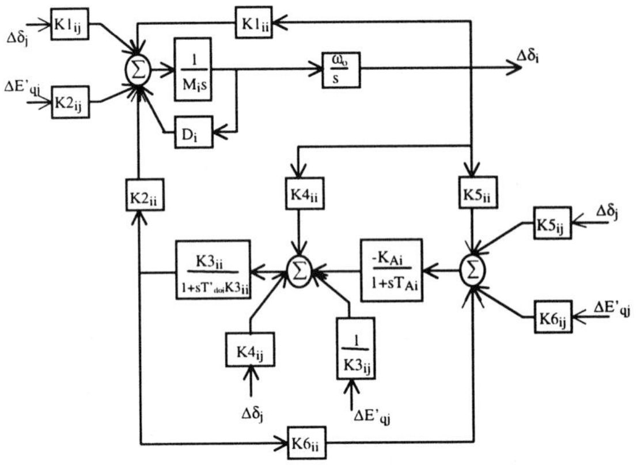

Fig. 1 Block diagram of a multimachine power system

The interraction damping and synchronizing torques of machine i with the other machines are:

$$T_{sij} = \sum_{j=1}^{n} \left(T_{seij} + T_{smij}\right)$$
$$T_{dij} = \sum_{j=1}^{n} \left(T_{deij} + T_{dmij}\right) \qquad (3)$$

2.2 New Stability Criterion

As the synchronizing and damping torques can not be determined in a closed form, a new criterion (torques criterion) is defined through necessary and sufficient conditions for the system to be stable as follows:

$$T_{sii} > T_{sij} \text{ and } T_{dii} > T_{dij} \qquad (4)$$

The coupling between the low frequency electromechanical modes is investigated by computing the contribution factors [7] which give a method of presenting the interaction between the eigenvectors associated with the eigenvalues or modes as one or more parameters of the system are varied. The synchronizing and damping torques analysis is correlated to the modal analysis to provide insight into the system stability.

2.3 Feature Selection

An acceptable simple criterion for selecting a variable as a feature is that it should provide more information for classification than those not selected [20]. The genetic algorithm solves the feature selection problem by representing each possible solution as a binary vector consisting of a number of bits equal to the number of features where a "0 bit in location i means the i-th parameter is not selected and a "1" bit means it is selected. An initial population of solution candidates, called "individuals", is generated, and the "fitness" F of the j-th individual is computed according to:

$$F(j) = \omega_1 F_1(j) + \omega_2 F_2(j) \qquad (5)$$

where F_1 and F_2 are the fitness components corresponding to neural network accuracy and cardinality, respectively, and ω_1 and ω_2 are their corresponding weights. We measure set sufficiency fitness (accuracy) of individual j as the reciprocal of the average absolute error computed by testying a neural network trained with inputs corresponding to the selected attributes of the individual. We measure cardinality fitness as $kn\ (j)$ where $n\ (j)$ is the cardinality level of member j (number of parameters selected) and:

$$\begin{cases} k = 1 & n(j) \le n_{desired} \\ k = 0 & otherwise \end{cases}$$

Here, $n_{desired}$ is the desired cardinality level. We also specify $\omega_1 = 1.0$ and ω_2 is large ($\omega_2 = 1000$).

After evaluating the fitness of all N_1 members of a population, a certain number of the most fit members are selected to undergo the so-called "crossover" procedure where a fraction of the selected individuals are randomly mated pairwise. In addition, a single bit of the chosen members may be "mutated" or toggled to try and improve their fitness. The new members of the population, called "children" (which evolved through either crossover or mutation), have their fitness evaluated, and again a certain chosen and the procedure repeated. The accuracy component, $F_1(j)$, for each j is computed by first randomly splitting the input data into train and test sets according to a user-specified ration. A neural network is then trained using the training data corresponding to feature set j. Finally, the average absolute error is obtained by testing the neural network using the test data. This error is given by:

$$F_1(j) = \frac{N_{test}}{\sum_{k=1}^{N_{test}} \left| R_{jk,nn} - R_{k,correct} \right|} \qquad (6)$$

where N_{test} is the number of test data points, $R_{jk,nn}$ is the neural network j's output for the k-th operating condition, and $R_{k,correct}$ is the correct performance measure for the k-th operating condition. The multilayer backpropagation neural network is chosen because of its computational simplicity and its ability to perform nonlinear functional approximation. To limit computation time, we use a simple neural network with only one neuron in one hidden layer. After using GANN to perform the feature selection, we then identify an optimal structure that we can use to perform the function approximation to be used in the visualization software. This step assumes that a feature set that is identified as "best" using the simple structure will be at least "one of the best" for more complicated structures. For different randomly perturbed loads, 200 patterns are generated for each contingency (transmission line outage). The set of features in terms of which input data will be represented is defined among the original input pattern composed by: the real and reactive powers of the n generators (PG_i , QG_i), their voltage regulator gains (K_{Ai}) and the real and reactive powers of the n_L load demands (PD_i , QD_i). The choice of these features is motivated by physical arguments and the selection process involves engineering judgement [13]. The feature extraction obtained from the original 15 dimensional input pattern, applied to the IEEE 9 bus test system, leads to the vector composed of 5 state variables, namely: PG_1, K_{A1} , PG_2 , QG_3 , PD_8.

3 Proposed Neural Network Architecture

3.1 Growing Hierarchical Self-Organizing Map

The choice of a clustering technique arises from the lack of a priori information of the problem. Unfortunately, SOMs require a previous knowledge of the problem to decide the shape of its grid. In order to overcome this limit, the GHSOM has been proposed [18]. The GHSOM has a hierarchical structure of multiple layers, where each layer consists of several independent growing SOMs. The map in layer 1 consists of 3 x 2 units and provides a rather rough organization of the main clusters in the input data. The six independent maps in the second layer offer a more detailed view of the data. For the control of the growth process, two different strategies can be used, using either the mean quantization error (*mqe*) of a unit or the absolute value namely, the quantization error (*qe*) of a unit.

The *mqe* of a unit i is calculated according to (7) as the mean Euclidean distance between its model vector m_i and the n_C input vectors x_j that are the set of input vectors C_i mapped onto this unit i.

$$mqe_i = \frac{1}{n_C} \sum_{x_j \in C_i} \left\| m_i - x_j \right\| \qquad n_C = \left| C_i \right|, C_i \neq \phi \tag{7}$$

The starting point for the GHSOM training process is the calculation of an mqe_0 of the unit forming the layer 0 map as provided in (8). With n_I we refer to the number of all input vectors x of the input data set I and m_0 denotes the mean of the input data

$$mqe_i = \frac{1}{n_I} \sum_{x_i \in I} \left\| m_0 - x_i \right\| \qquad n_I = \left| I \right| \tag{8}$$

The minimum quality of data representation of each unit will be specified as a fraction, indicated by a parameter τ, of mqe_0 to satisfy the global termination criterion:

$$mqe_i < \tau . mqe_0 \tag{9}$$

The growth process of a growing SOM can be described as follows. Let C_i be the subset of vectors x_j of the input data that is mapped onto unit i, and m_i the model vector of unit i. Then the error unit e is determined as the unit with the *mqe* as follows:

$$e = \arg\max_i \left(\sum_{x_j \in C_i} \left\| m_i - x_j \right\| \right) \qquad n_C = \left| C_i \right|, C_i \neq \phi \tag{10}$$

Following the selection of e, its most dissimilar neighbor d is determined as follows, where N_e is the set of neighboring units of e.

$$d = \arg\max_i \left(\left\| m_i - x_j \right\| \right) \qquad m_i \in N_e \tag{11}$$

A row or column of units is inserted between d and e. As more units are added to the growing SOM their *qes* decrease. The training process will continue until all units satisfy the global stopping criterion.

3.2 Problem Formulation

The task of dynamic security assessment of a large-scale power system using Neural Networks (NN) is an enormous computational exercise. One way of reducing this complexity is to divide the problem into smaller tasks at different levels and train dedicated NN classifiers to handle each task. The overall structure of the devised neural architecture is depicted in Fig. 2. The PNSI (Projection Network with Stochastic Interconnects) is a supervised three layer NN which models an input-output relationship by clustering both input and output spaces and establishing a correspondence between the clusters of the two spaces by viewing the training data only once [21]. The pre-fault state vector is the input of an array of PNSIs which work in a parallel way. Their purpose is the fast prediction of the post-fault state vector for each contingency. The learning phase of the PNSIs is performed on a main training set composed of pre-fault and post-fault state vector pairs of every outage for many operating points of the power system. The number of the hidden layer units is not to be determined a priori, but adapts automatically to the learnt data. One approach is to generate nonlinearities with stochastically chosen interconnects between the input and hidden neural layers with a sigmoidal nonlinearity at each hidden neuron. The hidden to output interconnects are chosen to be a trainable projection matrix. The input vectors are normalised between –1 and +1 according to their maximum values reached in the training and testing sets. The components of the weight vectors associated to each neuron have been randomly initialised in the interval [-0.1, 0.1]. The outputs of the PNSIs are then fed into a GHSOM which has been trained for classifying the post-fault vector in several bidimensionnal maps. The neighborhood function of each SOM is a unimodal function which is symmetric around the location of the winner neuron and monotonically decreasing with increasing distance from the winner. In this paper a Gaussian is used as a neighborhood function [17]. The GHSOM training set is composed of the post-fault state vectors for all the contingencies. The generalisation properties of the GHSOM when trained on a set of the post-fault state vectors, could fix the rough estimates obtained by the PNSIs.

4 Application to Dynamic Security Assessment

The proposed method uses combined supervised and unsupervised learning for evaluating power system dynamic security. The power system index is estimated by calling an output neuron where the estimated index is assigned. After an output neuron on the grid responds to an input pattern, the output calls the estimated index. It is assumed that the index is prepared by the off-line calculation and is defined by:

$$\begin{aligned} T_{crd} &= \min_i \left(T^i_{ds} - T^i_{dm} \right) \\ T_{crs} &= \min_i \left(T^i_{ss} - T^i_{sm} \right) \end{aligned} \qquad i=1,\ldots,n \tag{12}$$

As a result, power system dynamics are evaluated by the most critical value between the two critical synchronizing and damping torques:

$$T_{cr} = \min \left(T_{crd} \, , T_{crs} \right) \tag{13}$$

Let us assume that more than one learning data is assigned to each output neuron. The estimated index is calculated by the average value of the index corresponding to power system conditions classified into the same output neuron. Namely:

$$T^i_{cr} = \frac{1}{N_i} \sum_k T^i_{cr_k} \tag{14}$$

where :

T^i_{cr} : estimated index at output neuron i

N_i : Number of input patterns classified into output neuron i

$T^i_{cr_k}$: index of input-pattern k classified into output neuron i

The algorithm of the method is well described in [20].

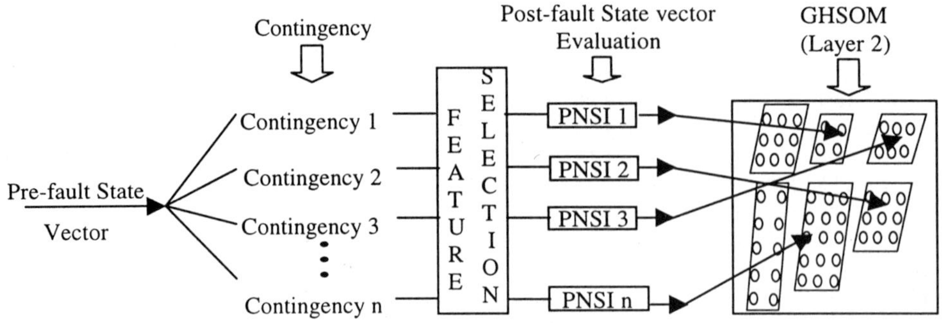

Fig. 2 Proposed mapping architecture

5. Numerical Results

5.1 Simulation Results

The proposed method has been applied to the IEEE 9 bus study system with 14 lines and 3 generators [4,5]. The main training set is composed of 200 different load scenarios. Therefore, a total amount of 2800 samples has been generated by load flow calculations. Once the key features are selected, some of the data are used for training the PNSIs (1700 samples) while some are used for testing the trained NN (1100 samples). The GHSOM has 7 inputs (state vector and security indexes) and the corresponding training set is composed of the true post-fault state vectors (the base case and the 14 contingencies). Based on the unit representing the mean of all data at layer 0, the GHSOM training algorithm starts with a 2 x 2 SOM at the first layer. The training process for this map continues with the additional units being added until the quantization error drops below a percentage τ_2 = 0.005 of the overall quantization error of the unit at layer 0, resulting in 3 x 2 units representing six major clusters depicted in Fig. 3. Further, the notation (*i*/*j*) is used to refer to the unit located in column *i* and row *j* of the map.

We find units (1/2), (1/3) and (2/1) to represent the contingencies cases corresponding to secure operating points. The main difference between these three clusters lies in the range of the values reached by both the synchronizing and damping torques, though they are both positive. For example, unit (2/1) classifies the secure cases characterized by high values of the critical synchronizing and damping torques $\left(T_{crd} \geq 0.25\,, T_{crs} \geq 4.25\right)$, whereas unit (1/2) groups the secure cases where the critical torques remain positive in the interval defined as: $\left(T_{crd} \leq 0.01\,, T_{crs} \leq 1.45\right)$. The non critical cases are presented by unit (2/2) where both synchronizing and damping torques are positive, corresponding to the outages that provoke major changes and cause severity conditions near the security contour, whereas the operating points remain inside this contour. The critical cases are placed in unit (2/3) representing insecure cases near to the security boundary. These unstable states are outside the security contour. Unit (1/1) classifies the insecure cases where the critical values of the damping and/or synchronizing torque are negative. Further maps are automatically trained to represent the various state vectors. This results in six individual maps on the second layer, each representing the data of the respective higher layer unit in more detail as QE_m reaches τ_l = 0.063 of the corresponding *qe*'s unit in the layer1. In particular, a 6x5 map represents the insecure patterns of unit (1/1) of the first-layer map.

Insecure cases (379) 6x5 SOM	Secure Cases III (29) 4x3 SOM
Secure Cases I (356) 6x4 SOM	Non critical cases (78) 5x5 SOM
Secure cases II (357) 4x7 SOM	Critical cases (491) 7x6 SOM

Fig. 3 GHSOM-layer1

When a post-fault state vector is presented, the corresponding unit of the corresponding map is activated. It is then possible to label each unit with security related information. The obtained six maps in the second layer provide a satisfactory and accurate enough pictorial view of static security assessment. The results show a better use of the neurons, avoiding the problem of the dead units (the neurons which never win the competitive process), and a smaller number of misclassifications.

For comparison purpose, a two-dimensional grid of Kohonen SOM is depicted in Fig. 4. This allows us to view the hierarchical structure of the data on the flat map which consists of 15x15 output neurons. In other words, 2800 power system conditions are classified into 225 patterns. The range of the estimated index in each cluster is given in the mapping, where the critical index is defined as the minimum value between the two critical torques: $T_{cr} = min\left(T_{crd}\,, T_{crs}\right)$. The white holes correspond to the output neurons where the estimated index is not assigned. As the holes colour becomes darker in the figure, the power system approaches insecure conditions. Five out of six clusters on the first layer of the GHSOM are recognized [20].

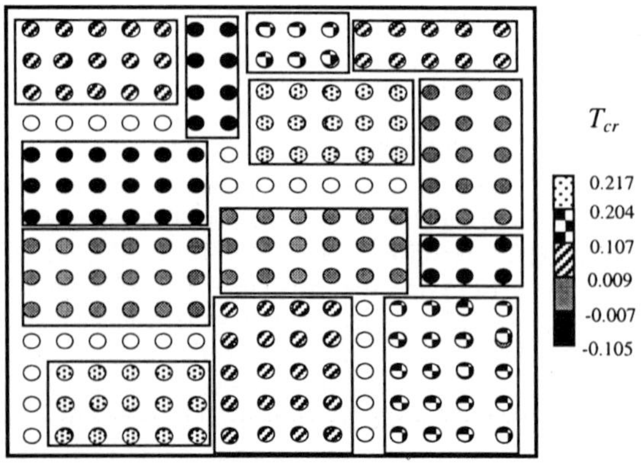

Fig. 4 Self-organizing map

5.2 Discussion

Simulations results show that the proposed method projects safe, critical and unsafe operation points to different clusters corresponding to different regions in the security space. Knowing which cluster relates to which security state and mapping an operational point to one of these immediately identifies whether the system is operating safely or not. The range of label attributes (T_{cr}) in each cluster indicates which associative actions should be taken in each instance. Different preventive control schemes may be advised for low and high critical values of the synchronizing and damping torques, depending on the installed system components, company policy, etc.

The following percentages are introduced to obtain a quantitative measure of the classification performance. The percentage of false alarms (FA), false dismissals (FD) and false classifications (FC) are calculated. The testing statistics are given in Table 1, comparing the use of the true or the predicted post-fault values in the security mapping.

Table 1 Classification Statistics

Testing statistics	True vectors	Predicted vectors
Testing data	1100	1100
True secure patterns	536	528
True insecure patterns	544	539
FA (%)	2.61	4.17
FD (%)	1.10	2.04
FC (%)	0.34	0.58

Note that identical proportions of secure and insecure data are used in both the training and testing sets. For larger τ_1 (τ_1 = 0.107), which serves as the control parameter for the (depth / shallowness) of the GHSOM architecture, more detailed data representation is delegated to additional maps resulting in 7 layers. The first layer map has grown to a size of 2x2 units with the remaining groups (1/1), (1/3), (2/2) and (2/3) of Fig. 3.

6. Conclusion

This paper has presented a new methodology to assess the dynamic security for large scale power systems of a possible contingency simply by considering the unit activated by the pre-fault state vector in an output map. A new dynamic stability index is developed. The torque components technique has an advantage that it allows us to understand the transition of the security margin on the two-dimensional grid since it is easy to visually understand both secure and insecure states close to the operational conditions. They can also provide a clear picture about the dynamic performance of the individual machines. The primary concern about whether this new concept can become useful as an on-line aid resides in the question of scaling. A feature selection technique based Genetic Algorithm is applied to reduce effective problem dimension. Using the GHSOM the overall training time is largely reduced since only the necessary number of units are developed to organise the state vectors and its architecture allows the user to understand and analyze large amounts of data in an explorative way. Parallel supervised neural network architecture and the adaptive capability of the growing hierarchical SOM can be combined to achieve high speeds of execution and good classification accuracy. This new approach significantly improves the efficiency and effectiveness of large-scale power systems security assessment.

References

[1] Anderson, P.M, Fouad, A.A.(1977) *Power system control and stability*, The Iowa State University Press

[2] Yu, Y.N.(1983) *Electric power system dynamics*, New York Academic Press

[3] Boudour, M., Bensenouci, A.(1994) Effect of load characteristics and PSS on the stability regions using damping and synchronizing torques. *Middle East Power Conference*, Assiut, Egypt

[4] Boudour, M., Bensenouci, A.(1998) Multimachine power system stability assessment using torques indexes and modal analysis.*AMSE Journal Modelling Measurement and Control* 69: 55-69

[5] Aggoune M.E.(1988) *Power System Security Assessment Using Artificial Neural Networks*, PHD Thesis, University of Washington

[6] Feilat E.A.(2000) Power system dynamic stability using least squares, Kalman filtering and genetic algorithms. Proceeding of the *IEEE South East Conference* 9: 489-492

[7] Mugwanya, D.K., Van Ness, J.E.(1987) Mode coupling in power systems. *IEEE Trans. Power System* 2: 264-270

[5] De Oliveira, S.E.M.(1994) Power system steady state stability as affected by static var compensators. *IEEE Trans. Power Systems* 9: 109-119

[6] Yoshimura, K., Uchida, N.(2001) Proposal of remote signal input PSS for improving power transfer capability considering damping and synchronizing torques. *IEEE Power Engineering Society Winter Meeting* 3: 1329-1334

[7] Pourbeik, P., Gibbard, M.J.(1994) Damping and synchronizing torques induced on generators by facts stabilizers in multimachine power systems. *IEEE Trans. Power Systems*11:1920-1925

[8] El-Sharkawi, M.A., et al.(1989) Dynamic security assessment of power systems using artificial neural networks. *Proc. of second Symposium on Expert Systems*, pp. 378-384

[9] Mansour, Y., et al., Dynamic security contingency screening and ranking using neural networks. *IEEE Trans. Neural Networks*, pp. 942-950

[10] Meyer, B., Nativel, G.(1999) New trends Requirements for dynamic security assessment. *Control Engineering Practice 7*: 375-380

[11] Srinivasan, D., et al.(1998) Power system security assessment and enhancement using artificial neural network . *Proceedings of the International Conference on Energy Management 2*:. 582-587

[12] Niebur, D., Germond, A.J.(1992) Power system security assessment using the Kohonen neural network classifier. *IEEE Trans. Power Systems* 7:865-872

[13] Jensen, C.A.(2001) Power system security assessment using neural networks : feature selection using fisher discrimination. *IEEE Trans. Power Systems* 16: 757-763

[14] Brandwain, V. et al.(1997) Severity Indices for Contingency Screening in Dynamic Security Assessment. *IEEE Trans. Power Systems* (12):1136-1142

[15] Vafai, H., Jong, K.(1992) Genetic algorithms as a tool for feature selection in machine learning. *Proc. offourth international Conference on tools with Artificial Intelligence*, Arlinton, VA, pp. 200-203

[16] Reed, R.D., Marks, R.J.(1999) *Neural smithing*: Supervised learning in feedforward ANN, Cambridge, MA : MIT Press

[17] Veerasooriya, S., El-Sharkawi, M.A.(1991) Use of Karhunen- Loe's expansion in training neural networks for static security assessment. *Proc. offirst internationalforum on Applications of Neural Netwoks to Power Systems*, Seattle, WA, pp. 59-64

[18] Kohonen, T.(1999) *Fast Evolutionary Learning With Batch-Type Self-Organizing Maps*, Neural Process Lett

[19] Rauber, A., Merkel, D., Dittenbach, M. (2002) The growing hierarchical self-organizing map: exploratory analysis of high-dimensional data. *IEEE Trans. Neural Networks*13: 1331-1341

[20] Boudour, M., Hellal, A.(2003) Self-organizing feature maps for power system dynamic security assessment using synchronizing and damping torques technique. *Proceedings of the 29th Conference of the IEEE Industrial Electronics Society, Roanoke, Virginia*, pp.752-758

[21] Marks II, R.J., et al.(1988) The effect of stochastic interconnects in artificial neural network classification. *IEEE International Conference on Neural Networks*

3D Self-organizing Convex Neural Network Architectures

F. Boudjemaï[1,2], P. Biela Enberg[1,2], J. G. Postaire[1]

[1] LAGIS UMR 8146 CNRS, University of Lille1, France
[2] ERASM, Hautes Etudes d'Ingénieurs, France
E-mail: farid.boudjemai@hei.fr, philippe.biela@hei.fr, jack-gerard.postaire@univ-lille1.fr

Abstract

Surface modeling and structure representation from unorganized sample points are key problems in many applications whose neural n
etworks are recently starting a gradual breakthrough. Our purpose is the development of innovative self-organizing neural network architecture for surface modeling.
We propose an original neural architecture and algorithm inspired by Kohonen's self-organizing maps, based on dynamic neighborhood propagation along with an adaptive learning and repulsion process applied to a generalized mesh structure that will lead to a topological definition of the surface given as an input.

1 Introduction

Surface modeling and structure representation from unorganized sample points (figure 1) are widely used in many applications such as data visualization, land surface modeling, medical imaging and reverse engineering. Usually in such applications the input data consist in set of sampling points collected on a 3D surface with no information available about their connectivity. That's why one of the main problems in surface reconstruction from unorganized data points is to build a coherent mesh structure that faithfully represents the sampled original surface. Classical methods [1, 2] tend to create connectivity among nearest points and finally build the mesh, but problems may appear with sparse sets of sample points, especially in under-sampled regions. To cope with this problem we propose a neural method based on an original convex neural network architecture inspired from Kohonen's self-organizing maps [3]. This neural mesh is a 3D-SOM with triangular architecture, which has the capacity, through totally unsupervised learning, to approximate the form of the surface to reconstruct. The learning phase is based on a local adaptation and repulsion processes which concern independent local areas of neurons with dynamic parameters for each neuron. Our method can be advantageous compared to certain traditional methods of reconstruction which have as severe precondition that the points cloud constitutes an optimal, non-noisy sampling of the surface which is less primordial in our approach.
The paper is organized as follows: Section 2 presents related works on surface reconstruction and representation using self-organizing maps and similar approaches including ours. In section 3 we present the general architecture and the learning algorithm associated to our 3D-SOM. Finally, the results presented in Section 4, demonstrate the efficiency of our 3D map. Section 5 brings some conclusions and perspectives.

2 State of the art

In the field of scientific visualization, Knopf and Sangole [4] have introduced a concept of spherical map based on Kohonen's self-organizing maps, however for the propagation process, their approach take into account a neighborhood restricted to the first level of vicinity, i.e. to the nearest neurons of the processed vertex. We believe that this restricted neighborhood limits the efficiency of the map adaptation. In respect of the original self-organizing maps philosophy, we use a neighborhood for adaptation using a large number of neurons, well beyond the first level of vicinity in the beginning of the adaptation processing. This improvement provides more flexibility to our structure and allows a better adaptive process.
A similar approach has been used by Yu [5] who proposed one of the first real rendering applications of the self-organizing maps in the field of surface reconstruction from unorganized sets of points. But the reconstruction process used a 2D-SOM architecture and needs some local mesh transformations and supervised forcing during the reconstruction.
Ito & Al introduced the concept of a 3D topologic map as a torus map for classification goals [6], but no visual representation of the latter was developed for visualization goals. According to the author this is due to the disadvantage that represents the absence of edges in the torus map. In [7] we have implemented a toric and spherical self-organizing map for 3D-reconstruction applications. Another approach using Growing Cell Network [8] for surface reconstruction was presented by Ivrissimtzis & Al in [9].

3 General architecture & learning rule

In our application, the Kohonen's learning procedure is used to carry a global arrangement in the mesh structure with respect to the topological repartition of the data.
Within the framework of our approach, we extend the concept of self-organizing map to any other triangular convex architecture. For example, with this method any convex triangular mesh structure can be used as a self organizing map. The main idea behind this generalization is the independence between the concept of self organizing map, as a learning process, and the topological architecture constraint of the map structure.
We added possibilities of local adaptation to the learning process, to enhance the topology learning efficiency. By those means, we obtain an original general dynamic self-organizing map architecture with extended adaptation capabilities.

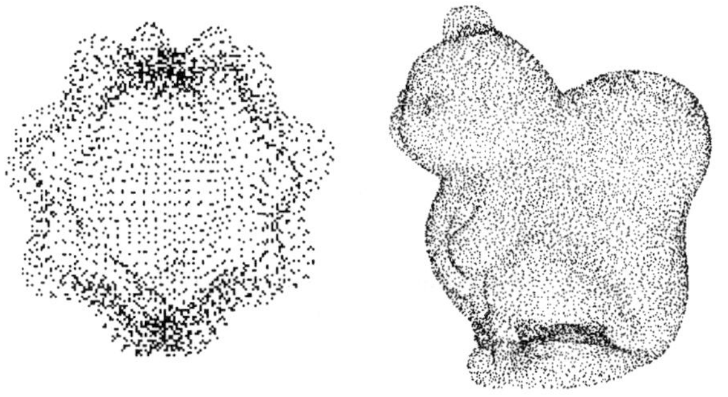

Fig. 1. spiky sphere 5042 points & Squirrel form 9995 points.

3.1 Generalized 3D-map architecture

In [7] we have defined the architecture of a spherical map, which was based primarily on the structure of an icosahedron. We have generalized the method for any convex triangular mesh structures like tetrahedron or cube (figures 2a, 2c, and 2e).
On these basis architectures, we apply a subdivision process to increase the density of the mesh, in order to represent more refined surfaces. We use the *butterfly* subdivision method [10] to increase the density of the mesh: examples are available on figures 2b, 2d, and 2f. With the subdivision method, a new vertex is inserted in the middle of each edge: connecting the new midpoints creates new edges. The old vertices are not modified and new vertices are inserted when splitting each edge in two sub-edges. Therefore, each triangle is subdivided into four triangles. Coordinates of the new vertex are obtained thanks to a *butterfly* mask that multiplies neighboring point coordinates by a set of coefficients and sums them up, to finally provide the new point coordinates.
The topological neighborhood in the map structure is obtained by using a graph representation of neighborhood links that enable a dynamical referencing of the topological neighborhood of the mesh hence the independence between the structure and the learning algorithm.

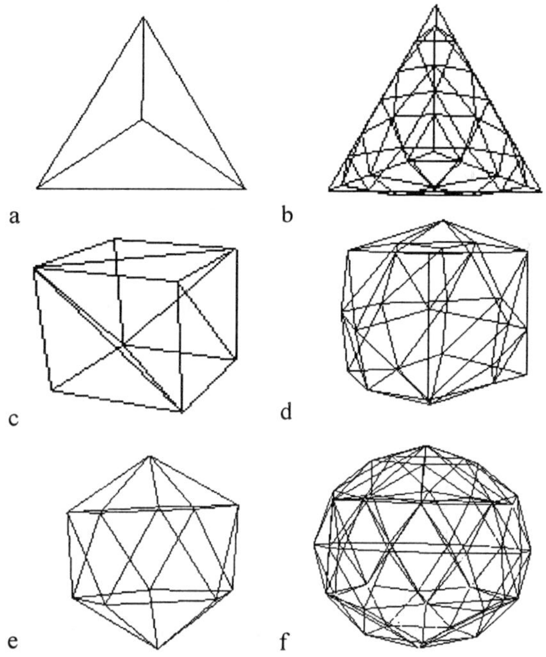

Fig. 2. (a) tetrahedron, (c) cube and (e) icosahedron after 2 levels of subdivision respectively (b), (d) and (f).

3.2 Adaptative learning algorithm

Our learning algorithm is inspired by the Fritzke's works on neural networks [8, 11] and Kohonen's works with the self-organizing maps [3]; we have adapted some of their ideas to our learning rule.
We use a principle of local adaptation inherent to each neuron, their parameters of adaptation are different and evolve in a dynamic way inside local regions. In addition, the efficiency of the adaptation process is locally improved according to a local parameter used as a statistical density estimation criterion; which consists in the accumulation of the error between the neuron and the input vector in the actual neighborhood of the winning neuron. More this local error is important, more the point density of the region represented by this neuron is significant, and needs more accurate and significant adaptation. This parameter is collected and used after a classical beginning phase of adaptation and repulsion process to allow the architecture during a processing time to get the pre-shape of the input data set.
Let us assume that the input data space is composed of a set of sampling points such as $\chi = \{ X_1, .., X_j, .., X_N \}$,

where X_j represents a real input data vector, with $X_j = [x_{j1}, x_{j2}, .., x_{jn}]^T \in R^n$. A weight vector $W_i = [w_{i1}, w_{i2}, .., w_{in}]^T \in R^n$ is associated with each neuron of the self-organizing map. In our case, the dimension of the input data space is n = 3.

1. Initialize the total number I of weight vectors W_i in the form of the initial structure.

   Set t = 0, t represents the discrete time reference of the learning rule procedure.

2. At time t, an input data vector X(t) is randomly selected from the input data space.

   X(t) is compared with all the weight vectors $W_i(t)$, assuming a distance measure (Euclidean distance in our case) denoted $d(X,W_i)$. The winning neuron, also called Best Matching Unit (BMU), with weight W_c, is defined to be the one with the smallest response and that has the index $c = Arg\min_i \{d(X,W_i)\}$.

3. The weight vectors W_c of the BMU and some of its neighboring neurons in the mesh are moved toward the input vector X(t),while the other neurons are subject to a weaker repulsion process, they are moved backward from X(t), equation 1 :

$$W_i(t+1) = W_i(t) + (\lambda(t).\mu_i(t) + \alpha(t))(X(t) - W_i(t)) \; ; \; For\ i \in N_c(t)$$

$$W_i(t+1) = W_i(t) \text{-} (\alpha(t)/\theta)(X(t) - W_i(t)) \; ; \; Otherwise$$

4. Do t = t+1, If the end-learning criterion (t > T) is not satisfied, return to step 2 and repeat the process. Otherwise stop.

Where the factor $\alpha(t) \in [0,1]$ represents the learning coefficient and $N_c(t)$ specifies the topological neighborhood for the neurons around the BMU inside the map architecture. Both $\alpha(t)$ and $N_c(t)$ are gradually decreasing as the training proceeds, following a uniform decreasing rule (1/t like).

The factor $\mu(t)$ represents a local error estimation criterion which consists in adding the square distance between each neuron W_i in the neighborhood $N_c(t)$ of the BMU, and the input vector X(t) :

$$\mu_i(t+1) = \mu_i(t) + \| X - Wi \|^2 \quad \text{for } i \in N_c(t).$$

The control parameter λ is equal to 0 all along the beginning of the learning procedure, until the process has reached a sufficient number of iteration T_λ for the network corresponding to a sufficient level of organization, after that $\lambda(t)=1$ when $t > T_\lambda$, this limit is fixed by the user. The factor θ is a modular reduction factor for learning coefficient $\alpha(t)$, this minimize the range of the repulsion process, in order to keep a coherent network structure.

4 Results

The spiky sphere (figure 4a) has been modeled with a network composed of 5K neurons.

The figurines in figure 4b and 4c have been modeled, starting from a tetrahedron structure composed of 2K neurons, obtained after 5 subdivision levels (figure 3). The learning rule applied to this architecture is the one defined in equation 1. The synthetic model of the spiky sphere on figure 4a contains 5042 points. The input data for the little pig figurine model [12] on figure 4b is composed of 15825 points, the squirrel model [13] on figure 4c of 9995 points. The convergence to those present results necessitates generally 300000 steps.

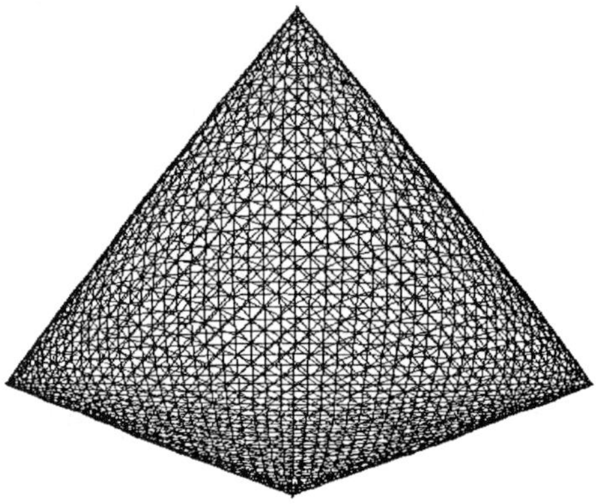

Fig. 3. Tetrahedron after 5 levels of subdivision.

Only a few minutes are necessary, even for large data sets, to complete the final shape, as the input data is not processed itself but is only sampled: one point for each iteration. Given that the sampling process is not the most time consuming process in our method, we can assume that the time performance is practically independent of the size of the input data set. For example, the time needed to achieve the complex shapes of the squirrel model is approximately 5 minutes on a machine with a Pentium 4 processor with 512MB of RAM and a 256MB graphic board. The quality and speed of the results is directly correlated with the algorithm parameters, which are the learning factor α and the initial neighborhood size N_c. A good choice of those parameters is crucial to allow a good behavior of the modeling process. A fast decreasing parameter or a small initial neighborhood width would lead to a distorted map. For example, the parameters for the squirrel model are: $\alpha(0)=0.5$; D=0.4 ; $N_c(0)=1000$; d=0.6 and the learning limit of T=300.000 iterations.

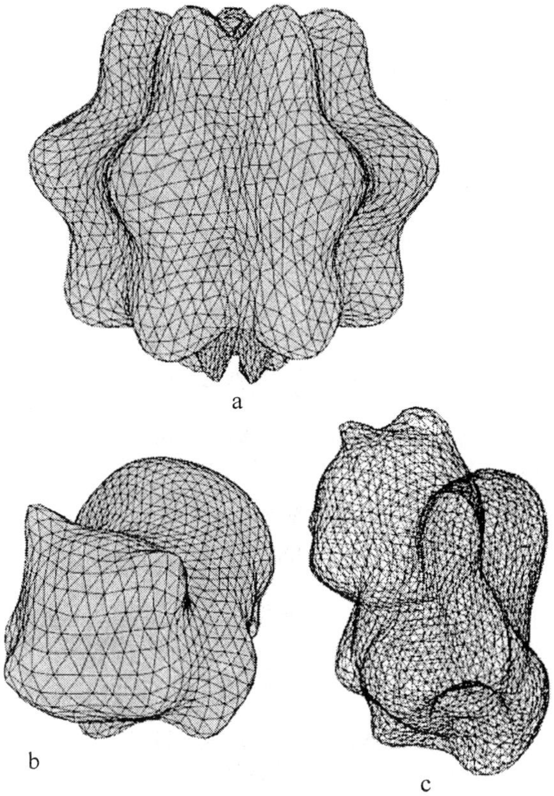

Fig. 4. Reconstruction results.

5 Conclusion and perspectives

In this work we have developed several new aspects of surface reconstruction using Kohonen's self organizing maps. We have shown the great potential of this approach for such graphical applications. The multiple applications of this method have appeared beyond our original purpose, surface reconstruction and representation, such as high dimensional data visualization, morphing and volume rendering. The strength and the efficiency of the learning process have been proved by our experiments. The interest of the architecture generalization is the potential use of the self-organizing maps in the morphing domain, using the same map structure to represent two different objects through a learning process, leading the mesh representing a certain object to model an other object. The next step in the development of a complete generalized self-organizing map is the implementation of a local subdivision feature.

Acknowledgements

The Region Nord – Pas de Calais, France, and the Norbert Ségard Fondation, Lille, France, have actively supported this work.

References

[1] Amenta, N., Bern, M., and Kamvysselis, M. (1998) A New Voronoi-Based Surface Reconstruction Algorithm. in Proc. of SIGGRAPH'98, pp.415-421

[2] Hoppe, H., DeRose, T., Duchamp, T., McDonald, J., and Stuetzle, W. (1992) Surface reconstruction from unorganized points. In Proc of SIGGRAPH'92, pp.71-78

[3] Kohonen, T. (2001) Self-Organizing Maps. Third Extended Edition Springer Series in Information Sciences, Berlin, Heidelberg, New York

[4] Knopf, G.K., Sangole, A. (2001) Scientific data visualization using three-dimensional self-organizing feature maps. IEEE Conference on SMC, pp. 759-764

[5] Yu, Y. (1999) Surface reconstruction from unorganized points using self-organizing neural networks, Proceedings of IEEE Visualization'99, pp. 61-64

[6] Ito, M., Miyoshi, T., Masuyama, H. (2000) The characteristics of the torus Self Organizing Map," Proceedings of 6th International Conference on Soft Computing (IIZUKA2000), Fukuoka, Vol. A-7-2, pp. 239-244

[7] Boudjemai, F., Biela, P., Postaire, J-G. (2003) Surface modeling by using self organizing maps of Kohonen. Systems, Man and Cybernetics, 2003. IEEE International Conference.Vol.3, pp. 2418 - 2423

[8] Fritzke, B. (1993) Growing cell structures - a self-organizing network for unsupervised and supervised learning. Technical Report, Ref. ICSTR-93-026, International Computer Science Institute, Berkeley

[9] Ivrissimtzis, P., Jeong, W-K., Seidel, H-P. (2003) Using Growing Cell Structures for Surface Reconstruction. In Proceedings of the Shape Modeling International, Seoul, IEEE, Los Alamitos, pp. 78-86

[10] Dyn, N., Levine, D., Gregory, A. (1990) A butterfly subdivision scheme for surface interpolation with tension control. ACM Transactions on Graphics (TOG)

[11] Fritzke, B. (1995) A growing neural gas network learns topologies G. Tesauro, D.S. Touretzky and T.K. Leen (eds.), Advances in Neural Information Processing Systems, MIT Press, Cambridge MA, pp. 625-632

[12] http://range.informatik.uni-stuttgart.de/

[13] http://www.mpisb.mpg.de/~shin/

Novel Learning Algorithm Aiming at Generating a Unique Units Distribution in Standard SOM

Kirmene Marzouki[1], Takeshi Yamakawa[2]
[1]Graduate School of Computer Science and Systems Engineering, Kyushu Institute of Technology, Japan
[2] Department of Brain Science and Engineering, Graduate School of Life Science and Systems Engineering, Kyushu Institute of Technology, Japan
E-mail: {kirmene-marzouki@edu.brain, yamakawa@brain}.kyutech.ac.jp

Abstract

Self-organizing maps, SOMs, are a data visualization technique developed to reduce the dimensions of data through the use of self-organizing neural networks. However, one of the limitations of Self Organizing Maps algorithm, is that every SOM is different and finds different similarities among the sample vectors each time the initial conditions are changed.

In this paper, we propose a modification of the SOM basic algorithm in order to make the resulted mapping invariant to the initial conditions. We extend the neighborhood concept to processing units, selected in a fashionable manner, other than those commonly selected relatively to the immediate surroundings of the best matching unit. We also introduce a new learning function for the newly introduced neighbors.

The modified algorithm was tested on a color classification application and performed very well in comparison with the traditional SOM.

1 Introduction

Kohonen formulated the Self Organizing Map algorithm as a mathematical model of the self-organization of topographic maps, which are found in brains of higher animals [1]. SOMs accomplish two things: they reduce dimensions and display similarities [2]. The way SOMs go about reducing dimensions is by producing a map of usually one or two dimensions plotting the similarities of the data by grouping similar data items together.

However, one of the limitations of the standard Self Organizing Map algorithm is that every SOM is different and finds different similarities among the sample vectors each time the initial conditions are changed. Consequently, a lot of maps need to be constructed in order to get one final good map, which is a tough and time consuming task.

On the other hand, in standard SOM, the acquired information by the net during the training process, although it gets more and more importance as learning goes on, is used only for finding the winner of the competitive layer.

The key idea introduced by Kohonen is the concept of neighborhood [3]. Each unit has a set of neighbors on the map. After finding the best matching unit, BMU, to the input vector, not only its weight is adjusted, but those of the neighbors are also changed.

However, updating is exclusively made to the selected winner and to those units determined by the neighborhood function. No other units can participate to the learning at the same time epoch.

Indeed, a standard SOM makes no difference between "close" or "far" neighbors, and assigns to both the same update. Thus, the output map provided by standard SOM depends highly on the initial conditions and the position of the winning unit on the net.

It leads us to think that SOM could be more effective not only if the neighborhood concept introduced by Kohonen is extended and used in a more sophisticated manner, but also if the information acquired by the net plays an active role in producing the final map.

2 Neighborhood Function

To improve SOM's classification capabilities we propose, at first, to not restrict the neighborhood concept only to units surrounding the BMU.

Inspired from the Neural Gas algorithm [4], we propose to use the concept of weights' vectors neighborhood in standard SOM, to generate a new group of neighbors.

Let *Nc*, be the traditional neighborhood function and *Nw* the neighborhood function relative to units' weights vectors as introduced in [4]. We note *df* the distance separating the BMU and the furthest unit belonging to *Nw*.

During training, when proceeding to updating, not only those units which have their position enclosed within *Nc* are updated, but also units located outside *Nc* and within *df* will also take part of the learning (Fig.1).

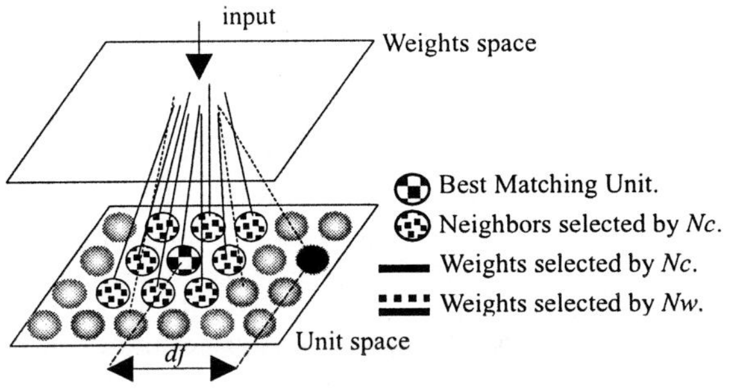

Fig.1. The determination of *df*.

As training goes on, both neighborhoods gradually shrink. At the end of training, both functions have shrunk to zero size.
Then, at a second stage, we try to extend the learning to the new type of neighbors, by introducing a new learning rule involving the information acquired by the net during the training so that it contributes to the formation of the resulted mapping.

3 Algorithm

3.1 Step 0

Input one sample vector of the training data set.

3.2 Step 1

Find the BMU relative to the input vector.

3.3 Step 2

Find those units to be updated. This is done by computing the two neighborhood functions. For a matter of simplicity, the two neighborhood functions we opted to use in this paper have the form

$$N = No(1- t/T) \tag{1}$$

where *No* is the value of *N* at t=0, t is the current time, e.g. current iteration number, and T is the total time, e.g. total number of iterations.

Applied to *Nc* and *Nw*, the two neighborhood functions are formulated by

$$Nc = Po(1-t/T) \tag{2}$$
$$Nw = do(1-t/T) \tag{3}$$

Initially, both *Nc* and *Nw* are chosen to cover all the map and the entire weights space respectively. Then, as learning proceeds they shrink at the same speed, but with different scales, since the former is relative to units' positions and the latter is relative to their weights vectors.

After determining *Nw*, we pick up *df* as explained in the previous section.
Then we proceed to updating.

3.4 Step 3

Proceed to weight adaptation.

3.4.1: The learning rule relative to the units selected by *Nc* is identical to the traditionally used adaptive rule of SOM, that is:

$$W_i(t) = W_i(t-1) + \alpha(t)[x(t)- W_i(t-1)] \tag{4}$$

for all units i such that $i \in Nc$, and where $W_i(t)$ is the newly updated weight, $W_i(t-1)$ is the old weight before update, $\alpha(t)$ is the learning rate and x(t) is the input.

3.4.2: The new learning rule we introduce in this algorithm is as follow:

$$W_j(t)=W_j(t-1)+\alpha(t)[x(t)-W_j(t-1)] + \alpha(t)(t/T)[A_j(t)-x(t)] \tag{5}$$

for all units j such that $j \in Nc \cup df$, and where $A_j(t)$ is the average of the weights of the neighboring units of unit j, t is the current time, and T is the total time. Eq.4 and Eq.5 can be used interchangeably, depending on even $df > Np$ or vice-versa.

3.5 Step 4

Input a new vector and go to step 1.

(End of Algorithm)

To obtain Eq.5, we start from the general learning rule of the SOM. From Eq4. we replace the updating term by a more sophisticated one, and we arrange the indices to obtain

$$W_j(t) = W_j(t-1) + \alpha(t)[\ \beta(t)[x(t)-W_j(t-1)] + \gamma(t)[\ A_j(t)-W_j(t-1)]\ ] \quad (6)$$

with $\beta(t)$ and $\gamma(t)$ are two discrete time-dependent weighing variables such that $\beta(t)+\gamma(t) = 1$, and $A_j(t)$ as defined above in Eq.5.

The main idea is that for those units located outside *Nc* and positioned inside *df*, learning is made by considering not only the newly input information, but also the already available one on the net.

From Eq.6, since $\beta(t)+\gamma(t) = 1$, we obtain

$$W_j(t) = W_j(t-1) + \alpha(t)[\ \beta(t)x(t) + \gamma(t)A_j(t)- W_j(t-1)] \quad (7)$$

considering $\gamma(t) = t/T$ and $\beta(t) = 1-(t/T)$, leads us to Eq.5.

In the proposed new learning rule, we replaced the learning term by a weighed combination of the input $x(t)$ and the average of the weight values of the surrounding units, $A_j(t)$.

In order to keep the spirit of the topographic mapping as introduced by Kohonen, we insisted to make our choice of the new neighbors based on their relative positions on the unit space and making their updates with respect to their surroundings.

Finally, to guarantee the convergence of the algorithm, $\gamma(t)$ and $\beta(t)$ are chosen to have antagonist roles.

In the beginning, $\gamma(t)$ starts with zero value, while $\beta(t)$ starts equal to 1. It means that learning starts identically as in the traditional SOM for both groups of neighbors, so that the new one can be considered as the extension of *Nc* beyond its borders.

Let *j* be one of those units located within *df* and positioned outside *Nc* on the map at a given time epoch *t*. In order to strengthen the neighborhood relationship, we update the weight $w_j(t)$ of this unit considering the information stocked in its immediate surroundings.

At the first stages of learning, this operation allows to put a maximum number of units in the context of the training data set. In a standard SOM, this operation is exclusive to the selected neighbors of the BMU. For example, considering the fact that the neighborhood function shrinks at each training epoch, if some region of the input space is more crowded than others and the initial density of weight vectors is too low in that region, specific regions of the units space will be updated consistently while others will be left apart.

There are number of approaches in the literature which tried to deal with this problem [5-7]. However all of these methods adapt the winning unit only, since they intend to bring solutions to underutilization of some units of the competitive layer.

In our algorithm, as learning goes on, the information stocked on the net gets more and more significance and importance, so that it participates actively in the training/learning process by keeping integrating a maximum number of units in their respective changing neighborhood environment making them "aware" of the evolution of the network.

In traditional SOM, every region of the map is being learned independently from the adjacent ones. There is no real consideration of the evolution of the network as a whole, but for finding the BMU.

We estimate that the information acquired by the map along the learning, should play a more significant role than just being used to select the BMU by making its selection more meaningful and consistent in the future steps of learning.

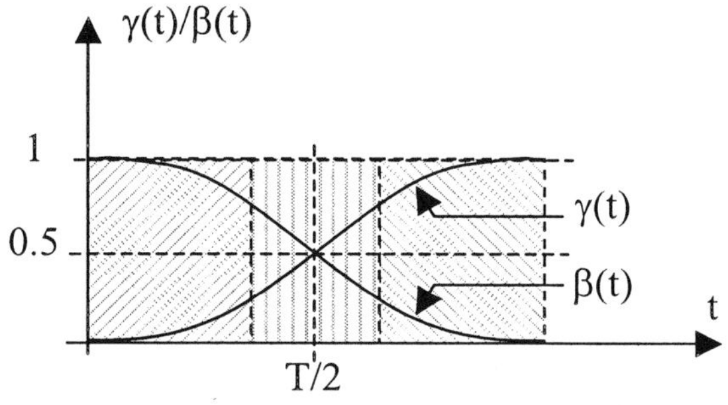

Fig.2. Different phases of the learning process. In the beginning of the algorithm, as $\gamma(t)$ is still small, learning is quite identical as in SOM. However, small updates aiming to put a maximum number of units in the context of the training data set are performed on the new group of neighbors. Then, as $\gamma(t)$ and $\beta(t)$ approach each other, the middle zone on the graph, the learned information will start participating actively to learning by integrating each unit of the new neighbors in its relative continuously changing environment. During this phase some conflicts may take place. Finally, the third phase of the learning will be rather a strengthening of the neighborhood relationship more than performing updates.

At the final stages of our algorithm when $\gamma(t)$ approaches the value 1 and $\beta(t)$ the value 0, the acquired information by the net will dominate the updates of the new neighbors. It means that the new learning rule strengthens the neighborhood relationship more than performing updating (Fig.2).

4 Experimental Results and discussion

For a matter of clarity, the chosen application to test the new algorithm consists of color mapping. As SOM produces a mapping of classified data, it is then easy to evaluate how good a map is and how strong the similarities between objects are. In the case of color classification, it is very easy to check the resulted map and make comparisons between the two algorithms.

Simulations were performed with different network topologies with different number of training data and different iteration numbers. The showed results were obtained using a 10x10 net lattice, 10000 training data set for 5000 iterations, which are the optimal parameters to implement color mapping in standard SOM. In fact, using our algorithm, we could obtain similar performance by using only 5000 training data in only 1000 iterations for the same net lattice size. The training data and initial states were three-dimensional vectors on the RGB scale having the form

$$X_1=(R_1,G_1,B_1)$$
$$X_2=(R_2,G_2,B_2)$$
$$\ldots$$
$$X_N=(R_n,G_n,B_n)$$

where $R_i \in [0.255]$, $G_i \in [0.255]$ and $B_i \in [0.255]$. The initial states were drawn at random (Fig.3). The showed initial distributions may appear highly localized, but these are only examples. We also used different initial distributions spreading on all the weight space and obtained similar results.

For the used parameter configuration, *Po* and *do* initial values of *Nc* (Eq.2) and *Nw* (Eq.3), are set to 12 and 400, respectively. These are the maximum Euclidian distances separating two units on the unit space (10x10), and two weights on the weight space (0-255), respectively.

Conversely to the resulted maps given by the traditional SOM, and as shown in Fig.4, the proposed algorithm is invariant to the initial conditions and in all cases we obtained the same final mapping.

As we explained above, every SOM finds different similarities among the sample data. In fact, SOMs organize sample data according to the position of the BMU on the map. If we re-initialize the net with the same set of initial states, but with different arrangements on the map, the obtained final classification will be definitely different of the one obtained before.

In our algorithm, the used new learning rule for the introduced new group of neighbors keeps adapting a maximum number of units to their continuously changing environment, so that initial position rearrangements of the units on the map have no effect on the final result.

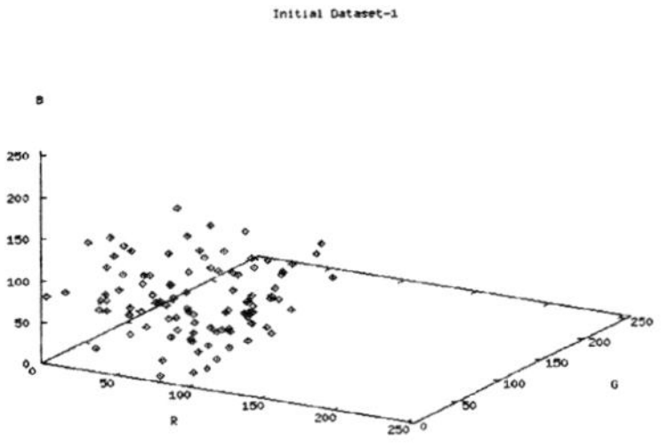

(a)

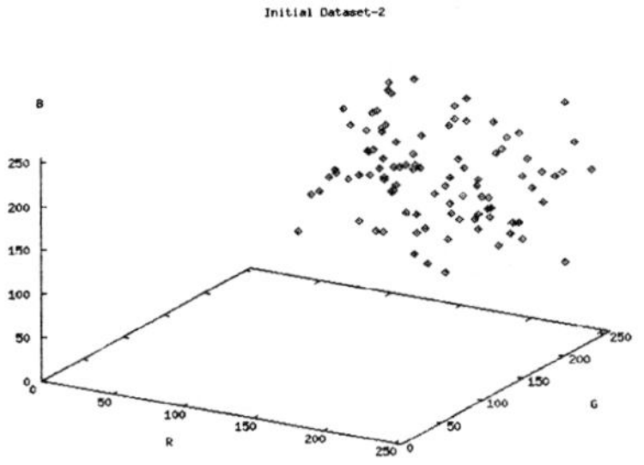

(b)

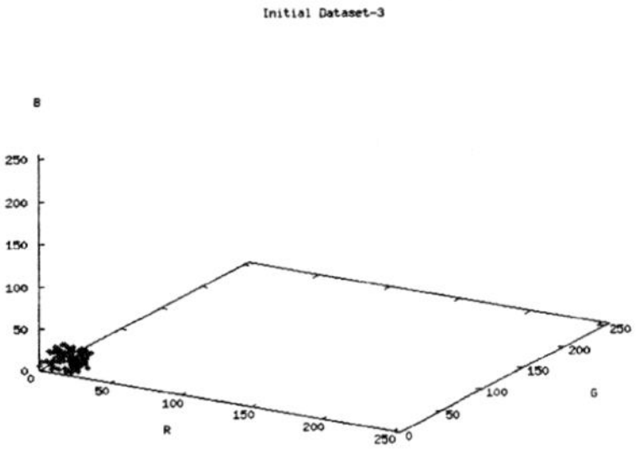

(c)

Fig.3. Different Initial distributions used for different sets of simulations (a), (b) and (c).

Another typical problem of SOM. SOMs organize sample data so that in the final product the same samples are usually surrounded by similar samples. However, similar samples are not always near each other. If we have a lot of shades of purple, not always we'll get one big cluster with all purples in it. Sometimes, the clusters will get split and there will be two groups of purples. As we can see on Fig.4, we have two groups of purple in two different locations, while they should be in the same region of the map. Unfortunately, the proposed algorithm, still suffers from this defect.

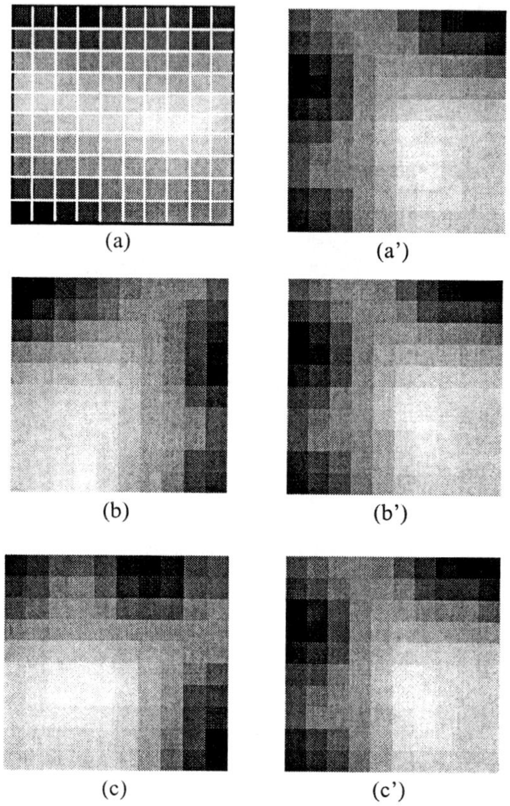

Fig.4. Representative results of three sets of simulations, (a,a'), (b,b') and (c,c'), performed on a 10x10 net matrix with a total number of 10000 training data for a total number of 5000 iterations. The same initial conditions and training data were used for each set, but different from one set to another. (a), (b), and (c) are maps given by SOM. (a'), (b') and (c') are the proposed algorithm's final maps. The grid placed on (a) illustrates the 10x10 competitive layer. Each square represents a unit on the map at its final state after learning is completed. Colors represent the different classes the network could classify.

5 Conclusion

In this paper we proposed some modifications on the traditional SOM algorithm, in order to obtain the same mapping independently from the initial conditions.

The modifications we proposed consist of introducing a new group of neighbors, based on weight neighborhood function, so that two kinds of neighbors are processed.

For this new type of neighbors we introduced, updating is made according to a new learning algorithm, in which not only the input information is considered, but also the information already existing on the map is used. At the beginning of the algorithm, the two learning rules are similar to the standard SOM. Then, more the learning goes on, more the contribution of the information learned by the net gets importance to the formation of the final map. At the final stages of the process, updating of the new group of neighbors will be more an operation of adaptation of the selected units to their changing environment, than a real updating to the input vector. The proposed algorithm exhibited very stable behavior in color classification application.

References

[1] T. Kohonen, Self-Organization and Associative memory, Springer-Verlag 1988.

[2] T.Kohonen, The Self-Organizing Map. Proceedings of the IEEE, 78:1464-1480, 1990.

[3] T.Kohonen, Self Organizing Maps. Springer-Verlag, Berlin, Germany, 1995

[4] Martinetz, T., &Schulten, K. (1991). "Neural Gas" network learns topologies. In T.Kohonen et al. (Eds). Artificial neural networks (Vol I,pp 397-402), Amsterdam, North Holland.

[5] R. Hecht-Nielsen, "Counterpropagation networks," *Appl. Opt.*, vol. 26, pp. 4979–4984, 1987.

[6] D. DeSieno, "Adding a conscience to competitive learning," in *IEEE Int. Conf. Neural Networks*, vol. I, San Diego, CA, 1988, pp. 117–124.

[7] G. Huerter, "Solution of the traveling salesman problem with an adaptive ring," in *IEEE Int. Conf. Neural Networks*, vol. I, pp. 85–92, San Diego, CA, 1988.

SOM-Based Estimation of Meteorological Profiles

T. Tambouratzis

Department of Industrial Management and Technology, University of Pireaus,
107 Delighiorgi St, Pireaus 185 34, Greece
e-mail tatiana@unipi.gr; tatiana@ipta.demokritos.gr

Abstract

The task of estimating the meteorological profile of any location of interest within a specified area is undertaken. Assuming that the meteorological profiles of a sufficient number of representative reference locations within the specified area are available, the proposed methodology is based on (a) the organisation of the meteorological profiles of the reference locations employing a self-organising map (SOM) and (b) the classification of the most salient morphological characteristics of the reference locations. Subsequently, the meteorological profile of any novel location of interest is approximated by a weighted average of the meteorological profiles represented on the SOM for those reference locations whose morphological characteristics most closely match the morphological characteristics of the location of interest. The proposed methodology is evaluated by comparing the accuracy of meteorological profile estimation with that of existing estimation techniques as well as with the actual meteorological profiles of the locations of interest.

1 Introduction

The estimation of the meteorological profile of any location of interest within a specified area is of obvious practical interest. Assuming that

- the morphology of the area is available, and
- the meteorological profiles of a sufficient number of representative reference locations within the specified area are known,

various methods can be employed for the estimation of the meteorological profile of the location of interest (e.g. "adoption-adaptation" of the meteorological profile of the reference location that is closest to the location of interest, interpolation between the nearest reference locations). In this piece of research, a novel methodology is proposed which combines the construction of a self-organising map (SOM) that orders the meteorological profiles of the reference locations with the classification of the reference locations according to their most salient morphological characteristics. Given a novel location of interest within the specified area, its meteorological profile is generated by:

- classifying the location of interest into one of the morphological classes and, subsequently,
- determining the meteorological profile represented on the SOM for those reference locations that most closely match the morphological characteristics of the location of interest.

This paper is structured as follows: the area specified for meteorological profile estimation is described in section 2; the SOM employed for organising the meteorological profiles of the reference locations within the specified area as well as morphological classification of the Greek territory are presented in section 3; the experimental results as well as a comparison with existing estimation techniques and the actual meteorological profiles of the locations of interest are given in section 4; finally, section 5 concludes the paper.

2 The specified area

The Greek territory has been used as the area on which the estimation of the meteorological profile of any location of interest is performed.

2.1 Meteorological parameters - profiles

The meteorological profiles of 130 locations in the Greek territory are available; these locations (shown in Fig. 1) correspond to the weather stations maintained by the National Meteorological Service (EMY) of Greece.

Each meteorological profile has been derived from 28 meteorological parameters[1] which have been collected over a period of 43 years (from 1955 to 1997) at a given weather station. A single value per meteorological parameter (equalling the numeric average of the collected values over the 43 years) has been utilised for the

---

[1] describing monthly and yearly maximum, minimum and average temperature, barometric pressure and relative humidity, monthly and yearly average number of sunny and cloudy hours per day, monthly and yearly average levels of precipitation and snow, monthly and yearly average wind speed

generation of the meteorological profile of the given weather station.

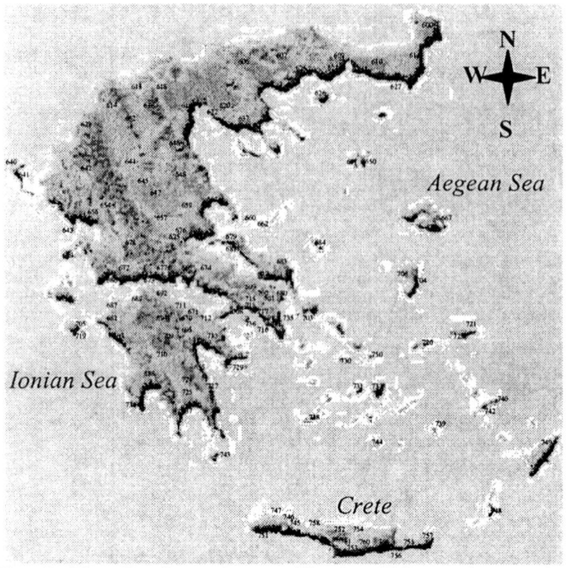

Fig. 1 The Greek territory with the 130 weather stations maintained by EMY; locations of the available meteorological profiles.

2.2 Morphological characteristics - classes

The morphological characteristics deemed as most salient for classifying any location of interest within the Greek territory are

- quantitative (and directly available), namely the altitude, amplitude and longitude of the location, and
- qualitative (derived), such as the location being on an island (also specifying the approximate island size and the proximity to other islands or the mainland) versus the location being on the mainland (also specifying the proximity to the sea), the proximity of the location to rivers and/or lakes, the location being on a plane versus the location being on a mountain (also specifying highland and lowland plateaus).

The classification of the Greek territory according exclusively to qualitative morphological characteristics results into 4,096 combinatorially possible classes, of which only 55 are observed in the 130 locations corresponding to the weather stations of Fig. 1. Classification according to both quantitative and qualitative morphological characteristics is detailed in section 3.

For reasons of clarity it should be mentioned that, while the morphological characteristics of any location in the Greek territory can be determined, the meteorological profiles of only 130 locations are available.

3 The Self-Organising Map (SOM)

The main sources of inspiration for the SOM [1] are the self-organisation of the neurons in biological brains and the emerging topology preservation such that similar stimuli excite (and, thus, are represented by) neighbouring neurons.

For meteorological profile estimation, two-dimensional SOMs arranged into hexagonal grids have been employed; the sizes of the grids range from 5x12 to 20x30 nodes. The SOMTOOLBOX for the MATLAB environment has been used [4]. The training procedure comprises 1,000 and 10,000 iterations of the entire training set during the rough-training and the fine-tuning phases, respectively. Following range normalisation, the 28-dimensional inputs (corresponding to the meteorological profiles of the locations of Fig. 1) have been used for training. Cross-validation of these 130 locations has been employed for meteorological profile estimation. A total of 130 tests have been performed where, in each test, 129 different locations constitute the reference locations[2].

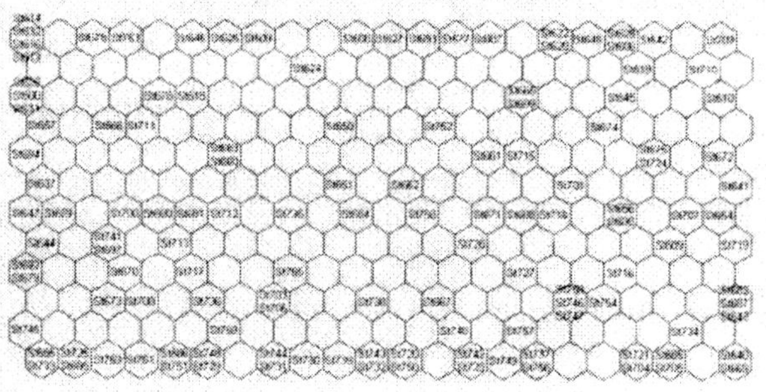

Fig. 2. One of the 11x22 SOMs employed for organizing the meteorological profiles of the Greek territory.

The 12x22 SOM (shown in Fig. 2) with range normalisation has been selected as the most consistent over the 130 trials as well as the best in preserving the distances of the original meteorological profiles, i.e. in ordering the meteorological profiles of the locations of Fig. 1 in a reasonable manner[3].

---

[2] the remaining location constitutes the location of interest

[3] large SOMs have been found preferable to smaller ones since they are more capable of exposing the proximity in terms of meteorological profiles between the 130 locations; smaller SOMs would be useful only if the aim were to ease computation (i.e. a class of meteorological profiles per node)

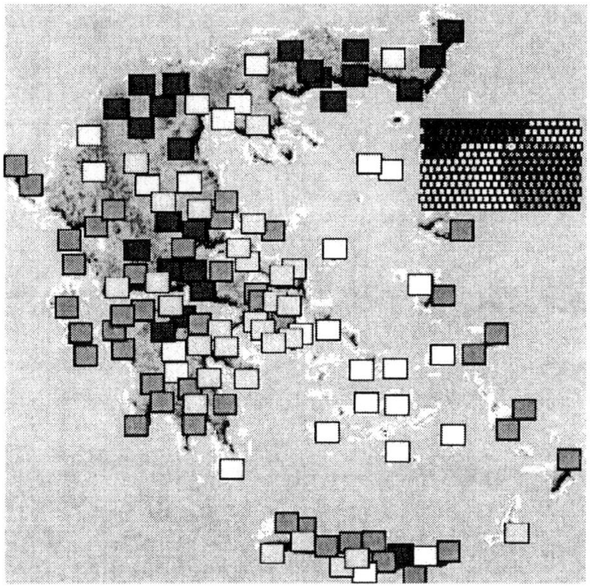

Fig. 3. The mapping of the 11x22 SOM on the Greek territory into five groups created by the U-matrix.

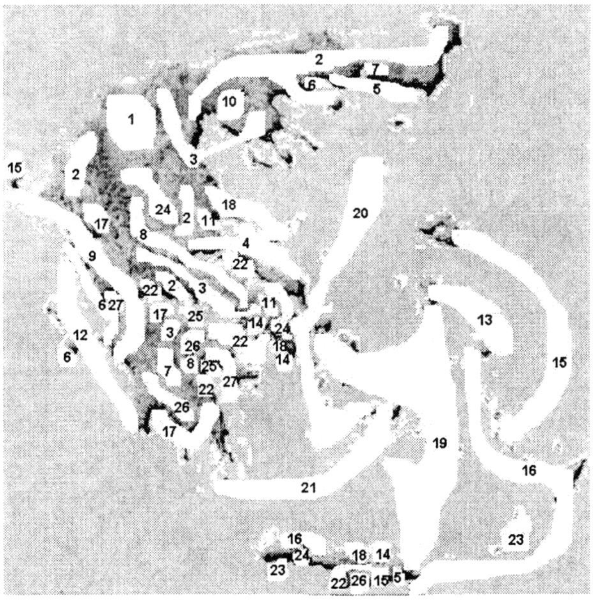

Fig. 4 The 27 morphological classes of the Greek territory.

The organisation accomplished by the SOM of Fig. 2 as well as the five groups created by its U-matrix (Fig. 3) have been employed in order to configure the morphological classification of the 130 locations of the Greek territory (according to both quantitative and qualitative morphological characteristics); the resulting 27 morphological classes are shown in Fig. 4. The relatively high degree of agreement between location-grouping and morphological classification (Figs. 3 and 4, respectively) underlies the fact that the morphological characteristics of any location largely affect its meteorological profile. The morphological classes are not as continuous as are the groups resulting from the U-matrix; this can also be seen in Fig. 5 which further illustrates the assignment of the SOM nodes into the 27 morphological classes. However, the similarity of the (range-normalised) meteorological profiles between locations assigned to the same class is high. Hence, some understanding of the relationship between meteorological parameters and morphological characteristics is established.

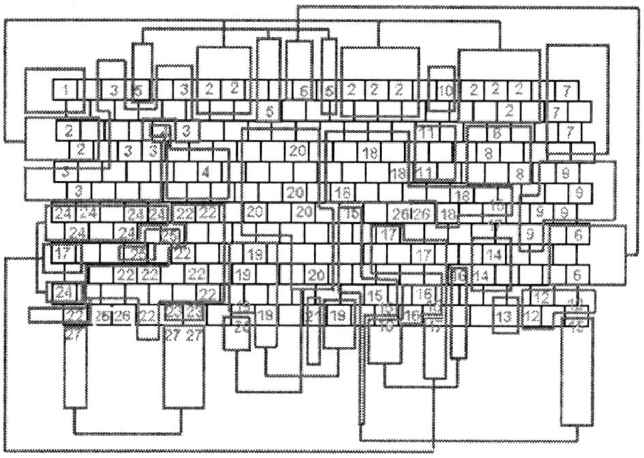

Fig. 5 The assignment of the nodes of the SOM of Fig. 2 into the 27 morphological classes of Fig. 4.

4 Experimental results - comparison with existing methods

In each of the 130 trials, the meteorological parameters – and hence the meteorological profile – of the remaining location (the location of interest) are assumed unknown. However, the morphological characteristics of the location of interest are available, whereby they can be employed in order to classify the location into one of the 27 morphological classes of Fig. 4. Subsequently, an estimation of the meteorological profile of the location of interest can be performed from the reference locations of the same morphological class that lie in the same closed area of Fig. 4. The estimation is based on the weighted average of the meteorological profiles of the reference locations involved, where the weights have been determined according to the similarity between reference location and location of interest in terms of the three quantitative morphological characteristics.

The proposed methodology for estimating the meteorological profiles of the locations of interest has been found superior to that of

- Simply adopting the meteorological profile of the nearest reference location. This approach ignores the fact that, although the two locations are near each other, the morphological characteristics of the reference location and the location of interest may differ significantly.
- Adopting and adapting the meteorological profile of the nearest reference location (e.g. adiabatic lapse rate normalization of temperature [5]). This approach takes into account the differentiation of the morphological characteristics between the reference location and the location of interest but is quite complicated to implement and not always accurate.
- Interpolating between the nearest reference locations. As for the simple adaptation, the fact that the morphological characteristics of one or more of the reference locations and the location of interest may differ significantly - although the locations are near each other - is ignored.

5 Conclusions

A novel methodology is proposed for the estimation of the meteorological profile of any location of interest within a specified area. Assuming that the morphology of the area is available, and that the meteorological profiles of a sufficient number of representative reference locations within the area are known, the methodology employs

- a self-organising map where the meteorological profiles of the reference locations have been ordered, and
- the morphological classification of the reference locations.

Subsequently, The meteorological profile of a location of interest within the specified area is generated by:

- classifying the location of interest into one of the morphological classes and, subsequently,
- determining a weighted average of the meteorological profiles represented on the SOM for those reference locations that belong to the same morphological class of Fig. 4.

The proposed methodology has been found superior to existing estimation techniques and consistently close to the actual meteorological profiles.

References

[1] Kohonen, T. (1997) Self-Organising Maps, 2nd ed. Springer-Verlag, Berlin.

[2] Kornaros, G. (1999) Climatic Data of the Stations of the Hellenic National Meteorological Service (period 1955-1997), Vols. 1 & 2. National Meteorological Service (E.M.Y.), Athens, Greece (in Greek).

[3] Tambouratzis, T., Tambouratzis G. (2003) Meteorological data mining employing Self-Organising Maps. In: Pearson, D.W., Steele, N.C., Albrecht R.F. (eds) Proceedings of the "International Conference on Artificial Neural Networks and Genetic Algorithms 2003", Roanne, France, April 23rd-25th, 2003, Artificial Neural Nets and Genetic Algorithms, Springer, Wien, pp. 149-152

[4] Vesanto, J., Himberg, J., Alhoniemi, E. & Parhankangas, J. (2000). "SOM Toolbox for Matlab 5". Report A57. SOM Toolbox Team, Helsinki University of Technology, Finland (available at http://www.cis.hut.fi/projects/somtoolbox).

[5] Watson, B.G., MacIver, D.C. (1995) Bioclimate Mapping of Ontario. Environment Canada and Ministry of Natural Resources of Ontario.

An Efficient Heuristic for the Traveling Salesman Problem Based on a Growing SOM-like Algorithm

Cristina García, José Alí Moreno *
*Laboratorio de Computacion Emergente, Facultades de Ciencias e Ingenieria, Universidad Central de Venezuela.
E-mail: {cgarcia,jose}@neurona.ciens.ucv.ve Web Page: http://neurona.ciens.ucv.ve/laboratorio

Abstract

A growing self-organizing (SOM) neural network, enhanced with a local search heuristic is proposed as an efficient traveling salesman problem solver. A ring structure of processing units is evolved in time with a Kohonen type adaptation dynamics together with a simple growing rule in the number of processing units. The result is a neural network heuristic for the TSP with a computational complexity of $O(n^2)$, comparable to other reported SOM-like networks. The tour emerging from the SOM network is enhanced by the application of a simple greedy 2-Opt local search. Experiments over a broad set of TSP instances are carried out. The experimental results show a solution accuracy equivalent to that of the best SOM based heuristics reported in the literature.

1 Introduction

The well known Traveling Salesman Problem (TSP) is the archetype of NP-complete problems, it has attracted considerably research interest and has been repeatedly used as benchmark in the comparison of different combinatorial optimization algorithms. The task in the TSP is to find a shortest close tour through a set of n cities with known inter-city distances such that each city is visited exactly once and the tour ends at the start city. In the literature [1, 2, 3, 4] many exact and heuristic algorithms for the TSP can be found. Several of them inspired from naturally occurring processes and with good performance figures. Typically the computation of exact solutions takes a time that grows exponentially with n. Hence in presence of large problems, often occurring in practical applications (e.g. VLSI routing, drilling machine steering and wall paper cutting) heuristic algorithms with considerably smaller complexity are preferred. It is argued [5, 6] that among the nature inspired heuristics, Neural Networks algorithms are the less competitive in overall performance. Nevertheless research in the area continues and recent results [7, 8] indicate that the Neural Network approach to the TSP remains as a promising alternative when all performance figures (problem scale, execution time and quality of the solution) are considered.

Considerable interest has attracted the self organizing map (SOM) or Kohonen [9] type approach to the TSP (e.g. see [7, 10] and references therein). The reasons are multiple: straightforward and efficient implementation, good scaling behavior with the size of the problem, relative simple parametrization and good solution quality. These SOM-like networks solve the TSP by simply inspecting, in an unsupervised way, the coordinate values of the cities (inputs) for patterns and regularities, adjusting themselves (self organizing) to fit the input data. The self organizing dynamics produces a gradual movement of the processing units (PUs) towards the cities and at the same time, due to the topology preserving property [9], paths between neighboring PUs (corresponding to neighboring cities) will be as short as possible. Since the PUs are logically organized in a ring shaped network a valid tour for the considered TSP can be easily extracted.

In this contribution we present a traveling salesman problem solver based on an enhanced SOM-like neural network with a growing structure [11]. Like all SOM-like networks for the TSP it consists of a ring of PUs over which an unsupervised learning scheme is applied. Initially the ring contains a small number of PUs, as time advances the PUs are updated in response to stimuli from the cities and successively new PUs are inserted. The organization process induces a topologically ordered description of the data that leads to a feasible tour of the considered TSP. This resulting tour is post processed with a straightforward 2-Opt local search procedure [12] in order to improve further its quality. A similar algorithm has been previously presented [13] but employing a more complicated insertion strategy and a different route post processing. Results from experiments carried out over a broad range of scales for the size of the TSP improve or are comparable, both in calculation time and quality of solution, those obtained with other more sophisticated SOM-like algorithms.

This paper is organized as follows. The growing SOM algorithm is briefly described in section 2. Comments on the complexity of the algorithms are presented in section 3. In section 4 aspects concerning the initialization and parametrization of the network are discussed. Experimental aspects and comparison of the results with other methods are exposed in section 5 and in section 6 the conclusions are presented.

2 The Growing SOM Algorithm

The proposed algorithm consists of a growing ring of PUs adapting themselves according to the stimuli coming from different cities. Initially the ring is composed of a small number of PUs and successively new PUs are inserted. The algorithm finds the winning PU (best matching unit BMU) nearest to a randomly chosen city. The weights of the BMU and those of its direct neighbors are updated. Each time a PU becomes a winner (BMU) an associated counter accumulating the number of wins is incremented. The weight update moves the position of the BMU and of its neighboring PUs a fraction in the direction of the stimulating city. Every fixed number of iterations the algorithm finds the most winning PU and inserts a new PU in the edge connecting it to the neighbor with the mayor number of wins. The counters of every PU are reset to zero and the process is repeated. This very simple growing dynamics is iterated until the number of PUs is twice the number of cities. Finally each city is assigned to the nearest available PU. The order of the PUs in the ring structure conforms a valid tour of the considered TSP. This resulting tour is post processed with a straightforward 2-Opt local search procedure in order to improve its quality. The algorithm consists then of two distinct stages: a PU weight update and insertion phase and at the end of network organization a tour improvement local search.

The PUs weight update is very similar to the well known Kohonen learning rule:

1. The position vector $\boldsymbol{r}(k)$ of a randomly selected city is presented to the network.

2. The BMU (PU with weight vector $\boldsymbol{w}_j$) nearest to $\boldsymbol{r}(k)$ is selected for update.

3. The weight vectors of the BMU and its two direct neighbors are updated according to:

$$\boldsymbol{w}_j \leftarrow \boldsymbol{w}_j + \eta(\boldsymbol{r}(k) - \boldsymbol{w}_j) \tag{1}$$

   where η is the learning rate, which remains constant during the process. The learning rate for the neighbors is a fraction of that for the BMU.

4. The counter win_j which accumulates the number of wins for the BMU is incremented.

The Growing_Rate_Parameter gives the number of iterations to be completed before the insertion procedure is applied:

1. The kth PU with the mayor number of wins is selected; $win_k > win_i$ for all i PUs.

2. From its direct neighbors the PU with most wins of the two is also selected, say the $k+1$ PU with win_{k+1}.

3. A new PU is inserted with weight vector given by

$$\boldsymbol{w}_{new} = p_1\, \boldsymbol{w}_k + p_2\, \boldsymbol{w}_{k+1} \tag{2}$$

   with $p_1 = win_k/(win_k + win_{k+1})$ and $p_2 = win_{k+1}/(win_k + win_{k+1})$.

4. Set the win counters to zero.

The update-insertion phase is iterated until the number of PUs in the ring network is equal to twice the number of cities in the TSP. The resulting organization of the ring network of PUs is such that every weight vector is in the immediate neighborhood of the position vector of a city so that each city can be assigned to its nearest available PU. At this stage the tour of the considered TSP is represented by the order of the PUs in the ring structure.

The tour improvement phase consists on the following local greedy heuristic which uses the Euclidean distance:

1. For every city i, a section of the tour with M cities is considered.

2. The distance $d(i, fc)$ between the city i and the following city (fc) in the considered section of the tour is evaluated.

3. For all other cities k in the tour section, the distance $d(i, fc)$ is compared with the distance $d(i, k)$.

4. If $d(i, fc) > d(i, k)$ then a 2-Opt inversion between the cities fc and k is proposed.

5. The proposed inversion is accepted if the length of the modified tour section is smaller than that of the original. In this case city k becomes city fc.

With a section of the tour containing $1/10$ of the number of cities (M) it is found that the above heuristic achieves a tour length improvement between 0 and 4 percent.

3 Algorithm Complexity

With an argumentation similar to that given in [13] it can be easily verified that the time complexity of the algorithm is at least of order $O(n^2)$ whereas its space complexity is of order $O(n)$. This complexity is typical of all SOM-like algorithms presented in the literature [7].

Nevertheless with some simple programming modifications the algorithm's complexity can be made linear, this aspect was not further pursued in this work. It is worth noting that in contrast, other nature inspired heuristics for the TSP (Genetic Algorithms, Ant Colony Optimization and Simulated Annealing) have much higher complexity.

4 Initialization and Parametrization of the Algorithm

The network is initialized as a ring of 8 PUs uniformly and randomly distributed on a rectangular frame containing all the cities. With this choice and as an effect of the first iterations of the learning dynamics, this initial configuration is organized in a ring oriented approximately along the centers of the main clusters of the cities distribution. The convenience of this simple initialization scheme has been discussed in [10]. Only a small number of parameters are involved in the algorithm: Growing_Rate_Parameter, learning rate of the BMU η, learning rate of the neighboring PUs and number of PUs to be grown. The adequate values for these parameters can be found after a small round of trial and error experimentation. It is experimentally found that of the four parameters only two of them (Growing_Rate_Parameter and learning rate η of the BMU) are important in overall performance and must be more or less carefully fine tuned. The final number of PUs can always be set to twice the number of cities in the problem and the learning rate of the neighboring PUs varied around $\eta/2$. In consequence a fine tuning of the algorithm's parameters is relatively easy to complete. In summary the best ranges and values used were:

- Growing_Rate_Parameter between 1000 and 4000, we used 2000 for all experiments.
- Learning rate of the BMU η between 0.01 and 0.06, we used 0.02 in all experiments.
- Learning rate of the neighboring PUs $0.6\,\eta$ was used.
- The number of PUs to be grown was set to twice the number n of cities in the TSP.

All experiments were carried out with the same set of parameters. The parameters of the algorithm were fine tuned with two instances of 100 and 500 cities and then kept fixed for the rest of the experiments. Nevertheless it was observed that the smaller instances ($n < 500$) were most sensible to parameter tuning than the bigger ones.

5 Simulation Experiments

The algorithm was implemented in C++ on a 2.4 GHz Pentium IV based machine. Experiments were made on a set of TSP instances available on the TSPLIB [14]. Sets of 10 experiments for each TSP instance were carried out, the best and average results were registered. In Table 1 the TSP instances studied are listed together with the best result obtained with our algorithm on each problem. In Table 2 the execution times averaged over 10 experiments are presented. It can be appreciated that the increase in time is slightly less than quadratic.

Table 1. Best results on 10 experiments for the TSP instances

Instance	Cities	Optimum	GSOM	Error(%)
ATT48	48	10628	10648	0.188
EIL51	51	426	427	0.235
EIL101	101	629	642	2.066
KROA150	150	26524	27229	2.657
KROA200	200	29368	30249	2.999
LK318	318	42029	43350	3.143
PCB442	442	50778	54351	7.03
ATT532	532	27686	28701	3.66
PR1002	1002	259045	269371	3.986
NRW1379	1379	56638	60431	6.96
PR2392	2392	378032	404055	6.884
PCB3038	3038	137694	149759	8.762
FNL4461	4461	182566	195582	7.13

Table 2. Average execution times for some TSP instances

No. Cities	Time (sec.)
101	4
200	12
532	88
1002	282
2392	1860
3038	3046
4461	5365

In order to evaluate the relevance of the experimental results these were compared with those presented in [7]. There the authors compare results obtained by a SOM-like network evolved with a Genetic Algorithm and those obtained by Simulated Annealing and other three SOM based heuristics including the FLEXMAP in [13]. In Table 3 this comparison is summarized. SA stands for Simulated Annealing, FMP for FLEXMAP, Bud for Budinich, ESOM and eiSOM are the author's algorithms (e.g. see [7]). The missing values in the table correspond to not reported results.

Table 3. Average solution quality of 5 enhanced SOMs

Instance	Average Error (%)					
	SA	FMP	Bud	ESOM	eiSOM	GSOM
ATT48	-	-	-	-	-	0.85
EIL51	2.33	1.88	2.48	0.93	1.97	2.05
EIL101	5.74	2.07	4.31	2.72	2.92	3.47
KROA150	4.31	2.90	2.23	1.69	1.26	3.125
KROA200	5.61	3.22	2.67	1.96	1.21	2.99
LK318	7.56	4.55	5.34	3.48	1.93	3.65
PCB442	9.15	5.31	6.88	5.11	5.67	7.41
ATT532	5.38	5.81	4.76	3.54	2.39	3.73
PR1002	7.32	6.99	7.44	5.07	4.01	5.29
NRW1379	-	-	-	-	-	7.05
PR2392	8.18	8.76	8.03	7.21	5.83	7.32
PCB3038	-	-	-	-	-	8.76
FNL4461	-	-	-	-	-	7.47

6 Discussion and Conclusions

In this contribution a TSP solver based on a growing SOM-like neural network enhanced with a greedy 2-Opt local search is presented. A Kohonen type learning rule together with a simple PU insertion strategy are combined in order to produce a SOM network with a growing structure. The parameters of the network are very easily fine tuned, no special procedure is needed. The experimental results over a wide spectrum of TSP instances show that the accuracy of the solution is, in most cases, equivalent to that of the best SOM based heuristics reported in the literature. Only the eiSOM [7] shows systematically better results than those of the present approach but for this case a Genetic Algorithm was necessary to evolve and parametrize the SOM network. Even for larger instances (3038, 4461 cities) the generated tours have a length with errors below 9% of the optimum values. This observed accuracy can still be further increased by a more careful tuning of the network parameters on each individual instance.

References

[1] E. L. Lawler, J. K. Lenstra, A.H.G.R.K., (editors), D.B.S.: (1990) The Traveling Salesman Problem: A Guided tour of Combinatorial Optimization. John Wiley & Sons, New York

[2] Reinelt, G.: (1994) The Traveling Salesman. Computational Solutions for TSP Applications. Springer, Berlin

[3] Johnson, D.S.: (1990) Local optimization and the traveling salesman problem. In: Proceedings of the 17th Colloquium on Automata, Languages and Programming. Springer, Berlin, New York

[4] T. Stuetzle, A. Gruen, S.L., Ruettger, M.: (2000) A comparison of nature inspired heuristics on the traveling salesman problem. In: Proceedings of Parallel Problem Solving from Nature $PPSN-VI$. Volume 1917. Springer, Berlin, pp. 687–692

[5] Johnson, D.S., McGeoch, L.A.: (1997) The traveling salesman: A case study in local optimization. In Aarts, E.H.L., Lenstra, J.K., eds.: Local Search in Combinatorial Optimization. John Wiley & Sons, Chichester, pp. 215–310

[6] Smith, K.: (1996) An argument for abandoning the traveling salesman problem as a neural-network benchmark. IEEE Transactions on Neural Networks **7:** 1542–1544

[7] Hui-Dong Jin, Kwong-Sak Leung, M.L.W., Zu, Z.B.: (2003) An efficient self organizing map designed by genetic algorithms for the traveling salesman problem. Transactions on Systems, Man and Cybernetics – Part B: Cybernetics **33:** 877–888

[8] Plebe, A.: (2002) An effective traveling salesman problem solver based on self organizing map. In Dorronsoro, J.R., ed.: Artificial Neural Networks - ICANN2002. Springer, Heidelberg, pp. 908–918

[9] Kohonen, T.: (1995) Self Organizing Maps. Springer, Berlin

[10] F. Carvalho Vieira, A.D.D.N., Costa, J.A.F.: (2003) An efficient approach to the traveling salesman problem using self-organizing maps. International Journal of Neural Systems **13:** 59–66

[11] Fritzke, B.: (1992) Wachsende zellstrukturen - ein selbstorganisierendes neuronales netzwerk. Technical report, Arbeits bericht des IMMD Universitaet Erlangen - Nuernberg

[12] Lin, S., Kernighan, B.W.: (1973) An effective heuristic algorithm for the traveling-salesman problem. Operations Research **21:** 498–516

[13] Fritzke, B., Wilke, P.: (1991) Flexmap - a neural network with linear time and space complexity for the traveling salesman problem. In: Proc. IJCNN-90 Int. Joint Conf. Neural Networks, Singapore, pp. 929–934

[14] Reinelt, G.: (1991) Tsplib - a traveling salesman problem library. ORSA Journal on Computing **3:** 376–384

Evolutionary Design and Evaluation of Modeling System for Forecasting Urban Airborne Maximum Pollutant Concentrations

H. Niska[1], T. Hiltunen[1], A. Karppinen[2], M. Kolehmainen[1]

[1]Department of Environmental Sciences, University of Kuopio, Finland
[2]Finnish Meteorological Institute, Helsinki, Finland
E-mail: harri.niska@uku.fi

Abstract

In this paper, an integrated modeling system based on a multi-layer perceptron model is developed and evaluated for the forecasting of urban airborne maximum pollutant concentrations. In the first phase, the multi-objective genetic algorithm (MOGA) and sensitivity analysis are used in combination for identifying feasible system inputs. In the second phase, the final evaluation of the developed system is performed for the concentrations of pollutants measured at an urban air quality station in central Helsinki, Finland. This study showed that the evolutionary design of neural network inputs is an efficient tool, which can help to improve the accuracy of the model. The evaluation work itself showed that the developed modeling system is capable of producing fairly good operational forecasts.

1 Introduction

A wide variety of neural network (NN) models such as multi-layer perceptron (MLP) [1, 2] models have been developed for the forecasting of air quality, as the main efforts have been focused on the forecasting of peak pollutant levels [3], i.e., episodes due to their negative health effects. The investigations have showed that NNs are fairly accurate and useful tools that can outperform other statistical and deterministic methods [4, 5]. However, up to date, most of the investigations have been focused on the so called "nowcasting" of air quality, in which the measured or pre-preprocessed meteorological data is used instead of using results from numerical weather prediction (NWP) models [4].

In this paper we develop and evaluate an integrated modeling system based on a MLP model for forecasting maximum concentrations of urban airborne pollutant concentrations; the corresponding study for the forecasting of hourly concentrations has been recently made by Niska et al. [6]. In the developed modeling system, the meteorological input data is derived from meteorological pre-processing and numerical weather prediction models, and the scheme corresponds to a realistic operational situation, i.e., actual forecasting. The system is designed through an evolutionary approach based on a multi-objective genetic algorithm (MOGA) [7, 8, 9] after which the final evaluation is performed.

2 Experimental data

The concentration data used in this study were obtained at the urban background station (Kallio) in central Helsinki. The data comprised hourly ambient airborne pollutant concentrations for the following species: NO_x, NO_2, O_3, PM_{10} and $PM_{2.5}$. Fairly small fractions (1-6%) of missing concentration values were detected and replaced using the combination of the Self-Organizing Map (SOM) and linear interpolation [10]. The reason for imputation was practical: it was employed to provide continuous data set to the model evaluation.

The pre-processed meteorological data, based on the combination of the data from synoptic stations at Helsinki-Vantaa airport (about 15 km north of monitoring sites) and Helsinki-Isosaari (an island about 20 km south of the monitoring site), were selected to be used in this study, as they were best representative for the whole of the urban area and contained relevant derived atmospheric turbulence parameters. The mixing height (the depth of the atmospheric layer where the pollutants emitted from the surface are effectively mixed) and turbulence parameters were evaluated using a meteorological pre-processing model MPP-FMI [11] which utilizes the sounding observations at Jokioinen (90 km northwest of Helsinki) and the routine meteorological observations. This pre-processor has also been adapted to better allow for urban meteorological conditions.

The numerical weather prediction data were derived from the Finnish variant of the HIRLAM [12] and its synoptic scale HIRLAM version 4.6.2 which was in operational use during the period 15 Nov. 1999 to 16 March 2003. In the calculations, HIRLAM data at a specified location (longitude 24.81°E and latitude 60.33°N) close to the air quality monitoring site were utilized. For this point, all the operational forecasts made within 6 hourly intervals (00, 06, 12, 18 UTC) were employed and the data from the model surface levels of 30 and 31 (the level 31 closest to ground) were used.

3 Methods

3.1 Modeling system based on MLP model

We address the forecasting of daily maximum concentrations of ambient airborne pollutants using a fully-connected MLP model [1], which was trained using scaled conjugate gradient back-propagation with 1000 training epochs. For controlling the over-fitting, we relied on the early stopping method of training. The training was stopped when the error of the validation set (20% of the training data) increased for 10 iterations, and consequently the weights and biases at the minimum of the validation error were utilised. The architectures of the MLP models were largely based on the previous studies in this domain [5], in which one hidden layer, sigmoid transfer functions for hidden units and linear transfer functions for output were found to be sufficient.

The aim was to model the relationship between airborne concentrations, pre-processed (MPP-FMI) and forecasted (HIRLAM) meteorological parameters applied at different time lags. The input data was selected from the operational point of view. The concentration and meteorological data were taken from the time lags T (current day) and T - 1 (previous day), and timing and the NWP forecast from the time T + 1 (next day; for which the forecast applies). Additionally, the meteorological and concentration variables were divided into 6-hourly daily averages: 1:00 a.m. to 6:00 a.m., 7:00 a.m. to 12:00 p.m., 13.00 p.m. to 18.00 p.m. and 19:00 p.m. to 24:00 p.m. The variables used in the modeling system are listed in Table 1.

3.2 Input selection using MOGA

In this study, the selection of input variables (predictors) was necessary due to the large data dimension (total number of variables 316); see Table 1. The selection of inputs was based on the combination of MOGA and the sensitivity analysis of the MLP model [14, 15]. In this approach, the number of inputs was minimized and the performance maximized, and the sensitivity of the MLP was used to estimate the relationship between inputs and performance [16].

In our experiments, we applied a structured genetic algorithm (GA) as the core of MOGA. The structured populations were implemented within the unconstrained migration model including five isolated subpopulations each having 20 individuals. The migration of best individuals was applied within 20 generation intervals. Inside the populations, we used the different mutation rates (0.5, 1, 1.5, 2.0 and 2.5 mutations per individual, on average) in order to maintain both rough and fine search capabilities.

The selection of genetic operators was based largely on the investigations made by Emmanoulidis [16]. Random sampling tournament selection and subset size oriented common features (SSOCF) recombination operators were utilized. The mutations were employed within the framework of binary mutations, since the problem (input selection) was represented as a bit string.

Table 1. The list of all variables used as inputs of the modeling system. A relevant input subset is selected from this set using MOGA for each pollutant respectively.

Input variable(s)	Unit(s)	Time Lags	N
Temporal variables			
Sine and cosine of year and week day	–	*T+ 1*	3
Concentration variables; all the daily 6-hourly averages			
NO_x, NO_2, O_3, PM_{10} and $PM_{2.5}$	μg/m^3	*T, T - 1*	40
Meteorological variables; all the 6-hourly averages			
Pressure, temperature and humidity	Pa, K, %	*T, T - 1*	25
State of ground and albedo	–	*T, T - 1*	16
Cloudiness	(0-8)/8	*T, T - 1*	8
Dewpoint, wetbulb and temp. scale	K	*T, T - 1*	24
Rain	mm	*T, T - 1*	8
Height of low clouds	m	*T, T - 1*	8
Sine and cosine of direction of flow	–	*T, T - 1*	16
Wind speed	m/s	*T, T - 1*	8
Sunshine duration and solar elevation	h, rad	*T, T - 1*	16
Solar and net radiations	W/m2	*T, T - 1*	16
Moisture parameter	–	*T, T - 1*	8
Monin-Obukhov length	m	*T, T - 1*	8
Friction and convective velocities	m/s	*T, T - 1*	16
Turb. and latent heat flux	W/m2	*T, T - 1*	16
Mixing height	m	*T, T - 1*	8
Gradient of potential temperature	K/m	*T, T - 1*	8
HIRLAM variables from the model surface levels 30 and 31; all the operational +24 hours forecasts (00, 06, 12, 18 UTC)			
U- and V-components of wind	m/s	*T+ 1*	16
Kinetic energy of turbulence	J/kg	*T+ 1*	8
Temperature	K	*T+ 1*	8
Specific humidity and cloud condensate	kg/kg	*T+ 1*	16
Total cloud cover	%	*T+ 1*	8
Pressure and temperature at 2m	Pa, K	*T+ 1*	8

The input time T+1 is the time (next day), for which the forecast applies; N is the number of variables.

The fitness assessment was performed here through the Pareto-ranking and goal attainment techniques [7, 8, 9]. The input subset fitness was estimated by applying the modified sensitivity analysis where the MLP was firstly trained on all the inputs, after which the sensitivity was assessed by calculating a statistical measure from the output of simulation. In this scheme, unselected inputs were replaced by their respective means computed on training data; data was variance scaled.

As the statistical measure, we used the well-known index of agreement [18] defined as follows:

$$d = 1 - \left[\frac{\sum_{i=1}^{N} (P_i - O_i)^2}{\sum_{i=1}^{N} \left(|P_i - \bar{O}| + |O_i - \bar{O}| \right)^2} \right] \quad (1)$$

where N is the number of observations, O_i the observed data point, P_i the predicted data point, $\bar{O}$ the mean of observed data and $\bar{P}$ mean of predicted data.

4 Experiments and results

4.1 Evaluation of MLP models

We have evaluated the modeling system for forecasting daily maximum values of NO_2, PM_{10}, $PM_{2.5}$ and O_3 concentrations obtained at the urban air quality station in central Helsinki. The evaluation work was performed through three main phases: (1) the selection of model-specific inputs, (2) the validation of input selection, and (3) the final evaluation of models.

In the phases 1 and 2, the data set was partitioned at random three data sets: a training data set (60% of all data rows), a test data set (20%) and a validation data set (20%). The training data set was used for constructing the initial MLP, the test data set for performing feature selection, and the validation data set for validating the achieved Pareto-optimal front. The validation step was made because it is known that the Pareto-optimal front by itself does not provide enough information to select best solution [14], i.e., does not have good generalization power on different data.

In the last phase, the best input subset was chosen according to the resulting Pareto-optimal fronts. The final evaluation was performed using the cross-validation method applying three different validation periods each containing a full year cycle of data. For each validation period, the general and episodic performances were assessed through statistical analysis, where the index of agreement (Eq. 1) was used for assessing the general performance of the model and the success index for assessing the episodic performance in the operational sense, i.e., whether an episodic threshold value is exceeded.

The episodic success index (SI) is calculated from the true positive rate (TPR), representing the sensitivity of the model, i.e., the fraction of correct alarms and the false positive rate (FPR), representing the specificity of the model, i.e., the relative fraction of false alarms.

$$TPR = A/M,\ 0 \le TPR \le 1, \quad (2)$$
$$FPR = (F-A)/(N-M),\ 0 \le FPR \le 1, \quad (3)$$

where A is the number of correctly predicted episodes, M is the number of all observed episodes, F is the number of all predicted episodes and N is the total number of observations. Sensitivity and specificity are combined into the SI as follows:

$$SI = TPR - FPR, \quad (4)$$

the value of SI ranging from -1 (TPR=0 and FPR=1) to 1 (TPR=1 and FPR=0).

The threshold values required in the computation of SI were defined for each airborne pollutant according to the 95th percentile of daily maximums or national guidelines (if available).

4.2 Numerical results

The model evaluation results are presented in Table 2. The results are presented as average values of the statistical indicators and their standard deviations (ten modeling runs) over the validation periods considered.

Table 2. Statistical model evaluation parameters of the forecasted and measured maximum concentrations.

	Index of agreement (d)					
	Period 1		Period 2		Period 3	
Pollutant	Mean	Std.	Mean	Std.	Mean	Std.
NO_2	0.91	0.03	0.85	0.05	0.78	0.02
O_3	0.94	0.02	0.94	0.01	0.89	0.01
PM_{10}	0.75	0.07	0.80	0.07	0.39	0.04
$PM_{2.5}$	0.76	0.08	0.76	0.04	0.55	0.05
	Succes index (SI)					
	Period 1		Period 2		Period 3	
Pollutant	Mean	Std.	Mean	Std.	Mean	Std.
NO_2	0.50	0.17	0.41	0.17	0.35	0.06
O_3	0.58	0.15	0.34	0.18	0.29	0.07
PM_{10}	0.15	0.11	0.54	0.09	0.41	0.09
$PM_{2.5}$	0.35	0.15	0.12	0.13	0.33	0.11

It can be seen that the best forecasting accuracies were achieved for the concentrations of NO_2 and O_3 according to index of agreement. This level of performance (d>0.9) is applicable to operational forecasting. However, the results obtained for particular matters: PM_{10} and $PM_{2.5}$ were slightly worse than the results obtained for NO_2 and O_3. Contrary to good general performance, only moderate episodic performance was achieved when considering success index. The episodic performance was especially degenerated in the case of particular matters.

5 Discussion and conclusions

In this paper, the modeling system based on the MLP neural network is evaluated for forecasting maximum ambient airborne pollutant concentrations at the urban air quality station in central Helsinki. Due to the large data dimension, the novel input selection method based on the MOGA and the sensitivity analysis of MLP is utilized for ensuring reliability of the evaluation process.

The results when considering the forecasts of MLP models showed good general agreement and moderate episodic performance; see Table 2. The results in general were slightly better than the results obtained for forecasting hourly concentrations in time [6].

Despite the modest capabilities of predicting episodic concentrations of particular matter, the proposed modeling system can be useful for assessing of maximum concentrations in urban areas. However, the performance of the proposed system depends ultimately on site-specific conditions such as climatic and topographic factors, and requires appropriate site- and time-specific data for training. Therefore, the results obtained in this study cannot be directly generalized to other areas.

Acknowledgements

This investigation was financially supported by the Academy of Finland (FORECAST, project no. 49946).

References

[1] Haykin, S. (1999) Neural Networks: A Comprehensive Foundation, 2nd Edition. Prentice Hall, Upper Saddle River, NJ

[2] Gardner, M. W., Dorling, S.R. (1999) Artificial neural networks (the multi-layer perceptron) – a review of applications in the atmospheric sciences. Atmospheric Environment 32:2627-2636

[3] Ziomas, I. C., Melas, D., Zerefos, C. S., Bais, A. F. (1995) Forecasting peak pollutant levels from meteorological variables. Atmospheric Environment 29:3703-3711

[4] Kukkonen, J., Partanen, L., Karppinen, A., Ruuskanen, J., Junninen, H., Kolehmainen, M., Niska, H., Dorling, S., Chatterton, T., Foxall, R., Cawley, G. (2003) Extensive evaluation of neural networks of NO_2 and PM_{10} concentrations, compared with a deterministic modelling system and measurements in central Helsinki, Atmospheric Environment 37:4539-4550

[5] Kolehmainen, M., Martikainen, H., Ruuskanen, J. (2001) Neural networks and periodic components used in air quality forecasting, Atmospheric Environment 35:815-825

[6] Niska, H., Rantamäki, M., Hiltunen, T., Karppinen, A., Kukkonen, J., Ruuskanen, J., Kolehmainen, M. (2004) Evaluation of the integrated modelling system containing a multi-layer perceptron model and the numerical weahter prediction model HIRLAM for forecasting urban airborne pollutant concentrations. Submitted to Atmospheric Environment.

[7] Goldberg, D. E. (1989) Genetic algorithms in search, optimization, and machine learning. Addison-Wesley Publishing Company, Inc.

[8] Fonseca, C. M., Fleming, P. J. (1993) Genetic algorithms for multi-objective optimization: formulation, discussion and generalization. In: Forrest, S. (eds.) 5th International Conference on Genetic Algorithms. Morgan Kauffman, California, pp. 416-423

[9] Deb, K. (1999) Evolutionary Algorithms for Multi-criterion Optimization in Engineering Design, Evolutionary Algorithms in Engineering and Computer Science. John Wiley & Sons Ltd, Chichester

[10] Junninen, H., Niska, H., Tuppurainen, K., Ruuskanen, J., Kolehmainen, M. (2004) Methods for imputation of missing values in air quality data sets. Atmospheric Environment 38:2895-2907

[11] Karppinen, A., Joffre, S., Vaajama, P. (1997) Boundary layer parametrization for Finnish regulatory dispersion models. International Journal of Environment and Pollution 8:557-564

[12] Eerola, K. (2002) The operational HIRLAM at the Finnish Meteorological Institute. HIRLAM Newsletter 41:19-24

[13] Emmanouilidis, C., Hunter, A., MacIntyre, J., Cox, C. (1999) Selecting features in neurofuzzy modelling using multiobjective genetic algorithms. In 9th International Conference on Artificial Neural Networks. Edinburgh, UK, pp. 749-754

[14] Oliveira L.S., Sabourin R., Bortolozzi F., and Suen C.Y. (2003) A methodology for feature selection using multi-objective genetic algorithms for handwritten digit string recognition. International Journal of Pattern Recognition and Artificial Intelligence 17:903-930

[15] Moody J., Utans, J. (1991) Principled architecture selection for neural networks: application to corporate bond rating predictions. Advances in Neural Information Processing Systems 4:683-690

[16] Emmanouilidis, C., Hunter, A., MacIntyre, J. (2000) A multiobjective evolutionary setting for feature selection and a commonality-based crossover operator. In Congress on Evolutionary Computation 2000. IEEE, New Jersey, pp. 309-316

[17] Willmott, C. J. (1981) On the validation of models. Phys. Geog. 2:184-194

Evolving Evolvability: Evolving both representations and operators

Grant W. Braught

Department of Mathematics and Computer Science
Dickinson College, Carlisle, Pennsylvania, USA
E-mail: braught@Dickinson.edu

Abstract

The behavior of an evolutionary system incorporating both an evolving genetic representation (a learning mechanism) and an evolving genetic operator (mutation) is explored. Simulations demonstrate the evolution of evolvability through the co-adaptation of these two mechanisms. It is also shown that this co-adaptation produces a transmission function that becomes more conservative as the strength of the learning mechanism increases.

1 Introduction

Altenberg [1] defines evolvability as the "ability of the genetic operator/representation scheme to produce offspring that are fitter than their parents." In addition to defining evolvability, this definition identifies two means by which evolvability can evolve. In particular, if either the genetic operators or the genetic representation[1] fall under evolutionary control then, by definition, it is possible for evolvability to evolve.

Although it is clear that the above conditions make it possible for evolvability to evolve, the dynamics by which evolvability evolves are far from clear. In attempts to gain insights into these dynamics, some researchers have studied systems with evolving genetic operators including mutation [2,3] and crossover [4]. Others have studied systems that incorporate evolving representations, including genetic programming [1], stochastic genotype-to-phenotype mappings [5] and simple learning mechanisms [6,7].

Each of the studies cited above considers either an evolving genetic operator or an evolving representation, but not both. Consequently, little is known about if and how these two mechanisms might interact – and interaction seems likely. Altenberg [1] argues that from the perspective of selection, there is a duality between representations and genetic operators. Meaning that in terms of their effect on fitness, any change in a genetic operator can be mimicked by a change in representation and vice versa. For example, an increase in mutation rate or a reduction of neutrality in the representation could have indistinguishable impacts on fitness. The existence of this duality strongly suggests that evolving representations and evolving genetic operators will interact. An understanding of the dynamics of these interactions may prove important in the creation of highly self-adaptive evolutionary algorithms and may generate insights into natural evolutionary phenomena as well. The work presented here begins an examination of such interactions through the study of a system that explicitly allows for both an evolving genetic operator and an evolving genetic representation.

2 Background

To talk about the dynamics of evolving evolvability it is helpful to introduce the notion of a transmission function [1, 8]. A transmission function describes the effect of the genetic operators and genetic representation on a population. In other words, a transmission function maps a given distribution of parent phenotypic fitness values into the distribution phenotypic fitness values for their offspring. It is also possible to talk about the conservatism of a transmission function. A more conservative transmission function will tend to produce offspring with phenotypic fitness closer to their parents' fitness than will a less conservative transmission function.

Two interesting dynamics related to the conservatism of evolving transmission functions, and relevant to the present work, have been reported. First, as the fitness of a population rises, the transmission function tends to become more conservative [1, 2, 9]. This dynamic arises because at high fitness, changes made during transmission tend to decrease fitness. Thus, a more conservative transmission function is favored and spreads through the population.

Second, for fitness levels above random quality, the weaker the selective force the more conservative the transmission function becomes [5, 10]. This dynamic arises from the effects of selection on the distribution of offspring fitness. When parents with less conservative transmission functions reproduce, a few of their offspring tend to have much higher fitness than the others. Under strong selection pressure, these few high-fitness offspring monopolize the reproductive opportunities, resulting in the spread of less conservative transmission functions. Conversely, when parents with more conservative transmission functions reproduce, they tend to produce offspring with closely clustered, but higher average fitness. Now, under weak selection pressure, reproductive opportunities are more evenly distributed among the offspring, leading to the spread of more conservative transmission functions.

---

[1] As used here the term genetic representation includes the effect of the genotype-to-phenotype mapping.

Through this second dynamic, the presence of learning will also influence the conservatism of the transmission function. By way of illustration, consider the view of learning as a process that moves a phenotype toward a nearby peak in the fitness landscape [see 11 for an illustration]. In this view, many individuals with distinct genotypes will have phenotypes that are approaching the same fitness peak. Thus, the fitness of these individuals will be similar, despite their underlying genetic differences. The net effect is that learning reduces the difference in selective pressure between distinct genotypes, leading to the evolution of a more conservative transmission function. Further, the more powerful the learning mechanism, the more pronounced this influence becomes.

3 Methods

To investigate the interaction between evolving genetic operators and evolving representations a model incorporating both of these mechanisms has been used. In this model, an organism's genotype consists of two parts: the genetic representation and a mutation rate gene. The genetic representation contains the genetic information from which the organism's phenotype is generated. This representation incorporates a simple learning mechanism that allows it to evolve. The mutation rate gene controls the frequency with which the genetic representation is mutated when the organism is reproduced. Encoding the mutation rate in the genotype permits the genetic operator to evolve. The following sub-sections provide further descriptions of this model and the genetic algorithm used for the simulations.

3.1 The Evolving Genetic Representation

The genetic representation that is used has been adopted from Mayley [7]. In this model, an organism's genetic representation consists of 20 loci, each of which may take on one of the alleles 1, 0 or ?. As in Hinton and Nowlan's seminal work [6], the ?'s represent traits that may be altered by learning. It is through the number and location of the ?'s (i.e. learning loci) within the genotype, that the genetic representation is thought of as evolving.

During every generation, each organism's genetic representation is converted into a phenotype by conducting a number of learning trials. In each learning trial a candidate phenotype, p_c, is generated as a copy of the organism's genotype. Every ? in the copy is then randomly replaced by either a 1 or a 0. The fitness of each candidate phenotype is then evaluated using an NK fitness landscape (see [7, 11, 12] for a complete description of NK landscapes). Finally, the organism is assigned the same fitness as the most fit candidate phenotype.

The precise method for calculating the fitness of an organism is shown in equation 1. In equation 1, T is the number of learning trials, and $F_{NK}(p_t)$ is the fitness of the tth candidate phenotype as determined by the NK landscape. F_{NKopt} is the maximum achievable fitness in the current NK landscape and normalizes the fitness so that comparisons can be made across multiple trials with different NK landscapes. N is the number of loci in the organism's genotype, m is the number of ?'s in the genotype and c is the cost associated with learning. Thus, the final term in equation 1 imposes a cost for learning that is proportional to the number of ? alleles in the organism's genotype. For an extensive discussion of learning costs see [7].

$$F(p) = \frac{MAX\{t = 1...T, F_{NK}(p_t)\}}{F_{NKopt}} - c\frac{m}{N} \quad (1)$$

3.2 The Evolving Genetic Operator

Self-adaptive mutation rates [2, 3] are used to model an evolving genetic operator. With self-adaptive mutation rates each organism's genotype contains a mutation rate gene. This gene indicates the probability that any given locus will be mutated during reproduction. When an organism is reproduced, the offspring receives mutated copies of its parent's genetic representation and mutation rate gene. The genetic representation is mutated at the rate specified by the parent's mutation rate gene. The mutation rate gene is mutated by adding a random number selected from a normal distribution with a mean of zero and a standard deviation of 0.001. The mutation rate gene is also confined to the range [0...1].

3.3 The Genetic Algorithm

The following parameters are common to all of the simulations performed. Each population contained 100 organisms and reproduced asexually using roulette wheel selection. Each allele in an organism's genetic representation was encoded using two bits with "00" representing 0, "01" representing 1, and both "10" and "11" representing ?. The initial populations were generated using random bit strings of length 40, resulting in an expected allele distribution of 25% 1's, 25% 0's and 50% ?. The mutation rate gene for every organism was initially set to be 0.25. The NK environments used to assign fitness were randomly generated and had N=20 loci and K=4. The cost of learning in equation 1 was set to 0.3.

4 Results

At the outset, it was unclear that evolution would occur reliably in the presence of both self-adaptive mutation rates and the learning mechanism employed. The selective force for self-adaptive mutation rates is already much weaker than that for primary traits, which directly affect fitness. The addition of learning further weakens this selection by hiding the effects of mutations from selection [11]. Despite this, it was found that evolution is in fact able to occur reliably in the system used here.

Figure 1 shows population averages for fitness, mutation rate, and the fraction of learning loci per genotype over the first 40,000 generations of a typical run using 10 learning trials. In this particular run, the evolving mutation rate begins to converge at about 15,000 generations and is fully converged by 22,000 generations. Although the rate of convergence varied between runs, the mutation rate converged before 75,000 generations in almost 90% of runs using 10 learning trials. Once the mutation rate nears convergence, the average population fitness rises quickly and the fraction of learning loci per genotype is dramatically reduced. This reduction is due to the Baldwin effect [6, 7, 13].

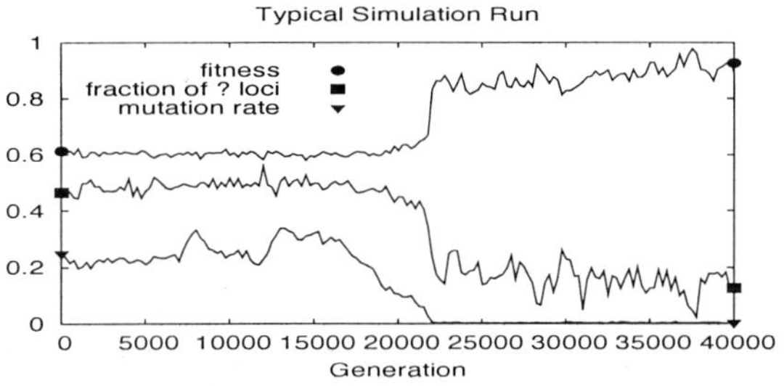

Fig. 1. Average population statistics for the first 40,000 generations of a typical simulation using 10 learning trials.

Having established that evolution will occur under the established conditions, the adaptive power of the representation was then varied. Specifically, the number of learning trials was varied from 2 to 30. For each number of learning trials, 50 runs were performed with 75,000 generations each. From those 50 runs, 30 in which the mutation rate converged prior to 40,000 generations were selected at random. For each of these 30 runs, the mutation rate and the percentage of learning loci were averaged across time from 50,000 to 75,000 generations. These time averages were then averaged across the 30 selected runs. These averages represent optimal values for mutation rate and the fraction of learning loci for each number of learning trials.

As shown in figures 2 and 3, both the mutation rate and the fraction of learning loci tend to rise as the number of learning trials increases. Increased mutation rates make the transmission function less conservative. Conversely, the combined increases in the fraction of learning loci and the number of learning trials make the transmission function more conservative.[2] Thus, the rises in these two curves represent opposite effects on the conservatism of the transmission function.

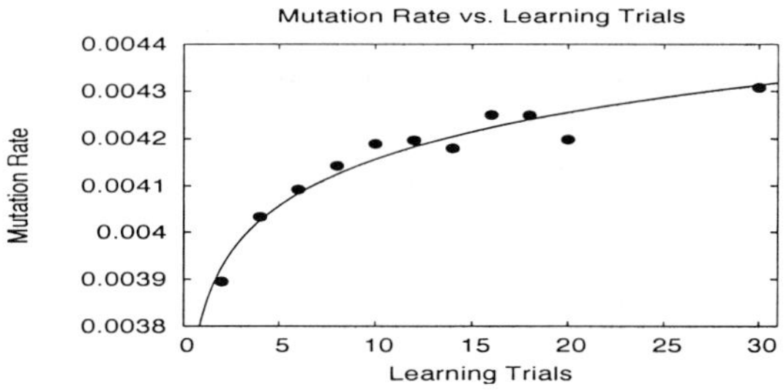

Fig. 2. The average mutation rates increase with the number of learning trials, decreasing the conservatism of the transmission function.

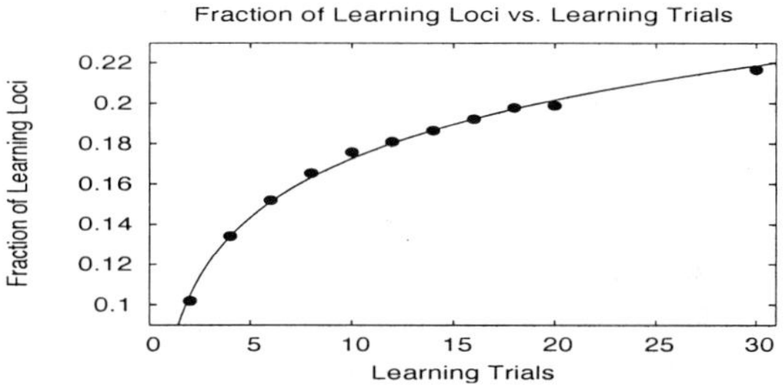

Fig. 3. The average fraction of learning loci increases with the number of learning trials, increasing the conservatism of the transmission function.

The opposing influences just described are interesting for two reasons. First, the influences oppose each other, making it impossible to immediately determine their net effect on the conservatism of the transmission function. Second, increasing mutation rates with decreasing selection pressure is exactly the opposite of what was found in earlier studies [2, 5, 9, 10]. However, these earlier studies used only self-adaptive mutation rates and did not allow the representation to evolve. Therefore, the expected conservatism-selection strength relationship may still hold if the increased conservatism in transmission due to the evolving representation exceeds the decrease due to the higher mutation rates.

To test this hypothesis, the conservatism of the transmission function was investigated directly. For each number of learning trials, 150 parent organisms were created with the optimal mutation rate and the expected number of learning loci. Each of these parents was reproduced 30 times and the root mean square difference in fitness between the parent and its offspring was computed. The average results from 10 repetitions of this process are shown in figure 4. As the number of learning trials increases, the root mean square difference in fitness between parent and child decreases, indicating a net increase in the conservatism of the transmission function.

From the results reviewed in the background section, this observed increase in conservatism could be

[2] This is true because as the number of learning trials increases, the number of learning loci that evolve allow for a higher percentage of all possible phenotypes to be guessed during the learning trials.

attributed to either increased fitness or to decreased selection pressure. However, the average, median and maximum fitness of the populations did not increase as the number of learning trials increased. Therefore, the observed increase in the conservatism of the transmission function must have occurred in response to the decreased selection pressure caused by the more powerful learning mechanism.

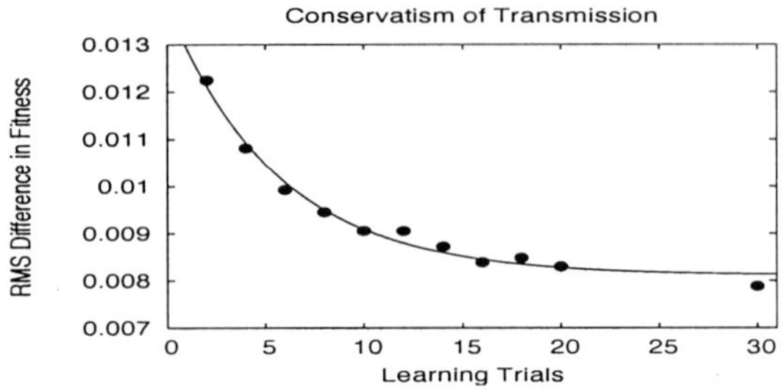

Fig. 4. The conservatism of the transmission function increases with increasing learning trials (i.e. decreasing selection pressure).

5 Discussion

A system allowing for the evolution of both a genetic operator (mutation) and a genetic representation (via learning) has been explored. Simulation results from this system have demonstrated the generality of earlier findings that decreased selection strength lead to the evolution of more conservative transmission functions. The results also show that the increase in conservatism can be achieved not just via evolution of the representation (as in [1, 5, 6, 7]) or the operator (as in [2, 5, 9, 10]), but can also be achieved through the co-adaptation of the representation and the operator.

In pursuing this work, two significant questions that will guide future study have arisen. First, what factors affect how the representation and operators co-adapt? In particular, why was the observed increase in the conservatism of transmission enacted by increases in both mutation and learning loci, as opposed to other seemingly equivalent alternatives? One intriguing possibility is a second-order conservatism in which the co-adaptation increases how reliably the conservatism of the transmission function is passed on to offspring. Second, how will representations and operators co-adapt in response to changes in the evolutionary environment? Recent results have shown that when an environmental change occurs (i.e. a change in the fitness function) a selective benefit exists for less conservative transmission functions [3, 13]. Thus, because both representation and operators affect the conservatism of the transmission function, it is presently unclear how they will respond to environmental changes.

References

[1] Altenberg, L. (1994) The Evolution of Evolvability in Genetic Programming. In: Kinnear, K. (ed.) Advances in Genetic Programming. MIT Press, Cambridge, pp. 48-74

[2] Bäck, T. (1992) Self-Adaptation in Genetic Algorithms. In: Varela, F. & Bourgine, P. (eds.) Proc. 1st ECAL. MIT Press, Cambridge, pp. 263-271

[3] Bedau, M., Packard, N. (2003) Evolution of evolvability via adaptation of mutation rates. Biosystems, 69(2,3) 143-162

[4] Angeline, P. (1996) Two Self-Adaptive Crossover Operations for Genetic Programming. In: Angeline, P., Kinnear, K. (eds.) Advances in Genetic Programming II. MIT Press, Cambridge, pp. 89-110

[5] Glickman, M., Sycara, K. (1999) Comparing Mechanisms for Evolving Evolvability. In: Wu. A. (ed.) Workshop Proc. of GECCO. Morgan Kaufmann, San Francisco, pp. 38-41

[6] Hinton, G., Nowlan, S. (1987) How Learning can Guide Evolution. Complex Systems 1, 496-502

[7] Mayley, G. (1996) The evolutionary cost of learning. In: Mayes, P., et al. (eds.) From Animals to Animats 4: Proc. 4th ICSAB. MIT Press, Cambridge, pp. 458-467

[8] Slatkin, M. (1970) Selection and polygenic characters. Proc. National Academy of sciences U.S.A. 66, 87-93

[9] Glickman, M., Sycara, K. (1998) Evolutionary Algorithms: Exploring the Dynamics of Self-Adaptation. In: Koza, J. et al. (eds.) Proc. 3rd Annual Conference on Genetic Programming. Morgan Kauffmann, San Francisco, pp. 762-769

[10] Glickman, M., Sycara, K. (2000) Reasons for Premature Convergence of Self-Adapting Mutation Rates. In Yao, X. (eds.) Proc. 2000 CEC. IEEE Press, Piscataway, pp. 62-69

[11] Mayley, G. (1997) Guiding or Hiding: Explorations into the Effects of Learning on the Rate of Evolution. In: Husbands, P., Harvey, I. (eds.) Proc. 4th ECAL. MIT Press, Cambridge, pp. 135-144

[12] Kauffman, S. (1993) The origins of order: Self-organization and selection in evolution, Oxford University Press, New York

[13] Buchanan, A., Triant, M., Bedau, M. (2004). The Flexible Balance of Evolutionary Novelty and Memory in the Face of Environmental Catastrophes." In: Pollack, J. et. al. (eds.) Artificial Life IX. MIT Press, Cambridge, MA, pp. 297-302.

A Multi-Objective Evolutionary Algorithm for Solving Traveling Salesman Problems: Application to the Design of Polymer Extruders

A. Gaspar-Cunha

IPC – Institute for Polymer and Composites, University of Minho, Guimarães, Portugal
E-mail: agc@dep.uminho.pt

Abstract

A Multi-Objective Evolutionary Algorithm (MOEA) for solving Traveling Salesman Problems (TSP) was developed and used in the design of screws for twin screw polymer extrusion. Besides the fact that MOEA for TSP have already been developed, this paper constitutes an important and original contribution, since in this case, they are applied in the design of machines. The Twin-Screw Configuration Problem (TSCP) can be formulated as a TSP. A different MOEA is developed, in order to take into account the discrete nature of the TSCP. The algorithm proposed was applied to some case studies where the practical usefulness of this approach was demonstrated. Finally, the computational results are confronted with experimental data showing the validity of the approach proposed.

1 Introduction

Co-rotating twin-screw extruders are widely used in the polymer industry for compounding and reactive processing applications, due to a number of interesting constructive and functional features [1-4]. As for other types of extruders, the performance of these machines is quite sensitive to the operating conditions (output, screw speed and barrel temperature profile) and screw configuration. Hence, both must be carefully defined for each polymer system.

A methodology based on an optimization approach to set automatically the operating point and satisfying the relevant process criteria was proposed [5]. The present work extends the method to the definition of the relative location of a set of modular screw elements (comprising conveying, staggering kneading discs and reverse elements) along the screw axis. This is a first step to deal with screw design in twin screw extrusion.

The method is applied to a case study where the objective is to determine the screw configuration yielding the best global performance in terms of melt temperature at the die exit, mechanical and thermal energy consumption, maximum pressure generated, viscous dissipation, degree of mixing and average residence time. The computational results were assessed experimentally using, for simplicity, two of these criteria, melt temperature at die exit and power consumption.

2 MOEA for TSP

The choice of an optimization methodology depends mainly of the type of problem to solve. In the present case, the aim is to select the best location of a fixed number of screw elements of a twin screw extruder that will yield the best performance for a particular application. The problem is illustrated in Figure 1: to determine the position along the screw of 5 transport elements, 4 kneading blocks (with different staggering angles) and one reverse element (considering that in both ends, other conveying elements will be used). This should be done in such a way that either single criterion (such as specific mechanical energy, SME), or a compromise between various criteria (SME, maximum pressure, viscous dissipation, etc) is/are satisfied. This will be denoted as Twin Screw Configuration Problem, TSCP.

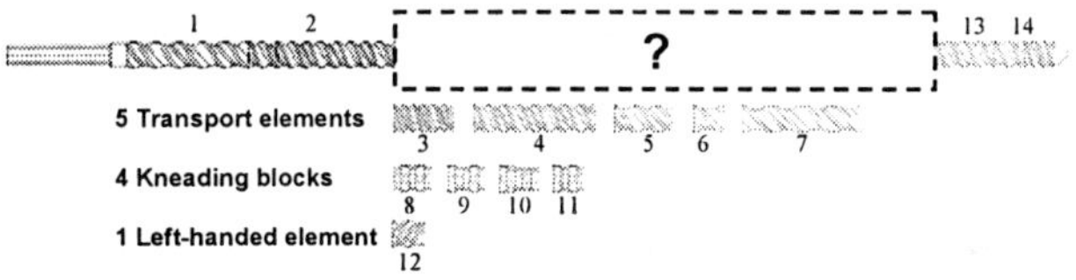

Fig. 1. Twin-screw extrusion configuration problem.

This is a complex optimization problem, due to the presence of discrete decision variables and the existence of a very wide search space. Also, one may face the eventual need to combine discrete and continuous decision variables (such as operating conditions and/or geometrical features of specific elements) [6]. To overcome these difficulties, the analogy between the well known Traveling Salesman Problem (TSP) [7] and TSCP was explored. This would be quite advantageous, given the methodologies available as a result of the on-going intensive research on TSPs [8-13].

In the TSP, the salesman aims to visit n cities but minimizing the distance traveled and/or the total cost (Figure 2-A). In the TSCP the polymer is the TSP's traveling salesman and the screw modules are the cities, the aim being to define the sequence of screw elements that maximize the process performance (Figure 2-B).

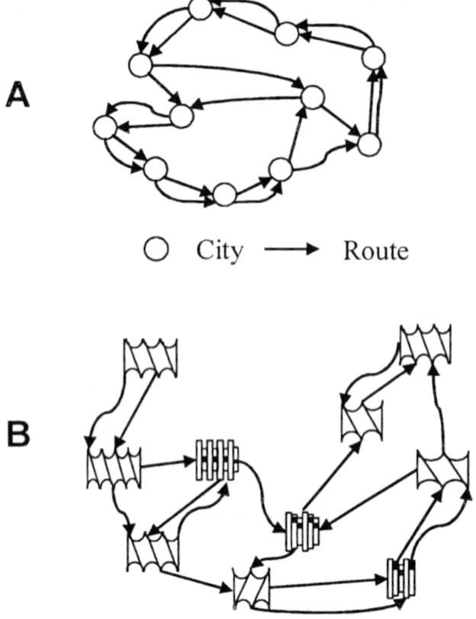

Fig. 2. Analogy between TSP (A) and TSCP (B).

Given the multi-objective nature of the TSCP, it makes sense to use a Multi-Objective Evolutionary Algorithm (MOEA) [6,14,15]. MOEAs have been used previously by the authors for the optimization of single screw extrusion, and for setting the operating conditions of twin screw extruders [5,6].

To solve the TSCP, a specific MOEA, developed previously by the authors [5,14] and denoted as Reduced Pareto Set Genetic Algorithm with elitism (RPSGAe), was modified in order to accommodate both the discrete representation of the chromosome and the adequate reproduction operator [6].

The formulation of the TSCP as TSP implies that each chromosome comprises a sequence of discrete numbers representing the positions of the screw elements. Consequently, the Algorithm 1 was implemented by modifying the Reduced Pareto Set Genetic Algorithm with elitism (RPSGAe) developed previously [6,14]. Two populations are considered, the internal (or main) population (P1) of size *N*, used to store the best individuals found in each generation, and an elitist population (P2), of size 2*N*, where the reproduction operator for TSP is accomplished. Also, a ranking scheme based on a clustering technique is employed to reduce the population size while maintaining intact their characteristics [6]. It is applied twice, first during the selection phase, in order to ensure that the best individuals are selected for reproduction, then after reproduction, to select the best *N* individuals of the elitist population who will integrate the main population.

Algorithm 1 (*MOEA for TSP*):

- Set P1 randomly;
- Evaluation of P1 (using the modeling routine);
- Set Gen = 0;
- Do (for each generation):
    - Non-domination test (P1);
    - Ranking of P1 (using the clustering technique);
    - Copy P1 to P2;
    - Apply the inver-over operator to P2;
    - Evaluation of P2;
    - Non-domination test (P2);
    - Ranking of P2;
    - Copy best *N* individuals from P2 to P1;
    - Gen = Gen +1;
- While (Gen < GenMax);
- Results (Pareto Plots);
- Apply the weighted sum;
- Best solution;

The algorithm starts with the random generation of the *N* individuals of the internal population (P1). These are evaluated using the modeling routine (see next section) and the generation counter is initialized (Gen = 0). At each generation, the following steps are performed: the non-dominated individuals of P1 were chosen; a clustering procedure (see Algorithm 2) is applied to rank the individuals of P1; the entire internal population (P1) is copied to the elitist population (P2); the remaining *N* individuals of the elitist population are generated by the application of the inver-over operator; the new individuals are evaluated; the non-domination test and the clustering procedure are applied to the elitist population, to rank its 2*N* individuals; the best *N* individuals of the elitist population are copied to the main population.

The algorithm ends when the maximum number of generations is reached, the solutions being the non-dominated individuals of the last main population. In practice, since the selection of a single solution from the pool of optimal points may be difficult (due to the most probable existence of Pareto plots consisting of clouds of points with feeble trade-offs), an overall simple objective function using the weighted sum of the individual criteria is applied to the final population:

$$F_i = \sum_{j=1}^{q} w_j \, f_j \qquad (1)$$

here, F_i is the fitness of individual i, q is the number of criteria, f_j is the objective function of criterion j and w_j is the corresponding weight ($0 \leq w_j \leq 1$ and $\sum_{j=1}^{q} w_j = 1$).

Thus, after defining the weight of each criterion, function (1) is applied to the non-dominated solutions and yields the best result.

Algorithm 2 starts with the definition of the number of ranks, N_{Ranks}, and the rank of each individual, *Rank*[*i*], is set to 0. For each rank, *r*, the population is reduced to *NR* individuals (where *NR* is the number of individuals of each rank), using the clustering technique proposed by Roseman and Gero [19]. Then, rank *r* is attributed to these *NR* individuals. The algorithm ends when the number of pre-defined ranks is reached. Finally, the fitness of individual *i* (F_i) is calculated using the following linear ranking function:

$$F_i = 2 - SP + \frac{2(SP-1)(N_{Ranks} + 1 - Rank[i])}{N_{Ranks}} \quad (2)$$

where *SP* is the selection pressure ($1<SP\leq 2$). Detailed information on these algorithms can be found elsewhere [6, 7].

Algorithm 2 (*Clustering*):

- Definition of N_{Ranks};
- *Rank*[*i*] = 0;
- *r* = 1;
- Do (for each rank):
  - $NR = r(N/N_{Ranks})$;
  - Reduce the population down to *NR* individuals;
  - *r* = *r* + 1;
- While ($r<N_{Ranks}$);
- Calculate fitness (equation 2);

3 Modeling of the Extrusion Process

The evaluation of the solutions is carried out using the LUDOVIC software [20]. The program assumes that melting occurs instantaneously, taking place just before the location of the first restrictive element upstream. From this location onwards, computations of melt flow are performed separately for each type of screw element (right-handed or left-handed screw elements, staggered kneading disks). Experimental studies have shown that the predictions of the software are within ± 10% of the experimental values [20].

Therefore, for each polymer / system geometry / operating conditions set, the software predicts with reasonable accuracy the evolution along the screw axis of variables such as temperature, melt pressure, shear rate, viscosity, residence time, specific energy and filling ratio and the values of relevant global performance parameters (e.g., average residence time, average strain, mechanical power consumption, maximum melt temperature, outlet temperature).

4 Results and Discussion

As an example, this section will tackle the resolution of the problem illustrated in Figure 1. Table 1 presents the criteria to be included in this exercise, the corresponding aim of the optimization and the individual range of variation. A modular Leistritz LSM 30.34 laboratorial extruder was used. The screw speed, output and barrel temperature were fixed at 100rpm, 10kg/hr and 200°C, respectively. It is also assumed that the machine processes a polypropylene homopolymer. The RPSGAe parameters chosen are the following. The main and elitist populations had 100 and 200 individuals, respectively, and 50 generations were to be studied. A roulette wheel selection strategy was adopted, a number of ranks of 30 and limits of indifference of the clustering technique of 0.01 (see [14] for more details) were adopted. Five different runs using different seed values were carried out in order to test the reproducibility of the algorithm.

Table 1- Optimization criteria, aim of optimization and prescribed range of variation.

Criteria	Aim	X_{min}	X_{max}
C1- Melt temperature at die exit – Texit (°C)	Minimize	180	240
C2- Specific mechanical energy – SME (MJ/kg)	Minimize	1	2
C3- Specific thermal energy – STE (MJ/kg)	Minimize	0.1	1
C4- Maximum pressure generated – Pmax (MPa)	Minimize	5	20
C5- Viscous dissipation – Tmax/Tb	Minimize	0.9	1.3
C6- Average strain – Avg. St.	Maximize	1000	15000
C7- Average residence time – Avg. RT (s)	Minimize	50	300

Figure 3 shows the Pareto curves in the criteria's domain (given the 7-dimensional character of the problem, only criteria C1 and C7 – see Table 1 - are represented). As anticipated, Pareto plots generate a cloud of data and not a well defined trade-off, because points that appear to be dominated in one Pareto plot (or dimension) are non-dominated in another. Thus, use of equation (1) provides a simple effective alternative to obtain the best solutions. Various individual weights were used to produce the six screws represented in Figure 4. First the aim was to select, from this pool of points, the best screw that satisfies each criterion individually (corresponding weight equal to 1). Then, the aim was to select the best screw that considers all the criteria with the same importance (all weights equal to 1/7).

When T_{exit} is the most important criterion, the screw proposed has the kneading blocks and the reverse element located quite upstream. This solution makes sense, as the earlier the restrictive elements are located, the earlier the viscous dissipation occurs and the bigger the opportunity for thermal relaxation during the remaining of the screw length.

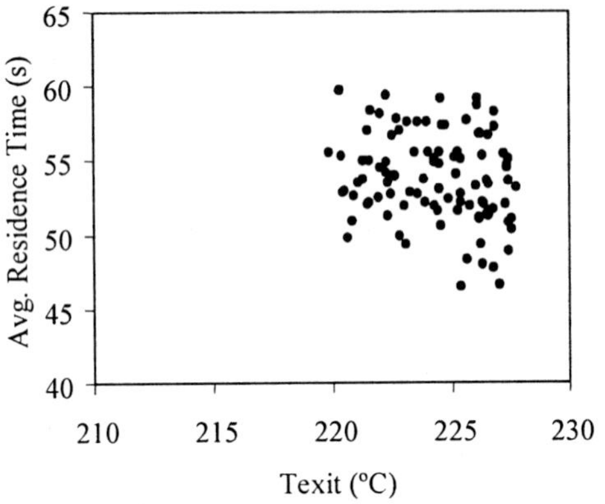

Fig. 3. Pareto plots in the criteria's domain (for simplicity, only C1 vs. C7 is represented).

One single configuration is proposed for minimizing energy consumption (SME and STE) and average residence time (Avg. RT). In this case, the material is made to melt much later and then flows along alternating restrictive and transport elements. Since SME and STE are computed only beyond polymer melting, this geometry involves much less screw length in melt flow and regular spacing of the elements minimizes the thermo-mechanical stresses induced to the polymer. In principle, flow towards the die is also easier, as less conveying elements will work fully filled.

The configurations proposed for decreasing maximum local pressure (Pmax), viscous dissipation (Tmax/Tb), or maximizing the average strain (Avg. Strain) maintain the distribution of the restrictive elements in between transport modules, but now the polymer melts as early as possible. Apparently, the algorithm attempts to use the available screw length to distribute shorter, less restrictive, mixing blocks, which create smaller local pressures and viscous dissipation, but maximize melt flow/mixing.

Finally, the screw at Figure 4 bottom reflects the necessary compromise between all criteria. Its geometry is very similar to that maximizing mixing.

5 Experimental Results

In order to assess experimentally the above proposed screw configurations, experiments were carried out considering the problem of the simultaneous minimization of the temperature at die exit (Texit) and the Specific Mechanical Energy (SME). These two criteria were selected since they are easy to obtain experimentally. Figure 5-A shows the evolution of the respective Pareto optimal front, from the randomly defined initial population to the final 50th generation. Each point represents a different screw profile. The improvement obtained with the optimization procedure is evident.

From this set of different screw configurations, six were selected in order to be replicated experimentally. As shown in the Figure, three were chosen from the optimal set and the other three from the initial population. Screw 1 has the lowest SME; screw 3 represents a solution yielding the minimal viscous dissipation; screw 2 results from a balance between the two criteria. Each experiment was repeated twice.

As can be seen in Figure 5-B, the global behavior of the computational and experimental results is very similar, the differences being within the expected values typical for LUDOVIC. More importantly, the relative location of the computational and experimental points is analogous, which validates the methodology.

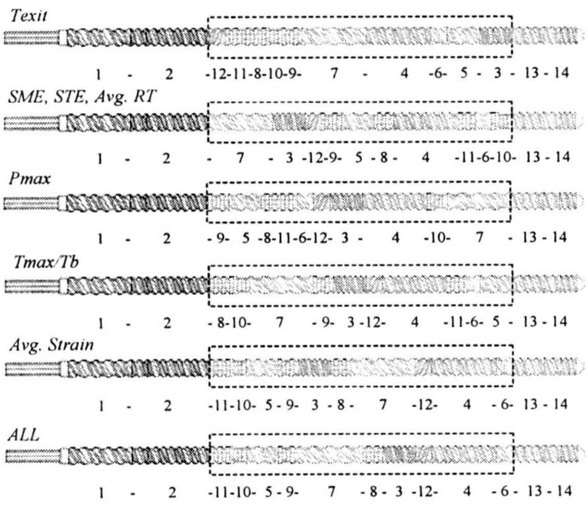

Fig. 4. Optimal screw configurations for various weights in equation 1 (Q=10 kg/hr, N=200 rpm, Tb= 200 °C).

6 Conclusions

An elitist multi-objective genetic algorithm, denoted as RPSGAe, was used to solve successfully the Twin-Screw Configuration Problem (TSCP), consisting of locating screw elements along the screw axis, which is an important step towards the design of co-rotating twin-screw screws.

The TSCP is a complex multi-objective, combinatorial, discrete and not always continuous problem. The

proposed optimization algorithm is sensitive to the type and relative importance of the individual criteria.
An TSP specialized algorithm was used in a difficult discrete and combinatorial problem on polymer extruders design.

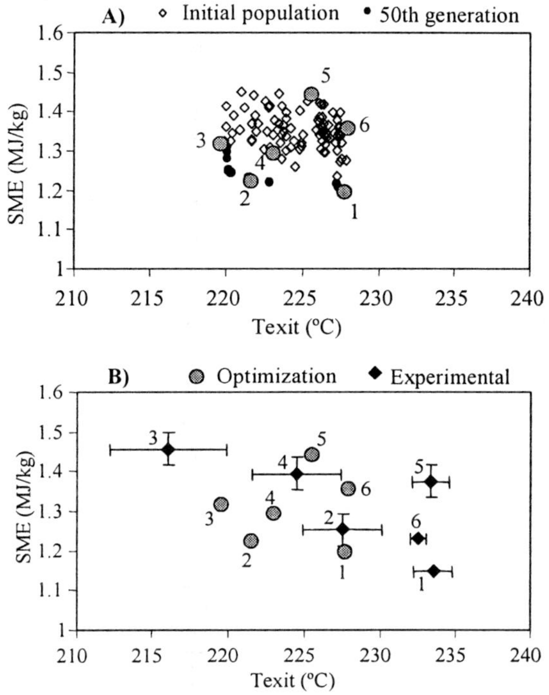

Fig. 5. Assessment of the optimization results: A) Pareto frontiers of the initial and final populations; B) Computational versus experimental results.

References

[1] White, J.L. (1990) Twin Screw Extrusion, Technology and Principles, Hanser, Munich

[2] White, J.L., Coran A.Y., Moet A. (2001) Polymer Mixing: Technology and Engineering, Hanser, Munich

[3] Potente, H., Ansahl, J., Wittemeier R. (1990) Throughput characteristics of Tightly Intermeshing Co-rotating Twin Screw Extruders. Intern. Polym. Proc., 5: 208-216

[4] Potente, H., Ansahl, J., Klarholtz, B. (1994) Design of Tightly Intermesing Co-Rotating Twin Screw Extruders. Intern. Polym. Proc., 9: 11-25

[5] Gaspar-Cunha, A., Poulesquen, A., Vergnes, B., Covas, J.A. (2002)Optimization of processing conditions for polymer twin-screw extrusion. Intern. Polym. Process., 17: 201-213

[6] Gaspar-Cunha, A., Vergnes, B., Covas, J.A., (2004) submitted to Polym. Eng. Sci.

[7] Baraglia, R., Hidalgo, J.I., Perego, R. (2001) A hybrid heuristic for the traveling salesman problem. IEEE Trans. On Evolut. Computation, 5: 613-622

[8] Padberg, M., Rinaldi, G. (1987) Optimization of a 532-city symmetric genetic salesman problem by branch and cut. Oper. Research Letters, 6: 1-7

[9] Johnson, D.S., MvGeoch, L.A. (1996) Local Search in Combinatorial Optimization, Wiley, New York

[10] Croes, G.A. (1958) A method for solving traveling salesman problems. Oper. Research, 6: 791-812

[11] Lin, S., Kernighan, B.W. (1973) An effective heuristic algorithm for the traveling salesman problem. Operational Research, 21: 498-516

[12] Lin, S., (1965) Computer solution of the traveling salesman problem. Bell Systems Technology Journal, 44: 2245-2269

[13] Applegate, D., Bixby, R., Chvátal, V., Cook, W. (1998) On the solution of traveling salesman problems. in Documenta Mathematica: Proc. Int. Cogr. Mathematicians, 3: 645-656

[14] Gaspar-Cunha, A., Covas, J.A, (2004) - RPSGAe - A Multiobjective Genetic Algorithm with Elitism: Application to Polymer Extrusion. In: Gandibleux, X.; Sevaux, M.; Sörensen, K.; T'kindt, V. (Eds.) Metaheuristics for Multiobjective Optimisation, Lecture Notes in Economics and Mathematical Systems, Springer, Wien, pp. 221-249

[15] Goldberg, D.E. (1989) Genetic Algorithms in Search, Optimization and Machine Learning, Addison-Wesley

[16] Tao, G., Michalewicz, Z. (1998) Inver-over operator for TSP. Proceedings of the 5th Parallel Problem Solving from Nature, Amsterdam

[17] Deb, K., (2001) Multi-Objective Optimization using Evolutionary Algorithms, Wiley

[18] Coello Coello, C.A., Van Veldhuizen, D.A., Lamont, G.B., (2002) Evolutionary Algorithms for Solving Multi-Objective Problems, Kluwer

[19] Roseman, M.A., Gero, J.S. (1985) Reducing the Pareto optimal set in multicriteria optimization. Eng. Optim. 8:189-206

[20] Vergnes, B., Della Valle, G., Delamare, L. (1998) A global computer software for polymer flows in corotating twin screw extruders. Polym. Eng. Sci., 38: 1781-1792

The PARETO-Box Problem for the Modelling of Evolutionary Multiobjective Optimization Algorithms

Mario Köppen, Raul Vicente-Garcia, Bertram Nickolay
Fraunhofer IPK, Pascalstr. 8-9, 10587 Berlin, Germany
{mario.koeppen|raul|nickolay}@ipk.fhg.de

Abstract

This paper presents the Pareto-Box problem for modelling evolutionary multi-objective search. The problem is to find the Pareto set of randomly selected points in the unit hypercube. While the Pareto set itself is only comprised of the point 0, this problem allows for a complete analysis of random search and demonstrates the fact that with increasing number of objectives, the probability of finding a dominated vector is decreasing exponentially. Since most nowadays evolutionary multi-objective optimization algorithms rely on the existence of dominated individuals, they show poor performance on this problem. However, the fuzzification of the Pareto-dominance is an example for an approach that does not need dominated individuals, thus it is able to solve the Pareto-Box problem even for a higher number of objectives.

1 Introduction

In multiobjective optimization, the optimization goal is given by more than one objective to be extreme. Formally, given a domain as subset of R^n, there are assigned m functions $f_1(x_1,\ldots,x_n),\ldots,f_m(x_1,\ldots,x_n)$. Usually, there is not a single optimum but rather the so-called PARETO set of *non-dominated* solutions.

Evolutionary Computation (EC) has been shown to be a powerful technique for multi-objective optimization [1][2][3] (EMO - Evolutionary Multi-Objective Optimization). This biologically inspired methodology offers both flexibility in goal specification and good performance in multimodal, nonlinear search spaces.

If we want to solve a highly complex multi-objective optimization problem, we might select one of the best ranked evolutionary approaches reviewed in the literature, like NSGA-II [4] or SPEA2 [5] and hopefully start reaching good results quickly. Sadly, that will be rarely the case when we face real-world problems with high number of objectives. Usually, the different algorithms are compared by measuring their performance indices in difficult test searches [1][2][3][6]. However, most of the benchmarks in the literature do not consider problems with high number of objectives. Moreover, very complex test functions should not be the only reference for the design of new approaches, as they prevent us from keeping track of the populations dynamics unambiguously (as already stated by Coello in [2]). In order to design a faster PARETO-dominance-based EC technique, we need an "easy" multi-objective test function that allows us to observe the search progress and that is yet easily scalable to higher number of objectives as well. The PARETO-Box Problem, which will be presented and studied in this paper, unifies these crucial properties. It will help us to know more about how the PARETO-front is searched for in EMO, and to measure the progress of the novel Fuzzy PARETO Dominance-Driven Genetic Algorithm (FDD-GA) approach in search problems with higher number of objectives. In the following section, we will introduce the PARETO-Box problem and its analysis for random search. These results will be used in an exemplary manner to study the dynamics of EMOs in section 3.

2 The Pareto-Box Problem

Given are m uniformly randomly selected n-dimensional points P_i in the n-dimensional unit hypercube ($1 \le i \le m$), with coordinates P_{ij} ($1 \le j \le n$). Thus, for each P_{ij} we have $0 \le P_{ij} \le 1$. The problem we state is:

PARETO-Box Problem: *What is the expectation value for the size of the* PARETO *set of these points?*

Here, we use the minimum version of PARETO dominance, so for two n-dimensional vectors $a = (a_i)$ and $b = (b_i)$ it is said that a dominates b (written as $a \prec b$) if and only if

$$\forall i:\ a_i \le b_i,\ \wedge\ \exists j:\ a_j < b_j\ \ (1 \le i,j \le n) \qquad (1)$$

For a set M of points, its PARETO set $P(M)$ is the subset for which none of its elements is dominated by any element of M. The PARETO set of the complete unit hypercube contains only one element, the point 0. The random sampling represents a random search in the unit hypercube, thus we are also going to answer the question if random search can find the PARETO set of the unit hypercube.

Obviously, the PARETO set of this problem is not hard to find, and there is also no conflict in the objectives. However, the following analysis will show that it is a hard problem for multi-objective optimization, once the dimension n of the problem is increased. Moreover, this problems allows for a precise analysis of the *progress* of algorithmic search, including the approach to the PARETO front and the entering of concave regions of the PARETO front.

In the following, $e_m(n)$ denotes the expectation value for the size of the PARETO set of m randomly selected points in the n-dimensional unit hypercube. Then, the following theorems hold:

Theorem 1. *Given are m randomly selected points in the n-dimensional hypercube. For the expectation value of the size of the* PARETO *set of these m points we have the recursive relation:*

$$\begin{aligned} e_1(n) &= 1 \\ e_m(1) &= 1 \\ e_m(n) &= e_{m-1}(n) + \frac{1}{m} e_m(n-1) \ (n, m \geq 2) \end{aligned} \tag{2}$$

Theorem 2. *The expectation value for the size of the* PARETO *set of $m \geq 1$ randomly selected points in the n-dimensional hypercube ($n \geq 1$) is*

$$e_m(n) = \sum_{k=1}^{m} \frac{(-1)^{k+1}}{k^{n-1}} \binom{m}{k} \tag{3}$$

Due to space limitations, the proofs of these theorems can not be given here.

Theorems 1 and 2 allow for the specification of the limiting behaviour of the expectation values for increasing number of points and increasing dimensions. This is stated in the following central theorem.

Theorem 3. *For fixed dimension $n > 1$ and the number of points $m \to \infty$, the expectation value $e_m(n) \to \infty$, the ratio of the non-dominated points $e_m(n)/m \to 0$ and for fixed $m > 1$ and dimension $n \to \infty$ it holds $e_m(n) \to m$.*

Proof. We see that

$$e_m(2) = \sum_{k=1}^{m} \frac{1}{k} = 1 + \frac{1}{2} + \frac{1}{3} + \frac{1}{4} + \ldots + \frac{1}{m} \tag{4}$$

which is the harmonic series and known to be divergent. Now, eq. (3) shows that for $n > 2$ always $e_m(n) \geq e_m(n-1) \geq \ldots \geq e_m(2)$, so for $m \to \infty$ $e_m(n) \to \infty$ as well. From the corresponding property of the harmonic series, $e_m(n)/m \to 0$ for $m \to \infty$ can be seen in a similar manner.

On the other hand, if $m > 1$ is fixed, all terms in eq. (3) but the one for $k = 1$ will go to 0 for $n \to \infty$, and the term for $k = 1$ itself computes to m. So, it is easy to see that $e_m(n) \to m$ for $n \to \infty$. □

We can express this result as follows: for increasing number of sample points in the hypercube, the number of non-dominated points will also increase, and never "shrink" to the PARETO set of the hypercube, which only contains the point 0. So, random search will not solve the problem to find the PARETO set of the hypercube in any dimension.

For increasing dimension, it will become more and more unlikely to find *any* dominated point in a population of random sampling points. In fact, the probability falls exponentially. The PARETO set of m points will contain nearly all m points.

We conclude this section by providing some special results:

$$\begin{aligned} (m,2): &\quad e_m(2) = \textstyle\sum_{k=1}^{m} \frac{1}{k} \\ (2,n): &\quad e_2(n) = 2 - \frac{1}{2^{n-1}} \\ (3,n): &\quad e_3(n) = 3 - \frac{3}{2^{n-1}} + \frac{1}{3^{n-1}} \\ (4,n): &\quad e_4(n) = 4 - \frac{6}{2^{n-1}} + \frac{4}{3^{n-1}} - \frac{1}{4^{n-1}} \end{aligned}$$

3 EMO Analysis

The remarkable point on the PARETO-Box problem is that it establishes the fact that the probability of finding dominated points in higher dimensions (i.e. increasing number of objectives) is falling exponentially with the dimension of the problem. Having a look on most prominent EMO algorithms like NSGA-II [5], SPEA2 [4] or PESA [7], it can be seen that they all need dominated points to perform their processing steps. For still yielding dominated points in the domain of higher number of objectives, these algorithms need an exponetially increasing *search effort*, be it by increasing the population size, or be it by increasing the number of generations.

Recently, the fuzzification of the PARETO dominance relation has been proposed [8], and a corresponding EMO has been presented. It will be shortly recalled here (see [9] for an alternative approach to introduce fuzzy logic in EMO). The underlying generic fuzzy ranking scheme for a set S of multivariate data (vectors) $\vec{a}_i$ with real-valued components a_{ij} and $1 \leq i \leq N$ is based on the provision of a *comparison function* $f_x(y) : R \times R \to [0,1]$ and a T-norm. Then, the following two steps are performed:

1. We compute the *comparison values* for any two vectors $\vec{a}_i = (a_{ik})$ and $\vec{a}_j = (a_{jk})$ by $c_{\vec{a}_i}(\vec{a}_j) = T(f_{a_{ik}}(a_{jk})\,|\,k = 1,\ldots,N)$ with N the number of components of each vector.

2. We compute the *ranking values* for any element $\vec{a}_i$ of S by $r_S(\vec{a}_i) = \max[c_{\vec{a}_i}(\vec{a}_j)|j \neq i]$.

Then, we consider vectors with lower numerical ranking values to be on a higher ranking position. For step 2, instead of max the min operator can be used as well, depending on the ranking to be favoured in increasing or decreasing order.

When using the comparison function bounded division and the algebraic (or product) norm as T-norm, the ranking scheme fulfills several useful properties like scale-independency in the data. The fuzzification of PARETO dominance relation can be written then as follows: It is said that vector $\vec{a}$ *dominates* vector $\vec{b}$ by degree μ_a with

$$\mu_a(\vec{a},\vec{b}) = \frac{\prod_i \min(a_i,b_i)}{\prod_i a_i} \tag{5}$$

and that vector $\vec{a}$ *is dominated by* vector $\vec{b}$ at degree μ_p with

$$\mu_p(\vec{a},\vec{b}) = \frac{\prod_i \min(a_i,b_i)}{\prod_i b_i} \tag{6}$$

For $\vec{a}$ PARETO-dominating $\vec{b}$, $\mu_a(\vec{a},\vec{b}) = 1$ and $\mu_p(\vec{b},\vec{a}) = 1$, but $\mu_p(\vec{a},\vec{b}) < 1$ and $\mu_a(\vec{b},\vec{a}) < 1$. Figure 1 gives a numerical example for the fuzzy PARETO dominance considered here.

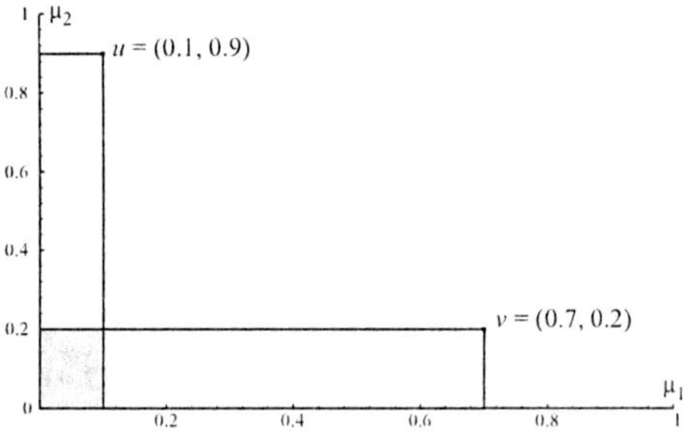

Fig. 1. Definition of Fuzzy-PARETO-Dominance. Here, u dominates v by degree $0.1 \cdot 0.2/0.1 \cdot 0.9 = 0.\bar{2}$ and is dominated by v by degree $0.1 \cdot 0.2/0.7 \cdot 0.2 \approx 0.143$.

The advantage of the FPD is that the problem of missing dominated points does not matter. This will be demonstrated by using the PARETO-Box problem. We also provide here a (Fuzzy-Dominance-Driven) FDD algorithm, a Genetic Algorithm (GA) variant that employs the FPD ordering of fitness values (represented as vectors in case of multiobjective optimization) for defining selection operators (see [10] for more details). As for nearly all newer EMOs, it also adds an *archive* for storing non-dominated individuals over the whole run of the algorithm.

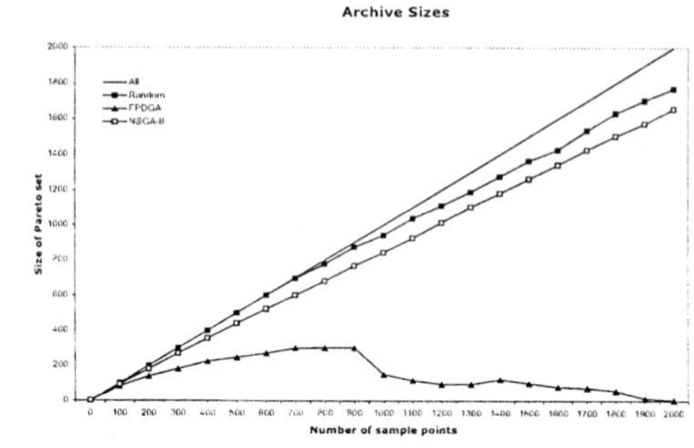

Fig. 2. Performances of NSGA-II and FDD-GA for the PARETO-Box problem.

Figure 2 compares the performance of a comparable set-up of NSGA-II and FDD-GA on the PARETO-Box problem for dimension $n = 20$. The NSGA-II implementation strictly followed [4]. For both algorithms, the population size was 10, and 200 generations were performed. Both algorithm used the same mutation probability and strength of 0.1. The selection scheme of FDD-GA was adapted due to having a known co-domain of the ranking values (aka fitness values). Tournament selection was performed using $\ln - r_i$ (with r_i the ranking values) as shared fitness values, and it was only selected among the non-dominated individuals. If all individuals got the same ranking values, it was randomly selected. The plot shows the size of the archive over the number of sample points (i.e. calls of the objective functions). Also given is the (numerically estimated) size of the PARETO-set for random sampling, and the total number of individuals (dominated and non-dominated). As established by Theorem 3, for random search the size of the PARETO set is close to the total number of points. However, also NSGA-II runs close to this curve, qualifying this search as more or less random as well. This is obvious from the fact that the probability of finding a dominated individual by applying randomized operators (mutation, crossover) is low.

For FDD-GA, we clearly see that even for dimension 20 it is able to find the single optimum of the PARETO-Box problem, and also stays strongly below the curve of random search all the time. To make this behaviour more clear, we considered the p.d.f. of the ranking values within a randomly created population (see fig. 3). This plot was obtained by 100 times creating a set of 20 random vector with 100 components from $[0, 1]$. Then, among these 20 vectors the ranking values r_i were computed, and the intervall frequencies for $-\ln r_i$ were derived. Thus, we can model the handling of randomly selected points by the FPD ranking scheme (as it happens when applied to the PARETO-Box problem). The distribution has a tail at the sider of smaller ranking values, so roulette-wheel selection will acknowledge the fact that such individuals gradually perform better (with respect to PARETO-dominance). Such a behaviour can not be achieved when an EMO is relying on the presence of dominated individuals alone.

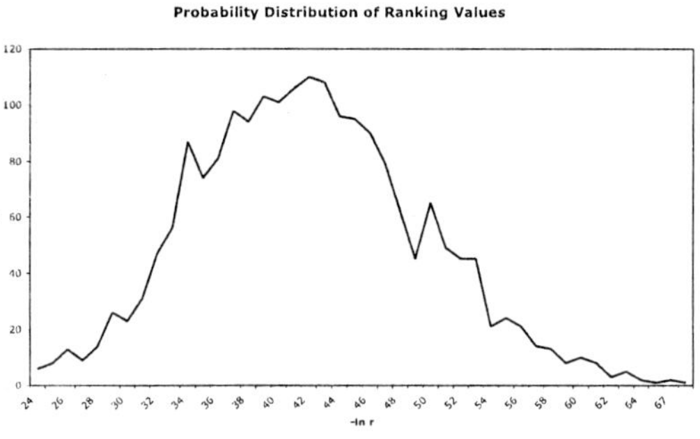

Fig. 3. Distribution of ranking values in FDD-GA algorithm.

It has to be noted (but can not be detailed here) that nevertheless NSGA-II, in this set-up, is also finding the optimum up to a problem dimension of 8. In low dimensions (2-3) the FDD-GA is also outperformed by NSGA-II.

Acknowledgment

The research presented in this paper was supported by German research project "VisionIC," funded by German Ministry of Research and Education (BMBF) as handled by "Deutsche Gesellschaft für Luft- und Raumfahrt" (DLR). The authors also would like to express their thanks to Frank Hoffmann from University of Dortmund and Anna Ukovich from University of Trieste for their valuable comments and inspiring discussions about the topics of this work.

References

[1] Carlos A. Coello Coello, "A short tutorial on evolutionary multiobjective optimization," in *First International Conference on Evolutionary Multi-Criterion Optimization*, Eckart Zitzler, Kalyanmoy Deb, Lothar Thiele, Carlos A. Coello Coello, and David Corne, Eds., vol. 1993 of *Lecture Notes in Computer Science*, pp. 21–40. Springer Verlag, 2001.

[2] Carlos A. Coello Coello, "An updated survey of GA-based multiobjective optimization techniques," Technical Report RD-98-08, Laboratorio Nacional de Informática Avanzada (LANIA), Xalapa, Veracruz, México, December 1998.

[3] Carlos M. Fonseca and Peter J. Fleming, "An overview of evolutionary algorithms in multiobjective optimization," *Evolutionary Computation*, vol. 3(1), no. 1, pp. 1–16, 1995.

[4] Kalayanmoy Deb, Samir Agrawal, Amrit Pratap, and T. Meyarivan, "A fast elitist non-dominated sorting genetic algorithm for multi-objective optimization: NSGA-II," in *Proceedings of the Parallel Problem Solving from Nature VI Conference*, Marc Schoenauer, Kalyanmoy Deb, Günter Rudolph, Xin Yao, Evelyne Lutton, J. J. Merelo, and Hans-Paul Schwefel, Eds., Paris, France, 2000, vol. 1917 of *Lecture Notes in Computer Science*, pp. 849–858, Springer.

[5] Eckart Zitzler, Marco Laumanns, and Lothar Thiele, "SPEA2: Improving the strength pareto evolutionary algorithm," Tech. Rep. 103, Computer Engineering and Networks Laboratory (TIK), Swiss Federal Institute of Technology, Zurich, ETH Zentrum, Gloriastr 35, CH-8092 Zurich, Switzerland, September 2001.

[6] Tatsuya Okabe, Yaouchu Jin, and Bernhard Sendhoff, "A critical survey of performance indices for multi-objective optimization," in *Proceedings of the 2003 Congress on Evolutionary Computation (CEC'2003)*, Canberra, Australia, December 2003, pp. 878–885, IEEE Press.

[7] David W. Corne, Joshua D. Knowles, and Martin J. Oates, "The pareto envelope-based selection algorithm for multiobjective optimization," in *Proceedings of the Parallel Problem Solving from Nature VI Conference*. 2000, vol. 1917 of *Lecture Notes in Computer Science*, pp. 839–848, Springer.

[8] Mario Köppen and Raul Vicente Garcia, "A fuzzy scheme for the ranking of multivariate data and its application," in *Proceedings of the 2004 Annual Meeting of the NAFIPS (CD-ROM)*, Banff, Alberta, Canada, 2004, NAFIPS, pp. 140–145.

[9] Marco Farina and Paolo Amato, "Fuzzy optimality and evolutionary multiobjective optimization," in *EMO 2003, LNCS 2632*, C.M. Fonseca et al., Ed., 2003, pp. 58–72.

[10] Mario Köppen, Katrin Franke, and Bertram Nickolay, "Fuzzy-pareto-dominance driven multiobjective genetic algorithm," in *Proceedings of the 10th IFSA World Congress*, 2003, pp. 450–453.

Implementation and Experimental Validation of the Population Learning Algorithm Applied to Solving QAP Instances

J. Jedrzejowicz[1], P. Jedrzejowicz[2]

[1]Institute of Mathematics, University of Gdansk, Poland
[2] Department of Information Systems, Gdynia Maritime University, Poland
E-mail: jj@math.univ.gda.pl , pj@am.gdynia.pl

Abstract

The paper proposes an implementation of the population learning algorithm designed to solve instances of the quadratic assignment problem. A short overview of the population-learning algorithm and a more detailed presentation of the proposed implementation is followed by the results of computational experiments carried. Particular attention is given to investigating performance characteristics and convergence of the PLA. Experiments have focused on identification of the probability distribution of solution time to a sub-optimal target value.

1 Introduction

The techniques used to solve difficult optimization problems have evolved from constructive algorithms to local search techniques, and finally to population-based meta-heuristics. Population based methods are optimization techniques inspired by natural evolution processes. Among best known population-based methods are evolutionary algorithms [1]. Despite the fact that evolutionary algorithms, in general, lack a strong theoretical background, the application results were more than encouraging. This success has led to emergence of numerous techniques, algorithms and their clones. Among well known population-based methods are: scatter search technique, ant colony algorithm, particle swarm optimization algorithm, cultural algorithm, memetic algorithm and greedy randomized adaptive search procedure.

Population learning algorithm (PLA) was proposed in ref. 2 as yet another population-based method, which can be applied to support solving difficult decision-making and optimization problems. Thus far PLA has been successfully applied to solving a variety of difficult scheduling problems [3], [4]. The algorithm proved also successful in the domain of ANN training [5], [6].

In this paper an implementation of the PLA designed to solve instances of the quadratic assignment problem (QAP) is studied and validated. The goal of this research was to investigate the PLA performance while dealing with one of the most difficult, widely researched and best-known combinatorial optimization problems. Section 2 of the paper offers a short overview of the population-learning algorithm. Section 3 gives details of the proposed implementation. In section 4 the results of computational experiments carried with a view to validate the implementation are discussed. Particular attention is paid to investigating performance characteristics and convergence of the PLA. Experiments have also focused on identification of the probability distribution of solution time to a sub-optimal target value.

2 Population-Learning Algorithm and Its Implementation

In the PLA an individual represents a coded solution of the considered problem. Initially, a number of individuals, known as the initial population, is generated. Once the initial population has been generated, individuals enter the first learning stage. It involves applying some, possibly basic and elementary, improvement schemes. These can be based, for example, on some simple local search procedures. The improved individuals are then evaluated and better ones pass to a subsequent stage. A strategy of selecting better or more promising individuals must be defined and applied. In the following stages the whole cycle is repeated. Individuals are subject to improvement and learning, either individually or through information exchange, and the selected ones are again promoted to a higher stage with the remaining ones dropped-out from the process. At the final stage the remaining individuals are reviewed and the best represents a solution to the problem at hand.

The quadratic assignment problem (QAP) is formulated as follows. Let $A = (a_{ij})_{n\times n}$ and $B = (b_{ij})_{n\times n}$ be two integer matrices for an integer value n. Find a permutation $x = (x(1),\ldots, x(n))$ that minimizes:

$$g(x) = \sum_{i=1}^{n}\sum_{j=1}^{n} a_{ij} b_{x(i)x(j)}$$

Since it was proved that QAP is NP-hard, heuristic approaches have been widely used.

The population learning algorithm applied to solving QAP makes use of several learning and improvement procedures, which, in turn, are based on four neighbourhood structures shown in Table 1. In what follows x denotes an individual encoded as a permutation of natural numbers representing an assignment, and $g(x)$ its fitness function.

All learning and improvement procedures denoted *LEARN* (i,P) operate on the population of individuals P, and perform local search algorithms shown in Table 2. A schema for the discussed procedure is shown below:

LEARN (i, P):
for each individual x in P **do**
    Local_ search (i,x)
end for

Table **1** Neighbourhood structures

Notation	Move	Neighbourhood space
$\mathcal{N}_1(x)$	Exchange of two consecutive elements in x	All possible exchanges
$\mathcal{N}_2(x)$	Exchange of two non-consecutive elements in x	All possible exchanges
$\mathcal{N}_3(x)$	Finding order of four consecutive elements in x by enumeration	All fourtuples of consecutive facilities
$\mathcal{N}_4(x)$	A single step rotation of three random elements in x	All possible triples

Table **2** Local search algorithms

i	Idea of the local search algorithm
1	Perform all moves from the neighborhood structure $\mathcal{N}_1(x)$; accept moves improving $g(x)$; stop when no further improvements of $g(x)$ are possible
2	Mutate x producing x' {mutation procedure is selected randomly from the two available ones – the two point random exchange or the rotation of all chromosomes between two random points}; perform all moves from the neighborhood structure $\mathcal{N}_1(x')$; accept moves improving $g(x')$; stop when no further improvements of $g(x')$ are possible
3	Repeat k' times {k' is a parameter set at the fine-tuning phase; in the reported experiment k' = 3 * initial population size}; generate offspring y and y' by a single point crossover of x and a random individual x'; perform all moves from the neighborhood structure $\mathcal{N}_1(y)$ and $\mathcal{N}_1(y')$; accept moves improving $g(y)$ and $g(y')$; stop when no further improvements of $g(y)$ and $g(y')$ are possible; adjust P by replacing x and x' with the two best individuals from $\{x, x', y, y'\}$
4	Perform all moves from the neighborhood structure $\mathcal{N}_2(x)$; accept moves improving $g(x)$; stop when no further improvements of $g(x)$ are possible
5	Perform all moves from the neighborhood structure $\mathcal{N}_3(x)$; accept moves improving $g(x)$; stop when no further improvements of $g(x)$ are possible
6	Perform *SIMULATED ANNEALING* (x) based on $\mathcal{N}_4(x)$ neighbourhood structure
7	Perform *TABU SEARCH* (x) based on $\mathcal{N}_2(x)$ neighbourhood structure

Simulated annealing is a metaheuristic introduced in [7]. Implementation of simulated annealing used within PLA as the 6th local search algorithm is based on random moves from the $\mathcal{N}_4(x)$ neighbourhood structure. If a better individual in $\mathcal{N}_4(x)$ is found, then it is chosen. Otherwise it is only accepted with some probability which decreases in the course of computation.

Tabu search is yet another metaheuristic proposed by Glover [8]. Here the robust tabu search algorithm, due to Taillard is used [9]. For a permutation x the following procedure is iterated. All neighbors from the neighborhood $\mathcal{N}_2(x)$ which are not tabu are considered and the one which minimizes g best, is chosen. The tabu criterion is dropped when a move is found which is better than the best found so far (so called aspiration criterion). Besides, an additional intensification mechanism is used at most once in r iterations (r is a parameter). That is, a local search around an element which improves the solution is performed.

The procedure *SELECT(P)* of selecting individuals from the population P for the consecutive stages of the PLA algorithm is based on the following criterion. For the population P, let var(P) stand for the variance, that is:

$$\text{var(P)} = \frac{\sqrt{\sum_{x\in P}(g(x)-m)^2}}{|P|-1}\;;\text{ where } m = \frac{\sum_{x\in P} g(x)}{|P|}$$

is the arithmetic mean. If, in the two consecutive stages, the variance increases from var(P) to var(P'), then all the individuals are promoted to the higher stage. Otherwise, only (1- var(P')/var(P)) ·100% best individuals from P' are selected. Now, PLA has the following structure:

begin
    generate randomly *initial_population*
    P := *initial_population*
    for i = 1 to 5 **do**
        LEARN (i, P)
    end for
    for i = 6,7 **do**
        SELECT (P)
        LEARN (i, P)
    end for
    output the best individual from P
end

3 Experimental Evaluation

To evaluate the proposed implementation of the population-learning algorithm several experiments have been carried out. All the reported experiments are based on the dataset of QAP instances contained in the OR-

LIBRARY. The dataset includes 136 instances with the respective solutions. Some of the solutions are known to be optimal. The remaining are the best known solutions (upper bounds) that have been obtained by numerous research teams during recent years. To assess the performance of the PLA all 136 instances have been solved, four times each, with four different initial population sizes. The results in terms of the mean relative error calculated over for the whole population of instances are shown in Table 3.

Table **3** Performance of the PLA with different initial population sizes

Population size	500	1000	1500	2000
MRE	0,2046%	0,0991%	0,0515%	0,0393%

To gain better understanding of the PLA performance the whole population of instances has been partitioned into 3 classes: small ($n < 20$), medium ($20 \leq n < 50$) and large ($n \geq 50$) with, respectively, 39, 57 and 40 instances. Estimated probabilities of finding a solution of the QAP instance in a single PLA run with the relative error equal to 0% are shown in Figure 1.

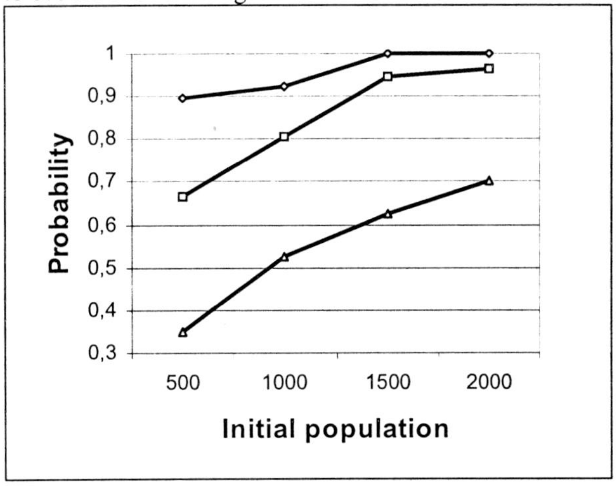

Figure 1 Estimated probabilities of finding a solution with the relative error = 0% for small, medium and large instances of the QAP problem (shown from top to down, respectively).

In order to check whether it is possible to achieve a speed-up by implementing QAP in parallel, the study of the probability distributions of the solution time to a sub-optimal target value has been carried out. Throughout the study the methodology suggested by Aiex et al. has been used [10].

The following experiments have been performed. For each of the QAP instance two different target values were defined. For each instance/target combination the following algorithm was run :

repeat
  generate randomly *initial_population*
  P := *initial_population*
  for *i* = 1 to 5 **do**
    LEARN (*i*, *P*)
  end for
  for *i* = 6,7 **do**
    SELECT (*P*)
    LEARN (*i*, *P*)
  end for
  xbest:= the best individual from P;
  Gbest:=g(x);
until Gbest <= target_value;

It was repeated $m = 200$ times and for each run the CPU time needed to achieve the target value was measured. Then for each instance/target pair the running times were sorted in increasing order. With the i-th sorted running time t_i, a probability $p_i = (i - 0.5)/m$ is associated. It indicates that $100p_i$% of data are below, or equal to the value t_i. In the next step the estimation of the parameters of the exponential distribution was performed. A quantile-quantile plot (or, *Q-Q* plot), was used. A theoretical *Q-Q* plot was obtained by plotting the quantiles of the empirical distribution against the theoretical quantiles. It allows to view whether there is a shift in going from one distribution to another.

The cumulative distribution function for the two-parameter exponential distribution is given by:

$$F(t)=1-\exp(-(t-\mu)/\lambda)$$

where λ is the mean of the distribution data and μ is the shift of the distribution. For each p_i, a p_i – quantile $Qt(p_i)$ is defined. Since

$$F(Qt(p_i)) = p_i,$$

thus

$$Qt(p_i) = F^{-1}(p_i), \text{ and}$$

$$Qt(p_i)) = -\lambda \ln(1 - p_i) + \mu.$$

And the empirical distribution is defined by (p_i, $Qt(p_i)$), for $i = 1,\dots, m$. The calculation of the estimated values of λ and μ is based on the following observation. If the theoretical distribution is a close approximation of the empirical distribution, then the points in the *Q-Q* plot follow the line $y = x$. Alternatively, in a plot of data against an exponential distribution with $\lambda = 1$ and $\mu = 0$ the points would tend to follow the line $y = \lambda x + \mu$. Therefore, the parameters λ and μ of theoretical distribution can be estimated from the line obtained in the *Q-Q* plot. Figure 2 presents example *Q-Q* plot with points ($-\ln(1 - p_i)$, t_i), that is the measured times (in the ordinate) against the quantiles of a two-parameter exponential distribution with $\lambda = 1$ and $\mu = 0$ (in the abscissa). To calculate λ and μ of the line $y = \lambda x + \mu$ which fits the *Q-Q* plot best, the upper quartile and lower quartiles are used. That is:

$$\lambda = \frac{z_u - z_l}{q_u - q_l}.$$

Since the number of measured times is m = 200, thus the upper quartile corresponds to the 150th value and the lower one to the 50th

$z_u = t_{150}$, $z_l = t_{50}$ and $q_u = -\ln(1 - p_{150})$, $q_l = -\ln(1 - p_{50})$

The parameter μ is calculated from the equation $y = \lambda x + \mu$, that is $\mu = z_u - \lambda q_u$.

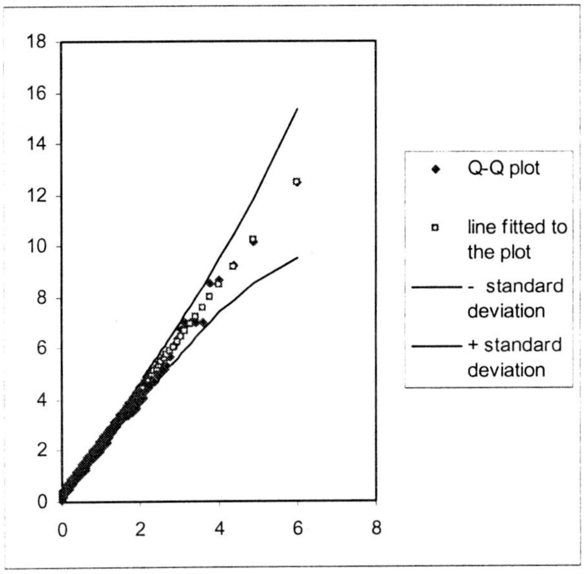

Figure 2 Q-Q plot with the variability information for example instances – tai17a with target 500000.

Subsequently, by superimposing empirical and theoretical distributions, it has been shown that distributing the PLA iterations evenly among parallel processors achieves linear speedup for total time needed to run all PLA iterations. This observation is derived from the fact that if time to a target solution value fits well a two-parameter exponential distribution, then the probability of finding a target solution in time ρt with a sequential process is equal to the probability of finding a solution at least as good as that target value in time t with ρ independent parallel processes.

4 Conclusion

Experiments have shown that the proposed implementation of the population-learning algorithm converges and performs well with respect to solving instances of the quadratic assignment problem known to be very hard from the computational point of view. It has been also shown that distributing the PLA iterations evenly among parallel processors achieves linear speedup for total time needed to run all PLA iterations. Showing that time to a target solution value, when using PLA, fits well a two-parameter exponential distribution may help in designing efficient population learning algorithms run on parallel machines.

References

[1] Michalewicz, Z. (1996) Genetic Algorithms + Data Structures = Evolution Programs, 3rd ed., Springer-Verlag, Berlin, Heidelberg, New York

[2] Jedrzejowicz, P., (1999) Social Learning Algorithm as a Tool for Solving Some Difficult Scheduling Problems. Foundation of Computing and Decision Sciences 24: 51-66

[3] Jedrzejowicz J., Jedrzejowicz P. (2003) PLA-based Permutation Scheduling. Foundation of Computing and Decision Sciences 28(3): 159 –177

[4] Jedrzejowicz J., Jedrzejowicz P. (2003) Population-based Approach to Multiprocessor Task Scheduling in Multistage Hybrid Flowshops. In: Palade V. & Howlett, R.J. & Jain, L. (eds.) Knowledge-Based Intelligent Information and Engineering Systems, Lecture Notes in Artificial Intelligence 2003, Springer, Berlin, pp. 279 – 286

[5] Czarnowski, I., Jedrzejowicz, P. (2002) Application of the Parallel Population Learning Algorithm to Training Feed-forward ANN. In: P.Sincak (ed.) Intelligent Technologies – Theory and Applications 2002, IOS Press, Amsterdam, pp. 10 – 16

[6] Czarnowski, I., Jedrzejowicz, P. (2002) An Approach to Artificial Neural Network Training. In: Bremer, M. & Preece, A. & Coenen, F. (eds.) Research and Development in Intelligent Systems XIX 2002, Springer, London, pp. 149-160

[7] Kirkpatrick, S., Gelatt Jr., C. D., Vecchi, M. P. (1983) Optimization by Simulated Annealing. Science 220: 671-680

[8] Glover, F. (1977) Heuristics for Integer Programming Using Surrogate Constraints. Decision Sciences 1(8): 156-166

[9] Taillard E. (1991) Robust taboo search for the QAP. Parallel Computing 17: 443-455

[10] Aiex, R. M., Resende, M. G. C., Ribeiro, C.C. (2000) Probability Distribution of Solution Time in GRASP: an Experimental Investigation, AT&T Labs Research Technical Report (00.7.1)

Estimating the distribution in an EDA

S. Shakya, J. McCall, D. F. Brown
School of Computing, The Robert Gordon University, Aberdeen, UK
E-mail: {ss, jm, db}@comp.rgu.ac.uk

Abstract

This paper presents an extension to our work on estimating the probability distribution by using a Markov Random Field (MRF) model in an Estimation of Distribution Algorithm (EDA) [1]. We propose a method that directly samples a MRF model to generate new population. We also present a new EDA, called the Distribution Estimation Using MRF with direct sampling (DEUM$_d$), that uses this method, and iteratively refines the probability distribution to generate better solutions. Our experiments show that the direct sampling of a MRF model as estimation of distribution provides a significant advantage over other techniques on problems where a univariate EDA is typically used.

1 Introduction

Estimation of Distribution Algorithms (EDAs) [2][3], also known as Probabilistic Model Building Genetic Algorithms (PMBGAs) [4], are a well-established topic in the field of evolutionary algorithms. EDAs are motivated by the idea of identifying important patterns or building blocks [4] from the population of promising solutions. A model of the probability distribution is used to preserve those patterns and is explicitly sampled to generate a child population. EDAs are classified as univariate, bivariate or multivariate according to the type of interaction between allele values that is allowed in the model of the probability distribution (see [3][4]).

An EDA regards a solution (chromosome) as a set of random variables (the alleles), each taking a particular value from a set of possible values. In particular, we represent a solution as $x = \{x_1, x_2, \ldots, x_n\}$ where each x_i is the value taken by the i-th random variable. Univariate EDAs do not consider dependencies between variables, i.e., they only model building blocks of order one. In this case, the joint probability distribution, $p(x)$, is simply the product of the univariate marginal probabilities of all variables in a solution x:

$$p(x) = \prod_{i=1}^{n} p(x_i) \tag{1}$$

Here, $p(x_i)$ is the marginal probability of the i-th variable having the value x_i. Population Based Incremental Learning (PBIL), the Univariate Marginal Distribution Algorithm (UMDA), and the Compact Genetic Algorithm (cGA) all use a univariate model of the probability distribution (see [3][4][5]).

In our recent work [1], we propose a different model of probability distribution for EDAs known as Markov Random Field (MRF) model [6]. As in PBIL, the algorithm proposed there, known as *Distribution Estimation Using MRFs* (DEUM), maintains a probability vector, however, uses a univariate MRF model to update it [1].

In this paper, we refine our use of MRF models for the estimation of distribution. We describe an updated version of DEUM called the *Distribution Estimation Using MRFs with direct sampling* (DEUM$_d$). DEUM$_d$ does not maintain a probability vector. Instead, it directly samples the MRF model to generate new population. The workflow of DEUM$_d$ is more similar to that of UMDA than PBIL. In UMDA, the marginal frequencies are directly sampled to generate successive populations. In DEUM$_d$, we replace these marginal frequencies with a MRF model that is also built from a selected subset of the population. This MRF model gives a maximum likelihood estimation of the optimal solution based on the selected set, and it is sampled to generate a successive population. The result of this, as we will show, is a significant improvement in learning on well-known univariate EDA problems.

The rest of the paper is constructed as follows. Section 2 introduces our univariate MRF model, and shows how we determine the model from a population. Section 3 describes the operation of DEUM$_d$ in detail, and Section 4 gives the results of several experiments that compare DEUM$_d$ with other univariate EDAs. Finally, Section 5 summarises and outlines further work.

2 A univariate MRF model of fitness

In [7], MRF theory was used to provide a formulation of the joint probability distribution that relates solution fitness to an energy function calculated from the values of the solution variables. To be precise:

$$p(x) = \frac{f(x)}{\sum_y f(y)} = \frac{e^{-U(x)/T}}{\sum_y e^{-U(y)/T}} \tag{2}$$

from which we can derive an equation for each solution x (see [7]):

$$-\ln(f(x)) = U(x) \tag{3}$$

Here, $f(x)$ is the fitness of an individual. $U(x)$ is an energy function derived from allele values and, T is a *temperature coefficient*. The summations are over all possible solutions y. $U(x)$ gives the full specification of the joint probability distribution, so it can be regarded as a probabilistic model of the fitness function. In particular, minimising $U(x)$ is equivalent to maximising $f(x)$. In general, the form of the energy function will involve interactions between the variables x_i. In DEUM$_d$, however, we use a simple form that assumes no such interactions. Instead each variable provides a contribution $\alpha_i x_i$ to the overall energy. From the above, we can derive an equation for each solution:

$$-\ln(f(x)) = \alpha_1 x_1 + \alpha_2 x_2 + \ldots + \alpha_n x_n \tag{4}$$

We refer to this as the *univariate MRF model*. The real-valued α_i are called the MRF parameters, and completely determine the probability distribution.

Each solution in a given population provides an equation satisfying the model. Selecting N promising solutions from a population therefore allows us to estimate the distribution by solving the system of equations:

$$A\alpha^T = F \tag{5}$$

Here, A is the $N \times n$-dimensional matrix of allele values in the selected set, α is the vector of MRF parameters $\alpha = (\alpha_1, \alpha_2, \ldots, \alpha_n)$, and F is the N-dimensional vector containing $-\ln(f(x))$ of the selected set of solutions x. Solving this system of linear equations, we get the set of MRF parameters α. Depending on the relationship between N and n, the system will be under-, over-, or precisely-specified. A standard fitting algorithm can be used to give a maximum likelihood estimation of the α_i. The α_i can then be used to provide a direct estimate of the probability of the value of x_i. For mathematical reasons, we use $\{-1, 1\}$ as the values of x_i in our model, rather than $\{0, 1\}$. This ensures arithmetical symmetry between the possible allele values. However, the following analysis generalises to any choice of distinct values.

Fixing the value of a particular allele divides the set Ω of all chromosomes into two disjoint sets, which we denote by A and B. More precisely, $A = \{x \in \Omega : x_i = 1\}$ and $B = \{x \in \Omega : x_i = -1\}$. We denote the probability that the allele value in position i is equal to 1 by $p(x_i = 1)$. Clearly, the probability that the allele value in position i is equal to -1 is $1 - p(x_i = 1)$. From (2), we obtain:

$$p(x_i = 1) = \sum_{x \in A} p(x) = \sum_{x \in A} \frac{e^{-U(x)/T}}{Z} \tag{6}$$

Here, $Z = \sum_y e^{-U(y)/T}$ is a (very large) normalising constant. Substituting for $U(x)$ from (4), and noting that $x_i = 1$ for all $x \in A$, we obtain:

$$p(x_i = 1) = e^{-\alpha_i/T}\frac{K}{Z} \tag{7}$$

where K is a large constant representing the sum over all chromosomes in A of contributions from alleles in positions other than i.

Similarly, summing over B we obtain the probability that the allele value in position i is equal to -1:

$$p(x_i = -1) = 1 - p(x_i = 1) = e^{\alpha_i/T}\frac{K}{Z} \tag{8}$$

Here, K is the same constant as in (7), because the chromosomes in A and B agree pairwise at allele positions other than i. Combining (7) and (8), the constants K and Z drop out, and we get the following expression as an estimate of the marginal probability for $x_i = 1$:

$$p(x_i = 1) = \frac{1}{1 + e^{\beta\alpha_i}} \tag{9}$$

where, $\beta = 2/T$.

Note that, as $T \to 0$, the value of β increases, and the value of $p(x_i = 1)$ tends to an extreme depending on the sign of α_i. If $\alpha_i > 0$, then $p(x_i = 1) \to 0$ as $T \to 0$. Conversely, if $\alpha_i < 0$, then $p(x_i = 1) \to 1$ as $T \to 0$. If $\alpha_i = 0$, then $p(x_i = 1) = 0.5$ regardless of the value of T. Therefore, the α_i are indicators of whether the allele value at the position i should be 1 or -1. This indication becomes stronger as the temperature is cooled towards zero.

This forms the basis for our estimation of distribution technique, which combines the univariate MRF model with a cooling scheme. We reduce T, i.e., increase β, as the population evolves, so the model becomes more exploitative rather than explorative as the evolution progresses.

3 DEUM$_d$: Distribution Estimation Using MRFs with direct sampling

The five step procedure of the algorithm for DEUM$_d$ are as follows:

1. Generate an initial population, P, of size M with uniform distribution.
2. Select the N fittest solutions from P, where $N \le M$.
3. Calculate the MRF parameters $\alpha = (\alpha_1, \alpha_2, \ldots, \alpha_n)$ by applying the univariate MRF model to the selected solutions and solving the system of linear equations.
4. Generate M new solutions using the following distribution:

$$p(x) = \prod_{i=1}^{n} p(x_i)$$

where, $p(x_i = 1) = \frac{1}{1+e^{\beta \alpha_i}}$ and $p(x_i = -1) = \frac{1}{1+e^{-\beta \alpha_i}}$. Here, β is defined as $\beta = g\tau$ where, g is the number of the current iteration and $\tau > 0$ is a *cooling rate* parameter chosen by the user.

5. Replace P by the new population, and go to Step 2 until the termination criterion is satisfied.

DEUM$_d$ uses the singular value decomposition (SVD) [8] technique to solve the system of linear equations. SVD proves to be the most stable technique, and can solve (in the sense of giving a useful numerical answer) systems of linear equations that are either under- or over-specified [8].

As described in Section 2, β has a direct effect on the convergence speed of DEUM$_d$. As the number of iterations (g) grows, the marginal probability ($p(x_i)$) gradually cools down to either 0 or 1. However, depending upon the type of problem, different cooling rate may be required. In particular, there is a trade-off between convergence speed of the algorithm and the exploration of the search space. Therefore, the cooling rate parameter, τ, has been introduced. τ gives the user explicit control over the convergence speed of DEUM$_d$. Decreasing τ slows the cooling, resulting in better exploration of the search space. However, it also slows the convergence of the algorithm. Increasing τ, on the other hand, makes the algorithm converge faster. However, the exploration of the search space will be reduced.

4 Experiments

In this section, we compare DEUM$_d$ with other univariate EDAs including DEUM and a GA, on two different problems. In order to compare best with best, we empirically determined the parameters for DEUM$_d$. For the rest of the algorithms, we used parameter settings from the literature or empirically determined parameters, depending on which proved best for particular problems. The performance of each algorithm was measured in terms of the number of fitness evaluation taken to find the optimal solution.

4.1 Onemax Problem

The Onemax problem [2] is a simple linear problem decomposable into building blocks of order one, and therefore is an ideal problem for univariate EDAs. It has been shown that UMDA works very well on this problem [2]. We compare the performance of DEUM$_d$ against a simple GA with uniform crossover (GA (uniform)), UMDA, and DEUM. 100 runs of each algorithm were executed for a series of Onemax problems with chromosomes ranging in size between 30 and 180. The number of fitness evaluations taken to find the optimal solution was recorded for each run. Uniform crossover with exchange probability of 0.5 was used for GA (uniform), crossover was applied all the time and mutation was not applied. Population size M ranged from 40 to 100 for GA (uniform), 50 to 170 UMDA, $1.5n$ for DEUM and was fixed at 40 for DEUM$_d$. λ for DEUM was from 0.5 to 0.6 and τ for DEUM$_d$ was from 5 to 4.

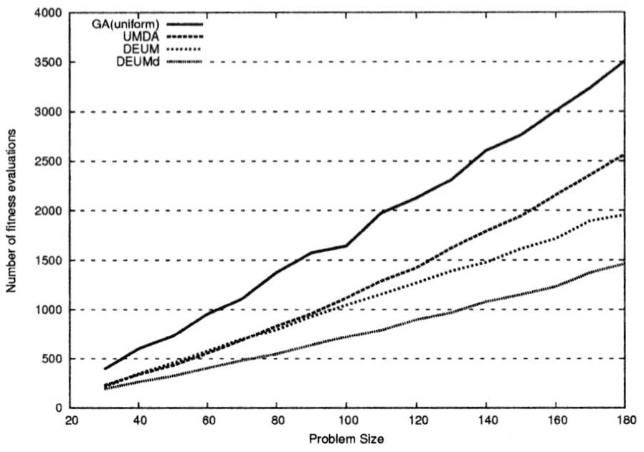

Fig. 1. Average number of fitness evaluations for 30 to 180 sized onemax problem where the population size was 40 to 100 for GA (uniform), 50 to 170 for UMDA, $1.5n$ for DEUM and 40 for DEUM$_d$.

Truncation selection was used where selection size N was $0.5M$ for GA (uniform), $0.3M$ for UMDA, $0.85M$ for DEUM and a fixed size of 10 was used for DEUM$_d$. No elitism was used and new populations were generated with complete replacement. Fig. 1 shows the average number of fitness evaluations for each algorithm over the range of onemax problems.

The success ratio of converging to the optimum was 93.5% for DEUM, 96% for DEUM$_d$, 98% for UMDA and 100% for GA (uniform). As we can see from Fig. 1, UMDA has an expected performance, better than that of GA (uniform) [2]. However, DEUM$_d$ performs better than all of the other algorithms, for all problem instances independent of their size.

4.2 Schaffer f6 function optimization

The Schaffer f6 function, described in [9] has been frequently used to evaluate the performance of GAs. An interesting feature of this function is that it has many local optima, but a single global optimal solution. So a hill-climbing algorithm will rapidly become trapped in one of the local optima. A simplified version of it is presented below:

$$f(x) = 1 + \left(\frac{\cos(x)}{1 + 0.001x^2} \right)$$

where $-300 \leq x \geq 300$.

The optimal solution is $f(x) = 2$ when $x = 0$. We performed experiments with a 20-bit Gray code representation of the f6 function.

Each algorithm with fixed parameter settings was run for total of 1000 runs. For each run the number of evaluations taken to find the optimum was recorded. For GA(uniform), the population size was 300, and truncation selection with a selection size $N = 0.5M$ was used. Crossover was applied all the time, mutation was set to 0.01 and 50% elitism was used. For PBIL, DEUM and DEUM$_d$, the population size was 500 and the selection size, N was 2. The learning rate, λ, for both PBIL and DEUM was 0.1 and the cooling rate, τ, for DEUM$_d$ was 1.5. Mutation shift was not applied in PBIL.

The experimental results are shown in Fig. 2, where the Run Length Distribution (RLD) [10] is plotted for each of the compared algorithms. We can see that, with DEUM$_d$, 80% of runs found the optimum within 4500 function evaluations, compared with 9000 function evaluations for PBIL. The success rate for finding a solution for DEUM$_d$ was 92% compared to 94% of DEUM, however the number of function evaluations needed to find the solution for DEUM$_d$ was significantly less than that of other algorithms.

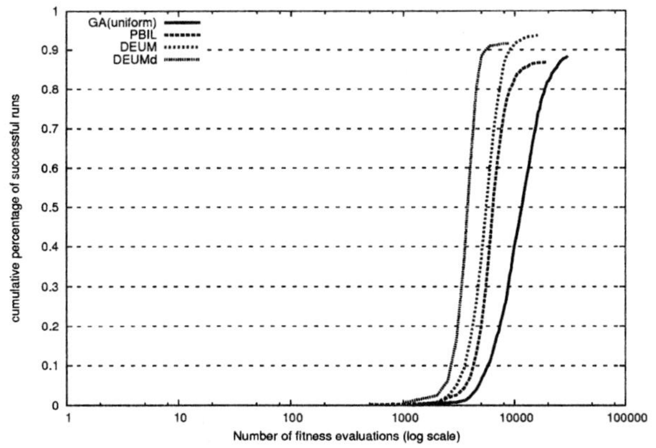

Fig. 2. Experimental results in the form of RLD showing, for each algorithm running on 20-bit Schaffer f6 function, the cumulative percentage of successful runs that terminated within a certain number of function evaluations.

5 Conclusions

In this paper, we have presented DEUM$_d$ as a novel EDA, which uses MRF modelling of fitness to estimate the probability distribution. The motivation behind DEUM$_d$ is to use a direct sampling of a MRF model to generate a new population in order to improve evolution. Our experiments shows that, for univariate problems, the use of MRF parameters instead of marginal probabilities does provide a better estimation of the distribution. This leads to better performance in terms of the number of function evaluations required for convergence to the global optimum. There are some penalties though. Calculating the MRF parameters is computationally more expensive than calculating marginal distributions, and so DEUM$_d$ will be particularly appropriate for problems where there is a positive trade-off in reducing the number of fitness evaluations.

A promising line of research in this area is to develop MRF models for bivariate and multivariate EDAs, where the extra computational costs are more likely to be compensated by a reduction in the number of fitness evaluations required to solve higher-order problems.

6 References

[1] Shakya, S., McCall, J., Brown D. (2004). Updating the probability vector using MRF technique for a Univariate EDA . In proceedings of STAIRS 2004, IOS press, pp. 15–25.

[2] Mühlenbein, H., Paass, G. (1996). From recombination of genes to the estimation of distributions I. Binary parameters. Voigt, H.-M et. El.(eds.) PPSN IV, LNCS 1141, Springer, pp. 178–187.

[3] Larrañaga P., and Lozano J. A. (2001) Estimation of Distribution Algorithms: A New Tool for Evolutionary Computation. Kluwer Academic Publishers, 2001.

[4] Pelikan, M., (2002). Bayesian optimization algorithm: From single level to hierarchy. Ph.D. thesis, University of Illinois at Urbana -Champaign, Urbana, IL.

[5] Baluja, S. (1994). Population based incremental learning: A method for integrating genetic search based function optimization and competitive learning. Technical Report CMUCS94163, Pittsburgh, PA.

[6] Li, S. Z. (1995). Markov random field modeling in computer vision. ISBN:4-431-70145-1, Springer-Verlag,1995.

[7] Brown D.F., Garmendia-Doval, A.B., McCall, J.A.W. (2001). Markov Random Field Modelling of Royal Road Genetic Algorithms. Evolution Artificielle 2001, LNCS 2310, pp. 65–78, Springer Verlag 2002.

[8] Press, W. H., Teukolsky, S. A., Vetterling, W. T., and Flannery, B. P (1993). Numerical Recipes in C: The Art of Scientific Computing. Cambridge University Press, 2nd edition, 1993

[9] Lawrence Davis (1991), editor. Handbook of Genetic Algorithms. Van Nostrand Reinhold, 1991.

[10] Hoos, H. H. and Stutzle, T. (1999). Towards a characterisation of the behaviour of stochastic local search algorithms for SAT. Artificial Intelligence, 112(1-2):213.232, 1999.

Modelling Chlorine Decay in Water Networks with Genetic Programming

Philip Jonkergouw, Ed Keedwell and Soon-Thiam Khu

Centre for Water Systems, School of Engineering, Computer Science and Maths,
University of Exeter, Exeter, UK, EX4 4QF

Abstract

The disinfection of water supplies for domestic consumption is often achieved with the use of chlorine. Aqueous chlorine reacts with many harmful micro-organisms and other aqueous constituents when added to the water supply, which causes the chlorine concentration to decay over time. Up to a certain extent, this decay can be modelled using various decay models that have been developed over the last 50+ years. Assuming an accurate prediction of the chlorine concentration over time, a measured deviation from the values provided by such a decay model could be used as an indicator of harmful (intentional) contamination. However, most current chlorine decay models have been based on assumptions that do not allow the modelling of another species, i.e. the species with which chlorine is reacting, thereby limiting their use for modelling the effect of a contaminant on chlorine. This paper investigates the use of genetic programming as a method for developing a mixed second-order chlorine decay model.

1 Introduction

Increasingly stringent government regulations and consumer satisfaction requirements force water utilities to supply "safe drinking water" to their consumers. The term "safe drinking water" implies that the water supplied does not carry pathogens at infectious doses, has few disinfectant by-products and good taste. Of these, harmful microorganisms are the main concern and in order to reduce the risk of infection, chlorine is typically used as a disinfectant. The general practice is to add large amounts of chlorine at the source so that the disinfectant properties at the extremities of the network are maintained. This practice has recently been brought into question [1], but remains the standard method of disinfection for water supplies.

Recent world developments have highlighted the risk of intentional contamination of water supply systems, which offer a method for conveying a biological or chemical agent to a large number of people with a minimal risk of detection. Most of these agents will react with the chlorine present in the water [2] but often not enough to neutralise the threat posed by the agent. As the chlorine reacts with the contaminant species to form disinfection by-products, the concentration of chlorine in the water reduces. Therefore, the introduction of a biological or chemical agent could produce a chlorine concentration profile different to that normally observed in the network. There is therefore scope for an automated system, based on the decay of chlorine, to differentiate between standard reactions and those associated with an unwanted contamination [2].

The disappearance, or decay, of chlorine from the water as a result of chemical reactions with various aqueous constituents has been modelled for over more than 50 years, with the first being a model developed by Feben and Taras [3]. The models that have been developed since have become increasingly complex [4, 5, 6], but all assume that the rate of decay of chlorine is constant throughout the reaction, whether chlorine reacts with one species or more (e.g. parallel first order). These methods have been fairly successful at predicting the decay of chlorine throughout a water distribution system under normal conditions, but often fail under re-chlorination conditions. In order to model the effect of varying contaminant concentrations on the decay of chlorine, a model would have to have the following two characteristics: (1) It should allow for the specification of the concentration of the contaminant and (2) It should be consistent with observed data when changing initial concentrations. To our knowledge, no current chlorine decay model meets these requirements.

This paper proposes the use of genetic programming to derive a modified mixed second order chlorine decay model from experimental data. In later sections, we will show that this approach can discover accurate solutions with a small computational outlay, which use only very simple operators and terminals. This is in contrast to previous approaches which are highly species-specific and can require long computation times.

2 Chlorine Decay Modelling

Chlorine decay is typically modelled using some form of the solution to the (pseudo-) first order rate law. The models based on this solution do not account for the decay of the species with which chlorine is reacting and can therefore not be used to model the effect of a contaminant on chlorine.

Chlorine decay based on a mixed reaction with one other species has the following chemical rate law,

$$-\frac{d[Cl]}{dt}=-\frac{d[X]}{dt}=-k[X][Cl] \qquad (1)$$

where $[Cl]$ is the molar concentration of chlorine, $[X]$ defines the concentration of the second species, t is time and k is the reaction rate constant. Since k is constant, this equation represents the rate of one particular reaction and this is not an adequate description of the decay of chlorine. Chlorine reacts concurrently with many species at many different rates. X should thus represent the *molar number of molecular sites* with which chlorine reacts. Some reactions will be quenched sooner than others, depending on the reaction rate and concentration of the sites. This implies a time-dependent average decay rate that is largest at t=0 and smallest when $t \rightarrow \infty$. In order to accommodate for these particulars, the mixed second order rate law is rewritten as,

$$-\frac{d[Cl]}{dt}=-\frac{d[X]}{dt}=-\lambda[X][Cl] \qquad (2)$$

where λ is an unknown function of time and

$$\int_0^t \lambda dt = k(t)t \qquad (3)$$

where $k(t)$ is an unknown function of time. The solution to this equation is given by,

$$[Cl]=[Cl]_0-[P] \text{ and } [X]=[X]_0-[P] \qquad (4)$$

and

$$[P]=\frac{\{\exp(k(t)t([Cl]_0-[X]_0))-1\}[Cl]_0[X]_0}{\exp(k(t)t([Cl]_0-[X]_0))[Cl]_0-[X]_0} \qquad (5)$$

where $[P]$ represents the molar number of reactive sites that have been quenched. The absolute behaviour of the function $k(t)$ is unknown, but it has to adhere to the following restrictions, due to the nature of the problem it represents:

> $k(0) \neq 0$. The rate of decay of chlorine at t=0 has a non-zero and defined value. Each reaction that chlorine will undergo has a certain rate of reaction (otherwise it would not react). Therefore, the overall reaction rate is defined and non-zero.
>
> $dk/dt < 0$. The reaction rate cannot increase with time. Reactions are being quenched and the overall rate of chlorine consumption will decrease over time. This behaviour has been shown in previous studies [7], particularly under re-chlorination conditions.
>
> $k(\infty) \neq 0$. The final rate of the reaction depends on the rate of the reaction of the last reactive site present, which will always be defined and non-zero.

In this paper we propose to use genetic programming to discover a function for k that fits the observed data. The function will be based on the initial concentrations of chlorine and the molar number of reactive sites (X), as well as time.

3 Genetic Programming

The genetic programming approach used in this paper is relatively standard. The solutions are evolved as trees which contain a number of operators and terminals. The evolution takes place by using genetic-programming specific crossover and mutation. Each of these are outlined below.

3.1 Genetic Parameters

3.1.1 Genetic Programming

The genetic programming uses a generational approach as its basis, with mutation rate and crossover rates of 0.9. The mutation rate is high because the terminals and operators are enumerated as integer variables. When a binary encoding is used, crossover can occur at any point through a variable, whereas this is not possible with integers. Therefore crossover on a binary string also implements some mutation. To compensate for the lack of this effect, the integer representation uses a large mutation rate.

To initialise the algorithm, a tree is "grown". This means that elements are selected at random from both the operator and terminal sets until such time as the initialisation limit is reached whereupon terminals are used to fill the remainder of the tree. The alternative "full" initialisation selects only from the operator list and results in substantially larger trees and therefore is not explored here.

3.1.2 Crossover

The standard genetic algorithm single-point crossover will not work with GP. Therefore a GP-specific operator is used here which selects a point on each tree and then swaps the sub-trees of each individual. This therefore ensures that the resulting tree is syntactically correct.

3.1.3 Mutation

The standard mutation operator used in genetic algorithms can be used in GP, but here a further GP-specific operator is used. A site is randomly selected on the tree and a new sub-tree grown using the same procedure as used to initialise the tree (in this case "grow").

3.1.4 Terminal Set

The terminal set of the genetic programming consists of three variables and ten constants as follows:

Table 1. Genetic Programming Terminal Set

InitCl	Initial molar chlorine concentration
InitX	Initial molar number of reactive sites
T	Time parameter
1-10	Constant integers from 1-10

It is anticipated that, if required, the genetic programming approach will derive its own set of floating point variables through the manipulation of these constants and the operators described below.

3.1.5 Operator Set

The operator set consists simply of multiplication, addition, subtraction and a protected division where division by zero is prevented. Exponentials and other operators are not included, as these can cause problems for genetic programming in nested expressions. Also, this approach ensures that simple expressions for the curves can be derived from this modest set of operators.

3.1.6 Objective Function

The objective function of the genetic programming is computed as the Root Mean Squared Error between the derived curves and the actual data. In addition to this, to implement the constraints 1 & 3 described in Section 2, the approach is heavily penalised if the k is less than $1*10^{-5}$ at timepoint 0, and a very large timepoint ($1*10^{31}$). These constraints approximate those real-world constraints as it is not computationally feasible to evaluate an expression at an infinite timepoint. The final constraint is on the size of the tree where the solution is penalised if a tree has a depth larger than that stipulated as an experimental parameter.

4 Results

The data used in this study have been taken from [8]. In this study, chlorine was added at three different initial concentrations (1, 3 and 5 mg/L) to an aqueous bacterial suspension (E. coli) with a concentration of approximately 107 cells/ml. The chlorine concentration was measured at various time intervals, shown in Figure 1. It should be noted that the original chlorine concentrations have been converted to mMol/L, so they can be used in this study. The molecular weight used for chlorine was 70.9 mg/mMol. The concentration of reactive sites (initX) was taken to be 0.034 mMol/L, which is equal to the difference between the initial chlorine concentration at 5 mg/L and the chlorine concentration after 30 minutes.

Genetic programming was used to determine a function for k, as described in previous sections. The GP was run a total of 12 times, with differing random seeds, initial depths and penalties for exceeding a depth 4, 6, 8, 10 and 12. Each run was conducted with a population size of 30 and for 5000 generations. In general, the results were very good, often with small RMSE values, but occasionally at the expense of the parsimony of the solution. In fact, we found there was a small effect with respect to the initial size of the trees, with smaller trees yielding more effective results after optimisation. Table 2 shows this effect.

Table 2. Average RMSE for 12 runs

Initial depth	Average RMSE
3 (5 runs)	0.001259
5 (4 runs)	0.001402
7 (3 runs)	0.001419

The best discovered result is shown in Figure 1. The experimental parameters and the discovered equation for the k-value over time are given in Table 3. As can be seen the solution is both succinct and accurate with respect to the data. In addition to this, a plot of k reveals that the second constraint (Section 2) that was not explicitly coded in the algorithm, was not violated by the discovered solution. This therefore shows that genetic programming has evolved a solution which fits all the criteria for this modelling task.

Table 3. Experimental parameters used for determination of best-fitting (lowest RMSE) k-function.

Random seed	2135
Population	30
Initial depth	3
Penalise depth	10
Iterations	5000
Time (sec)	74
RMSE	8.74E-04
k(t)	2/((InitCl+InitCl)*(1+InitX+InitX+T))+ 2+InitCl/2

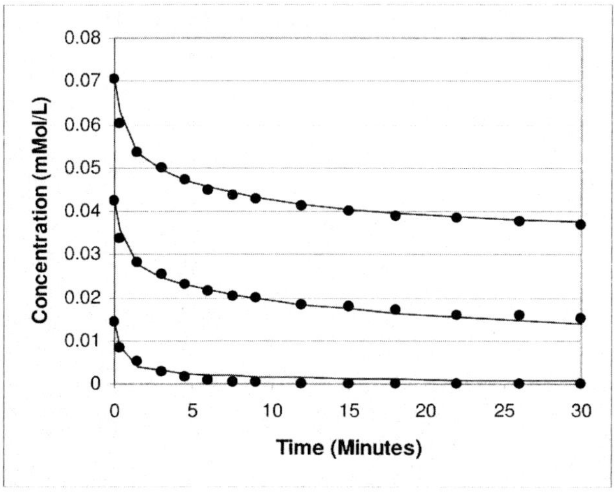

Fig. 1. Chlorine concentration versus time. The points are the measured chlorine concentrations taken from ref. 8. The GP-fitted curves are represented by the lines.

5 Conclusions

Previous attempts at modelling chlorine decay in water distribution networks has mostly been based on (pseudo-) first order reaction kinetics and has been limited to a description of one species only. Chlorine reacts with a large number of species in real-world water systems, each with different reaction rates. The average decay rate is likely to vary with time, whereas existing models use constant values to predict the decay of chlorine. In this paper genetic programming has been used to create an equation for k based on very few parameters, namely the initial dose of chlorine, the molar number of reactive sites and the time of the reaction. It has been shown that this approach is capable of closely matching experimental results. In addition to this, it has been shown that the approach discovers these solutions whilst subjected to constraints that ensure the results agree with the chemical situation the model represents. Furthermore, the genetic programming model has found good fit whilst only using very simple mathematical operators and obeyed a further constraint that was not made explicit to the algorithm. This ensures that when the approach is applied to large water distribution systems simulated over a large number of hours that the computation of chlorine concentrations can be completed efficiently. This accurate modelling of the decay of chlorine is anticipated to be the first step towards an early-warning contaminant detection system for water distribution systems.

6 References

[1] Devi Prasad T, Walters GA., and Savic D (2004) "Booster disinfection in water supply networks: A multi-objective approach", Journal of Water Resources Planning and Management ASCE, vol. 130, No.5.,2004.

[2] Jonkergouw PMR, Khu ST, Savic D (2004) Chlorine: An indicator of intentional chemical and biological contamination in a water distribution network? AutMoNet 2004, Proceedings of: The *2nd International IWA Conference on Automation in Water Quality Monitoring*, Vienna, Austria.

[3] Feben D, Taras MJ (1951) Studies on chlorine demand constants. *J. Am. Water Works Assoc.*, (**43**) 11, pp. 922–932.

[4] Vasconcelos JJ, Rossman LA, Grayman WM, Boulos PF, Clark RM (1997) Kinetics of chlorine decay. *J. Am. WaterWorks. Assoc.*, (**89**) 7, pp. 55–65.

[5] Clark RM, Sivaganesan M (2002) Predicting Chlorine Residuals in Drinking Water: Second Order Model. *J. Wat. Res. Plan. Man.* (**128**) 2, pp. 151–161.

[6] Boccelli DL, Tryby ME, Uber JG, Summers RS (2003) A reactive species model for chlorine decay and THM formation under rechlorination conditions. *Wat. Res.* (**37**), pp. 2654–2666.

[7] Powell JC, Hallam NB, West JR, Forster CF, Simms J (2000) Factors which control bulk chlorine decay rates. *Wat. Res.*, (**34**) 1, pp. 117–126.

[8] Shang C and Blatchley ER (2001) Chlorination of pure bacterial cultures in aqueous solution. *Wat. Res.* (**35**) 1, pp. 244 – 254.

Evolving Blackjack Strategies Using Cultural Learning

Dara Curran and Colm O'Riordan
Dept. of Information Technology
National University of Ireland, Galway

Abstract

This paper presents a new approach to the evolution of blackjack strategies, that of cultural learning. Populations of neural network agents are evolved using a genetic algorithm and at each generation the best performing agents are selected as teachers. Cultural learning is implemented through a hidden layer in each teacher's neural network that is used to produce utterances which are imitated by its pupils during many games of blackjack. Results show that the cultural learning approach outperforms previous work and equals the best known non-card counting human approaches.

1 Introduction

The game of blackjack has been the subject of much research, particularly in the reinforcement learning domain. However, given the nature of the game, perfect strategies do not exist. Thus, traditional methods of neural network evolution involving gradient descent techniques have no model to emulate. This paper introduces an alternative approach for the evolution of high-quality blackjack-playing agents. We evolve a population of neural network agents which play games of blackjack against an automated dealer. Cultural learning is introduced by taking a percentage of the population and allowing it to teach the following generation through specialised verbal output nodes. Two experiments are performed, one where agents may see the dealer's up-card and one where they cannot. We compare the performance of the evolved strategies with bench-marks obtained from a blackjack simulator.

The remainder of this paper is arranged as follows: Section 2 discusses related work. Section 3 presents the results of bench-marking several popular blackjack strategies. Section 4 introduces the artificial life simulator employed in the experiments. Section 5 presents the experimental results and Section 6 concludes and suggests future work.

2 Related Work

2.1 Cultural Learning

Culture can be succinctly described as a process of information transfer within a population that occurs without the use of genetic material. Culture can take many forms such as language, signals or artifactual materials. Such information exchange occurs during the lifetime of individuals in a population and can greatly enhance the behaviour of such species.

Artificial cultural evolution, or synthetic ethology[1], has been extensively researched. Approaches include fixed lexicons[2, 3], indexed memory[4], cultural artifacts[5, 6] and signal-situation tables[7]. The approach chosen in this work is the teacher/pupil scenario[8, 9, 3, 10] where a number of highly fit agents are selected from the population to act as teachers for the next generation of agents. These pupils learn from teachers by observing the teacher's verbal output and attempting to mimic it using their own verbal apparatus. As a result of these interactions, a lexicon of symbols evolves to describe situations within the population's environment.

2.2 The Game of BlackJack

Blackjack or twenty-one begins with the dealer dealing two cards face-up to each player and two to his/herself, with one card visible (the *up-card*) and the other face down. Cards are valued by their face value (10 for all picture cards) except for the ace which can be counted either as 11 or 1. The object of the game is to obtain a higher score (the sum of all card values) than that of the dealer's without exceeding 21. Each player can *draw* additional cards until they either *stand* or exceed 21 and go *bust*. Once all players have obtained their cards, the dealer turns over the hidden card and draws or stands as appropriate. Should the dealer's hand bust, all players win.

Several attempts have been made to develop high performing blackjack strategies with populations of neural networks using reinforcement learning techniques[11, 12]. The nature of the game means that there is no perfect set of neural network outputs from which to perform back-propagation. It is for this reason that we wish to show that the introduction of cultural learning can generate superior strategies than reinforcement learning methods and provides the learning framework required without knowledge of the perfect strategy.

3 Bench-marking

In order to assess the performance of any evolved strategy, a set of bench-marks were obtained for comparison purposes using a developed blackjack simulator.

Several strategies were considered including Dealer's (Stand on 17 or more, Draw on less), Random, Always stand, Hoyle's strategy and the Uribe Evolved Strategy.

The Hoyle strategy[13] is based on the dealer's up card and the possession of an ace. It can be summarised as:

```
if (dealer card < 6)
    if (ace is held)
        stand on 15
    else
        stand on 13
else
    stand on 17
```

The Uribe Evolved strategy is taken from the work of Uribe and Sanchez[12] and can be summarised as:

```
if (score> 9) OR
    [(score> 13)and(score< 19)
          AND (an ACE is held)]
    stand
else
    hit with 50% of probability
```

In order to produce statistically meaningful results, we performed 1000 runs of 1000 games for each strategy. The results presented in the table below are average wins for each strategy.

Strategy	% Wins	Standard Deviation
Hoyle	43.70	1.587
Dealer	41.55	1.576
Uribe et al	38.76	1.505
Always Stand	37.91	1.531
Random	30.41	1.511

We can see from these results that most strategies are quite poor against the dealer and that Hoyle's strategy performs best. This is most likely due to its inclusion of ace and dealer up-card information.

4 Simulator

Each agent in the population contains a neural network that allows it to play blackjack. Once cards are dealt to the agents the value of the hand is shown to the network using thermometer encoding[11]. Given the importance of the Ace, an additional input unit was created to signal the presence of an Ace to the network.

The agent's decision is determined by rounding the output of the single output node, where draw and stand are represented by 0 and 1 respectively. The number of hidden layers and nodes therein is unrestricted and is determined by the evolutionary process. The agents are evolved using a previously developed artificial life simulator[14, 15].

4.1 Simulating Cultural Evolution

In order to perform experiments related to cultural evolution, it was necessary to adapt the existing simulator architecture to allow agents to communicate with one another. This was implemented using an extended version of the approach adopted by Hutchins and Hazlehurst, where the last hidden layer of each agent's neural network functions as a verbal input/output layer[10].

At the end of each generation, a percentage of the population's fittest networks are selected and are allowed to become teachers for the next generation. Each pupil observes the teacher's verbal output as it interacts with its environment. A teaching cycle occurs when the pupil attempts to emulate its teacher's verbal output using backpropagation. Once the number of required teaching cycles is completed, the teachers die and new teachers are selected from the new generation.

5 Experiments

Each experiment allows 100 agents to evolve over 500 generations. At each generation, agents play 100 games against a dealer strategy and an agent's fitness is determined by the percentage of wins obtained scaled to [0.0,1.0]. Agents are linearly ranked and selected for reproduction using roulette wheel selection. Crossover and mutation are applied with probabilities 0.6 and 0.02 respectively. When cultural learning is applied, 10% of each generation are selected to act as teachers for the next. Teaching cycles are set at 2 and a noise value in the range [-0.5,0.5] is added to the each interaction with probability 0.01. Each of these settings were empirically determined to be suitable.

5.1 Experiment 1: Basic Experiment

The first experiment allowed agents to play hands of blackjack based on the contents of their cards. Cards were presented to each agent's neural network input layer. In addition, agents were also presented with a flag indicating the presence of an Ace card, reflecting its relative importance in the game.

The results for this experiment are presented in Figure 1. In addition to the average fitness of the population, both Dealer and Hoyle strategy levels are shown for comparison purposes.

The results illustrated in Figure 1 show that the populations are capable of attaining fitness levels above that of the Dealer strategy. However, statistical tests show (t-test with p value < 0.001) that cultural learning does not provide a significant improvement. This may be explained by the lack of information available and motivated our next experiment which takes account of dealer up-card information.

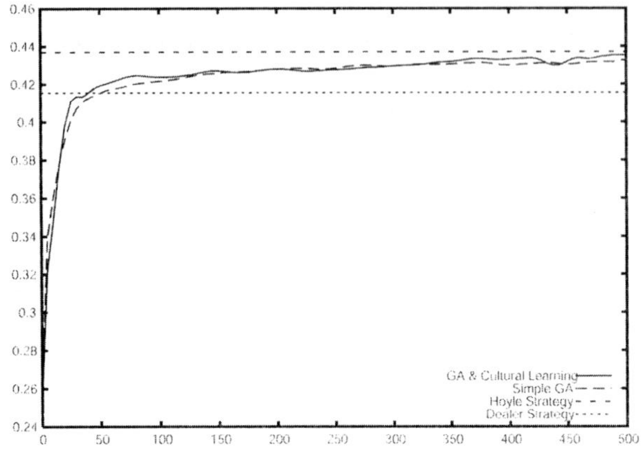

Fig. 1. *Experiment 1*

5.2 Experiment 2: Dealer Information

Most advanced strategies, including Hoyle's, take the dealer's up-card into account when determining a course of action. Therefore, we present the dealer's up-card to each agent by adding an additional 10 input nodes, one for each of the dealer's possible up-cards (2-9, 10 for all picture cards and the ten, Ace). All other variables remain equal as used in Experiment 2.

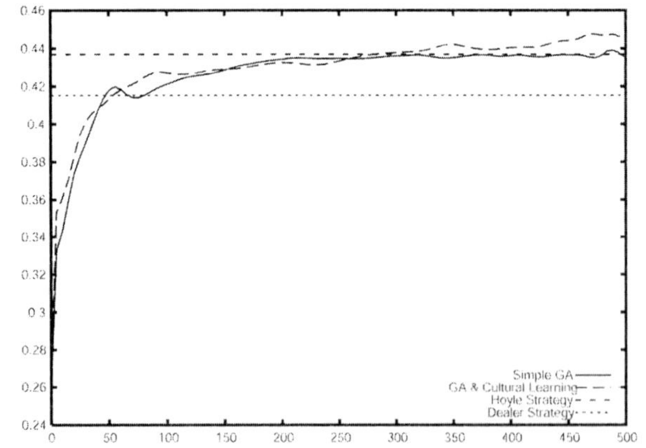

Fig. 2. *Experiment 2*

The graph in figure 2 shows that the addition of dealer information dramatically improves the performance of both populations but particularly that of the population employing cultural learning which achieved highs of nearly 0.45 (45% wins) versus 0.44 for population learning. There is strong evidence (p value < 0.0001, 95% confidence) that cultural learning is indeed improving the population's fitness. The following resulting strategy was extracted:

```
if (an Ace is held)
{
     if (dealer has a 6 or higher)
          stand on 16
     else
          stand on 17
}
else
{
     if (dealer has a 7 or higher)
          stand on 17
     else
          stand on 13
}
```

It is clear from the strategy that the evolved agents are employing the dealer information to the full extent and have identified a threshold value for the dealer up-card. The strategy is tested in the next experiment to ascertain its performance with respect to the bench-marked strategies.

5.2.1 Strategy Testing:

As a result of the inherent random nature of blackjack, it is necessary to test the strategies over a number of runs to observe whether they are successful. The final evolved strategy from Experiment 2 was taken and hard-coded into the blackjack simulator and 1000 runs of 1000 games were played. The averaged results are displayed in the table below:

Strategy	% Wins	Standard Deviation
Hoyle	43.69	1.573
Evolved	**43.67**	**1.582**
Dealer	41.52	1.571
Uribe et al	38.43	1.495
Always Stand	38.00	1.529
Random	30.67	1.507

There is strong evidence (p value <0.001, 95% confidence) to support the claim that the evolved strategy and Hoyle's strategy are equivalent in terms of performance, suggesting that the population has evolved an optimum strategy given the information available. It is likely that in order to out-perform Hoyle's strategy it is necessary to keep track of cards that have been played during a game, something which would only become truly useful if the number of players was increased.

6 Conclusion

We have shown that the addition of cultural learning to a population of neural network agents evolved using a genetic algorithm can produce robust blackjack strategies that out-perform those evolved thus far using reinforcement learning. As in previous work, the addition of dealer information to the population significantly improves performance. Through the bench-marking process we have shown that the evolved strategy is equivalent to the best human strategy which does not incorporate card-counting. Future work will introduce more players per game with the expectation of evolving agents capable of card counting strategies such as those developed by human experts.

References

[1] L. Steels. The synthetic modeling of language origins. In *Evolution of Communication*, pages 1–34, 1997.

[2] H. Yanco and L. Stein. An adaptive communication protocol for cooperating mobile robots. In *From Animals to Animats 2. Proceedings of the second International Conference on Simulation of Adaptive Behavior*, pages 478–485. MIT Press, Cambridge Ma., 1993.

[3] A. Cangelosi and D. Parisi. The emergence of a language in an evolving population of neural networks. *Technical Report NSAL–96004, National Research Council, Rome*, 1996.

[4] L. Spector and S. Luke. Culture enhances the evolvability of cognition. In *Cognitive Science (CogSci) 1996 Conference Proceedings*, 1996.

[5] E. Hutchins and B. Hazlehurst. Learning in the cultural process. In *Artificial Life II, ed. C. Langton et al.* MIT Press, 1991.

[6] A. Cangelosi. Evolution of communication using combination of grounded symbols in populations of neural networks. In *Proceedings of IJCNN99 International Joint Conference on Neural Networks (vol. 6)*, pages 4365–4368, Washington, DC, 1999.

[7] B. MacLennan and G. Burghardt. Synthetic ethology and the evolution of cooperative communication. In *Adaptive Behavior 2(2)*, pages 161–188, 1993.

[8] A. Billard and G. Hayes. Learning to communicate through imitation in autonomous robots. In *7th International Conference on Artificial Neural Networks*, pages 763–738, 1997.

[9] D. Denaro and D. Parisi. Cultural evolution in a population of neural networks. In *M.Marinaro and R.Tagliaferri (eds), Neural Nets Wirn-96.New York: Springer*, pages 100–111, 1996.

[10] E. Hutchins and B. Hazlehurst. How to invent a lexicon: The development of shared symbols in interaction. In N. Gilbert and R. Conte, editors, *Artificial Societies: The Computer Simulation of Social Life*, pages 157–189. UCL Press: London, 1995.

[11] D. K. Olson. *Learning to Play Games from Experience: An Application of Artificial Neural Networks and Temporal Difference Learning*. Pacific Lutheran University, 1993.

[12] Andrs Prez-Uribe and Eduardo Sanchez. Blackjack as a test bed for learning strategies in neural networks. In *International Joint Conference on Neural Networks, IJCNN'98*, pages 2022–2027, 1998.

[13] A. H. Morehead and G. M. Smith. *Hoyle's Rules of Games*. Plume, 1963.

[14] D. Curran and C. O'Riordan. On the design of an artificial life simulator. In V. Palade, R. J. Howlett, and L. C. Jain, editors, *Proceedings of the Seventh International Conference on Knowledge-Based Intelligent Information & Engineering Systems (KES 2003)*, pages 549–555, University of Oxford, United Kingdom, 2003.

[15] D. Curran and C. O'Riordan. Artificial life simulation using marker based encoding. In *Proceedings of the 2003 International Conference on Artificial Intelligence (IC-AI 2003)*, volume II, pages 665–668, Las Vegas, Nevada, USA, 2003.

Two-Criterion Optimization in State Assignment for Synchronous Finite State Machines using NSGA-II

Nitin Gupta[1], Vivek Kumar Agrawal[2]
[1]Department of Computer Science and Engineering, Indian Institute of Technology, Kanpur
[2]Department of Mechanical Engineering, Indian Institute of Technology, Kanpur, India
E-mail: ngupta@ucsd.edu

Abstract

One of the challenging problems in circuit implementations is finding the best state assignment for implementing a synchronous sequential circuit which are also represented as Finite State Machines. This problem, commonly known as State Assignment Problem (S.A.P.), has been studied extensively because of its importance in reducing the cost of implementation of circuits. The previous work on this problem assumes the number of coding bits as constant, making it a single objective problem with the only objective being to reduce the cumulative cost of transition between the connected states.

In this work, we add another dimension to this optimization problem by introducing a second objective of minimizing the number of bits used for assignment. This is desirable to reduce the complexity and the cost of a circuit. The two objectives are conflicting and thus the optimal solution requires a tradeoff. We present an evoluationary algorithms based approach to solve this multi-dimensional optimization problem. We compare the results from two algorithms, and find that an NSGA-II based approach, with some modifications to constraint handling, gives better results and running time than NSGA. We also gain some insights about the shape of the efficient frontier.

1 Introduction

A significant part of the digital circuits is constituted by sequential synchronous circuits behaviour of which can be presented by a finite state machine (FSM). Thus it is not strange that the FSM synthesis models are continually developed. One of the most crucial steps in FSM synthesis is the encoding of FSM states referred to as the state assignment problem (SAP). It consists of the unique assignment of bit strings to the states of sequential circuit (SC). This step of FSM synthesis is important because it affects the quality of realised SC (cost/area, maximum frequency, power consumption).

Attempts at solving the state assignment problem spreads over five decades. Amongst the early attempts are those due to Armstrong [1], Dolotta and McCluskey [2] and the decomposition technique due to Hartmanis and Stearns [3, 4]. Effective algorithms for the state encoding were developed, e.g. NOVA [5] for two-level implementation targeted to Programmable Logic Arrays (PLAs) or MUSTANG [6] and JEDI [7] for multilevel FSM implementation. However, state assignments generated by these methods, for FSMs implemented in modern programmable devices allowing efficient implementations of digital systems, are far from optimum [8]. Taking above into account, and considering other conditions (cf. [9, 10]) we decided to try to cope with SAP using evolutionary methods.

2 Multi-Objective Problem Formulation

We have a directed graph representing the FSM. Each vertex in the graph represents a state. Suppose there are m states. Each of these states is to be assigned a binary string of k bits. We have to find the assignment to these states that optimizes that objective functions.

2.1 Objective Functions

We have two conflicting objectives1 that we want to minimize: (1) k- Minimizing the number of bits used for state representation reduces the implementation costs and the size of the circuit. (2) Transition Costs- This represents the cumulative transition cost between states in the graph. Transition Cost for any pair of states i, j, such that j is connected to i is given by the Hamming distance between the string representation of states i and j, i.e. the number of bit flips required to move from i to j. Thus it is desirable to do the assignment of states in such a way that neighbouring states in the graph are assigned more "similar" binary strings than far-off states.

2.2 Constraints

Two constraints apply on the number of bits k used for assignment. The first constraint specifies a lower limit on

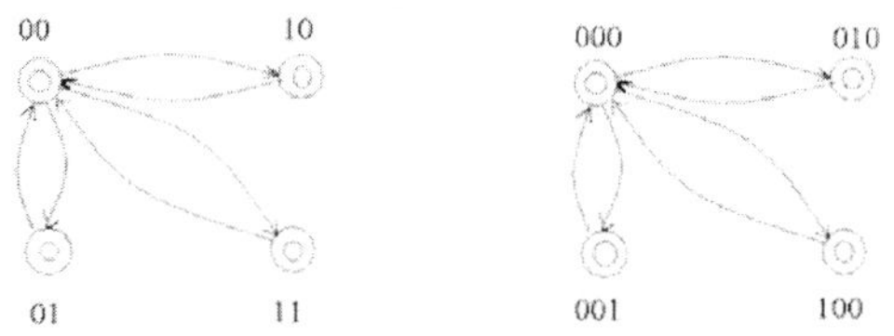

Fig. 1. Example of Tradeoff: The first figure shows the only possible two bit assignment (k=2) for the state, while the second figure shows one of the best possible three bit (k=3) assignments. We observe that number of flips in first image is 8, while only 6 in the second image. This example illustrates the significance of considering different k's.

the number of bits, that must be used to represent a set of m distinct states.

$$k \geq \lceil log_2(m) \rceil \quad (1)$$

The second constraint is an upper limit on the number of bits, which is same as the number of states. Any number of bits beyond this limit is redundant and can only increase the cost of transition.

$$k \leq m \quad (2)$$

There is one more constraint for the solutions to be valid: Each state should be given a distinct assignment. This is obviously required for unique identification of each state.

3 Methods

The first thing to do in Evolutionary Multiobjective Optimization(EMO) approach is to determine a feasible and solvable pair of representation and operators. The crossover and mutation operators are designed according to the chosen representation (genotype coding). We use a matrix representation for one solution, that is a set of assignments covering all the states of the given machine. In this representation, a certain genotype is represented by a 2-D matrix. The number of rows is equal to the number of states in the machine. The number of colums is equal to the upper limit on k. Our representation ensures that the constraint $k < m$ is followed. We need to check only for the remaining two constraints, that is the lowerbound on k, and that no two states get the same assignment.We do a two-point column-wise crossover. We use simple bitwise mutation with the mutation probability currently set at 0.002 which represents the small chance that each bit of the matrix can get changed.

4 Comparison of EMO Approaches and Results

We have experimented with two main EMO approaches, NSGA and NSGA-II, as discussed below. Note that all the simulations have been run on a personal PC with 800 MHz P-III processor and 128 MB memory.

4.1 NSGA

NSGA is a non-elitist evolutionary algorithms [12]. The basic principle of NSGA, like most other EMO algorithms is to stress on selection of dominating points, and avoid too much of crowding in any front. The individuals within a population are divided into diffrent fronts(ranks) and an individual in a lower front (in this minimization problem) will be given preference over an individual in a relatively higher front.

4.1.1 Test Machine: For testing the performance of the algorithm we have designed finite state machine M with 23 states. The first state of the machine is connected, both-ways, to the remaining 22 states. There are no connections within the 22 states. It is easy to observe that minimum transition cost will occur if the remaining 22 states are assigned strings which are different at only one position from the assignment of the first state. We have summarized the analytically computed minimum transition cost for each value of k in Table 1. Note that all the pairs mentioned in this table are pareto-optimal, except (23,44), as it is dominated by (22,44).

4.1.2 Results: The algorithm gives 7 near pareto-optimal points, of which only 5 are actually pareto-optimal. Two points (5,92) and (6,82) are not pareto-optimal, while the points (7,76), (8,72), (9,70), (10,68) and (11,66) are pareto optimal, as seen in Table 1. The initial population is quite far away from the efficient front which is obtained at the end of the simulations (see Figure 2). The algorithm is able to move towards the efficient frontier, and obtain five pareto-optimal points. The simulations were run for 75,000 generations, although the results did not change much after 25,000 generations. This indicated that the algorithm will not be able to detect the remaining 13 pareto-optimal points. All results are obtained for a population size of 500.

4.2 NSGA-II

NSGA-II [13] is an extension of NSGA. NSGA-II ensures that it has the following three properties:

1. It uses an elitist approach, i.e. the best individuals are preserved in each generation

2. It uses an explicit diversity preserving mechanism

k	5	6	7	8	9	10	11	12	13	14	15	16	17	18	19	20	21	22	23
Min Cost	92	78	76	72	70	68	66	64	62	60	58	56	54	52	50	48	46	44	44

Table 1. The minimum cost of transitions, calculated analytically, for the test machine M. Note that all the pairs mentioned in this table are pareto-optimal, except (23,44), as it is dominated by (22,44).

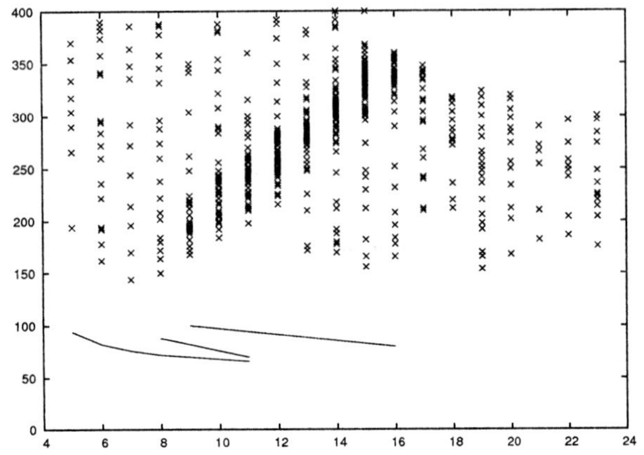

Fig. 2. Figure shows the distribution of the initial populations. The curves represent the best three fronts of individuals found by the algorithm.

3. It emphasizes the non-dominated solutions.

In this algorithm, the offspring population Q_t is first created from the pareant population P_t and the usual reproduction, crossover and mutation operators[14]. Thereafter, the two populations are combined together to form R_t of size 2N. Then, a non-dominated sorting is used to classify the entire population R_t. Once the non-dominated sorting is over, the new population is filled by solutions of different non-dominated fronts, one at a time. The filling starts with the best non-dominated front and continues with solution of the second non-dominated front, followed by the third non-dominated front, and so on. Since the overall population size of R_t is 2N, not all fronts may be accomodated in N slots available in the new population. All fronts which could not be accomodated are deleted. While the last front is under consideration, there may arise a situation when this front has more solutions than the number of free slots in the new population. In such situations the crowding measure is used, and the individuals with maximum crowding distance are chosen.

Here, we first used a penalty-function approach for handling constraints. A penalty equalling the number of clashes in state assignments was given to every individual. Feasible individuals, with no clash, were given a penalty zero. The domination criteria was adjusted to this change. If two given individuals are of different penalty value, the one with lower penalty is said to dominate the one with higher penalty value. In case of indiviuals with equal (or zero) penalty, the old domination criteria based on objective function value was used.

4.2.1 Results: The results are showed in Figure 3. We have obtained 16 points with rank1, and all of them actually correspond to pareto optimal points. The only two points that the method was unable to reach were (21,46) and (22,44), located at the far end of the solution space.

5 Discussions and Conclusion

In this study, we have attempted to solve the state assignment problem (S.A.P.), keeping the number of bits as variable. Thus, we have two objectives to be optimized: (1). Reducing the cost of transition between states and (2). Reducing the number of bits used. We have approached this two-dimensional optimization problem through evolutionary algorithms approach.

We observe that NSGA-II gave significantly better results than NSGA in lesser number of iterations, which can be attributed to the elitist approach. The inherent discreteness and discontinuity of the solution space increases the chance of losing a good solution during cross-over operators, and thus we find that an explicit mechanism of favoring the good solutions improves the qaulity of results. This is an evidence towards the discreteness of the efficient frontier of solutions.

References

[1] Armstrong, DB, "A programmed algorithm for assigning internal codes to sequential machines", IRE Trans, 1962, EC11, pp 466-472.

[2] Dolotta, TA, and McCluskey, E, "The coding of internal states of sequential machines", IEEE Trans, 1964, EC13, pp 549-562.

[3] Hartmanis, J, "On the state assignment problem for sequential machines I", IRE Trans, 1961, EC10, pp 157-165.

[4] Stearns, RE, and Hartmanis, J, "On the state assign-

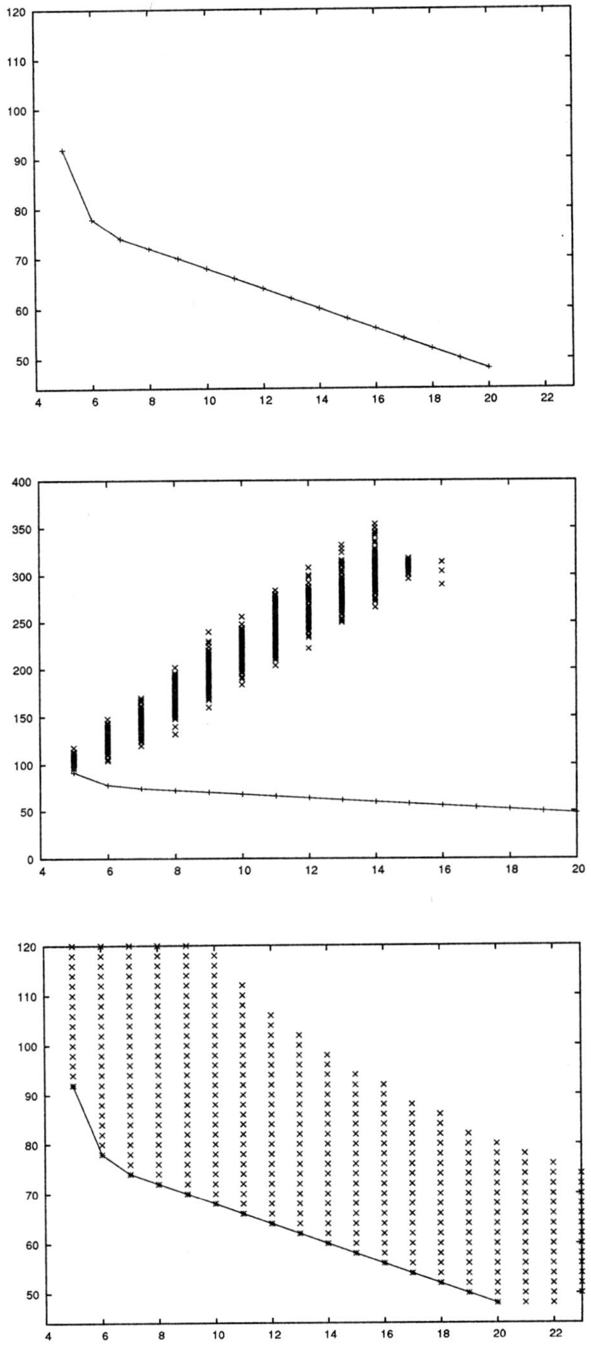

Fig. 3. The first figure shows the obtained efficient frontier, containing 16 points. The second figure shows the distribution of the initial population, and also the efficient frontier, to give a view of the distance that the population has to travel. The third figure shows the final population at the end of 3000 generations.

ment problem for sequential machines II", IRE Trans, 1961, EC10, pp 593-603.

[5] Villa T., Sangiovanni-Vincentelli A., NOVA: State Assignment of Finite State Machines for Optimal Two-Level Logic Implementation, IEEE Transactions on Computer- Aided Design, 1990, 9, (9), pp. 905-924.

[6] Devadas S., Ma H.-K., Newton R., Sangiovanni-Vincentelli A., MUSTANG: State Assignment of Finite State Machines Targeting Multilevel Logic Implementations, IEEE Transactions on Computer-Aided Design, 7, (12), (1988), pp. 1290-1300.

[7] Lin B., Newton R., Synthesis of Multiple Level Logic from Symbolic High-Level Description Languages, Proc. of the Int. Conf. on VLSI, Munchen, 1989, pp. 187-206.

[8] Deniziak S., Sapiecha K., A Comparison of State Encoding Algorithms for Different Reprogrammable Architectures, Proc. of the RUC Conference., Szczecin, Poland 1998.

[9] Gaat A., A system of data collection for small satellite systems with interfaces built with the use of FPGAs, Proc. of the RUC Conference, Szczecin, Poland 1999, pp.301-308.

[10] Michalewicz Z., Genetic Algorithms + Data Structures = Evolution Programs, Springer-Verlag, Berlin Heidelberg 1996, 3rd edition.

[11] De Micheli G., Synthesis and Optimization of Digital Circuits, McGraw-Hill , 1994.

[12] Srinivas N. and Deb K. Multi-objective function optimization using non-dominated sorting genetic algorithms. Evolutionary Computation Journal 2(3):221-248, 1994.

[13] Deb K., Agrawal S., Pratap K. and Meyarivan T. A fast and elitist multi-objective genetic algorithm: NSGA-II IEEE Transactions on Evolutionary Computation. 6(2):182-197, 2002.

[14] Goldberg D. E. Genetic Algorithms in search, optimization and machine learning. Addison-Wesley, Reading, Massachusetts, 1989.

Offspring Selection: A New Self-Adaptive Selection Scheme for Genetic Algorithms

M. Affenzeller, S. Wagner
Institute for Formal Models and Verification
Johannes Kepler University, Linz, Austria
E-mail: {michael, stefan}@heuristiclab.com

Abstract

In terms of goal orientedness, selection is the driving force of Genetic Algorithms (GAs). In contrast to crossover and mutation, selection is completely generic, i.e. independent of the actually employed problem and its representation. GA-selection is usually implemented as selection for reproduction (parent selection). In this paper we propose a second selection step after reproduction which is also absolutely problem independent. This self-adaptive selection mechanism, which will be referred to as offspring selection, is closely related to the general selection model of population genetics. As the problem- and representation-specific implementation of reproduction in GAs (crossover) is often critical in terms of preservation of essential genetic information, offspring selection has proven to be very suited for improving the global solution quality and robustness concerning parameter settings and operators of GAs in various fields of applications. The experimental part of the paper discusses the potential of the new selection model exemplarily on the basis of standardized real-valued test functions in high dimensions.

1 Introduction

By considering crossover as their main reproduction operator, Genetic Algorithms (GAs) take an approach which is fundamentally different to neighborhood-based search techniques. The most essential difference is given by the fact that recombination is able to combine properties of individuals from different regions of the search space. Therefore, provided that the problem representation and the operators are adequate, the advantage of applying GAs to hard optimization problems lies in their ability to search wider regions of the solution space than heuristic methods based upon neighborhood search do. Nevertheless, also GAs are frequently faced with a problem which, at least in its impact, is quite similar to the problem of stagnating in a local optimum. This drawback, called premature convergence in the terminology of GAs, occurs when the population of a GA reaches such a suboptimal state that the genetic operators can no longer produce offspring which outperform their parents (e.g. [1]).

The main aim of this paper is to find, analyze and improve new generic theoretical concepts for avoiding or at least retarding premature convergence in a generic way:

The basic approaches for retarding premature convergence discussed in GA literature aim to maintain genetic diversity. The most common techniques for this purpose are based upon preselection [2], crowding [3], or fitness-sharing [4]. The main idea of these techniques is to maintain genetic diversity by the preferred replacement of similar individuals [2], [3] or by the fitness-sharing of individuals which are located in densely-populated regions [4]. While methods based upon [3] or [4] require some kind of neighborhood measure depending on the problem representation, [4] is additionally quite restricted to proportional selection. Moreover, these techniques have the common goal to maintain genetic diversity which is very important in natural evolution where a rich gene pool is the guarantor in terms of adaptiveness w.r.t. changing environmental conditions. In case of artificial genetic search as being performed in Genetic Algorithm the optimization goal does not change during the run of a GA and the fixing of alleles of a global optimal solution is desirable in the same manner as the erasement of alleles which are definitely not part of of a good solution, i.e. we claim that pure diversity maintenance mechanisms as suggested in [2], [3], or [4] do not support goal-oriented genetic search w.r.t the locating of global optimal solutions.

A very essential question about the general performance of a GA is, whether or not good parents are able to produce children of comparable or even better fitness (the building block hypothesis implicitly relies on this). In natural evolution, this is almost always true. For Genetic Algorithms this property is not so easy to guarantee. The disillusioning fact is that the user has to take care of an appropriate coding in order to make this fundamental property hold.

In order to overcome this strong requirement we have developed an advanced selection mechanism which is based on the idea to consider not only the fitness of the parents, in order to produce a child for the ongoing evolutionary process. Additionally, the fitness value of the evenly produced offspring is compared with the fitness values of its own parents in order to decide whether or not the evenly produced offspring is accepted as a member of the next generation. The offspring is accepted as a candidate for the further evolutionary process if and only if the reproduction operator was able to produce an offspring that could outperform the fitness of its own parents. This strategy guarantees that evolution is presumed mainly with crossover results that were able to mix the properties of their parents in an advantageous way. Via these means we are already in a position to attack one of the reasons for premature convergence, namely the loss of relevant genetic information due to improper crossover operators. Furthermore, this strategy has proven to act as a precise mechanism for self-adaptive selection pressure steering.

2 Offspring Selection: A New Model for Self-Adaptive Selection Pressure Steering

In principle, the new selection strategy acts in the following way:

The first selection step chooses the parents for crossover either randomly or in any well-known way of Genetic Algorithms like roulette-wheel, linear-rank, or some kind of tournament selection strategy. After having created a new child from the selected parents by crossover and mutation, we introduce a further selection mechanism that considers the success of the apparently applied reproduction. In order to assure that genetic search proceeds mainly with successful offspring, it has to be attempted that the used crossover and mutation operators are able to create a sufficient number of children that surpass their parents' fitness. Therefore, a new parameter, called success ratio ($SuccRatio \in [0,1]$), is introduced. The success ratio gives the quotient of the next population members that have to be generated by successful mating in relation to the total population size. We define that a child is successful if its fitness is better than the fitness of its parents, whereby the meaning of 'better' has to be explained in more detail: is a child better than its parents, if it surpasses the fitness of the weaker, the better, or is it in fact some kind of mean value of both?

For this problem, we claim that an offspring only has to surpass the fitness value of the worse parent in order to be considered as 'successful' at the beginning, while as evolution proceeds the child has to be better than a fitness value continuously increasing between the fitness of the weaker and the better parent. As in the case of Simulated Annealing, this strategy gives a wider search at the beginning, whereas at the end of the search process this operator acts in a more and more directed way. Having filled up the claimed ratio ($SuccRatio$) of the next generation with successful individuals according to the above meaning, the rest of the next generation ($(1 - SuccRatio) \cdot |POP|$) is simply filled up with individuals randomly chosen from the pool of individuals that were also created by crossover but did not reach the success criterion. The actual selection pressure $ActSelPress$ at the end of a single generation is defined by the quotient of individuals that had to be considered until the success ratio was reached, and the number of individuals in the population in the following way:

$$ActSelPress = \frac{|POP_{i+1}| + |POOL|}{|POP|}$$

If it is more difficult to achieve improvements at a certain stage of genetic search, more selection pressure is needed and this GA variant is able to handle this in a self adaptive way by creating a greater number of offspring as candidates for the next generation. If, on the other hand, evolutionary progress is easier to achieve, this algorithm is able to operate under less selection pressure. The amount of required selection pressure also depends on the actually employed operators: If a certain crossover operator is appropriate for a concrete problem representation, it is easier to generate new offspring which are able to outperform their own parents and less selection pressure is needed. What is really rather remarkable is the fact that also provably worse operators [5] are able to achieve high-quality results as long as at least 'sometimes' advantageous crossover result are produced – the unprofitable crossover results are not considered for the ongoing evolutionary process anyway. Corresponding results for the path representation of the TSP, which show that MPX crossover for example is able to achieve results comparable to ERX crossover or OX crossover (when being equipped with our self-adaptive selection pressure model), are reported in [6].

An upper limit of selection pressure ($MaxSelPress$) defines the maximum number of offspring considered for the next generation (as a multiple of the actual population size) that may be produced in order to fulfill the success ratio. With a sufficiently high setting of ($MaxSelPress$), this new model also functions as a precise 'detector heuristics' for premature convergence:

If it is no longer possible to find a sufficient number ($SuccRatio \cdot |POP|$) of offspring outperforming their own parents even if ($MaxSelPress \cdot |POP|$) candidates have been generated, premature convergence has occurred.

3 Empirical Analysis on the Basis of Standardized Test Functions

All results of this section are achieved with the HeuristicLab prototyping and development environment (described in ref. 7). The programs are written in the C# programming language using the Microsoft .NET framework 1.1. All values presented in the following tables are the best resp. average values of thirty independent test runs executed for each test case. The average number of evaluated solutions gives a quite objective measure of the computational effort.

In addition to a detailed empirical analysis of standardized TSP benchmarks in higher dimensions (as stated in ref. 6), this paper discusses the performance of the newly introduced theoretical concepts for some commonly used real valued test functions in different dimensions. The main purpose of this additional analysis is to underpin the generality of the newly introduced concepts and methods. As no problem specific knowledge or local search is involved, the identical algorithm can be applied as in the tests for combinatorial optimization problems like the TSP - just the problem representation and the crossover and mutation operators have to be exchanged with standard operators known from GA literature for real valued encoding (e.g. ref. 8).

The test functions that are used in this section have been designed by several authors for analyzing and comparing different optimization algorithms. Most of the test functions have especially been designed to detect weak points of the different methods. In order to demonstrate the capability of offspring selection for achieving high quality solutions for very difficult optimization problems in a general way, we have selected the following test functions that are considered to be more difficult and whose degree of difficulty can by scaled up by increasing the dimension of the search space due to the exponentially increasing number of local minima (except the unimodal Rosenbrock function whose difficulty is given by the extremely flat region around the global optimum).

- **The n-dimensional Rosenbrock function:**

$$f(\vec{x}) = \sum_{i=1}^{n-1} 100 \cdot (x_{i+1} - x_i^2)^2 + (1 - x_i)^2$$

for $-2.048 \leq x(i) \leq 2.048$ with a global minimum of $f(\vec{x}) = 0$ at $\vec{x} = (1, 1, 1, \ldots\ldots 1)$.

- **The n-dimensional Rastrigin function:**

$$f(\vec{x}) = 10 \cdot n + \sum_{i=1}^{n} x_i^2 - 10 \cos(2 \cdot \pi \cdot x_i)$$

for $-5.12 \leq x(i) \leq 5.12$ with a global minimum of $f(\vec{x}) = 0$ at $\vec{x} = (0, 0, 0, \ldots\ldots 0)$.

- **The n-dimensional Schwefel function (Sine Root):**

$$f(\vec{x}) = 418.982887272433 \cdot n + \sum_{i=1}^{n} -x_i \sin(\sqrt{|x_i|})$$

for $-500 \leq x(i) \leq 500$ with a global minimum of $f(\vec{x}) = 0$ at $\vec{x} = (c, c, c, \ldots\ldots c)$ with c = 420.968746453712.

- **The n-dimensional Ackley function:**

$$f(\vec{x}) = 20 + e - 20 \cdot e^{-0.2\sqrt{\frac{1}{n}\sum_{i=1}^{n} x_i^2}} - e^{\frac{1}{n}\sum_{i=1}^{n}\cos(2\cdot\pi\cdot x_i)}$$

for $-32.768 \leq x(i) \leq 32.768$ with a global minimum of $f(\vec{x}) =$ 0 at $\vec{x} = (0, 0, 0, \ldots\ldots 0)$.

As all considered test functions have a unique global minimum of 0 we state the results of the following tables as the absolute difference between the achieved result and 0. Many contributions interpret a solution to be globally optimal if it is in a certain range (e.g. 10^{-5}) around the global optimum. For reasons of objectiveness we state the exact values of the achieved solutions as it frequently happens that it still takes plenty of iterations to improve the solution quality from a value of 10^{-5} to 10^{-20} for example. Consequently, a displayed result of 0.000000000 has to be interpreted to be smaller than the lower limit of the C# double data type.

Tab. 1 shows the parameter settings of the GA with offspring selection for the different benchmark test functions.

Table 1. Parameter values used in the test runs of the GA with offspring selection for the different test functions

Parameters for the n-dimensional Rosenbrock function	
Population Size	100
Elitism Rate	1
Mutation Rate	0.05
Selection Operator	Roulette
Crossover Operators	Discrete, Average, Heuristic
Mutation Operator	One-position Uniform Mutation
Success Ratio	0.6
Comparison Factor	1.0 (constantly)
Maximum Selection Pressure	30
Parameters for the n-dimensional Rastrigin, Schwefel (Sine Root), and Ackley functions	
Population Size	500
Elitism Rate	1
Mutation Rate	0.05
Selection Operator	Roulette
Crossover Operators	Discrete, Average, Heuristic
Mutation Operator	One-position Uniform Mutation
Success Ratio	0.9
Comparison Factor	0.9 - 1.0
Maximum Selection Pressure	25

Many new theoretical serial and parallel GA concepts are analyzed considering the highly multimodal Rastrigin, Ackley, Griewangk or Schwefel's Sine Root function (e.g. [9], [10], [11], etc.). However, these papers analyze rather low problem dimensions around $n = 10$ for the Rosenbrock function and about $n = 20$ or $n = 30$ for the other test functions. Under this perspective it is quite remarkable that the GA with offspring selection is

Table 2. Experimental results achieved for the benchmark test function in various dimensions

Problem Dimension	Best	Average	Evaluated Solutions
Results for the Rosenbrock function			
n=20	$1.24 \cdot 10^{-25}$	$5.70 \cdot 10^{-25}$	2'926'231
n=100	$1.37 \cdot 10^{-8}$	$2.23 \cdot 10^{-5}$	16'672'523
n=500	$2.31 \cdot 10^{-9}$	$1.02 \cdot 10^{-8}$	181'217'124
n=1000	$4.11 \cdot 10^{-8}$	$2.59 \cdot 10^{-7}$	493'559'814
Results for the Rastrigin function			
n=20	0.000000000	0.000000000	4'791'050
n=100	0.000000000	0.000000000	10'718'323
n=500	0.000000000	0.000000000	52'423'489
n=1000	$4.26 \cdot 10^{-9}$	$1.03 \cdot 10^{-7}$	67'145'078
n=5000	$4.35 \cdot 10^{-5}$	$4.32 \cdot 10^{-3}$	167'019'244
Results for the Schwefel (Sine Root) function			
n=20	$-1.34 \cdot 10^{-12}$	$-2.10 \cdot 10^{-11}$	735'411
n=100	0.000000000	0.000000000	3'321'411
n=500	$7.31 \cdot 10^{-9}$	$1.54 \cdot 10^{-4}$	17'311'622
n=1000	$5.21 \cdot 10^{-5}$	$2.76 \cdot 10^{-3}$	34'703'482
n=5000	$4.32 \cdot 10^{-4}$	$5.88 \cdot 10^{-3}$	146'235'943
Results for the Ackley function			
n=20	$4.51 \cdot 10^{-17}$	$7.83 \cdot 10^{-16}$	1'171'012
n=100	$3.11 \cdot 10^{-15}$	$7.16 \cdot 10^{-14}$	5'278'856
n=500	$2.85 \cdot 10^{-13}$	$8.78 \cdot 10^{-9}$	39'433'869
n=1000	$6.31 \cdot 10^{-8}$	$1.26 \cdot 10^{-7}$	95'327'512
n=5000	$3.21 \cdot 10^{-6}$	$7.65 \cdot 10^{-5}$	164'976'566

able to find the global optimum of all those test functions up to a problem dimension of $n = 1000$ for the Rosenbrock function, and even up to $n = 5000$ for Rastrigin, Ackley, Griewangk and Schwefel without any additional algorithmic adaptations, i.e. with absolutely the same algorithm as used for the combinatorial optimization problems like the TSP. Just the operators (crossover and mutation) have been replaced by standard crossover operators for real-valued encoding.

4 Conclusion

In this paper an enhanced GA selection concept has been presented and exemplarily tested on some real-valued test function with a scalable degree of difficulty. The proposed GA-based techniques couple aspects from Evolution Strategies (variable selection pressure) with established selection, crossover and mutation concepts known from GA theory in a general and self-adaptive way. Therefore established crossover and mutation concepts may be used analogously to the corresponding GA and offspring selection can be applied in various fields of GA applications.

The self-adaptive selection pressure steering makes sure that the actually used selection pressure is just high enough to successfully guide genetic search. Given an upper limit of selection pressure to be applied, this model also acts as a precise detector heuristics for premature convergence.

Furthermore, the potential of offspring selection for Genetic Programming (GP) applications is analyzed within the scope of a current research project [12]. First results for the identification of nonlinear structures in mechatronic systems by means of Genetic Programming indicate that offspring selection operates especially well in combination with genetic programming.

5 References

[1] Fogel, D.B. (1994) An introduction to simulated evolutionary optimization. IEEE Transactions on Neural Networks 5 (1): 3–14

[2] Cavicchio, D.J. (1970) Adaptive Search using Simulated Evolution. Unpublished doctoral dissertation, University of Michigan, Ann Arbor

[3] De Jong, K.A. (1975) An Analysis of the Behavior of a Class of Genetic Adaptive Systems. PhD. Thesis. University of Michigan

[4] Goldberg, D. E. (1989) Genetic Alogorithms in Search, Optimization and Machine Learning, Addison Wesley Longman

[5] Larranaga, P., Kuijpers, C.M.H., Murga, R.H., Inza, I., Dizdarevic, S. (1999): Genetic algorithms for the travelling salesman problem: A review of representations and operators. Artificial Intelligence Review 13: 129–170

[6] Affenzeller, M., Wagner S. (2004): SASEGASA: A new generic parallel evolutionary algorithm for achieving highest quality results. Journal of Heuristics - Special Issue on New Advances on Parallel Meta-Heuristics for Complex Problems, vol. 10: 239-263

[7] Wagner S., Affenzeller M. (2004): HeuristicLab - A generic and extensible optimization environment. Proceedings of ICANNGA 2005

[8] Dumitrescu, D., Lazzerini B, Jain L.C., Dumitrescu, A. (2000): Evolutionary computation. CRC Press

[9] Potter, M.A., De Jong K. (1994): A cooperative coevolutionary approach to function optimization. In: Parallel Problem Solving from Nature – PPSN III, pp. 249–257

[10] Hiroyasu, T., Miki M., Hamasaki M, Tanimura, Y. (2000): A new model of distributed genetic algorithm for cluster systems: Dual individual DGA. High Performance Computing, Lecture Notes in Computer Sicence 1940, pp.374–383

[11] Takahashi, O., Kita, H., Kobayashi S. (1999): A distance dependent alternation model on real-coded genetic algorithms. IEEE Transactions on Systems, Man, and Cybernetics 24(4): 619–624

[12] Winkler S., Affenzeller M., Wagner S. (2004): Identifying nonlinear model structures using genetic programming techniques. In: Proceedings of the European Meeting on Cybernetics and Systems Research - EMCSR 2004, pp. 689–694

Using Genetic Algorithms with Real-coded Binary Representation for Solving Non-stationary Problems

Jiří Kubalík
Department of Cybernetics, CTU Prague, Czech Republic
E-mail: kubalik@labe.felk.cvut.cz

Abstract

This paper presents *genetic algorithms with real-coded binary representation* - a novel approach to improve the performance of genetic algorithms. The algorithm is capable of maintaining the diversity of the evolved population during the whole run which protects it from the premature convergence. This is achieved by using a special encoding scheme, introducing a high redundancy, which is further supported by the so-called *gene-strength adaptation mechanism* for controlling the diversity. The mechanism for the population diversity self-regulation increases the robustness of the algorithm when solving non-stationary problems as was empirically proven on two test cases. The achieved results show the competitiveness of the proposed algorithm with other techniques designed for solving non-stationary problems.

1 Introduction

A weak point of conventional genetic algorithms (GAs) is that they often suffer from the so-called *premature convergence*, which is caused by an early homogenization of the genetic material in the population. The mechanisms for maintaining the population diversity become extremely important when GAs are used for the optimization in non-stationary environments. Thus GAs are extended to be capable of continuously adapting to changes in the environment when searching for optimal solutions. The simplest way of reacting to changes in the environment is to restart the genetic algorithm [1] or increase the mutation rate to re-introduce the diversity to the population [2] whenever a change of the system is detected. The drawback of such approaches is their poor ability to reuse the information gained in the past. Other approaches are based on using redundant representations. The dual GAs [3] extend standard binary representation by the so-called head-bit, which is used to determine the way the coding genes of the chromosome will be interpreted. A frequently used approach to redundant representations is the diploidy and dominance with multiple alleles for each gene position [4], [5] and a dominance switching mechanism [6]. However, results achieved with the multiploid representations published so far indicate that the approach is suitable for the non-stationary environments with only a few periodically changing states.

The next section introduces the proposed real-coded binary representation. Section 3 describes the test problems used for the experimental evaluation. The results are presented in section 4 and the final comments conclude the paper.

2 GAs with Real-coded Binary Representation

The motivation for the proposed real-coded representation is to introduce a redundancy into the genetic code that would make the genetic algorithm resistant against the premature convergence. This is achieved so that the chromosomes are not formed of direct zeros and ones as in the case of the conventional genetic algorithms with binary representation. Instead, each gene is expressed by a real number r from the interval $\langle 0.0, 1.0 \rangle$, which is interpreted as a binary gene 0 or 1 as follows:

$$\begin{aligned} interpretation(r) &= 1, \text{ when } r > 0.5 \\ &= 0, \text{ when } r < 0.5. \end{aligned}$$

The threshold value 0.5 is disabled in order to have unbiased interpretations of zeros and ones. The real-valued gene expressions represent a *gene strength*. The value $r = 0.0$ represents the strongest binary 0, and $r = 1.0$ represents the strongest 1. Then as the value of r changes from 1.0 to 0.5 the strength of a binary one decreases and vice versa. Similarly this applies to a binary zero as the value of r changes on the interval 0.0 to 0.5. Such a pseudo-binary representation provides many possibilities for expressing any arbitrary binary chromosome. For example the two chromosomes $C_1 = [0.9, 0.1, 0.2, 0.6]$ and $C_2 = [0.7, 0.2, 0.4, 0.9]$ are interpreted as the same binary chromosome $[1, 0, 0, 1]$.

The standard crossover operators can be used with this representation. First, the offspring is composed of parts taken from the parental chromosomes and then the gene strength of each gene is changed by the so-called *gene-*

strength adjustment mechanism. It plays a crutial role in maintaining the population diversity. The extent to which the genes will be adjusted depends (i) on their interpretation and (ii) on the relative frequency of ones at certain position in the chromosome. For this purpose a vector $P[]$ with relative frequency of binary ones at each position of the representation is maintained. For example the vector $P = [0.8, 0.17, 0.35, 68]$ shows that there is 80% of ones at the first, 17% of ones at the second, 35% of ones at the third, and 68% of ones at the last position of the chromosomes.

The gene at the i-th position is weakened proportionally to $P[i]$ if it interprets as the binary value that is more frequently sampled at the i-th position in the current population and vice versa, the gene at the i-th position is strengthened proportionally to $P[i]$ if it represents the binary value that is less frequently sampled at the i-th position. If the real-valued gene represents binary value that prevails in the population at the given position then the gene is weakened – the value is moved towards 0.5, so (1) is used for zero-valued genes (i.e. for $gene < 0.5$) and (2) is used for one-valued genes (i.e. for $gene > 0.5$), respectively. In opposite case the gene is strengthened – zero-valued genes are moved towards 0.0, so (2) is applied and one-valued genes are moved towards 1.0 using (1), respectively.

$$gene' = gene + c * (1.0 - P[i]) \tag{1}$$

$$gene' = gene - c * P[i] \tag{2}$$

Parameter c denotes the *maximal adjustment step* defining the maximal value each gene can be changed by in one adjustment step. It can take on values from the interval $(0.0, 0.5)$. However useful values are rather small ranging between 0.0 and 0.2.

The idea behind the introduced manipulation with the strength of the genes is as follows. In standard evolutionary algorithms the above-average building blocks are used more often for creating the new population that ends up in an undesirable homogenization of the population genotype. Here, the genes that are likely to prevail the population are weakened each time they are used in the newly created individual. Such genes are gradually approaching the value 0.5 and at some point of this weakening process they are adjusted so that they turn over the critical value 0.5 and change their interpretation from one to zero and vice versa, respectively. Even when some binary gene completely overwhelm the population at some position it is just a matter of time when the opposite value will emerge and reproduce again. In the opposite case, the binary genes that are sampled in the current population with below-average frequency are strengthened in the newly created individuals in order to make them more likely to survive and reproduce.

An important thing is that the weakening and strengthening of the genes is driven by the demand of the population for the needed gene-values at the given positions. There is no need for any explicit mutation operator, since *all the genes are* passed into the new population already *modified by the gene-strength adjustment mechanism*, which can be considered as an *implicit mutation*.

The gene-strength adjustment offers a possibility to distinguish between "promising" and "ordinary" individuals. Let us consider a newly generated individual that is better than both of its parents as a promising one. As such its genotype should be retained in the population with an unchanged interpretation for some time. In order not to change its interpretation by the adjustment too early all its genes are rescaled to be strong before it is inserted into the new population (Note that the generational evolutionary model is used in this work). This means that the genes interpreted as ones are set by random to be close to 1.0 and genes interpreted as zeros are set to be close to 0.0. For example the individual $o = (0.71, 0.45, 0.18, 0.57)$ would be rescaled to $o' = (0.97, 0.03, 0.02, 0.99)$. Obviously the latter one is more likely to survive unchanged for certain number of generations than the original one. This rescaling mechanism boosts up the exploitation ability of the algorithm.

3 Test Problems and Experimental Setup

The first non-stationary problem we used is the Ošmera's dynamic problem [7]. The basis of the fitness function is the function $g_1(x, t)$, where variable x is represented by a bit-string of length 31 and normalized to give a value in the range (0.0,2.0), $t \in \{1, 1000\}$ is the time-step and the function $c(t)$ specifies the changes in the optimal bit-string. The goal is to keep on tracking the optimal value of x, which minimizes the function $g_1(x, t)$. We used $c(t)$ of variant 2 introduced in [4], where the optimum changes every 20 generations and the changes are non-monotonous.

The second non-stationary problem is an oscillating version of the single knapsack problem. The goal is to fill a knapsack using a subset of objects from an available set of size n, such that the sum of object weights is as close as possible to the target weight t [5]. The problem uses 14 objects, each of them of the weight $w_i = 2^i$, where i ranges from 0 to 13. The optimum fitness is 1.0 and the fitness decreases towards 0.0 as the Hamming distance to the optimum solution string increases. The target weight oscillates between values 12643 and 2837 every 1500 generations as used in [5].

50 independent runs were carried out for each experiment and the results were averaged. The performance of GARB has been evaluated and compared to the standard genetic algorithm (SGA) and to the algorithms used in

[4] and [5]. The comparisons are based on the quality of the solutions achieved after the same number of fitness function evaluations, which is a commonly used condition when comparing different evolutionary algorithms. For the Ošmera's dynamic problem a *tracking error* calculated as the difference of the optimal fitness and the fitness of the best individual in each generation was used as the performance measure.

Parameters of GARB and SGA: population size 400 (Osmera's), 150 (Knapsack); tournament selection with $N = 3$; generational evolutionary model with elitism (the best individual is copied to the new population); crossover: 2-point (Osmera's), uniform (Knapsack) with the crossover rate 1.0; no mutation operator is used in GARB; a simple bit-swapping mutation is used in SGA; #fitness evaluations 400k (Osmera's), 30k (Knapsack).

4 Experiments and Results

Table 1 provides comparison of the GARB with the standard genetic algorithm (SGA) and the genetic algorithm extended with a case-based memory, a re-insertion strategy and a predictor (CBM-B) as presented in [4]. The performance is measured by the ability to track the optimum solution as it changes over the time. This is denoted as the overall mean % tracking error (MTE), which was calculated as an average tracking error over all generations and all experiments. A standard deviation ($StDev$) of the tracking error is calculated as well. Right after change measures MTE and $StDev$ show the tracking error observed in the generations in which the change of the optimum solution took place.

We can see that GARB's tracking capabilities are much better than those of the SGA and CBM-B. GARB performed best with $c = 0.225$ however, already with $c = 0.075$ it outperformed the other algorithms. This might be attributed to the fact that the GARB is able to maintain a diverse population even when the optimal solution had already been found. Due to such a permanent exploration activities the right after change MTE is already small.

Table 1. Comparison of GARB and other algorithms on Osmera's function.

Algorithm	Right after change MTE	Right after change $StDev$	Overall MTE	Overall $StDev$
GARB $c = 0.025$	83.3	30.6	50.4	25.2
GARB $c = 0.075$	25.6	34.6	2.4	7.4
GARB $c = 0.125$	12.8	22.4	1.0	3.9
GARB $c = 0.175$	10.2	19.7	0.7	3.0
GARB $c = 0.225$	**9.2**	**19.3**	**0.6**	**2.7**
SGA binary	N/A	N/A	57.3	43.61
SGA Gray	N/A	N/A	47.66	42.94
CBM-B	N/A	N/A	19.39	33.13

Table 2. Comparison of GARB and other algorithms on Oscillating knapsack problem. A number of runs (out of 50) in which optimum was achieved in each period is used as the performance measure.

Algorithm	Oscillation period 1	2	3	4	5	6	7	8
Oscillation period 10 generations								
GARB $C = 0.025$	46	0	7	0	17	6	0	0
GARB $C = 0.075$	45	19	5	11	9	16	7	10
GARB $C = 0.125$	**41**	**35**	**31**	**37**	**38**	**32**	**36**	**35**
GARB $C = 0.175$	32	28	25	28	29	33	35	21
Oscillation period 20 generations								
GARB $C = 0.025$	47	1	0	1	0	6	3	2
GARB $C = 0.075$	50	46	34	42	28	43	29	40
GARB $C = 0.125$	**49**	**49**	**49**	**47**	**49**	**49**	**50**	**50**
GARB $C = 0.175$	46	42	44	44	48	39	46	44
Haploid-Recover	45	44	33	45	33	44	29	43
Extended-Additive	43	29	44	42	39	40	45	37
Ng-Wong	32	21	41	25	34	27	32	26

Table 2 compares GARB with results obtained for the Haploid-Recover, Extended-Additive and Ng-Wong approaches as presented in [5]. The steady-state evolutionary model with the uniform crossover was running for 1500 generations between two subsequent target weight changes. This equals to either 1500 or 3000 fitness evaluations depending on whether the uniform crossover operator produces 1 or 2 offsprings. Our implementation of GARB uses generational model. Since we did not know the number of offsprings generated by the crossover in [5] we carried out experiments with two different oscillation periods – 10 and 20 generations – in order to change the target after 1500 and 3000 fitness evaluations, respectively. The results achieved with GARB for oscillation period 10 generations and $c = 0.125$ are better than those of Ng-Wong but slightly worse than those of Haploid-Recover and Extended-Additive. When the oscillation period 20 generations was used, GARB (with c set to 0.125) outperformed all other algorithms.

GARB's capability of maintaining the population diversity is best illustrated in Figure 4. Its capability of recovering from a completely homogeneous population is shown there on Knapsack problem. The knapsack problem was initialized by strings representing the target 12643 while the optimum target was 2837. The real gene values were picked from interval $\langle 0.0, 0.0 + c\rangle$ for zero-valued genes and from $\langle 1.0 - c, 1.0\rangle$ for one-valued

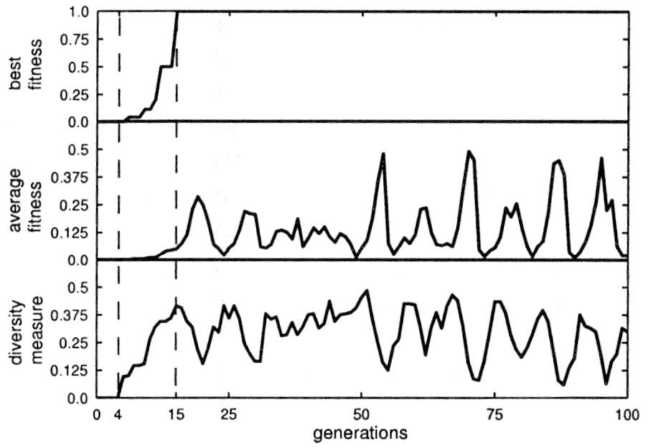

Fig. 1. Demonstration of GARB's ability to recover from the homogeneous population on knapsack problem.

genes where c was set to 0.1. The population diversity calculated as the relative frequency of unique binary chromosomes in the population is 0.0. We see that there nothing happened, regarding the observed measures, for the first four generations. The reason is that four generations is just the time needed to adjust the gene values to get close to the discrimination value 0.5. Then as some gene values change the interpretation the population starts to diversify which allows the algorithm to search for better solutions. The optimum solution was found in the 16th generation. The important observation is that the population remains diverse even after the optimum has been found. Notice the correlation of the diversity measure with the average fitness. When the average fitness reaches its peak the diversity drops down and vice versa. It illustrates the fact that the population can get homogeneous only to some extent. Due to the effect of the gene-strength adjustment mechanism on the gene values they become very weak at some point so that they are reverted to the opposite binary interpretations forcing the population to diversify again. The phases of homogenization and diversification repeat in cycles so that the evolution can never get stuck.

5 Conclusions and Future Work

This paper introduces a novel genetic algorithm well suited for solving non-stationary problems. It is based on a redundant pseudo-binary representation with the binary genes represented by real numbers where all the values greater than 0.5 are interpreted as binary ones and the values lower than 0.5 are interpreted as binary zeros, respectively. The *gene-strength adjustment mechanism* is proposed to control the diversity of the evolved population so that the real-valued gene corresponding to the binary value that prevails at the given position in the population is weakened and pushed towards the opposite binary interpretation. The real-valued genes interpreted as the minor binary value are strengthened making it more likely to remain in the population and reproduce.

The proposed algorithm was empirically evaluated on two non-stationary test problems proving that it represents a competitive alternative to the techniques designed for the non-stationary optimisation. The algorithm exhibits a strong capability of self-controlling of the population diversity that extends its explorative power and makes it capable of recovering even from a completely homogeneous population.

An analysis and a deep understanding of the algorithm's behavior will be the primary goal of the future work. The original version of the algorithm uses a generational model where the convergence characteristics of the population is updated after the whole population has been generated. The steady-state model could provide a faster response to the state of the population since the statistics could have been updated after inserting each single individual. The implementation and an analysis of the gene-strength adjustment mechanism with the steady-state model are subject to future research as well.

References

[1] Fukunaga, A. (1997) Restart scheduling for genetic algorithms. In Thomas Back, editor, Proceedings of ICGA'97, 1997.

[2] Morrison, R. W., De Jong, K. A. (2000) Triggered hypermutation revisited. In Proceedings of CEC'2000, pp. 1025-1032

[3] Collard, P., Gaspar, A. (1996) Royal-road landscapes for a dual genetic algorithm, in W. Wahlster, editor, ECAI 96: 12th European Conference on Artificial Intelligence, Wiley & Son, pp 214-217

[4] Eggermont, J., Lenaerts, T. (2002) Dynamic Optimization using Evolutionary Algorithms with a Case-based Memory, Proceedings of BNAIC'02, K.U.Leuven, Belgium, pages 107-114

[5] Lewis, J., Hart, E., Ritchie,G. (1998) A comparison of dominance mechanisms and simple mutation on non-stationary problems. In Eiben, A.E. et al. (Eds.). Proceedings of PPSN'98, LNCS 1498, pp. 139-148, Springer

[6] Ng, K. P., Wong, K. C. (1995) A new diploid scheme and dominance change mechanism for non-stationary function optimization. In Proceedings of ICGA'95, pp. 159-166. Morgan Kaufmann

[7] Ošmera, P., Kvasnička, V., Pospíchal, J. (1997) Genetic algorithms with diploid chromosomes. In Proceedings of Mendel'97, pp. 111-116

Dynamics in Proportionate Selection.

Abhishek Agrawal, Ian Mitchell, Peter Passmore, Ivan Litovski
[1]Middlesex University, Hendon, London, NW4 4BT, U.K.
{a.agrawal, i.mitchell, p.passmore, i.litovski} @ mdx.ac.uk

Abstract

This paper proposes a new selection method for Genetic Algorithms. The motivation behind the proposed method is to investigate the effect of different selection methods on the rate of convergence. The new method Dynamic Selection Method (DSM) is based on proportionate selection. DSM functions by continuously changing the criteria for parent selection (dynamic) based on the number of generations in a run and the current generation. Results show that by using DSM to maintain diversity in a population gives slower convergence, but, their overall performance was an improvement. Relationship between slower convergences, in GA runs, leading to better solutions, has been identified.

1 Introduction

Genetic Algorithms (GA) consists of many parameters, which may be optimised to a specific problem or problem domain. The parameters of most interest are those, which contribute to rapid development of a solution that is near-optimal. Analysis of a basic GA gives us five main parameters, which may be optimised to produce an algorithm optimal for generating a solution for a specific problem [1,2]. First, population size is important and can be changed throughout the GA. Second, Parent Selection mechanisms. Third, the Recombination techniques, which produce offspring. Fourth, replacement Ratio of Parents with Offspring, and fifth, Mutation [3]. This paper focuses only on Selection mechanisms as this is considered to be the main component which determines the character of the evolutionary search, whether it is volume orientated (convergence reliability) or path orientated (convergence velocity)[3].

There is an increasing interest in evolutionary algorithms that have dynamic parameters, which change with respect to time or generations. However most of our understanding is limited to selection mechanisms, which are static. The most commonly used algorithms for parent selection are Fitness Proportional Selection [4], and Linear Ranking [5] which are all static as their behaviour does not change with respect to generations. Tournament selection [4] is dynamic only in the sense that parents are selected by a new random function.

2 Features of Dynamic Selection Method (DSM)

This section will focus on a novel Dynamic Selection Method (DSM). Analysis of the DSM will provide insight into the effects of dynamic selection on the main properties of a GA such as rate of convergence. To test DSM an optimisation problem is used as the test bed. The optimisation problem chosen is the Traveling Salesman Problem [6] and instances are taken from TSPLIB [7].

Dynamic Selection Method is based on Proportional Selection. To characterise the probability distribution in DSM some assumptions have been made to simplify calculations. Primarily, all individuals that may occur in a population and are ordered with respect to fitness values

$$Pop(t) = \left\{ \vec{a}_1(t), \vec{a}_2(t), \cdots, \vec{a}_\mu(t) \right\}$$

$\Phi(\vec{a}_i(t))$ so that the first individual $\Phi(\vec{a}_1(t))$ should have the highest fitness and the last individual $\Phi(\vec{a}_\mu(t))$ the lowest. Also, each individual may be identified by the index $i \in \{1, 2 \ldots\ldots \mu\}$ such that,

$$p_s(\vec{a}_i(t)) = \mathrm{P}\{\vec{a}_i(t) \in s(Pop(t))\}$$

where p_s is probability of an individual being selected, $s(Pop(t))$ denotes a selection operator.

In order to make this algorithm dynamic we will use the concept of transposed fitness; $\Phi' = f(\Phi)$ where Φ' is the transposed fitness, f is the objective function and Φ is the original fitness. This approach will allow us to modify the algorithm itself by modifying the objective function.

As with all parent selection the sum of all probabilities should equal to 1. In DSM, the probability that an individual will be selected for mating (P_i) is equal to the fitness value $\Phi'(\vec{a}_i)$ divided by the sum of all fitness.

$$P_i = \frac{\Phi'(\vec{a}_i)}{\sum_{\forall a_j \in Pop} \Phi'(\vec{a}_j)}$$

As in proportionate selection an objective function (f) determines the selection probability of an individual P_i. In proportionate selection the objective is defined as $f(x) = x^C$ where the constant C may not be equal to zero.

In DSM, we use a non-constant, time dependent C to evaluate the objective function.

3 Dynamic Selection Method

Traditional methods of proportional selection rely on ranking the population (Pop) based on the individual's fitness. This gives a higher probability of selecting fitter individuals at the beginning of the run. Usually the implementation requires a random number being generated (0,1) representing the probability of selection. This then equates to the individual selected. The objective function responsible for calculating the individual's transposed fitness is in the equation below:

$$f(x) = x^{C(n)}$$

DSM tries to maintain diversity by changing the probability of selection of the individuals over generations. DSM works by periodically altering the selection criteria of the parents. This change happens every generation therefore C is defined as a function of the number of generations (n), where $C(n)$ for DSM may be defined as:

$$C(n) = \begin{cases} \frac{1}{N}n + \frac{1}{2}, & 0 < n < \frac{N}{2} \\ \frac{2n}{N}, & \frac{N}{2} \le n < N \end{cases}$$

Where N is the maximum number of generations.

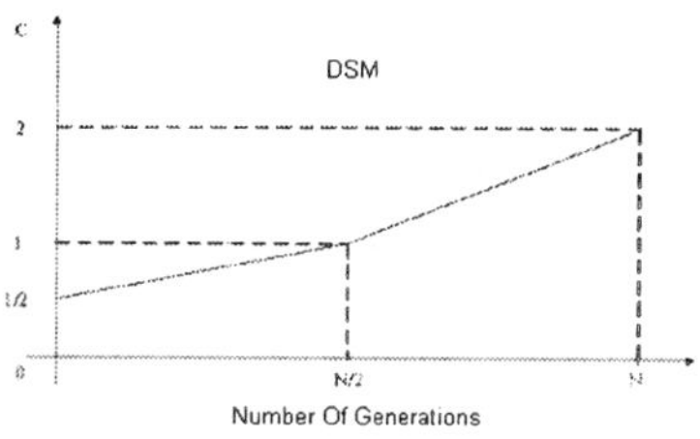

Figure 1. shows the increase in the function ($C(n)$) during a single GA run in DSM.

4 Inverse Dynamic Selection Method (IDSM)

It was also considered important to see the inverse effects of DSM where $C(n)$ follows a linear change from an exponent of 2 to 0.5. This is defined as IDSM and is shown below in Fig.2.

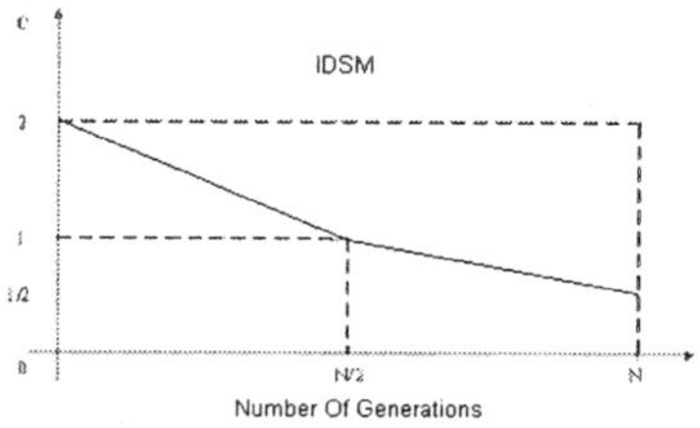

Figure 2. shows the decrease in the function ($C(n)$) during a single GA run in IDSM .

In essence, the function for IDSM changes as in Fig.3.

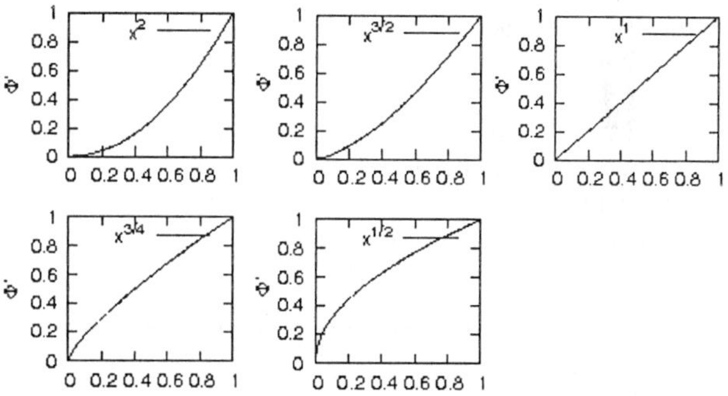

Figure 3. Shows changes in the objective function, Φ', during a single GA run for IDSM.

DSM follows the inverse of the objective function, Φ', shown in Figure 3.

5 Experimental Design

The effectiveness of DSM and IDSM were tested against the more traditional methods such as Proportional selection (PS), Ranking selection (RS) and Tournament

Selection (TS). The parameters that were used for the first genetic algorithm experiment are shown in Table 1.

Parameters	Value
Population size	2000
Replacement Ratio	100%, generational
Mutation rate	0.5%
Crossover	2 point Cyclic Crossover
Parent Selection	DSM,IDSM,RS,TS (5),PS
Max. Generations	2000

Table 1: GA Parameters.

For the experiments, Cyclic Crossover (CX) [8], was preferred over Partially mapped crossover (PMX) [4,9]. The reason for this preference is due to diversity. When using PMX without incest prevention, the selection of two identical individuals results in the same offspring and therefore diversity is reduced. Therefore CX was preferred which lowers the probability of identical offspring and hence increases the diversity.

TSP [7] was chosen as a "test bed" to test the results. The problems were selected from TSPLIB to see the effectiveness of the parent selection schemes on increasingly difficult problems.

6 Results

This paper is not preoccupied with the optimality of the results but rather the rate of convergence to near optimal solutions. Therefore, not all other parameters have been optimised for performance.

Table 2 presents the results achieved for a particular TSP instance. The results were verified by running each experiment 20 times. The numbers at the end of the TSP instance names indicate problem size, i.e. Atta48 is a 48 city problem.

EXP 1	DSM	IDSM	RS	TS	FPS
Atta48	**37878**	45200	41598	45535	42414
Eil51	540	539	550	563	**493**
Berlin52	9756	9953	**8829**	10562	9335
Eil76	846	824	844	**713**	818
Ch130	15147	14214	36641	**11611**	15890

Table 2. shows the average results over **2000** generations. The best results for each TSP instance are show in **Bold.**

As the optimality of results is not a focus of this work, it was considered more important to look at the approach by which the results were achieved by the various algorithms.

Other parent selection methods (especially Tournament selection) tend to have a much faster rate of convergence and reach stagnation very quickly. RS did not seem to perform very well under these parameters in comparison to other algorithms but has been shown to work effectively [10].

7 Further Results

Analysis of the data indicated that DSM could have much greater diversity (lower rate of convergence) and would reach the point of stagnation at a much later point, giving better results. To test this criteria all experiments were re-run with 3000 generations to see the effects of greater diversity produced by DSM. Table 3 presents the results achieved for a particular TSP instance. These results were also verified by running the experiment 20 times.

EXP 2	DSM	IDSM	RS	TS	FPS
Atta48	**32014**	44207	41590	42954	41418
Eil51	**447**	531	536	530	487
Berlin52	8910	9648	**8826**	9962	9238
Eil76	**646**	744	831	648	784
Ch130	**11350**	13996	35288	11357	15234

Table 3. shows the average results over **3000** generations. The best results for each TSP instance are show in **Bold.**

8 Conclusions

In this paper a novel selection method is proposed and tested on an optimisation problem. The dynamic properties of the proposed selection method assign different probability of selecting individuals with the same objective value, but, in different generations. Such a technique shows considerable favour with regards to DSM in its ability to maintain diversity in the population and thus produce better solutions.

In DSM the diversity is maintained due to selecting a higher percentage of weaker individuals in initial generations. This comes at the expense of the rate of convergence, however, the results shows that the point of stagnation is reached later. Therefore, slower rates of convergence and high diversity are needed to produce good solutions in GAs.

The data in tables 2 and 3 are summarised in table 4 and shows the effect of running the GA an extra 1000 generations. As can be seen many of the other methods have reached their optimality and have nearly converged and little improvement, if any in some cases, is shown.

The maximum reduction was 9% for TS and the minimum was none for RS. This should be compared with DSM that had not converged at 2000 generations since when run for extra 1000 iterations the optimality was improved significantly in all cases. The maximum reduction for DSM was 25% and the minimum was 9%.

% decrease	DSM	IDSM	RS	TS	FPS
Atta48	**15**	2	0	6	2
Eil51	**17**	1	3	6	1
Berlin52	**9**	3	0	6	1
Eil76	**24**	10	2	9	4
Ch130	**25**	2	4	2	4

Table 4. shows the difference in percentages between average results in Table 2 and 3.

Two new dynamic selection paradigms were tested; DSM and IDSM. Overall DSM performed better than IDSM. This is due to the fact that IDSM is biased towards selecting fitter individuals in earlier generations and thus affecting intermediate generations which soon become stagnated. However, DSM selects weaker individuals in early generations thus maintaining diversity and avoiding stagnation in intermediate generations.

EXP1	Best	EXP2	Best	Decrease
DSM	37878	***DSM***	32014	15%
FPS	493	***DSM***	447	9%
RS	8829	RS	8826	0%
TS	713	***DSM***	646	9%
TS	11611	***DSM***	11350	2%

Table 5. shows the decrease in TSP distance between the best values in experiment 1 and 2.

In table 5 a comparison was drawn between the best TSP routes achieved in 2000 (EXP1) generations and the best routes achieved in 3000 (EXP2) generations. DSM is not shown to be apparent in 2000 generations as it only produced one best result but after 3000 generations DSM has outperfomed other parent selection mechanisms in all but one case, where it achieved 8910 as opposed to RS's *8826* a 0.94% difference, indicates DSM to be a useful parent selection method.

We have tested the new dynamic parent selection methods DSM and IDSM against TS, RS, and FPS and shown that dynamic selection methods, i.e. TS, DSM and IDSM, tend to produce better results than static selection methods, i.e. RS and FPS when run over extra 1000 iterations.

The changes in DSM were controlled by objective functions that ranged from ½ to 2 further works in this area could reveal better functions and ranges. Also, dynamics in other areas of EA may provide improvements. Work is currently being carried out on the effects of dynamic mutation and replacement.

References

[1] Holland, J. (1975). Adaptation In Search, Optimization & Machine Learning. Addison-Wesley, 1989Natural and Artificial Systems. The University of Michigan Press, Ann Arbor.

[2] M. Mitchell. An Introduction to Genetic Algorithms. MIT Press, Cambridge, MA, 1996.

[3] Back T. (1996). Evolutionary Algorithms in Theory and Practice, Oxford University Press, NY.

[4] Goldberg, D. E. and Deb, K., "A Comparative Analysis of Selection Schemes Used in Genetic Algorithms," in FGA1, pp.69-93, 1991

[5] Baker, J. E.: Adaptive Selection Methods for Genetic Algorithms. Proceedings of an International Conference on Genetic Algorithms and their Application, pp. 101-111, Hillsdale, New Jersey, USA: Lawrence Erlbaum Associates, 1985

[6] E. Lawler, J. Lenstra, A. Rinnooy Kan, and D. Shmoys. The Traveling Salesman Problem. John Wiley, 1985.

[7] G. Reinelt. TSPLIB - a traveling salesman problem library. ORSA J. Computing, 3:376-384, 1991.

[8] I. Oliver, D. Smith, and J. Holland. A study of permutation crossover operators on the travelling salesman problem. In Grefenstette J.J.

[9] Larrañaga, P, C. Kuijpers, R. Murga, I. Inza and S. Dizdarevich. "Evolutionary algorithms for the travelling salesman problem: A review of representations and operators", Artificial Intelligence Review, Vol. 13 (1999), pp. 129-170.

[10] Whitley, D. (1989). The GENITOR Algorithm and Selective Pressure: Why Rank-Based Allocation of Reproductive Trials is Best. In Proc. 3th International Conf. on Genetic Algorithms. 116, D. Schaffer, ed., Morgan Kaufmann.

Generating grammatical plant models with genetic algorithms

Luis E. Da Costa, Jacques-André Landry
Laboratoire d'Imagerie, Vision, et Intelligence Artificielle (LIVIA)
École de Technologie Supérieure, Montréal, Québec (Canada)
E-mail: costa@livia.etsmtl.ca, jacques.landry@etsmtl.ca

Abstract

A method for synthesizing grammatical models of natural plants is presented. It is an attempt at solving the inverse problem of generating the model that best describes a plant growth process, presented in a set of 2D pictures. A geometric study is undertaken before translating it into grammatical meaning; a genetic algorithm, coupled with a deterministic rule generation algorithm, is then applied for navigating through the space of possible solutions. Preliminary results together with a detailed description of the method are presented.

1 Background and motivation

The detailed study of a set of plants from an agricultural field or a forest is a precious source of information about their health, the treatments that the plants have undergone and, consequently, about the appropriate management strategies. However, there is a physical impossibility in bringing to the field the specialized equipment needed to perform such a study. An approach to solve this constraint is to build a detailed model in order to obtain a-priory knowledge about the plants we wish to observe or to perform virtual computer studies. Our first attempt is to model dichotomous trees such as the great Maple.

Based on the self-similarity observed in its development (Mandelbrot [6], Barnsley [1]), a plant can be thought of as a natural ("*existing in nature*") representation of a fractal; therefore, it is modeled using tools created by the fractal research community. We are specially interested in a family of models called **L**-Systems (for **L**indenmayer Systems, [5]). With the addition of simple geometric features, **L**-Systems have been used extensively for visualization of natural developmental processes (see, for example, [7] and [8])[1].

However, *observing* a natural growth process and then *generating* a model that describes it is still more of an art than a structured approach more suitable to science. In section 2 we present this *inverse problem*; a solution is proposed in 3. Preliminary results are discussed in 4, and we conclude in 5.

[1] In this paper, we won't go into any deeper details concerning the basic concept and definition of **L**-Systems. These can be found in [9]

2 Evolving a model to describe plant growth: the inverse problem

2.1 The inverse problem for a fractal

As mentioned, a tree can be represented using its fractal characteristics. To represent a given tree, one must search for a grammar that generates a fractal that best represents the sought tree. The question to solve is then: how to find a model to generate a *specific* fractal figure?

Answers from different disciplines exist to related questions: in architecture, Coates *et al* ([2], chap. 14) combined **L**-Systems with genetic programming for exploring designs. In [10], Lindenmayer and Prusinkiewicz generated **FASS** (*space-***F***illing, self-***A***voiding,* **S***imple and self-***S***imilar*) curves with graph rewriting techniques. In [11], Vanyi *et al.* developed a system for graphically describing the circulation of blood in a human retine, coupling evolutionary methods coupled with a restricted form of **L**-Systems. And, finally, the *Collage Theorem* establishes a way of solving the inverse problem using *Iterated Function Systems* (**IFS**) ([1]).

2.2 The inverse problem for plant growth

The *growth process* of a plant can be described as a suite of fractal figures, each one representing a specific stage in time of this process; a *model* for this suite then describes the growth process of this plant. The question to answer at this stage is then: having *observed* a growth process of a plant (or tree), what is the model that best describes it?

The use of **L**-Systems (see [5, 7, 8, 9] for details) in an attempt to answer this question is quite appropriate as the recursive nature of the grammar usage makes them generate intermediate results in discrete time steps, much like observing a plant growth in predefined, equally spaced time intervals. Prusinkiewicz and Hanan, in [9], extensively demonstrate the usefulness of **L**-Systems for answering this question with what is known as *computational botany*. Somehow guided by a different objective, Koza, in [4], uses genetic programming for generating the (one and only) rule of an **L**-System.

The question that we address in this research project goes still a step further, adding a geometrical component: the object whose growth process is studied is three-dimensional, but the temporal information that we have is two-dimensional (a suite of 2D pictures). With this new restriction in mind, we propose in section 3 a method for generating a growth model for a tree (expressed as an **L-S**ystem grammar), having a set of 2D pictures. No hard restriction is set on the *shape* of the pictures (they don't have to be taken from any specific point in space) nor on their *age* (*when* they were taken is not important, as long as we have this information).

3 Proposal

This research problem, as stated, implies working with 2 parallel views of trees: the *geometric* representation, that defines how the tree is formed as a composition of shapes, and the *syntactic* representation: the **L-S**ystem string that is interpreted as the named geometric figure. We have to be able to handle both views at the same time, as the objective is to generate the same geometry, but this can only be done through the correct generation of a syntactic model.

In this section, a characterization of the kind of solutions we are looking for is described (3.1), and how this is represented in the syntactic domain is inferred (3.2).

3.1 Geometric considerations

Our goal being to model natural trees, the geometric nature of tree growth should be reflected in the chosen **L-S**ystem rules. We defined 3 ways in which a tree grows, exemplified with a series of drawings. If the shape of a given *young tree* is as in Fig. 1(a), the *natural* growth is a combination of **Y-*shape*** (Fig. 1(b)), **T-*shape*** (Fig. 1(c)) and **B-*shape*** (Fig. 1(d))

3.2 Grammatical translation

As shown in Fig. 1(b,c,d), the different types of growth amount to different ways of combining the "young tree" while forming a larger tree. We have defined 3 syntactic operators, named **B**, **T**, and **Y**, translating the geometric relations explained in Fig. 1 to **L-S**ystem notation. These are called the *growth* operators:

1. **Fig. 1(d)**, representing **B**, would be $S + (a)F(b)$. Extending it to a three dimensional space and specifying the contexts, the complete **L-System** rule would be:

   left ctx. $< B(a, b, c) >$ right ctx. $\rightarrow S+(a)\&(c)F(b)$

   This notation presents the advantage of having non-fixed contexts, allowing for a family of rules

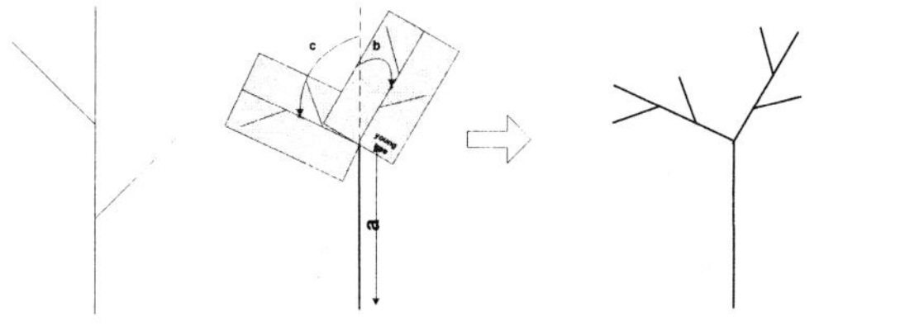

(a) Young tree

(b) **Growth in *Y-Shape***: a tree has a trunk of length a, and then the young tree appears rotated at angles b and c from the vertical

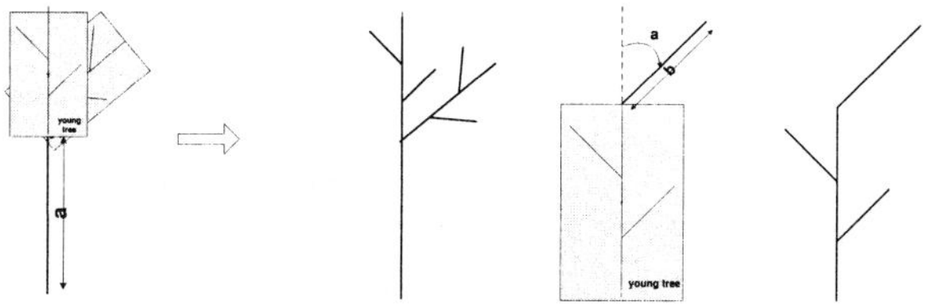

(c) **Growth in *T-Shape***: a tree has a trunk of length a, and then the young tree appears rotated an angle b from the vertical

(d) **Growth in *B-Shape***: from the young tree grows a line of length b rotated an angle a from the vertical

Fig. 1. Growth of a tree

   of this shape. Following this idea, we also derive a set of rules with the geometry of Fig. 1(b) and (c) (in three dimensions):

2. **Fig. 1(b)**:

$$Y(a, b, c, d) \rightarrow F(a)[+(b)\&(d)S][-(c)\&(d)S]$$

3. **Fig. 1(c)**:

$$T(a, b, c) \rightarrow F(a)[+(b)S]S$$

Now that we have defined the operators, we can take a closer look at the *primordial* components of the tree (Fig. 1(a)); this "young tree" has a specific meaning as part of an **L-S**ystem: it is the *axiom*, S. As mentioned, the operators only use combinations of this first tree for building the whole tree; so it is important to choose a large enough *family of axioms* for representing a large number of different type of trees, if we want our method to be general.

We have chosen this family to have the following syntactic form: $P_{ts}GP_{ts}GP_{ts}$, where:

- G is one of B, T, or Y (correspondingly parameterized) and
- P_{ts} is the **L-S**ystem string of a *primordial tree shape.*

This primordial shape is a generalization of the shape shown in Fig. 2. It is represented by the concatenation

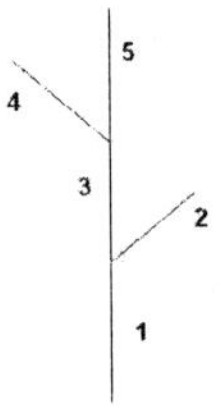

Fig. 2. General primordial tree shape

of 5 parameterized **L-S**ystem regular expressions: part 1, 3, and 5 in Fig. 2 are $F(l)^{0/1}$ (*"straight line of length l"*), and part 2 and 4 are $[\pm(\alpha_1)\&(\alpha_2)F(l)]^{0/1}$ (*"straight line of length l, growing from the vertical trunk at a 3D angle specified by α_1 and α_2"*). The exponent $0/1$ translates the possibility of presence/absence of the term.

3.3 The L-System family of solutions

These *translations* generate a *parametric context-sensitive* **L-**System family (Table 1, with rules representing objects in 3D)[2].

Table 1. The shape of the family of grammars

► 12 *reserved symbols*: $\mathcal{R}_S$ = {F,f,+,-,∧,&,\,/,(,),[,]}
► *Alphabet*: $\mathcal{T} = \mathcal{R}_S \cup \mathbb{R}$
► A *parameterized Axiom* $\mathcal{S} \in \mathcal{T}^+$
$\mathbf{lc}^b_1$ < B(a,b,c) > $\mathbf{rc}^b_1$ → S+(a)&(c)F(b)
⋮
$\mathbf{lc}^b_n$ < B(a,b,c) > $\mathbf{rc}^b_n$ → S+(a)&(c)F(b)
$\mathbf{lc}^y_1$ < Y(a,b,c,d) > $\mathbf{rc}^y_1$ → F(a)[+(b)&(d)S][-(c)&(d)S]
⋮
$\mathbf{lc}^y_m$ < Y(a,b,c,d) > $\mathbf{rc}^y_m$ → F(a)[+(b)&(d)S][-(c)&(d)S]
$\mathbf{lc}^t_1$ < T(a,b,c) > $\mathbf{rc}^t_1$ → F(a)[+(b)&(c)S]S
⋮
$\mathbf{lc}^t_k$ < T(a,b,c) > $\mathbf{rc}^t_k$ → F(a)[+(b)&(c)S]S

3.4 Method: generating a grammar

Now the problem is more specific:

> *Having a list of figures $\mathcal{L}_t$, find a grammar of the family defined in Table 1 whose graphical interpretation best approaches $\mathcal{L}_t$*

Our approach for solving this is shown in Alg. 1

[2] lc^r_i stands for *left context i for rule type r* and rc^r_j for *right context j for rule type r*

Algorithm 1 Grammar synthesis

```
1: procedure GENERATEGRAMMAR(L_t: target figures; P_t: spatial and temporal information about L_t)
2:   repeat
3:     Choose a parameterized grammar M_a from the axiom space   ▷
4:     Generate the best set of production rules (P_R) for the current axiom   ▷
5:     G = M_a ⊕ P_R   ▷ current grammar
6:   until G is good enough
7:   Return G
8: end procedure
```

The exploration of the axioms' space (line 1.3) is implemented as a genetic algorithm, described in 3.4.1. The production rules generation (line 1.4) is described in (3.4.2), and the fitness function for the whole process is presented in 3.4.3.

3.4.1 Choosing axioms: Each symbol of the axiom (as described in 3.2 and Table 1) is encoded as a gene in the chromosome, which is then 64 genes long. Straightforward 1-point crossover and mutation operators are implemented on such representation.

3.4.2 Navigating the production rules space: In this approach, we derive only the rules that could possibly expand a specific axiom. We use information from Alg.1: generation of the rules comes *after* choosing a possible axiom. The general method calculates each possible outcome of a string, starting from the axiom, and then derives the rule that would generate this string. For example, if the axiom is $\mathcal{S}: B(1,2,3)F(4)$, then its first derivation will be

$$\mathcal{S}{+}(1)\&(3)F(2)F(4) \qquad \textbf{or} \qquad B(1,2,3)F(4)$$

depending on whether there is a rule in the grammar G that replaces $B(1,2,3)$ or not. So G will be either G_1 or G_2:

$$G_1 = \begin{cases} \text{Axiom} = & \mathcal{S}: B(1,2,3)F(4) \\ \mathbf{B(a,b,c)} > F \rightarrow \mathcal{S} + (a)\&(c)F(b) \end{cases}$$

$$G_2 = \begin{cases} \text{Axiom} = & \mathcal{S}: B(1,2,3)F(4) \\ & \emptyset \end{cases}$$

Each possibility defines a current work string, that in turn could be replaced by the rules of a grammar; this iterative procedure is repeated for the number of generations indicated in the information about the target figures.

In that way, starting from the axiom, we build a graph in which each node is the state of the generation of a string, and whose leading transition is the rule that was applied in order to obtain the string (see Fig. 3). This type of graph does not have cycles, so it is a tree; each path from the root to each one of the leaves is a complete grammar, and a possible solution to the optimization problem. Finding the *best grammar* is now equivalent to *evaluating each grammar path with an appropriate fitness function (3.4.3) and choosing the one with better fitness*.

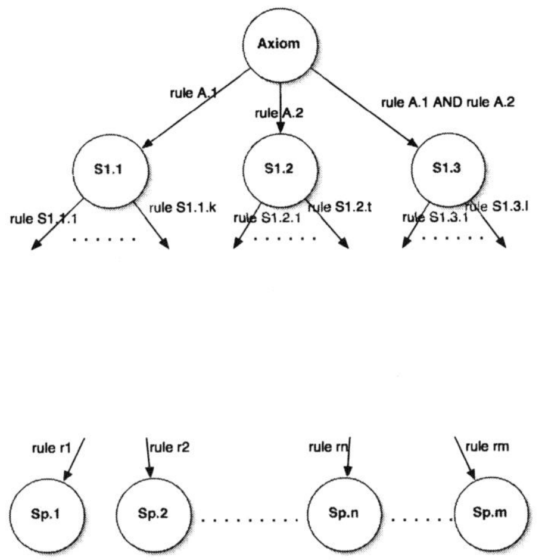

Fig. 3. Grammar evaluation graph

3.4.3 Fitness function: The fitness of a grammar G is a *distance* between a list of target figures $\mathcal{L}_t$ and the figures generated by G. The *distance* between 2 figures is evaluated by a comparison between their respective fractal dimensions (F, calculated using the box-counting algorithm) and their size (S):

$$\mathcal{D}(f_1, f_2) = \alpha*(F(f_1)-F(f_2))+(1-\alpha)*(S(f_1)-S(f_2))$$

α is a weighting estimator of the importance of each measure; for this paper, we fixed $\alpha = 0.5$

The fitness of a grammar G ($\mathcal{F}(G)$) is the sum of all the distances $\mathcal{D}$

4 Results and discussion

We implemented an environment to experiment with the visual appearance and with the formal definitions of **L-Systems** ([3]). We then built a synthetic database of 2D pictures of trees generated using 3D grammars. The proposed method is being evaluated for its ability to generate grammars that can evolve trees corresponding to the source synthetic pictures.

For this first set of tests we aimed at finding perfect fitness (fitness = 0, because *fitness* as in 3.4.3 is a sum of distances). Fitness being the ending condition in Alg. 1, this severe rule, associated with the cardinality of the axiom space being $\sim 10^{10}$, meant that the exploration of the whole space was unthinkable, even with very powerful machines. However, such a drastic constraint (fitness =0) was useful in evaluating the proposed approach when dealing with simple geometric figures and a subset of the grammar. Not only did this allowed to verify if we generated similar trees, but also similar grammars (remember that we are using synthetic trees).

To this end, we studied the geometry of grammars with simple axioms and one growth rule, either B, T, or Y. Under certain conditions, the geometry of each class of grammars is very distinctive, and is presented (in its 8th. iteration) in Fig. 4. We gave as inputs to our method

(a) Grammar with axiom: F[+F]F[-F]FB(30,1,45) and the B growth operator as the sole rule

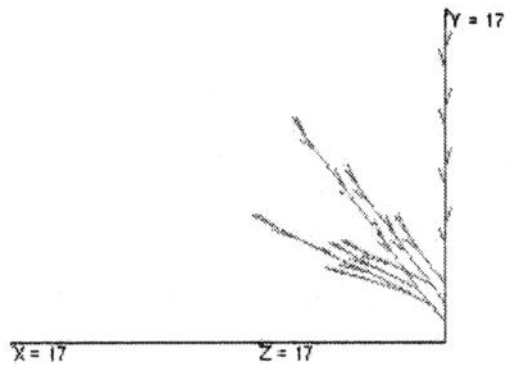

(b) Grammar with axiom: T(1,30,45)F[+F]F[-F]F and the T growth operator as the sole rule

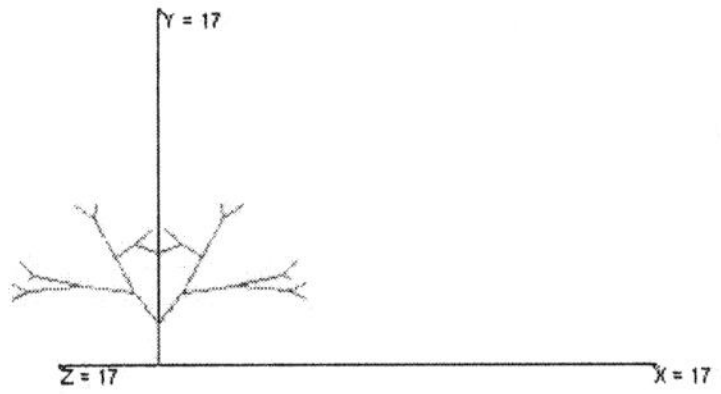

(c) Grammar with axiom: FY(1,45,45,45) and the Y growth operator as the sole rule

Fig. 4. Input to our experiences

the sequences generated using each of the growth rule, and hoped to obtain not only trees that were similar but also the same grammar as entered. We repeated the experiences 10 times with each sequence, and present here some of our observations:

1. for all replicates, the resulting grammar generated a tree that was similar to the input picture
2. in 8 replicates out of 10, the algorithm did not find the exact grammar as entered, and the search continued even though the solutions were very close and the fitness near 0.

3. in 2 replicates out of 10, the stochastic process quickly generated a perfect solution.

4. the solutions were generated in reasonable amounts of time. However, once again, the ending condition requiring a perfect fit (fitness = 0) did not allow the algorithm to stop when acceptable solutions were encountered.

5 Conclusion and future work

These first results are very promising, and demonstrate that the proposed approach is a solution to the inverse problem. We now have to improve the intuitiveness of our fitness function: make it closer to our expectations when comparing 2 natural trees. We, as humans, do not expect 2 trees to be *exactly* similar in 2 different pictures. However, we can recognize different pictures as coming from the same tree. To this end, additional criteria, such as comparing the number of segments, the compactness, or even the texture of the trees, would definitely improve the value returned by the fitness function.

Another important consideration is in devising a more objective way of evaluating our method. This is actually quite difficult to achieve, as we are evaluating the results as humans, in a subjective manner. However, we think that we would greatly benefit from a form of "mathematical evaluation" of our results.

We believe that the proposed approach for generating grammatical models from example is novel and very promising, even though the presented results are still preliminary. The next step will be to generate a consistent set of results, along with extending it to more complicated grammars (for example, stochastic). Finally, we would like to draw some conclusions about the correctness, usefulness and robustness of the proposed method.

References

[1] Michael F. Barnsley. *Fractals everywhere*. Academic Press Limited, London, 2 edition, 1993.

[2] P J Bentley. *Evolutionary Design By Computers*. Morgan Kaufmann, San Francisco, CA, 1999.

[3] Luis E Da Costa and Jacques-André Landry. A convivial visualization environment for l-systems. In *FRACTAL 2004: 8th International Multidisciplinary Conference*. Vancouver, BC, 2004.

[4] J.R. Koza. Discovery of rewrite rules in lindenmayer systems and state transition rules in cellular automata via genetic programming. In *Symposium on Pattern Formation (SPF-93)*. Claremont, California, 1993.

[5] A Lindenmayer. Mathematical models for cellular interaction in development. parts i and ii. *Journal of Theoretical Biology*, 18:280–299 and 300–315, 1968.

[6] B B Mandelbrot. *The fractal geometry of nature*. San Francisco, 1982.

[7] P Prusinkiewicz. Graphical applications of l-systems. In *Proceedings of Vision Interface '86*, pages 247–253. 1986.

[8] P Prusinkiewicz. Applications of l-systems to computer imagery. In H Ehrig, M Nagl, A Rosenfeld, and G Rozenberg, editors, *Graph grammars and their applications to computer science; Third international workshop*. Springer-Verlag, 1987.

[9] P Prusinkiewicz. Lindenmayer systems, fractals, and plants. *Lecture Notes in Biomathematics*, (79), 1989.

[10] P Prusinkiewicz and A Lindenmayer. *The Algorithmic Beauty of Plants*. Springer-Verlag, New York, 1990.

[11] Robert Vanyi, Gabriella Kokai, Zoltan Toth, and Tiinde Reto. Grammatical retina description with enhanced methods. In *Proceedings of Genetic Programming 2000*. 2000.

Golomb Rulers: Experiments with Marks Representation

Jorge Tavares[1], Francisco B. Pereira[1,2], Ernesto Costa[1]
[1]Centre for Informatics and Systems of the University of Coimbra,
Polo II - Pinhal de Marrocos, 3030 Coimbra, Portugal
[2]Instituto Superior de Engenharia de Coimbra,
Quinta da Nora, 3030 Coimbra, Portugal
E-mail: {jast, xico, ernesto}@dei.uc.pt

Abstract

We present a set of experiments regarding an evolutionary algorithm designed to efficiently search for Optimal Golomb Rulers. The approach uses a binary representation to codify the marks contained in a ruler and relies on standard genetic operators. Furthermore, during evaluation, an insertion and correction procedure is implemented in order to improve the algorithm performance. This method is successful in quickly identifying good solutions, outperforming previous evolutionary approaches by discovering optimal and near-optimal Golomb Rulers.

1 Introduction

A Golomb Ruler [1] is defined as a ruler that has marks unevenly spaced at integer locations in such a way that the distance between any two marks is unique. Unlike usual rulers, they have the ability to measure more discrete measures than the number of marks they possess. Additionally, they are not redundant since they do not measure the same distance twice. An Optimal Golomb Ruler (OGR) is defined as the shortest length ruler for a given number of marks [2]. There may exist multiple different OGRs for a specific number of marks. Golomb Rulers have application in a wide range of real world situations, mainly engineering [3].

For a small number of marks it is possible to construct OGRs by hand. As the number of marks increases, the problem becomes difficult and, for $n \geq 9$, computational approaches are required to find possible solutions. Currently, most of the techniques used to identify OGRs rely on parallel brute force algorithms. However, computation is prohibitively costly in terms of computer resources [3]. Evolutionary Computation (EC) approaches are a promising alternative to previous methods that usually need too much time to obtain an answer. To the best of our knowledge, there are some applications of EC to this problem [4], [5], [6] and [7]. In this paper, we present an aditional set of experiments using the approach presented in [6].

2 Problem Definition

A n-mark Golomb Ruler is an ordered set of n distinct nonnegative integers $\{a_1, a_2, \ldots, a_n\}$ such that all possible differences $|a_i - a_j|$, $i, j = 1, \ldots, n$ with $i \neq j$, are distinct. Values a_i correspond to positions where marks are placed. By convention, the first mark a_1 is placed on position 0, whereas the length of the ruler is given by the position of the rightmost mark a_n.

The length of a segment of a ruler is defined as the distance between two consecutive marks. This way, it is also possible to represent a Golomb Ruler with n marks through the specification of the length of the $n - 1$ segments that compose it. The Golomb Ruler $\{a_1, a_2, \ldots, a_n\}$ is an OGR if there exists no other n-mark ruler having a smaller largest mark a_n. In such a case a_n is called the length of the n-mark OGR (OGR-n, for short).

3 Evolution with Marks

We propose a representation where candidate solutions encode information about the maximum number of marks that can be placed on a ruler with a given length L [6]. A straightforward chromosome representation is adopted allowing us to evolve both the number and the position of the marks.

An individual is codified as a binary string with length L, where each bit is associated with an integer position in the ruler. If a given gene has the value 1 then it indicates that there is a mark in the corresponding position. This representation, where individuals represent sequences of marks in a ruler bounded by an upper length limit, enables us to search for rulers without a predetermined number of marks.

To generate descendants we selected variants of standard genetic operators usually applied to binary representations: two-point crossover and shift mutation. Shift mutation acts in the following way: it randomly selects a mark and shifts it left or right with equal probability. We selected this operator because it is somewhat sensitive to the structure of the solutions. Moving, deleting or

Table 1. Overall results achieved by Marks-EC approach with maximum ruler lengths between length 90 and 200

	CX = 0.25					CX = 0.5					CX = 0.75				
L	n	len	avg	std	runs	n	len	avg	std	runs	n	len	avg	std	runs
90	12	85	87.6	2.6	19	12	85	87.1	2.5	19	12	85	86.8	2.4	20
100	12	85	89.6	2.7	30	12	85	89.6	2.4	30	12	85	89.3	2.5	30
110	13	106	107.4	1.7	9	13	106	109.0	1.3	8	13	106	109.0	1.4	14
120	13	106	111.3	1.8	30	13	109	111.2	1.4	30	13	109	111.5	1.5	30
130	14	129	129.0	0.0	3	14	129	129.0	n/a	1	14	129	129.3	0.5	4
140	14	129	134.7	2.6	30	14	129	134.4	3.1	29	14	129	135.6	2.4	30
150	14	133	136.2	2.0	30	14	129	136.1	2.5	30	14	127	137.1	3.0	30
160	15	159	159.0	n/a	1	15	155	158.8	1.9	6	15	155	156.7	2.9	3
170	15	156	163.5	2.7	30	15	158	163.6	2.4	30	15	160	164.2	1.8	30
180	15	161	165.8	2.2	30	15	163	165.8	1.9	30	15	157	166.7	2.6	30
190	16	189	189.6	0.5	5	16	188	189.3	1.2	3	15	164	169.0	2.1	30
200	16	191	196.1	2.4	30	16	188	194.4	5.9	29	16	190	195.3	2.5	29

inserting a mark can cause a great disruption in a ruler. Shift mutation is applied to all individuals. The number of shift mutations applied to each chromosome is upper bounded by a small constant S.

To evaluate an individual we consider two criteria: ruler length and legality of the solution (i.e., whether it contains repeated measurements). The equation used to assign fitness to an individual x is the following:

$$fit(x) = \begin{cases} -\text{NR} & \text{, if } x \text{ is illegal} \\ \text{NM} \times \text{L} + (\text{L} - \text{M}) & \text{, if } x \text{ is legal} \end{cases} \quad (1)$$

Where *NR* is the number of repeated measurements, *NM* is the number of marks, L is the maximum length of the ruler and M is the length of the ruler encoded in the individual. This formula ensures that: the fitness of a legal solution is always higher than the fitness of an illegal solution; invalid solutions are ranked according to their illegality (number of repeated measurements); legal solutions are ranked according to their number of marks. Rulers with an equal number of marks are ranked according to their length.

A correction and insertion procedure is performed during the evaluation of an individual. The aim is twofold: to fix invalid rulers and to see if it is possible to add marks to legal solutions. The line of action is straightforward: a mark is randomly removed if, when checking the segments measured by the ruler, a duplicate measurement is found. A constant C gives the maximum number of rectifications allowed for a given solution. We use this upper bound so that the evaluation of an individual does not take too long. If a valid ruler is obtained at the end of the correction, then the second stage can proceed: an insertion operator tries to add marks in every possible position ensuring that the obtained ruler is still valid. The order in which the positions are selected for insertion is randomly selected.

4 Experimental Results

To evaluate our approach (which henceforth we will designate by Marks-EC) we performed a set of experiments with several ruler lengths. More precisely, we used the evolutionary algorithm to seek for good rulers with lengths between 90 and 200 (given the known OGRs we will be looking for rulers between 12 and 16 marks). The settings of the EC algorithm are the following: Number of generations: 20000; Population size: 200; Tournament selection with tourney size: 5; Elitist strategy; Two-point crossover with rates: $\{0.25, 0.5, 0.75\}$; Maximum number of shifts, $S = 3$; Maximum number of rectifications, $C = 4$.

For every ruler length L we performed 30 runs with the same initial conditions and with different random seeds. All initial populations were randomly generated, where each gene will have a value of 0 with probability of 0.9 (to increase the probability of having some legal rulers in the initial population).

In table 1, we present a brief summary of the attained results. Column "L" refers to the maximum ruler length, "n" is the number of marks of the best ruler found and "len" is the corresponding ruler length. Columns "avg" and "std" indicate the average and standard deviation of the best rulers found in each run. While the evolutionary algorithm performed 30 runs, not all of them were able to have their best rulers with the same number of marks. When this occurred, the averages presented in the table reflects only the runs with the highest number of marks

Table 2. Overall results achieved by Marks-EC approach with flip mutation (probability of 0.005)

	CX = 0.25					CX = 0.5					CX = 0.75				
L	n	len	avg	std	runs	n	len	avg	std	runs	n	len	avg	std	runs
90	12	85	87.0	2.5	23	12	85	87.5	2.6	20	12	85	87.1	2.5	19
100	12	85	90.1	2.1	30	12	85	89.7	2.5	30	12	85	90.2	2.1	30
110	13	106	108.6	1.7	15	13	106	108.9	1.3	14	13	106	108.9	1.4	13
120	13	109	111.3	2.0	30	13	109	111.8	1.4	30	13	109	111.8	1.6	30
130	14	128	128.5	0.6	4	14	128	128.6	0.5	5	14	129	129.3	0.6	3
140	14	129	135.2	2.2	30	14	129	136.2	2.3	30	14	127	136.2	2.7	30
150	14	129	137.0	2.6	30	14	136	138.3	1.5	30	14	128	138.3	2.5	30
160	15	155	158.0	2.6	3	15	155	158.8	2.5	4	15	158	158.5	0.7	2
170	15	155	164.9	2.6	30	15	160	164.9	2.3	30	15	158	164.8	2.5	30
180	15	163	167.1	1.7	30	15	163	167.3	2.0	30	15	163	167.3	2.0	30
190	16	177	185.8	5.9	4	16	184	184.0	n/a	1	15	165	170.8	2.2	30
200	16	187	196.5	2.6	26	16	192	196.2	2.0	26	16	189	196.5	7.7	28

(column "runs"). For example, with a crossover rate of 0.25 and length 110, only for 9 runs were found rulers with 13 marks (the average and standard deviation take into account only those 9 runs). It wouldn't be meaningful to compute all the runs.

A brief perusal of the results shows the effectiveness of our approach. In terms of best results, we found optimal or near-optimal solutions for the corresponding number of marks. In fact, below 15 marks, for all settings, the distance between the best solutions found by this approach and the optimal values never exceeds 2.8%. The single exception is the experiment with settings $L = 150$ and crossover rate 0.25, where this distance is 4.7%. Regarding 15 and 16 marks, the distance to the optimum is still reasonable, ranging between 2.7% and 8.6%. This observation can be reinforced by looking into the averages. The values are close to the best solutions discovered, showing that the algorithm can reliably find high quality rulers. The distances between the best solutions found and the averages never exceed 5.4% indicating that the algorithm is a robust approach.

Instances with 15 and 16 marks revealed to be the most difficult. This is an expected result since the search space topology defined by Golomb Rulers is very hard to sample. The number of interactions that occur between the genes that compose a chromosome is one of the reasons that contribute to the difficulty of the search. There is a high epistasis associated with this problem, since changing the position of a mark affects all other marks. As the number of marks increases, it is likely that there is a growth in the difficulty of controlling the effects of changes in the ruler.

The described approach does not have a standard mutation operator. To analyze if this option influences the search process, we performed a preliminary set of tests with the addition of flip mutation, a standard genetic operator for binary representations. The same experiments were run with a mutation probability of 0.005 and the results are presented in table 2.

Table 3. Best OGRs lengths for instances between 12 and 16 marks achieved by Marks-EC in comparison with previous approaches and optimal results

Instances	Optimal	S-EC	RKH-EC	Marks-EC
OGR-12	85	103	85	85
OGR-13	106	124	106	106
OGR-14	127	168	131	127
OGR-15	151	206	162	155
OGR-16	177	238	180	177

From the observation of this table is not possible to draw firm conclusions regarding the influence of flip mutation in the algorithm. There are not many significant alterations in the attained results. Further tests with alternative mutation operators and other rates are required. The most relevant result of this set of experiments was the discovery of the optimal value for the ruler with 16 marks. This solution was found with settings $L = 190$ and crossover rate 0.25. The study of the influence of the correction and insertion procedure which is performed during evaluation is done in another publication [6]. Results show that both correction and insertion help the EC algorithm to efficiently search for OGRs.

In table 3 we summarize the results achieved by our

Table 4. Averages of best runs comparison between Marks-EC approach with RKH-EC. Mutation rate = 0.

Instances	RKH-EC	CX 0.25	CX 0.5	CX 0.75
OGR-12	91.4	89.6	89.6	89.3
OGR-13	113.7	111.3	111.2	111.5
OGR-14	136.6	134.7	134.4	135.6
OGR-15	165.6	163.5	163.6	164.2
OGR-16	197.1	196.1	194.4	195.3

Table 5. Averages of best runs comparison between Marks-EC approach with RKH-EC. Mutation rate = 0.005.

Instances	RKH-EC	CX 0.25	CX 0.5	CX 0.75
OGR-12	91.4	90.1	89.7	90.2
OGR-13	113.7	111.3	111.8	111.8
OGR-14	136.6	135.2	136.2	136.2
OGR-15	165.6	164.9	164.9	164.8
OGR-16	197.1	196.5	196.2	196.5

algorithm and compare them with previous evolutionary approaches. We show the best solutions found for rulers containing from 12 to 16 marks.

Marks-EC was able to find the OGR for all tested instances with the single exception for 15 marks. Nevertheless, for OGR-15 the solution discovered is just 2.65% higher than the optimal value, whilst the previous best value found by an evolutionary approach is 7.28% higher. The overall results attained by our algorithm outperform previous approaches.

To conclude our analysis, it is important to understand how the global performance of the algorithm compares with the previous best EC approach, Random Keys with a Heuristic evolutionary algorithm. To achieve this goal, we will compare the averages of the best solutions found in the 30 runs. Comparing the two algorithms is a difficult task since we are not evolving rulers with a fixed number of marks (like the RKH-EC approach). To allow a fair interpretation, we will just consider experiments where all runs were able to reach the highest number of marks. Taking this into account, the experiments used for comparison are: OGR-12, $L = 100$; OGR-13, $L = 120$; OGR-14, $L = 140$; OGR-15, $L = 170$ and for OGR-16, $L = 200$. It is important to notice that, whenever possible we used the same settings as described in [5](e.g., number of generations, population size).

In table 4 and 5 we present the results. Although differences are not overwhelming, the averages achieved by the Marks-EC approach are always better (consult [5] for the RHK-EC averages). Moreover, in the experiments selected for comparison, the maximum ruler length is usually set to a value that constrains the individuals to have a larger final section of the ruler without marks. While it helps to find rulers with a higher number of marks, it may bring some difficulties when searching for shorter rulers. Nevertheless, this is an empirical comparison and conclusions must be regarded with care, since both algorithms rely on different representations influencing directly the evolutionary search.

5 Conclusions

Results achieved show that this evolutionary approach is effective since it was able to quickly discover good solutions. One of the main advantages of the representation used is that it does not require special precautions to build the initial population or specific genetic operators to ensure that feasible individuals are generated. Some additional scalability tests are required. Furthermore, we intend to extend our study in order to clarify the influence of the representation.

References

[1] Bloom, G., Golomb, S.: Applications of numbered undirected graphs. In: Proceedings of the IEEE. Volume 65. (1977) 562–570

[2] Dollas, A., Rankin, W., McCracken, D.: New algorithms for golomb ruler derivation and proof of the 19 mark ruler. IEEE Transactions on Information Theory **44** (1998) 379–382

[3] Rankin, W.T.: Optimal golomb rulers: An exhaustive parallel search implementation. Master's thesis, Duke University (1993)

[4] Soliday, S., Homaifar, A., G., L.: Genetic algorithm approach to the search for golomb rulers. In: Proceedings of the Sixth International Conference on Genetic Algorithms (ICGA-95). (1995) 528–535

[5] Pereira, F.B., Tavares, J., Costa, E.: Golomb rulers: The advantage of evolution. In: Proceedings of the 11th Portuguese Conference on Artificial Intelligence (EPIA'03). (2003) 27–33

[6] Tavares, J., Pereira, F.B., Costa, E.: Understanding the role of insertion and correction in the evolution of golomb rulers. In: Proceedings of the 2004 Congress on EC (CEC04). (2004) 69–76

[7] Cotta, C., Fernndez, A.J.: A hybrid grasp - evolutionary algorithm approach to golomb ruler search. In: Proceedings of Parallel Problem Solving from Nature (PPSNVIII). (2004)

Evolving Segments Length in Golomb Rulers

Jorge Tavares[1], Tiago Leitão[1], Francisco B. Pereira[1,2], Ernesto Costa[1]
[1]Centre for Informatics and Systems of the University of Coimbra,
Polo II - Pinhal de Marrocos, 3030 Coimbra, Portugal
[2]Instituto Superior de Engenharia de Coimbra,
Quinta da Nora, 3030 Coimbra, Portugal
E-mail: {jast, xico, ernesto}@dei.uc.pt, tleitao@student.dei.uc.pt

Abstract

An evolutionary algorithm based on Random Keys to represent Golomb Rulers segments, has been found to be a reliable option for finding Optimal Golomb Rulers in a short amount of time, when comparing with standard methods. This paper presents a modified version of this evolutionary algorithm where the maximum segment length for a Golomb Ruler is also part of the evolutionary process. Attained experimental results shows us that this alteration does not seems to provide significant benefits to the static version of the algorithm.

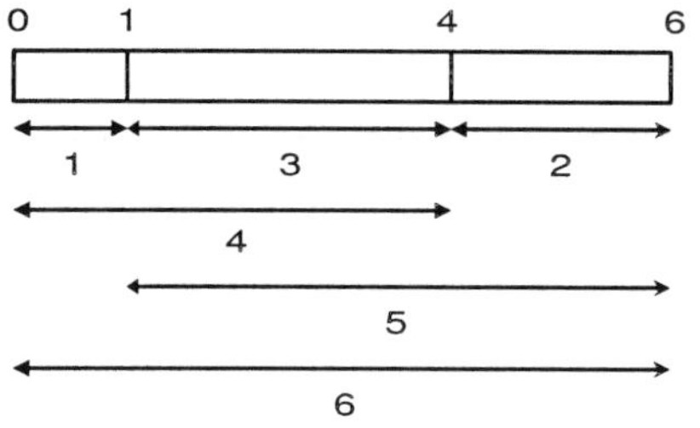

Fig. 1. A Golomb Ruler with 4 marks

1 Introduction

A Golomb ruler is defined as a ruler that has marks unevenly spaced at integer locations in such a way that the distance between any two marks is unique. They were named after the relevant work of the mathematician Solomon Golomb [1], and, unlike usual rulers, they have the ability to measure more discrete measures than the number of marks they carry. Also Golomb rulers are not redundant, since they do not measure the same distance twice. Figure 1 is an example of a Golomb Ruler.

Although the definition of a Golomb ruler does not place any restriction on the length of the ruler, researchers are usually interested in rulers with minimum length. An Optimal Golomb Ruler (OGR) is defined as the shortest length ruler for a given number of marks. There may exist multiple different OGRs for a specific number of marks. OGRs are used in a wide range of real world situations. For example, in the field of communications when setting up an interferometer for radio astronomy, placing the antennas on the marks of a Golomb ruler maximizes the recovery of information about the phases of the signal received [2].

Evolutionary Computation (EC) approaches are a promising alternative to brute force methods that usually need too much time to obtain an answer and so cannot be considered as a realistic option in real world situations. There have been some applications of EC to this problem [3], [4], [5] and [6]. These approaches, when searching for solutions evolve the length of a fixed number of segments. This way, during search EC algorithms try to discover good rulers for a specific number of marks. In [7] a different evolutionary approach is proposed. Prior to the application of the algorithm, a maximum ruler length is specified and then the search procedure tries to determine how many marks can be placed in such a ruler as well as where each one of the marks should be located. In this paper we continue the study presented in [5], by analyzing the influence of evolving the maximum length of a segment.

2 Golomb Rulers

In this section we present a formal definition of Golomb rulers. A n-mark Golomb ruler is an ordered set of n distinct nonnegative integers $\{a_1, a_2, \ldots, a_n\}$ such that all possible differences $|a_i - a_j|$, $i, j = 1, \ldots, n$ with $i \neq j$, are distinct. Values a_i correspond to positions where marks are placed. By convention, the first mark a_1 is placed on position 0, whereas the length of the ruler is given by the position of the rightmost mark an. The ruler from figure 1 can be defined as $\{0, 1, 4, 6\}$.

The length of a segment of a ruler is defined as the distance between two consecutive marks. This way, it is also possible to represent a Golomb ruler with n marks through the specification of the length of the $n - 1$ segments that compose it. According to this notation the example from figure 1 can be defined as $\{1, 3, 2\}$.

The Golomb ruler $\{a_1, a_2, \ldots, a_n\}$ is an OGR if there exists no other n-mark ruler having a smaller largest mark an. In such a case an is called the length of the n-mark OGR (OGR-n, for short).

Finding OGRs is a complex combinatorial optimization problem. Moreover, it has some specific features that differentiate it from other problems with similar characteristics, such as the Travelling Salesperson Problem (TSP). Whilst TSP can be classified as a complete ordered set (the goal is to find a permutation of the n cities that compose the problem), OGR can be considered as an incomplete ordered set [3]. Assume that we represent a ruler by a sequence composed by its segment's lengths. The OGR-n is a permutation of $n - 1$ elements taken from a set of m elements, where m is defined as the maximum distance between marks (usually $n \ll m$). The construction of such a solution poses several difficulties:

- Should a maximum value for m be pre-established or should it be adjusted during the construction of a ruler?
- How to select the $n - 1$ elements from a set of m values?
- How to build a valid permutation with the $n - 1$ elements selected?

3 An Evolutionary Approach with Random Keys

The representation chosen for individuals plays a crucial role on the performance of EC algorithms. In [5], an evolutionary approach based on the evolution of ruler segments is proposed. A candidate solution for an OGR-n instance is composed by a permutation of λ distinct values, where λ is the maximum segment length.

Even when λ is known, there are two crucial decisions to make when building a solution for OGR-n: how to select $n - 1$ distinct segments from the set of λ values and how to build a valid permutation with the selected elements. We adopted a representation that tries to deal efficiently with this situation:

- It provides a straightforward way to select which elements will compose the permutation;
- It finds a natural arrangement for the selected segments.

In our approach, the chromosome is composed by a permutation of λ distinct values. Encoding of the permutation is done with random keys (RK). RKs were introduced by Bean [8] in 1994 and obtained good results in situations where the relative order of the tasks is important [8], [9]. One of their main advantages is that it is possible to apply standard genetic operators to chromosomes (e.g., one point or uniform crossover) and still obtain feasible individuals. We will just present a brief overview of RKs. For a detailed description, consult [8] or [10]. RKs representation uses a sequence of N random numbers to encode a permutation of length N. These numbers are typically sampled from the real interval $[0, 1]$. Both the position and the value of the keys are important for the interpretation of the sequence. To obtain the permutation that corresponds to a given key sequence, all keys are sorted according to their values in decreasing order. Then, the original positions of the keys will be used to construct the permutation. For example, consider the following key sequence $r = \{0.5, 0.7, 0.3, 0.9, 0.4\}$. Position 4 contains the highest value of the key sequence (0.9), so 4 will be the first element of the resulting permutation. Then, the next highest value is at position 2. The ordering process continues in a similar way and at the end we get the permutation $\{4, 2, 1, 5, 3\}$. From a key sequence of length N we can always construct a permutation of N unique numbers between 1 and N (or between 0 and $N - 1$ if needed).

In [10], Rothlauf et. al. proposed NetKeys, an extension of RKs to problems dealing with tree network design. The situation addressed is that of the design of a minimum spanning tree over a fully connected graph with n nodes. In these circumstances, a NetKey sequence will be composed by $L = \frac{n(n-1)}{2}$ random numbers (the number of links in the graph). Positions are labelled and each one represents one possible link in the tree. The value of a particular key can be interpreted as the importance of the link it represents. The higher its value, the higher the probability that this link is used in the construction of the tree. From this sequence, a permutation of L numbers can be constructed in the same way as described for standard RKs. Then the construction of the tree begins: links are added to the tree in an order that is in accordance to the value of its key. If the insertion of a link would create a cycle, then it is skipped and construction continues with the next one. The process comes to an end when $n - 1$ links have been selected.

Our codification of a solution for an OGR-n follows the same principles as those expressed for NetKeys. Each one of the λ positions of the chromosome represents one possible segment. Without loss of generality, we assume that position i corresponds to a segment of length i $(i = 1, \ldots, \lambda)$. Also, just like with NetKeys, the value of a given key represents the importance of the related segment. If we compare both situations, there is nevertheless one additional difficulty associated with OGRs: the interpretation algorithm must determine, not only which segments will be part of the ruler, but also its

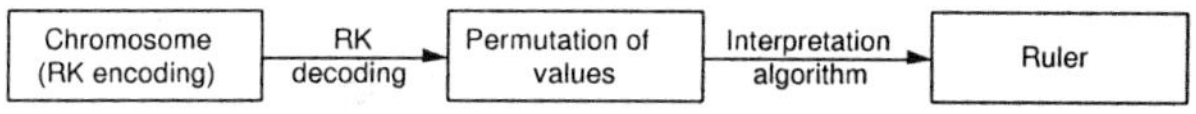

Fig. 2. Decoding and interpretation of the information contained in a chromosome

relative position. Figure 2 illustrates how decoding and interpretation (the two stages required to assign fitness to an individual) are related.

A step-by-step description helps to exemplify how the decoding of the permutation and subsequent interpretation of the information contained in a chromosome is performed. Consider that we are searching for OGR-5 and that λ is 10. Consider also that the chromosome encodes the following key sequence $\{0.87, 0.17, 0.67, 0.27, 0.86, 0.97, 0.71, 0.31, 0.38, 0.40\}$. After performing the RK decoding, the resulting permutation is $\{6, 1, 5, 7, 3, 10, 9, 8, 4, 2\}$.

During the interpretation phase, the first $n - 1$ valid segments from the permutation are used to build the ruler. The iterative algorithm used to build a valid ruler tries to ensure that segments are selected in such a way that no duplicate measurements exist. It is a deterministic process and segments on the left of the permutation have higher priority. Depending on the circumstances, it might happen that in a specific position all segments lead to duplicate measurements. If this situation arises, a random value between 1 and λ is chosen and a segment with this length is appended to the ruler.

Evaluation of an individual follows two criteria: ruler length and legality of the solution (i.e., whether it contains repeated measurements). The effect of the addition of a simple heuristic to the interpretation process is also analyzed [5]. Results presented show a small improvement in the performance of the EC algorithm.

In this paper, the value for λ is also evolved. This is accomplished in a simple manner. The value for λ is given by the chromosome length, this means that by having chromosomes with variable length each individual will have a different λ value. The only modification that is necessary to the previous approach, with fixed length, is made on the crossover operator, that for each individual, a different cut point is randomly selected. This will ensure the swap of genetic material of different lengths, thus providing individuals with variable length.

4 Experimental Results

To evaluate our approach we performed a set of experiments with several OGR instances. More precisely, we used the evolutionary algorithm to seek for good rulers with 10 to 17 marks. The settings of the EC algorithm are the following: Number of generations: 5000; Population size: 100; Tournament selection with tourney size: 5; Elitist strategy; One point crossover with rate: 0.75; Since we are mainly interest in comparing both random keys approaches, the number of total evalutions is smaller than the usual number needed for attaining the best results as presented in [5].

An evolutionary strategy like mutation operator is used. When undergoing mutation, the new value v_{new} for a given gene (i.e. a key in the chromosome) is obtained from the original value v_{old} in the following way:

$$v_{new} = v_{old} + \sigma \times N(0, 1) \tag{1}$$

Where $N(0, 1)$ represents a random value sampled from a standard normal distribution and σ is a parameter from the algorithm. In our experiments we used $\sigma = 0.1$. Mutation rate was set to 0.25 per gene. For every OGR instance we performed 30 runs with the same initial conditions and with different random seeds. All initial populations were randomly generated with values for keys selected from the real interval $[0, 1]$. Significance of the results was tested with a t-test with level of significance 0.05.

Table 1. Best rulers found for 30 runs without the use of a simple heuristic and comparing with optimal results.

Instances	Optimal	RK Fix λ	RK Evolve λ
OGR-10	55	55	55
OGR-11	72	72	72
OGR-12	85	91	91
OGR-13	106	111	114
OGR-14	127	131	131
OGR-15	151	167	167
OGR-16	177	200	202
OGR-17	199	236	230

The examination of table 1 shows that both RK approaches have found good quality solutions when comparing to the known optimal results. As expected, for a smaller number of marks, both RK approaches were able to find optimal solutions, while for the larger instances, the rulers found are of poorer quality. Nevertheless, it is interesting to observe that are not great differences between both RK approaches. As a matter of fact, the best solutions found are always the same with exception for the instances with 13, 16 and 17 marks. For OGR-12 and OGR-15, the static version of the algorithm provide better solutions, while for OGR-17 evolving λ has attained a better solution.

The observation of the best rulers found is not a sufficient indication if the introduction of the evalution of λ is a worthwhile addition to the algorithm. By looking

Table 2. Best rulers found and averages of the best rulers for 30 runs, for all the tested approaches without the use of a simple heuristic.

Instances	RK Fix λ		RK Evolve λ		t-test
	best	avg	best	avg	$p \leq 0.05$
OGR-10	55	60.1	55	**57.8**	**0.000159**
OGR-11	72	76.0	72	**74.9**	**0.031262**
OGR-12	91	95.6	91	**94.4**	**0.008301**
OGR-13	**111**	**117.2**	114	118.5	**0.017769**
OGR-14	131	**143.6**	131	144.4	0.672176
OGR-15	167	**172.8**	167	174.6	0.054341
OGR-16	**200**	**207.8**	202	210.5	**0.008406**
OGR-17	236	**245.5**	**230**	246.5	0.157805

Table 3. Best rulers found and averages of the best rulers for 30 runs, for all the tested approaches with the use of a simple heuristic.

Instances	RK Fix λ		RK Evolve λ		t-test
	best	avg	best	avg	$p \leq 0.05$
OGR-10	55	**58.0**	55	58.1	0.953500
OGR-11	**72**	**75.0**	74	75.5	0.130064
OGR-12	**92**	**94.5**	95	95.3	**0.001153**
OGR-13	113	**116.4**	**112**	118.0	**0.014721**
OGR-14	**137**	**142.1**	137	144.2	**0.001120**
OGR-15	**167**	**171.5**	169	173.3	**0.045272**
OGR-16	**197**	**205.0**	198	206.8	0.094827
OGR-17	**228**	**240.6**	234	243.5	**0.021833**

at table 2, we can also compare the averages of the best solutions found for the 30 runs (column avg). Results in bold indicates better results as a pattern is formed: for smaller instances, the evolution of λ consistently finds higher quality rulers while for a larger number of marks, not evolving λ attains overall better averages. Making a statistical analysis of the results, we can find significant differences for most of the OGR instances (column t-test, significant results in bold).

A final experiment was performed regarding this issue. Table 3 presents the attained results with both RK approaches with the usage of a simple heuristic (as in [5]). The use of an heuristic favouring small segments introduced another difficult factor for the evolution of the maximum segment length. In terms of averages, the static version is always better and for best rulers found (with the single exception of OGR-13).

5 Conclusions

In this paper we presented some experiments regarding the evolution of the maximum segment length of a Golomb Ruler. This effect was attained by introducing variable length cromosomes. Results presented in this paper suggest that there aren't significant gains in evolving the maximum segment length. Even so, it seems that there might be some advantages for lower instances.

References

[1] Golomb, S.: How to Number a Graph. In: Graph Theory and Computing. Academic Press (1972) 23–37

[2] Blum, E., Biraud, F., Ribes, J.: On optimal synthetic linear arrays with applications to radioastronomy. IEEE Transactions on Antennas and Propagation **AP-22** (1974) 108–109

[3] Soliday, S., Homaifar, A., G., L.: Genetic algorithm approach to the search for golomb rulers. In: Proceedings of the Sixth International Conference on Genetic Algorithms (ICGA-95), Morgan Kaufmann (1995) 528–535

[4] Feeney, B.: Determining optimum and near-optimum golomb rulers using genetic algorithms. Master's thesis, University College Cork (2003)

[5] Pereira, F.B., Tavares, J., Costa, E.: Golomb rulers: The advantage of evolution. In: Proceedings of the 11th Portuguese Conference on Artificial Intelligence, Workshop on Artificial Life and Evolutionary Algorithms (ALEA), EPIA'03. (2003) 27–33

[6] Cotta, C., Fernndez, A.J.: A hybrid grasp - evolutionary algorithm approach to golomb ruler search. In: Proceedings of Parallel Problem Solving from Nature (PPSNVIII). (2004)

[7] Tavares, J., Pereira, F.B., Costa, E.: Understanding the role of insertion and correction in the evolution of golomb rulers. In: Proceedings of the 2004 Congress on EC (CEC04). (2004) 69–76

[8] Bean, J.: Genetic algorithms and random keys for sequencing and optimization. ORSA Journal on Computing **6** (1994) 154–160

[9] Norman, B., Smith, A.: Random keys genetic algorithm with adaptive penalty function for optimization of constrained facility layout problems. In: Proceedings of the Fourth International Conference on Evolutionary Computation, IEEE (1997) 407–411

[10] Rothlauf, F., Goldberg, D., Heinzl, A.D.: Network random keys - a tree representation scheme for genetic and evolutionary algorithms. Evolutionary Computation **10** (2002) 75–97

Resource-Limited Genetic Programming: Replacing Tree Depth Limits

Sara Silva[1], Pedro J.N. Silva[2], Ernesto Costa[1]
[1]Centro de Informática e Sistemas da Univ. Coimbra, Dep. Engenharia Informática
Polo II – Pinhal de Marrocos, 3030 Coimbra, Portugal
[2]Centro de Genética e Biologia Molecular, Dep. Biologia Vegetal
Fac. Ciências Univ. Lisboa, Campo Grande, 1749-016 Lisboa, Portugal
E-mail: {sara,ernesto}@dei.uc.pt, pedro.silva@fc.ul.pt

Abstract

We propose replacing the traditional tree depth limit in Genetic Programming by a single limit on the amount of resources available to the whole population, where resources are the tree nodes. The resource-limited technique removes the disadvantages of using depth limits at the individual level, while introducing automatic population resizing, a natural side-effect of using an approach at the population level. The results show that the replacement of individual depth limits by a population resource limit can be done without impairing performance, thus validating this first and important step towards a new approach to improving the efficiency of GP.

1 Introduction

Genetic Programming (GP) solves complex problems by evolving populations of computer programs, using Darwinian evolution and Mendelian genetics as inspiration. Bloat is an excess of code growth caused by the genetic operators in search of better solutions, without a corresponding improvement in fitness. It is a serious problem in GP, often leading to the stagnation of the evolutionary process [1].

The traditional approach to maintaining code growth under control is by imposing a tree depth limit on the individuals accepted into the population, on a tree-based GP system [2]. Several other techniques have been used with various degrees of success (reviews and recent work in refs. 3–6), but none was ever as popular as the traditional depth limits.

This paper describes how to implement a simple technique to replace the traditional tree depth limits. It is based on a single limit imposed on the amount of resources available to the whole GP population, where resources are the tree nodes or other elements in non tree-based GP (*e.g.* code lines). The resource-limited technique removes most of the disadvantages of using depth limits at the individual level, while introducing automatic population resizing, a natural side-effect of using an approach at the population level.

Previous work, focused on financial time series prediction, also used limits on the total number of nodes in the population [7]. The results presented were, however, scarce, and the implications of the idea have not been explored any further.

Section 2 of this paper deals with several aspects of tree depth limits, while Sect. 3 introduces and explains the limited-resources technique. Section 4 describes the experiments made, Sect. 5 relates the results obtained, and finally Sect. 6 draws some conclusions and points towards future directions of this work.

2 Tree Depth Limits

Tree-based GP traditionally uses a depth limit to avoid excessive growth of its individuals [2]. When crossover creates an offspring that violates this limit, one of its parents is chosen for the new generation instead.

Traditional depth limits effectively avoid the growth of trees beyond a certain point, but they do nothing to control bloat until the limit is reached. They may also prevent the optimal solution to be found for problems of unsuspected high complexity. This may happen either because the number of possible nodes in a tree of maximum depth is not enough to represent the solution, or because a maximum depth may prove too hard a restriction to find the solution, regardless of the number of necessary nodes (for example, in symbolic regression of the quartic polynomial, $x^4 + x^3 + x^2 + x$, tree-based GP usually finds a solution of depth 7, whereas solutions of depth 6, 5, and even 4[1] are rarely found). Last but not least, depth limits cannot be used on non tree-based GP systems.

Various approaches have been tried in order to overcome the difficulties mentioned above. Some rely on choosing specialized genetic operators to keep tree

[1]The factored form of the polynomial, $(x^2 + 1)(x^2 + x)$, can be represented with a tree of depth 4.

growth under control, without imposing strict limits [8,9].

Recent work on *dynamic* limits has achieved promising results without the need for specific operators [10, 11]. It introduces a dynamic tree depth limit, initially set at a low value, but increased whenever needed to accommodate an offspring that, although deeper than the limit, is better than any other individual found during the run [10], and optionally decreased again whenever possible [11]. The idea has even been extended to create a dynamic *size* limit, where size is the number of tree nodes regardless of depth (which would enable its use in non tree-based GP), but this variation did not perform so well as the original idea [11]. All in all, new and better approaches to bloat control are still needed.

3 Limited Natural Resources in GP

Individuals in GP are built with small elements called functions and terminals. Each individual is made from a certain number of these elements, jointly designated as nodes. We can refer to the number of nodes of an individual as the amount of resources the individual uses (which is also directly related to the computational effort needed to evaluate it).

We propose limiting the total amount of resources the *population* can use in each generation. We can think of it as limiting the amount of natural resources available to a given biological population, where each individual competes with the others for its share, and the weakest individuals perish when resources are scarce. In GP, resources become scarce when the total number of nodes in the population exceeds the predefined limit. Beyond this point, not all offspring are guaranteed to be accepted into the new generation. The allocation of resources to individuals (ensuring their survival) is mainly based on fitness, with size playing a secondary role.

All the candidates to the new generation are queued by fitness, regardless of their size, and given the resources they need in a first come, first served basis. The individuals requiring more resources than the amount still available are skipped (do not survive) and the allocation continues until the end of the queue. Some resources may remain unused. A rule emerges from this procedure, promoting the survival of the best individuals and the rejection of 'not good enough for their size' individuals, where the relationship between size and fitness is not explicitly programmed, but a product of the evolutionary process itself.

The resource-limited approach is expected to cause a steady decrease of the population size (defined as the number of individuals) as long as code growth continues. According to recent work [12–15], this factor may bring some benefits in terms of convergence to good solutions. It is also expected that, after the resources have reached the exhaustion point, eventually some new generation will use them more sparingly, having all its individuals surviving and still leaving some resources available. We have considered two options on how to deal with this situation: (1) use the exceeding resources to allow the survival of the best individuals of the previous generation - the parents that would otherwise be replaced by their offspring - by applying the same allocation procedure described above (but ensuring that the population size never increases beyond its initial value), or (2) do not use the exceeding resources, thus never allowing the population size to increase again.

4 Experiments

The aim of these experiments is to see whether traditional tree depth limits can be replaced by resource limits as described above, without impairing performance in terms of fitness and computational effort. To perform a fair comparison, the resource limit for the new technique must be such that, during the entire run, the total amount of resources used is roughly the *same* for both approaches. We also want to study the differences in the evolution of population size and mean tree size inside the population.

A simple problem was used for the experiments: symbolic regression of the quartic polynomial ($x^4 + x^3 + x^2 + x$, with 21 equidistant points in the interval -1 to $+1$). An initial population of 500 individuals (Ramped Half-and-Half initialization [2] with maximum depth 6) was evolved for 50 generations, even if the optimal solution was found earlier. Tree crossover was the only genetic operator used, and reproduction rate was set at 0.1. The function and terminal sets were $\{+, -, \times, \div, sin, cos, log, exp\}$ (protected as in ref. 2) and $\{x\}$, respectively. Selection for reproduction used the lexicographic parsimony pressure tournament [5] and selection for survival used no elitism (in the traditional sense only, since the resource-limited approach can be considered highly elitist). A total of 50 runs were performed with each of the following techniques:

None → no limits
Depth → tree depth limit
R_{steady} → limited resources, forced steady usage
R_{low} → limited resources, possible low usage

The first technique (None) uses in fact no technique: nothing is done to control the growth of trees. The second technique (Depth) uses the traditional tree depth limit as described in Sect. 2, with the typical value 17. The following two techniques, R_{steady} and R_{low}, use limited resources as described in Sect. 3, set at 14500. The

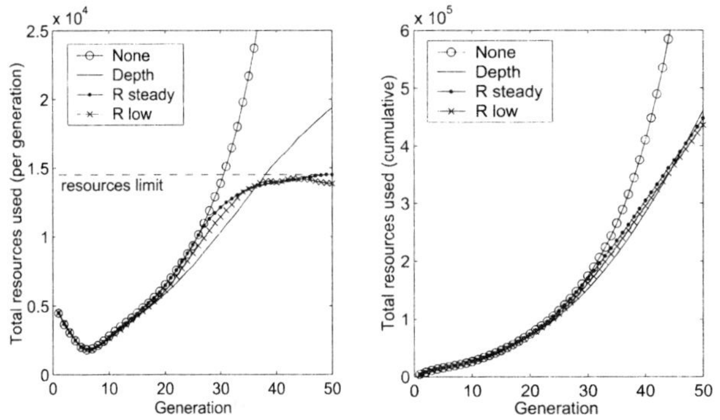

Fig. 1. Evolution of the amount of resources used by the population (left: per generation, right: cumulative).

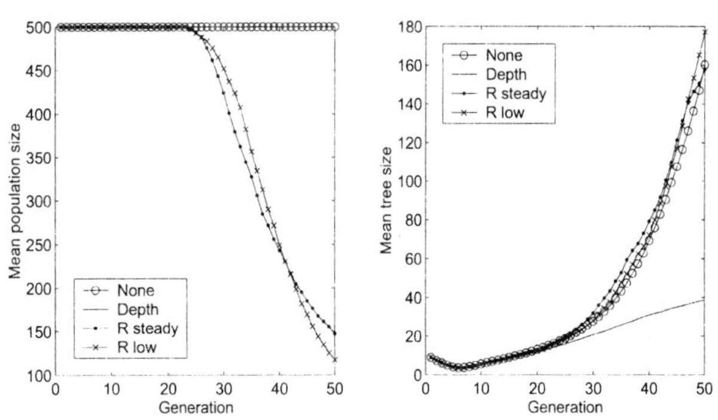

Fig. 2. Evolution of the population size (left plot) and mean tree size inside the population (right plot).

difference between them is that Rsteady forces the usage of all available resources (whenever possible), while Rlow leaves any exceeding resources unused (see Sect. 3 for details).

When looking for the ideal resources limit, 14500, we searched (in multiples of 500) for such a value that, when used to cap the amount of resources used per generation with the None technique, would result in an amount of *cumulative* resources (used during the entire run) similar to the cumulative amount used with the Depth technique. Capping at a lower value would have saved resources, but so would lowering the tree depth limit, and one must not forget we want to provide the same resources to both techniques.

All the experiments were performed using the GPLAB toolbox [16]. Statistical significance of the null hypothesis of no difference was determined with Kruskal-Wallis non-parametric ANOVAs at $p = 0.01$.

5 Results

The following results are based on the mean values over the 50 runs performed for each experiment.

Figure 1 shows the mean resources usage (per generation and cumulative) obtained with all techniques. Although the resources limit was chosen so that Depth and Rsteady would produce similar behavior (see Sect. 4), one could expect the Rsteady line to run close to None until it reached the resources limit (left plot, dashed line) and then stick to the limit until the end of the run. Instead, it diverts from None much sooner and barely reaches the limit on the last few generations, a result of the dynamics of the resource allocation procedure described in Sect. 3. The Rlow technique does not behave much differently from Rsteady. Consequently, the cumulative lines of Depth and both resource-limited techniques (Rsteady, Rlow) show very similar evolution (right plot), with no significant differences between them.

However similar the behavior of Depth, Rsteady and Rlow may look in Fig. 1 (except during the last 10 or 15 generations in the left plot), Fig. 2 shows a different reality, both in terms of population size and mean tree size inside the population (note that the left plot of Fig. 1 is obtained by "multiplying" both plots of Fig. 2). Regarding mean tree size, both Rsteady and Rlow run close to None during the entire run, while Depth diverts completely around generation 25 (right plot). In terms of population size, in both resource-limited techniques it starts dropping steeply (and not coincidentally) also by generation 25 (left plot).

Regardless of the relationship between population size and mean tree size, we are interested in looking at a different and important issue, fitness, and how it relates to the computational effort spent to obtain it. Effort can be roughly expressed as the total number of nodes evaluated – in other words, the total amount of resources used. Figure 3 shows best (lowest) fitness as a function of computational effort. Apart from the fact that the None technique proves to be a terrible waste of resources (which comes as no surprise), the remaining three techniques show similar behavior (note the logarithmic scale). The differences in the best fitness achieved during the run are not statistically significant between any two techniques.

6 Conclusions and Future Work

The replacement of the traditional tree depth limit in GP by a technique based on limited resources available to the entire population can be done without impairing performance, and at the same time removing most of the disadvantages of using depth limits.

The resource-limited technique could not be expected to perform better than tree depth limits in a simple problem like the one we have presented, where the resources available for both techniques were the same, and clearly more than needed to easily find the optimal solution. The

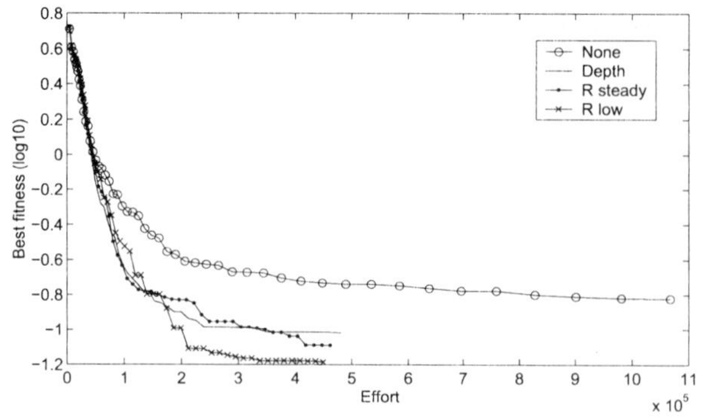

Fig. 3. Best fitness as a function of computational effort.

superiority of the new approach lies in its ability to automatically compensate higher tree size with lower population size, thus providing the necessary elements to cope with highly complex problems where tree depth techniques may fail, not for lack of resources, but for lack of flexibility. Unlike tree depth techniques, the resource-limited approach is also easily adaptable to non tree-based GP systems.

Still, the resource-limited technique may be criticized because, like the traditional depth limit, it does not prevent bloat from occurring freely until the limit is reached. The solution we propose for this problem is to adapt the idea of dynamic tree depth limits [10, 11] to the resource-limited approach: initially low, the limit is only raised if that results in improvement of the population fitness. This will be the next of a series of steps towards the achievement of more efficient GP systems.

Acknowledgements

We acknowledge grant SFRH/BD/14167/2003 from Fundação para a Ciência e a Tecnologia, Portugal. We also thank the Biomathematics Group at ITQB/UNL for allowing us the use of their computational resources.

References

[1] Banzhaf, W., Nordin, P., Keller, R. E., Francone, F. D. (1998). Genetic Programming - An Introduction. Morgan Kaufmann, San Francisco, CA.

[2] Koza, J. R. (1992). Genetic programming - on the programming of computers by means of natural selection. The MIT Press, Cambridge, MA.

[3] Soule, T., Foster, J. A. (1999). Effects of code growth and parsimony pressure on populations in genetic programming. Evolutionary Computation 6(4):293–309

[4] Poli, R. (2003). A simple but theoretically-motivated method to control bloat in genetic programming. In Ryan, C. *et al.* (eds.), Proceedings of EuroGP-2003. Springer, Berlin, pp. 204–217

[5] Luke, S., Panait, L. (2002). Lexicographic parsimony pressure. In Langdon, W. B. *et al.* (eds.), Proceedings of GECCO-2002. Morgan Kaufmann, San Francisco, CA, pp. 829–836

[6] Panait, L., Luke, S. (2004). Alternative bloat control methods. In Deb, K. *et al.* (eds.), Proceedings of GECCO-2004. Springer, Berlin, pp. 630–641

[7] Wagner, N., Michalewicz, Z. (2001). Genetic programming with efficient population control for financial time series prediction. In Goodman, E. D. (ed.), GECCO-2001 LBP, pp. 458–462

[8] Kennedy, C. J., Giraud-Carrier, C. (1999). A Depth Controlling Strategy for Strongly Typed Evolutionary Programming. In Banzhaf, W. *et al.* (eds.), Proceedings of GECCO-1999. Morgan Kaufman, San Francisco, CA, pp. 1–6

[9] Langdon, W. B. (2000). Size fair and homologous tree crossovers for tree genetic programming. Genetic Programming and Evolvable Machines 1:95–119

[10] Silva, S., Almeida, J. S. (2003). Dynamic maximum tree depth. In Cantú-Paz, E. *et al.* (eds.), Proceedings of GECCO-2003. Springer, Berlin, pp. 1776–1787

[11] Silva, S., Costa, E. (2004). Dynamic limits for bloat control. In Deb, K. *et al.* (eds.), Proceedings of GECCO-2004. Springer, Berlin, pp. 666–677

[12] Luke, S., Balan, G. C., Panait, L. (2003). Population implosion in genetic programming. In Cantú-Paz, E. *et al.* (eds.), Proceedings of GECCO-2003. Springer, Berlin, pp. 1729–1739

[13] Fernandez, F., Vanneschi, L., Tomassini, M. (2003). The effect of plagues in genetic programming: A study of variable-size populations. In Ryan, C. *et al.* (eds.), Proceedings of EuroGP-2003. Springer, Berlin, pp. 317–326

[14] Fernandez, F., Tomassini, M., Vanneschi, L. (2003). Saving computational effort in genetic programming by means of plagues. In Sarker, R. *et al.* (eds.), Proceedings of CEC-2003. IEEE Press, Piscataway, NJ, pp. 2042–2049

[15] Tomassini, M., Vanneschi, L., Cuendet, J., Fernandez, F. (2004). A new technique for dynamic size populations in genetic programming. In Proceedings of CEC-2004. IEEE Press, Piscataway, NJ, pp. 486–493

[16] Silva, S. (2004). GPLAB - a genetic programming toolbox for MATLAB. http://gplab.sourceforge.net

Treating Some Constraints as Hard Speeds up the ESG Local Search Algorithm

Y. kilani[1], A. MohdZin[2]

[1]Prince Hussein bin Abdullah Information Technology College, Al Al-bayt University, Jordan
[2]Faculty of Information Science and Technology, Universiti Kebangsaan Malaysia , Malaysia
E-mail: y_kilani@yahoo.com and amz@ftsm.ukm.my

Abstract

Local search (LS) methods for solving constraint satisfaction problems (CSP) such as GSAT, WalkSAT and DLM starts the search for a solution from a random assignment. LS then examines the neighbours of this assignment, using the penalty function to determine a better neighbour valuations to move to. It repeats this process until it finds a solution that satisfies all constraints. ICM considers some of the constraints as hard constraints that are always satisfied. In this way, the constraints reduce the possible neighbours in each move and hence the overall search space. We choose the hard constraints in such away that the space of valuations that satisfies these constraints is connected in order to guarantee that a local search can reach any solution from any valuation in this space. In this paper, we incorporate ICM into one of the most recent local search algorithm, ESG, and we show the improvement of the new algorithm.

1 Introduction

A *(CSP)* [1] is a tuple (Z, D, C), where Z is a finite set of variables, D defines a finite set D_x, called the domain of x, for each $x \in Z$, and C is a finite set of constraints restricting the combination of values that the variables can take [2]. A *solution* is an assignment of values from the domains to their respective variables so that all constraints are satisfied simultaneously [2]. CSPs are known to be NP-hard in general. LS techniques, for example GSAT [3], WalkSAT [4], DLM [5, 6], the min-conflicts heuristic [7] and ESG [8], have been successful in solving large CSPs [2]. LS first generates an initial variable assignment (or state). It then makes local adjustments (or repairs) to the assignment iteratively until a solution is reached. LS can be trapped in a local minimum, a non-solution state in which no further improvement can be made. To help escape from the local minimum, GSAT [3] and the min-conflicts heuristic [7] use random restart, while DLM [5,6] and ESG [10] modify the landscape of the search surface. LS traverses the search space to look for solutions using some heuristic function. Ref Schuurmans and Southey introduced three measures of LS performance: depth, mobility and coverage. Depth measures how many clauses remain unsatisfied as the search proceeds, mobility measures how rapidly a local search moves in the search space, and coverage measures how systematically the search explores the entire space. The efficiency of a LS algorithm depends on three things [2]: (1) the size of the search space (the number of variables and the size of the domain of each variable), (2) the search surface (the structure of each constraint and the topology of the constraint connection) and (3) the heuristic function (the definition of neighbourhood and how a ``good" neighbour is picked). The Island Confinement Method (ICM) aims to reduce the size of the search space [2]. In this paper, we incorporate ICM into ESG and we show how the new algorithm can improve ESG.

The rest of this paper is as follows. Section 2 gives the necessary background and definitions. Section 3 introduces local search algorithms. Section 4 shows the SAT translated from a CSP and the data of the experiments presented in this paper is this kind of SAT. Section 5 presents the notion of the island confinement method. Section 6 presents the ESGI algorithm and the results of this algorithm is in section 7. Last section gives conclusion remarks.

2 Background and Definitions

In this section, we illustrate some terminologies we use in this paper. Given a CSP(Z,D,C). We use *var(c)* to denote the set of variables that occur in constraint $c \in C$. If $|var(c)| = 2$ then c is a *binary constraint*. In a binary CSP, each constraint $c \in C$ is binary. A *valuation* for variable set $\{x_1,\ldots, x_n\} = Z$ is a mapping from variables to values denoted $\{x_1 \rightarrow a_1, \ldots, x_n \rightarrow a_n\}$, where $a_i \in Dx_i$.

A *state* of a CSP problem (Z,D,C) (or simply C) is a valuation for Z. A state s is a *solution* of a constraint c if s makes c true. A state s is a solution of a CSP (Z,D,C) if s is a solution to all constraints in C simultaneously. Let *unsat* to be the set of literals occurring in the unsatisfied clauses.

SAT problems are a special case of CSPs. A *propositional variable* can take the value of either 0 (false) or 1 (true). A literal is either a variable x or its complement x'. A literal *l* is *true* if *l* assumes the value 1; *l* is *false* otherwise.

A *clause* is a disjunction of literals, which is true when one of its literals is true. For simplicity we assume that no literal appears in a clause more than once and no literal and its negation appear in a clause. A *satisfiability problem* (SAT) consists of a finite set of clauses (treated as a conjunction). Let *l'* denote the complement of literal *l*: *l'* = *x'* if *l* = *x*, and *l'* = *x* if *l* = *x'*. Let *L'* = { *l'* | *l* ∈ *L*} for a literal set *L*. Since we are dealing with SAT problems we will often treat states as sets of literals. A state $\{x_1 \rightarrow a_1, \ldots, x_n \rightarrow a_n\}$ corresponds to the set of literals $\{ x_j \,/\, a_j = 1\} \cup \{x'_j \mid a_j = 0\}$.

3 Local Search

A LS solver moves from one state to another using a local move. The *neighbourhood* n(s) of a state *s* is the states that are reachable in a single move from state *s*. The neighbourhood states are the states reachable in one move from the current state regardless of the actual heuristic function used to choose the neighbour state to move to. The *Hamming distance* between states s_1 and s_2 measures the number of differences in variable assignment of s_1 and s_2. A *vector variable* vec(x) = $(x_1, \ldots, x_n)$. For the purpose of this paper, we are interested in SAT problems. We assume the neighbourhood function n(s) returns the states which are at a Hamming distance of 1 from the state *s*. In an abuse of terminology we will also refer to flipping a literal *l* which simply means flipping the variable occurring in the literal. A local move from state *s* is a transition, s ⇒ s', from s to s' ∈ n(s).

We will consider a SAT problem as a vector of clauses vec(c). The general LS algorithm starts the search from a random valuation. This valuation represents the current state. Some LS algorithms may start the search from a heuristically chosen valuation. LS then moves from the current state to a better neighbour. If there is no better neighbour then it is local minima, *trap*. It escapes this trap. Some LS algorithms may include a *restart* and/or *tabu list*. If the search could not find a solution within a number of flips it restarts the search. It uses tabu list to avoid flipping the same variable in the next coming number of steps.

4 Encoding CSP as SAT Problem

In this research, we focus on a specific class of SAT problems, namely those encoding from a binary CSP. We can encode any binary CSP (Z,D,C) to a SAT problem as follows. Every CSP variable x ∈ Z is mapped to a set of propositional variables $\{x_{a1}, \ldots, x_{an}\}$ where $D_x = \{a_1, \ldots, a_n\}$. For every x ∈ Z, SAT(Z,D,C) contains the *at-least-one-on* clause $x_{a1} \vee \ldots \vee x_{an}$ which ensures that any solution to the SAT problem gives a value to x. Each binary constraint c ∈ C with var(c) = {x,y} is mapped to a series of clauses. If {x → a, y → b} is not a solution of *c* we add the clause $x_a' \vee y_b'$ to SAT(Z,D,C), where x_a' and y_b' ∈ Z. This ensures that the constraint *c* holds in any solution to the SAT problem. We call these clauses problem clauses.

The above formulation allows the possibility that in a solution, some CSP variable *x* is assigned two values. Choosing either value is guaranteed to solve the original CSP. This method is used in the encoding of CSPs into SAT in the DIMACS archive. When a binary CSP (Z,D,C) is translated to a SAT problem SAT(Z,D,C), all the clause have the form x' ∨ y' except the *at-least-one-on* clauses.

5 The Island Confinement Method

The ICM is based on the observation: the solution space of any subset of constraints in *P* encloses all solutions of *P*. Solving a CSP thus amounts to locating this space to all the constraints in *P*, which could be either points or regions scattered around in the entire search space. The solution space of constraints *D* is connected if the search can move between any two solutions of *D* without violating any constraint in *D*. The idea of ICM works by finding a set of constraints which are connected, it starts the search from an assignment which satisfies all these constraints and finally restrict LS to search within this space instead of searching in the entire problem space. Therefore the search space becomes smaller and it contains all the solutions.

Let *sol(C)* denotes the set of all solutions to a set of constraints *C*, in other words the solution space of C. A set of constraints C is an *island* if, for any two states s_0, s_n ∈ sol(C), there exist states $s_1, \ldots, s_{n-1}$ ∈ sol(C) such that $s_i \Rightarrow s_{i+1}$ for all i ∈ {0, …,n-1}. That is we can move from any solution of C to any other solution using local moves that stay within the solution space of C.

Let *lit(c)* denote the set of all literals of a/set of clause *c*. A set *C* of clauses is *non-conflicting* if there does not exist a variable x such that *x, x'* ∈ *lit(C)*. A non-conflicting set C of clauses forms an island [2]. Therefore, the problem clauses are an island/*island clauses*. Given a SAT problem, we can incorporate ICM into any LS algorithm by the following steps: We split the clauses into c_i' and c_r', where c_i' and c_r' are the island and the at least-one-on clauses respectively. Make an initial valuation that satisfies c_i'; getting inside the island. $c_{i'}$ consists of clauses of the form x' ∨ y'. An arbitrary extension of lit(c_i') to all variables can always be such an initial valuation. Restricting the search to search within the at-least-one-on clauses while satisfying the problem (island) clauses. To do so, we exclude literals l from flipping when s' = s - l ∪ l' does not satisfy c_i. Hence we only examine states that are in n(s) and satisfy c_i'.

6 Incorporating ICM into ESG

The exponentiated subgradient algorithm (ESG) is a general-purpose Boolean linear program (BLP) search technique [10]. A BLP is a constrained optimization problem where one must choose a set of binary assignments to variables $vec(x) = x_1,\dots, x_n$ to satisfy a given set of m linear inequalities vec(d_1). vec(x) ≤ b_1,...,vec(d_m) vec(x) ≤ b_m, where vec(d_i) and b_i are constants, i=1, ..., m, while simultaneously optimizing a linear side objective vec(a). vec(x) [10]. Dale [10] describes how to encode a SAT problem as BLP. The LG for a SAT problem of m clauses vec(c) = c1,..., cm and k_i literals in each c_i can be given by the following equation:

$$L(s, vec(c)) = \sum_{i=1}^{m} y_i . ZO(c_i)$$

where y_i is the real valued Lagrangian multiplier (LM) associated with constraint c_i. c_i is the penalty for violating c_i. $ZO(c_i)$ = 1 or 0 when c_i is violated or satisfied by s respectively. s is the current assignment. The objective is to minimize *L(s, vec{c))*. The following is the ESG algorithm:

```
ESG(vec(c))
Let s be a random valuation
vec(y) := 1, tabulit := ∅
while (max number of tries not reached yet)
  while L(s, vec(c)) > 0 and (max n. flips is not over))
    unsat = n(s)
    if (unsat is empty) then it is a trap
      rand := choose a random number between 0-100
      var  := choose any variable randomly
      if ((rand) < noise) s - {var} ∪ {var'}
      else learn: increase the LMs for the unsat clauses
   else choose the best l from unsat
        s = s - { l } ∪ { l'}
   If (weight update condition holds) update vec(y)
 return s
```

ESG starts the search from a random assignment. It then searches for a solution and stops if it reaches a maximum number of flips. In each try, ESG restarts the search. ESG flips one of the best variables from the unsatisfied clauses of the current assignment to move to a better neighbour state. The better neighbour state is the state which has smaller LG than the LG of the current state. Note that flipping a variable from the unsatisfied clauses to move to a better neighbour is a common practice in local search algorithms like DLM and WalkSAT. When there is an ESG trap, with probability *noise*, where *noise* ≤ 100, ESG makes noise by flipping any variable randomly and with probability *100-noise,* ESG *learns* by increasing the LMs of the unsatisfied clauses. Note that a clause has higher LM means this clause has been involved in a higher number of traps traps. Therefore, ESG chooses to flip a variable from the unsatisfied clauses to satisfy the clauses of higher LMs in the next move in order to reduce LG. LG = 0 means all the clauses are satisfied. ESG updates the LMs after a cetain number of flips to prevent the clauses from having very high LMs.

The following is the EI algorithm, the ICM incorporating into ESG:

```
EI(vec(c))
  split vec(c) into vec(c_i) and vec(c_r)
  make an initial valuation s that satisfies vec(c_i)
  vec(y) := 1, tabulit := ∅
  while (max number of tries not reached yet)
   while L(s, vec{c}) > 0 and (max n. flips is not over))
     unsat = ∪ {l = lit(c) | c ∈ vec{c_r }, s ∉ sol(c) and
                 (s - l ∪ l') ∈ sol(c_i) }
      if (unsat is empty) then it is an island trap
        rand := choose a random number between 0-100
        vars := ∪ {var | if s-{var} ∪ {var'}} then s ∈
                                                sol({c_i})}
        var := choose a variable from vars randomly
        if ((rand) < noise) s - {var} ∪ {var'}
        else escapes this island trap in the same way
              mentioned in [2]
      else choose the best l from unsat
             s = s - { l } ∪ { l'}
return s
```

We implemented EI (The source code file can be taking by emailing the authors) by modifying the code of distribution of ESG. The input to EI is the set of SAT clauses. EI splits the clauses into island clauses and none island clauses. EI starts the search from a random valuation inside the island by initial assignment which satisfies all the island clauses. As mentioned in section 4 that an arbitrary extension of lit(c_i') can be such an initial assignment. A *free literal* is a literal once flipped the search will not violate any of the island clauses so that the search will remain searching inside the island. EI saves the free literals from the unsatisfied clauses into *unsat*. An *island trap* happens when flipping any of the literals in the unsatisfied clauses violates at least one of the island clauses and hence gets the search outside of the island. To escape this trap, EI flips the free literal *var* if the chosen probability is less than *noise*, where *noise* is a parameter. EI escapes the island trap using the same strategy used when escaping from the island trap in DLMI [2]. In this strategy, we heuristically choose to free one of the literals in the unsatisfied clauses by flipping some literals from the satisfied clauses. For instance, suppose we want to free the literal x from the unsatisfied clauses. We know that flipping x violates the island clauses, say, c_1 and c_2. Flipping x violates c_1 and c_2 means x is the only true literal in each of c_1 and c_2. Therefore, we try to make one more literal true in each of c_1 and c_2. By doing so we guarantee that there are two literals true, x is one of them, in each of c_1 and c_2. c_1 and c_2 still be satisfied after flipping x. If there is no island trap or after escaping from the island trap, EI flips the best literal *l* from *unsat* in order to move to a better neighbour.

7 Experiments

The ESG implementation has the following parameters: –*mf* : max flips before restarting, -*mr*: max restarts before aborting, -*cp* : number of reweights between corrections, -*alpha*: scaled reweight step size (1+alpha*n/m), -rho: rate of weight shrinkage to mean for sat clauses, -*noise*: probability of noise in a trap and -*rawalpha*: raw reweight step size, (never used with -*alpha* together). The -*cp*, -*alpha*, -*rho* and -*rawalpha* are the parameters used to update the LMs for the clauses. We have chosen to set the -*alpha* parameter instead of -*rawalpha* because of better performance. In all ESG experiments reported [10], the -*nr* flag is used to fix the random number generator seed to 0. We adopt the same practice.

The ESG distribution does not come with any recommended parameters sets. We tuned, with the help of the original authors, the parameter settings for each of the benchmark sets. The parameters –*rho* 0.99 and 0.99, -*alpha* 0.995 and 0.999, -*nose* 0.02 and 0.09, -*cp* 50 and 300, –*mr* 10 and 10 and –*mf* 500 and 10000 is for the *n* queens and random permutation generation problems respectively.

The following table gives a comparison of ESG and EI for *n* queens and random permutation generation problems, we give the P value for EI, the parameter needed when escaping from the island trap. The tables give the success ratio, average solution time (in seconds) and average flips on solved instances for ESG and EI.

The advantages of the ICM is evident in improving ESG. EI gives substantial and consistent improvement over ESG in term of both time and number of flips in all benchmark instances.

8 Conclusion

We have presented the EI algorithm which is the ICM incorporated into ESG. We have seen the improvement gained by EI. We believe that there is a plenty of scope for using the ICM concept to improve other LS algorithms, such as WalkSAT and the min-conflicts heuristic.

Instance	ESG			EI		
	Succ	Time	Flips	succ	time	Flips
N-queen problem and $P = 0.8$						
10queen	20/20	0.02	235	20/ 20	0.00	81
20queen	20/20	0.04	317	20/ 20	0.03	113
50queen	20/20	1.03	1424	20/ 20	0.90	179
100quen	20/20	15.23	7523	20/ 20	4.80	264
Random permutation generation problems and $P = 0.8$						
pp-50	20/20	0.83	2198	20/20	0.08	228
pp-60	20/20	1.39	2580	20/20	1.04	422
pp-70	20/20	5.22	6099	20/20	1.23	440
pp-80	20/20	5.10	4956	20/20	1.67	461
pp-90	20/20	4.23	5632	20/20	2.55	434
pp-100	20/20	9.25	7283	20/20	2.88	482

References

[1] Makworth, A. (1977) Consistency in networks of relations. AI 8(1): 99–118.

[2] Fang, H., Kilani, Y., Lee, J., Stucky., P. (2002) Reducing search space in local search for constraint satisfaction. In AAAI, AAAI press, pp. 200–207.

[3] Selman, B., Levesque, H., Mitchell, D. (1992). A new method for solving hard satisfiability problems. AAAI, AAAI press, pp. 440–446.

[4] Selman, B., Kauts, H. (1993) Domain-independent extensions to GSAT: Solving large structured satisfiability problems. IJCAI, pp. 290–295.

[5] Wu, Z., Wah, B. (1999) Trap escaping strategies in discrete lagrangian methods for solving hard satisfiability problems. AAAI, pp. 673–678.

[6] Wu, Z., Wah, B. (2000) An efficient global-search strategy in discrete lagrangian methods for solving hard satisfiability problems. AAAI, pp. 310–315.

[7] Minton, S., Johnston, M., Philips, A., Laird, P. (1992). Minimizing conflicts: a heuristic repair method for constraint satisfaction and scheduling problems. AI 58, pp. 161–205.

[8] Wu, Z., Wah, B. (2000). An efficient global-search strategy in discrete lagrangian methods for solving hard satisfiability problems. AAAI, AAAI press, pp. 310–315.

[9] Selman, B., Kauts, H., Cohen, B. (1994) Noise strategies for improving local search. AAAI, AAAI press, pp. 337–343.

[10] Schuurmans, D., Southey, F. (2000). Local search characteristics of incomplete sat procedures. AAAI, AAAI press, pp. 297–302.

Applications of PSO Algorithm and OIF Elman Neural Network to Assessment and Forecasting for Atmospheric Quality

L.M. Wang[1, 2], X.H. Shi[1], M.Li[2], G.J. Chen[3], H.W. Ge[1], H.P. Lee[4] and Y.C. Liang[1, 4*]

[1]College of Computer Science and Technology, Jilin University, Key Laboratory of Symbol Computation and Knowledge Engineering of Ministry of Education, Changchun 130012, P.R. China
[2]Department of Computer Science and Technology, Changchun Taxation College, Changchun 130021, P. R. China
[3]Department of Computer Science and Technology, Guangdong University of Finance, Guangzhou 510520, P.R.China
[4]Institute of High Performance Computing, Singapore 117528, Singapore
*Corresponding author, E-mail: liangyc@ihpc.a-star.edu.sg

Abstract

The assessment and forecast for atmospheric quality have become the key problem in the study of the quality of atmospheric environment. In order to evaluate the grade of the atmospheric pollution, a model based on the particle swarm optimization (PSO) algorithm is proposed in this paper. Experimental results show the advantages of the proposed models, such as pellucid principle and physical explication, predigested formula and low computation complexity. In addition, an improved Elman neural network, namely, the output-input feedback Elman (OIF Elman) neural network is also applied to forecast the atmospheric quality. Simulations show that the OIF Elman neural network has great potential in the field of forecasting the atmospheric quality.

1 Introduction

The air environment is closely linked with human health and life, whereas the atmospheric quality has been deteriorating with the quickening rhythm of economic growth and industrialized progress. The problems regarding the atmospheric pollution have attracted more and more attention.

In order to control the atmospheric pollution, many methods have been proposed and used to evaluate the atmospheric quality, such as algorithms based on grayer clustering, fuzzy mathematics and step analysis, and some others. But there is an obvious limitation that the assessment process and assessment results are lack of specific physical meaning when using these methods. In general, it is difficult to provide directly scientific basis for environmental renovation planning. A model based on PSO is proposed to evaluate the grade of atmospheric pollution in this paper.

The atmospheric quality forecasting is an important research subject. There are three kinds of commonly used atmospheric forecasting models, namely: latent, statistic, and numerical forecasting. These models are either excessively complicated or are lack of precision in forecasting[1]. Recently, some researches have explored some methods based on artificial neural networks (ANN) and applied them to the field of atmospheric pollution forecasting. In this paper, an improved output-input feedback Elman (OIF Elman) neural network is used to forecast the atmospheric pollution of a city in the Northeast of China. Experimental results show that the proposed method could provide an effective approach for the atmospheric pollution forecasting.

2 Application of PSO for atmospheric quality assessment

2.1 Brief introduction to PSO

Particle swarm optimization (PSO), originally developed by Kennedy and Elberhart[2], is a method for optimizing hard numerical functions on metaphor of social behavior of flocks of birds and schools of fish. A swarm consists of individuals, called particles, which change their positions over time. Each particle represents a potential solution to the problem.

Let the ith particle in a D-dimensional space be represented as $X_i= (x_{i1}, x_{i2},\ldots,x_{id})$ $(i=1,2,\ldots,m)$. The best previous position (which possesses the best fitness value) of the ith particle is denoted by $P_i=(P_{i1}, P_{i2},\ldots,P_{iD})$, which is also called P_{best}. The index of the best P_{best} among all the particles is represented by the symbol g. The location P_g is also called g_{best}. The velocity for the ith particle is represented as $V_i= (V_{i1}, V_{i2},\ldots,V_{iD})$. The concept of the particle swarm optimization consists of, at each time step, changing the velocity and location of each particle towards its P_{best} and g_{best} locations according to Eqs.(1) and (2), respectively:

$$v_{id} = wv_{id} + c_1 r_1 (p_{id} - x_{id}) + c_2 r_2 (p_{gd} - x_{id}) \quad (1)$$

$$x_{id} = x_{id} + d_{id} \quad (2)$$

where w is the inertia coefficient which is a constant in the interval [0,1] and can be adjusted in the direction of linear decrease; c_1 and c_2 are learning rates which are nonnegative constants; r_1 and r_2 are generated randomly in the interval [0,1]; v_{id} [$-v_{max}$, v_{max}], and v_{max} is a designated maximum velocity. The termination criterion for iterations is determined according to whether the maximum generation or a designated value of the fitness is reached.

Table 1 Limiting values of standard densities and benchmark values in "Environment Air Quality Standard (GB3095-1996)" and the relative values at each level

Atmospheric pollutants	0th grade		1st grade		2nd grade		3rd grade		4th grade	
	C_{i0}	X_{i0}	C_{i1}	X_{i1}	C_{i2}	X_{i2}	C_{i3}	X_{i3}	C_{i4}	X_{i4}
SO_2	0.020	1	0.05	2.50	0.15	7.50	0.25	12.50	0.50	25.0
NO_X	0.015	1	0.05	3.33	0.10	6.67	0.15	10.00	0.30	20.0
NO_2	0.015	1	0.04	2.67	0.08	5.33	0.12	8.00	0.25	16.7
FM_{14}	0.020	1	0.05	2.50	0.10	5.00	0.25	12.50	0.50	25.0
TSP	0.050	1	0.12	2.40	0.30	6.00	0.50	10.00	1.20	24.0
CO	0.500	1	1.50	3.00	4.00	8.00	6.00	12.00	10.00	20.0
Fallen dust	2.000	1	5.50	2.75	12.0	6.00	24.0	12.00	40.00	20.0

Table 2 Ranges of a, b and c for different PSO parameters

Ranges of PSO parameters	[-120, 120]	[-200, 200]	[-1000, 1000]	[-3000, 3000]
a, b, c	[119.996557, 0.354625, 0.824222]	[199.991384, 0.378879, 0.728762]	[999.984087, 0.460297, 0.533039]	[2999.993160, 0.518968, 0.450530]

2.2 A PSO-based model for assessment of atmospheric quality

2.2.1 Universal formula for calculating harm rate of atmospheric pollution

The harm rate of pollution of the ith atmospheric pollutant can be expressed as[3]:

$$R_i = 1/(1 + a_i e^{-b_i c_i}) \tag{3}$$

where a_i and b_i are parameters related to the ith pollutant, and c_i is the measured density of the ith pollutant. Parameter c_i can be replaced with a relative density[3], namely, $x_i = c_i / c_{i0}$. Then Eq.(3) can also be rewritten as:

$$R_i = 1/(1 + ae^{-bx_i}) \tag{4}$$

where c_{i0} is a predefined parameter of the ith pollutant, usually taken as the natural basic density of the ith pollutant. By using the relative density, a_i and b_i can be regarded as that they are irrelative with the characteristics of the pollutant. Thus Eq. (4) is also suitable to the cases of multi-pollutants.

To generalize the applicability of the model, a universal parameter c is introduced in Reference [4], which is needed to be ascertained unrelated to the ith pollutant in Eq. (4). Then it follows that

$$R_i = 1/(1 + ae^{-bx_i})^c \tag{5}$$

2.2.2 Optimization of parameters using PSO method

According to the limiting values of standard densities C_{i1}, C_{i2} and C_{i3}, the benchmark value C_{i0} enacted for seven special atmospheric pollutants and the limiting value of obvious harm C_{i4}, which are all regulated by the enactment "Environment Air Quality Standard (GB3095-1996)", the relative value at each level x_{ik} (k=0,1,2,3,4) can be worked out and listed in Table 1. In order to use the PSO method to optimize parameters in Eq. (5), the objective function is selected as follows

$$\min f(x) = \min\{(\sum_{k=0}^{K}\sum_{i=0}^{m} (R_{ik} - R_{ke})^2)/(Km)\} \tag{6}$$

where m is the number of the selected pollutants; K is the number of the atmospheric pollution levels; R_{ik} is the harm rate of atmospheric pollution of the ith pollutant for level k, and R_{ke} is the objective value of the harm rate of atmospheric pollution for level k, respectively. The values of m and K are taken as 7 and 5, respectively, in this paper.

According to the index grading principle of geometric proportion evaluating and arithmetical classification, the objective values of harm rate of atmospheric pollution for the five levels can be obtained as 0.01, 0.0463, 0.1284, 0.3562 and 0.99[3].

In the process of optimizing parameters in Eq.(6) by using the PSO, we find that the values of parameters are changed obviously with the differences of the intervals of the parameters in PSO as shown in Table 2. Moreover, the value of a is approached to the lower limit of the interval. Thus the functional relation between the harm rate of atmospheric pollution and the relative densities of pollutants can be set up by two parameters approximately.

The optimal values of parameters a and b obtained by using the PSO are 63.93 and 0.3401 respectively. Therefore the calculating formula of harm rate of atmospheric pollution suited for the cases of multi-pollutants can be formulized generally as:

$$R_i = 1/(1 + 63.93e^{-0.3401x_i}) \tag{7}$$

2.2.3 PSO-based assessment model for atmospheric pollution

The values of a and b could be calculated by using the PSO method and R_i can be obtained using Eq.(7). Then the assessment for the atmospheric quality could be performed. But how to integrate different R_i is a key problem. A comprehensive method is proposed in [3] for the assessment of atmospheric quality with multi-pollutants. Based on this method, the optimization is performed here by using PSO algorithm. Define a variable R to represent the comprehensive harm rate of atmospheric pollution. The final formula for the assessment of atmospheric quality can be illustrated as follows:

$$R = \sum_{i=1}^{m} w_i R_i = \sum_{i=1}^{m} w_i (1/(1+63.93e^{-0.3401x_i})) \quad (8)$$

where m is the number of the types of pollutants, R_i is the harm rate of atmospheric pollution of the ith pollutant, w_i is the normalized weight value of the ith pollutant, respectively. Parameter w_i can be defined by w'_k which is the relative importance of the corresponding level of k. The values of w'_k are[5]: $w'_0=0$, $w'_1=1$, $w'_2=1.67$, $w'_3=2.33$, $w'_4=3$. Then w_i can be formularized as follows:

$$w_i = v_i / \sum_{j=1}^{m} v_j \quad (9)$$

where v_j is the relative importance of the jth pollutant, namely, $v_j = w'_k$ if R_j ranked as k level.
The atmospheric quality could be assessed according to the value of R if it is known. The harm rate of pollution at each level of each atmospheric pollutant could be worked out according to Table 1 and Eq.(7). Taking the average of the same level of all atmospheric pollutants as the standard pollution harm rate value of the corresponding level, the ranges of R for all the five levels can be obtained as follows: [0,0.0215], [0.0215, 0.0383], [0.0383, 0.1245], [0.1245,0.4056], [0.4056, 0.939].

Table 3 Assessment results of a city of Northeast of China

Data	Harm rate R	Assessment results	Real results
2002.3.15	0.0273	good	good
2002.3.16	0.0305	good	good
2002.3.17	0.0427	slight	slight
2002.3.18	0.0302	good	good
2002.3.19	0.0272	good	good
2002.3.20	0.0249	good	good
2002.3.21	0.2355	moderate	serious
2002.3.22	0.1481	moderate	moderate
2002.3.23	0.0276	good	good
2002.3.24	0.0234	good	good
2002.3.25	0.0324	good	good
2002.3.26	0.0388	slight	slight
2002.3.27	0.0432	slight	slight
2002.3.28	0.0289	good	good
2002.3.29	0.0260	good	good
2002.3.30	0.0396	slight	good
2002.3.31	0.0285	good	good

2.3 An example for assessment atmospheric quality

The real monitoring density values of the three main pollutants (PM_{10}, SO_2 and NO_2) in a city of Northeast of China are selected as the sample data. Considering the characteristics of windy and sandy, and the notable pollution in spring in this city, we select the sample data from 15 March 2002 to 31 March 2002. The assessment results calculated by the proposed method are listed in Table 3. Satisfactory results in the table show that the proposed PSO-based method is effective and applicable for the assessment of atmospheric quality.

3 Application of OIF Elman network for atmospheric quality forecasting

3.1 OIF Elman neural network

Because of the strong nonlinear characteristics of the variation of the atmospheric pollution, we consider to use a neural network model to get more accurate forecasting. It has been demonstrated that the output-input feedback Elman neural network (OIF Elman NN) could be used successfully in the identification and forecasting for the nonlinear systems[6]. Accordingly, the OIF Elman NN is selected for forecasting of the atmospheric quality in this paper. The OIF Elman NN has two particular layers called context layer and context II layer besides the conventional input, hidden and output layers. The context layer and context II layer could memorize the former value of the hidden and output layer nodes, respectively, so they could be regarded as a time-delay operator. Therefore the feedforward part of the network could be modified in the learning process while the recurrent part is fixed. The structure of OIF Elman NN is depicted as Fig.1.The mathematic model of OIF Elman network is:

$$x(k) = f(W^{I1}x_C(k) + W^{I2}u(k-1) + W^{I4}y_C(k)) \quad (10)$$

$$x_C(k) = \alpha x_C(k-1) + x(k-1) \quad (11)$$

$$y_C(k) = \gamma y_C(k-1) + y(k-1) \quad (12)$$

$$y(k) = W^{I3}x(k) \quad (13)$$

where w^{I1}, w^{I2}, w^{I3}, w^{I4} are, respectively, the weights which are link the hidden and context layers, the hidden and input layers, the output and hidden layers and the hidden and context II layers. $x_c(k)$ and $x(k)$ are the outputs of context layer and hidden layer, $y_c(k)$ and $y(k)$ are the outputs of context II layer and output layer, α, γ are the feedback gains of the self-connections of context and context II layers, respectively.
The formulas of the learning algorithm of the OIF Elman network are as follows [6]

$$\Delta w_{js}^{I4} = \eta_4 \sum_{i=1}^{m} (\delta_i^0 w_{ij}^{I3}) \frac{\partial x_j(k)}{\partial w_{js}^{I4}} \quad (j=1,2,\cdots,n; s=1,2,\cdots,m) \tag{14}$$

$$\frac{\partial x_j(k)}{\partial w_{js}^{I4}} = f'_j(\cdot) y_s(k-1) + \gamma \frac{\partial x_j(k-1)}{\partial w_{js}^{I4}} \quad (j=1,2,\cdots,n; s=1,2,\cdots,m) \tag{15}$$

$$\delta_i^0 = (y_{dj}(k) - y_i(k)) \tag{16}$$

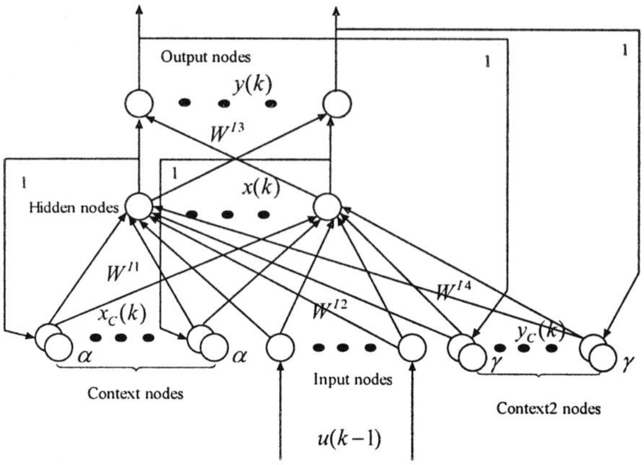

Fig. 1. Structure of OIF Elman

3.2 An example of the atmospheric quality forecasting

The real monitoring data from 1 December 2001 to 30 November 2002 in a city of Northeast of China are selected as samples. The atmospheric quality forecasting is performed using the OIF Elman NN. There are three main pollutants in the real data, namely the PM_{10}, SO_2 and NO_2. An OIF Elman NN with 3 input nodes, 10 hidden and context nodes, 1 output and context II node is used in this paper. The inputs of the network are the corresponding pollutant density values of three consecutive days, and the output is the forecasting density value of the next day.

The values of α and η are both set as 0.01 by trial and error. The maximum iterative number is set as 1000. The average errors of PM_{10}, SO_2 and NO_2 are 0.000252, 0.000203 and 000126, respectively. The forecasting curves of PM_{10} are shown in Fig.2, where the solid line represents the real density curve of the corresponding pollutant and the dot line represents the forecasting density curve, respectively. Numerical results show that the OIF Elman NN model is effective in the forecasting of atmospheric quality.

4 Conclusions

Because the values of parameters *a* and *b* in the equation to calculate the harm rate of the atmospheric pollution are uniform for different pollutants, the proposed method could have more advantageous when the number of pollutant types is large. Since the assessment of the atmospheric quality is performed according to the density values of multi-pollutants and the assessment criterion is according to the average density of multi-pollutants, the assessing process is more objective and has clear physical meaning.

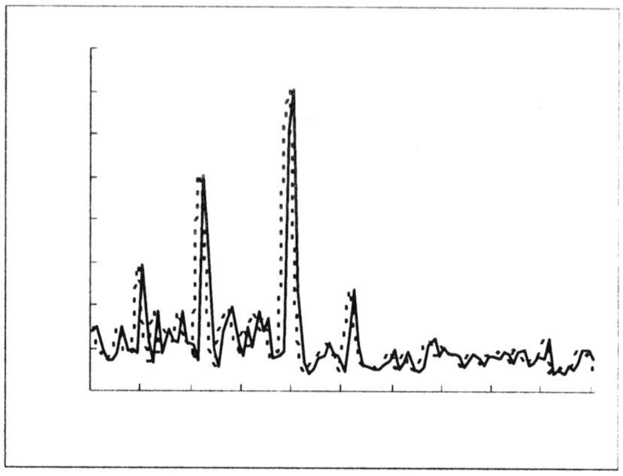

Fig. 2. Forecasting curves of PM_{10}

Acknowledgements

The support of the project of the National Natural Science Foundation of China (Grant No. 60433020), the science-technology development of Jilin Province of P. R. China (Grant No. 20030520) and the doctoral funds of the National Education Ministry of P. R. China (Grant No. 20030183060) are gratefully acknowledged.

References

[1] Wang, J., Hu, X.M., Zheng, L.X. and Liu, Z.S. (2002) Improvement and application of BP model in prediction of atmospheric pollution. Urban Environment & Urban Ecology 15: 17-19

[2] Kennedy, J., Eberhart, R. (1995) Particle swarm optimization. IEEE International Conference on Neural Networks 4: 1942-1948

[3] Li, Z.Y., Wang, Y. and Liu, G.D. (2001) Optimum estimation of parameters on assessment model of pollution loss rate of atmospheric environmental quality using genetic algorithm. Research of Environmental Sciences 14: 7-10

[4] Qian, L.W., Wu, C.Z., Hong, W., Chen, R. and Fan, H.L. (2003) Study on the improvement of index formula of pollution harm for atmosphere quality assessment. Journal of Fujian College of Forestry 23: 249-252

[5] Li, Z.Y., Zhang, X.L. and Ding, J. (2001) An universal formula of pollution harm index suited to water quality assessment. Advances in Water Science 12: 161-163

[6] Shi, X.H., Liang, Y.C., Lee, H.P., Lin, W.Z., Xu, X. and Lim, S.P. (2004) Improved Elman networks and applications for controlling ultrasonic motors. Applied Artificial Intelligence 18: 603-629

A hybrid particle swarm optimization model for the traveling salesman problem

Thiago R. Machado, Heitor S. Lopes[1]
Bioinformatics Lab./CPGEI,
Federal Center for Technological Education of Parana, Curitiba – Brazil
e-mails: {*tmachado, hslopes*}*@cpgei.cefetpr.br*

Abstract

This work presents a new hybrid model, based on Particle Swarm Optimization, Genetic Algorithms and Fast Local Search, for the symmetric blind traveling salesman problem. A detailed description of the model is provided. The implemented system was tested with instances from 76 to 2103 cities. For instances up to 439 cities, results were, in average, less than or around 1% in excess of the known optima. When considering all instances, results were 2.1498% in excess, in average. These excellent results encourage further research and improvement of the hybrid model.

1 Introduction

This paper presents a new hybrid model for the traveling salesman problem (TSP). The proposed model is based on the Particle Swarm Optimization (PSO) [1] heuristics and uses concepts of Genetic Algorithms (GA) [2] and Fast Local Search (FLS) [3]. From PSO, the model uses local maximum, global maximum and swarm movement. From GA, we used the representation of the solution and the order crossover for moving particles across the discrete search space. FLS is used to improve solutions found during the search, by evaluating points close to each particle. The motivation in developing a hybrid system is to achieve good performance for large instances.

2 Particle Swarm Optimization (PSO)

PSO is a heuristic method for optimization proposed by Eberhart and Kennedy [1], and is inspired in the behavior of social agents. In nature, this behavior can be observed in bird flocking, bee swarming, and fish shooling, for instance. The computational model is population-based, where agents, called particles, change their position (state) in the multidimensional search space of the problem, according to own experience and the influence of the neighboring particles. Each particle has a limited store capability, keeping track only of information about its current position, speed and quality (fitness regarding the other particles), as well as its best position ever visited (best local solution - BLS). Amongst the swarm of particles, the one with best quality is referred as "the best global solution" (BGS). At each time tick, particles move, influenced by both BLS and BGS, to a new position in the search space. This is an iterative process, repeated until a stop condition is met, usually a predefined number of iterations. BGS is updated whenever a better solution than the previous is found. This procedure is similar the principle of elitism, common to most GA applications since, throughout iterations, the best solution is conserved. However, there is a subtle difference: BGS is a reference for all particles in the same iteration (in GA, this would be similar to say that all individuals would mate with the best individuals). BLS is used only by a particle itself, not sharing this information with other particles. It is interesting that BLS would be a point with good fitness but it would be better if this point is far enough from the BGS so as to improve diversity. In population-based heuristics, diversity maintenance throughout iterations is frequently a challenge, but it is a necessary condition to assure a satisfactory exploration of the search space. In PSO, when many BLS's are somewhat close to the BGS, there will be a particle crowding and the search stagnates. A mechanism to avoid the consequences of this undesirable convergence will be described later. In the classical PSO model, the movement of a particle is defined by Equation 1, where its next position (X_{i+1}) is updated using the current position and a speed term (V_i). In fact, the speed term actually does not have the dimension of velocity. It could be better defined as ΔX_i but, for the sake of simplicity, it is called speed (V_i) [1].

$$X_{i+1} = X_i + V_i \tag{1}$$

The speed term is defined according to Equation 2:

$$V_i = c_1.r_1.dBLS + c_2.r_2.dBGS \tag{2}$$

[1]This work was partially supported by a research grant from the Brazilian National Research Council – CNPQ (process 350053/03-0)

where V_i is the current speed of particle i; r_1 and r_2 are random values in the range $[0..1]$; c_1 and c_2 are the weights of BLS and BGS, respectively (in percentage); dBLS and dBGS are the distance between the current position and BLS and the current position and BGS, respectively. The speed term (that is, the updating rate of the current position) is directly proportional to the distance between the current position to BLS and BGS. Therefore, within few iterations the particle will be attracted to either BLS or BGS. The speed term controls the amount of global and local exploration of the particle (that is, the balance between exploration and exploitation). A high speed facilitates global exploration, while a small speed will encourage local search. A user-defined upper bound (V_{max}) is established to limit the maximum speed of particles. As mentioned before, when particles agglutinate, a mechanism is necessary to avoid stagnation, and an explosion dissolves the crowd, repositioning all particles randomly in the search space, but not losing the previous references of BLS and BGS.

3 Genetic Algorithms (GA)

A GA is a search and optimization heuristic based on the Darwinian principle of the species evolution, where individuals better fitted to the environment are able to survive longer and propagate their genetic material to more descendants. A GA is also a population-based method with individuals that represent a possible solution for the problem. The genetic load is usually represented by a string of elements (genes). Individuals of a population are chosen according to their quality (fitness) to reproduce. Reproduction takes place by means of the application of genetic operators (crossover and mutation, for instance) to the selected pool of individuals, creating a new population. In particular, crossover is a genetic operator responsible for fostering local search, recombining pieces of two (or more) individuals. For permutation problems in combinatorial optimization, many crossover operators were proposed (see [2], for instance). In this work we used a strategy inspired in the Order crossover (OX). This operator works as follows: given two individuals, their genetic material is aligned and two random cut points are selected. The region of the chromosome between these two cut points is called matching section and will be exchanged between individuals. The remaining genes are mapped according to the matching section and are submitted to a sliding motion, so as to fill up the entire chromosome. Later, we shall explain how OX is used in our hybrid model.

4 Fast Local Search (FLS)

The FLS algorithm, also known as fast hill climbing, is an enhanced local search procedure first proposed by Voudouris and Tsang [3]. Although FLS can achieve similar results than the classic hill climbing (or neighborhood search), it is more efficient, concerning processing time. The main characteristic of FLS is the use of an activation bit for each position of the of the current solution vector. In the initial iteration, all bits of the binary activation vector are set to 1. Whenever two points are selected, the corresponding bits are reset. From the second iteration on, only the bits corresponding to permutations of higher fitness (than the previous one) are kept set. Consequently, the number of permutations is greatly reduced, since changes occur only for those positions where the corresponding bit is set.

5 Traveling Salesman Problem (TSP)

The TSP is a classical problem of combinatorial optimization and its modeling is of great interest for Engineering, Operations Research and Computer Science. For solving TSP, many methods have been proposed, including heuristic ones [2][3][4][5]. The simplest TSP considers a set of interconnected cities with symmetric distances between two points. The problem is to find the shortest path for visiting all cities passing only once at each point and returning to the initial city. There are many other variations of the problem, such as asymmetric distances between cities, capacitated vehicles, fuel/depot points with mandatory passing, time-windows restrictions, and so on. The blind TSP (BTSP) allows the starting point to be in any city and it is defined by a symmetric distance matrix between cities $D = [d_{ij}]$, which gives the distance between cities i and j. A tour (t) can be represented as a cyclic permutation [3], and its cost is the total distance travelled, according to the given permutation, passing just once by the NC cities and returning to the origin. In the TSP library (TSPLIB), at the Internet[1], a large collection of instances previously used as a benchmark is freely available. Those instances are used for evaluating performance of computational methods, since the optimal tours are always known.

6 Methodology

The hybrid model proposed in this work can be defined as a discrete PSO that uses explicit local search and concepts from GA. The original PSO was devised to cope with a continuous rather than discrete search space and, therefore, some adaptations were necessary. We also used a local search strategy so as to improve solutions globally found. Our model has a strongly coupled hybridism, since the GA characteristic appropriated to deal with the permutation problem (operator OX) is embodied in the model. This contrasts with loosely coupled

[1] http://www.iwr.uni-heidelberg.de/groups/comopt/software/TSPLIB95/

models that run independent systems and share only final results. We believe that the fusion of concepts from different paradigms in the same model is the most efficient way to tackle hard problems such as the TSP.

6.1 Initial swarm and particles

In our model, each particle represents a possible solution for the BTSP (a complete tour). A given particle is composed by the following information: current position (CP), BLS, current speed (V_i), current fitness of the particle ($Cfitness$) and fitness of BLS ($Mfitness$). Besides the particle-related parameters, other global parameters are defined: maximum number of iterations (N_{max}), number of particles (Np), minimum distance for computing fitness (D_{min}), number of cities of the problem (NC). Also, a matrix of distances between cities ($D = [d_{ij}]$), computed using Euclidean distance, is used as input data. Parameter NC determines the length of vectors used in CP, BLS and BGS.

Fitness is computed by dividing D_{min} by the cost of the tour represented by the current solution. Considering the instances used in this work, D_{min} is set to the optimal value of the tour and, therefore, fitness values represent the excess of distance relative to the known optimum for the instance. When the value of optimum tour is not known, D_{min} could be set to 1. Therefore, fitness is always inversely proportional to the cost of the tour, as in a minimization problem. The swarm represents a population of particles and, at the startup, Np particles are crated (usually $20 \leq Np \leq 50$). For each particle i, V_i is randomly initialized, respecting V_{max}, and the vector CP is set with a random, but valid, tour. For the first iteration, BLS receives the value of CP. All just generated particles will be in different points of the search space, and will have different values for CP. This represents a good diversity for the initial population. Next, the fitness of each particle is computed. The BGS and its fitness receives the corresponding values of the first generated particle and are updated as soon as any other particle is better than the previous stored value. The speed term of Equation 1 requests two parameters: dBLS and dBGS. In the classic PSO, those parameters are continuous. Therefore, the distance between particles is calculated as the value of BLS or BGS decreased by the CP value. However, for the BTSP, each position of the search space represents a complete tour (a permutation of cities), thus requesting a new way to compute the distance between particles, inspired in the Hamming distance. Given two particles "A" and "B" representing a tour, the distance between them is computed comparing vectors departing from a common city (point zero). Initially, the distance is null and, for each position of the vector, the corresponding values are compared. Whenever they are different, the distance is incremented by 1. In this way, the maximum distance possible will be exactly equal to NC. This computation may require a previous adjustment in one of the vectors: it will be slid circularly until the initial point of both vectors is the same. The new vector created does not change the encoded information about the tour (recall that it refers to a BTSP).

6.2 Diversity and movement

The average distance between particles and the BGS is calculated in each iteration. If this value is too close to NC (recall how the distance between particles is computed), agglutination is characterized and the swarm is exploded. This parameter is named diversity (δ). In the same way, it is possible to have agglutination around the BLS of a particle. In this particular case, all surrounding particles will have its BLS value changed to a random value. This procedure simulates a local explosion of the swarm. This diversity maintenance procedure does not affect the original number of iterations and it is essential for an efficient exploration of the search space.

The movement of particles is based on Equations 1 and 2 and the OX operator that recombines two possible solutions. The OX operator is adapted as follows: only one cut point ($P1_{OX}$) is randomly chosen and it is the same for the two particles. The second cut point ($P2_{OX}$), necessary to define the matching section, is found traversing circularly $NC - 1$ positions of the solution vector (tour represented by the particle). Operator OX is applied to two sets of vectors: CP and BLS, and CP and BGS. The number of positions of the matching section for each operation is given by Equations 3 and 4, where c_1 and c_2 are the same parameters of Equation 2.

$$S_{BLS} = c_1.V_i \tag{3}$$

$$S_{BGS} = c_2.V_i \tag{4}$$

This procedure generates new temporary solutions that are evaluated according to their fitness. The solution with best fitness will be considered the new CP of the particle. The concept of speed in this work is the same as in the classical PSO, and determines the level of exploration of the search space by the particles. As usual, after a particle has been moved in the search space, its BLS is updated accordingly. An iteration is defined by the movement of one particle and, after each iteration BGS is updated, if necessary.

6.3 Local search with FLS

Refinement of solutions takes place together with the search process. This refinement is accomplished by a local search, exploring positions of the search space surrounding to a given reference particle by means of the

FLS strategy. When two points of the activation vector are selected, the changes in the solution vector (tour) is done using the 2-opt heuristics [4]. That is, the sub-tour defined by these two points is inverted leading to a different tour, but preserving its structure. This newly created solution is stored in a temporary memory. The remaining bits set of the activation vector are browsed two-by-two and the same procedure as above is repeated. As result, the temporary memory holds several variations descending of the current solution. The best of them substitutes the current particle and the remaining are discarded. The 2-opt heuristics alone is not efficient, but combining its use with FLS in our hybrid model can enhance efficiency of the search, especially when the problem has many local maxima.

7 Results

Preliminary tests were done exhaustively to adjust the parameters of the model to maximum performance, considering accuracy and processing time. These tests used two instances (pr76 and pr299), and the parameters that produced the best results were: Np=20; N_{max}=1200; c_1=0.7; c_2=0.3; V_{max}=70% of NC; δ=0.4; $p(FLS)$=0.15 (probability of using local search in a given iteration). Further experiments were done using other instances of the BTSP found in the TSPLIB: rat195, pr439, d657, pr1002, rl1304 and d2103. For all instances, the suffix represents the number cities, that is, from 76 to 2103. The system was implemented in ANSI C programming language and tests were run in a PC-clone computer with Athlon XP 2.4 processor and 512 Mbytes of RAM. Table 1 presents the results obtained and values represent the amount of excess the solution is from the known optimal value (in percentage). Columns "Average" and "Best" are, respectively, the average of the best results in 10 independent runs, and the best solution found over all runs. Considering all instances, the average excess was 2,1498%.

Table 1. Results obtained for several instances of BTSP.

Problem	Average	Best
pr76	0.000	0.000
rat195	1.148	0.810
pr299	0.620	0.120
pr439	0.500	0.280
d657	3.193	2.114
pr1002	4.715	3.569
rl1304	2.498	1.454
d2103	4.524	3.433

8 Discussion and Conclusions

As expected, the quality of solutions found by the proposed method decreases as the size of the problem increases. However, for instances up to 439 cities, we obtained excellent results, less than or around 1% of excess, in average (less than 1% for the best). It is important to note that there are many models devised to solve TSP, including those that use PSO, such as [5] that relates the application of their model to quite small instances (up to 14 cities). Real-world problems usually have a much larger dimensionality, for which more efficient models must be used, such as the one here proposed.

Hybrid heuristic models are interesting for hard problems since they combine interesting features from several techniques in a single paradigm. The concepts of GA and FLS embodied in the PSO paradigm have lead to a robust and efficient model, therefore justifying the need for hybridism. Results can be considered excellent for a heuristic method, when compared with other similar methods in the recent literature. It is worth to emphasize the use of FLS with 2-opt. This strategy has enhanced solutions found by the algorithm, but at the expense of a larger, but acceptable, computational cost.

Results encourage further work that will comprise the study of a less expensive FLS, as well as other strategies that could improve the model, such as GA's concepts of niches and species [2] and breeding and subpopulations [6].

References

[1] Eberhart, R.C., Kennedy, J. (1995) Particle swarm optimization. In: Proc. IEEE Int. Conf. on Neural Networks, vol. 4, pp. 1942–1948

[2] Goldberg, D.E. (1989) Genetic Algorithms in Search, Optimization and Machine Learning, Addison Wesley, Reading

[3] Voudouris, C., Tsang, E. (1999) Guided local search and its application to the traveling salesman problem. Eur J Oper Res 113: 469–499

[4] Verhoeven, M.G.A., Aarts, E.H.L., Swinkels, P.C.J. (1995) A parallel 2-OPT algorithm for the traveling salesman problem. Fut Gen Comput Sys 11: 175–182

[5] Wang, K.P., Huang, L., Zhou, C.G., Pang, W. (2003) Particle swarm optimization for traveling salesman problem. In: Proc. 2nd IEEE Int. Conf. on Machine Learning and Cybernetics, pp. 1583–1585

[6] Løvbjerg, M., Rasmussen, T. K., Krink T. (2001) Hybrid particle swarm optimizer with breeding and subpopulations. In: Proc. Genetic and Evolutionary Computation Conf., pp.469–476

Perceptive Particle Swarm Optimisation

Boonserm Kaewkamnerdpong, Peter J. Bentley

Computer Science Department, University College London, UK
E-mail: {b.kaewkamnerdpong, p.bentley}@cs.ucl.ac.uk

Abstract

Conventional particle swarm optimisation relies on exchanging information through social interaction among individuals. However for real-world problems involving control of physical agents (i.e., robot control), such detailed social interaction is not always possible. In this study, we propose the Perceptive Particle Swarm Optimisation algorithm, in which both social interaction and environmental interaction are increased to mimic behaviours of social animals more closely.

1 Introduction

Collaborative behaviour among social animals exhibits a remarkable degree of intelligence. Often the behaviour of flocks, swarms and insect colonies, arises through interaction between the individuals in the collective and through interaction with their environment. For example, ants cooperate by modifying their environment (leaving pheromone trails) and have their behaviour modified by their environment (following pheromone trails) – a form of communication known as stigmergy.

Particle Swarm Optimisation – a "swarm intelligence" algorithm – was originally designed to simulate bird flocking in order to learn more about the human social behaviour [3]. However, the conventional particle swarm optimisation relies on social interaction among particles through exchanging detailed information on position and performance. In the physical world, this type of complex communication is not always possible. This is especially the case when each individual does not know or cannot communicate its own performance. To overcome this form of limitation, both social interaction and environmental interaction is required.

In traditional swarm intelligence, the environment of the swarm is the search space. We propose the Perceptive Particle Swarm Optimisation (PPSO) algorithm, which allows a finite perception range for each individual, in order to imitate social insects in the physical world. The particles in the PPSO algorithm can observe the search space and other neighbouring particles, as social insects observe the world and other individuals through senses.

Swarm intelligence and conventional particle swarm optimisation are described in section 2. The PPSO algorithm is discussed in comparison to conventional particle swarm optimisation in section 3. Section 4 describes experiments to investigate the performance of PPSO and conventional particle swarm optimisation. A discussion of the experimental results is provided in section 5.

2 Background

Swarm intelligence emerged from inspiration of the collective intelligence in social animals such as birds, ants, fish and termites. These social animals require no leader. Their collective behaviours emerge from interactions among individuals, in a process known as self-organisation. Each individual may not be intelligent, but together they perform complex collaborative behaviours. In swarm intelligence, the models of social animals are used to create software agents cooperating to solve complex problems [2]. Typical uses of swarm intelligence are to assist the study of human social behaviour by observing other social animals and to solve a variety of optimisation problems [1]. There are three main types of swarm intelligence techniques: models of bird flocking, the ant colony optimisation (ACO) algorithm, and the particle swarm optimisation (PSO) algorithm. Different techniques are suitable for different problems. We focus on the PSO algorithm in this paper.

The particle swarm optimisation algorithm was introduced by Kennedy and Eberhart in 1995 [3]. The algorithm consists of a swarm of particles flying through the search space. Each individual in the swarm contains parameters for position and velocity. The position of each particle represents a potential solution to the optimisation problem. The dynamic of the swarm is governed by a set of rules that modify the velocity of each particle according to the experience of the particle and that of its neighbours depending on the social network structure within the swarm. By adding a velocity to the current position, the position of each particle is modified. As the particles move around the space, different fitness values are given to the particles at different locations according to how the current positions of particles satisfy the objective. At each iteration, each particle keeps track of its personal best position, *pbest*. Depending on the social network structure of the swarm, the global best position, *gbest*, and/or the local best position, *lbest*, is used to influence the swarm dynamic. After a number of iterations, the particles will eventually cluster around the area where fittest solutions are. The particle swarm optimisation algorithm has been successfully employed to solve various optimisation problems [4, 5, 6]. The conventional particle swarm optimisation algorithm is summarised in Fig. 1.

The swarm behaviour in conventional particle swarm optimisation is influenced by the number of particles, the neighbourhood size, the inertia weight, the maximum velocity, and the acceleration calculation to modify the velocity. The larger the number of particles in the swarm, the more likely the swarm will converge on the global optimum, because the social information exchange is increased. (This is dependent on different neighbourhood types and the neighbourhood size.) The performance of PSO algorithm can be improved through other system parameters. The influence of the current velocity on the new velocity can be controlled by the inertia weight. A large inertia weight compels large exploration through the search space; a smaller inertia weight causes reduced exploration. The influence of the particle's experience and that of its neighbour is governed by the acceleration calculation (line 8, Fig. 1).

The further away the particle is from the best position from its own experience and its neighbour, the larger a change in velocity that is made in order to return to that best position. The acceleration limits the trajectory of the particle oscillation. The smaller the acceleration, the smoother the trajectory of the particle is. However, too small an acceleration may lead to slow convergence, whereas too large an acceleration drives the particles towards infinity. The new velocity is limited by the given maximum velocity to prevent particles from moving too fast in the space.

Line No.	Conventional Particle Swarm Optimisation
1	Initialise a population or a swarm of particles with a random position $\mathbf{x}_i(0)$ and a random velocity $\mathbf{v}_i(0)$, where $\mathbf{x}_i \in \mathbf{R}^n$, $\mathbf{v}_i \in \mathbf{R}^n$, and $i = \{1, 2, \ldots, m\}$ while n is the dimension of the search space and m is the number of particles in the swarm.
2	Loop for t = 1 to maximum iteration or until convergence
3	Set/update inertia weight, w
4	Loop for $i = 1$ to m
5	Get fitness value, $\mathbf{F}(\mathbf{x}_i(t))$
6	If $\mathbf{F}(\mathbf{x}_i(t)) \geq pbest_i$, $pbest_i = \mathbf{F}(\mathbf{x}_i(t))$ and $\mathbf{x}_{pbest,i} = \mathbf{x}_i(t)$
7	If $\mathbf{F}(\mathbf{x}_i(t)) \geq gbest$, $gbest = \mathbf{F}(\mathbf{x}_i(t))$ and $\mathbf{x}_{gbest} = \mathbf{x}_i(t)$ or If $\mathbf{F}(\mathbf{x}_i(t)) \geq lbest$, $lbest = \mathbf{F}(\mathbf{x}_i(t))$ and $\mathbf{x}_{lbest} = \mathbf{x}_i(t)$
8	Modify the velocity of particle i e.g. $\mathbf{v}_i(t+1) = w.\mathbf{v}_i(t) + c_1 r_1(\mathbf{x}_{pbest,i} - \mathbf{x}_i(t)) + c_2 r_2(\mathbf{x}_{gbest} - \mathbf{x}_i(t))$ where c is a constant number and r is a random number between 0 and 1 If $\mathbf{v}_i(t+1) > \mathbf{v}_{max}$, $\mathbf{v}_i(t+1)=\mathbf{v}_{max}.\text{sign}(\mathbf{v}_i(t+1)$
9	Modify the position of particle i as $\mathbf{x}_i(t+1) = \mathbf{x}_i(t) + \mathbf{v}_i(t+1)$
10	End loop I
11	Test for convergence
12	End loop t

Fig. 1. Algorithm of conventional particle swarm optimisation.

3 Perceptive Particle Swarm Optimisation

In particle swarm optimisation, all individuals in the swarm have the same behaviours and characteristics. It is assumed that the information on the position and the performance of particles can be exchanged during social interaction among particles in the neighbourhood. However, in the physical world, collective intelligence in social animals often cannot emerge from direct interaction among individuals. Instead, indirect social interaction (stigmergy) must be employed. The same constraints apply when swarm intelligence is used for applications such as robot control. Consequently, in order to imitate the physical collective intelligence in social insects, we propose the perceptive particle swarm optimisation (PPSO) algorithm.

The PPSO algorithm is illustrated in Fig. 2. Most of the processes are similar to the conventional particle swarm optimisation. However, instead of directly exchanging information among particles in their neighbourhoods, the PPSO algorithm allows each individual a finite range of *perception*. Each particle can observe the search space of the optimisation problem and the approximate positions of other individuals within its perception range. An n-dimensional objective function for the problem is treated as an (n+1)-dimensional physical landscape (the extra dimension being fitness), which particles are constrained to fly above, and never below.

Each particle in the swarm observes the search space within its perception range by sampling a fixed number of directions to observe and sampling a finite number of points along those directions. Fig 3 shows an example of a particle observing the search space for a landscape optimisation problem, where the goal is to find the highest position in the landscape.

For three-dimensional observations, the particle observes the search space in six directions along its +x,-x, +y, -y, +z and -z axes, where the +x axis is always in the direction of the particle velocity. (This allows particles to observe the search space at various angles and increases the chance that the swarm will find a good solution.) The particle attempts to perceive the landscape at several sampled distances from its position, in each direction. If the sampled point is within the landscape, the particle perceives the height of the landscape at that point. To be more realistic, the perception radius for observing the search space can be separated into an inner radius and an outer radius. Within the inner perception radius, the particle has excellent perception, while its perception is less reliable in the outer perception range. Particles can observe neighbouring particles in their perception range without sampling along specific directions. If there is any neighbour within the perception range, the particle perceives the approximate positions of neighbours and then chooses the local best position to move towards.

The performance of each particle in the neighbourhood is unknown to each other. Therefore, each neighbouring particle might be in either a better or worse position than its own position. The particle must randomly choose the neighbouring particles (line 8, Fig. 2), which will influence the particle to move towards them. The position of the chosen neighbour will be used as the local best position. If there is more than one neighbour chosen, the *lbest* position is the average position among those neighbours. The presence of the neighbouring particles influences the calculation of the new velocity for the next iteration in the same way as local social interaction, *lbest*, in the conventional PSO algorithm (line 9, Fig. 2). However, the particle will have no memory of the local best position from previous iterations. If the local best position at the current iteration does improve the performance of the particle, it will affect the personal best position in the next iteration because the *pbest* position is the position with maximum fitness value that the particle has ever been.

Perceptive Particle Swarm Optimisation	
Line No.	
1	Initialise a population or a swarm of particles with a random position $\mathbf{x}_i(0)$ and a random velocity $\mathbf{v}_i(0)$, where $\mathbf{x}_i \in \mathbf{R}^{n+1}$, $\mathbf{v}_i \in \mathbf{R}^{n+1}$, and $i = \{1, 2, \ldots, m\}$
2	Loop for $t = 1$ to maximum iteration or until convergence
3	Set/update inertia weight, w
4	Loop for $i = 1$ to m
5	$\mathbf{G}_j(\mathbf{x}_i(t))$ = Observation result of the search space within the perception range (e.g. heights of the landscape) by particle i in direction j, where $j = \{1,2, \ldots, d\}$ and d is the number of directions
6	$\mathbf{F}(\mathbf{x}_i(t)) = \dfrac{\sum_{\{j:\mathbf{G}_j(\mathbf{x}_i(t))>0\}} \mathbf{G}_j(\mathbf{x}_i(t))}{\text{\# landscape pts perceived}}$
7	If $\mathbf{F}(\mathbf{x}_i(t)) \geq pbest_i$, $pbest_i = \mathbf{F}(\mathbf{x}_i(t))$ and $\mathbf{x}_{pbest,i} = \mathbf{x}_i(t)$
8	$\mathbf{H}_k(\mathbf{x}_i(t)) = r_k$ where k is the index of the particle within the perception range, r is a random no. between 0 and 1; with probability p, $\mathbf{x}_{lbest} = \dfrac{\sum_{\{a:\mathbf{H}_a(\mathbf{x}_i(t))>p\}} \mathbf{x}_a(t) + s_a}{\text{\# neighbours}}$ where s is random no. between -1 and 1
9	Modify the velocity of particle i as $\mathbf{v}_i(t+1) = w.\mathbf{v}_i(t) + c.r(\mathbf{x}_{pbest,i} - \mathbf{x}_i(t))$ if no neighbouring particle is found, or $\mathbf{v}_i(t+1) = w.\mathbf{v}_i(t) + c_1 r_1(\mathbf{x}_{pbest,i} - \mathbf{x}_i(t)) + c_2 r_2(\mathbf{x}_{lbest} - \mathbf{x}_i(t))$ if otherwise. If $\mathbf{v}_i(t+1) > \mathbf{v}_{max}$, $\mathbf{v}_i(t+1) = \mathbf{v}_{max}.\text{sign}(\mathbf{v}_i(t+1))$
10	Modify the position of particle i as $\mathbf{x}_i(t+1) = \mathbf{x}_i(t) + \mathbf{v}_i(t+1)$
11	End loop i
12	Test for convergence
13	End loop t

Fig. 2. Algorithm of perceptive particle swarm optimisation.

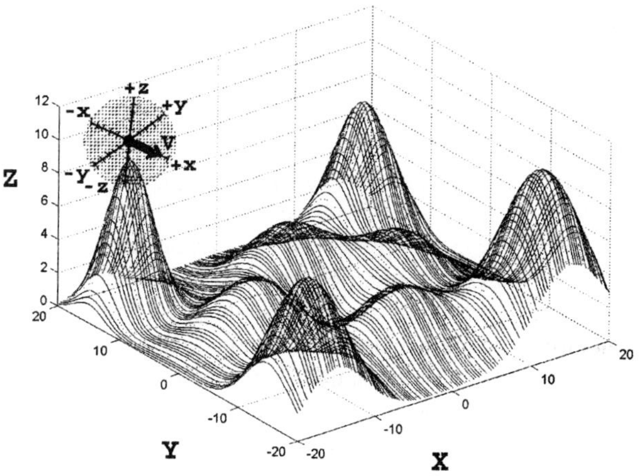

Fig. 3. The particle observing the search space in the problem of landscape optimisation

Apart from parameters in the conventional particle swarm optimisation, the main parameters of the perceptive particle swarm optimisation are: the perception radius, the number of observing directions and the number of points to observe along each observing direction. A larger perception radius allows more social interaction and encourages particles to explore the search space. This is because when there is no neighbouring particle within the perception range, the particle moves around its personal best position. However, the larger perception radius requires more computation time to observe the search space. A greater number of observing directions and a greater number of points to observe along each observing direction require more computation time as well. However, more observing directions allow a greater chance to obtain a good solution and the greater number of points offers more accuracy in observation.

4 Experiment and Result

The PPSO algorithm is designed for physically realistic optimisation problems (i.e. robot control) where the fitness function cannot accurately measure the performance of particles, i.e. the function is noisy or dynamic. In order to demonstrate the performance of the perceptive particle swarm optimisation, this experiment compares PPSO with conventional particle swarm optimisation on a landscape optimisation problem. The experiment on the conventional PSO is conducted with three types of social interactions; *pbest*, *lbest* (with a ring network) and *gbest* [7]. The particles operate in two dimensions, see landscape functions below. For the experiment on the PPSO algorithm, the particles operate in three dimensions. The fitness function for each particle is the average of the height of the landscape observed from all observation directions minus the distance between the particle and observed landscape.

Both experiments share the same number of particles and other settings. Both algorithms are applied on the same three landscapes, each with a different level of difficulty; single-peak, two-peak and multiple-peak landscape:

$$L_1(x,y) = 10 \cdot e^{(-x^2-y^2)}$$
$$L_2(x,y) = 10 \cdot e^{(-(x-4)^2-(y+3)^2)} + 5 \cdot e^{(-x^2-y^2)} - 2 \cdot e^{(-(x+3)^2-(y-4)^2)}$$
$$\begin{aligned} L_3(x,y) = {} & 10 \cdot e^{(-(x-15)^2-(y+15)^2)} + 9 \cdot e^{(-(x+15)^2-(y-15)^2)} - 2 \cdot e^{(-x^2-y^2)} \\ & + 8 \cdot e^{(-(x-15)^2-(y-15)^2)} + 8 \cdot e^{(-(x+15)^2-(y+15)^2)} + 4 \cdot e^{(-(x-9)^2-(y-1)^2)} \\ & + 4 \cdot e^{(-(x+9)^2-(y-1)^2)} + 4 \cdot e^{(-(x-1)^2-(y-9)^2)} + 4 \cdot e^{(-(x-1)^2-(y+9)^2)} \end{aligned}$$

where $x, y \in [-20, 20]$. Each experiment was run 20 times, with the result averaged in order to obtain a reliable result. The algorithms terminate when they reach the maximum iteration of 100,000 or when all particles move less than a distance of 0.2 units. The results we report here are the number of iterations used for each algorithm, the number of mis-optimisation and the optimisation error, which is the minimum distance between the maxima of the landscape and the final *pbest* position of each algorithm. The result of this experiment is shown in table 1, 2 and 3 for 10, 15 and 20 particles.

Table 1. Result for 10 particles

		Landscape 1		Landscape 2		Landscape 3	
		Error	Iter.	Error	Iter.	Error	Iter.
pbest	min	0.1233	14	0.0567	14	0.1396	17
	max	6.7466	53	7.3131	60	12.8083	73
	median	1.0958	20.5	1.9980	33	4.0390	27.5
	#miss	0		4		4	
lbest	min	0.0005	100000	0	100000	0.0090	44319
	max	0.0280	100000	0.0343	100000	29.1663	100000
	median	0.0066	100000	0.0109	100000	0.0273	100000
	#miss	0		0		2	
gbest	min	0.0001	51	0.0001	41	0.0073	69
	max	0.0031	119	4.9996	152	42.4193	6432
	median	0.0006	74	0.0006	74.5	0.0218	115
	#miss	0		1		7	
PPSO	min	0.0065	64	0.0387	90	0.04142	41
	max	13.6205	100000	8.6227	100000	23.5173	100000
	median	0.0657	72281.5	0.0698	11357	4.3769	173.5
	#miss	0		0		4	

Table 2. Result for 15 particles

		Landscape 1		Landscape 2		Landscape 3	
		Error	Iter.	Error	Iter.	Error	Iter.
pbest	min	0.0676	15	0.0682	15	0.0421	14
	max	3.6031	73	2.3845	50	13.2973	50
	median	0.5817	24	0.7654	31	1.7284	34
	#miss	0		0		3	
lbest	min	0.0012	100000	0	100000	0.0080	5748
	max	0.0122	100000	0.0184	100000	29.9703	100000
	median	0.0046	100000	0.0039	100000	0.0194	100000
	#miss	0		0		1	
gbest	min	0.0005	58	0	62	0.0026	61
	max	0.0013	107	0.0022	152	42.4193	5625
	median	0.0002	79.5	0.0001	91	0.0222	129
	#miss	0		0		8	
PPSO	min	0.0144	338	0.0064	12460	0.0224	150
	max	3.8519	100000	0.0744	100000	13.1444	100000
	median	0.0440	100000	0.0354	100000	0.0475	100000
	#miss	0		0		1	

Table 3. Result for 20 particles

		Landscape 1		Landscape 2		Landscape 3	
		Error	Iter.	Error	Iter.	Error	Iter.
pbest	min	0.0089	16	0.0328	16	0.0159	28
	max	2.2815	50	6.9141	70	4.6866	68
	median	0.2958	29	0.6425	34.5	0.2190	45.5
	#miss	0		0		0	
lbest	min	0.0007	100000	0.0004	100000	0.0067	100000
	max	0.0118	100000	0.0109	100000	0.1208	100000
	median	0.0042	100000	0.0034	100000	0.0171	100000
	#miss	0		0		0	
gbest	min	0.0000	51	0.0000	67	0.0066	70
	max	0.0009	117	0.0005	141	42.4111	24710
	median	0.0001	88	0.0002	88.5	0.0219	151.5
	#miss	0		0		3	
PPSO	min	0.0048	70239	0.0059	578	0.0257	139
	max	0.0629	100000	0.0772	100000	2.3208	100000
	median	0.0291	100000	0.0213	100000	0.0433	100000
	#miss	0		0		0	

5 Discussion

Despite the fact that particles can no longer communicate, they can only observe, the performance of the PPSO algorithm is comparable and sometimes better than the conventional one in terms of the optimisation error. For problems where reliable communication is impossible (or very difficult) such as robot control, the PPSO algorithm should be significantly beneficial. Also, the transformation of the objective function into a higher-dimensional landscape over which the particles fly, may enable the PPSO to perceive and move to optima in very complex or deceptive spaces. However, the observation mechanism requires more computation time to process, each iteration. The greater the number of particles in the swarm, the more the PPSO algorithm finds a good solution; however, at the expense of greater computation time. In contrast, having fewer particles might suffer from greater convergence error, but it requires less time. Alternative observation mechanisms can be investigated in future work in order to reduce the computation time. In addition, the PPSO algorithm can be adapted for other physical problems such as control or nanotechnology. In our future work, the PPSO algorithm for complex higher dimensional optimisation problem will be investigated.

References

[1] Eberhart, R., Shi, Y., Kennedy, J. (2001) Swarm intelligence, Morgan Kaufmann

[2] Bonabeau, B., Dorigo, M., Thraulaz, G. (1999) Swarm intelligence: from natural to artificial systems, Oxford University Press

[3] Kennedy, J., Eberhart, R. (1995) Particle swarm optimization, Proc. IEEE International Conference on Neural Networks, pp. 1942-1948

[4] Eberhart, R., Shi, Y. (1998) Evolving Artificial Neural Networks,

Proceedings of the 1998 International Conference on Neural Networks and Brain, pp. PL5 - PL13

[5] Fukuyama, Y., Yoshida, H. (2001) A Particle Swarm Optimization for Reactive Power and Voltage Control in Electric Power Systems, Proc. IEEE Congress on Evolutionary Computation

[6] Blackwell, T., Bentley, J. (2002) Improvised Music with Swarms, Proc. IEEE Congress on Evolutionary Computation

[7] Engelbrecht, A. P. (2002) Computational intelligence : an introduction, John Wiley & Sons.

Wasp swarm optimization of logistic systems

Pedro Pinto[1,2], Thomas A. Runkler[1], João M. Sousa[2]

[1]Siemens AG, Corporate Technology
Information and Communications, CT IC 4
81730 Munich - Germany

[2]Technical University of Lisbon, Instituto Superior Técnico
Dep. Mechanical Engineering - Control, Automation and Robotics Group
Av. Rovisco Pais, 1049-001 Lisbon - Portugal

E-mail: pedropinto@netcabo.pt, thomas.runkler@siemens.com, jmsousa@ist.utl.pt

Abstract

In this paper, we present the optimization of logistic processes in supply chains using the meta-heuristic algorithm known as wasp swarm, which draws parallels between the process to optimize and the way individuals in wasp colonies interact and allocate tasks to meet the demands of the nest.

1 Introduction

A supply chain is a modern organizational framework where goods are purchased from other companies instead of being produced in-house. Logistics can be defined as the sub-process of the supply chain that deals with the planning, handling and control of goods between the manufacturing and consumption points, with the goals of maintaining a flexible and economic storage structure and still deliver the goods (or orders) to the client in time. The scheduling algorithm has then to decide which goods are delivered to which customers at which time.

One way of achieving that goal is to assign individual agents to the orders and let the population of agents interactively find an optimal scheduling solution [1]. The interaction between the agents is performed by exchanging information about quantity, desired date and arriving date. These logistic problems are very complex, so efficient optimization methods are required. Natural systems often solve complex problems in a very efficient way by self organizing mechanisms. The idea here is to imitate this biological mechanisms.

Biologically motivated algorithms consist of a large variety of problem solving methods that in some way emulate a biological process observed either in specific organisms or in nature. They include, among others, genetic algorithms, which imitate the principle of natural evolution, social insect based algorithms, which emulate some aspects of the dynamic assignment of tasks in colonies, and swarm algorithms, based on the flock behavior of animals. Here, we present and apply a new algorithm based on wasp swarm optimization, tune the algorithm parameters and apply it to a simulation of a real-world logistic process at Fujitsu-Siemens Computers.

The paper is organized as follows. Firstly, we present an introduction to wasp swarm optimization, how it can be applied to a distributed dynamic scheduling problem and other optimization algorithms used in scheduling. Afterwards, we describe briefly the model of the studied logistic process and the obtained simulation results. The closing section concludes this paper and defines the future research work.

2 Wasp swarm optimization (WSO)

All termites, ants, many bees and some wasp species belong to the class of social insects. They live in organized communities (nests or colonies) where the members depend on each other to survive, unlike other kinds of insects where parents get together only to mate and resulting offsprings are abandoned. The dynamic assignment of tasks in those communities, such as foraging and brood care, to the community members, and the efficient way colonies achieve goals without a conscious planning effort from any of its agents has made them interesting study cases in the development of optimization algorithms.

In [2], a model for the organization characteristic of a wasp colony is presented. In addition to the tasks of foraging and brooding, also shared with the ant colonies, wasps colonies organize themselves in a hierarchy through interaction between the individuals. This hierarchy is an emergent social order resulting in a succession of wasps from the most dominant to the least dominant. The model of Theraulaz [2] describes the nature of interactions between individual wasps and their local environment with respect to task allocation.

Cicirello & Smith [3] have implemented this algorithm in order to optimize the job assignment problem in factories. Here we extend the wasp swarm optimization to a scheduling problem in logistic systems.

3 Wasp swarm optimization in scheduling

Our model is concerned with the allocation of available stock to incoming orders, with the goal of minimizing the overall cost. Basically, each order is assigned to a wasp in the beginning of the process. The WSO algorithm is a bidding algorithm, where the wasps take the role of bidders trying to acquire finite resources. Wasp strength is defined by the characteristics of the order: its urgency and its component requirements. After being assigned to an order, the wasps decide or not to bid for the available components in stock. The higher the urgency of the order, the higher the stimulus of the wasp to get the needed components to satisfy it. The probability of wasp A bidding for components in stock is given by:

$$p(\eta_i) = \frac{\eta_i^\alpha}{\eta_i^\alpha + \Theta_i^\alpha} \tag{1}$$

where η_i is the tardiness of the order, Θ_i is the bidding threshold and α is the bidding exponent. Both Θ_i and α can be adjusted to diminish or increase the number of actually competing wasps, with effects in the general speed of the optimization. Notice that all wasps use the same value of α.

The wasps that choose to bid will compete for the available resources, in order to satisfy their correspondent order (see figure 1).

The competing algorithm is based on a formation of a hierarchy within the wasps. The position of each wasp in the hierarchy is based on its strength. Each wasp can challenge the wasp that immediately precedes it. The probability that wasp A wins the tournament against wasp B (and thus switches places with the defeated wasp) is defined by:

$$p(A) = \frac{F_A^\beta}{F_A^\beta + F_B^\beta} \tag{2}$$

where β is the tournament exponent. The force F of each wasp can be defined as a combination of its order tardiness T and component availability A. Component availability is defined as the ratio between the needed components of an order, and their availability in stock. If the stock is running low in the required components, the strength of the wasp diminishes.

$$F(T, A) = W_t T - W_a A \tag{3}$$

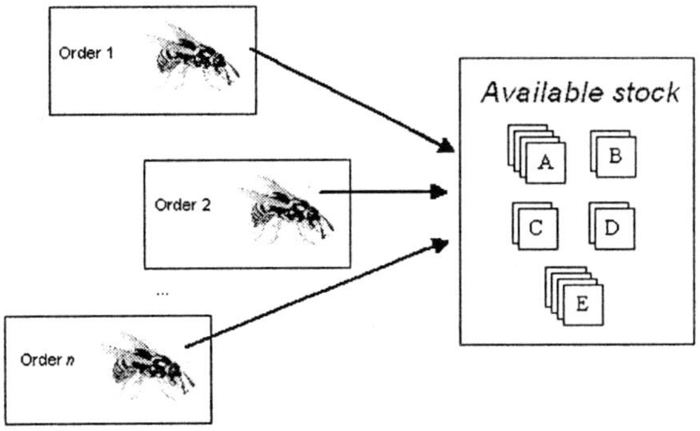

Fig. 1. order assignment to wasps

W_t and W_a are constants that weight the importance of the tardiness and the component availability, respectively. In each iteration of the tournament, the wasp on top of the hierarchy wins the right to get the components it requires, as long as there are enough of them in stock. If not, the wasp is eliminated from the process and the right of choice is passed to the next wasp in the hierarchy, and so on. For the next iteration, the force of each wasp is updated according to the components still in stock, there is a new tournament and the leading wasp can again be assigned the needed components. The procedure continues until all wasps are either removed from the tournament or satisfied.

4 Other optimization algorithms used in scheduling

In order to test the performance of the WSO, we compare its results with the results obtained using other commonly applied heuristic algorithms. Here, we consider the Ant Colony Optimization (ACO) and the Genetic Algorithms (GA).

- **Random Search(RS)**. The simplest optimization algorithm possible, it consists in randomly generating a large number of possible solutions and choose the one which results in the lowest cost.
- **Ant Colony Optimization(ACO)**. Ant algorithms are multi-agents systems where the behavior of each agent mimics the behavior of real life ants and on how they interact with each other. When ants travel from the nest to a food source and vice versa, they leave a chemical trail of pheromones marking the taken route. Other ants will catch the trail and follow it, leaving their own pheromones

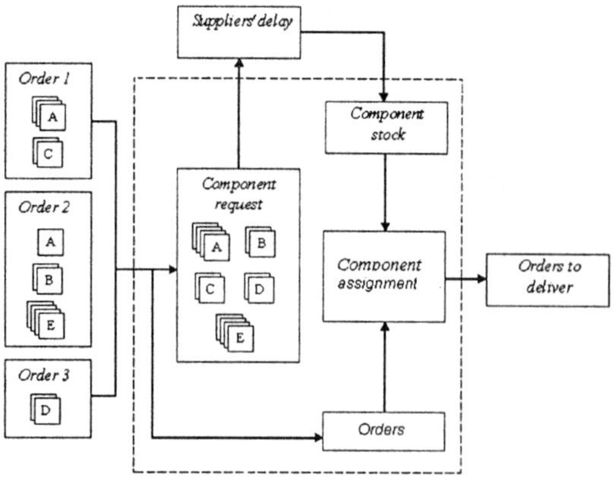

Fig. 2. Schematics of the scheduling process

behind. Since pheromones evaporate with time, the quickest route from the nest to the food source will eventually be the dominant route followed by most ants. The ACO was introduced by Dorigo [4] to solve different classes of NP-hard problems, in which dynamic scheduling is included. The algorithm used here and its parameters are exactly the same as presented in [5] and [6].

- **Genetic Algorithms (GAs)** GAs were introduced as a computational analogy of adaptive systems. They are modeled on the principles of the evolution via natural selection, employing a population of individuals that undergo selection in the presence of variation-inducing operators such as mutation and recombination (crossover). A fitness function is used to evaluate individuals, and reproductive success varies with fitness. Here, we used the algorithm and parameters presented in [7].

5 Model of the logistic process

Fig. 2 presents a schematic view of the analyzed logistic process, that can be described in probabilistic terms. The birth process of the system (arrival of new orders in a certain period of time) and the death process (delivery of orders per unit of time, or the time it took them to be processed), can be described by the classical theory of queuing processes [8].

For the process being studied, this theory asserts the Poisson distribution for the model of the birth process,

$$p(x, \lambda T) = \frac{(\lambda T)^x}{x!} e^{-\lambda T} \tag{4}$$

where x is the random variable *number of orders* and λT is the parameter indicating the probability of this event to occur on a certain time T. The death process is modeled by the exponential distribution

$$p(T, \mu) = \mu e^{-\mu T} \tag{5}$$

where the time T is the random variable and μ is the death rate. The process can be divided into five sequential steps: order arrival, component request, component arrival and component assignment.

The component assignment is the basic problem underneath a logistic process, since the company cannot influence the arrival rates of the orders (birth process), nor the suppliers delay. The service rates (death process) of the orders are the only control variable by considering the assignment of components to the orders. Ideally, all orders should be satisfied in time. In order to avoid large volumes of stocks and to minimize the number of delayed orders, the scheduling program assigns components to orders every day, not considering their order of arrival but their volumes, requirements and dates of delivery. The objective function used is presented in [7]. We impose as the most important objective that the highest number of orders is delivered at the correct date, although we consider also important to have a small tardiness variance (η_T) of the remaining orders. Thus, the objective function is given by

$$f(T) = \frac{1}{\eta_T + \frac{1}{\sum_{j=1}^{n} T_j}} \tag{6}$$

6 Simulation results and analysis of the algorithm

We applied the WSO algorithm described in sec. 3 to the logistic process described in sec. 5, and compared it with the RS, ACO and GA described in sec. 4.

In WSO, the bidding threshold is important when the number of orders far outweigh the available resources, as it reduces the number of competitors, thus diminishing the total computational cost. In the case of this simulation, computational cost was not a problem, therefore the bidding threshold had no significant influence in the overall result and in the end we picked $\alpha = 0$, thus making the probability of bidding equal to one. All the simulations were run with a tournament exponent of $\beta = 2$, This is the value used in the original wasp model in [3] and it appears to work well for this kind of problems, since any significant variation of that value generally resulted in poorer results.

The relative weights of the tardiness and the component availability, W_t and W_a, were tried with different values in order to obtain the a good solution. A ratio which gave good results was $\frac{W_a}{W_t} = 2$.

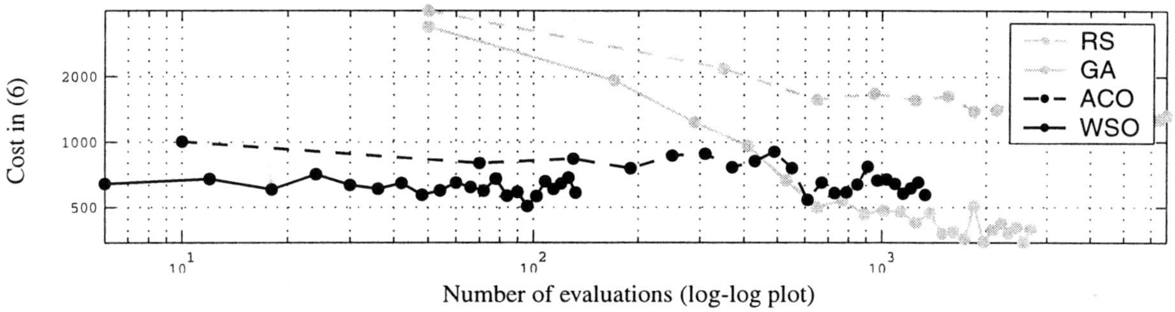

Fig. 3. Comparison of the WSO, ACO, GA and RS in terms of cost and number of evaluations

The RS and the GA were run with 50 individuals from 500 evaluations onwards (to have only reasonable results) and the ACO was run with five ants. The ACO parameters consist in the relative weight of the pheromone, set to 1, the relative weight of the heuristic matrix, set to 10, and the evaporation coefficient, set to 0.2.

Figure 3 shows the resulting cost for an increasing number of evaluations of the cost function.

Table 1 shows the cost and corresponding delivery delayed days for the different methods for a fixed number of evaluations (100). The results, although very similar for the ACO and the WSO, as can be expected from fig. 3, are better for the WSO.

Table 1. Comparison between the scheduling methods in number of orders (*dd* = days delayed)

Scheduling	Cost	$\#dd = 0$	$\#dd \leq 5$	$\#dd > 5$
RS	3730	326	118	17
GA	2780	332	114	19
ACO	817	335	127	3
WSO	543	350	104	11

For this problem the GA always achieves the best solution, given enough evaluations, but it needs 100 times more evaluations to even get to a "good" solution. Basically, the main advantage of using the WSO and ACO will be in problems were evaluating the cost function takes too much time.

7 Conclusions

This paper analyzes the performance of the wasp swarm optimization algorithm (WSO) for the optimization of logistic processes. For that we presented a simulation example, where the new algorithm is compared with the GA, ACO and RS. Even with a very low number of function evaluations, WSO produces good results when compared to the other algorithms. In finding those results, it is 100 times faster than GA.

8 References

[1] R. Palm and T. A. Runkler (2002) Multi-agent control of queuing processes. IFAC World Congress, Barcelona, Spain.

[2] G. Theraulaz, S. Goss, J. Gervet, and J. L. Deneubourg (1991) Task differentiation in polistes wasps colonies: A model for self-organizing groups of robots. From Animals to Animats: Proceedings of the First International Conference on Simulation of Adaptive Behavior. MIT Press, pp. 346-355

[3] V. A. Cicirello, and S. F. Smith (2004) Wasp-Like Agents for Distributed Factory Coordination Agents. Autonomous Agents and Multi-agent systems 8: 237–266

[4] M. Dorigo and V. Maniezzo and A. Colorni (1996) Ant System: Optimization by a colony of cooperating agents. EEE Transactions on Systems, Man, and Cybernetics-Part B 26(1): 29-41

[5] C. A. Silva and T. A. Runkler and J. M. Sousa and R. Palm (2002) Ant Colonies as Logistic Process Optimizers. Ant Algorithms, International Workshop ANTS 2002, Brussels, Belgium. Springer, pp. 76–87

[6] C. A. Silva and J. M. Sousa and R. Palm and T. A. Runkler (2002) Optimization of Logistic Processes Using Ant Colonies. Workshop Agent Based Simulation pp. 143–148

[7] C. A. Silva and T. A. Runkler and J. M. Sousa and J. M. Sá da Costa (2003) optimization of Logistic Processes in Supply–Chains using Meta–Heuristics. LNAI 2902, Progress in Artificial Intelligence, 11th Portuguese Conference on Artificial Intelligence, Beja, Portugal. Springer Verlag pp 9–23

[8] R. W. Wolff (1989) Stochastic Modeling and the Theory of Queues. Prentice -Hall

A Parallel Vector-Based Particle Swarm Optimizer

I. L Schoeman, A. P. Engelbrecht

Department of Computer Science, University of Pretoria, South Africa
E-mail: {lschoeman, engel}@cs.up.ac.za

Abstract

Several techniques have been employed to adapt particle swarm optimization to find multiple optimal solutions in a problem domain. Niching algorithms have to identify good candidate solutions among a population of particles in order to split the space into regions where an optimal solution may be found. Subsequently the swarm must be optimized so that particles contained inside the niches will converge on multiple optimal solutions.
This paper presents an improved vector-based particle swarm optimizer where subswarms contained in niches are optimized in parallel.

1 Introduction

Problem-solving forms a large part of many scientific and engineering applications. An optimal solution to a problem would be the solution where the output is an optimal value for a given set of inputs. A constant quest for more effective utilization of available resources has resulted in many techniques being developed to find optimal solutions to a large variety of problems.

Recently particle swarm optimization [1] have attracted a fair deal of interest as it proved to be an effective technique for solving more complex optimization problems. Most problems that have been investigated in this field, were solved by finding a single optimal solution in the search space. The original particle swarm optimizer has specifically been designed to avoid convergence on sub-optimal solutions. In some problems, however, the objective is to find all possible solutions in the search space. Techniques to accomplish this, entail division of the search space into regions containing candidate solutions that will all eventually converge on an optimal solution in that region. These techniques, called *niching*, have been studied in the fields of genetic algorithms and particle swarm optimization [2]. When adapting the original PSO algorithm to locate multiple optima, the most crucial tasks prove to be finding good candidate solutions where a niche may exist and then demarcate the boundaries of these niches to contain particles that will eventually converge on an optimal solution. Particles must be maintained inside these niches during optimization while niche boundaries are continuously updated.

In this paper an algorithm is presented where niches are identified using vector operations [3], but optimization takes place in parallel, giving a more effective result.

2 Particle Swarm Optimization

Particle swarm optimization is a relatively recent strategy for the optimization of functions that might be difficult to solve by means of other more established techniques. Conceptually similarities are to be found in the fields of genetic algorithms as well as artificial life [4]. Phenomena in nature involving populations of individuals interacting in order to improve their situation, were studied. In the case of flocks of birds and schools of fish, individuals controlled their own movements, but also seemed to share knowledge in order to move to a more profitable position.

The original particle swarm optimizer, introduced by Eberhart and Kennedy [1], consists of a population of particles, each of which is represented by a position in the search space where the optimal value of an objective function is to be found. The value of the function at a specific point is known as its *fitness*. The optimal value of the function will therefore be the position with the best fitness. A typical function used to illustrate the usefulness of this approach, may have many sub-optimal solutions. In the PSO, particle positions are updated by using a particle's best position found so far by the entire swarm, to guide the search for a new position. The stochastic nature of the algorithm will, however, also facilitate exploration of hitherto undiscovered regions in the search space, resulting in a fast and effective technique to find a single global optimum.

Updating of particles is formalized by the following equations:

$$v_{i,j}(t+1) = wv_{i,j}(t) + c_1 r_{1,j}(t)(y_{i,j}(t) - x_{i,j}(t)) + c_2 r_{2,j}(t)(\ _j(t) - x_{ij}(t)) \quad [1]$$

$$x_i(t+1) = x_i(t) + v_i(t+1) \quad [2]$$

where $0 \leq w < 1$ is an inertia weight incorporated to limit the contribution of the particle's previous velocity to the new velocity [5], c_1 and c_2 are two positive acceleration constants and $r_{1,j}$ and $r_{2,j}$ are random sequences between 0 and 1. y_i is the personal best position of the ith particle while is the best position found by any particle so far.

3 Locating Multiple Optima

The original PSO algorithm has purposely been designed to find a single optimum in a landscape often boasting many sub-optimal solutions. Popular benchmark functions like those of Rastrigrin or Griewank have precisely the topology enabling the user to conclusively test whether the PSO will indeed find the overall best solution.

Often, however, the purpose of the exercise entails finding all possible solutions in a search space. The PSO algorithm can be adapted for this purpose, but care must be taken that the exploratory nature of the original PSO is controlled so that candidate suboptimal solutions are maintained within a demarcated region. The process is therefore twofold: first all candidate solutions in the swarm of randomly placed particles must be identified and a portion of the search space demarcated where particles are expected to converge on that solution. Such a region is called a *niche* and its boundaries are set by calculating a niche radius. Secondly the particles must be contained in the niche by preventing them from moving outside the niche radius.

The particles in such a demarcated region can be considered to be a neighbourhood where the candidate solution is its neighbourhood best. The effect of neighbourhoods on PSO performance have been studied by Kennedy where a neighbourhood was defined as a subset of the swarm in which the particles are not necessarily adjacent to one another [6].

Several algorithms have been proposed to find multiple optima by means of particle swarm optimization. Parsopoulos and Vrahatis identified particles as potential solutions when $f(x_i) < \varepsilon$, a threshold value [7,8]. To prevent particles from moving away from the original target, "function stretching" is applied at the point where a candidate solution is found, entailing a transformation of the original function in order to eliminate local optima. During optimization, particles will converge on an optimum near the original candidate solution.

In their NichePSO algorithm Brits *et al* also use a threshold value to find initial possible solutions [9]. The process is refined by monitoring the fitness of the candidate particle while care is taken to prevent problem dependence of the threshold value. Subswarms are grown from the candidate solutions. These swarms are updated in parallel while a subswarm radius is maintained. While being updated, particles may be absorbed by subswarms. Swarms may also be merged when necessary.

4 A Vector-Based PSO

A new technique where vector operations were used to find niches in particle swarm optimization was proposed by Schoeman *et al* [3]. The main incentives driving the development of this algorithm comprised the following:

- to identify candidate solutions without prior knowledge of the problem domain
- to demarcate each niche in such a way that it would only contain particles that would eventually converge on the neighbourhood best of that niche.

The principle on which this technique is based, lies in the assumption that the dot-product of two vectors pointing in different directions will be negative while two vectors pointing in the same direction will have a positive dot product. As the original PSO already uses position vectors towards a particle's personal best and neighbourhood best positions, the dot product of these two vectors are calculated to determine whether the particle is moving towards or away from the neighbourhood best position. In addition, a niche radius is calculated by finding the distance from the neighbourhood best position to the nearest particle with a negative dot product. Particles inside the niche radius with positive dot products are marked as belonging to that niche.

The process to find all niches in a swarm, starts by setting the fittest particle in the entire swarm to the first neighbourhood best position. Particles belonging to the surrounding niche are identified as described above and marked. The process is repeated for the remaining particles until all particles have been processed.

In the original vector-based PSO, niches were optimized sequentially as they were found, by executing a fixed number of iterations. When niches are not symmetrical around each neighbourhood best, subsidiary niches may be formed between the genuine niches. To prevent these niches from becoming stationary, more particles were added when there were less than a specific number of particles in a niche. After sequential optimization of all the niches, it was found that particles from several niches have converged on the same optimum. The number of different optima comprised the true number of niches.

5 A Parallel Vector-Based PSO

The vector-based PSO described in the previous section is a sequential algorithm where each niche was optimized in turn. Because search spaces are not necessarily symmetrical, more than the required number of niches are initially identified. Although more than one niche may eventually have the same neighbourhood best position, and can be seen as finding the same niche, the performance of the algorithm could be increased by eliminating these duplicate niches.

A parallel vector-based PSO is therefore proposed where initial niches are identified similar to the vector-based PSO, but all particles are updated simultaneously. Convergence is guided by each particle's personal best position as well as the neighbourhood best of the niche it belongs to. Some subswarms may converge towards one another and will be merged once the distance between them becomes less than a certain size. Particles will be absorbed into the niche that has the fittest neighbourhood best. Only those particles in the subswarm that are near enough to the neighbourhood best of the other subswarm, will be merged.

To merge subswarms effectively, a problem-dependent parameter has to be introduced. Although one of the objectives of the vector-based PSO was to find all optima without any prior knowledge of the function, the use of this parameter is justified as the algorithm has been found to be faster.

An algorithm for the parallel vector-based PSO may now be presented.

1. Initialize the swarm.
   n particles are generated at random positions in the problem space. An initial personal best position for each particle is found by evaluating the fitness of a random position near the particle. The fittest of the two positions will be the particle's initial personal best position *pbest*. The vector v_{pi} is calculated where
   $$v_{pi}(t) = y_i(t) - x_i(t)$$
2. Identify initial niches by using vector operations.
   a. Set neighbourhood best *gbest* to *pbest* with the best fitness. Calculate v_{gi} where
   $$v_{gi}(t) = \quad (t) - x_i(t)$$
   b. Calculate the dot product λ of each particle.
   $$\lambda = v_{pi} * v_{gi}$$
   c. Calculate the niche radius – the distance between *gbest* and the nearest particle with a negative dot product.
   d. Identify all particles inside the niche radius having a positive dot product as belonging to the current niche.
   e. Repeat steps a to d for the remaining particles until all particles have been processed.
3. If a niche contains too few particle (less than 3 in this algorithm) new particles are spawned in the vicinity to prevent niches from becoming stationary.
4. Update particles
   a. For each particle: update its position by using *pbest* and its neighbourhood best *gbest*.
   b. Update *pbest* and *gbest* if better positions are found, as well as vectors v_{pi} , v_{gi} and their dot product λ. Repeat this step *k* times.
5. Update the niche radius.
6. Merge niches.
   If the distance between *gbest* of two niches becomes smaller than ε, particles are merged with the niche where *gbest* has the best fitness. Only particles nearer than ε to *gbest* of that niche will be merged.
7. Repeat steps 4 to 6 *m* times. The total number of iterations of the entire swarm will be *m* * *k*.

6 Experimental Results

A. Experimental setup

The parallel vector-based PSO was tested with four one-dimensional functions and three two-dimensional functions. The results were compared with those of the original vector-based PSO as well as the NichePSO algorithm.

$$F1(x) = \sin^6(5\pi x) \qquad [3]$$

$$F2(x) = \left(e^{-2\log(2)\times\left(\frac{x-0.1}{0.8}\right)^2} \right) \times \sin^6(5\pi x) \qquad [4]$$

$$F3(x) = \sin^6(5\pi(x^{3/4} - 0.05)) \qquad [5]$$

$$F4(x) = \left(e^{-2\log(2)\times\left(\frac{x-0.08}{0.854}\right)^2} \right) \times \sin^6(5\pi(x^{3/4} - 0.05)) \qquad [6]$$

$$F5(x, y) = 200 - (x^2 + y - 11)^2 - (x + y^2 - 7)^2 \qquad [7]$$

$$F6(x) = \left(\frac{1}{4000} \sum_{i=1}^{n} x_i^2 \right) - \left(\prod_{i=1}^{n} \cos\left(\frac{x_i}{\sqrt{i}} \right) \right) + 1 \qquad [8]$$

$$F7(x) = \sum_{i=1}^{n} \left[x_i^2 - 10\cos(2\pi x_i) + 10 \right] \qquad [9]$$

Functions *F*1 to *F*4 all have 5 maxima between 0 and 1.0, but the spacing between the maxima and the function values at the peaks differ.

For functions $F1$ to $F4$ the parallel vector-based PSO was randomly initialized with 20 particles each time. 30 iterations of the entire swarm were executed with $c_1 = c_2 = 1.0$ and $w = 0.8$. 100 experiments were done with each of the functions. As in the case of NichePSO and the original vector-based PSO good results were obtained as all maxima were successfully located each time.

The two-dimensional functions that were tested, were those of Himmelblau, Griewank for $-5.0 \leq x \leq 5.0$ and $-5.0 \leq y \leq 5.0$ and Rastrigrin for $-1.25 \leq x \leq 1.25$ and $-1.25 \leq y \leq 1.25$. Within these boundaries, Himmelblau has four, Griewank five and Rastrigrin nine optima. The parallel vector-based PSO was initialized with 40 particles in the case of the Himmelblau and Griewank functions and 60 particles for Rastrigrin. ε was set to 0.5 for Himmelblau and Griewank and 0.1 for Rastrigrin. c_1, c_2 and w were set as above.

B. Results

The results are summarized in table 1.

Table 1: Performance results

Function	% experiments locating all niches		
	Niche PSO	**Vector-based PSO**	**Parallel Vector-based PSO**
$F1$	100%	100%	100%
$F2$	100%	100%	100%
$F3$	100%	100%	100%
$F4$	100%	100%	100%
$F5$	100%	90%	98%
$F6$	100%	Not tested	95%
$F7$	100%	Not tested	91%

It will be noted that the original vector-based PSO has not been tested with functions $F6$ and $F7$. Although the principle used to identify niches were found to be sound, the algorithm became computationally expensive when dealing with more optima. An alternative was therefore investigated.

The performance results also show that the NichePSO algorithm outperforms both the original vector-based PSO and the parallel vector-based PSO. This might partly have resulted from a better initial distribution of particles, as NichePSO uses Faure sequences. NichePSO does, however, use more tunable parameters than the vector-based PSO algorithms. To create initial subswarms, a particle's fitness is tracked over a number of iterations. If the variance is less than a certain threshold, a subswarm is created with its nearest neighbour, assuming that it is also lies inside the niche. Subswarms are merged when the distance between them are less than a specific value. Setting these parameters is often problem-dependent assuming prior knowledge of the objective function.

Apart from the search space boundaries, the initial number of particles and the number of iterations, the parallel vector-based PSO uses only a value indicating the merging of two niches.

7 Conclusion and Future Work

It has been shown that the parallel vector-based PSO performs well on a number of one and two-dimensional test functions, but not as good as NichePSO. Further research is needed to determine whether more particles having a better initial distribution will give better results. An in-depth study comparing the three algorithms according to number of particles, parameter settings and scalability, is also envisaged.

References

[1] Kennedy, J., Eberhart, R.C. (1995) Particle swarm optimization. Proceedings of the IEEE International Conference on Neural Networks IV: 1942-1948.

[2] Brits, R. (2002) Niching strategies for particle swarm optimization. Master's thesis, Department of Computer Science, University of Pretoria, Pretoria, South Africa.

[3] Schoeman, I.L., Engelbrecht, A.P. (2004) Using vector operations to identify niches for particle swarm optimization. Proceedings of the conference on cybernetics and intelligent systems.

[4] Reynolds, C.W. (1987) Flocks, herds and schools: a distributed behavioural mode, Computer Graphics 21(4):25-34.

[5] Shi, Y., Eberhart, R.C. (1998) A modified particle swarm optimizer. Proceedings of the IEEE Congress on Evolutionary Computation : 69-73.

[6] Kennedy, J. (1999) Small worlds and mega-minds: Effects of neighborhood topology on particle swarm performance. Proceedings of the IEEE Congress on Evolutionary Computation :1931-1938.

[7] Parsopoulos, K.E., Vrahatis, M.N. (2001) Modification of the particle swarm optimimizer for locating all the global minima. In Kurkova, V., Steele, N.C., Neruda, R., Karny, N. (eds), Artificial Neural Networks and Genetic Algorithms, Springer, pp. 324-327

[8] Parsopoulos, K.E., Plagianakos V.P., Magoulas G.D., Vrahatis, M.N. (2001) Stretching technique for obtaining global minimizers through particle swarm optimization. Proceedings of the Particle Swarm Optimization Workshop : 22-29.

[9] Brits, R., Engelbrecht, A.P., Van den Bergh, F. (2002) A niching particle swarm optimizer. Proceedings of the 4th Asia-Pacific Conference on Simulated Evolution and Learning : 692-696.

Numerical Simulations of a Possible Hypercomputational Quantum Algorithm

Andrés Sicard, Juan Ospina, Mario Vélez
Logic and Computation Group, EAFIT University, A.A. 3300, Medellín, Colombia
E-mail: {asicard, jospina, mvelez}@eafit.edu.co

Abstract

The hypercomputers compute functions or numbers, or more generally solve problems or carry out tasks, that cannot be computed or solved by a Turing machine. Several numerical simulations of a possible hypercomputational algorithm based on quantum computations previously constructed by the authors are presented. The hypercomputability of our algorithm is based on the fact that this algorithm could solve a classically non-computable decision problem, the Hilbert's tenth problem. The numerical simulations were realized for three types of Diophantine equations: with and without solutions in non-negative integers, and without solutions by way of various traditional mathematical packages.

1 Introduction

Although there have been several conferences and various specialized publications dedicated to the topic of hypercomputation, this is still an area in its gestation and development process. The term 'hypercomputer' denotes any data processing device (theoretical, potentially realizable or that can be implemented) capable of carrying out tasks that cannot be performed by a Turing machine (TM) [1, 2]. Proposed models of hypercomputation surge from diverse disciplines, such as non-classical logics, computability, neural networks, or relativistic physics [1, 3, 4]. Notwithstanding the proliferation of theoretical hypercomputation models, the possibility of a *conceivable* construction of a hypermachine, continues to be controversial and held under study and analysis.

This work presents several numerical simulations of a possible hypercomputational algorithm based on quantum computations previously constructed by the authors [5]. The hypercomputability of our algorithm is based on the fact that this algorithm could solve a TM non-computable decision problem, Hilbert's tenth problem [6]. Our algorithm is based on the algorithm proposed by Tien D. Kieu [7, 8, 9], with the difference that we have selected the Infinite Square Well (ISW) instead of the (one-dimensional) Simple Harmonic Oscillator (SHO) as the underlying physical system. Since our model exploits the quantum adiabatic process for unbounded operators [10], and the characteristics of the infinite-dimensional irreducible representation of the dynamical algebra $\mathfrak{su}(1,1)$ associated to the ISW [11], the simulation is only able to partially illustrate our algorithm. In other words, this work does not pretend to affirm that the numerical simulations constitute our hypercomputational quantum algorithm.

2 The Hypercomputational Quantum Algorithm

The Hamiltonian operator H and the energy levels E_n, for a particle with mass m in an ISW with length πa are [11]

$$
\begin{aligned}
H &= i^2 \frac{\hbar^2}{2m}\frac{d^2}{dx^2} - \frac{\hbar^2}{2ma^2}, \\
E_n &= (\hbar^2/2ma^2)n(n+2),
\end{aligned} \tag{1}
$$

where the action of H on its eigenvectors

$$\{|\,n\rangle \mid n \in \mathbb{N}\}\,, \quad \mathbb{N} = \{0, 1, \dots\}, \tag{2}$$

is given by $H\,|\,n\rangle = E_n\,|\,n\rangle$. Due to the spectral structure of the ISW, the dynamical algebra associated with it, is the Lie algebra $\mathfrak{su}(1,1)$ [11]. This is a tridimensional algebra that satisfies the commutation relations $[K_-, K_+] = K_3$ and $[K_\pm, K_3] = \mp 2K_\pm$, where operators K_+, K_- and K_3 are called creation, annihilation and Cartan operators, respectively. The algebra $\mathfrak{su}(1,1)$ admits an infinite-dimensional irreducible representation where actions of K_+, K_- and K_3 on basis (2) are

$$
\begin{aligned}
K_+\,|\,n\rangle &= \sqrt{(n+1)(n+3)}\,|\,n+1\rangle\,, \\
K_-\,|\,n\rangle &= \sqrt{n(n+2)}\,|\,n-1\rangle\,, \\
K_3\,|\,n\rangle &= (2n+3)\,|\,n\rangle\,.
\end{aligned}
$$

With basis on the algebra $\mathfrak{su}(1,1)$, the Hamiltonian (1) is rewritten as $H = (\hbar^2/2ma^2)K_+K_-$, and a number operator N is given by

$$N = (1/2)(K_3 - 3). \tag{3}$$

Due to the dynamical algebra associated, the Barut-Girardello coherent states $|\,z\rangle\,, z\in\mathbb{C}$, for the ISW are eigenstates of annihilation operator K_- [12]

$$|\,z\rangle = \frac{|z|}{\sqrt{I_2(2\,|z|)}}\sum_{n=0}^{\infty}\frac{z^n}{\sqrt{n!(n+2)!}}\,|\,n\rangle\,.$$

On the other hand, a Diophantine equation is of the following form

$$D(x_1,\ldots,x_k)=0, \qquad (4)$$

where D is a polynomial with integer coefficients. In present terminology, Hilbert's tenth problem may be paraphrased as: Given a Diophantine equation of type (4), we should build a procedure to determine whether or not this equation has a solution in non-negative integers. From the concluding results obtained by Matiyasevich, Davis, Robinson, and Putnam we know that, in the general case, this problem is algorithmically insolvable or more precisely, it is TM incomputable [6].

With the basis for the ISW, our hypercomputational quantum algorithm for the Hilbert's tenth problem [5] is shown in Table (1). Our algorithm is probabilistic such are quantum algorithms in general. Unlike these however, our algorithm is of adiabatic quantum computation over infinite-dimensional spaces, given that parting from initial ground state (5) of (6), the final ground state (10) of (7), is obtained via the adiabatic theorem for unbounded operators [10]. As we have stated elsewhere [5], the construction of our algorithm, breaking from Kieu's algorithm, opens the possibility to construct similar algorithms based on physical references whose associated dynamic algebra possesses an infinite-dimensional irreducible representation.

3 Simulation Procedure

Once constructed (7) with (5) and (6) selected for a Diophantine equation of type (4), our problem revolves around solving (8) in search of a value of T which satisfies condition (11), whereby we can establish criteria (12). Given that (8) is formulated inside an infinite dimensional Fock space, the simulation consists in being solved numerically in a Fock space truncated from dimension $m+1$, selected such that [13]

$$|\,z\rangle^{m+1} = \frac{|z|}{\sqrt{I_2(2\,|z|)}}\sum_{n=0}^{m}\frac{z^n}{\sqrt{n!(n+2)!}}\,|\,n\rangle^m$$

has a norm less than one by some chosen ϵ

$$\left|1-\sqrt{{}^{m+1}\langle z\mid z\rangle^{m+1}}\right|\le\epsilon, \qquad (13)$$

where $|\,z\rangle^{m+1}$ represents the infinite-dimensional state $|\,z\rangle$ truncate to the $m+1$ dimension.

Table 1. Hypercomputational quantum algorithm.

Given a Diophantine equation with k unknowns of type (4), we provide the following quantum algorithm to decide whether this equation has any non-negative integer solution or not:

1. Construct a physical process in which a system initially starts with a direct product of k coherent states $|\,\psi(0)\rangle$, and in which the system is subject to a time-dependent Hamiltonian $H_{\mathrm{A}}(t)$ over the time interval $[0,T]$, for some time T, with the initial Hamiltonian H_{I} and the final Hamiltonian H_{D}, given by

$$|\,\psi(0)\rangle = \bigotimes_{i=1}^{k}|\,z_i\rangle\,, \qquad (5)$$

$$H_{\mathrm{A}}(t) = (1-t/T)H_{\mathrm{I}} + (t/T)H_{\mathrm{D}},$$

$$H_{\mathrm{I}} = \sum_{i=1}^{k}(K_{+_i}-z_i^*)(K_{-i}-z_i)\,, \qquad (6)$$

$$H_{\mathrm{D}} = (D(N_1,\ldots,N_k))^2\,. \qquad (7)$$

2. Measure through the time-dependent Schrödinger equation with the

$$i\partial_t\,|\,\psi(t)\rangle = H_{\mathrm{A}}(t)\,|\,\psi(t)\rangle\,,\ \text{for } t\in[0,T] \qquad (8)$$

the maximum probability to find the system in a particular number state at the chosen time T,

$$P_{\max}(T) = \max_{(n_1,\ldots,n_k)\in\mathbb{N}^k}|\langle\psi(T)\mid n_1,\ldots,n_k\rangle|^2\,,$$
$$= \left|\langle\psi(T)\mid n_1,\ldots,n_k\rangle_0\right|^2, \qquad (9)$$

where

$$|\,\{n\}\rangle_0 \equiv |\,n_1,\ldots,n_k\rangle_0\,. \qquad (10)$$

3. If $P_{\max}(T)\le 1/2$, increase T and repeat all the steps above.

4. If

$$P_{\max}(T) > 1/2 \qquad (11)$$

then $|\,\{n\}\rangle_0$ is the ground state of H_{D} (assuming no degeneracy) and we can terminate the algorithm and deduce a conclusion from the fact that

$$H_{\mathrm{D}}\,|\,\{n\}\rangle_0 = 0 \text{ iff (4) has a solution in } \mathbb{N}. \qquad (12)$$

The numerical solution involves a discretization procedure that consists in replacing (8) by a discrete equation, whose solver preserves the norm of the state vector in the evolution of time. This solver is unitary and it is given by the Cayley transform of H_A^{m+1} [13]

$$|\psi(t+1)\rangle^{m+1} = \frac{1 - \frac{i}{2} H_A^{m+1}(t)}{1 + \frac{i}{2} H_A^{m+1}(t)} |\psi(t)\rangle^{m+1}. \quad (14)$$

4 Simulation parameters and results

For the numerical simulations we have selected three Diophantine equations. Since we counted on all the information available for the simulation, the maximum probabilities (9) can be calculated directly. Of course, this situation would be different if we were measuring upon the physical reference, in which case these probabilities should be obtained parting from the expected values of the operator numbers (3).

4.1 Equation with solution

We now consider a simple example which nevertheless has all the interesting ingredients typical for a general simulation of our hypercomputing quantum algorithm. Let $D(x)$ be the Diophantine equation

$$D(x) \equiv x - 6 = 0. \quad (15)$$

In relation to the values of z, these should be greater than 1.6 to guarantee that the maximum peak of density of probability associated with the initial ground state (5) is less than 1/2 [5]. We choose $z = 4$ and $m = 9$ which satisfy (13) for $\epsilon = 10^{-4}$. The truncated Fock space has only ten dimensions, and is generated by $\{|0\rangle, \ldots, |9\rangle\}$. For (15), according to the algorithm pointed out in Table (1), we have that

$$|\psi(0)\rangle^{10} \approx$$
$$[0.16\ 0.36\ 0.51\ 0.53\ 0.43\ 0.29\ 0.17\ 0.08\ 0.04\ 0.02]^T,$$
$$H_D^{10\times10} = \left(N^{10\times10} - 6\mathbb{1}^{10\times10}\right)^2,$$
$$H_I^{10\times10} = \left(K_+^{10\times10} - 4\mathbb{1}^{10\times10}\right)\left(K_-^{10\times10} - 4\mathbb{1}^{10\times10}\right).$$

Applying (14) and incrementing T, figure (1) indicates that the maximum probability reaches the value of 1/2 for $T \approx 35$ and then the Fock state $|6\rangle^{10} = [0\ \ldots\ 1\ 0\ 0\ 0]^T$, is identified as the final ground state (10) according to (11). Now, given that $H_D^{10\times10}|6\rangle^{10} = 0$, we can conclude that (15) is solved in the non-negative integers in agreement with (12).

4.2 Equation without solution

Let $D(x)$ be the Diophantine Equation

$$D(x) \equiv x + 6 = 0, \quad (16)$$

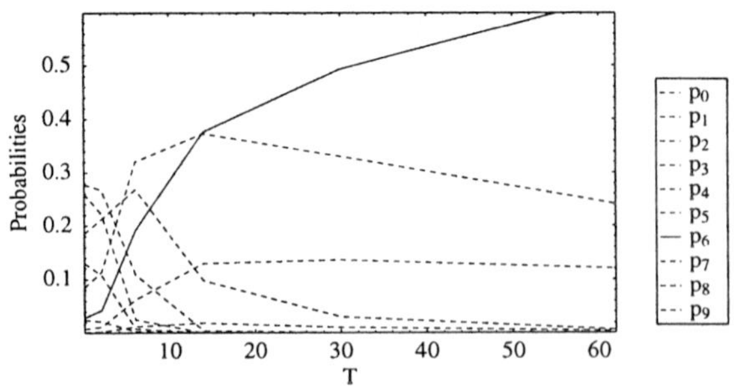

Figure 1. Simulation results for (15) with $(z, m) = (4, 9)$.

and let $z = 4$ and $m = 9$. The figure (2) shows the probabilities of states as a function of T. Below $T \approx 70$ none of the states have probabilities greater than 1/2, and in fact the excited states, clearly dominate in this regime. Eventually, we enter the quantum adiabatic regime upon which the dominant state raises its probability over the 1/2 value; indeed it corresponds to the Fock state $|0\rangle$. Parting from (12), it is concluded then that (16) cannot be solved in the non-negative integers.

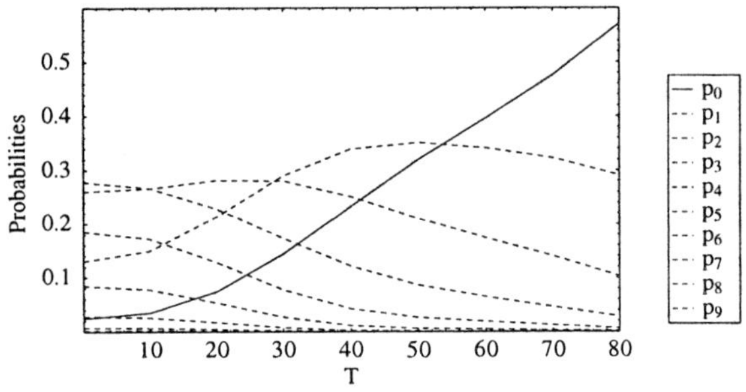

Figure 2. Simulation results for (16) with $(z, m) = (4, 9)$.

Unlike the previous example, this simulation does not give a conclusive result. In fact, it would be more precise to affirm that according to figure (2), equation (16) does not have a solution in $\{0, \ldots, m = 9\}$. Therefore, it would be necessary to increase the value of m and execute the algorithm once again. Unfortunately, different to what would occur if we could count on the physical reference, in the case of the simulation, a standing criteria to determine the maximum value of m from which we can obtain a conclusive result, does not exist.

4.3 Catalan Equation

Given that every exponential Diophantine equation can be reduced to a polynomial Diophantine equation [6], Hilbert's tenth problem also applies to these types of equations. It was recently proven that the Catalan Equation $x^p - y^q = 1$, with $x, y, p, q \geq 2$, has one and only one solution $(x = q = 3, y = p = 2)$ [14]. Although this solution could be found by simple inspection, different mathematical packages such as *Mathematica*^{TM} or *Maple*^{TM}, cannot find it. To put our algorithm to work in order to determine if the Catalan Equation can be solved in non-negative integers, we rewrite it as

$$(a+3)^{(b+2)} - (c+2)^{(d+3)} = 1, \qquad (17)$$

whose only solution is $(a = b = c = d = 0)$. The simulation results obtained are shown in figure (3). Upon verification of (12), we can observe that our algorithm determines the existence of a solution.

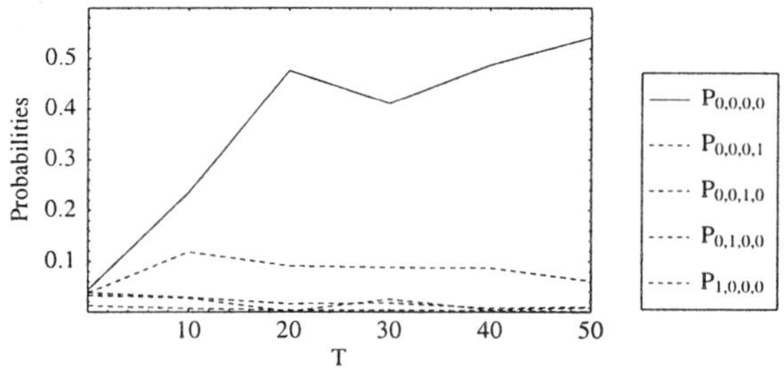

Figure 3. Simulation results for (17) with $(z, m) = (1.6, 3)$.

5 Conclusions

Although the numerical simulations cannot be equivalent to our possible hypercomputational quantum algorithm, due to the impossibility to simulate an infinite number of dimensions, their results give an idea of the behaviour that could be observed if we would have a real implementation and execution of our algorithm, supported in a determined physical process.

The nonequivalence is observed through an asymmetry between solvable and non-solvable Diophantine Equations in the non-negative integers, with respect to the conclusiveness of the numerical simulations. This asymmetry is usually a common characteristic of the Turing machine incomputable problems, for which, if the instance of the problem has a solution, a classic algorithm would find it; while if it doesn't have a solution, such algorithm will not halt.

Acknowledgments

We thank Tien D. Kieu for helpful discussions and feedback. This work was supported by COLCIENCIAS (grant #RC-284-2003) and by EAFIT University (grant #1216-05-13576).

References

[1] Copeland J. (2002) Hypercomputation. Minds and Machines 12: 461–502

[2] Burgin M., Klinger A. (2004). Three aspects of super-recursive algorithms and hypercomputation or finding black swans. Theoretical Computer Science 317: 1–11

[3] Stannett M. (2003) Hypercomputation models. In: Teuscher C. (ed.) Alan Turing: life and legaly of a great thinker 2003. Springer, Berlin, pp. 135–157

[4] Sicard A., Vélez M. (2001) Hipercomputación: la próxima generación de la computación (Spanish). Revista Universidad EAFIT, 123: 47–51

[5] Sicard A., Vélez M., Ospina J. (2004) Hypercomputation based on quantun computation. Eprint `arXiv.org/abs/quant-ph/0406137`

[6] Matiyasevich Y. V. (1993) Hilbert's tenth problem. The MIT Press, Cambridge, Massachusetts

[7] Kieu T. D. (2002) Quantum adiabatic algorithm for Hilbert´s tenth problem: I. The algorithm. Eprint `arXiv.org/abs/quant-ph/0310052`

[8] Kieu T. D. (2003) Computing the noncomputable. Contemporay Physics 44(1): 51-71

[9] Kieu T. D. (2003) Quantum algorithm for the Hilbert's tenth problem. Int. J. Theor. Phys. 42(7): 1461–1478

[10] Avron J. E., Elgart A. (1999) Adiabatic theorem without a gap condition. Commun. Math. Phys. 203: 444–463

[11] Antoine J. P., et al. (2001) Temporally stable coherent states for infinite well and Pöschl-Teller potentials. J. Math. Phys. 42(6): 2349–2387

[12] Wang X., Sanders B. C., Pan S. (2000) Entangled coherent states for systems with $SU(2)$ and $SU(1,1)$ symmetries. J. Phys. A: Math. Gen. 33(41): 7451–7467

[13] Kieu T. D. (2003) Numerical simulations of a quantum algorithm for Hilbert's tenth problem. In: Donkor E., Pirich A. R., Brandt H. E. (eds.) Quantum Information and Computation, vol. 5105 of Proc. of SPIE, SPIE, Bellingham, WA, pp. 89–95

[14] Waldschmidt M. (2004) Open diophantine problems. Moscow Mathematical Journal 4(1): 245–305

Simulated Fault Injection in Quantum Circuits with the Bubble Bit Technique

M. Udrescu [1], L. Prodan[1], M. Vlăduţiu[1]

[1]Advanced Computing Systems and Architectures Laboratory, University Politehnica of Timişoara, Romania
Web: www.acsa.utt.ro; E-mail: {mudrescu, lprodan, mvlad}@cs.utt.ro

Abstract

The simulation of quantum circuits is usually exponential. The Hardware Description Languages methodology is able to isolate the entanglement as source of simulation complexity. However, it was shown that this methodology is not efficient unless the bubble bit technique is employed [1]. In this paper, we present an extension of the HDL-bubble bit simulation methodology, which provides means for simulated fault injection – at the unitary level – in quantum circuits. The purpose is, just like in classical computer hardware design, to be able to verify the effectiveness of the considered quantum circuit fault tolerance methodologies.

1 Introduction

Simulating quantum computation (and quantum circuits in particular) is hard, even though there are many types of simulators [2] and promising new achievements [3], the high complexity involved by simulating quantum algorithms cannot be completely avoided by using symbolic representations. The main source of quantum circuit simulation complexity is the entanglement [4][5]. Nevertheless, efficient simulation is a prerequisite in order to approach CAD techniques for quantum circuits. The techniques described in this paper are based on the bubble bit insertions [1] in order to reduce the simulation runtimes. Other simulation frameworks are also considering the memory overhead as a parameter [3].

The gate-level simulation of quantum circuits [1] is based on the *circuit model* of quantum computation: the circuits process quantum register states, with the measurement being the only non-unitary operation. These *quantum circuits* are constrained networks of quantum gates, with no cloning and no feedback allowed [4].

With its paradigm suggested by Fig. 1, the gate-level HDL simulation methodology [1] acknowledges entanglement as the main source of simulation complexity. The use of Hardware Description Languages (HDLs) is motivated by the fact that they are able to describe a circuit, in a compact manner, with both structural and behavioral (functional) architectures. We show how the bubble bit technique provides, besides substantial runtime improvement, for simulated fault injection support.

2 HDL Simulation Methodology

When entanglement is not present in the processed quantum state, it is possible to describe the circuit and the states with structural architectures by employing only polynomial resources for simulation. By contrast, when entanglement is detected in the processed state, the circuit has to be described with a behavioral architecture. This case requires exponential resources, because the overall state cannot be represented correctly as a reunion of individual subsystem states [1].

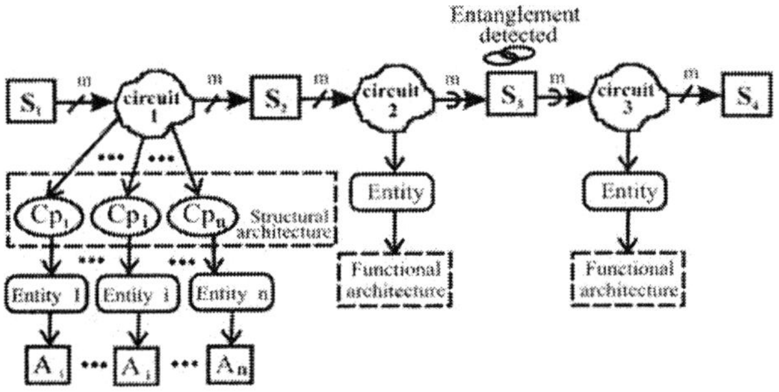

Fig. 1. HDL-based simulation approach example.

In Fig. 1 two quantum circuits (1 and 2, functionally described) guard the entangled quantum state (S_3). The first quantum circuit uses the structural description because it is guarded by 2 non-entangled states (S_1 and S_2). $Cp_1 \ldots Cp_n$ are the basic components of "circuit 1", while $A_1 \ldots A_n$ are their architectures. The non-entangled quantum register uses a '/' notation, while ')' is used for the entangled situation. Each circuit must be described both by structural and functional architectures. For a gate network, if entanglement is detected in the previous or next quantum state, then the functional architecture has to be selected to describe it; otherwise the structural architecture is chosen. For the non-entangled, structural situation, the quantum circuit can be reduced to a collection of 1-qubit quantum gates. In fact, the bubble bit technique is designed to avoid entangled representations and to encode the quantum state matrix representation so that it can be processed with only 1-qubit gates. Fig. 2 presents the model of bubble bit HDL simulation frameworks (including the possibility of fault injection).

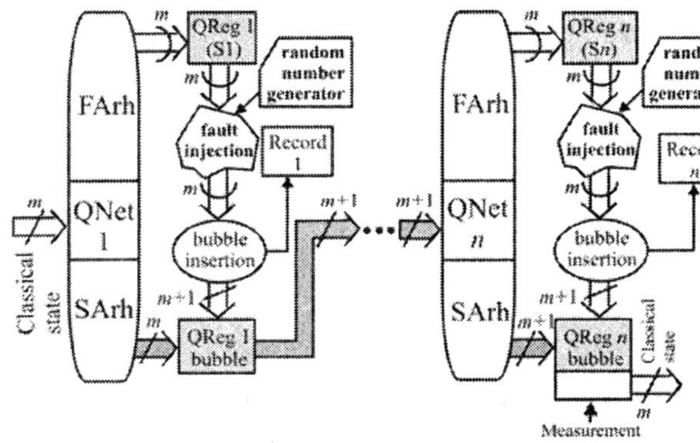

Fig. 2. HDL-based, bubble-bit simulation model.

3 Bubble Bit Insertion

The purpose of this technique is to avoid the $2^n \times 2^n$ matrix expression of the n-qubit register unitary operator. We take into consideration the specific pattern that characterizes the arithmetic states from Shor's algorithm and Grover's algorithm [4][5]:

$$\left|\psi_{\exp}\right\rangle = \xi\begin{bmatrix} b_0 & b_1 & \dots & b_{2^n-1}\end{bmatrix}^{\dagger} \quad (1)$$

where n is the number of qubits, $\xi \in \mathbb{C}$, and $b_i \in \{0,1,-1\}$. From this equation ξ will be ignored because all non-zero amplitudes are equal. We denote couple matrixes $\begin{bmatrix}0\\0\end{bmatrix} = \hat{0}$, $\begin{bmatrix}0\\1\end{bmatrix} = \hat{1}$, $\begin{bmatrix}1\\0\end{bmatrix} = \hat{2}$, $\begin{bmatrix}1\\1\end{bmatrix} = \hat{3}$ respectively.

The insertion algorithm is described by pseudocode:

```
Bubble step insertion algorithm
1. scan all the couples (b_k, b_{k+1}) from Eq.1;
1.1  memorize the first non-0̂ couple;
1.2 insert bubbles according to rules in Fig.3
and memorize their nature and position;
2. if the number of b_i elements is a power of 2
then go to step 4;
3. if the previous adjustement consisted of a 0s
padding then erase 0s so that the number of b_i
elements will be the closest power of 2;
4. extract the first detected non-0̂ couple as a
non-entangled qubit representation;
```

The oval from Fig. 3 is the first non-$\hat{0}$ couple detected, and the rectangle represents the current processed couple. The bubble bit is inserted between the bits shown in rectangles in Fig. 3. As a result, we will have a current processed couple (c) and a next couple (n). The '?' sign stands for a further matrix element, not processed in the current step. Several bubble steps must be performed until all qubits are extracted. Also, for maintaining a coherent matrix-form quantum state an extra-qubit will be added to the state representation. Thus, the number of b_i elements will be increased from 2^n to 2^{n+1} – at the first bubble step – by extra 0s padding. The next bubble steps will require erasure of 0s, so that the matrix-form representation further complies with the state coherence requirement. For every bubble-bit insertion, its position inside the vector is recorded: $\{b,\ pos\}$ is described by its nature (b=0/1/-1) and position in the resulted state (pos). Performing all the necessary bubble steps requires a total of $O(n^2)$ records be produced [1].

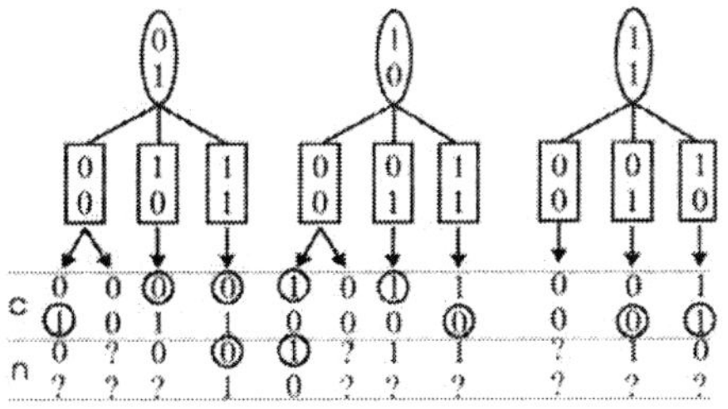

Fig. 3. Bubble bit insertion rules.

3.1 Fault Injection – The Setup Phase

The bubble bit framework can be adapted for simulated fault injection requirements. Due to the specific nature of the fault model in quantum computation [6], and the way bubble bit encoding is performed, fault injection must be applied before the bubble bit insertion algorithm. Also, fault injection is triggered only if the "random number generator" dictates so (see Fig. 2). The entire simulated fault injection process is inspired by the classical hardware design solutions [7] and consists of 3 cycles: initialization, simulation, and data computation. The setup phase corresponds to settling the nature of the fault and its triggering. The simulation phase is responsible for running the experiment according to the scenario provided by the setup phase, while the data processing phase uses the simulation signal traces results, in order to compute the appropriate reliability measure.

In our simulation framework [1], injecting a fault consists of accordingly modifying the quantum state matrix from equation (1). When the fault is a bit-flip, then the fault injection means the matrix elements will be rearranged, whereas for the phase shift some matrix elements will be multiplied with -1. In the bit flip case the elementary operation is exchanging values between two matrix positions: $b_i \leftrightarrow b_j$ for $i \neq j$. This allows building an exchange function that operates on blocks of matrix elements:

$$Exchange[(u_0,\dots u_{w-1}),(v_0,\dots v_{w-1})] \Leftrightarrow u_i \leftrightarrow v_i \quad (2)$$

for every $i = \overline{0, w-1}$.

Suppose we have a quantum state on n qubits, $\left|q_{n-1}\dots q_2 q_1\right\rangle$; if a fault occurs on qubit k, for the bit-flip and phase-shift faults we will execute one or both of the corresponding algorithms:

Bubble step insertion algorithm

For i:=0 to 2^{n-k}

$Exchange[(b_{i\cdot 2^k},\dots b_{(i+1)\cdot 2^k-1}),(b_{(i+1)\cdot 2^k},\dots b_{(i+2)\cdot 2^k-1})]$

End For

Phase-shift fault injection

For i:=0 to 2^n-1

If $i \bmod 2^k \ge 2^k - 1$ then $b_i := (-1)b_i$

End For

We have so far described the error injection process but it also needs a way to trigger it. According to the fault occurrence model [6] it has to be a random triggering. Therefore, we have to use a random number generator in order to find out if an error occurred. Then, we use the same generator for selecting one of the following fault types: bit-flip, phase-shift, both faults. When first used, the generator returns the number r_1. If $r_1 < n_\xi$ (for a n_ξ given by a fixed error rate), then we have a fault. We start the number generator again – yielding r_2 – and the selected fault nature is set by the following equation:

$$r_2 = \begin{cases} 0 \le r_2 \le \frac{1}{3} & \text{we have a bit-flip} \\ \frac{1}{3} < r_2 \le \frac{2}{3} & \text{we have a phase-shift} \\ \frac{2}{3} < r_2 \le 1 & \text{we have both faults} \end{cases} \tag{3}$$

3.2 The Data Processing Phase

The simulation phase is detailed in the example and experiment section. We now focus on how to automatically compute the most important parameter of quantum circuit reliability: the accuracy threshold [6].

Suppose that, at simulation time t we observe signals $\{s_0, s_1, \dots s_{n-1}\}$. Each such state has a bubble bit description. If s_i is on k_i qubits, the bubble-bit representation is given by the following equation:

$$s_i = |qb_0\rangle \otimes |qb_1\rangle \otimes \dots |qb_{k_i}\rangle + \boxed{rec_i}. \tag{4}$$

In our analysis, s_i is the state observed during non-faulty simulation, so for the same state in a faulty environment we will have the bubble expression given by:

$$s_i^* = |qb_0^*\rangle \otimes |qb_1^*\rangle \otimes \dots |qb_{k_i}^*\rangle + \boxed{rec_i^*}. \tag{5}$$

For validation of the quantum FTAMs (Fault Tolerant Algorithms and Methodologies), we need to compare s_i with s_i^*. This can be done with the operator presented in the following equation:

$$\mathrm{dif}(s_i, s_i^*) = \begin{cases} 1 & \text{if } |qb_i\rangle \ne |qb_i^*\rangle;\ \forall i = \overline{0, k_i} \text{ or } rec_i \ne rec_i^* \\ 0 & \text{otherwise} \end{cases} \tag{6}$$

Then, the total number of overall state errors at simulation time t is:

$$e_t = \sum_{i=0}^{n-1} \mathrm{dif}(s_i, s_i^*) \tag{7}$$

The error rate on the overall observed states at moments $t_0, t_1, \dots t_{m-1}$ will be given by:

$$\xi_{sim} = \frac{1}{m} \sum_{j=0}^{m-1} e_{t_j}. \tag{8}$$

As pointed out in ref. 6, the applied FTAMs are valid if the relationship between the experimental ξ_{sim} and the assumed singular error rate ξ is of the order $\xi_{sim} \sim \xi^2$.

4 Example and Experimental Results

In order to illustrate how the bubble bit technique works, we consider the simulation phase of fault injection, performed in bubble bit encoding conditions for the single error correcting quantum device in Fig. 4.

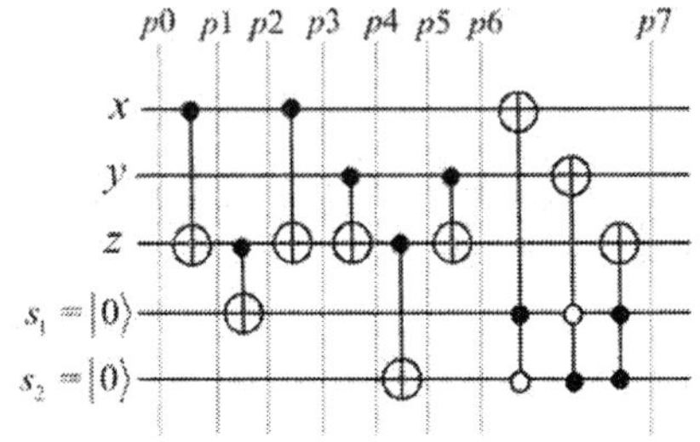

Fig. 4. Circuit for singular bit-flip error correction [6].

For this circuit we use a coding technique that replaces 1 qubit with a cluster of 3 qubits. The qubit state $|0\rangle$ is encoded as $|000\rangle$, and $|1\rangle$ as $|111\rangle$. For example, state $|\psi\rangle = \frac{1}{\sqrt{2}}(|0\rangle + |1\rangle)$ will become $|xyz\rangle = \frac{1}{\sqrt{2}}(|000\rangle + |111\rangle)$.

If a bit-flip occurs, then the error is indicated by the syndrome $|s_1 s_2\rangle$, where $s_1 = x \oplus z$ and $s_2 = y \oplus z$. The fault location corresponds to the following syndrome values: bit-flip on x by $|10\rangle$, on y by $|01\rangle$, and on z by $|11\rangle$. The syndrome $|00\rangle$ indicates that there is no error. In circuit from Fig. 4 we start with state $|p0\rangle_{correct} = \frac{1}{\sqrt{2}}(|000\rangle + |111\rangle)$, which is then affected by a bit-flip fault on qubit y: $|p0\rangle = \frac{1}{\sqrt{2}}(|010\rangle + |101\rangle)$.

The evolution of the bubble-bit quantum state representation throughout the simulation process (including the fault injection) is described by the following equations, which present the highlighted states from Fig. 4:

$$|p0\rangle = \left(\frac{1}{\sqrt{2}}\begin{bmatrix}1\\0\end{bmatrix} \otimes \begin{bmatrix}1\\1\end{bmatrix} \otimes \begin{bmatrix}0\\1\end{bmatrix} \otimes \begin{bmatrix}1\\0\end{bmatrix}\right) \otimes |0\rangle_{s_1} \otimes |0\rangle_{s_2} + \boxed{rec_0} \tag{9}$$

$$|p1\rangle = \left(\frac{1}{\sqrt{2}}\begin{bmatrix}1\\0\end{bmatrix} \otimes \begin{bmatrix}1\\1\end{bmatrix} \otimes \begin{bmatrix}0\\1\end{bmatrix} \otimes \begin{bmatrix}1\\0\end{bmatrix}\right) \otimes |0\rangle_{s_1} \otimes |0\rangle_{s_2} + \boxed{rec_1} \tag{10}$$

$$|p2\rangle \equiv |p1\rangle;\ |p3\rangle \equiv |p0\rangle \tag{11}$$

$$|p4\rangle = \left(\frac{1}{\sqrt{2}}\begin{bmatrix}1\\0\end{bmatrix}\otimes\begin{bmatrix}1\\1\end{bmatrix}\otimes\begin{bmatrix}0\\1\end{bmatrix}\otimes\begin{bmatrix}0\\1\end{bmatrix}\right)\otimes|0\rangle_{s_1}\otimes|0\rangle_{s_2} + \boxed{rec_2} \quad (12)$$

$$|p5\rangle = \left(\frac{1}{\sqrt{2}}\begin{bmatrix}1\\0\end{bmatrix}\otimes\begin{bmatrix}1\\1\end{bmatrix}\otimes\begin{bmatrix}0\\1\end{bmatrix}\otimes\begin{bmatrix}0\\1\end{bmatrix}\right)\otimes|0\rangle_{s_1}\otimes|1\rangle_{s_2} + \boxed{rec_2} \quad (13)$$

$$|p6\rangle = \left(\frac{1}{\sqrt{2}}\begin{bmatrix}1\\0\end{bmatrix}\otimes\begin{bmatrix}1\\1\end{bmatrix}\otimes\begin{bmatrix}0\\1\end{bmatrix}\otimes\begin{bmatrix}1\\0\end{bmatrix}\right)\otimes|0\rangle_{s_1}\otimes|1\rangle_{s_2} + \boxed{rec_0} \quad (14)$$

$$|p7\rangle = \left(\frac{1}{\sqrt{2}}\begin{bmatrix}1\\1\end{bmatrix}\otimes\begin{bmatrix}1\\0\end{bmatrix}\otimes\begin{bmatrix}1\\0\end{bmatrix}\otimes\begin{bmatrix}1\\0\end{bmatrix}\right)\otimes|0\rangle_{s_1}\otimes|1\rangle_{s_2} + \boxed{rec_3} \quad (10)$$

The corresponding records are presented in Fig. 5. This indicates the fact that state $|p7\rangle = \frac{1}{\sqrt{2}}(|000\rangle + |111\rangle)$, therefore the inflicted error has been corrected.

rec_0

step	bubble	zeros
1	{0,5}	+7
2	-	0

rec_1

step	bubble	zeros
1	-	0
2	{0,2}	+3

rec_2

step	bubble	zeros
1	-	0
2	{0,2}	+3

rec_3

step	bubble	zeros
1	{0,7}	+7
2	-	0

Fig. 5. Bubble records for simulating the circuit from Fig. 4.

The HDL bubble bit simulation of quantum circuits provides for substantial runtimes improvement, although it produces $\mathcal{O}(n^2)$ memory overhead because of the bubble records. Our results are due to VHDL simulations, carried on a Windows XP™, PENTIUM™ IV CPU 1,6GHz, 192MB RAM machine. We have performed the gate-level simulation of quantum arithmetic circuits (the full adder, modulo-k adder and multiplier [8]). The resulted runtimes are presented in Tab. 1 (we assumed $k=2^{size-1}$).

Grover's algorithm was simulated for an Oracle that marks just one basis state, like in ref. 3. The runtime evolution with the number of qubits in the data register is presented in Fig. 6. The measured simulation times are compared here with the runtime complexity reported in ref. 3, which is 0.22×1.44^n.

5 Conclusions

Efficient quantum gate-level simulation may be achieved by using the HDL simulation framework, at least for Grover iteration and arithmetic circuits [2][4]. Instead of using special algorithms for non-entangled qubit group extraction [2][8], the HDL simulation methodology relies on the *bubble bit* technique, introduced as a method of avoiding entangled representations. As experimented for Shor's algorithm arithmetic circuits and Grover's algorithm, the bubble bit technique allows for polynomial runtime simulation of unitary operators (a substantial improvement), at the expense of building records of size $\mathcal{O}(n^2)$. Furthermore, the bubble bit technique can be used in order to attain simulated fault injection of quantum circuits, which in turn creates incentive for experimental validation of quantum fault tolerant techniques. Our future work will focus on extending the HDL-based simulation framework according to the fault injection objective. All these will make up what we call the QUERIST (QUantum ERror Injection Simulation Tool) project.

Tab. 1. Runtimes for arithmetic circuits simulation.

size	full adder	modulo adder	modulo multiplier
4	2 sec	3.5 sec	9 sec
8	13 sec	17 sec	44.5 sec
16	41 sec	58.5 sec	2 min, 16 sec
32	3 min, 48 sec	5 min, 42 sec	16 min, 23 sec
64	16 min, 7 sec	21 min, 4 sec	53 min, 18 sec

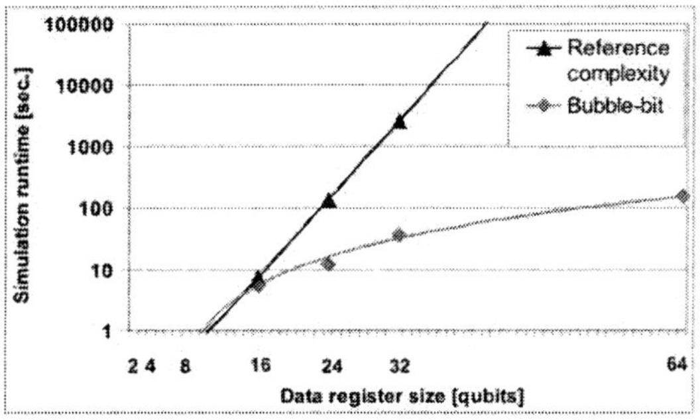

Fig. 6. Runtime evolution with the number of qubits.

References

[1] Udrescu, M., Prodan, L., Vlăduţiu, M. (2004) Using HDLs for Describing Quantum Circuits: A Framework for Efficient Quantum Algorithm Simulation. In Proc. 1st ACM Conf. on Computing Frontiers, Ischia, Italy, pp.96-110.

[2] Obenland, K.M., Despain, A. (1998) Simulating the Effect of Decoherence and Inaccuracies on a Quantum Computer. Proc. 1st NASA Conf. on Quantum Comp. and Comm.

[3] Viamontes, G.F., Markov, I.L., Hayes, J.P. (2004) High-performance QuIDD-based Simulation of Quantum Circuits. Proc. Design Autom. and Test in Europe, Paris, pp. 1354-1359.

[4] Nielsen, M.A, Chuang, I.L. (2000) Quantum Computation and Quantum Information. Cambridge University Press, UK.

[5] Parker, S., Plenio, M. (2002) Entanglement Simulations of Shor's Algorithm. J. Mod. Opt., Vol. 49, Nr. 8 pp. 1325-1353.

[6] Preskill, J. (1997) Fault-Tolerant Quantum Computation. Online preprint quant-ph/9712048.

[7] Rimen, M., Ohlsson, J., Karlsson, J., Jenn, E., Arlat, J. (1993) Design guidelines of a VHDL-based simulation tool for the validation of fault tolerance. Rapport LAAS No93170.

[8] Vedral, V., Barenco, A., Ekert, A. (1996) Quantum Networks for Elementary Arithmetic Operations. quant-ph/9511018, 1996.

Redundant Quantum Arithmetic

António Pereira, Rosália Rodrigues
Department of Mathematics, University of Aveiro, Portugal
E-mail: {antoniop, rosalia}@mat.ua.pt

Abstract

Redundant number systems have been widely used in the speedup of classical digital arithmetic. This work introduces the concept of redundancy in the quantum computation field. We show that a constant depth quantum adder circuit is attainable under this new framework.

1 Introduction

In the field of classical digital arithmetic, redundancy plays a central role in the speedup of computer arithmetic operations. Redundant number systems are at the core of several fast digital arithmetic algorithms [1].

Pioneer work on quantum arithmetic includes the carry-ripple quantum adder of Vedral et al [2] and the quantum carry-save adder proposed by Gossett [3]. Draper [4] developed a method for the addition on the quantum computer using the quantum Fourier transform. Recently Draper et al [5] developed a quantum carry-lookahead adder.

The best time complexity achieved so far is $\mathcal{O}(\log n)$. By bringing the concept of redundancy into the quantum arithmetic field we show that the bound $\mathcal{O}(1)$ is attainable.

2 Generalized Signed Digit Number Systems

Redundant number systems have been widely used in digital arithmetic algorithms. Parhami [6] made an important generalization and unification of these systems and their digital algorithms under a common framework named Generalized Signed Digit (GSD) number systems. The basic concepts are now introduced.

Definition 1 *Let $r \geq 2$ be an integer. A radix-r (signed) digit set is a set of consecutive integers, called (signed) digits, $\{-\alpha, \ldots, \beta\}$, such that $\alpha, \beta \geq 0$ and $\alpha+\beta+1 \geq r$.*

Definition 2 *A radix-r GSD number system is a positional number system such that any finite sequence of $n > 0$ (signed) digits from a digit set S associates with an integer by*

$$x_0 x_1 \cdots x_{n-1} \mapsto \sum_{j=0}^{n-1} x_j r^j \ . \tag{1}$$

The redundancy index *of a radix-r GSD number system with digit set $\{-\alpha, \ldots, \beta\}$ is*

$$\rho = \alpha + \beta + 1 - r \ . \tag{2}$$

A GSD number system is called redundant *if $\rho > 0$.*

Note that unlike standard practice in classical digital arithmetic, a least significant digit first notation is used, i.e., the most significant digit in $x_0 x_1 \cdots x_{n-1}$ is x_{n-1}.

3 Quantum GSD Representations

In this section GSD number systems are introduced in the field of quantum computation.

Consider some fixed radix-r GSD number system with a digit set of size $d \geq 2$, $S = \{-\alpha, \ldots, \beta\}$, $\alpha + \beta + 1 = d$.

Definition 3 *Let $d \geq 2$ be an integer. A* qudit *is a quantum system $\mathcal{Q}$ whose state is a unit vector in a d-dimensional Hilbert space.*

Without loss of generality, we assume that the underlying Hilbert space in Definition 3 is $\mathcal{H}_d = \mathbb{C}^d$. It has become common practice to identify the canonical computational basis of $\mathcal{H}_d$ with the set $\{|k\rangle, k \in \mathbb{Z}_d\}$, in Dirac notation. We will make use of the labelling scheme,

$$\{|k\rangle, k \in S\} . \tag{3}$$

The general state of a qudit thus is

$$|\psi\rangle = \sum_{k \in S} a_k |k\rangle \ , \tag{4}$$

where $a_k \in \mathbb{C}, k \in S$, satisfy the normalization condition $\sum_{k \in S} |a_k|^2 = 1$.

When the nonredundant binary system, $r = 2$, $S = \{0, 1\}$ is considered, (3) and (4) give the usual definition of the general state of a qubit, $|\psi\rangle = a|0\rangle + b|1\rangle$ with $|a|^2 + |b|^2 = 1$, $a, b \in \mathbb{C}$.

Let $\mathcal{Q}_1, \mathcal{Q}_2, \ldots, \mathcal{Q}_n$ be quantum systems with underlying spaces $\mathcal{H}_{d_1}, \mathcal{H}_{d_2}, \ldots, \mathcal{H}_{d_n}$. The underlying Hilbert space of the composite quantum system with components $\mathcal{Q}_1, \mathcal{Q}_2, \ldots, \mathcal{Q}_n$ is the tensor product space $\mathcal{H}_{d_1} \otimes \mathcal{H}_{d_2} \otimes \ldots \otimes \mathcal{H}_{d_n}$. Thus the state of a n-qudit quantum register is a normalized vector of $\mathcal{H}_d{}^{\otimes n}$.

Definition 4 *The computational basis for the underlying space* $\mathcal{H}_d^{\otimes n}$ *of a n-qudit quantum register is given by the* d^n *states*

$$|x_0\rangle \otimes |x_1\rangle \otimes \cdots \otimes |x_{n-1}\rangle \quad , \tag{5}$$

where $x_j \in S$, $j = 0, 1, \ldots, n-1$.

Each of these states naturally associates with the radix r representation of the integer $\sum_{j=0}^{n-1} x_j r^j$ by

$$|x_0\rangle \otimes |x_1\rangle \otimes \cdots \otimes |x_{n-1}\rangle \equiv |x_0 x_1 \cdots x_{n-1}\rangle \equiv \left| \sum_{j=0}^{n-1} x_j r^j \right\rangle . \tag{6}$$

In the case of a redundant GSD number system the same number can be represented by several basis states.

4 Carry Free Adder

In this section we outline the carry free adder algorithm developed by Parhami [6].

Let $S = \{-\alpha, \ldots, \beta\}$ be the digit set of some radix-r GSD number system where the integers x and y are represented by the sequences $x_0 x_1 \cdots x_{n-1}$ and $y_0 y_1 \cdots y_{n-1}$.

Adding x and y in conventional nonredundant digital arithmetic is basically a process of splitting $x_i + y_i$ into a sum digit z_i and a carry digit c_{i+1} to the next position. This operation is performed for each i and $z = x + y$ is encoded as $z_0 z_1 \cdots z_{n-1}$ plus an overflow digit c_n.

Generally, each carry c_{i+1} depends not only on x_i and y_i but also on the previous c_i. Therefore, the evaluation of c_{i+1} must wait for c_i to be computed. The generated carry propagation chains are responsible for the overall time complexity of the algorithm, tipically $\mathcal{O}(n)$.

The Carry Free Adder (CFA) algorithm starts by computing a partial sum w_i and a carry digit c_{i+1} from x_i and y_i, for every i. Then each of the final sum digits z_i is computed by adding the partial sum digit w_i and the carry c_i. The generated partial sum and carry digits must be such that the resulting $z_i = w_i + c_i$ still belongs to S.

An important property of the CFA algorithm is its full parallelization capability. As a consequence, the algorithm runs in $\mathcal{O}(1)$ (provided that sufficient hardware is available).

Algorithm 1 (Carry Free Adder)

```
FOR i FROM 0 TO n-1 DO
  compute c[i+1] and w[i]
  from x[i] and y[i]
END DO

FOR i FROM 0 TO n-1 DO
  compute z[i] from w[i] and c[i]
END DO
```

Parhami showed that the CFA algorithm is applicable to a large subclass of redundant GSD number systems [6].

Theorem 1 *Let* $S = \{-\alpha, \ldots, \beta\}$ *be the digit set of a radix-*r *GSD number system with redundancy index* ρ. *The CFA algorithm is applicable if and only if* ($\rho \geq 3$ *and* $r > 2$) *or* ($\rho = 2$, $\alpha \neq 1$ *and* $r > 2$) *or* ($\rho = 2$, $\beta \neq 1$ *and* $r > 2$).

Parhami also presented a limited carry propagation algorithm for the GSD number systems excluded by theorem 1. A quantum circuit implementing the limited carry propagation algorithm was already developed [7].

5 Quantum Carry Free Circuit

The quantum adder circuit will be represented by a unitary operator acting on the register space that stores the inputs x and y, the result z and the ancillary space.

The adder circuit belongs to the family of hybrid quantum systems [8] in the sense that is composed of qudits of several dimensions.

The general computing process, ignoring ancillae, is

$$U |x\rangle |y\rangle |z_{init}\rangle = |x\rangle |y\rangle |x + y\rangle \quad , \tag{7}$$

where $|z_{init}\rangle$ is the initial state of result register (similar to the initial state $|0\rangle$ in conventional qubit computations).

In what follows we will use the notation borrowed from [9]:

> The tensor product of an operator U acting on an ordered sequence of qudits $q_1, q_2, \ldots, q_m$ and the identity operator acting on the remaining qudits is denoted by $U[q_1, q_2, \ldots, q_m]$.

Generating each carry digit c_{i+1} can be viewed as the process of evaluating a function $c : S^2 \longrightarrow S'$ at the point (x_i, y_i), where $S' = \{-\lambda, \ldots, \mu\}$ is the set of possible carry digits. Function c depends only on the GSD number system in use and its definition can be found in Parhami [6]. Let $d' = \lambda + \mu + 1$ be the size of S' and $\widehat{c} : S^2 \times S' \longrightarrow S^2 \times S'$ be the extension of c defined by $\widehat{c}(x, y, f) = (x, y, -\lambda + (2\lambda + f + c(x, y)) \bmod d')$.

Since $\widehat{c}$ is just a permutation function, the operator $C \in \mathbf{U}(\mathcal{H}_d^{\otimes 2} \otimes \mathcal{H}_{d'})$ that implements the computation of each carry digit is given by

$$C |x\rangle |y\rangle |f\rangle = |x\rangle |y\rangle |\widehat{c}(x, y, f)\rangle . \tag{8}$$

This operator has action (see Fig. 1)

$$C |x_i\rangle |y_i\rangle |-\lambda\rangle = |x_i\rangle |y_i\rangle |c_{i+1}\rangle .$$

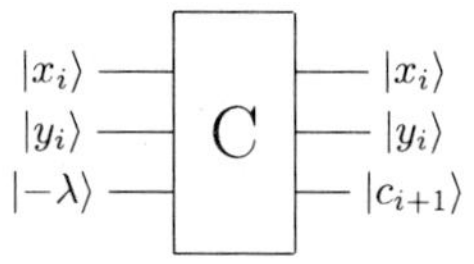

Fig. 1. Quantum gate C

Computation of the carry digits is performed in parallel by applying the unitary operator

$$L_1 = \prod_{i=0}^{n-1} C_{i+1} \equiv C_n C_{n-1} \cdots C_2 C_1 \ , \tag{9}$$

where $C_{i+1} = C[x_i, y_i, c_{i+1}]$ for each $i = 0, \ldots, n-1$.

Now let $w : S^2 \times S' \to S$ be defined by

$$w(x, y, c) = \begin{cases} x + y - r \cdot c & \text{if } c = c(x, y) \\ c & \text{otherwise} \end{cases} \tag{10}$$

and let $\widehat{w} : S^2 \times S' \times S \to S^2 \times S' \times S$ be the permutation function given by $\widehat{w}(x, y, c, g) = (x, y, c, -\alpha + (2\alpha + g + w(x, y, c)) \bmod d)$.

After all carry digits have been generated, computation of each partial sum w_i, $i = 0, \ldots, n-1$, is performed by applying the operator $W \in \mathbf{U}(\mathcal{H}_d^{\otimes 2} \otimes \mathcal{H}_{d'} \otimes \mathcal{H}_d)$ defined by

$$W |x\rangle |y\rangle |c\rangle |g\rangle = |x\rangle |y\rangle |c\rangle |\widehat{w}(x, y, c, g)\rangle \ . \tag{11}$$

The operator W has action $W |x_i\rangle |y_i\rangle |c_{i+1}\rangle |-\alpha\rangle = |x_i\rangle |y_i\rangle |c_{i+1}\rangle |w_i\rangle$, as illustrated in Fig. 2.

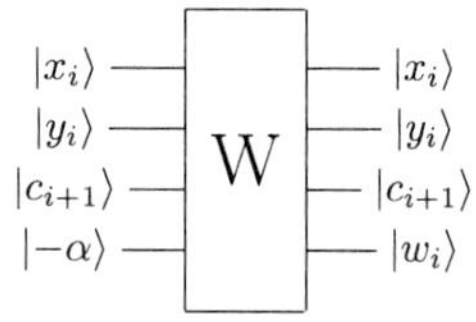

Fig. 2. Quantum gate W

The parallel computation of the partial sum digits consists on applying the following sequence of unitary operators acting on disjoint sets of qudits

$$L_2 = \prod_{i=0}^{n-1} W_i \equiv W_{n-1} W_{n-2} \cdots W_1 W_0 \ , \tag{12}$$

where $W_i = W[x_i, y_i, c_{i+1}, w_i]$, $i = 0, \ldots, n-1$.

The evaluation of each of the final sum digits, z_i, $i = 0, 1, \ldots, n-1$, is performed by the function $z : S \times S' \to S$ defined by $z(w, c) = w + c$. Associated to z is the unitary operator $Z \in \mathbf{U}(\mathcal{H}_d \otimes \mathcal{H}_{d'} \otimes \mathcal{H}_d)$ given by

$$\begin{aligned} Z |w\rangle |c\rangle |-\alpha\rangle = \\ |w\rangle |c\rangle |-\alpha + (2\alpha + z(w, c)) \bmod d\rangle \ . \end{aligned} \tag{13}$$

Figure 3 illustrates the action $Z |w_i\rangle |c_i\rangle |-\alpha\rangle = |w_i\rangle |c_i\rangle |z_i\rangle$.

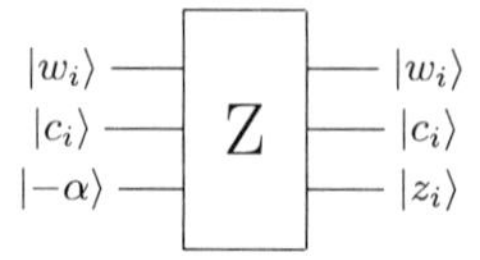

Fig. 3. Quantum gate Z

Let $Z_i = Z[w_i, c_i, z_i]$, $i = 0, \ldots, n-1$. The parallel computation of the final sum digits is provided by the unitary operator

$$L_3 = \prod_{i=0}^{n-1} Z_i \equiv Z_{n-1} Z_{n-2} \cdots Z_1 Z_0 \ . \tag{14}$$

Note that the most significant digit of the result, z_n, is just the last carry digit c_{n+1}.

We now address the problem of garbage disposal. Reverting the ancillary qudits to their initial states is achieved by applying the corresponding inverse operators in reverse order. This process, known as uncomputation of partial results, has two steps:

1. Reset the partial sums,

$$L_2^{-1} \equiv W_0^{-1} W_1^{-1} \cdots W_{n-2}^{-1} W_{n-1}^{-1} \ . \tag{15}$$

2. Reset the carry digits,

$$L_1'^{-1} \equiv C_1^{-1} C_2^{-1} \cdots C_{n-2}^{-1} C_{n-1}^{-1} \ . \tag{16}$$

Note that (16) does not reset the most significant digit of the result. A measurement is necessary before applying C_n^{-1}.

We are now in place to define the unitary operator U that implements the full CFA circuit.

Let $\mathcal{H}_{\text{work}} = \mathcal{H}_d^{\otimes n} \otimes \mathcal{H}_d^{\otimes n} \otimes \mathcal{H}_d^{\otimes n} \otimes \mathcal{H}_{d'}$ be the working register space and $\mathcal{H}_{\text{ancilla}} = \mathcal{H}_{d'}^{\otimes n} \otimes \mathcal{H}_d^{\otimes n}$ be the ancillary space. Then the quantum operation $V \in \mathbf{U}(\mathcal{H}_{\text{work}} \otimes \mathcal{H}_{\text{ancilla}})$ defined by

$$V = L_1'^{-1} L_2^{-1} L_3 L_2 L_1 \tag{17}$$

has action

$$V(|\xi\rangle \otimes |\zeta\rangle) = U(|\xi\rangle) \otimes |\zeta\rangle \ , \tag{18}$$

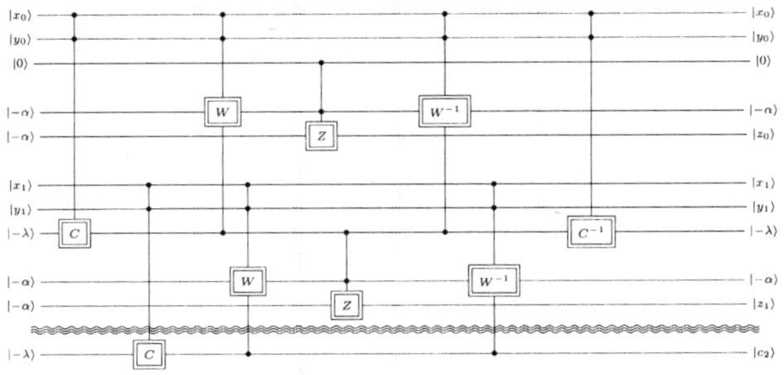

Fig. 4. Quantum CFA circuit for $n = 2$

where U is the adder operator (7) and $|\zeta\rangle \in \mathcal{H}_{\text{ancilla}}$ is the initial state of the ancillary space.

Each term of the product (17) can be computed in one-time slice since the gates act on distinct qudits. Thus the full circuit for the addition of two n-qudit numbers has constant depth , which gives the $\mathcal{O}(1)$ time complexity (see Fig. 4).

6 Applications

Quantum algorithms for multiplication and certain mathematical functions start to appear in the literature [2], [3]. The simple inclusion of a CFA circuit could noticeably improve the time requirements of such algorithms.

Quantum modular exponentiation is one of the most expensive parts of Shor's factoring algorithm [10]. This process could be made much faster, since it consists of a sequence of multiplications.

Quantum error-correcting codes allow for arbitrarily large quantum computations, provided the noise in individual quantum gates is below a certain constant threshold. Threshold theorems rely on assumptions about the type of noise and the architecture of the computer. In order that the techniques of fault-tolerant computation may be applied, quantum computer designers should develop architectures that are parallelizable [11]. The intrinsically parallel CFA here presented, when part of more complex redundant arithmetic algorithms, could provide efficient fault-tolerant quantum operations.

7 Conclusions and subsequent research

A new general framework for redundant quantum computations was introduced. This provided the building of a class of fully parallel circuits for the addition in constant time.

The use of redundancy in classical digital circuits has long been known to provide efficient algorithms for the evaluation of other arithmetic operations as well as elementary mathematical functions [12].

There remains to fully explore the combination of redundancy and quantum concepts in all arithmetics, evaluation of elementary mathematical functions and more complex computations.

Acknowledgments

This work was partially supported by the R&D unit Center for Research in Optimization and Control (CEOC) of the University of Aveiro.

References

[1] Phatak, D. S., Goff, T., and Koren, I. (2001) Constant-time addition and simultaneous format conversion based on redundant binary representations. IEEE Trans. Comput. 50: 1267–1278

[2] Vedral, V., Barenco, A., and Ekert, A. (1995) Quantum networks for elementary arithmetic operations. *http://xxx.lanl.gov/abs/quant-ph/9511018*

[3] Gossett, P. (1998) Quantum carry-save arithmetic. *http://arxiv.org/abs/quant-ph/9808061*

[4] Draper, T. G. (2000) Addition on a quantum computer. *http://xxx.lanl.gov/abs/quant-ph/0008033*

[5] Draper, T. G., et al (2004) A logarithmic-depth quantum carry-lookahead adder. *http://xxx.lanl.gov/abs/quant-ph/0406142*

[6] Parhami, B. (1990) Generalized signed-digit number systems: A unifying framework for redundant number representations. IEEE Trans. Comput. 39: 89–98

[7] Pereira, A. and Rodrigues, R. (2004) An $\mathcal{O}(1)$ adder for quantum arithmetic. Cadernos de Matemática, Dep. de Matemática da Univ. de Aveiro, CM04/I-13

[8] Daboul, J., Wang, X., and Sanders, B. C. (2003) Quantum gates on hybrid qudits. Journal of Physics A 36: 2525–2536

[9] Kitaev, A. Y., Shen, A., and Vyalyi, M. (2002) Classical and quantum computation. Graduate Studies in Mathematics 47, American Mathematical Society

[10] Shor, P. W. (1995) Polynomial-time algorithms for prime factorization and discrete logarithms on a quantum computer. *http://xxx.lanl.gov/abs/quant-ph/9508027*

[11] Nielsen, M. A. and Chuang, I. L. (2000) Quantum Computation and Quantum Information, Cambridge University Press, Cambridge, UK

[12] Rodrigues, M. R. D., Zurawski, J. H. P., and Gosling, J. B. (1981) Hardware evaluation of mathematical functions. IEE Proc. 128: 155–164

A Special Class of Additive Cyclic Codes for DNA Computing

Taher Abualrub[1], Ali Ghrayeb[2], and Xiang Nian Zeng[2]
[1]Department of Mathematics, American University of Sharjah, UAE
(e-mail: abualrub@ausharjah.edu).
[2]Department of Electrical and Computer Engineering, Concordia University,
Montreal, PQ, Canada (email: aghrayeb/x_zeng@ece.concordia.ca).

Abstract

In this paper, we study a special class of nonbinary additive cyclic codes over $GF(4)$ which we call *reversible complement cyclic codes*. Such codes are suitable for constructing codewords for DNA computing. We develop the theory behind constructing the set of generator polynomials for these codes. We study, as an example, all length-7 codes over $GF(4)$ and list those that have the largest minimum Hamming distance and largest number of codewords.

1 Introduction

Adleman's work [1] set up the stage for the possibility to use DNA to solve hard (NP-complete) computational problem such as the Hamiltonian Path problem by manipulations of DNA molecules in a test tube. His main idea was to use actual DNA sequences (strands) to solve problems that are unsolvable by conventional computers or require an enormous amount of computation. Because of the huge numbers of DNA molecules in a typical test tube, any method of computation based on DNA would seem to have potential massive parallelism, capacity, and power. This potential, however, is limited by the constraints imposed by the chemical structure of DNA.

A single DNA strand is a sequence of four possible nucleotides: adenine (A), guanine (G), cytosine (C) and thymine (T). The ends of a single DNA strand are chemically polar with the so called $5'$ end and the $3'$ end. Since there are 4^n possible single DNA strands of length n, and since DNA strands can be quickly and cheaply synthesized, DNA codewords can be used to store information at the molecular level, thus providing a basis for bimolecular computation. It is known that 1 gram of DNA contains 2.1×10^9 DNA bases. Since there are 4 DNA bases then each DNA bases can encode two bits which implies that 1 gram of DNA can store about 4.2×10^{21} bits. On the contrary, conventional storage technologies can store at most 10^9 bits.

These applications of DNA require success in achieving specific hybridization between a DNA codeword and its Watson-Crick complement. Each single string can be paired up with a complementary string to form a double helix. The Watson-Crick complement of a DNA strand is the strand obtained by replacing each A by T and vice versa, each C by G and vice verse, and switching the $3'$ and $5'$ ends. Specific hybridization is the process whereby a strand and its Watson-Crick complement bond to form a double helix. Nonspecific hybridization is the process whereby a strand bonds with another strand different that its complement. Nonspecific hybridization may occur between a DNA strand and the Watson-Crick complement of a distinct DNA strand. It may also occur between a DNA strand and the reverse of a distinct strand.

Several papers have proposed different techniques to construct a set of DNA codewords which are unlikely to form undesirable bonds with each other by hybridization [2]$-$[10]. For example, in [7], [8], [9] four different constraints on DNA codes are considered: The Hamming constraint for a distance d, the reverse-complement constraint, the reverse constraint and the fixed $GC-$content constraint. The purpose of the first three constraints is to make non-desirable hybridizations difficult to occur. The fixed $GC-$content constraint is used to make sure we have similar melting temperatures [9]. In [10], the authors have proposed cyclic codes over $GF(4)$ as a good field to construct DNA codes. They restricted their work only to linear reversible cyclic codes over $GF(4)$.

In this paper we will use the rich theory of cyclic codes to form codes that are suitable for DNA computing. In particular, we will construct additive cyclic codes over $GF(4)$ that are suitable to minimize non-specific hybridization. Our focus will be on constructing additive reversible complement cyclic codes over $GF(4)$ to generate codes that satisfy the first three constraints above. From these codes we then construct what we call DNA codes C_{DNA}. We give a list of all additive reversible complement cyclic codes of length 7 that can be used as DNA codes.

The rest of the paper is organized as follows. In section II we give some background and recall some basic results for DNA codes and additive cyclic codes over

$GF(4)$. In Section III, we construct the theory of additive reversible complement cyclic codes over $GF(4)$. In section IV, we list the best additive reversible complement cyclic codes of length 7 with their complete weight enumerators and their $GC-$ weight enumerators.

2 Background

We define a DNA code of length n to be a set of codewords $(x_1, \ldots, x_n)$ with $x_i \in \{A, C, G, T\}$. In this paper we shall identify codes over $\{A, C, G, T\}$ with codes over $GF(4) = \{0, 1, \omega, \varpi\}$ where $\varpi = \omega^2$, and $\omega^2 + \omega + 1 = 0$. For the letters A, C, G, T, over $GF(4)$, we designate $0, \omega, \varpi, 1$ respectively. The Watson-Crick complement is given by $A^c = T$, $T^c = A$, $C^c = G$, and $G^c = T$. An additive code $C(n, 2^k)$ over $GF(4)$ of length n and dimension k is an additive subgroup of $GF(4)^n$. Note that an additive code C over $GF(4)$ is closed under addition but not under scalar multiplication by elements from $GF(4)$. A linear code C over $GF(4)$ is called cyclic if it is invariant with respect to the shift operator that maps $(u_0, \ldots, u_{n-1})$ to $(u_{n-1}, u_0, \ldots, u_{n-2})$. For each codeword $(u_0, \ldots, u_{n-1})$, we associate the polynomial $u(x) = u_0 + u_1 x + \cdots u_{n-1} x^{n-1}$. For each codeword $u = (u_0, u_1, \cdots, u_{n-1})$, we define the reverse of u to be $u^r = (u_{n-1}, u_{n-2}, \ldots, u_0)$, the complement of u to be $u^c = (u_0^c, \ldots, u_{n-1}^c)$ and the reverse-complement to be $u^{rc} = (u_{n-1}^c, \ldots, u_0^c)$. The Hamming weight of a codeword $(u_0, \ldots, u_{n-1})$ is the number of nonzero coordinates. For example, the Hamming weight of (1002101) is 4. The Hamming distance $H(x, y)$ between two codewords is the number of coordinates in which they differ. The minimum (Hamming) distance of a code C is given by

$$d = \min\ H(x, y) \quad x, y \in C \text{ and } x \neq y.$$

If a code C is an additive code then the minimum distance of C is the minimum Hamming weight of any nonzero codeword. We denote the complete weight enumerator of a code C over $GF(4)$ by $CWE_C(x, y, z, t) = \sum_{c \in C} x^{n_0(c)} y^{n_1(c)} z^{n_\omega(c)} t^{n_\varpi(c)}$ where $n_k(c)$ is the number of occurrences of $k \in GF(4)$ in a codeword c. We denote by $GCW_C(x, y) = CWE_C(x, x, y, y)$ the GC-weight enumerator of C over $GF(4)$. The $GC-$weight enumerator of a code C over $GF(4)$ is the weight enumerator that counts the number of $\{0, 1\}$ and the number of $\{\omega, \varpi\}$ in C.

The trace map $Tr : GF(4) \to GF(2)$ is given by

$$Tr(x) = x + x^2.$$

In particular $Tr(0) = Tr(1) = 0$ and $Tr(\omega) = Tr(\varpi) = 1$.

For each polynomial $p(x) = p_0 + p_1 x + \cdots p_r x^r$, we define the reciprocal of $p(x)$ to be the polynomial $p^*(x) = x^r p(1/x) = p_r + p_{r-1} x^{r-1} + \cdots p_0 x^r$. Note that deg $p^*(x) \leq$ deg $p(x)$ and if $p_0 \neq 0$, then $p(x)$ and $p^*(x)$ always have the same degree. $p(x)$ is called self-reciprocal iff $p(x) = p^*(x)$.

Definition 1 *An additive code C of length n over $GF(4)$ will be called reversible if $\forall u \in C$, then $u^r \in C$.*

Definition 2 *An additive code C of length n over $GF(4)$ will be called complement if $\forall u \in C$, then $u^c \in C$.*

Definition 3 *An additive code C of minium distance d is called reversible complement cyclic if C is cyclic, reversible, and complement.*

We construct DNA codes C_{DNA} with distance δ from reversible complement cyclic codes by deleting u or u^{rc} from any pairs of codewords (u, u^{rc}) where $u \neq u^{rc}$. If the original code C has minimal distance d then $\delta \geq d$.

The following theorem classifies additive cyclic codes over $GF(4)$.

Theorem 1 *Let C be an $(n, 2^k)$ additive cyclic code of length n over $GF(4)$. Then $C = (\omega p(x) + q(x), r(x))$ where $p(x)$, $q(x)$, $r(x)$ are binary polynomials that divide $(x^n - 1) \bmod 2$, and $r(x)$ divides $q(x)\,(x^n - 1)\,/p(x)\,(\bmod 2)$, and $k = 2n - \deg p - \deg r$.*

Proof: See Theorem 14 in [12]. ∎

3 Cyclic Reversible Complement Codes over $GF(4)$

Our goal in this paper is to classify reversible complement cyclic codes over $GF(4)$. This class of codes is important because the set of codewords satisfy the following constraints.

- The Hamming constraint: For any two codewords w, x, we have $H(w, x) \geq d$. This constraint is intended to limit non-specific hybridization whereby the Watson-Crick complement of a codeword anneals to a distinct word w.
- The reverse-complement constraint: For any two codewords w, x (where w might equal x), we have $H(w^c, x^r) \geq d$. This constraint is intended to limit non-specific hybridization between a codeword and the reverse of another codeword.

Theorem 2 *Suppose $C = (r(x))$ is an additive cyclic code over $GF(4)$. C is reversible iff $r(x)$ is self-reciprocal.*

Proof: See Theorem 1in [13] ∎

Lemma 3 *An additive cyclic code* $C = (g(x))$ *over* $GF(4)$ *is complement iff* $(x-1)$ *does not divide* $g(x)$.

Proof: Let $C = (g(x))$ be a cyclic code over $GF(4)$. Let $c(x) = c_0 + c_1x + \cdots c_{n-1}x^{n-1}$ be a codeword in C. $c^c(x) = c_0^c + c_1^c x + \cdots c_{n-1}^c x^{n-1}$. Note that $a + a^c = 1$ for any a in $GF(4)$. This implies that $c(x) + c^c(x) = 1 + x + \cdots x^{n-1}$. Since C is linear then $c^c(x) \in C$ iff $c(x) + c^c(x) = 1 + x + \cdots x^{n-1} \in C$ iff $(x^n - 1)/(x-1) \in C$ iff $(x^n-1)/(x-1) = g(x)f(x)$ iff $(x^n - 1) = (x-1)g(x)f(x)$ for some polynomial $f(x)$. Since n is odd then $(x^n - 1)$ factors uniquely into a product of distinct irreducible polynomials. Hence $c^c(x) \in C$ iff $(x-1)$ does not divide $g(x)$. ■

Let $C = (\omega p(x) + q(x))$ be an additive cyclic code of length n over $GF(4)$ with $\deg(p(x)) = r$. Since C has no binary polynomial and since $c(x) + c^c(x) = (x^n - 1)/(x-1)$ is a binary polynomial for any codeword $c(x)$ then the following lemma holds

Lemma 4 *Let* $C = (\omega p(x) + q(x))$ *be an additive cyclic code of length* n *over* $GF(4)$ *with* $\deg(p(x)) = r$. *Then* C *cannot be a complement code.*

Lemma 5 *Let* C, A, B *be additive codes such that* $C = A + B$. *If* A *and* B *are reversible then* C *is reversible.*

Proof: See Lemma 5 in [15] ■

The next two theorems classify all additive reversible complement cyclic codes over $GF(4)$.

Theorem 6 *Let* $C = (\omega p(x), r(x))$ *or* $C = (\varpi p(x), r(x))$ *be an additive cyclic code of length* n *over* $GF(4)$. C *is reversible complement iff* $p(x)$, $r(x)$ *are self-reciprocal and* $r(x)$ *is not a multiple of* $(x-1)$.

Proof:

$\Leftarrow$Suppose $p(x)$, $r(x)$ are self-reciprocal and $r(x)$ is not a multiple of $(x-1)$. Then $A = (\omega p(x))$, $B = (r(x))$ are both reversible. By Lemma 5, C must be reversible. Since $r(x)$ is not a multiple of $(x-1)$, then C is also a complement cyclic code.

$\Rightarrow$Suppose C is reversible complement, then $\omega p^*(x)$, and $r^*(x)$ are in C. Since $\deg(\omega p^*(x)) = \deg(\omega p(x))$, and $\deg(r^*(x)) = \deg(r(x))$, then $\omega p^*(x) = \omega p(x)$, and $r^*(x) = r(x)$. Since C is complement then $r(x)$ cannot be multiple of $(x-1)$. Therefore, $p(x)$, $r(x)$ are self-reciprocal and $r(x)$ is not a multiple of $(x-1)$. ■

Theorem 7 *Let* $C = (\omega p(x) + q(x), r(x))$ *with* $q(x) \neq 0$ *be an additive cyclic code of length* n *over* $GF(4)$. C *is an additive reversible complement cyclic code iff* $p(x)$ *is self-reciprocal,* $r(x)$ *is self-reciprocal, not a multiple of* $(x-1)$, *and*

- *If* $\deg(q(x)) = \deg(p(x))$ *then* $q(x)$ *is self reciprocal*
- *If* $\deg(q(x)) < \deg(p(x))$ *then,* $r(x)$ *divides* $\left[x^{\deg(p)-\deg(q)}q^* + q(x)\right]$.
- *If* $\deg(q(x)) > \deg(p(x))$ *then,* $r(x)$ *divides* $\left[x^{\deg(q)-\deg(p)}q + q^*(x)\right]$.

Proof: See Theorem 8 in [15]

■

4 Additive Reversible complement cyclic codes of length 7

$(x^7 - 1)$ factors over $GF(2)$ as

$$x^7 - 1 = (x^3 + x + 1)(x + 1)(x^3 + x^2 + 1)$$

We have found 29 reversible complement cyclic codes of length 7 over $GF(4)$. The following list gives the ones with the largest minimum distance and maximum number of codewords:

$$C_1 = \begin{pmatrix} \omega + (x^4 + x^3), \\ (x^6 + x^5 + x^4 + x^3 + x^2 + x + 1) \end{pmatrix},$$
$$C_2 = \begin{pmatrix} \omega + (x^4 + x^3 + 1), \\ (x^6 + x^5 + x^4 + x^3 + x^2 + x + 1) \end{pmatrix},$$
$$C_3 = \begin{pmatrix} \omega + (x^5 + x^2), \\ (x^6 + x^5 + x^4 + x^3 + x^2 + x + 1) \end{pmatrix},$$
$$C_4 = \begin{pmatrix} \omega + (x^5 + x^4 + x^3 + x^2), \\ (x^6 + x^5 + x^4 + x^3 + x^2 + x + 1) \end{pmatrix},$$
$$C_5 = \begin{pmatrix} \omega + (x^5 + x^2 + 1), \\ (x^6 + x^5 + x^4 + x^3 + x^2 + x + 1) \end{pmatrix}$$
$$C_6 = \begin{pmatrix} \omega + (x^5 + x^4 + x^3 + x^2 + 1), \\ (x^6 + x^5 + x^4 + x^3 + x^2 + x + 1) \end{pmatrix}.$$

We have found that all the codes above have 256 codewords and Hamming distance 3. We also found that C_{DNA} for these codes have Hamming distance 3 and 208 codewords. In [16] the best $n = 7$, $d = 3$, DNA code has 180 codewords.

Codes C_2, C_3, C_4, C_5 above are the best in terms of constructing C_{DNA} with fixed $GC-$content. All these codes have the same $GC-$weight enumerator function given by:

$$GCW(x,y) = \begin{bmatrix} 70x^3y^4 + 14x^6y + 70x^4y^3 \\ +42x^5y^2 + 42x^2y^5 + 14xy^6 \\ +2x^7 + 2y^7. \end{bmatrix}.$$

This shows that we can construct codes with fixed $GC-$content (3/7) of size 70 codewords. C_2, C_4, and

C_5 have the same $CWE_C(x, y, z, t)$ given by

$$CWE_C(x,y,z,t) = \begin{bmatrix} y^7 + x^7 + 7x^2y^2z^3 + 7x^2z^2t^3 \\ +7y^2z^3t^2 + 7x^2z^3t^2 + 7y^2z^2t^3 \\ +7x^2yt^4 + 7xy^2z^4 + 7xy^4t^2 \\ +7x^4yz^2 + 7x^2y^3t^2 + t^7 \\ +7x^3y^2z^2 + 7x^4y^2t + 7x^2y^4z \\ +21x^2y^2z^2t + 21x^2y^2zt^2 + 7x^3y^2t^2 \\ +7x^2y^3z^2 + 7x^2y^2t^3 + z^7 \\ +7x^2z^4t + 7yz^4t^2 + 7xz^2t^4 \\ +7x^3z^2t^2 + 7y^3z^2t^2 + 21xy^2z^2t^2 \\ +7y^2zt^4 + 7x^4zt^2 + 7y^4z^2t \\ +21x^2yz^2t^2. \end{bmatrix}$$

C_3 has $CWE_C(x, y, z, t)$ given by:

$$CWEC_3(x,y,z,t) = \begin{bmatrix} y^7 + 7y^2z^4t + 7x^2zt^4 \\ +x^7 + 7x^2y^2z^3 + 7x^2y^3t^2 \\ +7x^3y^2z^2 + 7y^2z^2t^3 + 7x^2z^3t^2 \\ +7y^4zt^2 + t^7 + 21x^2y^2z^2t \\ +21x^2y^2zt^2 + 7x^4z^2t + z^7 \\ +7x^3y^2t^2 + 7x^2y^3z^2 + 7x^2y^2t^3 \\ +7xy^4z^2 + 7x^4yt^2 + 7x^4y^2z \\ +7x^2y^4t + 7xz^4t2 + 7x^2yz^4 \\ +7xy^2t^4 + 7x^3z^2t^2 + 7y^3z^2t^2 \\ +21xy^2z^2t^2 + 7yz^2t^4 + 21x^2yz^2t^2 \\ +7y^2z^3t^2 + 7x^2z^2t^3. \end{bmatrix}.$$

References

[1] L. Adleman (1994), "Molecular Computation of Solutions to Combinatorial Problems," *Science*, v. 266, 1021-1024.

[2] R. Deaton, R. Murphy, M. Garzon, D. R. Franceschetti, S. E. Stevens (1998), "Good encoding for DNA-based solutions to combinatorial problems," *Proceedings of DNA-Based Computers II*, Princeton. In AMS DIMACS Series, vol. 44, L. F. Landweber, E, Baum Eds., 247-258.

[3] P. Gaborit and O. King, "Linear constructions for DNA codes," Preprint.

[4] M. Garzon, P. Neathery, R. Deaton, M. Garzon, R. C. Murphy, D. R. Franceschetti, S. E. Stevens Jr. (1997), "A new metric for DNA computing," *Second Annual Genetic Programming Conference*, Stanford, CA, 472-478.

[5] M. Garzon, R. Deaton, L. F. Nino, S. E. Stevens Jr., M. Wittner (1998), "Genome encoding for DNA computing," *Proceedings of the Third Genetic Programming Conference*, Madison, WI, 684-690.

[6] L. Kari, R. Kitto, G. Thierrin (2003), "Codes, involutions, and DNA encoding," *Lecture Notes in Computer Science 2300*, Springer Verlag, 376-393.

[7] O. King (2003), "Bounds for DNA codes with constant GC-content," The Electronic Journal of Combinatorics, vol. 10, 1-13.

[8] A. Marathe, A. E. Condon and R. M. Corn (2001), "On Combinatorial DNA word design," *Journal of Computational Biology*, vol.8, 201-220.

[9] A. G. Frutos, Q. Liu, A. J. Thiel, A. M. W. Sanner, A. E. Condon, L. M. Smith and R. M. Corn (1997), "Demonstration of a word design strategy for DNA computing on surfaces," *Nucleic Acids Research*, vol. 25, 4748-4757.

[10] V. Rykov, A. J. Macula, D. Torney, P. White (2001), "DNA sequences and quaternary cyclic codes," *IEEE ISIT 2001*, Washington, DC, June 24-29, pp.248-248.

[11] F. J. MacWilliams and N. J. A. Sloane (1997), *The Theory of Error-Correcting Codes*, Ninth Impression, North-Holland, Amsterdam.

[12] A. R. Calderbank, E. M. Rains, P. W. Shor, and Neil J. A. Sloane (1998), "Quantum error correction via codes over GF(4)," *IEEE Trans. Info. Theory*, vol. 44, no. 4, 1369-1387.

[13] J. L. Massey (1964), "Reversible Codes," *Information and Control*, vol. 7, pp. 369-380.

[14] T. Abualrub and R. Oehmke (2003), "On the generators of Z_4 cyclic codes of Length 2^e," *IEEE Trans. Info. Theory*, vol. 49, no. 9, pp. 2126-2133.

[15] T. Abualrub and A. Ghrayeb (submitted 2004), "On the construction of Cyclic Codes for DNA Computing,".

[16] D. C. Tublan, "Tables of DNA codes," http://www.cs.ubs.ca/~dctulpan/papers/dna8/tables/index.html.

Evolutionary Algorithms for Static and Dynamic Optimization of Fed-batch Fermentation Processes [1]

Miguel Rocha†, José Neves†, Ana C.A. Veloso⋆‡,
Eugénio C. Ferreira⋆ and Isabel Rocha⋆
†Dep. Informática, University of Minho, Portugal
⋆Centro de Engenharia Biológica, University of Minho, Portugal
‡Esc. Superior Agrária de Bragança, Bragança, Portugal
{mrocha,jneves}@di.uminho.pt {anaveloso,ecferreira,irocha}@deb.uminho.pt

Abstract

In this work, *Evolutionary Algorithms (EAs)* are used to control a recombinant bacterial fed-batch fermentation process, that aims to produce a bio-pharmaceutical product. Initially, a novel *EA* was used to optimize the process, prior to its run, by simultaneously adjusting the feeding trajectory, the duration of the fermentation and the initial conditions of the process. Finally, dynamic optimization is proposed, where the *EA* is running simultaneously with the fermentation process, receiving information regarding the process, updating its internal model and reaching new solutions that will be used for online control.

Keywords: Evolutionary Computation, Fed-batch fermentation optimization, Online optimization, Variable size chromosomes, Real-valued representations

1 Introduction

Valuable products such as recombinant proteins, antibiotics and amino-acids are produced using fermentation techniques and thus there is an enormous economic incentive to optimize such processes. However, these are usually very complex, involving different transport phenomena, microbial components and biochemical reactions. Furthermore, the nonlinear behavior and time-varying properties make bioreactors difficult to control with traditional techniques. Thus, it is necessary to consider quantitative mathematical models, capable of describing the process dynamics and the interrelation among relevant variables. Additionally, robust optimization techniques must deal with the model's complexity, the environment constraints and the inherent noise of the experimental process.

Several optimization methods are described in literature. It has been shown that, for simple bioreactor systems, the problem can be solved analytically. Numerical methods, such as gradient algorithms based on the local sensitivities of the objective function for changes in the values of the control variables, have also been used [1]. One other popular method is dynamic programming which discretizes both time and control variables to a predefined number of values and uses a systematic backward search. However, those methodologies become too complex when the number of state variables increases. An alternative approach comes from *Evolutionary Algorithms (EAs)*, where previous work has reached interesting results in the optimization of feeding trajectories [2][3].

In this work, a fed-batch recombinant *E. coli* fermentation process to produce a bio-pharmaceutical product [4] was studied. In fed-batch fermentations there is an addition of certain nutrients along the process, in order to prevent the accumulation of toxic sub-products. The purpose is to apply real-valued representation based *EAs*, with variable sized chromosomes, in order to achieve both static and dynamic optimization. The former is conducted offline before the process starts and aims at setting some of the fermentation's variables, namely the substrate feeding trajectory, the initial conditions and also the duration of the fermentation. The latter is conducted in real time, interacting with the fermentation and optimizing the feeding trajectory by reacting to information about the real values of relevant state variables.

2 The fed-batch fermentation process

E. coli is the microorganism of choice for the production of bio-pharmaceuticals, usually grown under fed-batch mode due to the negative effect of acetic acid produced when glucose is present above certain concentrations. During this process the system states change considerably, from a low initial to a high biomass and product concentrations. This dynamic behavior motivates the development of optimization methods to find the optimal input feeding trajectories. The typical input is the sub-

[1] Ana Veloso is most grateful for the financial support provided by PRODEP.

strate inflow rate as a function of time. One way to evaluate the process performance is the productivity, defined as the units of product formed per unit of time, which is usually related with the final biomass obtained.

As previously presented [4], a white box mathematical model was developed, based on differential equations that represent the mass balances to the state variables, namely $F_{in,S}$ the substrate feeding rate (in kg/h), W the culture medium weight (in kg), S the glucose, X the biomass, O the dissolved oxygen, C the dissolved carbon dioxide and A the acetate. Real experiments were used as the basis for the model derivation, being conducted in a fermentation laboratory with a 5-L bioreactor and the experimental results were consistent with the model. This model will be used for the optimization described in the following, performing model simulation, by using the Euler numerical integration method, with a small step size d and a given duration for the process (T_f) measured in hours.

3 Evolutionary Algorithms for Static Optimization

3.1 Evolutionary Algorithms for Feeding Trajectory Optimization

The first approach to the problem was to develop an *EA* capable of optimizing the feeding trajectory, i.e., to determine the amount of substrate (glucose) to be fed into the bioreactor per time unit ($F_{in,S}$). Real-valued representations were used in order to encode the feeding amounts. Each gene will encode the amount of substrate to be introduced into the bioreactor in a given time unit and the genome will be given by the temporal sequence of such values.

In this case, the size of the genome is determined based on the final time of the process (T_f) and the discretization step (d) considered in the simulation, being given by the expression: $\frac{T_f}{d}$. However, this could produce a very large genome (a typical value would be 5000), which would difficult the *EA*'s convergence. Thus, feeding values are defined only at certain equally spaced points, and the remaining values are linearly interpolated. The size of the genome becomes $\frac{T_f}{dp} + 1$, where p stands for the number of points within each interpolation interval. The values used in the experiments are: $T_f = 25$, $d = 0.005$ and $p = 200$.

The amount of substrate that can be introduced per time unit is limited to a maximum of 0.4 kg/h due to the pump's capacity. Thus, there is the need to impose limits in the gene values that must be within the range $[0.0; 0.4]$. In the initial population, each individual is assigned, for each of its genes, a random value in this interval.

The evaluation process, for each individual in the population, measures the quality of the feeding trajectory in terms of the fermentation's productivity. This calculation is achieved by firstly running a simulation of the model, given as input the feeding values in the genome. In each simulation, the relevant state variables are initialized with the following values: $X_0 = 5$, $S_0 = 0$, $A_0 = 0$, $W_0 = 3$. The fitness value is then calculated from the final and initial values of the state variables X and W, by the expression

$$\frac{X_f * W_f - X_0 * W_0}{T_f}$$

Both mutation and crossover operators were taken into account. Two mutation operators were used, namely: *Random Mutation* (replaces one gene by a new randomly generated value, within the range $[min_i, max_i]$) and *Gaussian Mutation* (adds to a given gene a value taken from a Gaussian distribution, with a zero mean and a standard deviation given by $\frac{max_i - min_i}{4}$, where $[min_i; max_i]$ is the range of values allowed for gene i). In both cases, the operators were applied to a variable number of genes (a value that is randomly set between 1 and 10 in each application). On the other hand, the following crossover operators were chosen: *Two-Point*, *Uniform*, *Arithmetical* and *Sum* crossovers [5].

The population size was set to 200 and the selection procedure is done by converting the fitness value into a linear ranking in the population, and then applying a roulette wheel scheme. In each generation, 50% of the individuals were kept from the previous generation, and 50% were bred by the application of the genetic operators.

The implementation of the proposed *EA* was based on a general purpose package, developed by the authors in the *Java* programming language. All experiments reported were run on a *PC* with a *Pentium IV 2.4 GHz* processor.

A set of experiments was conducted in order to find the best set of genetic operators for this problem [4]. The best result was obtained using an alternative that contemplates the use of all genetic operators described above. In this case each crossover operator is responsible for breeding 12.5% of the offspring and each mutation operator 25%.

3.2 Optimization of Initial Conditions

The initial conditions of the experiments were set based on the practitioner's experience and wisdom. However, there is no guarantee that the initial values of the state variables are optimal. So, it was decided to incorporate the initial values of significant state variables in the optimization procedure.

Once each variable has different physical constraints it was necessary to define a genome where the limits are distinct in each position. The variables chosen to be optimized, additionally to the feeding trajectory, were the initial values of X, W, S and A, with their range of variability given by $X_0 \in [1;5]$, $W_0 \in [2;4]$, $S_0 \in [0;5]$ and $A_0 \in [0;5]$.

3.3 Optimization of the Final Time

The duration of the fermentation is not imposed by any theoretical result, yet is chosen by empirical knowledge, making it possible to optimize its value. In this section, an *EA* will be proposed for this task considering variable size chromosomes and new genetic operators.

The genetic operators defined in Section 3.1 were kept: the mutations were unchanged and the crossovers were updated to cope with parents of different sizes. In this case, each of the offspring keeps the size of one parent and for the genes where only one parent is defined (the one with greater length), their value is passed into the corresponding offspring. In the creation of the initial population the individuals are given chromosomes with distinct sizes, randomly selected in a range defined by two parameters: a minimum and a maximum size. Furthermore, two novel mutation operators were defined, in order to allow for the change of the size of individuals during the evolution process:

- *Grow*: consists in the introduction of a new gene into the genome, in a random position, being its value the average of the values of the two neighboring genes.
- *Shrink*: a randomly selected gene is removed from the genome.

Both operators are only applied when the maximum and minimum size constraints are obeyed. With the introduction of the new genetic operators, the probabilities used in the experiments are the following: each of the four crossover operators has a probability value of 10%, the random and gaussian mutation keep their probabilities of 25% and the new mutation operators have a probability of 5% each.

Two different experiments were conducted: in the first, only the final time and feeding trajectory are optimized, being the genome made out of the feeding trajectory; in the latter, the initial conditions are also considered a target of optimization, being the initial parameters encoded into the first group of four genes (fixed size), as before, and the remaining of the genome used to code the feeding trajectory (variable size). The minimum and maximum duration of the fermentation are set to 20 and 50 hours, respectively. The remaining parameters of the *EA* are kept unchanged.

Table 1. Comparison of the results obtained by the *EAs* for feeding trajectory (F), initial conditions (I) and duration (T) optimization.

Optim. aim	Mean and conf.interval	Best res.
F	8.98 ± 0.06	9.12
F+I	9.38 ± 0.06	9.46
F+T	9.16 ± 0.09	9.32
F+T+I	9.44 ± 0.05	9.50

3.4 Results

A set of experiments was conducted to test the previous approaches and the results obtained are displayed in Table 1. Each alternative was tested by 30 independent runs, and each run was stopped after 3000 generations. In the table, the first column indicates the purpose of the *EA*, where F stands for feeding trajectory, I for initial conditions and T for final time optimization. The results are given in terms of the defined fitness, being shown, in the second column, the mean of the runs and the confidence interval and in the third column the best result obtained over all the runs.

The optimization of both time, feeding trajectory and initial parameters had the best overall results, showing that the *EA* can simultaneously optimize all these aspects. An interpretation of the results led to the conclusion that the best results confirmed the findings of the practitioners and reached after years of practical experiments, although in some cases the results gave some interesting insights to the researchers.

4 Evolutionary Algorithms for Online Optimization

The offline optimization described previously makes use of a simulation model to evaluate each solution. Although this is a reliable model, validated by experimentation, in a real environment several sources of noise can contribute to changes in the observed values of the state values. This scenario has an impact on the experimental results that end up being worse than the ones predicted. During the fermentation process, some of the state variables can be measured, but its values tend to serve only for modeling purposes. However, it is possible to develop dynamic optimization algorithms that are capable of timely reacting to this new knowledge by updating its internal model and generating new solutions.

EAs make a promising approach to this real-time optimization task, since they keep a population of solutions that can be easily adapted to perform re-optimization. Indeed, the population can be evaluated under the new scenario and better adapted solutions created through the use of reproduction operators. The fact that a set of solutions is kept, and not only the best solution, makes a faster adaptation to new conditions possible, while taking advantage of previous work.

In this work, online optimization based on *EAs* is proposed, working in two stages: the former is the static optimization, conducted before the fermentation process and described in the previous section; the latter is the dynamic optimization, where the fermentation monitoring software sends information about the values of state variables to the *EA*, that reacts by updating its internal model and reaching a new optimal solution, that is sent back to the monitoring software.

The *EA* used to perform online optimization is similar to the ones described in the previous section. When new information is received, the *EA* determines a starting point (in time) for its optimization, by adding the time label of the received data with its predicted running time. Then it takes the last available population and adapts it by removing from the genome of each individual the genes that encode feeding values for elapsed time periods. Each of these individuals is re-evaluated taking the new knowledge into consideration and the normal process of the *EA* proceeds for a given number of iterations. The best solution (feed) obtained is then sent to the fermentation process and can be used.

In order to validate this method a set of experiments was conducted, where the online measurements are simulated by adding noise to relevant state variables (X and W were selected since they have a direct impact on the fitness values). These variables were disturbed in regular intervals in time (of 1 hour) by the following rules: $X_i = X_i + X_i.U(-p, p)$ and $W_i = W_i + W_i.U(-p, p)$, where $U(a, b)$ is a value taken from an uniform distribution in the range $[a, b]$. In the experiments, both in static and dynamic optimization, only feed trajectory optimization was considered and the experimental setup was kept from the previous section. The offline *EA* is stopped after 2000 generations, while the online one runs for 500 generations in each time step.

Table 2 summarizes the results obtained with different values of p ranging from 0.01 to 0.1. In the columns the result from the static optimization in shown, followed by the result obtained by the optimal feed with the added noise and finally the result of the online optimization method.

It is possible to observe that the added noise is enough to make an huge impact on the process even when the disturbance is small (p=0.01), a result that shows that the process is quite sensitive to small changes in state variables (a conclusion shared by the researchers with practical experience), and emphasizes the need for dynamic optimization. The online optimization is capable of results near the ones initially expected, therefore compensating the noise, even though there is a decrease in performance when p raises, as would be expected.

Table 2. Results obtained by the *EAs* for online optimization.

p	Initial opt.	Init.+ noise	Online Opt.
0.01	8.84 ± 0.06	4.58 ± 0.21	8.68 ± 0.07
0.02	8.81 ± 0.06	4.49 ± 0.17	8.35 ± 0.10
0.05	8.82 ± 0.09	4.26 ± 0.14	7.67 ± 0.12
0.1	8.68 ± 0.09	4.17 ± 0.18	7.44 ± 0.13

5 Conclusions and further work

In this work *EAs*, based on real-valued representations and variable size chromosomes were proposed in order to optimize relevant parameters for a fermentation process, both in offline and online modes. The results, although based on simulations, show that the offline *EA* is capable of simultaneously optimizing the feeding trajectory, the initial conditions and the duration of a fermentation process. On the other hand, the online optimization deals well with significant changes (up to 10%) in the state variables relevant to the fitness function.

Future work includes the validation of the system by further testing, including both simulations and real fermentation processes. An integration of the time and initial parameter optimization in the dynamic procedure will also be undertaken.

References

[1] A.E. Bryson and Y.C. Ho. *Applied Optimal Control - Optimization, Estimation and Control.* Hemisphere Publication Company, New York, 1975.

[2] J.A. Roubos, G. van Straten, and A.J. van Boxtel. An Evolutionary Strategy for Fed-batch Bioreactor Optimization: Concepts and Performance. *Journal of Biotechnology*, 67:173–187, 1999.

[3] R. Marteijn, O. Jurrius, J. Dhont, C. de Gooijer, J. Tramper, and D. Martens. Optimization of a Feed Medium for Fed-Batch Culture of Insect Cells Using a Genetic Algorithm. *Biotechnol Bioeng*, 81:269–278, 2003.

[4] M. Rocha, J. Neves, I. Rocha, and E. Ferreira. Evolutionary algorithms for optimal control in fed-batch fermentation processes. In G.Raidl et al., editor, *Proceedings of the Workshop on Evolutionary Bioinformatics - EvoWorkshops 2004, LNCS 3005*, pages pp.84–93. Springer, 2004.

[5] Z. Michalewicz. *Genetic Algorithms + Data Structures = Evolution Programs.* Springer-Verlag, USA, third edition, 1996.

Benchmark testing of simulated annealing, adaptive random search and genetic algorithms for the global optimization of bioprocesses

R. Oliveira[1] and R. Salcedo[2]

[1]REQUIMTE/CQFB Centro de Química Fina e Biotecnologia, Departamento de Química, Faculdade de Ciências e Tecnologia, Universidade Nova de Lisboa, P-2829-516 Caparica, Portugal, e-mail: rui.oliveira@dq.fct.unl.pt
[2]Department of Chemical Engineering, Faculty of Engineering, University of Porto, R.Dr. Roberto Frias, P-2781-901 Porto, Portugal, email: rsalcedo@fe.up.pt

Abstract

This paper studies the global optimisation of bioprocesses employing model-based dynamic programming schemes. Three stochastic optimisation algorithms were tested: simulated annealing, adaptive random search and genetic algorithms. The methods were employed for optimising two challenging optimal control problems of fed-batch bioreactors. The main results show that adaptive random search and genetic algorithms are superior at solving these problems than the simulated annealing based method, both in accuracy and in the number of function evaluations.

1 Introduction

Bioprocess optimisation is becoming an increasingly important tool in the biochemical industries. From the methodological point of view, bioprocess optimisation supported by mathematical models is still a challenging problem. Several optimisation strategies are possible. Deterministic gradient-based algorithms can certainly be used, although convergence is only guaranteed towards local optima. Many researchers have recently changed to stochastic algorithms for dynamic optimization, not only because discontinuities can be easily handled, but also because locating the global optimum becomes much easier [1-3]. The main objective in this paper is to perform a benchmark testing of common stochastic algorithms: simulated annealing, adaptive random search and genetic algorithms for the global optimization of bioprocesses. The methods are briefly described below and then applied to two benchmark problems for the optimal control of fed-batch bioreactors. Obviously, all these algorithms will produce results that may strongly depend on the discretization type used for the control variable [4]. For simplicity, a simple constant parameterization, which might however not be the best for these problems, was employed in all algorithms.

2 Bioprocess optimisation

The dynamic optimisation problem can be stated as to find the control inputs $\mathbf{u}(t)$ over a given time domain that maximise a given performance function that can be generally stated as follows [1]:

$$J(\mathbf{c},\mathbf{u}) = \Theta(\mathbf{c}(t_f)) + \int_{t_0}^{t_f} \Phi(\mathbf{c}(t),\mathbf{u}(t_f),t)dt \qquad (1)$$

with j the performance function, $\mathbf{c}$ a vector of state variables, $\mathbf{u}$ a vector of control inputs, t_f the batch time, Θ a terminal performance function and Φ a time varying performance function.

Adaptive random search methods are particularly attractive for the global optimization of non convex problems and are easily applicable to constrained functions and to optimal control problems [3]. Such an example is the MSGA algorithm [5,6].

Simulated annealing is a powerful technique for combinatorial optimization. Algorithms based on simulated annealing employ stochastic generation of solution vectors and share similarities between the physical process of annealing and a minimization problem. To apply simulated annealing to continuous spaces, Cardoso et al. [7] have coupled it with the deterministic nonlinear simplex [8], following a proposal for unconstrained minimization [9], while incorporating important additional features, such as the ability to handle constraints and MINLP problems [10]. This paper exploits both MSGA and MSIMPSA for the optimization of the test problems described below.

The genetic algorithm (GA) may be viewed as a stochastic algorithm based on the principle of survival of the fittest [11]. In short, a population of size n is initially randomly created. Each individual is characterised by the respective 'chromosome' consisting in a complete set of optimisation parameters. Then successive new generations are simulated implementing: selection, crossover and mutation, and the fittest individuals are selected for mating and propagated to the next generations. The children are subject to crossover and mutation operations. The genetic algorithm coded by Carrol et al. [12] was adopted in this work.

3 Results and discussion

3.1 Case study I

The optimisation of an heterologous protein process in a fed-batch bioreactor described in Park and Ramirez [13] was selected for the first study The mathematical model is as follows:

Material balance equations:

$$\frac{dX}{dt} = \mu(S)X - DX \quad (2)$$

$$\frac{dS}{dt} = -7.3\mu(S)X - D(S - S_o) \quad (3)$$

$$\frac{dP_t}{dt} = f_p(S)X - DP_t \quad (4)$$

$$\frac{dP_m}{dt} = \Phi(S)(P_t - P_{m)} - DP_m \quad (5)$$

$$\frac{dV}{dt} = F \quad (6)$$

Kinetic rates:

$$\mu(S) = \frac{21.87S}{(S+0.4)(S+62.5)} \quad (7)$$

$$f_p(S) = \frac{Se^{-5S}}{S+0.1} \quad (8)$$

$$\Phi(S) = \frac{4.75\mu(s)}{0.12+\mu(s)} \quad (9)$$

The process state variables are: the concentrations of biomass (X), glucose (S), total protein (Pt) and secreted protein (Pm). The optimisation goal is to find the optimal feeding strategy F(t) that maximizes

$$J = Pm(t_f)V(t_f) \quad (10)$$

where t_f is the batch time and V the medium volume in the bioreactor. The following initial conditions were assumed: PM=0g/L, PT=0g/L, X=1.0g/L, S=5g/L, V=1L. The batch time was fixed to t_f=15 h and the glucose concentration in the inlet feed was S_0=20 g/l. The Pontryagin's Minimum Principle (PMP) technique showed that the maximum performance is J=32.4g under the conditions described [13]. Since the proposed algorithms are stochastic, their behaviour was tested on a statistical basis, and 10 runs with different optimization trajectories were made, using 10 time intervals. The results for MSIMPSA are dependent on the upper bound for the control variable, and this were varied from 10 down to 3 in steps of one. Fig. 1a compares MSIMPSA with GA. The average number of function evaluations for MSIMPSA was ≈10^4 when the upper bound for **u** was set to 3, but decreased to 4500 otherwise, indicating premature ending of the optimization. The upper bound for **u** was kept at 10 with GA. It may be concluded that GA arrives near the global optimum with 10^5 function evaluations, and MSIMPSA only arrives at the global optimum with a high probability when the upper bound for **u** is kept below 4. Increasing the number of function evaluations for MSIMPSA did not improve the results.

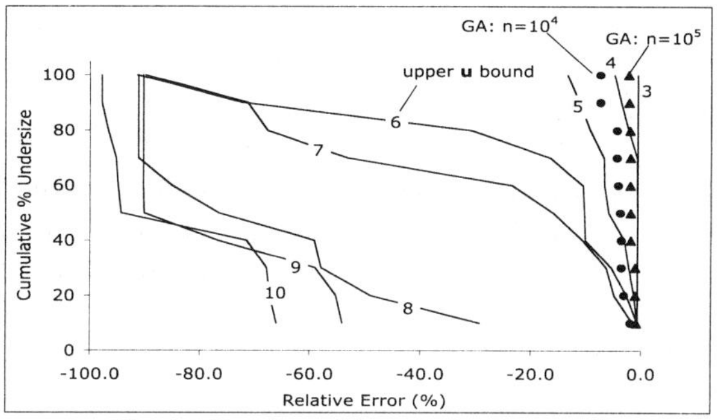

Fig. 1a. Benchmarking results for case I (MSIMPSA-lines *vs* GA-markers; n=function evaluations)

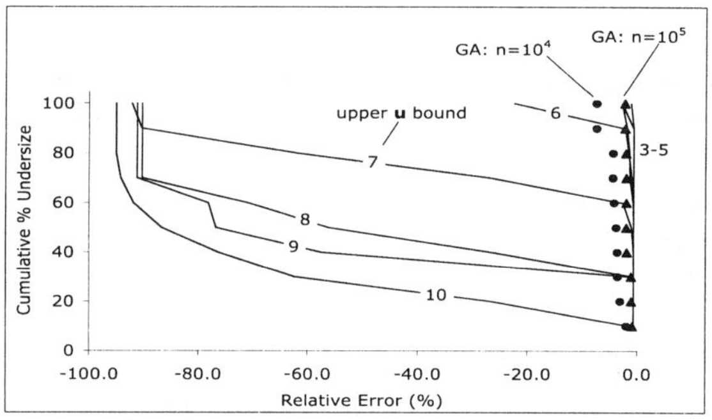

Fig. 1b. Benchmarking results for case I (MSGA-lines *vs* GA-markers; n=function evaluations)

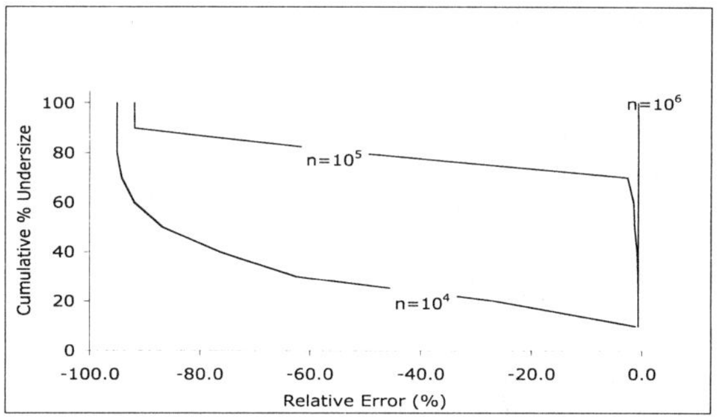

Fig. 1c. Effect of number of function evaluations for case I (MSGA with upper bounds on **u** set to 10)

Fig. 1b shows that MSGA, with 10^4 function evaluations, is more robust than MSIMPSA, and very good results are obtained with the upper bounds for the control variable set as high as 6. Fig. 1c further shows that MSGA can improve with the upper bounds on **u** set to the maximum value of 10, but at the expense of an increasing number of function evaluations. However, for this example, GA is somewhat superior, although less accurate.

3.2 Case study II

The maximisation of the profitability of a fed-batch fermentation for a protein production by recombinant bacteria described in [14-16], using 10 time intervals, was adopted for the second case. The material balance equations and corresponding kinetic rates are given below. The state of the process is defined by the biomass (X), glucose concentration (S), protein concentration (P), inducer concentration (I), schock factor (ISF) and recovery factor (IRF), and medium volume in the bioreactor (V). The model equations are:

Material balance equations:

$$\frac{dX}{dt} = \mu X - DX \qquad (11)$$

$$\frac{dS}{dt} = -Y_{x/s}^{-1}\mu X - DS + u_1 S_F / V \qquad (12)$$

$$\frac{dP}{dt} = R_{fp} X - DP \qquad (13)$$

$$\frac{dI}{dt} = -DI + u_2 I_F / V \qquad (14)$$

$$\frac{dISF}{dt} = -k_1 ISF \qquad (15)$$

$$\frac{dIRF}{dt} = k_2 (1 - IRF) \qquad (16)$$

$$\frac{dV}{dt} = u_1 + u_2 \qquad (17)$$

$$\text{with } D = (u_1 + u_2)/V \qquad (18)$$

Kinetic equations:

$$\phi = \frac{0.407S}{0.108 + S + S^2/14814.8} \qquad (19)$$

$$\mu = \phi\left(ISF + \frac{0.22IRF}{0.22 + I}\right) \qquad (20)$$

$$R_{fp} = \phi\frac{0.0005 + I}{0.022 + I} \qquad (21)$$

$$k_1 = k_2 = \frac{0.09I}{0.034 + I} \qquad (22)$$

Two streams are fed to the fermenter at different rates: glucose is fed at a volumetric feed rate $\mathbf{u}_1$ and concentration in the feed of S_F=100.0 g/l, while the inducer is fed with a different stream at a rate $\mathbf{u}_2$ with concentration in the stream of I_F=4.0 g/l. The degrees of freedom are the control inputs, i.e. the feeding rates, which in this case are constrained such that $0<u_1<1$ and $0<u_2<1$. The initial state is X(0) = 0.1 g/L, S(0)=40 g/L, P(0)=0.0 g/L, I(0)=0.0 g/L, V(0)=1L, ISF(0)=1, IRF(0)=0. The objective is to maximise the process profitability stated as follows:

$$J = P(t_f)V(t_f)/Q - \int_0^{t_f} u_2 I_F d\tau \qquad (23)$$

with t_f=10 hours and Q the ratio of protein value to inducer cost fixed to 1/5. In the original paper, this optimisation problem was solved employing the PMP theory, formulating the Hamiltonian and identifying singular control arcs, giving an optimal profitability of J=0.803. In [20] the same problem was solved using hybrid models, and the maximum profitability obtained was 0.740, viz. 7.8% lower than the optimum. The results obtained using the techniques presented before are shown in Figs. 2a-b. It may be concluded that MSGA gives better results than either MSIMPSA or GA, and that GA quickly deteriorates when the number of function evaluations decreases. However, all results are better than the ones given by the hybrid approach of Tholodur and Ramirez [15].

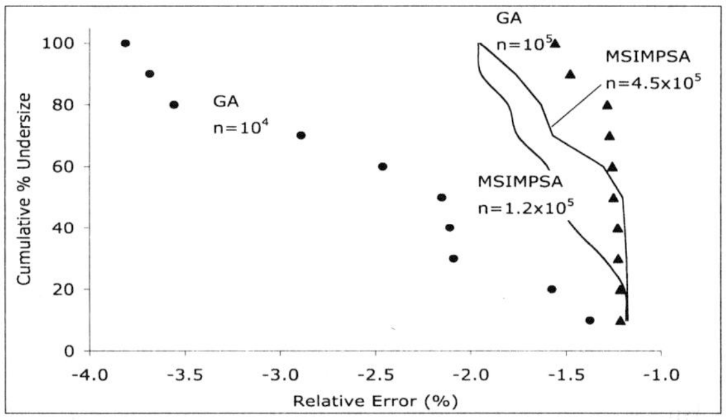

Fig. 2a. Benchmarking results for case II (MSIMPSA-lines *vs.* GA-markers; n=function evaluations)

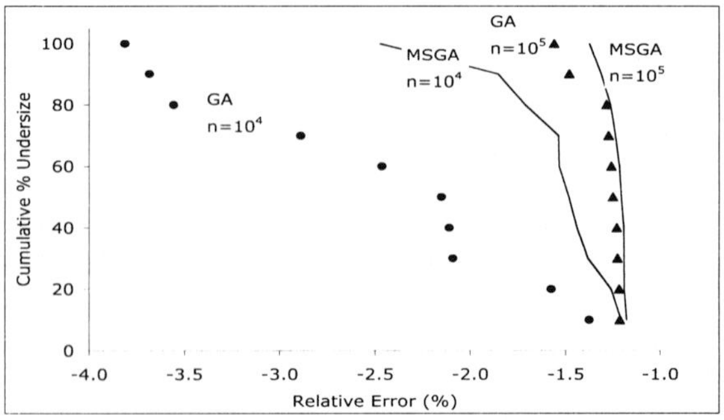

Fig. 2b. Benchmarking results for case II (MSGA-lines *vs.* GA-markers; n=function evaluations)

4. Conclusions

Bioprocesses optimisation is still a challenging problem in several respects. The resulting dynamic optimisation program is difficult to solve due to constraints and often due to the lack of sensitivity of the performance function over control inputs. This lack of sensitivity may be explained by the nature of kinetic behaviour characterised by saturation kinetics. Deterministic gradient-based algorithms tend to converge towards local optima and have difficulties handling constraints. The non linear nature and the occurrence of sudden metabolic switches difficult the evaluation of gradients. Stochastic algorithms are becoming more popular for dynamic optimization, not only because discontinuities can be easily handled, but also because locating the global optimum can be envisaged. In this paper the power of the stochastic approach is well illustrated with the two benchmark problems In particular it is demonstrated that adaptive random search and genetic algorithms seem superior at solving these problems than the simulated annealing based method, but adaptive random search and simulated annealing are more accurate. However, several other examples may be given where simulated annealing excels over either adaptive random search [4] or over genetic algorithms, especially for highly constrained functions [17]. Thus, more extensive tests have to be conducted before a definite conclusion can be taken.

Accronyms

GA – Genetic Algorithm
MSGA – Minlp SGA
MINLP – Mixed Integer Non-Linear Programming
MSIMPSA – Minlp Simplex Simulated Annealing algorithm
NLP – Non Linear Programming
SGA – Salcedo-Gonçalves-Azevedo algorithm

References

[1] Banga, J. R., E. Balsa-Canto, C. Moles and A. A. Alonso (2003). Dynamic optimization of bioreactors: a review, Proceedings INSA, Part A, nos. 3&4, May&July, 257.

[2] Martinez, E. C. And J. A. Wilson (2003). Evolutionary optimization of batch process systems using imperfect models, Proceedings INSA, Part A, nos. 3&4, May&July, 403.

[3] Salcedo, R. L., R. P. Lima and M. F. Cardoso (2003). Simulated annelaing for the global optimization of chemical processes, Proceedings INSA, Part A, nos. 3&4, May&July, 359.

[4] Lima, R. M., G. François, B. Srinivasan and R. L. Salcedo (2004). Dynamic optimization of batch emulsion polymerization using MSIMPSA, a simulated-annealing-based algorithm, *Ind. Eng. Chem. Res.*, 43(24), 7796.

[5] Salcedo, R., M. J. Gonçalves and S. Feyo de Azevedo (1990). An Improved Random-Search Algorithm for Non-Linear Optimization, *Comp. Chem. Engng.*, 14(10), 1111.

[6] Salcedo, R. L. (1992). Solving Non-Convex NLP and MINLP Problems with Adaptive Random-Search, *Ind. Eng. Chem. Res.*, 31(1), 262.

[7] Cardoso M.F., R.L. Salcedo and S.F. de Azevedo (1996). The simplex-simulated annealing approach to continuous non-linear optimization, *Computers Chem. Engng.*, 20(9), 1065.

[8] Nelder J.A. and R. Mead (1965). A simplex method for function minimization, *Comput. J.*, 7, 308.

[9] Press W.H. and S.A. Teukolsky (1991). Simulated annealing optimization over continuous spaces, *Computers in Physics*, 5(4), 426.

[10] Cardoso M.F.; Salcedo, R.L.; Feyo de Azevedo, S.; Barbosa, D. (1997). A Simulated Annealing Approach to the Solution of MINLP Problems, *Comp. Chem. Engng.* 21(12), 1349.

[11] Goldberg, D. (1989). Genetic Algorithms in Search, Optimization and Machine Learning, Addison-Wesley.

[12] Carroll, D. L. (1996). Chemical Laser Modeling with Genetic Algorithms, *AIAA J.*, Vol. 34(2), 338.

[13] Park, S. And Ramirez, W. F. (1988). Optimal Production of Secreted Protein in Fed-Batch Reactors, *aiche Journal*, 34(9), 1550.

[14] Lee, J. And Ramirez,W.F. (1994) Optimal fed-batch control of induced foreign protein-production by recombinant bacteria, *aiche Journal*, 40(5), 899.

[15] Tholudur, Arun and Ramirez, W. Fred (1996) Optimization of fed-batch bioreactors using neural network parameter function models, *Biotechenol. Prog.*, 12, 302.

[16] Carrasco E.F. and Banga J.R. (1997) Dynamic optimization of batch reactors using adaptive stochastic algorithms, *Ind. Eng. Chem. Res.*, 36 (11), 5047.

[17] Costa, L., P. Oliveira (2001). Evolutionary algorithms approach to the solution of mixed integer nonlinear programming problems. *Comput. Chem. Eng.*, 25, 257.

Dynamic modelling and optimisation of a mammalian cells process using hybrid grey-box systems

A. Teixeira[1], A. Cunha[2], J. Clemente[2], P.M. Alves[2], M. J. T Carrondo [1,2], R. Oliveira[1]

[1]REQUIMTE/CQFB Centro de Química Fina e Biotecnologia, Departamento de Química, Faculdade de Ciências e Tecnologia, Universidade Nova de Lisboa, P-2829-516 Caparica, Portugal

[2]IBET/ITQB Instituto de Biologia Experimental e Tecnologia/Instituto de Tecnologia Química e Biológica, Apartado 12, P-2781-901 Oeiras, Portugal

E-mail: {ana.teixeira, rui.oliveira}@dq.fct.unl.pt

E-mail: {cunha, clemente, marques, mjtc}@itqb.unl.pt

Abstract

In this work a model-based optimisation study of fed-batch BHK-21 cultures expressing the human fusion glycoprotein IgG1-IL2 was performed. Due to the complexity of the BHK metabolism it is rather difficult to develop an accurate kinetic model that could be used for optimisation studies. Many kinetic expressions and parameters are involved resulting in a complex identification problem. For this reason an alternative more cost-effective methodology was adopted, based on hybrid grey-box models. It was concluded that modulation particularities of BHK cultures were effectively captured by the hybrid model, this being of crucial importance for the successful optimisation of the process operation. From the optimisation study it was concluded that the glutamine and glucose concentrations should be maintained at low levels during the exponential growth phase and then glutamine feeding should be increased. In this way it is expected that both the cell density and final product titre can be considerably increased.

1 Introduction

The use of animal cells such as Baby hamster kidney (BHK) cell lines for the production of high valuable proteins has gained great importance in the last years. Low cell densities as well as high production costs are typical disadvantages associated with animal cultures. It is very important to improve the productivity and robustness in these processes by model-based optimisation and control strategies. There is not yet available in the literature a mathematical model describing accurately BHK cells metabolism. Several models have been proposed for other animal cell lines such as hybridomas or Chinese hamster ovary (CHO) cell lines [1-3]. In preliminary studies it was concluded that such models fail to describe the complete state of BHK cultures with sufficient accuracy for optimisation studies. Most likely due to the difficulty of synthesizing a good model, no quantitative optimisation studies supported by models have been reported in the literature so far. Therefore in a process control perspective BHK cultivation processes may be viewed as having high potential for improvement. Other modelling paradigms such as hybrid grey-box systems are currently available for the analysis of complex systems [4-8]. The main design concept is that the a priori mechanistic knowledge is not viewed as the only relevant source of knowledge, but also other sources, like heuristics or information hidden in databases, are considered as valuable complementary (not alternative) resources for model development. In this paper the main objective is to develop a hybrid grey-box model sufficiently accurate for optimization studies and then to perform a preliminary optimisation study of the process operation.

2 Proposed methodology

2.1 Hybrid model development

The hybrid model structure adopted is simultaneously parallel, in respect to the reaction kinetics model, and serial, in respect of the connection between reaction kinetics and material balance equations. The compensation of the reaction kinetics is performed by multiplying a mechanistic known term by a black-box kinetic term. The black-box compensator was a simple Multilayer Perceptron (feedforward neural network) with the tangent hyperbolic function in the hidden layers and linear output layer:

$$\mathbf{r}_{\text{nonp}} = \mathbf{w}_2 \tanh(\mathbf{w}_1\mathbf{x} + \mathbf{b}_1) + \mathbf{b}_2 \tag{1}$$

being $\mathbf{w}_1$ and $\mathbf{w}_2$ weight matrices in the connections between nodes of layers 1 and 2 and 2 and 3 respectively, whereas $\mathbf{b}_1$ and $\mathbf{b}_2$ are bias parameter vectors. This structure is in the statistical point of view most appropriate for non-linear regression problems [9].

2.2 Black-box model reliability monitoring

After the parameters of the black-box model have been estimated, the input space was subject to clustering using the k-mean algorithm [10-11]. The main idea is to represent in a compact way the input space, which defines the domain of experience of the black-box model. The clusters were defined as radial basis functions:

$$r(c, m_i, \sum_i) = e^{\{-1/2(c-m_i)^T \Sigma_i (c-m_i)\}} \quad (2)$$

with $\mathbf{m}_j$ the cluster centres and Σ_j diagonal covariance matrices (Σ_j = diag$\{\sigma_{ij}^2\}$). The calculation of the standard deviations is made on the basis of the interpolation tolerance. The resulting set of clusters forms a continuous density function f:$\mathbf{c} \rightarrow EM \in [0,1]$ that computes the neural network experience measure (EM) for given input vectors **c**.

$$EM(\mathbf{c}) = \max_j \{r(\mathbf{c}, \mathbf{m}_j, \mathbf{\Sigma}_j)\} \quad (3)$$

High EM values (higher than 0.5) indicate that the neural network outputs are reliable.

2.3 Process optimisation based on the hybrid model

The feeding strategies of glucose and glutamine were optimised for maximisation of the total amount of product produced using the best hybrid model obtained. The dynamic optimization problem can be generically stated as to find the control inputs **u**(t) over a given time domain $t \in [t_0, t_f]$ to maximize the performance index (**J**) subject to several constraints [12]. In the present case the optimisation criterion was to maximise the total mass of product obtained at the end of the run at time (t_f):

$$\max_u \quad J = C_{IgG1-IL2}(t_f)V(t_f) \quad (4)$$

The control input vector was the feeding rate of glucose and glutamine: $\mathbf{u} = [F_{glc}, F_{glut}]^T$.

The optimisation (4) is subject to several constraints:

- Differential equality constraints: the hybrid models equations form a set of ODE (ordinary differential equations) equality constraints that relate the performance function with the control inputs

- Upper/lower bounds inequality constraints:

$$0 \le V \le V_{max},\ 0 \le F_{glc} \le F_{glc,max},\ 0 \le F_{gln} \le F_{gln,max}$$

- Optimisation risk inequality constraint. In the present study we introduce a novel inequality constraint related with the use of hybrid models:

$$\frac{\int_{t_0}^{t_f} RISK[c(\tau)]d\tau}{t_f - t_0} \le RISK_{max} \quad (5)$$

with RISK=1-EM.

This constraint states that the average experience measure should be higher than a given acceptable experience EM_{min}, which is related with the risk of the model being unreliable. This will restrict the feasibility domain and is essential when employing hybrid models. As will be shown later the black-box estimates degrade in regions of the input space with sparse measurements. The constraint (5) prevents that the optimal process trajectories lie in those regions of the input space.

The optimisation algorithm was implemented in a Fortran program. The optimisation (4) was implemented with a micro-genetic algorithm [13]. The two control inputs were parameterised using a discretization technique.

3 Results and discussion

Several hybrid model structures were compared in terms of accuracy in describing the training data set, composed by 3 process runs, and a validation data set composed by 2 process runs. The criterion to stop the parameter estimation algorithm was the achievement of the minimum modelling error for the validation set. Since the final model is to be used for optimisation, the structure and size of the black-box compensator must be selected on the basis of improved predictive capacity.

3.1 Hybrid model results

The first hybrid model investigated was a simple serial structure. This structure combines the material balance equations with a feedforward neural network with 6 inputs (the full input space) and 6 outputs, which are directly the volumetric reaction rates of each of the state variables. The number of parameters in the hidden layer was tuned heuristically. The best result was obtained with NH = 2 hidden nodes corresponding to the overall error (sum of error of the training and validation set) of MSE = 0.0080. In subsequent studies the structure of the hybrid model increased in complexity by the incorporation of prior mechanistic knowledge in the reaction kinetics yielding simultaneously parallel and serial structures. The best result obtained was MSE = 0.0058 for a small network with 36 parameters. Despite the fact that all models could practically result in zero error in describing the training data set, the description of the validation set could be significantly improved by incorporating reliable mechanistic descriptions of the reaction kinetics in the serial/parallel hybrid structure. Other structures were tested, namely a multilinear description of the kinetic rates. The results were however worst probably because this structure is not able to capture the inherent non-linearities of the kinetic.

The final result using the best model is presented in Fig. 1. There is a remarkable agreement between model predictions and measurements. Interestingly the number of parameters is comparable to the mechanistic model first developed (23 parameters) and the mechanistic model has much higher error. Clearly the model can be improved, this requiring more experiments carefully planned being subject of future work.

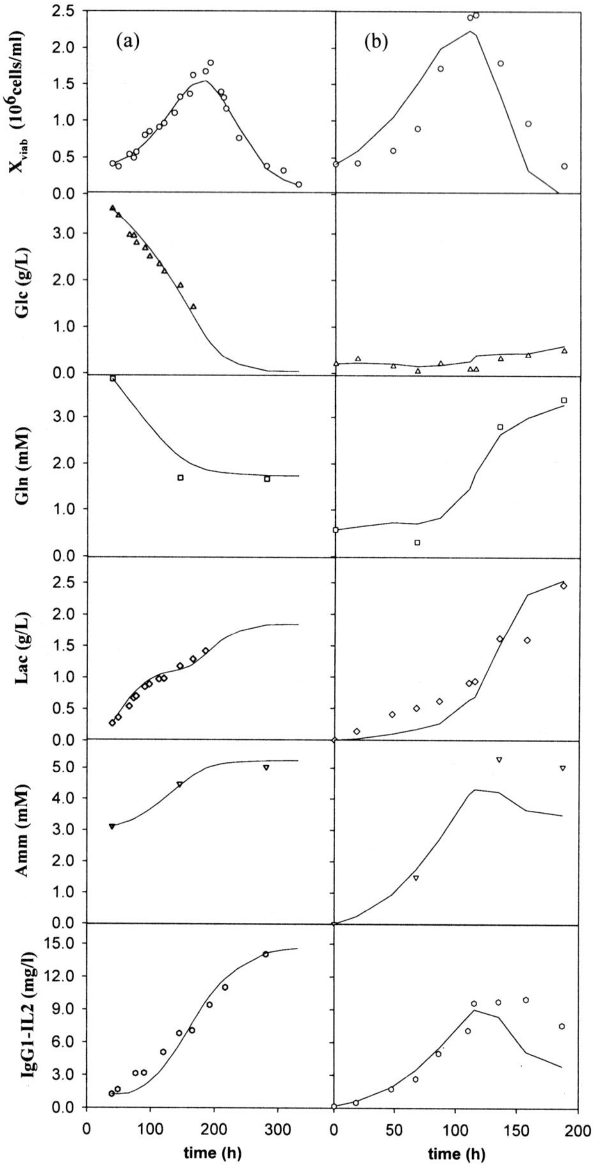

Fig. 1. Hybrid modelling results: (a) example of a batch run used for training; (b) example of fed-batch used for validation. Symbols represent measured values; full lines represent modelling results.

3.2 Metabolic significance

The reaction kinetics identified by the neural network embedded in the hybrid model were analysed in respect to their metabolic significance. The yield coefficients of lactate on glucose ($Y_{lac/glc}$) and ammonia on glutamine ($Y_{amm/gln}$) are in agreement with the so reported shift that occurs in BHK cell cultures [14]. The yield $Y_{lac/glc}$ increases with increasing glucose concentration reaching a relatively constant value of about 0.7 g/g for glucose concentrations higher than 0.4 g/L. The yield coefficient of ammonia on glutamine varies significantly with glutamine concentration. The metabolic shift is once more evident. For glutamine concentrations in the range 0.5-1.4 mM the yield $Y_{amm/gln}$ is constant and equal to 1.8 mol/mol, which is close to the theoretical value of 2 mol/mol. Above 1.4 mM of glutamine concentration this ratio decreases significantly approaching the theoretical value of 1 mol/mol. From this analysis it may be concluded that the hybrid model was able to properly describe the metabolic shifts, this being of crucial importance for the successful process optimisation.

3.3 Investigation of the optimal operating strategy

The feeding strategies of glucose and glutamine were optimised for maximising the total amount of IgG1-IL2 produced using the best hybrid model. As discussed previously, when using hybrid models for process optimization it may be important to constraint the risk of the model being unreliable. The possibility to explore novel process trajectories outside the trust region can be regulated through the maximum risk parameter (5). As expected the final IgG1-IL2 produced increases as higher risk levels are accepted. For risks higher than 25 % the optimization results do not comply with biological constraints. The optimisation results for the case of RISK $\leq$ 15 % may be interpreted as follows. Glucose and glutamine concentrations should be controlled to low levels while cells are growing at their maximum rate. During this phase the specific formation rate of IgG1-IL2 is affected by the low levels of glutamine but this is compensated by the increase in cell density. When cells start dying the best strategy seems to be increasing the glutamine feeding rate. This improves the IgG1-IL2 formation rate. A considerable improvement in productivity is expected when comparing with the fed-batch experiments performed so far. The final product titre is expected to be 31 mg/l against the 15 mg/l obtained so far in the fed-batch experiments.

4. Conclusions

The main objective in this work was to perform a detailed model-based optimisation study of fed-batch BHK-21 cultures expressing IgG1-IL2. It was first attempted to develop a mathematical model based on relatively simple unstructured and non-segregated kinetics. It was concluded that due to the complexity of the BHK metabolism it is rather difficult to develop such kinetic model sufficiently accurate for optimisation studies. An alternative more cost-effective methodology based on hybrid grey-box models was adopted. The best result was obtained with a serial/parallel structure in which a feedforward neural network compensates the inaccuracy of a simple first order kinetic description. The hybrid model was analyzed in respect to its metabolic significance. It was demonstrated that the yields are consistent with the well-known metabolic shifts occurring in BHK cells, this strengthening the potential application for process optimization.

The optimisation results show that the best operating strategy for maximising the volumetric productivity consists in controlling the glucose concentration at low levels during the whole process while glutamine should be controlled at low levels during the growth phase (0-0.5 mM) and then increased up to 1.8 mM in the production phase. By implementing such a strategy a considerable improvement on the final product titre is expected.

References

[1] Glacken, M. W., Adema, E., Sinskey, A. J. (1988) Mathematical descriptions of hybridoma culture kinetics: I. Initial metabolic rates. Biotechnol. Bioeng. 32: 491-506.

[2] Pörtner, R., Schäfer, T. (1996) Modeling of hybridoma cell growth and metabolism: A comparison of selected models and data. J. Biotechnol. 49: 119-135.

[3] Jang, J. D., Barford, J. P. (2000) An unstructured kinetic model of macromolecular metabolism in batch and fed-batch cultures of hybridoma cells producing monoclonal antibody. Biochemical Eng. J. 4: 153-168.

[4] Psichogios D.C., L. H. Ungar (1992). A hybrid neural network – first principles approach to process modelling, AIChE J., 38(10): 1499-1511.

[5] Thomson, M.L. and Kramer, M.A. (1994) Modeling chemical processes using prior knowledge and neural networks, AIChE J., 40: 1328-1340.

[6] Schubert, J., Simutis, R., Doors, M., Havlik, I., Lübbert, A. (1994) Hybrid Modelling of Yeast Production Processes. Chem. Eng. Technol., 17: 10-20.

[7] Preusting, H., Noordover, H., Simutis, R., Lübbert, A. (1996) The use of hybrid modeling for the optimization of the penicillin fermentation process. Chimia 50: 416–417.

[8] Braake, H.A., van Can, H.J., Verbruggen, H.B. (1998) Semi-mechanistic modeling of chemical processes with neural networks, Engineering Applications of Artificial Intelligence, 11(4): 507-515.

[9] Bishop, C.M., (1995) Neural Networks for Pattern recognition, Oxford University Press, Birmingham.

[10] Leonard, J.A., Kramer, M.A., Ungar, L.H. (1992) A neural network architecture that computes its own reliability. Comp. Chem. Eng. 16: 819–835.

[11] Simutis, R., Dors, M., Lübbert, A. (1995) Artificial neural networks of improved reliabiliity for industrial process supervision. In: Munack, A., Schügerl, K. (Eds.), Preprints of 6th International Conference on Computer Applications in Biotechnology, pp. 59–65.

[12] Banga, J.R., Irizarry-Rivera, R., Seider, W.D. (1998) Stochastic optimization for optimal and model-predictive control, Computers & Chemical Engineering, 22: 03-612.

[13] Goldberg, D. (1989) Genetic Algorithms in Search, Optimization and Machine Learning, Addison-Wesley.

[14] Cruz, H.J., Moreira, J.L., Carrondo, M.J.T. (1999) Metabolic shifts by nutrient manipulation in continuous cultures of BHK cells, Biotechnol. Bioeng. 66 (2), 104-113.

Adaptive DO-based control of substrate feeding in high cell density cultures operated under oxygen transfer limitation

R. Oliveira[1], A. Cunha[2], J. Clemente[2], M. J. T. Carrondo[1,2]
[1]REQUIMTE/CQFB - Centro de Química Fina e Biotecnologia, Faculdade de Ciências e Tecnologia, Universidade Nova de Lisboa, P-2829-516 Caparica, Portugal
[2]IBET - Instituto de Biologia Experimental e Tecnologica, Apartado 12 P-2781-901 Oeiras, Portugal
E-mail: {rui.oliveira}@dq.fct.unl.pt
E-mail: {cunha, clemente, mjtc}@itqb.unl.pt

Abstract

The carbon source feeding strategy is crucial for the productivity of many biochemical processes. In high density and shear sensitive cultures the feed of the carbon source is frequently constrained by the bioreactor maximum oxygen transfer capacity. In order to maximise the product formation, these processes should be operated at low dissolved oxygen (DO) concentrations close to the limitation level. This operating strategy may be realised with a closed-loop controller that regulates the DO concentration through the manipulation of the carbon source feed rate. The performance of this controller may have a significant influence on the final product production and should be as accurate as possible. In this work we study the application of adaptive control for solving this problem focusing not only on stability but also on accuracy. Whenever possible the convergence trajectories to the set point are characterised mathematically. Concerning the instrumentation, two situations are covered i) only the DO Tension (DOT) is measured, ii) both DOT and off-gas composition are measured on-line. The controllers are tested in a pilot plant recombinant *Pichia pastoris* process.

1 Introduction

The carbon source feeding strategy is normally determinant for bioprocess productivity because the availability of carbon source has a decisive influence on the cells metabolism and therefore on the final product production. In the biochemical industries the feeding strategy in fed-batch bioprocesses is commonly optimised off-line and then implemented on-line with open-loop controllers. In some cases the application of closed-loop control is justifiable and could introduce improvements in productivity. Ref. 1 demonstrated the advantage of closed-loop carbon source feed control based on direct carbon source measurements. This controller was implemented in a cultivation of Chinese hamster ovary cells in a high-density perfusion process and was shown to work reliably for more than 2.5 month.

In some aerobic high density processes, the feed of the carbon source can be controlled indirectly with a DO based closed-loop controller. In high-density cultures oxygen mass transfer limitation is likely to occur because beyond a given cell density the vessel oxygen transfer capacity is unable to sustain the oxygen metabolic demand. Ref. 2 studied this control problem in a recombinant *Escherichia coli* process with the general goal of avoiding acetate accumulation since this was determinant for the bioreactor productivity. The only sensor required was a standard dissolved oxygen probe. The controller managed to feed the maximum carbon source obeying to the oxygen transfer limitation constraint.

In targeting industrial application, issues such as reliability, sterility, and accuracy, are given high priority [1]. In this paper we study the application of adaptive control of glycerol addition under oxygen transfer limitation in a recombinant *P. pastoris* process employing reliable measurements. A major concern was given not only to the stability but also to the accuracy of the controller. The accuracy is in this case crucial because oscillations may lead to the irreversivel loss of culture productivity [3]. The accuracy is very much dependent on the tracking dynamics to the reference signal, hence the characterization of the tracking dynamics has been given a great deal of attention. Two of the most reliable measurements currently available in production facilities are DO (polarographic) probes and Mass spectrometers (MS). Hence adaptive controllers were designed that can utilise all this information but also the case of only DOT being available on-line is treated. Interestingly it was concluded that the availability of the MS does not represent a decisive advantage in context of the control problem studied.

2 Materials and methods

2.1 The process

A stable *Pichia pastoris* strain constitutively expressing a recombinant protein was used in this work. Fermentations were carried out in a 30 l working volume fermenter (Lab Pilot Fermenter Type LP351, 50 L, Bioengineering, Switzerland). Fermentation temperature was controlled at 30°C, and pH was controlled at 5.0 with addition of ammonium hydroxide

25%. Airflow was kept constant at 1800 l h^{-1}. Dissolved oxygen concentration was controlled in cascade mode: firstly by manipulating the overhead pressure between 80 and 500 mbar, controlling the DO level at 95% of saturation; then by manipulating the agitation rate between 320 and 1000 rpm maintaining the DO level at 50% of saturation. Finally the glycerol feeding rate was used to keep the dissolved oxygen concentration level at the desired set-point. The data acquisition and control algorithms were implemented in the Universal BIo-process CONtrol system – UBICON, Electronic System Design, Hannover, Germany. This supervision system, based on a VME architecture (industrial standard IEC 1014), includes hardware and software interfaces, connecting equipment via a CAN field-bus (ISO 11898), and enables real-time and multi-tasking operation.

2.2 Adaptive controller design

A process model for the fed-batch cultivation of the *P. pastoris* yeast was first developed with the purpose of supporting the design of the adaptive controllers. It is assumed that the process is run under carbon source limitation being the accumulation of the carbon source negligible at any instance. This is equivalent to state that all carbon source material fed to the bioreactor is instantly consumed and therefore the substrate consumption rate is equal to the rate of substrate addition to the vessel. In such circumstances the volumetric oxygen consumption rate can be related directly to the substrate consumption rate through the stoichiometry, i.e., the oxygen consumption rate is equal to the carbon source consumption rate (defined as the product $S_F D(t)$ where S_F is the carbon source concentration in the feed rate, and D is the dilution rate) multiplied by the yield of oxygen to substrate Y_{OS}. Since the intracellular reaction kinetics are complex and inherently dynamic it is necessary to assume that the yield coefficient is a time function. With these considerations and simplifications the following single-input single-output (SISO) process equation with D being the control input and Co the process output is obtained:

$$\frac{dC_o}{dt} = -Y_{OS} DS_F - DC_o + k_L a(C_o^* - C_o) \tag{1}$$

being $k_L a$ the volumetric oxygen mass transfer coefficient and C_o* the oxygen saturation concentration in the liquid phase. In Eq. (1) all parameters $Y_{OS}(t)$, Co* , S_F and $k_L a$ are considered to be badly known. The uncertainties around these parameters pose the important constraints for deriving a simple and robust controller directly from Eq. (1). It is however possible to design a stable direct or indirect adaptive scheme to cope with these uncertainties. Before we start designing the controllers we will transform the process Eq. (1) into a more convenient form. First the variable transformation Z=100-DOT is introduced with DOT being the dissolved oxygen tension defined as DOT=100Co/Co* being directly measured by oxygen probes after proper calibration. With this variable transformation and assuming that both the dynamics of Co* and the DO dilution term are negligible, the process equation is rewritten in the following simplified linear time-varying form:

$$\frac{dZ}{dt} = \theta D - k_L aZ \tag{2}$$

with $\theta = 100 Y_{OS} S_F / C_o^*$. In the process dynamical model (2) there are two possible unknowns. The term θ congregates several ill-defined process parameters and is considered always as being an unknown time varying 'parameter'. If no measurement device is available for the off-gas composition the $k_L a$ will be an additional unknown of the process. Though both cases were studied in detail, in the context of the present paper only the situation of unknown θ(t) and "measured" $k_L a$ (indirectly from the analysis of off-gas composition) is addressed.

A recursive least squares algorithm could be employed for estimating the unknown θ(t) on-line, however an even simpler alternative is to apply an observer-based estimator [4] that may be formulated as follows:

$$\frac{d\hat{Z}}{dt} = D\hat{\theta} - k_L a(100 - DOT) + \omega(100 - DOT - \hat{Z}) \tag{3a}$$

$$\frac{d\hat{\theta}}{dt} = \gamma(100 - DOT - \hat{Z}) \tag{3b}$$

If the design parameters ω, γ are defined as $w = 2\zeta\tau^{-1} - (dD/dt)D^{-1}$ and $\gamma = D^{-1}\tau^{-2}$, it may be shown that the estimates of θ(t) converge to the true process values with second order dynamics [5]. Subtracting Eq. (2) by eq (3a) and combining with (3b), after some manipulation yields the following result:

$$\tau^2 \frac{d^2\hat{\theta}}{dt^2} + 2\varsigma\tau \frac{d\hat{\theta}}{dt} + \hat{\theta} = \theta \tag{4}$$

that basically confirms that the convergence follows the well known second order equation. The advantage of this solution is that the user has full control on the speed of convergence and on the oscillatory behaviour of the estimator by setting the two parameters τ and ζ and hence trial and error tunings are avoided.

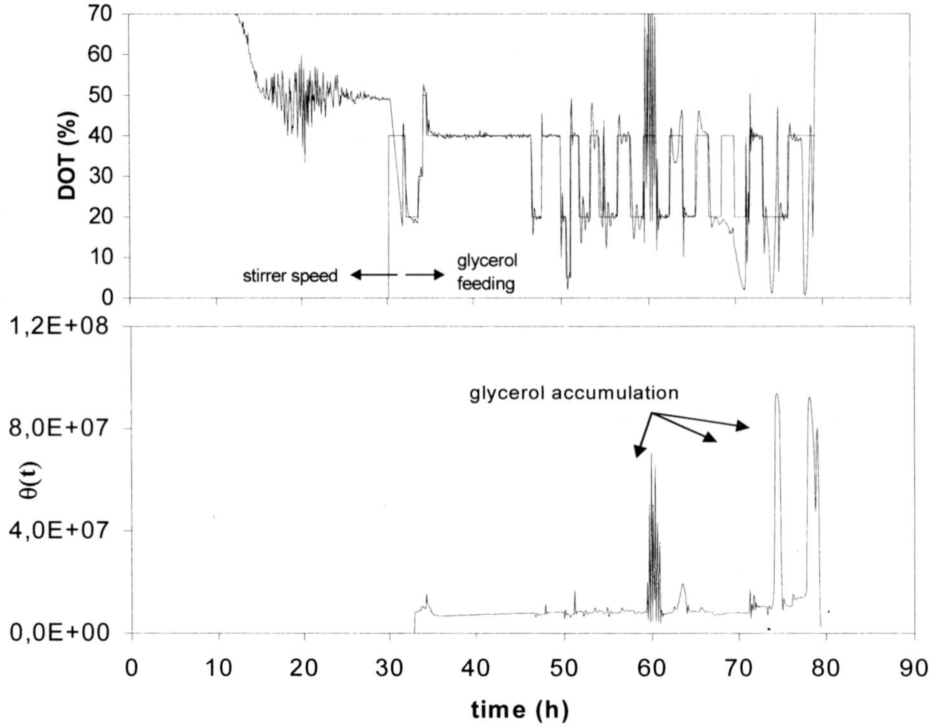

Fig. 1.DO control results for step changes in the set-point using 6 distinct adaptive control algorithms and different controller settings.

Povided that both θ(t) and $k_La(t)$ are known on-line it is straightforward to derive a control law for the control input D that stabilises system (2). Since the process is first order it is obvious to employ a first order reference model design:

$$\tau_C \frac{dDOT}{dt} + DOT = DOT^{SP} \tag{5}$$

being DOT^{SP} the set-point and τ_c the first order time constant. Combining equations (5) and (2) and substituting θ(t) and $k_La(t)$ by their estimates (or 'measurement' in the case of k_La), the following control law is obtained:

$$F = \frac{V\left[k_L aZ - e/\tau_c\right]}{\hat{\theta}(t)}, \qquad e = (DOT^{SP}\text{-}DOT) \tag{6}$$

3 Results and discussion

The controller equation (6) was implemented along with a discrete version of the estimator (3ab) employing direct Euler discretization with sampling time T=2 s. Also other controllers and estimation algorithms for the simultaneous θ(t) and k_La estimation were implemented and tested. The overall final results are plotted in Fig. 1. Besides the model reference adaptive controller described previously, an adaptive gain controller - with gain calculations that guarantee second order convergence to the set-point - and also a model reference controller employing a linear second order reference model were tested. The application of all control algorithms was rather successful. In all cases robust and accurate control was achieved. In Fig. (1) several results are presented for different controller tunings. It is important to recognise that the tuning is in this case intuitive: the convergence speed (τ) and damping coefficient (ζ) are the parameters that may be used to tune the dynamics of convergence. The divergence shown in Fig. (1) corresponded to a fast controller setting with τ=8 s, i.e close to the sampling time. The θ(t) could be estimated accurately in the whole time domain but was affected 3 times by glycerol accumulation. The two peaks observed at the end were caused by sudden loss of cell activity signalling the end of the batch. The results produced by the controller (6) with the estimator (3), which are representative of the other algorithms, can be assessed from Fig. (2). Three controller parameters must be properly set. The τ_1 and ζ are the time constant and damping coefficient respectively of a second order dynamic response for the estimation of the unknown function θ(t). They were set to 10 s and 1 respectively. The τ_C is the time constant for the exponential convergence of C_o to the desired set-point. Fig. (2a) shows the controller response to step changes in the set-point with τ_c=20, 40 and 60 s. It was observed that decreasing τ_c the response becomes faster and oscillations have higher amplitude. Though the controller design attempted to impose first order convergence, the controller response seems to be typical of a second order system. This was not unexpected and can be easily explained by probe dynamics as well as process time delays namely in what concerns the evaluation of the k_La and also delays in the pumping system. The estimator performance can be assessed from Fig (2b). The estimator was stable in the whole experiment. The oscillations observed in the estimation of θ(t) corresponded to glycerol accumulation that occurs whenever the set-point is changed.

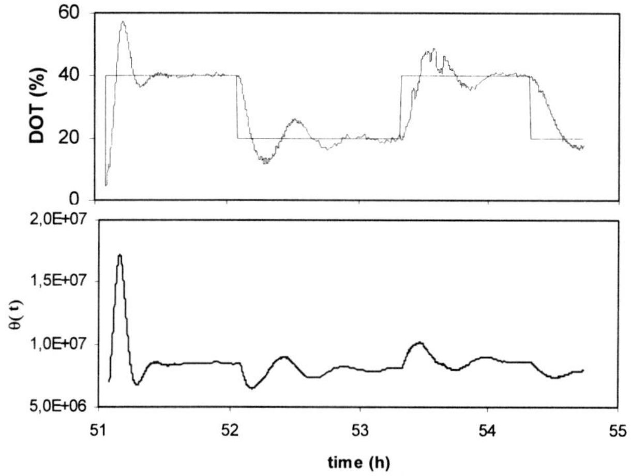

Fig. 2. Example of results obtained with controller (6) employing the adaptive estimator (3) (a) control results (b) θ estimator results.

4 Conclusions

In this work adaptive control algorithms were designed specifically for driving aerobic fermentations near maximum oxygen transfer capacity, were developed and tested in a recombinant *P. pastoris* process. This control problem is directly linked to economical issues and is of significant practical relevance. The control algorithms were derived on the basis of stability and accuracy and require two on-line measurements that are currently standard in most production facilities: DO electrodes and the MS or other device for measuring concentration of oxygen in the outlet gas stream.

References

[1] Konstantinov. K. B., Tsai, Y. S., Moles, D., Matanguihan, R. (1996) Control of long-term perfusion Chinese hamster ovary cell culture by glucose auxostat, Biotechnol Progress, 12 (1): 100-109.

[2] Akesson M., Hagander, P., Axelsson, J. P. (2001) Avoiding acetate accumulation in Escherichia coli cultures using feedback control of glucose feeding, Biotechnol Bioeng, 73 (3): 223-230.

[3] Chung JD (2000) Design of metabolic feed controllers: Application to high-density fermentations of Pichia pastoris, Biotechnol Bioeng, 68 (3): 298-307.

[4] Bastin, G., Dochain, D. (1990) On-Line Estimation and Adaptive Control of Bioreactors. Elsevier, Amsterdam.

[5] Oliveira R., Ferreira, E.C., Feyo de Azevedo, S. (2002) Stability and Dynamics of Convergence of Observer-Based Kinetics Estimators, J. Process Control, 12: 311-323.

Evolutionary Design of Neural Networks for Classification and Regression[1]

Miguel Rocha*, Paulo Cortez†, José Neves*
*Dep. Informática, University of Minho, Portugal
†Dep. Sistemas de Informação, University of Minho, Portugal
E-mail: mrocha@di.uminho.pt pcortez@dsi.uminho.pt jneves@di.uminho.pt

Abstract

The *Multilayer Perceptrons (MLPs)* are the most popular class of *Neural Networks*. When applying *MLPs*, the search for the ideal architecture is a crucial task, since it should should be complex enough to learn the input/output mapping, without overfitting the training data. Under this context, the use of *Evolutionary Computation* makes a promising global search approach for model selection. On the other hand, *ensembles* (combinations of models) have been boosting the performance of several *Machine Learning (ML)* algorithms. In this work, a novel evolutionary technique for *MLP* design is presented, being also used an ensemble based approach. A set of real world classification and regression tasks was used to test this strategy, comparing it with a heuristic model selection, as well as with other *ML* algorithms. The results favour the evolutionary *MLP* ensemble method.

Keywords: Supervised Machine Learning, Multilayer Perceptrons, Evolutionary Algorithms, Ensembles.

1 Introduction

Neural Networks (NNs) are important *Machine Learning (ML)* techniques, denoting a set of connectionist models inspired in the behavior of the human brain. In particular, the *Multilayer Perceptron (MLP)* is a popular architecture, where *neurons* are grouped in *layers* and only *forward connections* exist. This provides a powerful base-learner, capable of nonlinear mappings [1]. When compared to other *ML* methods, *MLPs* are known to behave well in terms of predictive knowledge [2], and there has also been research in terms of explanatory knowledge (e.g. extracting rules from *MLPs*) [3].

However, one of the major issues when applying *MLPs* is the *topology* (i.e. connectivity) design. This is a complex and crucial task, with a strong impact in performance (a small network may provide poor learning capabilities, while a large one will overfit the data). It is common to address this task by trial-and-error procedures (e.g. exploring different number of hidden nodes), in a *blind* search strategy, which only goes through a small set of possible configurations. More elaborated methods have been proposed, such as *pruning* [4] and *constructive* [5] algorithms, although these perform *hill-climbing*, being prone to local minima.

An alternative is offered by *Evolutionary Computation (EC)*, which performs a global multi-point (or *beam*) search, quickly locating areas of high quality, even when the search space is very large and complex. The combination of *EC* and *NN*, often called *Evolutionary Neural Networks (ENNs)*, is a better candidate for the topology design, due to the characteristics of the error surface [6]:

- the number of nodes/connections is unbounded;
- the mapping from the structure to its performance is indirect;
- changes are discrete and can provide discontinuous effects in the *NN* behaviour; and
- similar topologies may present different performances.

In addiction, this approach is biologically more plausible; i.e., living creatures have succesfully adapted to their environments as a result of the interaction of *evolution* and *learning*.

Another emergent *ML* research area is related to the use of *ensembles*, where a set of models are combined to produce an answer, being often more accurate than individual learners [7]. One interesting way to build *NN* ensembles is based on *heterogeneous topologies* [8]. This approach can be easily adapted to *ENNs* with no computacional effort increase, since *ENNs* already use a population of *NNs* with different connectivities.

The present work presents a novel *ENN* to the design of *MLP* topologies, where a direct *MLP* representation (closer to the phenotype) is used. This approach will be

[1]This work was supported by the *FCT* project POSI/ROBO/43904/2002, which is partially funded by FEDER.

tested in classification and regression tasks, using both single and ensemble based models. Finally, results will be compared with a heuristic *NN* selection procedure, as well with other *ML* methods.

2 Materials and Methods

2.1 Data Sets

Eight classification and eight regression data sets were selected from the UCI *ML* repository [9], and its main features are listed in Table 1, namely the number of numeric (**Nu**), binary (**Bi**) and nominal (**No**, i.e. discrete with 3 or more labels) input attributes, as well as the number of examples (**Ex**) and classes (**Cl**). The regression tasks are identified by the symbol $\Re$ in the **Cl** column (last eight rows).

Table 1. A summary of the data sets used.

Task	Inputs			Ex	Cl
	Nu	Bi	No		
Balance	4	0	0	625	3
Bupa	6	0	0	345	2
Car	0	0	6	1728	4
Cmc	5	3	1	1473	3
Dermatology	34	0	0	366	6
Ionosphere	34	0	0	351	2
Sonar	60	0	0	104	2
Yeast	7	1	0	1484	10
Abalone	7	0	1	4177	$\Re$
Auto-mpg	5	0	2	398	$\Re$
Autos	17	3	5	205	$\Re$
Breast-cancer	1	4	4	286	$\Re$
Heart-disease	6	3	4	303	$\Re$
Housing	12	1	0	506	$\Re$
Servo	2	0	2	167	$\Re$
WPBC	32	0	0	194	$\Re$

2.2 Neural Networks

Before feeding the *MLPs*, the data was preprocessed: a *1-of-C* encoding (one binary variable per class) was applied to the nominal attributes and all input values were rescaled within the range $[-1, 1]$. For example, the **safety** attribute from the task **car** was encoded according to: *low* $\rightarrow$ 1 -1 -1, *med* $\rightarrow$ -1 1 -1 and *high*$\rightarrow$ -1 -1 1.

Regarding the outputs, the discrete variables were normalized within the range $[0, 1]$ (using also a *1-of-C* encoding for the nominal attributes). Therefore, the predicted class is given by the nearest class value to the node's output, if one single node is used (binary variable), otherwise the node with the highest output value is considered. On the other hand, regression problems will be modeled by one real-valued output, which directly represents the *dependent* target variable.

The *MLPs* used make use of biases and sigmoid activation functions, with one hidden layer, containing a variable number of nodes. A different approach was followed for the regression tasks, since outputs may lie out of the logistic output range ($[0, 1]$). In this case, the *logistic* function was applied on hidden nodes, while the output ones used shortcut connections and *linear* functions, to scale the range of the outputs (Figure 1). This solution avoids the need of filtering procedures, which may give rise to information loss and has been successfully adopted in other regression applications, such as *Time Series Forecasting* [10].

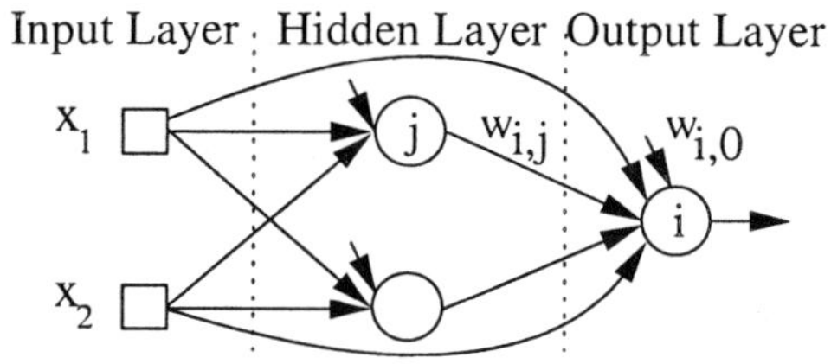

Fig. 1. A $2-2-1$ *MLP* topology with bias and shortcuts.

The initial weights will be randomly set within the range $[-1, 1]$. Then, the *RPROP* algorithm [11] is selected for training, due to its faster convergence and stability, being stopped after a maximum of 200 epochs or when the error slope is approaching zero.

Two distinct accuracy measures were adopted: the *Percentage of Correctly Classified Examples (PCCE)*, used in classification tasks; and the *Normalized Root Mean Squared Error (NRMSE)*, applied in the regression ones. These metrics are given by the equations:

$$
\begin{array}{l}
PCCE = \frac{\sum_{i=1}^{N} 1 \,,\, if(T_i=P_i)}{N} \times 100\,(\%) \\
NRMSE = \frac{\sqrt{\frac{\sum_{i=1}^{N} (T_i - P_i)^2}{N}}}{\frac{\sum_{i=1}^{N} T_i}{N}} \times 100\,(\%)
\end{array}
\quad (1)
$$

where N denotes the number of examples; P_i, T_i the predicted and target values for the i-th example.

An *Heuristic* approach (*HNN*) to model selection is defined by a simple trial-and-error procedure, where fully connected *MLPs*, with a number of hidden nodes ranging from 0 to 20, are trained. Then, the topology with the lowest validation error (computed over non training data) is selected. The trained *MLPs* will also be used to build an *Ensemble* (*HNNE*), where the output is given by the average over all 21 *MLPs*.

2.3 Evolutionary Neural Network

In this work, an *Evolutionary Algorithm* with a *direct* representation is embraced, where the genotype is the whole *MLP*. Each individual of the initial population is set by choosing a random number of hidden nodes (between 0 and 10). Then, each possible connection is set with probability of 50%. New individuals are bred by structural *mutation*, which works by adding or deleting a random number (from 1 to 5) of nodes or connections. The population size was set to 20, being the *selection* done by converting the fitness value (the error computed over a *validation* set) into its ranking, and then applying a roulette wheel scheme, being used a substitution rate of 50%. Finally, the *ENN* is stopped after 20 generations.

This scalable *ENN* is able to search through any kind of *MLP* connectivity, ranging from linear models to complex nonlinear *MLPs*. The *ENN Ensemble (ENNE)* will be built using the best 20 individuals (with lower validation error) obtained during the evolutionary process, being the output computed as the average of the *MLPs*.

3 Results

The *NN/EC* experiments were conducted using a software package developed in *Java* by the authors. The other techniques were computed using *WEKA ML* software (with its default parameters) [12]:

- *J48* – a classification decision tree based on the C4.5 algorithm;
- *M5P* – a regression decision tree (M5 algorithm);
- *IB5* – a *5-Nearest Neigboor*;
- *KStar* – an instance based algorithm; and
- *SVM* – a *Support Vector Machine*.

For each model, 10 runs of a 5-fold cross-validation process [13] (stratified in the classification tasks) were executed. This means that in each of these 50 experiments, 80% of the data is used for learning and 20% for testing. Regarding the *MLP* based approaches, the learning data was divided into training (50% of the original dataset) and validation sets (30%). Tables 2 and 3 show the average errors of the 10 runs for each learning model and classification/regression task. In both tables, the last row averages the global behaviour of each technique.

First, the classification results will be analyzed. The *NN* based approaches (last four columns) are competive when compared with the other *ML* algorithms. The few exceptions are the **dermatology** and **sonar** tasks, where the *SVM* and *KStar* get the best results. As expected, the *ENN* outperforms the *HNN*, with a 1% increase in performance. Furthermore, the ensemble approaches (*HNNE* and *ENNE*) obtain better results when compared with the single based methods (1.4% improvement in both cases). Indeed, the *ENNE* reveals the best overall behaviour, outperforming all other algorithms in 5 of the 8 tasks.

A similar scenario occurs in the regression tasks. In general, the *NN* methods are better than the other *ML* algorithms, although the *M5P* outperforms the *HNN* and *HNNE* approaches. When compared with the *ENN*, the *HNN* is outperformed by a wider difference (2.5%). As before, the ensembles behave better, although the impact is higher with the *ENNE* (1.5% improvement, being the best method in 5 tasks) than with the *HNNE* (0.7% improvement).

4 Conclusions

The surge of bio-inspired techniques, such as *Multilayer Perceptrons (MLPs)* and *Evolutionary Computation (EC)*, has created new exciting possibilities for the field of *Machine Learning (ML)*. Considering ensembles of learning models to improve its accuracy has also been a focus of attention by the research community. In this work, an evolutionary approach to *MLP* topology design was presented, considering single and ensemble combinations, being tested in supervised learning tasks (e.g., classification and regression).

The results obtained confirm than *Evolutionary Neural Network (ENN)* approach outperforms a heuristic trial-and-error *MLP* design procedure (*HNN*), as well as other *ML* algorithms (e.g. *K-Nearest Neigboor*). However, this improvement in performance has the handicap of increasing the computational complexity (the *HNN* requires only a tenth of the *ENN* computational effort).

On the other hand, the use of the *EC* population structure to construct ensembles is a recent research field. Indeed, the proposed *ENN Ensemble (ENNE)*, based on the average of the outputs from the best *EC* individuals (or *MLPs*), is competitive. The *ENNE* has the advantage of presenting the best overall performance while requiring the same computational effort, when compared with the single based *ENN*.

In future work, it is intended to explore similar approaches with different neural architectures (e.g., *Recurrent Neural Networks*). Moreover, more elaborated ensembles should be considered, by designing fitness functions which reward specialization [14].

References

[1] Haykin, S. (1999) Neural Networks - A Compreensive Foundation. 2nd ed. Prentice-Hall

[2] Quinlan, J.R. (1994) Comparing Connectionist and Symbolic Learning Methods. In: Hanson, S. et al. (eds.) Computation Learning: Theory and Natural Learning Systems, MIT Press, pp. 445-456

Table 2. The classification results (*PCCE* values, in %).

Task	J48	IB5	KStar	SVM	HNN	ENN	HNNE	ENNE
Balance	78.1	87.6	88.3	87.7	94.8	95.8	95.7	**96.5**
Bupa	64.8	60.7	65.9	58.0	68.4	69.0	68.5	**69.7**
Car	91.3	92.3	87.1	93.5	97.4	98.2	98.3	**98.7**
Cmc	51.2	47.2	49.6	48.4	50.6	54.2	52.1	55.9
Dermatology	95.7	96.6	94.5	**97.4**	94.9	95.5	95.9	96.8
Ionosphere	89.4	84.6	84.0	87.9	88.9	89.6	92.3	**92.6**
Sonar	72.5	80.5	**85.2**	76.7	79.9	79.0	80.9	81.2
Yeast	56.0	57.1	53.1	56.6	58.2	59.3	59.9	**60.2**
Mean	74.9	75.8	76.0	75.8	79.1	80.1	80.5	**81.5**

Table 3. The regression results (*NRMSE* values, in %).

Task	M5P	IB5	KStar	SVM	HNN	ENN	HNNE	ENNE
Abalone	24.3	25.3	24.8	25.0	**23.2**	**23.2**	**23.2**	**23.2**
Auto-mpg	**11.8**	15.1	14.6	13.5	14.2	13.0	12.2	12.2
Autos	**13.3**	21.4	21.6	13.8	14.6	16.1	14.2	14.9
Breast-cancer	40.8	40.8	43.6	44.0	42.6	40.6	47.4	**40.2**
Heart-disease	21.2	**21.0**	25.5	21.5	21.9	21.7	22.2	21.5
Housing	18.4	22.2	18.0	22.6	18.4	17.3	16.6	**16.0**
Servo	50.4	60.4	67.6	70.5	60.8	45.0	49.6	**39.4**
Wpbc	73.2	73.6	98.2	**71.7**	75.4	74.4	80.3	**71.7**
Mean	33.0	35.0	39.2	35.3	33.9	31.4	33.2	**29.9**

[3] Setiono, R. (2003) Techniques for Extracting Classification and Regression Rules from Artificial Neural Networks. In D. Fogel and C. Robinson, (eds.), Computational Intelligence: The Experts Speak, IEEE Press/Wiley, pp 99–114

[4] Thimm, G. and Fiesler, E. (1995) Evaluating pruning methods. In: Proc. of the Int. Symp. on Artificial Neural Networks, pp 20–25

[5] Kwok, T. and Yeung, D. (1997) Constructive algorithms for structure learning in feedforward neural networks for regression problems problems: A survey. IEEE Transactions on Neural Networks, 8(3):630–645

[6] Yao, X. (1999) Evolving Artificial Neural Networks. In: Proc. of the IEEE, 87(9): 1423-1447

[7] Dietterich, T. (1997) Machine Learning Research: Four Current Directions. AI Magazine, 18(4):97–136

[8] Rocha, M., Cortez, P. and Neves, J. Ensembles of Artificial Neural Networks with Heterogeneous Topologies. In: Proc. of the 4th Symposium on Engineering of Intelligent Systems (EIS2004). ICSC Academic Press

[9] Blake, C. and Merz, C. (1998) UCI Repository of Machine Learning Databases, University of California

[10] Cortez, P., Rocha, M. and Neves, J. (2001) Evolving Time Series Forecasting Neural Network Models. In: Proc. of the 3rd Int. Symposium on Adaptive Systems: Evolutionary Computation and Probabilistic Graphical Models (ISAS 2001), pp. 84–91

[11] Riedmiller, M. (1994) Supervised Learning in Multilayer Perceptrons - from Backpropagation to Adaptive Learning Techniques. Computer Standards and Interfaces 16

[12] Witten, I. and Frank, E. (2000) Data Mining: Practical Machine Learning Tools and Techniques with Java Implementations. Morgan Kaufmann

[13] Kohavi, R. (1995) A Study of Cross-Validation and Bootstrap for Accuracy Estimation and Model Selection. In: Proc. of the Int. Joint Conference on Artificial Intelligence (IJCAI)

[14] Liu, Y., Yao, X. and Higuchi, T. (2000) Evolutionary Ensembles with Negative Correlation Learning. IEEE Transactions on Evolutionary Computation, 4(4):380–387

Pelican – Protein-structure Alignment using Cellular Automaton models

Deepak K Gangadhar[1]

[1] IBM Software Labs (ISL), Golden Enclave, Airport Road, Bangalore, India
E-mail: deepakg@in.ibm.com

Abstract

With more than 23000 protein structures deposited in the Protein Data Bank (PDB) and more structures being discovered with each passing day, the experimental determination of the 3-dimensional structure of proteins is just the beginning of a journey in-silico. For a structural biologist, this enormous surge of structural data carries with it far greater computational challenges; Compare, align, classify, and categorize them under families, domains and functionally similar proteins already discovered. *Pelican* provides the structural biologist with a strong and easy technique that that will help him in facing these challenges. Pelican is a rapid way to align the backbones of two protein structures using 2-dimensional Cellular Automaton (CA) models. Breaking down the protein structure into distance matrices comprising of 5 peptide units, Pelican uses the differences of these matrices to construct the 2-dimensional CA grid. Starting from an initial unaligned state, the CA evolves through several generations according to a defined set of local rules. As the CA evolves through successive generations, the emergent patterns made by the live cells are the ones that contribute to the alignment. *Pelican is also an example of a system exhibiting emergent behavior. Each cell behaves in a strictly microscopic way, but each individual cell's behavior leads to a macroscopic long range behavior exhibited by the entire system which collectively gives the alignment.*

1 Introduction

Proteins have an uncanny ability to assume infinite shapes and structures and could be the reason why nature has chosen them to be the building blocks of life. This realization of the importance of proteins gave birth to a highly specialized field called Structural Genomics. *"Structural genomics will yield a large number of experimental protein structures (tens of thousands) and an even larger number of calculated comparative protein structure models (millions)..."* [1]. This mammoth surge in structure information has lead researchers to look out for efficient techniques to compare protein structures, detect motifs, classify proteins under specific families, etc. While the current dogma of genetics connecting sequences to structures (sequence→structure→function) suggests that it would suffice if sequences were studied in greater details to detect similarity in proteins and classify them, there are several instances when different sequences yield the same structure. As a result there is a concerted effort to study the 3-dimensional structure of proteins. [2, 3].

2 The Pelican algorithm

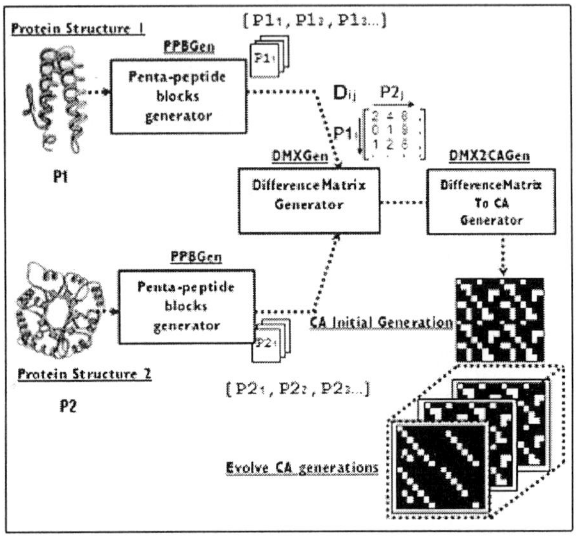

Fig. 1. The Pelican Algorithm

Fig. 1 illustrates the central idea of the Pelican algorithm:

- Breakdown large protein structures to smaller units... comprising of n successive C-alpha atoms of the primary amino acid sequence (n=5).
- Compute the distance matrix of each unit (Poly-peptide blocks).
- For the given 2 protein structures, compute the Difference Matrix (DMX)
- Using the DMX, generate the Initial Configuration of a 2-Dimensional CA.
- The CA evolves through several generations according to a pre-defined set of microscopic rules; the rules designed in a way such that only the cells corresponding to the longest alignments of the two structures survive.

3 Pelican's alignment principle

The central idea behind Pelican is that all alignments (significant or otherwise) between the two protein structures will be visible in the Difference Matrix. Fig. 2 shows the DMX (converted to the IC of a cellular automata [4]) of 2 protein structures; the A chains of 1a4f and 1c40. These two proteins are found to be similar according to DALI [5]. The DMX contains patterns that signify alignments and other similar motifs.

Fig. 2 Difference Matrix –1a4f_A and 1c40_A

Fig. 3 illustrates the principle of Pelican and how it can be used to identify these patterns. There are 2 protein structures whose PPBs are shown numbered ($P1_1 \ldots P1_{12}$) and ($P2_1 \ldots P2_8$). The shaded portions of the protein structures are similar. The Amino Acid sequences of the PPBs of the two structures are also shown. The DMX is computed and is shown in the figure. The similar structures correspond to the following.

A7 to A18 of protein structure 1 (corresponding to $P1_4$ to $P1_7$ PPBs)
A3 to A15 of protein structure 2. (corresponding to $P2_2$ to $P2_5$ PPBs)

Looking at the DMX, it is evident that there will be a contiguous **run of diagonal cells** of low value corresponding to the alignment because the distance matrices will be similar at these locations. The location of these low valued cells will be corresponding to the locations of $P1_4$ to $P1_7$ against $P2_2$ to $P2_5$. Fig. 2 shows the same principle applied to two similar protein structures obtained from the pdb servers (www.rcsb.org). The only difference is that they have been converted to the Initial Configuration of a CA. Notice from the DMX in Fig. 3 that alignment patterns can be formed as cross-diagonals running in either direction. It is evident from Fig. 2 that the long diagonal running from top-left to bottom-right comprises an alignment.

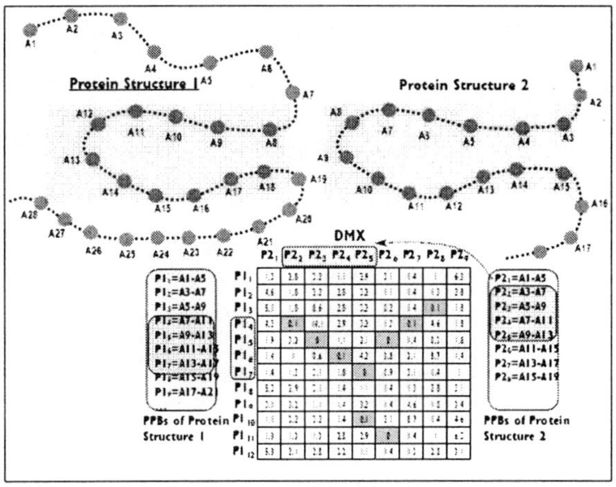

Fig. 3. The alignment principle behind Pelican

Detecting the significant contiguous alignments involves designing a set of rules using which the CA will evolve and in the process eliminate all other cells that are not part of the alignment. Instead of looking at all the eight neighbors of a cell for writing the rules, only the cross neighbors (cells at the diagonals) are of importance here.

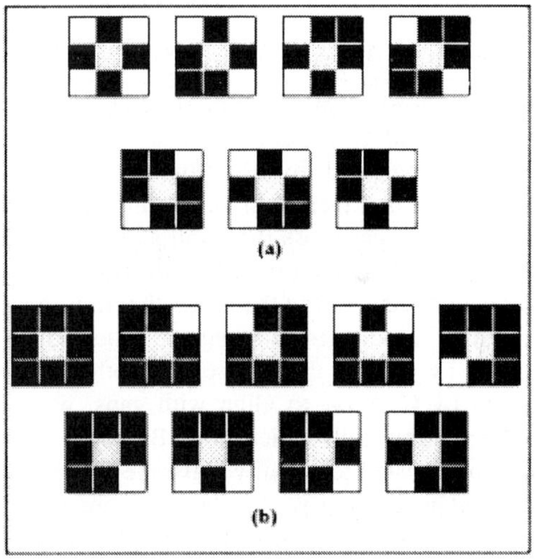

Fig. 4 (a) Neighborhood states for which the center cell lives and (b) Cells with states not found in (a) will die

The guiding rule is that a cell that is alive should remain alive into the next generation if and only if its diagonal neighbors are alive. Applying these rules successively to all the cells of a particular generation will eliminate all such cells that are not part of the alignment. Fig. 4(a) (b) shows all possible cross-neighbor combinations for which the center cell lives (a) and dies (b).

Applying the rules for the IC shown in Fig. 5(a), the next 4 generations of the CA are shown in Fig. 5(b)-(e).

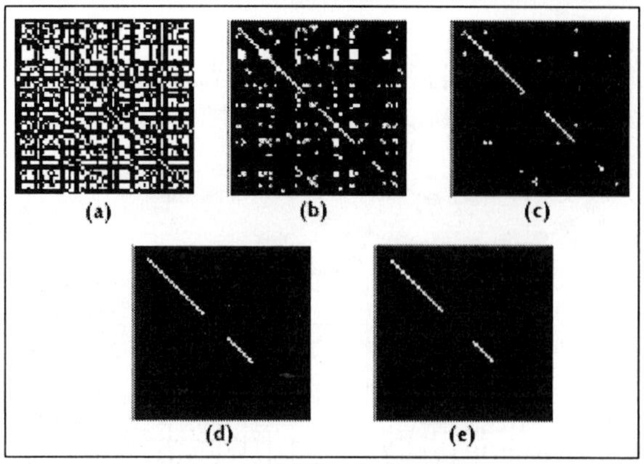

Fig. 5 Evolution of the (a) Initial configuration for 4 generations (b) to (e)

It can be observed that only those cells contributing to the alignment of similar structures survive in the final generations. At the same time, the live cells at the ends of the largest alignment are successively getting eliminated. However, this is not a cause of concern *because the survival of a lone live cell (a cell without any live neighbors) in the nth generation implies that it is part of an alignment that is at least of length 2n+1 (n cells on either direction plus itself).*

4 Aligning with gaps using Pelican

Fig. 6(a) shows two protein structures where the segments A and C are aligned with a gap in protein structure 1 indicated by B. Aligning with gaps in Pelican means that there will be continuous run of live cells in a particular cross direction and then there will be a gap in the horizontal/vertical direction followed by another continuous run in the cross direction. As shown in Fig. 6(b), A indicates an alignment followed by a gap indicated by B and then there is another alignment indicated by C. Now, to align with gaps, we need to consider the gray cells indicated by B and the rules also need to be updated to actually make the B cells alive in subsequent generations so that the entire alignment A+C is registered as a continuous alignment with a gap indicated by B.

5 Emergent behavior in Pelican

CA models are classic examples of systems exhibiting emergent behavior. Pelican goes on to corroborate this. Pelican is a grid of cells that evolve based on a set of locally applicable microscopic update rules. There is no long range communication between the cells.

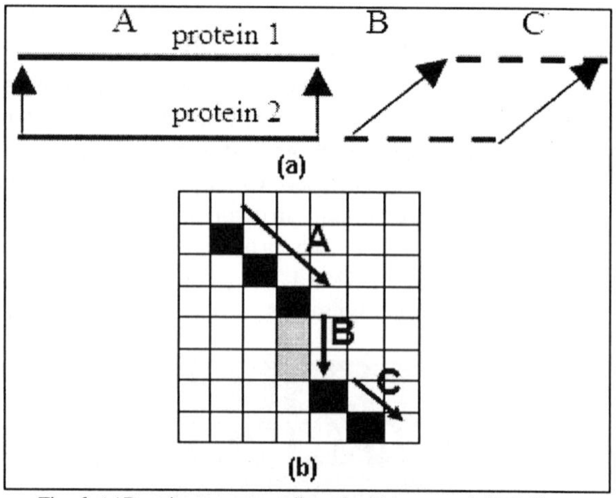

Fig. 6. (a)Protein structures aligned with gaps and (b) Pelican extended to detect alignments with gaps.

But these microscopic rules lead to large scale macroscopic behavior leading to detection of alignments of the two structures, thus making Pelican an emergent system.

6 Results

Pelican was tested with various sets of data. The data sets were obtained from www.rcsb.org. To test Pelican with similar proteins and dissimilar proteins, the DALI service for Fold Classification and Structural Alignments (FSSP) (http://www.ebi.ac.uk/dali/fssp) was used. The DALI server provides a percent similarity between two protein structures based on the amino acid sequences. The similarity is a measure of the percent of similar amino acids in the final alignment. On the other hand, Pelican is strictly a structure alignment algorithm. The scoring for Pelican uses only the structure information and no sequence related information. Pelican's scoring scheme is designed to yield a purely structure based numeric value of the alignment. In Pelican, the best alignment is when the smaller of the two structures is completely aligned with the larger of them. The score for a Pelican alignment calculates the percentage of cells that are contributing to the "largest alignment". Largest alignment here is a relative term since Pelican can be configured to detect alignments of any length. The results of all the runs shown below are for alignments that are greater than 5 continuous cells. Since each cell is comprised of 5 amino acids (with an overlap of two), the length of the alignments considered for scoring are greater than 13. If N_{MAX} be the number of cells (Penta-Peptide-Blocks) in the smaller of the two structures, and N_C be the number of cells of the final alignment, then Pelican score= N_{MAX}/N_C. Table 1-Table 2 and Fig. 7-Fig. 8 show the results of running Pelican on different data sets. *Apart from these data sets, Pelican was run on a* ***lot of different data sets and the results concurred with the results reported on the DALI server****. Where DALI reported low similarities,*

Pelican discovered a lot of small sized motifs, but no significant alignment.

7 Conclusion

The Pelican algorithm discussed till now considers aligning the two structures without gaps. With a little variation which involves considering the plus-neighbors and altering the update rules, the algorithm can very well be made to handle aligning structures with gaps as well. Pelican is not only useful for detecting alignments, but can also detect smaller motifs.

TABLE 1
RESULTS OF RUNNING PELICAN ON 1SCU_A AND 1EUD_A

Protein	# C-a atoms	Similarity DALI %	Pelican scoring(N_{MAX}/ N_C)*100		
			N_{MAX}	N_C	Pelican score
1SCU_A	288	66	143	87	61
1EUD_A	306				

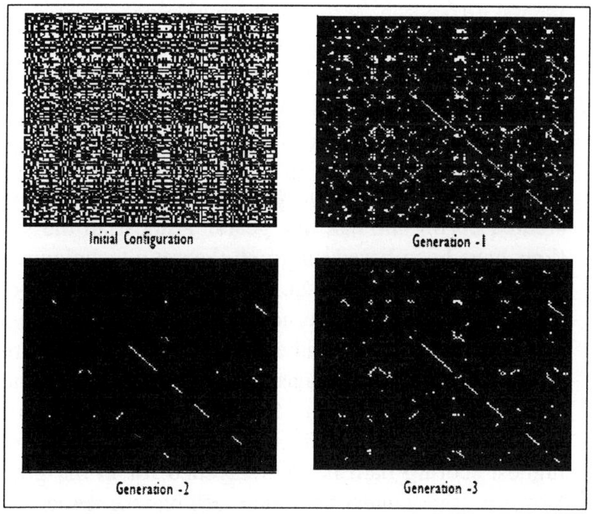

Fig. 7. Results of running Pelican on 1SCU_A and 1EUD_A

TABLE 2
RESULTS OF RUNNING PELICAN ON 1E6K_A AND 1M5T_A

Protein	# C-a atoms	Similarity DALI %	Pelican scoring(N_{MAX}/ N_C)*100		
			N_{MAX}	N_C	Pelican score
1E6K_A	129	23	61	18	29
1M5T_A	124				

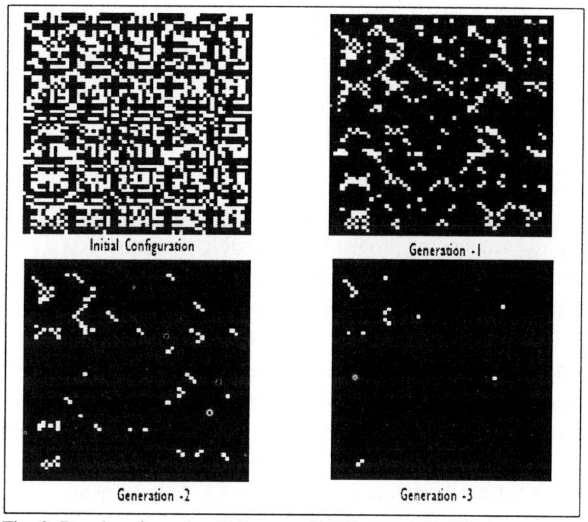

Fig. 8. Results of running Pelican on 1E6K_A and 1M5T_A

References

[1] Stephen K. Burley., (2000) An overview of structural genomics. Nature Structural Genomics supplement.

[2] Amit P. Singh and Douglas L. Brutlag., (2000) Protein Structure Alignment: A comparison of methods. http://citeseer.ist.psu.edu/390690.html

[3] Mark Gerstein & Michael Levitt., (1996) Using Iterative Dynamic Programming to Obtain Accurate Pairwise and Multiple Alignments of Protein Structures. Proceedings of the Fourth International Conference on Intelligent Systems in Molecular Biology (ISMB-96).

[4] S. Wolfram., (1982) Cellular Automata as Simple Self-Organizing Systems. Caltech preprint CALT-68-938, 1982.

[5] Liisa Holm and Chris Sander., (1993) Protein Structure Comparison by Alignment of Distance Matrices. J. Mol. Biol. 233, (1993), pp. 123-138.

An Efficient Algorithm for *De Novo* Peptide Sequencing

S. Brunetti[1], D. Dutta[2], S. Liberatori[3], E. Mori[1], D. Varrazzo[3],
[1]Dipartimento di Scienze Matematiche ed Informatiche "R. Magari", Università di Siena, Italy
E-mail: {sara.brunetti, morie}@unisi.it
[2] Department of Computer Science, University of Southern California, USA
E-mail: ddutta@catarina.usc.edu
[3]Dipartimento di Biologia Molecolare, Università di Siena, Italy
E-mail: {liberatori, varrazzo}@unisi.it

Abstract

In this paper we propose a new algorithm for the *de novo* peptide sequencing problem. This problem reconstructs a peptide sequence from a given tandem mass spectra data containing n peaks. We first build a directed acyclic graph $G = (V, E)$ in $O(n \log n)$ time, where $v \in V$ is a spectrum mass ion or a complementary mass to a spectrum ion. The solutions of this problem are then given by the paths in the graph between two designated vertices. Unlike previous approaches, the proposed algorithm does not use dynamic programming, but it builds the graph in a progressive fashion using a priority queue, thus obtaining an improvement over other methods [1, 2].

1 Introduction

Proteomics is defined as the study of many proteins simultaneously, and functional proteomics studies complex biological process as a whole, registering the behavior of several thousands of protein components at once. Analytical methods of protein separation, such as two-dimensional gel electrophoresis, are key techniques in proteomics as they allow the separation of complex protein mixtures, e.g. cells, tissues or whole organisms, in their components that can be afterwards characterized as individual proteins [3].

One of the key problems in Proteomics is the identification of peptides from mass spectrometry data. There are two fundamental approaches to this problem. One common approach is to search a database of mass spectra using various probabilistic methods. However, this approach has several limitations. For example, it is much harder to detect new peptides in this fashion and it is almost impossible to handle post translation modifications of the proteins. A relatively new technique is *de novo* sequencing. This technique reconstructs the amino acidic sequence of a peptide from a given tandem mass spectra data without the use of database search techniques.

In *de novo* sequencing, several copies of the "unknown" peptide (in term of amino acidic sequence) are broken up into two fragments, a prefix and a suffix sequence, by their random fragmentation due to collision. The masses of the resulting fragments are then measured by the mass spectrometer. Since we cannot distinguish if a mass peak corresponds to a prefix or a suffix, and additionally, some fragments may be lost due to experimental errors, the *de novo* sequencing problem takes in input a subset of the prefix and the suffix masses of a peptide, and determines an amino acidic sequence.

1.1 Related Work

The problem of *de novo* sequencing has been well studied during the last few years. Almost all the solutions have been based on graph theoretic techniques [1, 2]. For example in Dančik et. al [1], a directed acyclic graph is built, where the nodes of the graph are all the mass peaks and a labelled edge connects two nodes that differ by an amino acid mass. Then a mass peak is transformed into several nodes in the graph, each node representing a possible prefix sequence. The longest or the highest scoring path is then searched for in the graph. However, the longest path approach has the tendency to include multiple nodes associated with the same mass peak. Unlike this approach, we build a simpler graph based on the useful required peaks without the use of a scoring function.

The next major step was the dynamic programming approach of Chen [2]. This algorithm builds a graph in $O(n^2)$ time complexity, where n is the number of mass peaks. This algorithm finds only one solution in $O(n^2)$ time, and all the solutions in $O(pn+n^2)$ time, where p is the number of solutions. In [4], a more elaborate graph is built in $O(n^3)$ time, a scoring function is introduced, and all the solutions scoring more than a threshold are found in $O(pn^3)$ time. In [5] a new scoring function was introduced along with a dynamic programming algorithm, the time complexity of which is greater than that of the other papers mentioned above.

Another direction has been the application of the spectrum graph based algorithms that leverage scoring functions obtained from empirical biological knowledge [6] to *clean* the spectrum, and, use the method described in [2] to build the graph and obtain the solutions.

In another related work [7], the authors find only one solution in $O(n \log n)$ time, without building any graph, but they consider error-free spectrum, and so in this formalization the solution is unique.

Also, in a recent paper [8], the *de novo* peptide sequencing can be obtained using algorithms for finding substrings with particular submasses in a given weighted strings.

1.2 Our Contribution

In this paper we propose a simple and efficient algorithm for *de novo* sequencing problem by using a greedy strategy. We build a directed acyclic graph in $O(n \log n)$ time, from the given spectrum mass and the mass of the whole peptide. This graph is built using only the *useful* peaks in the spectrum (analogously for the edges). The solutions are obtained by a reading of the graph in $O(pn)$ time, where p is the number of solutions.

2 Background

In this section, we present the biological background and a brief description of the peptide sequencing. *Amino acids* are the building blocks of proteins. The amino acid form with missing hydrogen and hydroxyl group on the N-terminal and carboxyl sides respectively is called a *residue*. A *peptide* is formed by a chain of amino acid residues: a pair of amino acids undergoes a dehydration reaction where a loss of water leads to a *peptide bond* between the two molecules [9].

Peptides can fragment at three different places in the amide backbone (Fig. 1), denoted by a_i, b_i, c_i (if the fragment retains the N-terminal portion of the peptide) and x_i, y_i, z_i (if the carboxyl-terminal portion is retained). The subscripts indicate the number of amino acid residues contained in the ion. Thus if two adjacent ions with the same letter designation appear in a mass spectrum, their mass difference is equivalent to an amino acid residue mass.

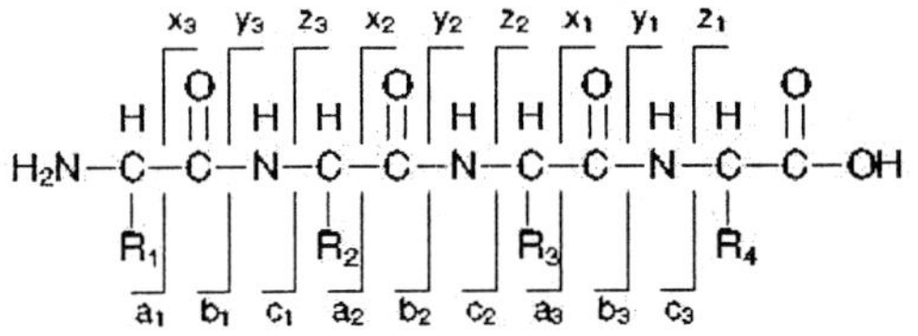

Fig. 1. Structural notation of a peptide backbone.

The *de novo* sequencing approach is based on the selective fragmentation of individual peptides by tandem mass spectrometry (MS/MS). The masses of the derived fragments are measured and plotted in a MS/MS spectrum. A MS/MS spectrum of peptides contains two levels of information: 1. the mass of the whole peptide; 2. a pattern of fragments (or ions) that can be associated to a given sequence basing on known fragmentation rules.

The b and y ions usually provide most of the information necessary to reconstruct the peptide analyte. All masses are measured in atomic mass units (abbrev. amu). The b-ion mass equals the sum of the residues plus 1 amu, due to the hydrogen atom on the N-terminal side of the backbone whereas the y-ion mass equals the sum of the residues plus 19 amu (17 amu from the carboxyl side OH group, 2 amu from the two hydrogen atoms). If a b-ion and a y-ion form a complementary pair, the sum of their masses equals the whole peptide mass plus 2 amu.

3 Our Model

In this section we give a formal characterization of the peptide sequencing problem.

We model a peptide as a string over the 20-letter alphabet $A = \{a_1, \dots a_{20}\}$, where $a_1, \dots a_{20}$ represent amino acid residues. We associate a mass to each residue denoted by a function $\mathrm{m} : A \longrightarrow \mathbb{R}$. We also introduce two special characters, b_0 and y_0, associated with masses $\mathrm{m}(b_0) = 1$ and $\mathrm{m}(y_0) = 19$, that will be used for the ions model. The definition of m can be naturally extended to $(A \cup \{b_0, y_0\})^*$ as follows:

$$\mathrm{m}(a_{i_1} \cdots a_{i_k}) = \sum_{j=1}^{k} \mathrm{m}(a_{i_j}) \text{ with } a_{i_j} \in A \cup \{b_0, y_0\}.$$

If we let any peptide P be represented by the string $a_{i_1} \cdots a_{i_k}$, then its b-ions and y-ions are represented by the strings

$$b_j = b_0 a_{i_1} \cdots a_{i_j} \qquad \text{for } j = 1, \dots, k$$
$$y_{k-j} = a_{i_{j+1}} \cdots a_{i_k} y_0 \qquad \text{for } j = 0, \dots, k-1$$

respectively. Note that b_k and y_k are not true peptide ions: such cleavage points are not detected in real experiments. They are instead a convenient notation. Similarly, we consider b_0 and y_0 as being ions too. Note that b_j and y_{k-j} are a *complementary* pair for $0 \le j \le k$. Thus the peptide mass is $\mathrm{m}_p = \mathrm{m}(P) + 18$, and we have $\mathrm{m}(b_j) + \mathrm{m}(y_{k-j}) = \mathrm{m}_p + 2$ for $i = 0, \dots, k$. More generally, we say that m_1 and m_2

are two *complementary masses* if $m_1 + m_2 = \mathrm{m}_p + 2$. We denote by m^c the complementary mass of m.

Since experimental spectra contain some background noise, we first perform some filtering operations to select only the most relevant peaks. For example, we can perform a threshold operation to select only those peaks with at least a given abundance.

We then model a spectrum by the set $M = \{m_{-1}, m_0, m_1, \ldots, m_n, m_{n+1}, m_{n+2}\}$, where $m_i < m_j, \forall i < j$. The elements $m_1, \ldots, m_n$ are the experimental peak filtered masses, with

$$\begin{aligned} m_{-1} &= \mathrm{m}(b_0) = 1, \\ m_0 &= \mathrm{m}(y_0) = 19, \\ m_{n+1} &= \mathrm{m}(b_k) = \mathrm{m}_p - 17, \\ m_{n+2} &= \mathrm{m}(y_k) = \mathrm{m}_p + 1. \end{aligned}$$

Furthermore, in the experimental data, we do not always obtain a complete peptide fragmentation. Instead of looking for mass peaks in the spectrum whose difference equals an amino acid residue mass, we look for mass differences equal to $s \in S$ where $S \subset \mathrm{m}[A^*]$. We can define S as

$$S = \{s \in \mathbb{R} | \exists r \in A^* \text{ s.t. } s = \mathrm{m}(r) \wedge |r| \leq \nu\},$$

where ν is the longest not-fragmentable sequence. In other words, S includes the masses of single amino acid residues and of small peptides. We notice that $|S|$ does not depend on the peptide mass nor on the spectrum size.

4 The *de novo* peptide Sequencing Problem

Using the notation introduced in the previous section, we can now state the *de novo* peptide sequencing problem as follows.

Input. A vector $M = \{m_{-1}, \ldots, m_{n+2}\}$ containing the spectrum mass peaks.

Output. All the strings $r_1 \cdots r_k \in A^*$ such that

i) $r_i \in A^*$, $\mathrm{m}(r_i) \in S$ for $i = 1, \ldots, k$

ii) $\mathrm{m}(r_1 \cdots r_k) = m_{n+1} - 1$

iii) $\mathrm{m}(b_0 r_1 \cdots r_j) \in M \vee \mathrm{m}(r_{j+1} \cdots r_k y_0) \in M$ for $j = 1, \ldots, k-1$.

We solve the problem in two steps. We first construct a spectrum graph from the relevant mass peaks and then we use this graph to obtain the solutions.

4.1 Graph building

In the following we do not distinguish between nodes and corresponding masses, with abuse notation.

We first build a directed acyclic graph $G = (V, E)$ in a progressive fashion starting from the initial node denoted by $v_0 = m_{n+1}$. At any step for each v already inserted in V we add a new node v' if $v - v' = s \in S$ and a new edge (v, v'). By proceeding in this way, we obtain a graph where

$$\begin{aligned} v \in V &\Rightarrow v \in M \vee v^c \in M \\ (v_1, v_2) \in E &\Rightarrow v_2 > v_1, v_2 - v_1 \in S. \end{aligned}$$

This is made, as described in Algorithm 1, by selecting the v elements of V in a mass decreasing order stored in a priority queue Q. We also have to look for already inserted elements in the queue, so we implement Q with a balanced binary search tree (such as a AVL-tree) [10].

Algorithm 1 *Solution graph building algorithm.*

Input: M;
Output: V, E;

$V = \{m_{n+1}\}$; $Q = \{m_{n+1}\}$; $E = \emptyset$;
while $Q \neq \emptyset$ **do**
  $v = ExtractMax[Q]$;
  for all $s \in S$ **do**
    if $\exists\, m \in M$ such that $v - m = s$ or $m - v^c = s$ **then**
      $v' = v - s$;
      if $v' \notin V$ **then**
        $V = V \cup \{v'\}$;
        $Q = Q \cup \{v'\}$;
      end if
      $E = E \cup \{(v', v)\}$.
    end if
  end for
end while

In order to deal with real experimental data, we also include a tolerance factor when seeking the spectrum. Given a tolerance δ, the mass $\tilde{m}$ is in the spectrum M if there exists $m \in M$ such that $|m - \tilde{m}| < \delta$.

When the algorithm terminates, we can have one of the following configurations:

1. if we inserted $v_e = m_{-1}$ in V, then we have at least a solution. The solutions are all the paths from v_e to v_0;

2. if we did not insert v_e in V, then we could not find any solution. This can happen for both missing spectrum peaks, or for unknown modifications in some amino acid.

In the latter case, we only built the terminal part of the path. We can then insert v_e in V and run the algorithm starting from v_e in a reverse fashion. This execution will surely stop before reaching v_0 too, giving us the beginning part of the path.

We could formally show that the solution of the *de-novo* peptide sequencing problem are equivalent to the paths in the graph constructed by our algorithm.

Proposition 1 *The spectrum graph is built in* $O(|M|\log|M|)$ *time.*

Proof: For each node in the queue Q (in step 2) and for each element in $|S|$ (step 4) Algorithm 1 checks if there exists an element in M such that the difference between the node and the element in M or between the element in M and the complementary mass to the node is the element of S (step 5). If it is true, that element is inserted in the queue Q and in the graph edge V when it was not already there (step 7), and the edge between the two peaks is inserted.

The loop on each element in Q is performed $|V|$ times, where $|V| < |M|$. For each iteration of the aforementioned loop, $|S|$ checks are executed; $|S|$ is constant with respect to the input size of our problem (i.e. the size of the mass spectra). The procedure to seek for an $m \in M$ in step 5 can be implemented by a binary search algorithm which takes a $O(\log|M|)$ time for each element extract from Q.

In step 7 we look for the presence of a node in the graph. We observe that we are looking for a node $v' < v$, where v was the maximum value in Q. Thus, if such node exists in the graph, it is still in Q, so we can search for it in the queue, taking a $O(\log|Q|)$ time. Finally, the obtained complex time is equal to $O(|M|\log(M|)$. ■

In [2], all the mass peaks have a corresponding node in the graph and an edge exists between nodes if the distance between them is an element of S, leading to $O(n^2)$ differences. Instead, we give an algorithm to built a graph not having all the mass peaks but only the peaks such that there exists a path starting from themselves and stopping to the first node, so reducing the number of edges with respect to [2].

4.2 Obtaining the Solutions

The solution are obtained trivially by performing a search in the spectrum graph, starting from the source node v_e. Hence, the time complexity is $O(n)$ to find one solution and $O(pn)$ for all the p solutions. We note that the way we built the graph, all such paths end in the v_0 node. Such a method, as described above, returns a list of sequences in S. Any of such sequences can be interpreted as a set of strings in A^*, expanding each sum into all the possible strings whose masses sum to it.

5 Conclusions and Future Work

In this paper, we presented a new algorithm for the *de novo* peptide sequencing problem. Our graph construction requires $O(n \log n)$ time and uses only the *useful* peaks. This is a significant improvement over existing algorithms.

This work is a first step in our plan. We have already developed two implementations, one each in C++ and Python. We are in the process of performing detailed performance tests on peptide data. One major future direction is to score the different solutions obtained by our algorithm.

Acknowledgments. We are grateful to Alberto Del Lungo for introducing us to the problem and for his support and valuable contribution.

References

[1] V. Dančík, T. A. Addona, K. R. Clauser, J. E. Vath, P. A. Pevzner (1999), De novo peptide sequencing via tandem mass spectrometry, Jour. Comp. Bio., vol. 6, pp. 327–342.

[2] T. Chen, M.-Y. Kao, M. Tepel, J. Rush, G. M. Church (2000), A dynamic programming approach to de novo peptide sequencing via tandem mass spectrometry, Proc. 11th ACM-SIAM symposium on Discrete algorithms, pp. 389–398.

[3] J. R. Yates, III(1998), Mass spectrometry and the age of the proteome, Jour. Mass Spec., no. 33, pp. 1–19.

[4] B. Lu, T. Chen (2003), A suboptimal algorithm for de novo peptide sequencing via tandem mass spectrometry, Jour. Comp. Bio., no. 10(1), pp. 1–12.

[5] V. Bafna, N. Edwards (2003), On de novo interpretation of tandem mass spectra for peptide identification, Proc. 7th conference on Computational Molecular Biology, ACM Press, pp. 9–18.

[6] S. Baginsky, M. Cieliebak, W. Gruissem, T. Klemann, Z. Lipták, M. Müller, P. Penna (2002), AuDeNS: A tool for automatic de novo peptide sequencing, tech. rep., ETH Zurich, Dept. of Computer Science.

[7] G. Pandurangan, H. Ramesh (2002), The restriction mapping problem revisited, Jour. Computer and System Sciences, no. 65, pp. 526–544.

[8] N. Bansal, M. Cieliebak, Z. Lipták (2004), Efficient algorithms for finding submasses in weighted strings, Combinatorial Pattern Matching 15th Annual Symposium, vol. 3109, CPM 2004.

[9] A. P. Snyder (2000), Interpreting Protein Mass Spectra: A Comprehensive Resource, American Chemical Society and Oxford University Press.

[10] T. H. Cormen, C. E. Leiserson, R. L. Rivest (1990), Introduction to Algorithms, MIT Press, Cambridge, MA.

Emergent Behavior of Interacting Groups of Communicative Agents

Alexander Bisler
Fraunhofer Institute for Computer Graphics
Darmstadt, Germany
Email: abisler@igd.fhg.de

Abstract

This paper presents a simulation of the behavior of different species of birds, which share the same habitat, but manage to use different times of the day to sing their songs. Therefore, they avoid a vocal competition and improve the conditions to find a mate. Communicative agents are used to model the birds and their behavior. A simple set of rules is used to make the decisions when and how to change the time for the search for a mate. By incorporating damping and amplifying feedback loops the collective behavior of each species led the system to a solution which was favorable to all agents.

1 Introduction

Emergent behavior can be seen in any type of system where behavior is determined not by a global control but instead by the interactions of decentralized entities [1] [2]. Many examples of emergent behavior can be found in biological systems, as for example in insect colonies like ants, bees, or termites [3] [4] [5].

To understand this collective behavior computer simulations are used to examine the parameters which are important for such a system [6]. Reynolds showed with his boids-simulation that a relatively simply behaving group of birds is capable of forming a flock without a centralized control [7].

In this paper the mating behavior of birds is examined. Male birds sing songs to attract females[1]. Most bird species do this at dawn [8]. But if all species, which share the same habitat, would choose the same time of day to sing their songs there would be likely a lot of vocal competition. This might be the reason why males of some species chose a different time of day to find a mate. Some species, like the nightingale, even use the night. In ref. [8] an even more sophisticated sharing of the time of day between two species is described:

> "Two birds of the Californian chaparral, the wrentit and the Bewick's wren, share the same habitat, and also have similar songs. But they manage to alternate their singing bouts so that they are never in direct conflict.
>
> The Bewick's wren starts the session in the morning and dominates the airways for the first hour. The wrentit stays quiet to avoid competition, but gets his turn sing in the second hour, after which the Bewick's wren starts again. And so this astonishing sharing of the stage goes on."

[1], defend their territory, and alarm others

To examine this obviously well coordinated behavior a simulation based on autonomous agents was used. Simple, communicative agents were used to imitate some of the behaviors of birds. They used a simple set of rules to make the decision when to change the time for their mating behavior. By incorporating damping and amplifying feedback loops the collective behavior of each species led the system to a solution which was favorable to all agents.

2 Simulation Setup

Different species of communicative agents were set into the 2-dimensional simulation environment, where they could fly around, search for food and mate. The simulation was frame based. It simulated one day within 4800 frames.

While programming a complex systems some specific properties have to be ensured [9]. E.g. the system has to be open. Therefore, old agents were periodically replaced by new agents. The rate of replacements was controlled.

It was ensured that no agent had a complete view of the system. All agents could only gather data and act within their near vicinity.

3 Attributes of the Agents

The agents are modelled as simple as possible and as complex as necessary.

Each agent can freely fly around, search for food, and rest at night. The male birds can emit signals (sing a song). The signals are the same within a species but have significant differences to signals of other species. Female birds know the signal of their species from birth on. A signal can be heard within a limited range by other agents. Agents can see other agents and places with food only in a short distance. This ensures that the relationships between agents are short-range.

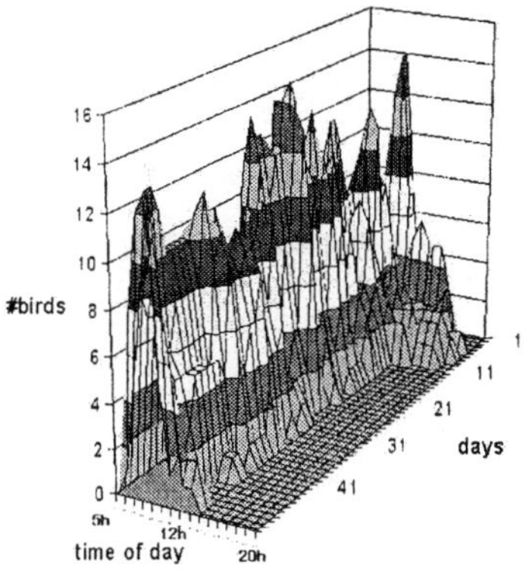

Fig. 1. Distribution of females of species A_2, which tried to fi nd a mate, over the day - observed for 50 days.

Each agent has a memory in which the signal of its species is stored. Additionally, an agent observes the activities of other (male) agents. It memorizes at what time of day how many signals of its own and of other species could be heard. The memory of these data fades out after a day and new observations have to be made.

The male birds have a value which defines their dominance. It ranges from 0 to 100 and is randomly chosen. This dominance is incorporated into their signals. Females can judge by the quality of the melody how dominant a male bird is. They tend to mate with more dominant males.

4 Behavior Model

Each communicative agent shows a specific behavior during the daytime. The behavior was simplified to three different phases: resting, finding food, and mating. A specific behavior was selected for each hour of the day. The resting phase could not be changed. The three hours for the mating behavior had to be in a row.

The duration for the mating behavior was set to three hours manually. In a more complex simulation it would depend on how much energy singing would cost and how much energy the food, found within a specific time, would provide.

At night all agents rest. During the day the agents would normally try to find food. At a specific time they would start to find a mate for three hours. After that, they would continue to find food again. Each agent was free to choose the time for mating. Why and when an agent decided to change its behavior is described in the next two subsections.

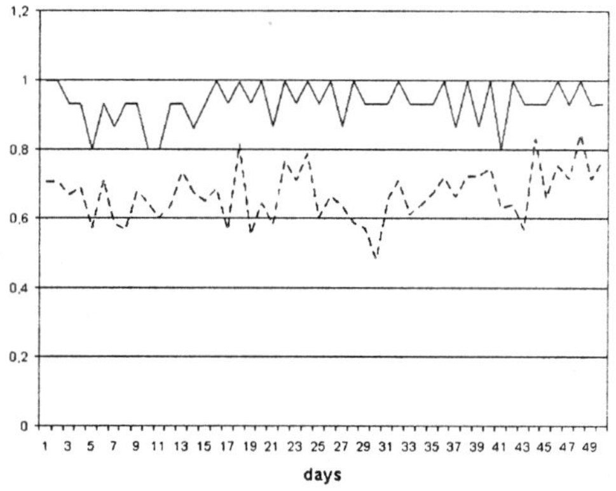

Fig. 2. Average success rate of male (dashed line) and female (straight line) birds of species A_2.

4.1 Behavior of Male Birds

During the period of mating a male bird flies around and periodically emits a signal (i.e. sings a songs) depending on a probability P_{send}. He waits for a response. If no female bird arrives, he moves on.

If his success decreases (no success over a time $T_{no\,success}$), he becomes less and less motivated and the probability P_{send} decreases by the factor $P_{decrease}$. Therefore, more successful birds sing more often.

If his success decreases over a longer period of time (D_{male}), he changes the time of day to find a mate. To find a better time of day he observes when males of his own species sing more often. He then imitates their behavior and starts his search for a mate just when they do it.

Another criterion to change the time for mating would be the vocal competition with other species. If during his mating time the number of foreign signals exceeds a threshold F_{max}, he would choose to change the time of day of his search. If he observed less foreign signals after his own mating time, he would try to find a mate even later; otherwise earlier.

The behavior of the male bird is kept quite simple: avoiding the time of day when too many birds of other species are active and imitating the more successful males of his own species.

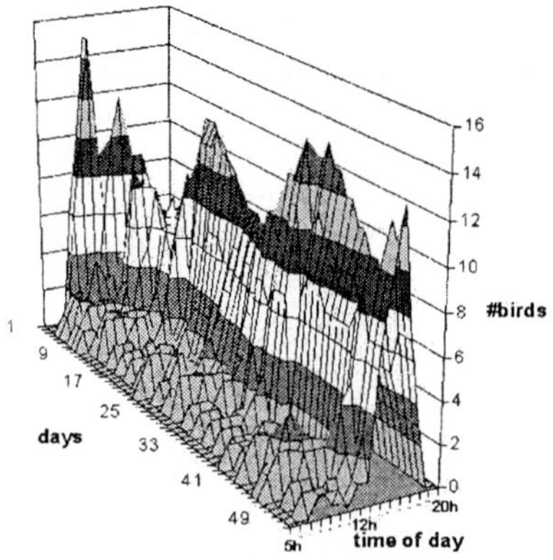

Fig. 3. Distribution of females of species B_2, which tried to fi nd a mate, over the day - observed for 50 days.

4.2 Behavior of Female Birds

During the daytime when female birds try to find a mate, they fly around and listen to signals. When they hear one, they compare it to the one they associate with their own species. If it is not similar, they just ignore it. But if it is similar they can decide to fly towards its source location. This decision is based on how dominant the male sounds. The more dominant he is, the more likely they choose to search for him. The minimal probability to fly to him is Q_{min}. If they find exactly one, they assume that he is the emitter and mate with him.

If females hear signals of their own species while they are not trying to find a mate, they just ignore them.

The criteria to change the time of day for their mating behavior are similar to those of the males. If they hear too many foreign signals ($> F_{max}$) during their mating time, they try to avoid this time of day and try to find a mate a little bit earlier or later.

If they have no success to find a male over a longer time (D_{female}), they observe when most signals of their own species can be heard and choose this time of day.

5 Simulation Scenarios

The absolute values used within these scenarios are not biologically founded. Most actions happen in a much shorter time than in nature. E.g. female birds tend to listen to a male's song for days instead of hours before deciding to mate with him.

Two different species A_2 and B_2 of 30 agents each (15 male, 15 female birds) were set into the simulated habitat. The simulation run lasted for 50 (simulated) days. By then the behavior of the two species had stabilized.

Every second day an agent was replaced with a new one to simulate an in-/out-flux of the system. If this frequency is increased too much, the system becomes instable, because the disturbances overweigh the time needed for the system to converge.

All agents of both species started with the same behavior pattern, to confront them with a situation which forced them to change their behavior. Resting from 9pm until 5am, then searching for food for most parts of the day. From 7am on, all agents started to search for a mate for three hours.

In a second scenario three different species A_3, B_3, and C_3 of 30 agents each (15 male, 15 female birds) were set into the habitat. The simulation run lasted for 50 days. All agents started again with the same behavior pattern.

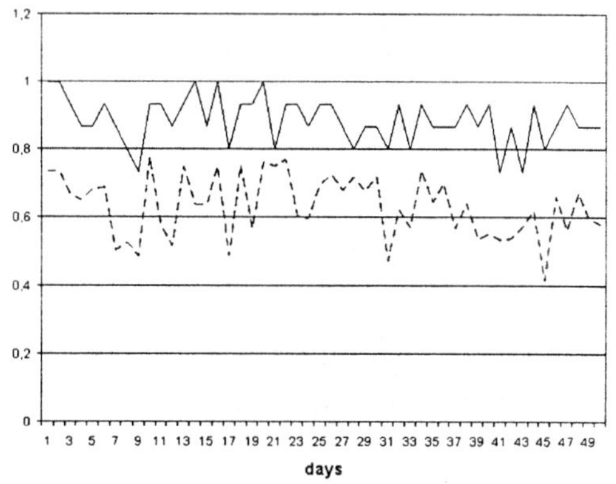

Fig. 4. Average success rate of male (dashed line) and female (straight line) birds of species B_2.

6 Examining the Parameter Space

The parameters used in the simulation were divided into two groups. Those which described the "physical reality" of the environment and those which influenced the behavior of the birds. The physical parameters were fixed.

There were too many parameters for the behavior, spanning a huge parameter space, to be analyzed analytically. Therefore, the parameter settings were examined with a genetic algorithm [10] [11].

These behavior parameters were grouped in a genome:

- Length of the time interval defining how long males would observe how many females wanted to mate with them ($T_{no\,success}$). If none arrived within this time, the male became less motivated to sing their song (decreasing P_{send})
- Number of days without success before changing behavior: D_{male}, D_{female} (separated parameters for females and males).

- Maximal number of foreign signals per hour before a bird thinks it is too noisy: F_{max}.
- Probability to send signals (initial & maximal value): P_{send}.
- Degradation of P_{send} if unsuccessful: $P_{decrease}$.
- Minimal value of P_{send}: P_{min}.
- Increase of P_{send} if successful: $P_{increase}$.
- Minimal probability that a female bird flies to the source of a signal: Q_{min}. Is added to dominance of the male bird divided by the maximal dominance.

As the fitness function the average mating success of each species was used. The population size[2] was set to 45 and the search lasted for 60 generations.

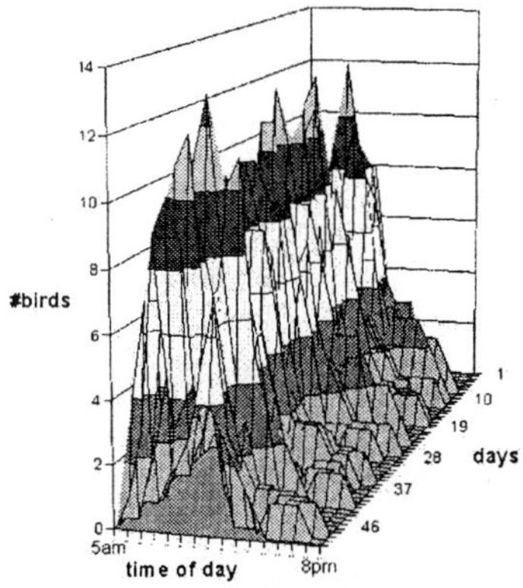

Fig. 5. Distribution of females of species A_3, which tried to fi nd a mate, over the day - observed for 50 days.

7 Results

7.1 Scenario with 2 Species

In the first scenario the females of both species started to change the time for their mating behavior after two days. Fig. 1 shows for species A_2 how many female birds at what time of day tried to find a mate over a period of 50 days (Fig. 3 for species B_2). They spread over several hours. But most females of species B_2 started to use the later hours of the day while species A_2 remained in the early hours. The males behaved in a similar way.

After day 10 the females of species A_2 started to reduce the span. And after day 13 a collective behavior emerged to only use the time from 6am to 1pm. The males followed them in a short distance.

[2]the number of simulations running parallel

Finally, species B_2 exploited the later hours while species A_2 tended to mate in the earlier hours of the day.

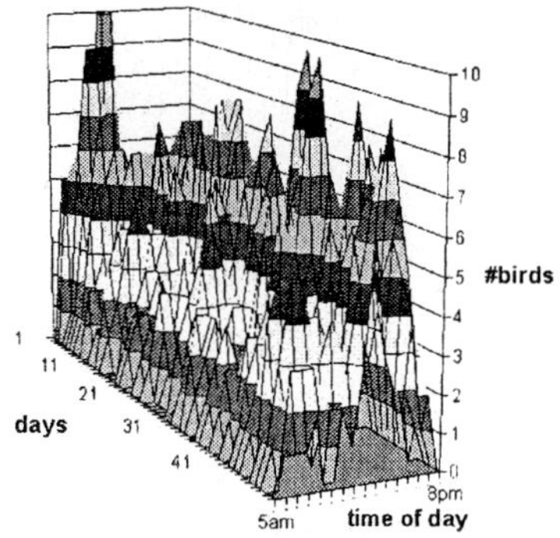

Fig. 6. Distribution of females of species B_3, which tried to fi nd a mate, over the day - observed for 50 days.

In Fig. 2 the average success rate to find a mate is shown for female and male birds of species A_2. A value of 1 means that a bird mated successfully within 24h. In the first days the rates dropped because the females started to search later in the day for a mate. But after the males followed them, their success rate increased again. Fig. 4 shows the success rates for species B_2 which developed in a similar way, but could not come back to the optimum like species A_2. Not all birds shifted their search to the later time of day as can be seen in Fig. 3.

The value of the male success rate could never reach 1 because the less dominant males would rarely find a mate.

7.2 Scenario with 3 Species

Fig. 5 shows the distribution of females of species A_3 over the day, Fig. 6 for species B_3, and Fig. 7 for C_3. In the first days all three species had to change their daily routine. After 16 days species A_3 established a maximum in the early hours of the day, while species B_3 gathered most birds around 4pm after 20 days. Species C_3 spread between A_3 and B_3, but without stabilizing in a specific time slot after 50 days. Its maximum rather alternated between the earlier and the later hours. Still, the birds of C_3 avoided the maxima of A_3 and B_3. The males behaved in a similar way.

In Fig. 8 the average success rate to find a mate is shown for female birds of species A_3, B_3, and C_3. Since three species had to share the daytime – instead of two as in the first scenario – the success rates were slightly worse and could not return to the optimum.

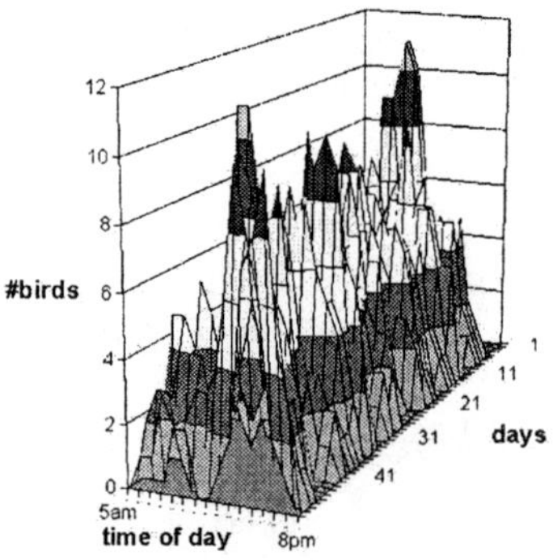

Fig. 7. Distribution of females of species C_3, which tried to fi nd a mate, over the day - observed for 50 days.

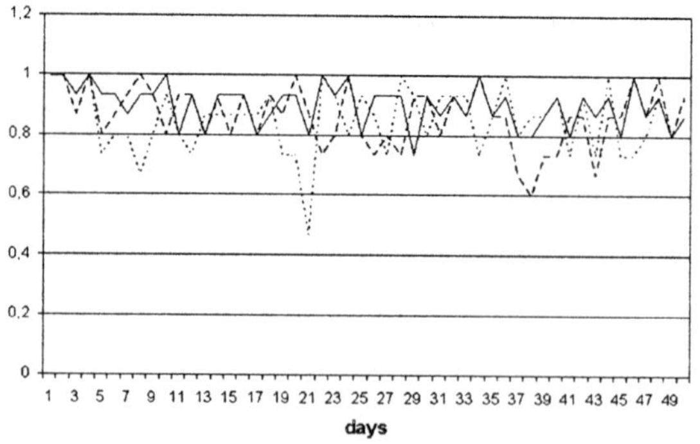

Fig. 8. Average success rate of female birds of species A_3 (straight line), B_3 (dashed line), and C_3 (dotted line).

8 Conclusion

The simulation does not claim to fit exactly for real birds. It is a simplified model to examine the idea of pushing groups which share the same habitat to use different time intervals for behaviors which would otherwise interfere.

Nevertheless, the simulation results provide an insight that under some conditions groups can adapt their behavior only by using feedback loops without any global control.

9 Outlook

The simulation could be made more complex. E.g. the duration for the mating behavior could depend on how much energy is used up during a day (while just flying around or singing) and how much energy the food, found within a specific time, provides. Since the fitness function of the genetic algorithm depends on the average mating success, the birds would have to search for an optimal solution, allowing them to search for a mate as long as possible, but still having enough time to find the needed food.

References

[1] M. Poggio and T. Poggio, "Cooperative physics of fly swarms: An emergent behavior," Tech. Rep., 1994.

[2] J. M. Epstein and R. Axtell, *Growing artificial societies: social science from the bottom up.* The Brookings Institution, 1996.

[3] C. Camazine, J. Deneubourg, N. Franks, J. Sneyd, G. Theraulaz, and E. Bonabeau, *Self-Organization in Biological Systems.* Princetown University Press, 2001.

[4] E. Bonabeau, M. Dorigo, and G. Theraulaz, *Swarm intelligence: from natural to artificial systems.* Oxford University Press, Inc., 1999.

[5] J. Kennedy, R. C. Eberhart, and Y. Shi, *Swarm Intelligence.* Morgan Kaufmann Publishers Inc., 2001.

[6] G. N. Gilbert and K. G. Troitzsch, *Simulation for the Social Scientist.* Taylor & Francis, Inc., 1999.

[7] C. W. Reynolds, "Flocks, herds, and schools: A distributed behavioral model," *Computer Graphics*, vol. 21, no. 4, pp. 25–34, 1987. [Online]. Available: citeseer.ist.psu.edu/reynolds87flocks.html

[8] G. H. Davies, "Bird songs." [Online]. Available: http://www.pbs.org/lifeofbirds/songs/index.html

[9] "Complex systems," Wikipedia. [Online]. Available: http://www.campusprogram.com/reference/en/wikipedia/c/co/complex_system.html

[10] J. H. Holland, *Adaptation in Natural and Artificial Systems.* Ann Arbor: University of Michigan Press, 1975.

[11] D. E. Goldberg, *Genetic Algorithms in Search, Optimization and Machine Learning.* Addison-Wesley Longman Publishing Co., Inc., 1989.

Integrating binding site predictions using meta classification methods

Y. Sun[1], M. Robinson[1], R. Adams[1], A. G. Rust[2], P. Kaye[1], N. Davey[1]
[1]Science and technology research school, University of Hertfordshire, United Kingdom
E-mail: { comrys, m.robinson, r.g.adams, p.kaye, n.davey}@herts.ac.uk
[2] Institute of Systems Biology, 1441 North 34th Street, Seattle, WA 98103, USA
E-mail:{arust@systemsbiology.org}

Abstract

Currently the best algorithms for transcription factor binding site prediction are severely limited in accuracy. There is good reason to believe that predictions from these different classes of algorithms could be used in conjunction to improve the quality of predictions. In this paper, we apply single layer networks and support vector machines on predictions from 12 key algorithms. Furthermore, we use a 'window' of consecutive results for the input vectors in order to contextualise the neighbouring results. Moreover, we improve the classification result with the aid of under- and over- sampling techniques. We find that by integrating 12 base algorithms, support vector machines and single layer networks can give better binding site predictions.

1 Introduction

In this paper, we address the problem of identifying transcription factor binding sites on sequences of DNA. There are many different algorithms in current use to search for binding sites [1, 2, 3, 4]. However, most of them produce a high rate of false positive predictions. The task addressed here is to reduce false positive predictions by means of classification techniques from the machine learning field.

To do this we first integrate the results from 12 different algorithms for identifying binding sites, using non-linear classification techniques. To further improve classification results, we employ windowed inputs where a fixed number of consecutive results are used as an input vector, so as to contextualise the neighbouring results. The datasets include two classes labeled as either binding sites or non-binding sites with about 93% being non-binding sites. We make use of sampling techniques, working with a traditional neural network: single layer networks (SLN) and a contemporary classification algorithm: support vector machines (SVM).

We expound the problem domain in the next section. In Section 3, we introduce the datasets used in this paper. We explain how we apply under- and over- sampling techniques in Section 4. Section 5 briefly introduces our experiments and gives all the experimental results. The paper ends in Section 6 with a conclusion.

2 Problem Domain

One of the most exciting and active areas of research in biology currently, is understanding how the exquisitely fine resolution of gene expression is achieved at the molecular level. It is clear that this is a highly non-trivial problem. While the mapping between the coding region of a gene and its protein product is straightforward and relatively well understood, the mapping between a gene's expression profile and the information contained in its non-coding region is neither so simple, nor well understood at present. It is estimated that as much as 50% of the human genome is cis-regulatory DNA [5] , undeciphered for the most part and tantalisingly full of regulatory instructions. Cis-regulatory elements form the nodes connecting the genes in the regulatory networks, controlling many important biological phenomena, and as such are an essential focus of research in this field [6].

It is known that many of the mechanisms of gene regulation act directly at the transcriptional or sequence level, for example in those genes known to play integral roles during embryogenesis [6]. One set of regulatory interactions are those between a class of DNA-binding proteins known as transcription factors and short sequences of DNA which are bound by the proteins by virtue of their three dimensional conformation. Transcription factors will bind to a number of different but related sequences.

The current state of the art algorithms for transcription factor binding site prediction are, in spite of recent advances, still severely limited in accuracy. We show that in a large sample of annotated yeast promoter sequences, a selection of 12 key algorithms were unable to reduce the false positive predictions below 80%, with between 20% and 65% of annotated binding sites re-

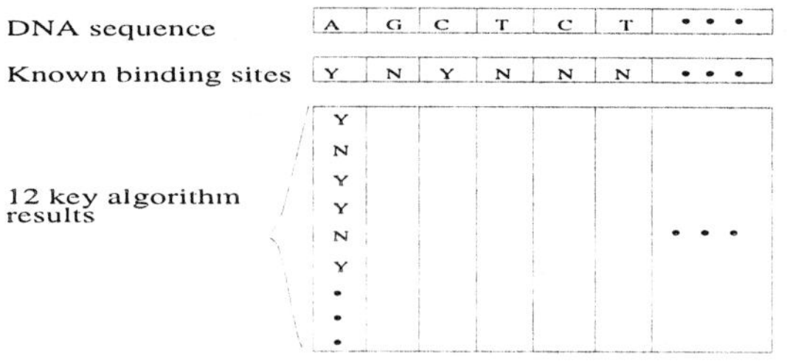

Fig. 1. The dataset has 68910 columns, each with a possible binding prediction (Y or N). The 12 algorithms give their own prediction for each sequence position and one such column is shown.

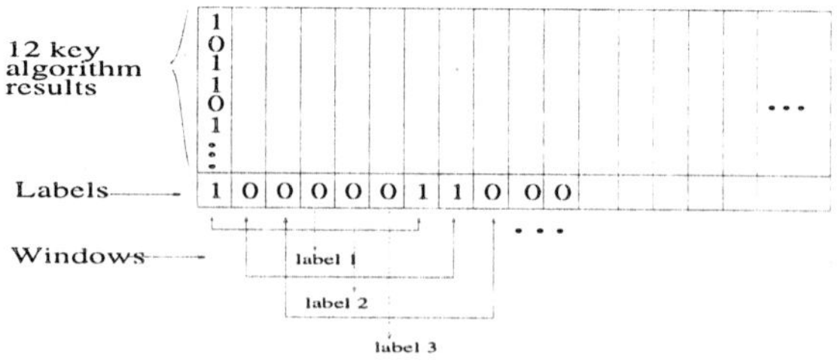

Fig. 2. The window size is set to 7 at the current study. The middle lable of 7 continuous possible binding sites is the label for a new windowed input. The length of each windowed input now is 12×7.

covered. These algorithms represent a wide variety of approaches to the problem of transcription factor binding site prediction, such as the use of regular expression searches, PWM scanning, statistical analysis, co-regulation and evolutionary comparisons. There is however good reason to believe that the predictions from these different classes of algorithms are complementary and could be integrated to improve the quality of predictions.

In the work described here we take the results from the 12 aforemention algorithms and combine them in 2 different ways, as shown in next section. We then investigate whether the integrated classification results of the algorithms can produce better binding site predictions than any one algorithm alone.

3 Description of The Dataset

The dataset is from a large sample of annotated yeast promoter sequences. It has 68910 possible binding positions and a prediction result for each of the 12 algorithms, see Figure 1. The 12 algorithms can be categorised into higher order groups as Single sequence algorithms (7) [1, 7, 8, 9]; Coregulatory algorithms (3) [2, 10]; Comparative algorithm (1) [3]; Evolutionary algorithm (1) [4].

The label information contains the best information we have been able to gather for the location of known binding sites in the sequences. Throughout we have used the following notation: 0 denotes the prediction that there is no binding site at this location; 1 the predictions that there is a binding site at this location, while 0.5 indicate that this algorithm is not able to analyse this sequence. Two datasets are generated based on the original set.

- *Single Inputs:* We take the results from the 12 algorithms and combine them into one 12-ary feature vector as a single input. All repeated and inconsistent data were removed and 1984 data points are left.

  We then randomly choose $2/3$ of the data points for training, while $1/3$ remained for testing.

- *Windowed Inputs:* To do this, we use a 'window' of consecutive results in the input vector in order to contextualise the neighbouring results, see Figure 2. In the current work, we set the window size $= 7$.

  We use the first $2/3$ of the data points for training. By using windowed inputs, after removing repeated and inconsistent data, 36247 out of 45933 data points are left. In this part, we do cross-validation first to choose a better classification algorithm, then train the selected classifier using the whole 36247 data points. For testing, we do experiments on the final third of the data points first without repeated and inconsistent data points, and second with full continuous set of rows for analysis by biologists.

4 Sampling techniques for Imbalanced Dataset Learning

In our dataset, there are less than 10% binding positions among the total vectors, so this is an *imbalanced* dataset [11]. Since the dataset is imbalanced, the supervised classification algorithms will be expected to tend to over predict the majority class, namely the non-binding site category. In this work, to cope with imbalanced data, we concentrate on the data-based method, i.e., using under-sampling of the majority class (negative samples) and over-sampling of the minority class (positive samples). We combine over-sampling and under-sampling methods in our experiments. For the under-sampling, we randomly selected a subset of data points from the majority class.

The over-sampling case is more complex. In [11], the author addresses an important issue that the class imbalance problem is only a problem when the minority class contains very small subclusters. This indicates that simply over sampling with replacements may not significantly improve minority class recognition. To overcome this problem, we apply a synthetic minority oversampling technique as proposed in [12]. For each patten in the minority class, we search for its $K-$nearest neighbours by computing Hamming distances. A new pattern belonging to the minority class will be generated by employing the majority voting principle to this pattern and its $K-$nearest neighbours in the feature vector space. For simplicity, in the following experiments, we set $K = 5$.

5 Experiments

The classification techniques we used are single layer network (SLN) [13], and support vector machine (SVM)[14].

The SVM experiments were completed using LIBSVM, which is available from the URL
`http://www.csie.ntu.edu.tw/~cjlin/libsvm`.
The others were implemented using the NETLAB toolbox, which is available from the URL
`http://www.ncrg.aston.ac.uk/netlab/`.

All the user-chosen parameters are attained by using cross-validation.

In the training dataset, the positive examples, namely binding sites, are double over-sampled. When combing with under-sampling, we selected a number of negative examples using cross-validation from the training dataset.

5.1 Common performance metrics

It is apparent that for a problem domain with an imbalanced dataset, classification accuracy rate is not sufficient as a standard performance measure. Instead *recall, precision and F-value* [15, 16] are calculated to understand the performance of the classification algorithm on the minority class. Based on the confusion matrix computed from the test results (see Table 1, where TN is the number of true negative samples; FP is false positive samples; FN is false negative samples; FP is true positive samples), several common performance metrics can be defined as follows:

$$R. = \text{Recall} = \text{TP} / (\text{TP} + \text{FN}), \quad (1)$$

$$P. = \text{Precision} = \text{TP} / (\text{TP} + \text{FP}), \quad (2)$$

$$F. = \text{F-score} = \frac{2 \cdot Recall \cdot Precision}{Recall + Precision}, \quad (3)$$

$$Accu. = \text{Accuracy} = \frac{TP + TN}{TP + FN + TN + FP}, \quad (4)$$

$$fp = \text{false positive rate} = \frac{FP}{FP + TN}. \quad (5)$$

Table 1. A confusion matrix.

TN	FP
FN	TP

5.2 Classification results

5.2.1 Classification results with Single Inputs: The results are shown in Table 2. Comparing with the best base algorithm, the SLN and SVM increase the F-score by 16.3%, and 36.6%, respectively; and decrease fp by 20.6%, and 48.8%, respectively. It can be seen that the SVM outperforms the SLN on this dataset. The integrated predictor is much better than the single best algorithm.

Table 2. Common performance metrics (%) calculated from confusion matrices on consistent single testing set.

Classifier	R.	P.	F.	Accu.	fp
best Alg.	74.2	20.4	32.0	70.4	30.1
SLN	75.8	24.7	37.3	76.1	23.9
SVM	69.4	31.9	43.7	83.2	15.4

5.2.2 Classification results with Windowed Inputs: All experimental results are presented in Table 3. The top half of the table gives the results for the consistent windowed data, as described in section 3. It can be seen that for both SLN and SVM, the F-scores are increased and fp are decreased when compared with the single best base algorithm on the consistent test set. And again, the SVM outperforms the SLN on this dataset.

For full test set, which includes inconsistent data points, it can also be seen that for the SVM the F-score is increased, whilst the SLN keeps the same value as the best base algorithm. On the other hand, the SVM reduces the fp by 40%, while the SLN does 15.5%.

Table 3. Common performance metrics (%) calculated from confusion matrices on consistent and full window test sets.

test set	Classifier	R.	P.	F.	Accu.	fp
consistent	best Alg.	39.2	14.4	21.0	86.0	11.6
	SLN	38.3	16.6	23.2	88.0	9.6
	SVM	36.8	21.6	27.3	90.7	6.6
full	best Alg.	37.2	17.2	23.5	86.1	11.0
	SLN	33.5	18.1	23.5	87.4	9.3
	SVM	32.5	23.3	27.1	89.9	6.6

6 Conclusions

The first significant result presented here is that by integrating the 12 algorithms we can considerably improve binding site prediction. In particular the SVM increases the F-score by 36.6% from the best base algorithm and decreases false positive rate by 48.8%. The second significant result is that windowing of the input vectors, thereby joining additional contextual information from the genome. Moreover the windowing of the data removes many of the inconsistent and repeated data points and thereby gives a much larger training and test set. As expected the SVM gave better classification results than the SLN.

Future work will investigate i) using real valued algorithm results in the input vector; ii) using algorithm based technologies to cope with the imbalanced dataset; iii) considering a range of supervised meta-classifiers or ensemble learning algorithms.

References

[1] *http://www.hgmp.mrc.ac.uk/Software/EMBOSS/.*

[2] Bailey, T.L. & Elkan, C. (1994) Fitting a mixture model by expectation maximization to discover motifs in biopolymers. Proceedings of the Second International Conference on Intelligent Systems for Molecular Biology: 28-36, AAAI Press.

[3] *http://family.caltech.edu/SeqComp/index.html.*

[4] Blanchette, M. & Tompa, M. (2003) FootPrinter: a program designed for phylogenetic footprinting. Nucleic Acids Research: Vol. 31, No. 13, 3840-3842.

[5] Markstein, M., Stathopoulos, A., Markstein, V., Markstein, P., Harafuji, N., Keys, D., Lee, B., Richardson, P., Rokshar, D., Levine, M. (2002) Decoding Noncoding Regulatory DNAs in Metazoan Genomes. Proceeding of 1st IEEE Computer Society Bioinformatics Conference (CSB 2002), Stanford, CA, USA.

[6] Arnone, M. I. and Davidson, E. H.(1997) The hardwiring of development: Organization and function of genomic regulatory systems. Development: 124, 1851-1864.

[7] Apostolico, A., Bock, M.E, Lonardi, S., & Xu, X.(2000) Efficient Detection of Unusual Words. Journal of Computational Biology: Vol.7, No.1/2.

[8] Rajewsky, N., Vergassola, M., Gaul, U. & Siggia, E.D. (2002) Computational detection of genomic cis regulatory modules, applied to body patterning in the early Drosophila embryo, BMC. Bioinformatics: **3**:30.

[9] Thijs, G., Marchal, K., Lescot, M., Rombauts, S., De Moor B, RouzP, & Moreau, Y. (2001) A Gibbs Sampling method to detect over-represented motifs in upstream regions of coexpressed genes, Proceedings Recomb'2001: pp. 305-312.

[10] Hughes, J.D., Estep, P.W., Tavazoie, S., & Church, G.M. (2000) Computational identification of cis-regulatory elements associated with groups of functionally related genes in Saccharomyces cerevisiae. Journal of Molecular Biology Mar 10;**296**(5): 1205-1214

[11] Japkowicz, N. (2003) Class imbalances: Are we focusing on the right issure? Workshop on learning from imbalanced datasets, II, ICML, Washington DC.

[12] Chawla, N. V., Bowyer, K. W., Hall, L. O. and Kegelmeyer, W. P. (2002) SMOTE: Synthetic minority over-sampling Technique. *Journal of Artificial Intelligence Research.* Vol. 16, pp. 321-357.

[13] Bishop, C.M. (1995) Neural Networks for Pattern Recognition. Oxford University Press, New York.

[14] Scholköpf, B and Smola, A. J. (2002) Learning with Kernels: Support Vector Machines, Regularization, Optimization, and Beyond, The MIT Press.

[15] Buckland, M. and Gey, F. (1994) The relationship between Recall and Precision. Journal of the American Society for Information Science: Vol. 45, No. 1, pp. 12–19.

[16] Joshi, M., Kumar, V., and Agarwal, R. (2001) Evaluating Boosting algorithms to classify rare classes: Comparison and improvements. First IEEE International Conference on Data Mining, San Jose, CA.

Reverse engineering gene networks with artificial neural networks

A.Krishna , A.Narayanan , E.C.Keedwell

Bioinformatics Lab, Washington Singer Laboratories, School of Biological and Chemical Sciences, School of Engineering, Computer Science and Mathematics
University of Exeter EX4 3PT, UK
E-mail: {A.Krishna, A.Narayanan, E.C.Keedwell}@exeter.ac.uk

Abstract

Temporal gene expression data is of particular interest to researchers as it can be used to create regulatory gene networks. Such gene networks represent the regulatory relationships between genes over time and provide insight into how genes up- and down-regulate each other from one time-point to the next (the Biological Motherboard). Reverse engineering gene networks from temporal gene expression data is considered an important step in the study of complex biological systems. This paper introduces sensitivity analysis of trained perceptrons to reverse engineer the gene networks from temporal gene expression data. It is shown that a trained neural network, with pruning (gene silencing), can also be described as a gene network with minimal re-interpretation, where the sensitivity between nodes reflects the probability of one gene affecting another gene in time. The methodology is known as the Neural Network System Biology Approach with Gene Silencing Simulations (NNSBAGSS). The methodology was applied to artificial temporal data and rat CNS development time-course data.

1) Introduction

Temporal gene expression analysis is an active area of research and is undergoing a transition in response to a shift from traditional biology to systems biology due to advances in high throughput technologies (see Figure 1). Initial approaches to the problem of temporal gene expression analysis include cluster analysis, principal component analysis (PCA) and singular value decomposition (SVD), as well as regression modeling and Bayesian Networks. All these methods perform well in identifying classificatory relationships but struggle to extract causal relationships (regulatory rules). Artificial neural networks (ANNs) [1] have been shown to be effective in extracting classificatory knowledge from non-temporal gene expression data, and an ANN-Genetic Algorithm hybrid [2] has also been applied to temporal data. One of the problems in the application of artificial neural networks in this research area

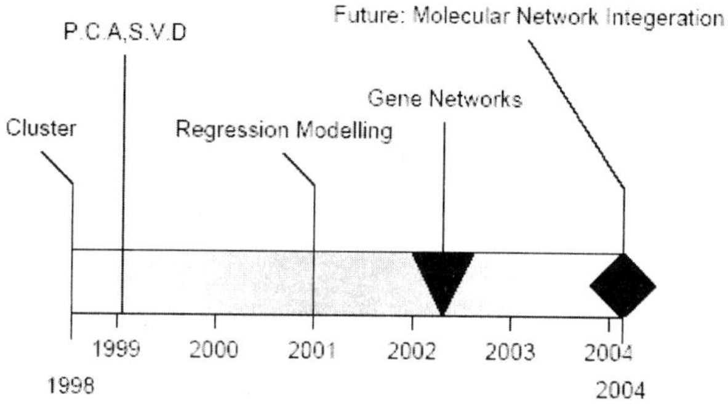

Figure 1 Time Line of Temporal gene expression analysis research

concerns the difficulty in creating complex regulatory networks from the analysis of high-dimensional weight matrices that represent the individual connections between genes over time. Narayanan et al [1] have demonstrated the applicability of single-layer ANNs (perceptrons) for static (non-temporal) gene expression data for classification purposes, where weight values were used to identify the importance of genes for classification. This paper extends this approach by introducing sensitivity analysis as a central theme of the NNSBAGSS methodology discussed here. Sensitivity analysis is more amenable than weight analysis and can be extended to more complex neural networks, such as those that result from the analysis of large temporal gene expression datasets.

2) NNSBAGSS Methodology

The NNSBAGSS methodology, represented as a flowchart in Figure 2, is a method for reverse engineering the regulatory rules from temporal gene expression data. At the heart of this methodology lies the sensitivity analysis of the trained, best neural network. Sensitivity analysis measures the effect of small changes in the input channel (gene expression at time t) on the output (gene expression at time t+1). Sensitivity analysis is performed to determine the effect

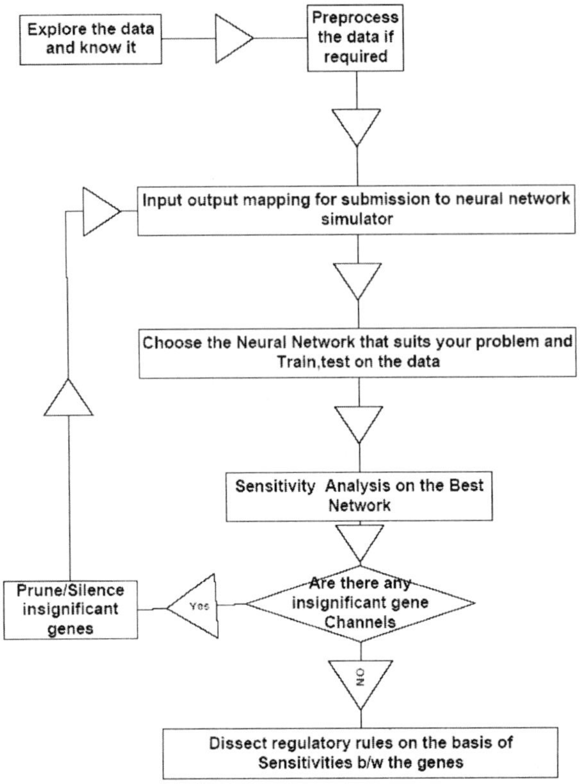

Figure 2 NNSBAGSS methodology

that each of the network inputs (gene expression values at one time step) has on the network output (gene expression at subsequent time step). This provides feedback as to which input channels are the most significant. From this, insignificant channels can be removed (pruned) from the input space (known as gene silencing). The first input is varied between its mean and ± 1 standard deviation while all other inputs are fixed at their respective means. The network output is computed for a number of steps above and below the mean. This process is repeated for each input. The variation of each output with respect to the variation in each input is analyzed and insignificant gene inputs can be silenced, resulting in regulatory rules being reverse engineered. Sensitivity analysis has an analogy with the systems biology approach described by Leroy Hood *et al* [3] in which they observe the effect of genetic and environmental perturbations in a system and then analyse the effect on other system components, thereby producing predictions and hypotheses about the system. A particular temporal gene expression experiment is regarded as a system. The algorithm perturbs the gene expression value of a gene at a previous time step to see the effect on the gene expression value of other genes at subsequent time steps, thereby revealing the inherent regulatory rules in the temporal data. All such *in silico* predictions must be checked by laboratory work. In the following experiments, the methodology was applied on artificial temporal Boolean data [1] and real world RT-PCR temporal data of rat CNS development [4].

2.1 Artificial Temporal Gene Expression data

2.1.1 Explore the data:

The data involved in this experiment is an artificial gene expression dataset forward engineered by a program which creates temporal Boolean Liang networks [5]. A Liang network consists of rules which are made of logical operators distributed randomly in a number of random fixed length logical expressions, but to a specific k value which determines the maximum number of genes that can affect others in the next time step. This data consists of only 2 k values. If k=2, then between 2 and 4 genes can be affected at time t+1 by genes at time t. When k=3, then anything between 3 and 9 genes are involved in the next time steps. The data contain 10 genes derived from Boolean rules with k values of 2 and 3. The data consists of 2^{10} (1024) records needed to specify every possible truth assignment for each dataset. The gene can either be on (1) or off (0). These 1024 records consist of the input values (10 gene expression values at time t) and the desired gene expression value (10 Gene expression values at time t+1). Thus our data consists of 1024 *20 expression values. This data was generated using a C^{++} programme and had the following 20 inherent rules between values at time t and the subsequent value at time t+1in the following format :3→1(If gene 3 is on then gene 1 goes on), 4&10→1 (if gene 4 and 10 are on then 1 is on)

Table 1 - Rules embedded in the artificial dataset

Rule	Arguments	Rule	Arguments
1	3→1	11	8→6
2	4 &10→1	12	5→6
3	6 &10→2	13	8 &7→7
4	4 &9→2	14	6 &10→7
5	5 &4→3	15	9 &3→8
6	9→3	16	4 &1→8
7	2→4	17	7→9
8	9→4	18	5 &4→9
9	6→5	19	5 &10→10
10	9 &6→5	20	1→10

If our method cannot reverse engineer approximately back to these rules for artificially constructed data, there is little chance of the method finding true rules in actual temporal gene expression data. Also note that a gene goes off at a time-step unless it is specifically switched on by some gene at the previous time-step.

2.1.2 Pre-process the data

The only pre-processing required for the artificial data is the randomisation of the input -desired pair and preparing the data for submitting to the neural network simulator[1] for learning the input-output function (the rules inherent in the data).

2.1.3 Choice of desired input-output mapping

The task that the ANN has to solve is a linear problem i.e. the artificial neural network has to learn the input (gene expression value at time t) to output (gene expression value at time t+1) mapping, where no non-linear relationships exist between input values and output value. There are 1024 exemplars of this input -output mapping.

2.1.4 Choice of neural architecture

Neural networks can be very powerful learning systems. However, it is very important to match the neural architecture to the problem. For the artificial temporal data we tried a number of single layer perceptrons (that is, full connectivity between the input nodes and output nodes, with no hidden layers), adopting different learning rule parameters and transfer functions. We followed the standard training, cross-validation and testing routine for perceptrons with different parameters. Out of 1024 randomised exemplars we used 60% (614) for training, 15% (154) for cross-validation and 25% (256) for testing. The best network with minimum Mean Squared Error for training as well as cross validation set was chosen. The best neural network architecture was a perceptron with the activation function as the hyperbolic Tanh function. This will squash the range of each neuron in the layer to between -1 and 1.

2.1.5 Sensitivity Analysis

For artificial temporal data a sensitivity analysis was performed on 624 training exemplars when the best network was trained on the data a number of times. Each input was varied between its mean and ± 1 standard deviation. The network output was computed for 50 steps below and above the mean and the sensitivity was calculated. The strongest weights during training were used, with the results represented as a graph. Figure 3 is a representation of the sensitivity matrix showing the sensitivity of all genes at previous time step over all other genes at next time steps over the 50 runs. It shows that the results remain more or less the same on repeated runs.

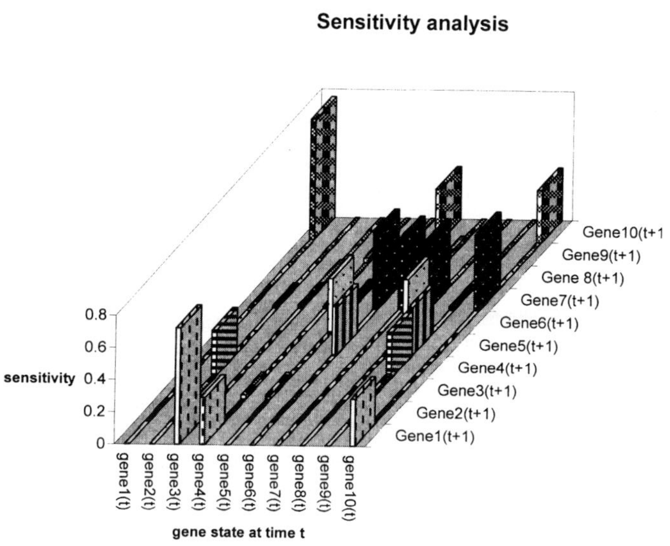

Figure 3: The x-axis represent gene state at time t, y- axis represents gene state at time t+1 and z-axis represents the sensitivity. For example, gene 1 has a strong sensitivity of (0.7) on gene 10 at next time step, thereby reverse engineering rule 20

2.1.6 Pruning/Reverse Engineering:

By analysing Figure 3 and the sensitivity matrix (not shown) all the embedded rules were reverse engineered by the approach. This can be seen most clearly in the first column of "Gene1(t+1)", where gene 3 has a large affect and genes 4 and 10 have a less marked affect. This corroborates the rules seen earlier as gene 3 can activate gene 1 by itself (Rule 1), whereas it requires the combined effort of gene 4 and gene 10 (Rule 2) to accomplish this. The sensitivity analysis therefore not only defines regulatory behaviour, but also the relative strengths of the regulation. Therefore, this demonstrates that our method is able to reverse engineer the embedded rules from artificial gene expression data.

2.2 Rat C.N.S development data:

The Rat (Sprague-Dawley albino rat) data is an RT-PCR study of 112 genes each measured on cervical spinal cord tissue in triplicate at nine different time points during the development of the rat central nervous system [4]. This gene expression data is accepted to be non-noisy, small and accurate, and is ideal for testing a new strategy because of previous work in literature. There are nine

[1] There are various neural network simulators available, Publicly available neural simulators are **SNNS,PDP^{++},Genesis,Joone** to name a few. Commercially available neural network simulators are **Neurosolutions, Matlab Neural Network Tool Box**.

different time points in the CNS development study namely E11, E13, E15, E18, E21, P0, P7, P14, A.
The input output mappings were E11 input -E13 output, E13 Input- E15 output, E15 input-E18 output,
E18 input-E21 output, E21 input-P0 output, P0 input-P7output, P7 input -P14 output, P14 input -A output.
There are eight exemplars for network to learn from. Each exemplar has 112 input gene expression values of the prior time step and 112 desired gene expression output values of the subsequent time step. The NNSBAGSS methodology was applied on this real world temporal data. The training, testing and crossvalidation regime was followed to choose the best network. The sensitivity analysis was performed on the perceptron trained on all eight exemplar pairs (9 time steps). Each input was varied between its mean and ±1 standard deviation. The network output was computed for 50 steps below and above the mean and the sensitivity was calculated . The best weights during the training were used and the results are represented as a graph. The graph in Figure 4 is the representation of sensitivity matrix showing the global sensitivity of all 112 genes at the previous time step affecting 112 genes at the next time step. Again these remain similar over the repeated runs.

3) Analysis

Some of the important rules from the real-world problem are as follows. The most strongly affecting gene according to sensitivities are *Ins1_RN1NS1*, which affects *Cellubrevin* (25.5), *Nestin* (15.9), *Grg1_RNgaba* (41.0243, the strongest signal in Fig4), *GRg2_RNGABA* (28). Gene *Ins1_RN1NS1,* involved in peptide signalling, has a very strong influence on the gene expression system. The gene expression of this gene is prominent only in the initial time points of E11-E13. It may be the case that the gene being switched off starts the process of cervical spinal cord development [4]. The other prominent cause-effect relationship as evident from the sensitivity analysis is *nAChRd_RNZCRD1* affecting nestin (21.77), thus corroborating previous suggestions that cholinergic transmission forms the leading edge in neurotransmitter signalling in cervical spinal cord [4]. Overall the sensitivity analysis is able to extract the regulatory importance of some of the genes described as activators of the CNS development programme and mentioned in the literature as the genes of study in spinal cord injuries because of their importance in activation of CNS development.

4) Conclusion

The reverse engineering of gene networks and causal relationships between genes is the major obstacle to extracting system knowledge from gene expression data. Our experiments demonstrate that a neural network approach, combined with sensitivity analysis, can reverse engineer both actual and biologically plausible rules from artificial and real-world data, respectively. While more work is required to determine why sensitivity analysis works in the way it does, the results presented here provide evidence of a novel, alternative approach to reverse engineering that can lead to automatic extraction of rules from temporal gene expression data. The process is simple and repeatable, with clear visualisations resulting from the method.

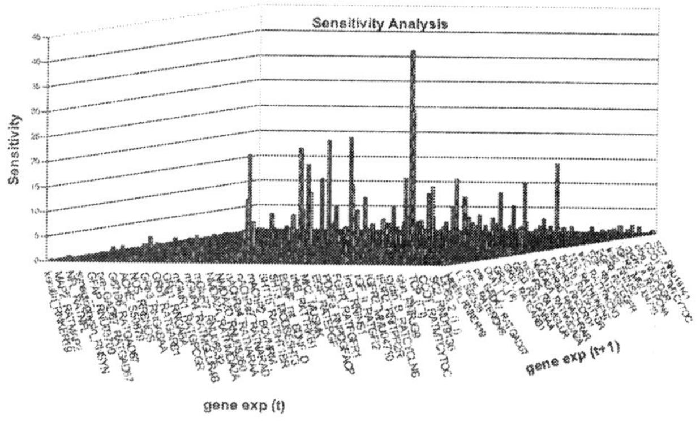

Figure 4: Sensitivity analysis for 112 genes in Rat C.N.S development. Only a small part of the full sensitivity analysis is provided here.

5) References

[1] A.Narayanan, E.C.keedwell, S.S.Tatineni and J.Gamalielsson, (2004) "Single-layer artificial neural networks for Gene expression analysis" *Neurocomputing*, Vol 61. pp 217-240.

[2] E.C.Keedwell and A.Narayanan, (2003) "Genetic algorithms for gene expression analysis." *Applications of Evolutionary Computation: Proceedings of the 1st European Workshop on Evolutionary Bioinformatics*, G. Raidl et al., Springer Verlag LNCS 2611, pp76-86.

[3] T.Ideker, V.Thorsson,J.A.Ranish, R.Christmas, J.Buhler, J.K.Eng, R.Bumgarner, D.R.Goodlett, R.Aebersold and L. Hood, (2001) "Integrated Genomic and Proteomic analysis of a systematically perturbed metabolic networks" *Science* 292 pp929-934.

[4] X.Wen, S.Fuhrman, G.D.Michaels, D.B.Carr, S.Smith, J.L.Barker, R.Somogyi, (1998) "Large-scale temporal Gene expression mapping of central nervous system development", *PNAS*, Vol95, pp. 334-339.

[5] S. Liang, S. Fuhrman, R. Somogyi, (1998) "REVEAL, a General reverse engineering algorithm for inference of Genetic network architectures," *Pacific Symposium on Biocomputing*, Vol. 3, Hawaii, USA, pp. 18–29.

A Meta-Level Architecture for Adaptive Applications

Fabrício J. Barth, Edson S. Gomi
Laboratory of Knowledge Engineering (KNOMA)
Polytechnic School. The University of São Paulo, Brazil
E-mail: {fabricio.barth, edson.gomi}@poli.usp.br

Abstract

The goal of this work is to investigate meta-level architectures for adaptive systems. The main application area is the user modeling for mobile and digital television systems. The results of a set of experiments performed on the proposed architecture showed that it is possible to reuse the components responsible for user modeling if they are designed as meta-level components.

1 Introduction

Currently most applications are developed for a large variety of users. An important issue is how to create adaptive applications that enable systems to satisfy heterogeneous needs. An adaptive system is a system with the ability of adapting its own behavior (with respect to performance and functionalities) to individual users needs. Some examples of system functions that can be adapted are: information retrieval, product recommendation, learning support, and user interfaces. Beyond the set of basic functions, systems may implement mechanisms that permit automatic development of a user profile, representing it in a user model. A user model is a knowledge representation of the preferences which determine the user's behavior. Preferences are the basic information necessary to provide system adaptability.

In traditional approaches, user modeling components have been created integrated with other system elements, without a specific component responsible for it. With the growing complexity of the adaptive systems, developers need new methodologies, concepts and techniques to overcome design difficulties and simplify the development process.

In this work we propose a meta-level architecture to separate the components responsible for the acquisition and manipulation of user modeling from other components of the adaptive system. This architecture has base-objects in its lower level and meta-objects in its upper level. Meta-objects can be defined as objects that describe the user model. The meta-objects can control a base-object, in order to modify its structure and behavior at execution time. This type of control is possible due to the reflective computing concept. Reflective computing is defined as the activities that a computational system execute over itself, in a different way from other processes, to solve their own problems and to search information about their own processes in real time.

To validate these architectures, experiments with interface adaptive systems will be conducted. The adaptation capacity, the reusability degree and the reduction in complexity of the implemented systems will be measured.

This paper is organized as follows. In section 2 the concepts of adaptive systems and user modeling are presented; in section 3 the concepts of meta-level architecture and reflective computing are described; in section 4 we propose an architecture that joins the meta-level concepts and the adaptive systems; in section 5 the implementation of a mobile system with adaptive interface is described. This system uses the architecture proposed in this paper. Finally, in section 6, the final considerations are presented.

2 Adaptive Systems

An adaptive system can be defined as a system with the ability to adapt its own behavior (i.e. functionalities) to the individual needs, expectations and abilities of the users [1]. This kind of system should be able to implement mechanisms that allow the creation and storage of the user model.

A user model is composed of the knowledge of the individual preferences that determine the user's behavior. Preferences are composed of information that is directly necessary to the adaptation of the system's behavior to the user's interests [2, 3]. For example, how many times, how frequent and for how long a user accesses one link in a web page.

Moreover, the user model may have personal information about the user such as age and profession. This information is not directly necessary for adapting a system to the user, but can be utilized to create user stereotypes. This allows the system to anticipate some of the user's behaviors [4, 5, 6]. Systems with that type of anticipation has shown benifits to both users and service providers [7]. However the development and maintenance of those personalizable systems are very complex and expensive, which suggests the need of a new approach for such undertaking [8].

In the majority of the related research works, user modeling is performed by the application system, and often there is no clear distinction between the system components that are used for user modeling purposes and the components that perform others tasks [8, 9].

In this context, we believe that the utilization of a meta-level architecture will help in the development of adaptive systems.

3 Meta-Level Architecture

Computational reflection is the activity executed by a computational system when the system computes (and possibly affects) its own computations. Reflection is defined as a way of introspection, in which the system tries to draw conclusions about its computations and eventually influence them [10].

Computational reflection defines a new software architecture paradigm. This architecture model is composed of a base-level, where we find the system objects that implement the system functionalities, and the meta-level, where we find the data structures and actions to be performed over the base-level objects [11]. The domain limits is the most interesting aspect of the reflexive architecture, not only because it permits the constructions of adaptive systems but also because it stimulates the reuse of components. The main point is to allow the application programmer to concentrate on the solution of a specific problem of the application domain. In this aspect, the meta-level architectures have been adopted to express non functional characteristics of the system, such as reliability and security, in an independent way from the application domain.

The basic concepts of computational reflection are [12]:

1. to separate the basic facilities from the non basic facilities through architectural levels;
2. the basic facilities must be satisfied by the objects of the application;
3. the non basic facilities must be satisfied by meta-objects;
4. the non functional capacities are added to an application object through its specific meta-objects, and;
5. the base-object can be structuraly and behavioraly changed at execution or compilation time.

In figure 1, we can see the reflection process in a computational system. The computational system is divided in two or more computational levels. The user sends a message to the computational system. The message is treated by the functional level, which is responsible for performing the work correctly, while the non functional level manages the work of the functional level.

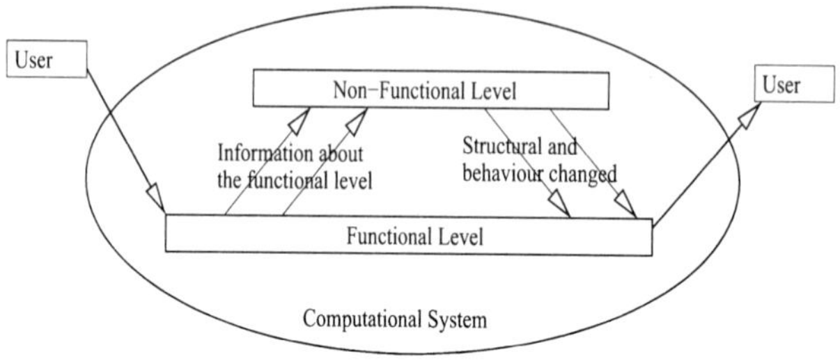

Fig. 1. Reflective Computing System Overview

With this ability a system is more capable of changing and adapting its structure and behavior. By separating the base-level functions from the meta-level functions we increase reusability, diminish complexity, and make the system more flexible overall.

Computational reflection can be implemented by functional, logic or imperative programming. However, it is with the natural flexibility of the object model that computational reflection has shown its effectiveness and elegance in the resolution of programming problems [13].

In computational reflection, classes, methods, attributes and objects representations are redefined by meta-classes and meta-objects. The level in which the meta-classes and meta-objects are available is called meta-level.

4 Proposed architecture

We propose an architecture for adaptive applications that use the meta-level and reflection concepts. The base-level contains all the basic functionalities (component 1 - figure 2) while the meta-level holds all the objects that control the creation and adjustment of the user models and stereotypes (component 2 - figure 2).

The meta-level has enough information to create and adapt the user models, stereotypes or any other structure necessary to adapt the system to the user. This is possible because the intersection concept (point (a) - figure 2) provides all the base-level system data (structure, behavior and user behavior) and observations of user bahavior. In this way, machine learning methods (mainly classification) process the data and offer support for resolution of decision-based problems.

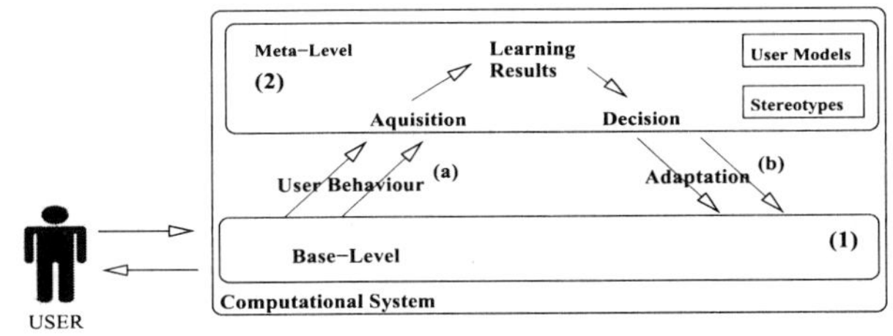

Fig. 2. Adaptive System Architecture

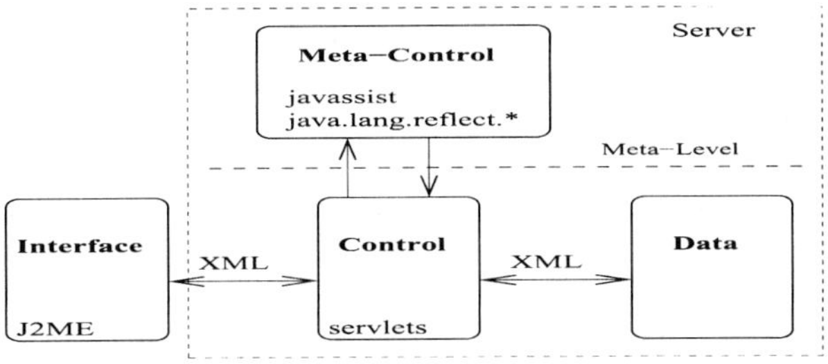

Fig. 3. Architecture of experiment

The decision can be used through the introspection ability (point (b) - figure 2) for changing the system functionalities (describes and executed by base-level) transforming the system and adapting it to the user's needs.

This architecture allows management of the functional aspects separately from the ones related to the acquisition and maintenance of the user's models. This is possible because the interface between base-level and meta-level is unique. Meta-level components can be removed and added to the architecture without modification to the base-level.

5 Application Example

To verify the adherence of the meta-levels architecture in adaptive systems, we implemented an adaptive system that uses the proposed architecture.

In this example a book store system (client and server) is implemented. The system is accessed by mobile equipments (cell phones) that use J2ME (Java Micro Edition) technology [14].

When the user is placing a book order using his cell phone, the following steps are executed:

1. he is identified by the cell phone number;
2. he searches the book by the book's name or by the author's name in the book's catalog;
3. he selects the desired book or books;
4. he adds the book to his "shopping cart";
5. he executes another search and adds more books to his "shopping cart";
6. he confirms the purchase, and;
7. he receives a message to confirm the purchase.

To attend the customers, the server side is implemented using the Java Servlets Technology [14]. Basically, the server is capable of keeping the "shopping carts" of the users, answering queries and finalize purchases. The book's database is written in XML. The system architecture can be visualized in figure 3.

5.1 Meta-Level

While the base-level is responsible for all functionalities described before, the meta-level is responsible for the acquisition of the user's behavior beyond the computational system. Through the use of that information the meta-level changes the behavior of the base-level with the intention of adapting the system to the user.

Basically, the implemented features for this prototype are: interface changes that adapt the presentation or the navigation path, and; a list of recommended items for each customer (like typically done by Amazon [15]).

During the adaptation of the interface, the text components on authors and books are changed - removing, inserting or ordering the components according to the users' preferences. In the adaptation of the navigation, links are shown or hidden according to users' preferences. The recommendation algorithms use input about a customer's interests to generate a list of recommended items.

In both cases, the acquisition and maintenance of the user model is carried out through the observation of the user behavior. This is determined by means of the user's clicks and purchases.

The experiment showed that it is possible to reuse the components (i.e. the presentation adaptation component, the navigation adaptation component and the recommendation algorithms). The only action that must be taken is to change the components in the meta-level.

6 Conclusions

This paper described a meta-level architecture to separate the components responsible for acquisition and manipulation of the users' models from the other components of an adaptive system.

The architecture proposed here can be used to add the ability of adaptation to stable computational systems. It also allows management of the functional aspects separately from the ones related to the acquisition and maintenance of the user's models.

To validate this architecture, a system with adaptive

interface was implemented. Through this experiment, it was verified that it is possible to reuse the components responsible for the user modeling only by adding them as stable meta-level components. Moreover, the need of tools to implement the computational reflection concept was identified. In the example, the meta-control component had to be built in the server side. This was because the technology used to implement the clients (J2ME) does not implement the computational reflection ability.

As future work, we plan to verify the applicability of this architecture in other kinds of adaptive systems, for example: adaptive information retrieval, recommendation systems and others.

Acknowledgments

The authors would like to thank Lavínia Boucault Napoleão and Luiz Carvalheira for all their remarks and cooperation in this paper.

References

[1] Palazzo, L.A.M. (2002) Sistemas de hipermídia adaptativa. In: JAI 2002 - XXI Jornada de Atualizacão em Informática, http://gpia.ucpel.tche.br/~lpalazzo/sha/

[2] Webb, G.I., Pazzani, M.J., Billsus, D. (2001) Machine learning for user modeling. User Modeling and User-Adapted Interaction 11: 19–29

[3] Danilowicz, C., Nguyen, H.C. (2002) Using user profiles in intelligent information retrieval. In Hacid, M.S., Rs, Z.W., Zighed, D.A., Kodratoff, Y., eds.: Foundations of Intelligent Systems. 13th International Symposium. Number LNAI 2366, Lyon, France, Springer-Verlag 223–231

[4] Rich, E. (1999) Users are individuals: Individualizing user models. International Journal of Man-Machine Studies 51: 323–338

[5] Papatheodorou, C. (2001) Machine learning in user modeling. In Paliouras, G., Karkaletsis, V., Spyropoulos, C.D., eds.: Machine Learning and Applications. Number LNAI 2049. Springer-Verlag Berlin Heidelberg, Berlin 286–294

[6] Orwant, J. (1995) Heterogeneous learning in the doppelganger user modeling system. User Modeling and User-Adapted Interaction 4: 107–130

[7] Fink, J., Kobsa, A. (2002) User modeling for personalized city tours. Artificial Intelligence Review 18: 33–74

[8] Kobsa, A. (2001) Generic user modeling systems. User Modeling and User-Adapted Interaction 11: 49–63

[9] Pohl, W., Nick, A. (1999) Machine learning and knowledge representation in the labour approach to user modeling. Proceedings of the Seventh International Conference on User Modeling. 179–188

[10] FERBER, J. (1989) Computational reflection in class based object-oriented languages. SIGPLAN Notices. 24: 317–326

[11] LISBÔA, M. (1997) Arquiteturas de meta-nível. Tutorial XI Simpósio Brasileiro de Engenharia de Software 1: 210-298

[12] WU, S. (1997) Reflective Java: making Java even more reflexible. http://www.ansa.co.uk, Cambridge, UK.

[13] KICZALES, J.G.R., BODROW, D. (1991) The art of the metaobjects protocol. MIT Press, Cambridge

[14] SUN (2004) Java Technology. http://java.sun.com

[15] Linden, G., Smith, B., York, J.: (2003) Amazon.com recommendations: Item-to-item collaborative filtering. IEEE Distributed Systems OnLine 1: 24-26

Adaptive Finite State Automata and Genetic Algorithms: Merging Individual Adaptation and Population Evolution

H. Pistori, P. S. Martins, A. A. de Castro Jr.
Department of Computer Engineering, Dom Bosco Catholic University, Brazil
E-mail: {pistori,martins,amaury}@ec.ucdb.br

Abstract

This paper presents adaptive finite state automata as an alternative formalism to model individuals in a genetic algorithm environment. Adaptive finite automata, which are basically finite state automata that can change their internal structures during operation, have proven to be an attractive way to represent simple learning strategies. We argue that the merging of adaptive finite state automata and GA results in an elegant and appropriate environment to explore the impact of individual adaptation, during lifetime, on population evolution.

1 Introduction

Most of the early works on genetic algorithms were based on the simplifying assumption that individuals do not change during lifetime. In contrast to this "rigid individuals" approach, some important trends on computational evolution, tackling the challenge of incorporating some kind of plasticity, into individuals modeling, started to emerge, in the last decade. Baldwinian [1] and Lamarckian [2] Evolution, are two of the major approaches, in which, individual phenotype is not a direct map from its genotype, but instead, the phenotype presents some flexibility in responding to the environment input.

Plasticity refers to the flexibility, or capacity for change [3], of a subject, and includes both reasonably simple traits, like the malleability of an amoeba cell membrane, and complex phenomena, like human learnable behaviour. It has been pointed out that individual plasticity is not always a benefi cial trait. Turney, for instance, presented ten dimensions of trade-offs related to the balance of phenotype rigidity and plasticity, like energy consumption, length of learning period and smoothness of the fi tness landscape [4].

Some simulation experiments involving plastic and learning individuals have been proposed. Most of these works are based on the use of artifi cial neural-networks with backpropagation [5] or some simpler hill-climbing strategy [2], to model the ability of an individual to perform a local search on the fi tness space, before a new population is produced by some standard genetic algorithm approach.

The concept of a plastic automaton, that can change its initial structure, during execution, have been initially proposed by Neto to address some problems on compiler construction design and implementation [6]. This formal device was named Adaptive Automaton. Independently, Shutt and Rubstein introduced a similar device, called Self-Modifying Automata [7]. More recently, Klein and Kutrib presented the Self-Assembling Automata [8], which share the same basic concepts created by Neto, Shutt and Rubstein, but with a new formalization. This paper presents the Adaptive Finite State Automata, or $\mathscr{A}$-FSA , as an alternative way to represent plastic individuals in a GA population. The $\mathscr{A}$-FSA formalism is heavily based on Adaptive Automata, but without some features that, despite being important in compiler construction design and some other application areas, would just add superflous complexity to the problem of represeting a GA genotype as a plastic automaton.

The three main reasons why $\mathscr{A}$-FSA is an interesting formalism to represent plastic individuals are: (1) plasticity is an inherent, but easily disabled, characteristic of $\mathscr{A}$-FSA ; (2) tradicional automata, like fi nite state [9], pushdown [10] and automata with multiplicities [11], are already being explored as individual modelling tool, in the context of "standard" genetic algorithms; (3) Adaptive fi nite state automata are related to two well stablished fi elds: formal language theory (FLT) and grammatical inference (GI) [12, 13]. The integration of $\mathscr{A}$-FSA and GA would bring these three fi elds (FLT, GI and GA) closer, facilitating information exchange, and tools reutilization.

The next section presents adaptive fi nite state automata and a graphical notation for their representation. Section 3 discuss the integration of Baldwinian and Lamarckian computational evolution with adaptive automata theory. Finally, conclusion and sugestions for future works are presented.

2 Adaptive Finite State Automata

An $\mathscr{A}$-FSA is a kind of finite state automaton that can change its transition relation during input reading. It can be seen as a simplified version of an adaptive automaton [6] and as a generalized self-assembling automata [8]. Each transition of an $\mathscr{A}$-FSA , besides operating as in a conventional FSA, can be attached to an *adaptive function*, which is executed just before the transition, removing or inserting new elements to the automaton's transition set.

Formally, an $\mathscr{A}$-FSA is a 10-uple $\mathrm{M} = \langle Q, \Sigma, q_0, F, \delta, \kappa, \Gamma, \Psi, \Phi, \Delta \rangle$. The first five elements define the *subjacent mechanism*, where:

Q is the state set.

Σ is the input alphabet, finite and non-empty.

$q_0 \in Q$ is the initial state of the automaton.

$F \subseteq Q$ is the final state set.

$\delta \subseteq Q \times \{\Sigma \cup \{\epsilon\}\} \times Q \times \{\kappa \cup \{\epsilon\}\} \times 2^{\Gamma \mapsto \{Q \cup \Sigma\}}$ is the non-deterministic transition relation.

The transition relation differs from the usual one by two new elements: an adaptive function label, taken from κ, and a set $P \in 2^{\Gamma \mapsto \{Q \cup \Sigma\}}$, of parameter assignments, where $\Gamma \mapsto \{Q \cup \Sigma\}$ is a partial function that maps formal parameters to states or input symbols. The adaptive function label may also be an epsilon symbol (ϵ), indicating that the transition is a regular one. The others five elements of M define the *adaptive mechanism*, where:

κ is the set of adaptive functions labels.

Γ is a set of formal parameters and variables.

Ψ is a set of generators.

$\Phi : \{\Gamma \cup \Psi\} \mapsto \kappa$ is a partial function, mapping formal parameters, variables and generators, to adaptive function labels.

$\Delta \subseteq \kappa \times \{?, +, -\} \times \{Q \cup \Gamma \cup \Psi\} \times \{\Sigma \cup \{\epsilon\} \cup \Gamma\} \times \{Q \cup \Gamma \cup \Psi\} \times \{\kappa \cup \{\epsilon\} \cup \Gamma\} \times 2^{\Gamma \mapsto \{Q \cup \Sigma \cup \Gamma \cup \Psi\}}$ is the set of adaptive actions.

The first element of each adaptive action just groups adaptive actions into adaptive functions. The second element defines the type of the adaptive action, which can be a query (?), a remove ($-$) or an insert action ($+$). The remaining elements represent the transitions to be queried, removed or inserted. These elements can be

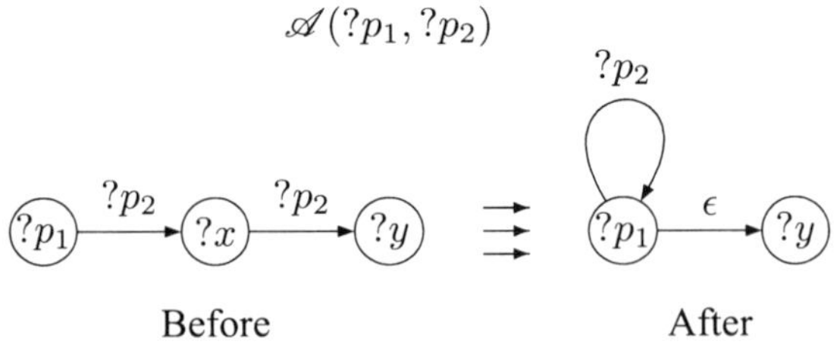

Fig. 1. Adaptive function Graphical Representation

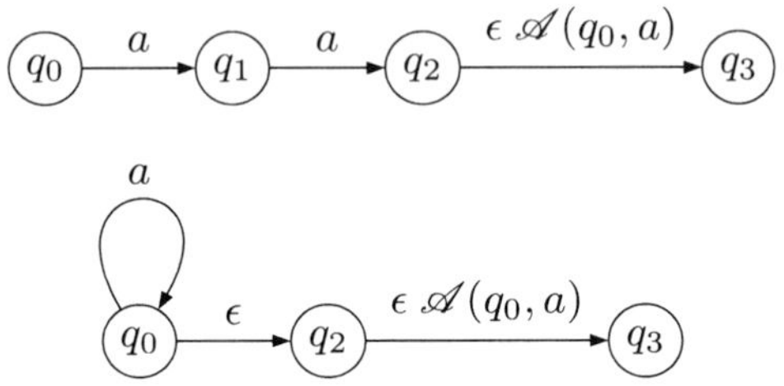

Fig. 2. Automaton structure before and after reading *aa*

replaced by a formal parameter, a variable or a generator. Formal parameters are mapped to the values defined in δ, during the adaptive function execution. Variables are used to indicate general transition patterns and generators indicate a state that is to be used for the first time inside the dynamically changing δ function. A detailed description of an $\mathscr{A}$-FSA is out of the scope of this paper, mainly for space restriction, but the following graphical representation proposal, can express, more intuitively, the operation of an adaptive function.

In the graphical representation, adaptive functions are illustrated by two prototypical automata (represented by the usual graphical notation, with circles and arrows), separated by a triple arrow. The triple arrow direction indicates an automaton sub-structure *before* and *after* the adaptive function execution. Variables and formal parameters are marked with the prefix "?", while generators are prefixed by "*". The adaptive function label and its formal parameters appear above the graphics, using standard notation for functions: the function label followed by a comma delimited, bracket enclosed, sequence of parameters.

Figure 1 shows a 2-parameter adaptive function that removes any two adjacent transitions, departing from state $?p_1$, that reads the same input symbol, $?p_2$; and inserts a loop, on state $?p_1$, reading $?p_2$. An empty transition is also created, from $?p_1$ to the state previously reached by the adjacent transitions. This adaptive function, if properly used, generalizes the language accept by the automaton, creating a loop, after reading two con-

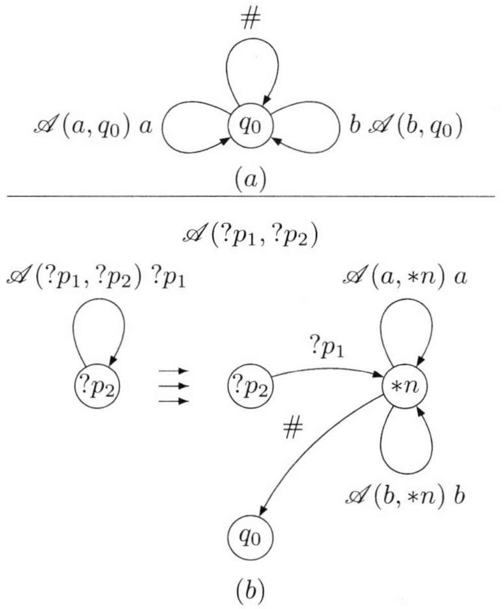

Fig. 3. Adaptive FSA for building Prefi x-Tree Acceptor

(a) Subjacent Mechanism (b) Adaptive Mechanism

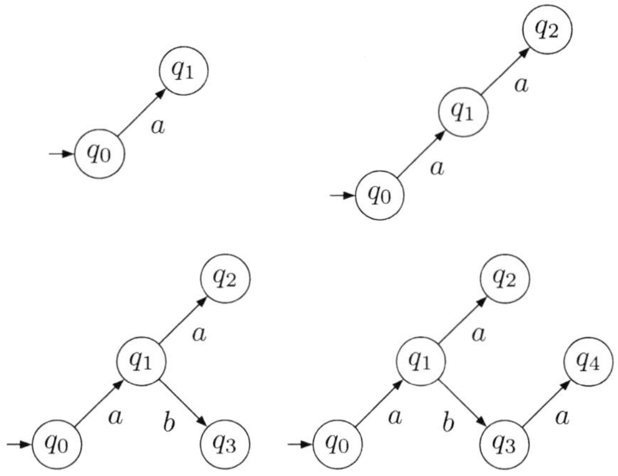

Fig. 4. Automaton structure as it reads *aa#aba*

secutive symbols. Figure 2 illustrates how a subjacent mechanism should look before and after the execution of this adaptive function.

Finally, its worth noting that an $\mathscr{A}$-FSA can be easily specialized to a self-assembling fi nite automata of degree k [8], just by: (1) restricting to k, the number of adaptive function parameters, (2) not accepting input symbols (just states) to be passed as parameters, (3) not allowing variables and (4) not allowing transitions to be removed. A formal proof of this result, which will imply that the class of languages accepted by self-assembling fi nite automata, is a subset of the one accepted by an $\mathscr{A}$-FSA , is under development.

2.1 An $\mathscr{A}$-FSA that builds a prefix-tree acceptor

Many grammar induction algorithms start from building a special kind of fi nite state automaton, called prefi x-tree acceptor, or simply, PTA [12]. Algorithms to build PTAs, from a set of positive strings, are straightforward. However, in the following example, both the PTA, and the algorithm that builds a PTA, from sample strings, will be modeled as an $\mathscr{A}$-FSA .

The automaton's subjacent layer, with alphabet $\Sigma = \{a, b, \#\}$, has one state, and is shown in fi gure 3.(a). The number sign, #, is just a string delimiter. Transitions (q_0, a, q_0) and (q_0, b, q_0) are both associated with the 2-parameters adaptive function, $\mathscr{A}$, presented in fi gure 3.(b). The adaptive function just breaks the loop (parameterized by $?p_1$ and $?p_2$), creating a new prefi x-tree edge, and two new loops, similar to the initial ones, that will allow the prefi x-tree to continue growing, if necessary. A string delimiter transition is also inserted to lead the automaton to its initial state whenever a number sign is read. Figure 4 illustrates the automaton's plasticity, as it reads the sample input *aa#aba*. Loops and delimiter transitions are omitted in the sake of clarity.

This automaton can be formalized as an $\mathscr{A}$-FSA $M = \langle Q, \Sigma, q_0, F, \delta, \kappa, \Gamma, \Psi, \Phi, \Delta \rangle$, where the subjacent mechanism has $Q = \{q_0\}$, $\Sigma = \{a, b, \#\}$, $q_0 = q_0$, $F = \{q_0\}$, $\delta = \{(q_0, a, q_0, \mathscr{A}(a, q_0)), (q_0, b, q_0, \mathscr{A}(b, q_0)), (q_0, \#, q_0, \epsilon)\}$ and the adaptive layer, containing just one adaptive function, $\mathscr{A}$, has $\kappa = \{\mathscr{A}\}$, $\Gamma = \{p_1, p_2\}$, $\Psi = \{n\}$, $\Phi = \{(p_1, \mathscr{A}), (p_2, \mathscr{A}), (n, \mathscr{A})\}$. The Δ relation, represented using a notation that emphasizes the adaptive action type (shown outside the brackets) and the association of adaptive function labels and parameters, is:

$$\begin{aligned} \mathscr{A}(?p_1, ?p_2) = \{ \quad & ?(?p_2, ?p_1, ?p_2, \mathscr{A}(?p_1, ?p_2)), \\ & -(?p_2, ?p_1, ?p_2, \mathscr{A}(?p_1, ?p_2)), \\ & +(?p_2, ?p_1, *n, \epsilon), \\ & +(*n, \#, q_0, \epsilon), \\ & +(*n, a, *n, \mathscr{A}(a, *n)), \\ & +(*n, b, *n, \mathscr{A}(b, *n))\} \end{aligned}$$

3 Baldwinian and Lamarckian Evolution with Adaptive Finite State Automata

Genetic Algorithms are very effi cient at exploring the entire search space; however, they are relatively poor at fi nding the precise local optimal solution in the region at which the algorithm converges. For make things better, there are hybrid algorithms which are the combination of improvement procedures, usually working as evaluation functions, and genetic algorithms. In order to improve the algorithms performances, local improvement procedures have been incorporated into GAs, through what could be called "learning" or "individual plasticity". There are two basic strategies in using hybrid GAs: Lamarckian and Baldwinian evolution. The Baldwin Effect, as utilized in genetic algorithms, was first investigated by Hinton and Nolan [14] by allowing an individual's fitness (phenotype) to be determined based on learning. Like in natural evolution, the result of the improvement does not change the genetic structure (genotype) of the individual. Although Lamarckian evolution has been universally rejected as a viable theory of genetic evolution in nature, using ideas inspired on it, in genetic algorithms, can improve their convergence speed [2]. In Lamarckian computational evolution, the genetic structure of an individual can be changed to reflect the results of learning.

In a GA environment where genotype are represented as an $\mathscr{A}$-FSA , the Baldwin effect could be explored by the appropriate utilization of adaptive functions to model plasticity or learning. The plasticity level could be controled by designing, or admitting the evolution, of different adaptive functions. A kind of "Lamarckian effect" could also be easily acchieved by retaining the structural changes suffered by the automaton, between generations.

4 Conclusion

Adaptive finite state automata have been described and an example illustrating their inherent ability to model plasticity, in a formal language and automata theory framework, has been presented. Using adaptive finite state automata to model individuals in GA opens a new path for investigations on the interaction of individual plasticity and evolution, which is an important topic in current computational evolution research. It may also strengths the links between grammar inference and computational evolution areas, providing a new environment for theory and results exchange. Some suggestions for future work include the integration of AdapTools [1], an environment for adaptive automata development, with some GA computational library. In-depth studies on genotype representation, performance and limits of hybrid GA-$\mathscr{A}$-FSA algorithms should also be conducted in the near future.

References

[1] Jones, M., Konstam, A. (1999) The use of genetic algorithms and neural networks to investigate the Baldwin effect. In: Proc. of the ACM symposium on Applied computing, pp. 275–279

[2] Wellock, C., Ross, B. J. (2001) An examination of Lamarckian genetic algorithms. In: Erik D. Goodman (ed.) Genetic and Evolutionary Computation Conference Late Breaking Papers 2001, San Francisco, California, USA, pp. 474–481

[3] Belew, R., Mitchell, M. (1996) Adaptive Individuals in Evolving Populations: Models and Algorithms - SFI Studies in the Sciences of Complexity - Vol.XXIII. Addison Wesley, Boston

[4] Turney, P. (1996) Myths and legends of the Baldwin effect. In: Proc. of the 13th Int. Conf. on Machine Learning, pp. 135–142

[5] Nolfi, N. (1999) How learning and evolution interact: The case of a learning task which differs from the evolutionary task. Adaptive Bahavior 7(2):231–236

[6] Neto, J. J. (1994) Adaptive automata for context-sensitive languages. SIGPLAN Notices 29(9):115–124

[7] Rubinstein, R., Shutt, J. N. (1994) Self-modifying finite automata. In: Proc. of the 13th IFIP World Computer Congress, pp. 493–498

[8] Klein, A., Kutrib, M. (2002) Self-assembling finite automata. In: Proc. of the 8th Annual Int. Conf. on Computing and Combinatorics, pp. 310-319

[9] Belz, A., Eskikaya, B. (1998) A genetic algorithm for finite state automata induction with an application to phonotactics. In: ESSLLI-98 Workshop on Automated Acquisition of Syntax and Parsing, pp. 9–17

[10] Lankhorst, M. M. (1995) A genetic algorithm for the induction of nondeterministic pushdown automata. In: Computing Science Report CS-R 9502, University of Groningen.

[11] Bertelle, C., Flouret, M., Jay, V., Olivier, D., Ponty, J. (2001) Genetic algorithms on automata with multiplicities for adaptive agent behaviour in emergenet organizations. In: Proc. of World Multiconference on Systemics, Cybernetics and Informatics 2001, pp. 22–25

[12] Cicchello, O., Kremer, S. C. (2003) Inducing grammars from sparse data sets: A survey of algorithms and results. Journal of Machine Learning Research 4:603–632

[13] Higuera, C. De La (2001) Current trends in grammatical inference. Lecture Notes in Computer Science 1876:28–30

[14] Hinton, G., Nolan, S. (1987) How learning can guide evolution. Complex Systems 1:495–502

[1] Freely available at http://www.ucdbnet.com.br/adaptools/

Modeling a tool for the generation of programming environments for adaptive formalisms

A.R. Camolesi

Departamento de Engenharia de Computação e Sistemas Digitais, Universidade de São Paulo, Brasil
Coordenadoria de Informática, Fundação Educacional do Município de Assis, Brasil
E-mail: camolesi@femanet.com.br

Abstract

This paper aims to present the logical model that makes up the structure of a tool for the definition of environments for rule-driven adaptive formalisms.

1 Introduction

Adaptive applications need resources to adapt themselves to the environment's momentary needs and to foresee the internal and external demands, thus making up for a complex, robust, and fault-tolerant structure, yet flexible and responsive. Such applications offer modern capacities that are very difficult to be modeled by using present techniques of software development.

In order to solve the adaptive applications' modeling, it was proposed in [1] a generic formalism that allows (underlying) rule-driven non-adaptive devices to be extended to concepts of adaptive mechanisms. Such formalism is based on an Adaptive Mechanism (AM) that involves the kernel of an underlying non-adaptive device (ND). This way, an Adaptive Device (AD) is formally defined by AD = (C, AR, S, c_0, A, NA, BA, AA).

In this formulation C is the set of all the possible configurations of ND and $c0 \in C$ means its initial configuration. S is the set of all possible events that make up AD's entry chain and set A represents the acceptance configurations for ND.

Sets BA and AA are adaptive actions' sets. NA is a set of all symbols that can be generated with exits by AD, in response to the application of adaptive rules.

AR is the set of adaptive rules that define the adaptive behavior of AD and is given by the relationship Ar $\subseteq BA \times C \times S \times C \times NA \times AA$ in which adaptive actions modify the current set of AR adaptive rules from AD to a new AR set by adding and/or deleting adaptive rules in AR.

Based on these definitions, it is proposed in this paper a logical model for the representation of the formal elements shown in [1]. Such model is fundamental to the developing of tools that support a design methodology for adaptive applications. This paper is organized as follows: in section 2, the stages of extensions for adaptive devices and its use will be described. In section 3, the logical representation for adaptive devices is shown, and finally, in section 4, some conclusions and future papers are discussed.

2 Stages of extensions for non adaptive rule-driven devices.

When extending a formalism of an underlying device to the concepts of adaptive rule-driven mechanisms, a specialist should involve the non adaptive device with an adaptive layer. In order to develop this job, the specialist should possess good knowledge both of the underlying formalism and of the concepts of adaptive mechanisms. On the other hand, a planner that uses a device extended by a specialist does not need a formal knowledge as deep as the one needed by the specialist in extension of devices. The planner needs to know the extended specification language and how to use it in the project of his applications.

When extending a non adaptive rule-driven device to support adaptive characteristics there is the need to accomplish 3 stages: the stage of extension of the formal (mathematical) model, the stage of definition of the logical model and the stage of definition of the physical model. Figure 1 illustrates the stages and the existent relationship among them.

The stage of extension of the formal model Figure 1(A), offers a view in which a specialist with good mathematical knowledge of underlying formalism accomplishes the conceptual definition of the extended device to the concepts of adaptive mechanisms. In [1] and [2], extensions of underlying devices are presented to the concepts of adaptive devices. In this phase, the junction of the formal concepts of both (underlying and adaptive) formalisms is achieved, thus obtaining a new underlying device extended to concepts of adaptive mechanisms.

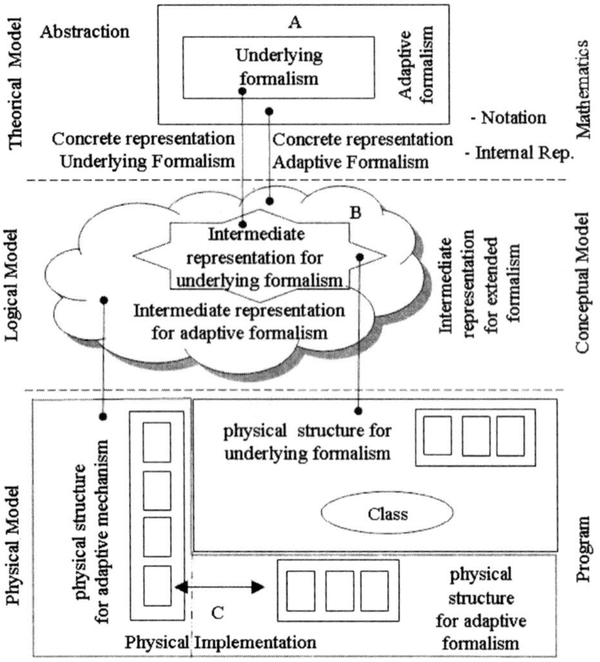

Figure 1. Stages of extension of non adaptive rule-driven devices.

After obtaining the adaptive formalism, it is necessary a mapping of its concepts for an intermediate representation, as shown in Figure 1(B). Such stage consists of the definition of the logical structure that represents the formal concepts of the new adaptive device. Such structure is of fundamental importance, because it is part of the information storage structure necessary for the development of tools that will help the planner in designing adaptive applications.

In the stage of physical definition, as shown in Figure 1(C), a planner with knowledge of the developed adaptive formalism accomplishes the specification of his application. At this stage, yielded specifications are to be later analyzed and implemented.

When performing his work the planner instances the defined objects in the logical stage and he defines the physical elements that represent the behavior of the application.

In this phase, it can be observed that the instantiated objects belong to two different classes, i.e., the objects that represent the behavior of the developed application and other objects that represent the adaptive functions and actions responsible for modifications in the behavior of the application in execution. Based on the set of the defined objects in this phase, the presentation, the simulation, the verification and the execution of the projected application are allowed. During the simulation and execution process of specification in the adaptive kernel, adaptive actions can be executed and rules can be added or removed from the behavior represented by specification, thus modifying its structure.

In [3], a methodology was proposed to give support to the project of adaptive applications by using concepts presented in this paper. In Figure 2, it is shown the design methodology for adaptive applications formed by the following phases: specification phase, transformation specification phase, and validation and specification simulation phase. In the specification phase the application is accomplished by using either a text or a graphic tool. Soon afterwards, the transformation of the produced specification to an intermediate representation (logical model) is accomplished and, based on the obtained representation the planner can inform entry string sequences and evaluate its specification. If mistakes or inconsistencies occur, the planner can make changes in the specification and restart the process.

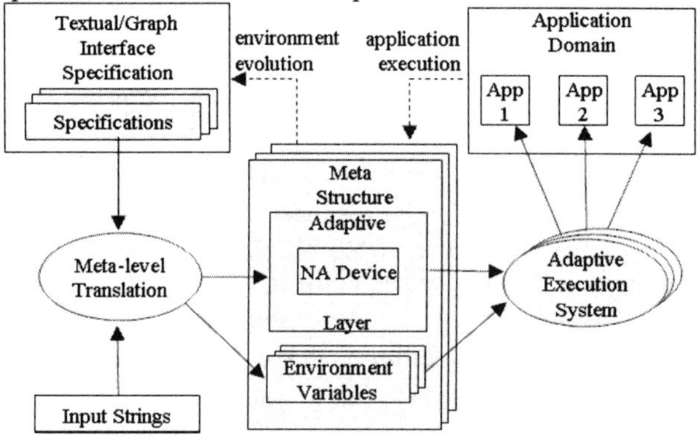

Figure 2. Methodology for Design of Adaptive Applications.

The proposed methodology is linked to the need to use tools for helping the planner in the performance of his job. During the specification phase there is the need of a text or graphic tool to aid the planner in the specification of an application. The phase of specification transformation of the application to an intermediate representation can be accomplished in two ways: automatic (generated by the editors at the moment of the edition), or through a translator that makes the transformation process after the specification process. And, finally, tools that allow the visualization, simulation and verification of the projected applications. In this phase, the planner, using an integrated environment informs the values regarding entry chain and submits their specification to the performer of the adaptive kernel. Initially, in case they exist, prior adaptive actions are performed, followed by elementary actions of the underlying device and finally the subsequent adaptive actions. This way, at each step the designer gets a new configuration (state of the system) and a new set of rules (behavior of the application) according to the adaptive actions that were performed. The obtained results should be displayed to the user, who can analyze them and, if necessary, make changes and restart the whole process.

3 Logical model for adaptive formalisms

Based on the concepts shown in [1], a logical model is proposed, so that it allows the construction of tools that help to plan adaptive applications. Such a model is represented by a data structure that gives support to the storage of the intermediate representation and allows the construction of a program that can manage the performance of the resulting specification by using the available facilities from adaptive devices.

In [3], a proposal was presented for the logical structuring of the formal definition of the concepts of adaptive devices. Figure 3 shows a diagram of entity relationship of the conceptual model for adaptive devices. Such a diagram is structured by objects of three types: Underlying Kernel (UK), Specification (S) and Adaptive Layer (AL) according to the characteristics they represent.

The objects of horizontal hachure (Device, Component Type, Connection Type and Attribute Type) are Underlying Kernel (UK) type and they correspond to the intermediate representation of the basic elements of an underlying device. In this structure, the conceptual elements of the underlying devices formally represented by set C are defined.

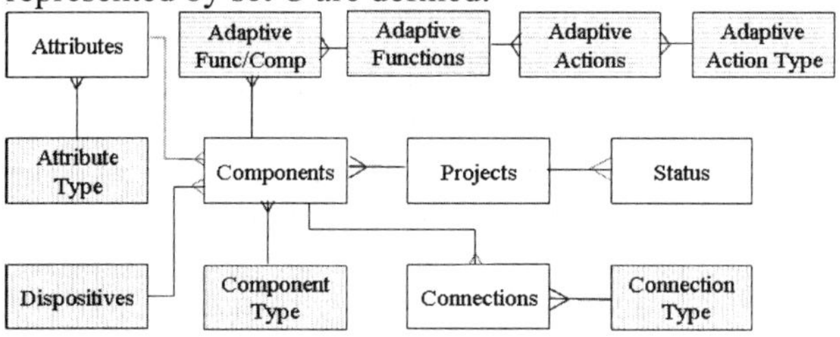

Figure 3. Diagram of entity relationship intermediate representation.

Solid color objects (Project, Attributes, Components, Connections and Status) are Specification (S) type and aim to represent the specifications yielded by a planner. Each object of this structure corresponds to elements of the formal definition, in which: each rule c that is part of the set of rules NR of an underlying device ND can be represented by the objects in S. The planner, when defining a specification, instances objects of the NS type (elements that constitute the underlying kernel) and defines the behavior of the application. This structure also stores the elements of set A that correspond to the rules of acceptance of an adaptive device and, furthermore, to the information on values of both the entry and exit chains in the Status object.

The objects with vertical hachure are Adaptive Layer (AL) type and they aim to provide the necessary resources to support the adaptive layer that involves the underlying kernel. The Adaptive Layer is structured in objects that correspond to the configuration of the adaptive device (Adaptive Action Type), and in objects that correspond to the AR conceptual elements that, in turn, correspond to adaptive functions and actions.

When defining the Underlying Kernel of a new device (Petri Nets, Automata, Grammar Free from Context, etc.) the specialist needs to store information related to the name of the device, the creation date and updating, etc.. Such information is stored in the Devices object.

Information on the types of components (places and transitions of a Petri Net, final states, and non-final states of Automata, etc.) that represent the behavior of an application and that is used by a planner when specifying their application, can be represented by the Component Type object.

When specifying a rule that represents the behavior of an application it is necessary to represent the form of the existing connection between its components. The Connection Type object represents the information on the connection type for a device: transition for Automata, Petri Net connections, etc., while the Attribute Type object contains information on the types of data that are available for attribution to a component of an application behavior.

When accomplishing the Specification of an application it is necessary to store information on the description of the specification, on the planner in charge, etc.. Such information is represented by the Projects object. At first, when defining the behavior of a project, one should define the components that constitute the application behavior. Such components are parts of the NR rules and they are represented by the Components object. One can mention the description of the states that constitute a specification of Automata or the description of the places and transitions of a Petri Net, etc. as examples of such components.

Following the definition of the components, one defines the rules (set c of the formal representation) that constitute the behavior of an application (formally acted by NR). Such structure establishes the relationship among the defined elements in both Component and Connection Type objects and defines the behavior of an application.

The value of each attribute associated to a (Component or Connection) object is represented by the Attributes object. The values of stimulus, and related information to the exit and other necessary information during execution are acted by the Status object.

The Adaptive Layer is associated to the elements of specification of an application. This results, at first, in the definition of the information on the type of adaptive action that can occur: consultation action, insert or removal. Such information is stored in the Adaptive Action Type object.

When the adaptive mechanism is joined to the underlying kernel it is necessary to define the adaptive functions (the conceptual elements BA and AA) that should be associated to the elements of the Components object. The Adaptive Functions object allows the extension of the underlying kernel to have the features of adaptive mechanisms and it makes the connection between the elements of the underlying kernel and their respective adaptive actions that are represented by the Adaptive Actions object.

The Adaptive Actions object represents the set of adaptive actions belonging to AR that has the function of accomplishing changes in the behavior of the projected application.

Based on the logical structure, a tool is being developed that will allow a specialist to configure the conceptual elements of a non adaptive device and to accomplish its extension for the adaptive mechanisms. Such tool will also allow a planner, by using a textual language (intermediate representation), to develop the project of their applications.

In a second stage other tools will be developed that will allow the specification and display of graphic elements of the extended adaptive devices. The tool development is being made in Java [3] due to the portability and reuse features inherent to this programming language. Figure 3 shows the interface of the tool that is responsible for the definition of the connections of a specification.

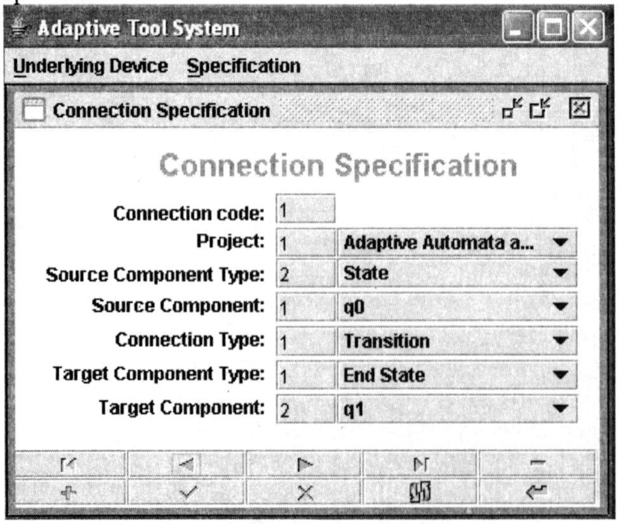

Figura 4. Interface of an Adaptive Tool System.

4 Conclusion.

This work aimed to present how to make the extension of a non adaptive device to support the characteristics of adaptive mechanisms.

Initially, the general structure of an adaptive mechanism was presented, followed by the stages for the extension of a non adaptive formalism to support the characteristics of adaptive mechanisms. Following, the methodology for the design of adaptive applications was shown by using these concepts. Finally, a logical model was presented for the construction of tools that will give support to a design methodology of adaptive applications.

The proposed methodology was used in [2] to modeling of applications that has support the use of graphic interface and tools are being implemented to facilitate specialists and planners in their job with adaptive technology.

In relationship the stages of definition of adaptive formalisms several works were accomplished in relation to formal definition and as resulted adaptive formalisms were developed. Such works served as base for the definition of the extension stages for adaptive formalisms and they were to base the proposal of a logical model that it seeks to represent adaptive rule-driven formalisms.

The defined logical structure represents the conceptual elements for adaptive formalisms and it constitutes an intermediate representation for the definition of tools that it will support the methodology of design of adaptive applications.

As a continuation to this work, it is suggested a deeper study for the validation of the proposed logical model and the definition of a physical model (computational) for the validation of the proposed structure.

References

[1] Neto J.J.(2001) Adaptive rule-driven devices - general formulation and case study, Sixth International Conference on Implementation and Application of Automata, Pretoria-South Africa.

[2] Camolesi, A.R. and Neto, J.J. (2004) Modelagem Adaptativa de Aplicações Complexas, XXX Conferencia Latino Americana de Informatica (CLEI), Arequipa, Peru.

[3] Camolesi, A.R. e Neto, J.J (2003) An adaptive model for modelling of distributed system, *Conference Argentina in la Ciência da Computacion (CACIC)*, La Plata, Argentina.

[4] Programming Language JAVA in http://www.sun.com (September 2004).

Adaptive Decision Tables
A Case Study of their Application to Decision-Taking Problems

T. C. Pedrazzi[1], A. H. Tchemra[1,2], R. L. A. Rocha[1]

[1]Department of Computing Engineering, Polytechnic School, University of São Paulo
[2]Department of Mathematics, Universidade Presbiteriana Mackenzie, São Paulo
E-mail: {thiago.pedrazzi, angela.tchemra, luis.rocha}@poli.usp.br

Abstract

Decision tables have been traditionally used for solving problems involving decision-taking tasks. In this paper, adaptive devices based on decision tables are used for the solution of decision-taking problems. The resulting adaptive decision tables have proved to be effective due to their generality and flexibility. They are helpful tools for automatically choosing an applicable alternative among several available at each stage in the decision-taking process. An illustrative example as well as an overall comparative evaluation is shown in the business management field.

1 Introduction

The evolution of information technology has contributed to transformations in several areas where its resources are applied. Computers and computational systems are very important support tools in the decision-taking process.

The decision processes can vary from the simplest to the most complex and dynamic, depending on the existing variables and the variables that can appear in the presented problems. Contingent on the complexity, the decision-taking process requires gathering of most information possible in order to reduce risks. The information can derive from past, present or inferred future facts.

Because of its features, decision processes are well suited as applications using the adaptive techniques. The use of methods based on the adaptive technology is an alternative to be considered in the resolution of complex problems, and those methods can be more efficient than traditional ones [1].

Adaptive devices are built of sets of rules that can be dynamically changed [2], which means that their internal procedures can be self-modified in order to face their input stimuli situations.

This paper shows how the adaptive decision tables' mechanism may be used in the decision-making processes, whose established criteria change at each decision-taking cycle. An application example of the adaptive decision table in the business area is shown in detail, in order to compare it to the traditional decision-making methods, and to make clear the use of adaptive technology.

2 Decision taking

A decision is a choice made between two or more available alternatives. Decision-taking is the process of choosing the best alternative to achieve the proposed goals [3].

A decision requires an individual, or a group of individuals, to choose one among several options. The options can vary from two to an unlimited number. The decisions can become too complex if the sequence of decisions taken affects the subsequent options [4].

In the decision process, the decision-taker usually must analyze the goals to be achieved by his actions, the situation within the problem, the available resources, and the consequence of the decision taken [5].

Hence, any problem whose solution is based on a decision-taking process can be planned using the experience and results of other similar processes. A database of adopted strategies in each case aids the decision process in many of its perspectives, and also improves information quality. In case there is no information in the database, there must have means to enter new data and, therefore, modify the database to improve the model.

Approaches about the decision-taking process can be found in several publications. Among them is the classic rational selection, where Ackoff and Sasieni (1968, apud [4]) state that the decision process model should gather all the data that can represent the control variables that will determine the alternative actions, the uncontrollable variables relevant to the problem, and the decision criteria that can lead to the best action. Hence, this model should show the selection outcome.

In management science, the use of decision process models implemented in computers may become a way to control and manage the consequences of a decision [4]. Those systems are based on the feedback concept. Depending on the availability of the data, and the performance of the model, the systems are fed back allowing the decision taker to gain control and improve his performance.

The computational tools currently available for use by organizations are designed to supply quantitative and qualitative information that assists in the decision-taking process.

As a rule, during the development of conventional systems, a previous analysis of the information in the actual context is performed. The system is developed based on the results of this analysis, allowing the information to be programmed and the situation to be simulated. If newer information emerges within the context, the system can no longer fit its answers, since its information is fixed.

With the adaptive techniques, however, the system is not only capable of determining the information needed for the decision-taking situation, but also it is capable of receiving new data, which will eventually appear in each cycle of the process. Therefore, the system changes itself to provide better conditions to such decisions. This system is called adaptive device. The main characteristic of the adaptive device is to dynamically modify its own procedures [2], in consequence of the inputs, without external action, such as, of the user.

3 Adaptive Technology

A formal device is said to be adaptive whenever its behavior changes dynamically, in a direct response to its input stimuli, without interference of external agents or even its users. In order to achieve this feature, adaptive devices have to be self-modifiable. In other words, any possible changes in the device's behavior must be known at their full extent, at any step of its operation in which the changes have to take place. Therefore, adaptive devices must be able to detect all situations causing possible modifications and adequately react by imposing corresponding changes to the device's behavior. In this work, devices are considered whose behavior is based on the operation of subjacent non-adaptive devices that can be fully described by some finite set of rules.

Application of adaptive technology is based on a formal model known as Adaptive Automata (AA) [6, 2], which is a Structured Pushdown Automata that, through the performance of predefined adaptive functions, change its behavior in response to its input stimuli.

Many projects have been developed using adaptive technology, which shows the versatility and applicability of these techniques in wide-range application.

The use of adaptive technology for solving computational problems is very interesting, since it presents compatible results with the most commonly used techniques with a cost-effective relationship even more interesting. We can list, e.g., Adaptive Statecharts, Adaptive Markov Chains, Adaptive Grammars, and Adaptive Decision Table, among others. Further information about these and other adaptive formalisms can be found at the Adaptive Technology Lab web site (www.pcs.usp.br/~lta).

4 Applying Adaptive Technology on Decision-Support Systems

After analyzing the operation, clarity and easiness of the learning process of each adaptive device available nowadays, we have concluded that the adaptive rule-driven device is the best choice for implementing decision-support systems.

In [2], the adaptive rule-driven devices can be seen as a two-layer system. The first layer is represented by a non-adaptive device, which is the basis for the system and is called underlying device. The second layer is represented by the set of adaptive actions associated to that underlying device. The addition of this second layer empowers a common decision table (or any other underlying device) to Turing's Machine level (the so-called Turing compatibility) at a minimum cost.

To operate such adaptive device, one should initially use the non-adaptive underlying decision table to determine the rule(s) that matches the current situation of the condition predicates. Then, the selected adaptive rule is performed by executing the indicated adaptive actions associated to that rule. The adaptive rule can change the underlying device rules, changing, therefore, the systems behavior.

material=	Mα	Mα	..	Mη
building=	B_a	B_a	..	B_b
supplier=	S_1	S_2		S_3
price=	BP	GP		GP
proximity=	NB	NB		FB
purchase:=		✓		

Fig 1 – non-adaptive decision table

The Adaptive Decision Table (ADT) was selected as the core of this decision-support system for reasons such as:

- The non-adaptive underlying device is the ordinary decision table, which is well-known among the information systems solution providers presenting, though, a higher commercial potential.
- The execution algorithm is quite straightforward, since it is as simple as the underlying device execution,
- Extending the non-adaptive underlying device to the adaptive one is very easy, as shown in [2], presenting extremely low-cost additions.

Further details about formal definition and implementation of ADT can be found in [2, 7].

As can be seen from the differences between the non-

adaptive decision table (fig 1) and the adaptive one (fig 2), the additional cost to 'upgrade' such device is really low. The underlying device is the same.
Attaching an adaptive layer to this device is a straightforward procedure, since it remains as a table, with increased size, though. The meaning of each extra column and row can be found in ref. 2.

As a table, just a few modifications need to be made for the runtime engine. However, ADT's computational power is much higher than its non-adaptive counterpart is. The last one can be used only for Finite State Automata simulation.

		Tag →	H	-	-	-	+	S	R	R	R	R	R	R	R	E
Subjacent decision table	Condition	material=		p1	p1	p1	p1		"μ"	M_α	M_α	M_α	M_β	M_β	M_χ	
		building=		p2	p2	p2	p2		"o"	B_a	B_b	B_b	B_a	B_b	B_a	
		supplier=					g1			S_1	S_1	S_2	S_1	S_2	S_2	
		price=		BP	BP	GP	GP			GP	GP	BP	GP	GP	BP	
		proximity=		NB	FB	FB	NB			NB	FB	FB	NB	NB	NB	
	Act	get(data)						✓								
Adaptive functions	Func	*F*	B													
	Other names	*p1*	P						"μ"							
		p2	P						"o"							
		g1	G													

Fig. 2 – ADT example before processing

5 Illustrative Example

The application of an ADT will be illustrated through a very simple example, such as the purchase of construction material for civil engineering business.
Ideal conditions for the purchase have been established, such as price has to be 'good' and the supplier has to be 'near' the building. These criteria (good and near) are pre-definite to simplify the example. At the moment of decision-taking, both conditions have to be true for the selected supplier. The possible non-deterministic solution will not be used, since the backtracking (or other techniques to simulate parallel processing) increases exponentially the running time.
Since this example must be short in length, some auxiliary functions such as the one that would perform price analysis, proximity criteria and insertion of new materials were not shown.
For inserting new materials, an adaptive function would receive material name / code to be inserted together its respective building name, and create new columns for it in the ADT. Note that the fields **supplier**, **price** and **proximity** would be empty.
A timely-started function (**bid**) would then replace the previous price analysis based on new market information, and also fulfill the just-inserted material's price and supplier field on those that does not have these fields fulfilled.
There may be another function that, based on GPS information and on some distance criteria, would analyze if the supplier are 'near' or 'far from' the building. Again, this function was also omitted for sake of simplicity.
An adaptive function (F) has been designed for this example. This function is composed of 4 elementary adaptive actions, as seen in fig. 2, which change the underlying device in the following manner:

- The first three rules (first, second and third column after the heading column) exclude the rules whose settings are not ideal; therefore, they are naturally excluded.
- The fourth rule add a rule whose settings are the ideal supplier.

A non-adaptive device rule is a 5-tuple (Mη, Bk, Sn, MPr, PO) format with:
•Mη – Material Name / Code to be purchased.
•Bk – Building name to which the material will be destined for.
•Sn – Supplier's Name / Code.
•MPr – Price of the material, which can be GP (good price) or BP (bad price).
•PO – Proximity to the Building, which can be NB (near the building) or FB (far from the building).
In the example, we will simulate a purchasing, where the material code is *Mβ* to be used at the *Bb* building.
Initially, the device is started by applying the starting rule, which is identified by a 'S' on the TAG row (as may be seen in Fig. 2)
Then, the action **get(data)**, which is checked in this rule, will be applied. This action actually acquires the material Name / Code and the building name
Now, we search the ADT for a rule that tests the conditions **material** and **building** against the values just

read (and, of course, with the ideal conditions for purchasing).
There is no eligible rule for this case, so the adaptive rule will call the adaptive function, passing as arguments **material** and **building** information just read.
The adaptive action will search for rules in the underlying device and delete any that has same material and building information and doesn't fulfill the ideal conditions (that could be {Mβ, Bb, ,GP,FB} or {Mβ, Bb, ,BP,FB} or else {Mβ, Bb, ,BP,NB}), and include a new supplier, which name will be generated by generator *g1*, with the desired settings {Mβ, Bb, *g1*,GP, NB}.
The system will then select the new supplier, just added. If there were already a rule with ideal conditions, it would be selected before the activation of any adaptive action. This is an intrinsic adaptive technology characteristic. The resulting ADT after this process can be seen in fig. 3.

		Tag →	H	-	-	-	+	S	R	R	R	R	R	R	E
Subjacent decision table	Condition	material=		p1	p1	p1	p1		"μ"	M_α	M_α	M_β	M_β	M_χ	
		building=		p2	p2	p2	p2		"o"	B_a	B_b	B_a	B_b	B_a	
		supplier=					g1			S_1	S_x	S_1	S_2	S_2	
		price=		BP	BP	GP	GP			GP	GP	GP	GP	BP	
		proximity=		NB	FB	FB	NB			NB	NB	NB	NB	NB	
	Act	get(data)						✓							
Adaptive functions	Func	*F*	B												
	Other names	*p1*	P						"μ"						
		p2	P						"o"						
		g1	G												

Fig. 3 – ADT example after processing

6 Conclusion

This study has shown that some real-world problems may be adequately modeled through adaptive techniques. The study has also shown that where conventional modeling may not be adequate, an adaptive modeling can fit the needs.
Among attractive characteristics of this project, two can be emphasized: reliability and affordability. Reliability is achieved by its formal developing process, which through mathematical definitions can 'predict' its behavior, preventing misleading. The production cost, however, as presented on this paper is very low.
In conclusion, this project is expected, as many others in this area, to help in the reduction of production cost and in the increasing application of adaptive solution to business problems in general.

References

[1] Neto, J. J., Pariente, C. A. B. (2002) Adaptive automata – a reduced complexity proposal. Conference on Implementation and Application of Automata – CIAA 2002, pages 161-170, July 2002.

[2] Neto, J. J. (2001) Adaptive rule-driven devices - general formulation and case study. In: International Conference on Implementation and Application of Automata - CIAA 2001, 6th, Pretoria, South Africa, July 2001. Lectures Notes on Computer Science 2494. Berlin: Springer-Verlag, p.234-250.

[3] Certo, S. C. (2000) Modern management: diversity, quality, ethics and the global environment. New Jersey: Prentice Hall

[4] Pidd, M. (1998) Modelagem empresarial: ferramentas para tomada de decisão. Trad. Gustavo Severo de Borba et al. Porto Alegre: Artes Médicas.

[5] Laudon, K., Laudon, J. (1999) Sistemas de Informação com Internet. LTC, Rio de Janeiro

[6] Neto, J. (1994) Adaptive automata for context-dependent languages. ACM SIGPLAN Notices, v.29, n.9, p.115-24, 1994.

[7] Rocha, R. L. A., Neto, J. J. (2003) Uma proposta de linguagem funcional com características adaptativas. CACIC 2003.

Robotic mapping and navigation in unknown environments using adaptive automata

Miguel Angelo de Abreu de Sousa and André Riyuiti Hirakawa

Escola Politécnica da Universidade de São Paulo,
Departamento de Engenharia de Computação e Sistemas Digitais,
Av. Prof. Luciano Gualberto, travessa 3, nº 158. São Paulo - SP - Brasil - CEP 05508-900
{miguel.sousa, andre.hirakawa}@poli.usp.br

Abstract

Real mobile robots should be able to build a virtual representation of the physical environment, in order to navigate and work in such environment. This paper presents an adaptive way to make such representation without any *a priori* information of the environment. The proposed system allows the robot to explore the entire environment and acquire the information incoming from the sensors while it travels and, due to the adaptability of the mapping method, the system is able to increase the memory usage according to the already mapped area. The map, built using the adaptive technique, is useful to provide navigation information for the robot, allowing it to move on the environment.

1 Introduction

Robotic mapping is one of the most important requirements for a truly autonomous mobile robot. Such robots should be able to build their own abstract representation of the physical environment instead of use a preliminary map stored in its memory. The main reasons are: First, the hard and repetitive task of manually mapping each house, office, factory, street and agricultural field which robots are intent to work and manually register this map in the their memory. The second reason is related to the possibility of robots work in hazardous, unstructured and unknown environments. Such places may be dangerous or impossible for humans to construct a map. Then, robots must build their own map before being able to execute others tasks [1].

Another question related to robotic mapping refers to the way of acquire such maps. Complete and geometrical representation of the environment increases the data amount and the computational complexity for the database searching during the robot localization and path planning process.

This work employs adaptive mechanisms to collect information of the world and to navigate on it. Such adaptive mechanisms allow the system to build maps without *a priori* information of the environment, and also allow the system to manage the memory space accordingly to the already mapped area.

This paper extends previous works on the representation of physical environments using adaptive automata [3], [8], [9]. The improvements are on the robot's motions, which are, movements during the exploring stage and the navigation process using the built map.

In order to present the general view of the system, the following section presents the information model used in this work, section 3 sketches the adopted formalism, section 4 describes the mapping structure, section 5 describes the automaton responsible for steer the robot during the exploration of the environment and section 6 presents the navigation mechanism using the built map.

2 The Information Model

The model proposed in this work is depicted in figure 1:

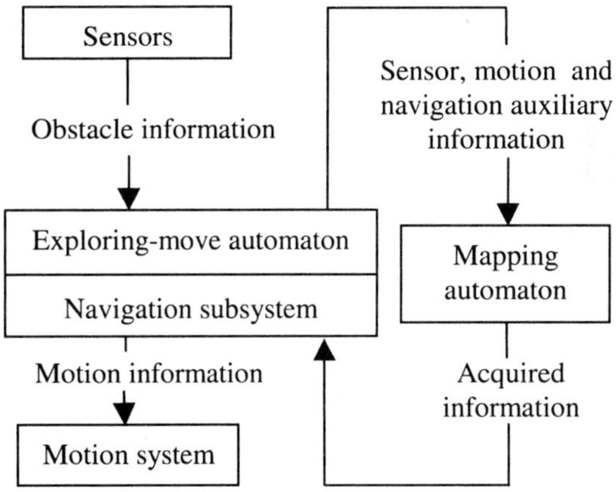

Fig. 1. System model.

In the model, the exploring-move automaton, receives data coming from the sensors and the current neighborhood information extracted from the mapping automaton. Its output contains information on the next move of the robot during its exploration's displacements. After the conclusion of the mapping process, decisions on displacements are performed by the navigation subsystem which also receives data from the sensors and the map. The mapping automaton stores all the sensor information on the presence or absence of obstacles close to the robot during its travel. Note that the environment, sensors and motions have been simulated in order to validate the proposed mechanisms.

3 Adaptive Automata

Adaptive automata extend the concept of finite automata by incorporating the feature of performing dynamic self-reconfiguration in response to external stimulus. As results, their behavior may be changed according to collected information. Such feature represents a trustful way for modeling physical environments and to conduct the robot, despite the complexity of the environment. It has been shown that adaptive automata have the same computational power as Turing devices [7]. Fig. 2 shows the graphic representation of adaptive automata, where:

e: current state before the transition;
e': current state after the transition;
a: input stimulus before the transition;
B: adaptive action before applying the transition;
A: adaptive action after applying the transition.

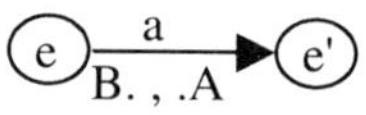

Fig. 2. Adaptive automata transition.

Adaptive actions are optional. Three different elementary adaptive actions are allowed: inspection – search the current state set for a given transition; deletion – erase a given transition from the current state set; and insertion – add a given transition to the current set of states. Such actions are denoted by preceding the desired transition by the signs ?, – and +, respectively. Reference [7] details the concept and notation of adaptive automata.

4 The Mapping Automaton

The proposed mapping automaton has the initial configuration of a square lattice (see figure 3a), composed by nine states connected by a set of transitions denoting areas to be mapped. The central state is the initial state of the automaton, and represents the starting point of the robot's exploring path. The length of each interconnection represents the distance covered by the robot size.

Figure 3b details the initial configuration of the automaton. The dot-marked state (•) represents the current position of the robot, special tags (X) are used for marking corner states and special transitions have function of support expansions in the lattice.

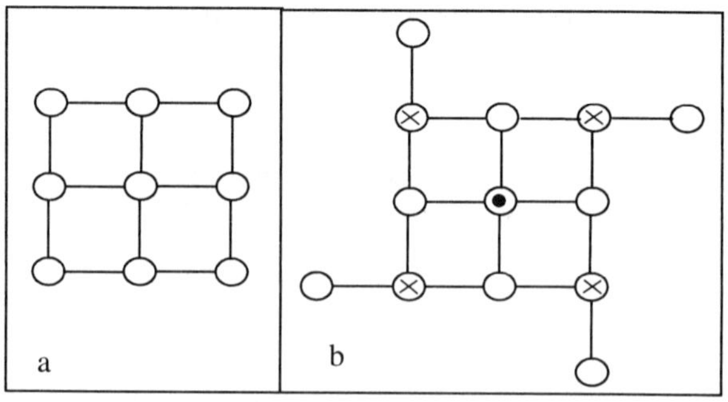

Fig. 3. (a) Initial lattice in an adaptive automaton. (b) Detailed initial automaton lattice.

Once the automaton is supplied with the data collected by the robot's sensors, the four adjacent non-filled transitions are properly replaced according to that information. The information collected by the sensors contain indications on the direction – north (N), south (S), east (E) or west (W) – and their respectively conditions – free to move or occupied (impossible to move). Figure 4a shows one example of the four-data information collected by sensors. In the figure, double arrows indicate non-obstructed areas (directions S and E) and bold lines denote obstructed ways (directions N and W).

Robot's displacements in the environment are represented by some consistent state change in the mapping automaton. As the robot moves, the current lattice is expanded in the direction of the movement. This expansion is performed by adding a line or a column to the existent lattice. Figure 4b shows the result of an N-move, after reconfiguring the vacant transitions.

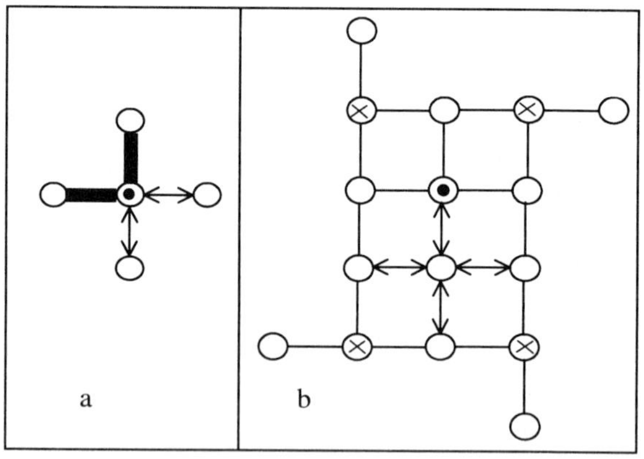

Fig. 4. (a) Example of information coming from the sensors: two directions obstructed and two free directions. (b) Expanded lattice after an N-move.

Figure 5 shows a complete map with the representation of the information acquired by the automaton after exploring an empty 'L' room. The robot has completed the exploration at the rightmost upper space of the room which is represented by the dot-marked state.

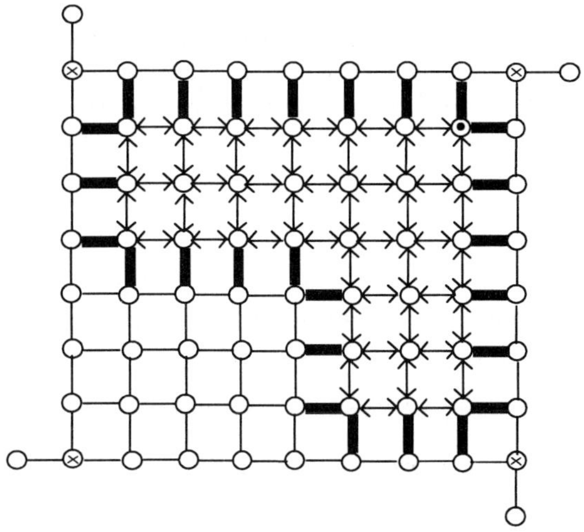

Fig. 5. Example of an empty 'L" room mapped.

As shown in the figure 5, the memory usage by the abstract representation of the real world, is proportional to the mapped area and all vacant transitions which fill the rectangular shape.

To keep the relation between the piece already built of the map and the real environment, the initial state of the automaton is adopted as representing the origin of the map. This state corresponds to the initial mapping place. Thus, one point in the map is linked to one point in the physical environment through the association of each transition in the automaton's representation with the corresponding displacement performed by the robot in the real world.

5 The Exploring-Motion Automaton

An adaptive automaton is employed for determining the robot's next move. For this purpose, it is supplied with four-data information collected by the sensors (figure 4a) and with neighborhood eight-data information previously modeled in the map (figure 6). Its operation allows the robot to cover the entire environment in a zigzag path.

The zigzag path, directed by the exploring-move automaton, may be adapted according the kind of environment detected and according the stage of exploration. To perform this adaptation, there are several branches of the exploring-move, one for each kind of possible situations. All of these branches may be connected to the initial state of the exploring-move automaton by a transition and, this connection is changed to the branch which should deal with the information from the sensor and from the map in a specific stage of exploration.

As described in section 3, this changing of connection between the initial state and the first state of the branch may be performed by adaptive actions, which allow insertion and deletion of transitions.

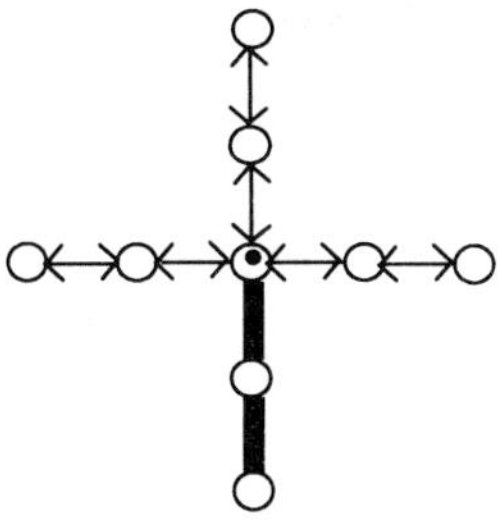

Fig. 6. Example of information extracted from the map representing two obstructed transitions and six free transitions.

While the environment is explored, the exploring-motion automaton may sign to the mapping automaton some special states, called landmarks, which are properly marked on the map. During the navigation process such landmarks are helpful for plan a trajectory.

If an obstacle is detected during the exploration's displacements, the proposed automaton marks the central state on the free space before and/or after the obstacle and it signs to the mapping automaton that this state is a landmark. Figure 7 depicts the landmarks defined for an environment-example. The mapping automaton performs this representation by properly adding a transition connecting the signed state to a special state. This special state indicates all the landmark states existent. Figure 8a shows an example of two close states signed as landmarks, in which the tag (X) marks the special state.

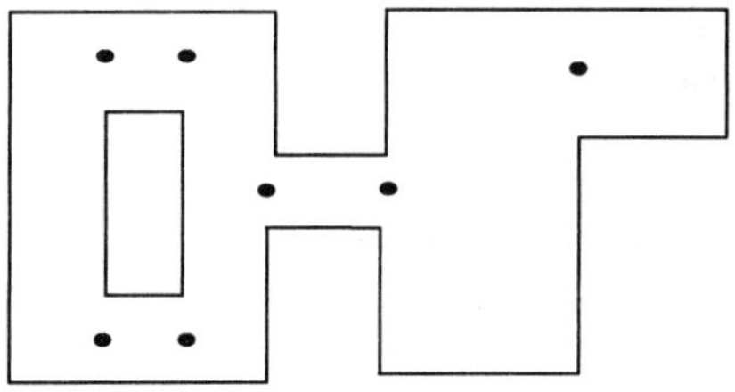

Fig. 7. Landmarks defined for an environment-example.

6 Navigation

During the navigation process the landmarks indicated by the exploring-move automaton are helpful for plan a trajectory from some initial position to a target position. The system calculates the path between such landmarks before the navigation and, during navigation, it must find which landmarks are nearest to the initial and to the target positions [4], [6]. Then, those landmarks may be viewed as sub-goals in the navigation process.

To perform the calculus of the path between two landmark-states in the mapping automaton a sequence generator is used. The sequences generated become the input of the mapping automaton, which is configured to present the landmark states as initial and final states. Then, the accepted sequences by the mapping automaton are the sequences which represent the path between two landmarks. The rejected sequences are discarded.

The four sequences firstly generated contain each one of the four main directions: N, S, E or W. Then, the following sequences generated are extensions of the previous one. For example, the three sequences generated from the N direction are NN, NE and NW (the sequence NS does not lead to a displacement, then it is not generated). Figure 8b shows the covered states by this example from the N direction (the triangle indicates the reached states and the dot-marked state (•) corresponding to the origin of searching).

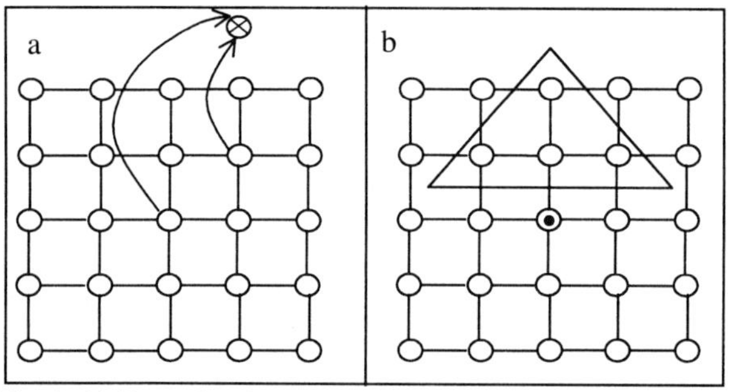

Fig. 8. (a) Example of two close states defined as landmarks. (b) Covered states by the first N-sequence generated.

7 Conclusions and Future Works

Robotic mapping is an important process for getting truly mobile robots and is also an essential feature to allow robots to complete certain tasks in hazardous, unstructured or unknown environments. The present work has shown an alternative to the classic robotic mapping approaches: adaptive mechanisms provide a new way to build maps and conduct the robot for the environment.

During the exploration, sensors attached to the robot scan the environment for the direction and presence or absence of close obstacles, and such information is collected into the model by enabling the mapping automaton to perform appropriate self-modifications.

The exploring-motion automaton allows the robot to cover the entire environment, which means that the robot is able to reach any target position during the exploration of the place. This automaton also provides special landmarks to the mapping automaton, which may be used as sub-goals during navigation. The navigation process calculates the path between two of these landmarks.

This purpose contrasts with some classic approaches (e.g., [2], [5], [10]) by presenting features of building a map without *a priori* knowledge of the space and memory space usage increasing with the actually mapped area.

Future works should deal with constructing landmarks with a serial identification attached in order to restrict the searching in the database during the calculus of the path between landmarks in the navigation process.

References

[1] Brooks, R. A. *"Visual Map Making for a Mobile Robot"*, Proc. IEEE International Conference Robotics and Automation., St. Louis, MO, 1985, pp. 824–829.

[2] Donald, B. R.; Howell J. *"Practical Mobile Robot Self-Localization"*, Proceedings International Conference on Robotics and Automation, San Francisco, CA, 2000.

[3] Hirakawa, A. R; Júnior, J. R. A.; Neto, J. J. *"Adaptive Automata for Independent Autonomous Navigation in Unknown Environment"*, International Conference on Applied Simulation and Modelling, Banff, Alberta, 2000.

[4] Innocenti, C. et al. *"Trajectory planning and real-time control of an autonomous mobile robot equipped with vision and ultrasonic sensors"*, Proc. of the International Conference on Intelligent Robots and Systems IROS'94, Munich, September 12-16, 1994, pp.1861-1866.

[5] Jennings, J.; Kirkwood-Watts, C.; Tanis, C. *"Distributed Map-Making and Navigation in Dynamic Environments"*, 1998.

[6] Ma, C.; Liu, L.; Li, W. *"Mobile Robot Motion by Integration of Low-Level Behavior Control and High-Level Global Planning"*, Proceedings of the IEEE International Conference on System, Man and Cybernetics, China, 1996, pp.310-315.

[7] Neto, J. J. "Adaptive Automata for Context-Dependent Languages", *ACM SIGPLAN Notices*, 1994, v.29, n.9, pp.115-24.

[8] Sousa, M. A. A.; Hirakawa, A. R.; Neto, J. J. *"Adaptive Automata for Mobile Robotic Mapping"*, Proceedings of 8th SBRN, São Luís, MA, Brazil. 2004.

[9] Sousa, M. A. A.; Hirakawa, A. R.; Neto, J. J. *"Adaptive Automata for Mapping Unknown Environments by Mobile Robots"*, Lecture Notes in Artificial Intelligence. Lemaître, C.; Reyes, C. A.; González, J. A. (Eds.): Advances in Artificial Intelligence – IBERAMIA 2004, Springer-Verlag, pp 562-571. 2004.

[10] Zimmer, U. R. *"Embedding local map patches in a globally consistent topological map"*, Proceedings of the Int. Symposium on Underwater Technology, Japan, 2000.

An Adaptive Framework for the Design of Software Specification Languages

João J. Neto, Paulo S. Muniz Silva

Department of Computer and Digital Systems Engineering,
Polytechnic School of the Sao Paulo University, Brazil
E-mail: {joao.jose, paulo.muniz}@poli.usp.br

Abstract

Software has been specified as domain theories. A useful strategy for building specifications is the incremental extension of an initial theory, in which increments add new terms and notions not considered in previous extensions. Given an increment, the corresponding theory is stated in a corresponding specification language. The next increment – or extension of the theory – typically requires a related language extension, which has been specified in a variety of ways, e.g. meta-computations, rewriting systems, etc. Adaptive devices naturally support such scheme, whose instances should reflect the impact of extension variations on the specification language. This paper describes an adaptive framework for the design of a class of software specification languages supporting the incremental process of elaborating software specifications.

1 Introduction

Software engineering strives to provide practitioners with principles that help building reliable software systems. One of the best-known principles concerns the elaboration of software specifications to bridge the gap between software requirements and their implementation in some programming language. It is not the case of discussing the rich debate about this issue, but it is worth mentioning some lessons learned from the use of specification languages in the traditional sequential program construction viewpoint[1]. Firstly, the use of a formal specification language, guided by some formal method, does not guarantee the correctness of the software system under construction [1], but specification languages with a formal basis increase our understanding of the specification by allowing the detection of inconsistencies and ambiguities. Secondly, a specification should support extensions, and should also exhibit operational capabilities. We are interested in these latter features as the main drivers of specification language design. The novelty of our approach is the use of an adaptive device [2] in the specification language design, supporting the realization of those desirable features. The motivation for using the adaptive device approach is that such devices naturally solve the realization problem. Section 2 presents the adopted viewpoint for specifications and specification languages. Sections 3, 4 and 5 describe our strategy to extend a specification (programming) language, and present simple examples. Section 6 draws some conclusions.

[1] In this paper, we are not considering the component based software construction approach, i.e., the reuse in-the-large.

2 Specifications and Specification Languages

Software specifications may be viewed as theories presentations elaborated for application domains. It is expected that the resulting working program be derived from its specification through a finite series of step-by-step transformations extending the base theory, i.e. the original specification [3]. Fundamental definitions underlying this approach may be found elsewhere [4, 5], to mention a few. We will briefly quote and synthesize the core definitions of [3], with respect to theory and language extensions.

- Since specifications and programs are linguistic constructs, they must be expressed in a defined linguistic system. A linguistic system consists of two parts: a collection of well-formed sentences, and a code of reasoning. Formally, $LS = <L_{LS}, \vdash_{LS}>$, where L_{LS} denotes the linguistic system language and $\vdash_{LS}$ denotes the rules of reasoning of the *LS*.
- A theory *T* in a linguistic system *LS* is a set of *LS* formulae, which is closed under $\vdash_{LS}$. A theory presentation is an axiomatization of the theory, i.e. a set of formulae from which all formulae of *T* can be derived by means of the derivability relation $\vdash_{LS}$. Formally, $T = <LS_T, A_T>$, where LS_T is the linguistic system of *T* and A_T is the presentation of *T*.
- An $LS' = <L_{LS'}, \vdash_{LS'}>$ is an extension of $LS = <L_{LS}, \vdash_{LS}>$, iff $L_{LS} \subseteq L_{LS'}$ e $\vdash_{LS} \subseteq \vdash_{LS'}$. If two theories are expressed in the same *LS*, their respective languages and presentations, leaving the underlying linguistic system implicit, can characterize them. Formally, $T = <L_T, A_T>$ e $T' = <L_{T'}, A_{T'}>$.
- A theory *T'* is an extension of *T*, i.e. $T \subseteq T'$, iff the properties defined by *T* for the symbols of L_T are still there in *T'*, and *T'* allows the proofs of some new properties.

- An extension $T \subseteq T'$ is conservative iff for all formulae A of L_T, if $A_{T'} \vdash_{SL'} A$, then $A_T \vdash_{SL} A$.

A program can also be viewed as a theory presentation with an underlying linguistic system, but due to the operational nature of programming languages there is a bias of the theory presentation towards a particular interpretation (a particular implementation) [3]. In other words, a program is a specification of another program written, say, in some machine language.

In brief, the program construction process is a series of conservative incremental extensions of an initial theory, in which increments add new terms and notions not considered in previous extensions, while preserving the properties included at each step. Given an increment, the corresponding theory is stated in a corresponding specification language. The next increment – or extension of the theory – typically requires a related language extension. To meet this requirement, the specification language should be *extensible.* Does this property imply writing a new compiler? Or would it be better using a purely syntactic extension based on the semantics of an appropriate existing language? We take the latter approach as the main strategy for the design of software specification languages. In the literature, a number of strategies are found that address this subject, e.g. using modular compilation [6].

In this paper, we illustrate this strategy with a very simple example. Let the following operation be a fragment of a theory presentation on integers, written in an imperative Pascal-like programming language:

```
FUNCTION divide (n: INTEGER; d: INTEGER): INTEGER;
        START
        divide:= n/d
        END;
```

Let us suppose we want to extend the theory with the notion of *preconditions* as of the *design by contract* approach [7]. This approach states that the function invocations at any point in a program must satisfy the precondition specified for the function. That means the programmer will not write explicit defensive code in order to check the precondition in the function body, because the respect for the precondition is an obligation for the client of certain service, not for the supplier of that service. The precondition specification for the above fragment may be written as: `PRE d <> 0`, where `PRE` denotes a precondition. This theory extension implies a language extension.

3 Extending Programming Languages

Let us sketch a proposal for extending programming languages as an intermediate goal towards the extension of software specifications. Our main intent is keeping the specifications executable. Starting from an available programming language, one is allowed to employ in the desired software specification strictly the abstractions provided by the host language's syntactic constructs. When this is a syntactically extensible language, users may create new syntactic constructs for representing abstractions not included in the original notation, so users may represent new abstractions in terms of existing ones. By proceeding in a hierarchical way, successive abstraction layers of increasing complexity may be built in order to bring the expressiveness of the language closer and closer to the domain of the particular software being specified. Since each layer's abstractions are stated strictly as combinations of previously defined abstractions, at each specification step any software will be easily translated into the immediately preceding layer's abstractions, until reaching the lowest level abstraction layer, represented by the host programming language.

With such a simple approach, specifying a software is reduced to specifying a sequence of successive abstractions, which are directly mapped into corresponding language syntactical extensions and their associate translating scheme into the abstractions defined in previously defined abstraction layers. This process proceeds until the set of available abstractions match the expressiveness requirements for defining the application software in terms of the application domain's abstractions.

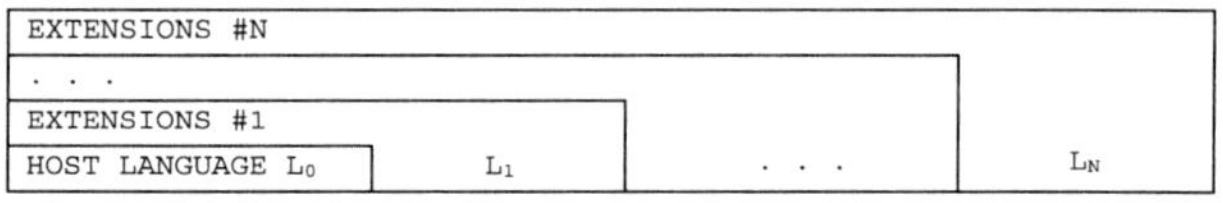

The main feature of this approach is keeping the specifications executable all along the development process. No semantic gaps are introduced between the specification and its implementation, since all specification steps may be immediately converted into executable code, and no abstraction is created unless it be strictly stated in terms of existing, executable ones.

By adequately choosing the host programming language, one may significantly reduce the need for supporting computing resources for implementing this approach. For instance, if L_0 is an already existing extensible language (e.g. LISP) all one must do is to define the abstractions needed as extensions of L_0 (in the case of LISP, the extensions take the form of either macros or functions). On the other hand, when L_0 is not already extensible, then an extension feature must be added in order to allow following the proposed approach.

For usual non-extensible procedural host languages, it is possible to implement a language extension additional layer in order to provide the needed facilities for programmers to provide their language extensions as definitions for new abstractions. A powerful way to allow the inclusion of extra abstractions in a given language is to offer some meta-linguistic feature for allowing to define new syntactical constructs. In our case, a modified Wirth's context-free extended BNF notation has been chosen [8]. That is enough for defining new syntax. Processing and incorporating such user-defined syntactical extensions as a preprocessor for the host language compiler is quite straightforward [9]. However, syntax is not all we need. It is necessary to state all extensions in terms of already existing language constructs. In order to overcome this problem, we adopted another well-known solution, e.g. typifying the extensions and using classical operational semantics for interpreting each new construct in terms of existing ones: each new extension is declared as a context-free grammar, and its meaning is also declared as a text stated as a program using the basic host language syntax enriched with previously declared syntactical constructs.

In this way, the compiler is informed on the new construct to be further accepted, as well as on how exactly it must be translated into lower-level abstractions. After accepting such an extension definition, the compiler extends its acceptor in order to recognize the new syntax, and associates the declared translation procedure to the syntactical recognition of the new construct. Whenever further input text contain excerpts that follow the new defined syntax, its related translation procedure is followed, converting the input text in the extended notation into another text written in the previous abstraction level.

4 A Very Simple Extension Layer

In this section we present a simplified proposal of a nucleus for an extension mechanism, to be used as a preprocessor for procedural, originally non-extensible languages. For space reasons, the host language has been reduced to a minimum: block structure has been simplified, declarations have been reduced to simple intege variables, and commands have been eliminated, except for if's, go to's and assignments of simple expressions.

The following context-free grammar, stated in modified Wirth's notation, defines, in its first part, the (non-extensible host language) nucleus L_0 we are going to use as the starting version of our extensible language, and in its second part, the proposed extension mechanism, represented by **EXT**. Each time the non-terminal **EXT** is instantiated, it extends the previous version of the language by adding a new non-terminal (**NEWNTERM**) to its grammar, and incorporates the corresponding abstraction to the language syntax. Note that **TERM** represents any terminal in the language, including identifiers (`id`), integers (`int`) and other elementary language components.

```
//* HOST LANGUAGE (EXTREMELY SIMPLIFIED) *//
PROG = "BEGIN" ( DECL \ ";" ) "START" ( COM \ ";" ) "END" .
DECL = "VAR" ( id \ "," ) ":" "INTEGER" | PROCEDURE | EXTENSION .
COM  =  LABEL ":" PROG  | id := EXPARIT | "GOTO" LABEL |
         "IF" EXPARIT ( ">" | "=" | "<" | "<>" ) EXPARIT
            "THEN" PROG ( "ELSE" PROG | ε ) | PREVIOUSNTERM .
EXPARIT = (( id | int | CALL ) \ ( "+" | "-" | "*" | "/" )) .
CALL = id "(" ( id | int | CALL  \ "," ) ")" .
PROCEDURE = "FUNCTION" id "(" ( id ":" "INTEGER" \ ";" ) ")"
                        ":" "INTEGER"  ";"
                        "START" ( COM \ ";" ) "END" ";" .
LABEL = id .
```

```
//* PROPOSED EXTENSION MECHANISM *//
PREVIOUSNTERM  =  Ø .
EXTENSION = "DEFINE" NEWNTERM ":" "NEW" NTERM "AS" WIRTHMOD
                         "MEANING" PREVIOUSWIRTHMOD "ENDDEFINE" .
NTERM = "PROG" | "DECL" | "COM" | "EXPARIT" | "EXTENSION"
      | "LABEL" | "NTERM" | "CALL" | "PROCEDURE" | "NEWNTERM"
      | "WIRTHMOD" | "PREVIOUSWIRTHMOD" | PREVIOUSNTERM .
NEWNTERM = id .
WIRTHMOD = ((( TERM | NTERM | NEWNTERM | "ε"
          | "(" WIRTHMOD ( "\" WIRTHMOD  | ε ) ")" )
            ( "#" int | ε ) \ ( "|" | ε  ) ) .
PREVIOUSWIRTHMOD = ((( TERM | NTERM | "ε"
        | "(" PREVIOUSWIRTHMOD ( "\" PREVIOUSWIRTHMOD | ε ) ")" )
          ( "#" int | ε ) \ ( "|" | ε  ) ) .
```

The interpretation of the above grammar is almost conventional, except for the meta-symbol Ø that refers to the empty set: initially there are no **PREVIOUSNTERM**s. After the full handling of the declaration of an **EXTENSION** the name corresponding to the **NEWNTERM** being declared is added to the **PREVIOUSNTERM**s set of already defined non-terminals of the grammar. So, **WIRTHMOD** refers to some syntactical definition involving any terminals or non-terminals, while **PREVIOUSWIRTHMOD** represents a syntax strictly stated in terms of the non-terminals representing abstractions known at the previous extension's abstraction layer. The extension pre-processor adequately updates the set of **PREVIOUSNTERM**s in order to keep the integrity of this mechanism.

5 Illustrating Case Studies

For illustration purposes, let us first work the small situation referred to at the end of section 2. Let us restrict the extension to including in the language the declaration of preconditions. The suggested syntax has been starting the construct with the word `"PRE"` followed by a condition (in our language, conditions are defined as relations between two arithmetic expressions. In order to explore the extensibility feature introduced by the preprocessor, the programmer should declare the desired new syntax as shown below.

In words, the denominator of the division will be automatically checked against zero every time the function `divide` is called, and an error report will be generated whenever that condition succeeds.

```
DEFINE PRECONDITION: NEW COM AS
   "PRE" EXPARIT # 1 ( "=" | "<" | ">" | "<>" ) EXPARIT # 2
MEANING
   "IF NOT (" EXPARIT # 1 ( "=" | "<" | ">" | "<>" ) EXPARIT # 2
         ") THEN ERROR() ELSE"
ENDDEFINE;
...
FUNCTION divide (n: INTEGER; d: INTEGER): INTEGER;
      START
      PRECONDITION d <> 0;
      divide:=n/d
      END;
...
```

Our preprocessor generates an expanded code as follows:

```
FUNCTION divide (n: INTEGER; d: INTEGER): INTEGER;
      START
      IF NOT ( d <> 0 ) THEN ERROR () ELSE divide:=n/d
      END;
```

The next is another simple illustrating example that creates a new command. By using such a feature, it is easy to create new abstractions from already existing ones. For instance, the next declaration adds a `WHILE` statement to our host language:

```
DEFINE WHILESTATEMENT: NEW COM AS
   "WHILE" EXPARIT # 1 ( "=" | "<" | ">" | "<>" ) EXPARIT # 2
   "REPEAT" PROG # 3
MEANING
 "LOOP#: IF"  EXPARIT # 1 ( "=" | "<" | ">" | "<>" ) EXPARIT # 2
       "THEN BEGIN" PROG # 3 "; GO TO LOOP# END"
ENDDEFINE;
```

In this case, note the meta-label `LOOP#` that must be instantiated each time a `WHILESTATEMENT` is called, in order to avoid label duplication in the resulting program. For instance, the `WHILESTATEMENT` below:

```
WHILE x < 3 REPEAT BEGIN x := x+1; y:= y-1 END
```

is expanded into the following equivalent program, according to the template previously defined:

```
LOOP0001: IF x < 3 THEN BEGIN
                        BEGIN x := x+1; y:= y-1 END;
                        GO TO LOOP0001
                        END
```

The host language must at least provide the full set of primitive constructs needed to specify all operations in the program we intend to build. If it is not the case, then it would not be possible to express the application program facts in terms of the available host language's basic syntactic constructs, unless some extra effort be made in order to provide the missing facilities for the host language before the desired extensions are created and used.

6 Conclusions

Obviously, there is much more to be said about using extensible features of a programming language in order to ease the specification of software projects than what one is allowed to fit into a four-page paper. Programming language extensibility and operational semantics have not received adequate attention in recent scientific works. However, as we tried to show in this paper, their features allow an easy way for implementing software directly from specifications, in a bottom-um fashion, always keeping the specifications executable, and eliminating the deep gap between specifications and implementation, usually found in current software engineering practices.
Using adaptive technology in the implementation of compilers and preprocessors substantially reduces the difficulty of implementing languages representing multi-layers of abstractions by allowing the user to interact with the kernel of the host language's compiler without opening its source code. Such great feature is accomplished by the unification got when using adaptive automata as the language's run-time abstract machine.
In this way, adaptive automata may be generated both as object-code and as a syntax recognizer, so programs may be compiled, their syntax may be modified and they also may be easily executed in the same unified framework provided by the underlying adaptive environment.

References

[1] Clarke, E.M., Wing, J.M. (1996) Formal methods: state of the art and future directions. ACM Computing Surveys 28(4): 626-643.

[2] Neto, J.J. (2001) Adaptive rule-driven devices – general formulation and case study. LNCS v.2494, Springer-Verlag, pp. 234-250.

[3] Turski, W.M., Maibaum, T.S.E. (1987) The specification of computer programs. Addison-Wesley, London, UK.

[4] Smith, D.R. (1999) Designware: software development by refinement. In Proc. 8th Internat. Conf. on Category Theory and Computer Science (CTCS '98), Edinburgh, UK.

[5] Maibaum, T.S.E. (2003) On what exactly goes on when software is developed step-by-step, II: the sequel. Information Processing Letters 88: 45-51.

[6] Harrison, W.L. and Kamin, S.N. (1998) Modular compilers based on monad transformers. In Proc. of the IEEE International Conference on Computer Languages, pp. 122-131.

[7] Meyer, B. (1997) Object-oriented software construction, 2nd. Ed. Prentice-Hall, New Jersey, USA.

[8] Wirth, N. (1977) What can we do with the unnecessary diversity of notation for syntactic definitions? CACM 20 (11): 822-823.

[9] Neto, J.J., Pariente, C.B., Leonardi, F. (1999) Compiler Construction – a Pedagogical Approach. V Int. Congress on Informatics Engineering – Buenos Aires.

Swarm Intelligence Clustering Algorithm based on Attractor

Qingyong Li, Zhiping Shi, Zhongzhi Shi

Institute of Computing Technology-Chinese Academy of Sciences – Beijing, China
E-mail: {liqy, shizp, shizz}@ics.ict.ac.cn

Abstract

Ant colonies behavior and their self-organizing capabilities have been popularly studied, and various swarm intelligence models and clustering algorithms also have been proposed. Unfortunately, the cluster number is often too high and convergence is also slow. We put forward a novel structure-attractor, which actively attracts and guides the ant's behavior, and implement an efficient strategy to adaptively control the clustering behavior. Our experiments show that swarm intelligence clustering algorithm based on attractor (**SICBA** for short) greatly improves the convergence speed and clustering quality compared with LF and also has many notable virtue such as flexibility, decentralization.

1 . Introduction

Swarm Intelligence emerged out of social insect collective behavior shows many interesting properties such as flexibility, robustness, decentralization and self-organization. Implementations of optimization and control algorithms based on swarm intelligence such as Ant Colony Optimization and Ant Colony Routing have been well known [1,2,3]. Clustering models and algorithms based on swarm intelligence, inspired by co-operative brood sorting by ants or other behavior, are also put forward, though they are still in a preliminary, proof-of -concept stage [4,5].

The swarm intelligence clustering models and algorithms have advantages in many aspects, such as no need of priori information, self-organization. However, the number of result cluster is often too high and the convergence is slow because of the ant's inefficient behaviors: randomly picking up items and dropping down items. Are there any methods to make ant to perform efficiently?

After some careful research, we believe that the algorithms inefficient performance is mainly because of the ant's inefficient moving. Especially in the first stage, the items is distributed sparsely, the probability for an ant to move to a place to pick up items or to drop down items is often small, so most of the ant's moving is inefficient. At the same time, the number of items loaded by an ant is an important factor for cluster number and accuracy.

In this paper, we put forward a new algorithm, named swarm intelligence clustering algorithm based on attractor (SICBA for short). SICBA attacks the problem in following two aspects:

Firstly, a novel, efficient structure, called attractor, is constructed. Simply, it is an item set converging the homogeneous items. Moreover, it contains not only local environment information, such as inner distance, but also global information, such as outer distance between attractors in the system. So it can actively attracts the ant to pick up dissimilar items or drop down similar items.

Secondly, an ant can pick up the farthest item or all items from an attractor controlled by a simple rule. Furthermore, a parameter is applied to form a strategy: first stage an ant is prior to pick up all items to coarsely but fast cluster; last stage the ant is mostly to pick up the farthest item to precisely partition. So it can distinctly improve the convergence speed and accuracy.

The paper is organized as follow: the following section introduces the related work; the next section describes the details of ***SICBA*** ; the experiments are showed in the section 4; at last we make our conclusion.

2. Related Work

Deneubourg et al [4] proposed an agent-based model to explain how ants manage to cluster the corpses of their dead nestmates. Artificial ants (or agents) are moving randomly on a square grid of cells on which some items are scattered. Each cell can only contain a single item. Whenever an unloaded ant encounters an item, this item is picked up with a probability which depends on an estimation of the density of items of the same type in the neighborhood. When a loaded ant encounters a free cell on the grid, the probability that this item is dropped also depends on an estimation of the local density of items of the same type.

Lumer and Faieta[5] (LF for short)extended the model of Deneubourg et al., using a dissimilarity-based evaluation of the local density, in order to make it suitable for data clustering. Unfortunately, the resulting number of clusters is often too high and convergence is slow. Therefore, a number of modifications were proposed, by Lumer and Faieta themselves as well as by others [6,7].

3. Swarm Intelligence Clustering Algorithm Based on Attractor

3.1 Basic Concept

Definition 1. Attractor is a data set which has similar items as a whole. The attractor can attract the ants to pick up the furthest item or all items from it according to a pick-attractive rate, and drop down an item or an item collect in it according to a drop-attractive rate. Furthermore, the attractor has it own status: active and inactive, if the attractor has not any item the attractor is considered as inactive, and it can't attract ant to pick up items or drop down items any longer; otherwise, it is considered as active and can attract any ant.

Definition 2. Inner distance is the average distance between the items and the attractor's centroid for an attractor.

Definition 3. Outer distance is the average distance with the other attractors in the system for an attractor.

Attractor's pick-attractive rate and drop-attractive rate are two very important features to control an ant to load items or unload items, differentiating the ant's randomly choosing method in [4, 5]. Pick-attractive rate (PAR) for an attractor is a numeric feature to indicate the attractor's affinity attracting the ant to pick up items, in other word, it indicate the probability for an ant to pick up a item or all the items from the attractor. If the PAR is greater, the probability for an ant to pick up items is also greater. The PAR value depends on the attractor's item number marked as C, if the number C is less than a threshold θ, the attractor is a small attractor whose PAR is determined by its inner distance, outer distance and item number, and if the inner distance, the outer distance and item number are less the PAR is greater; otherwise, if C is greater than θ the attractor is considered as big one and its PAR is effected by the maximal distance in the attractor and outer distance. The PAR is given by

$$PAR = \begin{cases} f_1(D_i)\times\alpha_1 + f_1(Do)\times\alpha_2 + T(C,\theta)\times\alpha_3 + \delta & if\ C<\theta \\ Dm\times\beta_1 + f_1(Do)\times\beta_2 & if\ C\geq\theta \end{cases}$$

(1)

where $D_i \in [0,1]$, $D_o \in [0,1]$ represent inner distance and outer distance, $f_1(x) = 1 - x$, and $T(C,\theta) = \dfrac{\theta}{C+\theta}$,

$0 \leq \alpha_1, \alpha_2, \alpha_3 \leq 1\ and\ \alpha_1 + \alpha_2 + \alpha_3 = 1$, δ is a constant parameter to control the choosing strategy: if δ>1, the attractor which has few items will have the absolute priority to be chosen comparing with the attractor which has many items. Dm is the maximal distance in the attractor, and $0 \leq \beta_1, \beta_2 \leq 1\ and\ \beta_1 + \beta_2 = 1$.

Drop-attractive rate (DAR) for an attractor is a contrast feature to indicate the attractor's affinity attracting the ant to drop down items. In the same way, The DAR value firstly depends on C, if the number C is less than the threshold θ the attractor is small one and the DAR is always small; if $C \geq \theta$ the attractor is a big one and its DAR is determined by the distance between the items loaded by a ant and the attractor's items, and the outer distance also has slight influence. The DAR is given by

$$DAR = \begin{cases} C/T & if \quad C < \theta \\ (1-D)\times\beta_1 + Do\times\beta_2 & if \quad C \geq \theta \end{cases} \qquad (2)$$

where T is the total number of items in the system, D is the distance between the loaded items by an ant and the attractor's items, $0 \leq \beta_1, \beta_2 \leq 1\ and\ \beta_1 + \beta_2 = 1$ β_1 is always greater than β_2.

3.2 Probability Conversion Function

Probability conversion function is a function which converts the attractor's quality and quantity characters into a pick up probability for an unloaded ant or drop down probability for an loaded ant. There are usually two related functions. One is for picking-up probability; another is for dropping probability.

Picking up stimulus An unloaded ant can perform the task: picking up an item or picking up all items. When the item number of the target attractor C is less than θ the ant picks up all the items. Obviously, the unloaded ant should pick up an entire attractor if the attractor is small, homogeneous and not isolated. The attractor is more homogeneous and coupling its inner distance and outer distance is less. So the $S_{pick-all}$ is given by

$$S_{pick-all} = f_1(D_i)\times\alpha_1 + f_1(Do)\times\alpha_2 \quad (4)$$

where $D_i \in [0,1]$, $D_o \in [0,1]$ represent inner distance and outer distance $0 \leq \alpha_1, \alpha_2 \leq 1\ and\ \alpha_1 + \alpha_2 = 1$.

While the item number C is greater than θ the attractor is considered as a big one, so only the most dissimilar item should be picked up. The stimulus $S_{pick-one}$ for picking up one item is mainly influenced by the furthest distance and slightly by the outer distance. $S_{pick-one}$ is given by

$$S_{pick-one} = Dm\times\beta_1 + f_1(Do)\times\beta_2 \quad (5)$$

where Dm is the maximal distance in the attractor, and $0 \leq \beta_1, \beta_2 \leq 1\ and\ \beta_1 + \beta_2 = 1$.

Drop down stimulus The stimulus for a loaded ant to drop its items L in an attractor is based on the local distance D between the loaded items and the attractor's items. If the ant loads an item collect the center of the loaded items is used to measure the similarity. Furthermore, we also consider the global factor – outer distance because the outer distance indicates the dependence among the attractors in the system. Stimulus for dropping down S_{drop} is given by

$$S_{drop} = (1-D)\times\beta_1 + Do\times\beta_2 \quad (6)$$

where $0 \le \beta_1, \beta_2 \le 1 \; and \; \beta_1 + \beta_2 = 1$ β_1 is always greater than β_2.

3.3 Algorithm Description

Based on the above description we have the following algorithm.

Algorithm (**SICBA**: Swarm Intelligence Clustering Algorithm Based on Attractor)

1)Initialize θ, δ, MAXCYCLENUMBER ATTRACTORNUMBER and other parameters;

2. Run k-means algorithm assigned with ATTRACTORNUMBER clusters to form the original ATTRACTORNUMBER attractors;
3. Give ants initial attractors, initial states of ants are unloaded;
4. **WHILE**(cycle_counter< MAXCYCLENUMBER and NotConvergent) {
5. **FOR** (number of ants) {
6. **IF** the ant is unloaded, **THEN**

{calculate P_p;

Compare P_p with a random probability P_r, **IF** $P_p < P_r$ **THEN**

{ Not pick up anything;

According to PAR values, the unloaded ant moves to the greatest attractor;

}**ELSE**

{ Pick up the most farther items or all items;

Update the attractor's state according to the picking task;

According to DAR values, the loaded ant moves to the greatest attractor;

}}**ELSE** the ant is loaded

{ Calculate P_d;

Compare P_d with a random probability P_r, **IF** $P_d < P_r$ **THEN**

{ Not drop down;

According to DAR values, the loaded ant moves to the greatest attractor;

}**ELSE**

{ Drop down the load;

Update the attractor's state according to the dropping task;

According to PAR values, the loaded ant moves to the greatest attractor;

}}}}

4. Experimental Results

In this section, we'll demonstrate the experimental results about the performance of **SICBA** and the influence of the parameter δ for clustering performance.

The dataset IRIS chosen from UCI machine learning repository (http://www.ics.uci.edu/~mlearn/MLRepository.html) are used in this paper. IRIS database has 150 records with 4 attributes. All experiments are performed on a 800-MHz Pentium machine which 512 megabytes main memory, running on Windows 2000 professional. Programs are written in Windows/Vision C++ 6.0.

4.1 Comparison with FL

In this experiment we compare SICBA with LF about the clustering performance. Because here accuracy is little important we measure the clustering result number when run N cycles. The result is shown in the figure 1.

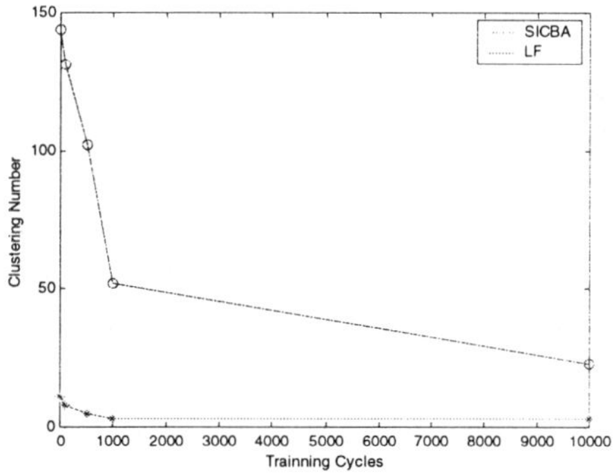

Fig. 1. Clustering Number with Training Cycle

Figure 1 shows that convergence speed of SICBA is faster than LF, especially in the first stage, SICBA quickly partitions dataset into a few clusters, however, the LF is very slow to be convergent. The last cluster number of SICBA is also better than LF. Figure 1 shows that the last number of SICBA is 4 a little greater than the real number 3, but LF is much greater than 3. Because SICBA initialize ATTRACTORNUMBER attractors with Kmeans, moreover, it picks up all items in the small attractor at first stage and picks up the farthest items in the big attractor at last stage, so SICBA can converge greatly faster than LF, furthermore, it also can get much better clustering result.

4.2 Influence of Parameter δ for the Clustering Performance

The parameter δ in formula (1) is a most important factor to affect the clustering performance, because it directly determines the picking up strategy. In this experiment the clustering performance is measured by clustering result number through 1000 times training. The influence shows in figure 2.

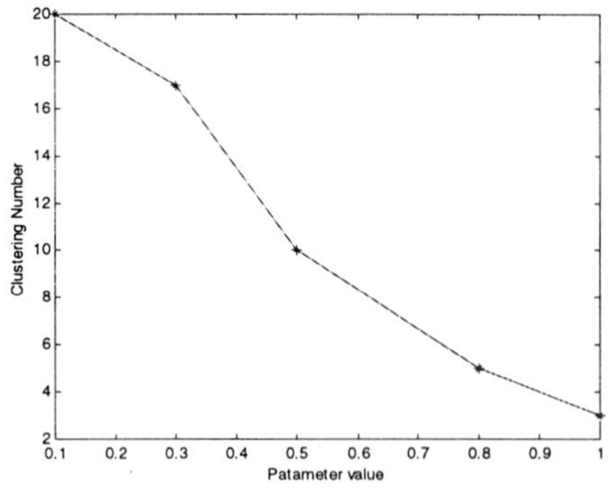

Fig.2. Influence of Parameter δ

Figure 2 shows that if value of δ is greater the clustering performance is better. From formula 1, we can easily get that δ directly effect the PAR value, that is to say, δ affects the probability for the small attractors to be chosen to pick up items. If δ is greater, the small attractor is more prior to be chosen than the big one, so SICBA is quicker to converge. As to last stage, most of the active attractors are all big one, and δ doesn't effect any longer. The probability of picking up item is mainly determine by the attractor's furthest item, so SICBA can precisely adjust the cluster and get better accuracy performance. Because the PAR values in the case of $C>\theta$ is always smaller than 1, obviously, δ greater than 1 is none meaning. In conclusion, when δ equals 1 SICBA can get the best performance and implement the efficient strategy: *small attractors prior.*

5. Conclusion

This paper put forward a novel structure in swarm intelligence clustering algorithm, named attractor, which contains not only the local information but also the global information, so it can actively attract the ant to pick up items or drop down items and avoid the ant aimless moving. Furthermore, we also implement an efficient strategy: *small attractors prior,* based on the attractor. The strategy makes SICBA pick up all items to coarsely but fast cluster in the first stage and pick up the furthest item to precisely fine-tune. The experiments prove that it can greatly improve the algorithm convergence speed and clustering quality compared with LF algorithm. Although SICBA has no advantages over classic kmeans algorithms on the aspect of space and time complexity, as a self-organization clustering algorithm, it has great advantages in robustness, visualization, flexibility and decentralization.

References

[1] Becker R., Holland O.E. and Deneubourg J.L. 'From local actions to global tasks: Stigmergy and collective robotics', in Brooks R. and Maes P. Artificial Life IV, MIT Press, 1994;

[2] E.Bonabeau, M.Dorigo,G.Theraulaz, Inspiration for optimization from social insect behaviour, Nature,vol 406,6 July 2000.

[3]Gianni Di Caro and Marco Dorigo, AntNet: Distributed Stigmergetic Control for Communications Networks, Journal of Artificial Intelligence Research 9(1998) 317-355;

[4] Deneubourg..J.L., Goss S.,Frank,N., Sendova-hanks, A.,Detrain C.,Chrerien L., The dynamics of collective sorting: robot-like ants and ant-like robots, in: Meyer J., Wilson S.W. (Eds.), Proceedings of the First International Conference on Simulation of Adaptive Behavior: From Animals to Animats, MIT Press/Bradford Books, Cambridge, MA, 1991, pp.356-363;

[5] E.Lumer, B.Faieta. Diversity and adaptation in populations of clustering ants . in J.-A.Meyer, S.W. Wilson(Eds.), Proceedings of the Third International Conference on Simulation of Adaptive Behavior: From Animals to Animats, Vol.3, MIT Press/ Bradford Books, Cambridge, MA, 1994, pp 501-508;

[6] J. Handl, B. Meyer. Improved Ant-Based Clustering and Sorting in a Document Retrieval Interface. Proc. of the 7th Int. Conf. on Parallel Problem Solving from Nature. 913-923 (2002).

[7] V. Ramos, F. Muge, P. Pina. Self-Organized Data and Image Retrieval as a Consequence of Inter-Dynamic Synergistic Relationships in Artificial Ant Colonies. Soft Computing Systems: Design, Management and Applications. 87, 500-509 (2002).

Ant-based distributed optimization for supply chain management[1]

Carlos. A. Silva[1], J. M. Sousa[1], T. Runkler[2], J.M.G Sá da Costa[1]
[1]Dep. Mechanical Engineering, Instituto Superior Técnico, Technical University of Lisbon
[2] Information and Communications, Siemens AG - Corporate Technology
E-mail: {csilva,j.sousa,sadacosta}@dem.ist.utl.pt

Abstract

Multi-agent systems are the best approach for an efficient supply chain management. However, the control of each sub-system in a supply-chain is a complex optimization problem and therefore the agents have to include powerful optimization resources along with the communication capacities. This paper presents a new methodology for supply-chain management, the distributed optimization based on ant colony optimization, where the concepts of multi-agent systems and metaheuristics are merged. A simulation example, with the logistic and the distribution sub-systems of a supply-chain, shows how the distributed optimization outperforms a centralized approach.

1 Introduction

In order to improve competitiveness and profitability, most of the companies today are organized as *supply-chains*: a world-wide network of external partners (suppliers, warehouses and distribution centers) through which raw materials are acquired, transformed into products and delivered to costumers [1]. The company's job is no longer to produce the goods, but to manage all the different partners in a coordinated manner such that in the end the costumer receives a quality product on a certain desired date.

The different partners in a supply-chain operate under different sets of constraints and objectives. However, the systems are highly interdependent and the optimization of objectives such as on-time deliveries or costs of one partner will influence the performance of the remaining partners. The supply-chain is a pure distributed system with several parallel and independent optimization problem and the coherence between the different decision making centers can be accomplished by a multi-agent based framework, based on explicit communication between constituent agents to control multiple systems [1, 5].

This paper introduces an innovative management methodology based on the description of the supply chain as a distributed optimization problem. The optimization problems are solved by the ant colonies metaheuristic that can also be used as a multi-agent framework.

2 Description of a supply-chain

A typical supply chain has at least two partners: the *logistic* system, that collects the orders from the customers, purchases the components from external suppliers and schedules the components gathered in cross-docking centers, e.g. airports, see [5]); and the *distribution* system, an external company that collects the components at the cross-docking centers and delivers them to the clients as orders. The task of each system can be modeled as an optimization problem.

2.1 Logistic process

The logistic system receives every day new orders requested by different clients, where an order o_j is a set of different types of components in certain quantities, with a certain due date d_j. The different components and their quantities are purchased from external suppliers, that deliver the components to the cross-docking centers after a certain period of time. The logistic process task is a scheduling problem that consists of observing the list of n orders and the list of components, and decides which orders are released at date r_j.

The difference between the release date and the due date is called the *lateness* $L_j = r_j - d_j$. The objective is to match the release date with the due date, i.e. to have for all orders $L_j = 0$. This decision step is done once per day. Two disturbances may influence the system: the fact that suppliers service may not be respected; and the fact that some clients ask for desired delivery dates not compatible to supplier services. The optimization

[1]This work is supported by the German Ministry of Education and Research (BMBF) under Contract no.13N7906 (project Nivelli) and by the Portuguese Foundation for Science and Technology (FCT) under Grant no. SFRH/BD/6366/2001 and "Programa de Financiamento Plurianual de Unidades de I&D (POCTI), do Quadro Comunitário de Apoio III".

objective is to minimize the cost function given by

$$f_L = \frac{\sum_{j \in O} |L_j| + \#\{j \notin O\} + \sum_{j \notin O} L_j}{\#\{j \in O : L_j = 0\}} \quad (1)$$

where $\sum_{j \in O} |L_j|$ accounts for the minimization of the lateness of the set of released orders O; $\#\{j \notin O\}$ refers to the minimization of the number of orders not released; $\sum_{j \notin O} L_j$ describes the minimization of the expected lateness of those orders that remain in the system and are delayed; and finally $\#\{j \in O, L_j = 0\}$ accounts for the maximization of the number of orders delivered at the correct date. This problem can be formally described by a disjunctive graph $G = \{V, A\}$, where the vertices V represent the n orders waiting to be released.

2.2 Distribution process

After the scheduling method has decided which orders will be delivered, a distribution company will pick-up the assigned components and deliver them to the different clients. There is a direct correspondence between clients and orders, but clients are described in this case by their geographical location.

In general, a distribution problem consists of determining how many trucks are necessary to transport the orders and which sequence should be followed in order to minimize the transportation costs. We consider here two constraints: the maximum load capacity Q and the maximum travel distance R of each truck. This distribution problem can be modeled as a *Vehicle Routing Problem* (VRP)[2]. In this case, the cost function to be minimized is the distance traveled by all the vehicles

$$f_D = \sum_{i=0}^{m} \sum_{j=0}^{m} \sum_{l=1}^{v} d_{ij} x_{ijl} \quad (2)$$

where $x_{ijl} = \{0, 1\}$ indicates if the vehicle l traveled the distance $d_{i,j}$ from client i to j: if yes, $x_{ijl} = 1$; if not $x_{ijl} = 0$. The problem can be represented by a disjunctive graph $G = \{V, A\}$, where the vertices V represent the location of the clients and the arcs A are associated with the traveling distance d_{ij} between the vertices.

3 Supply-chain management through distributed optimization

Two different strategies can be adopted for the supply chain management: the *centralized* or the *distributed* optimization. In the first case, the logistic partner is the dominant partner and defines the supply chain solution according only to the logistic objectives. The distribution partner can only optimize the static solution provided by the central system. In the distributed approach, both partners are equally important and the final solution is found after the systems agree about the solution that is better for both systems. This is achieved through a distributed optimization description of the supply chain management problem.

Consider that the supply-chain is a system $S = S_L \circ S_D$ consisting of an aggregation $\circ$ of the logistic system S_L and the distribution system S_D. Let $f = f_L \circ f_D$ be the cost function of the system S, where f_L and f_D are the expressions proposed in (1) and (2), respectively. Distributed optimization is a methodology where the two optimization processes are running in parallel and each of the processes is using the intermediate results of the other process. This can be defined as

$$\min f(t) = \min[f_L(t - l_1)] \circ \min[f_D(t - l_2)] \quad (3)$$

where t describes the actual optimization iteration and $t - l_1$ and $t - l_2$ describe previous optimization iterations. If $l_1 \neq l_2$, the distributed optimization is said to be *asynchronous*, which means that at every optimization iteration, the optimization method accesses information from previous and different iterations. The asynchronous method has been the most used method [6], in order to avoid convergence problems of the $\min[f_i]$ optimization methods, for example in cases where the computational effort of one iteration is different from method to method.

Next section shows how this framework is easily implemented using the Ant Colony Optimization methodology.

4 Ant Colony optimization

The Ant Colony Optimization (ACO) methodology [3] is an optimization method suited to find minimum cost paths in optimization problems described by graphs. Consider a problem with n nodes and a colony of g ants. Initially, the g ants are randomly placed in g different nodes. The probability that an ant k in node i chooses node j as the next node to visit is given by

$$p_{ij}^k(t) = \begin{cases} \frac{\tau_{ij}{}^{\alpha} \cdot \eta_{ij}{}^{\beta}}{\sum_{r \notin \Gamma}^{n} \tau_{ir}{}^{\alpha} \cdot \eta_{ir}{}^{\beta}} & \text{if } j \notin \Gamma \\ 0 & \text{otherwise} \end{cases} \quad (4)$$

where τ_{ij} and η_{ij} are the entries of the pheromone concentration matrix τ and heuristic function matrix η respectively, for the path (i, j). The pheromone matrix values are limited to $[\tau_{min}, \tau_{max}]$, with $\tau_{min} = 0$ and $\tau_{max} = 1$. Γ is the *tabu list*, which acts as the memory of the ants and contains all the trails that the ants have already passed and cannot be chosen again. The parameters α and β measure the relative importance of trail pheromone and heuristic knowledge, respectively.

After a complete tour, when all the g ants have visited all the n nodes, the pheromone concentration in the trails is updated by

$$\tau_{ij}(t+1) = \tau_{ij}(t) \times (1-\rho) + \Delta\tau_{ij}^{q} \quad (5)$$

where $\rho \in [0,1]$ expresses the pheromone evaporation phenomenon and $\Delta\tau_{ij}^{q}$ are pheromones deposited on the trails (i,j) followed by ant q that found the best solution $f^q(s)$ for this tour:

$$\Delta\tau_{ij}^{q} = \begin{cases} \frac{1}{f^q(s)} & \text{if arc } (i,j) \text{ is used by the ant } q \\ 0 & \text{otherwise} \end{cases} \quad (6)$$

The algorithm runs N times.

4.1 Implementation in the logistic process

In the scheduling problem of the logistic system, the orders waiting to be delivered are the nodes of the graph, and the role of the ants is to find the minimum cost path connecting the orders that should be delivered. We consider that each ant is traveling with a bag with the available stocks and is distributing the stocks between the orders that it is visiting. It only visits orders whose components it is able to deliver. In this way, the ACO only builds feasible solutions. When the stocks' bag is empty or the remaining components are not enough to deliver any missing order, the search for this ant is finished. In this case, the number of visited nodes may not be the same from one ant to another, while for the VRP the number of nodes to visit is fixed and equal to the number of clients to visit [2].

The heuristic function η is the order's lateness, as proposed in [4]: if an order has already a positive lateness, the ant will feel a stronger attraction to visit it, because the order is already delayed. We define the heuristic function as an exponential function in the interval $[0,1]$ where the value 0 is for the order that has the minimum lateness L_{min} and 1 is for the most delayed order L_{max} [4]. The objective is that the orders already delayed attract ants much more than the orders not yet delayed:

$$\eta_j = \frac{e^{\frac{L_j - Lmin}{L_{max} - L_{min}}} - 1}{e - 1} \quad (7)$$

Notice that in this case the heuristic information is only order dependent, therefore $\eta_j = \eta_{ij}$. The pheromone trails τ_{ij} are also restricted to the interval $[0,1]$, therefore $\alpha < \beta$ will indicate a higher relative weight of the pheromones trail. The Tabu list is the list of orders already delivered by the ant and also the orders which is not possible to visit, due to lack of stocks. The objective function to minimize by each ant k is f_L^k defined in (1) .

4.2 Implementation in the distribution process

To solve the VRP, the ACO algorithm constructs solutions by successively choosing clients to visit until all the orders have been delivered. The nodes are the locations of the clients and there is an extra node specifying the localization of the docking center. The heuristic information used in this case is the *saving function*, proposed in [2]:

$$w_{ij} = d_{i0} + d_{j0} - 2 \times d_{ij} + 2 \times |d_{i0} + d_{j0}| \quad (8)$$

where d_{ij} is the distance between clients i and j, and $d_{0i} = d_{i0}$ is the distance between client i and the docking center 0. The heuristic matrix η is a normalized version of this heuristic:

$$\eta_{ij} = \frac{w_{ij} - \min[w]}{\max[w] - \min[w]}. \quad (9)$$

Whenever the choice of a location will lead to infeasible solutions for reasons of vehicle capacity Q or total route length R, the cross-docking center is chosen as a final location to close the tour and a new tour with a new vehicle is started. On the next iteration, the algorithm will start from node 0 again and will repeat this procedure until all the clients are visited and all the orders are delivered. The objective function to be minimized is the one defined in (2).

4.3 Distributed optimization

In the logistic sub-system, the solution's search space is defined by the n orders that can be delivered today. The ACO algorithm uses $n \times n$ matrices τ_L and η_L to search for the optimal solution of f_L. In the distribution sub-system, the solution's search space is $O \cup 0$, i.e., it is equivalent to the search space of the logistic center plus the cross-docking center 0. The ACO algorithm uses $(n+1) \times (n+1)$ matrices τ_{D+0} and η_{D+0}.

The optimization problem $f = f_L o f_D$ is solved in an asynchronous way. The $\circ$ operator represents the composition of the individual pheromone matrices τ_L and τ_D. Note that they both represent a path connecting the clients, although based on different features: lateness and distance.

5 Simulation results

In this section, we compare the supply-chain performance using the centralized and distributed optimization approaches. We consider a simulation environment running one-day optimization problems during one fictive month, where each day a certain stochastic number of new orders enter the logistic system. The clients location follow a random distribution around the cross-docking center.

Table 1. Solutions for the one day problem

Optimization	f_L	f_D
Centralized	0.36	362
Distributed	0.36	355

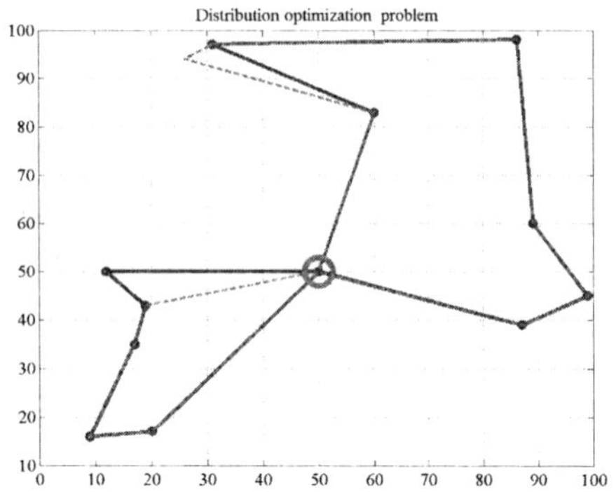

Fig. 1. Distribution problem solution: centralized (-) and distributed(..).

Table 1 shows the results of both logistic (f_L) and distribution (f_D) systems for the one day-problem. In both cases, there are 13 orders to be delivered: 11 orders are distributed at different clients using 2 vehicles, while 2 orders remained in the logistic system. However, when using the distributed optimization approach, the distribution system switched one of the orders that remained in the system by one that was delivered. In logistic terms the result is the same, but in routing terms, it is better, because the traveled distance is smaller. Figure 1 describes this result in detail by representing the cartesian coordinates of the clients' location and the routing solutions for both approaches.

The best cumulative results for the one month problem are presented in Table 2, as well as the mean and variance of the results for 10 different trials. The table presents also the *t-test* probabilities p_t, that indicate if there is a statistical difference between the results. It is clear that the results in terms of logistic optimization are very similar using both approaches, although the best result is obtained using a centralized approach, with a smaller mean result and a narrower variance. However, in terms of the distribution system, it is clear that the system performs better when using the distributed optimization, with lower traveled mean distances and narrower variance. This is confirmed by the t-test (considering a significance level of 0.05) that shows that the results for both approaches in the logistic system case can be considered the same, while for the distribution problem, it shows that the results are statistically different.

Table 2. Solutions for the one-month problem; *t-test* probabilities.

Optimization	f_L	f_D
Centralized	6.74	1511
	(6.86, 0.33)	(1586.8, 81.8)
Distributed	6.91	1422
	(7.28, 0.35)	(1463.2, 48.3)
t-test p_t	0.42	< 0.05

6 Conclusions

This paper introduces a new supply chain management technique, based on the distributed optimization paradigm solved by ant colony optimization. The results show that for a logistic-distribution partners supply chain, the distribution systems performance can be improved without compromising the logistic systems results, i.e. the global systems performance improved just through the exchange of information between the supply-chain partners.

References

[1] M. Barbuceanu and M. Fox. Coordinating multiple agents in the supply chain. In *Proceedings of the Fifth Workshops on Enabling Technology for Collaborative Enterprises, WET ICE'96*, pages 134–141. IEEE Computer Society Press, 1996.

[2] B. Bullnheimer, R.R. Hartl, and C. Strauss. Applying the ant system to the vehicle routing problem. In I.H. Osman, S. Vo, S. Martello, and C. Roucairol, editors, *Meta-heuristics: Advances and Trends in local search paradigms for optimization*, pages 109–120. Kluwer Academics, 1998.

[3] M. Dorigo and T. Stützle. *Ant Colony Optimization.* Cambridge, MA: MIT Press/Bradford Books, 2004.

[4] C. A. Silva, T. A. Runkler, J. M. Sousa, and J. M. Sá da Costa. Optimization of logistic processes in supply–chains using meta–heuristics. In *Proceedings of 11th Portuguese Conference on Artificial Intelligence*, pages 9–23. Springer Verlag, 2003.

[5] Jayashankar M. Swaminathan, Stephen F. Smith, and Norman M. Sadeh. Modeling supply chain dynamics: A multiagent approach. *Decision Sciences Journal*, 29(3):607–632, 1998.

[6] John N. Tsitsiklis, Dimitri P. Bertsekas, and Michael Athans. Distributed asynchronous deterministic and stochastic gradient optimization algorithms. *IEEE transactions on automatic control*, 9(31):803 – 812, 1986.

Comparison of nature inspired and deterministic scheduling heuristics considering optimal schedules

Udo Hönig and Wolfram Schiffmann
FernUniversität Hagen
Lehrgebiet Rechnerarchitektur
58084 Hagen, Germany
E-mail: {Udo.Hoenig, Wolfram.Schiffmann }@FernUni-Hagen.de

Abstract

We report about a performance evaluation of nature inspired stochastic vs. conventional deterministic scheduling algorithms. By means of a comprehensive test bench, that comprises task graphs with diverse properties, we determined the *absolute* performance of those algorithms with respect to the optimal solutions. Surprisingly, the nature inspired stochastic algorithms outperformed all the investigated deterministic algorithms.

1 Introduction

The analysis and comparison of scheduling heuristics is subject of many recent publications. The usually used test sets consist either of randomly generated task graphs or program traces of mathematical algorithms (e.g. Gauss-Seidel, Cholesky, ...). In very few cases, comparisons to optimal schedules were conducted. Although many authors use big task graphs to evaluate their heuristics, the number of test cases is rather small.

Unfortunately, most authors do not reveal their test sets, making it impossible for other scientists to compare their own algorithm to the analyzed ones. For our best knowledge, Kwok and Ahmad [1] are the only authors who published the used test set on their web-page. It consists of 350 task graphs of different sizes and is structured according to the method the task graphs were produced and the Computation to Communication Ratio (CCR) which represents the ratio of a task graph's node- and edge-weights. Since some of the test case classes consist of only five task graphs, it is possible that a single outlier could distort the analysis' results.

For these reasons, we developed a comprehensive structured test bench with 36000 test cases [2]. The task graphs were generated randomly and structured concerning the graph's size, its meshing degree, its average edge length and the node- and edge-weights. To emphasize a certain graph property (e.g. a high meshing degree), the random numbers were determined by a Gaussian distribution. Since this test bench should also provide test cases which are unbiased with respect to one or more graph attributes, test classes with uniform distributed attributes are provided as well.

2 Survey of the investigated algorithms

We used the current interim version[1] of this test bench to compare some well known scheduling algorithms, namely Dynamic Level Scheduling (DLS), Earliest Time First (ETF) and Modified Critical Path (MCP) with implementations of three nature inspired heuristics, Ant Colony Optimization Algorithm (ACO), Genetic Algorithm (GA) and Simulated Annealing (SA). Since Kwok and Ahmad [3] provide a detailed description of the observed deterministic algorithms, this section concentrates on some aspects of our nature inspired heuristics' implementations. An overview of these meta-heuristics' general properties is given by Blum and Roli [4]. Since nature inspired heuristics have many parameters for tuning their processing, their results can only represent a single implementation and its parameter settings but not the overall meta-heuristics it is based on.

To provide a fair comparison between the nature inspired and the deterministic heuristics, we decided not to make use of a heuristic for the determination of a starting point for the nature inspired algorithms. Otherwise it would be possible to select the best result of all the deterministic algorithms, implying that the nature inspired heuristics would perform better for every situation.

The process of scheduling a task graph can be subdivided into two phases: the selection of the task which will be mapped next and the mapping of this task to an idle processor. Since all of the observed deterministic heuristics perform a greedy mapping[2], we decided to reduce the complexity of the nature inspired heuristics' search spaces by reducing their search to the selection process, too. Therefore, the nature inspired heuristics generate task sequences which are forwarded to a

[1] Since the computation is still in progress, only 30511 optimal solution are currently available.

[2] This means, that a task is mapped to the processor where it can start as soon as possible.

greedy mapping mechanism that allocates the tasks to the processors. The resulting schedules are returned to the heuristics for evaluation and optimization of the search process.

Each individual of the genetic algorithm's population is a valid task sequence. Its fitness is computed by mapping the tasks in the given sequence to the considered target architecture. To keep the algorithm's runtime low, every run consists of only 36 generations with 32 individuals. While testing this algorithm, these settings, combined with a cross-over rate of 90% and a mutation rate of 10%, achieved the best results.

The starting temperature of the SA's cooling process is chosen with respect to the given task graph problem. The initial task order is generated randomly. To reduce the probability of getting stuck in local optima in the late phase of the cooling process, SA uses a proportional cooling strategy. A task sequence's neighborhood is defined by all sequences where exactly one task is placed at another position.

While SA and GA operate on complete task orders, ACO's virtual ants evaluate pairs of preceding and succeeding tasks. Every ant passes all tasks in a valid sequence and forwards this task sequence to the greedy mapper which returns the corresponding schedule's length. According to this schedule length a certain amount of pheromone is distributed equally to all predecessor/successor pairs belonging to the ant's path. Succeeding ants can use these information to select their path through the available tasks.

3 Results

The comprehensive structure of the test bench described above as well as the availability of optimal solutions is the base for a more thorough analysis than had ever been performed before. To point out the limits of the hitherto used methods, we will start this analysis by comparing the heuristics relatively to each other and to the best solution found by one of them. Next, we will show the additional information that can be achieved by using our comprehensive test bench: Firstly, by using the optimal schedules, the algorithms can be scored absolutely. In this way, one gets a more precise view of the algorithm's real performance. Secondly, the test bench's structure allows a more detailed examination of the observed heuristic's strengths and weaknesses.

The heuristics were pairwise compared considering the test bench's 36000 task graph problems[3]. The results are presented in table 1, where every cell contains the comparisons of the two heuristics which are assigned to the cell's column and row. The uppermost value represents the number of test cases, where the heuristics' results differ. The second value relates to the number of test cases, where the heuristic which belongs to the cell's row finds worse schedule lengths than the column's heuristic. The third value describes the inverse case.

Table 1: Relative comparison of the observed heuristics.

	SA	MCP	GA	ETF	DLS
ACO	1721 932 789	19522 16 19506	4576 367 4209	20614 39 20575	19220 15 19205
DLS	19232 19213 19	13447 6632 6815	19190 18377 813	9643 3539 6104	
ETF	20631 20604 27	16867 9523 7344	20505 19795 710		
GA	4543 4221 322	19425 853 18572			
MCP	19525 19509 16				

Obviously, the nature inspired algorithms find clearly better schedules than the deterministic heuristics. While SA and ACO are by far the most successful of the investigated algorithms, the GA performs still better than the deterministic algorithms from which DLS is the best, followed by MCP and ETF.

The degradation from the best known solution (degFromBest), is defined by means of the found schedule length SL and the best known schedule length SL_{best} as $degFromBest = 100 * \frac{SL - SL_{best}}{SL_{best}}$. This is another characteristic parameter which is widely used in scheduling literature [5]. A high degradation value indicates a strong deviation from the best solution and therefore poor results.

Table 2 shows the observed heuristics' degradation from the best found solution[4], averaged over all 36000 test cases. Again, SA and ACO perform much better than GA. The best deterministic algorithm is DLS, followed by MCP and ETF. Although this comparison gives a very clear view of the algorithms' results' relative quality, it has three significant disadvantages. Firstly, it still does not give any clue about the absolute quality of a heuristic's results. Secondly, since all comparisons are related to the best known solution, this analysis's results is strongly dependent of the heuristics selected for analysis. Choosing another set of algorithms could change the

[3] Note that this comparison is independent from the optimal solutions.

[4] This best solution was found by one of the observed heuristics and might therefore differ from the optimal solution.

results to a large extend. Thirdly, every time a heuristic is added for comparison, the whole analysis has to be repeated.

Table 2: Average degradation from the **best** solution (in %).

SA	MCP	GA	ETF	DLS	ACO
0,048	3,472	0,423	3,738	3,415	0,054

These disadvantages are eliminated by the knowledge of optimal schedule lengths. Table 3 shows the heuristic's average degradation from the optimal solutions of 30511 task graph problems. All heuristics perform worse than before, because for 5562 test cases (≈ 18,22%) none of the heuristics found the optimal solution. For the nature inspired algorithms, the difference between table 2 and 3 is larger than for the deterministic ones, because they found the best solution more often and therefore had only few suboptimal test cases which had to be considered in this analysis.

Table 3: Average degradation from the **optimal** solution (in %).

SA	MCP	GA	ETF	DLS	ACO
0,707	4,019	1,039	4,209	3,937	0,723

Figure 1 shows the percentage of test cases, to which the observed heuristics were able to find optimal solutions. Again SA and ACO perform much better than GA and the deterministic algorithms. Out of those, DLS found more optimal schedules than MCP and ETF. While the difference between the nature inspired and the deterministic algorithms is quite large, the last-mentioned behave nearly in the same manner and differ by only ≈2,8%.

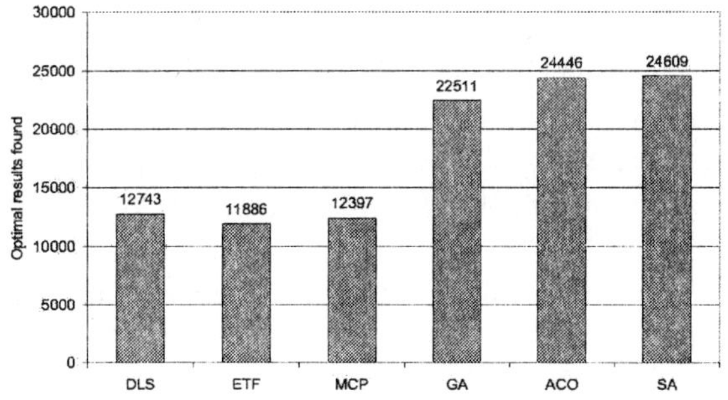

Fig. 1. Comparison considering the number of found optimal solutions.

As already mentioned above, the diverse structure of our test bench allows a thorough analysis of a heuristic's properties. The following results should give a first impression of possible examinations.

Figure 2 shows the effect of the target architecture's size on the quality of the heuristics' results. Obviously, SA and ACO perform better than all other heuristics. Except for target architectures with 2 processors, where SA finds ≈2,67% more optimal schedules. While ACO is almost unaffected by the target architecture's size, GA and the deterministic heuristics find optimal solutions more frequently, if the target architecture's size does not limit the parallelism of the schedule. Again, DLS performs better than MCP and ETF.

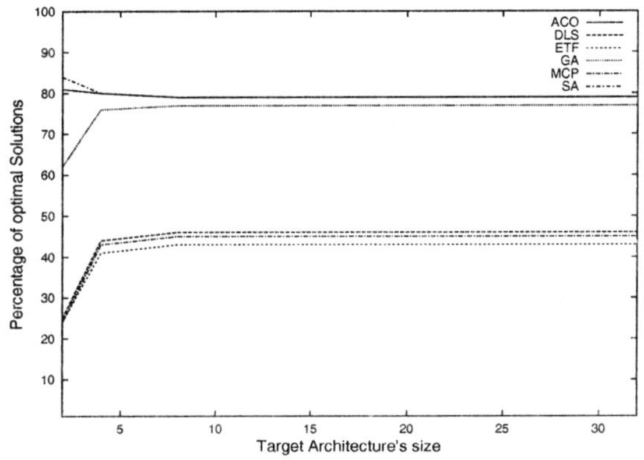

Fig. 2. Influence of the target architecture's size.

As can be seen in figure 2, SA is the only heuristic which finds more optimal schedules if the target architecture is small. Its success rate in finding an optimal solution drops from 84,21% for target systems with only two processors to approximately 80% for larger systems. Figure 3 shows, that for all target architecture sizes SA has clearly more difficulties in finding optimal schedules if the task graph has predominantly long edges. In contrast, DLS performs slightly better for task graphs with long edges if the target architecture is large.

Our investigation's next focus is the effect of the task graph's size on the quality of the found solutions. As can be seen in figure 4, the percentage of found optimal solutions decreases, when the task graphs' size increases. Nevertheless, SA and ACO scale better than GA which in turn performs much better than the deterministic algorithms. In contrast to SA, whose success rate drops from 87,42% to 72,58% (this is a difference of 14,84%), the MCP algorithm's success rate in finding optimal schedules drops by 52,36%.

With respect to the fact that most publications in literature use larger task graphs, with sometimes even more than 1000 tasks, the question arises, if the here presented

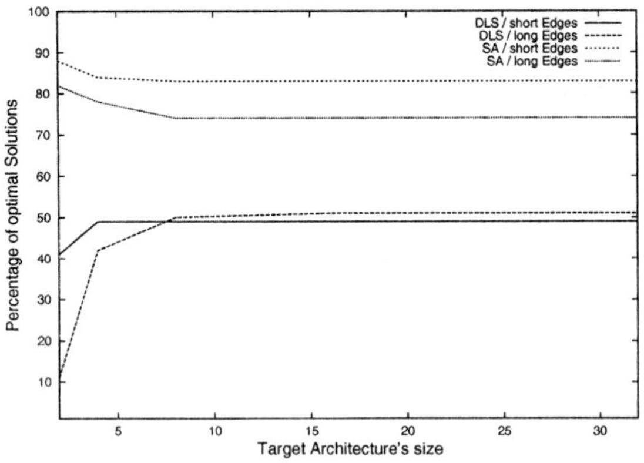

Fig. 3. Effect of the average edge length on SA's and DLS's success rate in finding optimal schedules.

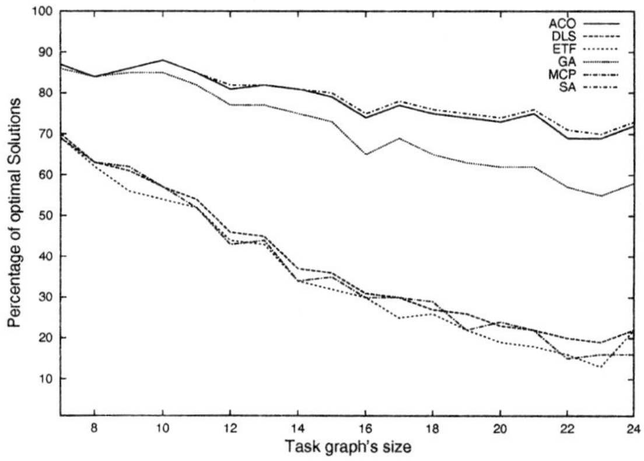

Fig. 4. Impact of the task graphs' size.

results are of any meaning to real world scheduling problems. In order to answer this question, we created another test bench which follows the same structure as the described one, but which consists of task graphs with up to 250 tasks. A representative test bench with larger task graphs would require too much hard disk space. First results with the large task graphs indicate, that the above observations will mainly hold for scheduling problems with larger task graphs (see table 4 for details).

4 Conclusion

In this paper, two different kinds of heuristic scheduling algorithms were compared: nature inspired stochastic and conventional deterministic algorithms. Our investigations are based on a comprehensive test bench that provides optimal schedules for 30511 test cases and therefore allows an elaborated comparison of heuristic algorithms. It could be clearly shown that the nature inspired algorithms outperform the conventional ones. Thus, future research in scheduling algorithms should pay more attention to this approach.

Table 4: Relative comparison with respect to larger task graph problems (up to 250 tasks).

	SA	MCP	GA	ETF	DLS
ACO	29056 20303 8753	34105 9167 24938	31904 5570 26334	34741 6901 27840	34194 8462 25732
DLS	34401 29612 4789	32645 17320 15325	34358 19361 14997	31746 11505 20241	
ETF	34750 30393 4357	33818 21120 12698	34652 22066 12586		
GA	32151 29047 3104	34294 15687 18607			
MCP	34354 28553 5801				

Acknowledgments: The authors would like to thank Mr. Markus Bank for developing the observed heuristics as part of his diploma thesis.

References

[1] Kwok, Y.-K., Ahmad, I. (1998) Benchmarking the Task Graph Scheduling Algorithms, Proceedings of the 12th International Parallel Processing Symposium, pp. 531–537

[2] Hönig, U., Schiffmann, W. (2004) A comprehensive Test Bench for the Evaluation of Scheduling Heuristics, Proceedings of the sixteenth IASTED International Conference on Parallel and Distributed Computing and Systems (PDCS 2004), pp. 437–442

[3] Kwok, Y.-K., Ahmad, I. (1999) Static scheduling algorithms for allocating directed task graphs to multiprocessors. ACM Computing Surveys, Vol. 31, No. 4, pp. 406–471

[4] Blum, C., Roli, A. (2003) Metaheuristics in Combinatorial Optimization: Overview and Conceptual Comparison, ACM Computing Surveys, Vol. 35, No. 3

[5] Dail, H., Casanova, H., Berman, F. (2002) A Decoupled Scheduling Approach for the GrADS Program Development Environment, Proceedings of the 2002 ACM/IEEE conference on Supercomputing, pp. 1–14

An External Memory Supported ACO for the Frequency Assignment Problem

Adnan Acan[1] and Akın Günay[1]
[1]Computer Engineering Department, Eastern Mediterranean University
Gazimağusa, T.R.N.C., Via Mersin 10, TURKEY
E-mail: adnan.acan@emu.edu.tr, akingunay@ieee.org

Abstract

Ant colony optimization algorithm is integrated with an external memory for the purpose of improving its efficiency for the solution of a well-known hard combinatorial optimization problem. The external memory keeps variable-size solution segments extracted from promising solutions of previous iterations. Each solution segment is associated with its parent's fitness value. In the construction of a solution, each ant retrieves a segment from the memory using tournament selection and constructs a complete solution by filling the absent components. The proposed approach is used for the solution of minimum span frequency assignment problem for which very promising results are obtained for provably difficult benchmark test problems that could not be handled by any other ACO-based approach so far.

1 Introduction

Ant colony optimization (ACO) is a general-purpose metaheuristic which can be applied to many kinds of optimization problems[1]. Among many efforts on the development of new variants of ACO algorithms toward improving their efficiency under different circumstances, recently the idea of knowledge incorporation from previous iterations became attractive and handled by a number of researchers. Mainly, these population- or memory-based approaches take their inspiration from studies in genetic algorithms (GAs). In memory-based GA implementations, information stored within a memory is used to adapt the GAs behavior either in problematic cases where the solution quality is not improved over a number of iterations, or a change in the problem environment is detected, or to provide further directions of exploration and exploitation. Memory in GAs can be provided externally (outside the population) or internally (within the population).

External memory implementations store specific information within a separate population (memory) and reintroduce that information into the main population at a later moment [2-4]. The most common approaches using internal memory are polyploidy structures in combination with dominance mechanisms, which imply the use of an internal memory through redundancy in genetic material by having more than one copy of each gene [5].

In ACO, the first internally implemented memory-based approach is the work of Montgomery et al. [6]. In their work, named as AEAC, they modified the characteristic element selection equations of ACO to incorporate a weighting term for the purpose of accumulated experience. This weighting is based on the characteristics of partial solutions generated within the current iteration. They aim to provide, in addition to normal pheromone and heuristic costs, a more immediate and objective feedback on the quality of the choices made. The authors suggested simple weight selection and update procedures and proposed two variations of their algorithm. They claimed that the achieved results for different TSP instances are either equally well or better than those achieved using normal ACS algorithm.

The work of Guntsch et al. is the first example of an external memory implementation within ACO [7]. Their approach, P-ACO, uses a population of previously best solutions from which the pheromone matrix can be derived. For every solution in the population, some amount of pheromone is added to the corresponding edges of the construction graph. Population is organized as a FIFO queue and it is simply updated by entering the best solution of an iteration into the population while the oldest one is removed. The authors proposed a simple pheromone update and a number of population management strategies in [8].

This paper introduces another population based external memory approach where the population includes variable-size solution segments taken from elite individuals of previous iterations. Each stored solution segment is associated with its parent's objective function value that will be used as a measure in segment selection and updating the memory. In order to construct a solution, a particular ant retrieves a segment from the external memory using tournament selection strategy and completes the rest of the solution. The details of the practical implementation are given in the following sections.

This paper is organized as follows. Current state of the art of using ACO for the solution of frequency assignment problem (FAP) is presented in Section 2. The proposed approach is described with its implementation details in Section 3. Section 4 covers the results and related discussions. Section 5 concludes the paper.

2 Ant Colony Optimization for the Frequency Assignment Problem

The frequency assignment problem (FAP) is an NP-hard combinatorial optimization problem and defined as the assignment of frequencies within a predefined bandwidth to radio transmitters in a mobile telecommunication network in such a way that certain interference and traffic constraints are all satisfied. In this paper, a well known minimum-order fixed-frequency assignment problem (MO-FFAP), namely the Philadelphia 21-cell MO-FAP is considered and solved using the proposed ACO strategy. This problem is handled by many authors in literature where two different demand vectors and several constraint matrices are used to define its different instances [9].

ACO for FAP is not yet widely studied in literature and two significant publications to be mentioned are published by Maniezzo et. al [10] and Montemanni et. al [11]. In the work of Maniezzo et.al, ANTS algorithm prepared for the solution of quadratic assignment problem is adapted for FAP [12]. Ants work on feasible solutions only, i.e. with no constraint violations, and an ant iteratively constructs a solution where partial solutions are considered as ant states. At each step t, each ant a_i computes a set $A^t_{a_i}$ of feasible expansions to its current state and moves to one of them probabilistically. The probability of moving from one state i to another j depends on the combination of the attractiveness η_{ij} of the move and the pheromone trail concentration τ_{ij} between the two states.

Montemanni et al. used the ANTS approach for the solution of minimum span frequency assignment problem. They first fix the available frequency span to a sufficiently high value and iteratively minimize the interference using the ANTS algorithm. Once the interference is reduced to zero, the frequency span is reduced by one and the ANTS algorithm is run again.

The results published in both of the studies demonstrate that the ANTS algorithm is quite effective for small size problems however it is outpeformed by other approaches for large and provably difficult problem instances.

3 The Proposed External Memory Strategy and Its Use in the Solution of FAP

The significant differences of our approach, compared to the above explained two ANTS-based algorithms, are its adaptation from the simple classical ACO approach, its use of a segment-based external memory for knowledge incorporation from previous iterations, and its efficiency in the solution of provably difficult benchmark problem instances that could not be solved by any previous ACO-based algorithm. In the proposed approach, variable-size solution segments from observably good solutions of previous iterations are used in the construction of new ones. These elite solution segments are stored in a dynamic library acting as the external memory for the ACO algorithm. Initially the memory is empty and a number of classical ACO iterations are performed to fill in the memory. To do this, at the end of each iteration, one randomly-positioned variable-size solution segment is cut from each of the best k solutions and stored into the memory. To construct a complete solution in this phase, ants use the demand vector, the constraint matrix, and the pheromone matrix as follows: starting from the first cell and for each frequency to be assigned within a particular cell, an ant first determines the set of available frequencies that can be assigned without violating the co-cell constraints such that the remaining demand can be satisfied with the remaining amount of frequencies. Then, a frequency from this set is selected using a pheromone-based probabilistic selection scheme as explained below. The frequency assignment procedure goes on until a solution with a number of co-channel constraints is obtained. In each iteration, all ants follow this simple frequency assignment procedure and form an assignment with no co-cell constraints. Consequently, the fitness value of a constructed solution S is computed as

$$Fitness(S) = L - \sum_{i=1}^{|C|} CCV(i) \qquad (1)$$

where $CCV(i)$ is the total number of co-channel constraint violations within cell i, C is the set of cells and L is a large integer constant.

The $|C|x|F|$ pheromone matrix H, where F is the set of frequencies, is initialized with a small constant value, h_0, and is updated after all ants complete their solution construction procedures. Only the best solution found in the current iteration is used for pheromone update and, assuming that the best solution found in iteration t is S_t, the amount pheromone to be deposited in $H(i,j)$ is updated as

$$H(i,j) = H(i,j) + \rho.\frac{Fitness(S_t)}{K} \qquad (2)$$

where ρ and K are experimentally determined constants. Note that, H(i,j) stores the amount of pheromone deposited on frequency f_j as a results of its assignments within cell C_i. At the beginning of each iteration, pheromone concentration of all frequencies within all cells are decreased by a constant factor as described in equation 3.

$$H(i,j) = max(h_0, (1-\rho).H(i,j)) \tag{3}$$

In the selection of frequencies for a particular assignment, ants follow a probabilistic selection scheme which favors those frequencies with higher pheromone concentration. Accordingly, the probability of assigning a frequency $f \in F$ to a cell $c \in C$ is computed using equation 4.

$$P_f^c = \frac{[H(c,f)]^\alpha.[n_f]^\beta}{\sum_{l \in F}[H(c,l)]^\alpha.[n_l]^\beta} \tag{4}$$

where α, β are constants and n_f for a particular frequency f is a heuristic value that is computed as $n_f = (|F| - f)/|F|$. The above formulation of ACO for frequency assignment problem is adapted from the work of Dorigo et al. [1].

When the initialization phase of the memory is completed, ACO algorithm starts to work in conjunction with the external memory. The size of memory M is fixed, there are m ants that build the solutions, and there is no particular assignment of ants over the problem space. In this respect, before an ant starts to construct an assignment, a solution segment is selected from the external memory using a tournament selection strategy and the ant constructs a complete solution over this partial assignment using the same construction procedure of the classical ACO algorithm as explained above. The only change is the usage of the new demand vector which is attached to the solution segment when it is extracted from its parent solution. This demand vector indicates the number of frequencies not yet assigned and needed to form a complete solution. After all ants complete their iterations, the external memory is updated using the resulting solutions. For this purpose, variable-length solution segments are extracted from the best n elite solutions to replace a number of low-fitness solution segments in the memory. In the implemented strategy, solutions contribute the memory update procedure in different amounts. In this respect, d, where $d > 1$, solution segments are cut from the best solution, whereas only one segment is taken from the elite solution with the lowest fitness. The number of segments to be taken from other elite solutions is determined by linear scaling. Each inserted segment replaces a non-updated library element with the lowest fitness value. This way, higher quality solutions have more characteristic components in the external memory compared to ones with lower fitness values.

As it is the case for many metaheuristic method, each ACO iteration is followed by a local search that aims to minimize the maximum constraint violations for each assigned frequency.

4 Experimental Results

In experimental evaluations, the number of ants m and the size of the external memory M are chosen as $|F|/2$. Constants α and β used in the computation of frequency assignment probabilities are set equal to 1 and 5, respectively. Initial pheromone concentration is 0.5 and constant ρ used in pheromone update procedures is taken as 0.02. In updating the memory, parameters n and d are selected as $0.1 * m$ and $d = 0.05 * m$, respectively. Constants L and K used in equations (1) and (2) are set equal to 10000 after a number of experimental evaluations.

Several commonly handled problem instances and the corresponding experimental results obtained for the Philadelphia system are listed in Table 1, where the proposed approach is named as EM-ACO. For these test problem instances, the most difficult ones are the second and sixth instances. For all the other problems, the optimal solutions are found within a few seconds and within one or two iterations. For the second and the sixth problem instances, best results are reported by Beckmann [9] and MCGA [13].

Table 1. Evaluation of EM-ACO with respect to published results

Problem	Approach and the best solution				
	EM-ACO	ACO	MCGA	[9]	[14]
1	381	381	381	381	381
2	433	450	427	427	433
3	533	533	533	533	533
4	533	533	533	533	533
5	221	221	221	221	221
6	256	271	253	253	263
7	309	309	309	309	309
8	309	309	309	309	309

Comparative convergence performances of EM-ACO and the classical ACO algorithm for the most difficult problem instances are given in Table 2 and Table 3. It can easily be seen that the proposed approach outperforms the classical ACO method for these two problem instances in convergence speed and solution quality. EM-ACO did not find the solutions with known best fitness values for these very difficult problem instances, however the obtained results are still significant from two point of views: firstly, this is the first time an ACO-based approach is used for the solution of these two difficult

Table 2. Evaluation of EM-ACO with respect to best published results for the second problem

2nd Problem				
$\|F\|$	EM-ACO		ACO	
	Perf.	Ave. Iter.	Perf.	Ave. Iter.
265	100%	192	30%	1343
264	100%	215	30%	1492
263	100%	498	0%	-
262	90%	436	0%	-
261	60%	588	0%	-
260	70%	674	0%	-
259	10%	1994	0%	-
258	20%	1504	0%	-
257	10%	1698	0%	-
256	20%	1392	0%	-

*$|F|$=Number of frequencies

Table 3. Evaluation of EM-ACO with respect to best published results for the sixth problem instance

6th Problem				
$\|F\|$	EM-ACO		ACO	
	P	Av. It	P	Av. It
442	100%	163	10%	628
441	100%	228	10%	928
440	100%	340	10%	432
439	70%	544	0%	-
438	60%	705	0%	-
437	40%	1250	0%	-
436	20%	979	0%	-
435	20%	1180	0%	-
434	50%	828	0%	-
433	10%	946	0%	-

*$|F|$=Number of frequencies

problem instances and the obtained results are still better than some GA-based algorithms, and secondly, they demonstrate the improvements gained with the use of of external memory in an ACO-based algorithm.

5 Conclusion

In this paper a novel ACO strategy using an external memory of solution segments from elite solutions of previous iterations is presented. The stored segments are used in the construction of solutions in the current iteration to provide further intensification around potentially promising solutions. Using partially constructed solutions also improve time spent in the construction of solutions since part of the solution is already available. The handled problem instances are well-known in literature and it is the first time that they are solved by an ACO-based approach and results with best known fitness values are obtained with most of the problem instances. As it is clearly seen from the obtained results, the use of a segment-based external memory has improved the convergence speed and solution quality of plain ACO; the obtained results are even better than those found by some GA-based algorithms. Additionally, the proposed ACO strategy is very simple to implement and it does not bring significantly additional computational or storage cost to the existing ACO algorithms.

6 References

[1] Dorigo, M., Caro, G.D., Gambardella, L.M. (1999) Ant algorithms for distributed discrete optimization, Artificial Life, Vol. 5, pp. 137-172.
[2] Louis, S. J., Johnson, J. (1997) Solving similar problems using genetic algorithms and case-based memory,. In: Back, T. (ed.) Proceedings of the Seventh International Conference on Genetic Algorithms - 1997. San Fransisco, CA, pp. 84-91.
[3] Simoes, A., Costa, E. (2002) Using genetic algorithms to deal with dynamical environments: comparative study of several approaches based on promoting diversity. In: Langton, W. B. et al. (eds.) Proceedings of the genetic and evolutionary computation conference - GECCO 2002. Morgan Kaufmann, New York, pp.698-105.
[4] Acan, A., Tekol, Y. (2003) Chromosome reuse in genetic algorithms. In: Cantu-Paz et al. (eds.) Genetic and Evolutionary Computation Conference - GECCO 2003. Springer-Verlag, Chicago, pp.695-705.
[5] Goldberg, D. E., Smith, R. E. (1987) Non-stationary function optimization using genetic algorithms and with dominance and diploidy, Genetic Algorithms and their Applcations: Proceedings of the Second International Conference on Genetic Algorithms, pp. 217-223.
[6] Montgomery, J., Randall, M. (20023) The accumulated experience ant colony for the travelling salesman problem, International Journal of Computational Intelligence and Applications, Vol.3, No. 2, World Scientific Publishing Company, pp.189-198.
[7] Guntsch, M., Middendorf, M. (2002) A population based approach for ACO. In: Cagnoni, S. et al. (eds.) Applications of Evolutionary Computing - EvoWorkshops 2002, Lecture Notes in Computer Science, No:2279, Springer Verlag, pp. 72-81.
[8] Guntsch, M., Middendorf, M. (2002) Applying population based ACO for dynamic optimization problems. In: Dorigo, M. et al. (eds.) Ant Algorithms - Third International Workshop ANTS 2002, Lecture Notes in Computer Science, No:2463, Springer Verlag, pp. 111-122.
[9] Beckmann, D., Killat, (1999) U. A new strategy for the application genetic algorithms to the channel assignment problem, IEEE Trans. On Vehicular Technology, Vol. 48, No. 4, pp. 1261-1269.
[10] Maniezzo, V., Carbonaro, A. (1998) An ant heuristic for the frequency assignment problem, Technical Report, University of Bologna.
[11] Montemanni, R., Smith, D.H., Allen, S.M. (2002) An ANTS algorithm for the minimum-span frequency assignment problem with multiple interference, IEEE Trans. on Vehicular Technology, Vol. 51, No. 5, pp. 949-953.
[12] Maniezzo, V.: (1999) Exact and approximate nondeterministic tree-search procedures for the quadratic assignment problem, Inform. Journal of Computing, Vol. 11, No. 4, pp. 358-369.
[13] Acan, A., Altincay, H., Tekol, Y., Unveren, A. (2003) A genetic algorithm with multiple crossover operators for optimal frequency assignment problem, Proceedings of IEEE Congress on Evolutionary Computation- CEC2003, Canberra, Australia, pp. 256-263.
[14] Battiti, R. (2001) A randomized saturation degree heuristic for channel assignment in cellular radio networks, IEEE Trans. On Vehicular Technology, Vol. 50, No.2, pp. 364-374.

Neural Networks for Extraction of Fuzzy Logic Rules with Application to EEG Data

Martin Holeňa*
*Institute of Computer Science, Academy of Sciences of the Czech Republic,
Pod vodárenskou věží 2, CZ-18207 Praha 8, e-mail: martin@cs.cas.cz

Abstract

The extraction of logical rules from data is a key application of artificial neural networks (ANNs) in data mining. However, most of the ANN-based rule extraction methods rely primarily on heuristics, and their underlying theoretical principles are not very deep. That is especially much true for methods extracting fuzzy logic rules, which usually allow to mix different logical connectives in such a way that extracted rules can not be correctly evaluated in any particular fuzzy logic model. This paper shows that mixing of connectives is not needed. A method for fuzzy rules extraction for which the evaluation of the extracted rules in a single model is the basic principle is outlined and illustrated on a case study with EEG data.

1 Introduction

One of the chief reasons why artificial neural networks are quite popular in practical applications is their ability to learn, from available data, very general nonlinear mappings. The learned mapping incorporates knowledge about the implications that certain values of the input variables have for the values of the output variables. That knowledge is represented through the network architecture and through numerical parameters of the mapping. Needless to say, such a representation is not much human-comprehensible. Therefore, methods allowing to extract from trained ANNs knowledge represented through *logical rules* have been developed since the early nineties [1, 2, 3].

Many ANN-based rule extraction methods have already been proposed, differing with respect to various aspects (cf. the classifications suggested in [1, 4, 5]). They can be divided into two main groups:

- *Boolean rules*, i.e., formulas of the Boolean logic, such as the propositional *if... then* rules or *M-of-N* rules [2, 3, 6].
- *Fuzzy rules*, i.e., formulas of some fuzzy logic, [1, 7, 8, 9] (a survey of many other methods can be found in [10]).

An important feature of ANN-based rule extraction methods is that most of them rely primarily on heuristics, and their underlying theoretical principles are not very deep. This is particularly much true for methods extracting fuzzy rules. Most of them allow to interpret different logical connectives by means of algebraic operations that do not belong to the same algebra. Typically, the conjunction is interpreted in a G-algebra, but the negation in a MV-algebra, due to the fact that the interpretation of the negation in a MV-algebra is involutive, whereas in a G-algebra, it is not [1, 10, 11]. Consequently, the extracted rules then can not be correctly evaluated in any single fuzzy logic model.

The objective of the present paper is to show that such a mixing of different interpretations is not needed. A method for the extraction of fuzzy rules in a disjunctive normal form (DNF) is outlined, for which the evaluation of the extracted rules in a single model is the basic principle, and its applicability is demonstrated on a case study with EEG data.

2 Extraction of Fuzzy DNF Rules from Data

In the Boolean logic, *DNF rules* are defined as rules $\Psi \equiv \bigvee_{i=1}^{d} \bigwedge_{j\in C_i} \varphi_{i,j}$ in which $C_1, \dots, C_d$ are nonempty finite sets, Ψ and $\varphi_{i,j}$, $i = 1, \dots, d$, $j \in C_i$ are atomic formulas, and $\vee, \wedge$ are connected via the *De Morgan law*

$$\alpha \vee \beta = \neg(\neg\alpha \wedge \neg\beta). \tag{1}$$

That definition can be in full generality transferred also to a fuzzy logic provided the meaning of the connectives $\wedge$ and $\neg$ is fixed. Recall some important properties of fuzzy logic connectives [12]:

- the meaning of a connective is given by its interpretation, i.e., an operation on truth values;
- each interpretation of each connective belongs to a BL-algebra, either a general one, or a BL-algebra of a specific kind (e.g., MV-algebra, G-algebra, PŁ-algebra, ŁΠ-algebra);

- to compute the truth value of a formula in a model, all connectives occurring in the formula have to be interpreted in the same algebra.

In this section, an ANN-based method for the extraction of fuzzy DNF rules will be outlined, which is based on the following principles:

(i) the ANN computes the truth value $\|\Psi\|_M$ of Ψ in some model M;

(ii) any atomic formulas $\varphi_{i_1,j}$, $\varphi_{i_2,j}$ are interpreted with the same kind of finitely-parametrizable fuzzy sets on a crisp domain D_j;

(iii) the ANN is trained with a sequence of pairs $(x^1, y^1), \ldots, (x^t, y^t)$ such that for $k = 1, \ldots, t$, $x^k = (x_1^k, \ldots, x_n^k) \in D_1 \times \cdots \times D_n$, and $y^k \in \langle 0, 1 \rangle$ is the desired $\|\Psi\|_M$ provided $\varphi_{i,j}$, $i = 1, \ldots, d$, $j \in C_i$, are interpreted with respect to x_j^k (i.e., $\|\varphi_{i,j}\|_M$ equals the membership of x_j^k in the fuzzy set $\tilde{\varphi}_{i,j}^M$ interpreting $\varphi_{i,j}$ in M);

(iv) M is chosen from some considered set of models $\mathcal{M}$ in such a way that the squared error over the training sequence is minimal.

Observe that the principle (i) implies that the underlying ANN has 1 output neuron, whereas from (ii) follows that it has n input neurons corresponding to the domains $D_1, \ldots, D_n$ and $|C_1| + \cdots + |C_d|$ hidden neurons corresponding to the considered atomic formulas, where $|C|$ denotes the cardinality of a set C, and $C_i \subset \{1, \ldots, n\}$ for $i = 1, \ldots, d$. Also, (ii) entails a finite-parametrizability of $\mathcal{M}$, and for each $M \in \mathcal{M}$, (iii) combined with (1) yields

$$M = ((\|\varphi\|_{i,j}^M)_{j \in C_i}^{i=1,\ldots,d}, T, ^c) = ((\tilde{\varphi}_{i,j}^M)_{j \in C_i}^{i=1,\ldots,d}, T, ^c), \tag{2}$$

where T and ^{c} are, respectively, the t-norm interpreting $\wedge$ and the precomplement interpreting $\neg$. Finally, (i)-(ii) together with (2) entail

$$\|\Psi\|_M = S_{i=1}^d (T_{j \in C_i} (\tilde{\varphi}_{i,j}^M)), \tag{3}$$

where S is an s-norm, connected with T, due to the De Morgan law, through $S(y, z) = T(y^c, z^c)^c$, and (iii)–(iv) together with (3) imply that training the neural network with $(x^1, y^1), \ldots, (x^t, y^t)$ leads to the optimization task

$$M = \arg \min_{M' \in \mathcal{M}} \sum_{k=1}^{t} (S_{i=1}^d (T_{j \in C_i} (\tilde{\varphi}_{i,j}^{M'} (x_j^k))) - y^k)^2. \tag{4}$$

The finite parametrizability of $\mathcal{M}$ turns (4) to a standard task of multidimensional optimization.

For any implementation of the method, the set of models $\mathcal{M}$ has to be specified. Taking into account (2) and the principle (ii), this means to specify:

- for $j = 1, \ldots, n$, the parametrization of the fuzzy sets on the domain D_j interpreting the formulas $\varphi_{i,j}$, $j \in C_i$, which is defined by a parametrizing mapping $\pi_j : \Re^{p_j} \to \mathcal{F}(D_j)$, where $p_j \in \mathcal{N}$, $\mathcal{F}(D_j)$ denotes the set of all fuzzy sets on D_j, and π_j fulfils $(\exists a_{i,j} \in \Re^{p_j})\ \tilde{\varphi}_{i,j} = \pi_j(a_{i,j})$ for $i = 1, \ldots, d$, $j \in C_i$;
- the particular fuzzy logic considered, which determines the t-norm T and the considered precomplement ^{c}.

In applications of fuzzy sets, various parametrizing mappings are encountered, for example, the Gaussian parametrization $\pi(a, b) = \gamma_{(a,b)}$ for $a \in \Re$, $b > 0$, where $\gamma_{(a,b)}(x) = e^{-\frac{(x-a)^2}{2b}}$, the triangular parametrization, $\pi(a, b, c) = \Delta_{(a,b,c)}$ for $a, b, c \in \Re, a < b < c$, where $\Delta_{(a,b,c)}(x) = \max(0, \min(\frac{x-a}{b-a}, \frac{c-x}{c-b}))$, or the bell-shaped parametrization $\pi(a, b, c) = B_{(a,b,c)}$ for $a, b, c \in \Re, a \neq 0$, where $B_{(a,b,c)}(x) = \frac{1}{1+(\frac{x-c}{a})^{2b}}$. Examples of particular fuzzy logics are the Łukasiewicz logic [12, 13], or the product-Łukasiewicz logic [14], entailing the t-norms $T_{\mathrm{L}}(x, y) = \max(x + y - 1, 0)$, and $T_{P\mathrm{L}}(x, y) = xy$, respectively, and the precomplements $x^{c_{\mathrm{L}}} = {}^{c_{P\mathrm{L}}} = 1 - x$.

3 A Case study with EEG data

In collaboration of neurophysiologists and transportation scientists, the specificity of EEG signals corresponding to somnolence has been investigated at the Czech Technical University Prague. Its ultimate objective is to provide an empiric knowledge base for a system automatically detecting impaired vigilance, a cause of severe traffic accidents. A description of that investigation and the collected data have been presented in [15] (due to space limitations, they are not reproduced here).

During data preprocessing, Gabor spectral analysis has been performed for EEG signals measured in 35 healthy volunteers and corresponding always to three vigilance levels – full vigilance, mental activity, and somnolence caused by sleep deprivation. The knowledge about the specificity of individual kinds of the signals was primarily obtained through visual inspection of the EEG records and of the corresponding spectrograms by expert physiologists. In addition, to 14 frequencies $f_1, \ldots, f_{14} = 1 - 14$ Hz of EEG spectra from two selected electrodes, several automated knowledge extraction methods were applied, including also the above method for the extraction of fuzzy-DNF rules. In that context, two important features of the way how human experts describe EEG spectra should be mentioned [15]:

- the spectra are usually not described in terms of idividual frequences, but instead in terms of differently looking bands of the spectrum, in particular, the considered frequencies 1–14 Hz are conventionally covered with three consecutive spectrum bands, called δ, θ, α bands;
- their descriptions are often vague and imprecise (e.g., "substantially higher", "a little higher", "clearly predominating").

4 Extracting fuzzy DNF rules about EEG spectra

Restricting knowledge about EEG spectra to knowledge about the δ, θ and α band has the following consequences for the extracted rules:

(i) Three ANNs need to be trained, for extracting fuzzy DNF rules corresponding to the three considered vigilance levels,

$$\Psi_\ell \equiv \bigvee_{i=1}^{d} \bigwedge_{j=1}^{2} \varphi_{i,\ell\delta_j} \wedge \varphi_{i,\ell\theta_j} \wedge \varphi_{i,\ell\alpha_j}, \quad (5)$$

where $\ell \in \{v,m,s\}$, Ψ_v, Ψ_m and Ψ_s states that the signal corresponds to full vigilance, to mental activity, and to somnolence, respectively, whereas $\varphi_{i,\ell\xi_j}$, $i = 1,\dots,d$, $\ell \in \{v,m,s\}$, $\xi \in C_i = \{\delta,\theta,\alpha\}$, and j=1,2, state approximate characterization of the intensity of the band ξ_j of the spectrum from the j-th electrode for the vigillance level ℓ,

(ii) For any training sample $(x^k, y^k), k = 1,\dots,t$, only $y^k \in \{0,1\}$ is known, $y^k = 1$ iff training the ANN for which the k-th sample is a positive example (e.g., $y^k = 1$ if that particular sample is a positive example of somnolence and the somnolence network is trained). On the other hand, the vector $x^k = (x^k_{\delta_1}, x^k_{\theta_1}, x^k_{\alpha_1}, x^k_{\delta_2}, x^k_{\theta_2}, x^k_{\alpha_2}) \in \langle 0,+\infty)^6$ of intensities corresponding to the δ, θ and α bands of the spectra obtained with the two electrodes is not known, but instead a vector $I^k = (I^k_{1,1},\dots,I^k_{1,14}, I^k_{2,1},\dots,I^k_{2,14}) \in \langle 0,+\infty)^{28}$ of intensities corresponding to the 14 considered frequencies $f_1,\dots,f_{14}$ is known.

(iii) The δ, θ and α bands of the spectrum from the j-th electrode, $j = 1,2$, are assumed to be described, respectively, with fuzzy sets $\tilde\delta_j$, $\tilde\theta_j$ and $\tilde\alpha_j$ on $\langle 0,+\infty)$, allowing to interconnect, for all (x^k,y^k), the intensities $x^k_{\delta_j}$, $x^k_{\theta_j}$, $x^k_{\alpha_j}$, with the intensities $I^k_{j,1},\dots,I^k_{j,14}$ through

$$(\forall (j,h,k) \in \{1,2\}\times\{1,\dots,14\}\times\{1,\dots,t\})\, I^k_{j,h} = \max\{x^k_{\delta_j}\tilde\delta_j(f_h), x^k_{\theta_j}\tilde\theta_j(f_h), x^k_{\alpha_j}\tilde\alpha_j(f_h)\} \quad (6)$$

Hence, for $j = 1,2$, $I^k_{j,1},\dots,I^k_{j,14}$ are membership grades of the frequencies $f_{j,1},\dots,f_{j,14}$ in the non-normalized fuzzy set

$$\tilde I^k_j = \max\{x^k_{\delta_j}\tilde\delta_j, x^k_{\theta_j}\tilde\theta_j, x^k_{\alpha_j}\tilde\alpha_j, x^k_{\beta_j}\tilde\beta_j\}. \quad (7)$$

(iv) The fuzzy sets $\tilde\delta_j$, $\tilde\theta_j$ and $\tilde\alpha_j$ for $j = 1,2$ are assumed to be finitely parametrizable independently of j, thus for each $\xi \in \{\delta_j,\theta_j,\alpha_j\}$, $\tilde\xi_j = \pi_\xi(a_{\xi_j})$, where $\pi_\xi : \Re^{p_\xi} \to \mathcal{F}(\langle 0,+\infty))$, $p_\xi \in \mathcal{N}$, $a_{\xi_j} \in \Re^{p_\xi}$, and the parametrizing mappings π_δ, π_θ and π_α can differ from each other, as well as from the mappings $\pi_{\ell\xi}$ that parametrize the fuzzy sets $\tilde\varphi_{i,\ell\xi_j}$, interpreting for $i = 1,\dots,d$, $\ell \in \{v,m,s\}$, $j = 1,2$ and $\xi = \delta_j,\theta_j,\alpha_j$ the atomic formulas $\varphi_{i,\ell\xi_j}$. Due to the finite parametrizability, the fuzzy sets $\tilde I^k_j$, $j = 1,2$ in (7) can not be arbitrary fuzzy sets on $\langle 0,+\infty)$, and the membership grades of $f_1,\dots,f_{14}$ in $\tilde I^k_j$ in general can not be arbitrary tuples of nonnegative numbers. Consequently, a solution to the system of equations (6) does not necessarily exist, and a more general task needs to be solved to get the band intensities $x^k_{\delta_j}$, $x^k_{\theta_j}$, $x^k_{\alpha_j}$, $k = 1,\dots,t$, together with the parameters a_{δ_j}, a_{θ_j}, a_{α_j} determining the fuzzy sets $\tilde\delta_j$, $\tilde\theta_j$ and $\tilde\alpha_j$ for $j = 1,2$. That task can be formulated as least-squares fitting of the fuzzy sets $\tilde I^k_j$ to given frequency intensities $I^k_{j,1},\dots,I^k_{j,14}$, $k = 1,\dots,t$, i.e.,

$$((x^k_{\delta_j}, x^k_{\theta_j}, x^k_{\alpha_j})^t_{k=1}, a_{\delta_j}, a_{\theta_j}, a_{\alpha_j}) =$$
$$\arg\min\{\sum_{k=1}^{t}\sum_{h=1}^{14}(\max\{x_{j,1}\tilde\delta_j(f_h), x_{j,2}\tilde\theta_j(f_h),$$
$$x_{j,3}\tilde\alpha_j(f_h)\} - I^k_{j,h})^2 : (x_{j,1},\dots,x_{j,3}) \in \langle 0,+\infty)^3 \;\&$$
$$a_{\delta_j}, a_{\theta_j}, a_{\alpha_j} \in \Re^{p_{\delta_j}} \times \Re^{p_{\theta_j}} \times \Re^{p_{\alpha_j}} \;\&\; \tilde\delta_j = \pi_\delta(a_{\delta_j})$$
$$\&\; \tilde\theta_j = \pi_\theta(a_{\theta_j}) \;\&\; \tilde\alpha_j = \pi_\alpha(a_{\alpha_j})\} \quad (8)$$

For illustration, Figure 1 shows the fitted fuzzy sets $\tilde I^k_1$ introduced in (7) for three particular training samples (one for each considered vigilance level), the Gaussian parametrization of the fuzzy set $\tilde\delta_1$, and the bell-shaped parametrization of the fuzzy sets $\tilde\theta_1$ and $\tilde\alpha_1$. The fit was constructed according to (8), and the obtained values $x^k_{\delta_1}$, $x^k_{\theta_1}$ and $x^k_{\alpha_1}$ of band intensities were used when solving the optimization task (4) for the rules (5) with $d = 5$, $\ell \in \{v,m,s\}$, Gaussian parametrization of the fuzzy sets $\tilde\varphi_{i,\ell\delta_1}$, $\tilde\varphi_{i,\ell\theta_1}$, $\tilde\varphi_{i,\ell\alpha_1}$, and the product-Łukasiewicz logic.

5 Conclusion

The paper proposed a neural-networks based method for the extraction of fuzzy DNF rules, requiring

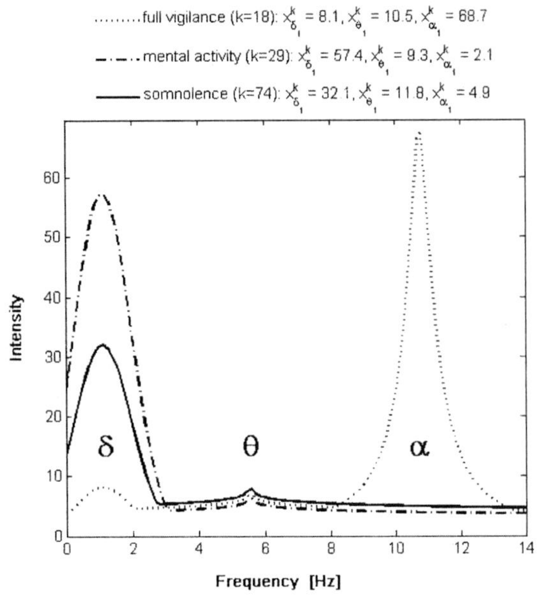

Fig. 1. Fitted fuzzy sets $\tilde{I}_1^k$ (7) for three particular training samples (one for each considered vigilance level), the Gaussian parametrization of the fuzzy set $\tilde{\delta}_1$, and the bell-shaped parametrization of the fuzzy sets $\tilde{\theta}_1$ and $\tilde{\alpha}_1$

all involved logical connectives to be interpreted in the same fuzzy logic model. This brings the advantage of assuring a correct interpretability of any extracted formula in the considered model, which is not the case for most existing fuzzy rule extraction methods. The method has been elaborated as far as to the formulation of a standard task of multidimensional optimization that needs to be solved to determine the parameters of all involved fuzzy sets. The feasibility of the method has been demonstrated on a case study with real data presented in [15], which concern the specificity of EEG signals corresponding to somnolence.

Acknowledgements

The research reported in this paper has been supported by the grant No. ME701, "Building Neuroinformation Bases, and Extracting Knowledge from them", of the Czech Ministry for Education. The author is indebted to Přemysl Posledník for implementing a substantial part of the method presented in Section 2 the specificity of EEG signals corresponding to somnolence.

References

[1] Duch, W., Adamczak, R., and Grabczewski, K. (2000). A new methodology of extraction, optimization and application of crisp and fuzzy logical rules. *IEEE Transactions on Neural Networks*, 11:277–306.

[2] Ishikawa, M. (2000). Rule extraction by successive regularization. *Neural Networks*, 13:1171–1183.

[3] Tsukimoto, H. (2000). Extracting rules from trained neural networks. *IEEE Transactions on Neural Networks*, 11:333–389.

[4] Andrews, R., Diederich, J., and Tickle, A. (1995). Survey and critique of techniques for extracting rules from trained artificical neural networks. *Knowledge Based Systems*, 8:378–389.

[5] Tickle, A., Andrews, R., Golea, M., and Diederich, J. (1998). The truth will come to light: Directions and challenges in extracting rules from trained artificial neural networks. *IEEE Transactions on Neural Networks*, 9:1057–1068.

[6] Alexander, J. and Mozer, M. (1999). Template-based procedures for neural network interpretation. *Neural Networks*, 12:479–498.

[7] Amato, P., Di Nola, A., and Gerla, B. (2005). Neural networks and rational Łukasiewicz logic. Accepted for Publication in Journal of Multiple-Valued Logic and Soft Computing.

[8] Amato, P. and Porto, M. (2000). An algorithm for the automatic generation of logical formula representing a control law. *Neural Network World*, 10:777–786.

[9] Nauck, D., Nauck, U., and Kruse, R. (1996). Generating classification rules with the neuro-fuzzy system NEFCLASS. In *Proceedings of the Biennial Conference of the North American Fuzzy Information Processing Society NAFIPS'96*, pp. 466–470.

[10] Mitra, S. and Hayashi, Y. (2000). Neuro-fuzzy rule generation: Survey in soft computing framework. *IEEE Transactions on Neural Networks*, 11:748–768.

[11] Finn, G. (1999). Learning fuzzy rules from data. *Neural Computing & Applications*, 8:9–24.

[12] Hájek, P. (1998). *Metamathematics of Fuzzy Logic*. Kluwer Academic Publishers, Dordrecht.

[13] Novák, V., Perfilieva, I., and Močkoř, J. (1999). *Mathematical Principles of Fuzzy Logic*. Kluwer Academic Publishers, Dordrecht.

[14] Horčík, R. and Cintula, P. (2004). Product Łukasiewicz logic. *Archive for Mathematical Logic*, 43:477–503.

[15] Faber, J., Novák, M., Svoboda, P., and Tatarinov, V. (2003). Electrical brain wave analysis during hypnagogium. *Neural Network World*, 13:41–54.

A New Neuro-Based Method for Short Term Load Forecasting of Iran National Power System

R. Barzamini[1, 2], M. B. Menhaj[1, 2], Sh. Kamalvand[2], M. A. Fasihi[1]

1 Department of Electrical Engineering, Amirkabir University of Technology, Tehran15914, IRAN
2 Department of Power System Operation, Niroo Research Institute (NRI), Tehran, IRAN
E-mail: {barzamini,menhaj}@cic.aut.ac.ir

Abstract

This paper presents a new neuro-based method for short term load forecasting of Iran national power system (INPS). A MultiLayer Perceptron (MLP) based Neural Network (NN) toolbox has been develeped to forecast 168 hours ahead. The proposed MLP has one hiden layer with 5 neurons. The effective inputs were selected through a peer investigation on historical data released from the INPS. To adjust the parameters of the MLP, the Levenberg-Marquardt Back Propagation (LMBP) training algorithm has been employed because of its remarkable fast speed of convergence. Most of papers dealt with 168-hour forecasting employed a hirachical method in the sense of monthly or seasonly provided that there are enough data. In the absence of rich data, forecasting error would increase. To remedy this problem, the proposed neuro-based approach uses only the weekly group data of concern while an extra input is added up to indicate the month. In other words for each weekly group, a unique MLP based neural network is designed for the purposed of load forecasting.

1 Introduction

Load forecasting has always been the essential part of an efficient power system planning and operation. Generally there are two groups of forecasting models, traditional models (model-based techniques) and modern techniques (known as model-free techniques) [1, 2]. Traditional load forecasting models are time series and regression analysis. In recent years, computational intelligent methods are more commonly used for load forecasting [1–10].

Multilayer feedforward neural networks, as universal approximation machines, are very suitable for load forecasting because they have remarkable ability to approximate nonlinear functions with any desired accuracy. Selection of the input-output training data and input vector of the neural network play a crucial role. Especially in our case (load forecasting problem) the MLP-based networks are greatly affected by selection of inputs, day type, month type, historical load data and weather information.

Importance of hourly load input data for each weekly group is more than other inputs (section II). This toolbox is based on a three-layered feedforward neural network building block. For training of this neural network instead of the conventional BP method, the LMBP method was employed in order to reach the (sub)optimum network's parameters faster; we need this especially for one hour ahead forecast (section III). Finally the paper presents the simulation results of this method (section IV) followed be section V that addresses future work.

2 Input Selection

Selection of proper and optimal number of inputs would result more accuracy and convergence speed in a multilayer feedforward neural network. Most of load forecasting methods use weather information, load power of the days before forecast day and day type for input variables. Day type is justified by correlation anlysis on the historical load information.

By correlation analysis on the historical load information for INPS, a week is divided into 4 groups: Saturdays (first day of the week), Sundays to Wednesdays (workdays), Thursdays and Fridays (weekends). For each group, most effective lags (load of the previous hours) on hourly load were selected by correlation analysis [11]. Table 1 shows the selected lags for each weekly group. For example first row in table 1 says that for Group "Saturdays" loads of one hour, two hours,.... and finally 169 hours earlier are used to predict the future load.

Table 1 - The selected load lags for weekly groups

Weekly Group	Selected Load Lags
Saturdays	1, 2, 3, 23, 24, 25, 167, 168, 169
Workdays	1, 2, 3, 23, 24, 25, 47, 48, 49, 147, 148, 149, 167, 168, 169
Thursdays	1, 2, 3, 23, 24, 25, 47, 48, 49, 167, 168, 169
Fridays	1, 2, 3, 23, 24, 25, 167, 168, 169, 191, 192, 193, 335, 336, 337

Iran has regions with various weather conditions. Temperature is the most effective weather information on hourly load. Instead of using temperatures of all cities in

Iran, temperatures of Ahvaz, Tehran and Tabriz were selected, by correlation analysis, as representatives of hot, moderate and cold region temperatures, respectively [11]. In this paper we also use an extra input that indicates which month of the year is under study.

3 Neural Network

3.1 Structure of Neural Network

Generally, neural networks with a hidden layer have the remarkable ability to approximate most nonlinear functions with a desired accuracy if there are enough hidden neurons. Therefore, the model shown in Fig. 1 is composed of three layers, and each layer has a feedforward connection. In this model, inputs for each weekly group are separately trained by a MLP._The input layer for hourly load forecast of each weekly group, Saturdays, workdays, Thursdays and Fridays, has respectively 13, 19, 16 and 19 neurons (consisting of selected load lags and 3 representatives' forecasted temperatures and 1 node to indicate month). Through a deep investigation we found that a hidden layer with 5 neurons works quite good. Of course, the network has one output neuron (the load of the forecasted hour).
In the proposed method, it is unnecessary to train the network using the data collected for one season or a month; this requires considering at least four previous years data that might make the set of training data huge. In this method, two previous years data is sufficient. Therefore, it is possible to reduce the size of the MLP and learning time as well.
To validate the quality of the developed MLP, we run it with year 2000 load data, in order to be able to judge the merit of the method.

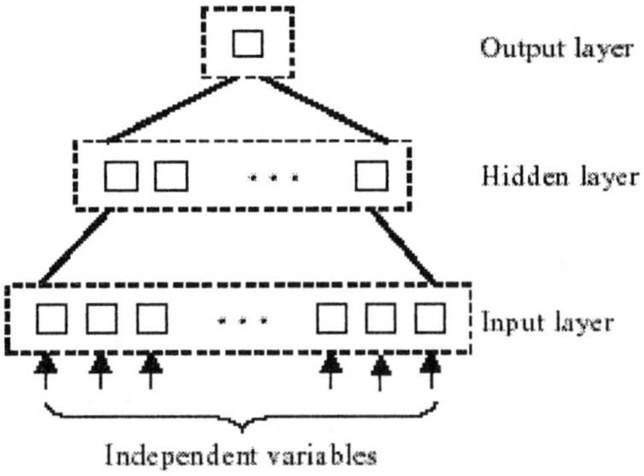

Fig1. The MLP architecture for each weekly group

3.2 Learning

For the purpose of training, we employed the Levenberg-Marquardt Back Propagation (MLBP) algorithm; this algorithm represents a network learning method that updates weight and baias values according to the Levenberg-Marquardt optimization. This is an improved Guass-Newton method that has an extra regularization term to deal with the additive noise in the training samples [16]. Conventional BP methods are often too slow for practical problems, but the LMBP method can converge from ten to one hundred times faster than the conventional BP method.
Neurons in the hidden and output layers have nonlinear transfer function known as the "tangent sigmoid" (tansig) function:

$$f(x) = \frac{2}{1+\exp(-2x)} - 1 \qquad (1)$$

The weighted inputs received by a tansig node are summed and passed through this non-linear function to produce an output. The tansig function generates outputs between –1 and +1 and its inputs should be in the same range. As a result, it is necessary to limit the MLP inputs and target outputs. Mean-standard deviation and minimum (min)-Maximum (max) normalization methods have been tested and min-max method has been selected:

$$X_{normlized} = \frac{X_{actual} - X_{\min}}{X_{\max} - X_{\min}} \times 2 - 1 \qquad (2)$$

This normalization method has also the advantage of mapping the target output to the non-saturated sector of tansig function. This process helps in improving the accuracy of both the learning and forecasting modes.
The MLPs can be trained for each weekly group of a year and the related weights and biases will be gained and used for forecasting.

3.3 Forecasting

The toolbox provides a user friendly environment for tuning and constructing the MLPs. It can be used for one hour up to a week load forecasting. The first hour load is forecasted and then it is used as one of the MLP inputs for the prediction of the next hours' load. Consequently, the error of each hour load's forecast will influence the prediction of next hours' load. However, the proposed method used in the toolbox eventually predicts the future loads effectively.

4 Simulation Result

To verify the predictive ability of this method, this section presents some examples of up to a week forecasting performance of this toolbox. The data used in the simiulations is the actual load power data of Iran National Power System (INPS) and temperature data from 1997 to 2000. we forecast the load power in 2000 by three different methods. Cases 1 to 3 are indicated as follows.

Case 1 presents a case in which a unique MLP was developed for each month in each weekly group. This method needs data of at least 4 years for learning.

Case 2 shows the results of load forecasting for a case in which for each weekly group of each season an MLP was designed.

Case 3 represents the results for the case where for each weekly group an MLP has been developed with adding an extra input indicating month of the corresponding data. This method only needs the data of two years for learning.

The mean absolute percentage error (MAPE) with the proposed technique is shown in Fig. 2. Figure 2 shows forecast results for Cases 1 to 3. The MAPE is defined as follows:

$$MAPE(\%) = \frac{1}{N}\sum_{i=1}^{N}\frac{\left|P_A^i - P_F^i\right|}{P_A^i}\times 100 \qquad (3)$$

Where P_A is the actual load, P_F is the forecast load, and N is the number of data. The mean absolute percentage error for Case 1 to Case 3 in the year 2000 are shown in Table 2. We can see from Fig. 2 and Table 2 that the prediction ability for Case 3 is better than the others (Case 1, Case 2). Therefore, less data has been used to train network (this cause's fast convergence) and at the same time error has reduced. Using some examples, performance of the proposed (used in Case 3) method is presented in Figs. 3 and 4.

5 Conclusions

One of the most successful applications of the MLP to real-world problems has been in the area of electric load forecasting. In this paper, a new method for load forecasting was presented, which is based on a three-layered feedforward neural network. The optimum inputs were selected for the MLP considering historical data of the INPS. The program has fast convergence, with the use of LMBP training method, and its performance is satisfactory for one-week load forecasting, noting that it can be extended for one-hour load forecasting too. Less data has been used to train network (this causes fast convergence) and at the same time error has reduced. To improve the accuracy of the program for peak load, special days' (holidays') load and also for sudden load changes caused by weather front, the authors are currently working on a peak load forecaster, an adaptive MLP and a fuzzy expert system.

Table 2. Mean Absolute Percentage Error in 2000.

2000	**MAPE(%)**
Case1	**4.59**
Case2	**2.53**
Case3	**1.79**

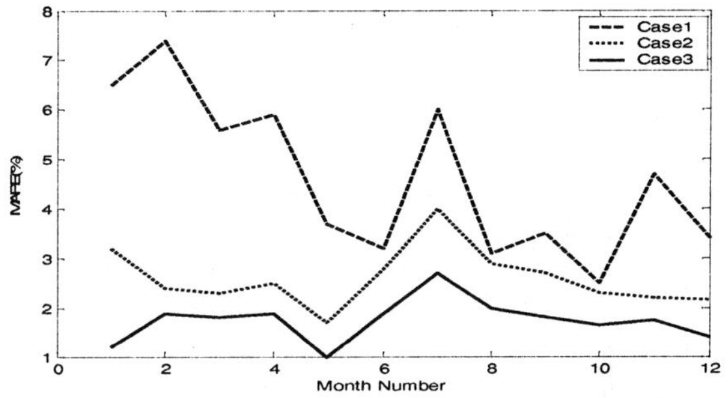

Fig. 2. Mean absolute percentage error corresponding to month

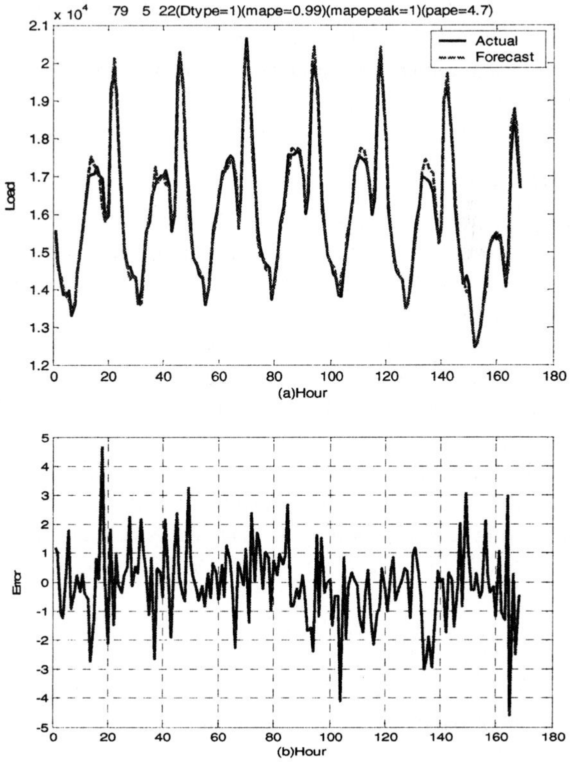

Fig. 3. Actual and forecast hourly loads from Aug. 12 to Aug. 18, 2000 (MAPE=0.99%)

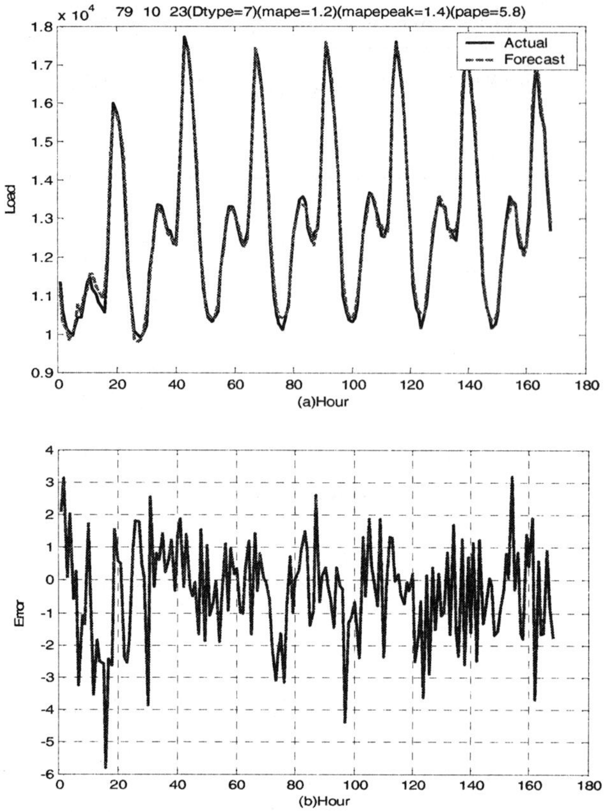

Fig 4. Actual and forecast hourly loads from Jan. 12 to Jan. 18, 2001 (MAPE=1.2 %)

References

[1] Nader S. Moharari, Atif S. Debs, "Short-Term Load Forecasting Using Artificial Neural Networks and Expert Systems", Neural Networks to Power Systems, Published 1993.

[2] A.Y. Varjani, P. Doulai, "Neural Network versus Time Series Methods for Short-Term Load Forecasting", Singapore International Power Engineering Conference, Feb. 1995.

[3] Shin-Tzo Chen, David C.Yu, A.R. Moghaddamjo, "Weather Sensitive Short-Term Load Forecasting Using Nonfully Connected Artificial Neural Network", IEEE Trans. on PWRS, Vol. 7, No. 3, Aug. 1992.

[4] A. Khotanzad, R. Afkhami-Rohani, Tsan-Liang Lu, A. Abaye, M.H. Davis, Dominic J. Maratukulam, "ANNSTLE--A Neural-Network Based Electric Load Forecasting System", IEEE Trans. on NN, Vol. 8, No. 4, July 1997.

[5] A. Khotanzad, Rey-Chune Hwang, A. Abaye, D. Maratakulam, "An Adaptive Modular Artificial Neural Network Hourly Load Forecaster and its Implementation at Electric Utilities", IEEE Trans. on PWRS, Vol. 10, No. 3, August 1995.

[6] A. Khotanzad, R. Afkhami-Rohani, D. Maratukulam, "ANNSTLF- Artificial Neural Network Short-Term Load Forecaster-Generation Three", IEEE Trans. on PWRS, 1998.

[7] B.S. Kermanshahi, C.H. Poskar, G. Swift, W. Buhr, A. Silk, "Artificial Neural Networks for Forcasting Daily Loads of a Canadian Electric Utility", 1993 IEEE.

[8] C.N. Lu, H.T. Wu, S. Vemuri, "Neural Network Based Short-Term Load Forecasting", IEEE Trans. on PWRS, Vol. 8, No. 1, Feb. 1993.

[9] T. Matsumoto, S. Kitamara, Y, Ueki, T Matsui, "Short Term Load Forecasting by Artificial Neural Networks Using Individual and Collective Data of Preceding Years", Neural Networks to Power Systems, 1993, ANNPS' 93, Page (s): 245-250.

[10] Nader S. Moharari, Atif S. Debs, "An Artificial Neural Network Based Short-Term Load Forecasting whit Special Tuning for Weekends and Seasonal Changes", Neural Networks to Power Systems, 1993, ANNPS' 93, Page (s): 279-283.

[11] NRI, "Short Term Load Forecasting of Iran National Power System Using Artificial Neural Networks" in Persian, Final Report of NRI Research Project PONVT04, Aug. 2000.

[12] S. Barghinia, P. Ansarimehr, N. Vafadar, "A New Approach to Bad Data Identification and Modification of Iran Power System Load Data Used in Short-Term Load Forecasting" in Persian, 15th Iran International Power System Conference, Nov. 2000.

[13] I. Drezga, S. Rahman, "Input Variable Selection for ANN-Based Short Term Load Forecasting ", IEEE Trans. on PWRS, Vol. 13, No. 4, Nov. 1998.

[14] H. Ma, A.A. El-Keib, X. Ma, "Training Data Sensitivity Problem of Artificial Neural Network-Based Power System Load Forecasting", System Theory, 1994, Proceeding of the 26th Southeastern Symposium on Published 1994.

[15] S.J. Kiartzis, C.E. Zoumas, A.G. Bakirtzis, V. Petridis, "Data Pre-Processing for Short-Term Load Forecasting in an Autonomous Power System Using Artificial Neural Networks", IEEE, 1996.

[16] M. T. Hagan, M. Menhaj, "Training Feed Forward Networks with the Marquardt Algorithm", IEEE Trans. on NN, Vol. 5, No. 6., 1994.

[17] A. Khotanzad, M.H. Davis, A. Abaye, D.J. Martukulam, "An Artificial Neural Network Hourly Temperature Forecaster with Applications in Load Forecasting", IEEE Trans. on PWRS, Vol. 10, Aug. 1995.

Approximating the Algebraic Solution of System of Interval Linear Equations with Use of Neural Networks

Nguyen Hoang Viet, Michał Kleiber *
*Institute of Fundamental Technological Research, Polish Academy of Sciences. E-mail: {viet,mkleiber}@ippt.gov.pl

Abstract

A new approach to approximate the algebraic solution of systems of interval linear equations (SILE) is proposed in this paper. The original SILE problem is first transformed into an optimization problem, which is in turn solved with use of artificial neural networks and gradient-based optimization techniques.

1 Introduction

Systems of linear equations play an important role in many theoretical as well as practical, real world problems. There are a number of techniques, such as the powerful Finite Element Methods (FEMs), which allow transforming the original problems into those of solving a system of linear equations. However, when it is about real world problems, it is inevitable to deal with uncertainties. As the result, the presence of uncertainties must properly be taken under consideration in the system of equations mentioned above. When all uncertainties can be considered as impreciseness and when they can be represented in form of interval numbers [1, 5], the system of linear equations becomes that of interval linear equations (SILE).

Solving SILE has been the topic of many researches for decades and various techniques have been successfully developed [2, 6, 8, 9]. Usually, a definition of the solution to such SILE must be stated first, and algorithms are provided in order to approximate the solution regarding that definition. The *exact* solution, when the size of the problem is considerably large, is almost unobtainable. It is very common to find only the so called *inner approximation* and the *outer approximation* (or the *hull approximation*) of the *difficult-to-find* exact solution [9].

In this paper the algebraic solution of SILE is considered. Finding the exact algebraic solution for a given SILE is usually a difficult task due to the characteristic properties of interval arithmetics. In many cases, such a solution may not exist at all. The aim of the technique proposed here is to approximate the *exact* algebraic solution, if any exists, or to find an *acceptable* solution in the other case.

One of the major problems in solving SILE lies in the specific properties of interval arithmetics. In this paper, this issue is dealt with use of artificial neural networks. Neural networks with its ability to learn from examples and to approximate a wide class of functions are applied here to replace some operations on interval numbers. Another advantage of using neural networks is that they allow parallelizing the computation process, which could increase the efficiency of the proposed approach and make it applicable in real-time systems.

In the next section, some basic notions regarding interval numbers and interval arithmetics together with the formulation of the SILE problem are provided. The SILE problem is then viewed as an optimization one in section 3. Numerical evaluation is provided in 4 and some final conclusions can be found in 5.

2 System of interval linear equations

2.1 Basic notions

We denote interval numbers in this paper by placing a bar over lowercase characters, i.e. $\overline{a}$, $\overline{b}$.... A real interval number $\overline{a}$ can be considered as a pair $[a^{\mathrm{L}}, a^{\mathrm{R}}]$, where $a^{\mathrm{L}}, a^{\mathrm{R}} \in \mathbb{R}$, $a^{\mathrm{L}} \le a^{\mathrm{R}}$ are called the left and the right endpoint of $\overline{a}$ respectively. The interval number $\overline{a}$ can be defined as follows:

$$\overline{a} = [a^{\mathrm{L}}, a^{\mathrm{R}}] \equiv \{x \in \mathbb{R} : a^{\mathrm{L}} \le x \le a^{\mathrm{R}}\}$$

The set of all real interval numbers is denoted by $\mathbb{IR}$. Operations in $\mathbb{R}$ can be extended to those in $\mathbb{IR}$. Namely, a binary operation $*$ in $\mathbb{IR}$ can be defined as follows:

$$\overline{a} * \overline{b} \equiv \{c = a * b : a \in \overline{a},\ b \in \overline{b}\} \qquad (1)$$

The concept of vector and matrix can be also extended to interval vector and interval matrix. Namely, an interval vector (an interval matrix) is a vector (matrix), whose elements are interval numbers. The set of all n-element real interval vectors is denoted by $\mathbb{IR}^n$, the set of all $n \times m$ element matrices is denote by $\mathbb{IR}^{n\times m}$. The concept of product of an interval matrix and an interval vector is essential in this paper. Let $\overline{\mathrm{A}} \in \mathbb{IR}^{m\times n}$, $\overline{\mathrm{x}} \in \mathbb{IR}^n$. The product of $\overline{\mathrm{A}}$ and $\overline{\mathrm{x}}$ can be defined in one of the following manners:

$$\overline{\mathrm{A}}\overline{\mathrm{x}} \equiv \left\{\mathrm{b} \in \mathbb{R}^m : \exists \mathrm{A} \in \overline{\mathrm{A}}, \exists \mathrm{x} \in \overline{\mathrm{x}}, \mathrm{b} = \mathrm{Ax} \right\} \quad (2)$$

or:

$$\overline{\mathrm{A}}\overline{\mathrm{x}} = \left[\overline{a}_{ij}\right]^{m\times n}\left[\overline{x}_j\right]^{n\times 1} \equiv \left[\sum_{k=1}^{n} \overline{a}_{ik} \times \overline{x}_k\right] \quad (3)$$

where the interval multiplication in (3) is that defined in (1).

Let us now define the distance between two vectors of interval numbers, namely for $\overline{\mathrm{a}}, \overline{b} \in \mathbb{IR}^n$:

$$d_2\left(\overline{\mathrm{a}}, \overline{\mathrm{b}}\right) \equiv \sqrt{\sum_{i=1}^{n}\left(\left(a_i^{\mathrm{L}} - b_i^{\mathrm{L}}\right)^2 + \left(a_i^{\mathrm{R}} - b_i^{\mathrm{R}}\right)^2\right)} \quad (4)$$

The last notion needed to be mentioned here is that of *interval hulls.* In general, an interval hull of a subset $\mathrm{A} \in \mathbb{R}^n$, denoted here by $\mathcal{H}(\mathrm{A})$ is said to be the *smallest* vector $\overline{\mathrm{h}} \in \mathbb{IR}^n$ that includes the set A. In other words:

$$\mathcal{H}(\mathrm{A}) \equiv \overline{\mathrm{h}} \in \mathbb{IR}^n \;:\; \mathrm{A} \subseteq \overline{\mathrm{h}} \wedge \nexists \overline{\mathrm{h}}' \in \mathbb{IR}^n,\ \mathrm{A} \subseteq \mathrm{h}' \subset \mathrm{h} \quad (5)$$

2.2 Algebraic solution for SILE

Let us now consider the system of interval linear equations:

$$\overline{\mathrm{A}}\overline{\mathrm{x}} = \overline{\mathrm{b}} \quad (6)$$

where $\overline{\mathrm{A}} \in \mathbb{IR}^{n\times n}$ is an interval matrix and $\overline{\mathrm{b}} \in \mathbb{IR}^n$ is an interval vector. The solution of (6) can be understood in different ways. The most common one is perhaps the so-called *united solution* which makes use of the interval matrix-vector product as provided in (2):

$$\mathrm{S}\left(\overline{\mathrm{A}}, \overline{\mathrm{b}}\right) \equiv \left\{\mathrm{x} \in \mathbb{R}^n : \exists \mathrm{A} \in \overline{\mathrm{A}}, \exists \mathrm{b} \in \overline{\mathrm{b}}, \mathrm{Ax} = \mathrm{b}\right\}$$

The solution set S in general is not an interval vector, it may not be convex or even not compact at all [8]. Usually instead of the exact solution S, $\mathcal{H}(\mathrm{S})$ is of major interest:

$$\overline{\mathrm{x}} \equiv \left[\overline{\mathrm{A}}, \overline{\mathrm{b}}\right]_{\exists\exists} \equiv \mathcal{H}\left(\mathrm{S}\left(\overline{\mathrm{A}}, \overline{\mathrm{b}}\right)\right)$$

A number of techniques has been developed so far to compute the interval hull of the solution set or to approximate it. Anther common approach is to find some inner approximation of the solution set S, i.e. an interval vector $\mathrm{S_I}$ such that $\mathrm{S_I} \subseteq \mathrm{S}$.

In this paper, the *algebraic solution* of (6) is taken under consideration [6]. The algebraic solution of (6) is an interval vector $\overline{\mathrm{x}}^* \in \mathbb{IR}^n$, for which the product $\overline{\mathrm{A}}\overline{\mathrm{x}}^*$, according to (3) is equal to $\overline{\mathrm{b}}$. Finding the exact solution $\overline{\mathrm{x}}^*$ is a complex problem since the interval arithmetic operations are defined in terms of min-max functions:

$$\overline{a}\,\overline{x} = \left[\min_{\alpha,\beta\in\{\mathrm{L,R}\}} a^\alpha x^\beta,\ \max_{\alpha,\beta\in\{\mathrm{L,R}\}} a^\alpha x^\beta\right] \quad (7)$$

Moreover, in many cases the algebraic solution may not exist at all.

The aim of this work is to find an interval vector approximating $\overline{\mathrm{x}}^*$, if such solution exists. In the other case, the approximating interval vector may give an idea of the solution for some solvable SILE, whose left-side matrix and/or right-side vector are slightly different from $\overline{\mathrm{A}}$ and $\overline{\mathrm{b}}$, respectively. This problem will be considered in an optimization context, which is the topic of the next section.

3 SILE as an optimization problem

3.1 Definition of the cost function

As mentioned above, the SILE problem (6) can be viewed in terms of an optimization (minimization) problem with the cost function F:

$$\mathrm{F}\left(\overline{\mathrm{x}}\right) = \frac{1}{2}d_2^2\left(\overline{\mathrm{A}}\overline{\mathrm{x}}, \overline{\mathrm{b}}\right) \quad (8)$$

where the interval matrix-vector product is defined as in (3). It can be easily observed that the algebraic solution $\overline{\mathrm{x}}^*$ for (6), if such exists, corresponds to the global minimum of the cost function $\mathrm{F}(\overline{\mathrm{x}})$. In (8), the cost function's domain is $\mathbb{IR}^n$. However, when endpoint notation is applied, $\mathrm{F}(\overline{\mathrm{x}})$ can be rewritten as $\mathrm{F}(\mathrm{X}) = \mathrm{F}\left(x_1^{\mathrm{L}}, x_1^{\mathrm{R}}, x_2^{\mathrm{L}}, x_2^{\mathrm{R}}, \ldots, x_n^{\mathrm{L}}, x_n^{\mathrm{R}}\right)$, which is a function in $\mathbb{R}^{2n}$. The cost function F may be treated hereby as either $\mathrm{F}(\overline{\mathrm{x}})$ or $\mathrm{F}(\mathrm{X})$, depending on the context in which it appears.

There are numerous techniques for real valued function optimization. The most popular ones are the gradient-based methods, which make use of the gradient of the optimized function. In our case, F is obviously not differentiable, therefore gradient-based optimization methods are inapplicable. It can be observed that the problem lies in the product of interval numbers:

$$\left[\overline{\mathrm{A}}\overline{\mathrm{x}}\right]_i = \left[\sum_{k=1}^{n} \min_{\alpha,\beta\in\{\mathrm{L,R}\}} a_{ik}^\alpha x_k^\beta,\ \sum_{k=1}^{n} \max_{\alpha,\beta\in\{\mathrm{L,R}\}} a_{ik}^\alpha x_k^\beta\right] \quad (9)$$

The cost function F can be minimized using genetic algorithms, or another idea is to replace interval number multiplication with an approximating differentiable function so that F becomes also differentiable. In our approach, interval numbers multiplication is approximated by use of a neural network. The details will be discussed in the next section.

3.2 Neural networks for approximation of interval number multiplication

Neural networks are well known for their ability to learn from examples and to be universal non-linear approximators. They can find various applications in numerous engineering problems. In this section, a feed-forward, multi-layer, sigmoidal neural network $\mathcal{S}$ is designed to approximate the interval number multiplication (7).

The network used here contains $\mathrm{N} = 4$ input neurons, $M = 2$ output neurons and one hidden layer with K hidden neurons, where K will be chosen later by experiments (Figure 1). The inputs to the network are $a^{\mathrm{L}}, a^{\mathrm{R}}, x^{\mathrm{L}}$ and x^{R}, whereas the two outputs represent the left and right endpoints of the result interval number. All neurons in the network are sigmoidal with the activation function μ.

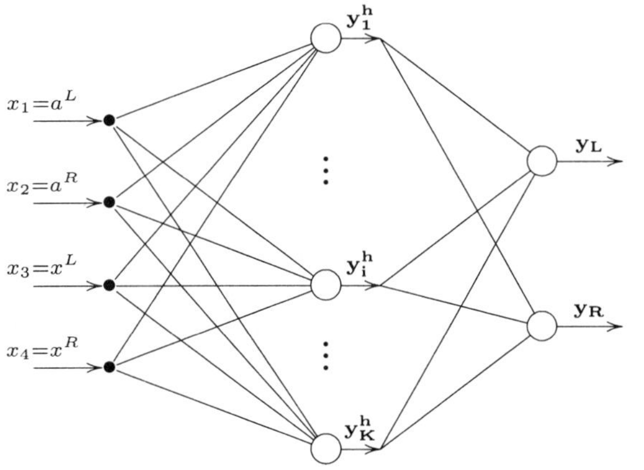

Fig. 1. The architecture of feed-forward neural network used to approximate interval number multiplication.

Let denote by $\mathrm{W} = [\mathrm{W}_{nk}]^{\mathrm{N}\times\mathrm{K}}$ and $\mathrm{V} = [\mathrm{V}_{km}]^{\mathrm{K}\times\mathrm{M}}$ the connection's weight matrices between neurons in the input layer and those in the hidden layer, and between hidden neurons and output neurons, resp.. It can be observed that the network described above can be interpreted as a function $\mathcal{N} : \mathbb{R}^{\mathrm{N}} \rightarrow \mathbb{R}^{\mathrm{M}}$, which is differentiable (since all neurons are sigmoidal). To train the given network, a set of random training samples $\mathbf{T} = \{\langle([a_i^{\mathrm{L}}, a_i^{\mathrm{R}}], [x_i^{\mathrm{L}}, x_i^{\mathrm{R}}]), [y_i^{\mathrm{L}}, y_i^{\mathrm{R}}]\rangle\}$ was generated so that $[a_i^{\mathrm{L}}, a_i^{\mathrm{R}}], [x_i^{\mathrm{L}}, x_i^{\mathrm{R}}] \subseteq [-1, 1]$. The network was then trained using the training set $\mathbf{T}$ and the scaled conjugate gradient algorithm developed by Möller [7].

It will later be in need of computing the gradient vector of $\mathcal{N}$ with respect to its input x_i:

$$\frac{\partial \mathcal{N}_m}{\partial x_i} = \mu'(u_m) \sum_{k=1}^{\mathrm{K}} \mathrm{V}_{km} \mu'\left(u_k^{\mathrm{h}}\right) \mathrm{W}_{ik} \quad (10)$$

where u_k^{h} and u_m are the aggregated input signals to the k-th hidden neuron and the m-th output neuron. It is worth noting that the partial derivatives $\partial \mathcal{N}_m / \partial x_i$ in (10) can be efficiently computed in term of an hybrid neural network $\mathcal{N}^{\partial}$ with 2K + M input units and M output units, which can be directly built from $\mathcal{N}$. In order to compute $\partial \mathcal{N}_m / \partial x_i$, one just only has to put the vector $\left[W_{ik} | \mu'\left(u_k^{\mathrm{h}}\right) | \mu'\left(u_j^{\mathrm{h}}\right)\right]$, $k = \overline{1,\mathrm{K}}, j = \overline{1,\mathrm{M}}$ $[W_{ik} |\, \mu'\left(u_k^{\mathrm{h}}\right)]$ as the input signal to $\mathcal{N}^{\partial}$.

3.3 The modified cost function

Let us now consider again the cost function (8) for our SILE problem. Let us replace the interval product $\left[\mathrm{A}\overline{\mathrm{x}}\right]$ as defined in (9) by a new one $\left[\mathrm{A} \times_{\mathcal{N}} \overline{\mathrm{x}}\right]$ using the approximating neural network $\mathcal{N}$ as previously described:

$$\left[\mathrm{A} \times_{\mathcal{N}} \overline{\mathrm{x}}\right]_i = \left[\sum_{k=1}^{n} \mathcal{N}_{\mathrm{L}}(a_{ik}, x_k),\ \sum_{k=1}^{n} \mathcal{N}_{\mathrm{R}}(a_{ik}, x_k)\right]$$

Consequently, the cost function F can be approximated by:

$$\mathrm{F}(\overline{x}) \simeq \mathrm{f}(\overline{x}) = \frac{1}{2} \sum_{i=1}^{n} \left[\left(\sum_{k=1}^{n} \mathcal{N}_{\mathrm{L}}(a_{ik}, x_k) - b_i^{\mathrm{L}}\right)^2 + \left(\sum_{k=1}^{n} \mathcal{N}_{\mathrm{R}}(a_{ik}, x_k) - b_i^{\mathrm{R}}\right)^2\right]$$

which is a differentiable function:

$$\frac{\partial f}{\partial x_j^{\alpha}} = \sum_{i=1}^{n} \left[\left(\sum_{k=1}^{n} \mathcal{N}_{\mathrm{L}}(a_{ik}, x_k) - b_i^{\mathrm{L}}\right) \frac{\partial \mathcal{N}_{\mathrm{L}}}{x_j^{\alpha}}(a_{ij}, x_j) + \left(\sum_{k=1}^{n} \mathcal{N}_{\mathrm{R}}(a_{ik}, x_k) - b_i^{\mathrm{R}}\right) \frac{\partial \mathcal{N}_{\mathrm{R}}}{x_j^{\alpha}}(a_{ij}, x_j)\right]$$

for $j = \overline{1, n}$, $\alpha \in \{\mathrm{L}, \mathrm{R}\}$. Now, instead of minimizing the original cost function F, its approximation f is

Table 1. The minimal, maximal and average cost function's values and Mean Absolute Square Error for $n = 2, 5, 10, 20$.

	$n = 2$	$n = 5$	$n = 10$	$n = 20$
Min.C	0	9.98e-30	7.04e-6	1.06e-2
Max.C	2.58e-4	4.53e-3	7.88e-2	3.25e-1
Min.E	0	1.27e-3	4.21e-5	1.88e-2
Max.E	7.36e-1	1.42e-1	2.32e-1	1.53e-1
Avg.C	7.57e-6	5.68e-4	1.55e-2	6.60e-2
Avg.E	4.61e-4	2.53e-2	5.70e-2	6.41e-2

minimized. In this paper, again, the scaled conjugate gradient algorithm is applied to minimize f.

4 Numerical evaluation

In order to test the performance of the proposed approach, a set of 50 neural networks for interval number multiplication with 5 neurons in the hidden layer were trained. These networks were trained with use of a set of 500 training and 300 validating samples, both randomly generated. Next, all those networks were tested against a test set composed of 2000 randomly generated testing samples. The network with the best performance was then chosen for further use in the task of solving the SILE problem.

The proposed technique was applied to approximate the algebraic solution of various systems of interval linear equations:

$$\overline{\mathrm{A}}\overline{\mathrm{x}} = \overline{\mathrm{b}}\,,\ \overline{\mathrm{A}} \in \mathbb{IR}^{n\times n},\ \overline{\mathrm{x}},\, \overline{\mathrm{b}} \in \mathbb{IR}^{n} \tag{11}$$

with various values of n. For each value of n, a set of 20 different configurations of $\overline{\mathrm{A}}$ and the exact algebraic solution $\overline{\mathrm{x}}^*$ were first randomly generated (each element of $\overline{\mathrm{A}}$ as well as of $\overline{\mathrm{x}}^*$ is an interval number in $[-1, 1]$). The interval vectors $\overline{\mathrm{b}}$ were then computed according to: $\overline{\mathrm{b}} = \overline{\mathrm{A}}\overline{\mathrm{x}}^*$.

For each configuration of $\{\overline{\mathrm{A}}, \overline{\mathrm{b}}\}$, the simulation was repeated 10 times (i.e. with 10 different randomly chosen starting points). The solution with the smallest value of the cost function (8) was chosen as the final solution for the system. For such a solution the Mean Absolute Square Error (E) was also computed. The results for $n = 2$, $n = 5$, $n = 10$ and $n = 20$, categorized according to the cost function value (C) as well as the error E are shown in Table 1.

5 Conclusions

In this paper, a neural network based method to approximate the algebraic solution of systems of interval linear equations was presented. The task of solving SILE was viewed as an optimization problem. Due to the specific properties of interval arithmetics, a simple neural network for interval number multiplication was used to facilitate the optimization problem. Both neural network training and cost function minimization was realized with use of the scaled conjugate gradient algorithm. A number of numerical simulations was also provided.

One important property of such an optimization based technique for SILE is that even when there is no analytical solution for a given SILE, a *possible solution* can still be found. The proposed approach can also be easily used in the more general case:

$$\overline{\mathrm{A}}\overline{\mathrm{x}} = \overline{\mathrm{b}}$$

where $\overline{\mathrm{A}} \in \mathbb{IR}^{n\times m}, \overline{\mathrm{b}} \in \mathbb{IR}^{n}$ and $n \neq m$.

Being able to solve the SILE problem would be the first step toward solving the more general problem, namely the problem of solving systems of fuzzy linear equations. This will be the topic for further works by the authors.

References

[1] G. Alefeld, J. Herzberger, *Introduction to Interval Computation*, Academic Press, London, 1983,

[2] G. Alefeld, V. Kreinovich, G. Mayer, *On the Solution Sets of Particular Classes of Linear Interval Systems*, Journal of Computaional and Applied Mathematics, vol. 152, 1-15, 2003,

[3] L.T. Fine, Feedforward Neural Network Methodology, Springer-Verlag: New York, 1999,

[4] F.M Ham, I. Kostanic, *Principles of Neurocomputing for Science and Engineering*, McGraw-Hill Education - Europe 2001,

[5] L. Jaulin, M. Kieffer, O. Didrit, E. Walter, *Applied Interval Analysis*, Springer-Verlag, London 2001,

[6] S. Markov, *An Iterative Method for Algebraic Solution to Interval Equations*, Applied Numerical Mathematics, vol. 30, 225-239, 1999,

[7] M.F. Möller, "A scaled conjugate gradient algorithm for fast supervised learning", Neural Networks, Vol. 6, 525-533, 1993,

[8] A. Neumaier, *Interval Methods for Systems of Linear Equations*, Cambridge University Press, 1990,

[9] S. P. Shary, *Algebraic Approach in the Outer Problem for Interval Linear Equations*, Reliable Computing, vol. 3, 103-135, 1997.

Comparing Diversity and Training Accuracy in Classifier Selection for Plurality Voting Based Fusion

H. Altınçay
Department of Computer Engineering, Eastern Mediterranean University
KKTC, Mersin 10, Turkey
E-mail: hakan.altincay@emu.edu.tr

Abstract

Selection of an optimal subset of classifiers in designing classifier ensembles is an important problem. The search algorithms used for this purpose maximize an objective function which may be the combined training accuracy or diversity of the selected classifiers. Taking into account the fact that there is no benefit in using multiple copies of the same classifier, it is generally argued that the classifiers should be diverse and several measures of diversity are proposed for this purpose. In this paper, the relative strengths of combined training accuracy and diversity based approaches are investigated for the plurality voting based combination rule. Moreover, we propose a diversity measure where the difference in classification behavior exploited by the plurality voting combination rule is taken into account.

1 Introduction

When a set of classifiers is available for the classification problem under concern, the simplest strategy would be to select the one providing best performance on the training data set. However, the best performing classifier is not guaranteed to provide best accuracy on the unseen data. Another reasonable approach would be to choose a subset of the available classifiers to build a multiple classifier system (MCS) which are shown to provide improved classification accuracies compared to the individual classifiers [1].

Given the rule of combination, an optimal subset can be selected by maximizing the combined accuracy. However, recent theoretical and experimental work have provided alternative criteria named as *diversity measures* for guiding classifier selection. It is argued that the selected classifiers should be diverse, which means that they should make errors on different input samples [2, 3]. There are several diversity measures proposed so far, many of which are not correlated with the combined accuracies provided by the popular combination methods such as majority voting. However, recent studies have shown that there is a strong relationship for the *double fault* measure proposed in Ref. [1] with the majority voting accuracy [4]. Krogh and Vedelsby have theoretically shown that the combined ensemble error is equal to the difference between the average error of the individual classifiers and the ambiguity (or, diversity) of the classifiers, $E_{ens} = E_{avg} - A$ [2]. In other words, average classification accuracy of the individual members and their diversity are two objective functions that should be simultaneously maximized in ensemble design.

The measure of diversity should depend on the method of combination. For instance, in the case of majority voting, the selected classifiers should not make coincident errors. In other words, the classifiers should generate errors on different input samples. In the case of plurality voting, the classifiers should generate different incorrect classifications for a given input sample [5].

In this paper, the use of combined accuracy and diversity are studied in classifier selection so as to investigate the strengths and weaknesses of each approach. A diversity measure is also proposed, aiming to satisfy the expectations of plurality voting combination rule. The proposed measure is based on quantifying the uncertainty in the incorrect classifier outputs. It is compared with two other diversity measures and also the combined accuracy on the training data. The experiments are conducted on two different classifier sets including strong and relatively weaker classifiers. The forward search algorithm which is an iterative approach is used to select optimal classifier sets by maximizing an objective function that may be either the combined training accuracy or any measure of diversity. The individual classification accuracies of the selected classifiers are also examined to clarify the relation between combined accuracy and individual classifier accuracies.

2 Diversity Measures

The correlation of a measure of diversity with the combined accuracy is generally considered to be important since the ultimate aim is maximized combined accuracy. Classifier selection based on making use of diversity measures quantifying *only* the differences in the classification behavior does not provide good results in general. On the other hand, in the cases when the combined error on training data is optimized, the difference in the selected classifiers is not explicitly examined.

In this study, we considered the *double fault* measure which is correlated with the majority voting accuracy and defined as,

$$DF = 1 - \frac{2}{M(M-1)} \sum_{i,j \in [1,M], i \neq j} \frac{N^{00}_{i,j}}{N} \quad (1)$$

and the *disagreement measure* defined as,

$$DA = \frac{2}{M(M-1)} \sum_{i,j \in [1,M], i \neq j} \frac{N^{01}_{i,j} + N^{10}_{i,j}}{N} \quad (2)$$

$N^{00}_{i,j}$ denotes the number of training samples that ith and jth classifiers misclassify simultaneously and similarly, $N^{01}_{i,j}$ denotes the number of training samples that ith and jth classifiers misclassify and correctly classify respectively. Taking into account the differences in classification behaviors, DA increases as the set of input samples where both classifiers generate either correct and incorrect classifications decreases. It should be noted that, DA does not have any explicit support on higher individual or combined accuracy during the classifier selection process. Moreover, both measures are mainly examining the correct/incorrect classification behavior of the classifiers but the labels of misclassifications are also important for plurality voting.

3 A Diversity Measure for Plurality Voting

In plurality voting, the joint decision is selected as the class getting the maximal number of votes. Hence, in order to increase the combined accuracy, the incorrect class labels generated by different classifiers should not be the same. Consider the case when an input sample x_i belonging to class ω_i is correctly classified by the first two classifiers, and misclassified by the other three classifiers in a five classifier subset. Then, the combined decision using plurality voting corresponding to the output set $\{\omega_i, \omega_i, \omega_j, \omega_k, \omega_n\}$ would be correct if $\omega_j \neq \omega_k \neq \omega_n$. However, if the last three classifiers agree on an another class as $\omega_j = \omega_k = \omega_n$, this would lead into a misclassification. This means that the output labels corresponding to incorrect classes should also be examined by the diversity measure in plurality voting rule instead of checking whether the input sample is correctly classified or not as in DA and DF.

Assume that we are looking for best M classifiers subset and C different classes are available. Let Y denote a random vector representing the number of *incorrect* labels. For instance, consider the above example where five classifiers are selected and there are $C = 4$ different classes. Let the outputs from the classifiers be $\{\omega_2, \omega_2, \omega_1, \omega_1, \omega_3\}$ respectively and the correct class is ω_1. Then, we obtain $Y = [2, 1, 0]$ which represents the fact that there 2 coincident incorrect labels and another incorrect label which is different from the other. The length of Y is equal to $\min\{M, C-1\}$. The length becomes M when $C > M$ and $C - 1$ when $C <= M$. In order to quantify the difference in the misclassifications, the vector is normalized to one and its entropy, $E(Y)$ is computed. For $Y = [y_1, y_2, \ldots, y_T]$, $E(Y) = \sum_{t=1}^{T} -y_t \log(y_t)$. Then, the diversity based on the entropy of the misclassifications is computed as,

$$ME = \frac{1}{N} \sum_{n=1}^{N} \begin{cases} 1 & combined = true \\ \frac{E(Y_n)}{\log(\min\{M,C-1\})} & otherwise \end{cases} \quad (3)$$

where N is the total number of training samples. As seen above, the diversity measure takes into account the combined accuracy. For the input samples where the plurality voting combination rule achieves correct classification, 1 is added to the diversity measure. For the misclassified input samples, the uncertainty in misclassification quantified by entropy is considered.

The scale factor $(1/\log(\min\{M, C-1\}))$ is used to normalize the entropy. In other words,

$$0 \leq \frac{E(Y_n)}{\log(\min\{M, C-1\})} \leq 1. \quad (4)$$

The above term is equal to zero when the incorrect labels coincide and to one when the incorrect labels are uniformly distributed among all possible incorrect labels. Since the contribution due to the second item is less than or equal to one, more support is given to a correct classification than uncertainty in misclassification during classifier selection. This is important for the measure to be correlated with the combined accuracy. The difference in the output labels of the selected classifiers

Table 1. The performance of the classifiers in SET1.

Data set	Oracle	Average	SB
iris	100.00	94.81	94.89
image-seg	99.28	92.58	95.17
vowel	98.48	79.63	83.23
satimage_CR	95.83	88.31	90.06
letters	95.13	81.15	88.07
vehicle	97.63	70.75	72.11
cone-torus	96.25	84.59	87.44
glass	94.53	63.31	61.02
Average	*97.14*	*81.89*	*84.00*

Table 2. The performance of the classifiers in SET2.

Data set	Oracle	Average	SB
iris	100.00	91.61	95.00
image-seg	99.28	83.04	90.82
vowel	97.31	55.90	81.16
satimage_CR	96.12	81.55	89.21
letters	94.34	64.60	79.88
vehicle	95.86	61.84	63.61
cone-torus	98.75	76.76	85.17
glass	100.00	54.13	61.48
Average	*97.71*	*71.18*	*80.79*

Table 3. The resultant accuracies (in %) using different diversity measures on SET1.

Data set	TA	DF	DA	ME
iris	94.89	95.11	95.33	95.33
image-seg	95.94	96.33	94.63	96.18
vowel	83.23	86.35	81.41	86.35
satimage_CR	90.06	89.99	88.38	90.23
letters	88.07	87.67	82.08	87.68
vehicle	72.11	76.51	71.70	76.61
cone-torus	86.98	86.48	85.92	86.69
glass	62.73	63.98	65.00	64.22
Average	*84.25*	*85.30*	*83.06*	*85.41*

is considered in the case of incorrect combined labels where, more support is given to the classifiers which are not generating same incorrect labels. These properties of this measure are in parallel with the diversity exploited by the plurality voting combination rule.

4 Data Sets and Classifier Sets

In the simulations performed on seven data sets, 40% of the available data is used for training and the rest for testing. For the 'letters' data set, 10% of the available data set is used for training. Two different classifier sets are implemented for the classifier selection problem. Eight classifiers based on Multilayer Perceptrons (MLPs) and seven Radial Basis Function neural networks (RBFs) with varying architectures and initial weights are trained on the same training set for the first classifier set, SET1. The second classifier set, SET2 involves fifteen classifiers generated using bagging algorithm. The classifier prototype is not the same for all classifiers and are randomly selected as either nearest mean classifier, Fisher's linear discriminant classifier, logistic regression classifier or parzen classifier and each classifier is trained on a different bootstrap replicate of the training set. For the training of the classifiers, the PRTOOLS toolbox for MATLAB is used [6].

The experiments are conducted ten times for each data set and the average performances are taken into account. The average accuracies of the classifiers over the test sets in ten different simulations on SET1 and SET2 are presented in Tables 1 and 2 respectively. The first column provides the percentage of test samples for which at least one of the classifiers generates the true label. The second column shows the average of the individual accuracies of all classifiers and the last column shows the accuracy provided by the single best classifier which is the test performance of the classifier providing the highest accuracy on the training data. As seen in the second columns of the tables, SET1 includes classifiers with better individual accuracies on the average.

4.1 Experimental Results

For the classifier selection, *forward search* method is used. The resultant accuracies on the test data obtained for SET1 and SET2 are presented in Tables 3 and 4. The second column presents the resultant accuracies when the combined accuracy on the training data is maximized (TA) whereas the rest columns show the accuracies for double fault (DF), disagreement measure (DA) and the proposed measure based on misclassification entropy (ME) respectively.

As seen in the tables, the ME based measure provides better results than the other diversity measures for both of the classifier sets. This is mainly due to the fact that the diversity measure should depend on the combination scheme. Consider the results on SET1 including classifiers with higher individual accuracies. As seen in Table 3, the diversity measures DF and ME which are correlated with the combined accuracy provide better results than TA. On the other hand, TA is advantageous on SET2 involving comparatively weaker classifiers. This means that, diversity of classifiers may not be a good criterion for classifier selection in the case of weak classifiers. This observation is also true in the case of DA which selects classifiers with non-coincident correct or incorrect classifications since its performance on SET2 is far below the one in SET1. Comparing the two tables, we can also conclude that availability of strong classifiers is important in ensemble design since SET1 provided better combined results than SET2 for all measures.

At this stage, the important question to be answered is: "Do better ensembles include only the classifiers with best individual accuracies?". In order to find the answer, the *selected* classifiers are analyzed in terms of individual accuracy and diversity. Since one classifier set may come out to be more diverse than another according to

Table 4. The resultant accuracies (in %) using different diversity measures on SET2.

Data set	TA	DF	DA	ME
iris	95.33	95.00	94.22	95.00
image-seg	92.45	92.49	87.83	92.19
vowel	83.38	78.30	65.29	77.90
satimage_CR	89.70	89.59	86.50	89.70
letters	82.14	81.18	63.70	81.38
vehicle	68.88	66.37	50.37	71.40
cone-torus	85.35	85.48	79.19	85.69
glass	63.13	61.25	55.63	62.58
Average	*82.54*	*81.21*	*72.84*	*81.98*

Table 5. Averages for mean and standard deviations (in parentheses) of the classifiers selected over ten simulations and eight data sets.

Obj. Funct	SET1	SET2
TA	96.86(0.40)	89.13(2.71)
DF	94.14(4.55)	87.74(4.26)
DA	88.08(8.89)	73.56(17.04)
ME	94.21(4.43)	86.02(7.21)

one particular diversity measure but the contrary may be true for another, we preferred to check the standard deviation of the individual accuracies of the selected classifiers to get some idea about the differences of the selected classifiers in terms of their classification behaviors. The mean and standard deviation of the individual accuracies of the selected classifiers are computed for each of the ten different simulations for each data set and the averages over eight data sets are computed for both mean and standard deviations. The experimental results are provided in Table 5. As seen in the table, TA selects classifiers with higher individual accuracies compared to the other measures. Since DA mainly concentrates on the differences, the selected classifiers have less average performance. Also, the difference in the individual classifier accuracies is smaller in TA. However, diversity based classifier selection approaches select classifiers with much different individual accuracies. The classification differences of the selected classifiers are also important for TA since, in SET2 where a larger difference is available among the selected classifiers compared to SET1, TA performs better than the other diversity measures. On SET1, similar individual classifier accuracy averages (DF and ME has 2-3% less average accuracy) but much larger differences in standard deviations (more than 10 times) mean that, both highly accurate and relatively weaker classifiers exist in the optimal sets selected by DF and ME. It can be argued that the contribution to the diversity is mainly from relatively weaker classifiers. Hence, not only strong classifier but the simultaneous use of both strong and relatively weaker classifiers are essential for good ensemble design.

5 Conclusions

It is observed that in the cases where the classifiers involved in the selection task are weak, the combined training accuracy based classifier selection performs better compared to the diversity measures. A new diversity measure is developed for plurality voting which takes into account the uncertainty in the output labels corresponding to incorrect classes. The proposed measure is observed to perform better compared to the double fault and disagreement measure. The mean and standard deviations of the individual accuracies of the selected classifiers are examined for both combined training accuracy and diversity based classifier selection. It is observed that the classifiers selected in the training accuracy based case have higher individual accuracies and less standard deviations compared to the diversity measures based cases. It is argued that both strong and weak classifiers should be considered in classifier fusion where the weak classifiers are expected to contribute the ensemble diversity.

References

[1] G. Giacinto and F. Roli. Design of effective neural network ensembles for image classification purposes. *Image, Vision and Computing Journal*, 19(9-10):669–707, 2001.

[2] A. Krogh and J. Vedelsby. Neural network ensembles, cross validation, and active learning. In *Advances in Neural Information Processing Systems*, volume 7, pages 231–238. The MIT Press, 1995.

[3] G. Zenobi and P. Cunningham. Using diversity in preparing ensembles of classifiers based on different feature subsets to minimize generalization error. *Lecture Notes in Computer Science*, 2167, 2001.

[4] L. I.Kuncheva and C. J. Whitaker. Measures of diversity in classifier ensembles and their relationship with the ensemble accuracy. *Machine Learning*, 51:181–207, 2003.

[5] D. Partridge et al. MCS diversity and classifier confidence: A Bayesian approach. *http://www.dcs.ex.ac.uk/academics/reverson/pubs/DataFusMCMC.pdf*, 2004.

[6] R. P. W. Duin. PRTOOLS (version 3.0). A Matlab toolbox for pattern recognition. *Pattern Recognition Group, Delft University, Netherlands*, January 2000.

Visualization of Meta-Reasoning in Multi-Agent Systems

D. Řehoř[1], J. Tožička[2], P. Slavík[1]
[1]Computer Graphics Group, Dept. of Computer Science and Engineering,
[2]Gerstner Laboratory, Agent Technology Group, Dept. of Cybernetics,
Czech Technical University in Prague, Czech Republic
E-mail: {rehor, slavik}@fel.cvut.cz, tozicka@labe.felk.cvut.cz

Abstract

This paper describes the advances of our research on visualization of multi-agent systems (MAS) for purposes of analysis, monitoring and debugging. MAS are getting more complex and widely used, such analysis tools are highly beneficial in order to achieve better understanding of agents' behaviour. Our solution is based on our originally offline visualization tools suite, which now uses a new realtime data acquisition framework. In this case we have focused on agent meta-reasoning in a MAS for planning of humanitarian relief operations. Previous tools were unable to deal with complex characteristics of these simulations. This paper describes our new approach, declares conditions and proposes visualization methods, which fulfil them.

1 Introduction

In our research two different areas meet. On one hand, it is the research of MAS, on the other hand it is visualization as a mean of presenting various issues present in the MAS. This paper is primarily oriented on methods of visualization based analysis of the MAS.

1.1 Visualization

From the point of view of visualization, our research can be characterized as visualization of dynamic changes in MAS, including detection of long term trends, hidden dependencies, mutual influence among various entities etc. Currently, there is a very little research directed similar way [1], [2]. We have already presented some results, mostly dealing with offline visualization tools for backward analysis of recorded simulations of MAS [3], [4]. In this paper we present our advances in this field of research. The main advance presented here is the fact, that we are now able to connect to a running simulation and perform visualization in realtime. This has brought some interesting problems and solutions which are described further.

1.2 Multi-Agent Systems

Multi-agent systems are collections of autonomous, heterogeneous agents with specialized functionalities. The agents are usually able to carry out collective decision making, share resources, integrate services or just collaboratively seek for specific information. A more detailed description can be found in [5] and is beyond the scope of this paper.

1.3 Application

Our designed visualization tools are general enough to cover a larger area of MAS. To validate it, we have chosen a special case application scenario, which has been motivated by real-life situations and was developed within InX (Inaccessibility in MAS) project [6]. Let us assume that several delivery vehicles are operating in an area without available communication infrastructure.In the scenario we have several kinds of actors:

- **villages** – locations in where the humanitarian aid is required or distributed from
- **transporters** – organizations owning trucks that compete for tasks
- **trucks** – vehicles that dispatch the humanitarian aid upon request from the ports to the villages
- **meta-agents with monitoring agents** – create appropriate models of community through observations (observers watch their neighbourhood) and intrusions (intruders break into coalitions)

These are represented in the $\mathcal{A}$-GLOBE system [6] by *containers* that include simulation agents. Requests in the villages are generated by the environment simulator.

Agents can create a *coalition*, a set of agents, which agreed to cooperate on a single, well-specified task. When an agent tries to form coalition with another agent, it checks its private restrictions with the properties of the other agent. Properties are e.g. tribe, language, location, aliance-name etc.

1.4 Meta-Reasoning and Meta-Agents

Meta-reasoning focuses on detection of these constrains by monitoring the community communication. Agent's meta-reasoning is a computational process that implements agent's capability to reason on a higher level about another agent or a community of agents. Meta-reasoning can be used for reconstructing agents' private knowledge, their mental states and for prediction of their future courses of action. Meta-agents should have the capability to reason about incomplete or imprecise information. A more detailed description is in [6]. The central point of the meta-agent's operation is an appropriate model of the community. This model has to be expressed in an appropriate language of adequate granularity. These models play an important role in visualization.

1.5 Visualization Challenges

$\mathcal{A}$-GLOBE platform is primarily directed towards MAS simulation and has very limited visualization possibilities. However, the platform can be connected to various visualization servers. Research of meta-reasoning involves many hidden relations in the system, which are hard to reveal from a set of tables. We were interested in a easy-to-understand complex visualization method, which will facilitate the folowing challenges:

- How geographical proximity of various agents influences their states.Visualize various agent properties to see important differences and similarities, which have influence on their state and also the global state of the MAS.
- Preview history of evolution of agents' properties to see temporal dependencies of consecutive events in the MAS. History of events is very important, because MAS evolves in time. In this case the state of an agent is closely related to its 2D position and its neighbourhood (e.g. observers watch their neighbourhood). So to analyze recent events in the MAS, we need also information from history, which can have direct consequences in the recent events.
- Realtime data capture with sufficient sampling resolution – i.e. capture realtime data from a running MAS simulation (has no central clock) with appropriate sampling frequency.
- Since we are in realtime, we cannot do preprocessing of the whole captured time interval; the framework has to be capable of capturing unspecified number of agents and their properties of various structure. Data structure of an agent can be viewed as a tree (e.g. XML document). However, not only the data change in time, but the structure of the tree changes too.

2 Our Solution

In this section we describe both the underlying capture framework and our specialized visualization techniques. The proposed visualization tool fits in the row of our previous tools [3][4] designed for offline analysis of recorded simulation. Their unsuitability to perform realtime visualization of a running simulation was the most limiting factor which has lead to development of the new solution.

2.1 Capture Framework

The capture framework is partially similar to a temporal database. The term *capture* has been adopted from the area of video capturing, because of similar behaviour of storing consecutive states of the MAS. Unlike temporal database, our capture framework does not use exact time stamps, but uses a preset sampling frequency. In the following text the term *time step* refers to time difference between two consecutive captured states of the simulated MAS. All events that happen during one time step are considered as if they happened at the same time. The loss of information in this case is irrelevant, because the simulation platform does not use any global clock, communication is via network environment (far from perfect) and the changes in the simulated MAS are relatively slow.

The capture framework is based on a circular buffer, which holds the captured history of states of the monitored MAS. Experiments have shown, that capturing up to 5 minutes of history with sampling frequency of 10 Hz is sufficient for most of the analysis needs.

2.1.1 Description of the Data Structures: Captured data are stored along with their structure. It was one of the main challenges to keep the memory cost as low as possible while being able to access the changing structure of recorded data throughout the whole captured history. Agents, containers, properties are created, deleted, modified and/or hierarchically organized during simulation. The state of the simulated MAS can be anytime serialized into a kind of tree-like structure, i.e. the system consists of a number of agents located in containers, they have set/unset properties of various types and the whole structure changes during simulation.

The data structures reflect the needs of the capture framework and visualization. Our current implementation keeps captured data in a form of a data stream. A stream represents the flow of time – history of one data variable. A variable can be a *data container* or a *name index*, which in fact is a conversion from a textual identifier into an index and serves as the main stone for building and keeping the tree structure of the captured data. The index can be used to select another data stream. Data streams are allocated and dealocated as necessary. When a change arrives, the corresponding data container is duplicated and filled with new value. Further time slices only referece this instance of data container. The same applies to name indices.

This solution has several advantages like (i) changes in one name index do not affect other name indices, (ii) can build and keep trees of any size and also (iii) memory cost is kept relatively low, because only changes are recorded.

2.2 Visualization Method

From the point of view of visualization, the MAS contains agents, located in containers moving in 2D environment. Chosen visualization techniques reflect this characteristics. The 2D environment is represented by a textured plane, which also serves as a cutting plane. It represents the current simulation time. The time flow axis is parallel with the normal vector of the plane facing up (anything below the plane is history). The texture of the environment provides geographical and proximity context (Fig. 2).

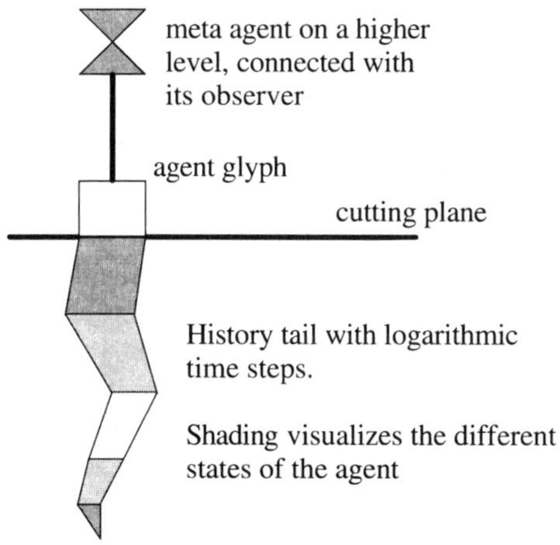

Fig. 1. Schematic view of the visualization technique.

Agents and containers are represented by glyphs [3]. There are several different types of agents in the MAS. Each type is assigned a unique glyph reflecting the purpose of the agent. Textual names of agents are drawn near the glyphs to distinguish particular agents. Containers are represented by glyphs too. If one container contains more than one agent, they are placed on an orbit around the container (Fig. 2). The orbit connects them and thus expresses that fact, that the agents belong to the particular container. Containers with only one agent do not use the orbit placement but keep the agent in the centre. Meta-agents are connected with their observers with a simple line (Fig. 1) and are placed on a different z-level to improve the readability of the connections.

To visualize the history of agents' properties, we have improved one method from our previous visualization approaches. As the time flows, each agent leaves a "path" behind (Fig. 1) – a sequence of agent's positions in the environment. The path reflects several important issues: history of proximity of agents with influence on agents' properties and/or environmental events. The technique of selecting a property to colour the steps of a history path has been adopted from our previous tools [4].

We have successfully tested non-linear time for the paths – i.e. while the simulation time is always linear, the paths are drawn with logarithmic decrease of distance between two corresponding time steps. Current time (at the cutting plane) is drawn with maximum distance between two time steps and fades with history. This approach has a very nice feature – it enables us to overview a much larger time interval while keeping important events visible.

Basically, agents contain two types of property values: numeric and textual. The framework recognizes them and uses two different methods of mapping them into colour palette. Minima and maxima are recorded along with numeric values and serve as scale for colouring. Textual properties are kept in a set which is used for colour selection. Basically, the algorithm assignes different colours to different strings. On a limited set of all strings (our case) the algorithm works well. Distinct colours emphasize the differences between agents.

3 Results

The proposed visualization tool already has significantly improved the efficiency of analysis of the simulated MAS. The realtime nature of the framework along with ability to present history brings a completely new view of the system. Meta-agents form their models of other agents. Through visualization of these model qualities and their history we are able to detect social structures inside the MAS, their relations to other agents, locations etc. or even explore the reasons or causing events very efficiently. This is very beneficial, because we can concentrate on the simulation itself or even perform some user intervention. For example, in particular areas, where a meta-agent is unable to provide some level of model qualities, the user can easily send an observer or an intruder there, because the visualization presents these issues in relation to geographical poswition.

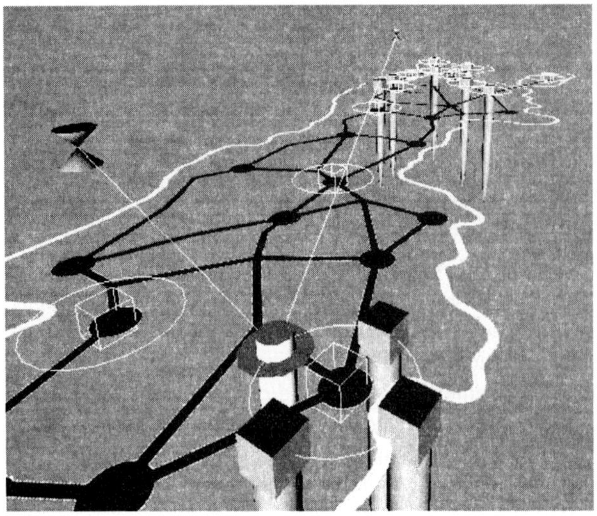

Fig. 2. Screenshot from a running visualization. In the foreground there are 4 agents in one container (placed on the orbit), textured cutting plane represents current simulation scenario with cities and roads.

We believe, our framework can be beneficial to monitoring of other agent platforms too. The capture framework itself is very modular. Input modules understand incoming data (parse communication protocol), convert it into internal form for the capture framework and build appropriate logical groups. In its current form it performs intrusive monitoring in such a way, that the MAS itself sends appropriate data into the visualization module. There exist other approaches, which collect data e.g. by monitoring communication among agents. The main contributions of our paper are as follows: (i) the ability to capture history of evolution of the agent system including the structure of logical groups and hierarchies and being able to interpret it – either in real-time visualization, or in a deeper, backward offline analysis; (ii) our visualization technique allows realtime preview of history of evolution of the MAS with many important contextual information (e.g. proximity issues).

4 Conclusion and Future Work

In this paper we have presented our advances on MAS visualization. Our visualization platform now captures and visualizes data from a realtime running MAS. Going realtime has brought new problems and challenges, some of them have been addressed in this paper.

In our further research we would like to concentrate on the following issues:

- Enlarge significantly the captured interval for tracking long term changes and dependencies and thus enable the user to walk through a much larger time interval. We expect to use some kind of disk cache for these purposes.
- Improve speed of the capture framework and colour selection and mapping mechanism.
- Test it with other MAS platforms to prove independence, flexibility and different dynamics of captured data.
- Introduce opposite direction of communication. Use visualization framework to analyse the MAS, stop the simulation if appropriate or perform other interaction with the simulation.

The first results are very promissing, the overview of the MAS has been significantly simplified and allows the user to explore and analyze many previously unseen issues. The visualized information helps the user to react efficiently and make particular decisions.

5 Acknowledgement

The project work has been co-funded by European Office for Aerospace Research and Development (EORD) Air Force Research Laboratory (AFRL) - contract number: FA8655-02-M-4057, Office of Naval Research (ONR) - award number: N00014-03-1-0292 and by the Grant No. LN00B096 of the Ministry of Education, Youth and Sports of the Czech Republic.

6 References

[1] Schroeder, M. and Noy, P. (2001) Multi-agent visualisation based on multivariate data. In: Proc. of the 5th International Conference on Autonomous Agents. ACM Press, Montreal

[2] Szekely, P., Rogers, C. M. Frank, M. (2001) Interfaces for understanding multi-agent behavior. In: Proc. of the 6th international conference on Intelligent user interfaces ACM Press, Santa Fe, New Mexico

[3] Nahodil, P., Kadleček, D., Řehoř, D. and Slavík, P. (2003) Dynamic Analysis of Agents' Behaviour – Combining ALife, Visualization and AI. In: Lecture Notes in Artificial Intelligence, Springer, Berlin.

[4] Kadleček, D., Řehoř, D., Nahodil, P. and Slavík, P. (2003) Analysis of Virtual Agent Communities by Means of AI Techniques and Visualization. In: Intelligent Virtual Agents. Springer, Heidelberg.

[5] Pěchouček, M., Štěpánková, O., Mařík, V. and Bárta, J. (2003) Abstract Architecture for Meta-reasoning in Multi-Agent Systems. In: Multi-Agent Systems and Applications III. Springer, Heidelberg.

[6] Šišlák, D., Rollo, M. and Pěchouček, M. (2004) A-globe: Agent Platform with Inaccessibility and Mobility Support. In: Cooperative Information Agents VIII. Springer, Heidelberg

Intelligent Agent-Inspired Genetic Algorithm

C. G. Wu[1], Y.C. Liang[*1, 2], H.P. Lee[2] and C. Lu[2]

[1] College of Computer Science and Technology, Jilin University, Key Laboratory of Symbol Computation and Knowledge Engineering of Ministry of Education, Changchun 130012, China

[2] Institute of High Performance Computing, Singapore 117528, Singapore

*Corresponding author, E-mail: liangyc@ihpc.a-star.edu.sg

Abstract

This paper presents an intelligent agent-inspired genetic algorithm (IAGA). Analogous to the intelligent agent, each individual in IAGA has its own properties, including crossover probability, mutation probability, etc. Numerical simulations demonstrate that, compared with the standard GA where all individuals in a population share the same crossover and mutation probabilities, the proposed algorithm is more flexible, efficient and effective.

1 Introduction

In the latest decades, genetic algorithm (GA), as a kind of heuristic algorithms, has been developed rapidly and applied to many fields including combinatorial optimization, function optimization, parameter selection, and many others [1-4]. Because of its high parallel properties and requiring minimal knowledge of the problems involved, GA has shown its powerful strength for solving problems in many fields. At the same time, intelligent agent (IA) is a distributive computing technique. It is far more flexible in handling problems[5]. Generally IA emphasizes on flexibility and responsiveness and GA pursues the global optimality of solutions. There are quite a few references relating to the combination of the two methods[6,7]. However, IA and GA still are applied relatively independently, for instance, the author of Ref. [5] presents an approach by combining GA and IA for manufacturing scheduling, in which GA provides optimized results to IA and IA schedules the limited resources according the scheme produced by GA. In this paper, we attempt to endow GA more distributive properties, similar to that possessed by IA, to improve the flexibilities of GA.

The remaining part of the paper is organized as follows. Section 2 gives the outline of the standard GA. Section 3 summarizes briefly the properties of the intelligent agent. Section 4 proposes a novel GA paradigm sharing some characteristics with IA. Section 5 shows the simulated experimental results. Conclusions and discussions are given in the last section.

2 Standard GA

The standard GA includes the following five primary elements in general: a) Problem coding, b) Population setting, c) Fitness function construction, d) Genetic operator design, and e) Control parameter selection.
The standard genetic algorithm can be formulated briefly as follows:

1. Select running parameters for the population, including population size, crossover probability, mutation probability and the max generation number.
2. Initialize a random population
3. Evaluate individuals in the population
4. Based on some rules, select parental individuals. If the crossover operation happens under uniform probability, cross the two parental individuals and then two child individuals are reproduced, or else, copy the parental individuals into child population directly.
5. For the two individuals newly entering the child population, perform mutation operation under uniform probability.
6. If the child population is full, replace the parental one with it, and then evaluate individuals in the parental population, or else, repeat from the 4th step.
7. If the number of max generation is reached, save the best solution in the final population and then terminate, or else repeat from the 4th step.

From the above description, it can be seen that the crossover and mutation operators use the same probabilities on all individuals, respectively, in the standard GA.

It has been demonstrated by practices that the selection of control parameter is very important to keep the population diversity and enable the population to converge efficiently. In contrast, if the individuals are almost the same with each other (lack of diversities) in the population, GA will be easily trapped into the local minimum and converge too early. In this paper, we will focus in the strategy of control parameter selection so that the control parameter can be distributive as possibly in the population, but not as that in the standard GA where all individuals share one parameter set which is determined before implementing the algorithm.

3 Characteristics of intelligent agent

Intelligent agent (IA), as a popular technique in distributive artificial intelligence field, has many desirable characteristics[5-7]. The following four items are its basic characteristics: a) autonomy, b) social ability, c) reactivity, d) pro-activeness. Autonomy enables agents to be more adaptive to the environment. Social ability ensures agents to collaborate with each other when implementing a given task. Reactivity enables agents to sense the surrounding environment and pro-activeness let agents go in the direction to finish their tasks. Based on the above traits, agents can adjust their properties to adapt to the changing environment as possibly, at the same time, collaborate mutually and try to finish their task actively and distributively.

4 Intelligent agent-inspired genetic algorithm

The flexibility of intelligent agent has been briefly introduced in Section 3. Compared with intelligent agent, GA should have more distributive properties. Motivated by IA, we attempt to endow GA some new characteristics, aiming at making it more parallel and more distributive. The main alterations focus on two aspects: one is that each individual is endowed a parameter set which includes crossover probability and mutation probability. The other is on selection method. In the proposed intelligent agent-inspired genetic algorithm (IAGA), we apply a method named local competition to select the second parental individuals. When it comes to selection, initially, one parental individual is selected according to the traditional proportion selection methods. And then randomly select a group of individuals (named as candidate set) in parental population. The individual with the highest fitness in the candidate set is taken as the second parental individual. This process strengthens the important principle "Survival of the fittest" in GA, since the individuals in the candidate set compete with each other to transmit its genetic traits.

Before implementing the algorithm, two genetic intervals $[P_{cL}, P_{cU}]$ and $[P_{mL}, P_{mU}]$ must be assigned. The first one is for adjusting the crossover probability and the second one for adjusting the mutation probability. If the crossover is performed on two selected individuals x and y, then new individuals x' and y' are reproduced. Taken x' for example, the crossover probability is changed according to

$$p_c(x') = p_c(x) + \mu\Delta f, \qquad (1)$$

and mutation probability is changed according to

$$p_m(x') = p_m(x) - \nu\Delta f, \qquad (2)$$

where $\Delta f = f(x') - f(x)$, $f(\cdot)$ is the fitness function, μ and ν are constants, which are predetermined to adjust the convergence speed. If the new crossover (or mutation) probability is larger than its corresponding upper bound, it is set to its upper bound P_{cU} (orP_{mU}); or smaller than corresponding lower bound, it is set to its lower bound P_{cL} (or P_{mL}).

The proposed IAGA can be described as follows:

1. Initialize the population. Randomly generate a group of feasible solutions and code them into individuals, then endow each individual a crossover probability and a mutation probability, which should be randomly set with their corresponding genetic intervals, i.e., $[P_{cL}, P_{cU}]$ and $[P_{mL}, P_{mU}]$.
2. Select parental individuals. According the traditional proportion selection method, select the first parental individual. And then randomly select some individuals to construct the candidate set. The individual with the highest fitness in the candidate set are taken as the second parental individual.
3. Reproduce child individuals. Crossover the parental individuals to reproduce child ones. The crossover mechanism includes single-point crossover, two-point crossover, etc.
4. Adjust genetic parameters. The crossover probability and mutation probability are adjusted after the reproduction of new individuals according to Eqs. (1) and (2). Both of them should be restricted within their genetic intervals. If any of them exceeds its corresponding genetic interval, the exceeding one will be set to be the nearest bound.
5. Check the size of child population. If it's smaller than the size of parental population, repeat from step 2; or else replace the parental population with the child one.
6. Check the generation number. If it reaches the maximum, terminate and save the best solution in

the final generation; or else repeat from the step 2 and continue.

From the above mentioned procedure of the proposed IAGA, it can be seen that if μ and ν are both taken as 0, the genetic probabilities (including crossover and mutation probabilities) of all individuals are set to be the same values, respectively, then the proposed algorithm is similar to the standard GA. In addition, if the size of candidate set is taken as 1, then the proposed IAGA is entirely the same as the standard GA. The flow chart of the proposed IAGA is shown in Fig. 1.

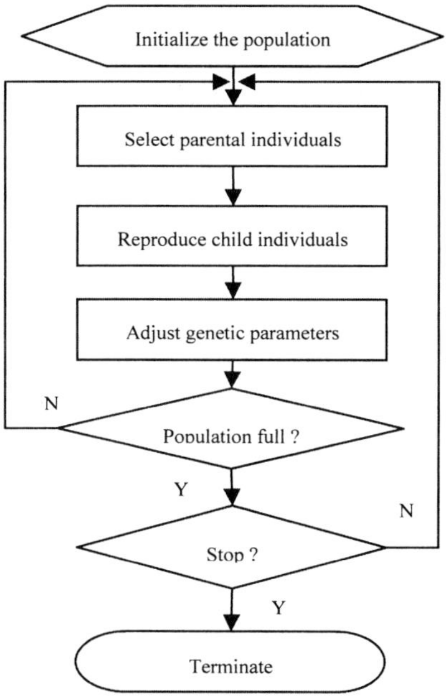

Fig. 1 Flow chart for proposed intelligent agent-inspired genetic algorithm

5 Simulation experiments

To demonstrate the validity of the proposed algorithm, we run it for some well-known traveling salesman problems[8], comparing it with the standard GA (SGA). For the SGA the population size, maximal generation number, crossover and mutation probabilities are taken as 200, 200, 0.98 and 0.005, respectively. For the proposed IAGA the corresponding running parameters are shown in table 1.

The results are listed in table 2, where the first column is for the instance names, the proceeding four columns for the tour lengths and run time corresponding to SGA and the proposed IAGA, respectively. And the last column is for the shortened tour lengths by the proposed algorithm. All the results are the averages of five independent runs. From table 2, it can be seen that the proposed IAGA can obtain the better solutions than those obtained by the SGA and shortens the tour lengths in different levels for all of the examined instances. The highest improved level is in 11.95% for the pr264 instance (in the sixth line of table 2).

Table 1. Running parameters for the proposed IAGA

Parameter	Value	Parameter	Value
Size of population	200	Maximal generation	200
Crossover Interval	[0.6, 0.98]	Mutation interval	[0.005, 0.3]
μ	0.01	ν	0.01
Size of candidate set	10		

Table 2. Comparison between standard GA and IAGA

Instance name	Tour length		Run time		Shortened tour length
	SGA	IAGA	SGA	IAGA	
KroA100	37192	36600	0.35	0.29	592
Bier127	138090	130449	0.48	0.42	7641
KroB150	53168	48012	0.70	0.45	5156
KroB200	75347	71509	0.59	0.41	3838
pr226	383389	375506	0.84	0.53	7883
pr264	248475	218792	1.02	0.83	29683
pr299	311084	294154	0.20	0.19	16930
rd400	62159	60434	0.45	0.37	1725
fl417	135790	135027	0.47	0.39	763
pcb442	222628	219825	0.52	0.37	2803

6 Conclusions and discussions

The proposed IAGA endows the GA more distributive properties and use a new selection mechanism to select the second parental individual. Numerical experiments show that the proposed algorithm improves the quality of solutions remarkably and is more efficient than the standard GA. The studies of distributive properties of GA could promote its applications in distributive artificial intelligence and enable it to run on internet grids.

Acknowledgements

The first two authors are grateful to the support of the National Natural Science Foundation of China (Grant No. 60433020), the Science-Technology Development Project of Jilin Province of China (Grant No. 20030520) and the doctoral funds of the National Education Ministry of China (Grant No. 20030183060).

References

[1] Reichert C., Magedanz T. (2004) A fast heuristic for genetic algorithms in link weight optimization. Lecture Notes in Computer Science, 3266: 144-153.

[2] Wu C.G., Xiang W., Liang Y.C., Lee H.P. and Zhou C.G. (2004) A modified integer-coding genetic algorithm for job shop scheduling problem. Lecture Notes in Computer Science, 3157: 373-380.

[3] Forman S.L. and Yue Y.D. (2003) Congressional districting using a TSP-based genetic algorithm. Lecture Notes in Computer Science 2724: 2072-2083.

[4] Varga V., Dumitrescu D. and Groşan C. (2004) Solving Stochastic Optimization in Distributed Databases Using Genetic Algorithms. Lecture Notes in Computer Science, 3255, 259-274.

[5] Shen W.M. (2002) Genetic algorithms in agent-based manufacturing scheduling systems. Integrated Computer-aided Engineering 9 (3): 207-217.

[6] Martin-Bautista M.J., Vila M.A. and Larsen H.L. (1999) A fuzzy genetic algorithm approach to an adaptive information retrieval agent. Journal of the American Society for Information Science 50 (9): 760-771.

[7] Gouveia P. and Ramos J. (2004) Multi-agent systems specification and certification: a situation and state calculus approach. Annals of Mathematics and Artificial Intelligence 41: 301–338.

[8] Reinelt G. (1991) TSPLIB-A traveling salesman problem library. ORSA Journal on Computing 3 (4): 376-384.

Combining Lazy Learning, Racing and Subsampling for Effective Feature Selection

Gianluca Bontempi, Mauro Birattari and Patrick E. Meyer
ULB, Université Libre de Bruxelles, Bruxelles - Belgium

Abstract

This paper presents a wrapper method for feature selection that combines Lazy Learning, racing and subsampling techniques. Lazy Learning (LL) is a local learning technique that, once a query is received, extracts a prediction by locally interpolating the neighboring examples of the query which are considered relevant according to a distance measure. Local learning techniques are often criticized for their limitations in dealing with problems with high number of features and large samples. Similarly wrapper methods are considered prohibitive for large number of features, due to the high cost of the evaluation step. The paper aims to show that a wrapper feature selection method based on LL can take advantage of two effective strategies: racing and subsampling. While the idea of racing was already proposed by Maron and Moore, this paper goes a step further by (i) proposing a multiple testing technique for less conservative racing (ii) combining racing with sub-sampling techniques.

1 Introduction

Lazy Learning (LL) is a local modeling technique which is *query-based* in the sense that the whole learning procedure (i.e. structural and parametric identification) is deferred until a prediction is required. In previous works we presented an original *Lazy Learning* algorithm [3, 1] that selects automatically on a query-by-query basis the optimal number of neighbors, and its application to data analysis, time series prediction, system identification and nonlinear control. Despite the large amount of applications to real tasks, local methods are often a target of criticism as far as computational requirements for dealing with large datasets (i.e. many variables and/or large samples) are taken into consideration.

This paper discusses how to effectively employ the LL algorithm for feature selection (for an up-to-date state of the art on feature selection see [7]). In particular, we will focus here on a wrapper method based on LL. In the wrapper approach [9] the feature subset selection algorithm exists as a wrapper around the learning algorithm, which is often considered as a black box able to return (e.g. via cross-validation) an evaluation of the quality of a feature subset. While most of the wrapper algorithms are defined independently of the learning machine, we propose here a method which takes advantage of the unique aspects of a local learning algorithm.

The usefulness of a local modeling approach for reducing the cost of feature selection was first presented in [10]. The idea consists in assessing a large number of feature subsets by performing cross-validation only on a reduced test set. On the basis of well-known statistical results, it is possible to show that families of good feature subsets can be rapidly found by quickly discarding the bad subsets and concentrating the computational effort on the better ones. This model selection technique was called the *Hoeffding race* by Maron and Moore [10], with reference to Hoeffding's formula which puts a bound on the accuracy of a sampled mean of N observations as an estimator of the expected value. Local modeling techniques fit well in this paradigm thanks to the property that modeling is performed only when a prediction is required. Then, by reducing the size of the test set, the computational cost required for assessment is reduced, too.

The aim of this work is: i) to speed up further the evaluation step of a wrapper method by reducing, together with the number of test predictions, also the size of the training set required for assessing the quality of a candidate feature set, ii) to improve the efficiency of the wrapper search by discarding more rapidly those candidates which, on the basis of the tests made till that moment, appear as significantly worse than others. In particular, the main two contributions of the paper are: the combination of racing and sub-sampling techniques with the Lazy Learning algorithm previously proposed by the authors and the use of a multiple testing criterion (the Friedman test) for performing a more statistically founded selection.

The use of sampling [8] enhances the performance of the racing approach to feature selection by making possible the definition of a number of lightweight experimental benchmarks to assess, compare and discriminate between candidate subsets. Each experimental benchmark relies on small training and test sets, which are obtained by randomly sub-sampling the available dataset. The idea is that, in order to discriminate between a very large number of candidates, the combination of multiple small and fast benchmarks can be more effective than a single big and time consuming validation procedure.

The second contribution of the paper lies in the use of a nonparametric multiple test, the Friedman test [4], to compare different configurations of input variables and to select the ones to be eliminated from the race. The use of the Friedman test for racing was proposed first by one of the authors in the context of a technique for comparing metaheuristics for combinatorial optimization problems [2]. This is the first time that the technique is used in a feature selection setting. Note that in some sense this method fills the gap between Hoeffding race [10] and BRACE [11]: similarly to Hoeffding race it performs a nonparametric test, and similarly to BRACE it considers a blocking design.

2 The RACSAM (racing+sub-sampling) algorithm

The RACSAM wrapper algorithm is based on the idea that once the evaluation is based on the LL algorithm,[1] this can be made faster by considering training and test sets of limited size.

The main steps of the RACSAM algorithm are: (i) the creation of a number of lightweight experimental benchmarks, each characterized by a small training and test set, (ii) the use of the experimental benchmarks to assess in parallel a large family of candidates, (iii) the discarding of those candidates resulting significantly worse than others.

2.1 The F-race algorithm

The racing algorithm proposed in this paper, F-Race in the following, takes as input a training set of size N, a test set of size N_{ts}, a set of M candidate feature sets $\mathcal{S}_m \subseteq \{1, \ldots, n\}$ (also called configurations) and a number $W \geq 1$ of expected winners. The expected output is a set of at most W configurations which are significantly better than the others.

Let $\Theta_0 = \{\mathcal{S}_m, m = 1, \ldots, M\}$ the set of configurations at the start of the algorithm. A racing algorithm proceeds by generating a sequence $\Theta_0 \supseteq \Theta_1 \supseteq \Theta_2 \supseteq \ldots$, of nested sets of candidate configurations. The step from a set Θ_{q-1} to Θ_q is obtained by possibly discarding some configurations that appear to be suboptimal on the basis of information available at step q. Let M_{q-1} the cardinality of Θ_{q-1}, that is the number of remaining configurations and $I_{q-1} = \{m : \mathcal{S}_m \in \Theta_{q-1}\}$ the indices of the configuration still competing before the execution of the qth step.

At the qth step, $1 \leq q \leq N_{ts}$, the racing algorithm computes for all $\mathcal{S}_m \in \Theta_{q-1}$

$$\hat{y}_{q,m} = LL(x_q, D_N, \mathcal{S}_m) \tag{1}$$

where $LL(\cdot)$ is the prediction returned by a LL algorithm where the feature set is $\mathcal{S}_m \in \Theta_{q-1}$ and the training set is made of N samples.

This value is used to fill an evaluation matrix E, sized $N_{ts} \times M$, whose generic $[q, m]$ term is

$$E[q, m] = |y_q - \hat{y}_{q,m}|, \quad q = 1, \ldots, N_{ts}, \quad m \in I_{q-1} \tag{2}$$

where y_q is the observed output for the input x_q. Step q terminates defining set Θ_q by dropping from Θ_{q-1} the configurations that appear to be suboptimal in the light of the statistical test described in the following section. Note that this test compares only the $M_{q-1} \leq M$ columns of the evaluation matrix, whose indices are in I_{q-1}, which represent the feature configurations still in the race.

The above described procedure is iterated and stops either when all configurations but W are discarded (i.e. $M_q \leq W$ for some q) or when $q = N_{ts}$. The advantage of racing is that in the first case, $M - W$ configurations are discarded by having recourse only to a subset of the N_{ts} samples available for testing.

The statistical test: the racing algorithm we propose in F-Race[2] is based on the Friedman test, a statistical method for hypothesis testing also known as Friedman two-way analysis of variance by ranks [4]. The null hypothesis of the test assumes that all remaining configurations in the race belong to the same error distribution. If at the qth step the null of the aggregate comparison is not rejected, all candidates in Θ_{q-1} pass to Θ_q. On the other hand, if the null is rejected, the configuration with the largest estimated mean-square error is discarded and the test repeated with $M_{q-1} - 1$ configurations, until the null hypothesis is not rejected.

The main merit of our nonparametric approach is that it does not require to formulate hypotheses on the distribution of the observations. A second role played by the Friedman test is to implement in a natural way a blocking design [5]. Blocking is an effective way for normalizing the costs observed on different conditions. By focusing only on the ranking of the different configurations within each condition, blocking eliminates the risks that the variation due to the difference among test samples washes out the variation due to the difference among configurations.

[1] We refer the reader to [3, 1] and to the publicly available R package `lazy` (http://cran.r-project.org/src/contrib/Descriptions/lazy.html) for a description of the main features of the LL algorithm.

[2] available at http://cran.r-project.org/src/contrib/Descriptions/race.html

2.2 The racing+sub-sampling combination

In spite of the improvement due to racing, evaluation based on LL may still appear computationally expensive if the training set is very large. It is sufficient to note that the complexity of the LL prediction in (1) is proportional to the number of samples in D_N.

The added value of sampling consists in using in (1) training sets $D_{\hat{N}}$, composed of $\hat{N} < N$ samples, which are a randomized subset of the original set D_N. The reduction of the training set size speeds up the LL prediction and consequently each step of the F-race algorithm.

Let us notice however that the computational gain occurs at the cost of the deterioration of the assessment of the quality of each single configuration. In terms of bias/variance trade-off sub-sampling implies variance increase. In general terms, the rationale for the RACSAM approach is that very bad models should be detected rapidly and with small effort by exploiting only a part of the information of the training set, reserving the intensive use of the entire information only to the most difficult cases. This is implemented in practice by increasing gradually the size $\hat{N}$ of the sub-sampled training set once the race moves forward.

The resulting RACSAM algorithm consists then in a modified version of the F-race algorithm, described in Section 2.1, where at the qth step only a subsample of the training set of size $\hat{N}^{(q)}$ is used for computing the accuracy of all the M_{q-1} configurations still in race. Equation (1) is then replaced by $\hat{y}_{q,m} = LL(x_q, D_{\hat{N}^{(q)}}, \mathcal{S}_m)$ where $\hat{N}^{(q)}$ is the size of the sub-sampled training set at the qth step. In our preliminary experiments we used the rule $\hat{N}^{(q+1)} = \hat{N}^{(q)} + 10$ with $\hat{N}^{(0)} = 50$ to update the size of the sub-sampled training set.

3 The exploration strategy

The RACSAM algorithm takes as input a set of M candidate feature sets and returns a set of W winners. While this approach makes possible an exhaustive exploration of the feature subspace (i.e $M = 2^n$) in the case of very small n, it requires some modifications if we intend to address problems with very large dimensionality. Our search strategy is quite simple. An initial set $\Theta_0^{(0)}$ of candidates is created either randomly or by adopting some filtering techniques (e.g. Pearson correlation or Gram-Schmidt orthogonalization [7]). This initial set is passed through the RACSAM algorithm which returns the W better candidates. Then an iterative procedure begins. The iterative search is composed of two steps: (i) generation (e.g. by neighborhood exploration) of a new set of M candidates $\Theta_0^{(i)}$ starting from the output $\Theta^{(i-1)}$ of the RACSAM algorithm (ii) racing of $\Theta_0^{(i)}$ by RACSAM. Alternative search strategies that could be easily combined with the RACSAM approach are discussed in [10].

4 Experimental results

Two experiments were carried out: the first one studies the ability of the algorithm to detect the subset of relevant variables in a very large set of irrelevant ones. The second one assesses the improvement in prediction accuracy that can be obtained by adopting the RACSAM approach.

Selection of relevant variables: we consider a problem of feature selection where the dimensionality of the input space is $n >> 10$ and the output is dependent only on 10 inputs according to the relation

$$y = 10\sin(\pi x_1 x_2) + 20(x_3 - 1/2)^2 + 10x_4 + 5x_5 + \\ + 10\sin(\pi x_6 x_7) + 20(x_8 - 1/2)^2 + 10x_9 + 5x_{10} + \varepsilon.$$

We suppose that the input vector x takes value in the hypercube $[0,1]^n$ and that ε is a standard Gaussian random variable. Note that this feature selection problem is a harder version of the problem proposed in [6]. We conduct two main experiments: the first with $n = 110$ and the second with $n = 210$ inputs. Each experiment has two variants: initialization by Gram Schmidt (GS) and by random generation (RN). For each experiment we generate 25 times a random training and test set by sampling uniformly the input hypercube. In the GS variant we first rank the variables according to the Gram-Schmidt orthogonalization procedure and we create an initial set of 5000 candidates $\Theta_0^{(0)}$ by combining the most relevant 50 variables according to the GS procedure. In the random variant the initial set $\Theta_0^{(0)}$ of 5000 candidates is generated randomly.

The RACSAM procedure performs a F-race where the p-value threshold is set to 0.01 and the number of winners to $W = 50$. The algorithm is stopped when at least 10000 models have been assessed. For each run, we define successful the RACSAM algorithm if it is able to discover and retain the best feature set (made of the first ten variables only) among the W winners. The success rate of the RACSAM algorithm is compared to that of a forward selection (FS) procedure using a LL algorithm with 270 training samples and 500 test samples.

Table 1 reports for each experiment (i) the RACSAM percentage of successes, (ii) the rate of success of the forward selection procedure, (iii) the average number of models (plus the standard deviation) which were examined by the RACSAM algorithm before discovering the correct set, (iv) the average number of training samples (plus the standard deviation) required for the assessment, (v) the average number of test samples (plus the standard deviation) required for the assessment. Note that an exhaustive search would have required the assessment of

n/Init	Racs	FS	Assessed	Train	Test
110/GS	92%	8%	7342 ± 572	259 ± 31	601 ± 307
110/RN	88%	8%	7054 ± 726	220 ± 21	213 ± 214
210/GS	88%	4%	7403 ± 595	265 ± 52	660 ± 522
210/RN	84%	4%	7017 ± 901	223 ± 37	240 ± 366

Table 1. Comparison RACSAM vs. Forward Selection.

2^n models. The experimental results show that the performance of the algorithm is independent of the initialization and quite robust to number of irrelevant variables.

Prediction accuracy: this experiment compares the performance accuracy of the LL algorithm enhanced by the RACSAM procedure to the the accuracy of two state-of-art algorithms, a SVM for regression and a regression tree (RTREE). These two algorithms are well-known and powerful examples of embedded techniques for feature selection. We use the implementations available in the `e1071` and `tree` R packages, respectively. The comparisons were carried out by performing a five-fold cross-validation on six real datasets[3] of high dimensionality: `Ailerons` ($N = 14308, n = 40$), `Pole` ($N = 15000, n = 48$), `Elevators` ($N = 16599, n = 18$), `Triazines` ($N = 186, n = 60$), `Wisconsin` ($N = 194, n = 32$) and `Census` ($N = 22784, n = 137$).

Two version of the RACSAM algorithm were tested: the first (LL-RAC1) takes as feature set the best one (in terms of estimate Mean absolute Error (MAE)) among the W winning candidates : the second (LL-RAC1) averages the predictions of W LL predictors, where each LL model takes as inputs one of the W sets returned by the RACSAM procedure. In both cases we set $W = 5$, and the p-value to 0.01. The selection procedure stops when at least 1000 different configurations have been analysed.

In Table 2 we present, for each learning method, the absolute prediction error averaged over the 5 cross-validation groups. Since the methods are tested on the same examples under the same conditions, we use the paired test of significance to perform an exhaustive paired comparison of all the methods for all the benchmarks. In what follows, by "significantly better" we mean better at least at a 5% significance level. As far as the comparison LL-RAC1 to LL-RAC2 is concerned, we obtained that LL-RAC2 is significantly better than LL-RAC1 3 times out of 6 and it is never significantly worse than LL-RAC1. As far as the comparison of LL-RAC2 to the other state-of-the-art techniques is concerned we obtain that the LL-RAC2 approach is never significantly worse than SVM and/or RTREE but that it performs 5 times out of 6 significantly better than SVM and 6 times out of 6 significantly better than RTREE.

[3] available at http://www.liacc.up.pt/~ltorgo/Regression/DataSets.html

Dataset	AIL	POL	ELE	TRI	WIS	CEN
LL-RAC1	9.7e-5	3.12	1.6e-3	0.21	27.39	0.17
LL-RAC2	9.0e-5	3.13	1.5e-3	0.12	27.41	0.16
SVM	1.3e-4	26.5	1.9e-3	0.11	29.91	0.21
RTREE	1.8e-4	8.80	3.1e-3	0.11	33.02	0.17

Table 2. Mean Absolute prediction errors.

5 Conclusions

Preliminary results show the effectiveness of the RACSAM approach in selecting relevant features and in improving the predictive accuracy, especially in the case of a combination of the predictors based on the feature sets returned by RACSAM. Future research will extend these preliminary results by combining the RACSAM assessment procedure with more sophisticate search strategies (e.g. the schemata search proposed in [10]) and by applying the technique to massive datasets with thousands of variables.

References

[1] M. Birattari, G. Bontempi, and H. Bersini. Lazy learning meets the recursive least-squares algorithm. In M. S. Kearns, S. A. Solla, and D. A. Cohn, editors, *NIPS 11*, pages 375–381, Cambridge, 1999. MIT Press.

[2] M. Birattari, T. Stützle, L. Paquete, and K. Varrentrapp. A racing algorithm for configuring metaheuristics. In W. B. Langdon, editor, *GECCO 2002*, pages 11–18. Morgan Kaufmann, 2002.

[3] G. Bontempi, M. Birattari, and H. Bersini. Lazy learning for modeling and control design. *International Journal of Control*, 72(7/8):643–658, 1999.

[4] W. J. Conover. *Practical Nonparametric Statistics*. John Wiley & Sons, New York, NY, USA, third edition, 1999.

[5] A. Dean and D. Voss. *Design and Analysis of Experiments*. Springer Verlag, New York, NY, USA, 1999.

[6] J. H. Friedamn. Multivariate adaptive regression splines. *The Annals of Statistics*, 19(1):1–141, 1991.

[7] I. Guyon and A. Elisseeff. An introduction to variable and feature selection. *Journal of Machine Learning Research*, 3:1157–1182, 2003.

[8] G. H. John and P. Langley. Static versus dynamic sampling for data mining. In *Proceedings of the Second International Conference on Knowledge Discovery in Databases and Data Mining*. AAAI/MIT Press, 1996.

[9] R. Kohavi and G. H. John. Wrappers for feature subset selection. *Artificial Intelligence*, 97(1-2):273–324, 1997.

[10] O. Maron and A. Moore. The racing algorithm: Model selection for lazy learners. *Artificial Intelligence Review*, 11(1–5):193–225, 1997.

[11] Andrew W. Moore and Mary S. Lee. Efficient algorithms for minimizing cross validation error. In *International Conference on Machine Learning*, pages 190–198. Morgan Kaufmann Publishers, Inc., 1994.

Personalized News Access

D. G. Kaklamanos, K. G. Margaritis

Parallel and Distributed Processing Laboratory, University of Macedonia, Greece
Email: {kaklaman, kmarg}@uom.gr

Abstract

PENA (Personalized News Access) is an adaptive system for the personalized access to news. The aims of the system are to collect news from predefined news sites, to select the sections and news in the server that are most relevant for each user and to present the selected news. In this paper are described the news collection process, the techniques adopted for structuring the news archive, the creation, maintenance and update of the user model and the generation of the personalized web pages. This is a preliminary work that is based on the system that is described in [1].

1 Introduction

Simple forms of personalization are implemented in many sites, based on preferences directly expressed by the users, while there are more complex approaches proposed in the literature, for example [2, 3]. In this paper, PENA is introduced, an adaptive web site for accessing news stories. PENA is a multi-agent system, which aims at personalizing both the selection of topics that are of interest to the user and the presentation of each news item.

This paper is organized as follows. In Section 2 we discuss the functionality requirements of the system. We then focus on three main aspects: the organization of content, described in Section 3, the adopted user modelling techniques, described in Section 4, and the news retrieval process, described in Section 5. Section 6 outlines the personalization strategy of PENA. Section 7 describes some testing experiments of the system and Section 8 gives an overview of the future goals.

2 Functionality, Requirements and Architecture of PENA.

In PENA we had two main functionality goals: Firstly, the system should be able to collect news from the web. Secondly, the system should present only the news that the user is interested in, and leave the others aside, as information which can be reached on demand.

As a first requirement, we imposed that these forms of personalization should be provided to both first time and frequent users. As a consequence, PENA must be able to generate an initial, possibly approximate, model of first time users.

Another important requirement is the ability of the system to revise the user model, in order to make it as close as possible to his actual characteristics and follow possible changes of interest.

Thirdly, the system should never impose its choices on the user. The user must be always able to modify system decisions. Moreover, being the user able to make changes supports the user modelling process, as the system can get information about the user's actual characteristics and interests. As a consequence, all the web pages generated by PENA contain buttons for modifying the system choices.

These functionality goals imposed some requirements on the design of PENA, which is organized as three tier architecture. The user can access the server using any web browser, first tier. The web browser interacts with a set of agents which are responsible for user-modelling and personalization activities, second tier, and which can access information maintained in a set of databases, third tier.

A user interaction agent collects user requests and returns web pages to the browser. User modelling is carried out by two agents: one is activated only when a user connects to the server for the first time and it creates the initial user model based on the data that are provided by the user in the registration form and a library of stereotypes. The second agent is activated at the start of each session and is responsible for revising the user model, based on the user's actions during his last session and on a set of user modelling rules. The user models are maintained in a database and thus can be retrieved each time the user connects to the site.

Two other agents are in charge of extracting, respectively, from the news database and from the user database, the specific pieces of information that are going to be presented to a user, given his model. All these pieces of information are then passed to the interaction agent, which generates the web pages and sends them to the user's browser.

3 Organization of Content.

The news database contains all the news stories that can be presented to the user, and is able to support the classification of news according to their topics. In PENA, we defined a hierarchy of sections, in which the news is classified. This hierarchy was decided based on the high level classification of news that is done to the web sites from where the news stories are downloaded. More specifically, we considered the high-level sections politics, economy, sport, technology, culture and world. As the news are retrieved from the web sites, the system, based on the web site news classification, stores them locally in the corresponding category.

There are many reasons that justify this choice. First of all, it worth's maintaining this high level classification since it is accurate. Second, the high level classification of news sections allows us to provide some form of personalization even when limited information about the user is available. This is very common for first time users. In this case, it is difficult to make a selection of topics more accurate than the one corresponding to high level sections in the hierarchy.

The news database consists of three Tables. The main table contains the article id, the source of the article and the high level category of the article. The other two tables contain the names of the categories and the name of the sources (news portal) of the articles. It must be mentioned that the retrieval part of PENA is responsible for populating this database.

4 User Modelling.

The user model plays a central role in the personalization process of PENA, because accurate personalization requires the modelling of different user features. For example, the selection of the sections to be displayed is related to the user's interests. In PENA, the user model is divided into a set of dimensions which represent the different level of user interest for a news section. In this case we considered four separate dimensions: high, medium, low, null.

We can turn now to discuss our user modelling approach more precisely. In our application, there are two separate phases. The first one regards the creation of an initial model for a user connecting to the server for the first time. The second one revises the user model, by tracking the user's actions, in order to identify his interests and, possibly, the changes about them during time. The creation of the initial model is based on the use of stereotypical knowledge. The second phase is based on user modelling rules that are activated based on the analysis of user's actions during the browsing the news.

4.1 Using Stereotypes for Initializing the user model.

The creation of an initial model for first-time users is based on the use of stereotypes [4], which represent the features of classes of readers. The data used in the classification are collected from the user in a registration form, which contains a small set of questions.

At this point, it must be mentioned that there are seven predefined stereotypes in the database. These stereotypes are twenty four fields that contain probabilities and are grouped in six logic groups. For each one of the six interests, which are presented in the registration form, the user is asked to choose one of the four possible interests, high, medium, low, null. Each one of the twenty four probabilities corresponds to each one of those possible user selections.

The stereotypes consist of two groups of slots: classification slots and prediction slots. The structure of the slots is the same: Each slot corresponds to a feature F_i of the user. For each feature F_i we have a set of linguistic values $(v_{i1}, , v_{ik})$. A numeric value x_{ij} E [0,1] is associated with each linguistic value v_{ij} of F_i . This number can be regarded as the probability that $F_i = v_{ij}$, given that the user belongs to the stereotype. This means that the numeric value x_{ij} measures the frequency of $F_i = v_{ij}$ for the individuals belonging to the stereotypical class and thus it is a measure of the compatibility of $F_i = v_{ij}$ with the stereotype.

The stereotypes use the data provided by the user in the registration form as classificatory information and make predictions on different features of the user. For each feature F_i, the corresponding slot provides the probability x_{ij} that $F_i = v_{ij}$, given the stereotype S_k. In order to compute the degree of match, we assume that the features are independent. Moreover, as a stereotype is the result of the conjunction of all the features, we compute the degree of match of the user with a stereotype as the product of the contributions of all the individual slots.

At this point, there are seven probabilities. The maximum probability is selected. In this way we decide about the stereotype which is closer to the user. All the probabilities of the selected stereotype are chosen and stored in the database, as the initial user profile.

It is worth noting that the information provided by the registration form and then used for classifying the user is very general. Thus, for example, we can only make predictions on the user's interests in high-level sections, such as sport, politics.

4.2 Dynamic revision of the user model.

Since the interests a user may change, we are interested in both tracking any changes and modifying the user model accordingly. This can be achieved only by adopting some techniques for tracking the user's behaviour and for dynamically revising the user model in accordance to user behaviour.

In order to do that: a set of events was isolated, corresponding to actions performed by the user during his browsing session. A set of dynamic user modelling rules for the regular revision of the user model was introduced. The rules are based on the monitored events and on the resulting statistics. In the following, these two aspects will be discussed in more detail.

4.3 Events monitored by the system.

PENA monitors and stores into the database the following events: Firstly, the news stories that the user reads and the amount of time devoted for reading by the user to each news story. Secondly, the sections that the system selected and the user maintained. Finally, the sections that were not selected by the system and that the user asked to display.

These events provide important feedback about the user actions to the system, in order to compute statistics about the user's behaviour. The reliable revision of the user model, evaluates as more important the whole history of events during time than a single event.

The computed statistics are: The percentage of stories that the user reads in each specific section. The percentage of news, the user does not read. For each section, the percentage of news not displayed by the system that the user asks to display and the percentage of news displayed by the system that the user suppresses. Finally, the amount of time the user spends in each news story he reads.

These statistics are used by a set of user modelling rules in order to revise the user models.

4.4 User modeling rules.

At this point two things must be mentioned about the rules. Firstly, each rule is activated independently of the others and is used to revise a specific portion of the user model. Secondly, the rules are organized as compositions of conditions and consequents and have the following format: the antecedents are formed by logical conditions on the statistics about the user's behaviour and the consequents specify changes to the probability distributions over some features of the user model.

The system contains three rules. The main goal of the first rule modifies the probabilities of the user model in a way that reassures the selection of the news category. Based on the fact that a news category is selected as relevant for a particular user if the sum of the low and null probabilities from this category is less than forty nine percent, the probabilities of a news category that was added is formulated as follows: the high probability of the news category is set to sixty percent, the medium probability is set to forty percent and the low and null probabilities of the news categories are set to zero.

The third rule is the opposite of the first one, as it deals with the removal of an entire news category from those that are presented to the user. Based again on the fact that a news category is selected as relevant for a particular user if the sum of the low and null probabilities from this category is less than forty nine percent, the probabilities of a news category that was removed is formulated as follows: the null probability of the news category is set to sixty percent, the low probability is set to forty percent and the high and medium probabilities of the news category are set to zero.

Finally, the second rule deals with the view of an article from the user. The main idea of this rule is to either reinforce the existing user profile, by increasing the high probability ten percent, if the user spent more than ten seconds in the article, or decrease the existing profile by decreasing the null probability of the news category by ten percent.

4.5 Rule Activation and Revision of the User Model.

Two aspects still have to be defined: When the rules are activated and how the user model is revised as a result of the rules activation.

In the current prototype, the rules are activated at the start of each session, based on the user actions that are stored in the database during his last session. Therefore, the user's behavior is monitored during a session. Any user changes are activated instantly, such as the addition of an entire section of news. At this point the user model does not change. It will change in the start of the next user session, when the user modeling rules for the revision of the user model are activated.

Our approach has several advantages concerning both the efficiency of the system and the coherency of its behavior. Firstly, the statistics derived from the user behavior analysis within a whole session are more reliable and more significant than those regarding shorter browsing periods. Secondly, since the analysis of the user's actions and the revision of the user model are carried out at the start of each user session, the system does delay due to the performance of the revision task.

The user model revision is based on the rules. Once a rule is activated, the probability distributions in the user model are first combined with the predictions expressed in the consequent of the rules, and then normalized obtaining the revised distributions. This new distribution is stored as the revised user model in the user's database.

5 News retrieval.

The main goal of the retrieval part of the personal news agent is to visit periodically some Hellenic news sites and e-newspapers, to retrieve new stories and store the plain text of the article in the news database of the system. The retrieval part contains five agents and is a thread. This feature allows us to arrange the time interval between two subsequent executions. The general idea for the retrieval agent is to be executed periodically and retrieve new articles.

6 Personalization of Content.

The personalisation task is performed by two different agents. The first agent personalizes the content of the presentation: given the pieces of information in the user model, it decides which sections and news have to be presented. The second agent generates dynamically the web pages and is also responsible for the personalization of the presentation. However, in the current prototype we decided to maintain a standard format for the presentation, as we focused on the personalization of the content. This means that all the users see the same layout of the web pages, but various forms of personalization can be adopted on this aspect.

All the pages are generated dynamically, including the home page, which contains a list of the high-level sections that are considered relevant for the specific user. After that, a list of all the articles of the relevant sections follows. Each node of this list contains a small portion of the article and a link to the full text article. In the case of a previously read article, there is a message that the specific article was read and a link that allows the user to read it again.

Whenever the user opens an article, the corresponding page is generated by the system. The page is divided into two parts: The upper part, which contains the links for getting back to the page with the list of all the articles and the lower part that contains the text of the article.

7 Experiments.

The testing was performed with a small set of users, who were selected as representatives of different categories of readers, as they differ in terms of education income and occupation. There were selected a high school student, a university student, a network administrator, a university professor, an accountant, a civil engineer, a doctor, a salesman, an economic manager and a PhD candidate.

They were presented with the complete news section list and were asked to express a measure of their interest: high, medium, low or null. After that, they were asked to register and use PENA. As a result, PENA classified the user based on the predefined stereotypes, generated predictions and produced personalized pages with news. The result of this personalization process has been compared with the preferences expressed a priori by the user.

The selection of the system was judged compatible with the preferences of the user, if the system included all the sections that the user indicated as of high interest and not included any section of no interest for the user. The selection was judged completely incompatible, if the system failed to include all sections indicated as of high interest and included at least a section of no interest. In 7 out of the 10 cases the system provided satisfactory results, while in 3 cases the system completely failed to satisfy the desired data.

Although the test set was very small and in many senses naive, the results are quite encouraging on the feasibility and on the practical applicability of the approach, since the system achieved high accuracy 70%, in the prediction of interesting news sections for the users.

8 Implementation details.

PENA is designed as a set of cooperating agents and implemented in a three -tier architecture. The system is implemented using Java, especially Servlets and JSP. The databases are implemented using MySQL and are accessed using JDBC.

9 Conclusions and Future Work.

The described system is in an early development and experimental stage and is part of a project that aims at the creation of an adaptive news site. There are three main goals of our future work. Firstly, alternative ways should be investigated for the initial user model. Secondly, news model adaptation can be achieved, by employing a further classification of the news, inside their high level classification that exists now. Finally, recommendations enhancement can be done by proposing articles based both on the user model and news stories similarity.

References

[1] Ardissono, L., Console, L., and Torre, I. (2001) An adaptive System for the personalized access to news. AI Communications 14: 129-147.

[2] Billsus, D., Pazzani, M., (1999) A personal news agent that talks, learns and explains. Autonomous Agents 1999, pp. 268-275.

[3] Joachims, T., Freitag, G., Mitchell, T. (1997) Web-Wacher: a tour guide for the world wide web. IJCAI, 1997, pp 770-775.

[4] Rich, E, (1989) User Models in Dialog Systems, chapter Stereotypes and user modeling, Springer, Wien.

A More Accurate Text Classifier for Positive and Unlabeled data

Rur Ming Xin[1], Wan li Zuo[2]

[1]Key Laboratory of Symbol Computation and Knowledge Engineering of the Ministry of Educating, Colloege of Computer Science, JiLin University of China

[2]Key Laboratory of Symbol Computation and Knowledge Engineering of the Ministry of Educating, College of Computer Science, JiLin University of China

E-mail: xinruiming_204@yahoo.com.cn

Abstract

Almost all *LPU* algorithms rely heavily on two steps: exploiting reliable negative dataset and supplementing positive dataset. For above two steps, this paper originally proposes a two-step approach, that is, *CoTrain-Active*. The first step, employing *CoTrain* algorithm, iterates to purify the unlabeled set with two individual *SVM* base classifiers. The second step, adopting active-learning algorithm, further expands the positive set effectively by request the true label for the "suspect positive" examples. Comprehensive experiments demonstrate that our approach is superior to *Biased-SVM* which is said to be previous best. Moreover, *CoTrain-Active* is especially suitable for those situations where the given positive dataset *P* is extremely insufficient.

1 Introduction

Traditionally, a general binary text classifier is built by employing some algorithm on the positive and the negative dataset. This kind of algorithms are termed supervised learning algorithm. However, in reality, hand-label data can be tedious and expensive to available. In practice, it is a common situation where only a small number of positive data is available. Since no negative data are provided, researchers turn to explore algorithms which are based on a small number of positive dataset and a comprehensive unlabeled dataset. Francois Denis originally proposes the framework for learning from only positive and unlabeled training data, abbreviated as *LPU* (Learning from Positive and Unlabeled data). A collection of *LPU* algorithms are detailed and compared in [2].

This paper proposed another *LPU* algorithm that contributes in two aspects. The one is the proposal of employing *CoTrain* for purifying unlabeled dataset. The other is supplementing the positive dataset by using of *Active-Learning* algorithm.

The remainder of the paper is organized as follows: section 2 gives a review of related works; section 3 proposes our *CoTrain-Active* approach. A number of comparative experiments have been made in section 4. Section 5 is conclusion.

2 Related Works

To tailor for *LPU* problem, F.Denis[10] introduced Naive Bayes algorithm for the positive and unlabeled examples (PNB). PEBL[3] is another two-step approach targeted at *LPU*. A detailed evaluation of several two-step *LPU* approaches, along with Biased-SVM, can be found in [2]. In experiment section, we list Biased-SVM result (previous best) as a comparison. Co-Training framework, which "assumes have two individual classifiers on different views "[10], is first proposed by Blum and Mitchell in 1998. Another comprehensive study using co-training method to tackle "labeled and unlabeled data" is conducted by Nigam and Ghani[11]. Active-Learning[13] intends to find the most informative instance to request the true label. A novel integration of active learning and EM can be found in [13].

3 CoTrain-Active method

The proposed algorithm falls into two sequential steps: *CoTrain* step and Active-Learning step. For short, we use *P* to denote positive set. Reliable negative set is expressed as *RN*. Unlabeled set can be denoted *U*.

3.1 CoTrain step

Unlike former *LPU* algorithms ([2],[3]) exploiting reliable negative dataset, *CoTrain* iterates to purify unlabeled dataset by filtering out some "suspect positive " examples. The rationality behind is that purifying the unlabeled data by filtering out a small fraction positive data is much easier than exploiting a small number of reliable negative data. Moreover, the "suspect positive" data being filtered out can also be made use of supplementing positive dataset.

Specifically, *CoTrain* consists of two individual SVM learners (base classifier) which are built on the same positive dataset and the different unlabeled datasets. The two base classifiers "help" or "co-purify" each other by filtering out "suspect positive" examples from the unlabeled dataset of the counterpart.

For presentation, d is an arbitrary document and the T is the set of "suspect positive" data. Classify(S, d) denote the result generated by classifier S on d. S_{2final} and S_{1final} are the classifiers when iteration converge.

Algorithm: *CoTrain*

Input: (P, U);
Output: (S_{2final} , S_{1final} , T , U_1 , U_2)
1. Randomly split U set into two set U_1 U_2 subject to $U_1 \cup U_2 == U$, $U_1 \cap U_2 == \{\}$
2. Label P as positive and U_1 as negative; build SVM classifier S_{10} with P and U_1;
3. Label P as positive and U_2 as negative; build SVM classifier S_{20} with P and U_2 ;
4. i=0;T={ };Q={ };R={ };
5. while(true)
6. $Q = \{d \mid$ Classify(S_{2i} , d)==positive}
7. $U_1 = U_1 - Q$;
8. $R = \{d \mid$ Classify(S_{1i} , d)==positive}
9. $U_2 = U_2 - R$;
10.   if (Q=={ } and R=={ }) then
11.   break;
12.   end if
13. $T = T \cup Q \cup R$;
14.   i=i+1;
15.   Build SVM classifier S_{1i} with P and U_1;
16.   Build SVM classifier S_{2i} with P and U_2;
17. end while

After *CoTrain* we got two robust classifiers and a purified unlabeled set, $U_1 \cup U_2$, which can be regard as reliable negative set *RN*. Another product of *CoTrain* is T, the doubtful positive set, which can be further utilized in later active learning step.

3.2 Active-learning step

Generally, active learning is usually employed for the purpose of reducing the necessary number of labeled training instance and enhancing classifier performance. We here employ active learning to expand positive set and therefore further improve classifier S_{final} .The key point of active learning is making decision on which unlabeled instance should be selected to query. Because each query is considered as an expensive operation, we attempt to better our classifier with as few queries as possible. We first classify the data in T, the set being removed from U, by S_{1final} and S_{2final}.If both of S_{1final} and S_{2final} consider the document in T is positive, we request the document for true label. Thus, we actually query the "most doubtful positive" document. Detailed about algorithm can be seen below.

Algorithm: Active-learning

Input: (S_{2final} S_{1final} T U_1 U_2 P)
Output: S_{final}
1. $O = \{d \mid$ Classify(S_{1final}, d)==positive and Classify(S_{2final} ,d)==positive }
2. for each d in O
3.   Request the true label for d;
4.   if (d labeled as positive)
5. $P = P \cup \{d\}$
6.   end if
7. end for
8. Rebuild final SVM classifier S_{final} with P and $U_1 \cup U_2$

Combining the algorithm *CoTrain* and Active-Learning is our final approach.

3.3 CoTrain-SVM

Since active learning needs some extra man efforts, we compare with other *LPU* method without active learning step as a fair comparison. After *CoTrain*, a linear SVM classifier is built on P as positive and $U_1 \cup U_2$ as negative.

4 Experiment

To compare with Biased-SVM, we set the same experimental environment as [2] did. Since our final classifier is enhanced by active learning involving extra expensive query, we list the result of CoTrain-SVM as a fair comparison

4.1 Experiment setup and Data preprocess

Here we use the Reuter-21578, the popular text collection in TC(text classification) experiment, which is made up of 21578 documents of 135 categories. Among 135 categories, only 10 commonly used categories, 9980 documents are selected to be our experiment set. Based on the documents of 10 categories, we choose each category as positive while the rest as negative. Thus we have 10 data sets to evaluate our approach. For each dataset, we randomly pick out 70% as training set and the rest 30% as test set. Further, we split the training set into P and U as follows: select the *Per* percent positive data as P and mixed the rest of positive data with negative data as U. In our experiments, we will vary *Per* to show the effect of the proposed algorithm with different positive mixed percentage in U.

Appropriate data preprocessing has been done, we employ stop-list to remove those common words but no feature selection and stemmer has been done. When indexing features, we choose itc-weighting[6]. For SVM, which is employed as basic classifier here, we use the SVM toolbox (developed by Anton Schwaighofer) with linear kernel.

4.2 How to measure our approach

We evaluate our approach from three aspects. So far as *CoTrain* step is concerned, the objective of this step is to purify U_1 and U_2. To measure the degree of unlabeled data being purified, we use q to denote the ratio of positive data being removed to total positive data being mixed in U_1 and U_2. Effectiveness of active learning can be measured by e, the ratio of positive data being added into P to query times. The performance of the final classifier can be measured by f, $f=2rp/(r+p)$, where p is the precision and r is recall.

4.3 Results analysis

To check the degree of being purified in *CoTrain* step, we calculate q, mentioned in section 4.2, to demonstrate the percentage of the removed positive instance mixed in U.

Table 1 list q scores of 10 Retuer data sets for several *Per* which is mentioned in section 4.1. For Table 1, we proved *CoTrain* method to be good at purifying U set .With the decrease of *Per*, the average of q scores declined synchronously. Such phenomenon can be interpreted as follows: the smaller *Per* is, the more positive data have been mixed with negative data and thus the two basic classifiers can be poorly constructed which results in low q score.

Table 2 list the average of 10 datasets f scores for several Per settings, the first column is given by [2], another column is calculated with CoTrain-SVM algorithm which doesn't employ active-learning step. An obvious comparison can be seen, even without the active-learning step, our approach CoTrain-SVM is superior to biased-SVM which is previous best one in a range of *Per* situation.

Tabel 2 Average f score of two algorithm

Per	CoTrain-SVM	Biased-SVM
0.7	0.8856	0.856
0.3	0.8569	0.785

Table 1 After *CoTrain* step, q scores of each dataset

Per	acq	corn	crude	earn	grain	interest	money	ship	trade	wheat	Average
0.7	0.988	0.934	0.992	0.992	0.984	0.878	0.933	0.806	0.921	0.862	0.929
0.6	0.994	0.969	0.910	0.994	0.939	0.892	0.928	0.878	0.920	0.901	0.932
0.5	0.992	0.679	0.924	0.978	0.937	0.884	0.928	0.824	0.889	0.893	0.893
0.3	0.973	0.727	0.857	0.971	0.928	0.878	0.871	0.689	0.859	0.750	0.850

Table 3 After Active learning, e scores of each dataset

Per	acq	corn	crude	earn	grain	interest	money	ship	trade	wheat	Average
0.7	0.718	0.518	0.654	0.921	0.685	0.424	0.564	0.769	0.639	0.466	0.636
0.6	0.852	0.696	0.707	0.939	0.823	0.532	0.618	0.797	0.690	0.534	0.719
0.5	0.843	0.729	0.768	0.968	0.877	0.600	0.688	0.850	0.786	0.582	0.769
0.3	0.921	0.868	0.855	0.984	0.907	0.707	0.818	0.885	0.883	0.760	0.859

Table 4 f of the final classifier on each dataset

Per	acq	corn	crude	earn	grain	interest	money	ship	trade	wheat	Average
0.7	0.950	0.851	0.863	0.984	0.960	0.816	0.874	0.866	0.863	0.845	0.887
0.6	0.951	0.822	0.893	0.986	0.905	0.833	0.873	0.883	0.925	0.836	0.891
0.5	0.946	0.853	0.912	0.981	0.920	0.816	0.889	0.863	0.913	0.845	0.894
0.3	0.960	0.852	0.903	0.980	0.921	0.772	0.888	0.858	0.922	0.807	0.886

The goal of active learning is to expand positive set effectively. It attempts to add new positive document as many as possible by querying users as few as possible. From Table 3, we can say, active-learning performs effectively especially when given positive dataset *P* is insufficient.

With *Cotrain* significantly purifying unlabeled data and active learning greatly expanding *P*, the enhancement of our algorithm has been proved by Table 4 experimentally. Notably, even with highly mixed unlabeled set, we can maintain a high *f* scores. Such phenomenon demonstrates that our method is still suitable for those situations where positive instances are deficient

5 Conclusion

This paper artfully solve the two difficulties, purifying unlabeled set and expanding positive set, in *LPU* problem.

Acknowledgement

This work is sponsored by the Natural Science Foundation of China under grant number 60373 and Science&Technology Development Program of JiLin Province under grant number 20020303-2

References

[1] Joachims,T.(1997).A probabilistic analysis of the Rocchio algorithm with TFIDF for text categorization. In Proceedings of international conference on Machine Learning(ICML).

[2] Liu, B., Dai, Y., Li, X., Lee, W.S. & Yu, P. (2003). Building Text Classifiers Using Positive and Unlabeled Examples. Proceedings ICDM-03

[3] H. Yu, J. Han, and K.C.-C. Chang, "PEBL: Positive-Example Based Learning for Web Page Classification Using SVM," Proc. Eighth Int'l Conf. Knowledge Discovery and Data Mining (KDD '02), pp. 239248, 2002.

[4] Rocchio, J. (1971) Relevance Feedback Information Retrieval. In Gerald Salton(Ed.), The Smart Retrieval System - Experiments in Automated Document Processing, 313-323. Prentice-Hall, Englewood Cliffs, NJ.

[5] Liu, B., Lee, W.S., Yu, P.S. and Li, X., Partially Supervised Classification of Text Documents, Proc. 19th Intl. Conf. on Machine Learning 2002, 387-394

[6] K. Aas and L. Eikvil. Text categorisation: A survey. Technical report, Norwegian Computing Center, June 1999

[7] Joachims, T. (1998). Text categorization with Support Vector Machines: Learning with many relevant features. In Machine Learning: ECML-98, Tenth European Conference on Machine Learning, pp. 137--142.

[8] Hua-Jun Zeng, Xuan-Hui Wang, Zheng Chen, Wei-Ying Ma. CBC: Clustering Based Text Classification Requiring Minimal Labeled Data. ICDM2003 page(s): 443- 450

[9] Porter,,M.(1980).An algorithm for suffix stripping. Program(Automated library and information Systems),14(3):130-137.

[10] F. Denis, R. Gilleron, A. Laurent, M. Tommasi, Text Classification and Co-Training from Positive and Unlabeled Examples, Proceedings of the ICML 2003

[11] Nigam, K., & Ghani, R. (2000). Analyzing the applicability and effectiveness of co-training. Proceedings of CIKM-00, 9th ACM International Conference on Information and Knowledge Management (pp. 86--93)..

[12] K. Nigam, A. McCallum, S. Thrun, and T. Mitchell. Learning to classify text from labeled and unlabeled documents. In Proceedings of the Fifteenth National Conference on Artificial Intelligence. AAAI Press, 1998.

[13] McCallum, A. K., & Nigam, K. (1998). Employing EM in pool-based active learning for text classification. In Machine Learning: Proceedings of the Fifteenth International Conference (ICML '98), pp. 350--358.

Efficiency Aspects of Neural Network Architecture Evolution Using Direct and Indirect Encoding

H. Kwasnicka, M. Paradowski
Department of Computer Science, Wroclaw University of Technology, Poland
E-mail: {halina.kwasnicka, mariusz.paradowski}@pwr.wroc.pl

Abstract

Using a GA as a NN designing tool deals with many aspects. We must decide, among others, about: coding schema, evaluation function, genetic operators, genetic parameters, etc. This paper focuses on an efficiency of NN architecture evolution. We use two main approaches for neural network representation in the form of chromosomes: direct and indirect encoding. Presented research is a part of our wider study of this problem [1, 2]. We present the influence of coding schemata on the possibilities of evolving optimal neural network.

1 Introduction

Genetic Algorithms (GAs) [3] and artificial Neural Networks (NNs) [4] were proposed as the imitation of natural process: biological evolution and real neural systems respectively. Designing architecture of networks causes some problems: How many neurons should be in the network? How they should be organized? How many connections we should establish and between which neurons? A designer of a NN architecture must relay on his experience or on informal heuristics [5].

Designing a NN is a problem of searching for an architecture, which performs some specific task with desired accuracy. This process can be seen as searching the surface defined by networks performance evaluation above the space of all possible neural networks architectures [6, 7, 8, 9, 10, 11]. The coding schema limits a search space for a GA. The smaller the search space is, the easier is to find the optimal, or near optimal network architecture. However, the smaller search space means the small number of different potential NN architectures.

A fitness function, used for evaluation of generated networks, strongly influences obtained results [2]. We developed a fitness function, which try to balance the size and an accuracy of an evaluated network.

2 Neural network representations

There are two major approaches of neural network architecture evolution: *direct encoding* [12] and *indirect encoding* [12, 13]. The main difference lies on a size of a set of representative networks (a size of a search space for a GA).

2.1 Direct encoding

Direct encoding represents neural networks as directional graphs. A very large subset of neural networks can be represented using this method. Direct encoding focuses on the ability to produce neural networks with the highest possible quality. The ability of representing a very large subset of neural networks and very few limitations on the size of the domain causes the time and memory requirements of this method to be very large. A network with n neurons is represented by a matrix and a vector. Matrix E is used to encode information if there are connections between neurons. For feed-forward neural networks only a half of the matrix is used. Vector V is used to encode information if neurons are active. This causes vector V to be a mask for matrix E. Both, vector and matrix encode input layer, hidden layer and output layer neurons and connections between neurons. There are no connections between neurons in input layer. The memory complexity of such approach is $O(n^2)$ for one neural network. An exemplary genotype with direct encoding and a decoded phenotype is shown on fig. 1.

$$V = [1 \quad 1 \quad 1 \quad 1 \quad 0 \quad 1]$$

$$E = \begin{bmatrix} 0 & 0 & 1 & 1 & 1 & 0 \\ 0 & 0 & 1 & 1 & 0 & 0 \\ 0 & 0 & 0 & 0 & 1 & 1 \\ 0 & 0 & 0 & 0 & 0 & 1 \\ 0 & 0 & 0 & 0 & 0 & 0 \\ 0 & 0 & 0 & 0 & 0 & 0 \end{bmatrix}$$

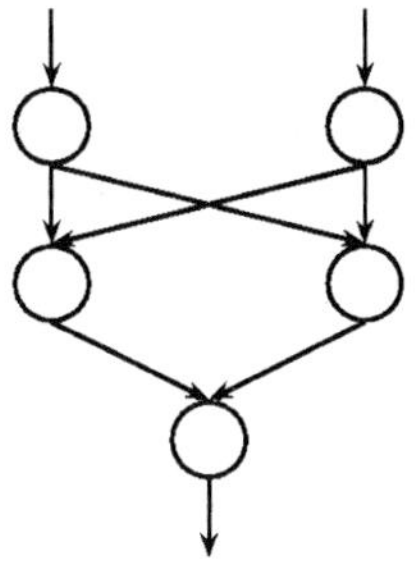

Fig. 1. Direct encoding genotype and the decoded phenotype

2.2 Indirect encoding

Indirect encoding comes as a compromise between time and quality. The most important feature of this approach is the limitation of the size of the domain of search. This causes the method to be very fast, but very often with suboptimal results. Memory requirements of this method are smaller than in direct encoding. Representation of a neural network with n neurons requires $O(n)$ memory. In used approach a neural network architecture is represented as a binary tree with a set of grammar rules for network architecture construction. Every node of the tree contains one of two non-terminal symbols (S and P) or a terminal symbol (E). All leaf nodes of the tree are terminal symbols. Non-terminal symbol S denotes division of the active neuron into two sequentially connected neurons. Non-terminal symbol P denotes division of the active neuron into two parallel neurons. An exemplary genotype with indirect encoding and a decoded phenotype is shown on fig. 2.

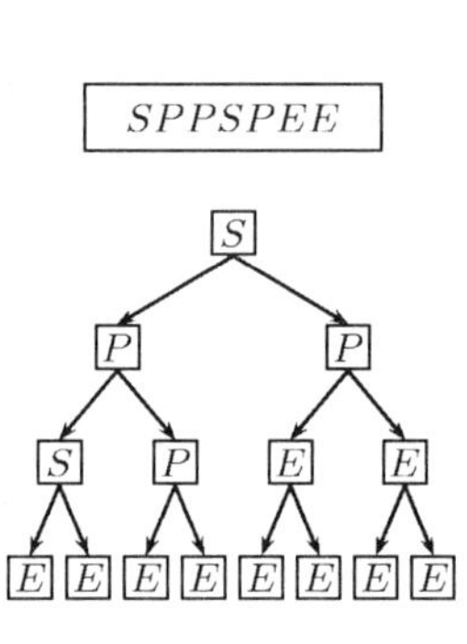

Fig. 2. Indirect encoding genotype and the decoded phenotype

2.3 Quality of a neural network

Quality of a neural network can be measured by two criteria: error rate during processing and size of the network itself. The smaller error, the better are the answers of a network during the processing of the test set. The smaller size, the better are the capabilities of a network to generalize during processing data which have not been seen previously.

The method of neural network evaluation used in all experiments has been presented in details in [2, 1]. Most of the experiments have been performed using data sets, especially HOUSING, IRIS and ZOO problems [14]. It is important to notice that used method allows to create neural networks with connections between more than one layer. A two phased evolution is used. The task of the first phase is to receive networks with small error rates, the task of the second phase is to reduce the size of neural networks and to keep the error rates small enough. The fitness function value consists of two elements: an error rate factor and a network size factor.

$$f = \begin{cases} f_t(1+f_s) & if \quad f_t > f_t^{min} \\ f_t \quad if & f_t \le f_t^{min} \end{cases} \tag{1}$$

Size factor f_s of a neural network is a function of number of connections between neurons. Error rate factor f_t is a function of error values on every output of a network during testing process. f_t^{min} is the maximum acceptable error factor. Detailed calculation of all factors have been presented in [2]. Evolved neural networks architectures are trained using Backpropagation method or by evolutionary approach.

3 Results of experiments

The difference of quality between direct and indirect encoding strongly depends on the discussed problem. The biggest difference is seen if the input vector is large. The natural ability of direct encoding of removing less or non important input features gives this method a big advantage. In case of such problems it has proven to be much more effective than the indirect approach in the aspect of quality. Indirect encoding is able to give very good solutions in very few generations, but evolved networks size is larger than in direct encoding (fig. 3).

Neural network evolved and trained for solving XOR problem is shown on fig. 1. This network architecture has been evolved using direct encoding and trained using evolutionary approach. It is impossible to receive this neural network using indirect encoding due it is outside the domain of search.

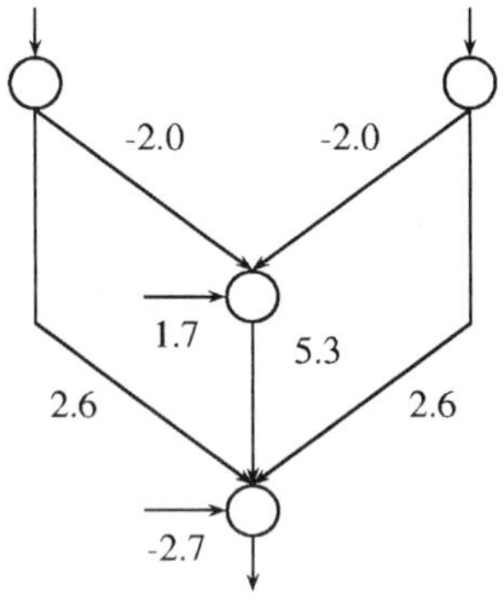

Fig. 3. Optimal neural network for XOR problem

Fig. 4. shows a neural network evolved for solving Parity of a number problem. It is evolved using direct encoding and trained using Backpropagation method. The smallest neural network for this problem evolved with indirect encoding contains 6 connections between neurons. This example shows the abilities of direct encoding for neural network pruning.

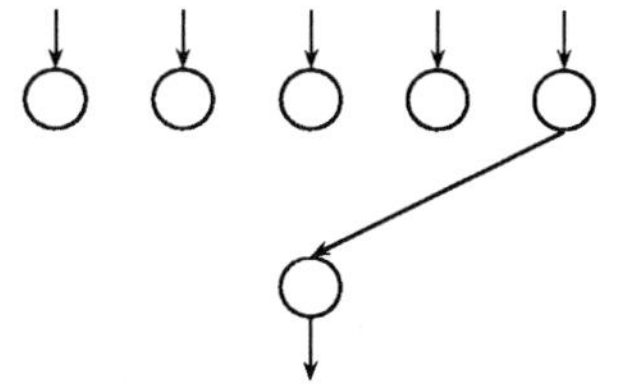

Fig. 4. Optimal neural network for Parity of a number problem

Fig. 6. shows the evolution results for the IRIS problem. In the experiment direct and indirect encoding give almost identical result. The neural network received using direct encoding is smaller only by 2 connection and evolution process took many more generations than for indirect encoding.

Exemplary results for ZOO, HOUSING and XOR problems using direct encoding have been shown on fig. 7. Evolution process requires a lot of computation time, but received neural networks are often very small and very well pruned. All neural networks have been trained using Backpropagation method.

4 Summary

Efficiency of direct encoding in generation of neural networks with very good quality has been proven by many experiments. A large part of experiments led to generation of very small networks with very good quality.

4.1 A size of networks and computation time

The biggest advantage of direct encoding is the ability of adding and removing single connections between neurons. The evolution process using direct encoding requires very long computation time (fig. 7.).

In every experiment indirect encoding has proven not to be able to reach global optimum. The optimal neural networks are outside the domain of search of this method. Received networks are often up to 6 times larger (fig. 6.) than in direct encoding. Indirect encoding has given very good neural networks in very few generations. It takes many more generations to reach the same solution by direct encoding.

Direct encoding requires much more computation time but is very effective in giving very small, very well pruned neural networks. Indirect encoding gives good results in very few generations, but it is not able to proceed further with optimization due to its limitations on the domain of search.

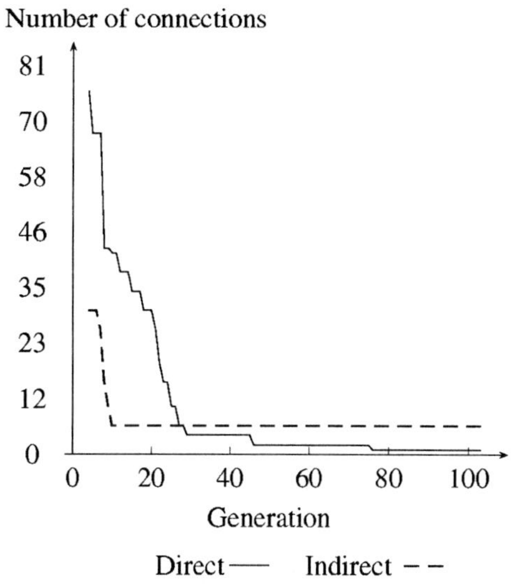

Fig. 5. Size of neural network in direct and indirect encoding, Parity of a number problem

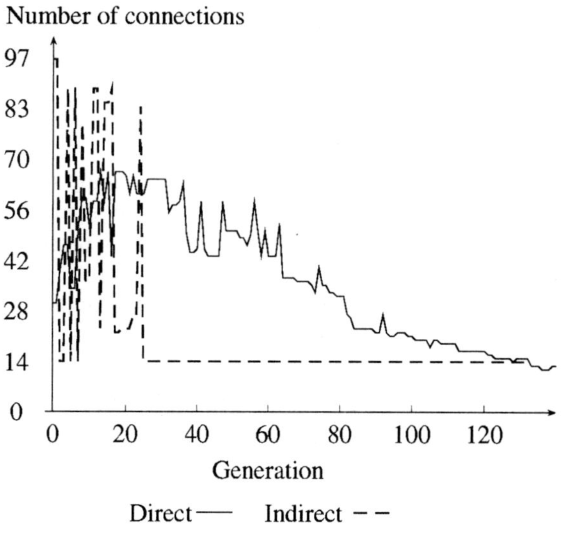

Fig. 6. Size of neural network in direct and indirect encoding, IRIS problem

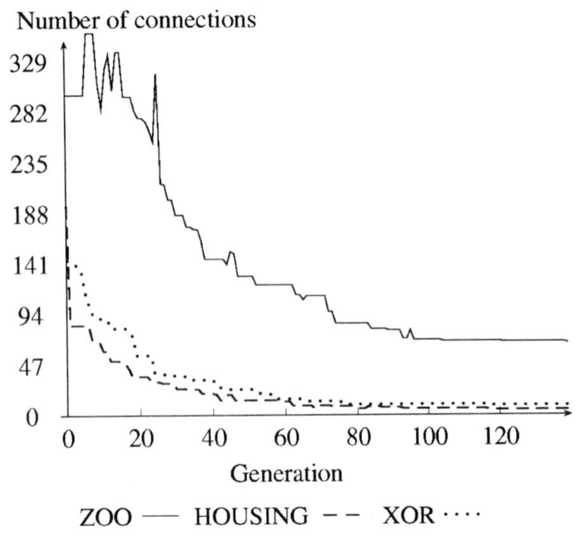

Fig. 7. Size of neural network in direct encoding

4.2 Size of the domain of search

Let us denote n as number of neurons in hidden layers, k as the size of input vector and l as the size of output vector. Used indirect encoding stores a neural network as a binary tree with two possible non-terminal symbols in each tree node. This gives the size of the domain equal to $2^n - 1$. Direct encoding uses a matrix (and a vector which is a mask) to encode a neural network and has the size of the domain equal to $2^{\sum_{x=1}^{n+l}(k+x-1)}$.

4.3 Further research

The further research will focus on combining the advantages of these two methods. Parallel process of evolution with two both types of encoding will be presented. Once, a network with indirect encoding is proven to be effective enough, it will be converted to a network with direct encoding and further evolved in the aspect of its quality optimization.

Other important problem concerns a learning method. As it is mentioned above, direct coding enables generation of small sized networks. Although, such a small network must learns its task. Our experiments showed that using Backpropagation method, a small network could not be able to learn desired task. Surprisingly, a GA can learn such a small network. An example is a XOR problem (see fig. 1). Backpropagation method starting from a set of random weights (from the range covering the proper values of weight) is not able to learn a network XOR problem while a GA does it. It seems that the optimum is very narrow and Backpropagation methods leads to the local optimum.

References

[1] Paradowski M. (2004) Evolution of neural network Architecture Using Genetic Algorithms (in Polish), Master Thesis, Wroclaw University of Technology

[2] Kwasnicka H., Paradowski M. (2004) Selection Pressure and an Efficiency of Neural Network Architecture Evolving, Lecture Notes on Artificial Intelligence, Springer-Verlag

[3] Goldberg D.E. (1989) Genetic Algorithms in Search, Optimization, and Machine Learning, Addison-Wesley Publishing Company, Inc.

[4] Tadeusiewicz R. (1993) Neural networks (in Polish), Akademicka Oficyna Wydawnicza RM

[5] Yao X. (1999) Evolving Artificial Neural Networks, School of Computer Science, The University of Birmingham

[6] Balakrishnan K., Honavar V. (1995) Properties of Genetic Representations of Neural Architectures, Proc. of the World Congress on Neural Networks (WCNN'95)

[7] Bornholdt S., Graudenz D. (1992) General Asymmetric Neural Networks and Structure Design by Genetic Algorithms, Neural Networks, Vol. 5, pp. 327-334

[8] Stanley K.O., Bryant B.D., Miikkulainen R (2003) Evolving Adaptive Neural Networks with and without Adaptive Synapses, 2003 IEEE Congress on Evolutionary Computation (CEC-2003)

[9] Chang S.O., Okurama A.M. (2004) Impedance-Reflecting Teleoperation with a Real-Time Evolving Neural Network Controller

[10] Garcia-Pedrajas N., Ortiz-Boyer D., Hervas-Martinez C. (2004) An alternative approach for neural network evolution with a genetic algorithm: Crossover by combinatorial optimization, Preprint submittet to Elsevier Science

[11] Dewri R. (2002) Evolutionary Neural Networks: Design metodologies, AI depot, http://ai-depot.com

[12] Whitley D. (1995) Genetic Algorithms and Neural Networks, Genetic Algorithms in Engineering and Computer Science

[13] Gruau F. (1994) Neural Network Synthesis Using Cellular Encoding And The Genetic Algorithm, Universite Claude Bernard, Lyon

[14] Perchelt L. (1994) Proben1 - A Set of Neural Network Benchmark Problems and Benchmarking Rules, Technical Report 21/94

Genetic Algorithm Optimization of an Artificial Neural Network for Financial Applications

Serge Hayward
Department of Finance
Ecole Supérieure de Commerce de Dijon, France
shayward@escdijon.com

Abstract

Model discovery and performance surface optimization with genetic algorithm demonstrate profitability improvement with an inconclusive effect on statistical criteria. The examination of relationships between statistics used for economic forecasts evaluation and profitability of investment decisions reveals that only the 'degree of improvement over efficient prediction' shows robust links with profitability. If profits are not observable, this measure is proposed as an evaluation criterion for an economic prediction. Also combined with directional accuracy, it could be used in an estimation technique for economic behavior, as an alternative to conventional least squares.

1 Introduction

Problems with applications of evolutionary computation (EC) in finance are often due to the lack of common methodology and statistical foundations of its numerous techniques. These deficiencies sometimes cast doubt on conjectured results and conclusions. At the same time, relationships between summary statistics used for predictions' evaluation and profitability of investment decisions based on these predictions are not straightforward in nature. The importance of the latter is particularly evident for applications of an evolutionary / artificial neural network (E/ANN) under supervised learning, where the process of network training is based on a chosen statistical criterion, but when economic performance is generally sought. Motivations for this paper come from the ongoing search for the foundation of EC in finance and a claim by [1] that traditional summary statistics are not closely related to a forecast's profit, with the exception of directional accuracy (DA).

Financial assets' prices often exhibit non-stationarity, autocovarience and frequent structural brakes, posing problems for their modeling. This paper also investigates how data mining benefits from genetic algorithm (GA) model discovery, performance surface optimization and pre/pro-processing, improving predictability or/and profitability.

2 Methodology

For our experiment we build ANN forecasts and generate a posterior optimal rule. The rule, using future information to determine the best current trading action, returns a buy/sell signal (B/S) today if prices tomorrow have increased/decreased. A posterior optimal rule signal (PORS) is then modeled with ANN forecasts, generating a trading B/S signal. Combining a trading signal with a strategy warrants a position to be taken. We consider a number of market timing strategies, appropriate for different strengths of the B/S signal. If we have a buy (sell) signal on the basis of prices expected to increase (decrease) than we enter a Long (Short) position. Note that our approach is different from standard B/S signal generation by a technical trading rule. In the latter it is only a signal from a technical trading rule that establishes that prices are expected to increase/decrease. In our model we collaborate signal's expectations of price change (given by PORS) with a time-series forecast.

To apply our methodology we develop the dual network structure, presented in Figure 1. The forecasting network feeds into the action network, from which the information set includes the output of the first network and PORS, as well as the inputs used for forecasting, in order to relate the forecast to the data upon which it was based.

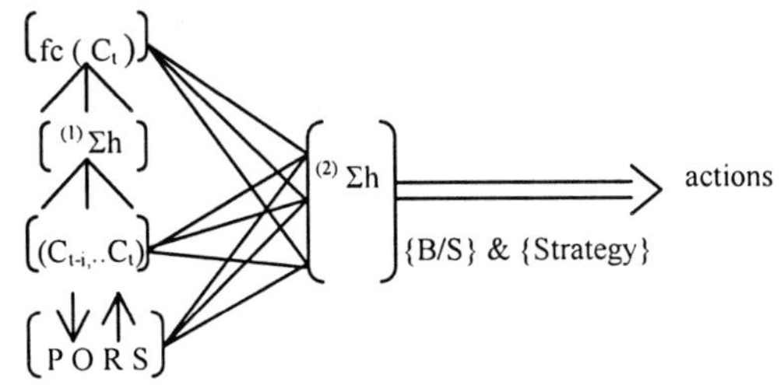

Fig. 1. Dual ANN: (1) forecasting network; (2) acting network

This structure is an effort to relate actions' profitability to forecasting quality, examining this relationship in computational settings. The model is evolutionary in the sense it considers a population of networks (individual agents facing identical problems/instances) that generate different solutions, which are assessed and selected on the basis of their fitness. Backpropagation is used in the forecasting net to learn to approximate the unknown conditional expectation function (without the need to make assumptions about data generating mechanism and beliefs formation). It is also employed in the action net to learn the relationship between forecasts' statistical and actions' economic characteristics. Lastly, agents discover their optimal models with GA; applying it for ANN model discovery makes technical decisions less arbitrary. The structure seems to be intuitive and simple to generate results independent from a chosen architecture. The results produced are sufficiently general, being stable for multiple independent runs with different random seeds for a dual forecasting/action net and a single forecasting net.

2.1. Generating Posterior Optimal Rule Signal

PORS is a function of a trading strategy adopted and based on the amount of minimum profit and the number of samples into the future. Stepping forward one sample at a time, the potential profit is examined. If the profit expected is enough to clear the minimum profit after transaction costs (TC), a PORS is generated. The direction of PORS is governed by the direction of the price movement. Normally, the strength of the signal reflects the size of underlying price changes, although, we also examine signals without this correlation to identify when profit generating conditions begin. Lastly, we consider PORS generated only at the points of highest profit to establish the maximum profit available.

3 Description of the Environment

Let Y be a random variable defined on a probability space $(\Omega, \mathcal{F}, P)$. Ω is a space of outcomes, $\mathcal{F}$ is a σ-field and P is a probability measure. For a space $(\Omega, \mathcal{F}, P)$ a conditional probability $P[A|\mathcal{F}]$ for a set A, defined with respect to a σ-field $\mathcal{F}$, is the conditional probability of the set A, being evaluated in light of the information available in the σ-field $\mathcal{F}$. Suppose economic agents' utility functions are given by a general form:

$$U(W_{t+s}) = g(Y_{t+s}, \delta(fc_{t+s})) \quad (1)$$

According to (1), agents' utility depends on: a target variable Y_{t+s}; a decision/strategy variable, $\delta(fc_{t+s})$, which is a function of the forecast, fc_{t+s}, where $s \geq 1$ is a forecasting horizon. Setting the horizon equal to 1, we examine the next period forecast (when this simplification does not undermine the results for $s \geq 1$). A reward variable W_{t+s} is sufficiently general to consider different types of economic agents and includes wealth, reputation, etc. $w_{t+1}(y_{t+1}, fc_{t+1})$ is the response function, stating that at time $t+1$ an agent's reward w_{t+1} depends on the realization of the target variable y_{t+1} and on the accuracy of the target's forecast, fc_{t+1}. Forecasting is regarded as a major factor of a decision rule, being close to the reality in financial markets. Also, it has a developed statistical foundation in econometrics allowing its application in evolutionary computation.

Let $fct_{t+1} = \theta' X_t$ to be a forecast of Y_{t+1} conditional on the information set $\mathcal{F}_t$, where unknown m-vector of parameters, $\theta \in \Theta$, with Θ to be compact in R^k and observable at time t n-vector of variables, X_t. X_t are $\mathcal{F}_t$-measurable and might include some exogenous variables, indicators, lags of Y_t, etc. An optimal forecast does not exclude model misspecification, which can be due to the form of fc_{t+1} or failure to include all relevant information in X_t. Under imperfect foresight, the response function and, therefore, the utility function are negatively correlated with forecast error, $e_{t+1} \equiv y_{t+1} - fc_{t+1}$; $|e_{t+1}| > 0$. A mapping of the forecast into a strategy rule, $\delta(fc_{t+1})$ (combined with elements of X_t) determines a predictive density g_y, which establishes agents' actions.

In this setting, maximizing expected utility requires us to find an optimal forecast, fc_{t+1} and to establish an optimal decision rule, $\delta(fc_{t+1})$. Note that optimality is with respect to a particular utility function, implemented through a loss function, in the sense that no loss for a correct decision and a positive loss for incorrect one. Given a utility function, expected utility maximization requires minimization of the expected value of a loss function, representing the relationship between the size of the forecast error and the economic loss incurred because of that error. A strategy development (mapping of the forecast into a decision rule) is another way to minimize the expected value of a loss function.

A loss function, $L: R \rightarrow R^+$, related to some economic criteria or a statistical measure of accuracy, takes a general form:

$$L(p, a, e) \equiv [a + (1-2a)1(e < 0)]e^p, \quad (2)$$

where p is a coefficient of risk aversion; e is the forecast error; $\alpha \in [0,1]$ is the degree of asymmetry in the forecaster's loss function. $L(p, \alpha, e)$ is $\mathcal{F}_t$-measurable. It could also be presented as:

$$L(p, a, \theta) \equiv [a + (1-2a)1(Y_{t+1} - fc_{t+1}(\theta) < 0)]|Y_{t+1} - fc_{t+1}(\theta)|^p, \quad (3)$$

where α and p are shape parameters and a vector of

unknown parameters, $\theta \in \Theta$. For given values of ρ and α an agent's optimal one-period forecast is

$$\min_{\theta \in \Theta} E[L(\rho,\ \alpha,\ \theta)] = E[L(Y_{t+1} - fc_{t+1})] = E[L(e_{t+1})]\ . \quad (4)$$

Training EANN under different criteria allows us to examine relationships between statistical measures and economic characteristics.

4 Experimental Design

We use ANN with GA optimization for the building/evolution of price forecast and trading strategy development/evolution upon relevant forecast. The mechanism appears to be an intuitive way to deal with agents' cognitive limits in forecasting and optimization, modeling the traders' learning process to approximate the unknown conditional expectation function. It also provides a natural procedure to consider decisions' heterogeneity by agents viewing similar information. A single hidden layer ANN seems to be sufficient for our problem, particularly considering the universal approximation property of feedforward nets. GA facilitates an optimal choice of network settings and adds additional explanatory power to the analysis.

4.1. Learning Law and Search Algorithm

Backpropagation is one of the most common algorithms in supervised learning. Although being simple and computationally efficient, the search here can get caught in local minima. Backpropagation is also often criticized for being noisy and slow to converge. To improve the original gradient learning, particularly its slowness of convergence, we consider a number of alternatives.

4.2. Performance Surface

The performance of ANN learning is monitored by observing how the cost changes over training iterations. The learning curve presents the internal error over each epoch of training, comparing the output of the ANN to the desired output. In price forecasting, the target is the next day closing price, where in signal modeling, the target is the current strategy. Achieving an accurate representation of the mapping between the input and the target might not necessarily lead to a forecast to be exploitable or a strategy using that forecast to be profitable.

We consider that evaluation criteria should measure not so much absolute effectiveness of the model with respect to the environment[1] but rather its relative effectiveness with respect to other models. Although we train ANN with the goal to minimize internal error function, we test and optimize its generalization ability by comparing its performance with the results of a benchmark, an efficient prediction (EP)[2]. In forecasting prices, EP is the last known value[3]. For predicting strategies, it is the buy/hold (B/H) strategy. The degree of improvement over efficient prediction (IEP) is calculated as an error from a de-normalized value of the ANN and a desired output, then normalizing the result with the difference between the target and EP value.

Making a prediction using a change or a percentage change, the value of IEP is particularly significant. IEP around 1, implying that the ANN predicted a change or a percentage change of zero, indicates that the network does not have adequate information to make a valid prediction. So, it ends up predicting the mean of all changes, zero. Predicting two samples or more in advance, one can have reduction in value of IEP (in comparison to one sample prediction). This does not mean that there is an improvement, since the change in the desired value is typically larger for a longer prediction. We classify our results using the following scale: IEP<0.8 $\Rightarrow$ excellent; IEP<0.85 $\Rightarrow$ very good; IEP<0.9 $\Rightarrow$ good; IEP<0.95 $\Rightarrow$ satisfactory; IEP$\geqslant$0.95 $\Rightarrow$ weak.

4.3. Profitability as Performance Measure

Similar to the performance evaluation criteria of investment managers (total realized returns adjusted for the riskness) the realized total continuously compounded returns or excess returns have been used to review trading rules developed under evolutionary learning. Unlike case-by-case evaluation of actions of portfolio managers, decisions of evolutionary agents are assessed on aggregate, over the entire trading period. Therefore, in computational modeling process/means used by agents need to be explicitly evaluated. Under continuously compounded reinvestment of realized returns, strategies with a higher number of trades and lower returns per trade receive greater fitness. [3] demonstrates that strategies with the lowest mean

---

[1] [2] found correlation between the Kolmogorov-Smirnov statistics and the length of validation period. Assuming that traders' beliefs with longer validation periods get closer to the true process in simulations and agents' accuracy increases, they consider the time horizon that agents use for validation as a representation of the accuracy of prediction.

[2] Note, market efficiency testing is not an objective of our studies per se. However, learning a profitable forecast/strategy is, in a way, discovering market inefficiency.

[3] If prices exhibit random walk behaviour, equally likely to change up or down, the average forecast has a change of zero from the last value. This makes the last value a good benchmark to determine if the prediction can improve on a random chance.

returns and variances per trade could be evaluated as best.

Simple aggregate realized returns overcome problems with frequent trading. Although the number of trades minimization favors infrequent but prolonged positions. More importantly, realized returns ignore opportunity costs (non-realized losses from missing profitable opportunities), incurred maintaining a certain market position. A proposed solution here is to use non-realized simple aggregate returns.

We examine the following forms of cumulative and individual trades' return measures: non-realized simple aggregate return; profit/loss factor; average, maximum gain/loss. In addition we estimate exit efficiency, measuring whether trades may have been held too long, relative to the maximum amount of profit to be made, as well as the frequency and the length of trades, including out of market position. To assess risk exposure we adopt the Sharpe ratio[4] and the maximum drawdown[5], as well as common 'primitive' statistics. To overcome the Fisher effect we consider trading positions with a one-day delay.

TC is assumed to be paid both when entering and exiting the market, as a percentage of the trade value. TC accounts for broker's fees, taxes, liquidity cost (bid-ask spread), as well as costs of collecting/analysis of information and opportunity costs. According to [4] large institutional investors achieve one-way TC about 0.1-0.2%. Often TC in this range is used in computational models. Since TC (defined above) would differ for heterogeneous agents, we report the break-even TC that offsets trading revenue with costs leading to zero profits.

Thus, in this paper profitability is a function of return, risk and transaction costs. The classification of the ANN output as different types of B/S signals determines the capability of the model to detect the key turning points of price movement. Evaluating the mapping of a forecast into a strategy, $\delta(fc_{t+1})$, assesses the success in establishing a predictive density, g_y that determines agents' actions.

4.4 Time Horizons and Trading Strategies Styles

Heterogeneous traders in the experiment use different lengths of past and forward time horizons to build their forecasts/strategies. We have run the experiment on stock indexes from a number of markets and found that 'optimal' length of training/validation period is a function of specific market conditions. In this paper we adopt three memory time horizons, [6; 5; 2½] years. We run the experiment with one year testing horizon, as it seems to be reasonable from the actual trading strategies perspective and supported by similar experiments.

Both long and short trades are allowed in the simulation. Investing total funds for the first trade, subsequent trades (during a year) are made by re-investing all of the money returned from the previous trades. If the account no longer has enough capital to cover TC, trading stops.

5. Genetic Algorithm Optimization

In this research EC is used for ANN model discovery, considering GA optimization for: network's topology; performance surface; learning rules; number of neurons and memory taps; weight update; step size and momentum rate. GA tests the performance of the following ANN models: Multilayer Perceptron (MLP), Jordan and Elman Networks (J/E), Time-Lag Recurrent Network (TLRN), Recurrent Network (RN), Modular Network (MN) and Support Vector Machine (SVM). We examine the performance surface optimized with GA for DA, discounting the least recent values and minimizing the number of large errors. For learning rule optimization we consider Steepest Descent; Conjugate Gradient; Quickprop; Delta Bar Delta and Momentum.

With GA optimization we test the integer interval [1, 20] for hidden layers' neurons, expecting that a higher number increases the network's learning ability, although at the expense of harder training and a tendency to overspecialization. GA optimization considers the range [1, 20] for the number of taps, affecting the memory of the net. The input layer, having access to the least modified data, has typically the highest number, decreasing in the hidden layers. GA optimization of the weight update for static networks considers whether the weights are updated following all data (batch) or after each piece of data (online) are presented. For dynamic networks GA determines a number of samples to be examined each time ANN updates weights during the training phase.

The step size, controlling the speed of weight adjustment, manages the trade-off between slow learning and a tendency to overreact. Usually the hidden layer has a larger step size than the output layer, and memory components generally have lower step size than other components of the same layer. GA optimizes the step size of the learning rates in the range [0, 1]. The momentum, using the recent weight update, speeds up the learning and helps to avoid local minima. GA

---

[4] Given by the average return divided by the standard deviation of that return.

[5] Accesses the size of the individual losses occurred while achieving given gains.

searches in the range [0, 1] for the value by which the most recent weight update is multiplied.

In terms of GA parameters, we apply the tournament selection with size 4, {prob=fitness/Σfitness}. Four types of mutation are considered in the experiment: uniform, non-uniform, boundary and Gaussian. Probability of mutation (PM) tested in the range [0, 0.05] and probability of uniform crossover is examined in the range [0.7, 0.95]. We test the effect of the increase in population size in the range [25, 200] on performance and computational time. The training optimization continues until a set of termination criteria is reached, given by maximum generations in the range [100, 500].

When a model lacks information, trading signals' predictions often stay near to the average. If ANN output remains too close to the mean to cross over the thresholds that differentiate entry/exit signals, post-processing is found to be useful (establishing thresholds within the range). Post-processing with GA optimization, examines a predicted signal with simulated trades after each training, searching for the thresholds against the values produced by ANN to generate maximum profit (see Appendix for details).

GA tests various settings from different initial conditions (in the absence of a priori knowledge and to avoid symmetry that can trap the search algorithm). Since the overall objective of financial forecasting is to make a trading decision, based on that forecast profitable, economic criteria rather than statistical qualities need to be employed for the final goal. We use GA optimization with the aim to minimize IEP value and profitability as a measure of overall success[6].

6. Empirical Application

6.1. Data

We consider daily closing prices for the MTMS (Moscow Times) share index obtained from Yahoo Finance. The time period under investigation is 01/01/97 to 23/01/04. There were altogether 1575 observations in row data sets. Examining the data graphically reveals that the stock prices exhibit a prominent upward, but non-linear trend, with pronounced and persistent fluctuations about it, which increase in variability as the level of the series increases. Asset prices look persistent and close to unit root or non-stationarity. Descriptive statistics confirm that the unit-root hypothesis cannot be rejected at any confidence level. The data also exhibits large and persistent price volatility with significant autocovarience even at high order lags.

Changes in prices increase in amplitude and exhibit clustering volatility. The daily return displays excess kurtosis and the null of no skewness is rejected at 5% critical level. The tests statistics lead to rejection of the Gaussian hypothesis for the distribution of the series. It confirms that high-frequency stock returns follow a leptokurtic and skewed distribution incompatible with normality assumed often in the analytical literature.

6.2. Experimental Result

ANN with GA optimization was programmed with various topologies[7]. Altogether we have generated and considered 93 forecasting and 143 trading strategies' settings. Effectiveness of search algorithm was examined with multiple trials for each setting. 92% of 10 individual runs produce identical results, confirming the replicability of our models. Efficiency of the search was assessed by the time it takes to find good results. The search with ANN unoptimized genetically took a few minutes, where the search with GA optimization lasted on average 120 minutes on a Pentium 4 processor.Over a one year testing period 19 trading strategies were able to outperform in economic terms the B/H strategy, with an investment of $10,000 and a TC of 2% of trade value. The average return improvement over B/H strategy was 20%, with the first five outperforming the benchmark by 50% and the last three by 2%. The primary strategy superiority over B/H strategy was 72%.

For the five best performing strategies, the break-even TC was estimated to be 2.75%, increasing to 3.5% for the first three and nearly 5% for the primary strategy. Thus, the break-even TC for at least primary strategy appears to be high enough to exceed actual TC. Profitability produced by our simple architecture supports computational model development based on economic and statistical foundations.

The experiment demonstrates that normalization reduces the effect of non-stationarity in the time series. The effect of persistency in prices diminishes with the use of

[6] Another possibility would be to use profit as the performance surface determinant. We leave this option out, since it wouldn't allow us to consider the questions proposed at the beginning. Setting the performance surface determined by the overall objective would not guarantee minimization of the underlying loss function used in forecasting and would not permit us to examine the relationship between statistical qualities and economic profitability.

[7] Programs in Visual C++, v. 6.0 are available upon request. We have run tests on TradingSolutons, v. 2.1, NeuroSolutions v. 4.22 and Matlab v. 6.

the 'percentage change' in values. Table 1, presenting the average effect of GA post-processing on performance, shows that it has generally improved (positive values) statistical characteristics. Although only accuracy[8] exhibits sizable change, the effects on IEP and correlation[9] were significantly smaller and not always positive.

Table 1. GA Post-Processing Effect

ΔStats./Sets	2000-2004	1998-2004	1997-2004
IEP	0.059	-0.838	0.001
Accuracy (%)	1.3	6.58	0.95
Correlation	0.016	0.011	0.001

The experiment with four types of GA mutation did not identify the dominance by a particular type. We have run simulations with different PM to test how the frequency of novel concepts' arrival affects modeling of the environment with structural brakes. The results, presented in Table 2, show that newcomers generally benefit the system. Although we have expected this outcome, its consistency among all (including short time) horizons was not anticipated. In economic terms, runs with a high probability of mutation {PM=0.05} have produced the highest returns. At the same time, this relationship is of non-linear character (e.g. {PM=0.001} consistently outperforms {PM=0.02}).
Some moderate, although consistent relationship between PM and strategies' risk exposure was found. Higher PM resulted in low riskness, given particularly by Sharpe ratio. We have also noticed some positive correlation between PM and annual trades' quantity, although this relationship appears to be of moderate significance and robustness. Trading frequency in simulations without mutation seems to be set at the beginning and stay until the end either at low or high values. The experiments without mutation have produced strong path-dependent dynamics, though not necessarily with sub-optimal outcome. It seems there exist some 'optimal' PM (in our experiment 0.05 and 0.001) and tinkering with this parameter can improve overall profitability. We have not found a robust relationship between the memory length and PM>0. Although, the memory length in simulations without mutation was on average 2.5 times shorter than in experiments with mutation. The relationship between PM and common statistical measures was inconclusive at acceptable significance or robustness.

GA model discovery reveals that MLP and TLRN with (Focus) Laguarre memory, with neurons number in the hidden layer in the range [5, 12] and Conjugate Gradient learning rule generate the best performance in statistical and economic terms for forecasting and acting nets. Generally models discovered with GA have lower trading frequencies, but without reduction in riskness. Annualized returns of those models were improved moderately. The effect of GA discovery on models' statistical performance was not conclusive, with a weak tendency towards accuracy amelioration. An increase in population size for GA optimization didn't lead to improvement in results. We explain this by the non multi-modal nature of our problem. Evidently, a higher population size has resulted in longer computational time. The relationship between statistical measures (accuracy, correlation, IEP) and trading strategies' profitability seems to be of a complicated nature. Among the ten statistically sound price forecasts, there is only one that was used in a trading strategy superior to B/H benchmark. The best five in economic terms strategies are among the worst 50% according to their accuracy. Three of the most accurate strategies are among the worst 25% in terms of their annualized return. Correlation of desired and ANN output characterizes one of the first five strategies with highest return among its best performers, another one among its worst results and the remaining are in the middle. IEP shows some robust relationships with annualized return. All five strategies with highest return have IEP<0.9. Furthermore, one of the first five profitable strategies has one of the three best IEP values. Therefore, if profits are not observable, IEP could be used as an evaluation criterion for an economic prediction
Regarding the performance surface optimization, two out of the three best strategies included an adjustment to treat directional information as more important than the raw error. We found that training ANN with the performance surface genetically optimized for DA, discounting least recent values or minimizing number of large errors generally improves profitability. Among 25% of the weak (in economic terms) strategies' annualized returns, there is none with learning criteria optimized. Our experiment has shown that among three optimizations of the performance surface considered, strategies trained on learning the sign of the desired output were generally superior to those trained to reduce the number of large errors or focusing learning on recent values. At the same time, the impact of optimization for DA on common statistical measures was insignificant, conforming that DA only weekly relates to conventional statistical criteria.

[8] Percentage of correct predictions.

[9] Correlation of desired and ANN output.

Table 2. Economic and Statistical Measures Under Different Probabilities of Mutation

Measures/PM	0	0.001	0.02	0.05	0	0.001	0.02	0.05	0	0.001	0.02	0.05
Return (%)	76.9	85.7	76.4	99.8	65.6	75.1	62.1	86.8	68.3	74.7	60.8	82
Sharpe Ratio	0.13	0.15	0.15	0.16	0.13	0.13	0.14	0.16	0.13	0.13	0.13	0.14
Trades (N°)	1	3	3	5	9	1	5	10	7	1	4	3
IEP	1.116	1.126	1.169	1.135	0.949	0.95	0.958	0.936	0.942	1.076	1.077	0.979
Accuracy (%)	51.5	32.9	37.66	54.98	41.2	45.92	40.77	42.06	32.38	32.9	32.9	32.4
Data Sets	2000-2004				1998-2004				1997-2004			

Our simulation generally supports a claim that DA relates to forecast profits more than mean squared or mean absolute errors criteria. At the same time, the experiment rejects an assertion that all other summary statistics are not related to forecast profit, as was demonstrated by the IEP relationship with profitability. As the results show that DA (alone or always) does not guarantee profitability of trading strategies trained with this criterion, it might be ineffective to base empirical estimates of economic relationships only on that measure. If conventional least squares are to be considered inadequate, an alternative estimation technique for economic behavior might use a combination of measures, demonstrated to have certain relationships with profitability; IEP and DA have been identified so far.

7. Conclusion

Profitability results produced by our simple architecture seem to be sufficiently general to support computational model development, based on economic and statistical foundations. The break-even TC, for at least primary strategy, appears to be high enough to exceed actual TC. GA post-processing has generally improved statistical characteristics. Novel concepts' arrival, determined by PM, benefits the system in economic terms, but is inconclusive statistically. It seems there exist some 'optimal' PM and tinkering with this parameter has a positive effect on profitability.

Models discovered with GA have moderately higher profitability, but the impact on their statistical characteristics was inconclusive. GA optimization of performance surface (particularly for DA) has a positive effect on strategies' profitability, though with little impact on their statistical characteristics. Since DA does not guarantee profitability of trading strategies trained with this criterion, it might be ineffective to base empirical estimates of economic relationships only on that measure.

When profits are not observable, IEP is proposed as an evaluation criterion for an economic prediction, due to its robust relationships with annualized returns. If conventional least squares are to be considered inadequate, an alternative estimation technique for economic behavior might use a combination of measures, demonstrated to have certain relationships with profitability; IEP and DA have been identified so far.

The performance surface set-up is viewed to be a crucial factor in search of a profitable prediction with an evolutionary model. Measures of trading strategies' predictive power might significantly differ from criteria leading to its profit maximization. The choice of evaluation criteria combining statistical qualities and economic profitability is viewed as essential for an adequate analysis of economic structures.

Presence of at least two objectives (statistical and economic) to be satisfied at the same time could be considered as a multiobjective optimization problem for further research. It seems, evolutionary algorithms, capable generating the Pareto optimal set in a single run, might be particularly appropriate for this task.

Appendix

Use of thresholds within the ranges against values produced by ANN ($\in [-1, 1]$) allows us to set different levels for predicted signals. For enter Long outputs the range is $\{\geq 0.5\}$, with scaling based on the distance between the enter Long and enter Short thresholds (enter Long and zero if thresholds are equal). Exit Short range is [0.2, 0.5], with scaling based on the distance between the exit Short and enter Long thresholds. For exit Long the range is [-0.2, -0.5], with scaling based on the distance between the exit Long and enter Short thresholds. Enter Short range is $\{\leq -0.5\}$, with scaling based on the distance between the enter Long and enter Short thresholds (enter Short and zero if thresholds are equal). For Hold outputs the range is [-0.2, 0.2], with scaling based on the distance between the exit Short and exit Long thresholds.

References

[1] G. Leitch and E. Tanner, "Economic Forecast Evaluation: Profits Versus the Conventional Error Measures," *American Economic Review*, vol. 81, pp. 580-590, 1991.

[2] S.-H. Chen and Y.-C. Huang, "Simulating the Evolution of Portfolio Behavior in a Multiple-Asset Agent-Based Artificial Stock Market," presented at the 9th International Conference on Computing in Economics and Finance, University of Washington, Seattle, USA, 2003.

[3] S. Bhattacharyya and K. Mehta, "Evolutionary Induction of Trading Models," in *Evolutionary Computation in Economics and Finance, Studies in Fuzziness and Soft Computing*, S.-H. Chen, Ed.: Physica-Verlag, 2002, pp. 311-331.

[4] R. J. Sweeney, "Some Filter Rule Tests: Methods and Results," *Journal of Financial and Quantitative Analysis*, vol. 23, pp. 285-301, 1988.

A Method to Improve Generalization of Neural Networks: Application to the Problem of Bankruptcy Prediction

Armando Vieira[1], João C. Neves [2], Bernardete Ribeiro[3]

[1]ISEP and Centro de Física Computacional, University of Coimbra, 3000 Coimbra, Portugal, asv@isep.ipp.pt
[2] ISEG, Rua Miguel Lupi 20, 1200 Lisboa, Portugal
[3]Department of Informatics Engineering, University of Coimbra, Portugal

Abstract

The Hidden Layer Learning Vector Quantization is used to correct the prediction of multilayer perceptrons in classification of high-dimensional data. Corrections are significant for problems with insufficient training data to constrain learning. Our method, HLVQ-C, allows the inclusion of a large number of attributes without compromising the generalization capabilities of the network. The method is applied to the problem of bankruptcy prediction with excellent results.

1 Introduction

Many actual machine learning problems, like image analysis, neurophysiology, remote detection and bioinformatics, involve classification and clustering of high-dimensional data. However, most classification methods have a poor performance when a large number of features are used. Since the search space increase exponentially with the number of features included, large training sets are usually required to obtain a reliable model. If the learning machine is an Artificial Neural Network, complex architectures may be needed with the risk of overfitting. Moreover, training is slower, due to the presence of many local minima, and often becomes an ill-conditioned problem [1].

Some classifiers, such as minimum distance, often perform relatively well on low dimensional data, but they show limited performance in high dimensional spaces. Dimensionality reduction through feature selection is a useful, and often necessary, preprocessing step. For difficult classification problems, with non-linear correlated inputs or samples containing many atypical examples, care must be taken during feature selection not to discharge information necessary for class separation.

In a previous work we showed that Hidden Layer Learning Vector Quantization (HLVQ) is an efficient method for classification [2]. HLVQ performs a non-linear feature selection that effectively reduce the dimensionality of the problem and improve class discrimination. In this work we apply HLVQ to correct the predictions of multi layer perceptron (MLP) in classification problems. The objective is to correct the outputs of the MLP, especially for points in poorly sampled regions in the training process. With this method we can handle more features without compromising the generalization capabilities of the MLP.

Next section presents the HLVQ algorithm and in section 3 it is used to implement the purposed correction algorithm. Section 4 presents brief description of Support Vector Machines (SVM), and section 5 describes the bankruptcy prediction problem. Section 6 presents the results of our method in comparison to the most conventional method of discrimination, the Multiple Discriminant Analysis (MDA) and more recent techniques such as Support Vector Machines.

2 The Hidden Layer Learning Vector Quantization (HLVQ)

The Hidden Layer Learning Vector Quantization (HLVQ) is an algorithm recently proposed for classification of high dimensional data [2, 3]. It is implemented in three steps. First, a multilayer perceptron is trained using back-propagation. Second, supervised Learning Vector Quantization is applied to the outputs of the last hidden layer to obtain the code-vectors $\vec{w}_{ci}$ corresponding to each class c_i in which data are to be classified. Each example, $\vec{x}_i$, is assigned to the class c_k having the smallest Euclidian distance to the respective code-vector:

$$k = \min_j \left\| \vec{w}_{c_j} - \vec{h}(\vec{x}) \right\| \tag{1}$$

where $\vec{h}$ is a vector containing the outputs of the hidden layer and $\|\cdot\|$ denotes the usual Euclidian distance. In the third step the MLP is retrained but with two differences regarding conventional multilayer training. First the error correction is not applied to the output layer but directly to the last hidden layer. The output layer is therefore ignored from now on. The second difference is in the error correction backpropagated to each hidden node:

$$E_1 = \frac{1}{2}\sum_{i=1}^{N_h} \left(\vec{w}_{ck} - \vec{h}(\vec{x}_i) \right)^2 \tag{2}$$

After retraining the MLP a new set of code-vectors,

$$\vec{w}_{c_i}^{\,new} = \vec{w}_{ci} + \Delta\vec{w}_{ci} \tag{3}$$

is obtained according to the following training scheme:

$$\Delta\vec{w}_{ci} = \alpha(n)(\vec{x} - \vec{w}_{c_i}) \text{ if } \vec{x} \in \text{class } c_i, \qquad \Delta\vec{w}_{c_i} = 0 \quad \text{if } x \notin \text{class } c_i \tag{4}$$

The parameter α is the learning rate, which should decrease with iteration n to guarantee convergence. Steps two and three are repeated following an iterative process. The stopping criterium is met when a minimum classification error is found.

The distance of given example x to each prototype is:

$$d_i = \left\| \vec{h}(\vec{x}) - \vec{w}_{c_i} \right\| \tag{5}$$

which is a proximity measure to each class.

3 HLVQ-C

One drawback of multilayer perceptrons is their poor performance in sparse regions not covered by training data, common in high dimensional datasets. To alleviate this situation we propose the following method to correct the MLP output. The objective is to identify these regions and evaluate the consistency of the perceptron results against the HLVQ prediction. After training the MLP and HLVQ, the algorithm runs according to the following steps.

Each test example, $\vec{x}^i$, is included in the training set and the neural network retrained. Since the class membership of this example is unknown, we first assign it to class "0" and determine the corresponding output $y_0(\vec{x}^i) = y_0^i$ as well as the respective distances to each class prototype obtained with HLVQ,

$$\vec{d}_0^i = (d_0^{c0}, d_0^{c1}) = \left(\left\| h_0(\vec{x}^i) - w_{c0} \right\|, \left\| h_0(\vec{x}^i) - w_{c1} \right\| \right) \tag{6}$$

In a second step the network is retrained considering the example as class "1". The new output $y_1(\vec{x}^i) = y_1^i$ and the respective distances to the prototypes are obtained in a similar way, thus:

$$\vec{d}_1^i = (d_1^{c0}, d_1^{c1}) = \left(\left\| h_1(\vec{x}^i) - w_{c0} \right\|, \left\| h_1(\vec{x}^i) - w_{c1} \right\| \right) \tag{7}$$

From these outputs, y_0^i and y_1^i, the choice is made following the heuristic rule:

$$y^i = y_0^i \text{ if } d_0^{c0} < d_0^{c1}$$
$$y^i = y_1^i \text{ if } d_1^{c1} < d_1^{c0}.$$

We call this method Hidden Layer Learning Vector Quantization Correction, HLVQ-C, and it corrects the class separation hyperplane.

Figure 1, show how the method works. Before correction, a new point "?", located outside the region covered by the training data, is classified as class 1 by the MLP, since it is above the separation hyperplane. However, the distance d_1 to class 0 code-vector (X) is smaller than the distance to the class 1 code-vector (Y) - d_2. HLVQ-C modifies the MLP prediction and this element is now correctly assigned to class 0.

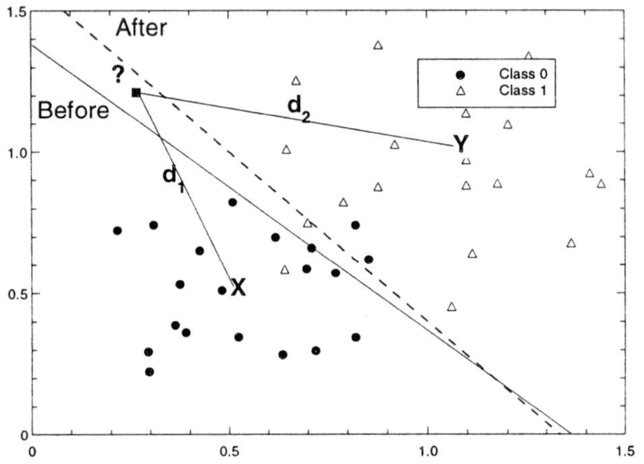

Figure 1: How HLVQ-C corrects the MLP predictions. Full line indicate the separation hyperplane before a new point "?" is included. Dashed line is the same but after retraining considering the new point as class 0.

4 Support Vector Machines (SVM)

Support vector machines (SVM) are a new learning-by-example paradigm spanning a broad range of classification, regression and density estimation problems [15]. Equipped with a sound mathematical background, SVM treat both the problem of complexit minimization while maximizing generalization. The advantage, regarding other approaches, is that SVM generalization error is not dependent on the dimensionality of the input data. The learning method uses input-output training examples from the data set

$$D = \{(\mathbf{x}_i, y_i) \in X \subseteq R^N \times Y \ : 1 \leq i \leq l\} \tag{8}$$

such that f classifies correctly test data $(\boldsymbol{x}, y)$ generated from the same underlying probability distribution $P(\boldsymbol{x}, y)$. Using the loss function defined by :

$$V(y_i, f(\mathbf{x}_i)) = |1 - y_i f(\mathbf{x}_i)|+ \quad (9)$$

The learning problem can be formulated minimizing the function (10) using

$$\frac{1}{l}\sum_{i=1}^{l} V(y_i, f(\mathbf{x}_i)) + \|f\|_{F_k}^2 \quad (10)$$

The minimizer of (10) has the form:

$$f(\mathbf{x}) = \sum_{i=1}^{l} \alpha_i k(\mathbf{x}, \mathbf{x}_i) + b \quad (11)$$

with $\alpha_i, b \in \boldsymbol{R}$. The equivalent quadratic programming problem originally proposed in [16] is:

$$\min_{f \in \mathbf{F}, \xi} \Phi(f, \xi) = \frac{C}{l}\sum_{i=1}^{l} \xi_i + \frac{1}{2}\|f\|_k^2 \quad (12)$$

subject to constraints:

$$y_i f(\mathbf{x}_i) \geq 1 - \xi_i \quad i = 1, \cdots, l \quad (13)$$

$$\xi_i \geq 0 \qquad i = 1, \cdots, l$$

where C is the penalty constant (regularization parameter) and ξ the slack variable.

Introducing Lagrange multipliers:

$$\max_{\boldsymbol{\alpha}} W(\boldsymbol{\alpha}) = \sum_{i=1}^{l} \alpha_i - \frac{1}{2}\sum_{i,j=1}^{l} \alpha_i \alpha_j y_i y_j k(\mathbf{x}_i, \mathbf{x}_j) \quad (14)$$

and solving with respect to α_i, under the constraints $0 \leq \alpha_i \leq \frac{C}{l}$, $i = 1 \ldots, l$ and $\sum_{i=1}^{l} \alpha_i y_i = 0$, the solution has again the form of Eq. (11). The empirical error measured by $\sum_{i=1}^{l} \xi_i$ is minimized while controlling the learning machine capacity.

5 Application to the Bankruptcy Problem

The bankruptcy prediction problem is to discriminate between healthy and distressed companies based on the record of several financial indicators from previous years [4]. Several modern approaches have been used to analyze this problem, ranging from Support Vector Machines, Genetic Algorithms and Neural Networks [4-10]. Although several authors have demonstrated that these methods are in general more accurate in signaling a distressed company, traditional Linear Discriminant Analysis (MDA) is still widely used. The success of MDA can be explained by its simplicity and the fact that, in most cases, the difference in performance are not substantial. Another reason to explain the analyst's resilience in using new approaches, concerns the quality of databases used to benchmark the predictions. The most common weakness are the small size of the database, limited historical records and the use of unbalanced samples, containing much more healthy companies than financially distressed ones.

We used a sample obtained from Diana, a database containing about 780,000 financial statements of French companies. The initial sample consisted of financial ratios on 2,800 non-financial French companies, for the years of 1998, 1999 and 2000, with at least 35 employees. From these companies, 311 were declared bankrupted in 2000 and 272 presented a restructuring plan ("Plan de redressement") to the court for creditors approval. We decided not to distinguish these two categories as both signal companies in financial distress.

The sample thus has 583 financial distressed firms, most of them of small and medium size, with a number of employees from 35 to 400. From the initial 30 financial ratios we select the 17 most relevant ratios excluding those with either a small average sensitivity or a small variance of the sensitivity. High correlated ratios were also excluded. All data were normalized to zero mean and unity variance.

Neural networks containing from 5 to 20 hidden nodes were tested. A hidden layer of 15 neurons, a learning rate of 0.1 and a momentum term of 0.25 were chosen.

The bankruptcy is a dynamic process. In order to capture the evolution and the trend of the financial position we also used three annual variances of the following ratios: debt ratio, value added per employees and margin before extra items and taxes, thus extending our previous study [3].

6. Results and conclusions

To test HLVQ-C, we used a balanced dataset containing 583 healthy and 583 distressed companies using ten-fold cross validation. We discriminate the accuracy of the classifiers for type I, type II error and the overall misclassification. Type I error is the percentage of undetected bankruptcies while type II error is the percentage of healthy companies predicted as bankrupt.

Table 1 summarizes the results with data from 1999, one-year prior to the announcement of bankruptcy. Our method surpasses all others, both in the overall accuracy, and, more important, on type I error. This term has a much higher cost for banks and insurance companies than type II error [4].

Support Vector Machines (SVM) reduced substantially error type I while keeping error type II at almost the same level as MLP. This may be due to the fact that the optimal compromise between minimization of the empirical risk and complexity has not been reached. However, the method could be improved using unlabeled data and the geometry of the separation hyperplane would allow less misclassifications of the type I.

Type I error is always higher than type II error since distressed companies has a more heterogeneous pattern and are therefore harder to classify. Finally, note that Multiple Discriminant Analysis (MDA) has a very small accuracy.

Table 1: Generalization errors (in percentage) for several machine learning algorithms.

Model	Error I	Error II	Total
MDA	26.4	21.0	23.7
SVM	17.6	12.2	14.8
MLP	25.7	13.1	19.4
HLVQ-C	11.1	10.6	10.8

We presented a technique for output correction of multilayer perceptrons, called HLVQ-C. With this technique a MLP can be trained with a large set of features without compromising the generalization. Corrections introduced by HLVQ-C can be substantial for high-dimensional data and small training datasets.

To our knowledge, results obtained with Hidden Layer Learning Vector Quantization Correction, represent the greatest improvement with respect to discriminant analysis. Support Vector Machines is also a very competitive approach.

References

[1] Bishop C. M. (1996) Neural Networks for Pattern Recognition, Oxford University Press, Oxford.

[2] Vieira A. and Barradas N. P. (2003) A training algorithm for classification of high dimensional data, Neurocomputing, 50C, 461-472.

[3] Vieira A. , Ribeiro B., Mukkamala S., Neves J. C. and Sung A. H. (2004) On the Performance of Learning Machines for Bankruptcy Detection, Second IEEE Int. Conf. on Computational Cybernetics, Vienna, Austria, August 30 – September 1, 223 – 227.

[4] Altman, E.I. (1989) Measuring Corporate Bond Mortality and Performance. Journal of Finance, **44**, 909-1022.

[5] G. Zhang, M. Y. Hu, B. E. Patuwo, D. C. Indro (1999) Artificial neural networks in bankruptcy prediction: General framework and cross-validation analysis, Europ. J. Op. Research. **116**, 16.

[6] Coats P.K., Fant L.F. (1993) Recognising Financial Distress Patterns Using a Neural Network Tool. Financial Management, 142-155.

[7] Jain, B., Nag B. (1997) Performance Evaluation of Neural Network Decision Models. Journal of Management Information Systems **14** (2), 201-216.

[8] Lee K.C., Han I., Known Y. (1996) Hybrid neural network models for bankruptcy predictions, Decision Support Systems 18, 63-72.

[9] Shah J.R., Murtaza, M. B. (2000) A Neural Network Based Clustering Procedure for Bankruptcy Prediction, American Business Review (June), 80-86.

[10] Varetto F. (1998) Genetic Algorithms Applications in the Analysis of Insolvency Risk, Journal of Banking and Finance, **22**, 1421-1439.

[11] Vieira A., P.A. Castillo and J.J. Merelo (2003) Comparison of HLVQ and GProp in the problem of bankruptcy prediction, IWANN03 - International Workshop on Artificial Neural Networks, LNCS 2687, Springer-Verlag, 655-662.

[12] J. S. Grice, M. T. Dugan (2001) The Limitations of Bankruptcy Prediction Models: Some Cautions for the Researcher Rev. of Quant. Finance and Account., **17** (2), 151.

[13] A. F. Atiya (2001) Bankruptcy Prediction for Credit Risk Using Neural Networks: A Survey and New Results, IEEE Trans. Neural. Net. 12 (4).

[15] C. Cortes and V. Vapnik, (1995) Support vector networks, Machine Learning, vol. 20, pp.273-297.

[16] V. Vapnik, (1995) The Nature of Statistical Learning Theory. New York, Springer.

An Adaptive Neural System for Financial Time Series Tracking

A. C. H. Dantas[1], J. M. Seixas[1]
[1]Signal Processing Lab
COPPE/EP - Federal University of Rio de Janeiro
C.P. 68504, Rio de Janeiro 21941-972, Brazil
E-mail: {augusto, seixas}@lps.ufrj.br

Abstract

In this paper, we present a neural network based system to generate an adaptive model for financial time series tracking. This kind of data is quite relevant for data quality monitoring in large databases. The proposed system uses the past samples of the series to indicate its future trend and to generate a corridor inside which the future samples should lie. This corridor is derived from an adaptive forecasting model, which makes use of the walk-forward method to take into account the most recent observations of the series and bring up to date the values of the neural model parameters. The model can serve also to manage other time series characteristics, such as the detection of irregularities.

1 Introduction

Specialists say that this century is surely the century of data [1]. In fact, the poor quality of customer data costs to U.S. business more than US$600 billion a year [2]. It is then easy to see that data are critical assets in the information economy, and that the quality of a company's data is a good predictor of its future success.

In this context, we propose a neural network based tool to treat one of the most important types of data involved in information systems: the financial time series. The development of time series forecasting methods is gaining importance as companies and research centers give more emphasis to a data based knowledge and make their investments in a data driven way. The idea in this paper is to generate an adaptive model from the series observations, and to use it to derive a monitoring "corridor" for the future samples of the series.

Note that our aim is not simply to develop a prediction algorithm; actually, what we propose is a system that monitors the quality of time series data by drawing reliability regions (the corridors) from the predicted values. These corridors may be estimated either for short-term or long-term prediction, and must be updated periodically.

In the next section, we give a brief description of some time series data quality issues. In Section 3, we summarize the neural system developed for time series modeling. The experimental data used in this work is presented in Section 4. Implementation and obtained results are detailed in Section 5, and conclusions are addressed in Section 6.

2 Time Series Data Quality Issues

Defining Data Quality (DQ) is a very context dependent task. Common sense says that it refers to conformity to the specifications. Concretely, what one is supposed to do is to set some attributes based on indicators and parameters, in order to get quantifiable metrics capable to inform the quality of a specific database [3].

For this work, since we are dealing with specific time series data, the considered DQ dimensions are *timeliness* (how up-to-date is your database?), *completeness* (does your database have missing values?) and *correctness* (how free-of-error are your data?).

2.1 Modeling Financial Time Series

Financial time series data have been traditionally decomposed in the following terms [4]:

- *Irregularities.* Sometimes a time series is affected by brusque changes that may not be predicted by any model. The September, 11 case is an example in civil aviation time series. So, in order to make data as "clean" as possible before they start to be processed, one must compensate for irregularities. The procedure proposed in this paper works towards furnishing to the system (or to the supervisor) the probabilities of the presence of irregularities in the time series;

- *Trend.* The first regular component to be removed of a time series before it can be efficiently modeled is its trend. Generally, the trend is a linear component that indicates the increase or the decrease of the series. It is important that the trend be calculated periodically, in order to permit the model to incorporate the most recent features introduced by new observations. Rarely, the trend may be modeled by nonlinear functions, such as exponential or polynomial ones [4];

- *Seasonal and cyclical components.* Besides the trend, time series generally contains cycles of regular nature. These cycles are due to the series seasonalities and ciclicities, and may be extracted by the removal of sinusoidal components (those which have greater spectral energy when compared to the others) [4].

Therefore, a time series Y_t (after the removal of the irregularities) may be expressed as:

$$Y_t = T_t + C_t + S_t + \varepsilon_t \quad (1)$$

where the right-hand terms are, respectively, the global trend, the cycles (longer period cyclical variations), the seasonal variations (weekly, monthly, annual) and a highly uncorrelated non-linear component ε_t, which we call the residual series, that is, the part of the series that remains after preprocessing is concluded. This residue may be generally viewed as a stationary stochastic process and is what should be processed (in our case, by neural models). The predicted values are then summed to the "deterministic" parts of the series in order to generate the desired forecasts.

3 The Adaptive Neural System

Since the 1990's, neural networks have been more and more applied in finance [5]. The number of financial applications such as pattern recognition, classification and time series forecasting have dramatically increased as a consequence of the fact that financial services organizations turned to be the second largest sponsors of neural network research [6].

Neural networks have an advantage over ARMA/ARIMA methods [7] because they can construct nonlinear models when mapping the input space to the output space, since they are universal function approximators. In this work, we use feedforward networks trained with the error backpropagation algorithm [8].

The proposed neural system makes use of two mechanisms that work together for performing the time series tracking task: classification and estimation. The first one is responsible to discover whether the next samples of the series should be increasing or decreasing (or even remaining within a stability region) the overall series values, and the second one gives an estimate of the value itself of the future samples. The aggregation of these two mechanisms gives us the requirements for tracking the series.

Figure 1 shows the architecture of the used network, for both classification and estimation tasks. The N most recent observations of the series form the network input vector. The hidden layer contains H biased neurons with the hyperbolic tangent as the activation function. For the output layer, we use either a hyperbolic tangent single neuron (for the classification problem) or a non-biased linear neuron (estimation). For both classification and estimation problems, the output stands for x_{t+F}, where F, the future lag, may be greater than one.

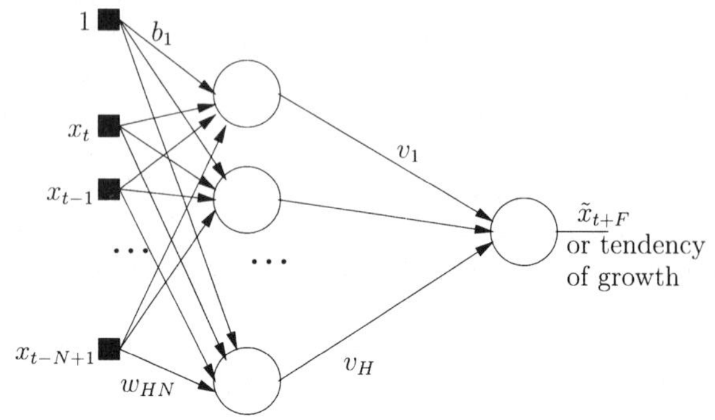

Fig. 1. Neural network architecture for time series tracking.

3.1 The Walk-forward Procedure

Time series data, due to their nature, demand efficient signal processing techniques for feature extraction and model adaption, in order to incorporate the changes of the series behaviour. For achieving this goal, the monitoring procedure will be continuously tuned to generate satisfactory predictions for the future values of the series.

The walk-forward procedure [6] is, in general, applied to an out-of-sample (offline) data set in order to simulate real-life trading and to test the robustness of the model through its frequent retraining. In our case, as it is assumed that we are continuously receiving new data, we may apply the procedure by substituting the older samples for the most recent ones, retraining the network and then predicting the future values in a up to date way. Figure 2 illustrates this process.

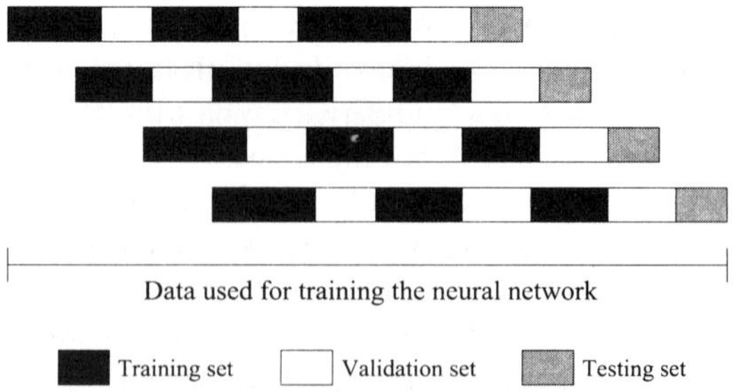

Fig. 2. Generic scheme for the walk-forward procedure.

Note that the size of the testing set is kept constant, in order to assure an independent check of the neural network; the training and the validation sets may be scaled together since the purpose of the validation set is to determine the ability of the network to generalize[1].

4 The Experimental Data

Financial market data may be bought from specialized companies (mainly if they are *intraday*, high-frequency data) or freely downloaded from websites that offer some amount of daily stock data for the past years. The latter alternative does not guarantee "total quality" data; this means that data may be noisy, corrupted, incomplete and so on. This reinforces the interest in DQ tool developments and introduces a new challenge to the prediction problem: if we are able to properly filter the noisy data out and then correctly predict the future values of the series, we can avoid the extra cost of certified stock data.

The procedure proposed in this work is suitable to operate in this noisy data environment: as the developed model generates "corridors" inside which future data are expected to lie, we may use these "certified" regions to check for the presence of noisy samples in non-certified data (this may be used to filter irregularities out, for example).

In this work, we used five years (1998-2002) of daily stock data for companies that appear in the SP500 index. Data are under the "open-high-low-close" (OHLC) format. In order to test the methodology, three companies were selected from the computer business segment: IBM, Sun and Microsoft.

5 Implementation and Results

The first step for implementing the procedure proposed here concerns the development of an automatic preprocessing block. In our system, the neural model is updated (and so the preprocessing of the series) for every new month of data[2]. The data lag used for the network training phase is initially set to 6 months, and the tune window (the number of observations N that comprises an input vector) is set to 10. The preprocessing phase aims at achieving a series decomposition according to Equation 1, which allows to obtain the residual term. Such residual term will be used to train the network.

We considered that trends are always linear, which may be a good approximation for financial time series in non-short periods of time. After the definition of the time lag, we fit a straight line to the series (by the least squares method) and define it as the series trend.

The resulting zero-trend series is now ready to have its cycles and seasons removed. This is done by computing the series spectrum and removing frequency components that exhibit higher energies. The number NC of components to be removed is defined by an energy threshold for surviving components[3].

Figure 3 shows an example of the series trend estimation, its compensation and the spectrum of the zero-trend series before removing cycles and seasonal components. Note that the component for $f = 0$ was zero even before the removal, because the detrended series has zero mean.

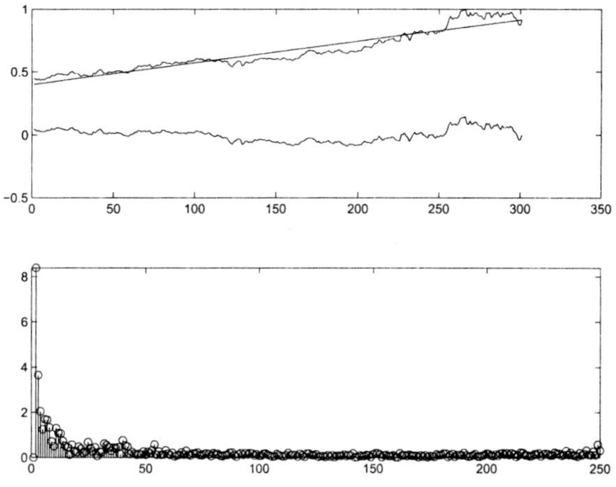

Fig. 3. The series trend compensation (top) and the spectrum of the zero-trend series (bottom).

Once the training, validation and testing sets are defined, a last normalization must be performed in order to flatten the distribution of the inputs according to the training set distribution. We calculate the mean μ and the standard deviation σ of the training set, and each input pattern x_i (for the training, validation and testing sets), becomes $x'_i = \frac{x_i - \mu}{S \times \sigma}$, where the parameter S is determined according to how wide the distribution of the training set is.

5.1 Generating the corridor

Once the networks (for classification and estimation) have converged, the validation corridor for the future values of the series may be computed. As shown in Figure 1, the network has a single output node, that is, it is optimized to predict the sample x_{t+F} having x_t to x_{t-N+1} as inputs.

For the prediction task, since there is an error in the estimation of x_{t+F} ($\tilde{x}_{t+F}$), that is, the output of the network may be seen as $x_{t+F} \pm \epsilon$, we will have access to upper and lower estimations that tend to be less accurate

[1] Usually, the training set correponds to 60-70% of the non-testing data.

[2] It may be shown that, for daily financial time series, one needs approximately one month to characterise a stable trend change. The user may, however, set another value for the retraining periodicity.

[3] We chose to remove cycles and seasons together, by looking at the spectrum of the detrended series; another option is to first eliminate seasons (when one knows their periods) and then remove cycles by spectrum analisys.

as we go further in time for the estimation process. These bounds form the estimation corridor, which is limited by the classification network output, that is, if the sample is expected to increase (decrease) the series value, the corridor may not go beyond a minimum (maximum) value, which is done by the series previous value[4]. The most accurate estimation we may produce is given by the one-day predictor ($F = 1$), in which the corridor is generated for the next series observation only. The bounds of the corridor are calculated from the last N estimations: for example, the corridor $\tilde{x}_{t+1} \pm \delta_N$ has

$$\delta_N = \frac{\sum_{i=0}^{N-1}(\tilde{x}_{t-i} - x_{t-i})}{N} \qquad (2)$$

that is, the average error from the last N predictions. The corridor interval also furnishes the value for the *correctness* metric (it is 1 if $x_{t+1} = \tilde{x}_{t+1}$). In general, this one-day predictor is sufficient to provide Data Quality Management, given that we are treating daily series and we can always apply a DQ test for a new sample. *Timeliness* and *completeness* have, in our case, the maximum value (1) if data are complete and up to date, and loses 0.1 for each missing or delayed sample. Thus, the system can always returns these DQ metrics to the user.

For the available experimental series, the topology of the best trained network was 10-4-1, with constant learning rate (0.2) and momentum term (0.1). For the treated series, the neural method achieved an average of less than 2% of estimation error for the reconstructed series (residue plus "deterministic" parts of the series), while an ARMA(10,1) process obtained an error of 2.24%. For the classification problem, we achieved more than 99% of efficiency for the classification problem when determining a one-day trend detection assuming 6% of stability region, that is, we were able to establish if the next future sample of the series would increase or decrease the series values (or remain within a 6% variation region).

6 Conclusions

A neural method for Data Quality assessment of time series data was developed. The neural system provides to the supervisor a tool for monitoring the quality of the next observed samples of the series: whenever a new sample falls outside the corridor, the system warns the supervisor and either apply a correction tool (when available) or label the unreliable sample. In both cases, an increase in the overall DQ is aimed. This monitoring system can be used for detecting irregularities (with posterior filtering) and in forecasting assessment, as discussed in the text.

The generation of long-term corridors through the feedback of short-term predictions may also be exploited. The main problem here is to maintain the stability of the network, which is not guaranteed due to the feedback error. We found that two kinds of difficulties may arise: the fading of the network output due to the feedback of negative errors, or the divergence of the output due to the feedback of positive errors. A new approach is being developed in which a mechanism to assure network stability is inserted in the recurrent prediction model. Preliminary results indicate that this new model will be able to furnish corridors longer than 10 days, enabling the supervisor to work with longer term forecasting.

Wavelet methods are also being tested in the preprocessing phase, in order to filter out irregularities from the training set and turn the network more sensitive to their presence in the testing set.

7 Acknowledgements

We would like to thank CNPq and FAPERJ/Brazil for funding, Insightful/USA for software support and Dr. Frank Block (Finscore/Switzerland) for his collaboration in this research.

8 References

[1] Donoho, D. (2000) High-Dimensional Data Analysis: The Curses and Blessings of Dimensionality. Lecture for the American Math. Society "Math Challenges of the 21st Century"

[2] Eckerson, W. (2001) Data Quality and the Bottom Line. Report from "The Data Warehousing Institute"

[3] Pipino, L., Lee, Y., Wang, R. (2002). Data Quality Assessment. Communications of the ACM, vol. 45, n. 4ve, pp. 211-218

[4] Chatfield, C. (1984). Analysis of Time Series. Chapman and Hall

[5] Poddig, T., Rehkugler, H. (1996). A 'world' model of integrated financial markets using artificial neural networks. In: Neurocomputing 10, Elsevier, pp. 251-273

[6] Kaastra, I., Boyd, M. (1996). Designing a neural network for forecasting financial and economic time series. In: Neurocomputing 10, Elsevier, pp. 215-236

[7] Box, G., Jenkins, G. (1970). Time Series Analisys, Forecasting and Control. Holden Day

[8] Haykin, S. (1999). Neural Networks - a Comprehensive Foundation, 2nd ed. Prentice-Hall

[4] A sample that falls very outside of this region should be considered as an irregularity.

Probabilistic Artificial Neural Networks for Malignant Melanoma Prognosis

R. Joshi[1], C. Reeves[2], C. Johnston[2]

[1]CTAC, Coventry University, UK
[2]School of Mathematical and Information Sciences, Coventry University, UK
E-mail: {r.joshi, c.reeves, c.johnston}@coventry.ac.uk

Abstract

Artificial Neural networks (ANNs) have found applications in a wide variety of medical problems and have proved successful for non-linear regression and classification. This paper details a novel and flexible probabilistic non-linear ANN model for the prediction of conditional survival probability of malignant melanoma patients. Hazard and probability density functions are also estimated. The model is trained using the log-likelihood function, and generalisation has been addressed. Unrestricted by assumptions that are unrealistic or parametric forms that are difficult to justify, the model thereby attains advantage over traditional statistical models. Furthermore, an estimate of the variance-covariance matrix is obtained using the asymptotic Fisher information matrix. Implemented in an Excel® spreadsheet, the model's user-friendly design further adds to its flexibility, with much potential for use by statisticians as well as researchers.

1 Introduction

Artificial Neural Network (ANN) research has become very popular in the analysis of medical data, as ANNs can incorporate non-linearities that are less restricted by unrealistic assumptions than conventional statistical models. One of the most common models is the multi-layer perceptron, which has found many applications in medicine, for example in diagnostic image analysis [1], clinical diagnosis [2]. They frequently give results comparable and in some cases superior to statistical models, for example in predictions of thalasemia minor in blood screening [3]. Indeed, theoretical work suggests that ANNs can match or outperform results obtained from statistical models [4]. Identification of prognostic factors of a particular disease is an important and widely studied aspect of medical and statistical research, useful for aiding clinicians to plan care pathways, and to predict survival characteristics, in addition to educating the patients and public.

Survival analysis describes the analysis of time-to event data, corresponding to the time from a defined time origin until the occurrence of a particular event of interest such as death or disease recurrence. In addition to incidence, corresponding clinical and pathological factors are recorded at times of diagnosis and treatment. The dependence of disease outcome on these factors and covariates can therefore be assessed. Standard statistical procedures are not suitable for the analysis of survival data, as they require particular modelling techniques due to the lack of symmetry, restriction to positive outcomes (survival times are not negative), and to *censoring* (i.e., the case has been lost to follow-up or the end point of interest has not occurred). Right censoring occurs when the survival time is greater than some defined time point. The analysis presented in this paper is based on data with right censored survival times. Much analysis of censored survival data uses Cox regression [5], or parametric models using (for example) a lognormal distribution. However, both types of models impose either distributional forms or assumptions which are not always tenable [6]. Therefore, there has been increasing interest in recent years in models that are flexible, and not restricted by unrealistic assumptions.

Recent research suggests that ANNs are suitable for modelling censored survival data and so may be used where regression models have been used traditionally. Studies proposing ANNs as extensions to standard regression models for survival data include Liestol *et al.* [7], who propose approaches for grouped time data, while Faraggi and Simon [8] generalise Cox regression, by replacing the linear functional in the Cox model with a non-linear ANN predictor. However, few ANN approaches in the published literature use continuous time. For the prediction of relapse from breast cancer, Ripley *et al.* [9] use 7 ANN models. Survival time is treated as a continuous variable in 4 of the models, and grouped and treated as a discrete variable in the remaining 3 models.

This paper investigates a novel flexible ANN approach for the analysis of censored survival data, set within a probabilistic framework. Censored observations are not omitted. Generalisation is addressed, in addition to an approximation to the variance-covariance matrix. The ANN's performance will be compared to Cox regression, as the latter is a method of choice for the analysis of censored data. In addition, comparisons to other popular survival modelling techniques, such as the empirical Kaplan-Meier estimate (KM) [10], and parametric log-normal model (LN) [11] will be performed.

2 Basics of Survival Analysis

When analysing survival data interest centres upon 2 main functions that are estimated from the observed survival times: the *survivor* and *hazard* functions. The probability density function of t, the actual survival time of an individual, is $f(t)$. The survivor function, $S(t)$, is the probability that the survival time is greater than or equal to t. The related hazard function $h(t)$ denotes the instantaneous death rate and represents the probability that the event occurs at time t, conditional on it not occurring prior to time t. The following relationship holds:

$$h(t)\,dt = \frac{f(t)dt}{S(t)} \tag{1}$$

The cumulative hazard function $H(t)$ is defined as

$$H(t) = \int_0^t h(u)du \tag{2}$$

so that

$$S(t) = e^{-H(t)} \tag{3}$$

3 The Data

The data set used for this study comprises 1946 malignant melanoma patients diagnosed in the West Midlands region from 1987-1996, of which 1160 are females and 786 are males. The survival time variable t is the number of months from date of diagnosis (entering the register) to the end of the study, due either to death or survival till the cut-off date (31st December 1999). Censored observations consist of patients that have not died. The data set is heavily censored, with 1628 censored cases. Significant prognostic factors used are summarised in Table 1. A logarithmic transformation was applied to the Pathological depth variable, as it is extremely positively skewed, resulting in the categories shown.

4 The ANN Model

4.1 The Model

A feed-forward multi-layer perceptron (MLP) was designed and implemented in an Excel® spreadsheet, using matrix multiplication. 24 patient subsets were created according to levels of the input variables defined in the table below:

Table 1: The Data

Factor	Description	Level
Townsend Score	**Measure of Social Deprivation**	
TQ=0	Townsend Score between -8.5 & -1.3	Affluent
TQ=1	Townsend Score Between -1.29 & 8.8	Deprived
Clarke Level	**Histological Stage of Cancer**	
Clar 12	Melanoma in situ (no invasion) or tumour invades papillary dermis	Least Severe
Clar 3	Tumour invades papillary reticular dermal interface	Medium
Clar 45	Tumour invades reticular dermal interface or subcutaneous tissue	Most Severe
Log-Pathological Depth	**Log of vertical thickness of tumour in mm**	
L.P.12	depth <= 0.75mm	Least Thick
L.P.3	depth between 0.75mm & 1.50mm	Medium Thick
L.P.4	depth between 1.51mm & 4.00mm	Medium Thick
L.P.5	depth > 4.00mm	Thickest

The source nodes in the input layer of the network comprise the covariates $\mathbf{X} = \{x_1, x_2, \ldots, x_{p-1}\}$, a bias, $x_0 \equiv 1$, and time, input x_p, coded as a prognostic variable by using it as a scaled normalized input (by dividing event time by the maximum value over the range of all event times). Multiplying these inputs by the synaptic weight parameters α_{jk}, for $j = 0, \ldots, p$ (for the connection from input node j to output node k), we have the input to the *hidden* layer. Next, a logistic *activation function* is applied to the weighted sum at each *hidden node*, such that

$$G(\vartheta) = \frac{1}{1 + exp(-a\,\vartheta)} \tag{4}$$

where a is the slope parameter of the sigmoid function. In this case, we set a=1. The use of such activation functions plays an important role in allowing the outputs to be given a probabilistic interpretation – it implies that the ANN model is in principle a mixture of logistic distributions. The inputs to the second layer, $G(y_k)$, for $k = 1, \ldots, m$, are then multiplied by weights β_k, and adding a bias parameter of unity, β_0, we have a single input to the output node. The same activation function is then applied, and the functional representation of the output from the ANN is

$$Z = S(t, \mathbf{X}) = G\left(\sum_{k=1}^{m} \beta_k G(y_k) + \beta_0\right) \tag{5}$$

Equation 5 is summed over all cases (the sample index i has been suppressed for clarity). As the batch method of learning is used, training and validation processes common to ANN studies are not required. The ANN model is illustrated in Figure 1.

4.2 Training

The back-propagation algorithm [13] is very popular for training feed-forward MLPs, but we have used the SOLVER® tool in Excel® as an alternative optimisation technique to make the ANN more user-friendly for explaining it to non-specialists. Although a common (cost) objective function is the sum of squared differences of the output and target values, by using the log-likelihood function additional valuable information may be obtained. This is discussed in section 4.4.

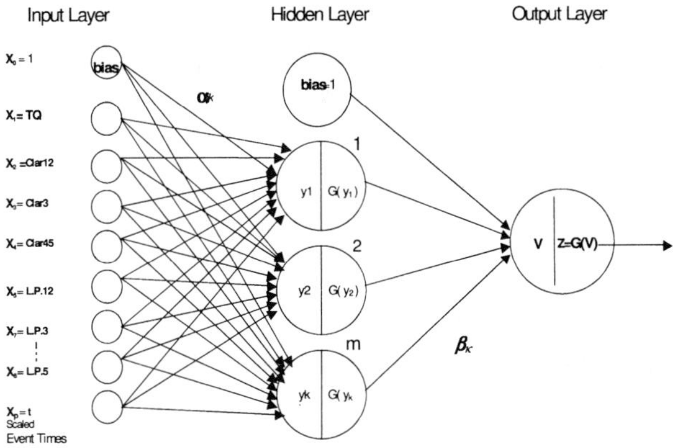

Fig. 1. The ANN Model

The log-likelihood for a data set of n exact or right censored survival times may be written as follows:

$$L = Log\ L = \sum_{i=1}^{n} \partial_i \ log\{f(x_i)\} + (1-\partial_i) log\{S(x_i)\} \qquad (6)$$

where ∂_i = Censoring Status (0 = Censored, 1 = Died). Upon differentiating Equation (5) with respect to time, the probability density function may be found as

$$-f(t) = \frac{dZ}{dt} = G'(V)\sum_{k=1}^{m} \beta_k G'(y_k)\ \alpha_{pk} \qquad (7)$$

where $G'(V) = G(V)(1-G(V))$. The log-likelihood is then maximised using the SOLVER® tool in Excel®. This procedure is stopped once convergence is achieved.

4.3 Monotonicity

In order to ensure monotonic survival functions it is necessary to constrain the α_{pk} and β_k weight parameters in Equation (7). It can be shown that a regularity condition of the ANN is

$$\alpha_{pk} \leq 0,\ \ \beta_k \geq 0\ \forall\ k \in \{1, \ldots, m\} \qquad (8)$$

This is a sufficient condition, and may over-constrain the parameters, but it is computationally convenient and easily incorporated in the SOLVER® tool.

4.4 Hazard Functions

As Equation (6) shows, a probability density estimate for each non-censored case is used to compute the log-likelihood. Furthermore, from Equation (1), estimates of the hazard function for each subset may also be obtained from the ANN by simply dividing the probability density function by the corresponding survivor function $S(t)$.

4.5 Regularisation and Generalisation

In order for the ANN to generalise well beyond the data set and thus avoid overfitting, the simplest and perhaps most effective form of regulariser, *weight decay*, was used to control model complexity. As over-fitted mappings with regions of large curvature tend to have large weights, penalizing the cost function by a multiple of the sum of squared weights, ω, encourages them to be small. ω was chosen by analysing the profile log-likelihood upon convergence of the ANN. A value of 0.01 was determined for the results in this paper. The significance of the weights was examined using the stepwise *saliency determination* procedure. The saliency of a weight is defined as the change in the cost function resulting from the deletion of that weight. The ANN was pruned using the optimal brain surgeon (OBS) [14] procedure, reducing the number of weight parameters in the model to 24. The OBS technique requires the Hessian, $\mathbf{H}$, and its inverse. Techniques used to compute both are described below.

4.6 Hessian and Variance-Covariance Matrices

Using the *Jacobian* of the log-likelihood function, the *outer product approximation* [4] (O.P) was used for generating a Hessian matrix, and is given by:

$$\mathbf{H}_N = \sum_{n=1}^{N} \mathbf{g}^n (\mathbf{g}^n)^T \qquad (9)$$

where $\mathbf{H}_N$ denotes the Hessian, n is the number of patterns in the data set and $\mathbf{g}$ is the gradient vector of the cost function (for ease of presentation we will not display the actual expression here). The inverse of the Hessian was computed using a computationally efficient procedure [4]

$$\mathbf{H}_{N+1}^{-1} = \mathbf{H}_N^{-1} - \frac{\mathbf{H}_N^{-1}\mathbf{g}^{N+1}(\mathbf{g}^{N+1})^T \mathbf{H}_N^{-1}}{1 + (\mathbf{g}^{N+1})^T \mathbf{H}_N^{-1} \mathbf{g}^{N+1}} \qquad (10)$$

which constructs the matrix one point at a time. The initial matrix $\mathbf{H}_0$ is $\alpha\mathbf{I}$, where α is a small quantity. Results are not sensitive to the precise value of α, and a value of 0.01 proved acceptable. Both matrices were computed using a specifically written Matlab® program.

Having computed the Hessian and its inverse, the large sample approximate variance-covariance matrix may be computed. The variance-covariance matrix is

$$\left(-E\left\{\frac{d^2 \log L(\boldsymbol{\beta})}{d\beta_j d\beta_k}\right\}\right)^{-1} = \left(E\left\{\frac{d \log L(\boldsymbol{\beta})}{d\beta_j d\beta_k}\right\}^2\right)^{-1} \qquad (11)$$

where E is the expectation operator and β_j is a parameter (not necessarily to be identified with the β_k weights of the network.). This expression is valid for censored data [16]. Using the law of large numbers, the sample average of the partial derivatives in Equation (11) converges to theexpectation of one term. Finally, the RHS of Equation (11) in its sample average form is the outer-product approximation of Equation (9).

5 Results

Figure 2 below shows a comparison of the ANN outputs with other survival modelling techniques discussed in section 1. The plot shows a representative example, corresponding to a patient residing in a deprived area, with Clarke level 3 (medium progressed stage) and L.P.4 (thick tumour depth). The ML estimate of the survival function fits the data very well, usually more closely than either a lognormal or a Cox model. Moreover, there is considerable doubt in this case as to the validity of the underlying assumptions of the Cox model [6].

Comparisons of similar plots and 10-year survival probability for other subsets confirmed the ANN's predictive accuracy.

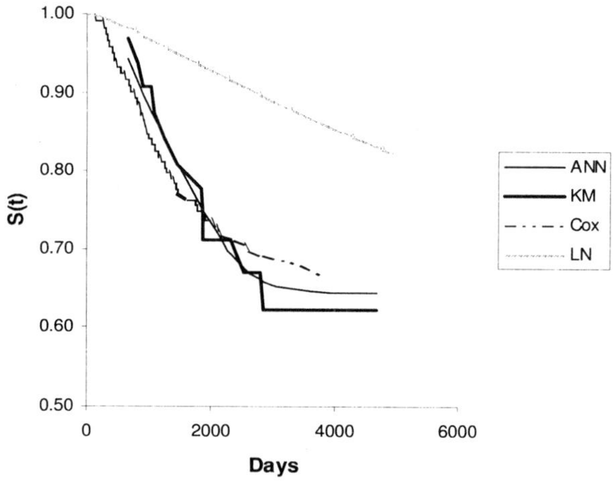

Fig. 2. The ANN Outputs With Traditional Models. LN denotes a log-normal model, and KM is the Kaplan-Meier estimate.

6 Discussion and Conclusion

A parsimonious non-linear ANN model for prediction of conditional survival probabilities, hazard and probability density functions, has been presented in this paper. The ANN is trained by maximising the log-likelihood function. Generalisation and model complexity have been addressed. Thus the model is expected to work well with a variety of different data sets. Survival predictions are at the very least as good as traditional modelling approaches.

Difficulty in interpreting the parameters forms the biggest disadvantage of the ANN in addition to the choice of regularisation parameter. However, our studies show that predictions are not too sensitive to the latter.

Using first derivatives of the log-likelihood, the sample approximation of the variance-covariance matrix may be calculated. Standard errors and intra-weight correlations may therefore be obtained, in addition to confidence intervals, detailed in [6]. The ANN therefore shares many desirable properties of statistical models and is a useful prognostic tool for the analysis of censored survival data. The model's implementation in an Excel® spreadsheet adds to its user-friendliness. As the only assumption involved is that the ANN is a mixture of logistic densities, the model provides huge potential for the analysis of survival data where a specified distribution is difficult to justify, or where conventional modelling assumptions are violated.

References

[1] Daponte, J. S., Sherman, P. (1991) Classification of ultrasonic image texture by statistical discriminant analysis of neural networks, Computerized Medical Imaging and Graphics 15: 3-9.

[2] Mann, N. H. I., and Brown, M. D. (1991) Artificial intelligence of low back pain, Orthopedic Clinics of North America 22: 303-314.

[3] Erler, B. S., Vitagliano, P., Lee, S. L. (1995) Superiority of neural networks over discriminant functions for thalasemia minor screening of red blood cell microcytosis", Archives of Pathology and Laboratory Medicine 119: 350-54.

[4] Bishop, C. M. (1995) Neural networks for pattern recognition, Oxford University Press.

[5] Cox, D.R. (1972), Regression models and life tables, Journal of the Royal Statistical Society, Series B, 34: 187-202.

[6] Joshi, R., PhD Thesis (2004), Coventry University.

[7] Liestol, K., Andersen, P. K., Andersen, U. (1994) Survival analysis and neural nets, Statistics in Medicine 13: 1189-1200.

[8] Faraggi, D. and Simon, R. (1995) A neural network model for survival data, Statistics in Medicine 14: 73-82.

[9] Ripley, R. M., Harris, A. L., Tarassenko, L. (2004) Non-linear survival analysis using neural networks, Statistics in Medicine 23: 825-842.

[10] Kaplan, E. L., and Meier, P. (1958) Nonparametric estimator from incomplete observations, Journal of the American Statistical Association 53: 457-481.

[11] Johnson, N. L., and Kotz, S. (1970) Distributions in Statistics: Continuous Univariate Distributions, 1, Boston.

[12] Johnson, N. L., Kotz, S., and Balakrishnan, N. (1995) Continuous Univariate Distributions, II, Wiley, Chichester.

[13] Rumelhart, D. E., and McClelland, J. L. (1986) (eds) Parallel distributed processing: explorations in the microstructure of cognition, Foundations, 1, MIT Press, Cambridge.

[14] Hassibi, B., Stork, D. G., and Wolff, G. J. (1993) Optimal brain surgeon and general network pruning, IEEE Int. Conf. Neural Networks, San Francisco: 293-299.

[15] Collett, D. (1994) Modelling Survival Data In Medical Research, Chapman and Hall.

[16] Escobar, L. A., and Meeker, W. Q. (1998) Fisher information matrices with censoring, truncation, and explanatory variables, Statistica Sinica 8: 221-237.

Boosting Kernel Discriminant Analysis with Adaptive Kernel Selection

Shinji Kita*, Satoshi Maekawa†, Seiichi Ozawa*, Shigeo Abe*
*Graduate School of Science and Technology, Kobe University, Japan
†National Institute of Information and Communications Technology, Japan
E-mail: {ozawasei,abe}@kobe-u.ac.jp, maekawa@nict.go.jp

Abstract

In this paper, we present a new method to enhance classification performance based on Boosting by introducing nonlinear discriminant analysis as feature selection. To reduce the dependency between hypotheses, each hypothesis is constructed in a different feature space formed by Kernel Discriminant Analysis (KDA). Then, these hypotheses are integrated based on AdaBoost. To conduct KDA in each Boosting iteration within realistic time, a new method of kernel selection is also proposed. Several experiments are carried out for the blood cell data and thyroid data to evaluate the proposed method. The result shows that it is almost the same as the best performance of Support Vector Machine without any time-consuming parameter search.

1 Introduction

Recently, kernel methods have been widely noticed as a powerful approach to solving difficult classification tasks. The Support Vector Machine (SVM) is a typical classifier based on the kernel method. The advantage of kernel methods originally comes from the nonlinear mapping to a high-dimensional feature space. If a proper kernel function is selected, the inputs in the same class can be completely isolated from the others and the class separability is maximized in the feature space. However, the features mapped to such a high-dimensional space often suffer from noise and outliers; hence dimensional reduction for the feature space such as Kernel Principal Component Analysis and Kernel Discriminant Analysis (KDA) have been often used [1][2].

KDA is a promising method of feature selection in which the class separability is maximized in a feature space. However, it is not easy to find a proper kernel function for a particular dataset. To find an optimal kernel, cross-validation has been often employed. However, an eigenvalue problem must be solved at every validation step in KDA; hence, the computation costs often become serious especially when a large training dataset is given. On the other hand, Boosting has been also widely known as a powerful method to realize a strong hypothesis by combining several weak hypotheses [3][4].

From the idea underlying in Boosting, we come upon a new idea for a practical KDA implementation without immense computations; that is, even if week hypotheses are constructed with low-performance features obtained by KDA using a small subset of training samples, we expect to construct a strong hypothesis by combining such week hypotheses based on the Boosting principle. In this framework, we apply KDA to a small number of training samples; hence, it is expected that the computation costs of kernel selection are greatly reduced. Unfortunately, however, if we adopt cross-validation as a kernel selection method, the computation costs are still high because we need to evaluate classification performance at every step in kernel selection and boosting. To overcome this problem, we can adopt another criterion in kernel selection: (between-class scatter)/(within-class scatter).

In this paper, we propose a novel boosting approach in which each weak learner is constructed based on a different feature space whose axes are obtained by KDA with a small subset of training samples. This approach provides a practical implementation for the combination of Boosting and KDA, which can lead to reducing classification dependency between constituent hypotheses

In the next section, we describe our strategy and the Boosting KDA algorithm. Then Section 3 shows some experimental results for two standard datasets.

2 Boosting Kernel Discriminant Analysis

Here, we briefly explain KDA and AdaBoost.M2 [3], and then we propose a novel method to select an appropriate parameter in KDA and a new Boosting strategy to increase the diversity of generated hypotheses. Finally we show the whole learning algorithm of the proposed method.

2.1 KDA and AdaBoost.M2

KDA [2], which is a nonlinear extension of LDA, is well known to give a subspace where the class separability is maximized in a high-dimensional feature space. In KDA, input data are projected into a high-dimensional feature space at first, then they are projected on a low-dimensional subspace called KDA subspace.

Suppose that a training set is given as $\{(\boldsymbol{x}_{ij}, y_{ij})_{j=1}^{C_i}\}_{i=1}^{C}$ where $\boldsymbol{x}_{ij}$ is an I dimensional column vector and y_{ij} is the class label of $\boldsymbol{x}_{ij}$. C is the number of classes and each class has C_i $(i = 1, \ldots, C)$ samples. Let $N = \sum_i C_i$ be the total number of training samples. We assume that observations have zero-mean in the feature space and this is achievable by adjusting 'Kernel Matrix' (see [5] for details). The inputs are mapped into a high dimensional feature space through a nonlinear mapping function $\phi : \boldsymbol{R}^I \rightarrow F$, where F is the feature space. A between-class scatter matrix B and a within-class scatter matrix W in F are given as follows:

$$B = \frac{1}{N} \sum_{i=1}^{C} C_i \boldsymbol{m}_i \boldsymbol{m}_i^t \qquad (1)$$

$$W = \frac{1}{N} \sum_{i=1}^{C} \sum_{j=1}^{C_i} (\phi(\boldsymbol{x}_{ij}) - \boldsymbol{m}_i)(\phi(\boldsymbol{x}_{ij}) - \boldsymbol{m}_i)^t \qquad (2)$$

where $\boldsymbol{m}_i = \sum_{j=1}^{C_i} \phi(\boldsymbol{x}_{ij})$ is the center of the class i in the feature space. Basis vectors v spanning a KDA subspace are obtained by solving the following eigenvalue problem: $\lambda W v = B v$.

To calculate B and W in the high dimensional space, we use a kernel function $K(\boldsymbol{x}, \boldsymbol{x}') = \phi(\boldsymbol{x})^t \phi(\boldsymbol{x}')$. This enable us to calculate B and W without treating $\phi(\boldsymbol{x})$. This method is well known as 'Kernel Trick' [1]. As one of various kernel functions, the Gaussian kernel

$$K(\boldsymbol{x}, \boldsymbol{x}') = \exp\left(-\frac{\|\boldsymbol{x} - \boldsymbol{x}'\|^2}{\sigma^2}\right) \qquad (3)$$

is often used, and we also use this Gaussian kernel here.

On the other hand, AdaBoost is a major boosting algorithm developed by Freund [3], which boosts the performance by the ensemble of weak learners whose performances are slightly better than random guessing. AdaBoost.M2 is one of the highly sophisticated multiclass extensions of AdaBoost, which has been proposed for two-class problems (see [3] for details). We use AdaBoost.M2 to integrate hypotheses.

2.2 Proposed Kernel Selection

In KDA, instead of finding an optimal parameter σ in Eq. (3) by cross-validation, we present a method to select σ such that the following criteria S is maximized:

$$S = \frac{\mathrm{trace}(B)}{\mathrm{trace}(W)}. \qquad (4)$$

This criterion is the same as used in the conventional LDA. Applying this kernel selection to several datasets,

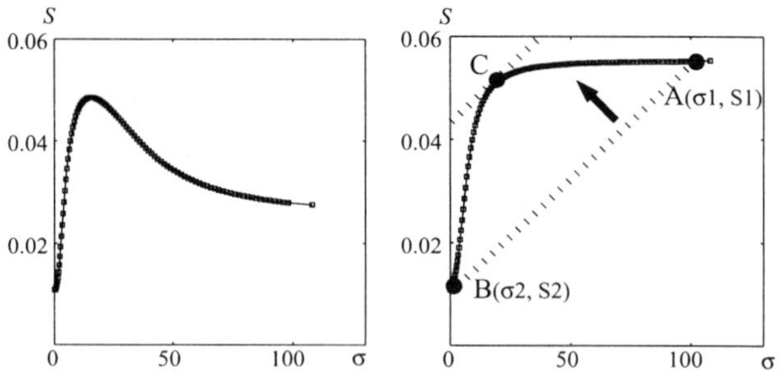

(a) Type 1 (b) Type 2

Fig. 1. σ-S curves for Gaussian kernel

we have observed two types of σ-S curves shown in Figs. 1(a),(b). As can be seen from Fig. 1(a), Type 1 has a single peak. In this case, we should select σ at this peak unless it is not too small. If σ is too small, some test samples might be projected to the complementary space of the feature space which is spanned by training data. Consequently, the test performance will get worse even though the training data are correctly classified. To avoid such an inappropriate situation, we should select σ with the largest S under the condition that σ is larger than a threshold value (here we set it to 0.01).

On the other hand, Type 2 in Fig. 1(b) does not have any peak and S monotonically increases to a certain value. Since the parameter σ that maximize S becomes so large, the Gaussian kernel has a similar value in any case; then the kernel matrix can be easily degenerated. Hence, it is preferable to select σ as small as possible under the condition that S is not too small compared with the maximum value. An ad hoc solution for this is to find σ ('C' in Fig. 1(b)) to maximize the following criteria H:

$$H = S - \frac{S_1 - S_2}{\sigma_1 - \sigma_2}\sigma \qquad (5)$$

where (σ_1, S_1) and (σ_2, S_2) correspond to the points with minimum and maximum σ, respectively ('A' and 'B' in Fig. 1(b)).

2.3 A Strategy to Generate Diverse Hypotheses

To obtain various KDA subspaces, a small subset of training samples is extracted from the whole training set, then a KDA subspace is constructed from it. Intuitively, this KDA subspace is regarded as a 'weak feature space' because it is constructed based on limited information on training samples. In such a weak feature space, the corresponding hypothesis must also have week performance. Therefore, it is expected that a strong hypothesis is constructed by combining these week hypotheses based on AdaBoost.

For each training subset, r% of training samples are

Input: Training set $\{(\boldsymbol{x}_{ij}, y_{ij})_{j=1}^{C_i}\}_{i=1}^{C}$ where $\boldsymbol{x}_{ij} \in \boldsymbol{R}^I$ and the class label $y_{ij} \in Y = \{1, \ldots, C\}$; the distribution probability of samples $D(i,j) = 1/N$ for $i = 1, \ldots, C,\ j = 1, \ldots, C_i$; the number of iterations T; kernel function K and the percentage r of selected data.

Initialize weight vectors: $w^1_{ij,y} = D(i,j)/(C-1)$ for $i = 1, \ldots, N,\ j = 1, \ldots, C_i,\ y \in Y - \{y_{ij}\}$

Do for $t = 1, 2, \ldots, T$

1. Set $W^t_{ij} = \sum_{y \neq y_{ij}} w^t_{ij,y}$; $q_t(i,j,y) = \frac{w^t_{ij,y}}{W^t_{ij}}$ for $y \in Y - \{y_{ij}\}$; and set $D_t(i,j) = \frac{W^t_{ij}}{\sum_{i=1}^{C} \sum_{j=1}^{C_i} W^t_{ij}}$
2. Choose r% of the training data randomly, then denote it as R_t.
3. Obtain a kernel parameter σ_t by applying the proposed kernel selection method to R_t.
4. Construct KDA subspace with (K, σ_t, R_t), and project all training data into this subspace, then denote a set of the feature vectors as F_t.
5. With (D_t, q_t, F_t), build the t-th hypothesis $h_t : \boldsymbol{R}^I \times Y \rightarrow [0, 1]$
6. Calculate the pseudo-loss of h_t :
$$\epsilon_t = \frac{1}{2} \sum_{i=1}^{C} \sum_{j=1}^{C_i} D_t(i,j)\Big(1 - h_t(\boldsymbol{x}_{ij}, y_{ij}) + \sum_{y \neq y_{ij}} q_t(i,j,y) h_t(\boldsymbol{x}_{ij}, y)\Big)$$
7. Set $\beta_t = \epsilon_t/(1 - \epsilon_t)$
8. Set a new weight vector to be $w^{t+1}_{ij,y} = w^t_{ij,y} \beta_t^{(1/2)(1 + h_t(\boldsymbol{x}_{ij}, y_{ij}) - h_t(\boldsymbol{x}_{ij}, y))}$ for $i = 1, \ldots, C,\ j = 1, \ldots, C_i,\ y \in Y - \{y_{ij}\}$

Output the final hypothesis

$$h_f(\boldsymbol{x}) = \arg\max_{y \in Y} \sum_{t=1}^{T} \left(\log \frac{1}{\beta_t}\right) h_t(\boldsymbol{x}, y)$$

Fig. 2. The Proposed Boosting KDA Algorithm

randomly chosen. Here, r is set to a small value to reduce the dependency between hypotheses generated in Boosting steps. If r is small, the total combinations of these subsets increase, then it will result in increasing the diversity of hypotheses that leads to high generalization ability [6].

In Fig. 2, we summarize the proposed Boosting KDA algorithm. Step **Initialize**, Steps 1, 5, 6, 7, 8, and Step **Output** correspond to the procedures in AdaBoost.M2. The percentage r of selecting training samples in Step 2 is set to 1, 3, 5, 10, 15 to study the effect of increasing the diversity of hypotheses. The kernel selection in Step 3 is carried out based on the procedure in 2.2 and KDA algorithm stated in 2.1 is conducted in Step 4.

3 Experiments

The proposed Boosting KDA is compared with AdaBoost [7] and SVM in terms of the recognition rate and training speed. In addition, the progress of recognition rates for different r are investigated to study the properties of Boosting KDA. Here, a blood cell dataset [1] and a thyroid dataset [8] are used for the evaluation. The blood cell dataset contains 3097 training data and 3100 test data. The number of classes and attributes are 12 and 13, respectively. The thyroid dataset contains 3772 training data and 3428 test data. The number of classes and attributes are 3 and 21, respectively.

We adopt RBF networks to create hypothesis, and the number of hidden units is set to twice the number of classes. Training of RBF is carried out based on the conjugate gradient descent, which optimizes the centers and widths of radial-basis functions, and modifies the weights to reduce training errors. The maximum iterations in the optimization is set to ten to avoid over-fitting.

Table 1 shows the test performance of the proposed method, AdaBoost, and SVM [1]. As seen from Table 1, the performance of SVM changes depending on the type of kernel and the implementation of multiclass problems. However, the performance of the proposed method is almost the same as the best performance of SVM. Figures 3(a),(b) show the progress of test performance in Boosting KDA as the boosting steps increase (r=1,3,5,10,15). As seen from Figs. 3(a),(b), the performance for larger r

Table 1. The comparison of test performance. For SVM, the best results are picked up from [1].

(a) Blood Cell Data

Algorithm	Kernel	Multiclass Ext.	rate(%)
Boosting KDA	Gauss	AdaBoost.M2	93.74
Boosting Only	—	AdaBoost.M2	91.42
L1/L2 SVM	Gauss	one-against-all	92.77
	Poly	one-against-all	93.58
DDAG SVM	Gauss	pairwise	92.41
	Poly	pairwise	93.00
ECOC SVM	Gauss	(63,10,27)	94.05
	Poly	one-against-all	92.84

(b) Thyroid Data

Algorithm	Kernel	Multiclass Ext.	rate(%)
Boosting KDA	Gauss	AdaBoost.M2	97.90
Boosting Only	—	AdaBoost.M2	96.03
L1/L2 SVM	Gauss	pairwise	97.29
	Poly	pairwise	97.72
DDAG SVM	Gauss	pairwise	97.40
	Poly	pairwise	97.86

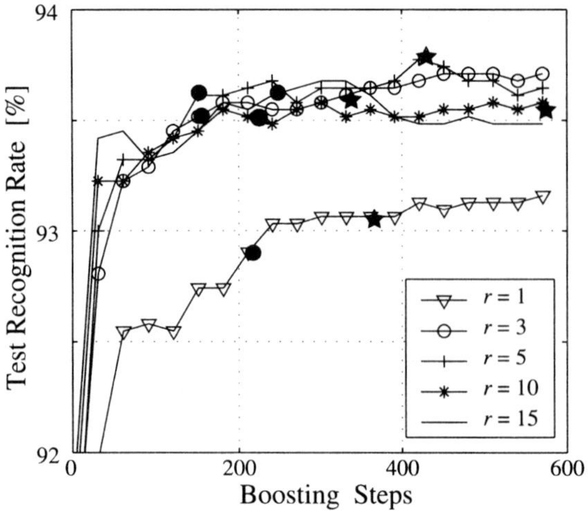

(a) Blood Cell Data

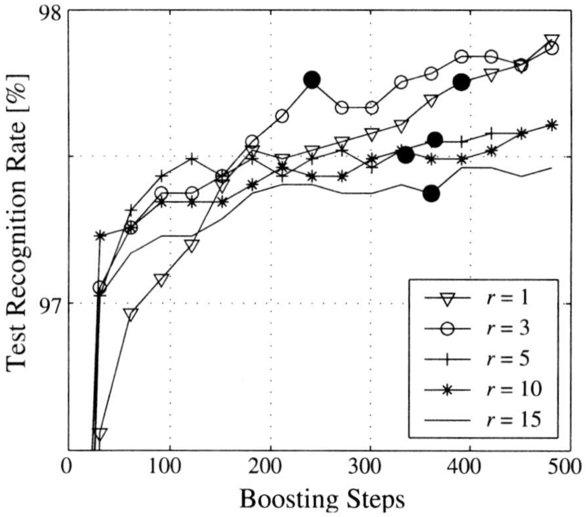

(b) Thyroid Data

Fig. 3. Test performance of Boosting KDA with r=1,3,5,10,15 for (a) blood cell data and (b) thyroid data. • and ⋆ mean the points where the training performance attains to 99% and 100%, respectively.

is higher than that for smaller r at early boosting steps, while this tendency reverses at late boosting steps. It should be noted that the performance for blood cell data is distinctively degraded when $r = 1$ even if there are no large differences in other cases. Considering that large r leads to slow convergence in training (see the result in Table 2), r should be selected as small as possible within an acceptable performance level.

When it comes to over-fitting, the test performance does not degrade even when the training performance attains to 100% except for the case of $r = 15$ in blood cell data.

Table 2. Average training time (sec.) per iteration in the proposed method (CPU: Intel Pentium-IV, 1.8GHz).

r [%]	1	3	5	10	15
blood cell	10.2	11.6	15.2	54.0	151.8
thyroid	3.5	6.5	16.7	88.9	232.6

4 Conclusions

We have developed a novel method to introduce Kernel Discriminant Analysis into Boosting and a method to choose an appropriate kernel parameter in KDA. The proposed method achieved fairly good performance without any time-consuming parameter tuning. But it takes a little longer time in training than SVM to achieve good recognition performance. To overcome this, a more effective method to choose subsets leading to fast convergence should be developed.

Acknowledgment

This research was partially supported by the Ministry of Education, Science, Sports and Culture, Grant-in-Aid for Scientific Research (B) and (C), and by the Okawa Foundation for Information and Telecommunications.

References

[1] Abe, S. (2005) Support vector machines for pattern classification. Springer

[2] Baudat, G., Anouar, F. (2000) Generalized discriminant analysis using a kernel approach. Neural Computation, Vol. 12: 2385-2404

[3] Freund, Y., Schapire, R. E. (1997) A decision-theoretic generalization of on-line learning and an application to boosting. Journal of Computer and System Sciences, Vol. 55, No. 1: 119-139

[4] Juwei, L., Plataniotis, K. N., Venetsanopoulos, A. N. (2003) Boosting linear discriminant analysis for face recognition. IEEE Int. Conf. on Image Processing: 14-17

[5] Schölkopf, B., Smola, A., Müller, K. -R. (1996) Nonlinear component analysis as a kernel eigenvalue problem. MPI Technical Report, No. 44

[6] Murua, A. (2002) Upper bounds for error rates of linear combinations of classifiers. IEEE Trans. on Pattern Analysis and Machine Intelligence, Vol. 24, No. 5: 591-602

[7] Rätsch, G., Onoda, T., Müller, K. -R. (2001) Soft margins for AdaBoost. Machine Learning, Vol. 42, No. 3: 287-320

[8] ftp://ftp.ics.uci.edu/pub/machine-learning-databases/

Product Kernel Regularization Networks

Kudová Petra, Šámalová Terezie
Institute of Computer Science, Academy of Sciences of the Czech Republic
Pod vodárenskou věží 2, P.0. Box 5, 182 07 Prague 8, Czech Republic
E-mail: {petra, terka}@cs.cas.cz

Abstract

We study approximation problems formulated as regularized minimization problems with kernel-based stabilizers. These approximation schemas exhibit easy derivation of solution to the problem in the shape of linear combination of kernel functions (one-hidden layer feed-forward neural network schemas). We prove uniqueness and existence of solution to the problem. We exploit the article by N. Aronszajn [1] on reproducing kernels and use his formulation of product of kernels and resulting kernel space to derive a new approximation schema – a Product Kernel Regularization Network. We present a concrete application of PKRN and compare it to classical Regularization Network and show that PKRN exhibit better approximation properties.

1 Reproducing Kernel Hilbert Spaces

Reproducing Kernel Hilbert Space (shortly RKHS) was defined by Aronszajn, 1950 ([1]) as Hilbert space $\mathcal{H}$ of functions (real or complex) defined over $\Omega \subset \mathbb{R}^d$ with the property, that for each $x \in \Omega$ the evaluation functional on $\mathcal{H}$ given by $\mathcal{F}_x : f \mapsto f(x)$ is bounded. This implies existence of a positive definite symmetric function $k : \Omega \times \Omega \rightarrow \mathbb{R}$ (so called *reproducing kernel*) corresponding to $\mathcal{H}$ such that

1. for any $f \in \mathcal{H}$ and $y \in \Omega$ the following reproducing property holds $f(y) = \langle f(x), k(x, y)\rangle$, where $\langle ., . \rangle$ is scalar product in $\mathcal{H}$ and

2. for every $y \in \Omega$, the function $k_y(x) = k(x, y)$ is an element of $\mathcal{H}$.

Note that the reproducing kernel for $\mathcal{H}$ is unique. On the other hand, every positive definite symmetric function is a reproducing kernel for exactly one Hilbert space, that can be described as $\mathrm{comp}\{\sum_{i=1}^{n} a_i k_{x_i}; x_i \in \Omega, a_i \in \mathbb{R}\}$, where comp means completion of the set.

Next we will consider product of Reproducing Kernel Hilbert Spaces. For $i = 1, 2$ let F_i be a RKHS of functions on Ω_i, let K_i be the corresponding kernel. Consider the following set of functions on $\Omega = \Omega_1 \times \Omega_2$ $F' = \{\sum_{i=1}^{n} f_{1,i}(x_1) f_{2,i}(x_2) \mid n \in \mathbb{N}, f_1 \in F_1, f_2 \in F_2\}$. Clearly, F' is a vector space, it is not complete though. For its completion, we first define a scalar product on F'. Let f, g be elements of F' expressed as $f(x_1, x_2) = \sum_{i=1}^{n} f_{1,i}(x_1) f_{2,i}(x_2)$, $g(x_1, x_2) = \sum_{j=1}^{m} g_{1,j}(x_1) g_{2,j}(x_2)$. We define $\langle f, g\rangle = \sum_{i=1}^{n} \sum_{j=1}^{m} \langle f_{1,i}, g_{1,j}\rangle_1 \langle f_{2,i}, g_{2,j}\rangle_2$, where $\langle \cdot, \cdot \rangle_i$ denotes the scalar product in F_i. It is a routine to check that this definition does not depend on the particular form in which f and g are expressed and that the properties of scalar product are satisfied. We define norm on F' by $\|f\| = \sqrt{\langle f, f\rangle}$. Finally, let F be the completion of F'. It can be shown ([1]) that the completion exists not only as an abstract Hilbert space but that F is in fact a space of functions on Ω. We call F the product of F_1 and F_2 and write $F = F_1 \otimes F_2$.

Theorem 1.1 *([1]) For $i = 1, 2$ let F_i be an RKHS on Ω_i with kernel K_i. Then the product $F = F_1 \otimes F_2$ on $\Omega_1 \times \Omega_2$ is an RKHS with kernel given by*

$$K((x_1, x_2), (y_1, y_2)) = K_1(x_1, y_1) K_2(x_2, y_2), \quad (1)$$

where $x_1, y_1 \in \Omega_1$, $x_2, y_2 \in \Omega_2$.

for proofs of the sketched properties we ask the reader to refer to [1] or to [12].

2 Learning from data as minimization of functionals

The task to find an optimal solution to the setting of approximating a data set $z = \{(u_i, v_i)\}_{i=1}^{N} \subseteq \mathbb{R}^d \times \mathbb{R}$ by a function from a general function space X (minimizing error) is ill-posed. Thus we impose additional (regularization) conditions on the solution ([5]). These are typically things like a-priori knowledge, or some smoothness constraints. The solution f_0 has to minimize a functional $\mathcal{F} : \Omega \rightarrow \mathbb{R}$ that is composed of the error part and the "smoothness" part: $\mathcal{F}(f) = \mathcal{E}_z(f) + \gamma\Phi(f)$, where $\mathcal{E}_z$ is the error functional depending on the data $z = \{(u_i, v_i)\}_{i=1}^{N} \subseteq \mathbb{R}^d \times \mathbb{R}$ and penalizing remoteness from the data, Φ is the regularization part — the so called stabilizer — penalizing "remoteness from the global property" and γ is the regularization parameter giving the trade-off between the two terms of the functional to be minimized.

To prove existence and uniqueness of solution to such a problem we will use some results from mathematical analysis. Error part of our functional doesn't exhibit sufficiently nice properties, so the regularization part has to do the job. We employ RKHS in such a way that we nicely and easily obtain existence, uniqueness and even form of the solution.

Let $\mathcal{H}$ be an RKHS over $\Omega \subseteq \mathbb{R}^d$ with kernel k and norm $\|.\|_k$. We construct the minimization functional composing of error part $\mathcal{E}_z(f)$ based on data $z = \{(u_i, v_i); i = 1, \ldots, N\} \subseteq \mathbb{R}^d \times \mathbb{R}$ and let the regularization part be $\Phi(f) = \|f\|_k^2$ forming

$$\mathcal{F}(f) = \mathcal{E}_z(f) + \gamma \|f\|_k^2, \tag{2}$$

where $\gamma \in \mathbb{R}^+$. (See section 2.1 for a more detailed construction.)

Now uniqueness of solution to such a problem comes clearly from strong quasiconvexity of the functional $\mathcal{F}$ composing of convex error part and strongly quasiconvex kernel part. To show existence of solution we need weak sequential lower semicontinuity of the functional which can be shown by computing second derivatives of the functional, for precise derivation see [11].

Derivation of the shape of the solution to the regularized minimization problem has been shown already in [5] but without taking advantage of RKHS, in [4], [8] and others known as Representer theorem, for the kernel case see [11]. All the proofs are based on the idea that minimum of a function can exist in an interior point only if first derivative equals zero.

Employing this theorem we obtain solution to the kernel-based minimization problem in the form of

$$f_0(x) = \sum_{i=1}^{N} c_i k(x, u_i), \tag{3}$$

where u_i are the data points and $k(\cdot, \cdot)$ the corresponding kernel.

2.1 Concrete minimization functionals and RKHS

An error functional is usually of the form $\mathcal{E}_z(f) = \sum_{i=1}^{N} V(f(u_i), v_i)$. A typical example of the empirical error functional is the classical mean square error: $\mathcal{E}_z(f) = \frac{1}{N}\sum_{i=1}^{N}(f(u_i) - v_i)^2$.

In [5] a special stabilizer based on the Fourier Transform was proposed: $\Phi_G(f) = \int_{\mathbb{R}^d} \frac{|\hat{f}(s)|^2}{\hat{G}(s)} dm_d(s)$, where $\hat{G} : \mathbb{R}^d \to \mathbb{R}_+$ is symmetric ($\hat{G}(s) = \hat{G}(-s)$) function tending to zero as $\|s\| \to \infty$ (the last holds for any $G \in \mathcal{L}_1$). That means $1/\hat{G}$ is a low-pass filter.

Thus the functional $\mathcal{F}_G$ to be minimized is of the form: $\mathcal{F}_G(f) = \mathcal{E}_z(f) + \gamma\Phi_G(f) = \frac{1}{N}\sum_{i=1}^{N}(f(u_i) - v_i)^2 + \gamma \int_{\mathbb{R}^d} \frac{|\hat{f}(s)|^2}{\hat{G}(s)} dm_d(s)$, where $\gamma \in \mathbb{R}^+$. Now we show how to build an RKHS corresponding to the regularization part of our functional:

Let us define $g(x, y) = G(x - y) = \int_{\mathbb{R}^d} \hat{G}(t)e^{it.x}e^{-it.y} dm_d(t)$. For $g \in \mathcal{S}(\mathbb{R}^{2d})$ symmetric positive definite we obtain an RKHS $\mathcal{H}$ (using the classical construction, see [4], [10],[13]). We put $\langle f, g\rangle_{\mathcal{H}} = \int_{\mathbb{R}^d} \frac{\hat{f}(s)\hat{g}^*(s)}{\hat{G}(s)} dm_n(s)$ and obtain the norm $\|f\|_{\mathcal{H}}^2 = \int_{\mathbb{R}^d} \frac{|\hat{f}(s)|^2}{\hat{G}(s)} dm_n(s)$, for $\mathcal{H} = \text{comp}\,\text{span}\{G^{\dagger}(x, .), x \in \mathbb{R}^d\}$, where $\text{comp}\{\ldots\}$ denotes completion of the set $\{\ldots\}$ and a^* means complex conjugate of a. It is easy to check the reproducing property of G on $\mathcal{H}$, that is $\langle f(x), G(x - y)\rangle_{\mathcal{H}} = f(y)$.

Special types of reproducing kernels and following RKHS are the well known Gaussian kernel $k_1(x, y) = e^{-\|x-y\|^2}$ with Fourier transform $\hat{k_1}(s) = e^{-\frac{\|s\|^2}{2}}$ or in one dimension kernel $k_2(x, y) = e^{-|x-y|}$ with Fourier transform $\hat{k_2}(s) = (1 + s^2)^{-1}$. The norm for this RKHS is of the form $\|f\|_k^2 = \int \frac{|\hat{f}|^2}{(1+s^2)^{-1}} = \|f\|_{\mathcal{L}_2}^2 + \|f'\|_{\mathcal{L}_2}^2$. So we see we obtain a Sobolev space W_2^1.

As a more general example we will consider the product of kernels introduced in section 1. Suppose that apriori knowledge of our data suggests to look for the solution as a member of product of two functional spaces. In one dimension the data may be clustered thus being suitable for approximation via Gaussian kernels. In the other dimension we have only information on smoothness of the data, hence we will use kernel resulting in Sobolev norm. Employing theorem 1.1 we obtain a kernel for the product space of the form: $K((x_1, x_2), (y_1, y_2)) = k_1(x_1, y_1) \cdot k_2(x_2 y_2) = e^{-\|x_1 - y_1\|^2} \cdot e^{-|x_2 - y_2|}$, where $x_1, y_1 \in \Omega_1, x_2, y_2 \in \Omega_2$.

Regularized minimization schema in this case is of the form:

$$\mathcal{F}_K(f) = \frac{1}{N}\sum_{i=1}^{N}(f(u_i) - v_i)^2 + \gamma \int_{\mathbb{R}^d} \frac{|\hat{f}(s)|^2}{\widehat{k_1 k_2}(s)} dm_n(s). \tag{4}$$

Taking advantage of this being an RKHS we have the form of the solution to such a type of minimization:

$$f_0(x_1, x_2) = \sum_{i=1}^{N} c_i e^{-\|x_1 - u_{i,1}\|^2} \cdot e^{-|x_2 - u_{i,2}|}. \tag{5}$$

Approximation schemas of this type exhibit so far nicer approximation properties since it can be better fitted to special types of data.

3 Learning algorithm

Now we present a learning algorithm based on the theoretical results from the previous sections. We as-

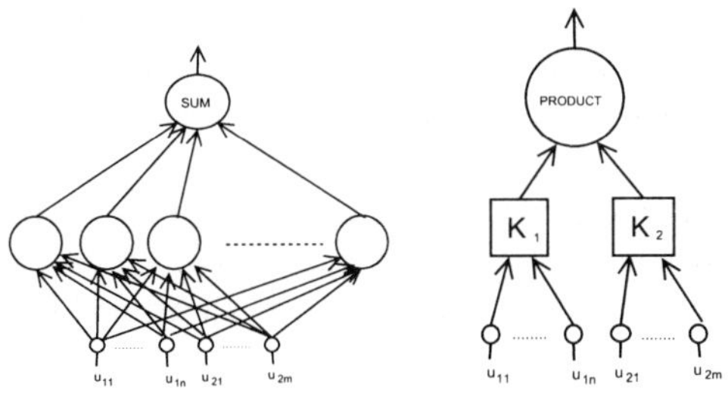

Fig. 1. a) Product Kernel Regularization Network b) Product Unit

sume that we have a data set $\{\vec{u_1}^i, \vec{u_2}^i, v^i\}_{i=1}^N$, where $\vec{u_1}^i \in \mathbb{R}^n$, $\vec{u_2}^i \in \mathbb{R}^m$, $v^i \in \mathbb{R}$ and N is a number of data samples. We will fit these data set using *Product Kernel Regularization Network* (PKRN) derived from the regularization schema 4.

It is a feed-forward neural network with one hidden layer of N *product units* and a linear output layer (see Fig. 1a). By a *product unit* (see Fig. 1b) we mean a unit with $(n+m)$ real inputs and one real output. It consists of two positive definite kernel functions $K_1(\vec{c_1}, \cdot)$, $K_2(\vec{c_2}, \cdot)$, one evaluating the first n inputs and one evaluating the other m inputs, the output of the product unit is computed as the product $K_1(\vec{c_1}, \vec{u_1}) \cdot K_2(\vec{c_2}, \vec{u_2})$.

The network then evaluates the function

$$f(\vec{u_1}, \vec{u_2}) = \sum_{i=1}^{k} w_i K_1(\vec{c_1}^i, \vec{u_1}) \cdot K_2(\vec{c_2}^i, \vec{u_2}), \quad (6)$$

where the parameters $\vec{c_1}^i$ and $\vec{c_2}^i$ are called *centers* and the coefficients of the linear combination w_i *weights*.

The learning algorithm for PKRN is sketched at Fig. 2. It is derived from Tikhonov regularization and for the case of Regularization Network was described in [8]. See also [6].

The algorithm is quite simple, setting the centers of kernels to the data points given by the training set and evaluating the values of output weights by solving linear system of equations. Parameter γ must be estimated in advance (cross-validation is usually used).

4 Experiments

We tested the performance of proposed Product Kernel Regularization Network on several experiments, including both benchmark and real life problems.

We always use two disjunct data sets, one for training and one for evaluating the error of the result, and compute the normalized error:

Input: Data set $\{\vec{u_1}^i, \vec{u_2}^i, v^i\}_{i=1}^k \subseteq \mathbb{R}^n \times \mathbb{R}^m \times \mathbb{R}$
Output: Product Kernel Regularization network.

1. Set the centers of kernels:
$$\forall i \in \{1, \ldots, k\}: \quad \vec{c_1}^i \leftarrow \vec{u_1}^i \qquad \vec{c_2}^i \leftarrow \vec{u_2}^i$$
2. Compute the values of weigths $w_1, \ldots, w_k$:
$$(k\gamma I + K)\vec{w} = \vec{v},$$
where I is the identity matrix,
$K_{i,j} = K_1(\vec{c_1}^i, \vec{u_1}^j) \cdot K_2(\vec{c_2}^i, \vec{u_2}^j)$,
and $\vec{v} = (v_1, \ldots, v_k)$, $\gamma > 0$.

Fig. 2. Learning algorithm for Product Kernel Regularization Network.

Table 1. Error values for PKRN and RN on Proben1 data sets.

	PKRN		**RN**	
	E_{train}	E_{test}	E_{train}	E_{test}
cancer1	2.739	**1.816**	2.658	1.875
cancer2	2.152	3.516	2.279	**3.199**
cancer3	2.374	**2.798**	2.348	2.873
glass1	6.141	8.590	4.899	**8.033**
glass2	5.269	**8.202**	4.570	8.317
glass3	3.691	**7.411**	4.837	7.691

$$E = 100 \frac{1}{N} \sum_{i=1}^{N} ||v^i - f(\vec{u_1}^i, \vec{u_2}^i)||^2, \quad (7)$$

where N is number of examples and f is the network output.

We have used the Gaussian function $e^{-\left(\frac{\|c-x\|}{b}\right)^2}$ for both kernel functions (K_1 and K_2), but the kernels differ in the width b of the Gaussian functions. All parameters γ, b_1 and b_2 were estimated by cross-validation. LAPACK library [7] was used for linear system solving.

The table 1 compares the resulting errors of PKRN and Regularization Networks (RN) on data sets selected from Proben1 [9] benchmark repository.

The applicability of PKRN on real life problems is demonstrated on the prediction of the flow rate on the Czech river Ploucnice. Our goal is to predict the current flow rate from the flow rate and total rainfall from the previous date, i.e. we are approximating function $f: \mathbb{R} \times \mathbb{R} \rightarrow \mathbb{R}$.

We have three different data sets for this task – called pl1, pl1s and pl2, each containing 1000 training samples and 367 testing samples. The results obtained on these data sets by PKRN are listed in the table 2.

Table 2. Error values of PKRN on training and testing sets of the Ploucnice data.

	pl1	pl1s	pl2
E_{train}	0.057180	0.057215	0.109477
E_{test}	0.048332	0.048475	0.097608

The table 3 shows that the PKRN overperformes the so called *conservative predictor*. Conservative prediction is a predictor saying that the value will be the same as it was yesterday, and in spite of its simplicity it is very successful on some tasks, including this one.

The prediction on the testing set made by PKRN is displayed at Fig. 3.

Table 3. Comparison of errors obtained by PKRN and conservative predictor (CP).

	PKRN	**CP**
E_{train}	0.057	0.093
E_{test}	0.048	0.054

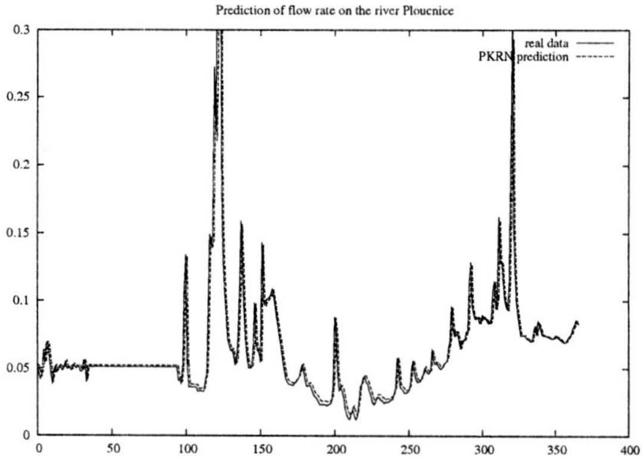

Fig. 3. Prediction of the flow rate on the river Ploucnice by Product Kernel Regularization Network.

5 Conclusion

We have shown how to employ RKHS in approximation theory and stressed advantages of this approach. Inspired by the article [1] we introduce kernel-product based approximation and derive the shape of Product Kernel Regularization Networks (PKRN).

We tested the performance of proposed PKRN on benchmark tasks from Proben1 repository and showed that its result are comparable to standard variant of Regularization Network. We demonstrated the applicability of PKRN on prediction of river flow rate, which is a real-life task.

We expect our algorithm to be useful particularly in situations where some prior knowledge of the character of data is available in the sense that we can expect that for some groups of inputs different kernel functions are suitable.

References

[1] Aronszajn N. (1950) *Theory of Reproducing Kernels*. Transactions of the AMS, **68**, 3, pp. 337-404.

[2] Cucker F., Smale S. (2001). *On the Mathematical Foundations of Learning*. Bulletin of the American Mathematical Society **39**, 1-49.

[3] Daniel J. W. (1971). *The Approximate Minimization of Functionals*. Prentice-Hall, Inc.

[4] Girossi F. (1998). *An Equivalence between Sparse Approximation and Support Vector Machines*. Neural Computation **10**, 1455-1480, MIT. (A.I. Memo No. 1606, MIT, 1997)

[5] Girossi F., Jones M., Poggio T. (1995). *Regularization Theory and Neural Networks Architectures*. Neural Computation, **7**, 219-269.

[6] Kudová P. (2004). *Kernel Based Regularization Networks and RBF Networks*. Doktorandský Den 2004, Paseky nad Jizerou.

[7] Lapack library. `http://www.netlib.org/lapack/`.

[8] Poggio T., Smale S. (2003). *The Mathematics of Learning: Delaing with Data*. Notices of the AMS **50**, 5, 536-544.

[9] Prechelt, L. (1994) *PROBEN1 – A Set of Benchmarks and Benchmarking Rules for Neural Network Training Algorithms*. Universitaet Karlsruhe, 21/94.

[10] Schölkopf B., Smola A. J. (2002). *Learning with Kernels*. MIT Press, Cambridge, Massachusetts.

[11] Šidlofová T. (2004). *Existence and Uniqueness of Minimization Problems with Fourier Based Stabilizers*. Compstat 2004, Prague.

[12] Šidlofová T. (2004). *Kernel Based Regularization and Neural Networks*. Doktorandský Den 2004, Paseky nad Jizerou.

[13] Wahba G. (1990). *Spline Models for Observational Data*. Series in Applied Mathematics, Vol. 59, SIAM, Philadelphia.

This work was supported by GA ČR grant 201/05/0557 and the National Research Program Information Society project T100300414.

Statistical Correlations and Machine Learning for Steganalysis

Qingzhong Liu[1], Andrew H. Sung[1], Bernardete M. Ribeiro[2]

[1]Department of Computer Science, New Mexico Tech, U.S.A
[2]Department of Informatics Engineering, University of Coimbra, Portugal
e-mail: [1]{liu, sung}@cs.nmt.edu, [2]bribeiro@dei.uc.pt

Abstract

In this paper, we present a scheme for steganalysis based on statistical correlations and machine learning. In general, digital images are highly correlated in the spatial domain and the wavelet domain; hiding data in images will affect the correlations. Different correlation features are chosen based on ANOVA (analysis of variance) in different steganographic systems. Several machine learning methods are applied to classify the extracted feature vectors. Experimental results indicate that our scheme in detecting the presence of hidden messages in several steganographic systems is highly effective.

1 Introduction

Steganography has recently received much attention due to its applications in copyright protection and covert communications. Unlike traditional watermark methods it does not affect the appearance of the image. With digital images (audios or videos) as carriers, detecting the presence of hidden messages poses significant challenges. Westfeld performed the blind steganalysis on the basis of statistical analysis of PoVs (pair of values). This method, so-called χ^2-statistical analysis [1], gave a successful result to a sequential LSB (Least Significant Bit) embedding steganography. Provos [2] extended this method by re-sampling the test interval and re-pairing values. Fridrich [3] introduced a RS steganalysis which is based on the partition of an image's pixels into three groups: Regular, Singular and Unusable and estimate the possible embedded message length of the LSB steganography. Farid and Lyu [4,5] described an approach to detect hidden messages in images that uses a wavelet-like decomposition to build higher-order statistical models of natural images. Support vector machines are then used to discriminate between untouched and adulterated images. In [6], Avcibas, et al. presented techniques for steganalysis of image based on image quality metrics. In [7], Rate-Distortion Curves are used for steganalysis.

On the side of steganography, Kawaguchi presented BPCS-steganography [8] to obtain a large hiding capacity; Westfeld proposed the F5 algorithm [9], which hides messages in the DCT (Discrete Cosine Transform) domain and can defeat χ^2-statistical analysis. Yu [10] proposed SES (Steganography Evading Statistical analyses) which can stand χ^2-statistical analysis and RS steganalysis. Meanwhile, many ordinary steganography tools can be downloaded from the Internet based on different hiding methods [11-14].

In this paper, we present a scheme for steganalysis based on statistical correlations and machine learning. In general, digital images are highly correlated in the spatial domain and the wavelet domain; hiding data will affect the statistics of images. Based on the correlation features extracted from images, ANOVA (analysis of variance) [15] is applied to choose the good measures and machine learning algorithms are applied to discriminate between untouched and adulterated images.

2 Correlation and Feature Extraction

2.1 Statistical Properties of Images

Several papers [16-19] described the statistical models of natural images such as probability models for images based on Markov Random Field models (MRFs) and the Gaussian Mixture Model (GMM). In general, natural images are highly correlated in adjacent pixels; as shown in the following.

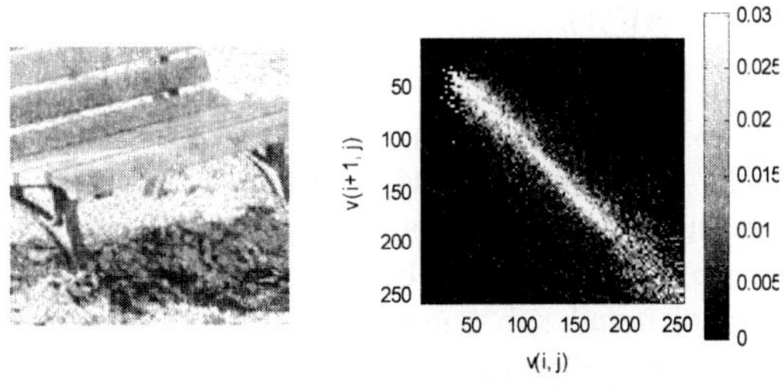

(a) (b)

Fig. 1. A grayscale image (a) and the joint probability of adjacent pixels (b)

Figure 1 (a) is a grayscale ([0 255]) image with size $m \times n$. $v(i,j)$ denotes the grayscale value at point (i,j), $v(i+1,j)$ denotes the grayscale value at the adjacent point $(i+1,j)$. $(v(i,j), v(i+1,j))$ is the grayscale value pair at the two adjacent points. Figure 1 (b) shows the joint probability distribution $\rho(v(i,j), v(i+1,j))$, which indicates that the adjacent pixels are highly correlated.

2.2 Statistical Correlations for Feature Extraction

Throughout plenty of experiments, there are also correlations in intra-bit plane and inter-bit plane of images. $M_1(1{:}m, 1{:}n)$ denotes the binary bits of the Least Significant Bit Plane and $M_2(1{:}m, 1{:}n)$ denotes the binary bits of the second Least Significant Bit Plane. We present the correlation coefficients C1, C2 and C3 as follows:

$$C1 = cor\,(M_1, M_2) \quad (1)$$

$$C2 = cor\,(X_5, X_6) \quad (2)$$

$$C3 = cor\,(X_7, X_8) \quad (3)$$

Where

$$X_5 = M_1(1{:}m-1, 1{:}n), \quad X_6 = M_1(2{:}m, 1{:}n)$$

$$X_7 = M_1(1{:}m, 1{:}n-1), \quad X_8 = M_1(1{:}m, 2{:}n)$$

Besides the correlations in the spatial domain, we consider the autocorrelation of the probability density in the histogram. ρ_k denotes the probability density of the histogram at grayscale sample k ($k = 0,1, \ldots, N-1$, for 8-bit grayscale image, $N = 256$).

$H = (\rho_0, \rho_1, \rho_2 \ldots \rho_{N-1})$ stands for the probability distribution in the histogram. H_e, H_o, H_{k1} and H_{k2} are defined as follows:

$$H_e = (\rho_0, \rho_2, \rho_4 \ldots \rho_{N-2}), \qquad H_o = (\rho_1, \rho_3, \rho_5 \ldots \rho_{N-1});$$

$$H_{k1} = (\rho_0, \rho_1, \rho_2 \ldots \rho_{N-1-k}), \quad H_{k2} = (\rho_k, \rho_{k+1}, \rho_{k+2} \ldots \rho_{N-1}).$$

The autocorrelation coefficients C4 and $C(k)$ are defined as follows:

$$C4 = cor\,(H_e, H_o) \quad (4)$$

$$C(k) = cor\,(H_{k1}, H_{k2}) \quad (5)$$

k is the lag distance in (5). Set k = 1, 2,

$$C5 = C(1) \quad (6)$$

$$C6 = C(2) \quad (7)$$

Meanwhile, wavelet decomposition is an analysis of scale- and location-dependence. There are high correlations in intra-subbands. cA, cH, cV and cD denote the approximate sub band, horizontal, vertical and diagonal detail sub bands with size $m' \times n'$, respectively. Define the following autocorrelation coefficients in the wavelet domain.

$$C7 = cor\,(cH_1, cH_2) \quad (8)$$

$$C8 = cor\,(cH_3, cH_4) \quad (9)$$

$$C9 = cor\,(cV_1, cV_2) \quad (10)$$

$$C10 = cor\,(cV_3, cV_4) \quad (11)$$

$$C11 = cor\,(cD_1, cD_2) \quad (12)$$

$$C12 = cor\,(cD_3, cD_4) \quad (13)$$

Where

$$cX_1 = cX(1{:}\, m'-1, 1{:}n'), \quad cX_2 = cX(2{:}\, m', 1{:}n'),$$

$$cX_3 = cX(1{:}\, m', 1{:}n'-1), \quad cX_4 = cX(1{:}\, m', 2{:}n'),$$

$$X \in \{H, V, D\}.$$

After extracting C1-C12 from the image, we apply ANOVA [15] to choose the good measures according to the steganographic system.

3 Experiments and Discussion

3.1 Experiments

Over 5000 images are taken from many different sources and cover several categories. Some are downloaded from http://www.freephoto.com and other websites. We store these images as 8-bit grayscales and hide messages in these carriers using the hiding methods SES [10], BPCS [9] and the hiding tools BMP Secrets [14], Invisible Secrets v4 [11] and Secure Engine 4.0 [12]. The corresponding hiding ratios are 12.5%, 25%, 25%, 12% and 12%, respectively. C1-C12 are extracted from the carriers and the steganograms, and ANOVA techniques are then applied to choose the good measures.

STPRtool and LS-SVMlab 1.5 are applied in our experiments. The core of the STPRtool comprises statistical pattern recognition algorithms [20] and the algorithms of LS-SVMlab 1.5 are described in [21]. Training sets are chosen at random and the remaining sets are tested. Classifiers are Fisher Linear Discriminant (FLD), Quadratic Classifier (QC), Support Vector Machines (SVM), Kernel Fisher Discriminant (KFD), and LSSVM [20-24]. RBF kernels are applied and the kernel parameters are 0.01 for SVM and KFD in STPRtool and 0.1 for LSSVM.

3.2 Results and Discussion

Table 1 lists the train accuracy and test accuracy for carriers and steganograms, using the five classifiers. The feature sets are {C1-C5} for SES, Invisible Secrets and Secure Engine; {C1, C4-C6} for BPCS and {C11, C12} for BMP Secrets. Table 1 shows kernel-based classifiers, SVM, KFD and LSSVM have better train accuracy than FLD and QC.

The ROC curves in Figures 2-6 indicate that the classification performance is best in the steganalysis of Invisible Secrets, followed by SES, BPCS and BMP Secrets; the classification performance in the steganalysis of Secure Engine is not as good as others. It is probably attributed to our lack of knowledge regarding its hiding methods, and so the feature set is likely not the best. Figure 6 also indicates that the kernel-based classifiers,

KFD and LSSVM are not as good as FLD and QC in steganalysis of Secure Engine, although kernel-based classifiers have a better train accuracy (table 1).

We note that the classification accuracy is related to feature set, kernel parameter, image file format and image type (gray or color). The details are presented in the expanded version of this paper.

Table 1. Accuracy comparison of train and classification using **F**(FLD), **Q**(QC), **S**(SVM), **K**(KFD) and **L**(LSSVM). The first row for each classifier gives train accuracy for carriers; the second row gives train accuracy for steganograms; the third gives test accuracy for carriers and the last row gives test accuracy for steganograms.

	SES	BPCS	BMP Secrets	Invisible Secrets	Secure Engine
F	63%	82.5	87.8	65.8	61
	99	54	84.4	100	87
	58.5	81.6	85.6	66.8	57.4
	98.9	51.7	85.8	100	90.1
Q	69.4	86.9	86.2	73.2	73
	99	70	88.8	100	86
	67.8	85.7	86.3	71.9	64.4
	98.9	63.8	89.5	100	90.1
S	89	100	94.7	98.7	98
	97.6	99.6	94.8	98.9	97
	85.4	69.7	86.1	97.8	80.2
	87.5	96	88	97.3	65.4
K	91.6	100	96.3	98.7	100
	96.6	100	95.8	99.0	97
	85.8	66.5	85.7	97.6	82.2
	85.6	97.7	86.9	97.2	50.5
L	87	98.9	92.0	96.8	97
	97.4	96.0	90.3	97.6	97
	85.4	88	88	95.2	80.2
	93	88.6	88.9	96.8	73.3

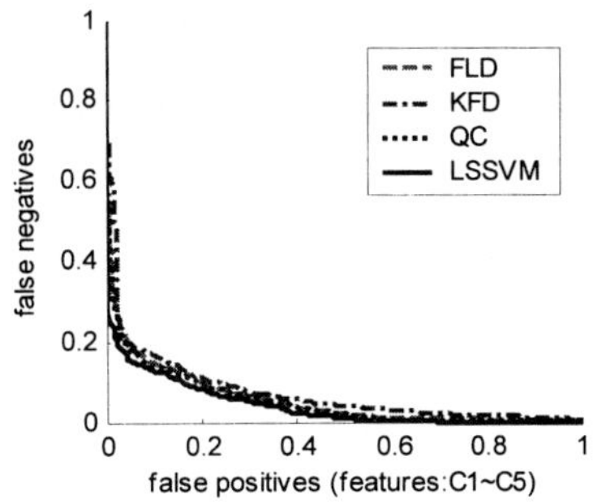

Fig. 2. ROC curves in steganalysis of SES

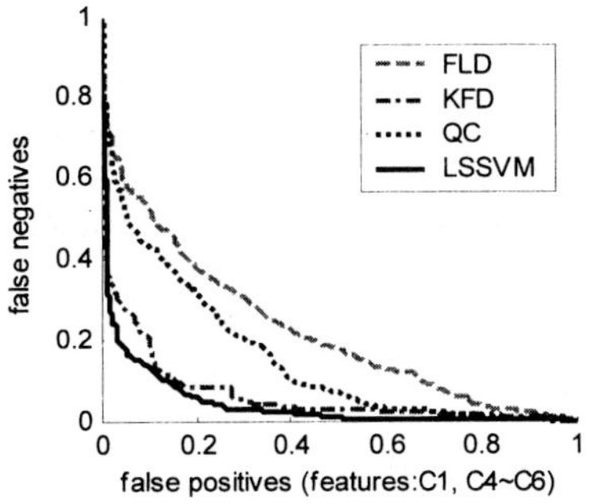

Fig. 3. ROC curves in steganalysis of BPCS

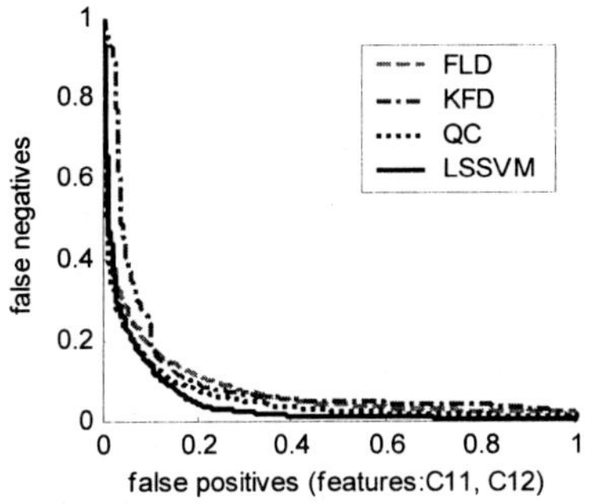

Fig. 4. ROC curves in steganalysis of BMP Secrets

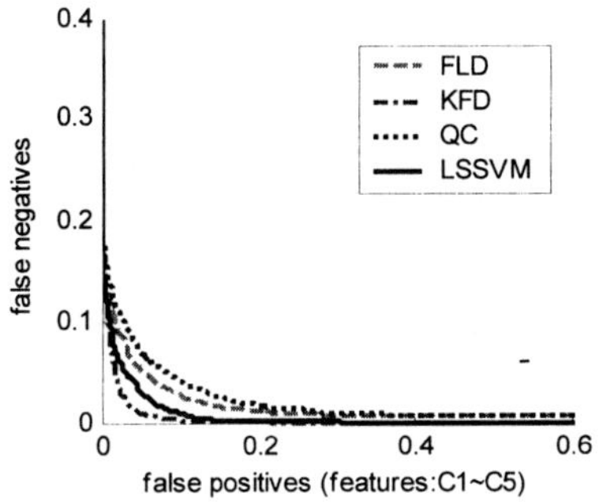

Fig. 5. ROC curves in steganalysis of Invisible Secrets

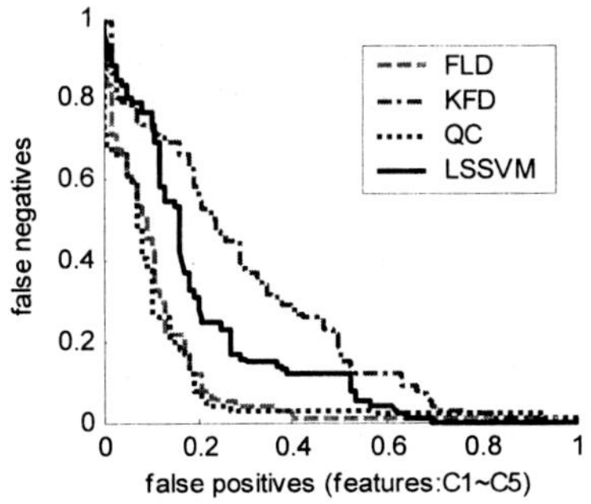

Fig. 6. ROC curves in steganalysis of Secure Engine

4 Conclusions

We presented a scheme for steganalysis based on statistical correlations and learning machine classifiers. Experimental results suggest that it can be applied successfully in the steganalysis of several steganographic systems. Overall, kernel-based classifiers give better train accuracy and test accuracy than the other classifiers, except in the steganalysis of images created using Secure Engine.

Steganalysis is a very challenging problem and, in our view, the successful development of a steganalytic tool will likely rely on multiple steganalytic algorithms and their independent decisions.

Acknowledgements

Partial support for this research received from ICASA (Institute for Complex Additive Systems Analysis, a division of New Mexico Tech), a DoD IASP Capacity Building grant, and an NSF SFS Capacity Building grant is gratefully acknowledged. We also thank S. Mukkamala, T. Xie, and J. Xu for numerous helpful discussions and the anonymous reviewers for their insightful comments and valuable suggestions. The collaborative work of the third author was performed during her sabbatical visit to New Mexico Tech in 2004.

References

[1] A. Westfeld, A. Pfitzmann (2000) Attacks on Steganographic Systems. LNCS 1768, Springer-Verlag, Berlin: 61-75

[2] N. Provos (2001) Defending against Statistical Steganalysis. Proceedings of the 10th USENIX Security Symposium: 323-335

[3] J. Fridrich, M. Goljan, and R. Du (2001) Detecting LSB steganography in color and gray-scale image, IEEE Multimedia: 22-28

[4] H. Farid (2002) Detecting hidden messages using higher-order statistical models. In ICIP 2002, Rochester, New York

[5] S. Lyu, H. Farid (2004) Steganalysis Using Color Wavelet Statistics and One-Class Support Vector Machines. Proc. SPIE: Security, Steganography, and Watermarking of Multimedia Contents VI, vol. 5306

[6] I. Avcibas, N. Memon and B. Sankur (2003) Steganalysis using image quality metrics. IEEE transactions on Image Processing, Vol. 12, No. 2

[7] M. U. Celik, G. Sharma, and A. M. Tekalp, (2004) Universal image steganalysis using rate-distortion curves. Proc. SPIE: Security, Steganography, and Watermarking of Multimedia Contents VI, vol. 5306

[8] E. Kawaguchi, R. O. Eason (1998) Principle and Applications of BPCS-Steganography. Proc. Of SPIE, Vol.3528: 464-473

[9] A. Westfeld (2001) High Capacity Despite Better Steganalysis (F5-A Steganographic Alogrithm). LNCS 2137, Springer-Verlag: 289-302

[10] J. Yu, J. Han, K. Lee, D. Ryu, S. Lee (2003) SES (Steganography Evading Statistical analyses). Pacific Rim Workshop on Digital Steganography 2003

[11] http://www.invisiblesecrets.com/

[12] http://securengine.isecurelabs.com/

[13] http://www.jjtc.com/stegoarchive/stego/

[14] http://www.pworlds.com/products/i_secrets.html

[15] Rencher A. C. (1995) Methods of Multivariate Analysis, John Wiley, New York

[16] J. Huang, D. Mumford (1999) Statistics of Natural Images and Models. Computer Vision and Pattern Recognition, Vol. 1

[17] A. Srivastava, A. Lee, E.P Simoncelli and S. Zhu (2003) on advances in statistical modeling of natural images, Journal of Mathematical Imaging and Vision 18(1): 17-33

[18] G. Winkler (1995) Image Analysis, Random Fields and Dynamic Monte Carlo Methods. Springer: Berlin

[19] M. J. Wainwright, E. P. Simoncelli (2000) Scale mixtures of Gaussians and the statistics of natural images. In: Advances in Neural Information Processing Systems, S. A. Solla, T. K. Leen and K.-R. Muller (Eds.): 855–861

[20] M. I. Schlesinger and V. Hlavac (2002) Ten lectures on statistical and structural pattern recognition. Kluwer Academic Publishers

[21] J. A. K. Suykens, T. Van Gestel, J. De Brabanter, B. De Moor, J. Vandewalle (2002) Least Squares Support Vector Machines, World Scientific, Singapore

[22] R. O. Duda, P. E. Hart, D. G. Stork (2001) Pattern Classification. John Wiley & Sons, 2nd edition

[23] B. Scholkopf, A. J. Smola (2002) Learning with Kernels, MIT Press

[24] S. Raudys (2001) Statistical and Neural Classifiers: An Integrated Approach to Design, Springer-Verlag

The Use of Multi-Criteria in Feature Selection to Enhance Text Categorization

Son Doan and Susumu Horiguchi
Japan Advance Institute of Science and Technology
1-1, Asahidai 1-8, Tatsunokuchi, Ishikawa, Japan
E-mail:{s-doan, hori}@jaist.ac.jp

Abstract

Feature selection has been an interesting issue in text categorization up to now. Previous works in feature selection often used filter model in which features, after ranked by a measure, are selected based on a given threshold. In this paper, we present a novel approach to feature selection based on multi-criteria of each feature. Instead of only one criterion, multi-criteria of a feature are used; and a procedure based on each threshold of feature selection is proposed. This framework seems to be suitable for text data and applied to text categorization. Experimental results on Reuters-21578 benchmark data show that our approach has a promising scheme and enhances the performance of a text categorization system.

1 Introduction

Feature selection is an interesting issue recently in machine learning and plays an important role in text categorization [1],[2],[3]. Up to now, there has been two most common approaches in feature selection: the filter and the wrapper [4],[5] approach. In the filter approach, features are chosen based on a given threshold of a specific measurements; the measurements should be one of four types: information, distance, dependence and consistency measures [6]. The wrapper model [5] choose features based on the accuracy of a classifier: features which achieve highest the accuracy of the classifier are selected. When the number of features increases, the wrapper method is impractical since the time of building the classifier; instead, the filtering approach is usually chosen because it is easy for implementing and independent of classifiers.

Feature selection for text categorization often applied the filter approach [2],[3]. Filter approach often uses one of the existing measurements, e.g. document frequency, information gain, mutual information as the measurement for filtering features [1],[2],[7],[8]. However, text has it own characteristics of linguistics such as semantics, syntax, thesaurus, etc; in other words, text has multi-criteria for evaluating its importance or weight in a document. Therefore, it seem to be suitable to apply several measurements for the same feature and the feature selection process is then implemented by the filter approach. In this paper, we propose a novel approach to the feature selection problem based on multi-criteria of features in text categorization and investigated its advantage on the benchmark text data.

The paper is organized as follows. Section 2 presents a novel framework for feature selection problem base on multi-criteria of features. The proposed framework is applied in text categorization in Section 3. Experimental results in Section 4 show the performance improvements of our approach. Discussions and conclusions are drawn in Section 5.

2 A Novel Framework for Feature Selection

2.1 The EFS procedure

The feature selection problem in text categorization can be stated as follows: Given a set $\mathcal{X}$ consisting of n features $x_1, x_2, \dots x_n$, the problem in feature selection is to choose the optimal subset S of $\mathcal{X}$ ($||S|| \ll ||\mathcal{X}||$) with highest effectiveness for the system.

To solve this problem, our basic idea is to filter features based on a procedure of multi-criteria ranking for terms (terms in text categorization are equivalent to features). Each feature, according to a criterion, will be weighted with a term weight; thus, with t criteria, we will have t ways of ordering features as follows. The feature selection problem can be stated mathematically as follows.

Choose a proper subset of $\mathcal{X}$, given a set of criteria $\theta_1, \dots, \theta_t$, within which each criterion determines a ranking of $\mathcal{X}$. Formally, we have:

$$\begin{array}{ll} \text{Criterion } \theta_1 : & x_{\sigma_1(1)} \preceq_{\theta_1} \cdots \preceq_{\theta_1} x_{\sigma_1(n)} \\ \dots & \dots \\ \text{Criterion } \theta_t : & x_{\sigma_t(1)} \preceq_{\theta_t} \cdots \preceq_{\theta_t} x_{\sigma_t(n)} \end{array}$$

where σ_i is a permutation of the set $\{1, \dots, n\}$, and $\preceq_{\theta_t}$

is the order relation based on criterion θ_i.

After ranked $\mathcal{X}$ according to a multiple criteria as above, for each criterion θ_i, we select a subset S_i of $\mathcal{X}$ based on a threshold τ_i. Then the set of selected terms is defined by

$$S = \bigcup_{i=1} S_i \quad (1)$$

Algorithmically, the EFS procedure for feature selection is depicted as in Figure 1.

Procedure EFS($\mathcal{X}$ - original feature set, S-optimal feature set, $\tau_1, \ldots, \tau_t$ - threshold values)

for i =1 **to** t **loop**
  $S_i \leftarrow \emptyset$;
  Step 1. Ranking all features based on criterion θ_i;
  Step 2. Choose the first features based on τ_i;
  Step 3. **Return** S_i;
end loop;
  $S \leftarrow S_1 \cup S_2 \cup \cdots \cup S_t$;

Return S

Fig. 1. The EFS procedure for selecting the optimal feature subset

2.2 How to choose the threshold for each criterion?

There is a question raising in the EFS procedure: How to choose the subsets of features based on threshold values ?. Intuitively, suppose that for each criterion θ_i, the optimal subset is S_{opt_i}, the final subset should be obtained by combination of optimal subsets S_{opt_i}, $S_{opt} = S_{opt_1} \cup \cdots \cup S_{opt_n}$ Finding the optimal subset for each criterion is also very difficult problem. In addition, it depends on the data type and characteristics of the problem under consideration. Thus, the threshold values should be approximated by experiments.

Among the criteria of feature in the problem under consideration, there must be a principal or primary criterion regards to the problem, denoted $\theta_p, \theta_p \in (\theta_1, \ldots, \theta_t)$. It should be reasonable to choose the optimal subset S_{opt_p} as the "core" set firstly and then other subsets are added increment ally to the core set. Finally the optimal subset obtained could be approximated as

$$S_{opt} = S_1 \cup \cdots \cup S_{opt_p} \cup \cdots \cup S_n. \quad (2)$$

3 Application to Text Categorization

3.1 Measurements of Text Data

Given a set of categories $\mathcal{C} = \{c_i\}_{i=1}^m$, features in text categorization usually are selected by one of the following measurements.

1. Document frequency criterion: Features are selected by their frequencies in document, with a threshold.
2. Class-based criterion: Select features based on their frequency in a class.
3. Information gain measure: The information gain of term x is given by [1],[3]:

$$\begin{aligned} IG(x) &= -\sum_{i=1}^m P(c_i)\log P(c_i) + \\ &+P(x)\sum_{i=1}^m P(c_i|x)\log(c_i|x) \\ &+P(x)\sum_{i=1}^m P(c_i|\bar{x})\log(c_i|\bar{x}). \quad (3) \end{aligned}$$

4. Mutual information measure: Mutual information of term t is given by [1],[3]:

$$MI(x) = \sum_{i=1}^m \log\frac{P(x \wedge c_i)}{P(x).P(c_i)}. \quad (4)$$

There are other measures for feature selection, for example, chi-square and odd-ratio ... [1],[2],[3]

3.2 Naive Bayes Classifier

The naive Bayes algorithm is viewed as the baseline algorithm for text categorization recently [2],[3]. It can be briefly described as follows.

Given m classes $\mathcal{C} = (c_1, c_2, \ldots, c_m)$, with a document d', our problem is to build a classifier σ that can assign the document d' to a class. Suppose a document d' consisting of terms $x_1, x_2, \ldots x_n$. The naive Bayes algorithm calculates the probability of a class belonging to each document by the following formulation:

$$\begin{aligned} P(c_i|d') \propto P(d'|c_i)P(c_i) &= P((x_1, \ldots, x_n)|c_i)P(c_i) \\ &= \prod_{j=1}^n P(x_j|c_i)P(c_i). \quad (5) \end{aligned}$$

Thus, the class of document d' is calculated by the following formula,

$$\sigma(d') = \arg\max_{i\in[1\ldots m]} P(c_i|d). \quad (6)$$

4 Experimental Results

4.1 Real-world Data Set

To examine our proposed method, we used a standard text data set Reuters-21578 [1]. The top 10 categories were

[1]This data set can be obtained from http://www.daviddlewis.com/resources/testcollection/reuters21578

Table 1. Details of top 10 categories of Reuters-21578 data set

Category	#training docs	#testing docs
Earn	2,877	1,083
Acq	1,650	719
Money-fx	538	179
Grain	433	149
Crude	389	189
Trade	368	117
Interest	347	131
Ship	197	89
Wheat	212	71
Corn	181	56
Total	7,769	3,019

chosen for implementation; they are described in Table 1. Reuters-21578 data set is preprocessed by removing common words such as *the*, *a*, *an*, etc in the stop list, words are stemmed by the Porter algorithm. After preprocessing, the number of vocabulary is 19,791 words.

In our experiments, we chose two standard methods in feature selection, all terms (that is method containing all terms in vocabulary) and feature selection based on mutual information measure. For easily understanding later, we called the first case all term method and the second case the baseline method.

In the baseline method, mutual information measure is considered as the most common measure used in feature selection [11],[1]. Thus, we take it as the principal criterion θ_p and S_{opt_p} is taken by a threshold of mutual information measure. The most threshold using this measure is $\simeq 1/10$ vocabulary [7], [11]. Thus we chose the number of vocabulary was $2,000$.

To compare our method with the baseline method and all term method, we used two criteria, the mutual information and class-based frequency. The principal criterion here was mutual information measurement. A threshold for the principal criterion was $\tau_1 = 2,000$ and two thresholds for class-based frequency measure were selected, $\tau_2 = 100$ and $\tau_2 = 200$, respectively. For convenience, we called the first case in the our proposed method the EFS-100 and the second the EFS-200. The number of terms in the EFS-100 and EFS-200 are $2,314$ and $2,619$ terms respectively.

Experiments are executed in SunOS 5.8 operating system, Perl, sed, awk, C programming languages and *libbow* library [12].

4.2 Performance Measures

Conventional performance measures in text categorization are precision/recall, F_1 measure and break-even point (BEP). They based on two basic measures pre-

Table 2. The contingency table for a category

Category c_i		Human assign YES	Human assign NO
Classifier predict	YES	a_i	b_i
Classifier predict	NO	c_i	d_i

cision and recall. Mathematically, they are expressed through the contingency table as in Table 2, for each category c_i, the precision P_i and the recall R_i are defined as follows, $P_i = a_i/(a_i + b_i)$ and $R_i = a_i/(a_i + c_i)$. Macro-averaging performances of the system are given by: macro-P $=$ $\sum_{i=1}^{k} P_i/k$ and macro-R $=$ $\sum_{i=1}^{k} R_i/k$. Microaveraging performances are calculated by, micro-P $=$ $\sum_{i=1}^{k} a_i/\sum_{i=1}^{k}(a_i + b_i)$, and micro-$R$ $=$ $\sum_{i=1}^{k} a_i/\sum_{i=1}^{k}(a_i + c_i)$. F_1 is defined as $F_1 = 2PR/(P + R)$. BEP measure is the point that $P = R$, this point is often calculated by taking the average of P and R. The macro- and micro-F_1 and BEP are calculated by replacing P, R with the corresponding macro and micro of P, R.

Macro-averaging and microaveraging of F_1 and BEP are treated as the measures to compare performances of text categorization systems.

4.3 Experimental Results

Table 3 shows the results of BEP and F_1. Results indicated that both two proposed methods the EFS-100 and the EFS-200 had always higher performances than the baseline and the all term methods.

The macroaveraging BEP for the EFS-100 is 69.26% vs. 68.13% when using the all term method and 68.69% when using the baseline method. In case of the EFS-200, the macroaveraging BEP is 69.14%; it is higher than both baseline and the all term methods but lower than the EFS-100. The microaveraging BEP for both proposed methods is the same (74.55%). It is also not different from that for the baseline method (74.54%) but higher than the all term method (72.74%).

In summary, our proposed method outperformed the baseline method and the all term method, especially for macro averaging measures. Furthermore, the results also showed that the EFS-100 has better performance than the EFS-200, it has been suggested that appropriate parameters τ_1 and τ_2 for our proposed method can be tuned for achieving better performance

In order to compare performances of the system, we compare macroaveraging measures of BEP and macroaveraging of F_1. Results shows that with all 10 categories, the proposed method have higher performances than using mutual information criterion and us-

Table 3. BEP and F_1 performance measures of Reuters-21578

	BEP				F_1			
Category	all terms	baseline	EFS-100	EFS-200	all terms	baseline	EFS-100	EFS-200
Earn	97.65	97.47	97.43	97.38	98.10	97.91	98.04	98.04
Acq	96.45	96.04	96.60	96.66	96.48	96.21	96.67	96.67
Money-fx	76.54	75.98	76.54	76.19	76.92	75.98	76.54	76.30
Grain	50.34	49.49	51.04	51.50	59.76	54.42	57.47	57.41
Crude	80.00	78.09	78.51	78.51	81.40	79.67	79.43	79.43
Trade	79.15	84.62	84.12	84.12	82.59	85.59	85.60	85.60
Interest	72.52	68.96	70.23	70.23	73.00	73.83	73.38	73.38
Ship	62.92	60.00	59.55	59.55	67.00	67.58	68.75	68.96
Wheat	31.76	40.85	41.13	39.72	39.82	49.24	48.39	47.83
Corn	33.93	35.40	37.50	37.50	35.89	46.40	44.02	44.30
macro ave	68.13	68.69	**69.26**	69.14	71.10	72.68	**72.83**	72.79
micro ave	72.31	74.54	**74.55**	74.55	73.34	73.86	**74.06**	74.03

ing all terms in both BEP and F_1. Macroaveraging of BEP with proposed method is 69.26% vs. 68.13% when using all terms and 68.69% when using mutual onformation criterion; macroaveraging of F_1 with proposed method is 72.83% vs. 71.10% when using all terms and 72.68% when using mutual information criterion.

5 Conclusions

This paper proposed a novel feature selection approach based on the multi-criteria ranking of features in text categorization problem. A general framework for feature selection was proposed and applied to Reuters-21578 data set. Experimental results shows that the proposed method using multi-criteria of features instead of using only one criterion as the baseline methods could enhance the performance of the system, especially for macroaveraging; in compared to methods using only one criterion and whole vocabulary. This results is significant and useful for feature selection in text categorization problem.

References

[1] Y. Yang and J.O. Pedersen. A comparative study on feature selection in text categorization. In *Proceeding of the 14th International Conference on Machine Learning (ICML97)*, pages 412–420, 1997.

[2] D. Mladenic. Feature subset selection in text learning. In *Proc of European Conference on Machine Learning(ECML)*, pages 95–100, 1998.

[3] F. Sebastiani. Machine learning in automated text categorization. *ACM computing survey*, 34(1):1–47, 2002.

[4] A. Blum and P. Langley. Selection of relevant features and examples in machine learning. *Artificial Intelligence*, 97(1-2):245–271, 1997.

[5] R. Kohavi and G. John. Wrappers for feature subset selection. *Artificial Intelligence*, 97(1-2):273–324, 1997.

[6] S. Huang. Dimensionality reduction in automatic knowledge acquisition: A simple greedy search approach. *IEEE Transactions on Knowledge and Data Engineering*, 15(6):1364–1373, 2003.

[7] M. Rogati and Y. Yang. High-performing feature selection for text classification. In *International Conference on Information and Knowledge Management-CIKM2002*, pages 659–661, 2002.

[8] P. Soucy and G.Mineau. A simple feature selection method for text classification. In *International Joint Conference of Artificial Intelligence (IJCAI)*, 2001.

[9] G. Salton, A. Wong, and C.S. Yang. A vector space model for automatic indexing. *Communications of the ACM*, 18(11):613–620, 1975.

[10] Y. Yang. An evaluation of statistical approaches to text categorization. *Information Retrieval Journal*, 1:69–90, 1999.

[11] F. Debole and F. Sebastiani. An analysis of the relative hardness of reuters-21578 subsets. *Journal of the American Society for Information Science and Technology (JASIST)*, 2004. Forthcoming.

[12] Andrew Kachites McCallum. Bow: A toolkit for statistical language modeling, text retrieval, classification and clustering, 1996. `http://www.cs.cmu.edu/~mccallum/bow`.

Text Classification from Partially Labeled Distributed Data

Catarina Silva†*, Bernardete Ribeiro*
*Departamento de Engenharia Informática, Centro de Informática e Sistemas, Universidade de Coimbra, Portugal
†Escola Superior de Tecnologia e Gestão - Instituto Politécnico de Leiria, Portugal

Abstract

One of the main problems with text classification systems is the lack of labeled data, as well as the cost of labeling unlabeled data [1]. Thus, there is a growing interest in exploring the combination of labeled and unlabeled data, i.e., partially labeled data [2], as a way to improve classification performance in text classification. The ready availability of this kind of data in most applications makes it an appealing source of information.

The distributed nature of the data, usually available online, makes it a very interesting problem suited to be solved with distributed computing tools, delivered by emerging GRID computing environments.

We evaluate the advantages obtained by blending supervised and unsupervised learning in a support vector machine automatic text classifier. We further evaluate the possibility of learning actively and propose a method for choosing the samples to be learned.

1 Introduction

Applications of text classification are ubiquitous, since almost 80% of the information available is stored as text. Thus, there is an effective interest in researching and developing applications that better help people handling text-based information. On the other hand, the wealth of text information has made the organization of that information into a complex and vitally important task.

Most text categorization methods, e.g., K-Nearest Neighbor, Naive Bayes, Neural Nets and Support Vector Machines, have their performance greatly defined by the training set available. This is one key difficulty with current text categorization algorithms, since they require manual labeling of more documents than a typical user can tolerate [3]. Thus, methods that need a small set of labeled examples are currently being explored.

Labeling data is expensive but, in most text categorization tasks, unlabeled data are often inexpensive, abundant and readily available. Therefore, to achieve this purpose, i.e., the use of relatively small training sets, the information that can be extracted from the testing set, or even unlabeled examples, is being investigated as a way to improve classification performance.

Seeger in [4] and Szummer in [2] present reports on learning with unlabeled data, comparing several approaches. Text is represented using the Bag of Words and text pre-processing tools are usually applied, such as, stopword removal, low frequency word removal and stemming. For a study on text pre-processing tools please refer to [5].

The rest of the paper is organized as follows. Section 2 focuses on the issues related to the use of partially labeled data on text classification and proposes two approaches. Section 3 presents the results obtained and Section 4 presents some conclusions and future work.

2 Using Partially Labeled Data

To achieve the best classification performance with a machine learning technique, there has to be enough labeled data. However these data are costly and sometimes difficult to gather. Therefore, using unlabeled data for text classification purposes has recently been actively investigated [6] [7].

In general, unlabeled examples are much less expensive and easier to gather than labeled ones. This is particularly true for text classification tasks involving online data sources, such as web pages, email and news stories, where large amounts of text are readily available. Collecting this text can frequently be done automatically, so it is feasible to collect a large set of unlabeled examples. If unlabeled examples can be integrated into supervised learning, then building text classification systems will be significantly faster, less expensive and more effective.

There is a catch however, because, at first glance, it might seem that nothing is to be gained from unlabeled data, since an unlabeled document does not contain the most important piece of information - its classification.

Consider the following example to give some insight of how unlabeled data can be useful. Suppose we are interested in recognizing web pages about conferences. We are given just a few conferences and non-conferences web pages, along with a large number of pages that are unlabeled. By looking at just the labeled data, we determine that pages containing the word *paper* tend to be about conferences. If we use this fact to estimate the classification of the many unlabeled web pages, we might find that the word *deadline* occurs frequently in the documents that are classified in the positive class.

This co-occurrence of the words *paper* and *deadline* over the large set of unlabeled training data can provide useful information to construct a more accurate classifier that considers both *paper* and *deadline* as indicators of positive examples.

Some authors [8] refer to unlabeled data as background knowledge, defining it as any unlabeled collection of text from any source that is related to the classification task.

Joachims presents in [9] a study on transductive SVM (TSVM) introduced by Vapnik [10]. TSVM make use of the testing set and extend inductive SVM, finding an optimal separating hyperplane not only of the training examples, but also of the testing examples [11].

The goal of active learning is to design and analyse learning algorithms that can effectively filter or choose the samples to be labeled by a supervisor. The incentive in using active learning is mainly to expedite the learning process and reduce the labeling efforts required by the teacher [12]. Schohn in [3] proposes a method to actively learn with SVM, exploring the examples that are orthogonal to the space spanned by the training set, in order to give to the classifier information about dimensions not yet explored.

2.1 Background Knowledge and Active Learning

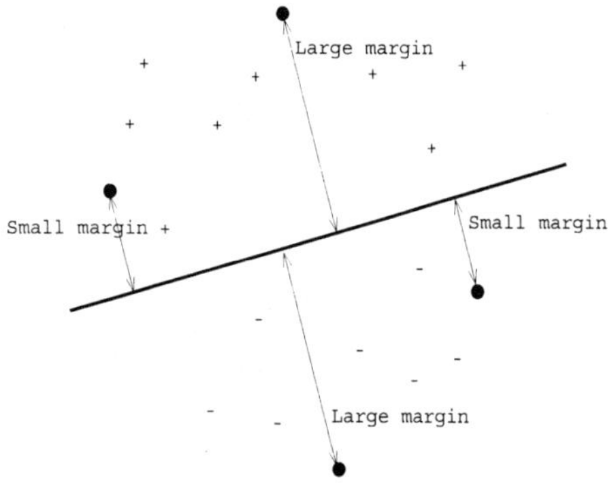

Fig. 1. Classifications task: positive examples marked with + signs; negative examples marked with − signs; and testing examples (with small and large margin) marked with black dots.

Figure 1 presents a classification example that we will use to illustrate both proposed approaches: background knowledge and active learning. Both approaches start with the inductive SVM trained with the training set available and add examples from the test set.

Background knowledge - Choose the testing examples classified by the SVM with more confidence (larger margin) and incorporate them, possibly incorrectly classified, in the training set.

Active learning - Choose the 20 testing examples, classified by the user/oracle/supervisor in which the SVM has less confidence (smaller margin) and integrate those correctly classified examples in the training set. Both approaches have advantages and disadvantages, and we expect to conjugate them to use the advantages and mitigate the disadvantages. For an objective comparison, we can use the following criteria:

1. User interaction: while the first approach is automated, the second approach needs some user interaction, since the selected items must be classified by the supervisor;
2. Correctness of training set: the first approach does not guarantee its correctness, since the added examples are classified by the inductive SVM, whereas in the second approach all examples in the training set are (correctly) classified by the supervisor;
3. Computational time: there is not a significant difference in the computational time used, however the first approach can take longer, because the examples are automatically classified and there is no limit on the number of examples added;
4. Performance measured as detailed in Section 3: the second approach has greater potential, since the information added is more reliable, but has limitations on the number of items the supervisor can tolerate/is able to classify (20 in our case).

3 Results

Reuters-21578 collection was used with the ModApte split publicly available at http://kdd.ics.uci.edu /databases/ reuters21578/ reuters21578.html.

Table 1. Number of positive training and testing documents for Reuters' most frequent categories.

Category	Training	Testing
Earn	2715	1044
Acquisitions	1547	680
Money-fx	496	161
Grain	395	138
Crude	358	176
Trade	346	113
Interest	313	121
Ship	186	89
Wheat	194	66
Corn	164	52

Table 2. F1 measures for ModApte split (9603/3299) and Small split (20/3299) on baseline SVM and on both proposed approaches: Background Knowledge (BK) and Active Learning (AL).

	ModApte Split			Small Split		
Category	**Baseline**	**BK**	**AL**	**Baseline**	**BK**	**AL**
Earn	93.99	94.42	94.98	86.24	87.40	90.41
Acquisitions	88.94	89.69	93.09	48.36	44.43	52.50
Money-fx	60.72	60.99	65.48	14.95	99.62	62.45
Grain	75.00	76.07	79.33	25.94	12.57	21.29
Crude	72.73	74.36	76.38	26.23	16.25	20.94
Trade	64.92	65.62	75.12	26.60	9.17	58.41
Interest	58.46	60.30	73.40	14.77	9.71	23.71
Ship	65.62	66.67	75.92	7.50	7.33	15.84
Wheat	68.47	69.03	80.68	41.29	4.67	63.81
Corn	68.29	81.87	82.22	33.77	3.88	52.80
Average	71.72	97.31	79.32	32.57	29.50	46.21

Reuters-21578 is a financial corpus with news articles averaging 200 words each. Example categories are trade, earn or crude. In this corpus there are about 12000 classified stories into 118 possible categories. The ModApte split was used, using 75% of the articles (9603 items) for training and 25% (3299 items) for testing. Table 1 presents the ten most frequent categories and the number of training and testing examples, comprising 75% of the items.

In addition to ModApte split, a Small split was also tested. The testing set was exactly the same for the sake of comparison, but the training set, instead of 9603 examples was randomly reduced to 10 positive examples and 10 negative examples. The idea was to reproduce a real situation in which a real user would be asked to provide these 20 examples.

The simulation results were evaluated using F1 measure, which combines both Precision and Recall. Precision is the percentage of predicted documents for the given topic that are correctly classified (1).

Recall is the percentage of total documents for the given topic that are correctly classified (2).

$$Precision = \frac{categories\ found\ and\ correct}{total\ categories\ found} \quad (1)$$

$$Recall = \frac{categories\ found\ and\ correct}{total\ categories\ correct} \quad (2)$$

$$F1 = \frac{2 * precision * recall}{precision + recall} \quad (3)$$

Table 2 presents F1 measures for ModApte split (9603 training examples and 3299 testing examples) and for Small split (20 training examples and 3299 testing examples) on baseline SVM and on both proposed approaches: Background Knowledge and Active Learning. F1 values were preferred to analyse the methods' performance, since they fuse precision and recall values.

Where Backgroung Knowledge is concerned, there is an improvement of 3% (from 71,72% to 73,86%) for the ModApte split, but not for the Small split, where there is a decrease (from 32,57% to 29,50%). For this approach to be successful the baseline classifier can not be too weak, since it will be responsible for classifying the testing examples. That is not the case with Small split. With only 20 examples the initial classifier is not accurate enough to determine new training examples.

With active learning the improvement is more relevant (improvement of 40%, from 32.57% to 46.21%) on the Small split than on the ModApte split, a predictable outcome, since the training set was doubled (20 initial examples plus 20 examples actively chosen to be classified by the supervisor).

In what ModApte split is concerned this active approach improves 10% the baseline results (from 71,72% to 79,32%).

4 Conclusions and Future Work

The results presented in the previous section are encouraging to the improvement achieved by the introduction of unlabeled documents' information in the learning procedure. The introduction of background knowledge by the SVM classified testing examples should not be used with small training sets, however it can constitute a slight improvement when the baseline classifier is not too weak.

The proposed margin-based active learning method has potential to substantially improve performance when small training sets are available. This conclusion is very important in text classification tasks, since usually there is a small number of classified examples and a huge number of unlabeled ones. Further testing in the hybrid conjugation of both proposed methods is also foreseen.

Future work will deal with the theoretical foundations of these experiments and research on use of data grid technology to make use of physically distributed data sources.

The motivation behind the use of unlabeled or partially labeled data has to do with its readily online availability. To effectively explore the wealth of data a grid infrastructure has to be taken advantage of. Not only the most obvious data grid infrastructure, but also the crucial resource grid. Yet another appealing issue to be analysed has to do with the possibility of distributing the learning methods computational load. Using a service grid it would be possible to not only to exploit the unused computational power in an intranet, but also to interchange and compare methods with other researchers on the internet.

Acknowledgments

CISUC - Center of Informatics and Systems of University of Coimbra and Project POSI/SRI/41234/2001 are gratefully acknowledged for partial financing support.

References

[1] S.Kiritchenko and S.Matwin, "Email Classification with Co-Training", *2001 Conference of the Centre for Advanced Studies on Collaborative Research*, pp. 8.

[2] M.Szummer, "Learning from Partially Labeled Data", *PhD Thesis*, Massachusetts Institute of Technology, 2002.

[3] G.Schohn and D.Cohn, "Less is more: Active Learning with Support Vector Machines", *ICML*, 2000, pp. 839-846.

[4] M.Seeger, "Learning with Labeled and Unlabeled Data", *Technical Report*, Institute for Adaptive and Neural Computation, Univ. Edinburgh, 2001.

[5] C.Silva and B.Ribeiro, "On the Evaluation of Text Processing in Inductive Categorization", *ICMLA*, 2003, pp. 121-127.

[6] J.Hong and S.Cho, "Incremental Support Vector Machine for Unlabeled Data Classification", *ICONIP*, 2002, pp. 1403-1407.

[7] B. Liu, Y. Dai, X. Li, W. Lee and P. Yu, "Building Text Classifiers Using Positive and Unlabeled Examples", *ICDM*, 2003, pp. 179.

[8] S.Zelikovitz and H.Hirsh, "Improving Text Classification with LSI Using Background Knowledge", *ICIKM*, 2001.

[9] T.Joachims, "Transductive Inference for Text Classification using Support Vector Machines", *ICML*, 1999, pp. 200-209.

[10] V.Vapnik, *The Nature of Statistical Learning Theory*, Springer, 1995.

[11] C.Silva and B.Ribeiro, "Labeled and Unlabeled Data in Text Categorization", *IEEE IJCNN*, 2004.

[12] Y.Baram, R.El-Yaniv and K.Luz, "Online Choice of Active Learning Algorithms", *ICML*, 2003, pp. 255-291.

Recovering the Cyclic-Code of Generated Polynomial by Using Evolutionary Computation

Kangshun Li [1 2 3], Yuanxiang Li [1 2], Haifang Mo [2]

[1] State Key Laboratory of Software Engineering, Wuhan University, Wuhan, 430072, China.
[2] Computer School of Wuhan University, Wuhan, 430072, China
[3] School of Information Engineering, Jiangxi University of Science & Technology, Jiangxi, 341000, China.
Li Kangshun,lks@public1.gzptt.jx.cn

Abstract

The data integrity in computer security is a key component of what we call trustworthy computing, and one of the most important issues in data integrity is to detect and correct error codes, which is also a crucial step in software and hardware design. Numerous methods have been recently proposed to solve legal-codes of the cyclic-code generated polynomial g(x). We think that a better approach for this purpose is to solve the legal-codes by finding the roots of the cyclic-code generated polynomial. However, as it is well known, finding roots of polynomials of high degree in the modulo-q space GF(q) is very difficult. In this paper we propose a method to solve the roots of cyclic-code generated polynomial by using evolutionary computation, which makes use of randomized searching method from biological natural selection and natural genetic system.

1 Introduction

With the fast development of computer, there are more and more demand on computer application in all fields, and the trustworthy computing play more and more important role in the software design and network communication. The computer detecting error codes and correcting error codes is the basic demand of trustworthy computing, therefore, the computer detecting error codes and correcting error codes is of significance to trustworthy computing and to computer security. The precise bits of detecting error codes and correcting error codes[1] are 64 bits nowadays in MILKWAY-II computer while there was only 1 bit in the past, and in the future, the spaceflight technology of shipping persons and GPS will all request higher capacity of error-correcting and error-detecting. Therefore, we need more scientific and more complete trusted computing means. This paper introduces evolutionary algorithm with simulative biologic genetic evolution through population crossover and mutation to attain the solutions of problem, and to solve the roots of exponential cyclic-code generated polynomial $g(x)$ in modulo-q space $GF(q)$.

2 The Theory of Evolutionary Computation

The method of evolutionary computation[2] is a random searching method referencing biological natural selection and natural genetic mechanism. It includes genetic algorithms (GAs), evolution strategies (ESs), evolutionary programming (EP) and genetic programming (GP). It is also a new method to solve problems through combining nature genetics with computer science.

If we consider an optimizing problem as following:

$$\min\{f(x) \mid x \in X\}$$

Foundation item: This work was supported by the National Natural Science Key Foundation of China with the Grant No.60133010 and the National Research Foundation for the Doctoral Program of Higher Education of China with the Grant No.20030486049.

Biography: Li Kangshun (1962-), male, Ph. D candidate, senior engineer, research direction: Evolutionary computation, computer security.

subject to

$$g_i(x) = 0, \qquad i = 1,2,\cdots,n$$

where $f(x)$ is an objective function in domain X, $\forall x \in X, f(x) \geq 0$ 。 X is a feasible set of its solutions. It can be a finite set (for example, a combination optimization problem), it can also be a subset of real space R^n (for example, continuous optimization problem) etc.

3 Steps of Evolutionary Computation Algorithm

Step 1: Producing the initial population $P(t) = \{x_1(t), x_2(t), \ldots\ldots, x_n(t)\}$ and computing the corresponding function fitness values. The basic genetic unit of biological individual $x_i(t)$ is gene, and the genes are ranked in sequence to form chromosomes. The initial population[3] consists of a certain amount of chromosomes at the beginning of the search. In fact fitness function is the objective function of optimization problem, the evolutionary algorithm will search for the chromosome with maximum fitness.

Step 2: Selecting, crossing and mutating the individuals. The individuals with the better fitness replace the worse individuals, namely, the selection operation embodies the biological genetic process of fitted survival. Crossover operation is the exchanging of gene pieces of the male parent chromosome and female parent chromosome. Some parts of the genes are changed in mutation operation

Step 3: Computing the fitness values of the new individuals: the sub-generation generated by the male parent and the female parent forms a new generation (population), and then we compute the all fitness values of a new generation to prepare a new evolutionary operation.

Step 4: Stopping the operation when the most optimal fitness solutions are solved or stop criterions are satisfied, and then output the solutions of the chromosome; otherwise, return to step 1.

4 The Theory of Cyclic-Code Generated Polynomial

4.1 Field[1,4]

If *F* is a set included at least one element, and satisfy the follow criterions:

a. All elements of *F* constitute a plus Abel-group, and the plus unit element is marked 0.
b. All non-0 elements of *F* constitute a multiple Abel-group, and the multiple unit element is marked 1.
c. The elements of *F* satisfy the distributive and associative regulations between plus and multiple as follows:
   $a(b+c)=ab+ac$
   $(b+c)a=ba+ca$

where $a,b,c \in F$, so we call *F* a field.

4.2 $GF(q)$-Field

$GF(q)$ is a field which satisfies the following criterions

a. The $GF(q)$ includes q elements
b. All results calculated by modulo-q plus and modulo-q multiple with the elements of $GF(q)$ belong to $GF(q)$.

Example 1: Assume $GF(9)$ ={0,1,2,3,4,5,6,7,8}, {8} and {7} belong to $GF(9)$, then (8+7) modulo 9=6 belongs to $GF(9)$, 8*7 modulo 9=2 belongs to $GF(9)$ too.

Example 2: Assume $p(x) = x^3 + x + 1$ is a polynomial that cannot be divided by other polynomial of $GF(2)$, then the set consisting of 8 leaves of polynomials

$$\{\overline{0}, \overline{1}, \ \overline{x}, \overline{x^2}, \overline{x+1}, \overline{x^2+x}, \overline{x^2+1}, \overline{x^2+x+1}\}$$

produced by modulo $p(x)$ form a finite set $GF(2^3)$ of 8 degrees.

Therefore, we know that the $GF(q)$ is a plus cyclic group and a multiple cyclic group.

4.3 Cyclic codes[1,6]

In N-dimensional linear space V_n in field $GF(q)$, if $V_{n,k}$ is a k-dimension subspace, and to any $C_i = (C_{n-1}, C_{n-2}, ..., C_0) \in V_{n,k}$ there is $\overline{C_i} = (C_{n-2}, ..., C_0, C_{n-1}) \in V_{n,k}$, so we call $V_{n,k}$ a cyclic subspace or cyclic code, where $GF(q)$ is finite field of q exponentials, it's clear that $V_{n,k}$ is a k-dimension linear subspace of n-dimension linear space.

From above, we know that cyclic codes in field $GF(q)$ are linear grouping codes which every code has cyclic feature. Therefore, we can get another definition of cyclic codes:

Assume that C is a linear and $a = (a_0, a_1, ..., a_{n-1})$ is a legal code of C, a new code is formed by shifting the elements along one place and taking one off the end and putting it on to the beginning, then we get $b = aT = (a_1, a_2, ..., a_{n-1}, a_0)$, and b is a legal code of C too, so we call C a cyclic code.

For example, suppose a cyclic-code generated matrix and a checkout matrix of Hamming code C are as follow:

$$G=\begin{bmatrix}1000101\\0100111\\0010110\\0001011\end{bmatrix} \qquad H=\begin{bmatrix}c_1\\c_2\\c_3\end{bmatrix}=\begin{bmatrix}1110100\\0111010\\1101001\end{bmatrix}$$

From the first line of generated matrix based on left moving a place, we get 16 codes (1000101) (0001011) (0010110) (0101100) (1011000) (0110001) (1100010)

(0100111) (1001110) (0011101) (0111010) (1110100) (1101001) (1010011) (1111111) (0000000)

From these codes we can easily see that if C_i is the code of C, moving cyclic code leftward or rightward a place, we can also get the code of C. So this group of linear grouping code is cyclic code.

Assume $C = (C_{n-1}, C_{n-2}, ..., C_0)$ is a code of $[n,k]$ cyclic codes, the corresponding polynomial is $C(x) = C_{n-1}x^{n-1} + C_{n-2}x^{n-2} + \cdots + C_0$. It is called the polynomial of code C (or code polynomial).

4.4 The cyclic-code generated polynomial [3,4,6,7]

Assume $f(x)$ an n -exponential polynomial on $GF(q)$, and $g(x)$ the first 1-coefficient factor of $f(x)$, namely, $f(x) = g(x)h(x)$.

A set: $I = \{m(x)g(x) \mid \partial m(x) \leq n-1-\partial g(x)\}$, where ∂ is exponential numbers of polynomial. Then, we call $g(x)$ the generated polynomial of I, namely, I is generated by $g(x)$.

Assume $g(x)$ a cyclic code (polynomial), then we call it the cyclic-code generated polynomial. There are two important theories of solving legal codes as follows:

Theory 1: $C(x)$ is the cyclic legal code if and only if $C(\alpha_i) = 0, i = 1,2,\cdots,n-k$. Where α_i is the root of $g(x)$ $(i = 1,2,\cdots,n-k)$

Theory 2: Assume $C(x)$ is the cyclic code, $c_{n-1}, c_{n-2}, \cdots, c_0$ are the corresponding coefficients of $C(x)$, and $C(\alpha_i) = 0, i = 1,2,\cdots,n-k$. Where α_i is the root of $g(x)$ ($i = 1,2,\cdots,n-k$), and $\alpha_i \neq \alpha_j$ $(i \neq j)$. Then, we can solve the values of $c_{n-1}, c_{n-2}, \cdots, c_0$ through solving

$$\begin{bmatrix}\alpha_1^{n-1}, \alpha_1^{n-2}, \cdots, \alpha_1, 1\\ \alpha_2^{n-1}, \alpha_2^{n-2}, \cdots, \alpha_2, 1\\ \vdots \qquad\qquad \vdots\ \vdots\\ \alpha_{n-k}^{n-1}, \alpha_{n-k}^{n-2}, \cdots, \alpha_{n-k}, 1\end{bmatrix}\begin{bmatrix}c_{n-1}\\c_{n-2}\\ \vdots\\c_0\end{bmatrix}=0,$$

and the cyclic legal code is $(c_0, c_1, \cdots, c_{n-1})$.

Therefore, we can use the roots of cyclic-code generated polynomial $g(x)$ to construct the cyclic legal code C. But the difficulty in constructing cyclic legal code C is how to solve the roots of cyclic-code generated polynomial. But general method to solve the roots is to factor cyclic-code generated polynomial, as we know, this method is infeasible for solving the roots of high-exponential polynomial. To resolve this difficulty, in the next section we will propose a feasible method on how to use evolutionary computation to solve the roots of cyclic-code generated high-exponential polynomial

5 The Algorithm of Solving the roots of Cyclic-Code Generated polynomial by Evolutionary Computation

Assume $g(x)$ is a cyclic-code generated polynomial, then we convert the solving problem to a single-objective optimal problem as following,

$$\min g(x)$$

subject to $x \in GF(q)$ and $g(x) \geq 0$

Because the field $GF(q)$ is the modulo-q field, the method of finding optimal solutions is more difficult than classical method of evolutionary computation. Therefore, we have to improve the algorithm of evolutionary computation as following:

Step 1, initialize population P_0, set $t = 0$, produce n integral individuals $P_0 = \{\alpha^0{}_i\}$, $i = 1,2,\cdots N$ at random, and calculate the function fitness values of the corresponding individual, and sort the modulo-q values in the order from small to large.

Step 2, Select some individuals to form a reproduction pool

$P_1 = \{\alpha^1{}_i\}$, $i = 1,2,\cdots m$

Step 3, crossover the individuals of P_1

$$\alpha^2 = \text{int}(\sum_{j=1}^{m} \alpha_j{}^1 \beta_j)$$

where $\sum_{l=1}^{m} \beta_l = q$, $0 < \beta_l < q$ are random number, $l = 1,2,\cdots m$

Step 4, use new individual to mutate the parents. If the function fitness value of independent α^2 is better than the worst modulo-q value calculated by step 1, then, use α^2 to replace the corresponding individual, otherwise, return to step 3.

Step 5, If the fitness value of function $\|G(x)\| < 10^{-10}$ or $t > T$ is reached, output the modulo-q individuals, and stop the running. Otherwise, return to step 3.

6 Numerical Experiments

In this section, we apply evolutionary computation to solving the roots of cyclic-code generated polynomial in a special modulo-q field $GF(q)$ to indicate that we can use this method to replace traditional method in which factor the polynomial by artificial operation to solve the roots in this complex modulo-q field $GF(q)$.

Experiment 1: Suppose the length of a code 14, namely, n=14, and the places of checking-code is 4,so n-k equals 10, the cyclic-code generated polynomial is as following:

$$g(x) = x^{14} + 4x^{12} + 7x^{10} + 6x^9 + 4x^3 + x + 1 = 0$$

subject to $g(x) \geq 0$ and $1 \leq x \leq 1000$, $x \in GF(2^3)$

we set maximum iteration times T=20000, after running 543 times the 10 roots are attained as in the following table

No.	Root	No.	Root
1	49	6	681
2	985	7	41
3	129	8	993
4	689	9	681
5	33	10	105

Experiment 2: Suppose the length of a code 14, namely, n=14, and the places of checking-code is 2,so n-k equals 12, the cyclic-code generated polynomial is as following:

$$g(x) = x^8 + 3x^7 + x^6 + 5x^3 + 3x^2 + 2x + 1 = 0$$

subject to $g(x) \geq 0$ and $1 \leq x \leq 1000$, $x \in GF(2^3)$

we set the maximum iteration times T=15000, after running 148 times the 12 roots are attained as in the following table

No.	Root	No.	Root
1	993	7	25
2	757	8	997
3	989	9	785
4	65	10	21
5	13	11	13
6	73	12	77

Through the analysis of the above experiments, we conclude that this method has at least two advantages, compared with the method of traditional manual factorizing polynomial. The first is that using evolutionary algorithm to solve the roots of cyclic-code generated polynomial can save much time, it only takes some minutes to solve the roots, which is impossible by factorizing the polynomial in such short time; the second is that this method can solve all the roots of the polynomial, while using the method of factorizing the polynomial is difficult to solve all the roots of the cyclic-code generated polynomial, sometimes not even a single one.

7 Conclusions

This paper proposed an innovative method to solve the roots of the cyclic-code generated polynomial in the complex modulo-q field $GF(q)$ by using evolutionary computation. Through the experiments and the theoretical analysis, this method is the best method to solve the roots of this type of polynomials up to now, and authors haven't found anyone who proposed such high performance method. Therefore this method will change the method that only factor the cyclic-code generated polynomial of high exponential to find the roots manually in modulo-q field $GF(q)$, and it will also resolve the difficult problem of having roots but could not be solved. So, it extended the application of evolutionary computation in computer security, it also enhanced the reliability of the trustworthy computing at the same time.

References

[1] Wang Xingmei, Zhang Huanguo, Ma Jianfeng, Tan Zhongping, 1999. "Error-Correction Code Technology in Computer". People's Post Press.

[2]. Pan Zhengjun. Kang Lishan, Chen Yuping, Evolutionary Computation, Tsinghua University Press, 1998.

[3] Ye Dacheng, 1996, 24(12). "The solution to genetic algorithm of route selection and volume allocation problem in computer communication network". Electronic Journal.

[4] Wang Xinmei, 1991, 280 ~ 284. "Error-Correcting Code—Theory and Method". Xian: Xidian University Press.

[5] Holland J H. 1975, "Adaption in Natural and Artificial Systems". Ann Arbor: University of Michigan Press.

[6] Spillman R. 1993, 17(4), "Cryptanalysis of Knapsack Chipers Using Genetic Algorithms". Cryptologyia,

[7] Zhang Muxiang, 1994, 76(3), "Simulated Annealing Approach to the Minimum Distance of Error-Correcting Codes". Int J Electronics.

Intrusion Detection System Based on a Cooperative Topology Preserving Method

Emilio Corchado, Álvaro Herrero, Bruno Baruque, José Manuel Sáiz

Department of Civil Engineering, University of Burgos, Spain.
E-mail: escorchardo@ubu.es

Abstract

This work describes ongoing multidisciplinary research which aims to analyse and to apply connectionist architectures to the interesting field of computer security. In this paper, we present a novel approach for Intrusion Detection Systems (IDS) based on an unsupervised connectionist model used as a method for classifying data. It is used in this special case, as a method to analyse the traffic which travels along the analysed network, detecting anomalous traffic patterns related to SNMP (Simple Network Management Protocol). Once the data has been collected and pre-processed, we use a novel connectionist topology preserving model to analyse the traffic data. It is an extension of the negative feedback network characterised by the use of lateral connections on the output layer. These lateral connections have been derived from the Rectified Gaussian distribution.

1 Introduction

The identification of intrusions is a difficult problem due to the dynamic nature of systems and networks, the creativity of attackers, the wide range of computer hardware and operating systems and so on.

This complexity increases if we talk about distributed network-based systems and insecure networks as Internet. An attack and intrusion to a network would end up affecting any of the three computer security principles: availability, integrity and confidentiality, exploiting for example the Denial of Service, Modification and Destruction vulnerabilities [1]. Further, network intruders are constantly updating their attack technology.

For these reasons, intrusion detection systems have become a required element in addition to the computer security infrastructure of most organizations. Intrusion Detection (ID) is a field focused on the identification of attempted or ongoing attacks on a computer system or network. The accurate detection in real-time of computer and network system intrusions has always been a complicated and interesting problem for system administrators and information security researchers.

Intrusion Detection Systems (IDS) are software or hardware systems that speed up and automate the process of monitoring the events which take place in a computer system or network, analyzing them to identify security attacks.

There are two main models to analyze events for detecting attacks: Anomaly detection (identifies activities that diverge from known patterns for users) and Misuse detection (based on the comparison of a user's activities with the known behaviors of attackers attempting to penetrate a system) [2].

2 The Connectionist Analyzer Model

The Data Classification step used by this IDS model is based on the use of the neural architecture called Cooperative Maximum Likelihood Hebbian Learning [3, 4, 5]. It is based on the Negative Feedback Network [6]. Consider an N-dimensional input vector, $\mathbf{x}$, and a M-dimensional output vector, $\mathbf{y}$, with W_{ij} being the weight linking input j to output i and let η be the learning rate.

It can be expressed as:

$$y_i = \sum_{j=1}^{\mathbf{N}} W_{ij} x_j, \forall i \qquad (1)$$

The activation is fed back through the same weights and subtracted from the inputs.

$$e_j = x_j - \sum_{i=1}^{M} W_{ij} y_i, \forall j, \qquad (2)$$

After that simple Hebbian learning is performed between input and outputs.

Lateral connections [3, 4, 5] have been derived from the Rectified Gaussian Distribution [7] and applied to the negative feedback network. The net result [3, 4, 5, 8] will be shown to be a network which can find the independent factors of a data set but do so in a way which captures some type of global ordering in the data set.

We use the standard Maximum-Likelihood Network [3, 5, 9, 10] but now with a lateral connection (which acts after the feed forward but before the feedback). Thus we have: a feed forward step (Eq. 1) follows by:

Lateral Activation Passing: $$y_i(t+1) = [y_i(t) + \tau(b - Ay)]^+ \quad (3)$$

Feedback: $$e_j = x_j - \sum_{i=1}^{M} W_{ij} y_i \quad (4)$$

Weight change: $$\Delta W_{ij} = \eta . y_i . sign(e_j) | e_j |^{p-1} \quad (5)$$

3 IDS Model

The aim of this work is the design of a layered system capable of detecting anomalous situations for a computer network. The information analysed by our system is obtained from the packets which travel along the network. So, it is a Network-Based IDS [2]. The necessary data for the traffic analysis is contained on the captured packets headers. This data can be obtained using a network analyser.

When we talk about anomaly detection model we refer to IDS which detect intrusions by looking for abnormal network traffic. Anomaly detection is based on the assumption that misuse or intrusive behaviour deviates from normal system use [11, 12, 13]. In many cases, as in the case of the attacker who breaks into a legitimate user's account, this is a right assumption. The attacker may behave differently than the regular user, so if the IDS has established what the user normally does during a session, it can determine that the user is not behaving in a usual way and detect the attack.

So in summary, we have developed a system for detecting anomalous traffic patterns, this includes proper attacks and dangerous situations without being an attack. Examples of these ones are management actions performed by the network administrator, so in those cases, the administrator will know that is a real attack or just a false alarm in the case that he has performed it.

3.1 Structure of the Model

The structure of this novel layered IDS model is showed in Fig.1 and it is described as follows:

- **First step**. - Network Traffic Capture: one of the network interfaces is set up as "promiscuous" mode, in such a manner that it is capable of capture all the packets which are travelling along the network.
- **Second step**.- Data Pre-processing: the captured data is pre-processed and used as an input data to the following stage. We only select traffic based on UDP (User Datagram Protocol) as it is explained later. This means that in terms of TCP/IP (Transmission Control Protocol/Internet Protocol) protocol stack, the model analyses only the packets which use UDP at transport layer and IP protocol at network layer.
- **Third step**.- Data Classification: once the data has been captured and pre-processed, the connectionist model presented in section 2 is used to analyse the data and identify the anomalous pattern.
- **Fourth step**.– Result Display: the last step is related to the visualization stage. Finally the output of the network is presented to the administrator or person in charge of the network security. Up to the actual research state, this visualization tool displays data projections highlighting anomalous situations clearly enough to alert the network administrator as we show in Fig.2, taking into account aspects as the traffic density or "anomalous" traffic directions.

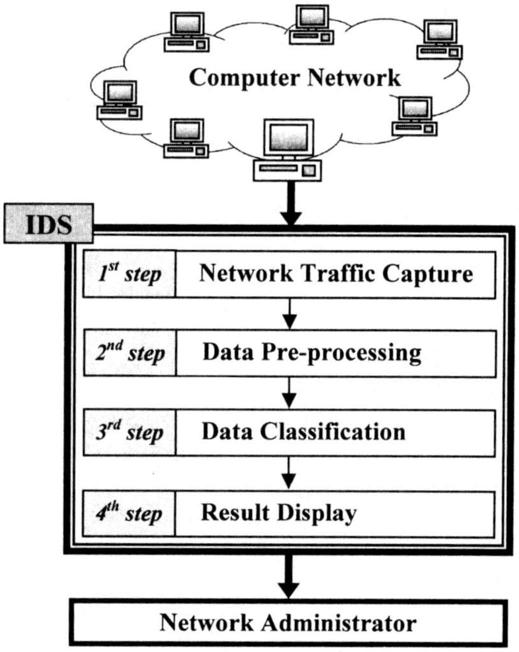

Fig. 1. Model Structure of the Layered IDS.

4 Real Data Set

The data pre-processing step is performed taking into account the following:

There are several ports/protocols that can be considered dangerous for the network security: SNMP, ICMP, TFTP and so on. Among those, we have actually focused our effort in the study of SNMP because an attack based on this protocol may severely compromise the systems security. CISCO [14] found the top five most vulnerable services in order of importance, and SNMP was one of them.

The study of SNMP protocol is the reason why the system selects packets based on UDP during the data pre-processing layer.

This research will continue trying to extend the model to cover several different situations, including other SNMP anomalous situations and protocols, until to cover all of them.

Data selection: we used only 7 variables extracted from the packet headers among all the information captured for each one:

- Timestamp: the time difference in relation to the first captured packet.
- Source Port: the port of the source host from where the packet is sent.
- Destination Port: the port of the destination host to where the packet is sent.
- Size: total packet size (in Bytes).
- Protocol: in this case we have used values between 1 and 35 to identify the packet protocol.
- Source IP: numeric value which codifies the source host IP address.
- Destination IP: numeric value which codifies the destination host IP address.

In terms of IP address, we have fixed numeric values to addresses included in each range in which the network is divided, given special values to the multicast and broadcast addresses.

This specific data set contains a scanning of network computers for the SNMP (Simple Network Management Protocol) port using sniffing methods. The aim is to make a systematic sweep in a group of hosts to verify if SNMP protocol is active in one of the following ports: 161, 162 and 3750. The sweep has been done using these port numbers because:

- 161 and 162 are the default port numbers for SNMP, as RFC 1157 [15] says: protocol entity receives messages at UDP port 161, and messages which report traps should be received on UDP port 162.
- We have also included a random port (3750) in the sweep as a test random element.

Some features of the analysed traffic along the network are the following:

- The SNMP packets are generated and sent inside the own network, this is, it is an internal protocol and any host out of the network can not introduce any packets of this type in the network. This is mainly warranted by the external security implemented through the firewall.
- We have taken into account all the traffic to ensure the existence of both, anomalous and non-anomalous situations. These have similar behaviours so the differences are difficult to identify making it an interesting problem to investigate.

5 Results

Fig. 2 shows traffic based on several protocols such as BOOTP, NETBIOS, DNS, TIMED and SNMP.

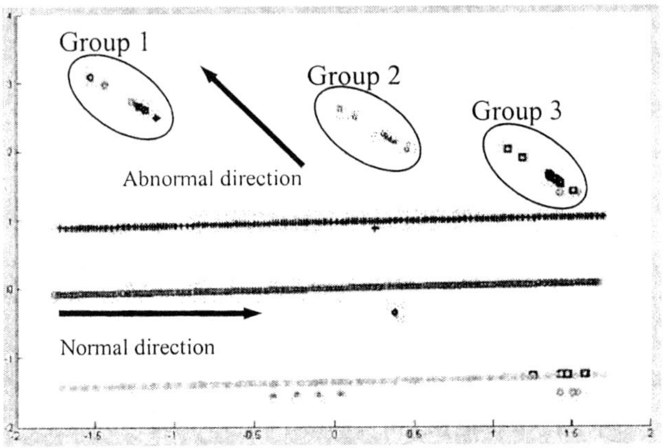

Fig.2. Data projections displayed by the model identifying anomalous situations.

Through a simple visual analysis of Fig.2 we can see that while most of the traffic evolves in the same direction, it is easy to identify three groups (Groups 1, 2 and 3.- Fig. 2) progressing in a different direction. We have study this matter (identifying every represented point) and we have concluded that these groups are related to the SNMP sweep mentioned above. Each group identified in Fig. 2 contains points that represent packets sent to each port included in the sweep (161, 162 and 3750) which is embedded in data set introduced to the model. All the packets belonging to SNMP protocol are contained in one of these three groups and there are no packets belonging to another protocol. In terms of performance results, our model has identified the three anomalous situations existing in the data set, as we known.

These graphical features allow the Network-Administrator to identify the sweep anomalous situation just by looking. The main feature that allows identifying the anomalous patterns is the growth direction. It can be seen that it is not parallel to the normal traffic direction.

6 Conclusions

We have developed a novel layered system for detecting anomalous traffic patterns including proper attacks and dangerous situations without being an attack, which can be considered an IDS. This work is actually focused on the study of SNMP because an attack based on this protocol may severely compromise the systems security.
We have applied different methods such as Principal Component Analysis [6, 16] or Maximum Likelihood Hebbian Learning for the classification step. Cooperative Maximum Likelihood Hebbian Learning provides more sparse projections than the others [5].
This is an ongoing research with the aim of showing the viability of the system developed. Later on it will be extended to cover a greater variety of anomalous situations as dictionary attacks or spoofing.
Future work will be based on the study of different distributions and learning rules to improve the whole architecture and to improve the system in such a way that it can be able to capture, process, classify and display the data in real time.

References

[1] Myerson, J.M. (2002) Identifying enterprise network vulnerabilities. International Journal of Network Management. 12(3).

[2] Planquart, J-P. (2002) Application of Neural Networks to Intrusion Detection. Information Security Reading Room - SANS (SysAdmin, Audit, Network, Security) Institute.

[3] Corchado, E., Fyfe, C. (2003) Orientation Selection Using Maximum Likelihood Hebbian Learning. International Journal of Knowledge-Based Intelligent Engineering Systems. 7(2).

[4] Corchado, E., Han, Y., Fyfe, C. (2003) Structuring global responses of local filters using lateral connections. J. Exp. Theor. Artif. Intell. 15(4): 473-487.

[5] Corchado, E., Fyfe, C. (2003) Connectionist Techniques for the Identification and Suppression of Interfering Underlying Factors. International Journal of Pattern Recognition and Artificial Intelligence. 17(8): 1447-1466.

[6] Fyfe, C. (1996) A Neural Network for PCA and Beyond. Neural Processing Letters. 6:33-41.

[7] Seung, H.S., Socci, N.D., Lee, D. (1998) The Rectified Gaussian Distribution. Advances in Neural Information Processing Systems, 10: 350.

[8] Corchado, E., Corchado, J.M., Sáiz, L., Lara, A. (2004) Constructing a Global and Integral Model of Business Management Using a CBR system. First International Conference on Cooperative Design, Visualization and Engineering (CDVE 04).

[9] Corchado, E., MacDonald, D., Fyfe, C. (2004) Maximum and Minimum Likelihood Hebbian Learning for Exploratory Projection Pursuit, Data mining and Knowledge Discovery. Kluwer Academic Publishing. 8(3): 203-225.

[10] Fyfe, C., Corchado, E. (2002) Maximum Likelihood Hebbian Rules. European Symposium on Artificial Neural Networks.

[11] Lunt, T., Tamaru, A., Gilham, F., Jaganathan, R., Neuman, P., Jalali, C. (1990) IDES: A Progress Report. Sixth Annual Computer Security Applications Conference.

[12] Denning, D. (1987) An Intrusion Detection Model. IEEE Transactions on Software Engineering. SE-13(2).

[13] Debar, H., Becker, M., Siboni, D. (1992) A Neural Network Component for an Intrusion Detection System. IEEE Symposium on Research in Computer Security and Privacy.

[14] Cisco Secure Consulting. (2000) Vulnerability Statistics Report.

[15] Case, J., Fedor, M.S., Schoffstall, M.L., Davin, C. (1990) Simple Network Management (SNMP). RFC-1157.

[16] Oja, E. (1989) Neural Networks, Principal Components and Subspaces. International Journal of Neural Systems. 1: 61-68.

Model Selection for Kernel Based Intrusion Detection Systems

Srinvas Mukkamala[1], A. H. Sung[1], B. M. Ribeiro[2]

[1]Department of Computer Science, New Mexico Tech, Socorro NM 87801, U.S.A.
[2]Department of Informatics Engineering, University of Coimbra, Portugal
Email: {srinivas|sung|bmr@cs.nmt.edu}

Abstract

This paper describes results concerning the robustness and generalization capabilities of a supervised machine learning method in detecting intrusions using network audit trails. We also evaluate the impact of kernel type and parameter values on the accuracy with which a support vector machine (SVM) performs intrusion classification. We show that classification accuracy varies with the kernel type and the parameter values; thus, with appropriately chosen parameter values, intrusions can be detected by SVMs with higher accuracy and lower rates of false alarms.
Feature selection is as important for intrusion detection as it is for many other problems. We present support vector decision feature selection method for intrusion detection. It is demonstrated that, with appropriately chosen features, intrusions can be detected in real time or near real time.

1 Introduction

Intrusion detection is a problem of great importance to protecting information systems security, especially in view of the worldwide increasing incidents of cyber attacks. Since the ability of an Intrusion Detection System (IDS) to identify a large variety of intrusions in real time with accuracy is of primary concern, we will in this paper consider performance of SVM-based IDSs with respect to classification accuracy and false alarm rates, and their relation to parameter selection and kernel type.
AI techniques have been used to automate the intrusion detection process; they include neural networks, fuzzy inference systems, evolutionary computation, machine learning, etc. Several research groups recently have used SVMs to build IDSs. However, most groups that studied SVMs for IDS considered only a small set of kernels and parameters [1-5]. Although several groups have extensively considered model selection in SVMs, optimal parameters are usually domain specific. In this paper, we present a methodology to evaluate the impact of model selection (kernel types and parameter values) on the performance of a SVM to detect intrusions.
Data mining techniques have been introduced to identify key features that characterize intrusions [6-8]. We performed experiments to rank the importance of input features using support vector decision function for each of the five classes (normal, probe, denial of service, user to super-user, and remote to local) of network traffic patterns in the DARPA data. It is shown that using only the important features for classification gives better performance.
Intrusion detection data used for experiments is briefly explained in section 2. A brief introduction to model selection using SVMs for intrusion detection is given in section 3. In section 4, we analyze classification accuracies of SVMs using ROC curves. A brief introduction to feature selection and SVM-specific feature identification is given in section 5.

2 Data Used for Analysis

A subset of the DARPA intrusion detection data set is used for offline analysis. In the DARPA intrusion detection evaluation program, an environment was set up to acquire raw TCP/IP dump data for a network by simulating a typical U.S. Air Force LAN. The LAN was operated like a real environment, but being blasted with multiple attacks [9,10]. For each TCP/IP connection, 41 various quantitative and qualitative features were extracted [6] for intrusion analysis. Attacks are classified into the following types.
Attack types fall into four main categories:

1. DOS: denial of service
2. R2L: unauthorized access from a remote machine
3. U2Su: unauthorized access to local super user (root) privileges
4. Probing: surveillance and other probing

3 Model Selection

In any predictive learning task, such as classification, both a model and a parameter estimation method should be selected in order to achieve a high level of performance of the learning machine. Recent approaches allow a wide class of models of varying complexity to be chosen. Then the task of learning amounts to selecting the sought-after model of optimal complexity and estimating parameters from training data [11,12].
Within the SVMs approach, usually parameters to be chosen are (i) the penalty term C which determines

the trade-off between the complexity of the decision function and the number of training examples misclassified; (ii) the mapping function Φ; and (iii) the kernel function such that $K(\mathbf{x}_i, \mathbf{x}_j) = \Phi(\mathbf{x}_i) \cdot \Phi(\mathbf{x}_j)$. In the case of RBF kernel, the width, which implicitly defines the high dimensional feature space, is the other parameter to be selected [13].

We performed a grid search using 10-fold cross validation for each of the five faults in our data set.

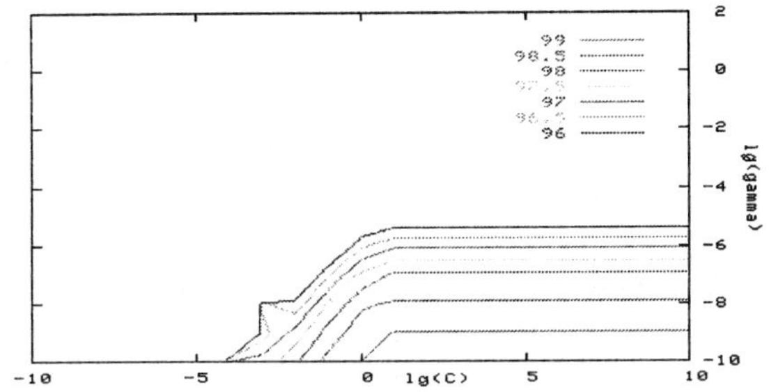

Fig. 1. SVM model for Normal.

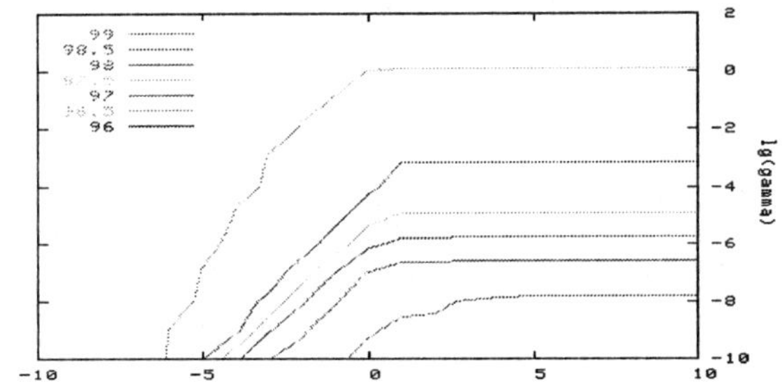

Fig. 2. SVM model for Probe.

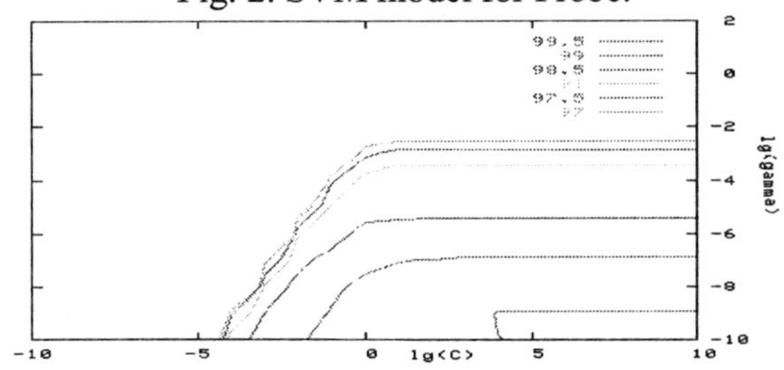

Fig. 3. SVM model for DoS.

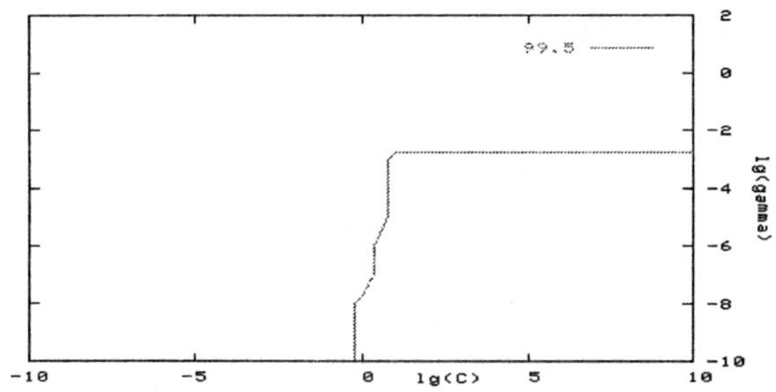

Fig. 4. SVM model for U2Su.

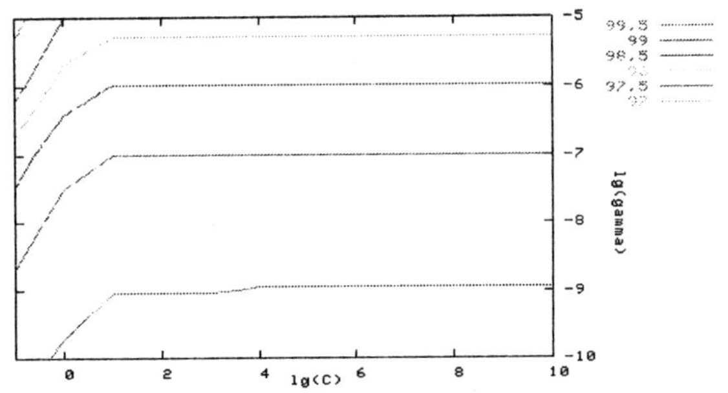

Fig. 5. SVM model for R2L.

First, we achieved the search of parameters C and γ in a coarse scale and then we carried through a fine tunning into the five detection faults proper space. Model selection results obtained through grid search are given in figures 1 to 5 for normal, probe, DoS, U2Su, and R2L, respectively.

4 ROC Curves

The Receiver Operating Characteristic (ROC) curves are generated by considering the rate at which true positives accumulate versus the rate at which false positives accumulate with each one corresponding, respectively, to the vertical axis and the horizontal axis in Figures 6 to 10.

The point (0,1) is the perfect classifier, since it classifies all positive cases and negative cases correctly. Thus an ideal system will initiate by identifying all the positive examples and so the curve will rise to (0,1) immediately, having a zero rate of false positives, and then continue along to (1,1).

Detection rates and false alarms are evaluated for the five-class pattern in the DARPA data set and the obtained results are used to form the ROC curves.

Figures 6 to 10 show the ROC curves of the detection models by attack categories as well as on all intrusions. In each of these ROC plots, the x-axis is the false alarm rate, calculated as the percentage of normal connections considered as intrusions; the y-axis is the detection rate, calculated as the percentage of intrusions detected. A data point in the upper left corner corresponds to optimal high performance, i.e, high detection rate with low false alarm rate [14].

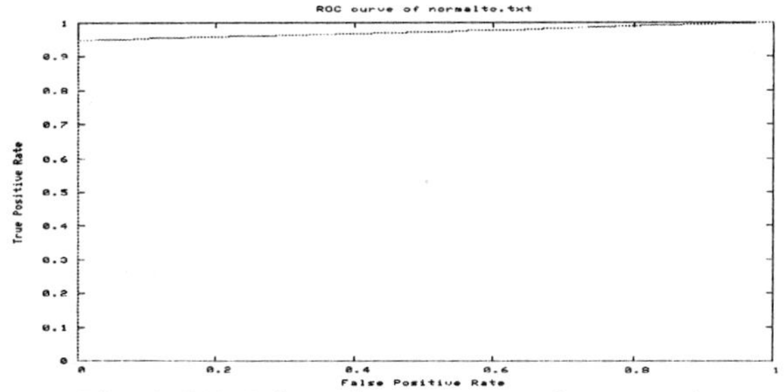

Fig. 6. SVM detection accuracy for normal.

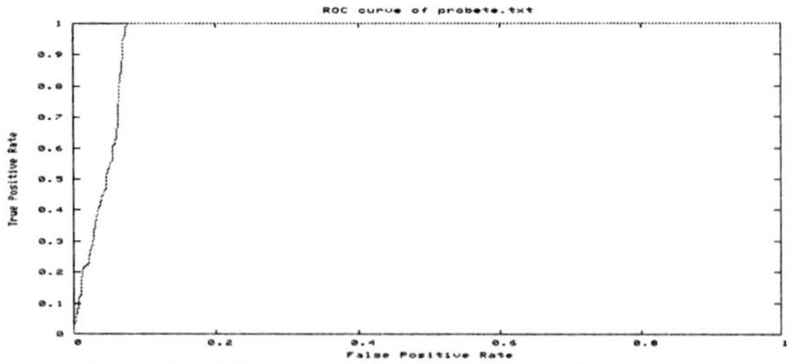

Fig. 7. SVM detection accuracy for probe.

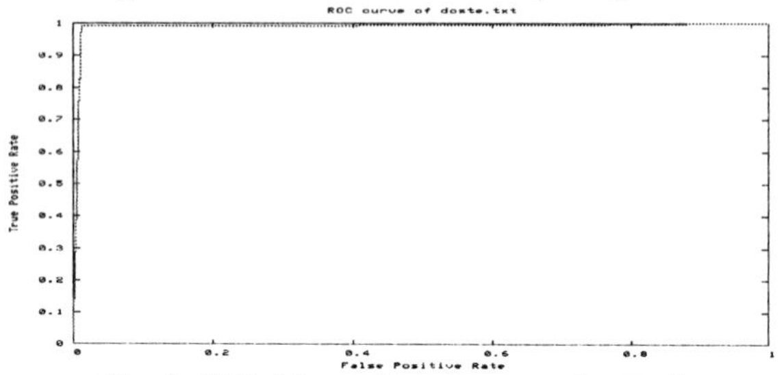

Fig. 8. SVM detection accuracy for DoS.

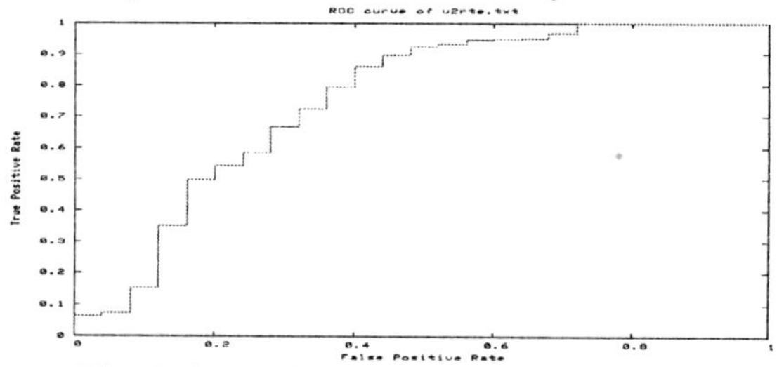

Fig. 9. SVM detection accuracy for U2Su.

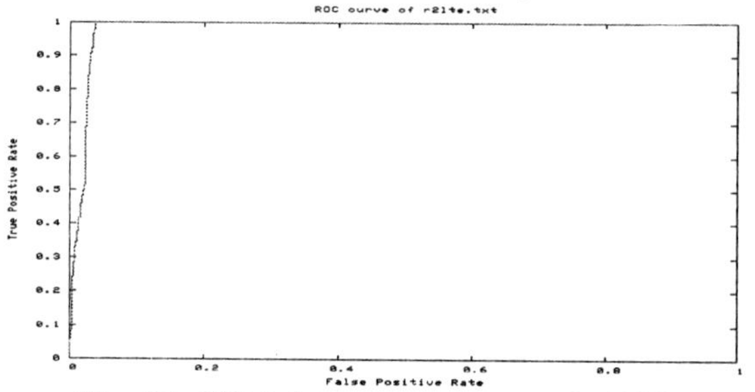

Fig. 10. SVM detection accuracy for R2L.

5 Feature Ranking and Selection

Feature selection is an important issue in intrusion detection. Of the large number of features that can be monitored for intrusion detection purpose, which are truly useful, which are less significant, and which may be useless? The question is relevant because the elimination of useless features (the so-called audit trail reduction) enhances the accuracy of detection while speeding up the computation, thus improving the overall performance of IDS. In cases where there are no useless features, by concentrating on the most important ones one may well improve the time performance of an IDS without affecting the accuracy of detection in statistically significant ways.

The feature selection problem for intrusion detection is similar in nature to various engineering problems that are characterized by:

- Having a large number of input variables $\mathbf{x} = (x_1, x_2, \ldots, x_n)$ of varying degrees of importance to the output **y**; i.e., some elements of **x** are essential, some are less important, some of them may not be mutually independent, and some may be useless or noise (in determining the value of **y**)
- Lacking an analytical model that provides the basis for a mathematical formula that precisely describes the input-output relationship, $y = F(x)$
- Having available a finite set of experimental data, based on which a model (e.g. intelligent systems) can be built for simulation and prediction purposes

5.1 SVM-specific Feature Ranking Method

Information about the features and their contribution towards classification is hidden in the support vector decision function. Using this information one can rank their significance, i.e., in the equation

$$F(X) = \Sigma W_i X_i + b$$

The point X belongs to the positive class if F(X) is a positive value. The point X belongs to the negative class if F(X) is negative. The value of F(X) depends on the contribution of each value of X and W_i. The absolute value of W_i measures the strength of the classification. If W_i is a large positive value then the ith feature is a key factor for positive class. If W_i is a large negative value then the ith feature is a key factor for negative class. If W_i is a value close to zero on either the positive or the negative side, then the ith feature does not contribute significantly to the classification. Thus, a ranking can be done by considering the support vector decision function.
We validate the ranking by comparing the performance of the classifier using all input features to that using the important features; and we also compare the performance of a classifier using the union of the important features for all fives classes.

Table 1 SVM detection accuracies

Class	Classifier Accuracy (%)	
	SVMs (41 features)	SVMs (6 features)
Normal	99.55	99.23
Probe	99.70	99.16
DoS	99.25	99.16
U2Su	99.87	99.87
R2L	99.78	99.78

Table 2 Most important feature descriptions

Class	Feature Description
6 Most Important Features	▪ source bytes: number of bytes sent from the host system to the destination system ▪ dst_host_srv_count: : number of connections from the same host with same service to the destination host during a specified time window ▪ count: number of connections made to the same host system in a given interval of time ▪ protocol type: type of protocol used to connect (e.g. tcp, udp, icmp, etc.) ▪ srv_count: number of connections to the same service as the current connection during a specified time window ▪ flag: normal or error status of the connection

6. Conclusions

A number of observations and conclusions are drawn from the results reported in this paper:

SVMs easily achieve high detection accuracy (higher than 99%) for each of the 5 classes of DARPA data, regardless of whether all 41 features are used, or only the important features for each class are used. Using the important features for each class gives the most accurate performance.

A grid search for intrusion detection (Figures 1 to 5) which seeks the optimal values of the constraint penalty for method solution and the kernel width (C,γ) has been performed. We demonstrate that the ability with which SVMs can classify intrusions is highly dependent upon both the kernel type and the parameter settings.

We note, however, that the difference in accuracy figures tend to be small and may not be statistically significant, especially in view of the fact that the 5 classes of patterns differ tremendously in their sizes. More definitive conclusions perhaps can only be drawn after analyzing more comprehensive sets of network data.

Acknowledgements

Partial support for this research received from ICASA (Institute for Complex Additive Systems Analysis, a division of New Mexico Tech), a DoD IASP, and an NSF SFS Capacity Building grants are gratefully acknowledged. We would also like to acknowledge many insightful discussions with Dr. Jean-Louis Lassez that helped clarify our ideas. The collaborative work of the third author was performed during a sabbatical visit to New Mexico Tech in 2004.

References

[1] Mukkamala, S., Janowski, G., Sung. A.H. (2002) Intrusion Detection Using Neural Networks and Support Vector Machines. Proceedings of IEEE International Joint Conference on Neural Networks 2002, IEEE press, pp. 1702-1707

[2] Fugate, M., Gattiker, J.R. (2003) Computer Intrusion Detection with Classification and Anomaly Detection, Using SVMs. International Journal of Pattern Recognition and Artificial Intelligence 17(3): 441-458

[3] Hu, W., Liao, Y., Vemuri., V.R. (2003) Robust Support Vector Machines for Anamoly Detection in Computer Security. International Conference on Machine Learning, pp. 168-174

[4] Heller, K.A., Svore, K.M., Keromytis, A.D., Stolfo, S. J. (2003) One Class Support Vector Machines for Detecting Anomalous Window Registry Accesses. *In* 3rd IEEE Conference Data Mining Workshop on Data Mining for Computer Security

[5] Lazarevic, A., Ertoz, L., Ozgur, A., Srivastava, J., Kumar, V. (2003) A Comparative Study of Anomaly Detection Schemes in Network Intrusion Detection. In Third SIAM Conference on Data Mining

[6] Stolfo, J., Wei, F., Lee, W., Prodromidis, A., Chan, P.K. (1999) Cost-based Modeling and Evaluation for Data Mining with Application to Fraud and Intrusion Detection. Results from the JAM Project

[7] Mukkamala, S., Sung, A.H., (2003) Feature Selection for Intrusion Detection Using Neural Networks and Support Vector Machines. Journal of the Transportation Research Board of the National Academics, Transportation Research Record No 1822: 33-39

[8] Mukkamala, S., Sung, A.H. (2003) Identifying Significant Features for Network Forensic Analysis Using Artificial Intelligence Techniques. In International Journal on Digital Evidence, IJDE 3

[9] Kendall, K. (1998) A Database of Computer Attacks for the Evaluation of Intrusion Detection Systems. Master's Thesis, Massachusetts Institute of Technology (MIT)

[10] Webster, S.E. (1998) The Development and Analysis of Intrusion Detection Algorithms. Master's Thesis, MIT

[11] Chapelle, O., Vapnik, V. (1999) Model selection for support vector machines. Advances in Neural Information Processing Systems 12

[12] Cherkassy, V. (2002) Model complexity control and statistical learning theory. Journal of natural computing 1: 109–133

[13] Cristianini, N., Taylor, J.S. (2000) Support Vector Machines and Other Kernel-based Learning Algorithms. Cambridge, UK: Cambridge University Press

[14] Egan, J.P. (1975) Signal detection theory and ROC analysis. New York: Academic Press

A Comparison of Three Genetic Algorithms for Locking-Cache Contents Selection in Real-Time Systems

E. Tamura[1], J.V. Busquets-Mataix[2], J. J. Serrano Martín[2], A. Martí Campoy[2]
[1]Grupo de Automática y Robótica, Pontificia Universidad Javeriana - Cali, Colombia.
[2]Departamento de Informática de Sistemas y Computadores*,
Universidad Politécnica de Valencia, Spain
E-mail: eutamo@doctor.upv.es, {vbusque, jserrano, amarti}@disca.upv.es

Abstract

Locking caches, providing full determinism and good performance, are a very interesting solution to replacing conventional caches in real-time systems. In such systems, temporal correctness must be guaranteed. The use of predictable components, like locking caches, helps the system designer to determine if all the tasks will meet its deadlines. However, when locking caches are used in a static manner, the system performance depends on the instructions loaded and locked in cache. The selection of these instructions may be accomplished through a genetic algorithm.

This paper shows the impact of the fitness function in the final performance provided by the real-time system. Three fitness functions have been evaluated, showing differences in the utilisation and performance obtained.

1 Introduction

To cope with the processing power demanded by today complex engineering systems, real-time system designers are resorting to high-performance contemporary processors. These processors are designed with a major goal: to provide good average execution times over a great variety of computing needs. Unfortunately, in real-time systems, what the designer needs is to guarantee that the tasks must execute on time even under adverse circumstances. To do so, it is necessary to estimate the worst-case response time of every task in the system.

Given the speed disparity between the memory system and contemporary processors, computer architects introduce a cache memory in between. However, cache operation makes hard to predict the execution times since cache contents change dynamically, adapting to the execution path in each moment. Furthermore, during its execution, the tasks working sets interferes with each other. Therefore, to ease the analysis, it is feasible to lock the contents of the cache memory with some predetermined instructions such that the system provides not just better performance but also predictability as shown in ref. 1 and ref. 2. The problem is that, as it is proved in ref. 3, determining an optimal placement of cache contents by trying to maximise the number of times that the referenced datum is in the cache memory is NP-hard. Hence, it may be a good idea to apply some heuristics. Previous works published in ref. 4 show that using a genetic algorithm represents a good solution for this particular problem.

However, performance of locking cache is highly dependant on the genetic algorithm characteristics. In addition, performance of real-time systems must be quantified with several metrics: system utilisation, task slack (distance from end of task execution to its deadline), or average task slacks in the system.

This paper explores three different fitness functions, and evaluates the performance they provide in terms of system utilisation and slack time, two metrics commonly used in evaluating real-time systems. The results will illustrate which is the most adequate fitness function to get the best value for each metric.

The remainder of the paper is organised as follows. Section 2 introduces the genetic algorithm and its three fitness functions. Section 3 illustrates the experimental procedure. Finally, in Section 4 some conclusions and future work are summarised.

2 Genetic Algorithm

2.1 Representation

Each individual is modelled as a binary vector of dimension n, where n is the number of memory blocks occupied by all of the tasks. If a bit is set to one, the associated memory block is to be loaded and locked in the cache memory.

2.2 Fitness functions

The first fitness function will attempt to allocate more blocks for those real-time tasks with lower priorities, try-

*This work has been supported in part by the Spanish *Comisión Interministerial de Ciencia y Tecnología* under project CICYT-TIC2003-08106-C02-01 ing to compensate the time lost due to the execution of higher priority real-time tasks. Equation 1 shows the resulting fitness function:

$$f_A = \frac{R_1 + \sum_{i=2}^{N} 2^{i-2} R_i}{2^{N-1}} \qquad (1)$$

where N is the number of tasks in the system, R_i is the response time of each task, τ_i, computed using CRTA as given in ref. 5.

The second fitness function will try to provide a lower processor utilisation at the system level.

$$f_B = \sum_{i=1}^{N} \frac{C'_i}{T_i} \qquad (2)$$

where C'_i, the computation time of τ_i includes all cache effects.

The third fitness function tries to improve the average slack time at the system level.

$$f_C = \frac{\sum_{i=1}^{N} S_i}{N} \qquad (3)$$

with

$$S_i = 1 - \frac{R_i}{D_i} \qquad (4)$$

where R_i and D_i are respectively the response time and deadline of τ_i.

2.3 Selection Criteria

Given any of the previous fitness function there are two possible outcomes:

- The number of locked blocks required by the task set is less than or equal to the cache memory size. This is a valid individual.
- The number of locked blocks required by the task set is greater than the cache memory size. This is a non-valid individual, since even though the system may provide acceptable response times, it does not satisfy the intended requirements.

A rank-based selection is used, in which the individuals are sorted according to two basic rules. If the individual is valid, it is ranked according to its fitness value. If the individual is not valid, it is ranked according to the number of blocks used; the lesser the number of blocks, the better the individual.

2.4 Crossover and Mutation

For crossover, two individuals are randomly selected from the previous ranking. Both individuals are divided in two ends, randomly selecting the splitting point. Then, by exchanging the two portions to the right of the cut-off point, two new individuals are created. Once the crossover is done, the resulting new individuals may use more cache lines than the cache memory has available. However, to make a broader exploration of the search space, those individuals are not discarded. Therefore, mutation is applied by following one of three schemas:

- For individuals with a number of locked blocks greater than cache size, the mutation procedure selects at random a set of locked blocks and marks them as unlocked, reducing the number of locked blocks. The resulting individual may have a number of locked blocks that are greater, equal or lower than the cache size.
- For individuals with a number of locked blocks lower than cache size, mutation randomly selects a set of unlocked blocks and mark them as locked, increasing the number of locked blocks. The resulting individual may have a number of locked blocks greater, equal or lower than the cache size.
- For individuals with a number of locked blocks equal than cache size, mutation randomly selects a set of pairs, each pair with one locked block and one unlocked block, and exchanges them, leaving unchanged the number of locked blocks.

This policy allows the existence of non-valid individual, but also helps to keep its number low.

2.5 Initial population and tuning parameters

Although a genetic algorithm can explore all the search space through crossover and mutation, selecting adequately the initial population may help the algorithm to find a sub-optimal solution with a lower number of iterations. Due to the structure of the tasks, the ideal solution is an individual with a number of 1's equal to the cache size; hence, the best solution includes a large sequence of consecutive 1's. The population is initialised with sequences of 1's, randomly selecting the beginning.

Other parameter settings are: Population size: 200; Number of generations: 5000; Probability of crossover: 0.6; Probability of mutation for individual with number of locked blocks equal to cache size: 0.01; Probability of mutation for individual with number of locked blocks distinct to cache size: 0.001; Probability of selection of the highest ranked individual: 0.1. The parameter settings are based on results of several preliminary runs.

They are comparable to the typical values mentioned in ref. 6.

3 Experimental Procedure

There were 26 different setups, each with 3 to 8 tasks. The code for each task is synthetic; it does nothing useful but it has a mix of instructions such that it is easy to generate different programs, which is adequate for the purpose. Each experiment was tested using seven different cache memories ranging from 64 lines to 4096 lines, for 182 experiments comprising 770 tasks. In some of the experiments, the footprint (the amount of memory required) of the task set was smaller than the cache size, which means that they will run as fast as possible, since there will be no interference at all. Because of this, they were discarded; the final number of valid experiments is 146 and the number of tasks is 610.

For each of the 146 experiments, three runs of the genetic algorithm -one run for each fitness function- were accomplished. For each run, the overall system utilisation, the per-task slack times, and, finally, the overall slack time, were estimated.

3.1 Overall System Utilisation

Let N be the number of tasks in the system, the computation time, C'_i, of each task, τ_i, is calculated using CRTA. Assuming that task τ_i has higher priority than task τ_j whenever $i < j$, the calculation of the computation times, are given by:

$$C'_1 = R_1 \tag{5}$$

$$C'_i = R_i - \sum_{\forall\, 0<j<i} C'_j \left\lceil \frac{R_i}{T_j} \right\rceil, \forall\, i \mid 1 < i \leq N \tag{6}$$

which takes into account the execution times of those tasks whose priority is higher than the priority of the current task.

The system utilisation, U_l, is then given by:

$$U_l = \sum_{i=1}^{N} \frac{C'_i}{T_i} \tag{7}$$

3.2 Slack Time

Let N be the number of tasks in the system, the per task slack time, S_i, is given by:

$$S_i = 1 - \frac{R_i}{D_i} \tag{8}$$

where D_i is the deadline of task τ_i.

Then, the system average slack time is calculated by means of:

$$S_{avg} = \frac{1}{N} \sum_{i=1}^{N} S_i \tag{9}$$

3.3 Statistical Analysis

With just the bare results attained in the experiments it is not possible to declare, which fitness function is better, so statistical analyses will be used. First, a comparison of the statistical summaries is done; this is followed by a paired-sample analysis, a procedure designed to test for significant differences between two data samples where the data is collected as pairs. Several statistical values have been used to determine if there exist differences between samples. In addition, three null hypothesis tests have been done for each comparison. These are t-test, sign test, and signed rank test.

3.3.1 Analysis of System Utilisation: Figure 1 shows a comparison of the average, median, lower and upper quartile for the utilisation obtained from each fitness function. It can be observed that function f_A provides the largest (the worst) utilisation, while functions f_B and f_C offers similar results. In order to determine if there are significant differences between the utilisation provided by f_B and f_C, a paired-sample analysis has been performed. The analysis shows that the difference between f_B and f_C is small but statistically significant, giving lower utilisation (better performance) for the results obtained when using the fitness function f_B.

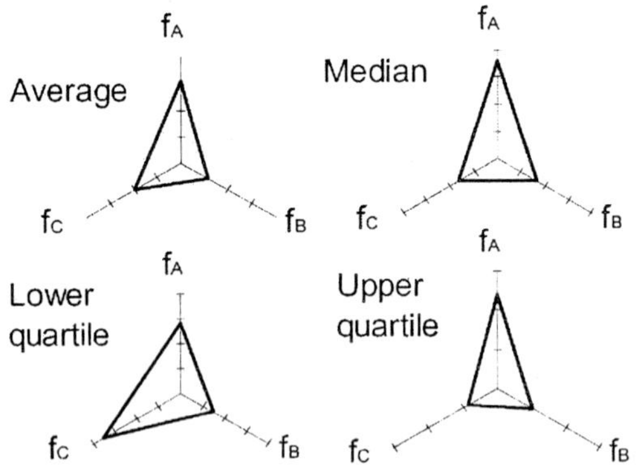

Fig. 1. Utilisation from each fitness function

3.3.2 Overall Slack Time: From Figure 2 it can be clearly recognised than f_C offers a large (better) average slack than f_A and f_B. Regarding the comparison between f_A and f_B, Figure 2 presents contradictory results. Besides that, the paired-sample analysis for the data coming from f_A and f_B does not help to decide which of the two fitness functions provides better average slack. Albeit, it may be reasonable to say that both f_A and f_B provide the same average slack.

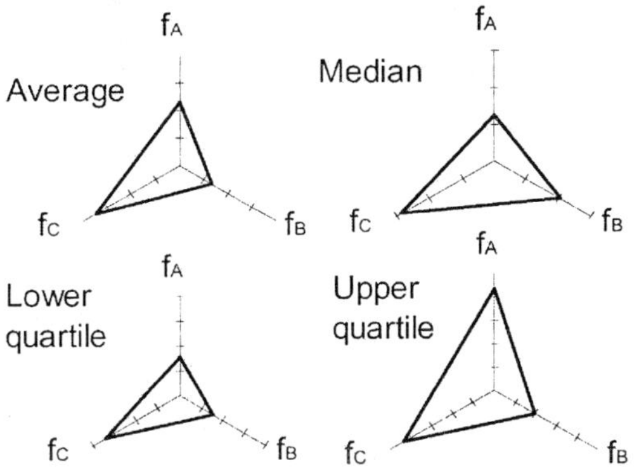

Fig. 2. Average slack from each fitness function

4 Conclusions and Future Work

This paper showed the impact of three different fitness functions over the results provided by a genetic algorithm that selects the contents of a locking cache memory, which is used in a static way in a real-time system. The first fitness function, f_A, tries to minimise the average response time of the tasks. The second one, f_B, tries to minimise the global system utilisation. Finally, the third one, f_C, tries to minimise the average slack time.

The evaluation is based upon two metrics: global system utilisation, U, and Overall Average Slack Time, S.

- Fitness function f_B is statistically better than f_C whenever it is required to optimise the overall utilisation. Fitness function f_A presents worst utilisation than the other two functions.

- Regarding overall slack time, fitness function f_C offers the better performance, while there is no difference between f_B and f_A.

From the results it can be seen that none of the fitness functions perform well in the two basic metrics, although fitness function f_C may be the best option to get the optimal average slack and quasi-optimal utilisation. However, if the performance required for one of the two metrics is a critical parameter, the system designer should choose between f_B and f_C the one that it is more appropriate to its particular optimisation interests. Table 1 gives the proposed selection criteria of fitness function.

Table 1. Function selection criteria versus optimising parameter

Rank	Optimising for utilisation	Optimising for slack
First option	f_B	f_C
Second option	f_C	f_B or f_A
First option	f_A	f_B or f_A

The statistically significant but very small difference between f_B and f_C concerning utilisation allows to expect that the combined use of the two fitness functions in the same genetic algorithm, may bring a trade-off solution for both metrics. Therefore, future work will involve the development of a selection operator that considers two or more metrics for different fitness functions in order to sort the individuals.

References

[1] Martí, A., Perles, A., Busquets Mataix, J. V. (2001) Static Use of Locking Caches in Multitask Preemptive Real-Time Systems. In: Proceedings of the IEEE/IEE Real-Time Embedded Systems Workshop (Satellite of the 22nd IEEE Real-Time Systems Symposium), London, UK.

[2] Martí, A., Perles, A., Busquets-Mataix, J.V. (2002) Dynamic Use Of Locking Caches In Multitask, Preemptive Real-Time Systems. In: Proceedings of the 15th World Congress of the International Federation of Automatic Control, Elsevier Science, Barcelona, Spain.

[3] Petrank, E., Rawitz, D. (2002) The harness of cache conscious data placement. In: Proceedings of the 29th ACM SIGPLAN-SIGACT Symposium on Principles of Programming Languages, pp. 101-102, Portland, U.S.A.

[4] Martí Campoy, A., Pérez Jiménez, A., Perles Ivars, A., Busquets Mataix, J.V. (2001) Using Genetic Algorithms in Content Selection for Locking-Caches. In: Proceedings of the IASTED International Symposia Applied Informatics. pp. 271-276. Acta Press. Innsbruck, Austria.

[5] Busquets-Mataix, J.V., Wellings, A.J., Serrano, J.J., Ors, R., Gil, P. (1996) Adding Instruction Cache Effect to an Exact Schedulability Analysis of Preemptive Real-Time Systems. In: Proceedings of the 8th Euromicro Workshop on Real-Time Systems, pp. 8-15, L'Aquila, Italy.

[6] Mitchell, M. (1996) An Introduction to Genetic Algorithms, MIT Press, Cambridge

A Binary Digital Watermarking Scheme Based On The Orthogonal Vector And ICA-SCS Denoising

Han dongfeng , Li wenhui

Department of Computer Science, University of JiLin, China

E-mail: {hhhhhhdf,handongfenggood}@sohu.com

Abstract

This paper proposed a new perceptual digital watermarking scheme based on ICA, SCS, the human visual system (HVS), discrete wavelet transform (DWT) and the orthogonal vector. The original gray image first is divided into 8×8 blocks, and then permuted. A 1-level DWT is applied to each 8×8 block. Each watermark bit is modulated by orthogonal vector, then the watermark is add to the original image. Finally the IDWT is performed to form the watermarked image. In the watermarking detection process the independent component analysis (ICA)-based sparse code shrinkage (SCS) technique is employed to denoise, and make using of the orthogonal vector character. By hypothetical testing, the watermark can be extracted exactly. The experimental results show that the proposed technique successfully survives image processing operations, image cropping, noise adding and the JPEG lossy compression. Especially, the scheme is robust towards image sharping and image enhancement.

1. Introduction

The digital watermarking has been proposed as a solution to the problem of copyright protection of multimedia data in a networked environment. By the term "digital watermark," we mean a signal which is superimposed on the digital image, in such a way that the following hold.

(1) The visual perception of the image remains unaltered, and the watermark is unnoticed.

(2) We are in a position to detect a certain digital watermark by examining the alterations caused by the superposition.

(3) A great number of different digital watermarks, all distinguishable from each other, can be produced.

(4) Distortion or removal of the digital watermark through general image operations and manipulations should be extremely difficult and, preferably, impossible.

The satisfaction of the above-mentioned demands provides a way to superimpose an "invisible" watermark on images. On the other hand, we must know that no watermarking scheme can survive all attacks, and the watermarking scheme should associate with the cryptography. A certain watermarking scheme should be used under some circumstances. When it comes to copyright protection, there are many problems to solve. Because some issuses should be solved before it applies to actual application. We think for the copyright authorization a image should first registers in the watermark authentication center, then it is regarded as the protected image. When verifying a image, the watermark authentication center should extract the watermark based on the registration information, or it is regarded as a counterfeit. Generally this watermark model can applied to copyright protection, or the watermark authentication center.

There are several techniques have been developed for watermarking mainly in spatial domain and frequency domain. In [1], three coding methods for hiding electronic marking in document were proposed. In [2]–[4], the watermarks are applied on the spatial domain. The major advantage of the spatial domain method is that the algorithm is sample and the the major disadvantage of spatial domain watermarking is that a common picture cropping operation may eliminate the watermark. On the other hand the frequency domain method is more robust than the spatial domain method. So many researchers focus their attentions on this method

The proposed scheme of this paper is based on ICA, SCS, the human visual system (HVS), discrete wavelet transform (DWT) and the orthogonal vector. This method can exactly extract the watermark. Especially, it can resist the signal enhancement operation.

This paper is organized as follows: section 2, section 3 and section 4 describes the HVS, the watermark embedding and detection algorithm. In section 5 experimental results are shown. Finally, in section 6 we draw the conclusion of this paper.

2. The Human Visual System (HVS)

The wavelet transform finds its way into the field of

signal analysis. Compared with the traditional transforms, the Fourier transform for instance, the wavelet transform has an advantage of achieving both spatial and frequency localization.
The DWT is very suitable to the human visual system. In [9] the authors proposed a method calculating the image masking and the JND (just noticeable differences). It can be calculate through these terms:

$$q_l^{\theta}(i,j) = \Theta(l,\theta)\Lambda(l,i,j)\Xi(l,i,j)^{0.2} \quad (1)$$

$$\Theta(l,\theta) = \begin{cases} \sqrt{2}, & \text{if } \theta = 1 \\ 1, & \text{otherwise} \end{cases} \bullet \begin{cases} 1.00, & \text{if } l = 0 \\ 0.32, & \text{if } l = 1 \\ 0.16, & \text{if } l = 2 \\ 0.10, & \text{if } l = 3 \end{cases} \quad (2)$$

$$\Lambda(l,i,j) = 1 + L(l,i,j) \quad (3)$$

$$L(l,i,j) = \frac{1}{256} I_3^3\left(1 + \left\lfloor \frac{i}{2^{3-l}} \right\rfloor, 1 + \left\lfloor \frac{j}{2^{3-l}} \right\rfloor\right) \quad (4)$$

$$\Xi(l,i,j) = \sum_{k=0}^{3-l} \frac{1}{16^k} \sum_{\theta=0}^{2} \sum_{x=0}^{1} \sum_{y=0}^{1} \left[I_{k+l}^{\theta}\left(y + \frac{i}{2^k}, x + \frac{j}{2^k}\right)\right]^2 \bullet \operatorname{Var}\left\{ I_3^3\left(1 + y + \frac{i}{2^{3-l}}, 1 + x + \frac{j}{2^{3-l}}\right)\right\}_{\substack{x=0,1 \\ y=0,1}} \quad (5)$$

l and θ are the resolution level where l=1, 2, 3 and the frequency orientation where θ =1, 2, 3. From above we can get the JND :

$$JND_l^{\theta}(i,j) = \frac{1}{2} \bullet q_l^{\theta}(i,j) \quad (6)$$

The detail of the calculation can refer [5].

3. Watermark Permution And Embedding

We use a binary meaningful image of size 64×64. In order to resist cropping operation, a fast two-dimensional (2-D) pseudorandom number traversing method is used to permute the watermark to disperse its spatial relationship, i.e.

$$W' = \text{Permute}\ (W, \text{Key}\ (K)) \quad (7)$$

During the detection process we can extract W', and then reverse W' to W using :

$$W = \text{InvPermute}\ (W', \text{Key}\ (K)) \quad (8)$$

First we set two orthogonal vector S0 and S1 like below:

$$S0 = [0\,1\,0\,0\,1\,1\,1\,1\,0\,0\,0\,1\,1\,1\,0\,0]$$

$$S1 = [1\,0\,1\,1\,0\,0\,0\,0\,1\,1\,1\,0\,0\,0\,1\,1]$$

S0 , S1 are selected by the Zig-zig scan and then projecting to one dimension, which can warrant this method have a good virtue towards JPEG compression. Patterns S0 and S1 are so selected as to be symmetric and orthogonal to each other. The embedding process follow:
(1) For each 8×8 block we apply 1-level DWT, then we get three detail subbands HL,LH,HH and one approximate subband LL.
(2)select one detail subband of size 4×4, for example HL. The coefficiens are modified :

$$\overline{W}(i,j) = \begin{cases} HL(i,j) + alpha * JND(i,j) * S0((i-1)*4+j) & \text{if } W' = 0; \\ HL(i,j) + alpha * JND(i,j) * S1((i-1)*4+j) & \text{if } W' = 1; \end{cases} \quad i,j = 1..4 \quad (9)$$

Where alpha is a positive number controlling the trad-off between perception and the robust of the watermark. For a image of size 512×512 we can embed 64×64 bits information. So three versions watermark are embedded.
(3)Performing the IDWT and forming the watermarked image $\overline{I}$.

4. Watermark Detection

The watermark detection process need the original image I .In the watermark detection process two-setp detection is used.
(1) denoising the possibly corrupted image $\tilde{I}$.
(2) extracting the watermark.
(1) First step the PSNR of a given possibly corrupted image $\tilde{I}$ is caculated. If PSNR is smaller than a given threshold $PSNR_0$, then the independent component analysis (ICA)-based sparse code shrinkage (SCS) technique [7]] [8] is employed to model the denoising problem. Therefore, the noisy image $\tilde{I}$ can be denoted As:

$$\tilde{I} = I + N = As + N \quad (10)$$

Suppose only the observed data is given; the basis matrix and the ICs can be obtained by first finding a separating matrix W (with $W^{-1} = A$) via sparse coding [7]. Then, can be determined by $s = WX$. After sparse coding, the noisy image $\tilde{I}$ can be transformed by means of W, and a noisy independent component, $s + \tilde{N}$ (in the ICA transformed domain), can finally be derived as follows:

$$W\tilde{I} = WI + WN = WAs + WN = s + \tilde{N} \qquad (11)$$

Then, one can shrink $s + \tilde{N}$ by means of g and then get the cleaned version of s, which is represented as $\bar{s}$, where

$$\bar{s} = g\left(s + \tilde{N}\right) \qquad (12)$$

In general, the shrinkage function, is explicitly defined [8] based on the sparse density distribution of noisy independent components to have the effects that small arguments are set to zero and the absolute value of large arguments are reduced by an amount depending on the noise level. In the third step, the approximated host image $\bar{I}$ can be derived by an inverse ICA transformation: $\bar{I} = A\bar{s}$. After the estimated host image is determined, it can be used for watermark detection.

(2) After the pre-processing, we then can extract the watermark from image $\bar{I}$, for each bit the algorithm includes the following steps:

1) For the original image I and the watermarked image $\bar{I}$ each 8×8 block we apply 1-level DWT, we get three detail subbands HL, LH, HH, $\overline{HL}, \overline{LH}, \overline{HH}$ and one approximate subband LL, $\overline{LL}$.

2) using (13)

$$\begin{aligned} PH0 &= ABS\left(SUM\left(\left(HL - \overline{HL}\right).*S0\right)\right) \\ PH1 &= ABS\left(SUM\left(\left(HL - \overline{HL}\right).*S1\right)\right) \\ \tilde{W}'(i,j) &= 0; \quad \text{if} \quad PH0 - PH1 > 0 \\ \tilde{W}'(i,j) &= 1; \quad \text{if} \quad PH1 - PH0 > 0 \end{aligned} \qquad (13)$$

one bit watermark can be extracted.

(3)Then we can extract 64×64 bits watermark. So the extracted watermark $\overline{W}'$ can obtain by

$$\overline{W}' = InvPermute\left(\tilde{W}, Key\left(K\right)\right) \qquad (14)$$

We define the similarity measurement between the referenced watermark W and extracted watermark $\overline{W}'$ as

$$NC = \frac{\sum_i \sum_j \left(w(i,j) \oplus \overline{W}'(i,j)\right)}{M \times N} \qquad (15)$$

Where $M\times N$ is the size of the watermark.

5. Experimental Results

We have performed experiments with the test images Lena, Baboon, and Peppers of size 512×512. The watermark "JLU" image of size 64×64. All the test images and the watermark image are shown in fig.2. 18 commonly used attacks are adopted to test the robustness of our method. These attacks included 1) median filtering; 2) Wiener filtering; 3) sharpening 3times; 4) sharpening 4times; 5) Possion noise addition; 6) Salt noise addition; 7) enhanced brightness +50; 8) enhanced brightness +100; 9) enhanced darkness +50; 10) enhanced darkness +100; 11)-14) JPEG compression with quality factors of 90%, 80%, 70% and 60%; 15) dithering; 16) rotation 1 degree; 17) scale from 512×512 to 400×400; 18) scale from 512×512 to 128×128. It can be seen from the experience our method receives good robustness under signal enhancement operations, for example sharpening 3 times the PSNR is only 13.285 and the NC value is 1. When sharpening 4 times the PSNR is 8.699 and the NC can also get a high value 0.9031. Another signal enhancement experience is enhanced darkness +50 the NC is 1. When enhanced darkness +100 the NC is also 1. So from this results we can see our method is very robust towards signal enhancement. Additionally, we also test the robust under noise adding attack. Because our method uses sparse code shrinkage (SCS) technique, the experiences show it has a good performance in denoising. The corresponding results can also be seen from fig.1.

We compare our method with the method proposed in [6]. The strongly enhanced contrast experiences of our method can get the value of NC is 1, and the method proposed in [6] only get the value of NC is 0.97. When sharpening experiences, our method the value of NC is 1, and the method proposed in [6] not mentioned. It can be seen from table.1 our method have a better performance than the method proposed in [6], especially under signal enhancement attacks.

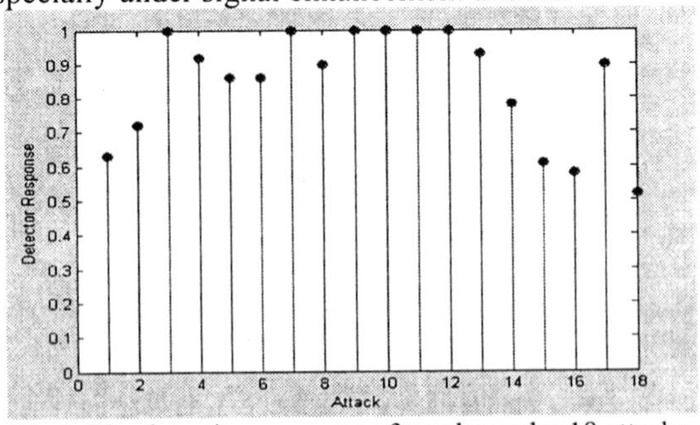

Fig.1. The detection response of results under 18 attacks.

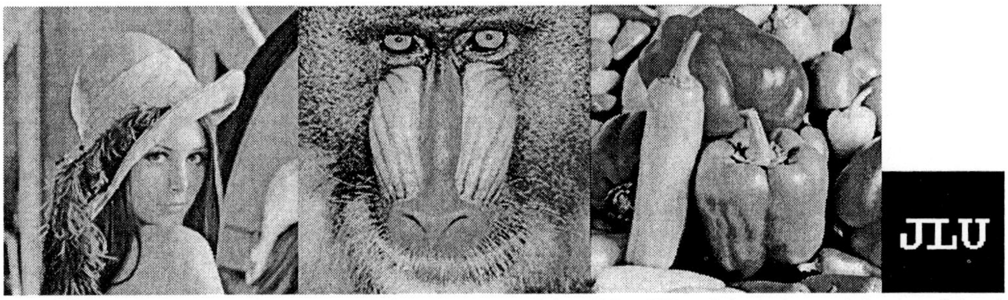

Fig.2. The test images are respectively Lena, Baboo, Peppers of size 512 × 512 and the watermark image of size 64 × 64.

Table.1.

Comparison robustness to JPEG lossy compression, PSNR, slightly enhanced contrast, strongly enhanced contrast, sharpening 3 times and cropping (1/4) of our method and the method prorosed in [6].

Image	Method	JPEG (ratio3.3)		PSNR	Slightly enhanced contrast	Strongly enhanced contrast	Sharpening 3 times	Cropping (1/4)
Lena	Proposed in[6]	PSNR	33.8	39.8	0.9985	0.973	No	0.7623
		NC	0.99					
	Our method	PSNR	34.1	40.4	1	1	1	0.7942
		NC	0.999					
Baboon	Proposed in[6]	PSNR	34.8	40.2	0.9992	0.980	No	0.7746
		NC	0.992					
	Our method	PSNR	35.2	40.2	1	1	1	0.7931
		NC	0.9995					
Peppers	Proposed in[6]	PSNR	34.1	38.9	0.9982	0.968	No	0.7741
		NC	0.99					
	Our method	PSNR	33.9	37.6	1	1	1	0.7602
		NC	0.999					

6. Conclusion

In this paper, we proposed an adaptive watermarking algorithm for binary image watermark. The proposed algorithm is evaluated from the transparency point of view and the robustness against some common attacks, such as a JPEG compression, filtering, noise corruption and cropping. The results show the desirable features of proposed algorithm, especially it shows robustness against signal enhancement. The performance of the proposed algorithm is compared with the algorithm proposed in [6]. The comparison results show the advantage of the proposed algorithm over the algorithm in [6].

7. References

[1] J. T. Brassil, S. Low, N. F. Maxemchuk, and L. O'Gorman. (1995) Electronic marking and identification techniques to discourage document copying. IEEE J. Select. Areas Commum,vol. 13, pp. 1495–1504, Oct.

[2] I. Pitas and T. H. Kaskalis. (1995) Applying signatures on digital images. in Proc. IEEE Nonlinear Signal and Image Processing, pp. 460–463.

[3] O. Bruyndonckx, J. J. Quisquater, and B. Macq. (1995) Spatial method for copyright labeling of digital images. in Proc. IEEE Nonlinear Signal and Image Processing, pp. 456–459.

[4] S. Walton. (1995) Image authentication for a slippery new age. Dr. Dobb's J, pp. 18–26.

[5] Barni M, Bartolini F, Piva A. (2001) Improved wavelet-based watermarking through pixel-wise masking. IEEE Transactions on Image Processing, 10(5):783~791.

[6] Hsu,Wu JL. (1999) Hidden Digital watermarks in images[J]. IEEE Trans Image Processing 8(1):58-68.

[7] A. Hyvarinen. (1999) Sparse code shrinkage: Denoising of nongaussian data by maximum likelihood estimation. Neural Comput, vol. 11, pp.1739–1768.

[8] A. Hyvarinen, P. Hoyer, and E. Oja. (2001) Image denoising by sparse code shrinkage. in Intelligent Signal Processing, S. Haykin and B. Kosko,Eds. New York: IEEE Press.

Simulating binocular eye movements based on 3-D short-term memory image in reading

Satoru Morita
Faculty of Engineering, Yamaguchi University
E-mail: smorita@yamaguchi-u.ac.jp

Abstract

We simulate binocular eye movements in reading. We introduce the 3-D edge features reconstructed from the binocular foveated vision to determine the next fixation point in reading. The next fixation point is determined statistically from the feature points in the 3-D short-term memory edge image. We show the effectiveness of simulating eyes movement based on 3-D short-term memory image to realize humanlike robots.

1 Introduction

It is important to determine eye movement to realize the vision of autonomous robot such as human vision. It is proposed to determine tasks and the viewpoint according to it in the field of computer vision[1][2]. On the other hand, it is based on the psychological experiments[3]that the viewpoint determine according to the given tasks. On the other hand, it is one of important problem that many psychologist studies[4]. It aims at realizing the viewpoint movement in the task of reading. The vision system used CCD device is realized based on the foveated vision that the resolution in the retina center is high and the resolution in the retina periphery is low[5].

But it is not argued for the binocular eye movements in reading. Human understands the 3-D world by computing the depth from the binocular foveated vision. Thus, we realize the viewpoint movement in reading based on the 3-D world derived from binocular eye movements. In this paper, we realize binocular eye movements that viewpoint moves in the wide region such as to search the next line and word. We use the 3-D edge features reconstructed from the foveated vision of binocular eyes to determine the next fixation point in reading. It is argued that human behaviour is related to human memory[6]. In this paper, we introduce the 3-D short-term memory related to binocular vision. The next fixation point is determined statistically from the feature points saved on the short-term memory. We show the effectiveness of simulating eyes movement based on 3-D short-term memory image to realize humanlike robots .

2 Foveated Vision

The center of the retina is called the fovea. Vision in which resolution is low at the periphery of the retina and high at the center of it is called foveated vision. Because the log-polar mapping model varies its scale in rotation, it is used widely as a image sampling model. Wilson proposed the arrangement of the receptive field according to the mapping model and explained the human sensing facility related to the contrast[7]. The receptive field is located on circles whose center is the center of the retina. In order to realize the types of vision, it is necessary that the resolution diminishes as the radius of the circle grows larger. The eccentricity R_n of the nth circle is defined in the following:

$$R_n = R_0(1 + \frac{2(1 - Ov)Cm}{2 - (1 - Ov)Cm})^n \qquad (1)$$

R_0 is the radius of foveated vision, and C_m is the rate between the eccentricity and the distance from the retina center to the center of the receptive field. Ov is the overlapping rate between the receptive fields in the neighbor.

3 Calculation of determining the camera directions based on stereo vision

We describe the necessary technique to reconstruct the 3-D world based on binocular eyes.

3.1 Calculation of camera matrix

Camera matrix M is the translation matrix used to translate from world coordinates to camera display. It is defined using camera position, direction and focus and image size. Camera matrix is calculated as the products of the following four matrix.

$$M = T \cdot M_1 \cdot M_2 \cdot M_3 \qquad (2)$$

T is the translation matrix used to translate from the point coordinate to the homogeneous coordinate . M_1 is the perspective translation matrix used to translate from the homogeneous coordinate to the display in the 3-D coordinate on the display. M_2 is the matrix used to translate from the 3-D coordinate on 3-D display to the 2-D coordinate on the display. M_3 is the matrix used to translate from the 2-D coordinate to the 2-D display. Camera matrix of left camera and right camera is calculated in the following.

$$M_L = T_L \cdot M_{L1} \cdot M_{L2} \cdot M_{L3} \quad (3)$$

$$M_R = T_R \cdot M_{R1} \cdot M_{R2} \cdot M_{R3} \quad (4)$$

We calibrate using the method of Tsai[9] in the case of simulating the eye movement using the real pan-tilt camera.

3.2 Calculation of the points in 3-D space from binocular images

The point of the 3-D space P is represented using the homogeneous coordinate as $P[p] = P[p_1, p_2, p_3, 1]$. If h_L and h_R are real and the coordinate on left image is $P[p_L] = P[p_{L1}, p_{L2}, 1]$ and the coordinate on right image is $P[p_L] = P[p_{L1}, p_{L2}, 1]$,

$$h_L < p_{L1}, p_{L2}, 1 >= [< p_1, p_2, p_3, 1 > M_L] \quad (5)$$

$$h_R < p_{R1}, p_{R2}, 1 >= [< p_1, p_2, p_3, 1 > M_R] \quad (6)$$

. Thus, the coordinate (p_1, p_2, p_3) is calculated from the right image $< p_{R1}, p_{R2}, 1 >$, the left image $< p_{L1}, p_{L2}, 1 >$ and the camera matrix M_L, M_R.

3.3 Determining of the camera direction from the fixation point

The pan and tilt rotation angle $< \theta_l, \phi_l >$ and $< \theta_r, \phi_r >$ for the left and right camera are calculated from the fixation point (p_1, p_2, p_3) in the 3-D world. The rotation center of the left camera is (cl_1, cl_2, cl_3) and the rotation center of the right camera is (cr_1, cr_2, cr_3). The direction of the left camera is $X' < i_{L1}, i_{L2}, i_{L3} >$ and the direction of the right camera is $X' < i_{R1}, i_{R2}, i_{R3} >$. The direction of the left camera Y' required to look at the fixation point is calculated in the following:

$$A =< p_1 - cl_1, p_2 - cl_2, p_3 - cl_3 > \quad (7)$$

$$Y' =< \frac{p_1 - cl_1}{|A|}, \frac{p_2 - cl_2}{|A|}, \frac{p_3 - cl_3}{|A|} > \quad (8)$$

. The rotation matrix R_L is calculated using $X'Y'$ as

$$Y' = R_L X' \quad (9)$$

$$Y'(X'X'^T)^{-1}X'^T = R_L \quad (10)$$

. After the pan tilt camera rotates as the rotation axis is y axis, rotates as the rotation axis is x axis. The rotation around the y axis shows T_{ry}, and the rotation around the x axis shows T_{rx}. R_L is represented as

$$R_L = T_{rx} \cdot T_{ry} = \begin{bmatrix} r_{11} & r_{12} & r_{13} \\ r_{21} & r_{22} & r_{23} \\ r_{31} & r_{32} & r_{33} \end{bmatrix} \quad (11)$$

, T_{ry} and T_{ry} are calculated using ϕ_l and θ_l as

$$T_{ry} = \begin{bmatrix} cos\phi_l & 0 & sin\phi_l \\ 0 & 1 & 0 \\ -sin\phi_l & 0 & cos\phi_l \end{bmatrix} \quad (12)$$

$$T_{rx} = \begin{bmatrix} 1 & 0 & 0 \\ 0 & cos\theta_l & sin\theta_l \\ 0 & -sin\theta_l & cos\theta_l \end{bmatrix} \quad (13)$$

. Therefore, we can get R_L from previous equations:

$$sin\phi_l = r_{13} \quad (14)$$

$$cos\theta_l = r_{22} \quad (15)$$

. In the case of

$$X' =< i_{L1}, i_{L2}, i_{L3} >=< 0, 0, 1 >$$

, as the following equation is gotten:

$$sin\phi_l = \frac{p_1 - cl_1}{|A|} \quad (16)$$

$$cos\theta_l = \frac{p_2 - cl_2}{|A|} / (1 - sin^2\phi_l) \quad (17)$$

, ϕ_l, θ_l can be calculated. In the similar, the right camera direction Y'' required to look at the point (p_1, p_2, p_3) is calculated :

$$Y'' =< \frac{p_1 - cr_1}{|A|}, \frac{p_2 - cr_2}{|A|}, \frac{p_3 - cr_3}{|A|} > \quad (18)$$

.

$$A =< p_1 - cl_1, p_2 - cl_2, p_3 - cl_3 > \quad (19)$$

The rotation matrixR_R is calculated :

$$Y''(X''X''^T)^{-1}X''^T = R_R \quad (20)$$

$$Y'' = R_R X'' \quad (21)$$

In the similar, ϕ_r, θ_r are calculated. Thus, left camera direction $< \phi_l, \theta_l >$ and right camera direction $< \phi_r, \theta_r >$ are calculated from the fixation point (p_1, p_2, p_3).

4 Binocular eye movements in reading based on 3-D short-term memory image

We describe the eye movement in reading based on short-term memory image and foveated vision.

4.1 3-D short-term memory image

Short-term memory saves the information for about 20 seconds[6]. Human does not have the consciousness while the fxation point moves quickly. This is reason why the image saved for the short term does not change while the foveated image of the viewpoint changes quickly. Thus, we realize the observed image that does not change suddenly though the fixation point moves in the wide region. We take the attention to the short-term memory related to vision. In especially, the vision system should not save 2-D image but 3-D image in short-term mermoy image to determine the motion of binocular eye balls.

The feature is saved in the 3-D short-term memory after 3-D point is recovered from these feature in the case that the feature of right image is same as the feature of left image and the feature is on the epipolar line of left camera corresponding to a feature of right camera. The next fixation point is statistically determined from the feature points of the short-term memory. Thus, 3-D coordinates (x, y, z) are recovered from the two feature points in right and left camera.

4.2 Generating 3D short-term memory feature image

We explains the algorithm of 3-D short-term memory edge feature image.

1) A pixel is selected from the xy coordinate of right foveated vision. A plane determined from the center of two cameras and a pixel on the image plane of the right camera. The epipolar line is generated by intersecting the plane with the left image plane.

2) We determine the rotation that the sum of absolute difference between RGB values is minimum using the correlation method for each pixel, we get the corresponding point on the epipolar line[8].

3) The 3-D point coordinate is recovered from xy coordinate gotten from the foveated vision of right and left camera ((3.1)(3.2)).

4) The color and the time value are saved in the coordinate of the 3-D point in the 3-D short-term memory image. We generate 3-D short-term memory image using voronoi algorithm from 3-D point sets in 3-D short-term memory.

5) The sequences 1)2)3)4) are applied for the xy coordinate corresponding to each pixel of foveated vision and the 3-D points are reconstructed.

6) The next fixation point is determined from the edge feature image included in the attention region in 3-D short-term memory edge feature image(4.3). The motion of the right camera and left camera is determined(3.3).

7) The time value is incremented. The difference between the current time and the time on the short-term memory image is calculated. If the difference is over the constant value, the point is deleted in the short-term memory image.

4.3 Binocular eye movements in reading

In the case that the fixation point is on the line in reading, we use the square mask. The direction that the most edges exist in the square mask is determined while the square mask is rotated. The attention region is composed of the regions to the space from the fixation point backward and forward for the reading direction. If the edge number is not over the constant value, the next fixation point is in the attention region. If the edge number is over the constant value, the next fixation point is in the region from the next space following forward the current attention region to the space after the space. Thus, it is simulated to determine the next fixation point both backward and forward. In the case that a fixation point is on the end of the line, the fixation point moves to the top of the next line through the line saved in the 3-D short-term memory image. At first, the direction that edges are many is determined using the square. In this time, the attention region is selected in the opposite of the reading direction. The top of the line is detected by repeating this. After the space is found under the current line, the next fixation point is determined from the attention region that the edges are many under the space.

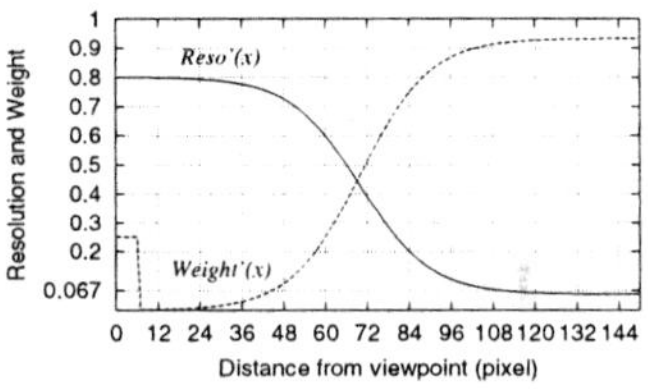

Fig. 1. The change of resolution and weight on the foveated vision. : $Reso'(x)$ and $Weight'(x)$ in the case of $\alpha = 0.8$, $\beta = 0.25$.

α makes the resolution $Reso'(x)$ change. The resolution becomes low, if α decreases. β makes the weight $Weight'(x)$ change. The frequency that the next fixation point is determined to the feature points where the distance between the fixation ooint and the feature point is less than 6 pixels is high, if β is big.

Figure 1 shows the change of resolution and weight on the foveated vision. $Reso'(x)$ and $Weight'(x)$ is determined as $\alpha = 0.8$,

$\beta = 0.25$. The resolution $Reso'(x)$ calculated the feature point number per unit volume from the short-term memory feature image. In this paper, the edge length of a cube is 24 pixels in a unit volume.

$$Reso(x) = \frac{1 + \exp(-ab)}{1 + \exp(a(x - b))}(1.0 - 0.067) + 0.067 \quad (22)$$

$$Weight(x) = 1.0 - Reso(x) \quad (23)$$

$$Reso'(x) = \alpha Reso(x)$$

$$Weight'(x) = \begin{cases} 1.0 - Reso(x) & if\ x > 6 \\ \beta & otherwise \end{cases} \quad (24)$$

If the edge number including in the attention region is N, the probability f_i to determine the next fixation point to the ith edge is defined in the following.

$$f_i = \frac{Weight(x_i)}{\sum_{j=1}^{N} Weight(x_j)} \quad (25)$$

5 Simulating binocular eye movements in reading

We simulate binocular eye movements in reading. The pan-tilt cameras that the focus length is $f = 519.615$ are arranged in $(-5, 0, 0)$ and $(5, 0, 0)$. The document size is $512 * 256$ pixels. The center position of the document is fixed in $(0, 0, 700)$. We use the camera that the focus distance f is 519.615. The parameter requiered to generate the foveated image is set in $R_0 = 7$,$C_m = 0.5$,$O_v = 0.9$ and $length = 170$. Figure2(a) shows edges detected on the right foveated vision at the middle of the second line. Figure2(b) shows The 3-D edge recovered from the foveated vision of the right and left camera. Figure2 (c) shows the viewpoint movement. Figure2(d) shows the 3-D short-term memory edge image when the fixation point is on the end of second line. Figure3(a) shows the 3-D short-term memory edge image for the curved document when the fixation point is on the end of second line. Figure3 (b) shows the edge image generated from a foveated vision when the fixation point is on the end of second line. The 3-D short-term memory edge is recovered adequately. It is found that the viewpoint movement to read the document located in the space is realized from these results. Binocular eye movements in reading the curved document in the bound book was simulated using this model.

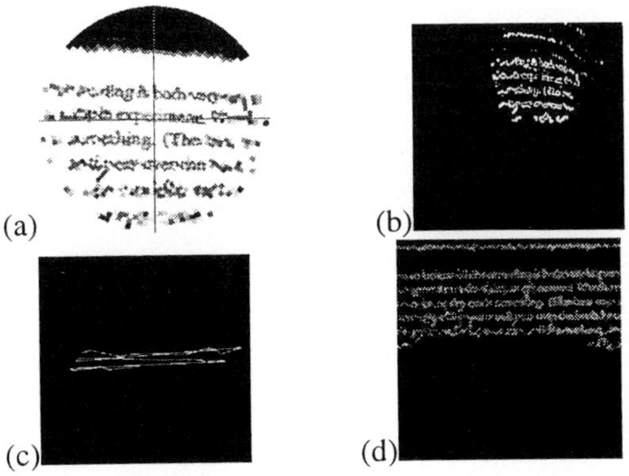

Fig. 2. (a)Edges detected on the right foveated vision at the middle of the second line. (b)The 3-D edge recovered from the foveated vision of the right and left camera(c)The viewpoint movement.(d)The 3-D short-term memory edge image when the fixation point is on the end of second line.

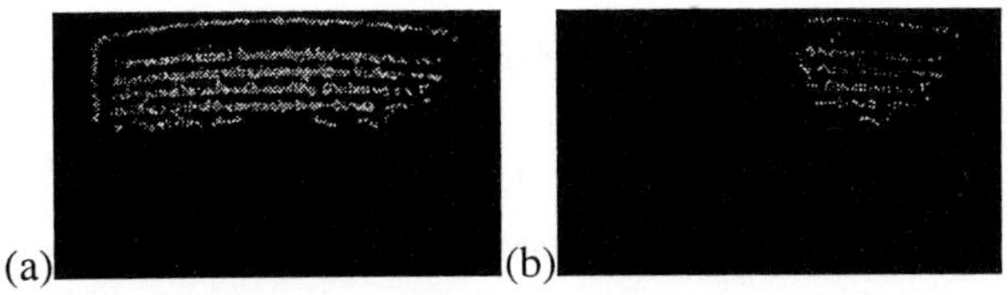

Fig. 3. (a) The 3-D short-term memory edge image for the curved document when the fixation point is on the end of second line.(b)The edge image generated from a foveated vision when the fixation point is on the end of second line

6 Conclusions

The binocular eye movements in reading based on the 3-D short-term memory are simulated by controlling two pan-tilt cameras.

References

[1] M. J. Swain, R.E. Kahn, and D. H. Ballard (1992) Low resolution cues for guiding saccadic eye movements. CVPR, pp. 737-740

[2] L. Birnbaum, M. Brand, and P. Cooper (1993) Looking for trouble: Using causal semantics to direct focus of attention, ICCV, pp. 49-56

[3] A. Yarbus (1967) Eye movements and vision. Plenum Press

[4] Levy-Schoen, A and ORegan, K. (1979) The control of eye movements in reading, Proc. visible language, Plenum Press, pp. 7-36

[5] G. Sandini, P. Dario and F. Fantini (1990) A RETINA LIKE SPACE VARIANT CCD SENSOR," SPIE 1242, pp. 133-140

[6] R. C. Atkinson and R. M. Shiffrin (1968) Human memory: A proposed system and its control process. The Psychology of Learning and Motivation, Vol. 2, Academic Press

[7] S. W. Wilson, (1983) On the retina-cortial mapping. Int. J. Man-Machine Stud. 18, pp. 361-389

[8] T. Kanade and M. Okutomi, (1994) A stereo matching algorithm with an adaptive window: Theory and experiment, IEEE Trans, PAMI, Vol. 16, No. 9, pp. 920-932

[9] R. Y. Tsai, (1986) An efficeint and accurate camera calibration technique for 3D machine vision, CVPR, pp. 364-374

An Algorithm For Face Pose Adjustment Based On Gray-scale Static Image

Wenming Cao , Shoujue Wang

Institute of Intelligent Information System, Infornation College, Zhejiang University of Technology , Hangzhou 310014 , China

(Email: csann@zjut.edu.cn)

Abstract:

Face pose adjustment, as a loop of human face location, is very important in computer face recognition. In this paper, we present a new approach to automatic face pose adjustment on gray-scale static images with a single face. In the first stage, with the degree of mediacy make every little image, then ask for one piece of image including two eyes using match degree. And continue the nose and mouth part horizontal gray projection in small scope. Finally adjust this piece correctly. In the second stage, based on the location and the symmetry feature of eyes, the inclination angle is calculated and the face position is redressed. The experimentations show that the algorithm performs very well both in terms of rate and of efficiency. What's more, due to the precise location of eyes, the apples of the eyes are detected.

1 Introduction

With the development of computer technology pattern recognition[4] has emerged as an active research area with the characteristic of friendliness and convenience. Human face recognition technique[1][5] embodies human intelligence and has general applications including identity validation and human- computer interaction. Face pose location is one of the important parts in human face recognition. Whether or not detecting the correct human face pose has crucial influence on the performance of the whole human face recognition system. In human face recognition systems the facial features need to be located after detecting the human face. Due to the behavior habit and physical factors, the facial images are not upright but have an inclination angle in the vertical plane sometimes. Thus increase the difficulty of facial features location. Therefore the human face location can be divided into two parts: face pose adjustment and facial features location. Face pose adjustment is the precondition of facial features location and an important part in face recognition system.

Considering the variation of gray scale, geometrical characteristics and symmetry features of the eyes, this paper presents a novel approach for face pose adjustment based on eye location in gray scale image with single face. In the first stage, the face or the eyes search field is detected according to vertical variance projection and horizontal integral gray projection of the image. Then make the desired field at middle with the degree of mediacy. Thereafter ask for one piece of image including two eyes using match degree. And continue the nose and mouth part horizontal gray projection in small scope. In the second stage, based on the location and vertical complexity of the image block and black block eye model the eyes are located. Based on the location and the symmetry feature of eyes, the inclination angle is calculated and the face pose is redressed. In this paper we use gray scale images with single face in simple background. In condition of color images, they will be transformed to gray scale images by deriving the red component(The reason is that the base color of human skin is red, the red component contain large amount of human face information and its SNR is higher than the other components. Thus it is more effective.)

2 The algorithm parameter design

The gray level variance is the main feature to represent face characteristic. Face area is complex contrast to the simple background. Meanwhile, the gray difference between eyes and its surrounding skin is greater than other features in the face. Furthermore, the eyes geometrical shape is around round or ellipse and the eyes are symmetrical to the face axis. Based on the previous analysis, this paper design and utilize the parameters presented below to the images of horizontal size of N pixels and vertical size of M pixels.(Its gray matrix is G^{MxN} , the gray value is $B_{i,j}$ in row i and column j.)

2.1 Vertical variance projection σ_V^2

σ_V^2 indicates the gray variance intensity in vertical direction in an image. Its expression is:

$$\sigma_V^2(j) = \frac{1}{M}\sum_{i=1}^{M}[B_{i,j} - V(j)]^2 \quad (1)$$

Where V is the vertical integral projection of the image block and can be obtained by

$$V(j) = \frac{1}{M}\sum_{i=1}^{M} B_{i,j} \quad (2)$$

2.2 Horizontal integral projection H

H indicates the gray variance in vertical direction. Its expression is

$$H(i) = \frac{1}{N}\sum_{j=1}^{N} B_{i,j} \quad (3)$$

2.3 Vertical complexity of image block C

C is designed according to the characteristic that the vertical gray variance of eyes is greater than that of the other features in a face image. Its expression is defined as follows:

$$C = \sum_{i=1}^{M-1}\sum_{j=1}^{N} | B_{i+1,j} - B_{i,j} | \qquad (4)$$

2.4 The dark block eye model F

F is designed base on the fact that the geometrical shape of eye is approximate round or ellipse and the eye gray variance is greater than the surrounding skin. Its value indicates the match degree between the image block and an eye so that it can be used to accurate eye location. Its expression is below.

F = min

$$\Big(\sum_{i=1}^{\frac{M}{2}-1}\sum_{j=1}^{\frac{N}{2}}(B_{i,j}-B_{i+1,j})(\frac{M}{2}-i)+\alpha\sum_{i=1}^{\frac{M}{2}}\sum_{j=1}^{\frac{N}{2}-1}(B_{i,j}-B_{i,j+1})(\frac{N}{2}-j),\ \sum_{i=1}^{\frac{M}{2}-1}\sum_{j=\frac{N}{2}+1}^{N}(B_{i,j}-B_{i+1,j})$$

$$(\frac{M}{2}-i)+\alpha\sum_{i=1}^{\frac{M}{2}}\sum_{j=\frac{N}{2}+1}^{N-1}(B_{i,j+1}-B_{i,j})(j-\frac{N}{2}),\ \sum_{i=\frac{M}{2}+1}^{M-1}\sum_{j=1}^{\frac{N}{2}}(B_{i+1,j}-B_{i,j})$$

$$(i-\frac{M}{2})+\alpha\sum_{i=\frac{M}{2}+1}^{M}\sum_{j=1}^{\frac{N}{2}-1}(B_{i,j}-B_{i,j+1})(\frac{N}{2}-j),\ \sum_{i=\frac{M}{2}+1}^{M-1}\sum_{j=\frac{N}{2}+1}^{N}(B_{i+1,j}-B_{i,j})$$

$$(i-\frac{M}{2})+\alpha\sum_{i=\frac{M}{2}+1}^{M}\sum_{j=\frac{N}{2}+1}^{N-1}(B_{i,j+1}-B_{i,j})(j-\frac{N}{2})\Big) \qquad (5)$$

Where α is a constant and is used to modulate the influence of horizontal gray variance on the black block eye model.

3. Implementation of the face pose adjustment algorithm.

3.1 Extraction of the face region(the eye search region)

The face region detection algorithm based on σ_V^2 and H is described as below:

Analyses the approximate face location in the gray scale image with a single face and decide the face detection region. This step can speed up face location and decrease the disturbance of the redundant information(mainly the clothes).

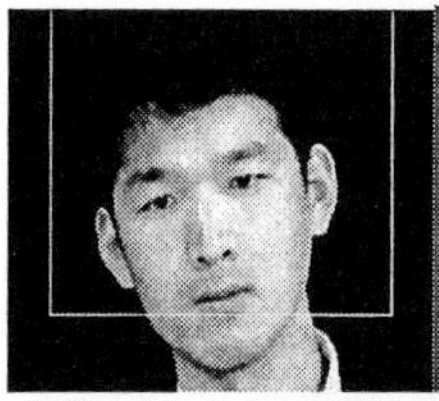

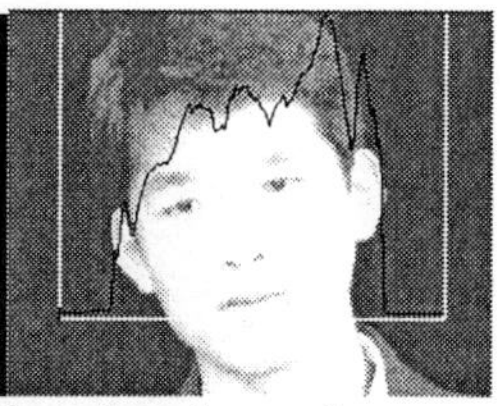

Fig 1(left: Decided gray scale face detection region

Fig 2(right).The relationship between the vertical variance projection curve and the face image

In this paper we use gray scale images of 256 degree against black background and the face occupy a relatively large region in each image. The course face location is to determine the approximate region of the face in an image, so it's the precondition of the fine facial features extraction. As to the images examined in this paper two steps can be performed.

First, determine the left and right boundary by the vertical gray projection and the upper region by the horizontal gray projection in an image. The algorithm [2] is described as follows.

Suppose the face gray image is I(x, y) (sized by n*m). The vertical gray projection of the image is below.

$$P(x) = \sum_{y=1}^{m} I(x, y) \qquad (x = 1,2,3 \ldots n)$$

The curve corresponding to this function is called as vertical gray projection curve. It indicates the total gray variance in X axis.

The horizontal gray variance is defined as

$$P(y) = \sum_{x=1}^{n} I(x, y) \qquad (y = 1,2,3 \ldots m)$$

The curve this function corresponding to is called horizontal gray projection curve. It denotes the total vertical gray variance in Y axis.

Take the vertical gray projection curve into account. It can be find in general conditions the face gray value is higher than that of the background and it vary obviously in vertical direction, thus the vertical gray projection curve has a peak in the corresponding face area. The left and right limit of the peak is also that of the face. Therefore as long as the left and right limit of the main peak of the vertical gray projection curve is decided, the relatively precise left and right limit of the peak can be obtained.

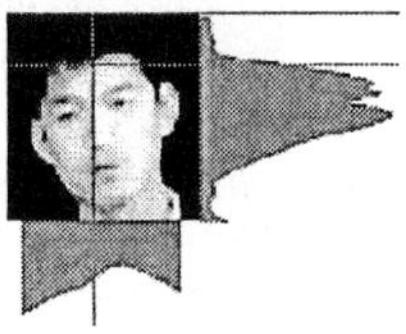

Fig 3 (left): The face left and right limit location

Fig 4(right) The vertical and horizontal gray projection

Let the rectangle region deciding by the left right boundary and upper boundary be the final result of the detected gray face region. In fact, because the eyes generally locate in the upper half region of the rectangle, the bottom boundary can be raised approximately considering the eye detection speed and computational burden(Fig 5).

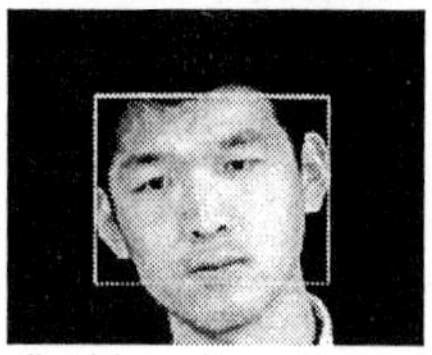

Fig 5: The final location of gray face region

3.2 The eyes location

The eyes location can be divided into two parts: detecting the image block of maximum complexity and fine locating the eyes by the black block eye model.

3.2.1 Detecting the image block of maximum complexity

Firstly, Divide the eyes region obtained from above into little image blocks averagely. The width of the little image block is 1/10 that of the eyes detection region, a little larger than the apples of the eyes in the gray scale face image with single face. The height of the little image block is 1/5 that of the eyes detection region. Secondly, calculate and record the vertical complexity of all little images blocks. Thereafter shift all the little image blocks rightwards by 1/2 unit width and upwards by 1/2 unit height respectively, then calculate and record the vertical complexity of the little image blocks. Finally, sort the complexity obtained from above decreasingly and select 10 little image blocks of the largest complexity.

The experiments demonstrate these blocks generally concentrate on the significantly gray variant area such as the eye, eyebrow, bang and temple. Many of these 10 little image blocks adjoin or overlap each other. For these little image blocks of the same height that adjoin or overlap each other, only the little image blocks of the maximum complexity is reserved and the other are deleted. Slightly slide the remaining little image blocks over its surrounding area until the vertical complexity of the little image blocks reach the maximum. Delete the overlapped little image blocks according to reserving the little image blocks of the maximum vertical complexity. Up to now the remaining little image blocks is little, there are must two little image blocks containing the eye and eyebrow simultaneity.

Fig 6(left) 10 little image blocks with the maximum vertical complexity

Fig 7(right) The remaining little image blocks after slightly adjusted according to vertical complexity

3.2.2 Accurate eye location according to the black block eye model

The results from 3.2.1 can locate the eyes roughly. In order to decrease the disturbance of the eyebrow to accurate eye location, select the lower half part of the little image blocks as the processing object for fine eye location based on black block eye model. In this way, these new little image block are all square, its sideline being 1/10 the width of the eye detection region.

Slightly slide these new little image blocks in their surrounding region and calculate their black block eye model value until reaching the optimal location of these new little image blocks. Sort these new little image blocks by descending order according the values obtained from above. The experimental results(Fig 8 for example) demonstrate the new little image block of the maximum black block eye model value must contain an eye and this block lies in the optimal eye location(The optimal eye location here is the result that the eye locate in the middle of the image block and the middle of the image block is the apple of an eye). Select the new little image block of the maximum value as the final location of one eye.

Before the final location of the other eye is determined delete three kinds of new little image blocks that obviously don't contain an eye.

(1)The new little image blocks of the negative black block eye model value because the geometrical figure and gray variance in these images differ a lot from those of an eye.

(2) The new little image blocks that closely locate in the boundary of the eye detection region. These blocks are disturbance blocks locating in such area as the bang, temple, dividing region of the face from the background.

(3) The new little image blocks that locate below the block of the maximum black block eye model value besides the absolute value of the incline rate is larger than 1(namely the absolute value of the inclination angle is bigger than 45°).

Slightly shift the remaining new little image blocks again in the nearby small region(excluding the little block that has been considered as the final location of one eye). Get the optimal location of each little block according to the output of the black block eye model. Thereafter select the little block of the maximum black block model value among these little blocks as the final location of the other eye. The reason that these new little image blocks need to be adjusted twice lies in that owing to the limit of slightly shifting scope, slightly adjustment once can't guarantee both the two new little image blocks that contain the eyes locate in the final location in condition that the inclination angle is relatively big in the vertical plane.

3.3 . Face pose adjustment according to the symmetry of the face image

The human face image is a symmetric geometric configuration and the eyes lie in the two sides of the face symmetry axis symmetrically. This kind of symmetry does not change while the face tilts. Therefore the face pose can be redressed by select the face symmetry axis through the accurate eye location.

Two little image blocks can be found by the eye location technique mentioned in 3.2. Subsequently calculate the face inclination angle in the vertical plane according to these two little image blocks. Let the perpendicular bisector of the line between the centers of the two little image blocks (namely the apples of the eyes location) be the face symmetry axis and redress the face pose. The redressed face image illuminates in Fig 9.

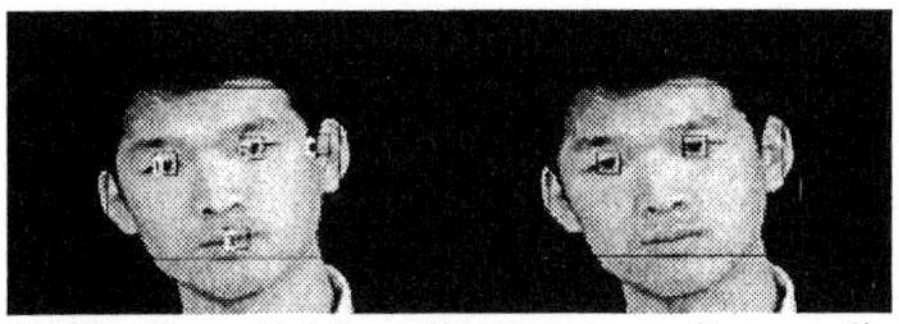

Fig 8(left) The first slightly adjustment result according to the black block eye model.
Fig 9 (right) The final image blocks location containing the eyes

4. Experimental results analysis

This experiments use 100 color images of 20 people (5 images per people, 300x200 pixels,256 gray scale) captured by digital camera in simple background. Take the gray scale images obtained by deriving the red component from the above images as the initial sample database. The face inclination angle doesn't exceed positive or negative 40° in vertical plane. The programs using VC++ are implemented on PIII 450 PC. Experimental results show that our algorithm can perform face pose adjustment successfully. Face pose is successfully redressed in 99 images and the correct rate is 99%. Partial face images processing results is illustrated in Fig 10.

It has been experimentally verified that the proposed algorithm is the same effective even in condition that the face rotates to a large degree rightward or leftward as long as the two eyes can be found. Merely the face symmetry axis deduced by the two eyes location runs parallel with the real face symmetry line. This shows the robustness of the algorithm. The adjustment result of this kind of images is shown in Fig 11.

The precondition of valid face pose adjustment in this algorithm is the accurate eye location. In other words this algorithm can firstly detect the eyes location in gray scale images with a single face which has important significance on the subsequent face feature(eye) location in face recognition.

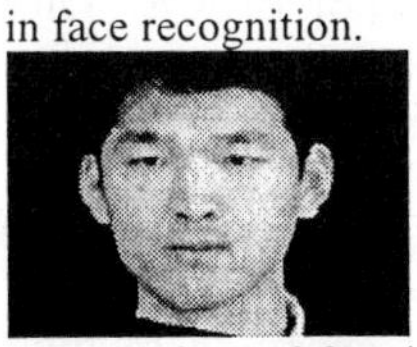

Fig 10: The redressed face image

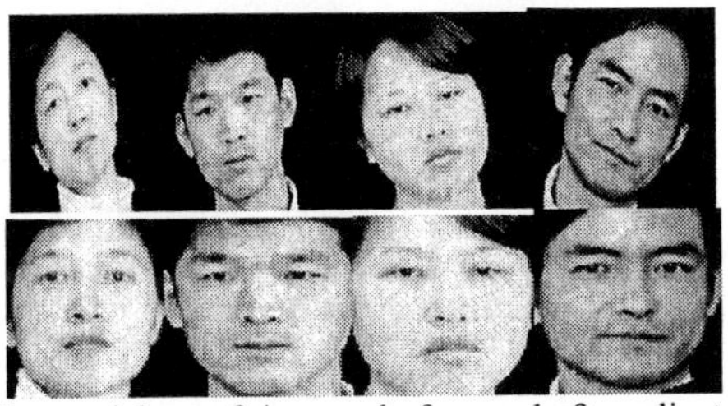

Fig 11 The partial set of images before and after adjusted in the face set

Fig 12: The images rotated greatly in depth before and after adjusted

5. Conclusion

This work presents an approach for face pose adjustment based on eye location in gray scale images with a single face which has fully concerned the variation of gray scale variance, geometrical figure and symmetry features of the eyes. The concepts of vertical complexity of image block and black block eye model are proposed. After the gray scale face region is detected, the face pose is adjusted according to the accurate eye location and symmetry of the eyes. This algorithm has high correct rate and is very robust thus suitable to face pose adjustment in simple background specifically. Furthermore, this algorithm can accurately locate the eyes so that it has leading significance in the subsequent feature location in face recognition.

References

[1] Jiezhou, Chunyu lu, Changshui Zhang, Yanda Li. The review of the automatic human face recognition technique. Electronic Journal. 2004.4: 102-106.

[2] Zhaohan Xu, Xingang Wang, Hao Feng, Shoujue Wang. An rapid algorithm for finding the symmetry of images configuration in human face recognition. Neural networks and intelligent computing .Zhejiang University Publisher,2003, 245-248.

[3] Xin Zhang,Yanjun Xu, LiMin Du. Automatic detection and location of main facial features in color images[j]. Chinese Graphics and Image Processing Journal. 2000.5(A2): 222~241.

[4] Cao, Wenming; Hao, Feng; Wang, Shoujue. The application of DBF neural networks for object recognition. Information Sciences, Volume 160, Issue:1-4,March 22, 2004, Pages 153-160

[5] Frank Y. Shih and Chao-Fa Chuang, Automatic extraction of head and face boundaries and facial features, Information Sciences, Volume 158, January 2004, Pages 117-130

Learning Image Filtering from a Gold Sample Based on Genetic Optimization of Morphological Processing

S. Rahnamayan[1], H.R. Tizhoosh[1], M.M.A. Salama[2]
[1] Department of Systems Design Engineering, University of Waterloo, Canada
[2]Department of Electrical and Computer Engineering, University of Waterloo, Canada
shahryar@pami.uwaterloo.ca, tizhoosh@uwaterloo.ca, msalama@hivolt1.uwaterloo.ca

Abstract

This paper deals with the design of a semi-automated noise filtering approach, which receives just original noisy image and corresponding gold (user manipulated) image to learn filtering task. It tries to generate an optimized mathematical morphology procedure for image filtering by applying a genetic algorithm as an optimizer. After training and generating a morphological procedure, the approach is ready to apply the learned procedure on new noisy images. The main advantage of this approach is that it takes just one gold sample to learn filtering and does not need any prior context knowledge. Using the morphological operators makes the filtering procedure robust, effective, and computationally efficient. Furthermore, the proposed filter shows little distortion on the noise free parts of an image and it can extract objects from heavily noisy environments. Architecture of the system and details of implementation are presented. The approach feasibility is tested by well-prepared synthetic noisy images and results are given and discussed.

Keywords: Image Filtering, Mathematical Morphology, Genetic Algorithm, Learning, Morphological Filtering, Dilation, Erosion, Opening-Closing

1 Introduction

Image filtering and enhancement are crucial tasks in image processing because for other related tasks, such as segmentation or object tracking, these preprocessing tasks are vital to get desirable results. Many different filtering methods are presented in literature [1, 2, 3, 4]. Learning by just one gold sample image without depending on any prior knowledge has not been a major research field, and is the main focus of our approach. Indeed, in any scientific or engineering application, preparing one sample does not require much effort. Our approach uses morphological operators to build image filtering procedure. In the next section a short review of binary mathematical morphology will be presented.

This paper is organized as follows: section 2 is about binary morphology; section 3 presents proposed architecture and methodology; experiments and results are given in section 4; and finally discussion and conclusions are presented in section 5.

2 Binary Mathematical Morphology(MM)

Mathematical morphology was developed based on the work of Serra and Sternberg [5, 6]. Morphology is a shape-based approach to image processing. The value of each pixel in the output image is based on a comparison of the corresponding pixel in the input image with its neighbors [5, 7]. The MM provides techniques for pre- or post processing, such as filtering, and also for segmentation and representation of the objects. Dilation and erosion are fundamental operators of mathematical morphology . Let A be the image and B the structuring element (SE), then the basic operations can be defined as follows [7, 8]:

Dilation - The dilation of A by B is defined as:

$$A \oplus B = \{z | (\hat{B})_z \cap A \neq \phi\}, \tag{1}$$

where $\hat{B}$ is the reflection of B about its origin and $(\hat{B})_z$ is the shifted version of $\hat{B}$ by z. Dilation expands the boundaries of the object.

Erosion - The erosion of A by B is defined analogously as:

$$A \ominus B = \{z | (B)_z \subseteq A\}. \tag{2}$$

Erosion, as a dual operation to dilation, shrinks the boundaries of the object.

Opening - It is defined as an erosion followed by a dilation:

$$A \circ B = (A \ominus B) \oplus B. \tag{3}$$

Objects and connections between them can be eliminated by opening with suitable structuring elements.

Closing - It is defined as a dilation followed by an erosion:

$$A \bullet B = (A \oplus B) \ominus B. \tag{4}$$

Closing removes small holes on the foreground, which are smaller than the chosen SE. The combination of closing and opening is also known as morphological filtering[7].

3 Architecture and Methodology

Figure 1 shows the main architecture of the proposed approach. It has two key units, namely, Genetic Optimizer of Mathematical Morphology Procedure (GOMMP) and Mathematical Morphology Procedure Applier (MMPA).

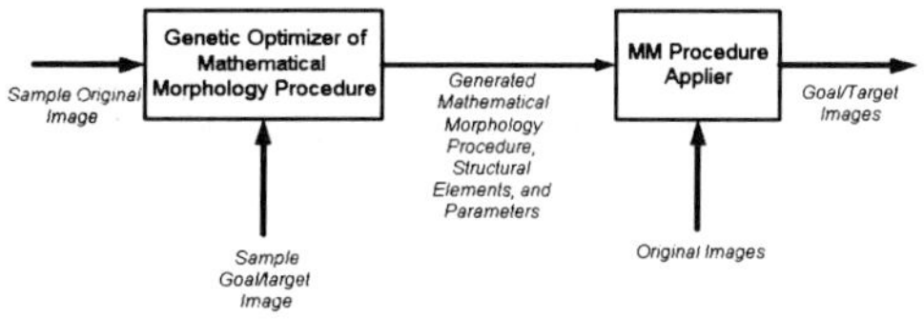

Fig. 1. The main structure of the proposed approach

The GOMMP receives the pair of images, the original and corresponding gold sample, and then generates the desirable mathematical morphology procedure to achieve the filtering effect illustrated in the gold sample. The MMPA applies the generated procedure on noisy images to filter new input images. Morphological operators are selected to build the image processing procedure because they are computationally efficient robust shape-based image processing tools. The MM procedure uses three fundamental operators, namely, dilation, erosion, and opening-closing (as a filtering operator). These operators can be applied more than once ($K1, K2$, and $K3$ times); and any operator uses its own 5×5 structuring element. One of six different combinations of these three operators can be a chain as follows:

$$\begin{aligned} K3 \times O(SE1) - C(SE2) \to \\ K1 \times E(SE3) \to K2 \times D(SE4) \end{aligned} \tag{5}$$

Now we must choose the desirable MM procedure (one of six possible combinations of operator chains) and find the optimal structuring elements ($SE1, SE2, SE3$ and $SE4$) and their repetition time ($K1, K2$, and $K3$). Genetic Algorithm (GA) is a good candidate to determine these variables to optimize the MM procedure. It is a suitable tool for those function optimization, especially those functions with complex searching space [9]. In order to generate the ideal MM procedure, a function with 104 variables should be optimized (100 variables for elements of four 5×5 structuring elements, 3 variables for operator repetition parameters, and one variable for chain indicator, to point out one of six possible combinations of operator chains).

The following steps show how GOMMP optimizes the MM processing chain:

A. Population Initialization: Producing 80 randomly generated chromosomes as an initial population. Any chromosome is built by concatenating binary coded strings of 104 decision variables.

B. Applying Mathematical Morphology Procedures: Applying MM procedure related to any individual; parameters of that procedure are taken by decoding and mapping of any chromosome to decision variables.

C. Evaluation of Similarity: After applying MM procedure related to any chromosome, the similarity of goal image and processed image is measured by a normalized correlation coefficient. This measure quantifies the fitness value of any corresponding chromosome.

D. Stopping Criteria: The number of produced generations is considered if that exceeds a pre-specified threshold, the algorithm terminates and shows the individuals with the higher fitness value in the population; otherwise it goes to the next step.

E. Selection: Selecting a pre-specified number of individuals to produce offspring. The Roulette Wheel method is used to select candidates from the current population[9].

F. Crossover: Applying single point crossover for candidate chromosomes to produce offspring[9].

G. Mutation: Applying mutation as a background operator with low probability ($p = 0.01$) to generate new chromosomes resulted by randomly filliping of their bits[9]. Go to step B.

For aforementioned GA, the fitness function obviously can be defined based on the similarity between the gold sample and generated image by MM procedure, normalized correlation coefficient, $S \in [0, 1]$, is chosen to measure this similarity [10]:

$$S = \frac{\sum_{i=1}^{N}\sum_{j=1}^{N} C_{ij}\hat{C}_{ij}}{\sqrt{\sum_{i=1}^{N}\sum_{j=1}^{N} C_{ij}C_{ij}}\sqrt{\sum_{i=1}^{N}\sum_{j=1}^{N} \hat{C}_{ij}\hat{C}_{ij}}}. \tag{6}$$

In this measure, C is the gold sample and $\hat{C}$ is the image generated by the MM procedure. The similarity between these two images should be maximized by GA.

4 Experiments and Results

The training phase of approach is done by introducing the binary original and the gold samples to the Genetic Optimizer. A heavily noised synthetic image which includes four objects is used as the original image, as shown in Figure 2a. Figure 2b shows the gold sample as the user manipulated image with filtering effect. Our aim is to remove the noise and keep the objects. The optimal structuring elements, the optimized chain of mathematical morphology operations (MM procedure), and the iteration time for any operation are generated by the training (genetic optimization) process. The results are as follows:

$$SE1 = \begin{bmatrix} 0 & 1 & 1 & 1 & 1 \\ 0 & 1 & 1 & 1 & 1 \\ 0 & 0 & 0 & 0 & 0 \\ 1 & 1 & 1 & 1 & 0 \\ 1 & 1 & 1 & 0 & 0 \end{bmatrix} \quad (7)$$

$$SE2 = \begin{bmatrix} 1 & 0 & 1 & 0 & 0 \\ 0 & 1 & 1 & 1 & 0 \\ 0 & 0 & 0 & 0 & 1 \\ 1 & 0 & 0 & 0 & 1 \\ 0 & 0 & 1 & 0 & 0 \end{bmatrix} \quad (8)$$

$$SE3 = \begin{bmatrix} 1 & 1 & 1 & 1 & 0 \\ 0 & 1 & 1 & 0 & 0 \\ 0 & 1 & 0 & 0 & 0 \\ 0 & 0 & 1 & 0 & 0 \\ 1 & 1 & 1 & 0 & 1 \end{bmatrix} \quad (9)$$

$$SE4 = \begin{bmatrix} 1 & 1 & 1 & 1 & 0 \\ 1 & 0 & 0 & 1 & 1 \\ 0 & 1 & 1 & 1 & 1 \\ 1 & 1 & 1 & 1 & 0 \\ 1 & 1 & 1 & 1 & 1 \end{bmatrix} \quad (10)$$

The selected ideal morphological operation chain by GA based on these structuring elements is as (applying left to right):

$$K3 \times (O(SE1) - C(SE2)) \rightarrow K2 \times D(SE4) \rightarrow K1 \times E(SE3), \quad (11)$$

where $K1 = 2, K2 = 2$, and $K3 = 1$.

The result of applying the generated MM filtering procedure on a sample noisy image is shown in Figure 2c, with 96.77% of similarity to the gold sample. Now we have everything of MM procedure to apply it on a test image set.

In our test set, three kinds of noise, namely random, uniform, and gaussian noise, are used to corrupt the test images. The different noise levels and the combined forms of them are added to the test images to generate the test data. Furthermore, some thin lines, thick lines, and speckle noise are added to images to investigate the performance of the approach in the presence of irrelevant structures in noisy images (these irrelevant components are not present in the gold sample). The testing set contains 19 noisy images. Some results of applying the generated MM filtering procedure on noisy test images are shown in Figure 3. In the conducted tests, the matching rate η between the result image and the gold sample is calculated as following:

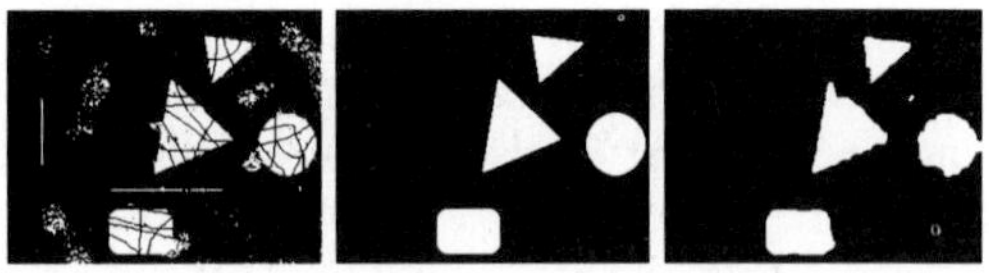

Fig. 2. Example for noise filtering. From left to right: sample noisy image, gold sample provided by the user, and filtered image by generated MM procedure. The result has a 96.77% similarity to the gold sample.

$$\eta = \frac{N - 8N_{UP}}{N} \times 100\%, \quad (12)$$

where N is the number of pixels in the noisy image, and N_{UP} is the number of unmatched pixels when the gold sample and the result image are compared. The number of unmatched pixels is multiplied by 8 to magnify the dissimilarity between the result image and the gold sample. The level of magnification was determined empirically to match the visual expectation more realistically.

5 Discussion and Conclusion

The learning/optimization process uses the sample noisy image and the corresponding gold sample to learn noise filtering. After optimization by GA, the MM filtering procedure according to Eq.11 is generated, where $K1 = 2, K2 = 2$, and $K3 = 1$. Hence, the optimal morphological processing consists of opening-closing followed by two times dilation and two times erosion (any operator has its own optimal structuring element).

After finding the ideal MM filtering procedure, it is applied on the test set which includes 19 noisy images. The overall average matching rate was 94.81% with standard deviation of 7.80%. In cases number 4 and 5 the filtering task can be interpreted as object extraction because the filter removed irrelevant objects, thin lines, thick lines, and speckles, and correctly extracted four objects of interest. Other filters such as median and mean filters cannot deal with these cases.

Our approach takes just one gold sample to learn

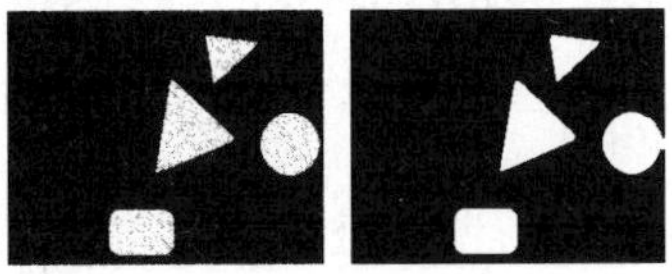

(a) corrupted by uniform noise

(b) $\eta = 98.46\%$

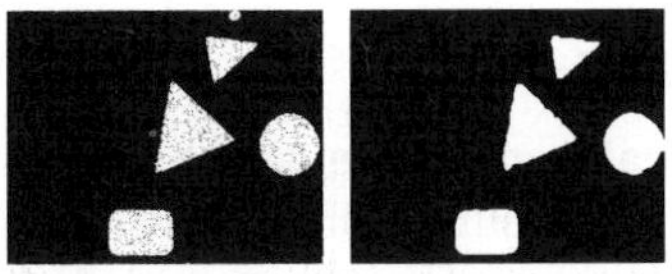

(c) corrupted by gaussian noise

(d) $\eta = 95.99\%$

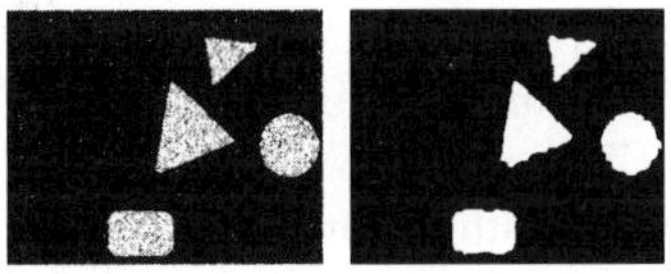

(e) corrupted by combined (uniform, gaussian, and random) noise

(f) $\eta = 93.05\%$

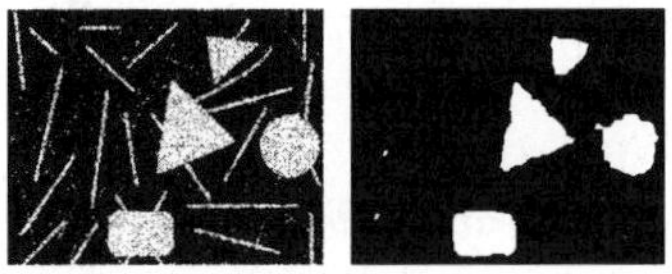

(g) added thin and thick lines to (e)

(h) $\eta = 89.55\%$

(i) added speckle noise to (g)

(j) $\eta = 86.36\%$

Fig. 3. Some results of applying generated MM filtering procedure on noisy test images: the overall average matching rate is 94.81% with standard deviation of 7.80% for test set with 19 images. (a), (c), (e), (g), and (i) are the noisy images and (b), (d), (f), (h), and (j) are the result images. The different amount of three noise (random, gaussian, and uniform noise), thin and thick lines, and speckle noise are used to corrupt the test images.

the filtering task and does not need any prior context knowledge: that is the main advantage of this approach. Using the morphological operators for image filtering makes the filtering procedure robust and computationally efficient. Overall results are very promising.

The proposed approach can be extended to gray level images in order to cover a wider range of image processing applications with the need of segmentation, object extraction, and classification. This will be the subject of future work.

References

[1] L. Gorman and J.V. Nickerson, (1989) An approach to fingerprint filter design, Pattern Recognition, Vol. 22, No. 1, p.p. 29-38

[2] T. Law, H. Itoh, H. Seki, (1996) Image filtering, edge detection, and edge tracing using fuzzy reasoning, IEEE Transaction on Pattern Analysis and Machine Intelligence, Vol. 18, No. 5

[3] T. Sun, M. Gabbouji, Y. Neuvo, (1994) Center weighted median filters: some properties and and their applications in image processing, Signal Processing , Vol. 35, p.p. 213-229

[4] K.N. Plataniotis, D. Androutos, A.N. Venetsanopoulos, (1996) Fuzzy adaptive filters for multichannel image processing, Signal Processing, Vol. 55, no.1, pp. 93-106

[5] J.Serra, (1982) Image analysis and mathematical morphology, Academic Press Inc.

[6] S.R. Sternberg, (1986) Grayscale morphology, Computer Vision, Graphics, and Image Processing, Vol. 35, pp. 333-355

[7] Rafael C. Gonzalez and Richard E. Woods, (2002) Digital Image Processing, Prentice Hall, Second Edition, pp. 519-566

[8] J. Goutsias, L. Vincent, Dan S. Bloomberg (Editors), (2000) Mathematical Morphology and its Applications to Image and Signal Processing, Kluwer Academic Publishers, Vol. 18

[9] Darrell Whitley, A Genetic Algorithm Tutorial, Computer Science Department, Colorado State University.

[10] G. Hamarneh, (2001) Towards intelligent deformable models for medical image analysis, Thesis submitted to Department of Signals and Systems, Chalmers University of Technology, Gteborg, Sweden

Associative Memories and Diagnostic Classification of EMG Signals

C. Shirota[1], M. Y. Barretto[1], C. Itiki[1]

[1]Biomedical Engineering Laboratory, Escola Politécnica da Universidade de São Paulo, Brazil
E-mail: {camila, mateus, cinthia}@leb.usp.br

Abstract

In this work, associative memories are used for diagnostic classification of needle EMG signals. Vectors containing 44 autoregressive coefficients represent each signal and are presented as stimuli to associative memories. As the number of training stimuli increases, the method recursively updates associative memories. The obtained classification results are equivalent to the ones provided by the traditional Fisher's discriminant, indicating the feasibility of the proposed method.

1 Introduction

Adaptive techniques have been applied to the study of electromyographic (EMG) signals. Artificial neural networks were used to control functional electrical stimulation in paraplegics, by adapting to changing environments and allowing patient's interaction with the network's operation [1]. An adaptive feature extraction algorithm was used for the classification of motion commands used in prosthetic arms [2]. Adaptive classification of motor unit action potentials was designed to deal with shape changes due to muscular fatigue [3].

EMG signals have also been used for diagnostic purposes [4]. Clinicians diagnose patients, based on the knowledge or experience that has been acquired over the years. An automatic classification technique should mimic a physician, by storing previous classification results and improving its diagnostic ability. However, most papers that deal with automatic EMG classification for diagnostic purposes present non-adaptive techniques [5], [6], [7].

In this work, we propose an adaptive EMG classification technique for diagnostic purposes and study its feasibility. EMG signals are represented by autoregressive models, since encouraging classification results have been presented in the literature [6], [7]. Associative memories are used as classifiers, for they provide reliable results [8] and they can be computed by a recursive algorithm. In order to validate the obtained classification results, they are compared to the ones obtained by the traditional Fisher's discriminant.

2 Methodology

2.1 Signal Acquisition and Autoregressive Modeling

A data bank with needle EMG signals was used in this work. These signals were recorded at 50% MVC, from the *biceps brachii* muscle, using a 10kHz low-pass filter, at the Duke University Medical Center.

Signals were acquired at the Biomedical Engineering Laboratory of the University of São Paulo with a 12-bit A/D converter and a sampling rate of 25kHz. We selected 800-ms EMG signals, recorded from 6 normal patients (72 signals), 6 myopathic patients (56 signals) and 5 neuropathic patients (43 signals). All signals were classified as stationary, both in mean and variance, by the run test [9], for a 5% significance level and 20 segments. EMG signals were modeled as outputs of autoregressive models. Burg's method [10] was used to estimate the autoregressive coefficients. Order 44 was used in this work, since previous studies [8] showed that this order provided the best classification results for the same data bank.

2.2 Classification

The use of a two-step classification procedure is reported in the literature [5], [6]. We used the two-step classification, since it provided better results than a single-step classification, in a previous research using the same data bank [11]. The first classification step separated EMG signals into normal and pathological classes. The second step separated the signals previously classified as pathological into neuropathic and myopathic classes.

The classification procedure was implemented by two linear classifiers—associative memories and Fisher's discriminant—, which are described in the following subsections.

2.2.1 Linear Associative Memories

According to psychologists, an associative memory (AM) often results from learning the relationship between a stimulus and a response (operant conditioning) [12]. The same terminology is used in the mathematical formulation of associative memories

[13]. A response r is associated to a stimulus s, through an associative memory M, according to equation 1:

$$r = M \cdot s\,. \tag{1}$$

The conditioning process is called training. Among all the signals from the data bank, a training set of signals is created. Autoregressive coefficients of each training signal form a column of the training stimuli matrix. The known classification of the signal is represented in a column of the training response vector. For each signal, an integer number represents the classification. In the first classification step, '1' stands for normal and '2' for pathological signal. In the second step, '1' stands for neuropathic, while '2' stands for myopathic. An associative memory vector M_{j+1} is estimated, by using the training stimuli matrix S_j and the training response vector R_j [13]

$$M_{j+1} = R_j \cdot S_j^+\,, \tag{2}$$

where S^+_j indicates the Moore-Penrose inverse of the training stimuli matrix S_j [14].

As the number j of stimuli in the training matrix S_j increases, the computation of the generalized inverse becomes more burdensome. The stimuli matrix at the j-th iteration can be expressed as:

$$S_j = \begin{bmatrix} S_{j-1} & s_j \end{bmatrix}, \tag{3}$$

where S_{j-1} is the previous stimuli matrix and s_j is the newest stimulus vector to be included in the training group. As one can observe, the number of columns in the stimuli matrix increases with the number of training signals. As a consequence, the direct computation of the generalized inverse could be a hindrance to the use of associative memories as classifiers. However, Greville's recursion [14] enables us to update the generalized inverse by calculating:

$$S_j^+ = \begin{bmatrix} S_{j-1}^+ - S_{j-1}^+ \, s_j \, b_j^+ \\ b_j^+ \end{bmatrix}, \tag{4}$$

where b^+_j is the generalized inverse of vector b_j, which is defined by

$$b_j = s_j - S_{j-1} \, S_{j-1}^+ \, s_j\,. \tag{5}$$

If b_j given by equation 5 is equal to zero, then we must use another formula to compute b_j

$$b_j = \frac{1 + s_j^t \left(S_{j-1} S_{j-1}^t\right)^+ s_j}{s_j^t \left(S_{j-1} S_{j-1}^t\right)^+ \left(S_{j-1} S_{j-1}^t\right)^+ s_j} \left(S_{j-1} S_{j-1}^t\right)^+ s_j\,, \tag{6}$$

where the superscript 't' indicates the transpose matrix. In our Matlab® implementation, we calculated the norm of vector b_j given by equation 5. Whenever this norm was smaller than 10^{-7}, we substituted b_j by the value provided by equation 6. This procedure avoided numerical errors.

The generalized inverse of matrix S_j is obtained by computing the generalized inverse of column vector b_j, which has length 44 (the number of autoregressive coefficients). Vector b_j itself is either a function of the previously computed S^+_{j-1} or a function of the generalized inverse of the product ($S_{j-1}\, S^t_{j-1}$), which is square matrix of size (44 x 44). In this way, we can compute S^+_j by inverting smaller matrices or vectors that do not increase in size, as the number of training signals increases.

We also provide the recursion to compute the associative memory vector:

$$M_{j+1} = M_j - M_j \, s_j \, b_j^+ + r_j \, b_j^+\,, \tag{7}$$

where r_j is the response corresponding to the newest training stimulus vector s_j, and M_j is the previous memory vector.

In order to evaluate the performance of these adaptive associative memories, we must use test stimuli s_k of known classification responses. The response r_k associated to each test stimulus is estimated by

$$\hat{r}_k = M_j \cdot s_k\,, \tag{8}$$

for j varying from 1 to the number of training stimuli and for k varying from 1 to the number of test stimuli. These responses are real values and do not correspond exactly to the integer values associated to the classes. As a consequence, we must use a criterion to separate the obtained responses into the classes. In the first classification step, responses r_j below 1.5 were classified as normal signals (class 1), while responses r_j greater or equal to 1.5 were classified as pathological signals (class 2). This same value was used in the second classification step, in order to separate neuropathic (class 1) from myopathic signals (class 2).

2.2.2 Fisher's Linear Discriminant

Fisher's linear discriminant (FLD) classifies a signal associated to a stimulus vector s_j, by estimating the response [15]

$$\hat{r}_j = v^T s_j\,. \tag{9}$$

In order to compute vector v, training stimuli must be separated into classes 1 and 2. For class 1, the mean stimulus vector u_1 and the covariance matrix Σ_1 are calculated. Similarly, for class 2, the mean stimulus vector u_2 and the covariance matrix Σ_2 are computed. The optimal projection direction that separates classes 1 and 2 is given by vector v:

$$v = \left(\frac{(n_1 - 1)\Sigma_1 + (n_2 - 1)\Sigma_2}{n_1 + n_2 - 2} \right)^{-1} \cdot (u_1 - u_2)\,, \tag{10}$$

where n_1 and n_2 are the number of stimuli in classes 1 and 2 respectively. The stimulus s_j is separated into

class 1, if the response $r_j^{'}$ is above a threshold value. Otherwise, it is separated into class 2.
In this work, both AM and FLD were implemented in Matlab®.

2.2.3 Training group

In a real system, the diagnosis of a patient is based on previous knowledge. Similarly, an initial set of subjects was used to compose the initial training group. Three subjects were chosen to form the initial training group, providing 11 normal, 10 neuropathic and 10 myopathic signals.
Eight additional subjects were later added to the training group, at a random order. This random order of subjects was used because physicians cannot choose the order in which patients arrive at the clinic. All the signals of a given subject were inserted one-by-one into the training process. The following order of insertion was used: 12 normal, 15 normal, 7 myopathic, 9 neuropathic, 9 myopathic, 6 neuropathic, 16 normal and 12 myopathic signals.
At each training step, one signal was classified using all the information stored in the previous training group. Then, autoregressive coefficients of this signal were included in the next training stimuli group. The response associated to each stimulus varied according to the type of training: unsupervised or supervised [16]. Unsupervised training was based on the results given by the classifier. The estimated class for each new signal was inserted as the next training response. Supervised training was based on the known signal classification, which was provided by a teacher. For supervised training, the known classification of the subject was included as response, regardless of the estimated class. The known classification inserted in supervised training could represent the diagnosis provided by a physician, based on clinical evidence and exam results, other than EMG.

2.2.4 Test Group

In order to test the variation of classification rate with the increasing number of training stimuli, we needed signals that were not used previously in training. All six of the remaining subjects had already been separated for the test group, providing 18 signals of each type. All test signals were classified several times, as each new training signal was included in the training group.

3 Results

Classification rates obtained by AM are shown in Figure 1, while results obtained by FLD are illustrated in Figure 2. The horizontal axis presents the number of training signals used at each iteration and ranges from 31 (initial training group) to 117 (final training group). The vertical axis represents the correct classification rate in percentage. It is equivalent to the ratio between the number of correct classifications and the total number of test signals (54). Dashed lines represent the classification rate obtained for unsupervised training, while continuous lines represent supervised training.

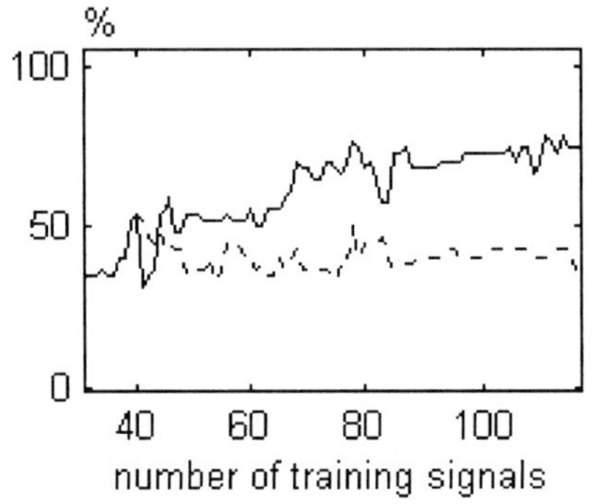

Fig. 1. Classification rates obtained by AM, for supervised (continuous line) and unsupervised training (dashed line).

According to figures 1 and 2, supervised training (continuous line) clearly showed better results than unsupervised training (dashed line), for both classifiers (AM and FLD). This result was expected, since the correct training is fundamental for the classifier's performance.

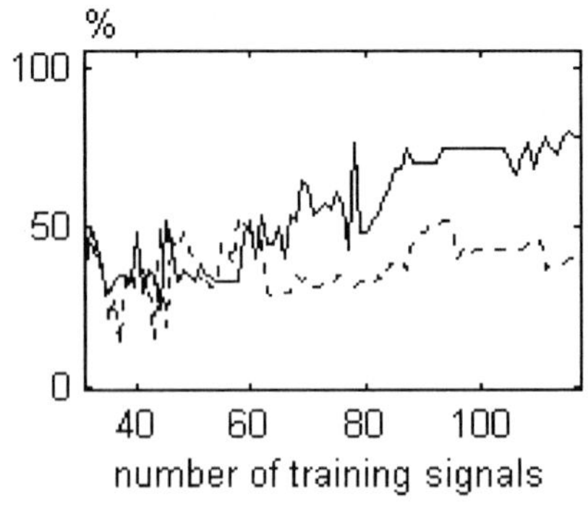

Fig. 2. Classification rates obtained by FLD, for supervised (continuous line) and unsupervised training (dashed line).

Unsupervised training seemed to stabilize at values close to 40%, which is similar to the classification rate obtained for the initial training group. This result shows that both classifiers did not 'learn' from

unsupervised training. On the other hand, supervised training presented increasing classification rates that reached values above 75%. The performance of both supervised classifiers improved as the size of the training group increased. This result shows that the supervised classifiers were able not only to store but also to use previous training information, in order to improve classification rates.
Comparing FLD and AM results, one can observe that FLD provides a more erratic behavior than AM, whenever less than 60 signals are used in training.
The method presented in this work uses simple update equations, whose computation does not increase in complexity, as additional training signals are included. So, the method can be applied to a larger group of signals and subjects. If we had more subjects in our data bank, we could have provided the classification rates in percentage of subjects. However, since we had only six test subjects, we provided our final classification rates (above 75%) in percentage of signals. The classification rates presented in the literature for FLD (60% to 87.5%) [7], [6] and neural networks (47.5% to 90%) [17], [7] are given in percentage of subjects and were obtained for the same parameters of our study—autoregressive coefficients of EMG signals.

4 Conclusion

This work presented a recursive method to update associative memories. These memories 'adapted' as new signals and their known classification were included in the training process. The method was applied to 117 training signals and 54 test signals. The obtained 75% classification rate showed the method's feasibility, since it is within the range of 47.5% to 90% presented by other methods in the literature [6], [7], [17].
Training signals were included one-by-one in the training group and associative memories were updated by recursive equations. As a consequence, the computational load did not increase as new information was added to the training process. This is one of the advantages of associative memories in comparison to other methods.
These results indicate that recursive associative memories could be applied, in the near future, to a large number of signals in a clinical setting, providing high classification rates.

5 Acknowledgements

This work was partially supported by FAPESP and CNPq. We thank Prof. Sanders from the Duke University Medical Center, who kindly provided the EMG data bank used in this study.

References

[1] Graupe, D., Kordylewski, H. (1995) Artificial Neural Network Control of FES in Paraplegics for Patient Responsive Ambulation. IEEE Trans. Biom. Eng. 42(7): 699-707.
[2] Lee, S., Kim, J., Park, S. (1996) An Enhanced Feature Extraction Algorithm for EMG Pattern Classification. IEEE Trans. on Rehabilitation Engineering 4(4): 439-443.
[3] Gazzoni, M., Farina, D., Merletti, R. (2004) A new method for the extraction and classification of single motor unit action potentials from surface EMG signals. Journal of Neuroscience Methods 136(2): 165-177.
[4] Desmedt, J. E. (1989) Computer-Aided Electromyography and Expert Systems. Elsevier-Science, Amsterdam.
[5] Berzuini, C., Maranzana-Figini, M., Bernardinelli, L. (1982) Effective use of EMG parameters in the assessment of neuromuscular diseases. International Journal of Bio-Medical Computing 13: 481-499.
[6] Inbar, G. F., Noujaim, A. E. (1984) On Surface EMG spectral characterization and its application to diagnostic classification. IEEE Trans. on Biomedical Engineering 31(9): 597-604.
[7] Abou-Chadi, F. E., Nashar, A., Saad, M. (2001) Automatic analysis and classification of surface electromyography. Frontiers of Medical and Biological Engineering 11: 13-29.
[8] Shirota, C., Barretto, M. Y., Itiki, C. (2004) Classificação de sinais eletromiográficos de agulha por memórias associativas e modelagem auto-regressiva. Proceedings of the International Federation for Medical and Biological Engineering 5(1): 959-962.
[9] Bendat, J. S., Piersol, A. G. (1986) Random Data - Analysis and Measurement Procedures, 2nd ed. John Wiley & Sons, New York.
[10] Marple, S. L. (1987) Digital Spectral Analysis: with Applications. Prentice-Hall, Englewood Cliffs.
[11] Barretto, M. Y., Kohn, A. F., Itiki, C. (2004) Modelagem auto-regressiva e discriminante de Fisher na classificação de sinais eletromiográficos de agulha. Proceedings of the International Federation for Medical and Biological Engineering 5(1): 939-942.
[12] Kandel, E. R., Schwartz, J. H., Jessell, T. M. (1991) Principles of Neural Science, 3rd ed. Elsevier Science, New York.
[13] Kohonen, T. (1984) Self-Organization and Associative Memory. Springer-Verlag, Berlin.
[14] Graybill, F. A. (1983) Matrices with Applications in Statistics. Wadsworth, Pacific Grove.
[15] Nadler, M., Smith, E. P. (1993) Pattern Recognition Engineering. John Wiley, New York.
[16] Haykin, S. (1999) Neural Networks: a Comprehensive Foundation, 2nd ed. Prentice Hall, Upper Saddle River.
[17] Pattichis, C., Elia, A. G. (1999) Autoregressive and cepstral analyses of motor unit action potentials. Med. Eng. Phys. 21: 405-419.

Discretization of Series of Communication Signals in Noisy Environment by Reinforcement Learning

Katsunari Shibata
Department of Electrical and Electronic Engineering, Oita University, Japan
E-mail: shibata@cc.oita-u.ac.jp

Abstract

Thinking about the "Symbol Grounding Problem" and the brain structure of living things, the author believes that it is the best solution for generating communication in robot-like systems to use a neural network that is trained based on reinforcement learning. As the first step of the research of symbol emergence using neural network, it was examined that parallel analog communication signals are binarized in some degree by noise addition in reinforcement learning-based communication acquisition. In this paper, it is shown that two consecutive analog communication signals are binarized by noise addition using recurrent neural networks. Furthermore, when the noise ratio becomes larger, the degree of the binarization becomes larger.

1 Introduction

We humans can communicate complicated information skillfully using symbols. It has been thought that artificial neural networks(ANNs) are good at continuous nonlinear approximation, but are not good at symbol handling or logical processing. In living things, the functional difference has been pointed out between the left-brain and the right-brain[1]. Based on these, the idea of the specialization that the ANN corresponding to the right-brain is used for pattern processing, and a digital computer corresponding to the left-brain is used for logical processing has been accepted generally. However, there is no general idea about what signals should be transferred between the ANN and the computer, and that causes the "symbol grounding problem". Furthermore, the left-brain and the right-brain looks almost the same in the real brain compared with the difference between the ANN and the digital computer.

The author believes that in order to solve the serious "symbol grounding problem", the pattern processing and the logical processing should not be distinguished. Accordingly it is expected for the ANN to perform the both without any discriminations. For this reason, it is very significant to show that symbols emerge in the ANN only by applying reinforcement learning inspired by the learning of living things. Then, there appears a question "is it true that ANN is not good at symbol processing?"

Here, for simplicity, symbols are considered as discretized signals. Considering from the point of necessity, the reason why the communication signals are discretized can be either "logical thinking" or "elimination of noise effect". Considering from the point of structure, associative memory, in other words, fixed-point dynamics can be a solution to realize the discretization. As the first step of the research of symbol emergence, the necessity of "elimination of noise effect" has been focused on, and it was examined whether parallel analog communication signals are discretized by noise addition in reinforcement learning-based communication acquisition[2][3].

In this paper, like a "word", a series of signals are communicated on behalf of the parallel signals. Same as the previous work, binarization of the signals by reinforcement learning in noisy environment is examined. It seems more difficult than the case of parallel signals because memory is necessary to generate and recognize the communication signals.

2 Learning and Task

A simple communication environment in which only two agents exist is assumed here. Referring to [4], one of them can transmit a communication signal to the other. They are put on an one-dimensional space as shown in Fig. 1. When the both agents touch together, they get a reward. The transmitting agent (transmitter) cannot

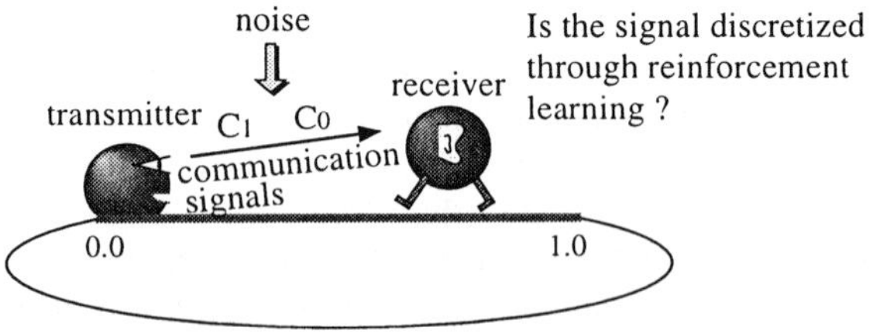

Fig. 1. Communication task employed in this paper.

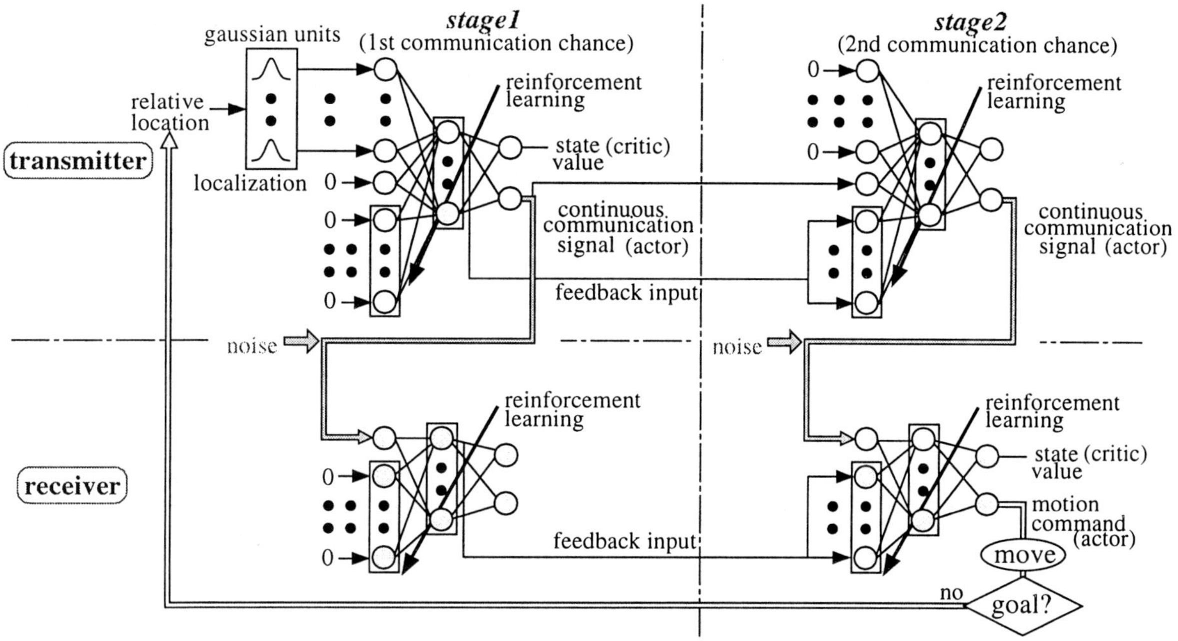

Fig. 2. Architecture of each agent and signal flow in one time step.

move, but can observe the relative location of the opponent, and generates a series of communication signals by its own recurrent neural network (RNN). The receiving agent (receiver) interprets the communication signals and generates its motion command also by its own RNN. It can move according to the motion command. It cannot observe anything except for the communication signal, and cannot transmit anything. Both agents are trained based on reinforcement learning independently.

The transmitter is fixed at the left edge on a one-dimensional space where the left edge is linked to the right edge. The length of the space is 1.0. The receiver is located randomly at every trial. When the motion command is positive, it moves to the right, and when negative, it moves to the left. The absolute value decides the moving distance. When the distance between the transmitter and receiver is less than 0.11, they can touch each other and get a reward. However, if the motion command is too large for the distance, it goes past the transmitter, and they cannot get the reward. Accordingly, the receiver's motion should be in a range, and the range is gradually sifted according to the receiver's location.

Fig. 2 shows the architecture of each agent and the signal flow in one time step. Each agent has an Elman-type RNN that enables memorization of some necessary information. There are two stages in one time step. In the first stage, the transmitter observes the receiver's location, and generates the first communication signal by computing its RNN. The receiver receives the signal, and computes its RNN. The outputs are not used at this stage, but the hidden outputs are used as the feedback inputs at the stage 2. In the stage 2, the transmitter receives the first communication signal as input, and generates the second communication signal. The receiver receives the signal, and generates the motion command.

The information of receiver's location is localized by N Gaussian units. This helps the neural network to learn a strong nonlinear transformation. The center of each Gaussian is arranged between 0.0 and 1.0 with the constant interval of $1.0/(N-1)$, where N is the number of Gaussian units. The size of each Gaussian σ is the same as the interval. The output is described as

$$GS_i(loc) = exp\left(-\frac{1}{2}\left((N-1)\cdot loc - i\right)^2\right), \quad (1)$$

where i is the index of each Gaussian unit ($i = 0, 1, 2, .., N-1$), loc is the receiver's location. Here, $N = 30$.

As a reinforcement learning architecture, actor-critic is employed for each agent. Here, the transmitter deals with the communication signal as its action, while the receiver deals with the communication signal as its state. One of the outputs of each network is used as critic, and the other is used as actor. The hidden neurons are used in common by the both types of outputs. The training signals are generated based on reinforcement learning, and the network is trained by BPTT (Back Propagation Through Time). TD (Temporal Difference) error $\hat{r}$ is calculated as

$$\hat{r}_{t-1} = r_t + \gamma P_t - P_{t-1} \tag{2}$$

where r_t is the reward, P_t is the critic output at t time step, and γ is a discount factor. The critic output is trained by the training signal as

$$P_{s,t-1} = P_{t-1} + \hat{r}_{t-1} = r_t + \gamma P_t. \tag{3}$$

As the critic output P_t, the output at the stage 1 is used for the transmitter, while the output at the stage 2 is used for the receiver. The training signal is also given to the output at the stage 1 for the transmitter, while to the output at the stage 2 for the receiver. The actual motion M_t is calculated as

$$M_t = \alpha(2.5A_t + rnd_t + n_t) \tag{4}$$

where A_t is the actor output, rnd is the random number for trial and error, and n is the noise factor that is not added in the case of the receiver's motion, but is added in the case of the communication signal. α is a constant. The actor output is trained by the training signal as

$$A_{s,t-1} = A_{t-1} + \beta \hat{r}_{t-1} rnd_{t-1} \tag{5}$$

where β is a constant, and it is 0.5 here. The training signal is given to the output at each stage for the transmitter, while to the output at the stage 2 for the receiver. When the training signal is given to the output at the stage 2, the learning traces back also to the stage 1 based on BPTT.

The output function of each hidden or output neuron is a sigmoid function that ranges from -0.5 to 0.5. All the training signals are limited from -0.4 to 0.4 to avoid the saturation area of the sigmoid function. In Eq. (4), by multiplying 2.5 to each actor output, the range becomes from -1.0 to 1.0, and after that, the trial and error factor and noise are added. When the value becomes larger than 1.0 or less than -1.0, it is returned to 1.0 or -1.0 respectively. Here, the trial and error factor is cubed uniform random number whose level, in other words, whose amplitude is ± 0.4. The noise factor is a uniform random number whose level is varied from ± 0.0 to ± 1.6 with the interval of 0.2 in the following simulations. Even in the case that the noise factor is always zero, the random number for the trial and error factor of the transmitter is received as a noise for the receiver.

For the critic computation based on TD learning in Eq. (2) and (3), 0.5 is added to the critic output actually. The reward that is given to each agent is 0.9. To generate the communication signal, α in Eq. 4 is 1.0 in the transmitter. For the motion command, α is 0.4 in the receiver so as that the receiver can touch the transmitter in one step from any locations by an appropriate motion. If the receiver's motion is discretized completely, no less than 4 levels of output is required. The number of layers is 3, and the number of neurons in the hidden layer is 10 for both agents. All the initial connection weights from the hidden layer to the output layer are 0.0, and those from the input layer including feedback inputs to the hidden layer are decided randomly in the range from -1.0 to 1.0.

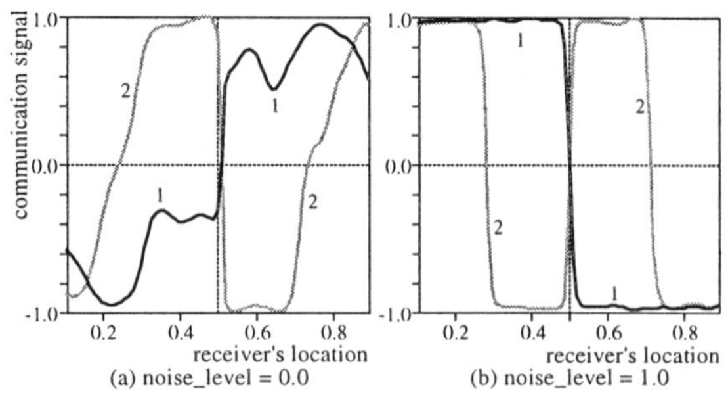

Fig. 3. The communication signals as a function of the receiver's location. The random number level is 0.4.

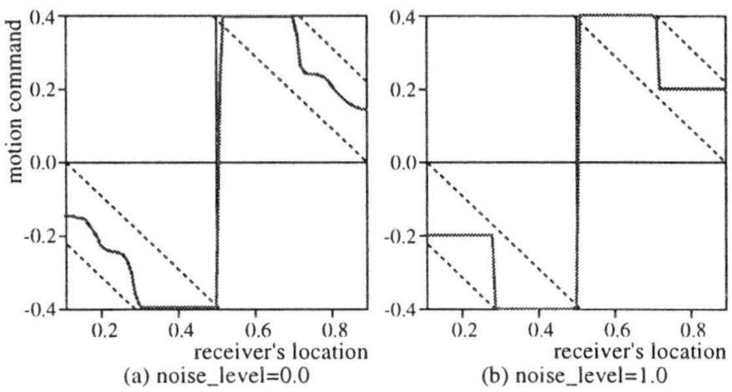

Fig. 4. The receiver's motion command as a function of the receiver's location. The random number level is 0.4 and the noise level is 1.0.

3 Result

It was observed whether the transmitted signals became discrete when the noises were added to the communication signals during the learning. The consecutive two communication signals and the motion command after learning with no noise are shown in Fig. 3(a) and Fig. 4(a), and those with some noise (level=1.0) are shown in Fig. 3(b) and Fig. 4(b). The initial connection weights are the same between the two cases. The sloping lines in Fig. 4 indicate the maximum and minimum limit values of the motion for the receiver to touch the transmitter in one time step by the motion as a function of the receiver's location.

Roughly, the results are similar to the case of the parallel communication signals[2][3]. In the both cases, the transmitter generated the first communication signal from the receiver's location and generated the second one from the first communication signal and the feedback inputs. Then the receiver received the first communication signal, kept the information through the feedback inputs, and could acquire appropriate motions from the stored information and the second signal at the stage 2. After learning with noise, each communication signal was almost binarized, and only around the boundary of the binary values, the signal took a medium value. However, it is clear that the degree of binarization is larger than in the case of no noise. The receiver's motion is discretized into four levels in the optimal range by the combination of the two consecutive binary communication signals. The motion is more clearly discretized than the communication signals. The reason might be that the receiver also learned to binarize the received signal utilizing non-linear transformation of the neural network. When the number of communication chances was increased to three times, the assignment of information to the signal of each chance could not be done well, and the optimal motion command could not be acquired. That is different from the case of parallel signals.

The degree of the binarization according to the noise level was also observed. The degree of binarization that means how the signal is close to the maximum value 1.0 or the minimum value -1.0 is defined as

$$bin = \sum_{i}^{Nc} \sum_{j}^{Nd} |com_{i,j}| / (Nc \cdot Nd) \qquad (6)$$

where Nc is the number of communication chances, Nd is the number of sampled receiver's locations, and com is the communication signal without the random number and noise. If the communication signal is always -1.0 or 1.0, the degree becomes the maximum of 1.0, while if the signal is always 0.0, it becomes the minimum of 0.0. The degree of binarization according to the noise level is shown in Fig. 5(a). Each small circle shows the average degree over 50 simulations, and the vertical line shows the standard deviation. It can be seen that when the noise level becomes larger, the degree becomes larger and the deviation becomes smaller. However, when the noise level becomes larger than 1.0, the degree decreases slightly according to the noise level. When the noise level is 1.6, the both signals encode the same information as the signal 1 in Fig. 3(b) in most cases. That results in the high degree of binarization.

The noise tolerance was also examined. Fig. 5(b) shows the average time steps to the goal as a function of the noise level in the learning phase for each noise level in the test phase after learning. It can be seen that when some noise is added in the test phase, the performance is the best when the noise level in the learning phase is 1.0. It is interesting that under the same condition, the degree of binarization is the maximum. However, when

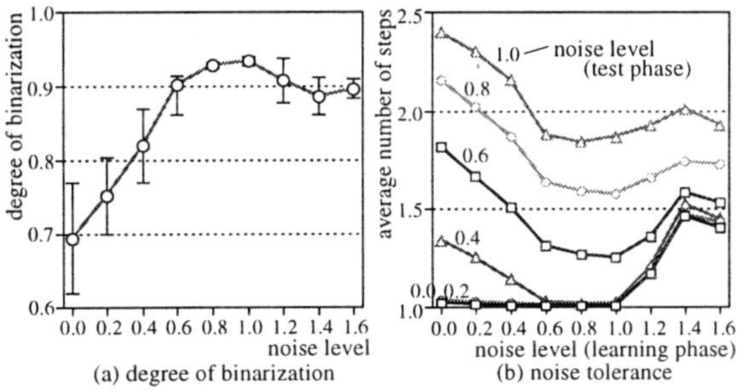

Fig. 5. Noise Effect. (a)Degree of binarization according to the noise level in the learning phase. (b)Noise tolerance according to the noise level in the learning phase.

the noise level in the learning phase is larger than 1.0, the performance becomes worse even if the noise level in the test phase is 0.0. This means that the large noise in the learning phase disturbed the proper learning even for the case of no noise.

4 Conclusion

It was shown that as well as the case of the parallel communication signals, noise addition has the effect to promote the binarization of the signal only by reinforcement learning in the case of a series of signals. The reason why appropriate signals cannot be obtained in the case of more than two signals should be examined.

Acknowledgment

This research was supported by the Japan Society for the Promotion of Science, Grants-in-Aid for Scientific Research, #14350227 and #15300064. The author also would like to say thanks to Dr. J. Yoshimoto at Nara Inst. of Science and Technology, Japan who advised the interest of series of communication signals.

References

[1] Sperry, R.W. (1968) Hemisphere Deconnection and Unity in Conscious Awareness, *American Psychologist*, **23**, pp. 723–733

[2] Shibata, K. and Sugisaka, M. (2004) Discretization of Analog Communication Signals by Noise Addition in Communication Learning, *Proc. of AROB 9th*, **2**, pp. 351-354

[3] Shibata, K (2004) Discretization of Analog Communication Signals by Noise Addition in Reinforcement Learning of Communication, *Technical Report of IEICE*, **103** (734), NC2003-203, pp. 55-60 (in Japanese)

[4] Ono, N, Ohira, T. and Rahmani, A.T. (1995) Emergent Organization of Interspecies Communication in Q-Learning Artificial Organs, *Advances in Artificial Life*, pp.396–405

The Research of Speaker-Independent Continuous Mandarin Digits Speech-Recognition Based on the Dynamic Search Method of High-Dimension Space Vertex Cover

Wenming Cao [1] ,Xiaoxia Pan [1] , Shoujue Wang [2]

[1]Institute of Intelligent Information System, Information College, Zhejiang University of Technology, Hangzhou 310032, China

csann@zjut.edu.cn

[2]Institute of Semiconductors, Chinese Academy of Science, Beijing 100083, China

Abstract.

In this paper, we present a novel algorithm of speaker-independent continuous Mandarin digits speech-recognition, which is based on the dynamic searching method of high-dimension space vertex cover. It doesn't need endpoint detecting and segmenting. We construct a coverage area for every class of digits firstly, and then we put every numeric string into these coverage-areas, and the numeric string is recognized directly by the dynamic search method. Finally, there are 32 people in experiment, 16 female and 16 male, and 256 digits all together. All these digits are not learned. The correct recognition result is 218, and error recognition result is 26. Correct recognition rate is 85%

1 Introduction

With the development of the technology of speaker-independent small vocabulary isolating speech recognition, its application has been prospected in many fields. But the applied continuous speech recognition system is still rare. Therefore, the research of speech recognition was mainly concentrated on improving the performance of speaker-independent large vocabulary continuous speech recognition since 1990s.

The traditional algorithm of the speaker-independent continuous speech-recognition is mainly depended on the endpoint detection. First, the speech was segmented into small units, syllables or phonemes, and then these units were recognized by the method of Hidden Markov Models (HMMs)[2] or Dynamic Time Warping (DTW). The advantages of these methods are that they can convert the continuous speech into isolated speech and reduce the difficulties of the recognition process. But their method is fragile in the noise, limited in their ability to handle pronunciation variation, and costly for large vocabulary spontaneous speech transcription. That is to say, their ability to represent dynamic behavior is limited [10].

In order to resolve these problems of continuous speech recognition, in this paper, we present a novel algorithm which is based on the Theory of High-Dimension Space Vertex Cover[5],[8]. We classify the corpora firstly and then analyze the point property of the feature space by this means. Because of CASSAN-II neural computer have been widely applied the Theory of High-Dimension Space Geometry to Artificial Neural Networks [3],[7],[6],[4], and soon it succeeds in many fields, such as pattern recognition[9], face recognition[12] and adaptive controller[1],Finally we implement it on the CASSAN-II neural computer[11] invented by Wang Shoujue.

2 The characteristic of continuous speech

(1) Co-articulation exists in the speech units (e.g. phoneme, syllable).

(2) Because there are not clear endpoints between two units, the phone can not be segmented completely or separated from each other.

(3) Both the rate-of-speech[5] and the duration-of-speech are variable, so it is difficult to find an uniform module;

Because of the three characteristics mentioned above, the endpoint detecting and the speech segmentation become difficult. In addition, the environment noise makes the acoustic model establishment even more difficult. Therefore, the decision tree and many phonemic models were used in the traditional Hidden Markov Models method, such as Word-Dependent (WD), Left-Context-Dependent (LCD), Right-Context-Dependent (RCD), context-Dependent (CD, also called TRIPHONE model, triple phone), and so on. This makes the models which appear seldom can't be trained sufficiently, so the recognition rate is lower than those which appeared more.

(4) There are some characters of the nature speech, which is more optional and has a few stochastic phenomena, such as hesitation, pause, filled pauses etc.

3 Feature extraction of the continuous digits speech recognition and the constructing of High-Dimension Space coverage areas

3.1 The collection and establishment of the speech corpora

There are two speech patterns. One is spontaneous speech that is the utterance we speak in our daily life. It is unbending. At least, it has no special prepare in terms of speech pattern. It is always slack, and goes with random events (filled pauses etc.). The other is the reading speech. It is always changeless. Its speech pattern and speech context should be prepared beforehand and accord with the grammar as well.

The continuous speech we adopted in this paper is between the reading and the spontaneous speech. The context of our corpora is the phone numbers. The read pattern is similar to the spontaneous speech. According to the experiment requires, we remove some complex noise but some background noise (e.g. stir of cars in road) is left.

There are two corpora. As for the first corpora, we segment the continuous speech into syllables by hand and then select the better result as "the learning corpora". We must point out that these syllable samples are different from the isolated samples. They have many characteristics of continuous speech. As for the second corpora, we regard all these articulate continuous phone number as "the recognition corpora". Both these corpora are collected in 8000Hz (the Sample Frequency) and 16bits (the Bit Depth).

3.2 Feature extraction method of the learning corpora

There are three steps in this process.

Step 1. Change the wave samples into Mel Frequency Cepstrum Coefficient (MFCC).

First, pre-emphasis processing:

$$x'(n)=x(n)-0.9375x(n-1).$$

Second, hamming windows:

$$x'(n)=[0.54-0.46\cos(2\pi n/255)]x(n)$$

the width of this window is 256, and its offset is 64.

Third, Mel Frequency Cepstrum conversion: the number of Mel filters is 24. We remove the first and the last 7 values, and remain 16 values as feature parameters.

Step 2. Remove the redundant data.

First, suppose that every 16 feature parameters consist of one 16 dimension-vector Ci, i=1,2,3...n, shown as fig.1.

Fig.1. As for a digit sample, here are n 16-dimension-vectors. Every vector has 16 dimensions which are 16 MFCC after step 1.

Second, compute the angle of the two adjacent vectors:

$$\theta_j = ar\cos(\frac{C_j \bullet C_{j+1}}{|C_j| \cdot |C_{j+1}|})$$

When the angle θ_j is below the experiment statistic number 0.13rad, remove Cj or Cj+1 until all $\theta_j \geq 0.13 rad$. In this way, we can not only remain more than 8 vectors but also compress data in a certain extent.

Step 3. Normalize a digit sample into a certain length.

First, select one shortest sample from each class of MFCC vectors. These vectors have already been compressed. Choose the optimal 8 vectors in hearing by hand. And regard these 128 values, a new feature vector of High-Dimension Space, as a standard of this class of MFCC vectors.

Second, compare all MFCC vectors of each class with the standard, and choose sequential 8 16-dimension-vectors whose angle with the standard is smallest. And regard this 128-dimension-vector as the feature of this sample. The coverage areas of feature space are constructed by these 128-dimension feature vectors. This process is shown as figure 2.

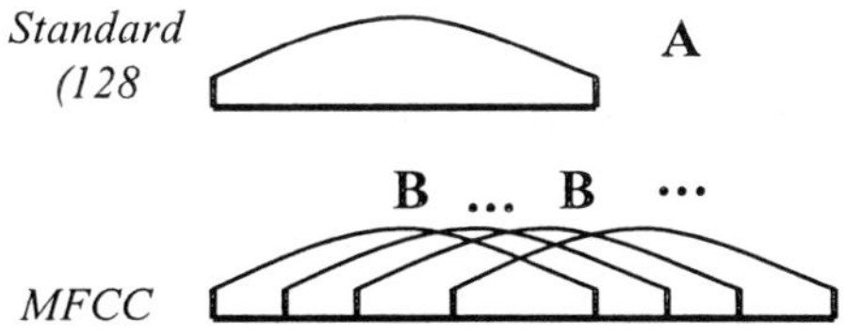

Fig. 2. Choose the most similar 8 vectors (consist of a vector of 128-dimension) from the MFCC vectors.

Suppose,

$$\theta_k = ar\cos(\frac{A \bullet B_k}{|A| \cdot |B_k|})$$

If,

$$\theta_{\min} = \theta_p = \min\{\theta_k, k = 1,2,\cdots n\}$$

Then regard this 128-dimension vector B_p as the most similar one in these vectors $B_k (k = 1,2,\cdots n)$.

Therefore, select B_p as this sample's feature vector (128-dimension). This feature vector is one of the samples, which construct the coverage areas of this class of feature space.

3.3 Put the feature vectors into neural networks and construct the coverage areas of every syllable (we see every syllable as one class) in the high-dimension feature space

From 0 to 9 except 1, there are 9 classes of samples. The digit "1" is divided into two classes: "yi" and "yao". Then there are 11 classes of samples totally. Regard each sample class as one set Si（i=1,2,...11）,

$$S_i=\{x_{ij} \mid x_{ij}\in S_i ,\ j\in N\}$$

Si is from the learning corpora. And select n samples (every sample xij is one of the point of High Dimension Space) from every sample class as a new sample set S'i. Put the new sample set S'i into neural networks, and adopt the Theory of High-Dimension Space Vertex Cover[4],[8],[11],[12] to construct the coverage areas of the ith class in feature space.

4 The novel algorithm and its implementation of Speaker-Independent Continuous Digit Speech Recognition Based on the Dynamic Search Theory of High-Dimension Space Vertex Cover

4.1 Feature extraction method of recognition corpora

According to the same procedure of the step 1 and step 2 in section 3.2, change wave samples into 16 dimension MFCC vectors and remove the redundant data. And then extract 8 16-dimension vectors as a 128-dimension vector from the beginning of the speech. Extract next 8 16-dimension as the next 128-dimension from the second 16-dimension vector. The rest may be deduced by analogy. Form a series of 128-dimension vectors whose length (the number of 128-dimension vectors) varies with the length of the speech. In this experiment, it varies from 207 to 465.In this procedure, the frame width of Hamming window is 256 sample points, the offset length of every frame is 64 sample points, the number of Mel filter is 24, and the threshold θ=0.13 radian.

4.2 Dynamic search and recognition method of High-Dimension Space Vertex Cover

Regard n 128-dimension feature vectors of continuous speech to be recognized as n points of high dimension space. Here, take the sentence "san wu ba er qi ling ling san" as an example. The length of this sentence is 433, that is n=433. Compute the distance between these n points and every class of the coverage area. The result curve is shown as 11 Figures, we select 2 figures fig.3a-3c. (The x-coordinate is the points' serial number and it is a time-axis. And y-coordinate is the distance between the present point and the present coverage area. The distance will be changed by time). In fig.3a, two distinct minimum points are shown in 79th and 357th points respectively. Since it is in the coverage area of "san", this result indicates that the two positions in time-axis are the pronunciation of "san". Fig.3b indicates the distance between this sentence and the coverage area of "wu". There is only one minimum point, the 109th point. This point stands for "wu" in this sentence. The curves in fig. 3c, 3d and 3e indicate the distance between this sentence and the coverage areas of " ba", "er" and "qi" respectively. The minimum points correspond to the pronunciation of "ba", "er" and "qi" of this sentence respectively.

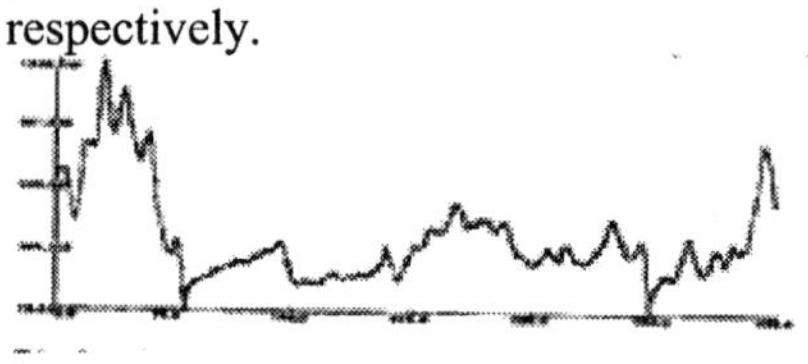

Fig.3a. the distance from "san wu ba er qi ling ling san" to the coverage area of "san"

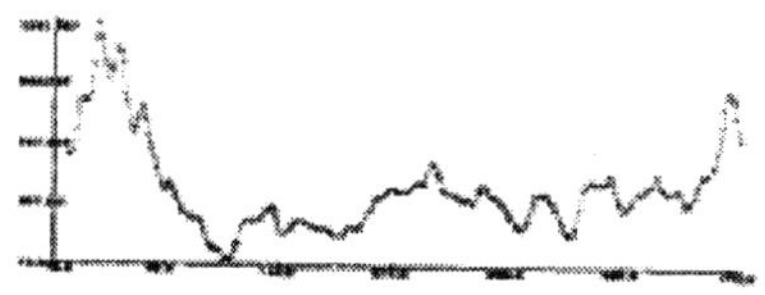

Fig.3b. the distance from "san wu ba er qi ling ling san" to the coverage area of "wu"

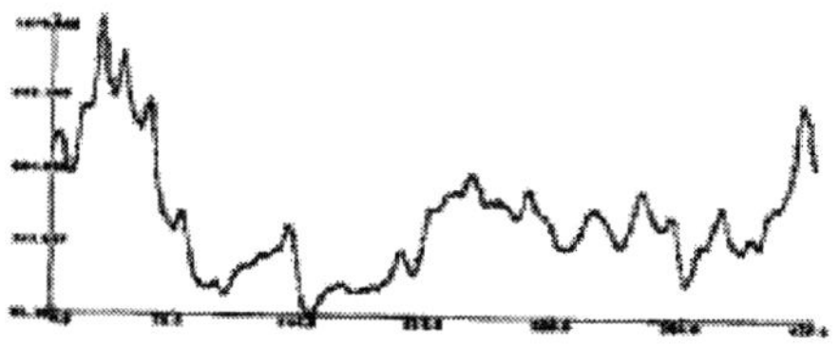

Fig.3c. the distance from "san wu ba er qi ling ling san" to the coverage area of "ba"

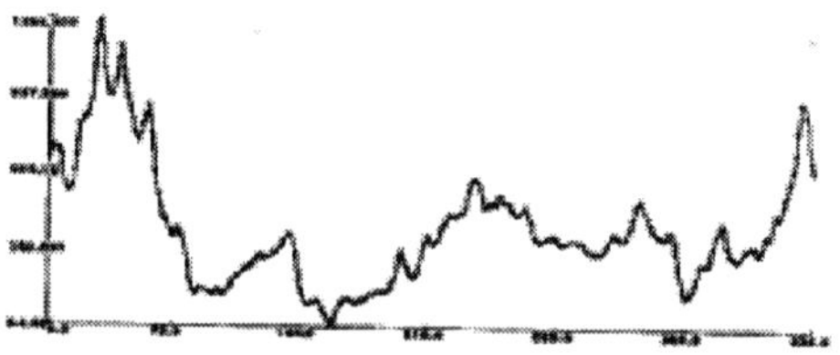

Fig.3d. the distance from "san wu ba er qi ling ling san" to the coverage area of "er"

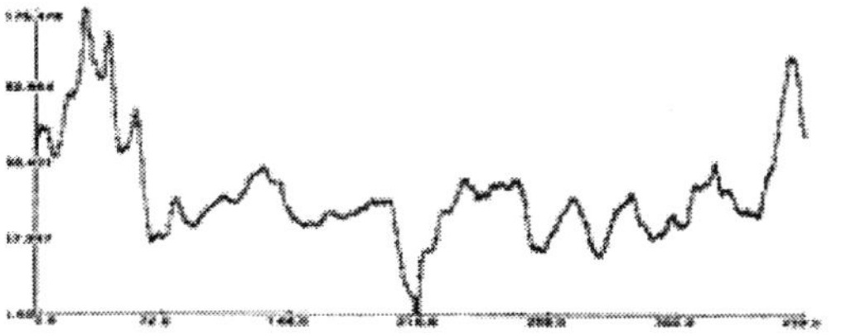

Fig.3e. the distance from "san wu ba er qi ling ling san" to the coverage area of "qi"

All these 11 curves form a mesh surface shown as fig.6.

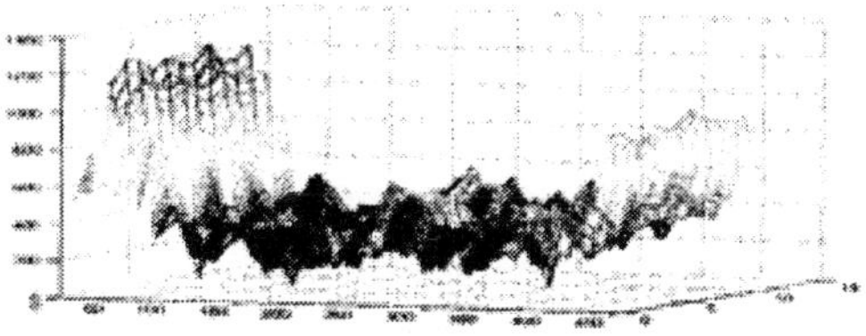

Fig.4. the 3D display of the recognition result

5 Result and discussion

5.1 Statistic result:

There are 32 people in this experiment, 16 female and 16 male, and 256 digits all together. All these digits are not learned. The correct recognition result is 218. and Error result is 26.

5.2 Discussion

We discussed a new method of the Speaker-Independent Continuous Speech-Recognition in this paper. As it dose not need to detect the endpoint and segment speech, it can be used in the recognition of the continuous nature speech. Although the result is not perfect, but we can see the performance of robustness is good. It will be a promising new research direction in the continuous nature speech recognition.

References

[1] Cao, Wenming , Feng, Hao; Zhang, Dongmei; Wang, Shoujue (2002) : An adaptive controller for a class of nonlinear system using direction basis function. Chinese Journal of Electronics, v 11, n 3, July, p 303-306

[2]L. R. Rabiner (1989): A tutorial on hidden Markov models and selected applications in speech recognition. Proceedings of the IEEE, vol. 77, 257-286.

[3] Shoujue Wang et al (1998).: The Sequential Learning Ahead Masking (SLAM) model of neural networks for pattern classification, Proceedings of JCIS98, vol. IV, October 1998, RTP, NC, USA, 199–202

[4] Shoujue Wang (2003): A new development on ANN in China - Biomimetic pattern recognition and multi weight vector neurons. Rough sets, fuzzy sets, data mining, and granular computing lecture notes in artificial intelligence, LECTURE NOTES IN ARTIFICIAL INTELLIGENCE, vol. 2639, 35-43

[5]Jing Zheng , Horacio Franco, Andreas Stolcke (2003): Modeling word-level rate-of-speech variation in large vocabulary conversational speech recognition. Speech Communication, Vol.41, 273-285

[6] Shoujue Wang(1999): Direction-Basis-Function neural networks, Proceedings of IJCNN'99, 10–16 July, Washington, DC, USA, 1251–2171

[7] Shoujue Wang, Priority ordered neural networks with better similarity to human knowledge representation. Chinese Journal Of Electronics,Vol.8, (1999)1–4.

[8] Shoujue Wang, Wang Bainan(2002): Analysis and Theory of High-Dimension Space Geometry for Artificial Neural Networks, Chinese Journal of Electronics, Vol.30 1-4.

[9] Wenming Cao, Feng Hao and Shoujue Wang (2004): The application of DBF neural networks for object recognition, Information Sciences Vol.160, Issues 1-4 , 22 March (2004)153-160

[10]Martin J. Russell Jeff A. Bilmes (2003): Introduction to the special issue on new computational paradigms for acoustic modeling in speech recognition, Computer Speech and Language, Vol.17, 107 - 112

[11] Wang Shoujue, LiZhao zhou, Chen Xiangdong, WangBainan (2001): Discussion on the Basic Mathematical Models of Neurons in General Purpose Neuro Computer, Acta Electronica Sinica, Vol.29, 577-580

[12] Wang Shoujue Xu Jian Wang Xianbao Qin Hong (2003): Multi-Camera Human-Face Personal Identification System Based on the biomimetic pattern recognition, Acta Electronica Sinica, Vol.31, 1-3

A Connectionist Model of Finding Partial Groups in Music Recordings with Application to Music Transcription

Matija Marolt

Faculty of Computer and Information Science, University of Ljubljana, Slovenia
E-mail: matija.marolt@fri.uni-lj.si

Abstract

In this paper, we present a technique for tracking groups of partials in musical signals, based on networks of adaptive oscillators. We show how synchronization of adaptive oscillators can be utilized to detect periodic patterns in outputs of a human auditory model and thus track stable frequency components (partials) in musical signals. We present the integration of the partial tracking model into a connectionist system for transcription of polyphonic piano music. We provide a short overview of our transcription system and present its performance on transcriptions of several real piano recordings.

1 Introduction

Music transcription could be defined as a process of converting an audio signal into a note-level (parametric) representation, where notes (pitches), their starting times and durations are extracted from the signal. Transcription is a challenging problem for current computer systems; separating notes from a mixture of other sounds, which may include other notes played by the same or different instruments or simply background noise, requires robust algorithms with performance that should degrade gracefully when noise increases.

Automatic transcription of polyphonic music would be useful in a variety of applications, ranging from content-based retrieval of music (i.e. query by example systems) and music analysis systems to accompaniment systems and musicological studies.

In recent years, several transcription systems have been developed [1-4]. All authors, except for Bello [2], base their systems on frequency domain analysis of the musical signal. Cues, such as local energy maxima, are extracted from the time-frequency representation of the signal and used in subsequent processing stages to find notes that are present in the signal. Various techniques, such as statistical frameworks, blackboard architectures, distance metrics or ICA are used in the process of grouping the found cues into notes, relying on information such as harmonicity and common onset/offset times. To reduce the complexity of the TF representation, to reduce noise and to incorporate some kind of temporal processing, partial tracking has been used in some systems to locate stable frequency components in the audio signal [4,5].

In this paper, we present a connectionist approach to music transcription. Transcription is a challenging task, so we limited the domain of our system to transcription of polyphonic piano music. The paper focuses on our approach to partial tracking with networks of adaptive oscillators, provides a short description of our entire transcription system, and presents some results obtained on transcriptions of real piano recordings.

2 Partial tracking with networks of adaptive oscillators

A melodic sound can be roughly described as a sum of components with relatively stable frequencies and time-varying amplitudes, called partials. By finding partials in a signal, one isolates the stable frequency components most likely belonging to tones, and discards noisy components. This is especially desirable in transcription systems, where the goal is to find all the tones present in the audio signal. Currently, most partial trackers used in transcription systems are based on a procedure similar to the tracking phase vocoder [6], where peaks are computed in each frame of the time-frequency representation. Detected peaks are then linked over time according to intuitive criteria such as proximity in frequency and amplitude to form partial tracks. Such approach is quite susceptible to errors in the peak peaking procedure, where missed or spurious peaks can lead to fragmented or spurious partial tracks.

We propose an alternative partial tracking approach that is not based on the standard peak-picking/peak connecting paradigm, but on connectionist principles. It is composed of two parts: an auditory model, and adaptive oscillators that extract partials from outputs of the auditory model.

2.1 Auditory Model

The auditory model emulates the functionality of human ear and transforms the audio signal into a probabilistic representation of firing activity in the auditory nerve. Amongst the several auditory models available, we chose to use a combination of the Patterson-Hodsworth gammatone filterbank [7] and Meddis' model of hair cell transduction [8], as their implementations are readily available. The gammatone filterbank emulates the movement of basilar membrane in the inner ear. Its outputs are processed by the hair cell model, which converts each output into a probabilistic representation of firing activity in the auditory nerve. Its operations are based on a biological model of the hair cell and it simulates several of the cell's characteristics, most notably half-wave rectification, saturation and adaptation. Saturation and adaptation are very important to our model, as they reduce the dynamic range of the signal, and in turn enable our partial tracking model to track partials with low amplitude. These characteristics can be observed in Fig. 1, displaying outputs of three gammatone filters and the hair cell model on the 1., 2., and 4. partial of piano tone F3 (pitch 174 Hz).

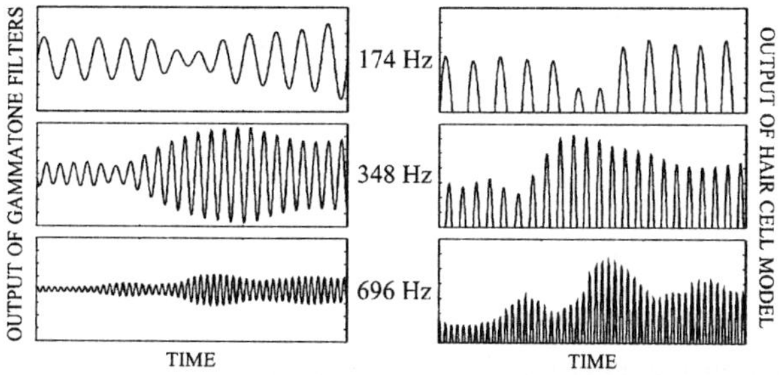

Fig. 1. Auditory analysis of three partials of piano tone F3.

2.2 Partial Tracking

Output of the auditory model consists of a set of quasi-periodic functions describing firing activities of inner hair cells in different parts of the basilar membrane (Fig. 1). Temporal models of pitch perception are based on the assumption that detection of periodicity in output channels of the auditory model forms the basis of human pitch perception. Periodicity is usually calculated with autocorrelation, resulting in a three-dimensional time-frequency representation of the signal called autocorrelogram, with time, channel center frequency and autocorrelation lag represented on orthogonal axes. In contrast, we propose to use a set of adaptive oscillators to estimate periodicity in output channels of the auditory model.

An oscillator is a system with periodic behavior. It oscillates in time according to its two internal parameters: phase and frequency. An adaptive oscillator adapts its phase and frequency in response to its input (driving) signal. When a periodic signal is presented to an adaptive oscillator, it adjusts its phase and frequency to match that of the driving signal and thus synchronizes with the signal. By observing the frequency and phase of a synchronized oscillator, an accurate estimate of the frequency and phase of its driving signal can be made. After reviewing several models, we decided to use a modified version of the Large-Kolen adaptive oscillator [9] in our partial tracking model.

The rationale behind the use of adaptive oscillators for partial tracking is simple. As periodicity in an output channel of the auditory model points to the presence of a frequency component (partial) in the input signal, analysis of periodicity in the channel indicates the exact frequency of the partial. In our model, periodicity is detected by a set of adaptive oscillators. If these synchronize with their stimuli (outputs of the auditory model), this indicates that the stimuli are periodic, and consequently that partials are present in the input signal. Frequencies of partials can be estimated by observing the frequencies of synchronized oscillators. Such a model has several advantages, when compared to standard approaches: it produces a continuous estimate of partials in a signal; because oscillators constantly adapt to their stimuli, partials with slowly changing frequencies (vibrato...) can be tracked; and as the auditory model reduces the dynamic range of the input signal and thus boosts partials with low amplitudes, these can be tracked as well.

2.3 Oscillator networks

As most tones are harmonic, we extended the model of tracking individual partials to a model of tracking groups of harmonically related partials by joining adaptive oscillators into fully-connected networks. Each network contains oscillators that track a series of harmonically related partials, so the frequencies of oscillators in a network are set to integer multiples of the frequency of the first oscillator (Fig. 2). As each oscillator in the network tracks a single partial close to its initial frequency, a network of oscillators tracks a group of harmonically related partials, which may belong to one tone with pitch equal to the frequency of the first oscillator.

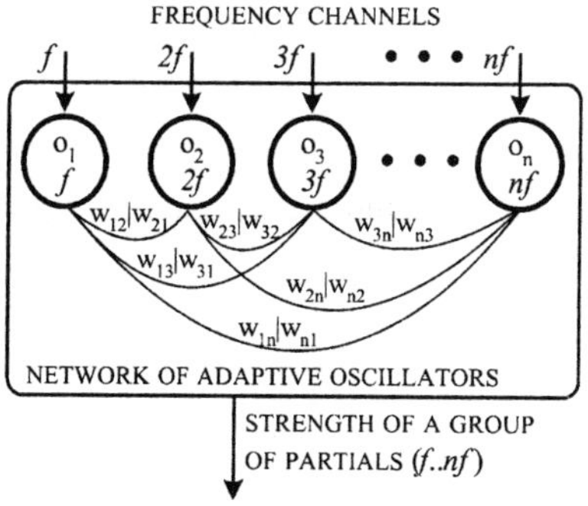

Fig. 2. A network of adaptive oscillators.

Within a network, each oscillator is connected to all other oscillators with excitatory connections. These connections are used to adjust frequencies and outputs of non-synchronized oscillators in the network with the goal of speeding up their synchronization. Only a synchronized oscillator can affect frequencies and outputs of other oscillators in the network. Output of a network is calculated as a weighted sum of outputs of individual oscillators in the network and represents the strength of a group of partials tracked by oscillators in the network.

Connecting oscillators into networks has several advantages if the goal is to obtain a compact representation of a signal, suitable for transcription. Output of a network represents the strength of a group of harmonically related partials tracked by its oscillators. Such output provides a better indication of presence of a harmonic tone in the input signal than do outputs of individual oscillators. Noise usually doesn't appear in the form of harmonically related frequency components, so networks are more resistant to noise and provide a clearer time-frequency representation. Network connections are used by synchronized oscillators to speed up synchronization of non-synchronized oscillators, leading to a faster network response and faster discovery of a group of partials. Missing partials (even missing fundamental) are tolerated, if enough partials are found by other oscillators in the network.

An example is given in Fig. 3, which displays slices taken from three time-frequency representations of piano chord C3E3B4, calculated 100 ms after the onset: representation with uncoupled oscillators, repre-sentation with networks of adaptive oscillators and short-time Fourier transform. The representation with uncoupled oscillators was calculated with 88 oscillators tuned to pitches of piano tones A0-C8. Oscillator outputs (independent of partial amplitudes) are presented in Fig. 3A. Fig. 3B shows outputs of 88 oscillator networks, tuned to the same pitches. Product of networks' outputs and amplitudes of partials is shown in Fig. 3C. Fig. 3D displays the first 440 frequency bins of the Fourier transform calculated with a 100 ms Hamming window.

Individual oscillators have no difficulty in finding the first few partials of all tones (A). Some of the higher partials are not found, as they are masked by louder partials of other tones (we use only one oscillator per semitone). Oscillator networks (B) produce a clearer representation of the signal; the first two or three partial groups of each tone stand out. Networks coinciding with tones E3 and B4 produce the highest outputs, because almost all partials in the networks are found. When amplitudes are combined with network outputs (Fig. 3C), only four partial groups stand out, corresponding to first partials of all three tones (C3, E3, B4) and the second partial of tone E3. If we compare Fig. 3C with the Fourier transform in 3D, advantages of partial group tracking for transcription are obvious.

Overall, oscillator networks produce a compact and clear representation of partial groups in a musical signal. The main problem of this representation lies in occasional slow synchronization of oscillators in networks, which can lead to delayed discovery of partial groups. This is especially true at lower frequencies, where delays of 40-50 ms are quite common, because synchronization only occurs once per oscillator cycle; an oscillator at 100 Hz synchronizes with the signal every 10 ms, so several 10s of milliseconds are needed for synchronization. Closely spaced partials may also slow down synchronization, although it is quite rare for a group of partials not to be found.

3 Transcription of piano music

The described partial tracking model has been incorporated into our system for transcription of piano music, called SONIC [10]. Next to partial tracking, the system also includes a note recognition module, an onset detector based on a network of integrate-and-fire neurons, a module for resolving repeated notes, based on multilayer perceptrons and simple algorithms for estimation of tuning, note length and loudness.

A note recognition module is the central part of every transcription system. Its input consists of a set of cues extracted from the time-frequency representation of the input signal and its task is to associate the found cues with notes. Statistical methods are frequently used for this task; in our transcription system the task is performed by a set of neural networks. Inputs of each network are taken from outputs of the partial tracking module presented in

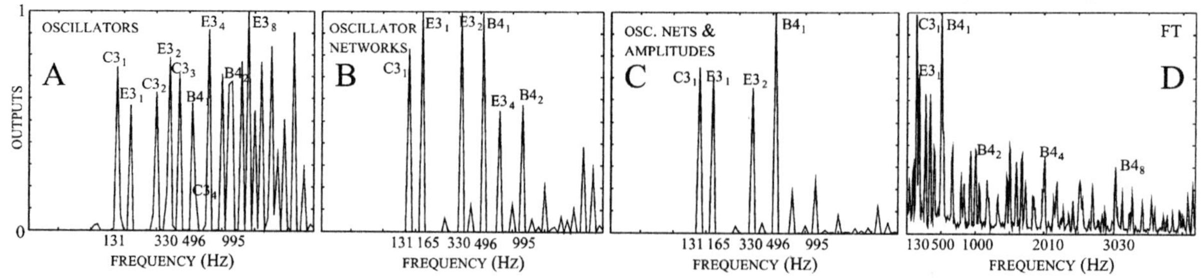

Fig. 3. Representations of piano chord C3E3B4

previous sections. Each network is trained to recognize one piano note in its input; i.e. one network is trained to recognize note A4, another network recognizes note G4... After extensive testing of several neural network models, we decided to use time-delay neural networks (TDNNs) in our system, as they provided the best performance. Networks were trained and tested on a database of approx. 150 synthesized piano pieces of various styles, combined with randomly generated chords. Tests showed that the incorporation of the partial tracking module significantly improved overall accuracy of transcription, halving the number of spurious notes [10].

4 Performance analysis

To analyze the performance of our transcription system, we tested it on a number of synthetisized and real recordings. The real recordings were transcribed by hand with the help of the original score. Originals and transcriptions of several pieces can be found on http://lgm.fri.uni-lj.si/SONIC. Table 1 lists performance statistics of three real piano performances: percentages of correctly found and spurious notes in transcriptions, as well as percentages of octave errors for missed and spurious notes are given.

	corr. notes	spur. notes	missed octave err.	spurious octave err.
1	88.5	15.5	35.1	80.5
2	68.3	13.6	30.3	79
3	85.9	15.2	70.3	87.4

Table 1. Performance statistics on real recordings

The transcribed recordings are: (1) J.S. Bach, English suite no. 5, 1st mvt, performer Murray Perahia, Sony Classical SK 60277; (2) F. Chopin, Nocturne no. 2, Op. 9/2, perf. Artur Rubinstein, RCA 60822; (3) S. Joplin, The Entertainer, performer unknown, MCA 11836.

Overall, most transcription errors are either due to octave errors or errors related to missed or spurious repeated notes. In Bach's English Suite, next to octave and repeated note errors, most of the missed notes are either quiet low pitched notes or notes in arpeggios and thrills. Chopin's Nocturne is a good example of very expressive playing, where a distinctive melody is accompanied by quiet, sometimes barely audible left hand chords. The system therefore misses over 30% of all notes, but even so the resynthesized transcription sounds quite similar to the original (listen to the example on the aforementioned URL address). When we compared transcriptions of the real and a synthesized version of The Entertainer, both turned out to be very similar. Transcription of the real recording contains more spurious notes, mostly occurring because of pedaling, which was not used in the synthesized version. The number of correctly found notes is almost the same in both pieces. Octave errors are the main cause of both missed and spurious notes. For a more detailed analysis, see [10].

5 Conclusion

In this paper, we presented a connectionist approach to partial tracking in musical signals. Our approach is based on a human auditory model and on adaptive oscillators for discovery and tracking of partial groups. By using a connectionist approach, we avoided some of the pitfalls of classical partial tracking approaches. We presented a brief overview of our transcription system and presented performance statistics on transcriptions of several real piano recordings. Overall, results are very promising and we believe that connectionist approaches to transcription should be further studied.

References

[1] Abdallah S., Plumbley M. (2004). Polyphonic transcription by non-negative sparse coding of power spectra. Proceedings of ISMIR'2004, Barcelona, Spain.

[2] Bello, J. P., Daudet, L. & Sandler, M. B. (2002). Time-Domain Polyphonic Transcription using Self-Generating Databases. In Proceedings of the 112th AES Convention. Munich, Germany.

[3] Klapuri A., Virtanen, T., Eronen, A. & Seppänen, J. (2001). Automatic transcription of musical recordings. Proceedings of Consistent & Reliable Acoustic Cues Workshop, CRAC-01, Aalborg, Denmark.

[4] Sterian. A.D. (1999). Model-based Segmentation of Time-Frequency Images for Musical Transcription. Ph.D. Thesis, Univesity of Michigan.

[5] Marchand, S. (2001). An efficient pitch-tracking algorithm using a combination of Fourier transforms, Proceedings of DAFX-01, Limerick, Ireland.

[6] Roads. C. (1996). The Computer Music Tutorial. Cambridge, MA: MIT Press.

[7] Patterson, R. D. & Hodsworth J. (1990). A functional model of neural activity patterns and auditory images. Advances in speech, hearing and auditory images 3, W.A. Ainsworth (ed.), London: JAI Press.

[8] Meddis. R. (1986). Simulations of mechanical to neural transduction in the auditory receptor. J.Acoust.Soc.Amer., vol. 79, no. 3, 702-711.

[9] Large E.W., Kolen, J.F. (1994). Resonance and the perception of musical meter. Connection Science, vol. 2, no. 6, 177-208.

[10] Marolt M. (2004). A Connectionist Approach to Transcription of Polyphonic Piano Music. IEEE Transactions on Multimedia, June 2004, Vol.6, Issue 3.

Adaptive ICA Algorithm Based on Asymmetric Generalized Gaussian Density Model

Fasong Wang[1], Hongwei Li[1]
[1]Department of Mathematics and Physics, China University of Geosciences, Wuhan, China
E-mail: wfslirui@sohu.com, hwli@cug.edu.cn

Abstract

A novel Independent Component Analysis(ICA) algorithm is achieved, which enable to separate mixtures of symmetric and asymmetric sources with self adaptive nonlinear score functions. It is derived by using the parameterized asymmetric generalized Gaussian density(AGGD) model. Compared with conventional ICA algorithm, the proposed AGGD-ICA method can separate a wide range of signals including skewed sources. Simulations confirm the effectiveness and performance of the approach.

1 Introduction

The independent component analysis(ICA) technique is to recover source signals which are considered statistically independent given only the output of a number of sensor mixtures[1]. In this paper we consider the linear, instantaneous, noiseless mixtures model with the form $\mathbf{x}(t) = \mathbf{A}\mathbf{s}(t)$, where $\mathbf{s}(t) \in R^M$ denotes the vector of zero-mean, stationary, statistically independent source signals, $\mathbf{x}(t) \in R^N$ denotes the vector of observable mixtures and $\mathbf{A} \in R^{N\times M}$ is a full-rank constant mixing matrix. We assume that there are as many sources as mixtures i.e. $M = N$. These observed signals are processed by a de-mixing matrix $\mathbf{W}$ that

$$\mathbf{y}(t) = \mathbf{W}\mathbf{x}(t) = \mathbf{W}\mathbf{A}\mathbf{s}(t), \tag{1}$$

where $\mathbf{y}(t) \in R^N$ is an estimate of $\mathbf{s}(t)$ to within the permutation and scaling ambiguities. Then we get the performance matrix $\mathbf{P}$ and $\mathbf{P} = \mathbf{WA}$.

A useful algorithm that has been developed is the important contribution of Amari et.al.according to ref.2(independently by Cardoso, which was termed the relative gradient algorithm[3])

$$\mathbf{W}^{k+1} = \mathbf{W}^k + \eta[\mathbf{I} - \varphi(\mathbf{y})\mathbf{y}^T]\mathbf{W}^k, \tag{2}$$

where $\eta > 0$ is the learning rate which can be chosen by the perfect idea in ref.4 and $\varphi(\cdot)$ is the vector of score functions whose optimal components are

$$\varphi_i(y_i) = -\frac{d}{dy_i}\log p_i(y_i) = -\frac{p_i'(y_i)}{p_i(y_i)}. \tag{3}$$

The optimal score function depends on the probability distribution of the sources heavily. In order to achieve a good separation result, the ICA algorithm must adapt to different source distributions. Various methods have been proposed to separate mixtures of sub- and super-Gaussian distribution signals[5], but they did not consider the condition that the sources have asymmetric distributions. Here, we present an adaptive algorithm which is derived by using an AGGD model for symmetric and asymmetric source signals.

This paper is organized as follows: Section 1 introduces briefly the ICA model and natural gradient algorithm; then in section 2, we use two parametric density models: generalized Gaussian density(GGD) model and AGGD model to cover an extensive range of source distributions, at the end of this part the new algorithm is proposed; Simulations illustrating the good performance of the proposed method are given in section 3; Finally, section 4 concludes the paper.

2 Generalized Gaussian Density(GGD), AGGD Model and AGGD-ICA Algorithm

Firstly, let us introduce two definitions which can measure the skewness and sharpness of a random variable's distribution so called skewness and kurtosis. They are defined as

$$\gamma_1 = skew(y) = \frac{m_3(y)}{\sqrt{m_2^3(y)}};$$

$$\gamma_2 = kurt(y) = \frac{m_4(y)}{m_2^2(y)} - 3,$$

where $m_2(y)$, $m_3(y)$ and $m_4(y)$ denote the 2nd-, 3rd-, 4th-order moments. The skewness with zero and nonzero values corresponds to symmetric and asymmetric distribution. Similarly the kurtosis with positive, zero and negative values corresponds to super-Gaussian, Gaussian and sub-Gaussian distribution respectively.

The relationship between γ_1, γ_2 and freedom degree parameters in GGD and AGGD model will be determined in the next two subsections.

2.1 GGD Model

The PDF of GGD[5-7] with shaping factor c and scaling factor γ is represented by

$$p_{gg}(y) = \frac{\gamma c}{2\Gamma(1/c)} \exp[-|\gamma(y - \mu_y)|^c],$$

where $\Gamma(\cdot)$ is the Gamma function defined by $\Gamma(x) = \int_0^\infty t^{x-1}e^{-t}dt$. By changing the values $c(c > 0)$, a family of distributions with different sharpness will be given. When $c = 2$, the distribution reduced to Gaussian. If $c > 2$ the distribution becomes sub-Gaussian, that is $\gamma_2 < 0$, otherwise $c < 2$, it becomes super-Gaussian PDF, that is $\gamma_2 > 0$. Figure 1 shows a family of GGD.

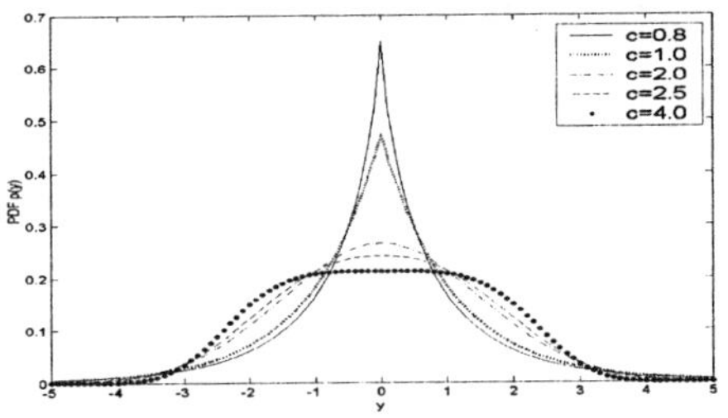

Fig.1. Some Densities of GGD PDF($\mu_y = 0, \sigma_y^2 = 1$).

Because the parameter c can not be estimated directly from data samples, but the kurtosis γ_2 can be, so we should find the relationship between c and kurtosis γ_2. From the definition, the 2nd- and 4th-order moment can be obtained easily

$$m_2 = \int_{-\infty}^{\infty} y^2 p_{gg}(y)dy = \frac{\Gamma(\frac{3}{c})}{\gamma^2\Gamma(\frac{1}{c})},$$

$$m_4 = \int_{-\infty}^{\infty} y^4 p_{gg}(y)dy = \frac{\Gamma(\frac{5}{c})}{\gamma^4\Gamma(\frac{1}{c})}.$$

The scaling factor γ can be derived by using one of above equations, e.g. from the 2nd-order moment

$$\gamma = \sqrt{\frac{\Gamma(\frac{3}{c})}{m_2\Gamma(\frac{1}{c})}}.$$

Then using the definition of γ_2, we can get

$$\gamma_2 = \frac{m_4}{m_2^2} - 3 = \frac{\Gamma(\frac{5}{c})\Gamma(\frac{1}{c})}{\Gamma^2(\frac{3}{c})} - 3. \tag{4}$$

So given a value of γ_2, the parameter c can be determined from (4). But, because of the $\Gamma(\cdot)$ function's definition, it is impossible to express c in terms of γ_2 by an analytically exact expression. Hence an approximation is required. A good approximation was found in ref.8 by applying the least squares method(LSM), the LSM estimate result follows

$$c \doteq \sqrt{\frac{5}{\gamma_2 - 1.865}} - 0.12. \tag{5}$$

From above analysis, the score function based on GGD PDF in the algorithm (2) can be derived from (3)

$$\varphi_{i,gg}(y_i) = c\gamma^c|y_i - \mu_{y_i}|^{c-1}sgn(y_i - \mu_{y_i}). \tag{6}$$

For different value of parameter c, the score functions are shown in figure 2.

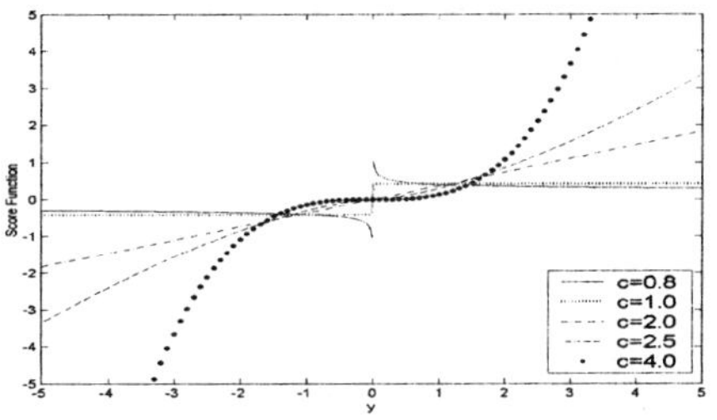

Fig.2. Some Score Function Based on GGD PDF Family($\mu_y = 0, \sigma_y^2 = 1$).

Then for any symmetric source signals, we can substitute (6) to (2), the ICA algorithm is obtained.

2.2 AGGD Model

Choi et.al.[7] present the flexible ICA algorithm where the score function was derived from the GGD model. They had shown that the algorithm can separate the mixtures of sub- and super-Gaussian signals with self-adaptive score function which is controlled by Gaussian exponent c. But they did not consider the condition that the source signals have asymmetric distributions. The main limitation affecting the GGD model is its symmetry. So we will introduce an asymmetric PDF model-AGGD model[6]. It depends on two second-order parameters:σ_l^2, σ_r^2 called left and right variance respectively and defined as: $\hat{\sigma}_l^2 = \frac{1}{N_l-1}\sum_{i=1,y_i<\hat{m}_y}^{N_l}(y_i - \hat{m}_y)^2$, $\hat{\sigma}_r^2 = \frac{1}{N_r-1}\sum_{i=1,y_i>\hat{m}_y}^{N_r}(y_i - \hat{m}_y)^2$, where $\hat{m}_y$ is the estimated mode(which is not coincide with the mean in the case of asymmetry distribution) and $N_l(N_r)$ is the number of y_i that is less than $\hat{m}_y$(or greater than $\hat{m}_y$).

Hence, replacing the variance of GGD PDF with σ_l^2 and σ_r^2, we can get the AGGD distribution model

$$p_{agg}(y) = \begin{cases} \frac{c\gamma}{\Gamma(\frac{1}{c})}\exp\{-\gamma_l^c[-(y - m_y)]^c\}, & y < m_y \\ \frac{c\gamma}{\Gamma(\frac{1}{c})}\exp\{-\gamma_r^c[-(y - m_y)]^c\}, & y \geq m_y \end{cases} \tag{7}$$

where $\gamma = \frac{1}{\sigma_l+\sigma_r}\sqrt{\frac{\Gamma(\frac{3}{c})}{\Gamma(\frac{1}{c})}}$, $\gamma_l = \frac{1}{\sigma_l}\sqrt{\frac{\Gamma(\frac{3}{c})}{\Gamma(\frac{1}{c})}}$, $\gamma_r = \frac{1}{\sigma_r}\sqrt{\frac{\Gamma(\frac{3}{c})}{\Gamma(\frac{1}{c})}}$ and m_y is the mode. From the definition

(7) we can prove that AGGD model satisfies the required properties of density function that is $p_{agg}(y) > 0$, $\int_{-\infty}^{\infty} p_{agg}(y)dy = 1$, at the same time, $p_{agg}(y)$ is continuous when $y = m_y$.

From this definition, it is easy to notice that when $\sigma_l^2 = \sigma_r^2$, the PDF coincides with the GGD model, that is to say it has symmetric distribution; Furthermore, if $\sigma_l^2 = \sigma_r^2$ and $\gamma_2 = 0$, the AGGD coincides with the Gaussian model, so if $\gamma_2 > 0$, it coincides with super-Gaussian distribution, otherwise, $\gamma_2 < 0$, it degrades to sub-Gaussian PDF; then if $\sigma_l^2 \neq \sigma_r^2$ and $\gamma_2 = 0$, we can get general asymmetric Gaussian class. Figure 3 shows a family of AGGD as c varies.

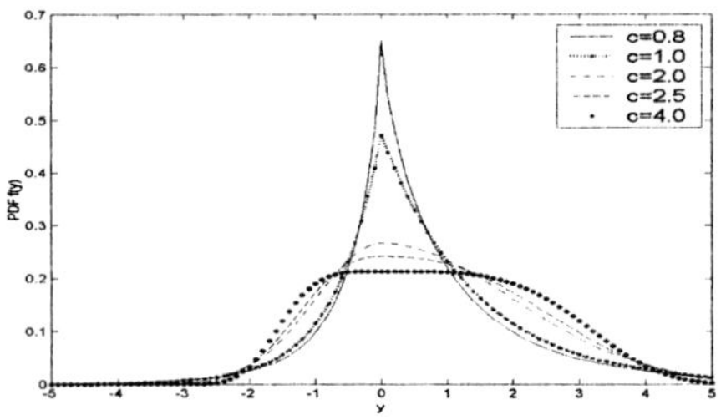

Fig.3. Some Densities of AGGD PDF Family for Different $c(\mu_y = 0, \sigma_l^2 = 1, \sigma_r^2 = 4)$.

From the definition, the 2nd- and 4th-order moment of AGGD can be obtained

$$m_2 = (\sigma_r - \sigma_l)^2 \left(1 - \frac{(\Gamma(\frac{2}{c}))^2}{\Gamma(\frac{1}{c})\Gamma(\frac{3}{c})}\right) + \sigma_l\sigma_r; \quad (8)$$

$$\begin{aligned} m_4 = & \frac{\Gamma(\frac{5}{c})\Gamma(\frac{1}{c})}{(\Gamma(\frac{3}{c}))^2}(\sigma_r^4 - \sigma_r^3\sigma_l + \sigma_r^2\sigma_l^2 - \sigma_r\sigma_l^3 + \sigma_l^4) \\ & -4\frac{\Gamma(\frac{2}{c})\Gamma(\frac{4}{c})}{(\Gamma(\frac{3}{c}))^2}(\sigma_r^2 + \sigma_l^2)(\sigma_r - \sigma_l)^2 \\ & +6\frac{(\Gamma(\frac{2}{c}))^2}{\Gamma(\frac{3}{c})\Gamma(\frac{1}{c})}(\sigma_r^2 - \sigma_r\sigma_l + \sigma_l^2)(\sigma_r - \sigma_l)^2 \\ & -3\frac{(\Gamma(\frac{2}{c}))^4}{(\Gamma(\frac{3}{c})\Gamma(\frac{1}{c}))^2}(\sigma_r - \sigma_l)^4. \end{aligned} \quad (9)$$

Then the relationship between γ_2 and c can be obtained from (4), (8-9). The relationship between γ_2 and c is given in figure 4 for different combinations of σ_l^2 and σ_r^2.

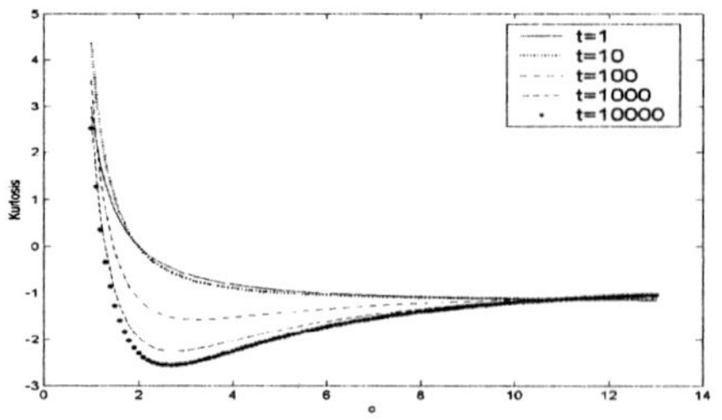

Fig.4. Relationship Between γ_2 and c for Different Left and Right Variance($\mu_y = 0$, $\sigma_l^2 = 1$, $t = \sigma_r^2$).

From (3), we can get score function of AGGD easily

$$\varphi_{i,agg}(y_i) = \begin{cases} -c\gamma_l^c\left(-(y_i - m_{y_i})\right)^{c-1}, & y_i < m_{y_i} \\ c\gamma_r^c\left(y_i - m_{y_i}\right)^{c-1}. & y_i \geq m_{y_i} \end{cases} \quad (10)$$

Substitute (10) to (2), we can get the separation algorithm based on AGGD model—AGGD-ICA. For different value of parameter c, the score functions are shown in figure 5.

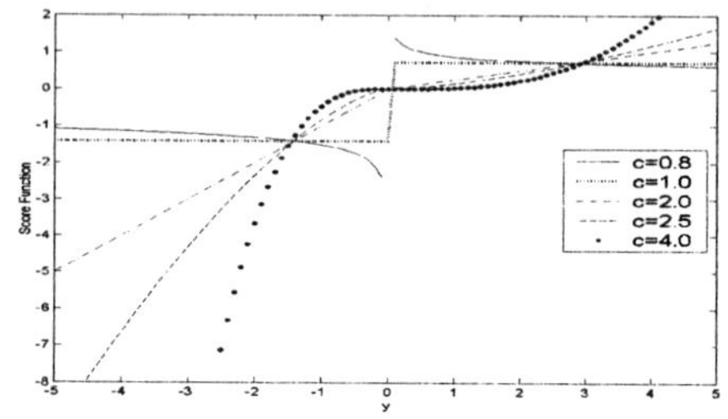

Fig.5. Some Score Function Based on AGGD PDF Family($\mu_y = 0$, $\sigma_l^2 = 1$, $\sigma_r^2 = 2$).

2.3 AGGD-ICA Algorithm

As discussed above, we can use a parametric PDF model AGGD to estimate symmetric and asymmetric source signals. Under the framework of natural gradient, we derive a novel AGGD-ICA separated algorithm. The proposed AGGD-ICA algorithm have a wide application range compared with classic ICA algorithm, It paves the way to wider applications of ICA methods to real world signal processing. The procedure of proposed AGGD-ICA algorithm is simply summarized as:

Step1 Calculating the output signals $\mathbf{y}(t)$ by observations $\mathbf{x}(t)$ and the given initial separating matrix $\mathbf{W}$;

Step2 Estimating the skewness and kurtosis of the extracted signals $\mathbf{y}(t)$;

Step3 If skewness equals to zero or very close to zero(< 0.1), then $\sigma_l^2 \doteq \sigma_r^2$. We seek c from (5) using (4) according to the values of $\hat{\gamma}_2$. Else if γ_1 far away zero(> 0.1), we will seek c from (5) using (4),(8) and (9) according to the values of $\hat{\gamma}_2$;

Step4 Calculate the score function (6) or (10);

Step5 Updating $\mathbf{W}$ using (2), if $\|\mathbf{W}^{k+1} - \mathbf{W}^k\| > \varepsilon$, $k = k + 1$ and go to step1, otherwise, end the algorithm.

3 Simulations

In order to confirm the validity and performance of our proposed AGGD-ICA algorithm, simulation experiment is given below. Five source signals which have different waveforms are used, they are one Gaussian signal, one super-Gaussian signal, one sub-Gaussian signal and two asymmetric source signals. The waveform of them are shown in figure 6(a). The source signals are mixed through a mixing matrix which is randomly generated.

The initial condition of $\mathbf{W}$ is an identity matrix, the learning rate η is chosen by the variable step-size method, which Chamber et.al. in[4] has proved this procedure can make the algorithm convergence fast and robust. We can choose the learning rate η as follows: at the first stage of the separation process(first 50 iterations),$\eta = 0.01$, then it reduces to $\eta = 0.001$ at the last stage of the optimization. In the experiment we find that the AGGD-ICA algorithm can convergent before 100 iterations. The performance matrix $\mathbf{P} = \mathbf{WA}$ can be obtained

$$\mathbf{P} = \begin{bmatrix} \underline{-0.960} & -0.030 & -0.006 & 0.043 & 0.014 \\ 0.005 & 0.030 & -0.025 & 0.094 & \underline{1.046} \\ 0.002 & \underline{1.000} & 0.000 & 0.001 & 0.001 \\ -0.000 & -0.001 & \underline{-0.990} & 0.004 & -0.005 \\ 0.177 & 0.059 & -0.032 & \underline{2.373} & -0.023 \end{bmatrix}.$$

The recovered signals $\mathbf{y}(t)$ are plotted in figure 6(b). It can be seen from the result that the source signals are well separated by using the AGGD-ICA algorithm.

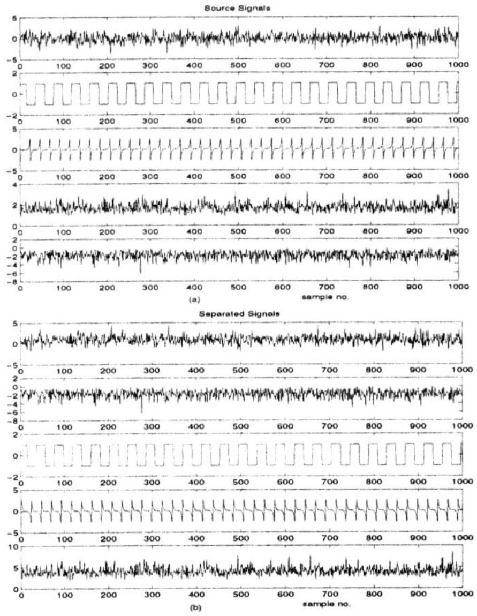

Fig.6. Experiment Results Showing the Separation of Symmetric and Asymmetric Sources.

The separation performance can be evaluated in terms of median signal-to-interference ratio(SIR), defined as

$$SIR(y_j) = 10\log_{10}\left(\frac{\sum_{i=1}^{n} s_{ij}^2}{\sum_{i=1}^{n}(y_{ij} - s_{ij})^2}\right),$$

where s_j is the sources and y_j is the corresponding reconstructed signals, $j = 1, \cdots, N$. So, if we want to get good separation, we must make SIR as big as possible.

The SIR of the recovered signals are: SIR$(y_1) = 19.1867$; SIR$(y_2) = 28.4527$; SIR$(y_3) = 22.2379$; SIR$(y_4) = 26.3291$; SIR$(y_5) = 23.4001$. The proposed AGGD-ICA algorithm can separate the mixed signals efficiently.

4 Conclusion

Under the natural gradient framework, using a single parametric non-linearity model–AGGD model as the density estimation of source signals, a novel AGGD PDF based adaptive ICA algorithm has been gained. The AGGD-ICA algorithm can adaptively choose the parameter by estimating the high order sample statistics γ_1, γ_2. As a result it can be applied to more extensive range of source distributions including skewed sources. From above simulation we can see that the mixed signals achieve good separation and the AGGD-ICA algorithm has perfect convergent properties.

Acknowledgements: This work was supported by National Natural Science Foundation of China(Grant No.60472062) and Natural Science Foundation of Hubei Province, China(Grant No.2004ABA038).

5 References

[1] Comon, P.(1994)Independent component analysis, A new concept?. Signal Processing 36: 287-314

[2] Amari, S.(1998)Natural gradient works efficiently in learning. Neural Computation 10: 251-276

[3] Cardoso, J.F., Laheld, B.(1996)Equivariant adaptive source separation. IEEE Trans. on Signal Processing 44: 3017-3030

[4] Chamber, J.A., Jafari, M.G., McLaughlin, S.(2004)Variable step-size EASI algorithm for sequential blind source separation. Electronics Letters 40: 393-394

[5] Choi, S., Cichocki, A., Amari, S.(2000)Flexible Independent Component Analysis. J. VLSI Signal Process 20: 25C38

[6] Tesei, A., Regazzoni, C.S.(1998)HOS-based generalized noise pdf models for signal detection optimization. Signal Processing 65: 267-281

[7] Cao, J., Murata, N., Amari, S., Cichocki, A., Takeda, T. (2003)A Robust Approach to Independent Component Analysis of Signals With High-Level Noise Measurements. IEEE Tans. on Neural Network 14: 631-645

[8] Tesei, A., Regazzoni, C.S.(1996)Use of four-order statistics for non-Gaussian noise modeling: the generalized Gaussian Pdf in terms of kurtosis. In: Ramponi, G., Sicuranza, G.L., Carrato, S., Marsi, S.(eds.) Proc. EUSIPCO'96. EdizioniLINT, pp. 671-674

Toward an On-Line Handwriting Recognition System Based on Visual Coding and Genetic Algorithm

M. Kherallah, F. Bouri, A.M. Alimi

REGIM: Research Group on Intelligent Machines,
Department of Electrical Engineering. University of Sfax, ENIS, Tunisia.
E-mail: {monji.kherallah, fatma.bouri, adel.alimi}@ ieee.org

Abstract

One of the most promising methods of interacting with small portable computing devices, such as personal digital assistants, is the use of handwriting. In order to make this communication method more natural, we proposed to visually observe the writing process on ordinary paper and to automatically recover the pen trajectory from numerical tablet sequences. On the basis of this work we developed handwriting recognition system based on visual coding and genetic algorithm. The system was applied on Arabic script. In this paper we will present the different steps of the handwriting recognition system. We focus our contribution on genetic algorithm method.

1 Introduction

In the last few years a number of pen-based computers and personal digital assistants (PDA's) have been released. Most of the PDA's have difficulty to recognize handwritten text. Since handwriting is one of the most familiar communication media, pen-based interfaces combined with automatic handwriting recognition offer a very easy and natural input method. We notice the great increase of interest in the development of on-line handwriting recognition systems [1,2,3,4]. The user writes on the surface of a digitizing tablet with an electronic pen. The digitizing tablet captures the dynamic information about handwriting, such as number of stokes, stroke order, writing speed, etc, all in real time.

In recent years, some researches have made a lot of efforts in the field of the cognitive psychology and the linguistics to conceive the model of perception and reading by men [1,5,6]. The encoding of Arabic handwritten word consists of representing a word from its original sequence of 2D coordinating it to sequence of visual codes. In fact, our approach is based on the psychology of the visual perception to extract primitive description of the word, which is useful in the stage of on-line handwritten recognition. For this reason, some algorithms are used to extract automatically the visual codes from the cursive script. The extraction rate of this experiment is 72 % [1]. We correct the weakness of this heuristic strategy chosen for encoding the Arabic handwritten word by developing a basic concept of the genetic approach in order to select the best combination of characters that represent the correct Arabic word.

In this paper, we developed a system that recognizes on-line Arabic cursive handwriting. Previous research in the field of calligraphic characteristics and visual coding of the Arabic handwriting is reviewed in section 2. Section 3 is devoted to present our proposed recognition system which consists of the genetic algorithm concept. Experimental results and discussions were given in section 4.

2 Calligraphic characteristics and primitive extraction of the Arabic Handwriting

The Arabic is a consonantal and cursive writing. This property is met in two forms printed or handwritten. The Arabic alphabet is composed of 28 different letters. The difference between these letters is their positions in the word, the number and the position of the diacritic dots, the presence of the "hamza" and Vowels. In fact, the majority of letters change slightly in the shape of their character according to their position in the word (initial, medium or final) if they are joined to another letter either if they are isolated. Therefore we have 58 Arabic letters without diacritics as shown in the following figure.

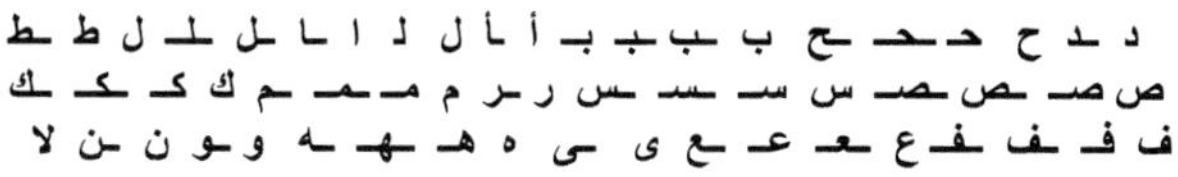

Fig.1. The 58 Arabic letters in their different positions

2.1 Diacritical symbols influence

Some Arabic letters have the same shape; however, they are distinguished from each other by the addition of dots in different positions relative to the main stroke. Some Arabic characters use special marks to modify the character accent. When diacritical symbols (dots, specials marks) are used, they appear above or below the characters and they are drawn as isolated entities as shown in fig. 2. Diacritical symbols are positioned at a certain distance from the character. In fact, this makes some difficulties in separating the border of a text line.

Indeed, diacritical symbols can generate some redundant separate lines [1]. We count 15 among the 28 letters of the alphabet, which contain dots. Some letters present a zigzag shape said 'Hamza'. It is considered as accent "vowel" in the Arabic alphabet. All diacritics are not considered in our work.

2.2 Primitive extraction

The word has generally two main zones: the median zone and the striking zone. Every zone is composed of its own visual indications as it is presented in the tables (1,2) [1].

Table.1 Visual codes information of the prominent zone

Notation	Visual code	Shape
Alif	Al	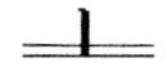
Shaft	Ha	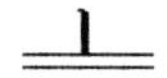
Left oblique shaft	Hog	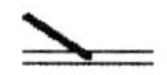
Left oblique shaft	Hod	
Leg	Ja	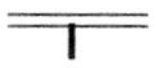
Pocket	Po	

Table.2 Visual codes information of the median zone

Notation	Visual code	Shape
Valley	Va	
Occlusion	Oc	
Curve open right	Cod	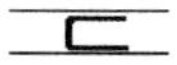
Left open curve	Cog	
Ain	Ain	
Sad	Sad	
Space inter tracing	#	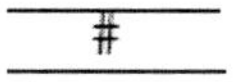

We attribute for every visual code a label as shown in table 3.

Table.3 Visual codes labels

Visual code	Va	Oc	Po	Al	Ja	Ha	CoD
Label	*1*	*2*	*3*	*4*	*5*	*6*	*7*

CoG	HoG	HoD	Ain	Sad	#
8	*9*	*10*	*11*	*12*	*13*

2.3 Features extraction of the Arabic script

We represent an arabic word by 2 strings: the first string represents the visual indices and the second one represents the different letters which are called "chromosomes". The number of strokes is determined by calculating the number of the velocity maxima of the handwriting trajectory [2].
The following table shows the visual indices and the visual codes of some Arabic letters. These features are used in the preliminary operations of genetic algorithm.

Table.4 Example of some letters with their visual features and stroke number

Arabic Letters	Indices	Stroke Number	Visual Codes
ا	1	1-2	Al
ـا	2	2-3	Al / Va
ٮ	3	3-4	Va
ٮ	4	3-4-5	Va / Va / Va

3 Genetic algorithms

Genetic algorithms are a class of optimization and search methods that use randomness to avoid local extrema. They are capable of adaptive and robust search over a wide range of space topologies [3]. Genetic algorithms are distinguished from other techniques by a principal characteristic: they search in intrinsically parallel fashion from a population of solutions and not from a single solution. A genetic algorithm is too an iterative algorithm that depends on the generation-by-generation development of possible solutions, with selection schemes permitting the elimination of bad solutions and the replication of good solutions that can be modified. The robustness of a genetic algorithm is also due to the fact that it manipulates according to parameters rather than the parameters themselves. There are three stages in a genetic search process: selection, crossover and mutation.

3.1 Gene structure

A handwritten word is represented by a continuation of visual codes of letters. In this case the order of these letters is considered. We attribute N-the number of basic letters extracted from a cursive word. Therefore, every gene of the population has N chromosomes and every chromosome has one of the 58 possible values (1 to 57 for the basic Arabian characters and the value 0 for characters with more than one visual indication) numbered from the right to the left.

For example, if the letter ▯ has 6 strokes, we code it as follow. Visual codes: va/ain/0/0/0/va

Label codes: 1/11/0/0/0/1

An other example for the complete word: علم

The visual codes are: Cod/Va/Va/Al/Va/Va/Occ/Ja

The chromosomes are : 27 / 40 / 42.

The following figure shows us the different codes:

Letters	0	0	0	0	م	0	0	ـلـ	0	0	ع
Visual indices	5	0	0	2	1	1	4	1	1	0	7
Chromosomes	0	0	0	0	42	0	0	40	0	0	27
Stroke order	11	10	9	8	7	6	5	4	3	2	1

Fig.2. Example of an Arabic word representation

3.2. Fitness function

The value returned from the fitness function for one chromosome represents the degree of match between the word represented by that chromosome and the real handwritten word. The calculation of the fitness value is based on sum of the resemblance measurement between two visual indices strings. The following table (Table 5) presents the different values of fitness function. The genetic algorithm will find the best combination of characters to reconstruct the analyzed word. In the selection-reproduction stage, the solutions with high

Table.5 Fitness values matrix

VI	1	2	3	4	5	6	7	8	9	10	11	12	13
1	0	1	1	1	1	1	1	1	1	1	1	1	1
2	1	0	1	1	1	1	0.5	0.5	1	1	0.5	0.5	1
3	1	1	0	1	0.5	1	1	1	1	1	1	1	1
4	1	1	1	0	1	0.5	1	1	0.5	0.5	1	1	1
5	1	1	0.5	1	0	1	1	1	1	1	1	1	1
6	1	1	1	0.5	1	0	1	1	0.5	0.5	1	1	1
7	1	0.5	1	1	1	1	0	1	1	1	0.5	0.5	1
8	1	0.5	1	1	1	1	1	0	1	1	0.5	0.5	1
9	1	1	1	0.5	1	0.5	1	1	0	0.5	1	1	1
10	1	1	1	0.5	1	0.5	1	1	0.5	0	1	1	1
11	1	0.5	1	1	1	1	0.5	0.5	1	1	0	0.5	1
12	1	0.5	1	1	1	1	0.5	0.5	1	1	0.5	0	1
13	1	1	1	1	1	1	1	1	1	1	1	1	0

"**VI**": Visual indices

fitness values are given more chance to reproduce, which gives them more effect on the population change. In this stage, the best chromosomes will dominate while the worse ones are discarded.

3.3 Crossover operation

The crossover operation consists on the selection of two parent solutions, with a higher probability for chromosomes possessing high fitness values. During crossover, portions of each parent pair are selected and resulting chromosomes are created by concatenation of the exchanged substrings. The main goal of the crossover stage is to generate new solutions, which will contain the useful parts of both parent solutions and which will have better fitness values. The number of crossover positions should not be before a zero, i.e. it should not break a character between two blocs. (See figure 3)

Word	0	0	ں	ـ	‘ ’	0	ـر	0	0	0	ـهـ	ـ
IV	0	3	1	1	13	5	1	1	0	2	1	1
CH	0	0	46	4	10	0	13	0	0	0	44	4
Order stroke	12	11	10	9	8	7	6	5	4	3	2	1

Word	0	ر	‘ ’	0	0	0	ـو	0	0	0	حـ
IV	0	5	13	5	0	2	1	1	0	0	8
CH	0	14	10	0	0	0	51	0	0	0	9
Order strokes	11	10	9	8	7	6	5	4	3	2	1

Fig.3. Example of crossover operation between 2 words

The crossover will be done on the fourth stroke. Two generations have been reproduced. (See figure 4)

Word1 « حزين »

Word2« تمور »

Word	0	0	ں	0	ـ	‘ ’	0	ـر	0	0	0	حـ
VI	0	3	1	0	1	13	5	1	1	0	0	8
CH	0	0	46	0	4	10	0	13	0	0	0	9
Order stroke	12	11	10	9	8	7	6	5	4	3	2	1

Word	0	ر	‘ ’	0	0	0	ـو	0	0	0	ـهـ	0	ـ
VI	0	5	13	5	0	2	1	1	0	2	1	0	1
CH	0	14	10	0	0	0	51	0	0	0	44	0	4
Order stroke	13	12	11	10	9	8	7	6	5	4	3	2	1

Fig.4. Crossover results

"CH": Chromosome

3.4 Mutation operation

Just as in nature, some individuals will have random mutations occur in their genes. The mutation rate specifies the odds that a given gene in an individual will be mutated. If a gene is selected for mutation then its value will be changed. In the case of bit representation, the gene will simply be flipped. (See figure5)

Word	0	0	م	‘ ’	0	0	0	و	0	0	ل	0	0	ع
IV	5	0	2	13	5	0	2	1	1	4	1	1	0	7
CH	0	0	41	10	0	0	0	51	0	0	40	0	0	27
Order stroke	14	13	12	11	10	9	8	7	6	5	4	3	2	1

Word	0	0	ى	‘ ’	0	0	0	و	0	0	ل	0	0	ع
IV	3	0	7	13	5	0	2	1	1	4	1	1	0	7
CH	0	0	53	10	0	0	0	51	0	0	40	0	0	27
Order stroke	14	13	12	11	10	9	8	7	6	5	4	3	2	1

Fig.5. Mutation result of the word « علوم »

4 Experimental results and discussions

In order to evaluate the performance of our system, 24 participants were invited to contribute to the handwriting data construction for our recognition experiment. The data set of words of each participant was stored in one data file. When producing the data file, each participant was asked to write some Arabic words. We collected 500 words written by different writers. 200 words we used as data prototypes for the selection of the initial population of the genetic algorithm, the others used for testing our system. In the first stage where the heuristic strategy chosen for encoding the Arabic handwritten word, The extraction rate obtained is about 72 %, but in the second stage which consist to correct the weakness of the previous method, we developed a genetic algorithm in order to select the best combination of visual codes extracted by a heuristic method from a word. The evolutionary approach here permits the recognition of cursive handwriting without limitation of a lexical dictionary. The number of generations (500) and the fitness value (0.5) were fixed as a convergence condition criterion. The recognition rate obtained is about 89%. These results obtained were encouraging. Compared to the on line recognition system of the cursive handwriting in [3] which have a manually segmentation system of the handwritten words, our system contains an automatically segmentation process of the handwritten words. This segmentation approach is based on a combination of the visual extractor and the evolutionary genetic algorithm system. Some errors were due to the style of the writer handwriting and were difficult to be avoided even by a human reader. For example there is confusion between the two Arabic words shown in the following figure.

Maksabon Maktabon

Fig.6. The confusion between two Arabic words

As a future work, we can remove this confusion and improve the recognition rate obtained by adding to our system a matching template algorithm and a large lexical Arabic dictionary.

Acknowledgement

The authors would like to acknowledge the financial support of this work by grants from the General Direction of Scientific Research and Technological Renovation (DGRST), Tunisia, under the ARUB program 01/UR/11/02.

References

[1] Jouini, B., Kherallah, M., Alimi, M. A., (2003) A new approach for on-line visual encoding and recognition of handwriting script by using neural network system. In: David W. Person. Ed. Artificial Neural Nets and Genetic Algorithms. Springer, Wien, pp. 161–167

[2] Kherallah, M., Haddad, L., Mitiche, A., Alimi, M.A., (2004) Towards the design of handwriting recognition system by neuro-fuzzy and Beta-Elliptical approaches. Proc. AIAI'2004. 18th IFIP World Computer Congress, pp. 187–196

[3] Alimi, M. A . (2002) Evolutionary Computation for the Recognition of On-Line Cursive Handwriting, IETE Journal of Research, Special Issue on "Evolutionary Computation in Engineering Sciences" edited by S.K. Pal et al.

[4] Plamondon, R., Srihari, S. N. (2000) On-line and off-line handwriting recognition: A comprehensive survey. IEEE Trans. Pattern Anal. Mach. Intell., 22(1):63–84

[5] Coté, M., Cherie, M., Leconet, E., Suen, C.Y., (1995). Building a perception Based Model for Reading Cursive Script, ICDAR, vol I, pp 898-901

[6] Foorster, K., (1994) Computational modeling and elementary process analysis in visual word recognition. Journal of Experimental Psychology, "Human Perception and Performance", Special Section, Modeling Visual Word Recognition, 20, no 6, pp. 1292-1310

Multi-objective genetic algorithm applied to the structure selection of RBFNN temperature estimators

C. A. Teixeira[1], W. C. A. Pereira[2], A. E. Ruano[1], M. Graça Ruano[1]
[1]Centre for Intelligent Systems, University of Algarve, Portugal
[2]Biomedical Eng. Program - COPPE/ Federal University of Rio de Janeiro, Brazil
E-mail: cateixeira@ualg.pt

Abstract

Temperature modelling of a homogeneous medium, when this medium is radiated by therapeutic ultrasound, is a fundamental step in order to analyse the performance of estimators for in-vivo modelling. In this paper punctual and invasive temperature estimation in a homogeneous medium is employed. Radial Basis Functions Neural Networks (RBFNNs) are used as estimators. The best fitted RBFNNs are selected using a Multi-objective Genetic Algorithm (MOGA). An absolute average error of $0.0084^oC$was attained with these estimators.

1 Introduction

The use of ultrasound for therapeutic proposes, specially for hyperthermia and surgery, is based on the thermal effect induced by the ultrasound. Accurate knowledge of tissue temperature evolution with respect to time and space is envisage for the development of therapeutic hyperthermia models. As for as known, neither precise nor effective spatial-temporal temperature systems are available, disabling an even more frequent use of ultrasonic instrumentation in therapy. To be also mentioned the possibility of broadening the ultrasound instrumentation to other specific and well delimited regions of applications if an accurate temperature model is achieved. Previous work in the area considers that these spatial-temporal temperature patterns are linearly related with the changes of sound velocity and medium expansion. The temperature range considered by [1] was between $\approx 20^oC$and $\approx 25^oC$. The researchers obtained a mean square error of 0.03 $(^oC)^2$, an average error of -0.02 ^oC, and a maximum error of 0.44^oC.

In this paper, punctual and invasive temperature estimation in a homogeneous medium is proposed, when the medium is radiated by therapeutic ultrasound. The temperature range employed on the experiments included the normal human temperature, therefore a more realistic modelling is proposed. The temperature in the point of interest is considered nonlinearly related with the spectral intensity information at that point. For this propose Radial Basis Functions Neural Networks (RBFNN) were used. The best fitted RBFNN structures were searched by a Multi-objective Genetic Algorithm (MOGA). Since the selection procedure consumes a high computational time, a distributed computing approach was used, as in [2].

Table 1. Temperature ranges.

Intensity	**Temperature** (^oC)		
$(Watt/cm^2)$	**Initial**	**Max.**	**Final**
1	29	34.5	28.5
1.5	30	37.2	28
2	31	38.8	31

2 Experimental setup

The real data used in this work are temperature and acoustic intensity signals, in a point 48 mm distant (axial distance) from an ultrasonic therapeutic transducer, in a glycerin (homogeneous medium) tank. Data was acquired during approximately 110 min. At each 10 seconds, a temperature value was acquired from the developing waveform using a thermocouple. Also, a window of $5\mu s$ of the acoustic intensity signal, corresponding to a window of 2000 points of the intensity waveform was recorded. Mechanical energy was supplied only in the first 60 min, by the ultrasonic therapeutic device (Ibramed Sonopulse (São Paulo)). During the remaining 50 minutes the acoustical energy was maintained at zero level, while temperature variations were observed and recorded. Three sets of signals were acquired at 3 MHz in continuous operating mode, at three different intensities: $1Watt/cm^2$, $1.5Watt/cm^2$ and $2Watt/cm^2$. The initial, maximum, and final temperature of the glycerin medium in the point of interest are summarised in Table 1.

3 Methods

3.1 Processing the experimental data

From the intensity signals collected from the experiment, some spectral features were extracted aiming the development of temperature models. After a *Fast Fourier Transform* computation the amplitude of the fundamental component (3 MHz), of the first harmonic (≈6MHz), and of the second harmonic (≈9MHz) were saved for future use in the RBFNNs training, test and validation.
Afterwards, the features extracted and the measured temperature signals were normalised to values between 0 and 1, in order to enable a correct training of the neural networks (NNs). The conventions used are the following ones: normalised amplitude of the fundamental component - I_{fc}, normalised amplitude of the first harmonic - I_{1h}, normalised amplitude of the second harmonic - I_{2h}, and normalised temperature - T.
At the end of this processing phase the data collected at $1Watt/cm^2$ was selected for training, the data collected at $1.5Watt/cm^2$ was selected for test, and the data collected at $2Watt/cm^2$ was selected for model validation. The training, test, and validation sets are formed by 429, 427, and 400 patterns, respectively.

3.2 RBFNN

A RBF (Fig. 1) is a three fully connected layers neural network. The first layer is a set of inputs, the second is formed by a set of processing elements, called neurons, that performs a nonlinear transformation on the input data. The outputs of the hidden layer are linearly combined at the last layer to calculate the overall network output. The input/output relation for a RBF is given by:

$$f(x_j) = b + \sum_{i=1}^{n} \alpha_i \varphi\left(||x_j - c_i||\right) \qquad (1)$$

where n is the number of neurons in the hidden layer, b is the bias term, $||.||$ is a norm (an Euclidean norm was employed), and $\varphi\left(||x_j - c_i||\right)$ is a set of nonlinear radial basis functions weighted by $\{\alpha_i\}_{i=1}^{n}$. The basis functions are centred at $\{c_i\}_{i=1}^{n}$ (centres) and are evaluated at points x_j. Normally these functions are Gaussian:

$$\varphi_i = e^{\frac{1}{2\sigma_i^2}||x_j - c_i||^2} \qquad (2)$$

3.3 Multi-objective Genetic Algorithm (MOGA)

Manual selection of the best RBFNN structure for a particular problem is a hard process according to the enormous number of possible structures. In order to perform an automatic and intelligent search, MOGA[3] was

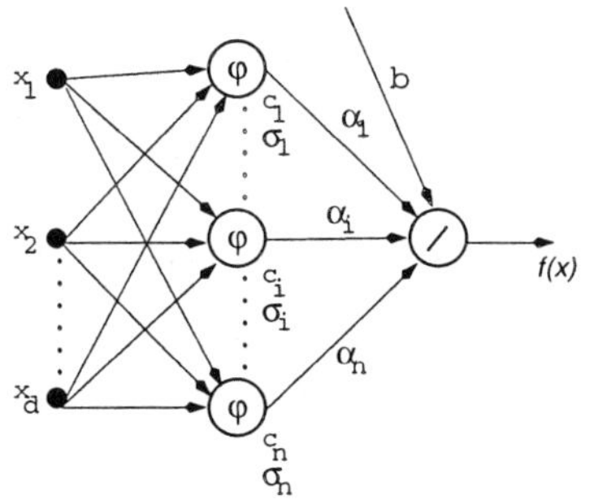

Fig. 1. RBFNN structure.

applied to the problem under study.
In this work MOGA was assigned to select the important inputs, and the best number of neurons. The lags of the variables: I_{fc}, I_{1h}, I_{2h}, and T were selected as NNs inputs. In this particular problem MOGA runs during 100 generations, of 100 RBF (individuals) each. The crossover and mutations probabilities were defined as 0.7 and 0.5 respectively. To maintain population diversity in each generation 10% of the population was changed by a set of randomly generated individuals. The previously referred values are the ones that reproduce the best MOGA results, and are selected after several tests, considering different parameters arrangements.

3.3.1 Search space: In order to define the MOGA search space, the number of inputs was restricted to the interval [2,20]. In the same way, the number of neurons was restricted to the interval [2, 8], and the maximum admissible lag (MLAG) was defined as 20. These values were also selected after several runs, testing different parameters arrangements.
Considering NV=4 the number of variables that can contribute with information to the RBFNN inputs, then the total number of inputs available is: TI = NV×MLAG = 80. The total number of possible input combinations is given by: $\sum_{n=2}^{20} {}^{NV}C_n = 5.2E18$. Considering the restrictions on the number of neurons, the total number of possible models is: $8 \times \sum_{n=2}^{20} {}^{NV}C_n = 4.1E19$, an extremely high value to consider an exhaustive search.

3.3.2 Computational architecture: In order to accelerate MOGA execution, a distribute computing scheme was used (Fig. 2). The main computer places the training and test data in the database, together with the initial population. In a next step each one of the workers queries the database for NNs to be trained, and the features of each trained NN are placed in the database. When all the individuals in the initial population are trained, the master computer recombines and mutate the

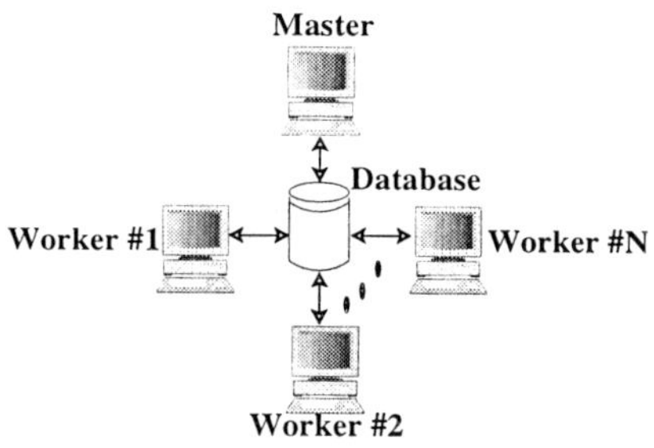

Fig. 2. Distributed computational architecture.

trained population. A portion of 10% of the population is replaced by a set of randomly generated individuals, and the changed population is stored in the database for future training of the workers. After the training of this population the master mutates, recombines, and introduces random individuals, and a new stage begins. The process continues until the specified number of generations is met.

3.3.3 Individual performance assessment: The training of each individual was performed using the Levenberg Marquardt (LM) algorithm with the "early-stopping" termination criteria[5]. In each iteration the LM optimises only the values of the centres and spreads, while the linear weights ($\{\alpha_i\}_{i=1}^{n}$) were obtained using the "Least Squares"(LS) strategy[5]. The initial values of the centres and spreads were determined using the optimal adaptive k-means algorithm [6].
After training, the performance of each individual was accessed according the the following measures: Root Mean Square error in the TRaining set (RMSTR), Root Mean Square error in the TEst set (RMSTE), Model-validity tests, and Model ComPlexity (MCP). The model-validity tests considered in this work are explained in [7], and used in [2]. These tests involve the computation of auto-correlation, cross-correlation and higher correlations functions involving model residuals, inputs, and outputs. If the fitted model is adequate, the following conditions should hold:

$$\begin{aligned} R_{ee}(\tau) &= \delta(\tau) \\ R_{ue}(\tau) &= 0 \quad ,\forall\tau \\ R_{u^{2'}e}(\tau) &= 0 \quad ,\forall\tau \\ R_{u^{2'}e^2}(\tau) &= 0 \quad ,\forall\tau \\ R_{e(eu)}(\tau) &= 0 \quad ,\tau \geq 0 \\ R_{e^2e^2}(\tau) &= \delta(\tau) \\ R_{(ye)e^2}(\tau) &= k\delta(\tau) \\ R_{(ye)u^2}(\tau) &= 0 \quad ,\forall\tau \end{aligned} \tag{3}$$

In reality, the correlations presented in eq. 3 will never be precisely zero for all lags. This way the model is considered adequate if the normalised correlation tests lie within 95% confidence limits defined as:

$$CI = 1.96/\sqrt{N}, \tag{4}$$

where N is the number of training patterns. In the same way, the autocorrelations of the residuals never equals perfectly the delta function, but will be considered as adequate if the autocorrelation plot enters the 95% confidence interval before lag one.
The MCP was computed as the total number of parameters for a particular NN structure:

$$\text{MCP} = NC \times NE + NS + NW, \tag{5}$$

where NC is the number of centres, NE is the number of inputs, NS is the number of spreads, and $NW = n$ is the number of weights.

3.3.4 MOGA objectives: From the MOGA point of view, the RMSTR, RMSTE, maximum of correlation tests, and MCP are objectives to minimise. Having in mind the attainment of models with a higher generalisation capacity, the RMSTE was defined as a goal of priority 2, and with goal value of 0.003. The maximum of the correlation tests are defined with a goal of value $CI = 1.96/\sqrt{429 - \text{MLAG}} = 0.097$, and with priority 1. In order to discard large models of heavy computation, MCP was defined as a goal with priority 1, and value 70. This value was selected having in mind the MOGA search space, defined in sub-section 3.3.1.

4 Results and discussion

In this section the selection of the best individual from the MOGA preferable set, was based on Root Mean Square Error in the Validation set (RMSEV). This means that the best individual is the one that presents the smallest RMSEV, when compared to the other preferable individuals.
After 100 generations, MOGA yields a non-dominated set of 2598 individuals, and a preferable set of 20 individuals. From the preferable set, the best individual presents a RMSEV of 0.0015, a maximum absolute error of 0.0748^oC, a mean squared error of 0.0037 $(^oC)^2$, and an average error of -0.0084 ^oC. In addition, this model has 6 neurons, and a weights norm of 4.2044. Comparing the previously referred error values with the ones presented in [1], it can be said that the maximum absolute error was reduced from 0.44$^oC$to 0.0748^oC, the mean squared error was reduced from 0.03 $(^oC)^2$ to 0.0037 $(^oC)^2$, ie. one magnitude order, and the average error was reduced from -0.02$^oC$to -0.0084^oC. These increases in performance show that probably temperature

Table 2. Performance of the best model

	RMSTR	RMSTE	MCP	R_{ee}	$R_{e^2e^2}$	$R_{(ye)e^2}$	$R_{(ye)u^2}$	R_{ue}	$R_{u^2e^2}$	R_{u^2e}	$R_{e(eu)}$
	0.0008	**0.0016**	**61**	**0.0782**	**0.0138**	**0.0933**	**0.0767**	**0.0863**	0.1040	**0.0757**	0.1213
G	-	0.003	70	0.097	0.097	0.097	0.097	0.097	0.097	0.097	0.097
P	-	2	1	1	1	1	1	1	1	1	1

G - MOGA Goals; **P** - MOGA Priorities

modelling in the same conditions of [1] would be better performed with RBFNN.
Table 2 presents MOGA optimisation objectives for the best RBFNN. Looking to that table, it can be stated that this individual fulfils 8 out of 10 objectives defined. The goals that are not fulfilled are close to the desired value.

Table 3. Inputs of the best model

I_{fc}	I_{h1}	I_{h2}	T
-	4,15	1,6,7,15,18	1

Table 3 presents the inputs of the best model. From the analysis of this table it can be said that the information of I_{fc} is completely discarded, as well as in the majority of the preferable individuals. The relevance of the information of I_{h1} is marked by the presence of two lags ($I_{h1}(k-4)$ and $I_{h1}(k-15)$). The information of I_{h2} is of great importance for the model, given the presence of 5 lags related with this variable. This crescent relevance from I_{fc} to I_{h2} is because harmonics at higher frequencies are more attenuated, and more sensitive to temperature variations. The relevance of the past memory of the system is marked by the presence of the 1st lag of T. The presence of this lag proves the physical validity of the model, since the actual temperature $T(k)$ is obviously dependent on the temperature in the past 10 seconds.

5 Conclusions & future work

The work hereby presented deals with RBFNN structure selection for punctual and invasive temperature modelling in a homogeneous medium. The results obtained with the best selected models are promising, and points that coherent results can be obtained in a non-invasive spatial-temporal environment, that is, in the same conditions of [1]. A reduction of one magnitude order is attained in terms of mean squared error.
For future work the use of B-Splines NNs is suggested. These NNs use different basis functions through the input space, adjusting itself locally to the problem, and better results are expected. It will be tested a non-invasive spatial-temporal temperature estimation, using these methods. This is proposed as a fundamental step towards ultrasonic therapeutic instrumentation feedback control.

Acknowledgments

The authors gratefully acknowledge the financial support of: Fundação para a Ciência e a Tecnologia (scholarship SFRH/BD/1461/2003), and Conselho Nacional de Desenvolvimento Científico e Tecnológico (CNPq/CYTED 490.013/03-1), Brazil.

References

[1] Simon, C., VanBaren, P., Ebbini, E. S. (1998) Two-dimensional temperature estimation using diagnostic ultrasound. IEEE Transactions on Ultrasonics, Ferroelectrics, and Frequency Control 45: 1088-1099.

[2] Ferreira, P. M., Ruano, A. E., Fonseca, C. M. (2003) Genetic assisted selection of RBF model structures for greenhouse inside air temperature prediction. In Proc. IEEE Conference on Control Applications. Vol 1 and 2. Instanbul, Turkey. pp. 576-581.

[3] Fonseca, C. M., Fleming P. J. (1993) Genetic algorithms for multi-objective optimization: Formulation, discution and generalization. In: Proc. 5th Int. Conf. Genetic Algorithms, Forrest, S. (eds.), pp. 416-423.

[4] Ferreira, P. M., Ruano, A. E. (2004) Predicting solar radiation with RBF neural networks. In: Proc. 6th Portuguese Conf. on Automatic Control, Vol. One, pp. 31-36.

[5] Ferreira, P. M., Faria, E. A., Ruano, A. E. (2002) Neural Network Models in Greenhouse Air Temperature Prediction. Neurocomputing 43: 51-75.

[6] Chinrungrueng, C., Séquin, C. H. (1995) Optimal Adaptive K-Means Algorithm with Dynamic Adjustment of Learning Rate. IEEE Transactions on Neural Networks 6: 157-169.

[7] Billings, S., Voon, W. (1986) Correlation based model validity tests for non-linear models. International Journal of Control 44: 235-244.

Assessing the Reliability of Complex Networks through Hybrid Intelligent Systems

D.E. Torres D., C. M. Rocco S.

Facultad de Ingeniería, Universidad Central de Venezuela, Venezuela
E-mail: douglastd@cantv.net , crocco@reacciun.ve

Abstract

This paper describes the application of Hybrid Intelligent Systems in a new domain: reliability of complex networks. The reliability is assessed by employing two algorithms (TREPAN and Adaptive Neuro-Fuzzy Inference Systems (ANFIS)), both belonging to the Hybrid Intelligent Systems paradigm. TREPAN is a technique to extract linguistic rules from a trained Neural Network, whereas ANFIS is a method that combines fuzzy inference systems and neural networks. In the experiment presented, the structure function of the complex network analyzed is properly emulated by training both models on a subset of possible system configurations, generated by a Monte Carlo simulation and an appropriate Evaluation Function. Both approaches are able to successfully describe the network status through a set of rules, which allows the reliability assessment

1 Introduction

The central role played by systems in most real world situations requires an adequate design phase. In communication networks, for example, besides basic constraints (connection speed, throughput, etc.), an important issue to be addressed regards the reliability of the system [1].

A convenient way of modelling any system is to adopt an undirected or a directed connected graph, called Reliability Block Diagram (RBD), in which every block or link is associated with a system component. Each block in an RBD can be considered to assume one of two possible states, operating or failed. In this way, the whole system can be described by a Boolean vector $\boldsymbol{x}$, having as many components as the number of edges in the RBD.

The state of the whole system, uniquely determined by the Boolean vector $\boldsymbol{x}$, is also described by a binary variable y [2]. The Boolean mapping that associate every input vector $\boldsymbol{x}$ to its corresponding output y is called Structure Function (*SF*) [3]. The procedure employed to retrieve the value of y that corresponds to a given $\boldsymbol{x}$ is usually referred to as an Evaluation Function (*EF*) [4]. The most widely studied reliability measure (*s-t* reliability) assumes that the system is operating if there exists at least one working path from the source node s to the terminal node t. In this case a depth-first procedure [5-6] can be employed as an *EF*. Of course, there are also situations where the reliability assessment is more complicated (e.g. electric power system), and more complex *EF* are required [7]. An important issue is that the *SF* and therefore the reliability assessment require the solution of an NP-hard problem [8]. A possible way to reduce the computational burden is to employ Monte Carlo techniques, which attempt to produce an estimate of the network reliability by analyzing a subset of possible system states $\boldsymbol{x}$.

Generally, Monte Carlo techniques require a large number of *EF* evaluations to establish the reliability of a system; therefore, it seems to be convenient to employ a machine learning method for approximating the reliability expression through a reduced collection of *EF* values. To this aim several Machine Learning approaches have been considered in the literature [9-12].

However, in search of more comprehensible models, novel investigation areas are developed, by integrating several intelligent systems. This operative synergy, called Hybrid Intelligent Systems (HIS) [13], seeks to improve the efficiency, reasoning power and comprehensibility of the integrand systems.

This paper presents, under the integrative perspective of HIS, an approach for the reliability assessment of complex networks.

To this aim empirical models induced by two techniques (TREPAN [15] and ANFIS [16]) are compared, when applied to the samples generated by a Monte Carlo simulation for a given *EF*. To our best knowledge, this approach, based on HIS paradigm, has not been yet used to assess the reliability of complex systems.

The paper is organized as follows: In Sec. 2 some definitions are presented. Sec. 3 introduces the machine learning methods considered for approximating the reliability of a network, while Sec. 4 presents the proposed approach to assess the system reliability. Sec. 5 compares the results obtained on a example related to a complex network with 21 links. Finally, Sec. 6 contains the conclusions.

2 Definitions

It is assumed that the system components have two states (operating and failed) and that component failures are independent events. The state x_i of the ith component is defined as [2]:

$$x_i = \begin{cases} 1 \text{ (operating state)} & \text{with probability } P_i \\ 0 \text{ (failed state)} & \text{with probability } Q_i = 1 - P_i \end{cases} \quad (1)$$

where P_i is the probability of success of component i.
The state of a system containing d components is expressed by a vector $\boldsymbol{x}=(x_1,x_2,\ldots, x_d)$. To establish if $\boldsymbol{x}$ is an operating or a failed state for the network, a proper Evaluation Function (*EF*) is defined:

$$y = EF(x) = \begin{cases} 1 & \text{if the system is operating in this state} \\ 0 & \text{if the system is failed in this state} \end{cases}$$

For example, a depth-first procedure [5-6] can be employed as an *EF*, if the criterion to be used for establishing reliability is simple connectivity. For other metrics, special *EF* may be used.

3 Hybrid Intelligent Systems Models

3.1 Extraction of knowledge from trained NN

The Extraction of Knowledge from Neural Networks consists of the development of techniques that allow the comprehensible representation of the knowledge acquired by a trained network. This can be expressed in diverse ways, through symbolic rules, rules fuzzy or decision trees. The Extraction of Knowledge allows the validation and refinement of the neural networks, as well as the integration of connectionist and symbolic systems. TREPAN [15] is a technique to extract decision trees from a trained neural network and contains aspects, which differ from other algorithms used for extraction of information from neural networks [15]. TREPAN requires as input the weights and biases of the trained neural network (NN) and a training data set. As output it produces a decision tree that provides an approximation to the function represented by the network.

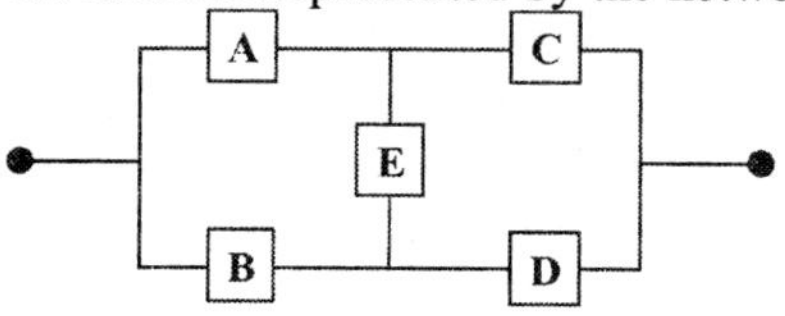

Fig. 1. A Complex Reliability Network

Figure 2 present a decision tree extracted by TREPAN, from the system network the figure 1. In general the first node in the tree refers to the most important component of the network Note that from the extracted tree, it is easy to obtain rules.

3.2 Fuzzy Model Identification

Fuzzy system identification is the process of identifying the structure and the parameters of a fuzzy model. The construction of the fuzzy model requires two phases. The first phase is model structure identification that is, the identification of the input variables and rules. As a result, the first phase produces a set of If/Then rules [17]. The second phase corresponds to the adaptation of the parameters (membership functions and coefficients).
The first phase was performed through the construction of knowledge-based neural networks (KBNN) [18], using the procedure Neural Fuzzy Networks (FuNN) [19]. This procedure combines elements of fuzzy modeling and neural network computations into single connectionist architecture. The procedure employed (FuNN) [20-21] consists on five layers: input variable layer, condition elements layer, rule layer, action elements layer, and Output layer. The rules obtained by FuNN are of the linguistic type. For example, referred to the system shown in Figure 1:**If** (B is Operating) and (C is Operating) and (E is Operating) **then** (System is Operating)

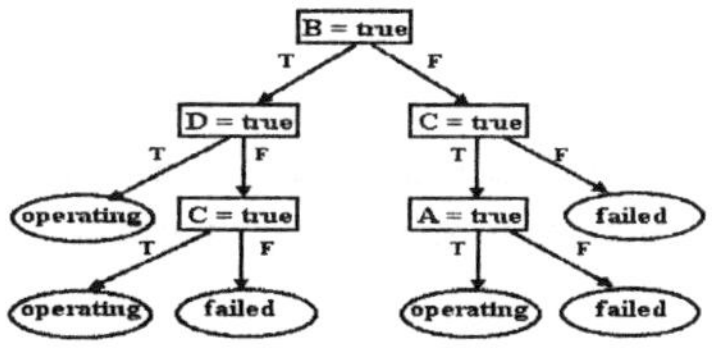

Fig. 2. Tree extracted by TREPAN from the network in Fig. 1

3.3 Parameters optimization

The fuzzy inference system described in 3.2 provides the knowledge expressed through rules of the type **If/Then**, which represent the structure and a first approach to the *SF* to be estimated. An additional phase is required for tuning the parameters of a preliminary fuzzy system and then to carry out the evaluation of the final system. ANFIS [16-17] uses a hybrid learning algorithm to identify parameters of Sugeno-type fuzzy inference systems. The ANFIS procedure can construct an input-output mapping based on both human knowledge (in the form of fuzzy if-then rules) and input-output data pairs. The parameters that define membership functions are adjusted through the learning process by a back-propagation algorithm.

4 The Proposed Approach

To evaluate the performance of the methods presented in the previous section, the network shown in Fig. 3 has been considered [22]. It is assumed that each link has reliability P_i and the goal is to obtain models that approximate the *s-t* reliability metrics. In order to apply the HIS paradigm, such as TREPAN or ANFIS, it is first necessary to collect a set of examples $(\boldsymbol{x},y)$, where $y=EF(\boldsymbol{x})$, to be used in the training phase and in the subsequent performance evaluation of the resulting models. To this aim, N_T system states have been randomly selected without replacement. In the case to be analyzed, only connectivity is checked to assess if a selected state $\boldsymbol{x}$ corresponds to an operating or to a failed state. To select the appropriate models a 10-fold cross-validation (CV) was performed. The performance of each method is measured using sensitivity, specificity and accuracy indexes as suggested in [23].

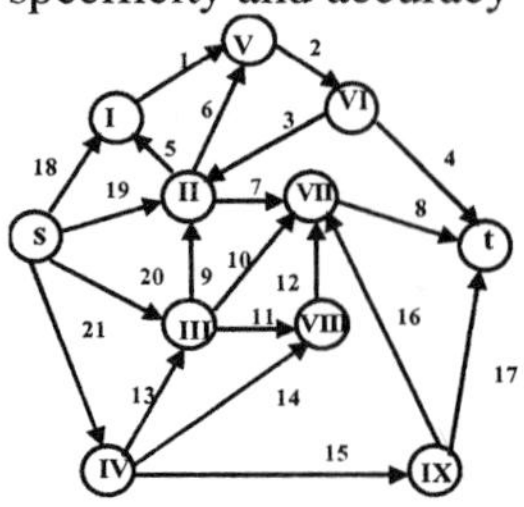

Fig. 3. Complex Network to be evaluated [22]

The average accuracy across the CV is computed, whereas the model with the highest value of accuracy and with the lower complexity is selected for evaluating the system reliability. To this aim, a system state $\boldsymbol{x}^*_i$ is generated at random and it is decided if it is a failed state or not using the induced models. The process is repeated by analyzing N_M system states. System reliability is estimated as the ratio of operating states to N_M.

Models determination

The state space associated to the system shown in figure 3 is randomly sampled and a data set with 2000 different $(\boldsymbol{x},y)$ pairs is generated. First the data set is used to train an MLP network composed of one input layer, one hidden layer and one output layer. The architecture of the network is denoted by '*i:h:o*' indicating *i* neurons in input layer, *h* neurons in hidden layer and *o* neurons in output layer. The activation function used was sigmoid function. Different numbers of neurons in the hidden layer were evaluated. The network was trained using the Levenberg-Marquardt algorithm with Bayesian optimization of the regularization parameters. The aim of regularization is to avoid over-fitting of the model by minimizing the sum of squares of errors and the sum of squares of model parameters. [24]. The best average accuracy during testing (95.2 %) was obtained with 67 units. The average accuracy during training was 100 %. The Neural Network Toolbox of Matlab was used to train the network [25]. The trained neural network (21:67:1) is then integrated with the TREPAN model. The fidelity, (percentage of predictions made by the extracted tree that agree with the predictions made by the network) was 97.46% during the training and 94.35% during the testing. In order to obtain the ANFIS model, a preliminary fuzzy system was induced, using the FuNN model [19]. Some rules extracted by FuNN are:

a) **If** x_{16} is Operating and x_{17} is Failed and x_{21} is Operating **then** System is Operating
b) **If** x_8 is Failed and x_{17} is Failed and x_{18} is Failed **then** System is Failed
c) **If** x_{15} is Operating and x_{17} is Operating and x_{21} is Operating **then** System is Operating

Some of the rules generated have a physical meaning, related to the minimal path (e.g. rule c) and cut (e.g. rule b) sets of the network [2]. Finally, the optimization of the previous fuzzy model was performed by ANFIS. The Fuzzy Logic Toolbox of Matlab [25] was used for training. Table 1 shows the average number of rules along with the average performance results obtained using the 10-fold cross-validation for the algorithms compared, for the training and testing phases. It is interesting to note that although the NN model presents the highest performance indexes during the training, the performance indexes of the TREPAN model during the testing, are superior to the NN and ANFIS models, even if the complexity of the induced ANFIS model (rules generated) is lower. Table 1 also shows that, during the testing phase, TREPAN performs better that the ANFIS model.

Once TREPAN and ANFIS are trained, their models are used to estimate the network reliability. A random data set with N_M = 10000 data pairs $(\boldsymbol{x}_i,y_i)$ was generated using P_i=0.90. Each system state is evaluated using a depth-first procedure and both trained models. The average system reliability based on *EF* was 0.9943. TREPAN and ANFIS produced the same average system reliability.

6 Conclusions

This paper has presented the reliability assessment of a complex system based on two methodological approaches (TREPAN and ANFIS) that uses the hybridization of different soft computing techniques. For the case analyzed, both models, built from a small sample of the state space, produce approximations with a satisfactory accuracy but, in average, the TREPAN model outperforms the best ANFIS model. Nevertheless, from a complexity point of view, the ANFIS model is better since it produces a smaller number of rules. Future researches, such as parameterizations, different NN training methods, different numbers of layers and nodes among others, are required to obtain more concrete conclusions about the performance of the different techniques. However, the hybridization of intelligent systems is a promising approach for assessing the reliability of complex networks. It not only improves the efficiency of the integrand systems, but also increases the capacity of understanding, since it produces useful topological information about the network, such as minimal path and/or cut sets.

Table 1. Average performance results for models

Model	Rules	Sensitivity %		Specificity %		Accuracy %	
		Train	Test	Train	Test	Train	Test
NN	67[1]	100	94.4	100	96.0	100	95.2
TREPAN	96	96.5	94.1	98.4	96.9	97.57	95.6
ANFIS	58	93.5	91.9	94.4	93.5	94.06	92.8

[1] number of neurons

References

[1]Lynn N, Singpurwalla, N(1998): Bayesian assessment of network reliability. SIAM Review, 40: 202-227.

[2] Billinton R., Allan R. (1992) Reliability Evaluation of Engineering Systems, Concepts and Techniques, 2nd ed. Plenum Press, New York

[3] Dubi A. (2001) Modeling of realistic system with the Monte Carlo method: A unified system engineering approach. Proc. Annual Reliability and Maintainability Symposium – Tutorial Notes.

[4] Grimaldi R.P., Shier D.R. Redundancy and reliability of communication networks. www.math.clemson.edu/~shierd/ Shier/abstracts/randr.html

[5] Reingold E, Nievergelt J, Deo N(1977): Combinatorial Algorithms. Prentice Hall, New Jersey

[6] Papadimitriou C. H., Steiglitz K. (1982) Combinatorial Optimization: Algorithms and Complexity, Prentice Hall, New Jersey

[7] Aggarwal K.K., Chopra Y.C., Bajwa J.S. (1982) Capacity consideration in reliability analysis of communication systems. IEEE Trans on Reliability, 31

[8] Billinton R. Allan R.N. (1992) Reliability Evaluation of Engineering Systems, Concepts and Techniques, 2nd Ed. Plenum Press, New York

[9] Rocco C.M., Moreno J.A. (2002) Machine Learning Models for On-Line Dynamic Security Assessment of Electric Power Systems. In: Lecture Notes in Artificial Intelligence, Vol. 2527, Springer-Verlag, Berlin

[10] Rocco C.M. (2003) A rule induction approach to improve Monte Carlo system reliability assessment. Reliability Engineering and System Safety, 82:87-94.

[11] Rocco C.M., Moreno J.M. (2002) Reliability evaluation using Monte Carlo simulation and support vector machine. Lecture Notes in Computer Science, Vol. 2329, Springer-Verlag, Berlin

[12] Rocco C., Muselli M (2004) Empirical models based on machine learning techniques for determining approximate reliability expressions. Reliability Engineering and System Safety, 83:301-309

[13] Jacobsen H.A. (1998) A Generic Architecture for Hybrid Intelligent Systems. IEEE Fuzzy Systems IEEE Fuzzy Systems. Anchorage, Alaska

[14] Jain L.C, Martin N. M.(1998) Fusion of Neural Networks, Fuzzy Sets, and Genetic Algorithms, Industrial Applications. Ed. CRC Press

[15] Craven M.W. (1996) Extracting Comprehensible Models from Trained Neural Networks. Ph.D.Thesis. University of Wisconsin- Madison

[16] Jang J.S, Sun C.T., Mizutani E. (1997) Neuro-Fuzzy and Soft Computing. A Computational Approach to Learning and Machine Intelligence. Prentice Hall, NJ

[17] Jang J.S.(1993) ANFIS Adaptive-Network-based Fuzzy Inference Systems. IEEE Transactions on Systems, Man and Cybernetics, 23:665-685.

[18] Lin C.T, Lee C.S.G. (1996) Neural Fuzzy Systems: A Neuro-Fuzzy Synergism to Intelligent Systems. Prentice Hall, NJ

[19] http://divcom.otago.ac.nz/infoscie/kel/cbiis.htm

[20] Kasabov N. (1996) Investigating the Adaptation and Forgetting in Fuzzy Neural Networks Through a Method of Training. *Proc.of ICNN'96.* Vol 1996: 118-123

[21] Kasabov N. (2001) On-line learning, reasoning, rule extraction and aggregation in locally optimized evolving fuzzy neural networks. Neurocomput. Issues 1-4: 25-45

[22] Yoo Y.B, Deo N. (1988) A Comparison of Algorithm for Terminal-Pair Reliability, IEEE Transaction on Reliability, 37(2): 210-215

[23] Veropoulos K, Camppbell, Cristianini N. (1999) Controlling the Sensitivity of Support Vector Machines. Proceedings of the IJCAI99

[24] Foresee F.D., Hagan M.T. (1997) Gauss-Newton Approximation to Bayesian Regularization Proc. of IJCNN′97

[25] MathWorks (2002) *MatLab 6.5* R13

Autonomous Behavior of Computational Agents

Roman Vaculín[1], Roman Neruda[2]
[1]Faculty of Mathematics and Physics, Charles University, Malostranské nám. 25, Prague
[2]Institute of Computer Science, ASCR, P.O. Box 5, 18207 Prague, Czech Republic
Email: vaculin@cs.cas.cz, roman@cs.cas.cz

Abstract

In this paper we present an architecture for decision making of software agents that allows the agent to behave autonomously. Our target area is computational agents — encapsulating various neural networks, genetic algorithms, and similar methods — that are expected to solve problems of different nature within an environment of a hybrid computational multi-agent system. The architecture is based on the vertically-layered and belief-desire-intention architectures. Several experiments with computational agents were conducted to demonstrate the benefits of the architecture.

1 Introduction

Software agents can be seen as small self-contained programs that can solve simple problems in a well defined domain [6]. In order to solve complex problems agents have to cooperate and exhibit some level of autonomy. Autonomy, adaptivity, cooperation ability, and several other properties distinguish agents from "conventional" programs.

In this paper we present an architecture that allows simple design of adaptive, or intelligent, agents. The architecture enables the agent to solve problems of different nature within an environment of a computational multi-agent system, and thus increase its autonomy, adaptivity and the performance of the whole system. The architecture is implemented within a distributed multi-agent system *Bang3* that provides a platform for an easy creation of hybrid artificial intelligence models by means of autonomous agents (see [4]).

2 Computational agents

An *agent* is a computer system that is *situated* in some *environment*, and that is capable of *autonomous action* in this environment in order to meet its design objectives [9, Ch. 1]. *Autonomy* is used to express that agents are able to act (to perform actions) without the intervention of humans or other system.

[2]This research has been supported by the National Research Program Information Society project no. 1ET100300419.

An *intelligent agent* is one that is capable of *flexible* autonomous action in order to meet its design objectives, where flexibility means three things: *proactiveness* (goal-directed behavior), *reactivity* (response to changes), and *social ability* (interaction with other agents). Building purely *goal-directed* or *purely reactive* agents — one that continually responds to its environment — is not difficult in some environments. The problem is to build a system that achieves an effective balance between the goal-directed and reactive behavior, which strongly depends on the characteristics of the environment.

A *computational agent* is a highly encapsulated object realizing a particular computational method [5], such as a neural network, a genetic algorithm, or a fuzzy logic controller. The main objective of our architecture is to allow a simple design of adaptive autonomous agents within an environment of a computational multi-agent system. In order to act autonomously, an agent should be able to cope with three different kind of problems [8]: cooperation of agents, a computation processing support, and an optimization of the partner choice. The architecture we present is general in the sense that it can be easily extended to cope with different problems than those mentioned, nevertheless, we present its capabilities in these three areas.

Cooperation of agents: An intelligent agent should be able to answer the questions about its willingness to participate with particular agent or on a particular task. The following subproblems follow: (1) deciding whether two agents are able to cooperate, (2) evaluating the agents (according to reliability, speed, availability, etc.), (3) reasoning about its own state of affairs (state of an agent, load, etc.), (4) reasoning about tasks (identification of a task, distinguishing task types, etc.).

Computations processing: The agent should be able to recognize what it can solve and whether it is good at it, to decide whether it should persist in the started task, and whether it should wait for the result of task assigned to another agent. This implies the following new subproblems: (1) learning (remembering) tasks the agent has computed in the past (we use the principles of case-based

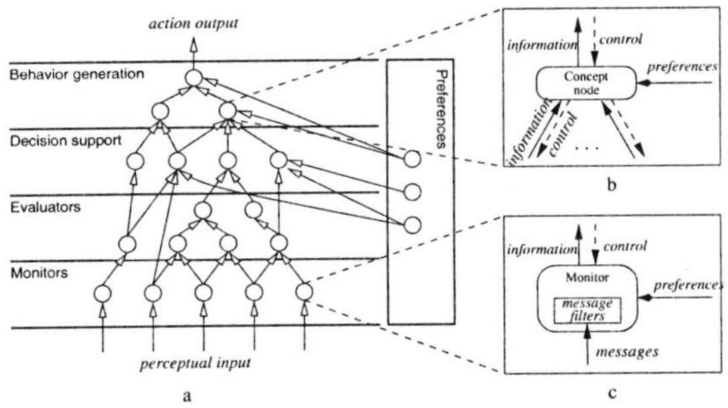

Fig. 1. Architecture — network of concepts (a); Concept node (b); Monitor (c)

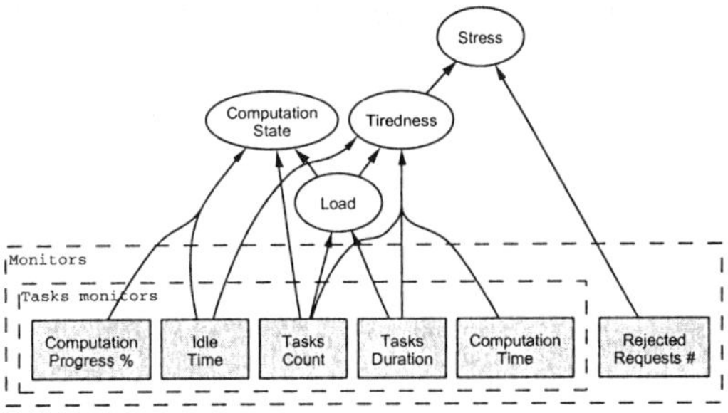

Fig. 2. Modeling state of an agent

learning and reasoning — see [2], [1] — to remember task cases), (2) monitoring and evaluation of task parameters (duration, progress, count, etc.), (3) evaluating tasks according to different criteria (duration, error, etc.).

Optimization of the partner choice: An intelligent agent should be able to distinguish good partners from unsuitable ones. The resulting subproblems follow: (1) recognizing a suitable (admissible) partner for a particular task, (2) increasing the quality of an evaluation with growing experience.

So, the architecture must support *reasoning*, *descriptions* of agents and tasks (we use ontologies in descriptions logics - see, e.g., [3]), *monitoring* and *evaluation* of various parameters, and *learning*.

3 Network of concepts

The architecture is organized into layers. Its logic is similar to the vertically-layered architecture with one-pass control (see [9, p. 36]). The lowest layer takes perceptual inputs from the environment, while the topmost layer is responsible for the execution of actions.

The architecture consists of four layers (see Figure 1): the *monitors* layer, the *evaluators modeling* layer, the layer for *decision support*, and the *behavior generation* layer. All layers are influenced by global *preferences*.

Global preferences allow us to model different flavors of an agent's behavior, namely, we can set an agent's pro-activity regime, its cooperation regime and its approach to reconsideration. *The monitors layer* interfaces directly with the environment. It works in a purely reactive way. It consists of rules of the form *condition* $\longrightarrow$ *action*. *Evaluators modeling layer* is used to model more aggregate concepts on top of already defined concepts (either monitors or other evaluators). *Decision support layer* enables an agent to solve concrete problems. *Behavior generation layer* generates appropriate actions that the agent should perform, and thus controls the agent's behavior. The mechanisms for action generation and selection are provided by the BDI model (see [9, pages 55–61]).

The basic element of our architecture is a *concept node*. We can imagine a concept node as a class in some common object-oriented programming language, which defines explicitly its *dependences* on other concept nodes. The concept node is dependent on some other concept node if it needs some services provided by this other concept node in order to provide its own services. Each part of the architecture is defined as a concept node.

The network of concept nodes is a directed acyclic graph of concept nodes — see Figure 1 (a). Edges express dependences between concept nodes. This graph respects described layers. Figure 1 (b) shows a detailed view of a common concept node and Figure 1 (c) depicts a detailed view of a monitor. Each monitor can define several *filters* which represent rules as described above.

Explicitly defined dependences allow each agent to use only those those concept nodes that it really needs.

4 Modeling in the network of concepts

Evaluators are used to describe an agent's state, and to estimate services' quality of partner agents. They usually perform aggregations of several simpler concept nodes, typically the monitors. Typically, evaluators have the form of a non-linear real function that may differ in individual evaluators.

In order to describe an agent's state, we have defined four evaluators — *Load*, *Tiredness*, *Stress*, and *Computation state* (see Figure 2). For example, the *Load* evaluator depends on the count of currently running tasks and on their demandingness (complexity, etc.). We approximate the complexity of tasks by the average duration of past tasks. The load grows proportionally with the count of tasks and the average duration of tasks. The other evaluators can be described in a similar way.

The decision support concept nodes are used to represent particular decision problems and provide suggestions of how to solve these problems.

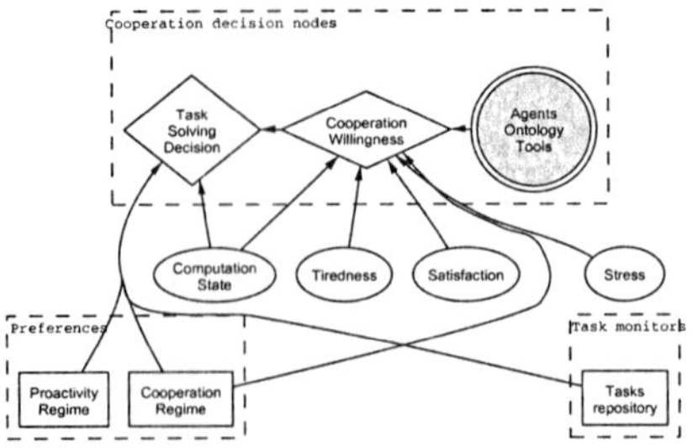

Fig. 3. Support for cooperation

Figure 3 depicts concept nodes that solve decision problems common in the area of agents cooperation. The *Agents ontology tools* concept node encapsulates reasoning services about agents' capabilities. The *Cooperation willingness* (CW) concept node suggests whether to cooperate with a particular agent or not. The *Task solving decision* concept node suggests whether to solve a particular task for a particular agent. For details about other areas of decision support see [7].

In the behavior generation layer, we use the BDI model (see [9]) to generate and choose the appropriate actions. The purpose of computational agents is to solve assigned tasks in an effective way. We distinguish two different situations:

1. If an agent does not use services of other agents in order to solve its task, it can perform the following basic actions: (a) *Accept / postpone / reject a new task,* (b) *Finish / interrupt a started task,* (c) *Evaluate task* (if there are some tasks in the tasks repository with incomplete information), d) *Find and solve new tasks.*
2. If an agent uses services of other computational agents in order to solve its own task, it acts as a simple task manager (an agent that assigns tasks to other agents), and it can further perform the following actions: (a) *Search for suitable partners,* (b) *Test and evaluate possible partners,* (c) *Distribute / redistribute task to partners.*

We base our implementation of the BDI model on the algorithm described in [9, pages 55–61]. According to it, we have to specify the implementation of a belief revision function (*brf*), an options generation function (*options*), a *filter* function, and an *execute* function. The monitors layer and the evaluators modeling layer represent the agent's knowledge about its environment, and thus stand for its beliefs. Beliefs are updated automatically by filters of monitors, which can be seen as the *brf* function.

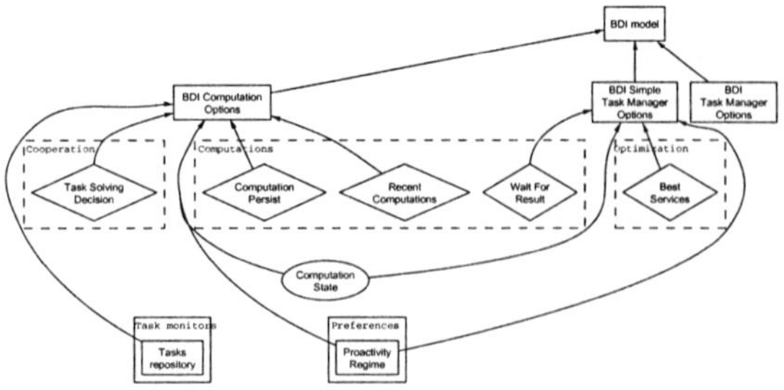

Fig. 4. BDI architecture within the network of concepts

Figure 4 shows concept nodes implementing the BDI architecture. The *BDI model* encapsulates the basic logic of the algorithm. We have further defined several concept nodes that are responsible for option generation. Each such a concept node implements its own *options* function and its own *filter* function which is responsible for filtering desires and intentions. The *action* function is implemented in the *BDI model* as defined by the pseudocode in the algorithm 1. B denotes the set of beliefs, I the set of intentions, D the set of desires, D_x the set of desires generated by the x-th option generation node, $ONodes$ a vector of all agent's option generation nodes.

Algorithm 1 *Action* function of the BDI agent.

```
weight_max = 0
for i = 1 to ONodes.length do
    ⟨D_i, weight_i⟩ = ONodes[i].options(B, I)
    if weight_i > weight_max then
        weight_max = weight_i
        D = D_i
        node_max = ONodes[i]
    end if
end for
return node_max.filter(B, D, I)
```

5 Experiments

We have adapted two existing computational agents embedding the multi-layer perceptron (MLP) and the radial basis function (RBF) neural network. These agents represent two different computational methods for the solution of similar categories of tasks.

Overheads of the architecture are summarized in Table 1. The creation of the agent takes 2-3 times longer since all the structures must be initialized. The communication overhead is around 30% when dealing with message delivering. However, in real-life scenario of task solving, the overhead is only about 10%.

	Without the arch.	With the arch.
Agent creation time	3604 μs	9890 μs
Message delivery time	2056 μs	2672 μs
Total computation time	8994681 μs	9820032 μs

Table 1. Comparison of the agent with and without the autonomous support architecture

	Error	Duration
Random choice	11.70	208710ms
Best speed	1.35	123259ms
Best Accuracy	1.08	274482ms
Best services	1.17	102247ms

Table 2. Optimization of the partner choice. Comparison of choices made by different criteria.

Table 2 summarizes the measured results of *optimization of the partner choice.* We simulated a usual scenario when an agent needs to assign some tasks to one of admissible partners. This agent uses a collection of different tasks and assigns them to the computational agents successively. The total duration of the computation and the average error of computed tasks were measured. A significant improvement of the efficiency can be seen.

Experiments with *optimization by reusing results* are summarized in Table 3. We have constructed several collections of tasks with different ratios of repeated tasks (quite a usual situation when, e.g., evaluating the population in genetic algorithms). We compared the total computation-times of the whole collection with and without the optimization enabled. We can see that the optimization is advantageous when the ratio of repeated tasks is higher than 20%. When more than 40% are repeated the results are significant.

Repeated tasks	Standard	Optimized
0 %	135777097	121712748
20%	94151838	90964553
40%	50704363	91406591
60%	47682940	90804052

Table 3. Optimization by reusing the results of previously-computed tasks (duration in milliseconds).

6 Conclusions

In this paper, we have described a general architecture that allows a simple design of adaptive software agents. It supports both agents' decision making and the generation of autonomous behavior. The architecture incorporates learning capabilities and support for reasoning based on ontologies which allows reasoning about agents' capabilities and activities and optimization of the performance.

The experiments have demonstrated that (1) it allows faster and more precise execution of tasks; (2) it supports a better cooperation of agents; (3) the performance drawbacks are not high.

The realized architecture provides several challenges for future work. The exchange and sharing of task cases can be a useful extension of the current implementation. We plan to perform more exhaustive experiments with groups / ensembles of cooperative computational agents. Finally, we plan to experiment with algorithms (e.g., by genetic algorithms) for automatic learning (generation) of global and local preferences of the architecture suitable for a particular situation (task).

References

[1] Agnar Aamodt and Enric Plaza. Case-based reasoning : Foundational issues, methodological variations, and system approaches. *AICom — Artificial Intelligence Communications*, 7(1):39–59, 1994.

[2] David W. Aha and Dietrich Wettschereck. Case-based learning: Beyond classification of feature vectors. 1997.

[3] Franz Baader, Diego Calvanese, Deborah McGuinness, Daniele Nardi, and Peter Patel-Schneider, editors. *The Description Logic Handbook.* Cambridge University Press, 2003.

[4] Bang3 web page. http://www.cs.cas.cz/bang3/.

[5] Roman Neruda, Pavel Krušina, and Zuzana Petrova. Towards soft computing agents. *Neural Network World*, 10(5):859–868, 2000.

[6] H. S. Nwana. Software agents: An overview. *Knowledge Engineering Review*, 11(2):205–204, 1995.

[7] Roman Vaculin. Artificial intelligence models in adaptive agents. Master's thesis, Faculty of Mathematics and Physics, Charles University, Prague, 2003.

[8] Roman Vaculin and Roman Neruda. Concept nodes architecture within the Bang3 system. Technical report, Institute of Computer Science, Academy of Science of the Czech Republic, 2004.

[9] Gerhard Weiss, editor. *Multiagents Systems.* The MIT Press, 1999.

Neural Network Generating Hidden Markov Chain

J. Koutník[1], M. Šnorek[1]
[1]Neural Computing Group,
Department of Computer Science and Engineering,
Czech Technical University, Prague, Czech Republic

Abstract

In this paper we introduce technique how a neural network can generate a Hidden Markov Chain. We use neural network called Temporal Information Categorizing and Learning Map. The network is an enhanced version of standard Categorizing and Learning Module (CALM). Our modifications include Euclidean metrics instead of weighted sum formerly used for categorization of the input space. Construction of the Hidden Markov Chain is provided by turning steady weight internal synapses to associative learning synapses. Result obtained from testing on simple artificial data promises applicability in a real problem domain. We present a visualization technique of the obtained Hidden Markov Chain and the method how the results can be validated. Experiments are being performed.

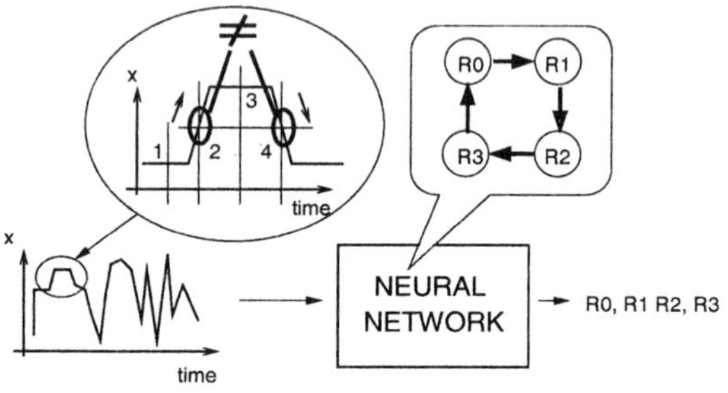

Fig. 1. Neural network can distinguish between rising and falling edge in a sequence even if they have the same values. It generated Hidden Markov Chain. Output of the network is a sequence of states of that chain as well as the chain as a temporal model of the sequence.

1 Introduction

Many signals that are processed in the field of artificial intelligence have temporal components. Of course, that is because we live in temporal environment. If we want to process such signals with neural networks, we have to take the temporal information into account.

Feed-forward neural networks could not be used with a great success in sequential data processing. There are, of course, applications which employ such neural networks by e.g. passing several consequent signal values to the network input at the same time. But such state-free neural network could not discover relationship between consequent signal samples. Another drawback is that such application of a neural network is limited to one-dimensional signals only.

Processing of temporal sequences and signals is domain of recurrent neural networks. Analogically in the field of logical circuits static signals are processed by combinatorial logic and sequences are handled by sequential logic. In the field of artificial neural networks we can name Elman networks [1] as a basic networks for sequence processing. Multilayered perceptron networks were enhanced with time-delay connections which can represent sequential memory in a feed-forward neural network. There are spiking neural networks [2] which cover temporal behavior with leaky integration of an input and transmission of temporal activation pulses (spikes). The most precise and complicated structure for sequence processing is Long Short-Term Memory [3].

Hidden Markov Models (HMM) [4] are widely used tools for processing and recognition of temporal sequences. HMM is a tool for statistical modeling of sequences. In short, the HMM is based on a stochastic automaton - Hidden Markov Chain (HMC), hidden behind a probabilistic observation function. For each sequence we can create the corresponding stochastic automaton which corresponds to the sequence by the probabilistic function. HMC is based on Markov condition which says that future state of the automaton depends only on the present state, not on the past states.

There are also combinations of HMMs and neural networks [5]. Some of them use neural networks instead of the probabilistic observation function. Some of them use small neural networks for computation of state to state transition probability etc.

Since our objective are neural networks and we need to measure their performance, the HMM are suitable tools for comparison. We did not create so popular hybrids of HMM and neural networks but we developed a neural network that can somehow act as a HMM. Our focus is to make the comparison to HMM and to discover mutual consequence and relation of HMMs and neural networks in general.

2 Theoretical Part

Based on previous experiment in information categorization, we found that a suitable neural network that can facilitate this task is Murre's Categorizing and Learning Module (CALM) [6] developed in Leiden in the year 1992. Original CALM neural network has many pleasant features such as:

- Categorization of the input vectors. CALM processes input vectors and sorts them into categories.
- Module is strongly inspired by live neural network. It can be taken as a simple model of a cortical mini-column.
- *Winner takes all* principle is applied within the network. After several iterations one R-V neuron pair is chosen.
- Plasticity/stability dilemma is gently controlled by the arousal process.
- Modularity. The CALM is an autonomous module with built-in learning algorithm (in contradiction to supervised neural networks with external teacher). Learning phase does not need to be separated from the recalling phase.
- Degree of the module (amount of R-V pairs) does not depend on dimension of input vector. It depends on desired number of categories.
- CALM is recurrent neural network. This is the main reason why we have chosen it.
- Module has leaky integrating behavior in the arousal subsystem and built-in activation decay of the neurons.

Function of CALM can be described by three main processes (see Fig.4 where additional fourth process for temporal memory occurred as well). Excitatory process is performed by R-neurons. Each R-neuron reacts to the input signal based on the inter-synapse weights (which are being modified by learning algorithm). Each R-neuron excites its matching V-neuron. V-neurons realize competition layer. When the new input vector is presented, one V-neuron wins the competition and suppress activation of other V- and R-neurons. CALM contains additional arousal process that controls learning rate and

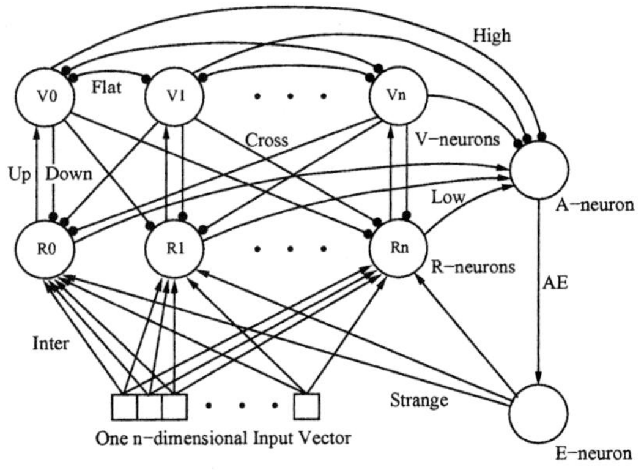

Fig. 2. Original Categorizing and Learning module can process isolated input patterns. Neurons are depicted as circles. Synapses are either excitatory (arrow) or inhibitory (dot). Each synapse has its weights. Inter-synapses only are being learned by learning algorithm other synapse weights are steady.

selects new representants (R-neurons) for new categories of input vectors.

Original CALM module has pleasant features in static pattern recognition. It has fast convergence (iteration time decreases with increasing number of presentation of one input vector) but it is not suitable for recognition of sequences and it needs certain improvements.

We made such improvements that were partly introduced in [7]. Namely it is the improved measuring of the input space and autonomous mapping of the temporal consequences in cross-synapses that we consider as a tool for construction of a HMC.

First, we replaced weighted sum by Euclidean metrics in the R-neuron inputs. The scalar product is a good solution for high dimensional patterns. But it could not separate low dimensional patterns precisely. The major problem is with one-dimensional input vector where all values are treated as one category because all of them are linear combinations. Euclidean metrics allows categorization of low dimensional patterns and is suitable for processing of sequences.

Second improvement overcomes another drawback of the original CALM module that is a need of a reset of all activations in the module between presentations of two different patters. Original CALM forgets its state after the recognition thus it is not usable in sequence recognition.

We found that cross-synapses (in Fig.3 denoted by thick lines with dots) in combination with a temporal memory can provide historically recognized information to recognition of future input vectors. We preserve activation of the V-neurons and following input vector is being categorized based on its position in the input space (by the meaning of Euclidean metrics) and also by previously selected category that influents R-neurons using matching cross-synapse.

Original CALM has cross-synapses as well. Certain setup of their weights makes the CALM to become self organizing map (proximate vectors in the input space form categories which are represented by neighboring R-neurons). Our improvement makes the cross-weights become *learning*. With help of the temporal memory in R-neurons cross-weights maps often consequent vectors to neighboring categories. Learning cross-synapses use associative Hebb's learning rule. If the present input vector activates one R-neuron, then the cross-synapse between this R-neuron and V-neuron activated by previous input vector is reinforced, others are diminished.

Thus there are two mappings in the enhanced CALM. First one maps input space to the categories. Second one maps temporal consequences of input patterns to cross-synapses. Enhanced CALM was named *Temporal Information Categorizing and Learning Map* (TICALM).

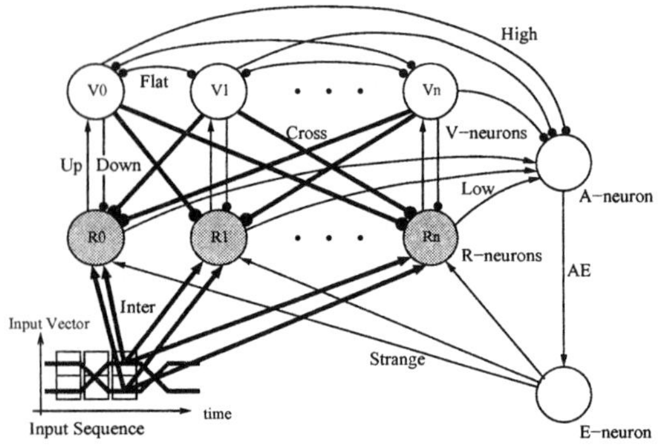

Fig. 3. TICALM does not process isolated input vectors but it takes temporal nature of pattern sequences into account.

In Fig.3 we can see that TICALM no longer process isolated patterns. In contradiction to commonly used processing of the temporal sequences by presenting a subsequence (several continuous samples of the sequence at the same time) to a neural network input the TICALM processes input vectors one by one. This approach does not limit the sequence to be one-dimensional. In Fig.3 the input sequence is two-dimensional and is passing through network input from left to right.

For easier consideration of the results we use a visualization technique of the HMC graph where states are depicted as circles and transition probability is drawn by an arrow with arbitrary strength. Narrow and dashed arrow means low probability, black solid arrow means high probability, see Fig.3.

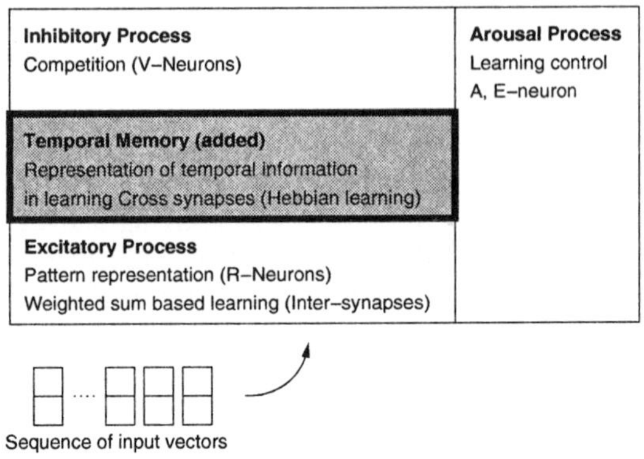

Fig. 4. Block diagram of the TICALM. There is the temporal memory added besides excitatory, inhibitory and arousal processes included in original CALM.

Now we can treat categories in the TICALM as *states*. Transition from one state to another is influenced by cross-synapse weight. One property of the HMC is transition matrix that denotes probability of a transition from one state to another. In our network the probability is expressed by the cross-weight. As the HMC is driven by probabilities during transitions, TICALM is driven by cross-synapse weights.

In experiments we proof it the cross-synapses are capable to incorporate sequential and stateful behavior in the TICALM. The cross-synapse weights could be filled into the matrix. Strong inhibitory weights denote low probability of transition to another state, weak inhibition by cross-synapses causes high probability of the transition. This is the reason why we say that TICALM can construct Hidden Markov Chain.

3 Experiments

First of all we test the network on a special artificial and simple data set. Since the original CALM is a state-free network (it treats the same input vectors in the same category), TICALM should separate even same input patterns based on their context or position between other input patterns respectively. Let us have a simple sequence of two dimensional testing vectors $((0\ 1)(0.5\ 0.5)(1\ 0)(0.5\ 0.5))$. If we present this sequence to original CALM, the module will assign three categories to vectors $(0\ 1)$, $(0.5\ 0.5)$ and $(1\ 0)$. One can expect that TICALM should assign vector $(0.5\ 0.5)$ two categories. One for the vector between $(0\ 1)$ and $(1\ 0)$ and another one for the vector between $(1\ 0)$ and $(0\ 1)$.

In the other words TICALM will differ the same input vector into two transitions, one forward and one backward.

In Fig.3 we can see four images that show progress of the HMC being created within the network. Cross-synapses were rewritten to a matrix which is depicted in a series of graphs.

Each iteration in module ends with one activated R-V-pair (network state). Cross-synapses indicate how probable transitions between network states are - the network simulation can be displayed as a stochastic automaton. At the set of automata we can see how the consequence of clusters is being created within the network during the learning.

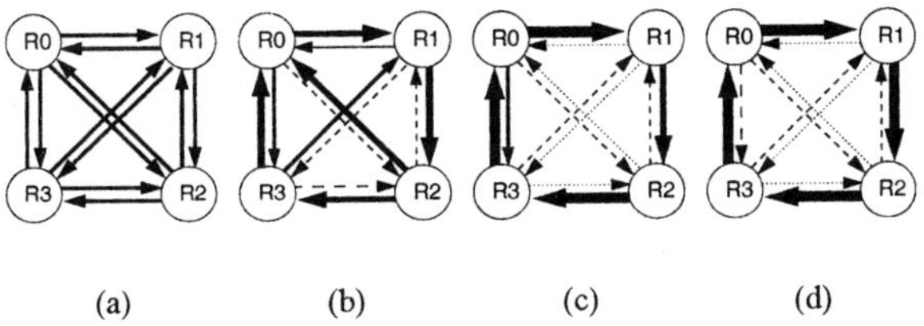

Fig. 5. Progress of construction of the HMC. Transitions with high probability are represented by thick arrows. Transitions with low probability are represented by thin, dashed and dotted arrows.

Next experiment uses artificial data as well but the data are not as simple as in previous experiment where the result is easily predictable.

We will compare the temporal model created by TICALM to constructed within the HMM from the same data sequences. Schema of the experiment is depicted in Fig.6 Let us create the HMM respectively its transition matrix manually. Then we will use particular HMM for generating sequences. The sequences will be used as training data for another HMM and for TICALM. Afterwards, we will display HMC created within the HMM and HMC created by TICALM.

We could not directly compare HMC created by TICALM with HMC that created the testing data sequence because one sequence can be generated by more than one HMM. Our visualization technique can help us to check whether the created models are the same in several repeated experiments.

4 Conclusion

We introduced a technique how the neural network can generate a probabilistic automaton - Hidden Markov Chain, that describes temporal characteristics of the input sequence. We have built in Euclidean metrics instead of scalar sum previously embedded in the network. Main

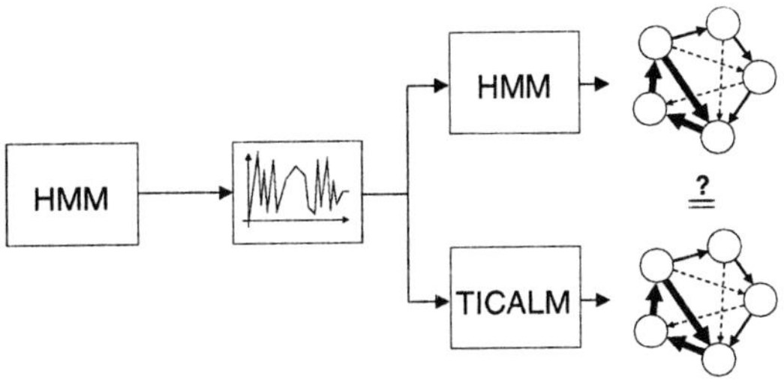

Fig. 6. Schema of comparison of HMM and TICALM generated models.

contribution is an improvement of the internal cross-synapses by a learning algorithm that adds a possibility of construction of the Hidden Markov Chain by the network. Functionality of the network was proven by simple experiments. Currently running experiments will prove that generated chains are reasonable and comparable to Hidden Markov Models on more complex and non-artificial data.

References

[1] Elman J.L., Finding structure in time. *Cognitive Science*, 14, pp. 179-211. 1990

[2] Maass W., Networks of Spiking Neurons: The Third Generation of Neural Network Models, Neural Networks 10(9), pp. 1659-1671, 1997

[3] Gers F.A., Schmidhuber J., LSTM recurrent networks learn simple context free and context sensitive languages. *IEEE Transactions on Neural Networks*, 12(6), pp. 1333-1340, 2001.

[4] Rabiner L.R., A Tutorial on Hidden Markov Models and Selected Applications in Speech Recognition, in Proceedings of the IEEE, vol.77,No.2,Feb 1989.

[5] Morita M., Oliveira L. S., Sabourin R., Bortolozzi F., Suen C. Y., An HMM-MLP Hybrid System to Recognize Handwritten Dates. International Joint Conference on Neural Networks, (ICJNN'02), pp. 867-872, Honolulu-USA, May 12-17, 2002.

[6] Murre J.M.J., Phaf R.H., Wolters G., CALM: Categorizing and Learning Module, *Neural Networks*, Vol. 5, pp. 55-82, Pergamon Press 1992

[7] Koutník J., Šnorek M., Single Categorizing and Learning Module for Temporal Sequences, in Proceedings of the International Joint Conference on Neural Networks (IJCNN'04), Budapest, July 25-29, 2004.

Datamining in Grid Environment

M. Ciglarič, M. Pančur, B. Šter, and A. Dobnikar

Faculty of Computer and Information Science, University of Ljubljana, Slovenia
E-mail: {mojca.ciglaric, matjaz.pancur, branko.ster, andrej.dobnikar}@fri.uni-lj.si

Abstract

The paper deals with assessing performance improvements and some implementation issues of two well-known data mining algorithms, Apriori and FP-growth, in Alchemi grid environment. We compare execution times and speed-up of two parallel implementations: pure Apriori and hybrid FP-growth - Apriori version on grid with one to six processors. As expected, the latter shows superior performances. We also discuss the effects of database characteristics on overall performance, and give directions for proper choice of execution parameters and suitable number of executors.

1 Introduction

Since the introduction of association rule mining [1], the frequent itemset mining has received a great deal of attention. Within the last decade, a large number of algorithms have been proposed for mining either all frequent itemsets [2], closed frequent itemsets [3] or only maximal frequent itemsets [4] within large databases consisting of item subsets. An important issue is to perform either of the tasks as fast as possible.

However, there is an alternative approach to performance improvement, namely concurrent implementation of existing algorithms. We use GRID technology to speed up mining procedures and present the results of our experimental setup. Two popular mining algorithms, Apriori and FP-growth, are used on Alchemi platform, which is selected as the implementation of GRID environment.

The paper is organized as follows. In Sec. 2, we present the background on association rule mining and main features of the two chosen mining algorithms. Sec. 3 briefly describes GRID environment, Alchemi platform and our experimental setup, together with some implementation issues. The databases and experiments are described, followed by the results of mining procedures executed in GRID environment in Sec. 4. The performance of the two algorithms is evaluated in GRID environment with different number of processors, for the two databases and different support values. We conclude the paper with discussion of results and directions for future work.

2 Background Theory and Related Work

Association rule mining deals with large databases, where it is impossible for human to extract relevant or existing features. The use of automated procedures and computer technology is therefore imperative.

2.1 Frequent Itemsets

Let us assume a database D with large number of patterns or itemsets, taken from a set of items or attributes I. A row in D is called transaction $T = (tid, I)$, where tid is the transaction identifier and I is an itemset, $I \subseteq \mathrm{I}$. A transaction $T = (tid, I)$ is said to support an itemset X, if $X \subseteq I$. Each transaction has a unique identifier. The *cover* of an itemset X in D consists of the set of transaction identifiers of transactions in D that support X:

$$cover(X, D) := \{tid \mid (tid, I) \in D, X \subseteq I\}$$

The *support* of an itemset X in database D is the number of transactions in the cover of X and D:

$$support(X, D) := |cover(X, D)|.$$

An itemset is called *frequent* in D if its *support* in D exceeds a given minimal support threshold σ. The goal is usually to find *all* frequent itemsets (AFI) in a database, given a minimal support threshold. As the search space of all subsets of I is huge, there are two well known sub-problems that are looking only for *closed* frequent itemsets (CFI) or *maximal* frequent itemsets (MFI).

A frequent itemset I is called *closed*, if it has no frequent superset with the same *support*, i.e., if

$$I = \bigcap_{(tid, J) \in cover(I)} J$$

A frequent itemset is called *maximal* if it has no frequent superset. It is easy to see that MFI $\subseteq$ FCI $\subseteq$ AFI, which influences the speed of corresponding mining algorithms.

Among a large number of sequential algorithms for discovering frequent itemsets, we shall briefly describe two of the most popular: Apriori and FP-growth [6, 7], followed by a short comment on parallel versions.

2.2 Apriori algorithm

The Apriori algorithm uses the following theorem to reduce the search space: if an itemset is large, then all of its subsets are large as well. This means that it is possible to generate the potentially large (i+1) itemset using large i itemset. All subsets of candidate i+1 itemset must be large itemset. It is therefore possible to find all large itemsets using database scan repeatedly. During the i-th scan, Apriori counts the occurrence of the i-itemset and at the end of the pass i, it generates the candidates with i+1 items. Its main disadvantage is multiple database scan, however some solutions exist which overcome this issue [6].

2.3 FP-growth algorithm

The FP-growth (Frequent Pattern) is an alternative to Apriori-like algorithms since it doesn't generate time consuming candidate lists, but rather adopts a pattern fragment growth method, based on a compressed prefix-tree data structure. There are basically two steps: 1) FP-tree construction and 2) mining frequent patterns with FP-tree by pattern fragment growth. The first step requires one scan of database to identify the set of frequent items followed by the supplement of the list of ordered frequent items to the database transactions, while the second scan is sufficient for the FP-tree creation. The second step is a recursive procedure that analyses FP-tree in a bottom-up manner for each frequent item and constructs its conditional FP-trees that call the same procedure recursively. As it performs mining on a compressed (tree) data structure, it is very fast.

2.4 Parallel algorithms

Among Apriori-like algorithms, three parallel versions are popular: Count Distribution, Data Distribution and Intelligent Data Distribution algorithm [8]. Since Alchemi does not support direct communication among threads or executors, we could not take advantage of these algorithms. So we adopted an alternative approach, described in Section 3.2.

3 GRID Environment and Alchemi Platform

Grid environment represents a kind of distributed platform which enables dynamic sharing of processing and storage capacities on (possibly) geographically dispersed end-computers, taking into account their availability, capabilities and resources. Grid can be considered either as a single giant virtual computer or as an interconnected cooperative environment. The term grid comes from analogy with power grid, which acts as a constant and reliable energy source and its users do not need to care where electricity comes from, how it is produced and distributed etc. Similarly, grid users do not have to know how many and what kind of computers execute the submitted task, they only want to get a correct solution within a reasonable time.

3.1 Platform of choice

Our platform of choice Alchemi [9] is an open source project from Melbourne University and part of the Gridbus project. It is developed in C# for Microsoft .NET framework. It offers an OO model with a multithreading paradigm. The smallest unit of parallel processing is a grid thread. Grid application consists of several grid threads that can be executed in parallel. Although tightly integrated with .NET platform, Alchemi can also run on other platforms using web services, or even use alternative open source environment (Mono, dotGNU) on platforms other than MS Windows. Alchemi grid consists of three components: manager, executor and owner, as shown in Fig. 1 [10].

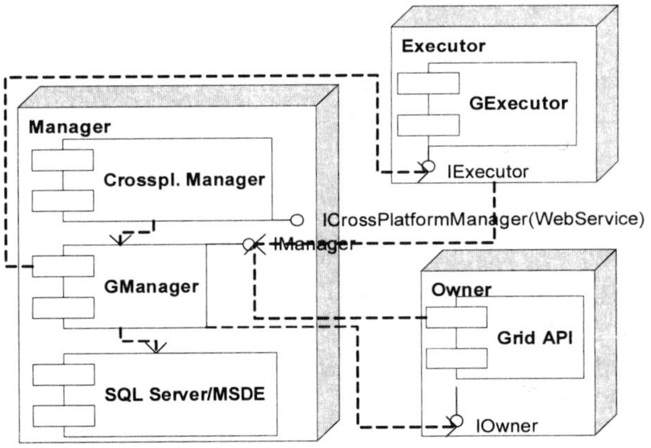

Fig. 1. Alchemi grid components.

The manager manages grid application execution and thread execution. Executors sign up with manager. Owner submits grid threads to manager who adds them into the thread pool and from there, schedules threads for execution on available executors. Executors afterwards return the finished threads or results to the manager, who passes them on to the owner. Owner may afterwards perform some final processing if necessary.

Although simple, Alchemi represents an appropriate environment for experimental evaluation of intraorganisational grid benefits. Its major drawbacks are absence of security mechanisms and no inter-thread communication. In our case, security is not an issue (yet), since our experimental grid is situated within laboratory environment, while the absence of inter-thread communication limits the choice of algorithms appropriate for implementation.

3.2 Problem and experimental setup

Our experimental grid consists of six personal computers with Pentium IV 2 GHz processors and 500 MB memory, connected with 100 Mbps links over Ethernet switch and running Windows 2000 Advanced Server operating system. The experiments are always performed on the most recent version of Alchemi software, which is at the moment of writing 0.8. All of them function as executors, while one of them additionally acts as a grid manager and application owner, as shown in Fig. 2.

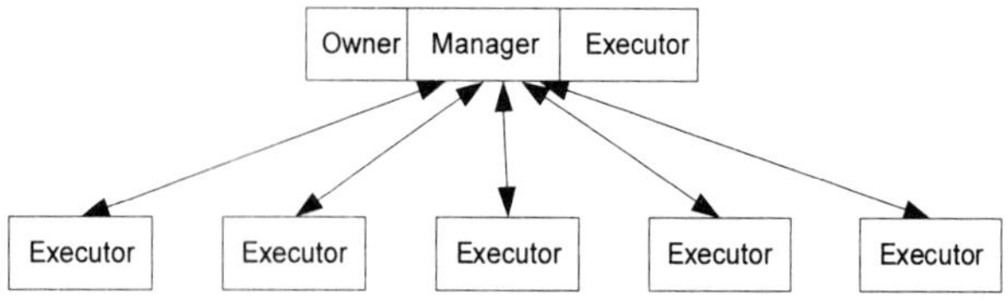

Fig. 2. Experimental grid architecture.

We study the problem of mining for association rules in large databases. More precisely, we investigate how much we can shorten the time needed for completion of known algorithms like Apriori when using a grid environment.

We have rewritten a basic Apriori implementation to run on Alchemi grid. When the owner specifies the desired number of total threads, the dataset is partitioned into a corresponding number of horizontal partitions. Each partition becomes a parameter which is sent out to the executor together with a thread to be executed. The executors search for frequent itemsets in the received partitions with additional parameter - support s_1. The result itemsets are returned to the owner, who accomplishes the aggregation by searching for frequent itemsets in the whole database with (usually higher) support s_2 and taking as candidate itemsets only those returned from the executors.

We also experimented with FP-growth algorithm, which is supposed to return the same results as Apriori in shorter time. We took an existing FP-growth implementation [11] in C++ and wrapped it with .NET managed code to suit the distributed nature of our grid environment. An alternative model to grid thread is used - a basic unit of distributed processing is called grid job. The algorithm ran on the same database partitions as Apriori in the former experiment. Although a job-oriented model is used as opposed to the thread-oriented one in our Apriori implementation, the amounts of overhead in both cases are comparable. Since FP-growth cannot make use of candidate itemsets, we employed Apriori in the aggregation (reduction) phase just like in the first experiment. The cost of aggregation is achieved centrally and is therefore constant. We rather focus on the total amount of processing time on executors when employing different number of executors. With only one executor, the execution is centralized, while with more executors, the level of parallelism is increased.

Our hypothesis is that the more executors we occupy, the shorter will be the overall execution time. However, depending on the database size and characteristics, after a certain point adding new executors would probably not bring further reduction due to higher amount of overhead.

4 Results and discussion

To show the benefits of running datamining applications in grid environment, we performed the two experiments described in Section 3.2, on two different databases, Sava [12] and UCI-Adult [5], each of them six times, with one to six executors. Since the databases have different characteristics and sizes, the results obtained also slightly differ.

Figs. 3 and 4 show the execution times versus number of executors - processors, obtained with different values of relative partition support parameter s_1, which was always taken as the third of the aggregation support, $s_1 = s_2/3$. (when the support is lower, the algorithm has to check more candidates). It is obvious that with higher number of executors we get shorter execution times, at least up to six executors on UCI-Adult database, and up to five on Sava database.

On both databases, FP-growth runs considerably faster than Apriori. However, there are a few points worth mentioning. If number of computers is further increased, the overhead of communication becomes excessive, thereby limiting the benefits of parallelism. On the other hand, our databases are relatively small. With the largest known databases used for datamining, we could still benefit from as much as hundred and more executors. Next, when database partitions consist of randomly chosen transactions, execution times on different partitions are rather similar.

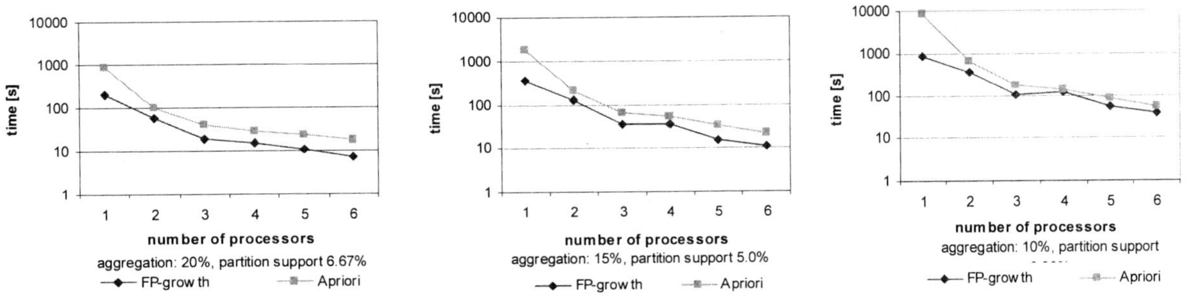

Fig. 3: The results on UCI-Adult database.

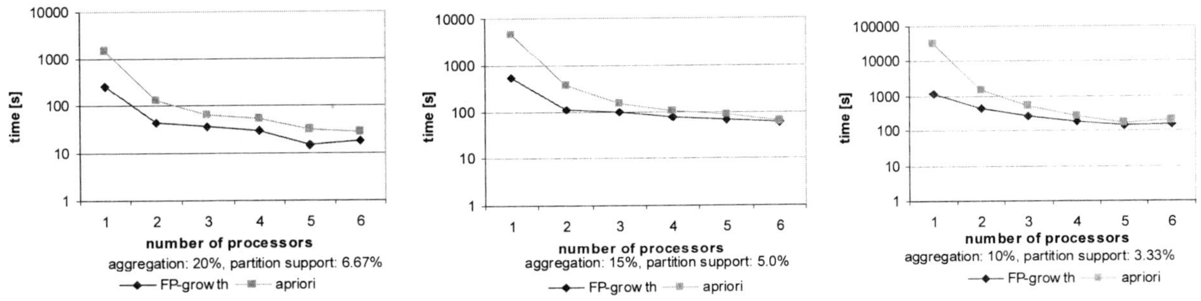

Fig 4: The results on Sava database.

When this is not the case, e.g. when partitions on different machines consist of differently distributed data, individual execution times may be drastically different - even larger than on the entire dataset (due to smaller absolute minimal support values).

5 Conclusions

The paper deals with datamining in grid technology. We show its benefits when using Apriori and FP-growth algorithms with Alchemi. The best results are obtained with hybrid combination, where FP-growth is used in parallel on partition databases and Apriori for the final aggregation. In our future work we shall try to implement the different parallel versions of mining algorithms on the latest grid platforms, as well as to explore more thoroughly issues of effectiveness and robustness.

Acknowledgements

The described research is partly funded by Slovenian Ministry of Education, Science and Sport and Ministry of Information Society, by grants V2-0886, L2-6460 and L2-6143.

References

[1] R. Agrawal, T. Imielinski, A. Swami, Mining association rules between sets of items in large databases. Proc. 1993 ACM SIGMOD Int. Conf. on Management of Data, 207-216, ACM Press, 1993.

[2] M.J. Zaki et al. New algorithms for fast discovery of association rules. 3. Int. Conf. on Knowledge Discovery and Data Mining, 1997

[3] N. Pasquir, Y. Bastide, R. Taouil, L. Lakhal. Discovering frequent closed itemsets for association rules. 7. Int. Conf. on Database Theory, Jan., 1999

[4] K. Gouda, J. Zaki. Efficiently mining maximal frequent itemsets. 1. IEEE Int. Conf. on Data Mining, Nov. 2001

[5] UCI Machine Learning Database Repository, http://www.ics.uci.edu/~mlearn/MLRepository.html

[6] M. H. Dunham, Data mining, Introductory and Advanced Topics, Prentice Hall, 2003

[7] J. Han, J. Pei, Y. Yin. Mining Frequent Patterns without Candidate Generation. In ACM SIGMOD Int. Conf. on Management of Data, May, 2000

[8] E. Hong Ham, G. Karypis, V. Kumar. Scalable Parallel Data Mining for Association Rules. In IEEE Tr. on Knowledge and Data Engineering, 1999.

[9] Alchemi, http://www.alchemi.net, 2004.

[10] A. Luther, R. Buyya, R. Ranjan, S. Venugopal, Alchemi: A .NET-based Grid Computing Framework and its Integration Into Global Grids, TR, GRIDS-TR-2003-8, University of Melbourne, Australia, 2003.

[11] B. Goethals. FP-growth implementation. http://www.cs.helsinki.fi/u/goethals/software/index.html

[12] M. Trebar, U. Lotric. Predictive data mining on rubber compound database. ICANNGA 2005, Coimbra, Portugal.

Parallel Placement Procedure based on Distributed Genetic Algorithms

Masaya Yoshikawa[1], Takeshi Fujino[1], Hidekazu Terai[1]

[1]Department of VLSI System Design, Ritsumeikan University, Japan
E-mail: yosikawa@se.ritsumei.ac.jp

Abstract

This paper discusses a novel performance driven placement technique based on distributed Genetic Algorithms, and focuses particularly on the following points:(1) The algorithm has two-level hierarchical structure consisting of outline placement and detail placement. (2) For selection control, which is one of the genetic operations, new multi-objective functions are introduced. (3) In order to reduce the computation time, a parallel processing is introduced. Results show improvement of 22.5% for worst path delay, 11.7% for power consumption, 15.9% for wire congestion and 10.7% for chip area.

1 Introduction

Deep-Sub-Micron (DSM) technology of 0.18 micron and below enable the integration of logical circuits having more than 10 million gates. In such a DSM technology, it is important to optimize timing constraint, power consumption, wire congestion and chip area at initial phase of layout design. In this paper, we proposed a novel performance driven placement procedure based on Genetic Algorithm (GA)[1]. GA, which is population-based algorithm, is effective to such a multi-objective optimization problems. An essential difficulty exits in GA, However, with regard to large amount of computation time. In order to overcome the difficulty of GA, a parallel processing technique, which divides the population into several sub-population as shown in Fig.1, is introduced. Studies on analog module placement[2], floorplanning [3],[4] and routing[5],[6] have been reported as being applications of GA to the VLSI layout problem. However, no studies have ever seen the effect of applying GA in consideration of timing constraint, power consumption, wire congestion and chip area.

2 Base Algorithm

The placement model we assume in this paper is for CMOS standard cells. The cells have the same height and various widths, and are placed as arrays in several lines. The algorithm proposed herein has two-level hierarchical structure consisting of outline placement and detail placement. GA is applied the basic algorithm of each placement process. In outline placement, the placement area is divided into four partitions, and cells are allocated to the partitioned areas. This process repeated sequentially in order to determine cell positions in stages as shown in Fig.2 (1). In detail placement, two-dimensional areas of the LSI, as shown in Fig.2 (2), are searched to determine optimal cell positions in a partitioned area. The genetic coding in outline placement is as explained below. An example of an intermediate stage during outline placement is shown in Fig.3 (1), where eight cells are partitioned into four areas, denoted A through D. The result represents an individual (that is; a solution). The gene locus provides cell IDs and assigns area-names to genes. As Fig.3 (1) shows, cells 2 and 8 are assigned in Area-A, and cells 5 and 6 are assigned in Area-B. Thus, an individual indicates the status of a partition. Detail placement is executed after outline placement. In detail placement, the gene locus corresponds cell IDs and each gene gives cell location (grid number). An example of coding in detail placement and individual are presented in Fig.3 (2). In this example, five cells are placed in a 4 *2 placement mesh.

3 Selection Operators

For GAs, it's important to set suitable evaluation parameters for controlling the selection of individuals. In order to realize high quality placement, the proposed algorithm uses a new multi-objective function that consists of three evaluations besides chip area.

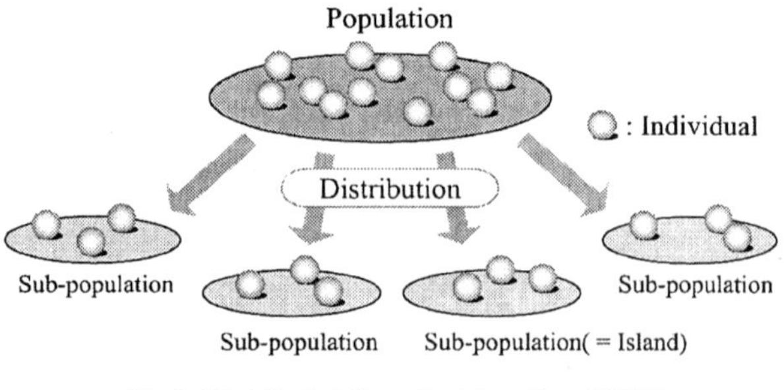

Fig.1. Distributed Genetic Algorithm (DGA)

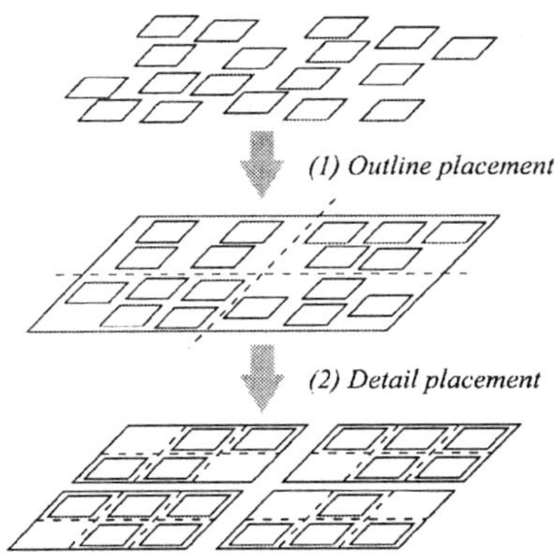

Fig.2. Hierarchical placement

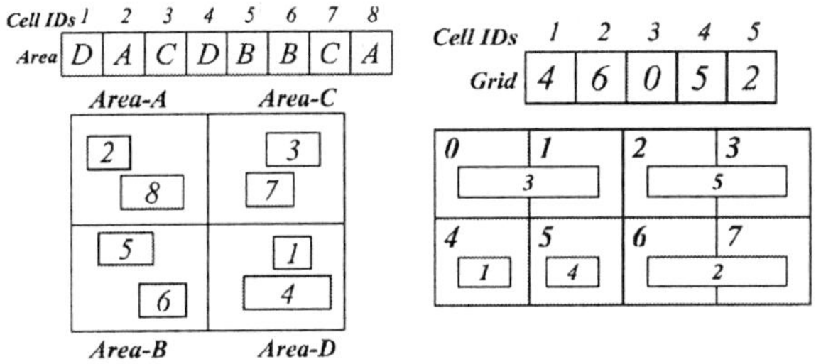

(1) Outline placement (2) Detail placement

Fig.3. Coding

3.1 Timing Constraint

We assume that the interconnections are done using two layers, the first metal layer (M1) and the second metal layer (M2). Wire length of a net is calculated from the circumscribed rectangle lengths of the net as shown in Fig.4. In this paper, The wiring delay of del_{ij} for a net n from the source i to the load j is as follows.

$$del_{ij} = (c_1 L_v + c_2 L_h + \sum_j C_j) \times (R_0 + r_1 l_v + r_2 l_h) \quad \text{----} \ (1)$$

Where, $L_{v(h)}$ is the total wire length of the circumscribed rectangle for the net n routed on M1(2), $l_{v(h)}$ is the length of the circumscribed rectangle from the source to the load j routed on M1(2), $c_{1(2)}$ is the capacitance of M1(2) per unit length, $r_{1(2)}$ is the resistance of M1(2) per unit length, $\sum_j C_j$ is the sum of the load capacitance for the net n, and R_0 is the equivalent output resistance of the source for the net n. We define the timing constraint. The proposed algorithm doesn't calculate the delay time of all paths on every generation. Concretely, it is only calculated the delay of all paths on every T generation and we build up *calculation path list*. The *calculation path list* consists of 10 worst paths in all paths. The proposed algorithm only calculates the delay of paths in *calculation path list* from T+1 generation to 2T generation as shown Fig.5. Thus, we save run-time keeping quality of solutions.

3.2 Power Consumption

The dominant source of power consumption in a digital CMOS circuit synchronized by a clock is the switching of signals. It can occupy about 70% of total power consumption. Therefore, it is effective to reduce the power consumption of switching of signals. The power consumption of switching of signals in CMOS is given by

$$P = a \times C \times f \times V^2 \quad \text{----} \ (2)$$

Where "C" is the load capacitance, "V" is the supply voltage, "f" is the clock frequency and "a" is the switching ratio. The load capacitance "C" consists of input gate capacitance for the next stage, output gate capacitance and routing capacitance. So, we reduce the capacitance (routing capacitance) of cell with the high ratio of switching.

3.3 Wire Congetion

Wire congestion may cause un-connect nets or the chip size may grow if the wiring capacity is increased. In order to avoid wire congestion, the new function called cut-line deviation is introduced. Cut-line deviation refers to the balance of the maximum number of cross-border cut-line in the process of iterative partitioning minus the minimum number of cut-lines. The example shown in Fig.6, the cut-line deviation of (1) is 3 and that of (2) is 0. Preferably, this value should be small. Thus, the example Fig.6 (2) gives a better result than (1).

4 Crossover

This algorithm adopts general uniform crossover for outline placement. While according to the coding in outline placement, a sequence of genes is not so essential. One-point or two-point crossover operations tend to save the genetic sequence, whereas uniform crossover operations are expected to hold a gene and gene locus pair across generations and thus maintain population versatility.

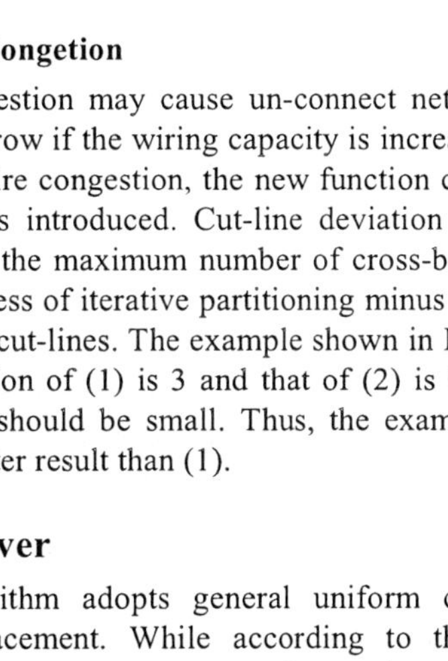

Fig.4. Virtual Wire Length

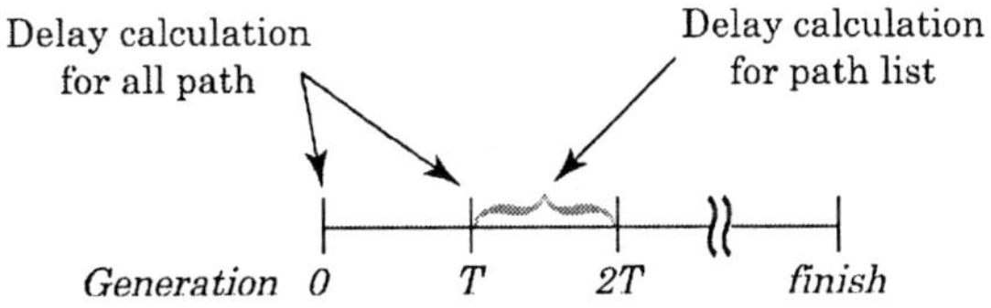

Fig.5. Timing of calculation for path delay

Therefore, this algorithm adopts general uniform crossover for outline placement. However, the crossover cannot adjust the number of cells contained in an area for newly generated individuals. This deviation of cell count does not produce a lethal gene but remarkably reduces the processing speed. In order to prevent this deviation, the heuristic adjustment technique shown in Fig.7 is used for new individuals generated by crossover. If the deviation of cell count exceeds the threshold, cells migrate from the most populous area to the least populous area. The number of migrating cells is calculated in order to minimize the virtual wiring length. Uniform crossover was used in outline placement, however this generates many lethal genes if used in detail placement, because crossover cause several cells coexist in a single grid or overlap each other. In order to supper lethal genes, we used the crossover that modified PMX (Partially Mapped Crossover)[1] for corresponding to the proposed coding.

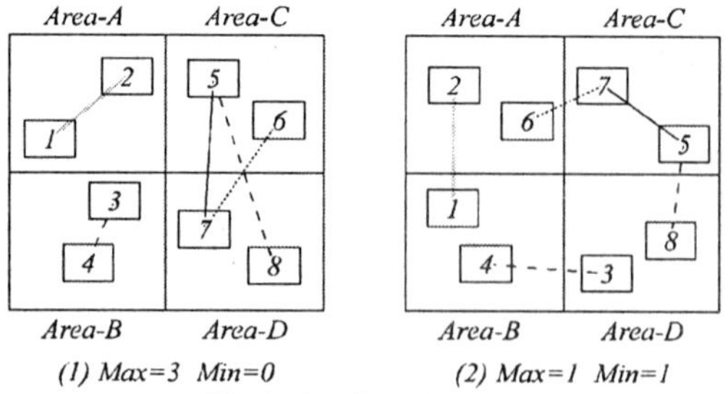

Fig.6. Cut-line deviation

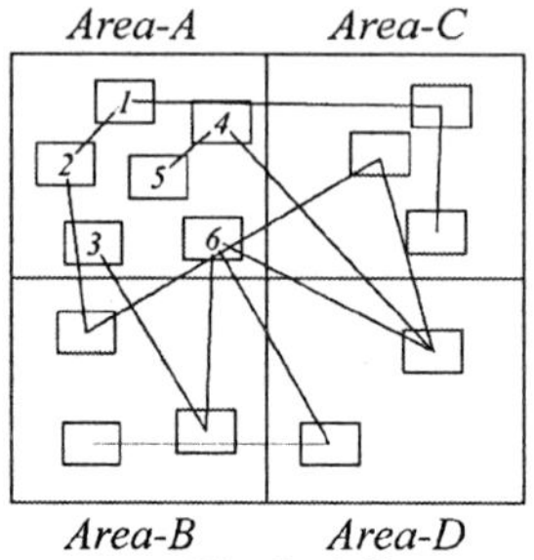

Cell	Gain
Cell-1	-1
Cell-2	-1
Cell-3	0
Cell-4	0
Cell-5	0
Cell-6	2

Fig.7. Adjustment technique

5 Implementation on Parallel Computer

5.1 Parallel Processing of the proposed Algorithm

When parallel processing in placement is considered, it can be considered that the partial areas of LSI is assigned to Processing Unit (PU). The proposed algorithm takes processing time most to the 1st division processing, because gene length is the longest. In early division processing in which many processing time is required, a sufficient number (at least the number of PUs) of partial areas are not generated yet. Therefore, the parallel processing system, which assigns the partial areas to PUs, is not effective. In order to realize efficient parallel processing, we adopt Distributed GA.

5.2 Hierarchical structure ring topology

The new ring topology of two steps of hierarchical structures was introduced into reference for biological niche as shown in Fig.8. The hierarchical ring topology (H ring topology) consists of lower layer, which means the neighboring sub-population, and upper layer, which means the distant sub-population. In lower layer, the migration interval is made high. By contraries, in upper layer, the migration interval is made low.

6 Result and Discussion

6.1 Experimental Set-up

In order to evaluate the effectiveness of the techniques proposed in this paper (hereafter, called GAP), we compared with commercial EDA tool. The experimental data is 0.35 micron COMS technology.

6.2 Experimental Result I

The result of the experiment is shown in Table.1. Routing capacitance can be reduced if the value of "Length of the nets with high SW" rate in Table 1 is small. That is, power consumption is reducible. The proposed algorithm improved power consumption, timing constraint, wire congestion and chip area in comparison with EDA. However, our runtime spends more than one thousand times in comparison with EDA.

6.3 Tuning of The Parameter of DGA

Migration of individuals is controlled by several parameters such as: (1) a migration interval that affects the frequency of migration, (2) a migration rate that controls how many individuals migrate. In this work, the preliminary experiments determined the optimal migration interval and rate. In case of the migration interval, the preliminary experiment was conducted about the case of the migrate interval of 20, 30, 40, 50 or 60 generations.

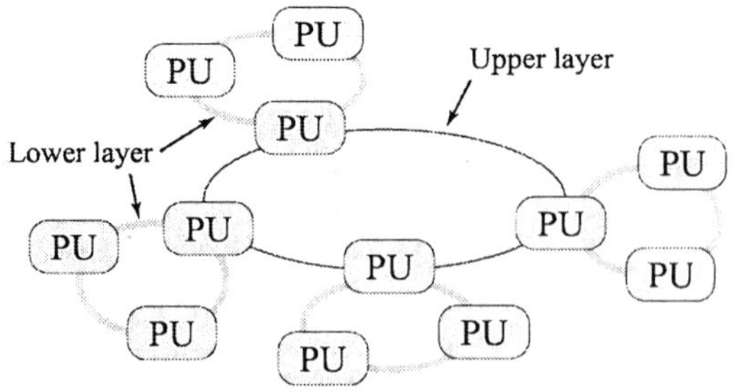

Fig.8 Hierarchical Ring Topology

The experiment results are shown in Fig. 9. The migrate interval was the best result in the case of 50 generations irrespective of the number of processors from Fig. 9. In case of the migration rate, the preliminary experiment was also conducted about the case of the migrate rate of 5%, 10%, 20%, 30% or 40%. The experiment results are shown in Fig. 10. The migrate rate was the best result in the case of 10% irrespective of the number of processors from Fig. 10. Thus, in this work, interval was made into every 50 generations and migrate rate was made into 10%.

6.4 Experimental Result II

The results of H ring topology and ring topology, which use 12 processors, and that of a single processor are shown in Fig.11. Fig. 11 shows the relative value which set the value of a single processor to 1. In the case of 12 processors, the ring and the H ring realized 11 times as many improvement in speed as this compared with the single processor. Moreover, parallel processing has been improved as the quality of placement. When the H ring topology was compared with ring topology, in all evaluation criteria, H ring topology was made a good result.

7 Conclusion

In this paper, a novel parallel placement procedure was presented. The experiment verified that the proposed multi-objective functions are effective. It also became clear that Distributed Genetic Algorithm could improve placement performance reducing processing time.

References

[1] D.E.Goldberg (1989) Genetic Algorithms in search, Optimization and Machine Learning, Addison-Wesley.

[2] L.Zhang, et al. (2004) Analog Module Placement Realizing Symmetry Constraints Based on a Radiation Decoder , Proc. IEEE MWSCAS., pp. I-481-484.

[3] S.Nakaya, et al.(2000) An adaptive genetic algorithm for VLSI floor planning based on sequence-pair, Proc. IEEE ISCAS, Vol.3, pp.65-68.

[4] K.Hatta, et al.(1999) Solving the rectangular packing problem by an adaptive GA based on Sequence-pair, Proc ASP-DAC, pp.181-184.

[5] S,Wakabayashi, et al.(2001). A genetic algorithm for generating a Steiner tree with wire sizing and buffer insertion, Proc.GECCO, pp. 1431-1438

[6] A. Kanasugi, et al. (2001) A Greedy Router based on Maze Router and Genetic Algorithm, Proc. Int. Conf. on Electronics Packaging, pp.427-430.

Table.1 Placement Result

	GAP	EDA	Improvement
Worst path delay (Timing constraint)	34.59ns	44.65ns	22.5%
Length of the nets with high SW rate	33317μm	37752μm	11.7%
Wire congestion	1571grid	1869grid	15.9%
Total wire length (Chip area)	138085μm	154714μm	10.7%
Run time	55680sec	58sec	—

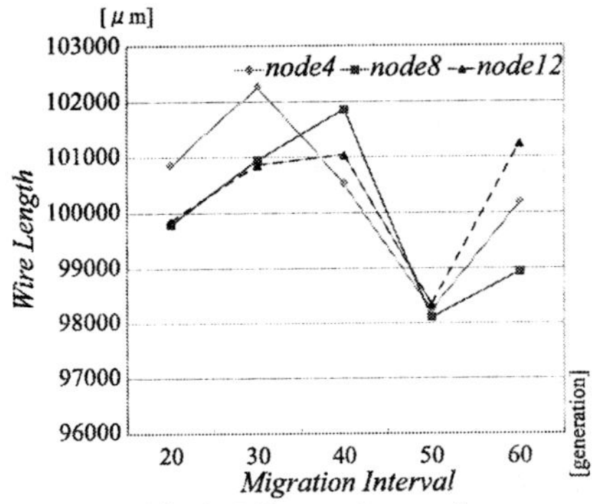

Fig.9. Migrate interval

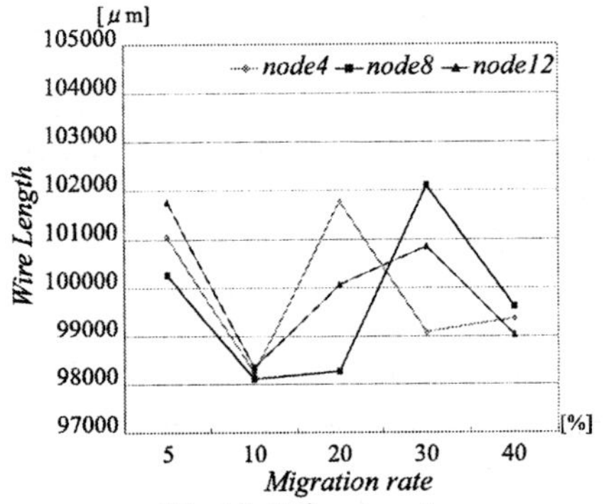

Fig.10. Migrate rate

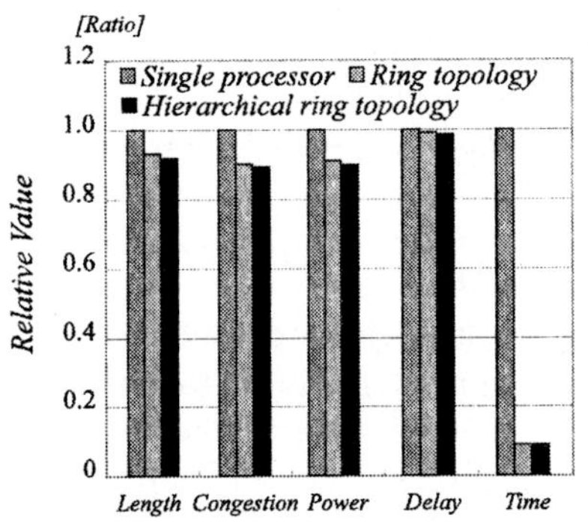

Fig.11. Result of Hierarchical ring topology

Massive parallelization of the compact genetic algorithm

Fernando G. Lobo, Cláudio F. Lima, Hugo Mártires
DEEI-FCT, Universidade do Algarve
Campus de Gambelas, 8000-062 Faro, Portugal
E-mail: {flobo,clima}@ualg.pt, hmartires@myrealbox.com

Abstract

This paper presents an architecture which is suitable for a massive parallelization of the compact genetic algorithm. The resulting scheme has three major advantages. First, it has low synchronization costs. Second, it is fault tolerant, and third, it is scalable.

The paper argues that the benefits that can be obtained with the proposed approach is potentially higher than those obtained with traditional parallel genetic algorithms.

1 Introduction

One of the efficiency enhancement techniques that has been investigated in the field of evolutionary computation, both in theory and in practice, is the topic of parallelization [1]. With a traditional parallel genetic algorithm (GA) implementation, population members need to be sent over a computer network, and that imposes a limit on how fast they can be. This paper addresses the parallelization of the compact genetic algorithm (cGA) [2], and take advantage of its compact representation of the population do develop a scheme which significantly reduces the communication overhead.

The paper is organized as follows. The next section presents background material on parallel GAs and section 3 reviews the cGA. Section 4 shows the motivation for parallelizing the cGA and presents an architecture that allows its massive parallelization. In section 5 computer experiments are conducted and its results are discussed. Finally, a number of extensions are outlined, and we finish with a brief summary and the main conclusions of this work.

2 Parallel GAs

An important efficiency question that people are faced with in problem solving is the following: Given a fixed computational time, what is the best way to allocate computer resources in order to have as good a solution as possible. Under such a challenge, the idea of parallelization stands out naturally as a way of improving the efficiency of the problem solving task. By using multiple computers in parallel, there is an opportunity for delivering better solutions in a shorter period of time.

Several researchers have investigated the topic of parallel GAs and the major design issues are in choices such as using one or more populations, and in the case of using multiple populations, decide when, with whom, and how often do individuals communicate with other individuals of other populations.

Although implementing parallel genetic algorithms is relatively simple, the answers to the questions raised above are not straightforward and traditionally have only been answered by means of empirical experimentation. One exception to that has been the work of Cantú-Paz [1] who has built theoretical models that lead to rational decisions for setting the different parameters involved in parallelizing GAs. There are two major ways of implementing parallel GAs: (1) using a single population, and (2) using multiple populations.

In single population parallel GAs, also called Master-Slave parallel GAs, one computer (the master) executes the GA operations and distributes individuals to be evaluated by other computers (the slaves). After evaluating the individuals, the slaves return the results back to the master. There can be significant benefits with such a scheme because the slaves can work in parallel, independently of one another. On the other hand, there is an extra overhead in communication costs that must be paid in order to communicate individuals and fitness values back and forth.

In multiple population parallel GAs, what would be a whole population in a regular non-parallel GA, becomes several smaller populations (usually called demes), each of which is located in a different computer. Each computer executes a regular GA and occasionally, individuals may be exchanged with individuals from other populations. Multiple population parallel GAs are much harder to design because there are more degrees of freedom to explore. Specifically, four main things need to be chosen: (1) the size of each population, (2) the topology of the connection between the populations, (3) the number of individuals that are exchanged, and (4) how often do the individuals exchange.

Cantú-Paz investigated both approaches and concluded that for the case of the Master-Slave architecture, the benefits of parallelization occur mainly on problems with long function evaluation times because it needs constant communication. Multiple population parallel GAs have less communication costs but do not avoid completely the communication scalability problem. In other words, in either approach, communication costs impose a limit on how fast parallel GAs can be.

The next section gives an overview of the cGA, and after that, its parallelization is discussed.

3 The Compact Genetic Algorithm

Harik et al. [2] noticed that it was possible to mimic the behavior of a simple GA without storing the population explicitly. Such observation came from the fact that during the course of a regular GA run, alleles compete with each other at every gene position. At the beginning, scanning the population column-wise, we should expect to observe that roughly 50% of the alleles have value 0 and 50% of the alleles have value 1. As the search progresses, for each column, either the zeros take over the ones, or vice-versa. Harik et al. built an algorithm that explicitly simulates the random walk that takes place on the allele frequency makeup for every gene position. The resulting algorithm, the cGA, was shown to be operationally equivalent to a simple GA that does not assume any linkage between genes.

Under the cGA, the population is represented by a probability vector. The elements of the vector are the relative frequency counts of the number of 1's for the different gene positions. The cGA manipulates the population in an indirect way through an update step in each allele frequency of the probability vector. Notice that each allele frequency value is a member of a finite set of $N + 1$ possible values, and can be stored with $\log_2(N + 1)$ bits. (N denotes the population size of a regular GA). Instead, a regular GA would require N bits to represent each bit position. Further details about the algorithm can be found in the original source [2].

4 Massive parallelization of the compact GA

The main motivation for parallelizing the cGA comes from the observation that the probability vector is a compact representation of the population, and it is possible to communicate the vector rather than individuals themselves. Communication costs can be reduced this way because the probability vector needs significant less storage than the whole population. This observation has first been made by Harik [3] when the cGA was developed.

Since communication costs can be drastically reduced, it makes sense to clone the probability vector to several computers, and each computer can work independently on solving a problem by running a separate cGA. Then, the different probability vectors would have to be consolidated (or mixed) once in a while.

We have developed an asynchronous parallelization scheme which consists of a manager processor, and an arbitrary number of worker processors (see Figure 1). Initially, the manager starts with a probability vector with 0.5 in all positions, just like in a regular cGA. After that, it sends the vector to all workers who are willing to contribute with CPU time.

Each worker processor runs a cGA on its own based on a local copy of the probability vector. Workers do their job independently and only interrupt the manager once in a while, after a predefined number of m fitness function evaluations have elapsed.

During the interruption period, a worker sends the accumulated results of the last m function evaluations as a vector of probability fluxes with respect to the original probability vector. Subsequently, the manager adds the probability fluxes (values are truncated so that they never exceed 1.0 and never go below 0.0) to its own probability vector, and resends the resulting vector back to the worker. Notice that the manager's probability vector not only incorporates the results of the m function evaluations performed by that particular worker, but it also incorporates the results of the evaluations conducted by the other workers. That is, while a particular worker is working, other workers might be updating the manager's probability vector. Thus, at a given point in time, workers are working with a slightly outdated probability vector. Although this might seem a disadvantage at first sight, the error that is committed by working with a slightly outdated probability vector is likely to be negligible for the overall search because an iteration of the cGA represents only a small step in the action of the GA (this is especially true for large population sizes).

One could think of different ways of parallelizing the cGA but the scheme that we are proposing is particularly attractive because once the manager starts, there can be an arbitrary number of workers, each of which can start and finish at any given point in time making the whole system fault tolerant. When a worker starts, it receives a copy of the manager's probability vector, which already contains the accumulated results of the other cGA workers. On the other hand, when a worker quits, we simply loose a maximum of m function evaluations, which is not a big problem.

The proposed parallelization scheme has several advantages: (1) it has low synchronization costs, (2) it is fault tolerant, and (3) it is scalable.

All the communication that takes place consists of short transactions. Workers do their job independently and only interrupt the manager once in a while. During the interruption period, the manager communicates

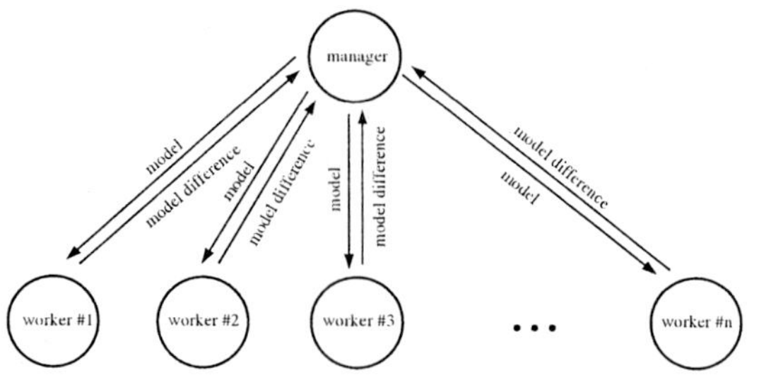

Fig. 1. Manager-worker architecture.

with a single worker, and the other workers can continue working non-stop.

The architecture is fault tolerant because workers can go up or down at any given point in time. This makes it suitable for massive parallelization using the Internet. It is scalable because potentially there is no limit on the number of workers.

5 Computer simulations

This section presents computer simulations that were done to validate the proposed approach. In order to simplify both the implementation and the interpretation of the results, we decided to do a serial implementation of the parallel cGA architecture. The serial implementation simulates that there are a number of P worker processors and 1 manager processor. The P worker processors start running at the same time and they all execute at the same speed. In addition, it is assumed that the communication cost associated with a manager-worker transaction takes a constant time which is proportional to the probability vector's size. Such a scheme can be implemented by having a collection of P regular compact GAs, each one with its own probability vector, and iterating through all of them, doing a small step of the cGA main loop, one at a time. After a particular cGA worker completes m fitness function evaluations, the worker-manager communication is simulated as described in section 4.

We present experiments on a bounded deceptive function consisting of the concatenation of 10 copies of a 3-bit trap function with deceptive-to-optimal ratio of 0.7 [4]. This same function has been used in the original cGA work. We simulate a selection rate of $s = 8$ and did tests with a population size of $N = 100000$ individuals (each worker processor runs a cGA that simulates a 100000 population size). We chose this population size because we wanted to use a size large enough to solve all the building blocks correctly. We use $s = 8$ following the recommendation given by Harik et al. in the original cGA paper for this type of problem. Finally, we chose this problem as a test function because, even

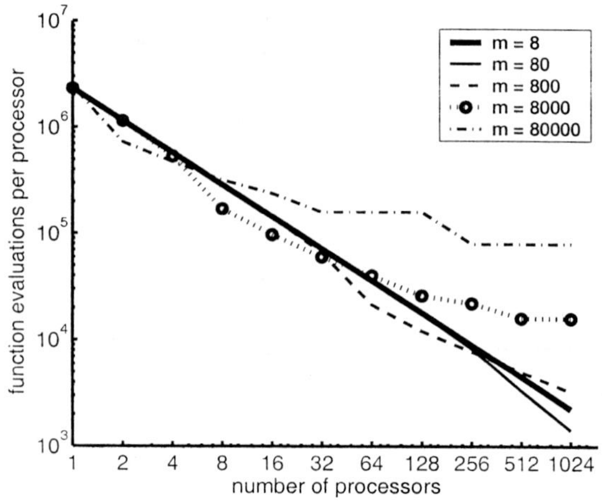

Fig. 2. Function evaluations per processor.

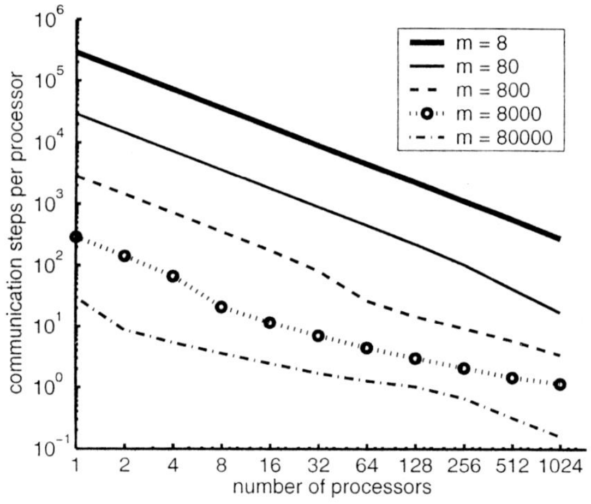

Fig. 3. Communication steps per processor.

though the cGA is a poor algorithm in solving the problem, we wanted to use a function that requires a large population size because those are the situations where the benefits from parallelization are more pronounced.

Having fixed both the population size and the selection rate, we decided to systematically vary the number of worker processors P, as well as the m parameter which has an effect on the rate of communication that occurs between the manager and a worker. We did experiments for P in $\{1, 2, 4, 8, 16, 32, 64, 128, 256, 512, 1024\}$, and for a particular P, we varied the parameter m in $\{8, 80, 800, 8000, 80000\}$. This totalled 55 different configurations, each of which was run 30 independent times.

The m parameter is important because it is the one that affects communication costs. Smaller m values imply an increase in communication costs. On the other hand, for very large m values, performance degrades because the cGA workers start sampling individuals from outdated probability vectors.

Figures 2 and 3 show the results. In terms of fitness function evaluations per processor, we observe a linear speedup for low m values. For instance, for $m = 8$ we observe a straight line on the log-log plot. Using the data directly, we calculated the slope of the line and obtained an approximate value of -0.3. In order to take into account the different logarithm bases, we need to multiply it by $\log_2 10$ (y-axis is $\log_{10}$, x-axis is $\log_2$) yielding a slope of approximately -1. This means that the number of function evaluations per processor decreases linearly with a growing number of processors. That is, whenever we double the number of processors, the average number of fitness function evaluations per processor gets cut by a half. Likewise, in terms of communication costs, as we raise the parameter m, the average number of communication steps between manager and worker decreases in the same proportion as expected. For instance, for $m = 80$, communication costs are reduced 10 times when compared with $m = 8$. Notice that there is a degradation in terms of speedup for the larger m values. For instance, for $m = 8000$ and $m = 80000$ (which is about the same order of the population size), the speedup obtained goes away from the idealized case. This can be explained by the fact that in this case (and especially with a large number of processors), the average number of communication steps per processor approaches zero. That means that a large fraction of processors were actually doing some work but never communicated their results back to the manager because the problem was solved before they had a chance to do so.

6 Extensions

It would be interesting to do a mathematical analysis of the proposed parallel cGA. A number of questions come to mind. For instance, what is the effect of the m parameter? What about the number of workers P? Should m be adjusted automatically as a function of P and N? Our experiments suggest that there is an "optimal" m that depends on the number of cGA workers P, and most likely depends on the population size N as well.

Another extension that could be done is to compare the proposed parallel architecture with those used more often in traditional parallel GAs, either master-slave and multiple deme GAs. Again, our experiments suggest that the parallel cGA is likely to be on top of regular parallel GAs due to lower communication costs.

The model structure of the cGA never changes, every gene is always treated independently. There are other probabilistic model building genetic algorithms (PMBGAs) [5] which are able to learn a more complex structure dynamically as the search progresses. One could think of using some of the ideas presented here for parallelizing these more complex PMBGAs.

Finally, it would be interesting to have a parallel cGA implementation based on the Internet infrastructure, where computers around the world could contribute with some processing power when they are idle. Similar schemes have been done with other projects, one of the most well known is the SETI@home project [6].

7 Summary and conclusions

This paper reviewed the compact GA and presented an architecture that allows its massive parallelization. The motivation for doing so has been discussed and a serial implementation of the parallel architecture was simulated. Computer experiments were done under idealized conditions and we have verified an almost linear speedup with a growing number of processors.

The paper presented a novel way of parallelizing GAs. This was possible due to the different operational mechanisms of the cGA when compared with a more traditional GA. By taking advantage of the compact representation of the population, it becomes possible do distribute its representation to different computers without the associated cost of sending it individual by individual.

Acknowledgements

This work was sponsored by FCT/MCES under grant POSI/SRI/42065/2001.

References

[1] Cantú-Paz, E. (2000) Efficient and accurate parallel genetic algorithms. Kluwer Academic Publishers, Boston, MA.

[2] Harik, G.R., Lobo, F.G., Goldberg, D.E. (1999) The compact genetic algorithm. IEEE Transactions on Evolutionary Computation **3**, pp. 287–297

[3] Harik, G.R. (1997) Personal communication.

[4] Deb, K., Goldberg, D.E. (1993) Analyzing deception in trap functions. In Whitley, L.D., ed.: Foundations of Genetic Algorithms 2, San Mateo, CA, Morgan Kaufmann, pp. 93–108

[5] Pelikan, M., Goldberg, D.E., Lobo, F. (2002) A survey of optimization by building and using probabilistic models. Computational Optimization and Applications **21**, pp. 5–20

[6] Korpela, E., Werthimer, D., Anderson, D., Cobb, J., Lebofsky, M. (2001) SETI@home - massively distributed computing for SETI. Computing in Science and Engineering **3** pp. 79

Parallel implementations of feed-forward neural network using MPI and C# on .NET platform

U. Lotrič, A. Dobnikar[1]
Faculty of Computer and Information Science, University of Ljubljana, Slovenia
E-mail: {uros.lotric, andrej.dobnikar}@fri.uni-lj.si

Abstract

The parallelization of gradient descent training algorithm with momentum and the Levenberg-Marquardt algorithm is implemented using C# and Message Passing Interface (MPI) on .NET platform. The turnaround times of both algorithms are analyzed on cluster of homogeneous computers. It is shown that the optimal number of cluster nodes is a compromise between the decrease of computational time due to parallelization and corresponding increase of time needed for communication.

1 Introduction

A neural network model training itself is computationally expensive. If large datasets are considered, the modelling becomes time consuming. A practical solution to the problem is to parallelize the algorithms. The basic idea of concurrency on which the architecture of feed-forward neural network is built can be efficiently exploited for computational purposes. However, this can only be effectively realized on SIMD (single instruction multiple data) computers [1]. Today, clusters of loosely coupled desktop computers present extremely popular infrastructure for development of parallel algorithms. The processes, running on computers in cluster, communicate with each other through messages. Message Passing Interface (MPI) is standardized and portable implementation of this concept, providing several abstractions that simplify the use of parallel computers with distributed memory [2]. Thus, each process in MPI program in MIMD (multiple instruction multiple data) architecture executes its own code on its own data.

Although, the majority of today clusters run on Linux operating systems, Microsoft Windows operating systems coupled with the .NET platform is also becoming an interesting alternative. The .NET platform, designed to simplify the connection of information, people, systems and devices has two important parts [3]: (i) Common Language Infrastructure (CLI), a layer built upon operating system, allowing development of operating system independent applications and (ii) new programming language C# – simple, safe, object-oriented, network centered high performance language.

In this paper parallel implementation of neural network training algorithms in aspect of linking Message Passing Interface to the C# and .NET framework is considered. In next section parallelization of training algorithms for feed-forward neural networks is presented. Furthermore, important implementation details with emphasis on technology are exposed in section three. In section four the experimental setup and results in terms of computational times are given. The main findings are summarized in the last section.

2 Parallelization of feedforward neural network training

Feedforward neural networks [4] are designed to find nonlinear relations between specified input-output pairs. Suppose Q input-output pairs $\{\mathbf{p}(q), \mathbf{d}(q)\}$, $q = 1, \ldots, Q$ are given. Each n_0-dimensional input sample $\mathbf{p}(q)$ is propagated through the neural network layer by layer in order to obtain n_L dimensional output $\mathbf{y}^L(q)$ in the L-th layer. Output of n-th neuron in the l-th layer is calculated as

$$y_n^l(q) = \varphi^l(s_n^l(q)) \, , \; s_n^l(q) = \sum_{i=0}^{N_{l-1}} \omega_{n,i}^l y_i^{l-1}(q) \, , \qquad (1)$$

where $\varphi^l(s) = 1/(1 + \mathrm{e}^{-s})$ is sigmoid activation function, $y_i^0 = p_i(q)$ are neural network inputs, $y_0^{l-1}(q) = 1$ bias inputs and $\omega_{n,i}^l$ neural network weights.

A training algorithm objective is to find such set of weights that minimize the error function $\mathcal{E} = \frac{1}{2}\sum_{q=1}^{Q}\sum_{n=1}^{N_L} e_n(q)^2$, with $e_n(q) = d_n(q) - y_n^L(q)$ being the difference between calculated and desired output on the n-th neuron in the output layer.

In the following we will consider parallelization of two gradient based training algorithms: the classical gradient descent algorithm with momentum and the second-order derivative based Levenberg-Marquardt algorithm. In both cases the weights are iteratively updated in batch mode, i.e., only after the entire set of input-output pairs has been applied to the network. Having in mind the

[1]The work is sponsored in part by Slovenian Ministry of Education, Science and Sport by grants V2-0886, L2-6460 and L2-6143.

MIMD architecture of the MPI cluster, the parallelization can be implemented in terms of parallel execution of the processes running the same neural network model with different input-output pairs. The parts of algorithm which can be parallelized are divided among slave processes, while other tasks are handled by the master process. By this division the need for communication between the master and slave processes arises.

2.1 Gradient descent algorithm with momentum

The idea of the gradient descent algorithm [4] is to update the network weights in the opposite direction of gradient in which the error function decreases most rapidly,

$$\Delta\omega_{n,i}^{l} \leftarrow -\eta \frac{1}{Q} \sum_{q=1}^{Q} \frac{\partial \mathcal{E}}{\partial \omega_{n,i}^{l}} + \alpha \Delta\omega_{n,i}^{l} \,. \tag{2}$$

The first term with learning parameter η is determined using a technique called backpropagation [4], which involves computations backwards through the neural network. The second term, with parameter α, enables a network to ignore small features in the error function surface and respond more efficiently to the general trends.

The parallelization flowchart is presented in Fig. 1. Within the initialization phase input-output pairs are divided among P slaves processes. During the training phase each slave process needs to obtain weight updates for its group of approximately Q/P input-output pairs.

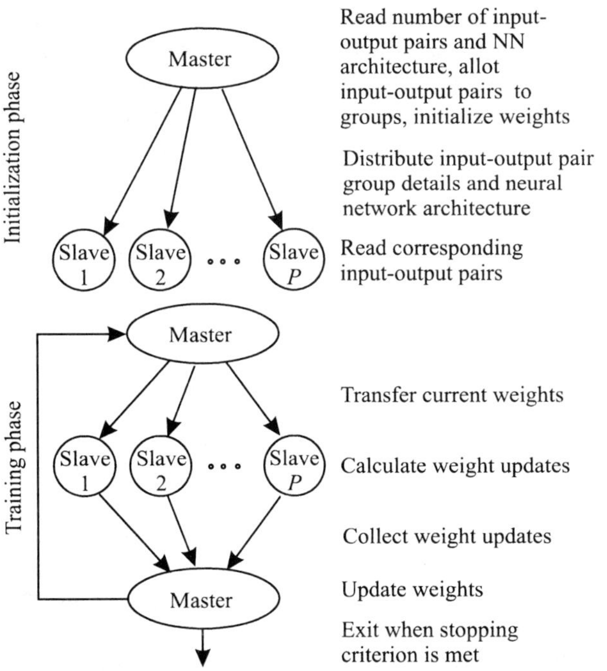

Fig. 1. Parallelization of the gradient descent algorithm with momentum.

As a result, the computational burden of Eq. 2, is reduced by a factor of P. However, the processes need to communicate with each other in order to exchange weights and weights updates, which requires some additional computational resources and reduces overall performance.

2.2 Levenberg-Marquardt algorithm

The Levenberg-Marquardt algorithm was designed to approach second-order training speed without having to compute the Hessian matrix. With error function $\mathcal{E}$ the Hessian matrix can be approximated as $\mathbf{H} \approx \mathbf{J}^{\mathrm{T}}\mathbf{J}$, where $\mathbf{J}$ is the Jacobian matrix that contains first derivatives of the errors $e_m(q)$ with respect to the weights $\omega_{n,i}^{l}$. For the calculation of Jacobian matrix $\mathbf{J}$ the weight vector $\mathbf{\Omega} = (\omega_{1,0}^{1}, \dots \omega_{1,N_0}^{1}, \omega_{2,0}^{1}, \dots, \omega_{N_1,N_0}^{1}, \omega_{1,0}^{2}, \dots \omega_{N_L,N_{L-1}}^{L})^{\mathrm{T}}$ having N_Ω elements and the error vector $\epsilon = (e_1(1), \dots, e_{N_L}(1), e_1(2), \dots, e_{N_L}(Q))^{\mathrm{T}}$ with $N_\epsilon = N_L Q$ elements are introduced. Assuming that $e_m(q)$ is the r-th element of the vector ϵ and $\omega_{n,i}^{l}$ is the c-th element of vector $\mathbf{\Omega}$, the elements of the Jacobian matrix having N_ϵ rows and N_Ω columns are given as $J_{r,c} = \partial\epsilon_r/\partial\Omega_c = \partial e_m(q)/\partial\omega_{n,i}^{l}$. The elements of Jacobian matrix can be computed through a slightly modified backpropagation technique [5], i.e., instead of one backward pass N_L passes are required, one for each neuron in the output layer. When the Jacobian matrix is calculated, the weight update is given as

$$\Delta\mathbf{\Omega} = -[\mathbf{J}^{\mathrm{T}}\mathbf{J} + \mu\mathbf{I}]^{-1}\mathbf{J}^{\mathrm{T}}\epsilon \,. \tag{3}$$

Marquardt has proposed to set learning rate μ to some small value $\mu = 0.001$ and then update it according to the following rules until the stopping criterion is met:

1. Following the Eq. 3 the weight update $\Delta\mathbf{\Omega}$ is obtained and error function $\mathcal{E}_{\mathbf{\Omega}+\Delta\mathbf{\Omega}}$ for the updated weights $\mathbf{\Omega} + \Delta\mathbf{\Omega}$ is calculated.

2. When the error function $\mathcal{E}_{\mathbf{\Omega}+\Delta\mathbf{\Omega}}$ is grater or equal to the error function $\mathcal{E}_{\mathbf{\Omega}}$ of the weights $\Delta\mathbf{\Omega}$, increase the learning rate, $\mu \leftarrow 10\mu$, and return to item 1.

3. If the case $\mathcal{E}_{\mathbf{\Omega}+\Delta\mathbf{\Omega}} < \mathcal{E}_{\mathbf{\Omega}}$, reduce the learning rate, $\mu \leftarrow 0.1\mu$, and continue with item 1.

Parallel implementation of Levenberg-Marquardt training algorithm follows the ideas applied in parallelization of the gradient descent algorithm with momentum. The calculation of the Hessian approximation from Jacobian matrix having N_ϵ rows and N_Ω involves $N_\Omega^2 N_\epsilon$ multiplications, which represents a huge computational and storage burden. Fortunately, it turns out that the Jacobian matrix does not have to be computed and stored as a whole. It can be divided into P matrices $\mathbf{J}_1, \ldots, \mathbf{J}_P$ with approximately N_ϵ/P rows and N_Ω columns and each of these matrices is obtained from some predetermined group of input-output pairs. In this case the approximation to the Hessian matrix is calculated as

$$\mathbf{H}_a \approx \mathbf{J}^{\mathrm{T}}\mathbf{J} = [\mathbf{J}_1^{\mathrm{T}} \ldots \mathbf{J}_P^{\mathrm{T}}] \begin{bmatrix} \mathbf{J}_1 \\ \vdots \\ \mathbf{J}_P \end{bmatrix} = \mathbf{J}_1^{\mathrm{T}}\mathbf{J}_1 + \ldots + \mathbf{J}_P^{\mathrm{T}}\mathbf{J}_P \,. \quad (4)$$

Similarly, the vector ϵ can be divided into P sub-vectors of length approximately N_ϵ/P, $\epsilon^{\mathrm{T}} = (\epsilon_1^{\mathrm{T}}, \ldots, \epsilon_P^{\mathrm{T}})^{\mathrm{T}}$, and the product $\mathbf{J}^T\epsilon$ can be calculated as

$$\mathbf{J}^T\epsilon = \mathbf{J}_1^{\mathrm{T}}\epsilon_1 + \ldots + \mathbf{J}_P^{\mathrm{T}}\epsilon_P \,. \quad (5)$$

While the initialization phase of the algorithm is equal to the gradient descent algorithm with momentum, presented in Fig. 1, the training phase differs substantially. The computational burden of each slave is increased due to more complex derivation of gradients using backpropagation technique and the fact that slaves also calculate their contributions to the Hessian approximation and to the product $\mathbf{J}^T\epsilon$ (Fig. 2). Further, the Levenberg-

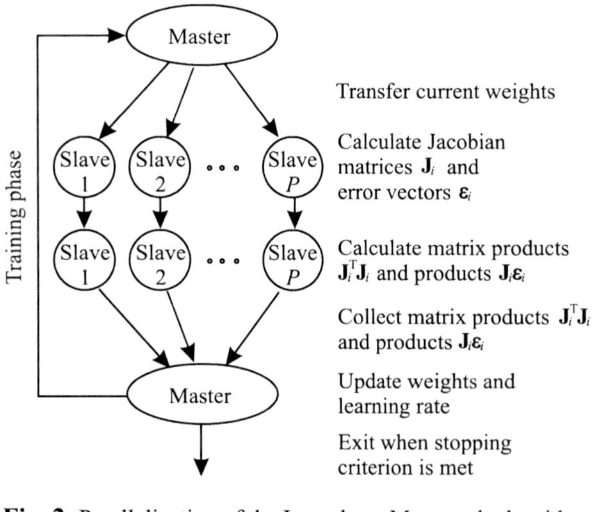

Fig. 2. Parallelization of the Levenberg-Marquardt algorithm.

Marquardt algorithm also needs additional communication resources. While in the gradient descent algorithm master process only collects N_Ω weight updates from each slave process, the master process in Levenberg-Marquardt algorithm collects the N_Ω^2 elements of the Hessian approximation and N_Ω elements of the product $\mathbf{J}^{\mathrm{T}}\epsilon$.

3 Binding C# and MPI on .NET platform

The langauge C# is object oriented language with bounds checking and garbage collection. Thus it helps writing safe code by protecting from dangerous pointer and memory-management errors, such as accessing the element of array out of its bounds or problems connected to creation and deletion of objects. The meta code, produced by C# compiler is then executed by the CLI interpreter, available for Windows systems and also for Linux systems [6].

Due to the fact that MPI standard only requires source compatibility and that current MPI implementations do not support .NET platform, the final code, using MPI libraries written in C language is not platform independent. On Windows systems the freely available MPICH library [2] is mostly used.

Besides the compatibility issues, there are also some problems regarding the binding of the MPI libraries to the C# language. First, the objects in .NET can be arbitrarily moved by garbage collector, and this must be prevented when they are in use by MPI functions. The solution, which still generates safe code, is to use special C# class to pin the object in some memory location and than obtain pointer to that object needed by MPI library functions. Special care is needed to unpin the object when it is not needed anymore. Secondly, MPI data types and constants are defined in C++ header files, which cannot be directly imported into C#. Therefore, MPI constants need to be represented as functions, which can return the value of particular constant on startup. Similarly special functions are written to create, access and delete special MPI data types. In order to put the binding problems out of the C# programmer's sight, a wrapper was written in C# and partially in C [7], providing an interface to the MPICH library that looks more like a normal C# class. It is reported that with careful pinning and unpinning of objects the performance of the MPI is only slightly affected [7].

4 Results

Performance of both algorithms was tested on two problem domains: (i) the character recognition problem and (ii) the rubber testing data set. In the first case 7800 samples of numbers 0-9 represented as 8×8 black and white pixels were generated. Additionally, noise was introduced as a 0.15 probability of a wrong pixel being generated. The neural network with 64 inputs, 10 neurons in the hidden layer and 10 neurons in the output layer was used to classify the samples. In the second case, rubber hardness was related to its rheological properties for 25000 samples of different rubber compounds. The neural network having 7 inputs, 21 neurons in the hidden layer and one neuron in the output layer was used.

In all runs neural networks were randomly initialized using the same seed. The training was stopped when the specified error was reached or after 20000 iterations.

A cluster of 8 computers with processor Intel Pentium IV 2.0 GHz, 512 MB RAM, running Windows 2000 was used. One computer was reserved for the master process and the remaining for slave processes.

The cluster utilizing from 1 to 7 slaves was considered in different trials to run both algorithms and the turnaround times of those trials were recorded. The performance of the gradient descent algorithm with momentum on both problem domains is presented in Fig 3. It can be observed that the time of the tasks which

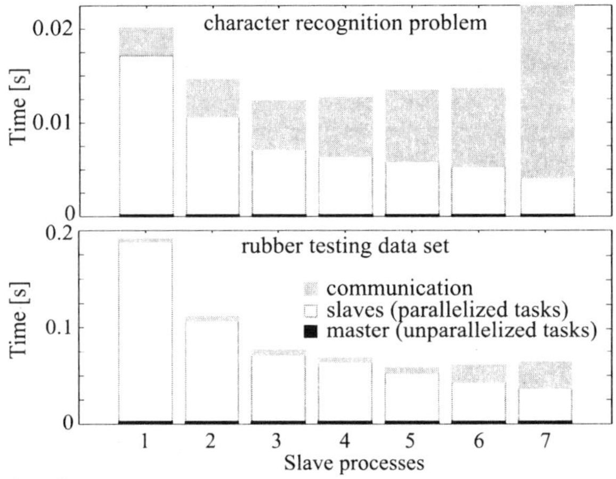

Fig. 3. Gradient descent learning algorithm with momentum: computational time per iteration depending on the number of slave processes.

are parallelized (white) is reduced with the increasing number of slave processes. On the other hand, the increasing number of slaves also increases the time needed for weights transfer and weight update collection (gray). The best overall performance is thus reached with 4 slave processes for the first problem and 5 for the second one.

Performance of the Levenberg-Marquardt learning algorithm measured in the same way is presented in Fig. 4. In this case the contribution of the non-parallelized processes is notably greater (black) due to the fact that solving of equation 3 was not parallelized. Although, $N_\Omega + 1$ times more data is exchanged during the collection phase, compared to the gradient descent algorithm with momentum, its relative contribution to the required computational time is small. Therefore, the turning point was not reached within the cluster of 8 computers.

5 Conclusion

The parallelization of two different feed-forward neural network training algorithms was considered in C# on .NET platform using Message Passing Interface (MPI).

The parallelization of certain tasks reduces the total

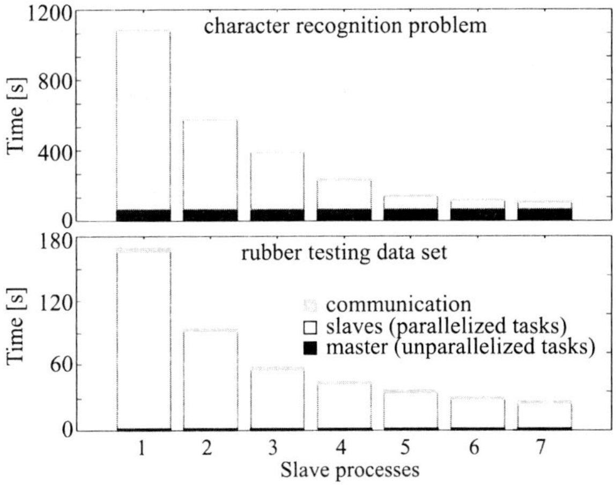

Fig. 4. Levenberg-Marquardt algorithm: computational time per iteration depending on the number of slave processes.

computational time, however the ratio between the parallelized tasks and unavoidable inter-processes communication is decisive for the optimal number of slaves. In terms of speedups the gain is considerably greater in the case of Levenberg-Marquardt algorithm, since the relative contribution of communication to the iteration times is small.

References

[1] Strey, A. (2003) On the suitability of SIMD extensions for neural network simulation. Microprocessors and Microsystems 27: 341-351

[2] Gropp, W., Lusk, E., Skjellum, A. (1999) Using MPI: portable parallel programming with the message-passing interface, MIT, Cambridge

[3] ECMA (2001) Common language infrastructure (CLI), C# language specification, ECMA, http://www.ecma.ch

[4] Haykin, S. (1999) Neural networks: a comprehensive foundation, 2nd ed., Prentice-Hall, New Jersey

[5] Hagan, M. T., Menhaj, M. B. (1994) Training feed-forward networks with the marquardt algorithm, IEEE Trans. Neural Netw 5(6): 989-993

[6] Mono project (2004) Open source platform based on .NET, http://www.mono-project.com

[7] Willcock, J., et. al. (2002) Using MPI with C# and the Common language infrastructure, Technical report TR570, Indiana University, Bloomington

[8] Ranga Suri, N. N. R., et. al. (2002) Parallel Levenberg-Marquardt-based neural network training on Linux clusters - a case study, 3rd Indian conference on computer vision, graphics and image processing.

HeuristicLab: A Generic and Extensible Optimization Environment

S. Wagner, M. Affenzeller
Institute for Formal Models and Verification
Johannes Kepler University, Linz, Austria
E-mail: {stefan, michael}@heuristiclab.com

Abstract

Today numerous variants of heuristic optimization algorithms are used to solve different kinds of optimization problems. This huge variety makes it very difficult to reuse already implemented algorithms or problems. In this paper the authors describe a generic, extensible, and paradigm-independent optimization environment that strongly abstracts the process of heuristic optimization. By providing a well organized and strictly separated class structure and by introducing a generic operator concept for the interaction between algorithms and problems, HeuristicLab makes it possible to reuse an algorithm implementation for the attacking of lots of different kinds of problems and vice versa. Consequently HeuristicLab is very well suited for rapid prototyping of new algorithms and is also useful for educational support due to its state-of-the-art user interface, its self-explanatory API and the use of modern programming concepts.

1 Introduction

Naturally inspired optimization heuristics have already been used to attack various kinds of problems. They were applied to e.g. routing problems, scheduling problems, function optimization, or graph problems. As a basis for the different optimization algorithms different natural archetypes were considered: E.g. the process of natural evolution inspired Evolutionary Computation, the foraging behavior of ants provided a basis for Ant Colony Optimization or the movement and social behavior of bird flocks or fish schools led to Particle Swarm Optimization. Due to their generality a lot of these heuristics are able to attack very different kinds of problems. However, to obtain very high quality results, various problem specific adaptations and hybrid variants were also developed. This leads to the situation that not only the number of problems but also the variety of different algorithms is enormous today.

So scientists working in the area of heuristic optimization have to deal with this variety of optimization algorithms and problems. On the one hand it is necessary to test newly developed algorithms with different problems and on the other hand different optimization algorithms have to be evaluated concerning their applicability for an optimization problem. Consequently a paradigm-independent optimization environment is favorable that separates the algorithm and problem implementation from each other, making it possible to exchange both of them freely. Furthermore, it is preferable to have an environment at hand that supports the users not only concerning the implementation of algorithms and problems but also offers a very high degree of usability by providing e.g. a state-of-the-art user interface, a self-explanatory installer, and a concise documentation, making it also useful for educational purposes in student courses.

In this paper the authors present a new generic, flexible, and extensible optimization environment named HeuristicLab that is designed to deal with very different kinds of optimization algorithms and problems. Due to the strict separation of algorithms and problems (see section 4) one of the most important aspects is the development of a new generic concept to define abstract operators for all interactions between algorithms and problems (described in section 5).

2 Requirements Analysis

One of the basic features of each generic optimization environment is to enable the user to add new optimization algorithms as well as new optimization problem types in a very fast and easy way. Furthermore, the program should not be focused on a specific kind of optimization problem or algorithm. The environment should provide a framework with a well structured and self-explanatory application programmer's interface (API) that enables the user to implement all different kinds of algorithms and problems. This framework should especially take care of all not optimization specific parts like threading, saving and loading settings, reading and writing files, or the graphical user interface (GUI). Moreover, it should also provide a plug-in architecture that allows adding, removing and using algorithm and problem implementations at any time.

Concerning the implementation of various different kinds of algorithms and problems the most important aspect should be to follow one of the fundamental principles of object oriented programming: the exchangeability of modules. Consequently it should be possible to attack all different kinds of implemented problems with *one* single algorithm implementation (e.g. a generic Genetic Algorithm (GA)) without the need to change any code of the algorithm plug-in. Vice versa *one* problem implementation (e.g. an implementation of the Traveling Salesman Problem (TSP)) should be useable for all available algorithms. So it would be of crucial importance to provide an abstract way of communication between algorithms and problems to be able to exchange both parts independently.

Additionally to an usable API the whole program itself should be very intuitional in order to use it not only for algorithm development and problem solving but also for educational aspects. It should provide a state-of-the-art installer and a GUI to enable students or other interested users to experiment with different optimization techniques and problems out of the box.

3 Other Available Optimization Software Packages

Of course the idea of a generic and extensible optimization environment is not new. Especially in the area of Evolutionary Computation there are numerous more or less mature libraries (a comprehensive list can be found e.g. on the EvoWeb homepage[1]). Among them are a lot of small projects which have just been developed to solve a specific optimization problem, some libraries which are a little bit outdated or have not yet reached a state of broad application (like GAlib [1], OpenBEAGLE [2], TEA, GENOM, JDEAL), and a few highly-developed frameworks (like EO [3], JEO [4], ParadiseEO [5], ECJ). Especially the Evolving Objects (EO) project and its further developments JEO and the DREAM project [6] have to be mentioned as very ambitious and successful approaches towards a generic and flexible framework for (parallel) Evolutionary Computation. However, all these libraries suffer from some weaknesses concerning the requirements stated in the previous section:

Almost all of the existing libraries are dedicated to a specific optimization paradigm like Evolutionary Algorithms, Local Search or Swarm Algorithms. It is a very hard task to break up the libraries in such a way that optimization algorithms of another paradigm can be equally used. However, for practical applications the ability to compare results of different paradigms is very essential.

All the existing libraries are - as the name suggests - libraries. Therefore, their usability cannot really be compared with state-of-the-art software products. If at all GUIs are only rudimentally available, self-explanatory installers are missing altogether. None of the existing packages can be used out of the box without deeper programming knowledge. Consequently, the usability for e.g. university courses is rather limited as the students should not spend their time with installing but with using the software.

So it can be said that there definitively exist some powerful optimization libraries, but all of them seem not to be able to fulfill the desires of the authors: a generic and extensible optimization environment, open for all different kinds of optimization paradigms, comparable to state-of-the-art software packages, equipped with a GUI and a self-explanatory installer, usable out of the box not only for rapid prototyping and comparison of new optimization algorithms but also for educational purposes.

4 Modeling the Optimization Process

After having thought of all the different requirements for the project the first point to start with is the modeling of the optimization process itself. As already stated in the introduction there are manifold natural archetypes that inspired heuristic optimization techniques. Consequently the resulting algorithms have almost no structural similarities. So due to the strict demand for generality and openness the process has to be abstracted to a very high level to assure that all different kinds of algorithms and problems can be integrated into the framework.

Primarily a heuristic optimization process can be stated in the following way: An *algorithm* iteratively modifies one or more *solutions* of a *problem* in a specific way in order to increase their quality and generates *results* to inform others about the progress of the optimization. Obviously the four modules Algorithm, Problem, Solution, and Result form the main actors and consequently they have been modeled as base classes in the HeuristicLab framework. Furthermore, also the EvaluationOperator delegate is included as the calculation of the quality value of the different solutions is one of the main issues in every heuristic optimization algorithm. More details on the operator concept are given in section 5. In Fig. 1 an UML class diagram shows these classes and their most important properties and methods.

Due to the very high level of abstraction the base classes only provide very rudimental functionality. The three classes Algorithm, Problem, and Result contain basic parameter properties (e.g. Rounds, RandomSeed, Is-

[1] http://evonet.lri.fr/evoweb/resources/software

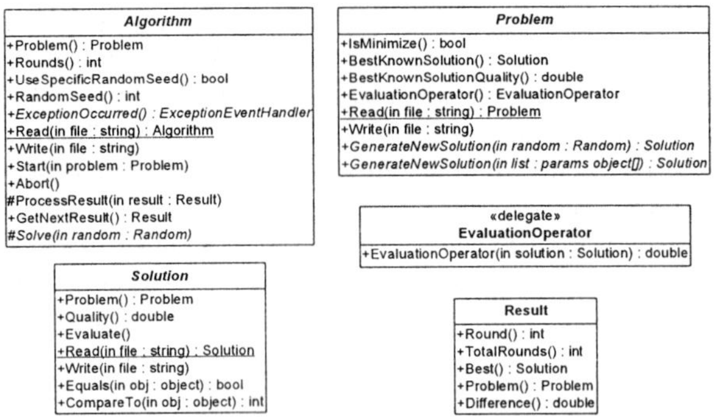

Fig. 1. Class diagram of the main HeuristicLab framework classes.

Minimize, etc.) which are reasonable in every kind of optimization task. Furthermore, they provide a reading and writing mechanism to store and retrieve the current object (binary serialization). Additionally Algorithm also takes care about threading by performing the algorithm execution in an own thread, separating this computational heavy task from the representation and administration issues.

Based upon these four base classes the HeuristicLab framework can be used to implement very different kinds of optimization algorithms and problems by inheriting from them and by implementing their abstract methods. However, HeuristicLab is not just one more class library offering some more assistance for solving optimization problems. HeuristicLab is more thought of as a fully functional optimization environment which can be extended by own algorithm and problem plug-ins. So a layer is needed above the framework base classes that implements a GUI front end and is therefore responsible for all user interactions, for the presentation of various results generated by the algorithms and for administrative tasks like installing or removing plug-ins or changing global settings. The propagation of results from the Algorithm to the front end is done via Result objects. Fig. 2 visualizes the interplay between the different HeuristicLab classes.

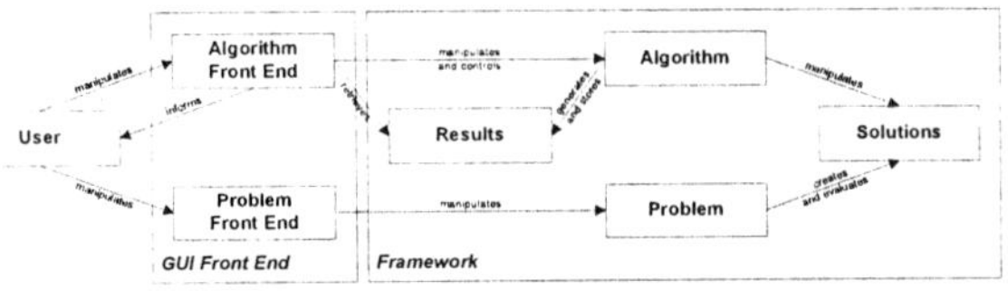

Fig. 2. Diagram showing the interaction between the various HeuristicLab classes.

5 Communication between Algorithms and Problems – The Operator Concept

Due to the high demand for generality and extensibility a fundamental problem occurs concerning the interaction between algorithms and problems. It has to be possible to use a specific algorithm with any optimization problem that is available and the other way round to attack a concrete problem implementation with any algorithm. Consequently both sides can only use the base classes Algorithm, Problem, and Solution to work with each other as they are the least common denominator of all algorithms and problems. However, e.g. in the case of GAs the algorithm has to manipulate the solution candidates via the genetic operators crossover and mutation. These operators are problem specific as they depend on the problem encoding. But the algorithm itself has no idea about the encoding used for the problem implementation and furthermore also cannot provide any crossover or mutation operator on its own.

This general dilemma occurs not only in the case of GAs. In some way any heuristic optimization algorithm has to manipulate solution candidates in order to find better solutions. So there has to be a way that the algorithm can perform problem specific operations without having to know details about the problem's implementation.

To handle this problem the operator concept was developed. This programming technique is based on a main idea of object oriented programming (OOP): the separation of the declaration and the implementation of objects as it is done in OOP by interfaces and classes. However, in the case of HeuristicLab this separation doesn't take place on the level of classes but on the level of methods. An algorithm declares the parameters and the return type of the operators it needs (i.e. the method interface) and a problem has to provide concrete methods that match exactly with the declared interfaces and represent those operators. So in the case of GAs the algorithm declares e.g. a crossover operator as a method that takes two solutions as parameters (the parents) and returns a new solution (the child).

From a more software development oriented point of view this can be realized by using delegates. A delegate is a type that represents methods. When a delegate is defined the parameter types and the return type of the method are specified. Then any concrete method that has the same parameter types and return type can be assigned to a variable of the delegate type. This variable can be used like any other method. So e.g. the Algorithm class in a GA implementation has to have two properties that are used to store the two genetic operators crossover and mutation. When a problem instance is selected for optimization a value is assigned to these properties that references to the methods implemented by the problem that should be used as crossover or mutation operators. Furthermore, as the algorithm and problem implementations are located in different HeuristicLab plug-ins meta data is used to label those methods that implement specific operators. So an algorithm can read this meta data via reflection at runtime to initialize the needed delegates properly.

By this operator concept no specific information about the problem implementation is needed and so all problems that provide the necessary operators can be attacked by an algorithm. On the other hand a problem just has to provide some methods implementing certain operators and to label them correctly. So e.g. if a problem implements a neighborhood operator, it can be solved with any neighborhood-based optimization technique that only needs to compute the neighborhood of a solution (like Simulated Annealing, etc.).

6 Current State of Development

As reflection, meta data and delegates are needed to realize the above discussed architecture and concepts, the Microsoft .NET Framework 1.1 and C# were used for the implementation of HeuristicLab. At the moment HeuristicLab 1.0.0 is freely available for non-commercial and educational use. More details are available at http://www.heuristiclab.com.

Up to now several different optimization algorithms and problems have been implemented for HeuristicLab and lots of them are already included in the actual version. The following plug-ins are currently available (plug-ins printed in italic are currently still under development): GA (Generic Genetic Algorithm), SSGA (Steady-State Genetic Algorithm), IslandGA (Coarse Grained Parallel Genetic Algorithm), *SASEGASA* [7], GP (Genetic Programming), ES (Evolution Strategy), SA (Simulated Annealing), PSO (Particle Swarm Optimization), *ACO* (Ant Colony Optimization), TS (Tabu Search), TSP (Traveling Salesman Problem), JSSP (Job Shop Scheduling Problem), MPSP (Multiprocessor Scheduling Problem), *GCP* (Graph Coloring Problem), *BP* (Bin Packing Problem), TestFunctionsND (n-Dimensional Real-Valued Test Functions). The architecture of the HeuristicLab framework as well as the operator concept have proven themselves to be extremely flexible and very well suited for all different kinds of algorithms and problems. Besides its potential for the rapid prototyping of new optimization heuristics HeuristicLab was also already used as supporting tool in university courses and is also used as development environment for real-world applications ([8],[9]).

7 Conclusion

In this paper we have presented the basic architecture of HeuristicLab - a generic, extensible, and paradigm-independent optimization environment. A clear and well organized class structure, the operator concept, a self-explanatory API and the use of modern programming concepts make it possible to easily use HeuristicLab for the development of different kinds of heuristic optimization algorithms and problems as well as to exchange the different implementations freely. Consequently especially for rapid prototyping of new optimization algorithms HeuristicLab turned out the be a powerful tool. Moreover, HeuristicLab is not just an optimization class library. It also provides an easy to use and state-of-the-art GUI and installer and is therefore also suitable as an educational tool for university courses.

8 References

[1] Wall, M. (1996) GAlib: A C++ Library of Genetic Algorithm Components. Mechanical Engineering Dept. MIT

[2] Gagné, C., Parizeau, M. (2002) Open BEAGLE: A New Versatile C++ Framework for Evolutionary Computations. Proceedings of the Genetic and Evolutionary Computation Conference (GECCO) 2002

[3] Keijzer, M. et al. (2001) Evolving Objects: A General Purpose Evolutionary Computation Library. Proceedings of the 5th International Conference on Evolutionary Algorithms (EA 01)

[4] Arenas, M. G. et al. (2002) JEO: Java Evolving Objects. Proceedings of the Genetic and Evolutionary Computation Conference (GECCO) 2002

[5] Cahon, S. et al. (2004) ParadisEO: A Framework for the Reusable Design of Parallel and Distributed Metaheuristics. Journal of Heuristics – Special Issue on New Advances on Parallel Meta-Heuristics for Complex Problems 10:357–380

[6] Arenas, M. G. et al. (2002) A Framework for Distributed Evolutionary Algorithms. Lecture Notes in Computer Science 2439:665–675

[7] Affenzeller, M., Wagner, S. (2004) SASEGASA: A New Generic Parallel Evolutionary Algorithm for Achieving Highest Quality Results. Journal of Heuristics – Special Issue on New Advances on Parallel Meta-Heuristics for Complex Problems 10:239–263

[8] Winkler, S., Affenzeller, M., Wagner, S. (2004) Identifying Nonlinear Structures Using Genetic Programming Techniques. Cybernetics and Systems 2004 689–694

[9] Braune, R., Wagner, S., Affenzeller, M. (2004) Applying Genetic Algorithms to the Optimization of Production Planning in a Real-World Manufacturing Environment. Cybernetics and Systems 2004 79–84

THE SATELLITE LIST
A Reversible Doubly-Linked List

C. Osterman[1], C. Rego[1], D. Gamboa[2]

[1]School of Business Administration, University of Mississippi, University, MS 38677, USA,
[2]Escola Superior de Tecnologia e Gestão de Felgueiras, Instituto Politécnico do Porto, Apt. 205, 4610-156, Felgueiras, Portugal,
E-mail: {crego, costerman}@bus.olemiss.edu, dgamboa@estgf.ipp.pt

Abstract

Subpath reversals are common operations in graph-based structures arising in a wide range of applications in combinatorial optimization. We describe the satellite list, a variation on the doubly-linked list that is symmetric, efficient, and can be reversed or reverse subsections in constant time.

1 Introduction

Linked lists and doubly-linked lists are among the first topics in any elementary data structures text. Though these lists are often stored in arrays, they are dynamic list structures in the purest definitions. The list size can be changed dynamically, and large splicing operations can be performed in constant time, because pieces of the list can be deleted or inserted by reassigning pointers in the endlinks of the spliced subsections.

The general-purpose operation that reverses a doubly-linked list (or a subsection) is a common pedagogic programming exercise, and the fastest solution is known to be $O(n)$. This paper describes an efficient approach to coding a list that abandons a fixed orientation in favor of a symmetric design, so that subsections appear the same backwards and forwards. This structure, a *satellite list*, can be reversed or have a subsection reversed in constant time. The reverse operation is equivalent to deleting and re-inserting the subsection with the head and tail switched. This allows algorithms that require the use of the "reverse" operation to become more efficient. All other list operations remain unchanged and the new design requires the same number of pointers to store the list as a doubly-linked list. However, algorithms that rely on a list's fixed global orientation will not benefit from using the satellite list design.

A well-known variant of the doubly linked list, sometimes called the *xor-list*, is symmetric and can be reversed in constant time. This ingenious space-saving representation, discussed in one of the exercises by Knuth [1] (550), combines the successor and predecessor of each node in a single pointer field using the 'exclusive or' operation, or some other operation that yields equivalent functionality. Thus a pointer to either adjacent list element can be found with knowledge of the other. Clever as it may be, the problem with this representation is that one cannot jump to the middle of the list and start a traversal, or perform any other operation. It seems to give up the dynamic properties that would make a linked list preferable to a static array. We have not yet found an application for which this representation would be appropriate.

Another method for implementing reversible lists is described by Tatham [5]. This method uses a doubly-linked list, but adds additional computation to the traversal queries (e.g. *previous()* and *next()*). Rather than designating one pointer 'next' and the other 'previous', Tatham's traversal uses '*if*' structures to manually check at each node which link refers to the node just visited. This allows the list to be treated symmetrically, and requires no more memory than a traditional doubly-linked list. Like the satellite list, this method abandons the notion of a fixed global orientation.

In the following sections, we discuss the issue of subpath reversal with a doubly-linked list, describe the satellite list, and discuss logical and physical representation, memory and time efficiency, and subpath reversal with the satellite list.

1.1 Subpath Reversal with a Doubly-Linked List

Subpath reversal is the alteration of a graph such that some subpath in the induced graph is reversed relative to the path that contains it. Subpath reversal is often equated to the removal two node-links, or *arcs*, (a,b) and (c,d) and the addition of two others (a,c) and (b,d) in the path $(a,b,\ldots,c,d)$. A problem that arises when performing a subpath reversal in a path or cycle represented by a doubly-linked list motivates the desirability of a reversible list structure. The difficulty is that the linked list makes the task of performing a subpath reversal while maintaining a readable list too computationally taxing.

A simple example can best describe how the problem occurs. The subpath reversal operation deletes two arcs (a,b) and (c,d) in favor of another two arcs (a,c)

and *(b,d)* from a cycle contained in a circular, doubly-linked list. If the resulting cycle is traversed from any point in either direction, exactly one of the subpaths between the new arcs (e.g. the *b-c* path) will be traversed in reverse order. The obvious part of the operation is updating the arcs among *a,b,c*, and *d* by assigning *c* to follow *a*, *a* to precede *c*, *d* to follow *b*, and *b* to precede *d*. Now, however, the "Next" and "Previous" pointers of a node chosen from the path between *b* and *c* no longer reference the correct nodes. For example, *c* now follows *a*, but the node that now follows *c* is the node that previously preceded *c*; so the pointer must be updated. Completing the move and maintaining a readable tour involves swapping the "Next" and "Previous" pointers in each node of the reversed subpath.

With a linked list structure, the expense of the subpath reversal grows proportionally with the size of the list (*O(n)*). Hence the challenge is to find a data structure that allows a given algorithm to make the move and restore a readable structure efficiently.

2 The Satellite List

The satellite list can operate in the same capacity as a doubly-linked list with the following key difference: the satellite list does not depend on a fixed orientation. The "next" node in the list depends on the *current* orientation, information that is naturally preserved in traversing the satellite list but otherwise neglected by the doubly-linked list. If a global orientation can be avoided altogether, a primary consequence is that the subpath reversal operation is performed easily and in constant time. In addition, the satellite list retains the same efficiency as the doubly-linked list in terms of the number of memory fields it occupies and the commands it requires to perform other list operations.

To obtain a suitable structure that lacks the structural weakness of the linked list but retains its simplicity, its representation is disassembled. A linked list node contains two data members: a pointer to the previous client node and a pointer to the next client node. Pointers should operate in a symmetric fashion, suggesting that they point toward "adjacent" nodes, rather than "next" or "previous" ones. To accomplish this, the pointers are removed from the structure and given their own structures, called *satellites*, whose sole purpose is handling the links among clients in the list. Each satellite points not to an adjacent list node, as in the linked list representation, but rather to the list node's satellite. To read the tour starting from a client, one of its satellites is chosen to begin the traversal. The traversal operation then follows the satellite's pointer to the next satellite, the client is noted, and the process continues. Therefore, the list is traversed on one of two distinct singularly linked lists of satellites.

2.1 Logical Representation

Figure 1 depicts a doubly-linked list node and a satellite list node. The distinguishing characteristic of the satellite list node is its avoidance of direct links with adjacent list nodes. Instead, its satellites link the list nodes indirectly. The dashed lines in the figure emphasize the parts of the structure that indicate relationships among components of the list node. From a satellite, there exists some means to immediately access the *complement satellite*. A similar relationship associates a satellite with the client ID or index of the list node.

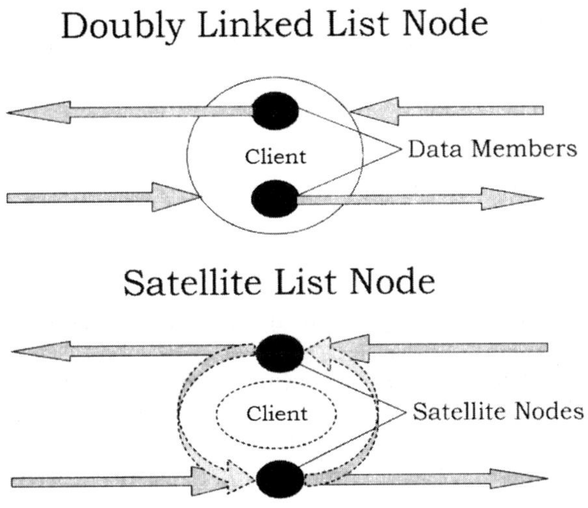

Fig. 1. The Doubly-Linked List vs. the Satellite List

Figure 1 discloses the functional differences between the two types of nodes. The city represented by the doubly-linked list node is constructed in memory in such a way that its data members (typically "Next" and "Previous") are not referenced directly. The data members are not interchangeable and necessitate an orientation. Also, since the data members store references to entire list nodes, the node from which a reference was obtained is not implicit. For example, it is not correct to assume that a node has been obtained from the "Next" pointer of the preceding node because it is equally likely to have been obtained from the "Previous" pointer of the following node. On the other hand, this information remains evident when using the satellite list because each satellite is reached from a distinct source. Furthermore, two satellites sharing a list node operate as independent nodes in separate singly-linked lists, allowing them to be interchangeable. This symmetry allows for a "current direction" during traversal without incorporating any of the costly, explicit decision making that would otherwise be necessary to overcome the asymmetry of a doubly linked list.

2.2 Efficient Physical Representation

A satellite list may be constructed and used without specifically declaring each satellite's links (pointers or indices) to its associated complement satellite or its client. This property reduces the number of pointers needed to represent a list node from four to two, cutting the memory requirement in half.

The list indices are maintained in a single one-dimensional array of length twice n (where n is the number of list nodes). Each element in the array is a satellite node and contains the value of (or pointer to) its adjacent satellite. For each client list node, there exists two physical array positions (satellites), which, together, are considered a logical position. The client itself needs no physical position since it can be uniquely identified by the indices of its satellites. Figure 2 diagrams an example satellite list stored in an integer array representing the cycle (0,1,2,3,4,0).

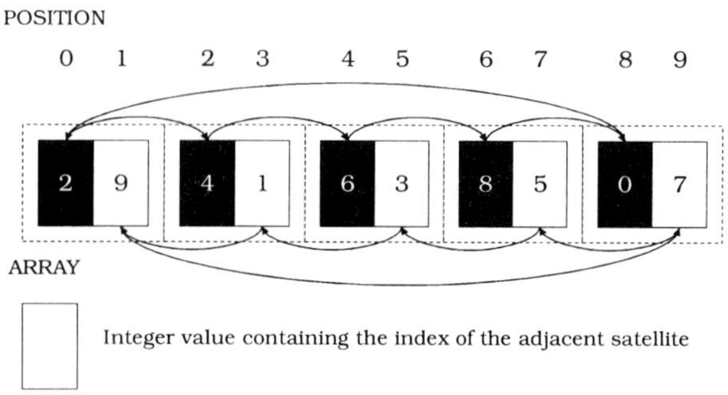

	In General	Bitwise Ops in C
Index of the Satellite:	i	i
ID number of the node:	$i/2$	$i>>1$
Index of the complement satellite:	$i+1-2(i \text{ Mod } 2)$	i^1

(The dashed line indicates that the boundary is logical)

Fig. 2. The Doubly-Linked List vs. the Satellite List

A key factor that makes the use of the efficient implementation desirable is the ease with which the implied relationships mentioned in the previous section can be found. Given some satellite, the queries are to determine its complement satellite and its client. The general formulas to compute these queries are given in the figure along with the equivalent bitwise operations in the C language. A satellite's client can be computed by integer-dividing the index by 2 (no remainder). An identical result is achieved with the bitwise shift operator, ">>". To find a satellite's complement, the index should be incremented if it is even and decremented if it is odd. A bitwise operator is also available for computing a satellite's complement satellite—the exclusive or, "^". These and other bitwise operators are extremely fast because they don't request any arithmetical or logical computation from the processor; thus, the measure of overhead they contribute to routine operations is insignificant. For languages that do not support bitwise operations, using the general formulas may be too costly. In this case, the computational overhead can be avoided by storing "client" and "complement" explicitly, although doing so doubles the memory requirement for the structure.

Although Figure 2 shows all odd satellites pointing to odd satellites, the list is perfectly functional when some odd satellites point to even satellites, or vice versa. Indeed, this becomes the case for the endpoints of a reversed subpath. Interestingly, when a satellite list is organized so that the entire list can be read from just the even or odd satellites, the satellite list array can be unioned with a doubly-linked list structure and the list can be read correctly from either.

2.3 Memory and Time Efficiency

In practice, the satellite list does not impose an additional memory requirement. The number of pointer fields required to store the tour are the same for the two lists (two per client). Of course, this is not to say that additional fields cannot be added to a list node of either representation for use with specialized algorithms. As a point of academic interest, however, the representation of the satellite list in the previous section clearly uses one extra bit per field, since the index of a satellite node is twice that of the city. So, if ever the size of the list being dealt with nears a billion elements, the ceiling for using 32-bit fields in a satellite list will be half that of the doubly linked list.

The satellite list does not impose additional computational effort over the doubly-linked list for the traversal procedures, practically speaking. Enumerating across several nodes is simply a matter of following a singly-linked list of satellite nodes, and the cost of using the shift operator is marginal.

2.4 Subpath Reversal with a Satellite List

Because of the list's symmetric design, the subpath reversal operation is a natural one and is performed easily in constant time, (*O(1)*). The operation is illustrated in Figure 3.

Figure 3 depicts isolated subpaths for a doubly-linked list (top) and a satellite list (bottom), where straight arcs point to the following nodes in the list, and the curved arcs point to preceding nodes. The subpath in the satellite list remains feasible when rotated 180 degrees—

only the end-links need to be changed. The subpath in the doubly-linked list, however, does not match correctly when rotated. Additional time must be taken to correct the intermediate arcs in the reversed subpath.

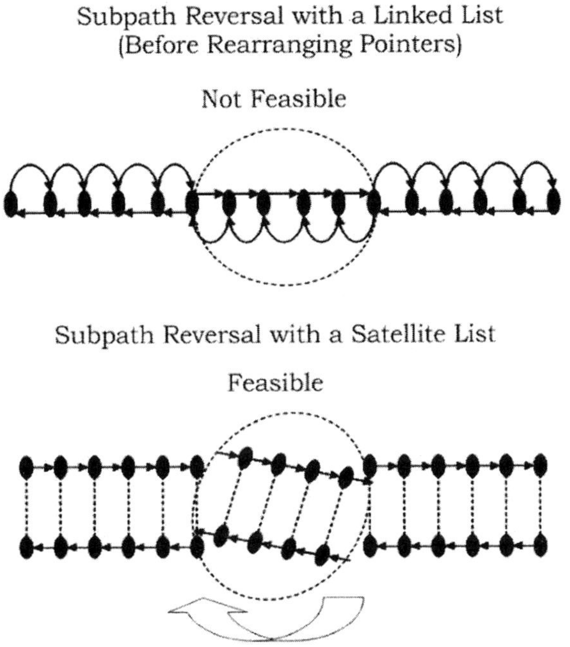

Fig. 3. The Doubly-Linked List vs. the Satellite List

3 Remarks

The strengths of the satellite list are symmetry and subpath reversal. However, if a global orientation is essential to the algorithm, the *O(1)* subpath reversal property cannot be maintained. The satellite list seems to have tremendous potential for application. The development of this list was motivated by our research on data structures for the Traveling Salesman Problem (TSP), for which an efficient method of handling paths and cycles is needed for a number of operations. While the satellite list proved successful in speeding up TSP algorithms in comparison to the doubly-linked list, it proved even more useful in replacing the doubly-linked list component of the *k*-level tree [3]—our variable-level generalization of the 2-level tree structure for the TSP by Chrobak et al. [2]. In addition, the structure was found particularly relevant in the implementation of adaptive neighborhood search procedures based on ejection chain reference structures that have proved effective in a wide range of applications in combinatorial optimization. Likewise, ejection chains have often been fundamental components in tabu search adaptive memory procedures and in a number of evolutionary algorithms based on scatter search and path-relinking models. Preliminary results for the efficiency of these structures in the context of TSP algorithms can be found in Osterman, Rego, and Gamboa [4]. The successful results obtained with the use of the satellite list in the TSP setting greatly motivates the application of similar algorithm designs to more general routing problems as well as a variety of other challenging problems commonly cast into TSP models such as those arising in computational biology (Agarwala et al. [6], Avner et al. [7]).

References

[1] Knuth, D. (1968) The Art of Computer Programming, v.1: Fundamental Algorithms. Addison-Wesley, Reading, MA

[2] Chrobak, M., Szymacha, T., Krawczyk, A. (1995) A Data Structure Useful for Finding Hamiltonian Cycles. Theoretical Computer Science, 71: 419-424

[3] Osterman, C., Rego, C. (2003) New Data Structures for Symmetric Traveling Salesman Problems. Hearin Center for Enterprise Science, Research Report HCES-06-03, University of Mississippi

[4] Osterman, C., Gamboa, D., Rego, C. (2003) On the Performance of Data Structures for the Traveling Salesman Problem. In: CE: The Vision for the Future Generation in Research and Applications 365-371. Swets & Zeitlinger, Lisse

[5] Tatham, S. (2004) Reversible Linked Lists. http://www.chiark.greenend.org.uk/~sgtatham/algorithms/revlist.html

[6] Agarwala, R., Applegate, D., Maglott, D., Schuler, G., Schäffer, A. (2000).,A Fast and Scalable Radiation Hybrid Map Construction and Integration Strategy. Genome Research 10:350-364.

[7] Avner, P., Bruls, T., Poras, I., Eley, L., Gas, S., Ruiz, P., Wiles, M., Sousa-Nunes, R., Kettleborough, R., Rana, A., Morissette, J., Bentley, L., Goldsworthy, M., Haynes, A., Herbert, E., Southam, L., Lehrach, H., Weissenbach, J., Manenti, G., Rodriguez-Tome, P., Beddington11, R., Dunwoodie, S., Cox, R. (2001) A Radiation Hybrid Transcript May of the Mouse Genome, Nature Genetics 29, 194-200.

Springer and the Environment